D. Readey
6/98

YO-AWJ-281

Ceramic Microstructures
Control at the Atomic Level

Ceramic Microstructures
Control at the Atomic Level

Edited by
Antoni P. Tomsia
Lawrence Berkeley National Laboratory
Berkeley, California

and

Andreas M. Glaeser
University of California, Berkeley
Berkeley, California

Plenum Press • New York and London

Library of Congress Cataloging-in-Publication Data

Ceramic microstructures : control at the atomic level / edited by
  Antoni P. Tomsia and Andreas M. Glaeser.
     p.   cm.
  "Proceedings of the International Materials Symposium on Ceramic
Microstructures: Control at the Atomic Level, held June 24-27, 1996,
in Berkeley, California"--T.p. verso.
  Includes bibliographical references and index.
  ISBN 0-306-45817-9
  1. Ceramic materials--Congresses. 2. Microstructure--Congresses.
I. Tomsia, Antoni P. II. Glaeser, Andreas M. III. International
Materials Symposium on Ceramic Microstructures: Control at the
Atomic Level (1996 : Berkeley, Calif.)
TA455.C43C464 1998
620.1'404299--dc21   98-16169
                                                             CIP

Proceedings of the International Materials Symposium on Ceramic Microstructures '96: Control at the Atomic Level, held June 24 – 27, 1996, in Berkeley, California

ISBN 0-306-45817-9

© 1998 Plenum Press, New York
A Division of Plenum Publishing Corporation
233 Spring Street, New York, N.Y. 10013

http://www.plenum.com

10 9 8 7 6 5 4 3 2 1

All rights reserved

No part of this book may be reproduced, stored in a retrieval system, or transmitted in any form
or by any means, electronic, mechanical, photocopying, microfilming, recording, or otherwise,
without written permission from the Publisher

Printed in the United States of America

MEMBERS OF INTERNATIONAL ADVISORY COMMITTEE

Salvador De Aza, CSIC, Madrid, Spain
Hisao Banno, NGK Spark Plug Co., Nagoya, Japan
Dario Beruto, University of Genoa, Italy
Phillippe Boch, ESPCI, Paris, France
Richard C. Bradt, University of Alabama, USA
Richard J. Brook, University of Oxford, UK
Nils E. Claussen, Technical University, Hamburg, Germany
Anthony G. Evans, Harvard University, USA
H. Eckart Exner, University of Darmstadt, Germany
Carol Handwerker, NIST, USA
David W. Johnson, Jr., AT&T Bell Labs, USA
Mitsue Koizumi, Ryukoku University, Japan
Fred F. Lange, University of California at Santa Barbara, USA
Ronald E. Loehman, Sandia National Laboratory, USA
Gary L. Messing, Pennsylvania State University, USA
Kenji Morinaga, Kyushu University, Japan
Koichi Niihara, Osaka University, Japan
Koichi Niwa, Fujitsu Laboratories Ltd., Japan
Janusz Nowotny, ANSTO, Australia
William H. Rhodes, Osram Sylvania, Inc., USA
Manfred Rühle, Max-Planck-Institut, Germany
Hartmut Schneider, German Aerospace Res. Establishment, Germany

PROGRAM CHAIRMAN

Antoni P. Tomsia, Lawrence Berkeley National Laboratory

PROGRAM CO-CHAIRMAN

Andreas M. Glaeser, University of California, Berkeley

HONORARY CHAIRMAN

Joseph A. Pask, University of California, Berkeley

CONFERENCE COORDINATOR

Mollie Field, Lawrence Berkeley National Laboratory

SUPPORT

The financial support of the following organizations for the Ceramic Microstructures '96 Conference is greatly appreciated.

- Fujitsu Laboratories, Ltd.
- Matsushita Electric Industrial Co., Ltd.
- Murata Manufacturing Co., Ltd.
- NGK Spark Plug Co., Ltd.
- Shoei Chemical Inc.
- TDK Corp.
- TOTO, Ltd.

◆

- National Science Foundation
- Wright Patterson Air Force Base
- Corning Glass
- Osram Sylvania, Inc.

◆

- Ernest Orlando Lawrence Berkeley National Laboratory
- Materials Sciences Division
- Department of Materials Science and Mineral Engineering
- University of California, Berkeley, California

SPONSORING SOCIETIES

- The American Ceramic Society
- The Ceramic Society of Japan
- The European Ceramic Society

PREFACE

This volume, titled Proceedings of the International Materials Symposium on *Ceramic Microstructures: Control at the Atomic Level* summarizes the progress that has been achieved during the past decade in understanding and controlling microstructures in ceramics. A particular emphasis of the symposium, and therefore of this volume, is advances in the characterization, understanding, and control of microstructures at the atomic or near-atomic level.

This symposium is the fourth in a series of meetings, held every ten years, devoted to ceramic microstructures. The inaugural meeting took place in 1966, and focussed on the analysis, significance, and production of microstructure; the symposium emphasized the need for, and importance of characterization in achieving a more complete understanding of the physical and chemical characteristics of ceramics. A consensus emerged at that meeting on the critical importance of characterization in achieving a more complete understanding of ceramic properties. That point of view became widely accepted in the ensuing decade.

The second meeting took place in 1976 at a time of world-wide energy shortages and thus emphasized energy-related applications of ceramics, and more specifically, microstructure-property relationships of those materials. The third meeting, held in 1986, was devoted to the role that interfaces played both during processing, and in influencing the ultimate properties of single and polyphase ceramics, and ceramic-metal systems.

The past decade has been marked by the development and refinement of new analytical tools and procedures that allow experimental determination of the structural and chemical characteristics of materials at an atomic or near-atomic level. Concurrently, the decade has seen remarkable increases in raw computational power, and equally remarkable decreases in the cost of this capability. These trends have allowed and encouraged efforts to model structure, properties, and processes at an atomic or near-atomic scale. The emergence of computational materials science is an outcome.

This proceedings includes accounts of our progress in both the experimental and theoretical domains. Presented here are results of research where the atomic-scale or near-atomic length scale is of particular relevance to ceramic microstructures. These topics include: the nature of thin intergranular films in structural and electronic ceramics, characterization of physical and chemical grain boundary widths, the role of interphases and atomic-level interactions in composites and ceramic-metal systems, the growth of textured thin films on both solid substrates and on two-dimensionally assembled organic molecules.

There has also been substantial progress in other areas, where the relevant length scale is somewhat larger. For example, the decade has witnessed an increased interest in the development of materials in which properties depend strongly on direction within the sample or vary with position in a controlled way. The former category of materials relies on seeding and texturing to produce anisometric and oriented microstructures; the latter category has been labelled functionally graded materials (FGMs). The use of nonequilibrium structures that are carefully designed at the micron or submicron scale provides new routes to processing and joining materials. Exciting new tools for materials characterization have also emerged, that provide information on particular aspects of microstructure that previously were difficult to quantify; atomic force microscopy for characterizing surfaces, orientation imaging microscopy for determining texture and misorientation relationships, and a liquid immersion technique for identifying flaws in green bodies are examples of such new research tools. Finally, new approaches to synthesizing powders and processing materials have been developed, and their characteristics are being studied and quantified. This volume provides an account of our progress in these areas as well.

Symposium volumes such as this serve two purposes. They are intended to be a summary of progress in an area, but more importantly they are also intended to stimulate new research that leads to important breakthroughs and further progress. Over a thirty year period, this symposium series, under the guidance of Professor Joseph A. Pask, has charted our ability to understand and control ceramic microstructures. It is hoped that this volume, dedicated to Joe, will prove to be as enduring and as influential as those earlier ones. Those past symposia considered ceramic microstructures in various contexts and at progressively finer length scales. Perhaps our challenge for the next decade is to bridge the length scales, and to develop the connections between continuum level treatments and atomistic treatments of the fundamental processes that control microstructure and define the properties of advanced ceramics. We look forward to Ceramic Microstructures 2006, where our progress will again be assessed.

CONTENTS

INTRODUCTION AND OVERVIEWS

Structure and Composition of Interfaces in Ceramics and Ceramic Composites 1
   M. Rühle, G. Dehm and C. Scheu

Making the Connection between Atomistic Modelling of Interfaces and Real
     Materials 13
   J. H. Harding, A. H. Harker, A. L. Shluger and A. M. Stoneham

Local Chemical Bonding at Grain Boundary of Si_3N_4 Ceramics 23
   I. Tanaka, H. Adachi, T. Nakayasu and T. Yamada

Design and Life Prediction Issues for High Temperature Engineering Ceramics and
     Their Composites 35
   A. G. Evans

Wetting and Work of Adhesion in Oxide/Metal Systems 65
   E. Saiz, A. P. Tomsia and R. M. Cannon, Jr.

Interface Materials for Oxide Composites 83
   D. B. Marshall and P. E. D. Morgan

MICROSTRUCTURES AND INTERFACES

Near Atomic Scale Nanochemistry and Structure: Ceramic Grain Boundaries and
     Interfaces 95
   R. W. Carpenter and W. Braue

Anion Segregation at Si_3N_4 Interfaces Studied by High-Resolution Transmission
     Microscopy and Internal Friction Measurements: A Model System 107
   H.-J. Kleebe and G. Pezzotti

Testing Induced Nanoscale Instablities of Secondary Phases in Si_3N_4: The TEM
     Approach to High-Temperature Microstructures 115
   W. Braue and H.-J. Kleebe

Grain-Boundary Films in a Silicon Nitride Ceramic at High Temperatures 123
   M. K. Cinibulk and H.-J. Kleebe

Microstructure and Intergranular Phase Distribution in Bi_2O_3-Doped ZnO 131
Y.-M. Chiang, J-R. Lee and H. Wang

Thin Intergranular Films in Ceramics: Thermodynamic Calculations and Model
　　Experiments in the System Titania-Silica 149
H. D. Ackler and Y.-M. Chiang

Atomic Structure of the Σ5 (210)/[001] Symmetric Tilt Grain Boundary in Yttrium
　　Aluminum Garnet 161
G. H. Campbell and W. E. King

Synthesis and Microstructure of Mullite Fibers Grown from Deeply Undercooled
　　Melts 169
W. M. Kriven, M. H. Jivali, D. Zhu, J. K. R. Weber, B. Cho, J. Felton and
P. C. Nordine

The Cubic-to-Hexagonal Transformation to Toughen SiC 177
W. J. MoberlyChan, J. J. Cao and L. C. DeJonghe

Morphology and Microstructure of AlN Single Crystals on Si (111):
　　A Combination of Surface Electron Spectroscopies and Transmission Electron
　　Microscopies 191
F. Malengreau, S. Hagège, R. Sporken, M. Vermeersch, R. Caudano and
D. Imhoff

Nanometer Level Characterization of Rapidly Densified Ceramics and
　　Glass-Semiconductor Composites 199
S. H. Risbud and V. J. Leppert

Accommodation of Volume Changes During Diffusion-Controlled Metal
　　Precipitation Inside an Oxide Matrix 209
M. Backhaus-Ricoult

Control of the Microstructure of Polycrystalline Diamond and Related Materials
　　via an Enhanced CVD Process 221
M. L. Carasso, S. S. Staehle, P. A. Demkowicz, D. J. Gilbert, R. K. Singh and
J. H. Adair

SINTERING AND GRAIN GROWTH

Microdesigned Interfaces: New Opportunities for Studies of Surfaces and Grain
　　Boundaries 229
M. Kitayama, J. D. Powers, L. Kulinsky and A. M. Glaeser

Effect of Packing Structure of Powder Particles in a Ceramic Green Body on
　　Microstructure Development During Densification 239
K. Uematsu and N. Uchida

Influences of Interparticle Stresses on Intermediate-Stage Densification and Grain
　　Growth 247
A. W. Searcy, J. W. Bullard and D. Beruto

The Al$_2$O$_3$-SiO$_2$ System: Logical Analysis of Phenomenological Experimental Data .. 255
　J. A. Pask

Sintering Behavior, Microstructure and Properties of Fine Silicon Nitride Ceramics .. 263
　A. Bellosi, F. Monteverde, C. Melandri and S. Guicciardi

Simulation of Anisotropic Grain Growth by Ostwald Ripening 277
　V. Tikare and J. D. Cawley

Mullitization Behavior of Alpha Alumina/Silica Microcomposite Powders 285
　M. D. Sacks, K. Wang, G. W. Scheiffele and N. Bozkurt

Processing of Textured Ceramics by Templated Grain Growth 303
　M. M. Seabaugh, S. H. Hong and G. L. Messing

Characterization of Second Phases in Translucent Alumina by Analytical
　Transmission Electron Microscopy 311
　G. C. Wei, S. J. Jeon, C. Sung and W. Rhodes

Origin and Control of Abnormal Grain Growth in Alumina 323
　J.-C. Nam, I.-J. Bae and S. Baik

The Origin and Growth Kinetics of Plate-Like Abnormal Grains in Liquid Phase
　Sintered Barium Titanate 331
　D. Kolar, A. Recnik and M. Ceh

Microstructural Control of Zinc Oxide Varistor Ceramics 339
　R. C. Bradt and S. L. Burkett

CERAMIC-METAL INTERFACES

Metal/Oxide Interfaces: Chemistry, Wetting, Adhesion and Oxygen Activity 349
　D. Chatain, V. Ghetta and J. Fouletier

Ceramic/Metal Reactions and Microstructures in Ceramic Joints 359
　S. D. Peteves, M. Paulasto, G. Ceccone and M. G. Nicholas

Joining Technology for Ceramic/Metal Composite Structures 369
　K. Suganuma

Microanalysis of Buried Metal/Ceramic Interfaces Using Neutron Reflection 377
　P. Xiao, B. Derby, J. Webster and J. Penfold

Recent Progress in Surface Activated Bonding 385
　T. Suga

Ceramic-Metal Interfaces in Electronic Ceramics – Interface between AlN
　Ceramics and Conductors 391
　K. Niwa, K. Omote, Y. Goto, N. Kamehara

Densification of Tungsten Conductors in Cofired Aluminum Nitride Multilayer
　Substrates by the Addition of Manganese Oxide 399
　J. Monma, T. Yasumoto, T. Takahashi and N. Iwase

Wettability of Cu-Based Alloys on Alumina and Joining of Alumina with
  Microdesigned Nickel-Chromium Alloy Interlayer 407
  K. Nakashima, K. Mori and A.M. Glaeser

Interface Nanostructure of Brazed Silicon Nitride 415
  C. Iwamoto and S.-I. Tanaka

SPECIAL TECHNIQUES/NOVEL PROCESSES

The Impact of Interface Nonstoichiometry on Gas/Solid Kinetics 421
  J. Nowotny

Reactions at Nanometric Solid-Liquid Interfaces Influenced by Low Frequency
  Electromagnetic Fields and Microstructure Development of the Precipitates 429
  D. Beruto

Growth of Inorganic Crystals on the Surfaces of Two-Dimensionally Assembled
  Organic Molecules 437
  K. Koumoto and H. Lin

Reactive Coatings on Ceramic Substrates 447
  J. S. Moya, J. Requena, H. P. Steier, A. H. De Aza, P. Pena and R. Torrecillas

Multiphase Ceramics from Combustion-Derived Powders 463
  R. Pampuch

Absence of Microwave Effect in Ceramics: Precise Temperature, Thermal
  Gradient, and Densification Determination in a Proportional Power Microwave
  Furnace 471
  C. Sorrell, N. Ehsani, A. J. Ruys and O.C. Standard

State and Structure of Free Carbon in Dense SiC Developed by Smart Plasma
  Sintering 487
  K. Kijima and Y. Ikuhara

COLLOIDAL PROCESSING

Colloidal Stability in Complex Fluids 495
  J. A. Lewis

High Temperature Colloidal Processing for Glass-Metal and Glass/Ceramic FGMs .. 503
  L. Esposito, E. Saiz, A. P. Tomsia and R. M. Cannon

GLASSES AND CERAMIC COATINGS

Multicomponent Oxide Coatings via Sol-Gel Process 513
  L. Bonhomme-Coury, M. Najman, F. Babonneau and P. Boch

Modeling the Structure of Glasses and Ceramic Grain Boundaries (An Interstitial
  Approach) 527
  J. F. Shackelford

Interactions between Al$_2$O$_3$ Substrate and Glass Melts 535
   K. Morinaga, H. Takebe and Y. Kuromitsu

Bioactive Coatings on Ti and Ti-6Al-4V Alloys for Medical Applications 543
   A. Pazo, E. Saiz and A. P. Tomsia

NANOSTRUCTURED MATERIALS

Characterization of Alumina/Silicon Carbide Ceramic Nanocomposites 551
   B. Derby and M. Sternitzke

Composite Powder Synthesis 559
   L. C. De Jonghe and Y. He

Synthesis and Characterization of Ceramic Powders by Dual Irradiation of CO$_2$ and
   Excimer Lasers 567
   T. Yamada, Y. Tanaka, T. Suematsu and Y. Kohtoku

Ionic Transport to Produce Diverse Morphologies in MgAl$_2$O$_4$ Formations Started
   from Powders of Varied Physical Nature 577
   T. Kume, T. Kuwabara, O. Sakurada, M. Hashiba, Y. Nurishi, J. Requena and
   J. S. Moya

Microstructure Evolution During Oxidation of MgO-SiC Composites 589
   D. J. Aldrich, M. E. F. Camey and D. W. Readey

Synthesis Tailored Microstructures and "Colossal" Magnetoresistance in Oxide
   Thin Films 597
   K. M. Krishnan, A. R. Modak, H. Ju and P. Bandaru

Magnetic and Interface Characterization of rf-Sputter Deposited Multilayered and
   Granular Fe/AlN Films 605
   S. Kikkawa, M. Fujiki, M. Takahashi and F. Kanamaru

Nano-Sized Oxide Composite Powders, Bulk Materials and Superplastic Behavior
   at Room Temperature 613
   D. S. Yan, J. L. Shi and Y. S. Zheng

Electron Beam Control of Alumina Nanostructure and Al Bonding 623
   S.-I. Tanaka and B. Xu

ELECTRICAL PROPERTIES

Microstructure and Properties of SiC/SiC and SiC/III-V Nitride Thin Film
   Heterostructural Assemblies 629
   R. F. Davis, S. Tanaka, S. Kern, M. Bremser, K. S. Ailey, W. Perry and
   T. Zheleva

Correlating Simulation Defect Energy Calculations with HREM Images of Planar
   Defects in Electroceramics 637
   W. E. Lee, M. A. McCoy and R. W. Grimes

Microstructural Studies of Strontium Titanate Dielectric Ceramics 645
   Z. Mao and K. M. Knowles

Electrical and Optical Properties at the Interfaces of n-BaTiO$_3$/p-Si 653
   S. Sugihara, T. Ebina, S. Suzuki and T. Andoh

Bulk and Thin Film Properties of (Pb,La)TiO$_3$ Ferroelectrics 659
   K. Okazaki and H. Maiwa

Preparation and Properties of PZT/PbTiO$_3$ Ceramic Composite 669
   H. Banno, N. Sugimoto and T. Hayashi

Growth of Oriented SnO$_2$ Thin Films from Organotin Compounds on Glass
   Substrates by Spray Pyrolysis 679
   S. Kaneko, I. Yagi, T. Kosugi and K. Murakami

The Energy Barrier at the Interface of ZnO Contact 687
   S. Fujitsu, K. Nakanish, S. Hidaka and Y. Awakura

CERAMIC COMPOSITES

Fiber Coating Design Parameters for Ceramic Composites as Implied by
   Considerations of Debond Crack Roughness 695
   R. J. Kerans and T. A. Parthasarathy

Interfacial Design and Properties of Layered BN(+C) Coated Nicalon Fiber
   Reinforced Glass-Ceramic Matrix Composites 705
   J. J. Brennan

Stability of Carbon Interphases and Their Effect on the Deformation of
   Fiber-Reinforced Ceramic Matrix Composites 713
   M. M. Stackpoole, R. K. Bordia, C. H. Henager, Jr., C. F. Windisch and
   R. H. Jones

Thermal Stability of Nextel 720 Alumino-Silicate Fibers 721
   H. Schneider, J. Göring, M. Schmücker and F. Flucht

Textured Calcium Hexaluminate Fiber Matrix Interphase for Ceramic-Matrix
   Composites 731
   M. K. Cinibulk, R. S. Hay and D. E. Dutton

Microstructure of the Particulate Composites in the (Y)TZP WC System 741
   K. Haberko, Z. Pedzich, J. Dutkiewicz, M. Faryna and A. Kowal

Reaction Mechanisms and Microstructures of Ceramic-Metal Composites Made by
   Reactive Metal Penetration 749
   W. F. Fahrenholtz, K. G. Ewsuk, A. P. Tomsia and R. E. Loehman

Eutectic Structures that Mimic Porous Human Bone 761
   P. N. de Aza, F. Guitian and S. de Aza

Mullite - A Modern Ceramic Material 771
   S. Somiya, H. Ohira and T. Akiba

MECHANICAL PROPERTIES

Control of Interface Fracture in Silicon Nitride Ceramics: Influence of Different
    Rare Earth Elements 779
  E. Y. Sun, P. F. Becher, S. B. Waters, C.-H. Hsueh, K. P. Plucknett and
  M. J. Hoffmann

Microstructural Evaluation of Deformation Mechanisms in Silicon Nitride
    Ceramics 787
  J. A. Schneider and A. K. Mukherjee

Microstructurally Induced Internal Stresses in the Silicon Nitride Layered
    Composites 795
  P. Šajgalik and Z. Lenčéš

Characterization of Microstructure and Crack Propagation in Alumina Using
    Orientation Imaging Microscopy (OIM) 803
  S. J. Glass, J. R. Michael, M. J. Readey, S. I. Wright and D. P. Field

Grain Boundary Chemistry and Creep Resistance of Alumina 815
  Y. Z. Li, M. P. Harmer, H. M. Chan and J. M. Rickman

Fracture of Copper/Alumina Interfaces: The Role of Microstructure and Chemistry .. 823
  I. E. Reimanis and K. P. Trumble

Effect of Microstructure and Internal Stress on Mechanical Properties of WC-Co
    Doped Al_2O_3/TiC/Ni FGMs 833
  J. Lin, Y. Miyamoto, K. Tanihata, M. Yamamoto and R. Tanaka

Index 843

STRUCTURE AND COMPOSITION OF INTERFACES IN CERAMICS AND CERAMIC COMPOSITES

Manfred Rühle, Gerhard Dehm and Christina Scheu

Max-Planck-Institut für Metallforschung
Seestr. 92, D-70174 Stuttgart, Germany

INTRODUCTION

The inherently brittle nature of ceramics, arising from their lack of plastic deformation mechanisms at ambient temperatures, subjects their macroscopic mechanical properties to microscopic flaws, such as processing voids, secondary phases, and/or grain boundaries. Advances in ceramic processing have curtailed the detrimental aspects of the former two flaw types. Utilizing recent advances in microstructural characterization techniques, the structure and chemistry of grain boundaries, and/or interfaces in ceramics composites, can be engineered to actually enhance mechanical properties.[1-3] Thus, future improvements in ceramic properties via the control of ceramic microstructure requires a comprehensive knowledge of atomic structure and chemical composition of interfaces.

The determination of the atomic structure of the grain boundaries is not as easily done as for surfaces. Only two techniques are available for the study of internal interfaces, namely, X-ray studies[4] and high-resolution transmission electron microscopy.[5,6] This paper concentrates on studies of transmission electron microscopy with emphasis on advanced techniques such as high-resolution electron microscopy and analytical electron microscopy. There exist basically two different types of grain boundaries, those which are atomically discrete and those which comprised a thin amorphous grain boundary layer. Alumina ceramics belong to the first group of materials.[7] Silicon nitride would be a typical representative of the second group.[8,9] In this paper results on specific grain boundaries in alumina will be reported. Also metal/ceramic interfaces between Cu/Al_2O_3 will be discussed.

TRANSMISSION ELECTRON MICROSCOPY

Within the last decade astonishing advances in the TEM instrumentation have been developed.[10,11] Besides conventional TEM, new techniques are emerging which in many cases are based on specialized instruments.[12,13] With these advanced methods dynamical processes can be investigated, atomic structures can be studied, and the chemical composition can be determined with extremely high lateral resolution. Table 1 presents the different TEM techniques, with each column representing a specialized area of microstructural characterization. The instruments required for optimum application of the different techniques are becoming more and more specialized. The ultimate microscope would enable the analysis of the atomic structure as well as the chemical composition in each region of a material.

A TEM micrograph usually includes a wealth of information which should be evaluated and from which quantitative data on physical parameters should be extracted. In many cases this requires extensive image processing and frequently computer simulation to retrieve all information. Whereas in many fields these new quantitative techniques are only at their beginning, in the field of materials science a number of methods are already well established. A summary will be provided in this paper.

Table 1. Summary of the different transmission electron microscopy techniques

Transmission Electron Microscopy (TEM)			
CTEM	**In situ HVEM**	**AEM**	**HRTEM**
-Microstructure -Morphology -Phase distribution -Defect analysis (point defects, dislocations, grain boundaries)	-Irradiation experiments -Deformation experiments -Environmental cells (corrosion)	· Chemical composition -EDX -EELS -EXELFS -ELNES · CBED -lattice strain -bonding -charge density	-Lattice imaging -Structure of complex materials -Atomic structure of defects (interfaces, disclocations)

Conventional Transmission Electron Microscopy (CTEM)

In the classification of the different techniques, CTEM stands for all imaging modes which only use one beam, either primary the beam or one of the diffracted beams.[14-16] Furthermore, CTEM includes diffraction studies with parallel illumination of a specific area, the selected area diffraction (SAD). CTEM still plays the most important role in materials research concerning the microstructural evaluation of applied and advanced materials.

Analytical Electron Microscopy (AEM)

AEM allows the determination of the chemical composition and gradients of the chemical composition with extremely high spatial resolution. In AEM, signals resulting from inelastic scattering processes are used to identify the chemical composition.[17,18]

A prerequisite for analysis with high spatial resolution is the formation of a very small probe (typical diameter < 1 nm). This is best realized in a dedicated scanning transmission electron microscope (STEM) working in ultra high vacuum (UHV). CTEM with field emission guns are also now capable of forming those small probes.

A completely different approach is realized in the new energy filtering TEMs (EFTEMs). In these instruments a large area on the specimen is illuminated and spatially-resolved, analytical information is obtained by forming images with inelastically scattered electrons which have suffered a particular energy-loss. These electrons can be selected by means of the energy filter and the resulting images are projected onto a CCD camera for effective two-dimensional recording.

Another major progress in analytical electron microscopy has been the improvement of detectors. The diode arrays in energy-loss spectrometers are noise-free and all electrons can be detected. Energy dispersive X-ray spectrometers now operate digitally, allowing the detection of high intensities of x-rays.

In an inelastic scattering event the primary process is the loss of energy. Electron energy-loss spectra, where the intensity of the inelastic scattering is plotted against the energy loss, provide a wide range of material information. Two different electron energy-loss regions are commonly used for analyses. From the valence loss regions (energy loss < 100 eV), optical properties and electronic structures can be deduced. Core loss spectroscopy (~50 eV to 3000 eV) yields information on the concentration of different elements. By studying the specific behaviour in the near-core-loss regions, information on bonding, coordination numbers and distances to nearest neighbors can be obtained. The energy-loss near-edge structure (ELNES) influences the spectra from each edge onset up to ~30 eV additional loss, whereas the extended energy-loss fine structure (EXELFS) is responsible for intensity modulations from there to several hundred eV past the edge onset.[17]

ELNES and EXELFS are explained as follows. One can view the excited electron as a wave which is scattered in the crystal at neighboring atoms. The interference at the location of the excited atom determines the intensity measured in a spectrum. The mean free path for the excited electrons within the solid are short. For EXELFS this means that only paths with single scattering events have to be considered. The measured spectrum can be analysed to provide a radial distribution function. However, this method is not used very often in the electron microscope because the achievable signal-to noise ratio is not sufficient. For ELNES the mean free path becomes longer, so that multiple scattering has to be considered. ELNES is therefore generated by complex processes and cannot be analysed directly, however, the intensity of the ELNES structure is sufficiently high that the spectrum can be determined with accuracy. Egerton[17] showed that the ELNES depends on coordination number, bond distances and bond angles. Multiple scattering calculations for clusters yield the expected ELNES for a given arrangement of atoms. This approach is amenable to interfaces,[19] where the more traditional band structure calculations are unfeasible, because the low symmetry requires large supercells that are currently beyond the scope of computing power.

Changes in the ELNES are observed for different oxidation states (chemical shifts, white line intensities) and coordination.[18,20,21] The first is a more localized atomic effect whereas the latter depends on the environment of the probed atom. The ELNES can be used to determine the bonding, electronic structure and real space structure. The interpretation makes use of reference spectra which for some classes of material can be used as fingerprints, i.e. specific structures in the ELNES correspond to specific structural units.[22] A more complex interpretation compares ELNES calculated for model structures with experimental data. This yields valuable insight into the correspondence of spectral features and real space structures. The important point about ELNES is that it contains information about the three dimensional atomic arrangement and therefore complements the information obtained from imaging, which is limited to two dimensional projections.

Convergent Beam Electron Diffraction (CBED)

For CBED a convergent beam enters the specimen and the multiple diffraction events can be directly observed in the diffraction planes for the different directions of the incoming beam (included in the convergent beam). However, two important problems arise in electron diffraction owing to the strong interaction of the electron with matter: i) inelastic scattering processes lead to a diffuse background which for thicker specimens may dominate the whole diffraction pattern, and ii) the kinematical approximation loses its validity for much lower specimen thicknesses. The first problem has been overcome by the introduction of energy filtering in TEM. The second problem has been solved by the introduction of the dynamical theory of electron diffraction.[23,24]

High-Resolution Transmission Electron Microscopy (HRTEM)

An image of a specimen (following the description of Abbé) is formed by the interference of the transmitted beam with diffracted beams. The interpretation of HRTEM micrographs of crystals and crystal defects with respect to the atomic structure is very difficult.[25,26] Direct interpretation is restricted to very thin specimens, where imaging conditions are free of phase reversals or low strain fields. In practice, dynamical electron diffraction and aberrations of electron optics lead to a complex relationship between the image and the structure of the object. For example, in the regions of a crystal defect, the image patterns representing projected atom columns may appear displaced relative to their true position. To interpret HRTEM images 'safely', therefore, one compares them with simulated images of model structures. Formerly, it was common practice to compare experimental and simulated images by visual inspection, however, recent procedures have been developed to evaluate image intensities *quantitatively,* through the use of digital image processing.[27-31] This image comparison is, however, very difficult if the valid structure retrieval under dynamical electron diffraction, and non-linear image conditions require detailed numerical image simulation and image matching to the experimental image. For the processing of the digital images powerful computer hardware and software is required to handle large data fields on disk and in memory and to run time-consuming programs for off-line image restoration, structure model generation and image simulation.

STRUCTURE OF GRAIN BOUNDARIES IN ALUMINA

Alumina is one of the most widely utilized oxide ceramics.[32] However, it is quite amazing that many of its basic properties are not yet understood. This is true not only for experimental studies but also for theoretical predictions. It was only recently that the theoretical modeling of the most stable phase led to the actual structure of α-Al_2O_3. *Ab initio* calculations of Gillan et al.[33,34] were able to show that α-Al_2O_3 is the most stable phase of many different metastable configurations. Recently, Wilson et al.[35] showed by introducing a new compressible ion model which includes both dipole and quadrupolar polarisability of the O^{2-} ions, that a shell model describes α-Al_2O_3 as the most stable phase.

Another problem exists for the interpretation of high-resolution micrographs of pure α-Al_2O_3. Whereas the HRTEM micrographs are well understood for certain directions of the incoming electron beam with respect to the crystal structure, there is at least one direction (electron beam direction parallel to [10$\bar{1}$0]) where a discrepancy between the simulated image and the experimental micrograph still exists. HRTEM micrographs include details that suggest the (0003) reflection is excited. However, this reflection should be forbidden in pure α-Al_2O_3. Subtle changes to the composition in the alumina, via the introduction of small amounts of impurities, lead to internal strains which modify the intensity distribution of a HRTEM micrograph. For nearly all projections of alumina the projected distance between Al^{3+} ions and O^{2-} ions is very small, thus even the highest resolution microscopes are too limited to resolve this spacing.

Failure of a material often starts at grain boundaries. Therefore, it is important to understand the structure and composition of grain boundaries and regions close to the grain boundaries. These studies can be performed at the atomic level only by electron microscopy and high-resolution electron microscopy (HRTEM).[7,37] Analytical electron microscopy (AEM) yields information on the composition, bonding, and local coordination.

A bulk of experimental studies of grain boundaries in alumina have been confined to commercial grade polycrystalline alumina and have been concerned mainly with the crystallographic classification of grain boundaries and the description of grain boundaries structures. It is quite interesting that small additions of impurities such as Y lead to a change in the distribution of different grain boundary classes.[38-40]

Only specific grain boundaries, mainly tilt boundaries, can be studied by HRTEM.[41] The condition has to be fulfilled that the direction of the incoming electron beam is parallel to strict periodicity within the specimen.[42] (The electron beam has to be parallel to atom columns of the material). Studies were performed experimentally for different grain boundaries (near $\Sigma 11$ and $1\bar{1}04$ twin).[7,37] Those grain boundaries were not selected in a technical polycrystalline pure material but from bicrystalline specimens produced by diffusion bonding.[43] By this method, the grain boundary misorientation can be accurately adjusted.

The interpretation of HRTEM micrographs requires a comparison between experimental micrographs and simulated images, so that the actual structure is retrieved. This structure can be compared then to results from static lattice calculations. The experimental work by Höche et al.[7,37] resulted in quite interesting high-resolution micrographs for the $N\Sigma 11$ and the $(1\bar{1}04)$ twin. Hoche et al.[7,37] showed that reasonably good agreement exists between the experimentally determined structure and the simulated structure. For the interpretation of HRTEM micrographs, it was possible to differentiate between different metastable structures predicted by theoretical modelling.[44] It was found that the agreement between the simulated images of a specific relaxed structure showed the best agreement with experiments.[7] The agreement was studied quantitatively. It was found that an accuracy of ±0.02 nm was obtained for the positions of columns of ions. However, the agreement is not quite satisfactory at a grain boundary. It is not yet established what causes the difference between the experimental micrograph and the calculated image. This could be caused either by deficiencies in the knowledge of the interatomic potentials or by experimental errors (inhomogeneous TEM specimen, nonequilibrium state of the diffusion-bonded grain boundary).

Recently *ab-initio* calculations of the $(1\bar{1}04)$ twin were performed.[45] The atomic structure in the regions close to the grain boundary is slightly modified, and the volume change at the grain boundary now corresponds exactly to the experimental observations. However, the agreement of the atomic structure at the grain boundary still remains unsatisfactory. This could be caused either by the rather limited high-resolution microscopy of these grain boundaries or by specific experimental changes, especially, preferential ion thinning at grain boundaries.

Studies of Segregated Grain Boundaries in α-Al_2O_3

Alumina possesses a very small solubility in the bulk for almost all anions and cations.[46-48] Therefore, small additions of impurities usually segregate at grain boundaries, and these segregated atoms influence the bonding between grains.

It is well established that small additions of MgO dramatically change the microstructure of alumina. MgO allows complete densification of alumina, inhibits abnormal grain growth of α-Al$_2$O$_3$ grains, and is therefore essential as a sintering aid. However, there exist different models to explain the influence of MgO on the sintering behaviour. Harmer recently developed a model[49,50] which seems to explain the different mechanisms leading to a change in sintering behaviour. However, thus far Mg has not been detected by analytical electron microscopy. The concentration along a grain boundary should be < 0.001 per nm^2.

Detailed experimental studies have been done for the segregation of Ca doping for a specific grain boundary. These boundaries were produced using ultra-high vacuum diffusion bonding of two highly pure single crystals.[52] Two distinct boundary configurations (each having mirror and glide mirror symmetry) were studied in the Ca-doped bicrystals. Thus far only the mirror related configuration has been identified for the undoped boundary. To estimate the atomic structure of the grain boundary, simulated images of regions of the crystals with different Ca contents were compared to experimental HRTEM micrographs. The results indicated that the amount of Ca found at this specific twin boundary corresponded to a replacement of ~0.5 monolayers (ML) of Al by Ca. These results compare well to those obtained by analytical techniques.

Recently, great interest has been directed toward studies of the influence of Y on the microstructure, segregation behavior, and properties of grain boundaries. It has been established that Y increases the number of special boundaries in bulk α-Al$_2$O$_3$.[38,39] An α-Al$_2$O$_3$ scale formed by oxidation of an Al-containing metallic alloy also developed a specific interfacial relationship and a columnar grain morphology.[53] Y segregates at grain boundaries and interfaces. Harmer et al.[49,50] showed that Y decreases the creep rate of alumina by about one order of magnitude. The strong influence of these impurities are not yet understood.

THE Cu/Al$_2$O$_3$ INTERFACE: A MODEL SYSTEM FOR METAL/CERAMIC INTERFACES

Geometrical Structure of the Interface. Thin films of Cu were grown by molecular beam epitaxy (MBE) on (0001) α-Al$_2$O$_3$ substrates. The deposition was carried out at a substrate temperature of 200°C in UHV conditions. During the growth of copper *in situ* reflection high-energy electron diffraction (RHEED) was performed to investigate the growth mechanism and the crystallographic orientation relationship between Cu and α-Al$_2$O$_3$.[54] Initial RHEED images exhibit a ring pattern, which started to fade at a Cu thickness of about 10 nm. With further deposition sharp RHEED streaks develop, characteristic of single crystal films, and persist throughout the growth of a 1.1 µm thick Cu film. Cross-section and plan-view TEM investigations confirm the well-defined orientation relationship indicated by *in situ* RHEED streaks, whereby the close-packed planes and close-packed directions between Cu and α-Al$_2$O$_3$ are parallel:

$$(\bar{1}11)_{Cu}[110]_{Cu} \| (0001)_{Al_2O_3}[10\bar{1}0]_{Al_2O_3} \tag{1}$$

Plan-view TEM studies reveal a homogeneous copper film of near-epitaxial quality. However, a few small pores (< 30nm in size) are observed at the interface by cross-sectional TEM. These pores show that the Cu films grown at 200°C initially nucleate as three dimensional clusters on the (0001) α-Al$_2$O$_3$ substrate.

HRTEM micrographs of the edge-on Cu/Al$_2$O$_3$ interface illustrate that both the close-packed planes and close-packed directions of Cu and Al$_2$O$_3$ are parallel to one another. The lattice image shown in Fig. 1 was taken along the $[211]_{Cu}$ and $[2\bar{1}\bar{1}0]_{Al_2O_3}$ zone axes with the JEOL-ARM 1250. The $(0\bar{2}2)$ Cu planes, with a spacing of 0.128 nm, are clearly resolved. The $(\bar{1}11)$ Cu planes and (0006) Al_2O_3 planes are parallel to the interface. The Cu/Al_2O_3 interface is atomically flat and reveals a sharp transition between the α-Al_2O_3 substrate and the copper overgrowth. The interface is devoid of possible reaction phases between Cu and α-Al_2O_3. This is ascribed to the low substrate temperature of 200°C and the UHV conditions used during growth.

Figure 1. Lattice image of the atomically flat Cu/Al_2O_3 interface taken along the $[211]_{Cu}$ and $[2\bar{1}\bar{1}0]_{Al_2O_3}$ zone axes reveals no misfit dislocations. The clearly resolved $(0\bar{2}2)$ Cu planes have a spacing of 0.128 nm.

The epitaxial orientation relationship leads to a mismatch of about 7% between the corresponding spacings of the adjacent Cu and α-Al_2O_3 lattices. In order to minimize the interfacial energy, either misfit dislocations can accommodate the lattice disregistry or the two lattices form a strain free incoherent interface. For the case of a semicoherent interface, two different hexagonal misfit dislocation networks are possible. One network consists of edge dislocations with $a_{Cu}/2$ <110>Cu Burgers vectors and <211>Cu line directions. Thus a network would be visible by imaging the Cu/Al_2O_3 interface edge-on in the $[211]_{Cu}$ direction. The resolved $(0\bar{2}2)_{Cu}$ planes in the lattice image (Fig. 1) do not reveal any strain field indicative of a dislocation network. The other possible misfit dislocation network consists of 60° dislocations with <110>$_{Cu}$ line vectors and would be observed edge-on in a lattice image, by tilting the sample 30° in $[10\bar{1}0]_{Al_2O_3}$ and $[110]_{Cu}$ zone axes. The main periodicities in this projection of the copper lattice correspond to the spacings of $(\bar{1}11)_{Cu}$, $(1\bar{1}1)_{Cu}$ and $(002)_{Cu}$ planes. Since the experimental image is devoid of strain fields caused by such misfit dislocations, we discard the possibility of the 60° misfit dislocation network[54], also. Additionally, no periodic correlation between the terminating copper and α-Al_2O_3 planes can be observed, which is consistent with an incoherent interface.

Atomic Structure of the Interface. The HRTEM investigations described above demonstrate that the interface is atomically flat. The (0001) plane of α-Al$_2$O$_3$, which consists of an alternating stacking sequence of oxygen and aluminum layers, could be terminated by either an oxygen or aluminum layer at the interface. Either all Cu-O bonds or all Cu-Al bonds are expected to exist across the Cu/Al$_2$O$_3$ interface. The type of bonding can be determined by EELS studies of the fine structure of the Al, O, and Cu ionization edges.[20] The ELNES of an ionization edge contains the information about local coordination and electronic structure. Thus, local changes of the oxidation state of Cu at the interface, due to bonding with the oxygen sublattice of α-Al$_2$O$_3$ or due to metallic bonding between Cu and the Al-sublattice, will modify the ELNES.

The ELNES studies were carried out on the VG HB501 dedicated STEM equipped with a parallel EEL spectrometer (Gatan 666 PEELS). The spectra were recorded by applying the spatial difference technique.[20] The interface was aligned edge-on and spectra recorded in the substrate, in the film and at the interface with the beam scanning an area of 10x12 nm^2. All spectra were corrected for dark current and read-out pattern of the parallel detector. The pre-edge background was extrapolated by a power law and subtracted from the raw data. The EEL spectrum of the interface was corrected following the method described in ref. 20 for both the bulk substrate and film contributions, which are present due to the width of the measured area. The remaining difference spectrum represents the ELNES of interfacial atoms possessing a different oxidation state and/or environment compared to the bulk.

The Al$_{L2,3}$ and Cu$_{M2,3}$ edge spectra recorded from the α-Al$_2$O$_3$ substrate, the Cu/Al$_2$O$_3$ interface and the Cu film lead to a difference spectrum which remains zero within the detection limit. This means that no interface specific component exists within this energy-loss region. Since the Al$_{L2,3}$ ELNES is known to be sensitive to subtle changes in local Al coordination[20] (e.g. tetrahedrally and octahedrally coordinated Al in oxides have been distinguished by ELNES studies[23]), this result indicates that Al does not change its local coordination and is therefore not involved in bonding at the interface.

In Fig. 2 the measured Cu$_{L2,3}$ spectra of the Cu-film (I$_{Cu}$) and the interface (I$_{IF}$) are presented. The difference spectrum of the interface (ΔI$_{IF}$) is shown in Fig. 2(c). The spectrum was obtained by ΔI$_{IF}$ = I$_{IF}$ - aI$_{Cu}$, with the scaling factor a = 0.4. Since no electron transitions occur in α-Al$_2$O$_3$ within that energy-loss region, contributions of bulk α-Al$_2$O$_3$ need not be considered. The difference spectrum of the interface reveals a Cu$_{L2,3}$ ELNES which is clearly different from bulk Cu. At an energy-loss of 933 eV, a L$_3$ white line is observed for interfacial Cu-atoms. White lines are typical for transition metals and their oxides and arise from electron transitions from 2p-states into empty 3d-states.[21] In metallic Cu° all 3d states are occupied and no white line exists in the ELNES (see Fig. 2a). This means Cu atoms at the interface must possess empty 3d-states. Taking into account that no interface specific component exists for the Al$_{L2,3}$ edge, the empty 3d-states in Cu must result from a charge transfer between copper and oxygen at the Cu/Al$_2$O$_3$ interface and/or a hybridisation of Cu-3d and O-2p states. Comparing the interfacial Cu$_{L2,3}$ ELNES (Fig. 2c) to reference spectra of Cu$_2$O (nominally Cu^{1+}) and CuO (nominally Cu^{2+}, Fig. 2d,e) indicates that the nominal oxidation state of Cu at the interface is Cu^{1+}, since no chemical shift, which is typical for Cu^{2+}, is detected.

Examination of the O$_K$ interface difference ELNES indicates that oxygen has changed its local environment at the interface. A direct comparison of the difference spectrum with the bulk α-Al$_2$O$_3$ shows a broadening of the main peak at 540 eV and a shift to slightly lower energy-losses. At approximately 532 eV electron transitions cause a shoulder in the difference spectrum which is not present in bulk α-Al$_2$O$_3$. At the same energy-loss a peak is observed in the O$_K$ ELNES of bulk Cu$_2$O due to transitions into a hybridized Cu-3d-O-2p state. Furthermore, the peaks above 550 eV change their shape compared to those observed in the O$_K$ ELNES of α-Al$_2$O$_3$.

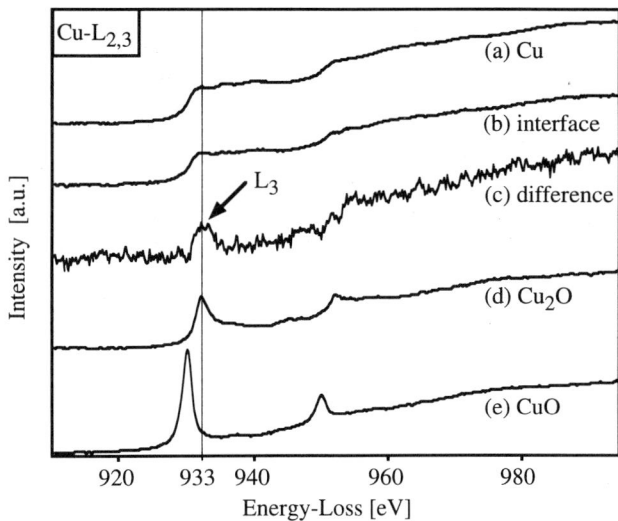

Figure 2. $Cu_{L2,3}$ ELNES acquired in a region of 10x12 nm² in (a) the Cu-film and (b) at the Cu/Al_2O_3 interface. The (c) difference spectrum, which contains the interface specific component shows a white line at 933 eV for interfacial Cu atoms due to Cu-O bonds at the interface. Reference spectra of (d) Cu_2O and (e) CuO indicate a nominal Cu^{1+} oxidation state for Cu atoms at the Cu/Al_2O_3 interface.

The interface specific components found in $Cu_{L2,3}$ and O_K edge imply the presence of ionic-covalent Cu-O bonds across the interface. Image simulations to match experimental HRTEM micrographs were then obtained under the assumption of Cu-O bonds at the interface, as established by the ELNES studies. A reasonable atomic model of the interfacial region must result in a realistic interatomic spacing across the interface. The best agreement between the experimental and simulated interface image (Fig. 3) was achieved for an average projected bonding distance of 0.2 nm ± 0.03 nm between the terminating oxygen layer and copper layer. The individual atomic columns fluctuate by the denoted error bar of ± 0.03 nm around the mean monolayer position. The projected bonding distance of 0.2 nm between the terminating oxygen layer and copper layer at the interface appears real considering the minimum interatomic spacing in Cu_2O to be 0.18 nm. The experimental HRTEM micrograph is presented in Fig. 3a and the simulated image is shown in Fig. 3b.

CONCLUSIONS

The different techniques of advanced transmission electron microscopy give access to a broad variety of complimentary information with respect to structure and composition of materials. The different TEM techniques enable us to perform the characterization of crystalline materials at atomic dimensions. For α-Al_2O_3, however, the access is still limited. The characterization comprises the structure, chemistry and bonding at the atomic level. In order to gain optimum information, constant advancements of the techniques are required.

The increasing resolution with which structural and compositional information can be obtained also means that the information is extracted from smaller and smaller volumes. Hence, a good correlation between macroscopic properties and the underlying structures on the atomic level can only be obtained for model systems, like planar bimaterial interfaces or layered structures with different components. An extrapolation to a general grain boundary and the properties of a general polycrystalline material is still impractical. As an example, we have discussed the results obtained on grain boundaries in alumina and on the Cu/Al_2O_3 interface. Information was not only obtained on the local arrangement of the atoms at the

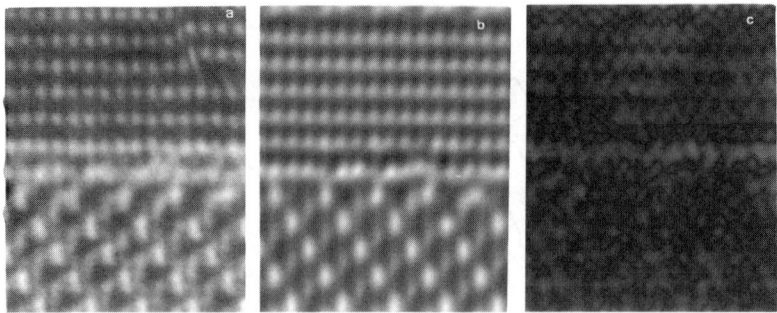

Figure 3. (a) Experimental and (b) simulated image of the Cu/Al$_2$O$_3$ interface. The position of the atomic columns are indicated in the simulation. (c) shows the difference image between (a) and (b).

interface, but also on the interfacial chemistry and the type of chemical bonding at the interface. An inherent problem arises as advances in TEM are made, that more and more information is gained on less and less material. For real materials with complex microstructures it may be difficult to infer predictions on macroscopic properties from only a few local areas with nm-size dimensions. Another unresolved question is how much the preparation of the thin TEM specimens may introduce artifacts or provide non-representative sections of the overall microstructure. Both limitations of the TEM technique must be kept in mind when carefully interpreting advanced TEM results.

REFERENCES

1. J. A. Pask, A. G. Evans (Eds.), *Ceramic Microstructures '86: Role of Interfaces,* Plenum Press, New York, London (1987).
2. M. Rühle and A. G. Evans, *Prog Mat. Sci.,* **33**:85 (1989).
3. A. G. Evans, *J. Am. Ceram. Soc.,* **73**:187 (1990).
4. I. Majid, P. D. Bristowe and R. W. Balluffi, *Phys. Rev. B,* **40**:2779 (1989).
5. M. W. Finnis and M. Ruhle, in: *Materials Science and Technology,* R. Cahn, P. Haasen and E. Kramer, eds., Vol. 1: *Structure of Solids,* V. Gerold, ed., VCH-Verlag, Weinheim, 533-605 (1992).
6. A. P. Sutton and R. W. Balluffi, *Interfaces in Crystalline Materials,* Clarendon Press, Oxford (1995).
7. T. Hoche, P. R. Kenway, H. J. Kleebe and M. Rühle, *J. Am. Ceram. Soc.,* **77**:339 (1994).
8. D. R. Clarke, *J. Am. Ceram. Soc.* 70: 15 (1987).
9. H. J. Kleebe, M. K. Cinibulk, R. M. Cannon and M. Rühle, *J. Am. Ceram. Soc.,* **76**:1969 (1993).
10. *High-Voltage and High-Resolution Electron-Microscopy,* M. Ruhle, F. Phillipp, A. Seeger, J. Heidenreich, eds., *Ultramicroscopy,* **56**:1-232 (1994).
11. *Proceedings of the Second International Workshop on Electron Energy Loss Spectroscopy and Imaging (EELSI),* O. L. Krivanek, ed., *Microsc. Microanal. Microstruct.,* **6**:1-157 (1995).
12. *French-Japanese Seminar on In Situ Electron Microscopy,* F. Louchet, H. Saka, eds., *Microsc. Microanal. Microstruct.,* **4**:101-346 (1993).
13. S. J. Pennycook and D. E. Jesson, *Ultramicroscopy,* **37**:14 (1991).
14. M. H. Loretto, *Electron Beam Analysis of Materials,* Chapman and Hall, London (1984).

15. P. Hirsch, A. Howie, R. Nicholson, D. W. Pashley and M. J. Whelan, *Electron Microscopy of Thin Crystals,* Butterworths, London (1965).
16. E. Hornbogen, B. Skrotzki, *Werstoffe Mikroskopie,* 2nd. ed, Springer Verlage, Berlin (1993).
17. R. F. Egerton, *Electron Energy Loss Spectroscopy in the Electron Microscope,* 2nd. ed., Plenum Press, New York (1996).
18. P. Rez, in: *Microbeam Analysis,* R. Gooley, ed., San Francisco Press, San Francisco (1983).
19. R. Brydson, J. Bruley, H. Mullejans, C. Scheu and M. Rühle, *Ultramicroscopy,* **59**:81 (1995).
20. C. Scheu, G. Dehm, H. Mullejans, R. Brydson and M. Rühle, *Microsc. Microanal. Microstruct.,* **6**:19 (1995).
21. R. D. Leapman, L. A. Gruncs, P. L. Fejes, *Phys. Rev. B,* **26**:614 (1982).
22. R. Brydson, H. Sauer and W. Engel, in: *Transmission Electron Energy Loss Spectrometry in Materials Science,* M. M. Disko, C. C. Ahn, B. Fultz, eds., TMS Monograph Series, Warrendale, USA, **131** 154(1992).
23. P. Buseck, J. Cowley, L. Eyring, *High Resolution Transmission Electron Microscopy and Associated Techniques,* Oxford University Press, Oxford (1992).
24. L. Reimer, *Transmission Electron Microscopy: Physics of Image Formation and Microanalysis,* Springer, Berlin (1989).
25. J. C. H. Spence, *Experimental High-Resolution Electron Microscopy,* 2nd ed., Oxford University Press, New York, Oxford (1988).
26. S. Horiuchi, *Fundamentals of High-Resolution Transmission Electron Microscopy,* Elsevier Science B.V., Amsterdam (1994).
27. D. Hofmann and F. Ernst, *Ultramicroscopy,* **53**:205 (1994).
28. D. Hoffmann and F. Ernst, *Interface Science,* **2**:210 (1994).
29. F. Phillipp, R. Hoeschen, G. Möbus, M. Osaki and M. Rühle, *Ultramicroscopy,* **56**:1 (1994).
30. K. Nadarzinski and F. Ernst, in: *Intergranular and Interphase Boundaries in Materials,* A. C. Ferro, E. P. Conde, E. A. Fortes, eds., Trans Tech Publications, Zurich, 309-312 (1996).
31. K. Nadarzinski and F. Ernst, *PhiL Mag A,* (1996) in press.
32. J. D. Cawley, W. E. Lee, *Oxide Ceramics, in: Materials Science and Technology,* R.W. Cahn, P. Haasen, E. J. Kramer, eds., *Vol. 11: Structure and Properties of Ceramics,* M.V. Swarz, ed., VCH, Weinheim (1994).
33. I. Manassidis, A. DeVita and M. J. Gillan, *Surface Science Letter* **285**:L517 (1993).
34. I. Manassidis, M. J. Gillan, *J. Am. Ceram. Soc.,* **77**:335 (1994).
35. M. Wilson, M. Exner, Y.-M.Huang and M.W. Finnis, unpublished research.
36. T. Gemming, G. Möbus and M. Rühle, unpublished research.
37. T. Hoche, P. R. Kenway, H. J. Kleebe, M. W. Finnis and M. Rühle, *J. Phys. Chem. Solids,* **55**:1067 (1994).
38. H. Grimmer, R. Bonnet, S. Lartigue and L. Priester, *Phil. Mag.,* **61**:493 (1990).
39. D. Bouchet, F. Dupar and S. Lartigue-Korinek, *Microsc. Microanal. Microstr.,* **4**:561 (1993).
40. W. Swiatuicki, S. Lartigue-Korinek, A. Dubon and J. Y. Laval, *Mater. Sci. Forum,* 126-128:193 (1993).
41. M. Rühle, *Fresenius Z. Anal Chem.,* **349**:49 (1994).
42. M. Rühle, G. Necker and W. Mader, *Ceramic Transactions,* **5**:340 (1989).
43. H. F. Fischmeister, G. Elssner, B. Gibbesch, K.-H. Kadow D. Korn. F. Kawa. W. Mader and M. Turwitt, *Rev. Sci Instrument,* **64**:234 (1992).
44. P. R. Kenway. *Phil. Mag.,* **B68**:171 (1993).

45. Y. M. Huang, M. W. Finnis, G. Möbus and M. Rühle, unpublished research.
46. W. H. Gibson (Ed.), *Alumina as a Ceramic Material,* The American Ceramic Soc., Inc., Columbus/Ohio (1970).
47. E. Dorre, H. Hubner, *Alumina,* Springer-Verlag, Berlin (1984).
48. W. E. Lee, W. M. Rainforth, *Ceramic Microstructures,* Chapman and Hall, London (1994).
49. J. Fang, A. M. Thompson, M. P. Harmer and H. M. Chan, *J. Am. Ceram. Soc.,* **79** (1996), in print.
50. S. J. Bennison and M. P. Harmer, *Ceramic Transactions,* **7**:13 (1990).
51. J. Bruley, T. Hoche, H.-J. Kleebe and M. Rühle, *J. Am. Ceram. Soc.,* **77**:2273 (1994).
52. T. Hoche and M. Ruhle, *J. Am. Ceram. Soc.,* **79**:1961 (1996).
53. E. Schumann, J. C. Yang, M. Rühle and M. J. Graham, *MRS Symp. Proc.,* **364**:1291 (1995).
54. G. Dehm M. Rühle, G. Dinz and R. Rai, *Phil. Mag.,* **B71**:1111 (1995).

MAKING THE CONNECTION BETWEEN ATOMISTIC MODELLING OF INTERFACES AND REAL MATERIALS

J. H. Harding, A. H. Harker, A. L. Shluger and A. M. Stoneham
Materials Research Centre, Dept. Physics and Astronomy,
University College London,
London, WC1E 6BT
UK

INTRODUCTION

Why is anyone interested in control at the atomic level? One obvious answer is that control at the atomic level enables control at larger scales. Some examples of this are clear enough; nano-engineering is precisely about this. However, in many cases the question of how control at the atomic level results in control at other length scales is far from clear. This is particularly the case in complex materials where many mechanisms are active at once. Putting reactive elements into an alloy to control the growth of the resulting scale is an attempt to control atomistic processes, but the details of how it works are often obscure. Yet control implies that we can identify what the important mechanisms are and then find a procedure that will enhance (or inhibit) the mechanism we want to affect without causing problems elsewhere. The traditional way of controlling complex systems is trial and error. The equally traditional difficulty is the large error to trial ratio. In the hope of reducing this a bit, people have turned to modelling complex processes. However, unless we are in the (unusual) position of being able to model everything the problem of deciding what the important mechanisms are remains. The strategy of *mesoscopic modelling* is to attempt to identify the individual mechanisms and model these at the appropriate length scale (which will often be atomistic but need not be) and then to build a model of how these mechanisms interact to produce the process. If the result of a process is a material, a final goal will often be to predict the properties of the material as a function of the conditions under which it was made. 'Properties' may be of two main kinds. First, we may require the effective bulk properties (elastic constants, thermal conductivity). These depend on the averaged behaviour of the microstructure but are unlikely to be dependent on individual features of the microstructure. Other properties such as the probability of fracture may well depend on such features.

Clearly a major problem of such modelling is validation. This needs to be done at least two levels. First, is our modelling of the individual mechanisms correct (and have we chosen the right ones)? Second, is our modelling of the interaction of those mechanisms correct? This may require validation both of the microstructures themselves and the properties predicted. In the rest of this paper we consider a number of examples of the links between various scales of modelling. In the three examples that follow we shall concentrate on the problems that arise when one attempts to deal with problems at different levels of description; from atomistic to mesoscopic or from mesoscopic to macroscopic. (These issues have also been reviewed by King et al[1] but they only discuss metals and to a lesser extent semiconductors. Ceramics are not metals; in particular, there are effects from long-range coulombic forces that can never be neglected.)

ATOMIC FORCE MICROSCOPY; FROM ATOMISTIC TO MESO-SCALE

The atomic force microscope[2] is a tool for investigating the structure and behaviour of surfaces by probing the microscopic forces present. The instrument probes the surface with a microscopic tip. The displacements of this tip are transmitted through a rigid cantilever and measured at the top of the instrument. They are often measured by optical means. Since the atomic force microscope does not require the surface to be conducting, it was hoped that it would repeat the success of scanning tunnelling microscopy in obtaining atomic-scale resolution in those cases where that technique could not be used. Images have been produced that show a regular surface lattice, even periodicities characteristic of the crystal structure. However, it has become clear that the physics of the tip-sample contact is a complex matter involving the interaction of large numbers of atoms. Moreover, the long-range (Van der Waals) forces produce an instability in the system as the tip approaches the surface; the 'jump to contact'. Scanning the tip over the surface in the subsequent 'hard contact' regime disrupts both tip and surface. The interaction is akin to those involved in friction or wear. This hard contact can be prevented by inserting liquid between the tip and sample.[3] This reduces the effect of the long-range forces and enables images to be obtained with the tip a few Ångstroms above the surface. This, together with the use of ultra-sharp tips and sensitive cantilevers, permits forces as small as a few pico-Newtons to be measured. However, the nature of the forces between tip and surface is an even more complex matter than before and the debate continues as to whether the periodicities observed in the images are to be interpreted as atomic-scale features on the surface. One thing is clear; the original model of the tip as a hard-sphere atom rolling across a surface of hard-sphere atoms is far too simple.

Shluger and co-workers[4-7] have performed a series of calculations on the tip surface interaction for ionic systems. In the most recent calculations[7] they model the force microscope as a sphere connected to a support by a spring. The nano-asperity at the end of the tip is a cube of thirty-two cells of MgO. This represents a hard oxide tip. Two cases are considered; a neutral tip where the cube has the [111] axis perpendicular to the surface with an oxide ion at the base, and a hydrated tip where the basal oxide is replaced by a hydroxide ion and another hydroxide ion is added to the opposite (Mg^{2+}) corner to maintain charge neutrality. Previous work[6] had investigated the effects of a protonated silica tip; the choice of MgO simplifies the calculations without altering the qualitative behaviour. The calculation considers the effect of the mesoscopic tip using the standard Hamaker constants and superimposes the chemical interactions of the nano-asperity with the surface by explicit calculation using a pair-potential model of the interactions. The effect of liquid between the tip and the surface is considered by taking a range of Hamaker constants and also through the possibility of protonation of the tip. The calculations show that, for distances of 5Å or more, the interaction is dominated by long-range forces; Van der Waals and static multipole interactions. At closer distances, the short-range 'chemical' interactions become important and different surface ions become distinguishable. In the absence of a liquid to moderate the effect of the long-range forces the surface ions become unstable; they break away and begin to coat the tip. At about 1.5Å the tip encounters a repulsive wall that prevents it digging further into the substrate. This behaviour is summarised in Figure 1. If the tip is pulled away from the substrate, the spring must do work against the adhesion between the tip and the surface. 'Jump off' occurs at about 2Å. The ions of the substrate close to the probe form a neck that breaks, the upper part coating the probe, the lower part returning to the substrate. This is shown in the force-distance curve by a series of excursions in the force on the tip. Not surprisingly, this behaviour results in hysteresis.

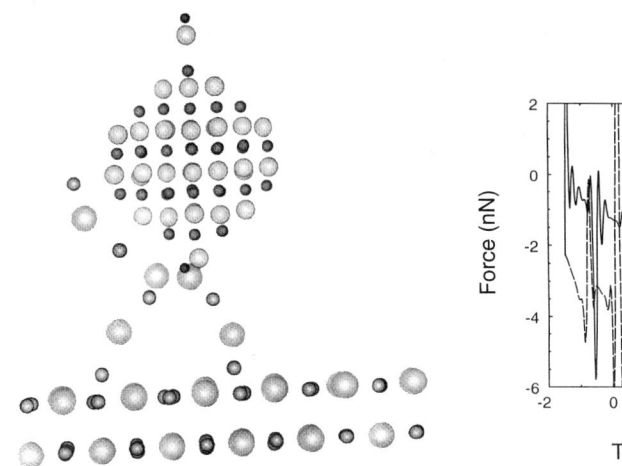
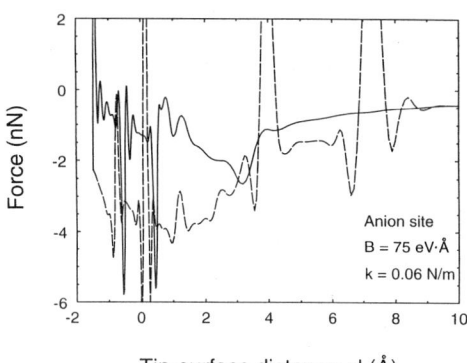

(a) Atomic force microscope tip pulling away from the surface; the neck is breaking away

(b) Force distance curve. Full line is the tip approach; broken line is the tip pulling away.

Figure 1. The interaction of an atomic force microscope tip with the surface.

The scanning of the tip at constant force is modelled by moving the probe across the surface in 0.3-0.4Å steps parallel to the ideal surface plane. At each step the force-distance curve is calculated and a scan-line (the trajectory of the tip) obtained. If the system is arranged so that the height of the scan-line is 3-4Å above the surface, a periodic image is obtained. The surface distorts elastically but is not violently disrupted. However, although a periodic image is obtained; there is no simple rule such as the tip always being displaced upwards by the larger ion. What happens depends on the detailed chemistry between the tip and the surface close to it. The atomic force microscope can produce atomic-scale resolution but it does so as a mesoscopic object, combining a variety of atomistic processes to produce a force whose meaning can only be unravelled by careful simulation. Although it produces a periodicity on the scale of the unit cell, it does so because a mesoscopic process repeats on an atomic length-scale. If the tip-surface distance is less, say 1.5Å, the tip is still deflected but the periodicity may disappear. The interaction is much stronger and ions move from the tip to the surface and back again continually altering the chemical nature of the tip. If the mesoscopic process producing the periodicity is not reversible, the periodicity will vanish.

The calculations support the idea that, in the right conditions, atomic force microscopy can give atomic-scale resolution. In principle surface defects should be observed. However there are two caveats. First, the lifetime of the defect must be long since the mean time for a scan-line is about 10^{-3} sec. Second, the defect may be affected by the presence of the probe and so the image may only give an average of the possible states produced by the defect-tip interaction. Recent work[8] monitoring the lateral forces on the tip has shown how the instability of the ions near the tip can give rise to stick-slip behaviour. One asperity moving across a surface is a microscope; enough of them moving together is friction.

HETEROEPITAXY AND STRESS; FROM MACRO TO ATOMISTIC

When a layer of a different material is grown on a substrate there will be stresses. Provided that the new layer is thin enough and the bonding of layer and substrate is similar, a pseudomorphic layer may grow. The structure and lattice parameters are the same and any lattice mismatch appears as stress. Beyond a critical thickness, a relaxed epilayer with misfit dislocations is favoured. Attempts have been made to predict values for this critical thickness (see [9] for a review). The general problem of the calculation of stress and strain in epilayers has recently been reviewed in detail.[10,11]

Consider a stripe of material of width $2l$ and height h on a substrate of width l_s and height h_s. There are three main mechanisms of stress relaxation (shown in Figure 2).

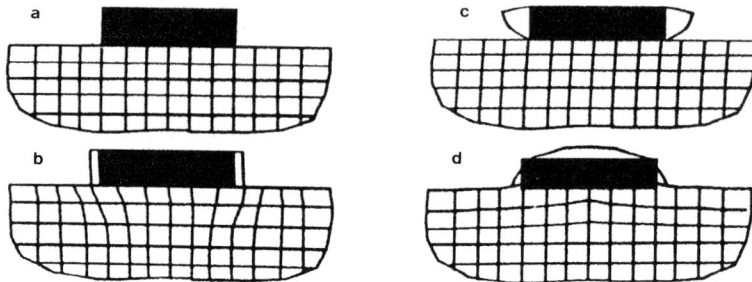

Figure 2. Stress relaxation modes in epilayers: (a) unrelaxed stripe; (b) outward movement of stripe; (c) bowing outwards; (d) convexity (after [10])

- Outward movement of the stripe, dragging the lattice planes under the stripe along. This relaxes the stress in the stripe at the price of creating it in the substrate.
- Matching at the interface but the outer layers of the stripe bow outwards. This avoids creating stress in the substrate; all the stress is confined to the stripe
- The lattice planes in the stripe and the substrate bend down at the edges and bow up in the centre. This produces a convex stripe.

In general, all three mechanisms contribute but their relative importance depends on the ratio l/h. This problem has been considered in detail by Harker and co-workers.[12-14] Provided that we are far from the edges of the stripe, for $l/h>10$ the first mechanism dominates. For $l/h<7$ the second mechanism becomes important and for $l/h<3$ even the third mechanism can no longer be ignored. For $l/h<1$ the third mechanism dominates and the stress in the top surface is everywhere tensile due to the curvature of the lattice planes. Near the edges, matters are more complex. As Figure 3 shows, the first mechanism is never correct. Also, the width of the substrate is important; stress relaxation is greater if $l=l_s$. Comparison of the calculated stress relaxation with Raman shift measurements confirm the full finite element calculation but demonstrate that various analytic approximations that have been used are unreliable.[10,11]

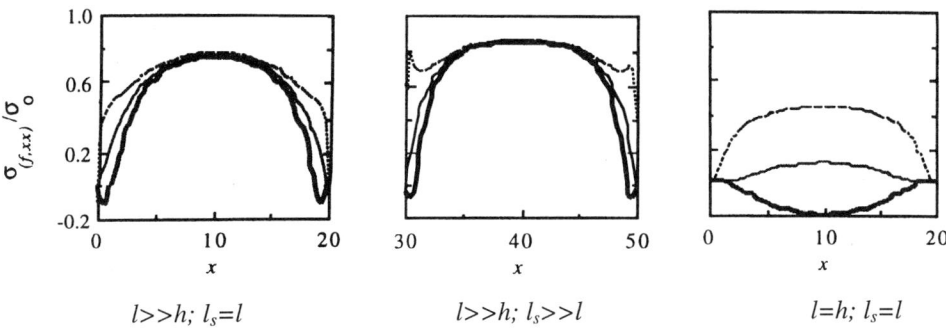

Figure 3. Stresses in epilayers; the effects of the edge. Thick line gives stresses at the surface of the stripe, thin line gives stresses at the midpoint, dotted line gives stresses at the stripe/substrate interface (after [10,12]).

All this, however, says nothing of the structure of the interface. For this we must turn to atomistic simulation. Most simulations of grain boundaries have assumed a pair potential model for the interactions between the ions (see [15] for a review of the method and results), although a few density functional calculations have recently been performed (see [16] for an example). So far, these confirm the results of the pair potential simulations. We consider two examples; the BaO(100)/MgO(100) and the $CeO_2(111)/\alpha$-$Al_2O_3(0001)$ interfaces.[17-18]

The simulations we will discuss have been performed with the MIDAS[19] code. This divides the problem into an inner block where the ions are relaxed explicitly and outer blocks which are only allowed to move rigidly. The purpose of the outer blocks is to ensure that the electrostatics of the inner block is correctly accounted for. Atomistic models of hetero-interfaces have usually generated the interfacial structure using near-coincidence site lattice theory. This is a special case of the coincidence site lattice theory;[20] we follow the approach of Mykura *et al*[21]. Exact coincidence would be obtained if $a_1/a_2=(m^2+n^2)/(k^2+l^2)$ where a_1, a_2 are the lattice parameters of the materials and m,n,k,l are integers. The rotational angle needed to bring these lattices into coincidence is then $\theta=\tan^{-1}(n/m)+\tan^{-1}(l/k)$. The planar reciprocal density of the two surfaces that form the interface is $\Sigma_1^p = m^2+n^2$ and $\Sigma_2^p = k^2+l^2$. Since the ratio of lattice parameters is most unlikely to be a simple fraction this condition is not met in practice. There must therefore be a misfit, F, defined as

$$F = 2\left|a_1\sqrt{\Sigma_1^p} - a_2\sqrt{\Sigma_2^p}\right| / \left|a_1\sqrt{\Sigma_1^p} + a_2\sqrt{\Sigma_2^p}\right| \qquad (1)$$

It is always possible to decrease F by increasing Σ^p, but this will decrease the density of interplanar sites which may itself decrease interplanar stability.[22] (Note that there are further complexities in the formulation for non-cubic systems [23]). Calculations have been performed for a variety of layers from one to six for the two interfaces. The results are shown in Table 1.

The near-coincidence site lattice is a poor predictor for very thin layers. This is because, when the BaO (or CeO_2) film is only a few atomic layers thick, the system can accommodate the misfit by rumpling the film. As the film gets thicker, it approximates to the bulk lattice and so the interface is constrained. The interface has one further way of accommodating the mismatch without resorting to misfit dislocations and that is to incorporate point defects. This phenomenon has been seen in grain boundaries.[24,25] A similar effect is seen in the $CeO_2(111)/Al_2O_3(0001)$ interface.[25,26] In this case, it is favourable for oxide vacancies to migrate from the surface to the interfacial plane.

Table 1: Stability of hetero-epitaxial interfaces (Note that misfit is defined with respect to a planar block containing the same number of layers but with the lattice parameter minimised); after [17].

Layer	Lowest misfit (misfit%)	Lowest energy (misfit%)	Binding energy (J/m^2)
BaO(100) layers on MgO(100); the ratios are Σ^p (BaO)/Σ^p (MgO)			
1	17/26 (0.9)	13/17 (-6.9)	0.132
2	8/13 (-0.2)	5/8 (-0.9)	0.189
3	10/17 (1.0)	5/8 (-2.0)	0.116
6	10/17 (-0.03)	10/17 (-0.03)	0.117
Stability of CeO$_2$(111) on Al$_2$O$_3$ (0001): the ratios are Σ^p (CeO$_2$)/Σ^p (Al$_2$O$_3$).			
1	7/4 (-0.06)	21/13 (3.9)	1.453
2	61/37 (0.2)	21/13 (1.2)	1.419
3	31/19 (-0.04)	21/13 (0.5)	1.244
6	21/13 (-0.3)	21/13 (-0.3)	1.560

Most simulations of the interface cannot consider longer-range stress relaxation in the interfacial plane since they assume constant interfacial area. We[8] have evaded this problem by 'turning the calculation on its side'. The results show the same behaviour as the third mechanism in the finite element calculations discussed above. The next stage is to incorporate the effect of dislocations at the interface. Some trial calculations have already been performed.[27] Up to now, most atomistic simulation has been directed to the detailed structure of the interface. However, this confirms that the simple, pseudomorphic picture that the finite element calculations assume is a special case. In many ceramics, when the films are thin, the misfit that the system chooses depends on details of the construction of the interface that are difficult to predict without detailed calculation.

The problems of stress are also of interest in growing films. A common growth mode is Stranski-Krasnov growth (for a general review of growth modes see [28]). Here the film begins with layer growth but the strain in the growing layer increases the effective interfacial energy sufficiently for the system to convert to island growth. The effect of stress on the growth of layers can be seen in a series of calculations for the growth of BaO on MgO.[18] If BaO units are added one by one to a surface, they prefer to remain as single units rather than trimers or tetramers. At higher loadings the BaO forms clusters and eventually three-dimensional islands. If the system is forced to form a monolayer, this has a hexagonal structure rather than the simple cubic structure. This general sequence is supported by results from molecular beam epitaxy.[29] Similar distortions of island structure from perfect lattice structure are seen for NaCl on MgO.[30] However, the effects of the hetero-epitaxial stress stretch beyond this. There is experimental data (and considerable interest in) the self-ordering of islands due to strain.[31-33] This opens up the possibility of a new kind of control.

FILMS, STRUCTURES AND PROPERTIES; FROM MESO TO MACRO

There is a wide variety of methods for growing films on substrates. These produce characteristic microstructures and hence properties of the film. If one can understand (and control) the way that the microstructure is produced, one can control the properties of the resulting film.

Consider a film made up of material, pores and cracks. A common method for evaluating the bulk properties of such a solid is effective medium theory. The essential idea is to calculate the average response <r> to a stimulus <f> using the idea of a generalised susceptibility, S. In general, <r> and <f> are vectors and so S is a tensor. Effective medium theory calculates these tensors for inhomogeneous materials. A general review of this type of approach is given by Batchelor,[34] more recent developments are discussed by Ferrari.[35] An adaptation of these methods to porous coatings has been developed by Harding and Harker.[36] The effective thermal conductivities and elastic properties of materials containing ellipsoidal inclusions may be expressed in terms of tensors which are generalisations of the depolarisation factors used in the Clausius-Mosotti theory of Bragg and Pippard.[37] For a porous material, the effective thermal conductivity is given by

$$<\kappa> = \kappa\{1 - \phi[[(1-\phi)T + \phi] - 1]^{-1}\}^{-1} \qquad (2)$$

where T is Eshelby's tensor, ϕ is the volume fraction of the material removed and κ is the thermal conductivity of the pure material. The same form of expression holds for the elastic properties. We now need to add the cracks. If ϕ_p is the volume fraction of pores and ϕ_c is the volume fraction of cracks, a simple intuitive method is to use the proposed scheme first to calculate the properties of a material of which $(1 - \phi_p - \phi_c)$ is solid and ϕ_p is void. This medium is then taken as the solid phase for a second calculation where the cracks are introduced. The cracks differ from the voids only in their Eshelby tensor. Thus the whole problem is contained in the matrices that described the properties and distribution of the pores and cracks. If we are to proceed any further we need a description of the distributions of their sizes and shapes. This can only come from exhaustive (and exhausting) experimental work or a model.

We shall briefly consider the example of plasma-sprayed coatings. These are produced by projecting micron-sized particles of ceramic onto a substrate using a plasma torch. The particles, accelerate, melt and splash and the resulting splats build up the coating. Thus the natural units are not atoms, but the individual particles. The behaviour of the individual particles can be calculated[38] provided that the behaviour of the plasma is known. This in turn can be measured, or calculated using fluid dynamics (see for example, [39] for a discussion of the properties of torches, plasmas and the interaction between the particles and the plasma). The coating is then built from the individual splats. Initially, the material flows out from the point of impact, filling cracks in the surface. However, the material rapidly cools, the splats form and curl up. Cracking may occur at the intersplat boundaries or within individual splats. The core of the particle may not melt at all, and becomes embedded in the growing coating. All this is modelled by a set of physically-based rules. These are discussed in detail in reference [40]. The resulting simulations of porous, cracked coatings are then compared with experiment. Comparison is made both with experimental microstructures and with measured experimental properties. The result of a simulation using a standard set of parameters is shown in Figure 4 together with a calculation of the effective Young's modulus as a fraction of the bulk modulus (for a typical thermal barrier coating the bulk modulus is about 250GPa). The strong difference in behaviour between the Young's modulus for a force applied parallel to the interface between film and substrate (Y(p)) and one applied normal to this interface (Y(n)) shows the effect of the structure of the film; the splats that make it up are largely parallel to the interface. These large reductions are typical of what is observed in experimental coatings. The diagrams show that the porosity of the coatings varies with the configuration of the apparatus and hence the properties vary also.

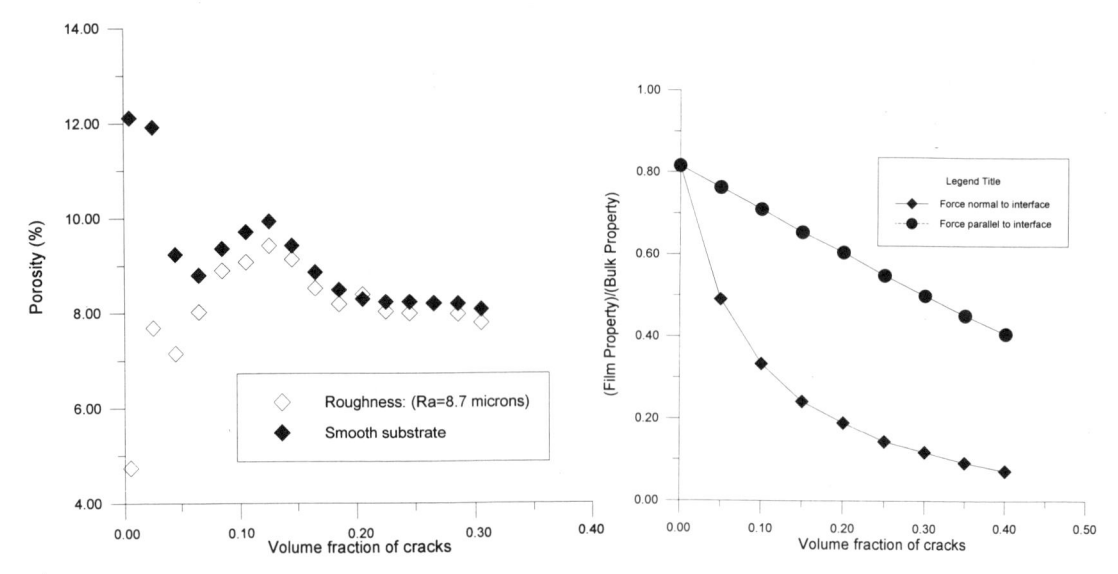

Figure 4. Porosity curves and effective properties in plasma-sprayed coatings. Y(n) is the Young's modulus for a force applied normal to the interface; Y(p) the Young's modulus for a force parallel to the interface and Y(b) is the bulk modulus.

In this case control is to be exercised through altering the parameters of the process; the type of gun, the distance from the gun to the substrate, the size of the particles. Here the purpose of simulation is to guide the experimental process by suggesting how the microstructure, and hence the properties, are related to the experimental variables that can be controlled.

CONCLUSIONS

In this paper we have discussed a number of examples of problems looked at from two different angles. There is the 'bottom up' approach where one attempts to study the system from the point of view of the individual constituents. These are often atoms or ions but need not be so. Here the hope of control is based on manipulating these constituents to produce the desired effect. However, the desired effect is often not best described in terms of atomistic variables. People wish to control meso-scale effects such as microstructures either because they are interested in phenomena on those scale or as a step towards controlling macro-scale effects such as elastic properties, stresses, corrosion. An example where understanding (and control) is sought at a number of length scales is the growth of oxide scales.[41] There is therefore a boundary between the atomistic and the mesoscopic behaviour (or equivalently between the mesoscopic and macroscopic behaviour). Such a boundary is, in a sense, arbitrary; it is a function of how we choose to describe things. However, the problems of obtaining a description that is both correct and useful are very real, and need to be solved if effective control is to be possible.

ACKNOWLEDGEMENTS

The authors would like to thank EPSRC for a grant under which the work on atomic force microscopy was performed; AEA Technology and EPSRC for support for the work on oxide/oxide interfaces; the Commission of the European Communities for funding (under the BRITE/EURAM programme) and the AEA Technology Corporate Research Programme for the work on plasma sprayed coatings. The Royal Society is also thanked for the provision of a travel grant.

REFERENCES

1. W. E. King, G. Campbell, T. Gonis, G. Henshall, D. Lesuer, E. Zywicz, and S. Foiles, Theory, simulation, and modeling of interfaces in materials-bridging the length-scale gap: a workshop report, *Mater. Sci. Eng.,* A**191**:1 (1995).
2. G. Binnig, C. F. Quate and C. Gerber, Atomic force microscope *Phys. Rev. Lett.,* **56**:930 (1986).
3. F. Ohnesorge and G. Binnig, True atomic resolution by atomic force microscopy through atomic and repulsive forces, *Science,* **260**:1451 (1993).
4. A. Shluger, C. Pisani, C. Roetti and R. Orlando, Ab initio simulation of the interaction between ionic crystal surfaces and the atomic force microscope tip, *J. Vac. Sci.,* A**8**:3967 (1990).
5. A. L. Shluger, A. L. Rohl, D. H. Gay, and R. T. Williams, Atomistic theory of the interaction between AFM tips and ionic surfaces, *J. Phys. Cond. Mater.,* **6**:1825 (1994).
6. A. L. Shluger, R. M. Wilson and R. T. Williams, Theoretical and experimental investigation of force imaging at the atomic scale on alkali halide crystals, *Phys. Rev.,* B**49**:4915 (1994).
7. A. L. Shluger, A. L. Rohl, R. T. Williams and R. M. Wilson, Model of a scanning force microscope on ionic surfaces, *Phys. Rev.,* B**52**:11398 (1995).
8. A. L. Shluger, R. T. Williams and A. L. Rohl, Lateral and friction forces originating during force force microscopy scanning of ionic surfaces; *Surf Sci.,* **343**:273 (1996).
9. S. C. Jain, T. J. Gosling, J. R. Willis, R. Bullough, and P. Balk, A new study of critical layer thickness, stability and strain relaxation in pseudomorphic Ge_xSi_{1-x} strained epilayers, *Phil Mag.,* A**65**:1151 (1992).
10. S. C. Jain, M. Willander and H. Maes, Stresses and strains in epilayers, stripes and quantum structures of III-V compound semiconductors, *Semicond. Sci. Tech.,* **11**:461 (1996).
11. S. C. Jain, A. H. Harker and R. A. Cowley, Misfit strains and misfit dislocations in lattice-mismatched epitaxial layers and other systems, *Adv. in Phys.* in press.
12. A. H. Harker, K. Pinardi, S. C. Jain, A. Atkinson and R. Bullough, Two-dimensional finite-element calculation of stress and strain in a stripe epilayer and substrate, *Phil, Mag.,* A**71**:871 (1995).
13. S. C. Jain, A. H. Harker, A. Atkinson and K. Pinardi, Edge induced stress and strain in stripe films and substrates: A 2D finite element calculation, *J. Appl. Phys.,* **78**:1630 (1995).
14. S. C. Jain, B. Dietrich, H. Richter, A. Atkinson and A. H. Harker, Stresses in strained GeSi stripes: Calculation and determination from Raman measurements, *Phys. Rev.* B**52**:6247 (1995).
15. J. H. Harding, Computer simulation of defects in ionic solids; *Rep. Prog. Phys.,* **53**:1403 (1990).
16. I. Dawson, P. D. Bristowe, M-H Lee, M. C. Payne, M. D. Segall and J. A. White. A first principles study of a tilt grain boundary in rutile, *J. Phys. Cond. Mater.* in press.
17. T. X. T. Sayle, C. R. A. Catlow, D. C. Sayle, S. C. Parker and J. H. Harding, Computer simulation of thin film heteroepitaxial ceramic interfaces using a near-coincidence-site lattice theory, *Phil. Mag.,* A**68**:565 (1993).
18. D.C. Sayle, S. C. Parker and J. H. Harding, Accommodation of the misfit strain energy in the BaO(100)/MgO(100) heteroepitaxial ceramic interface using computer simulation techniques, *J. Mater. Sci.,* **4**:1883 (1994).
19. P. W. Tasker, The surface energy, surface tensions and surface structure of the alkali halide crystals, *Phil. Mag.,* A**39**:119 (1979).

20. H. Grimmer, W. Bollman, and D. H. Warrington, Coincidence site lattices and complete pattern shift lattices in cubic crystals, *Acta Cryst.* 30:197 (1974).
21. H. Mykura, P. S. Bansal and M. H. Lewis, Coincidence-site-lattice relations for MgO-CdO interfaces, *Phil. Mag.*, A**42**:225 (1980).
22. D. G. Brandon, B. Ralph, S. Ranganathan and M. S. Wald, A field ion microscope study of atomic configuration at grain boundaries, *Acta Metall.*, **12**:813 (1964).
23. A. H. King and A. Singh, Generalising the coincidence site lattice model to non-cubic systems, *J. Phys. Chem. Solids,* **55**:1023 (1994).
24. D. M. Duffy, Grain boundaries in ionic crystals, *J. Phys.*, C**19**:4393 (1986).
25. D. C. Sayle, T. X. T. Sayle, S. C. Parker, C. R. A. Catlow and J. H. Harding, The effect of defects on the stability of heteroepitaxial ceramic interfaces as studied by computer simulation, *Phys. Rev.*, B**50**:14498 (1994).
26. D. C. Sayle, T. X. T. Sayle, S. C. Parker, J. H. Harding and C. R. A. Catlow, The stability of defects in the ceramic interfaces MgO/MgO and CeO_2/Al_2O_3, *Surf. Sci.,* **334**:170 (1994).
27. G. W. Watson, E. T. Kelsey, N. H. de Leeuw, D. J. Harris and S. C. Parker, Atomistic simulation of dislocations, surfaces and interfaces in MgO, *J. Phys. Cond. Mater.* in press.
28. J. A. Venables, Atomic processes in crystal growth, *Surf: Sci.*, **299/300**:798 (1994).
29. R. A. McKee, F. J. Walker, E. D. Specht, G. E. Jellison, L. A. Boatner and J. H. Harding, Interface stability and the growth of optical quality perovskites on MgO, *Phys. Rev. Lett.,* **72**:2741 (1994).
30. A L. Shluger, A. L. Rohl and D. H. Gay, Properties of small clusters at ionic surfaces: (NaCl)n clusters (n=1-48) at the (100)MgO surface, *Phys. Rev.*, B**51**:13631 (1995).
31. J. Tersoff, A. W. D. van der Gon.and R. M. Tromp, Critical ion size for layer-by-layer growth *Phys. Rev. Lett.,* **72**:266 (1994).
32. J. Tersoff, C. Tiechert and M. G. Lagally, Self organisation in quamum dot superlattices, *Phys. Rev. Lett.,* **76**:1675 (1996).
33. Q. Xie, A. Madhukar, P.Chen and N. P. Kobayashi, Vertically self-organised InAs quantum box islands on GaAs (100), *Phys. Rev. Lett.,* **47**:1459 (1981).
34. K. Batchelor, Transport properties of two-phase materials with random structure, *Ann. Rev. Fluid Mech.,* **6**:227 (1974)
35. M. Ferrari, Composite homogenisation via the equivalent poly-inclusion approach, *Comp. Eng.,* **4**:37 (1994).
36. J. H. Harding and A. H. Harker, Effective medium theories for cracked and porous media, unpublished work.
37. W. L. Bragg, and A. B. Pippard, The form birefringence of macromolecules, *Acta Cryst.,* **6**:865. (1953).
38. M. Bertagnolli, M. Marchese, G. Jacucci, I. St Doltsinis and S.Nolting, Finite element thermo-mechanical simulation of droplets impacting on a rigid substrate, *Materials and Design Technology* (ed T.J Kozik) **62**:199, ASME (1994).
39. P. Fauchais A. Grimaud, A. Vardelle, and M. Vardelle, La projection par plasma: une revue; *Ann. Phys. Fr.,* **14**:261 (1989)
40. S. Cirolini, M. Marchese, G. Jacucci, J. H. Harding, and P. A. Mulheran, Modelling the microstructure of thermal barrier coatings, *Materials and Design Technology* (ed T.J Kozik) **62**:189 ASME (1994).
41. M. Schutze, An approach to a global model of the mechanical behaviour of oxide scales, *Mater. at High Temp.,* **12**:237 (1994).

LOCAL CHEMICAL BONDING AT GRAIN BOUNDARY OF Si$_3$N$_4$ CERAMICS

Isao TANAKA[1], Hirohiko ADACHI[1]
Tetsuo NAKAYASU[2] and Tetsuo YAMADA[2]

[1]Department of Materials Science and Engineering,
Kyoto University, Sakyo, Kyoto 606-01 Japan
[2]Ube Research Laboratory, UBE Industries, Ltd.,
Ube, Yamaguchi 755, Japan

ABSTRACT

We perform a first principles molecular orbital calculation with special interest on the local chemical bonding at the grain boundary of Si$_3$N$_4$. The computational result is twofold: 1) On the basis of our calculations using model clusters, the Si-L$_{23}$ edge ELNES of the intergranular glassy film reported in literature is interpreted; The N/(O+N) ratio of the intergranular glassy film is found to be in the range of 30 to 40%. A model of atomic arrangement at the interface between the prismatic plane of β-Si$_3$N$_4$ and the glassy film is proposed which has no broken bond at the interface. 2) Magnitude of chemical bond-strength around rare-earth ions at an interstitial sites in bulk Si$_3$N$_4$ is examined by the calculation of bond overlap populations. The difference in solubility of rare-earth ions between α- and β-Si$_3$N$_4$ matrix is well explained. Finally the bond-strength around rare-earth ions at the interface between β-Si$_3$N$_4$ and intergranular glassy film is examined. The rare-earth ions are implied to be more stable at this interface than at the interstitial site of the bulk crystal. The grain-boundary bond-strength is suggested to be weakened when large rare-earth ions such as La^{3+} are present at this interface. This computational result explains why the interfacial bond strength varies with ionic radius of rare-earth ions in doped β-Si$_3$N$_4$ ceramics.

INTRODUCTION

A glassy film of 1 nm in thickness is usually present between two grains in silicon nitride ceramics. Since character and behavior of this film determine mechanical performance of the sintered bodies, great efforts have been made to understand the property/structure/chemistry relationships of the glassy film. However, our knowledge of the chemical bondings around the glassy film is still limited: Atomic models of the grain boundary film has yet to be revealed.

Recent progress of computer performance enables detailed examination of the electronic states around grain boundaries through quantum mechanical calculations, without including any adjustable parameters, in other words, using first principles method. The present authors have employed a first principles molecular orbital method for grain boundary problems in silicon nitride ceramics. The computation we have used can be categorized into two parts: 1) Interpretation of the ELNES, i.e., the near edge-structure of electron energy-loss spectrum (EELS), that is experimentally obtained from the intergranular film using a sub-nanometer sized electron probe in a field-emission type transmission electron microscope. 2) Calculation of the chemical bondings around impurity atoms in Si_3N_4. Some elements were found to significantly reduce the covalent bond strength when they were present in the Si_3N_4 matrix or at the grain boundary of Si_3N_4. Good correlations were found between the bond strength and solid-solubility of the impurities.[1-4] All of the computations were made by the discrete variational Xa (DV-Xα) method which has been widely used and successful for solving a number of problems in ceramic systems. In the present manuscript, we serve an overview of our recent theoretical works on Si_3N_4.

THE DV-Xα CALCULATIONS USING MODEL CLUSTERS

The first principles molecular orbital (MO) calculations by the DV-Xα method were made for model clusters. The computational code was originally developed by Ellis, Adachi and Averill.[5] We have been using a substantially modified version of this program made by Adachi, Tsukada and Satoko.[6] The advantage of this computational procedure for the first principles MO calculations is summarized as follows: (1) Since we use numerical atomic orbital (AO) basis-set which is obtained by solving the radial part of the Schrödinger equation for a given chemical environment of all atoms in a cluster, precise electronic calculations for all elements in the periodic table can be made using a minimal number of basic functions. This is extremely useful for ceramics systems since we have to treat various types of elements from the periodic table. (2) Core electrons are included throughout the calculation. They have often been ignored in other types of calculations. However, core orbitals should be included when we want to calculate precisely the electronic transition associated with the electron energy loss (EELS) process, because the EELS measures the electric dipole transition from a core orbital to the unoccupied band. (3) All the calculations were done in real space and no periodicity was assumed for the electronic states. This is especially useful for the calculation of localized imperfections in solids. Information on chemical bonding can be readily and intuitively understood by the real space calculation. In the present study the magnitude of bond overlap population obtained by the Mulliken's population analysis was often used as a measure of the covalent bond strength.

In order to obtain the theoretical ELNES, an extra procedure in addition to the normal ground state calculations was made. These self-consistent calculations were made under the constraint condition that a half electron was removed from the core orbital to fill the unoccupied MO. This is the Slater's transition state for the energy-loss process.[7] Slater found that the total energy difference between initial and final states of the transition is well approximated by the difference in one-electron MO energies for the transition state configuration. The absolute transition energy was evaluated by taking account of the temporary spin-polarization associated with this electronic transition. Details of the computational procedure have been described elsewhere.[8,9]

INTERPRETATION OF ELNES FROM HIGH PURITY Si_3N_4-SiO_2 SYSTEM

High purity Si_3N_4 ceramics without additives can be fabricated by the glass encapsulation technique of hot isostatic pressing (HIP).[10] This material contains only a few mol% of SiO_2 that is derived from the starting powder. Since this is a good reference system for liquid-phase sintered Si_3N_4 ceramics, its mechanical performance as well as its microstructures have been extensively studied.[11,12] Regarding the glassy film between two grains, high resolution electron microscopy (HREM) studies were carefully done in order to measure the film thickness.[13,14]: It was found to be 1.0 ± 0.1 nm, independent of the grain misorientation and the total volume fraction of the glassy phase. A typical HREM image is shown in Fig. 1. Spatially resolved EELS investigation of the glassy film was done by Gu and coworkers.[15] They found that the chemical composition of the glassy film is approximately $O_{0.7}N_{0.3}$ except for Si atoms from the analysis of spectral intensities of O- and N-K edge EELS spectra. No other elements were detected in the film. This means that the film is composed of Si-oxynitride glass. They have also recorded ELNES at various edges from the film.[16] The spectra from the film and Si_3N_4 matrix are most clearly discriminated at the Si-L_{23} edge as shown in Fig. 2 (left). The first peak located at around 105 eV was shifted to higher energy by 0.9 eV in the film. The ELNES structures around 110 - 120 eV were also notably different.

The coordination number of Si in all polytypes of SiO_2, Si_3N_4 and Si_2N_2O is four except for the high-pressure phase, stishovite (SiO_2). The Si-O bond length in these crystals is in the range of 1.61 to 1.62 Å. The Si-N bond is in the range of 1.70 to 1.75 Å. We performed model calculations for two tetrahedral clusters in T_d symmetry, i.e., $(SiO_4)^{4-}$ and $(SiN_4)^{8-}$ keeping the Si-O and Si-N distance at 1.61 Å and 1.73 Å. Regarding unoccupied levels, we obtained a singly degenerated $6a_1$ orbital (mainly Si-3s), a triply degenerated $6t_2$ orbital (mainly Si-3p), a doubly degenerated $2e$ orbital (mainly Si-3d) and a triply degenerated $7t_2$ orbital (mainly Si-3d). In our previous study,[8] we reported that the Si-L_{23} edge ELNES within 30 eV from the edge can be classified into four groups, i.e., A, T, E, and TT, which originate from the $6a_1$, $6t_2$, $2e$ and $7t_2$ MOs for the tetrahedral unit, respectively. For example, the $6a_1$ orbitals in the neighboring tetrahedra interact to form more complicated ELNES of A-group in the solid system. However, the center of gravity of the MO energies in the A-group is well approximated by the MO energy of the $6a_1$ orbital. The transition energy of the $6a_1$ orbital from the Si-2p orbital was obtained to be 105.56 eV for $(SiN_4)^{8-}$ and 107.08 eV for $(SiO_4)^{4-}$ cluster. A chemical shift of 1.52 eV can be expected for the first peak between Si_3N_4 and SiO_2 at the Si-L_{23} edge. The chemical shift of 0.9 eV was experimentally found in the gb film relative to β-Si_3N_4 as shown in Fig. 2 (left). Assuming linear dependence of the chemical shift on the amount of $(SiN_4)^{8-}$ unit in the film as shown in Fig. 3, the chemical shift of 0.9 eV corresponds to 41% of N/(N+O) in the gb film. As described previously, the areal intensity analysis at N and O-K edges by Gu et al.[16] found 30% of N in the film. Since these two values are obtained from independent analysis of the EELS at different edges, the agreement is satisfactory. The present results show that the gb film of the high purity Si_3N_4-SiO_2 ceramics is made of silicon oxynitride glass whose N/(N+O) ratio is in the range of 30 to 40%. It should be noted that such high N solubility has never been reported in a bulk Si-O-N glass system when cation impurities are absent. A high amount of N may have to be present in the intergranular film because of a topological requirement of the glassy film.

A simple cluster, $(Si_3O_9N_1)^{9-}$, that is composed of Si, O, and N with saturated Si and N bonds is shown in Fig. 4. In the proposed Si-O-N glass cluster, nine N ions at the surface of the $(Si_3N_{10})^{18-}$ cluster taken from the β-Si_3N_4 crystal were replaced by O ions, and the Si-O distance was changed to be 1.61 Å, which is the Si-O distance in SiO_2. Since the electric dipole selection rule allows transition from Si-2p core-orbital to the unoccupied Si-s and d orbitals, the sum of Si-3s and 3d PDOS (partial density of states) for the $(Si_3O_9N_1)^{9-}$ is compared in Fig. 2 (right) with that of a model cluster $(Si_9N_{25})^{39-}$ obtained from the β-Si_3N_4 crystal. The first peak, A_1, is steeper in the $(Si_9N_{25})^{39-}$ cluster. The peak A_2 seemingly disappears in the theoretical spectrum of the $(Si_3O_9N_1)^{9-}$ cluster. The intensity of the higher energy shoulder of the peak E is also reduced in the $(Si_3O_9N_1)^{9-}$: These features are in good agreement with the experimental spectra shown in Fig. 2 (left). The present model satisfies the necessary conditions to reproduce the Si-L_{23} edge ELNES spectrum of the glassy film.

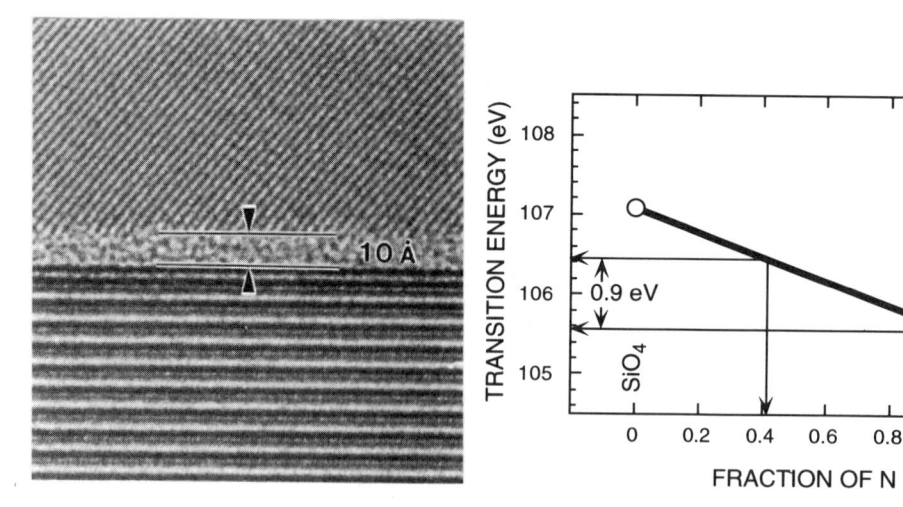

Figure 1. A HREM image of intergranular glassy film in high purity Si_3N_4-SiO_2 ceramics.

Figure 3. Calculated energy of the first peak of Si-L_{23} edge ELNES for $(SiN_4)^{8-}$ and $(SiO_4)^{4-}$ clusters. Assuming linear dependence of the peak energy on the N content between $(SiN_4)^{8-}$ and $(SiO_4)^{4-}$, the chemical shift of 0.9 eV corresponds to N/(O+N) of 0.41.

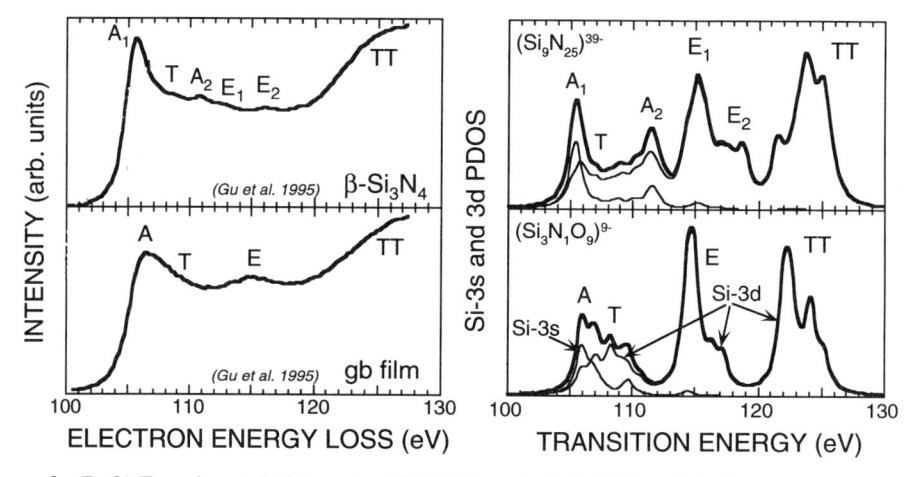

Figure 2. (Left) Experimental Si-L_{23} edge ELNES from bulk β-Si_3N_4 and the intergranular glassy film reported by Gu et al. (Ref. 15). (Right) The sum of Si-3s and Si-3d partial density of states (PDOS) obtained by the present cluster calculations for the Slater's transition states.

The β-Si_3N_4 grains often exhibit an elongated shape, and the surface area of the prismatic plane (1 0 - 1 0) of β-Si_3N_4 is greatest. Figure 5 shows an example of the β-Si_3N_4 (prism) // Si-O-N glass interface without any unsaturated bonds at the interface. Because the coordination number decreases from 3 to 2 when N is substituted by O at the interface, generation of a chemically abrupt interface may be very difficult without dangling bonds. If N and O are mixed at the interface as exemplified in Fig. 5, the structurally abrupt interface can be generated without any unsaturated bonds. In this model, one third of the N at the surface of the prismatic plane of β-Si_3N_4 are substituted by O.

Clarke[17] was the first to propose the presence of the formation of an epitaxial-like partially-ordered monolayer of silica at the interface between Si_3N_4 and SiO_2. According to his model, the silica glass network must conform to the partially-ordered monolayer of silica. The conformation becomes severe when the distance between two crystalline surface is only a few times the size of the basic tetrahedral unit. He proposed that the distortional energy of the glass network is the origin of the disjoining pressure which balances the van der Waals attraction force to provide an equilibrium thickness of the glassy film. Our model is consistent with the Clarke's model; The atomistic modeling suggested that the mixture of N and O at the surface of β-Si_3N_4 is a more favorable configuration than the partially ordered monolayer of pure silica at the interface. However, we need more efforts to model the interface in detail.

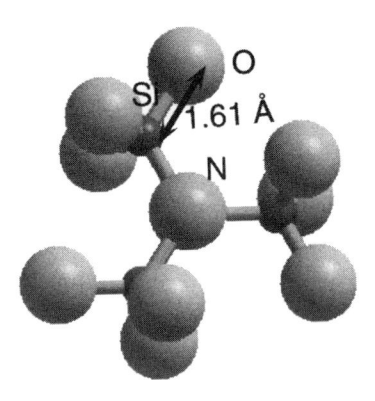

Figure 4. A hypothetical Si-O-N glass structure that is used for the calculation shown in Fig. 2 (right).

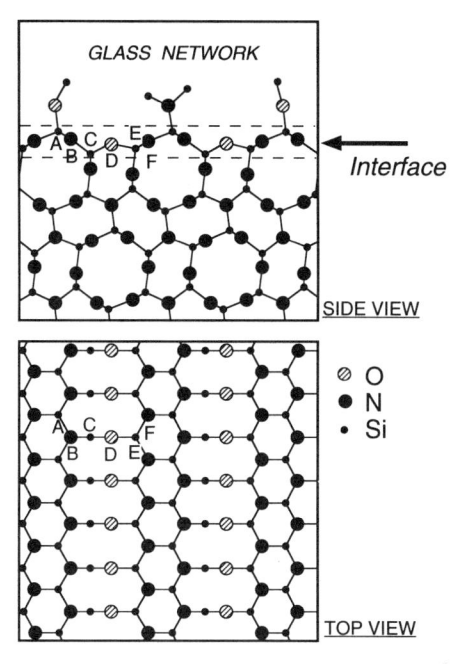

Figure 5. An example of atomic arrangement at the interface between β-Si_3N_4 (prism plane) and the intergranular Si-O-N glassy phase. Note that no bonds are broken at this interface.

SOLUBILITY OF RARE-EARTH IONS IN Si₃N₄

Rare-earth oxides are often used as sintering aids for Si₃N₄ ceramics, and a number of works have been reported on these systems. The α-Si₃N₄ shows high solubility of rare-earth ions (hereafter denoted by Ln^{3+}) with concurrent substitution of Si by Al and N by O. The solid solution is called Ln-α-sialon. On the other hand, no notable solubility has been reported for any Ln^{3+} into the β-Si₃N₄. Atomic arrangements of α- and β-Si₃N₄ are shown in Fig. 6. They are the clusters used in the present calculations. At the center of each cluster, there is a cage-like interstitial site in α-Si₃N₄, and a tunnel-like interstitial site in β-Si₃N₄. The Ln^{3+} is believed to occupy these interstitial sites.[18] Radial distribution of host ions measured from the center of the interstitial site is shown together in Fig. 6. The nearest neighbor distance from the interstitial site is smaller in α-Si₃N₄ than in β-Si₃N₄. This means that the different solubility of Ln^{3+} in the α- and the β-Si₃N₄ cannot be explained simply by the size effect.

Total overlap population which is the sum of the bond overlap-population over the whole cluster, was computed by means of the DV-Xα methods using two types of clusters shown in Fig. 6. Eight kinds of Ln^{3+} were inserted into the interstitial site. Calculations were made also for undoped clusters. Figure 7 shows that the total overlap populations for any Ln^{3+} doped cluster is smaller than the value for the undoped cluster in the case of β-

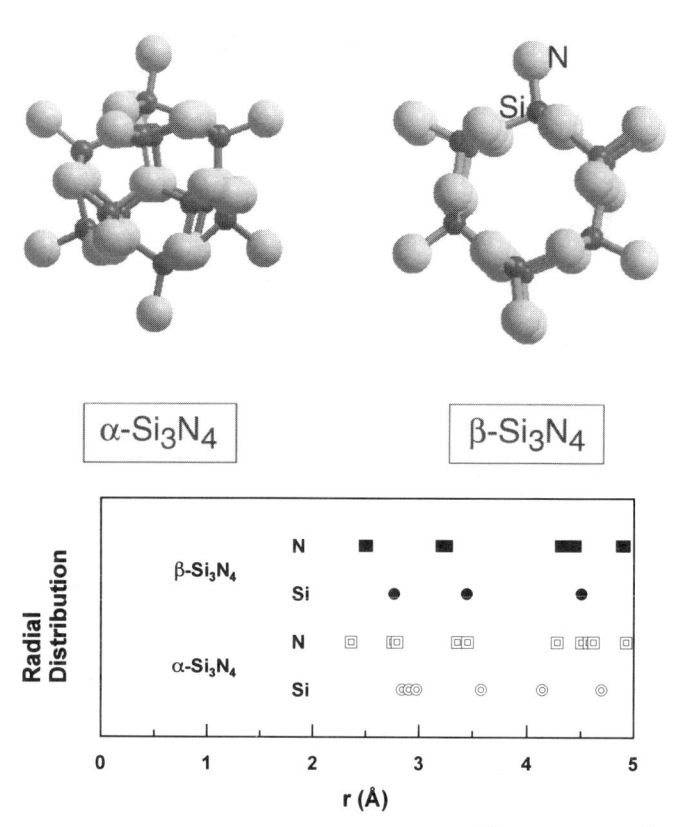

Figure 6. (Top) Model clusters for α- and β-Si₃N₄, i.e., $(Si_{12}N_{29})^{39-}$ and $(Si_{12}N_{30})^{42-}$ clusters, respectively. (Bottom) Radial distribution of ions measured from the center of the interstitial sites are shown together.

Si$_3$N$_4$. On the other hand, Yb^{3+}, Tm^{3+}, Er^{3+}, Ho^{3+}, Dy^{3+}, and Gd^{3+} -doped clusters exhibit greater total overlap population than the undoped cluster in the case of α-Si$_3$N$_4$. Smaller total overlap population for the doped cluster than the undoped one suggests strong anti-bonding between Ln^{3+} and Si$_3$N$_4$ host. These Ln^{3+} ions are suggested to show little solubility into the Si$_3$N$_4$ host. The correlation between the experimental solubility and the total overlap population is satisfactory. No notable solid solubility of Ln^{3+} has been reported for β-Si$_3$N$_4$. In α-Si$_3$N$_4$, Yamada et al.[19] reported that Yb^{3+}, Tm^{3+}, Er^{3+}, Ho^{3+}, Dy^{3+}, and Gd^{3+} showed solubility of x = 0.30 - 0.34 in the formula of Ln$_x$(Si,Al)$_{12}$(O,N)$_{16}$. They found La^{3+} was insoluble in α-Si$_3$N$_4$.

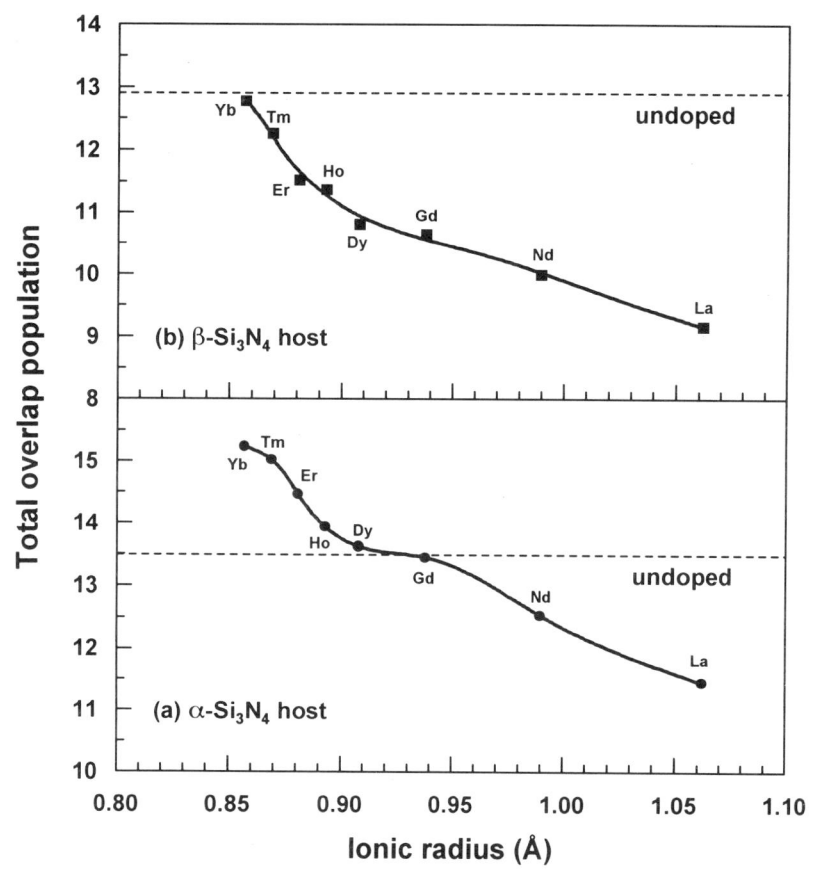

Figure 7. Total overlap population over the whole cluster as a function of the ionic radius of Ln^{3+}. Values for undoped clusters are indicated by broken lines for comparison.

Figure 8 displays contour maps of the total charge density on the basal plane of an α-Si$_3$N$_4$ cluster. The difference in charge density between Yb^{3+}-doped and an undoped cluster implies that the charge distribution at the Si-N bond is significantly altered by the presence of the Yb^{3+} ion: The charge that contributes to the Si-N bonding is polarized due to the electric field generated by the positive charge at the interstitial site. The total overlap population can be divided into three parts, i.e., 1) the sum of bond overlap populations between Ln^{3+} and all other ions in the cluster, 2) the sum of Si-N nearest neighbor (nn) bond overlap populations in the cluster, and 3) all other bond overlap populations. Quantitative results for 1) and 2) are shown in Fig. 9. The Ln^{3+} and all other ions are found to be antibonding, and the magnitude of the anti-bonding is almost proportional to the ionic radius of Ln^{3+}. No significant difference can be found between α- and β-Si_3N_4 matrix regarding this component. The Si-N (nn) bond is polarized due to the electric field associated with the Ln^{3+} ion. It is strengthened when the Ln-Si-N angle is more than 90 deg., and otherwise weakened as schematically shown in Fig. 10. The sum of the Si-N (nn) overlap populations over the whole cluster is found to be greater than that of the undoped cluster. It means the Si-N bond is reinforced by the presence of Ln^{3+} at the interstitial site of both α- and β-Si_3N_4.

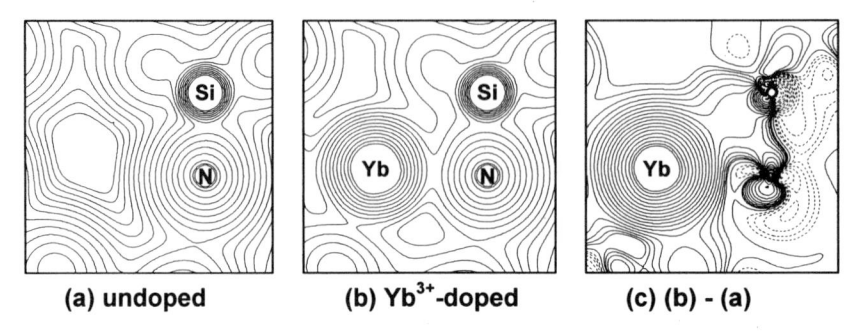

Figure 8. Contour maps of charge density in model clusters of α-Si_3N_4. a) Total charge density of the $(Si_{12}N_{29})^{39-}$ cluster. b) Total charge density of $(Yb_1Si_{12}N_{29})^{36-}$ cluster. c) The difference between a) and b).

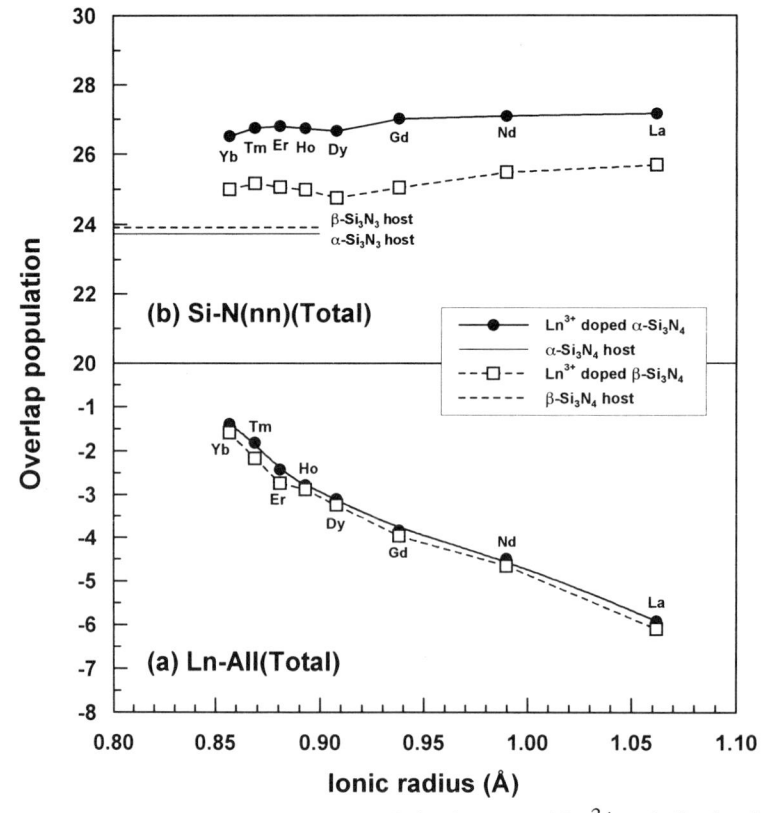

Figure 9. Ionic size dependence of overlap population between a) Ln^{3+} and all other ions in the clusters, and b) nearest neighbor (nn) Si-N ions.

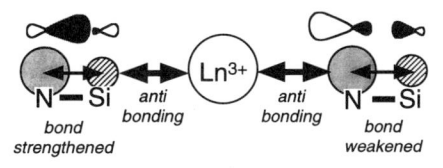

Figure 10. Schematic picture of chemical bondings around an Ln^{3+} ion at the interstitial sites of α- and β-Si$_3$N$_4$.

However, the magnitude of the reinforcement is much smaller for the β-Si$_3$N$_4$ matrix. This component should therefore be responsible for the higher stability of Ln^{3+} ions in the α-matrix than in the β-matrix.

GRAIN-BOUNDARY BOND-STRENGTH OF RARE-EARTH DOPED β-Si$_3$N$_4$

As discussed in the previous section, rare-earth ions show little solubility in the β-Si$_3$N$_4$ matrix. When rare-earth oxides are used as additives in β-Si$_3$N$_4$, there are two scenarios for the dopants: one is the segregation to the intergranular glassy film between grains. The second possibility is the evacuation from the sintered material to the external surface or to glass pockets at the triple grain-junctions. Wang et al.[20] recently measured the intergranular film thickness of β-Si$_3$N$_4$ ceramics sintered with Al$_2$O$_3$ and Ln$_2$O$_3$. They found that the film thickness is dependent on the ionic radius of Ln^{3+} ions. This result implies that the Ln^{3+} ions are localized at the intergranular glassy film. They proposed that these Ln^{3+} ions are adsorbed on the surface of β-Si$_3$N$_4$ grains.

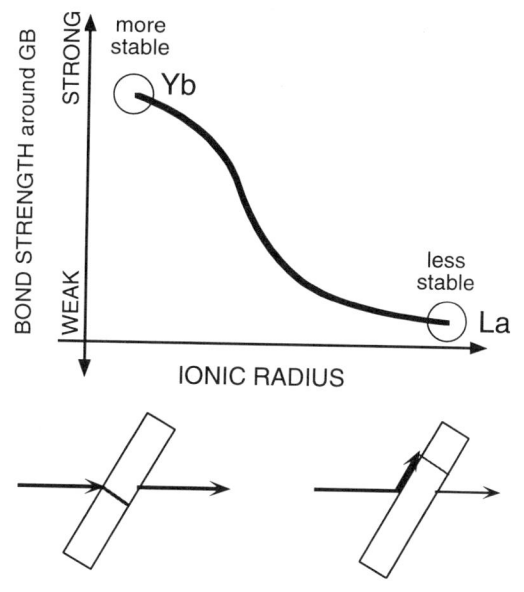

Figure 11. Schematic picture of the dependence of intergranular bond strength in rare-earth doped β-Si$_3$N$_4$ ceramics on the ionic size of Ln^{3+}.

In the previous chapter, we found that the stability of Ln^{3+} ions at the interstitial site of Si$_3$N$_4$ is determined by the balance of two contributions: 1) strong antibonding between Ln^{3+} and matrix Si$_3$N$_4$. This contribution is significantly dependent on the ionic size of Ln^{3+} ion, and 2) reinforcement of the nearest neighbor Si-N bond due to the electric field induced by the trivalent charge. This contribution depends weakly on the ionic radius of Ln^{3+}. The situation of the chemical bonding may not be significantly altered when the Ln^{3+} is located at the interstitial site of Si_3N_4 adjacent to the intergranular glassy film. When larger Ln^{3+} is present, the interface bond strength is weaker than the case for smaller Ln^{3+} ions, because the larger ions generate strong antibonding with surrounding ions. Therefore, the ionic size dependence of the interface bond strength may be schematically drawn in Fig. 11.

Kanamaru and Hoffmann[21,22] investigated the crack propagation behavior of rare-earth doped aluminosilicate glass dispersed with a small amount of β-Si_3N_4 crystals. They found that the crack propagation behavior changed from transgranular to intergranular with the increase of the ionic radius of rare earth ions. Sun, Becher and coworkers[23,24] have recently investigated the intergranular fracture behavior of β-Si_3N_4 whiskers embedded in Si-Al-Ln-O-N glasses. The interfacial fracture behavior of Si_3N_4 ceramics has often been discussed from the viewpoint of a thermal expansion coefficient mismatch between the glassy phase and Si_3N_4 grain. Peterson and Tien[25] found a good correlation between the thermal expansion coefficient of glass and the mean crack deflection angle in Si_3N_4 sintered with SiO_2-Y_2O_3-Al_2O_3-MgO. However, in the rare-earth doped system, Sun, Becher and coworkers[24] reported that the thermal expansion mismatch stress does not make a major influence on the interfacial fracture behavior on the basis of their systematic experiments on thermal expansion coefficient and crack propagation analysis. They have concluded that the interfacial bond strength is determined by the interfacial microstructure/chemistry.

In order to evaluate the chemical bonding around Ln^{3+} ions at the interface between the prismatic plane of β-Si_3N_4 and the intergranular Si-O-N glassy film, a set of model calculations were made. The cluster used to model the interface is shown in Fig. 12. The Madelung potential was taken into account only below the surface of β-Si_3N_4. The effect of adsorbed $Si(O,N)_4$ molecule on the surface of the β-Si_3N_4 was included in the simplest manner; a set of point charges were put at the Si site above the N-terminated surface. Only the Coulombic potential due to the adsorbed tetrahedra can be taken into account in this way.

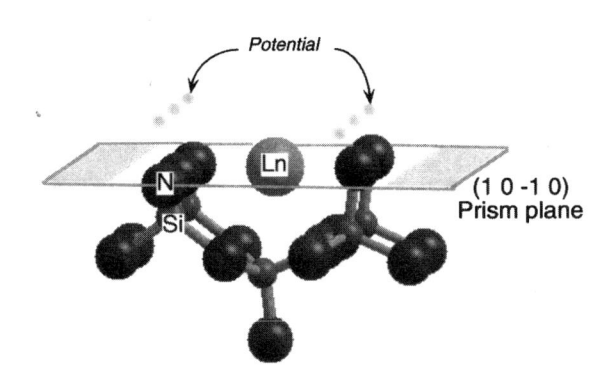

Figure 12. A cluster used to model the interface between β-Si_3N_4 (prism plane) and the intergranular Si-O-N glassy phase.

The total overlap populations for the undoped and Ln^{3+}-doped interface model-cluster are compared with that of the bulk model-cluster in Fig. 13. Contrary to the bulk model, the total overlap population of the Ln^{3+} doped cluster is found to be greater than that of the undoped cluster. Although the present interface model is simple, it is plausible that most of these Ln^{3+} ions are stable at the interface, while they are unstable in the β-Si_3N_4 matrix. We

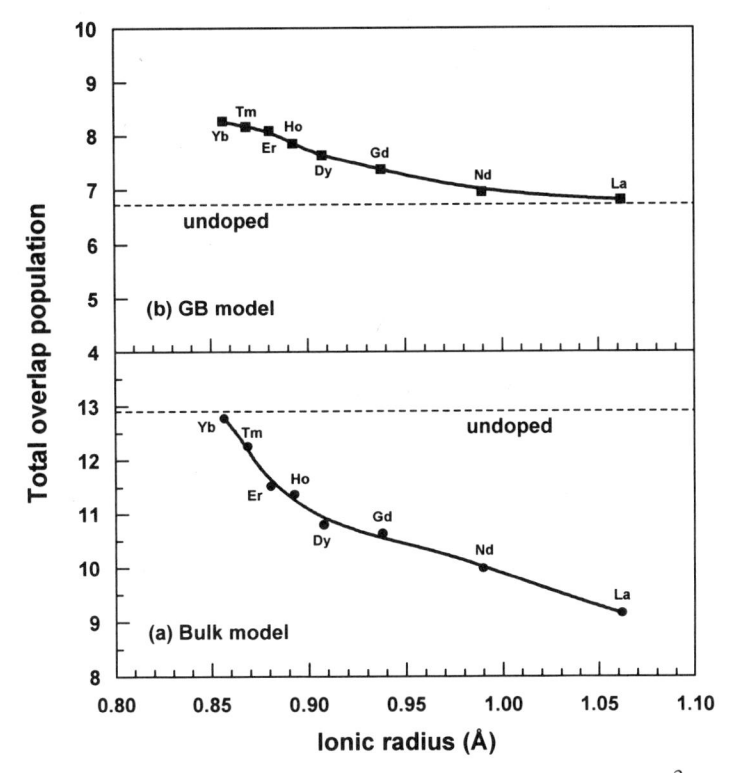

Figure 13. The dependence of total overlap populations on the ionic size of Ln^{3+} for bulk model and grain boundary (GB) model.

can also conclude that the interface bond strength is greater for smaller Ln^{3+} ions, as we have predicted with Fig. 11.

The present computational result will be consistent with the experimental observations by Sun, Becher and coworkers[23] and Wang et al.[20], if highly concentrated Ln^{3+} ions are detected at the interface between β-Si$_3$N$_4$ and the intergranular glassy film. If verified, the adsorbed Ln^{3+} has a dual role for the intergranular glassy films in Si$_3$N$_4$ ceramics: (1) weakening the interface bond strength, especially by larger-sized Ln^{3+}, such as La^{3+}, and (2) changing the interplanar forces across the intergranular film to alter the equilibrium film thickness.

ACKNOWLEDGMENTS

This work is supported by a grant-in-aid for Scientific Research from The Ministry of Education, Sports, Science and Culture of Japan. Helpful discussions with M. Kanamaru, E. Sun, H. Gu, D. R. Clarke and R. M. Cannon are gratefully acknowledged.

REFERENCES

1. I. Tanaka, K. Niihara, S. Nasu and H. Adachi, Calculation of the electronic structure of sp elements in β-Si$_3$N$_4$ with correlation to solubility and solution effects, *J. Am. Ceram. Soc.*, **76**, 2833-38 (1993).
2. I. Tanaka and H. Adachi, Electronic structure of 3d transition elements in β-Si$_3$N$_4$, *Phil. Mag. B*, **72**, 459-473 (1995).

3. T. Nakayasu, T. Yamada, I. Tanaka, H. Adachi and S. Goto, Electronic structures of Ln^{3+} α-Sialons with correlations to solubility and solution effects, *J. Am. Ceram. Soc.*, in press.
4. T. Nakayasu, T. Yamada, I. Tanaka, H. Adachi and S. Goto, Local chemical bonding around rare earth ions in α- and β-Si_3N_4, *J. Am. Ceram. Soc.*, submitted.
5. D. E. Ellis, H. Adachi and F. W. Averill, Molecular cluster theory for chemisorption of first row atoms on nickel (100) surfaces, *Surf. Sci.*, **58**, 497-510 (1976).
6. H. Adachi, M. Tsukada, and C. Satoko, Discrete variational Xa cluster calculatios. I. Application to metal clusters, *J. Phys. Soc. Jpn.*, **45**, 875-83 (1978).
7. J. C. Slater, Quantum theory of molecules and solids, McGraw-Hill: NY, Vol. 4, 1974.
8. I. Tanaka, J. Kawai and H. Adachi Near edge X-ray absorption fine structure of crystalline silicon dioxides, *Phys. Rev.* B, **52**, 11733-11739 (1995).
9. I. Tanaka and H. Adachi, First principles molecular orbital calculation of electron energy loss near edge structures of α-quartz, *J. Phys.* D. (1996) in press.
10. I. Tanaka, G. Pezzotti, T. Okamoto, Y. Miyamoto and M. Koizumi, Hot isostatic press sintering and properties of silicon nitride without additives, *J. Am. Ceram. Soc.*, 72, (1989), 1656-1660.
11. I. Tanaka, G. Pezzotti, K. Matsushita, Y. Miyamoto and T. Okamoto, Impurity enhanced cavity formation in Si_3N_4 at elevated temperatures, *J. Am. Ceram. Soc.*, **74** (1991) 752-59.
12. I. Tanaka and G. Pezzotti: "Delayed failure resistance of high-purity Si_3N_4 without additives at 1400°C", *J. Am. Ceram. Soc.*, **75**, 1023-1025 (1992).
13. I. Tanaka, H-J. Kleebe, M. K. Cinibulk, J. Bruley, D. R. Clarke and M. Rühle, Calcium concentration dependence of the intergranular film thickness in silicon nitride, *J. Am. Ceram. Soc.*, **77**, 911-914 (1994).
14. H-J. Kleebe, M. K. Cinibulk, I. Tanaka, J. Bruley, J. S. Vetrano and M. Rühle, High resolution electron microscopy studies on silicon nitride ceramics, Tailoring of High Temperature Properties of Si_3N_4 Ceramics, pp.259-274, Edited by M. J. Hoffmann and G. Petzow, Kluwer Academic Publishers, Dordrecht, Netherland, 1994.
15. H. Gu, M. Ceh, S. Stemmer, H. Müllejans, and M. Rühle, A quantitative approach for spatially-resolved electron energy-loss spectroscopy of grain boundaries and planar defects on a subnanometer scale, *Ultramicroscopy*, **59** (1995) 215-227.
16. H. Gu, R. M. Cannon, H. Müllejans, M. J. Hoffmann and M. Rühle, unpublished.
17. D. R. Clarke, Grain boundaries in polyphase ceramics, *J. Phys.*, Paris C4 (1985) 51-59.
18. S. Hampshire, H. K. Park, D. P Thompson and K. H. Jack, α'-Sialon ceramics, *Nature* (London), **274**, 880-82 (1978).
19. T. Yamada, T. Nakayasu, T. Takahashi, T. Yamao and Y. Kohtoku, Preparation of Ln-SiAlON powders and sintering behavior, pp. 672-79 in Proceedings of Fourth International Symposium on Ceramic Materials and Components for Engine. Edited by R. Carlsson, T. Johansson, and L. Kahlman. Elsevier Applied Science, London, U.K., 1992.
20. C. M. Wang, X. Pan, M. J. Hoffmann, R. M. Cannon and M. Rühle, Grain boundary films in rare-earth-glass-based silicon nitride, *J. Am. Ceram. Soc.*, **79**, 788-92 (1996).
21. M. Kanamaru, Microstructure of silicon nitride densified with rare-earth oxides, Ph.D thesis, University of Stuttgart, Germany 1994.
22. M. J. Hoffmann, Analysis of microstructural development and mechanical properties of Si_3N_4 ceramics, pp.59-72 in Tailoring of Mechanical properties of Si_3N_4 ceramics, ed. by M. J. Hoffmann and G. Petzow, Kluwer Academic Publishers, Dordrecht, Netherland, 1994.
23. E. Y. Sun, P. F. Becher, S. B. Waters, C. H. Hsueh, K. P. Plucknett and M. J. Hoffmann, Control of interface fracture in silicon nitride ceramics: influence of different rare earth elements, this issue.
24. P. F. Becher, E. Y. Sun, C. H. Hsueh, K. B. Alexander, S. L. Hwang, S. B. Waters and C. G. Westmoreland, Debonding of Interfaces between beta-silicon nitride whiskers and Si-Al-Y oxynitride glasses, *Acta Mater.* in press.
25. I. M. Peterson and T. Y. Tien, Effect of the grain boundary thermal expansion coefficient on the fracture toughness of silicon nitride, *J. Am. Ceram. Soc.*, **78**, 2345-52 (1995).

DESIGN AND LIFE PREDICTION ISSUES FOR HIGH-TEMPERATURE ENGINEERING CERAMICS AND THEIR COMPOSITES

Anthony G. Evans

Division of Applied Science
Harvard University
Cambridge, Massachusetts 02138

ABSTRACT

Perspectives are presented on ceramics and ceramic matrix composites (CMCs) as high temperature materials. The emphasis is on design and life prediction requirements and their role in directing research. Important themes include the relative roles of fracture toughness and inelastic strain (ductility), as well as scaling and stochastic effects caused by manufacturing defects. Ceramics with high toughness have been developed. But, because they are inductile, design with such materials is based on *elastic stresses,* combined with weakest link scaling and extreme value statistics. Procedures that ensure reliable performance under these circumstances are inherently constrained. Opportunities to mitigate these restrictions by matching mechanisms to design are explored. By contrast CMCs exhibit inelastic strain mechanisms that provide an efficient means of redistributing stress. These mechanisms eliminate stress concentrations and suppress scaling effects, enabling design procedures similar to those used with metals. The sources and mechanisms of inelastic strain are described, as well as the ensuing constitutive models. Examples of their finite element implementation in design are presented. A life prediction methodology requires a robust procedure for characterizing fatigue effects in conjunction with manufacturing and machining flaws. A lifting approach is described having commonalty between ceramics and CMCs.

1. INTRODUCTION

A goal for many decades has been the creation of affordable structural materials capable of reliable operation at 1200-1400°C, subject to oxidizing conditions as well as *tensile loading*. This goal has yet to be realized. The strategy has been to begin with stable materials, inherently deformation resistant at these temperatures. Such materials are typically ceramics or intermetallics. Because of brittleness, components made from these materials have not exhibited acceptable reliability upon tensile loading. The addition of *reinforcements* has provided damage tolerance. Various approaches have been successful, though each has degraded *stability and deformation resistance.* To exploit this promise, coherent engineering objectives are essential. Guidelines needed for this purpose have yet to be clarified. The consequence has been minimal engineering *experience,* and few applications. The intent of this overview is to identify the technical problems, as well as explore strategies which might enable such materials to be implemented extensively in commercial systems. To facilitate this objective, methodologies that have close analogies with those currently practiced on metals are emphasized. Two parallel themes address *mechanism-based concepts* in engineering design and life prediction. (i) The relative roles of ductility and toughness. (ii) Engineering design and life prediction methodologies.

The earliest goal was to create ductility by inducing dislocation *plasticity*.[1-17] But, even when plasticity was achieved, *the materials still had a low fracture toughness* and were prone to catastrophic rupture. This paradox arose because cracks in ceramics do not exhibit plastic blunting.[11-17] Subsequently, the research realigned into two conceptual directions: one concerned with the development of *toughness*, the other emphasized inelastic deformation mechanisms and *"ductility"*. These research efforts have resulted in major discoveries of mechanisms with theories capable of characterizing the principal effects. An excellent overview is given in the book by Wachtman.[18] However, the technological ramifications have still been minimal. One reason has been an ineffectual translation of the physics and mechanics concepts into design and life prediction methodologies.

The central technical problem for "toughened" materials, is their lack of ductility. That is, these materials have no mechanism for redistributing stress: therefore, *strain concentration sites are also regions of high stress.* There is an associated *notch sensitivity,* coupled with a *scale dependence* caused by weakest link behavior. These factors impose stringent demands on design and engineering exploitation has been seriously impeded. Conversely, materials designed to induce "ductility", through inelastic deformation mechanisms, enable stress redistribution and, in some cases, exhibit *notch insensitivity.*[19] Equally important, the ultimate tensile strengths become *scale insensitive*, because local stress concentrations are alleviated.[10-23] These characteristics allow a design strategy similar to that used with metals. Such merits have given a stimulus to the continued development of materials that exhibit "ductility", particularly ceramic matrix composites (CMCs).

Having a life prediction capability is an equally important requirement for the reliable implementation of new materials. Fatigue, creep, stress corrosion and their interactions are all involved. Here, some issues that arise in fatigue and its interaction with stress corrosion are presented with a focus on approaches for realizing a life prediction methodology.

2. DESIGN APPROACHES

Design procedures used in practice are straightforward and robust. They are based on rigorous mechanics principles, but simplified to allow ready implementation at acceptable precision. They use the theories of elasticity and plasticity, combined with Gaussian statistics. The engineering challenge for ceramics and their composites is to provide analogous strategies that can be applied with similar confidence. Central to the challenge is the approach used to address the roles of inelastic strain and fracture toughness.

In the absence of macroscopic inelastic strain, stress concentrations persist at slots, attachments, impact sites, etc. The consequent design procedure requires that the "strength" be compared with the concentrated stresses.[26-28] This design strategy has deficiencies, associated with the acquisition of high-fidelity strength data. The problem is caused by weakest link scaling and the extreme value nature of the stochastics. It can be alleviated, but not obviated, by using higher toughness materials. A more robust design strategy can be implemented if the material has a capacity to exhibit inelastic strain. Such strain diminishes stress concentrations, leading to a decreased sensitivity to manufacturing flaws, notches and impacts.[19,29] The effects are analogous to plasticity in metals. Further confidence is achieved *if weakest link scaling can be suppressed* by imparting damage tolerance to the material, through mechanisms that stabilize sub-critical cracks. The following discussion summarizes these design issues and makes a preliminary attempt to understand the opportunities for enhancing the "design friendliness" by optimizing toughness, by inducing inelastic strain and by suppressing weakest link scaling.

2.1 Weakest Link Design

A consequence of material linearity is that stress concentrations are overwhelmingly important. This difficulty is exacerbated by weakest link scaling of the strength, as well as the extreme value stochastics. These issues are elaborated. Designing a load bearing component using a linear elastic material involves two steps. (1) An elastic analysis is performed in order to obtain the *stresses*. (2) The *survival probability* of each element is evaluated using the principles of weakest link statistics.[23-25,27-33] The procedure is understood and has been implemented in software programs compatible with finite element codes. The codes calculate the survival probability, ϕ_S of volume elements δV using the weakest link expression,[27]

$$\phi_S = 1 - \delta V \int_0^{\sigma_D} g_S(S) dS \quad (1a)$$

$$\equiv 1 - \delta V G_S(\sigma_D)$$

where $g_S(S)dS$ is the number of flaws in unit volume having strength between S and $S + dS$, σ_D is the average stress within the element *at the design load* and G_S is the cumulative distribution. There is a corresponding formula for the surface elements. The survival probability of the component is the product of the survival probabilities of all elements (volume plus surface), $\Pi(\phi_S)$.[25,27] Usually, G_S is approximated by a power law (Weibull distribution) with a scale parameter, S_0, and a shape parameter, m,

$$G_S(S) \approx (S/S_0)^m \quad (1b)$$

The practical utilization of these codes is limited by the procedures used to obtain data. The integration of Eqn. (1a) is *between zero and the design stress*. However, strength data are normally obtained using flexure tests or tension tests on small specimens. These provide information about the flaw populations at stresses *above* σ_D, because of *weakest link scaling*[3] (Fig. 1). A robust procedure for *extrapolating* has not been provided. It is often *asserted* that the scale and shape parameters obtained from a power law fit to laboratory data at high strength levels apply at stresses below σ_D. Such assertions are not justified. A polynomial fit to the data could be made with equal fidelity, resulting in large differences in the predicted survival probability upon data extrapolation. A key question is whether it might *ever* be feasible to establish a robust, affordable design procedure for systems having these characteristics. A corollary is whether the situation can be ameliorated by creating materials with relatively high toughness. The following two approaches address these problems.

(1) Provide data which assure acceptable confidence in the scale and shape parameters at the design stress. *For this purpose, tests must be performed on relatively large tension specimens in order to obtain data close to* σ_D (Fig. 1). But, the gathering of such data is costly. These costs might be justified when the processing and machining have been standardized and subjected to a *rigorous control regimen,* such that the flaw populations are stable and consistent. Otherwise, batch-to-batch variations and deviations among machining

runs result in population changes that have to be recalibrated in order to provide the level of confidence required for design. The associated costs are usually prohibitive.

(2) With these problems, diminishing the sensitivity of the strength to processing and machining flaws appears to be the preferred alternative. In principle, improving the toughness should result in these attributes. However, the resistance-curve behavior *inherent in the toughening mechanisms* limits exploitation, because toughening cannot be utilized by the small flaws typically introduced upon processing.[34-36] Nevertheless, the scaling difficulties can be diminished by matching the design to the crack growth resistance.

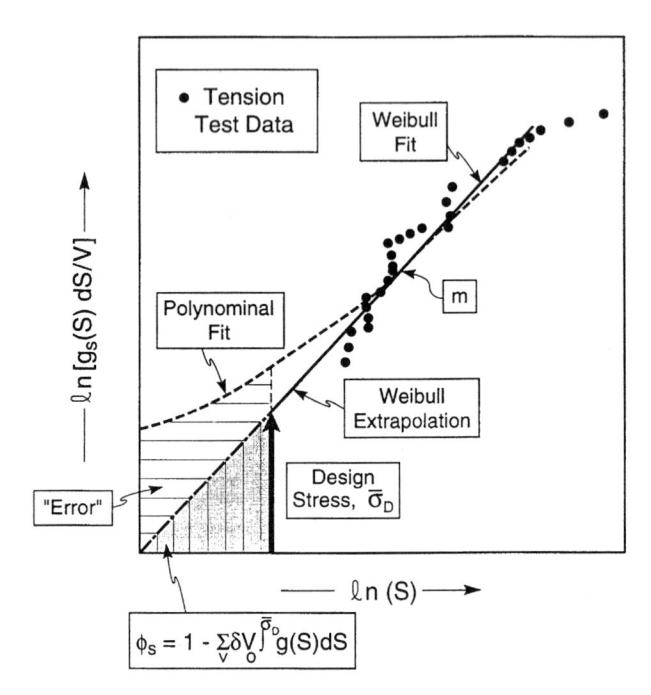

Figure 1. Typical statistical data obtained upon tension testing of ceramics. The extrapolation from the (linear) Weibull fit is shown, as well as the extrapolation from a polynomial fit. These indicate the large difference in the projected survival probability (note that the axes are logarithmic).

2.2 Stress Redistribution and Design

The design situation is completely different for "ductile" materials exhibiting *inelastic deformation modes* such as CMCs (Fig. 2). Such deformations redistribute stress.[26,37] Two fundamentally different stress redistribution mechanisms operate. (Fig. 3) (a) One derives from the inelastic strain produced by multiple matrix cracking accompanied by interface debonding and friction. This mechanism is subject to strain hardening.[19,26] (b) Another involves fiber pull-out, beyond the Ultimate Tensile Strength (UTS) after localization.[19,38] This mechanism is subject to strain softening. It is associated with frictional tractions enabled by the pull-out of the failed fibers. These mechanisms are so efficient at dissipating energy by internal friction that the materials become notch insensitive.[37] Inelastic constitutive laws associated of these mechanisms have been developed and implemented in finite element codes.[26] Stress redistribution is illustrated by two representative design calculations: one for tensile plates containing holes or slots (Fig. 4b): another for Pin-loaded holes (Fig. 4b), which simulate mechanical attachments. These calculations demonstrate that the inelastic strain spreads out the stress concentrations and diminishes their peak magnitudes.[26] In design these peak stresses are compared with the ultimate tensile strength.[19,26] The UTS is *scale insensitive* and its distribution is Gaussian, because it is controlled by stable damage modes.[20,21]

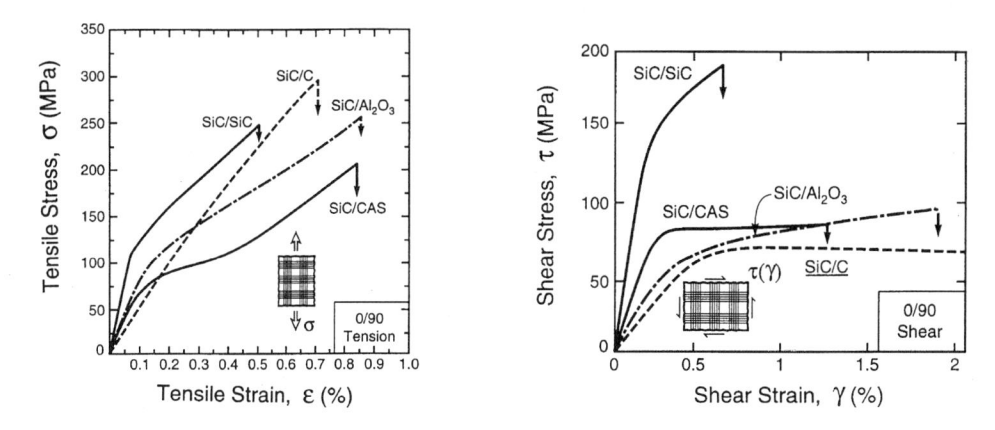

Figure 2. Typical stress/strain curves for CMCs in tension and shear.

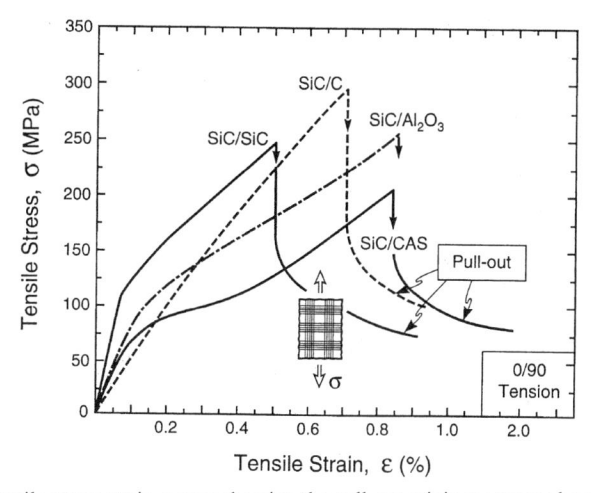

Figure 3. Tensile stress-strain curves showing the pull-out origin at stresses beyond the UTS.

The consequence of this stress redistribution capability is that components can be designed with small holes and slots, without concern for stress concentrations. That is, a net-section stress calculation is often adequate. This feature has been used to advantage in the design of combustors and other thermally-loaded components. Another consequence is that mechanical attachments can be used, with the loads transferred through holes drilled in the material, again without concern for stress concentrations. A dramatic illustration of this behavior is notch insensitivity (Fig. 5) and the immunity of the strength to impact damage, etc.

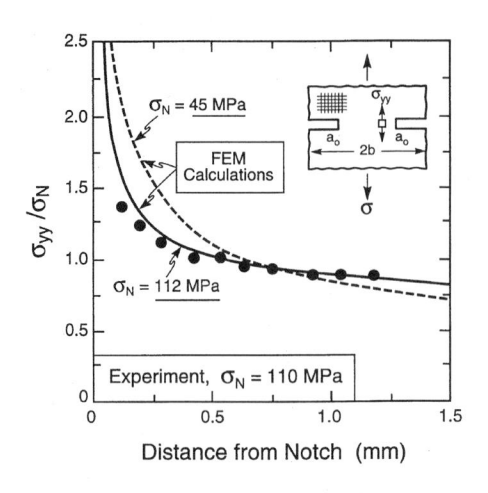

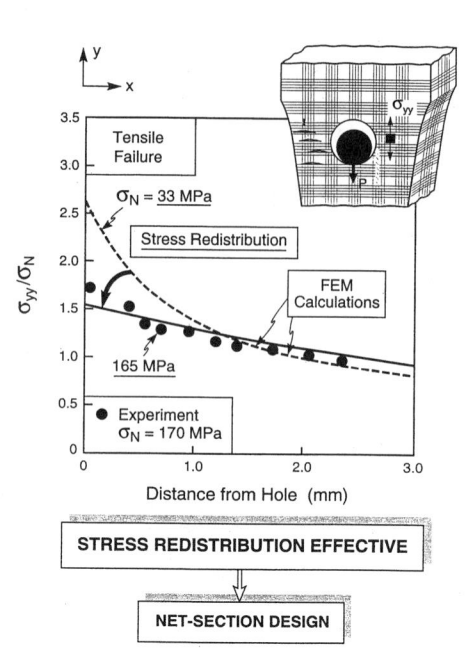

Figure 4a. A notched tension test showing the stress ahead of the notch. The measured values are indicated as well as those calculated by FEM using eqn. (28)

Figure 4b. A pin-loaded hole test performed on SiC/CAS showing the stress across the net section. The measured values are compared with those calculated by FEM using the inelastic constitutive law (eqn. 28).

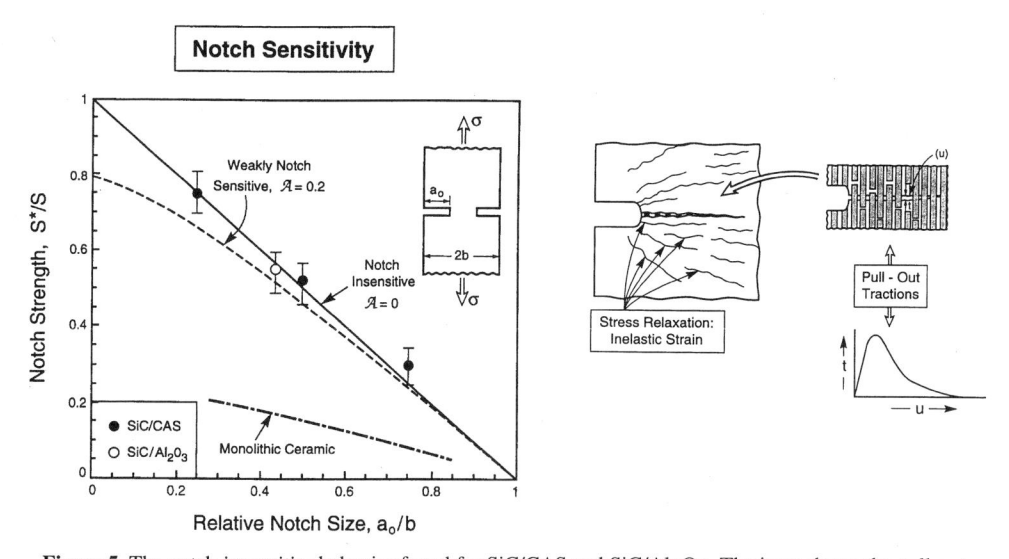

Figure 5. The notch *insensitive* behavior found for SiC/CAS and SiC/Al$_2$O$_3$. The inset shows the pull-out mechanism.

2.3 Illustration of Differences

The consequence of weakest link scaling is that the strength S of a component, at specified survival probability, decreases as its stressed volume, V, increases.[30] Large components have much lower strength than laboratory test specimens, even when the flaw populations are calibrated and invariant. Moreover, if a strain concentration is introduced, such as a small hole or attachment, the stress concentration interacts with the flaws and further diminishes the strength. A large component with an attachment thus has a tensile strength about one-quarter that of a smooth tensile bar and one-eighth that of a flexure beam. The corresponding strength requirements for a "ductile" material are considerably less stringent, since there are minimal scaling and stress concentration effects.[22] That is, a large component with a small hole has about the same load capacity as a laboratory tensile coupon. Moreover, most ceramic and CMC components are designed primarily to withstand *thermal loads*, with minimal pressure. In such cases, design is strain-based, such that the failure strain of the material is most relevant. For CMCs, these strains are typically 0.8%[19-20,43] well in excess of design strains which are typically < 0.2%.

3. THE ROLE OF TOUGHENING
3.1 Reliability Through Enhanced Toughness

The fracture toughness has no direct affect on the design of elastic materials. Its influence is reflected in the magnitudes of the stochastic *strength parameters*, m and S_0, through their effect on strengthening and reliability. An approach that provides insights about these effects is developed and presented. Toughening is manifest as a fracture resistance, Γ_R, that increases with crack extension, Δa (Fig. 6)[44-51]. Concepts of *initiation toughness*, Γ_0, *steady-state toughness*, $\Delta \Gamma_S$, and *tearing index*, λ are needed to characterize this behavior. To bring out the concepts in the most straightforward manner, the simplest relation for transient toughening ($\Delta a \leq L$) having a sound theoretical basis is used. It is given by,[51]

$$\Delta \Gamma_R = \Gamma_R - \Gamma_0 \approx \Delta \Gamma_S \sqrt{(\Delta a / L)} \tag{2a}$$

and for steady-state ($\Delta a \geq L$),

$$\Delta \Gamma_R = \Delta \Gamma_S \tag{2b}$$

where L is a reference length related to the inelastic zone size. The tearing index is a measure of the slope of the resistance curve and is defined as

$$\lambda = (\Delta \Gamma_S / \Gamma_0)^2 / L \tag{3}$$

It has units of reciprocal length. *Reliability is dictated by the product of L with the size a_0 of the flaws responsible for component survival at the design stress.*

To develop understanding, the energy release rate for the small processing and machining flaws that typically control strength is used

$$G = 4\sigma^2 a / \pi E \tag{4}$$

where a is the crack radius, ($a = a_0 + \Delta a$), with a_0 being the initial crack radius and σ the applied stress. The ultimate tensile strength S is obtained from these formulae (2 and 4) by requiring that $G = \Gamma_R$ and $dG/da = d\Gamma_R/d\Delta a$. Accordingly, S is determined to vary with the tearing index in accordance with,

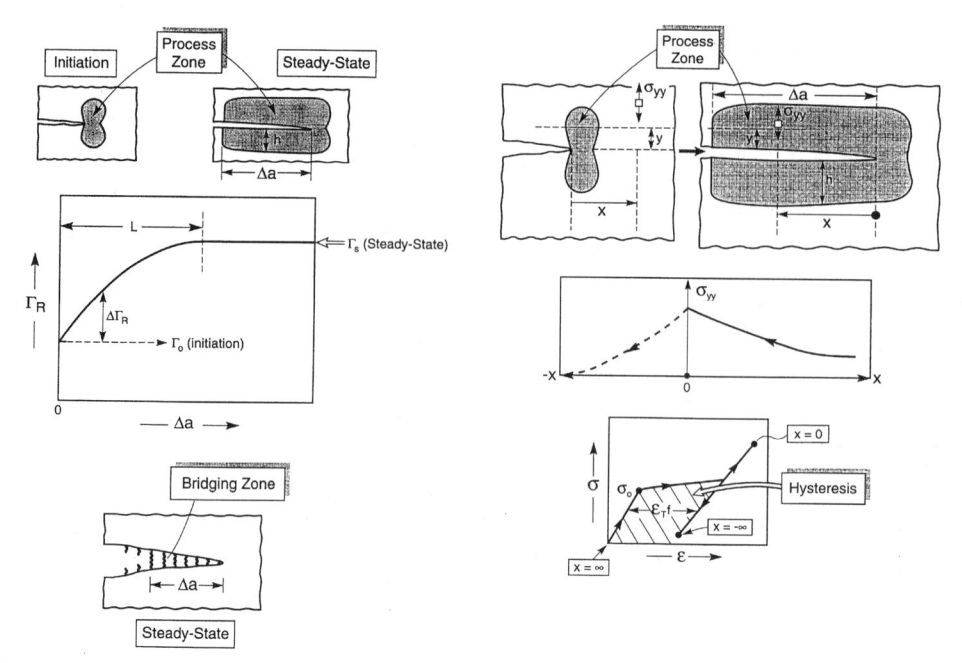

Figure 6a. A schematic of a resistance curve showing the evolution of the inelastic zones and the parameters that characterize the behavior.

Figure 6b. A schematic of crack extension in the presence of a process zone and the strain/stress history experienced by an element of material within the zone.

$$S/S^* = [\Lambda/2(\sqrt{1+\Lambda}-1)]^{1/2} \quad (5)$$

where Λ is related to the tearing index by

$$\Lambda = \lambda a_0 \quad (6)$$

and S^* is the UTS of the untoughened material: that is,

$$S^* = (\sqrt{\pi}/2)\sqrt{E\Gamma_0/a_0} \quad (7)$$

The corresponding crack extension Δa_c when it becomes critical is,

$$\Delta a_c/a_0 = [\sqrt{1+\Lambda}-1]^2/\Lambda \quad (8)$$

The larger Λ, the more beneficial the effect of the toughening on the strength and reliability. The inference is that the attainment of "beneficial" toughening requires an optimization based on L, as well as $\Delta\Gamma_S$, through λ and Λ.

The key importance of Λ is illustrated in Fig. 7. When the flaws a_0 are very small ($\Lambda \ll 1$), fracture occurs unstably and the strength stochastics are unaffected by the toughening. Conversely, larger flaws are able to experience stable crack growth such that the strength exceeds the untoughened magnitude,[34,35] and the Weibull shape parameter increases ($m^* > m$, Fig. 7). That is, the material has a diminished sensitivity to processing and machining flaws and size scaling effects are alleviated. Toughening mechanisms that achieve a reliability enhancement require: $\Lambda = \lambda a_0 \geq 1$. The means for attaining this objective are addressed next.

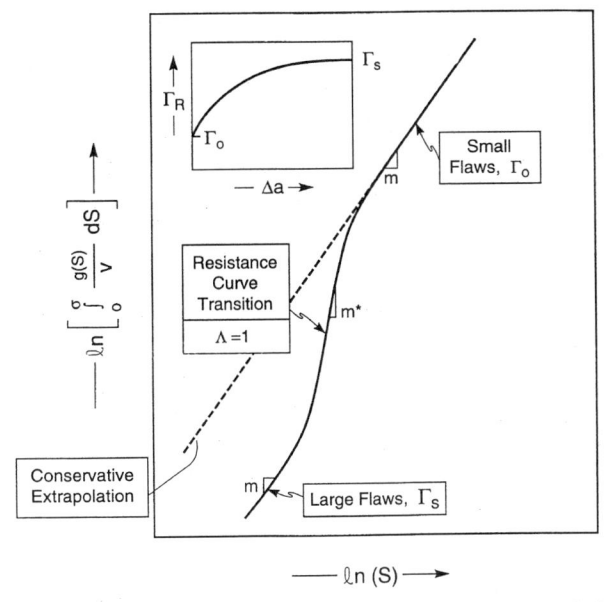

Figure 7. Superior confidence associated with extrapolation when a toughened ceramic is used at the optimum value of the strengthening index, $\Lambda \approx 1$. That is, the slope which governs the shape parameter increases above m when the tearing behavior becomes significant.

3.2 Toughening Configurations

The known toughening mechanisms involve inelastic deformation zones having two distinct configurations (Fig. 6). (i) Zones that concentrate in a thin region around the crack plane which rupture as the crack extends: referred to as either bridging or Dugdale zones. (ii) Regions that extend normal to the crack plane and remain in a deformed but intact state after the crack has propagated through the material: these are designated process zones. Process and bridging zone mechanisms may operate simultaneously and synergistically.[50-53]

Toughening by *bridging* is caused by tractions *along the crack surface* induced by intact inelastic material ligaments. Usually, the tractions soften as the crack extends. Then, in steady-state, the toughening $\Delta\Gamma_S$, is[50,51]

$$\Delta\Gamma_S = b\,[t_0\,u_c] \tag{9}$$

where u_c is the crack opening at the edge of the bridging zone, t_0 is the peak traction and $b \approx 1/2$. The inelastic zone length is[51]

$$L \approx c\,[Eu_c/t_0] \tag{10}$$

where $c \approx 0.12\,\pi$. The tearing index is thus

$$\lambda = (b^2 c)^{-1}[t_0^3 u_c / E\Gamma_0^2] \tag{11}$$

Note the particularly strong influence of the traction t_0. Ductile reinforcements, as well as brittle fibers and anisotropic grains, toughen by means of bridging tractions. Some typical magnitudes are described below.

Process zone toughening may be characterized by the product of the critical stress for activating the inelastic strain mechanism, σ_0, the associated stress-free strain, ε_T, and the zone height, h, in accordance with the stress-strain hysteresis of material elements within the process zone (Fig. 6b),[48,52-54]

$$\Delta\Gamma_S = 2h\sigma_0\varepsilon_T \tag{12}$$

The inelastic zone length is governed by the zone height,

$$L = c_0 h \tag{13}$$

where $c_0 \approx 3$. Moreover, the zone height is related to the critical stress by

$$h = d[E\Gamma_0 / \sigma_0^2] \tag{14}$$

where d is a mechanism dependent coefficient of order, 0.1.[54] The tearing index is thus,

$$\lambda = (4d/c)[\varepsilon_T^2 E / \Gamma_0] \tag{15}$$

Note that at this level of simplification, the only inelastic material property affecting the tearing and strengthening is the transformation strain, ε_T. The critical stress and the zone height are of secondary importance, in contrast to their primary influence on the steady-state toughening. Because of this, the *strengthening flexibility for process zone toughening is appreciably less than that for the bridging mechanisms*. Moreover, strengthening predictions are often optimistic, because there are important *short crack effects*. For crack lengths, $a_0 < 2L$, interactions between the inelastic zones at the opposite crack tips, *reduce the toughening*. Some appreciation for the effect is provided by results obtained for transformation toughening[54] (Fig. 8).

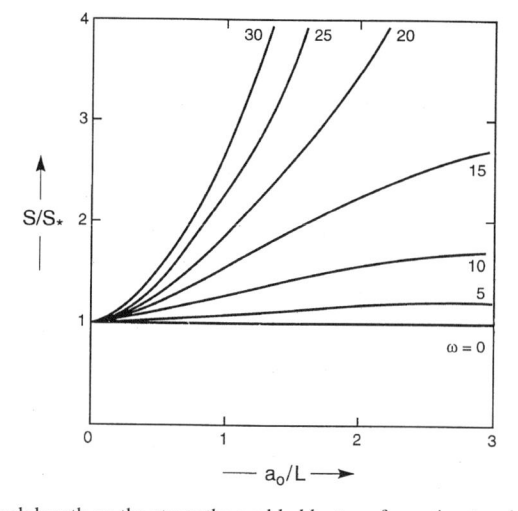

Figure 8. The effect of crack length on the strengths enabled by transformation toughening: ω is a parameter related to the tearing index.

3.3 Bridging Mechanisms

Reinforcing elements can be either ductile or brittle. The former rely on plasticity to create ligaments and dissipate energy[51,55-56] (Fig. 9). When the elements are *brittle*, bridging requires either microstructural residual stresses or weak interfaces. Residual stresses caused by thermal expansion mismatch can suppress local crack propagation and, thereby, allow intact ligaments to exist behind the crack front.[57] When these ligaments eventually fail in the crack wake, energy is dissipated through acoustic waves and causes toughening.[36] *Low fracture energy interfaces are more effective.* They cause the crack to deflect and debond

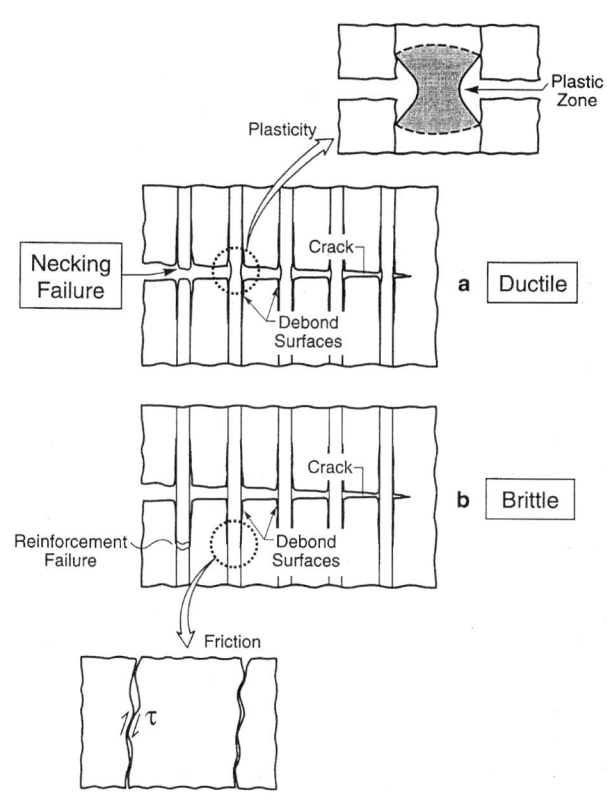

Figure 9. A schematic indicating the dissipation that occurs with a bridging zone: (a) ductile ligaments, (b) brittle reinforcements.

the interfaces.[58] The debonds acquire mode II (shear) characteristics, leading to friction, stability, and intact ligaments.[19] As the crack extends, further debonding occurs, subject to friction.[19] Eventually, the bridging material fails, either by debonding around the ends or by fracture. Following reinforcement failure, additional friction may occur along the debonded surfaces. The dissipation thus includes terms from the debonded interfaces, the acoustic energy upon reinforcement failure, and friction[36] (Fig. 9). The latter is typically dominant. Moreover, the internal friction can become exceptionally large, resulting in toughnesses approaching those for ductile metals (section 4).[19]

3.3.1. Ductile Phases

There are three distinct ductile phase microstructures: (a) isolated ductile reinforcements in an elastic matrix, (b) interpenetrating ductile/elastic networks and (c) a ductile matrix with a dispersed elastic phase.[36,50,55,59,60] An important difference between the first two microstructures and the third concerns the potential for macroscopic plastic strain. Plastic strain in the former is limited by the elastic network, such that the only ductile material experiencing extensive strain is that stretching *between the crack surfaces in the bridging zone*.[55,56,60] The latter develop an additional *plastic zone* which enables further, often substantial, inelastic dissipation.[61,62]

Ductile phase toughening in an otherwise elastic material is contingent upon the ligament failure mechanism, through the stress/stretch relation. Such toughening can be reexpressed by noting that the traction scales with the uniaxial yield strength, Y, of the ligaments and that the plastic stretch is proportional to the radius of the cross section of the reinforcements, R. Consequently, the asymptotic toughening is

$$\Delta\Gamma_s = \chi f R Y \tag{16a}$$

and the zone length is

$$L = \beta R E / Y f \tag{16b}$$

where f is the volume fraction of the reinforcement: χ and β are "ductility" parameters, which also depend on the extent of the interface debonding d. Without debonding, the traction attains high levels, because of the elastic constraint of the matrix, but then decreases as the crack opens because of necking.[55,60] Debonding reduces the constraint, but increases the plastic stretch to failure.[55,63] The latter dominates, causing the dissipation to increase as the debond length increases. Values of χ and β have been obtained both by calculation and by experiment.[55-63] For well-bonded interfaces with ductile phases that fail by necking to a point, $\chi \approx 0.5$ and $\beta \approx 0.1$, both increasing as the strain hardening increases. Less-ductile ligaments that rupture prematurely have correspondingly smaller χ and β. With debonding, χ and β depend on the reinforcement morphology. For interpenetrating or continuous reinforcing phases, χ and β are both increased by debonding, because the dissipation is spread laterally away from the crack plane: χ approaches 8 for large d/R. For discontinuous phases, debonding around the ends of the reinforcement diminishes χ and β.

The tearing index for such materials is obtained from (3) and (16) as

$$\lambda = [\chi^2/\beta] [R/E\Gamma_o^2] (Yf)^3 \tag{17}$$

The important implication is that the scaling that controls strength and reliability is most strongly affected by the yield strength of the reinforcing material Y and its ductility, through χ. Other factors are relatively unimportant. A plot of λ against Y for various R (Fig. 10) indicates the range in toughening parameters needed as a function of the flaw size. For example, when the manufacturing and machining flaws responsible for component performance at the design stress are typically ~10 μm, and the reinforcements are in the 1 μm range, Y should exceed 800 MPa if effective strengthening is to be achieved. Such results guide material design for engineering performance.

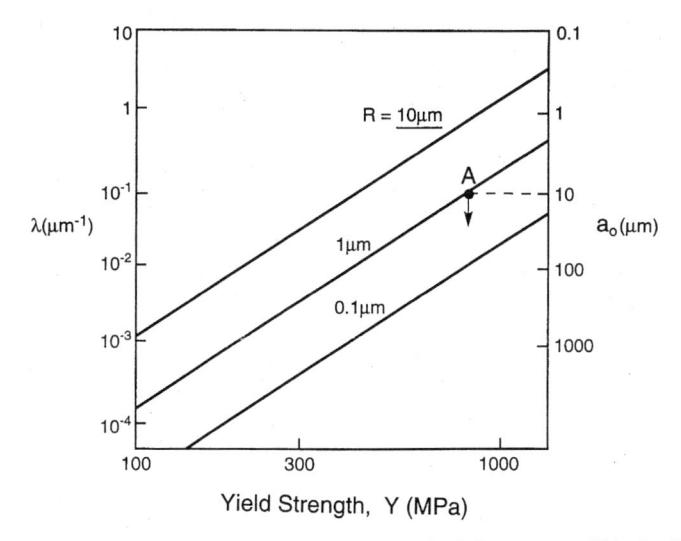

Figure 10. Effect of yield strength on the tearing index for several reinforcement radii in ductile phase toughened materials, (with $f = 0.2$ and $\Gamma_0 = 20$ Jm^{-2}). For a 10 μm initial crack and 1 μm reinforcements (A), the yield strength needed to achieve appreciable strengthening is, $Y \approx 800$ MPa

3.3.2 Brittle Reinforcements

The highest toughnesses are achievable in systems containing "weak" interfaces, which enable debonding and allow dissipation by internal friction.[19,35] There are major effects of the morphology of the debond planes (roughness), the residual stresses, the failure stochastics of the reinforcements and the stiffnesses of the constituent phases.[64] A prerequisite to toughening is that the debond criterion be satisfied. That is, Γ_i/Γ_f be small enough to lie within the debond zone depicted on Fig. 11 (typically $\leq 1/4$).[58] The extent of initial debonding is small. But, further debonding is induced in the crack wake, as the crack extends, having extent governed largely by the residual field, the debond surface roughness and the friction coefficient.[65] Reinforcement failure involves stochastics, subject to a friction stress τ.[20-21,66-69] Large τ causes the stress to vary rapidly and induces reinforcement failure close to the crack, leading to a small pullout length, p, and vice versa.[66,67,69] The consequent tractions are relatively complex. Insight is gained from solutions for short, strong, aligned reinforcements, subject to friction. For this case,[36,39]

$$\Delta\Gamma_s = f\tau \, (p/R)^2 \, R \tag{18}$$

$$L = (c/2)ER/f\tau$$

where R is the reinforcement radius, p is their length, such that p/R is the aspect ratio. The corresponding tearing index is

$$\lambda = (2/c) \, [R/E\Gamma_o^2](f\tau)^3 (p/R)^4 \tag{19}$$

Note the major influences of the aspect ratio and the friction stress. This toughening and strengthening is only attained provided that the reinforcement strength, S, satisfies[69]

$$S > \tau \, (p/R) \tag{20}$$

A plot of λ as a function of the aspect ratio for various friction stresses (Fig. 12) establishes the potential for strengthening. A comparison with Fig 10 indicates the relatively greater efficiency of the internal friction mechanism with aligned, brittle reinforcements than the plasticity mechanism with ductile reinforcements. That is, strengthening by friction is attained with τ ~ 10 MPa, whereas the comparable effect achievable with yielding requires, Y~ 500 MPa. The difference is attributable to the enhanced influence zone achieved by spreading of the friction normal to the crack, resulting in greater dissipation. The implication is that frictional toughening is more effective then ductile phase toughening.

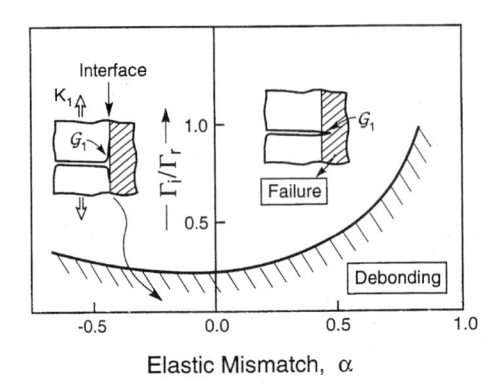

Figure 11. A debond diagram for brittle reinforcements.[58]

Figure 12. Effect of friction stress on the tearing index at several pull out lengths for reinforcements with radii, $R=10$ μm (for $f=0.4$ with $\Gamma_0 = 20$ Jm^{-2}). Note the relatively small values of τ needed to achieve strengthening with 100 μm flaws.

When the reinforcements are long and aligned, but susceptible to fracturing as the crack extends, the tractions are more complex[19]:

$$t/f\, S_L = \sqrt{w}\, \exp[-\alpha\, w^{(m+1)/2}] \qquad (21)$$

where m is the Weibull shape parameter associated with the reinforcements, S_L is their average strength at length L_r, and[19]

$$w = 8\zeta^2(u/L_r)$$
$$\alpha = [\Gamma(1+1/m)]^m /[m+1]$$
$$\zeta = [E/2E_m(1-f)](E_r/S_L)^{1/2}$$
$$L_r = S_L R/\tau$$

with the subscripts m and r referring to the matrix and reinforcements, respectively, and here Γ is the gamma function. Even in this simplified case, the friction stress and the reinforcement strength have interactive effects on toughening and strengthening which are not evident without detailed analysis. Consequently, for practical implementation, the integrated effect of these variables on the *notch sensitivity*, is more useful. The notch behavior predicted by (21) for short, aligned reinforcements is plotted on Fig. 13. It demonstrates how variability in the reinforcement strength (low m) alleviates the notch sensitivity.

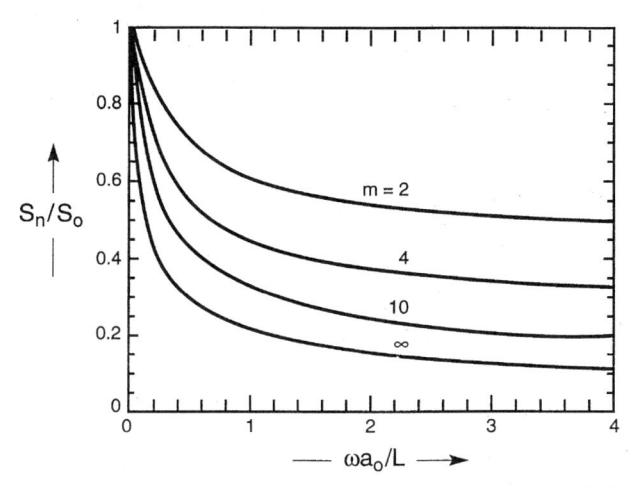

Figure 13. The notch properties of materials reinforced with short aligned, brittle reinforcements: m is the Weibull shape parameter.[69]

3.3.3 Anisotropic Grains

Low fracture energy planes or grain boundaries can allow debonding, as in reinforced materials with weak interfaces, such that toughening involves the same considerations. Certain anisotropic ceramics with elongated grains exhibit such toughening (particularly alumina and silicon nitride[47,49,70]). The dominant effect is the friction that operates along the rough, non-aligned, debonded grain boundaries. The trends are broadly consistent with the above results for short, strong reinforcements. That is, the toughening and the strengthening increase as the grain radius R and their aspect ratio, p/R, increase[70]. Intermetallics such as TiAl also toughen in this manner,[71,72] because of extreme anisotropy in the cleavage energies. But, in this case, the energy of the additional surface created by debonding and ligament formation appears to be more important than the friction. The relevant toughening and strengthening parameters are controlled entirely by the extra surface energy and given by,[72]

$$\Delta\Gamma_S / \Gamma_L = \omega_L^* \qquad (22)$$

and

$$L = 2cb_1\omega_L^* / \chi^2$$

such that

$$\lambda = (\Gamma_L / \Gamma_0)^2 \omega_L^* \chi^2 / 2cb_1$$

where Γ_L is the cleavage energy on the "tough", transverse planes, b_1 is the ligament width (Fig. 14) and ω_L^*, χ are non-dimensional quantities that depend only on the aspect ratio and orientation of the ligaments formed by debonding (Fig. 14). The inverse dependence of λ on size b_1, contrasts with that for the other mechanisms. This dependence arises because more surface is created per unit area of crack growth as b_1 decreases. The consequence is a strengthening effect that can become very large as b_1 becomes smaller. A plot of the strengthening achievable as the anisotropy increases for different b_1 (Fig 15) quantifies the importance of the size scale.

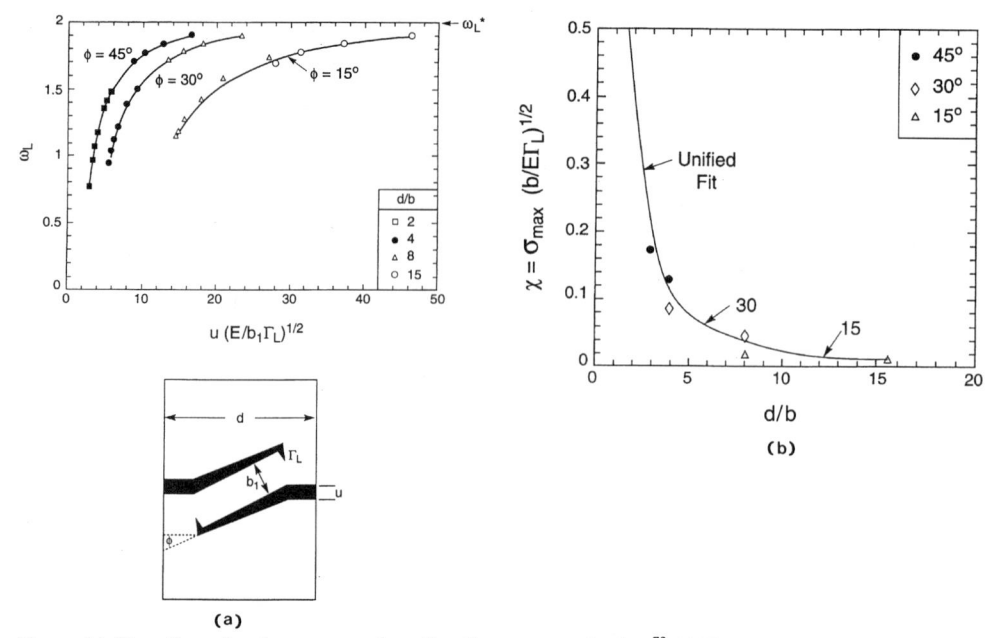

Figure 14. Non-dimensional parameters that affect ligament toughening.[72] (a) The toughening as a function of opening u, with ω_L being steady-state. (b) A tearing coefficient and its dependence on ligament dimensions.

In summary, when such a mechanism prevails, there can be no effect of the ligament width on the steady-strain toughening. The only quantities affecting this behavior are the aspect ratio and orientations of the ligaments that form, as well as the cleavage energies. Conversely, the strengthening *is* influenced by the ligament size, with the opposite trend to that associated with frictional toughening. That is, *the strengthening increases as the microstructure is refined* (Fig.15).

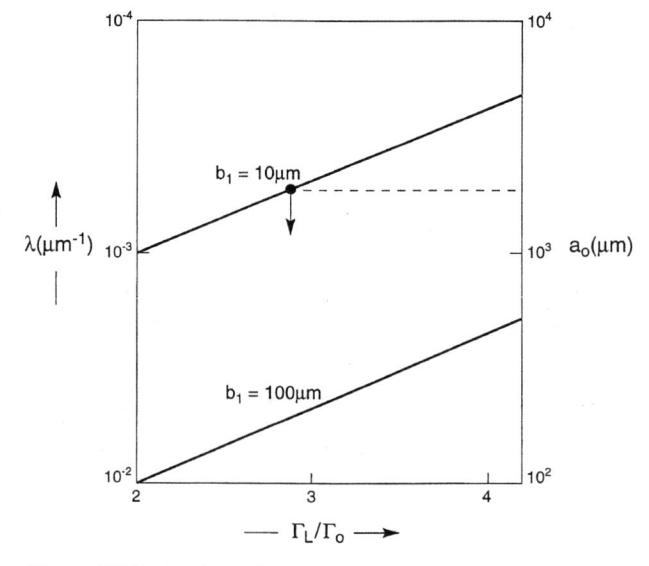

Figure 15. Tearing index for elastic ligament toughening ($d/b_1 = 4$).

4 THE ROLE OF "DUCTILITY"
4.1 Inelastic Strains

A mechanism-based strategy for calculating stress redistribution is illustrated using results for CMCs. These composites exhibit inelastic deformations when matrix microcracks are stabilized.[19,58,73] This is achieved by using either fiber coatings[42,74,75] or porous matrices[73] that deviate cracks toward the loading axis (Fig. 16). The resultant composite microstructure and the ensuing mechanical responses resemble those found in various naturally occurring materials. The fundamental requirement is that the crack deviating *sites* be homogeneously dispersed throughout the body and that these regions have a low debond energy relative to the fracture energy of the reinforcements.[58,76] Once the deviation criterion

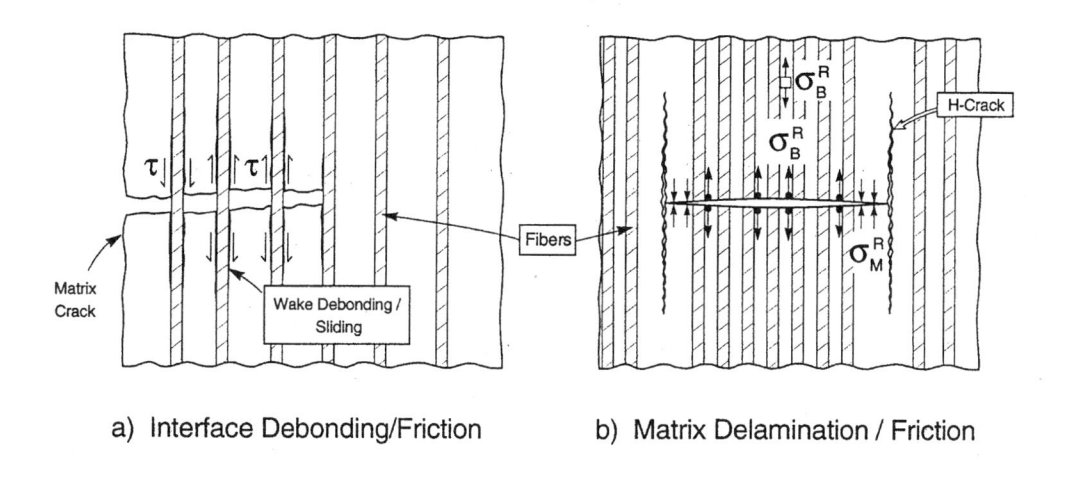

Figure 16. The two concepts for deviating cracks and inducing frictional dissipation along debonded surfaces. (a) Debonding within fiber coatings. (b) Debonding within the (porous) matrix.

has been satisfied, the inelastic strain is governed primarily by the number density of cracking sites and the friction stress that operates along the debonded crack surfaces.[64]

Debonding and friction are commonly encountered in thin brittle layers. They occur by the formation and eventual coalescence of microcracks *en echelon* within a cohesive zone[77]. The associated fracture energy is about four times the mode I fracture toughness of the material in the layer. When debonding occurs, the microcracks coalesce behind the debond and fragment the layer. This process results in a slip zone subject to a friction stress, τ. Models and experiments indicate that τ is affected primarily by the roughness amplitude behind the debond.[64] The residual stress, the elastic compliance of the circumventing material and the friction coefficient are also important.[19]

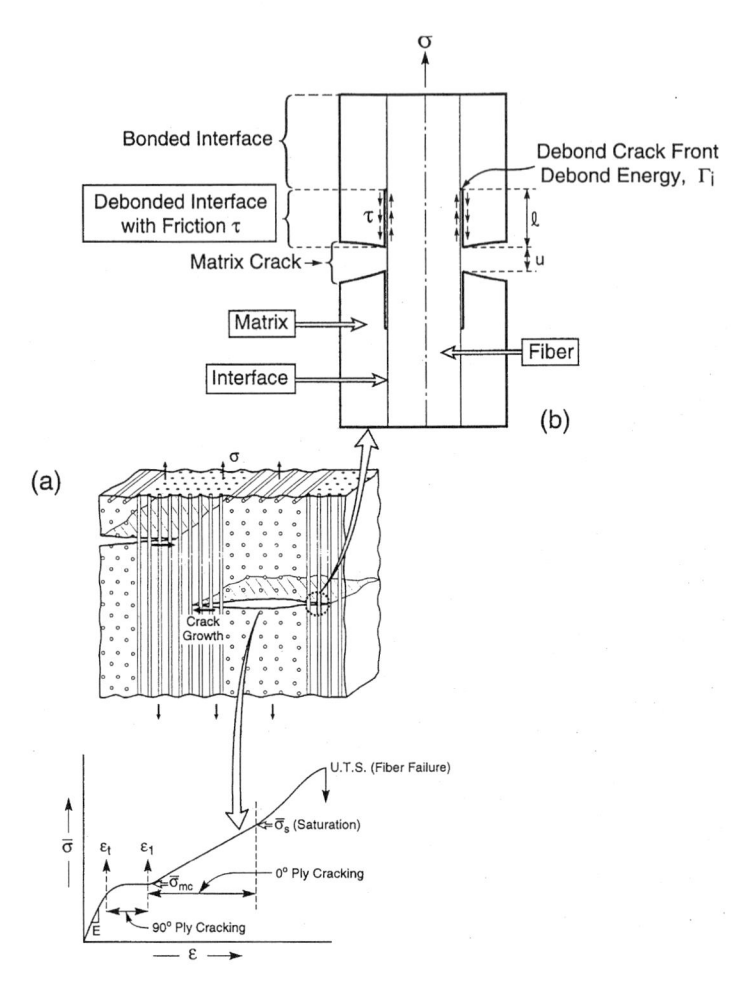

Figure 17. (a) The ply cracking model and the associated stress/strain curve. (b) The cell model for inelastic strains in the 0° plies.

4.2 Stress/Strain Curves

Two cell models represent most of the important physical relationships between the inelastic strains and the mechanisms of matrix cracking, friction and debonding. One model applies to cracks that first form on the 90° plies (Fig. 17a).[78-80] The other represents cracks that penetrate the 0° plies[81] (Fig. 17b). Cracks form first in the 90° plies by tunneling with an associated inelastic strain[19,80] (Fig. 17a). Upon subsequent loading, as the cracks penetrate the 0° plies, they interact with the fibers and the coatings.[79] When debonding and slip occur *within the coating,* two stresses characterize the inelastic strain: a friction stress τ, and a debond stress σ_i[81] (Fig. 17b). The latter is related to the debond toughness for the coating and the residual stress. The consequent inelastic tensile strain ε depends on the stress σ acting on the 0° plies. It has linear and parabolic terms, given by[81,82]

$$\varepsilon = \left(1+\Sigma^T\right)\sigma/E^* + 2\mathcal{L}\sigma^2\left(1-\Sigma_i\right)x\left(1+\Sigma_i+2\Sigma^T\right)-\sigma^T/E \tag{23}$$

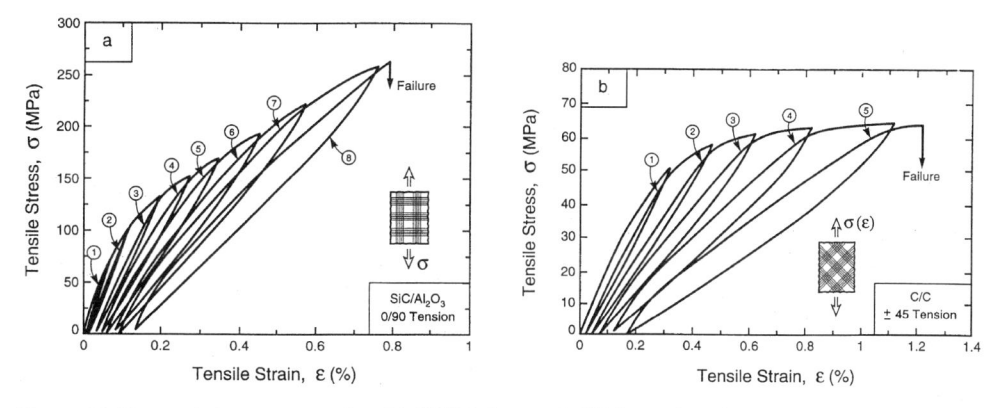

Figure 18. Hysteresis loops measured on (a) SiC/Al$_2$O$_3$ tested 0/90 in tension, (b) C/C tested in ± 45° tension.

where Σ_i is the non-dimensional debond stress, ($\Sigma_i = \sigma_i/\sigma$), E* is the diminished elastic modulus caused by matrix cracking, σ^T is the residual stress ($\Sigma^T = \sigma^T/\sigma$) and $\mathcal{L}$ is an interface friction index, given by[82]

$$\mathcal{L} = \frac{(1-f)^2 R}{4f^2 \tau d E_m} \tag{24}$$

with d being the crack spacing. The parameters E*, $\mathcal{L}$ and Σ_i, evaluated from hysteresis loops[82] (Fig. 18) provide understanding about the separate influences of debonding, friction and matrix cracking on the inelastic strain. They have also provided the insight needed to develop a constitutive law compatible with finite element codes. Simulations, as well as hysteresis measurements have established the influences of friction and debonding on the stress/strain behavior of CMCs.[65,82] The principal effects associated with (23) are as follows. The inelastic strain and the hysteresis loop width increase as the friction stress τ decreases (or $\mathcal{L}$ increases). Small τ accounts for the wide hysteresis loops in SiC/CAS. A large debond stress σ_i limits the role of friction and diminishes both the inelastic strain and the hysteresis. Large σ_i accounts for the narrow hysteresis loops in SiC/SiC and SiC/MAS.

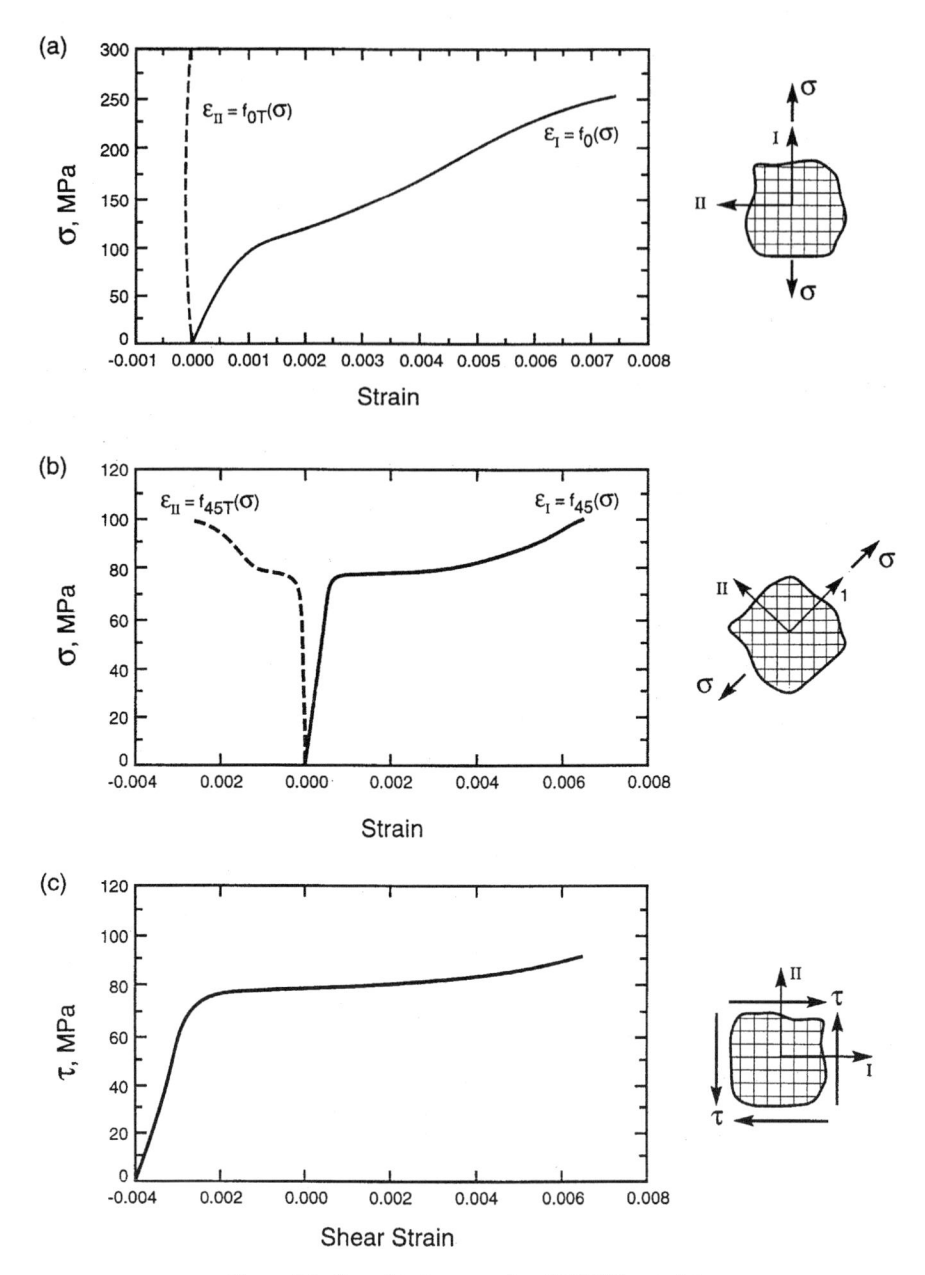

Figure 19. Stress/strain curves for SiC/CAS materials
(a) 0/90 : longitudinal and transverse strains:
(b) ± 45°, axial and transverse strains:
(c) shear stress/strain.

These insights have led to the formulation of a constitutive law. The plane stress relation for the stresses and strains in (1,2) obtained by expressing the inelasticity through the stress drops that occur upon matrix cracking is (Fig. 19),[26]

$$\sigma_1 = \frac{E_0}{1-v_0^2}(\varepsilon_1 + v_0\varepsilon_2) + \Delta\sigma_1 \cos^2\theta + \Delta\sigma_{11}\sin^2\theta$$

$$\sigma_2 = \frac{E_0}{1-v_0^2}(\varepsilon_2 + v_0\varepsilon_1) + \Delta\sigma_1 \sin^2\theta + \Delta\sigma_{11}\cos^2\theta \tag{25}$$

$$\tau = \frac{E_{45}}{2(1-v_{45})}\gamma_{12} - (\Delta\sigma_1 - \Delta\sigma_{11})\sin\theta\cos\theta$$

where τ is now the macroscopic shear stress, $\varepsilon_{1,2}$ the normal strains and γ_{12} the shear strain:
- $\sigma_{1,11}$ are the stress drops upon matrix cracking at fixed strain, parallel and normal to the fiber directions, θ is the angle between (1) and the fiber direction. Some calculations of stress redistribution in a pin loaded configuration have been summarized on Fig 4. These calculations use (25) with the data for SiC/CAS on Fig. 19. A comparison with experiment is also shown. These results establish that appreciable stress reduction is enabled by the inelastic strain, resulting in a response insensitive to the strain concentration caused by the hole. Other configurations have also been analyzed, in some cases demonstrating notch insensitivity.[26,83,84]

4.3 The Ultimate Tensile Strength

The UTS of CMCs is not subject to weakest link behavior, because of the stress redistribution enabled by internal friction. The important consequence is that these materials are scale insensitive and subject to Gaussian statistics.[20,21,68] Design practice is thus essentially the same as that used in metals. Namely, the peak tensile stress obtained through FEM calculations is equated to the UTS.

The underlying phenomena governing the UTS reside in the effects of internal friction on the dimension, δ, over which the stress concentrations in failed fibers are eliminated. This length becomes an *internal* scale parameter, enabling the UTS to be independent of the actual size of the body.[22] The stress evolution of fiber failures still satisfy extreme value statistics and are characterized by a shape parameter, m. Hence, m appears in the expression representing the UTS.[22] For CMCs in which the fiber/matrix interfaces enable stress redistribution through debonding and friction, fiber failures occur in a spatially uncorrelated manner, resulting in global load sharing (GLS) characteristics. In consequence, beyond a slip transfer length, the stress in the fibers is unaffected by the existence of the failure.[22] This length dictates the dimension δ governing the fiber bundle strength. For materials of infinite size, the UTS is given by[22]

$$S_u = fS_c F(m) \tag{26}$$

where

$$F(m) = [2/(m+1)]^{1/m+1}[(m+1)/(m+2)]$$

and

$$S_c = S_0\left[\frac{\tau\delta}{S_0 R}\right]^{1/m}$$

where

$$\delta = \left[L_0(S_0 R/\tau)^m\right]^{1/(m+1)}$$

with L_0 being a reference gauge length (normally 25mm). The main features described by this result are that the UTS is affected by friction, which sets the "internal gauge length", there is a weak dependence on the shape parameter m, but a strong effect of the reinforcement scale parameter, S_0 (or equivalently, the mean reinforcement strength, S_L).

Even when the friction is less effective and some concentrated stresses persist upon fiber failure, resulting in local load sharing (LLS), the effects on the UTS are not especially deleterious. Actual composites having finite size, subject to LLS, exhibit strengths slightly smaller than that predicted by (26). They also exhibit a moderate dependence of the UTS on the gauge length characterized by a Weibull shape parameter over an order of magnitude larger than that for the fibers alone.[21,68] Moreover, the strength distribution is essentially Gaussian, rather than extreme value. The "design friendliness" is thus retained.

Large *manufacturing flaws* may have a more profound effect on the UTS. These flaws induce a stress concentrations governed by the following index,[54,85]

$$\eta = \frac{\pi f^2 E_f E a \tau}{(1-f)^2 E_m^2 RSA} \quad (27)$$

such that the composite strength S becomes,[85]

$$S/S_u = \left(1 - \eta^{2/3}\right)^{-1/2} \quad (28)$$

where a is the size of the *largest* manufacturing flaw, f is again the fiber volume fraction, A is an anisotropy coefficient of order unity. When $\eta \to 0$, there is no effect of manufacturing flaws. Otherwise, such flaws would degrade the UTS. Note that materials with a high friction stress and relatively weak fibers are most prone to degradation by flaws.

Usually, the damaging flaws are limited in size by the ply dimension, h.[19,79,80] For typical fibers strengths, this results in a *maximum acceptable* τ that ensures flaw insensitivity and enables attainment of the GLS strength. This value is, τ_{max} • 50 MPa. Otherwise, the UTS would be lower than the GLS magnitude and the material would have a diminished inelastic tensile strain capacity. Decisions regarding the behavior of "ductile" materials need to be made subject to this level of understanding.

5. FATIGUE LIFE PREDICTION

Cyclic degradation acts upon the mechanisms that provide the toughening and the inelastic deformation, causing the materials to embrittle. An understanding of the degradation mechanisms and of their roles in diminishing either toughness or inelastic strain enables development of a fatigue life prediction methodology. In contrast to metals, there is no cyclic growth mechanism operating *at the crack front,* because there is no plastic blunting.[86] Instead, fatigue operates in the *crack wake*. For materials toughened by bridging, two principal mechanisms are involved. (i) The friction stress τ diminishes by a "wear" phenomenon acting along the debonded interfaces.[86-93] (ii) The reinforcement strength decreases by an abrasion mechanism, exacerbated by environmental interactions. Diagnosis of these two factors is achieved with hysteresis and retained strength measurements (Figs. 20-22). The method is particularly well-developed for CMCs.[86-93] By measuring and analyzing hysteresis strains, changes in the friction stress and the debond energy upon cycling can be determined.[86] Changes in the reinforcement strengths are obtained by measuring the UTS *retained after cycling*.

Typically, τ decays after cycling from the initial value τ_0 to a steady-state value τ_S (of order 5–10 MPa). An empirical function that represents the experimental measurements is (Fig. 22),

$$(\tau - \tau_S)/(\tau_0 - \tau_S) = (1 + b_0)(1 + b_0 N^j)^{-1} \quad (29)$$

where b_0 is a coefficient: j is an exponent which determines the rate at which τ drops with the number of cycles N. It is dependent upon the properties of the interface. Both b_0 and j must be obtained by experiment. The reasons for steady-state friction, τ_S are not

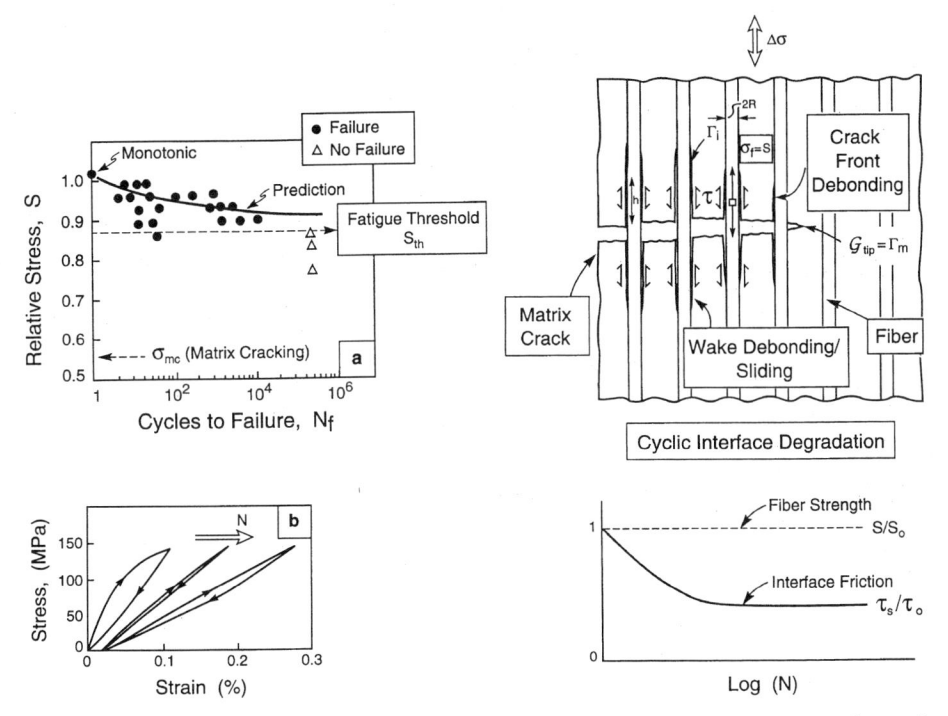

Figure 20. (a) The S/N curve for a SiC/SiC composite. (b) Hysteresis loop measurements.

Figure 21. The phenomena accompanying cyclic loading of CMCs at maximum stresses above that needed to introduce matrix cracks.

understood, but its existence is crucial to the presence of a threshold stress, S_{th} (Fig. 20). When there is no reinforcement degradation, the ambient temperature fatigue threshold for CMCs becomes,[86,93]

$$S_{th}/S_u = (\tau_S/\tau_0)^{1/(m+1)} \quad (30)$$

Because the power law exponent, $1/(m+1)$, is small, S_{th} is a large fraction of the UTS: typically 0.7 to 0.8 (Fig. 20). Consequently, for lower temperature applications, where the fibers retain their strength, fatigue does not present life prediction challenges for CMCs. Design to the threshold provides a robust strategy.

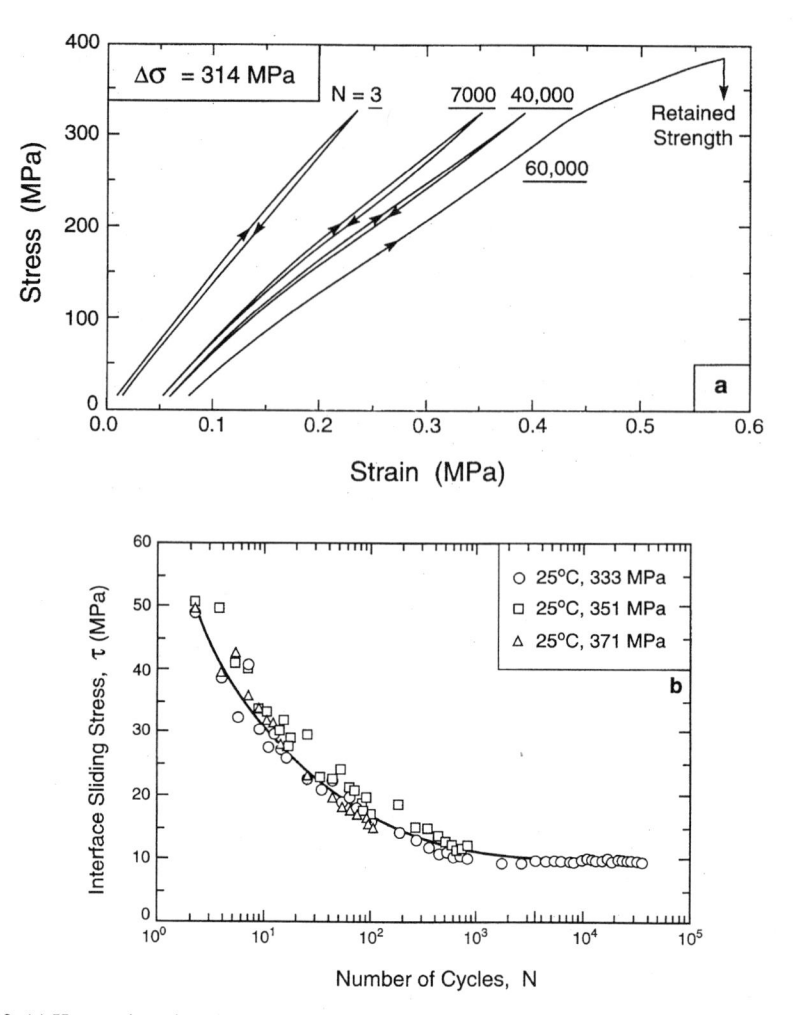

Figure 22. (a) Hysteresis and retained strength measurements upon fatigue testing of SiC/MAS composites. (b) The change in friction stress upon cycling evaluated from the hysteresis measurements.

Similar characteristics arise in ceramics toughened by a bridging zone. Changes in the bridging tractions upon cycling lead to crack growth. *A simple model again identifies the predominant characteristics.* In the model, cyclic degradation proceeds uniformly in accordance with (29), whereupon it will be shown that $\Lambda(N)$ fully characterizes cyclic crack growth. The energy release rate at the *maximum stress* is always equal to the current fracture resistance Γ_R and the tip energy release rate remains at Γ_0. That is, for the crack to extend,

$$G_{max}/\Gamma_0 - 1 = (\Lambda/a_0)^{1/2} \tag{31}$$

But Λ is now cycle dependent and given by (19) and (16) with τ as a variable:

$$\Lambda = \Lambda_0(\tau/\tau_0)^3 [1 + \Delta a/a_0] \tag{32}$$

with Λ_0 being the magnitude of Λ after the first cycle, when $\tau = \tau_0$. Fracture is considered to occur in accordance with the same requirements used for monotonic loading: that is, when Δa reaches a critical magnitude Δa_c, still given by Eqn. (8). Combining (31) with (5) and (7), then gives the fracture relation,

$$(S/S*)^2 [1 + \Delta a_c / a_0] = (\Lambda \Delta a_c / a_0)^{1/2} + 1 \tag{33}$$

With (8), this becomes,

$$\Lambda = 4(S/S*)^4 [1 - (S*/S)^2] \tag{34}$$

The cycle dependence of Λ is now invoked, by combining (32) with (29),

$$\Lambda = \Lambda_0 [(1 + b_0)(1 + N_f^j)]^{-1} \tag{35}$$

Combining Eqns. (34) and (35), the cycles to failure N_f becomes,

$$N_f = \left\{ (1 + 1/b_0) \frac{(S_{th}/S)^{4/3} [1 - \Omega^2]^{1/3}}{\Omega^{4/3} [1 - (S_{th}/S)^2]^{1/3}} - 1 \right\}^{1/j} \tag{36}$$

There is a threshold stress, S_{th}, as $N_f \to \infty$. The parameter, $\Omega = S_{th}/S_u$, is a reference stress ratio, with S_u being the UTS. The S/N behavior associated with these results is summarized on Fig. 23. The straightforward dependence on a few parameters that can be calibrated by selected experiments provides a basis for a life prediction methodology. A relatively few test data calibrate the key quantities, particularly the threshold stress parameter, Ω. This life prediction approach is analogous to that developed for stress corrosion cracking.[94-102] More complexity could be added to the model, but the methodology then becomes less tractable.

Cyclic tests are performed at three or more peak stress S levels, and the cycles to failure measured. The results at each S are *ordered* to obtain the survival probability Φ_s as a function of N_f, by using order statistics. Then, the data at given Φ_s are correlated and plotted as a function of S to obtain the slope of the S/N plot. Specimens that exhibit run-out (no failure after a designated number of cycles) provide particularly important information. *These data give the threshold stress*, at the associated Φ_s. These S_{th} may be compared with the initial strengths S_u at the same Φ_s to determine the *reference threshold stress*, Ω (Φ_s). This is the crucial quantity, because it *scales all other data* required for life prediction.

Most non-oxide materials are susceptible to stress oxidation:[19,86,103] a high-temperature manifestation of stress corrosion (Fig. 24). Again, the phenomenon degrades the toughening. It is *particularly debilitating for very tough non-oxide CMCs*, especially at intermediate temperature (700–900°C); that is, "pest" behavior applies. Matrix cracks created upon loading become pathways for the relatively rapid ingress of oxygen. The oxygen reacts to form both solid and gaseous products. There is a threshold stress below which the phenomenon does not occur, given by[104]

$$S_{th} = \left(\frac{2 E_m \Gamma_0}{\pi h} \right)^{1/2} \tag{37}$$

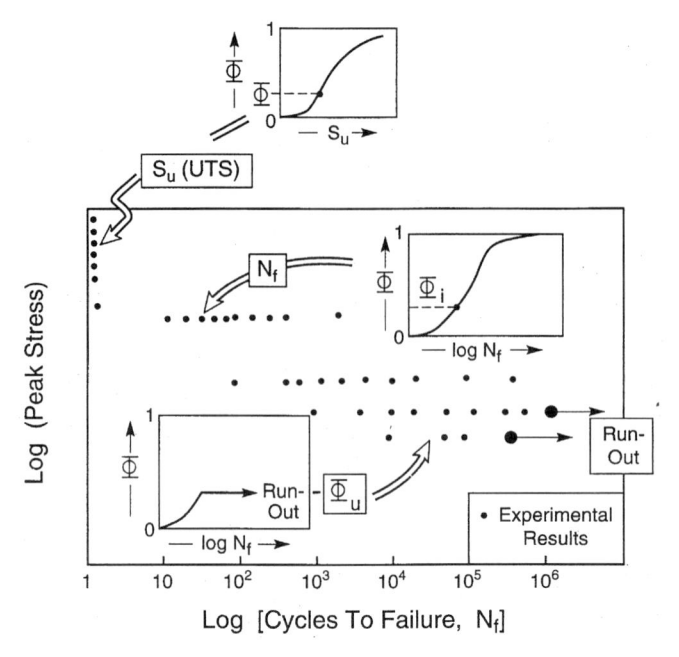

Figure 23. An approach for measuring the fatigue threshold stress for ceramics.

For CMCs, this stress is typically small, of order 10–50 MPa, and too low to enable efficient, lightweight design. Degradation when $\sigma > S_{th}$ occurs according to two rate-limiting phenomena.[104] (i) When the oxygen flow within the cracks is relatively rapid, all reinforcements bridging surface-connected cracks oxidize and weaken *simultaneously*. When they have degraded sufficiently to fail, the surface cracks extend across the weakened zone and form *new crack segments* bridged by pristine reinforcements (Fig. 24). This new bridged region again gradually weakens and fails. The process continues in a manner resembling reaction-controlled, stress corrosion cracking. This is the more important regime, because it *governs the rupture life at long times*. (ii) When the matrix crack opening is narrow, oxygen gradients develop along the crack, resulting in a degradation front that progresses into the material. This process is similar to diffusion controlled, stress corrosion cracking (Fig. 24).

In order to provide a life prediction methodology, three principal phenomena must be addressed[104]: (a) The reduced strength of the reinforcements with an oxide reaction layer. (b) The stress concentration on the reinforcements at the perimeters of unbridged crack segments, (28). (c) The oxygen concentration within the matrix cracks, which is coupled with the thickness of the oxide reaction product on the reinforcements. Analysis of these effects leads to an expression for the failure time • t_c,[104]

$$\Delta t_c \approx t_0 \left[\frac{S_0^4 \sqrt{E\Gamma_0} R^{2/3}}{(\sigma^{13}\tau^{1/3}) h^{7/6}} \right] \tag{38}$$

The features emphasized by (38) are the explicit effects of the applied stress and the friction stress as well as the fiber strength. But, the major issue relates to the chemistry, which determines the reference time t_0. The basic kinetic factors are presently unresolved. An alternative approach is to implement all-oxide systems, which use a porous matrix to stabilize cracks and to induce internal friction (Fig. 16b).

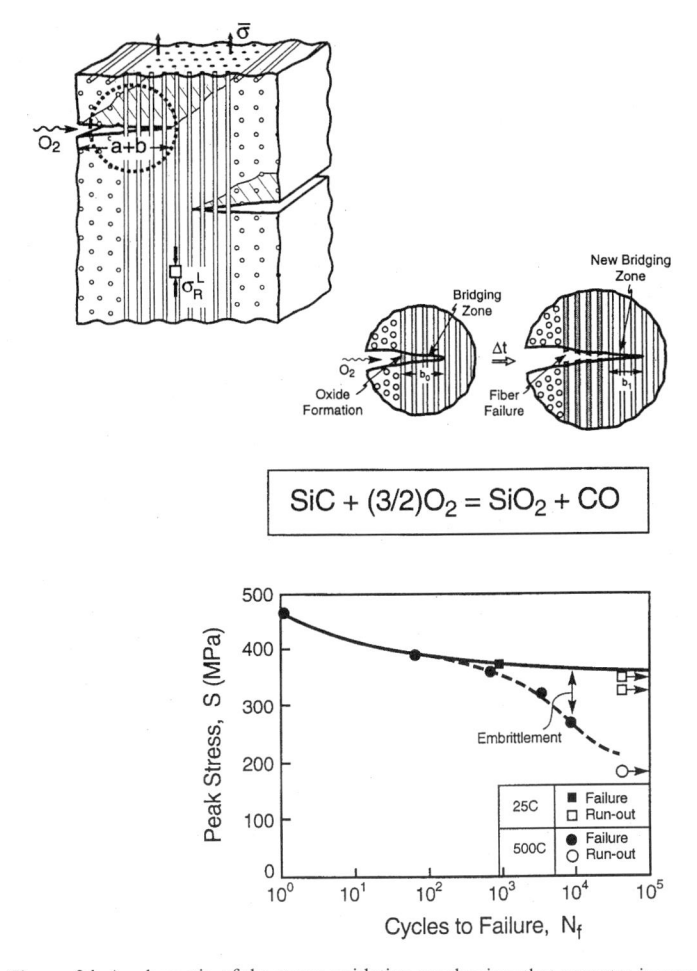

Figure 24. A schematic of the stress oxidation mechanism that operates in non-oxide CMCs.

6 CHALLENGES AND OPPORTUNITIES

Ceramics have the disadvantage that weakest link statistics must be used for design. The associated size scaling causes practical difficulties in the acquisition of design data, since the survival probabilities at design stress levels must be specified with high confidence. Such testing becomes prohibitively costly unless either proof testing can be performed on every component or the manufacturing is so consistent that batch-to-batch variability in the stochastic parameters is within a narrow range. The latter would require a processing regimen that *regulates* the stochastic strength parameters. The stringency of the testing and control regimen can be ameliorated by using reinforcing schemes that toughen the material. The toughening goals are set by requirements for diminished variability in the stochastic parameters governing component survival probabilities at the design stresses. These goals are manifest in the tearing index for the toughening process. This index has explicit forms for each important toughening mechanism. Derivation and discussion of these indices has demonstrated preferences for two mechanisms. (i) Internal friction, achievable through the use of aligned, strong reinforcements with "weak" interfaces. Materials that optimize this mechanism can become sufficiently tough that ductility is achieved. Then the thermomechanical phenomena that dictate design and reliability change, as elaborated next. (ii) Microstructures having fine scale cleavage anisotropy sufficient to form elastic ligaments, such as lamellar TiAl, have a tearing index that increases as the microstructure is refined, resulting in potent strengthening. There are opportunities to exploit these toughening and strengthening mechanisms in a manner that enhances "design friendliness".

Materials that develop inelastic deformation have the advantage that these strains enable stress redistribution and the implementation of familiar design rules used for metals. Such materials are exemplified by ceramic matrix composites reinforced with continuous, aligned fibers. Moreover, the damage stabilization mechanisms that operate in such materials lead to an "internal" scaling, which causes the UTS to be insensitive to size and to be distributed in a Gaussian manner. The remaining problems with these materials concern high-temperature degradation and cost. For non-oxide CMCs, oxidation embrittlement is debilitating. Retarding the degradation requires research on multiple coating concepts, as well as on the development of fibers having high purity, at acceptable manufacturing costs.

Implementation requires a robust life prediction methodology in addition to design strategies. One component of such a methodology involves fatigue. Some of the principles governing development of fatigue life prediction methods have been discussed, through phenomenon common to toughened ceramics and CMCs, especially those that rely on internal friction to impart toughness and ductility. It is predicated on the interplay between cyclic friction and reinforcement degradation. However, much additional analysis and modeling is needed to establish a comprehensive lifing methodology. The phenomena still to be addressed include stress oxidation, cyclic creep and thermomechanical fatigue.

REFERENCES

1. R. J. Stokes, *Trans. Met. Soc. of AIME,* **222** (1962) 1227–37.
2. R. J. Stokes, *J. Am. Ceram. Soc.,* **48** (1965) 60–67.
3. R. J. Stokes, *J. Am. Ceram. Soc.,* **49** (1966a) 39–41.
4. R. J. Stokes, in *Ceramic Microstructures,"* pp. 379–386, J. M. Fulrath and J. A. Pask, editors. John Wiley & Sons, New York, 1968.
5. R. J. Stokes and C. H. Li, *Acta Metall.,* **10** (1962) 535–42.
6. R. J. Stokes and C. H. Li, in *Fracture of Solids,* pp. 289–297, D. C. Drucker and J. J. Gilman, eds. Interscience Publishers, John Wiley & Sons, New York, 1963.
7. J. B. Wachtman, Jr., and L. H. Maxwell, *J. Am. Ceram. Soc.,* **37** (1954) 291–99.
8. J. B. Wachtman, Jr., and L. H. Maxwell, *J. Am. Ceram. Soc.,* **40** (1954) 377–85.
9. J. B. Wachtman, Jr., and L. H. Maxwell, *J. Am. Ceram. Soc.,* **42** (1959) 432–33.
10. S. M. Wiederhorn, *Ann. Rev. Mater. Sci.,* **14** (1984) 374–403.
11. F. J. P. Clarke, R. A. J. Sambell and H. G. Tattersall, *Phil. Mag.,* **7** (1962) 393–413.
12. P. B. Hirsch and S. G. Roberts, *Phil. Mag.,* A**64** (1991) 55.
13. B. J. Hockey, *J. Am. Ceram. Soc.,* **54** (1971) 223-31.
14. M. L. Kronberg, *Acta Metall.,* **5** (1957) 507–24.
15. F. R. N. Nabarro, editor, *Dislocations in Solids,* **1-8**, 1979-1989.
16. J. R. Rice and G. E. Beltz, *J. Mech. Phys. Solids,* **42** (1994) 333.
17. G. Xu, A. S. Argon and M. Ortiz, *Phil. Mag.,* (1995) in press.
18. J. B. Wachtman, Mechanical Properties of Ceramics, to be published.
19. A. G. Evans and F. W. Zok, *J. Mater. Sci.,* **29** (1994) 3857–96.
20. W. A. Curtin, *Appl Phys. Lett.,* **58** (1991) 1155.
21. R. B. Henstenburg and S. L. Phoenix, *Polym. Compos.,* **10** (1989) 389.
22. W. Curtin, *J. Am. Ceram. Soc.,* **74** (1991) 2837.
23. J. Lamon, *J. Am. Ceram. Soc.,* **71** (1988) 106–12.
24. A. G. Evans, *J. Am. Ceram. Soc.,* **61** (1978) 302–308.
25. J. R. Matthews, W. J. Shack and F. A. McClintock, *J. Am. Ceram. Soc.,* **59** (1976) 304.
26. G. Genin and J.W. Hutchinson, *J. Am. Ceram. Soc.,* **80** (1997) 1245-55.
27. A. Freudenthal, *Fracture* (ed. H. Liebowitz) pp. 341–45, Academic Press, New York, 1967.
28. W. Weibull, *Ingeniorsvetenskapakademiens,* p. 153. Handlingar, Nr, 1939.
29. S. B. Batdorf and J. G. Crose, *J. Appl. Mech.* (June 1974) 459–64.
30. S. B. Batdorf and D. J. Chang, *Int. J. Frac.,* **15** (1979) 191–99.

31. S. B. Batdorf and H. L. Heinisch, Jr., *J. Am. Ceram. Soc.,* **61** (1978) 355–58.
32. S. B. Batdorf and G. Sines, *J. Am. Ceram. Soc.*, **63** (1980) 214-18.
33. CARES, Software, NASA Lewis Research Center.
34. K. Kendall, N. McNalford, S.R. Tan and J.D. Birchall, *J. Mater. Sci.,* **1** (1986) 120.
35. R. F. Cooke and D.R. Clarke, *Acta Metall.,* **36** (1988) 555.
36. A. G. Evans, *J. Am. Ceram. Soc.,* **73** (1990) 187–206.
37. C. Cady, T.J. Mackin and A.G. Evans, *J. Am. Ceram. Soc.,* **78** (1995) 77–82.
38. G. Bao and Z. Suo, *Appl. Mech. Rev.,* **45** (1992) 355–66.
39. C. Cady, F.E. Heredia and A.G. Evans, *J. Am. Ceram. Soc.*, **78** (1995) 2065-78.
40. K. M. Prewo, *J. Mater. Sci.,* **22** (1987) 2595
41. D. C. Phillips, *J. Mater. Sci.,* **9** (1974) 1874.
42. J. J. Brennan, *Tailoring of Multiphase Ceramics* (ed. R. Tressler), Plenum, New York: 1986, p. 549
43. J. Aveston, G.A. Cooper and A. Kelly, in *Properties of Fiber Composites: Conf. Proc. National Physical Laboratories,* pp. 15-26. Surrey, UK: IPC Science and Technology Press, 1971.
44. P. Becher, *J. Am. Ceram. Soc.,* **74** (1991) 255–69.
45. R. M. McMeeking and A. G. Evans, *J. Am. Ceram. Soc.,* **65** (1982) 242–47.
46. D. J. Green, R. H. Hanninck and M. V. Swain, *Transformation Toughening of Ceramics,* Boca Raton, FL: CRC Press, 1989.
47. P. R. Becher and T. N. Tiegs, *J. Am. Ceram. Soc.,* **70** (1987) 651–54.
48. A.G. Evans and R.M. Cannon, *Acta Metall.,* **34** (1986) 761–800.
49. C.-W. Li and J. Yamanis, *Ceram. Eng. Sci. Proc.,* **10** (1989) 632–45.
50. G. Vekinis, M. F. Ashby and P. W. R. Beaumont, *Acta Metall. Mater.,* **38** (1990) 1151–62.
51. G. Bao and C. Y. Hui, *Int. J. Solids & Structures*, **26** (1990) 631.
52. B. Budiansky, and J. C. Amazigo, *Int. J. Solids & Structures*, **24** (1988), 7.
53. B. Budiansky and J. C. Amazigo, *J. Mech. Phys. Solids,* **36** (1988), 5.
54. B. Budiansky and D. M. Stump, *Acta Metall.*, **37** (1989) 12.
55. M. F. Ashby, F. J. Blunt, and M. Bannister, *Acta Metall.*, **37** (1989) 1947-57.
56. A. G. Evans and R. M. McMeeking, *Acta Metall.*, **34** (1986) 2435-2441.
57. J. W. Hutchinson and D. K. M. Shum, *Mechanics of Materials,* **9** (1990).
58. M. Y. He and J. W. Hutchinson, *J. Applied Mechanicsl*(Trans. ASME), **56** (1989) 270-278.
59. B. D. Flinn, C. S. Lo., F. W. Zok, and A. G. Evans, *J. Am. Ceram. Soc.*, **76** (1993) 369-76.
60. P. A. Mataga, *Acta Metall.*, **37** (1989) 3349-59.
61. J. W. Hutchinson, and V.Tvergaard, *J. Mech. Phys. Solids*, **40** (1992) 6.
62. J. W. Hutchinson, and V.Tvergaard, *J. Mech. Phys. Solids,* **41** (1993) 1119-1135.
63. H. Deve, A. G., Evans, and R. Mehrabian, *Mat .Res. Soc. Symp Proc.*, **170** (1990) 33-38.
64. R. Kerans, *Scripta Metall. Mater.*, **32** (1994) 1075.
65. J. W. Hutchinson, M. Y. He, B.-X. Wu, and A. G. Evans, *Mechanics of Materials*, **18** (1994) 213-229.
66. M. D. Thouless, and A. G. Evans, *Acta Metall.*, **36** (1988) 517.
67. M.Sutcu, *Acta Metall.*, **37** (1989) 651.
68. W. A. Curtin, and S. J. Zhou, *J. Mech. Phys. Solids*, **43** (1995) 343.
69. B. Budiansky, and J. C. Amazigo, IUTAM Symposium on Nonlinear Analysis of Fracture, Cambridge, Sept. (1995).
70. P. F. Becher, *J. Am. Ceram. Soc.,* **73** (1991) 255-269.
71. K. S. Chan, and Y. W. Kim, *Met. Trans.,* **24A** (1993) 113.
72. D. J. Wissuchek, M. Y. He, and A. G. Evans, *Acta Mater.,* **45** (1997) 2813-2820
73. W. Tu, F.F. Lange and A.G. Evans, *J. Am. Ceram. Soc.,* **79** (1996) 417-24.
74. P. E. D. Morgan, and D. B. Marshall, *J. Amer. Ceramic Soc.*, **78** (1995) 113.
75. J. B. Davis, J . P. A. Löfvander, and A. G. Evans, *J. Amer. Cer. Soc.*, **76** (1993) 1249-1257.
76. J. E. Gordon, *The New Science of Strong Materials*, London: Penguin, 1968.
77. J. W. Hutchinson and Z. C. Xia, *Int. J. Solids Structures,* **31** (1994) 1133-1148.
78. C. Xia, J. W. Hutchinson, B. Budiansky and A. G. Evans, *J. Mech. Phys. Solids,* **42** (1994) 1139–58.
79. C. Xia and J. W. Hutchinson, *Acta Metall. Mater.,* **42** (1994) 1935–45.
80. C. Xia, R. R. Carr and J. W. Hutchinson, *Acta Metall. Mater.,* **41** (1993) 2365.
81. J. W. Hutchinson and H. Jensen, *Mech. of Mater.*, **9** (1990) 139.
82. J.-M. Domergue, E. Vagaggini, A. G. Evans and J. Parenteau, *J. Am. Ceram. Soc.,* **78** (1995) 2721.
83. C. M. Cady, T. J. Mackin and A. G. Evans, *J. Amer. Ceram. Soc.,* **78** (1995) 1.
84. F. A. Heredia, A. G. Evans and C. E. Anderson, *J. Amer. Ceram. Soc.,* **78** (1995) 2790.
85. B. Budiansky and L. Cui, *J. Mech. Phys. Solids,* **42** (1994) 1–19.
86. A. G. Evans, F.W. Zok and R.M. McMeeking, *Acta Metall. Mater.* **43** (1995) 859.
87. J. E. Ritter, Jr., K. Jakus, A. Batakis, and N. Bandyopadhyay, *J. Non-Crystalline Solids,* **38–39** (1980) 419–24.
88. C. J. Gilbert, R. H. Dauskardt and R. O. Ritchie, *J. Am. Ceram. Soc.,* **78** (1995) 2291–300.
89. S. Lathabai, J. Rödel and B. Lawn, *J. Am. Ceram. Soc.,* **74** (1991) 1360–48.

90. R. H. Dauskardt, *Acta Metall. Mater.,* **41** (1993) 2765–81.
91. H. C. Cao, E. Bischoff, O. Sbaizero, M. Rühle, A. G. Evans, D. B. Marshall and J. J. Brennan, *J. Am. Ceram. Soc.,* **73** (1990) 1691–99.
92. C. J. Gilbert, R. H. Dauskardt. R. W. Steinbrech, R. N. Petrany and R. O. Ritchie, *J. Mater. Sci.,* **30** (1995) 643–54.
93. D. Rouby and P. Reynaud, *Compos. Sci. Technol.,* **48** (1993) 109–18.
94. A. G. Evans and S. M. Wiederhorn, *Int. J. Frac.,* **10** (1974) 379–92.
95. A. G. Evans and E. R. Fuller, *Metall. Trans.,* **5** (1974) 27–33.
96. K. Jakus, D. C. Coyne and J. E. Ritter, Jr., *J. Mater. Sci.,* **13** (1978) 2071–80.
97. K. Jakus and J. E. Ritter, Jr., *Res. Mechanica,* **2** (1981) 39–52.
98. K. Jakus, J. E. Ritter, Jr., T. Service and D. Sonderman, *J. Am. Ceram. Soc.,* **64** (1981) C-174 to C-175.
99. K. Jakus, J. E. Ritter, Jr. and J. M. Sullivan *J. Am. Ceram. Soc.,* **64** (1981) 372–74.
100. J. E. Ritter, Jr., pp. 667–86 in *Fracture Mechanics of Ceramics,* Vol. 4, (ed. R.C. Bradt, D. P. H. Hasselman and F. F. Lange) Plenum, New York, 1978.
101. J. E. Ritter, Jr., N. Bandyopadhyay and K. Jakus, *J. Am. Ceram. Soc.,* **62** (1979) 542–43.
102. J. E. Ritter, Jr., N. Bandyopadhyay and K. Jakus, *Bulletin of the Am. Ceram. Soc.,* **60** (1981) 7989–806.
103. F. A. Heredia, J. C. McNulty, F. W. Zok and A. G. Evans, *J. Am. Ceram. Soc.,* **78** (1995) 2097.
104. A. G. Evans, F. W. Zok, R. M. McMeeking and Z. Z. Du, *J. Am. Ceram. Soc.,* **79** (1996) 2345-52.

WETTING AND WORK OF ADHESION IN OXIDE/METAL SYSTEMS

E. Saiz, A. P. Tomsia and R. M. Cannon

Lawrence Berkeley National Laboratory,
Berkeley, California

ABSTRACT

This paper evaluates the inherent influence of $p(O_2)$ on contact angle, θ, and work of adhesion W_{ad}, of liquid metal/solid oxide systems. It also outlines factors that can alter macroscopic contact angles from their equilibrium values.

For a metal/metal-oxide system, the two phases are compatible over a certain range of $p(O_2)$, within which interfacial compositions and energies may change. The dependencies of W_{ad} and θ on oxygen activity are addressed in terms of proposed, mutual adsorption models for each of the three interfaces. Favorable adsorption energies can induce changes especially near the high and low $p(O_2)$ limits of the coexistence range that often mimic bulk oxidation or reduction reactions. However, in some systems, an intermediate range of $p(O_2)$ exists wherein all the interfaces are stoichiometric, and so θ and W_{ad} are independent of $p(O_2)$ (denoted "plateau values"); these are characteristics of the quasi-binary metal/oxide system. A critique of several pure metal/Al_2O_3 systems affirms predicted trends and reveals that θ is 110-130° at the plateau and drops to or below 90° at the high and low $p(O_2)$ limits.

Potentially confusing effects on spreading kinetics and θ values frequently arise from triple point ridging, developed to achieve two dimensional equilibrium at junctions, and from transient oxide skin on the metal. A key finding is that a time scale exists in which triple point ridging can control the spreading rate, yet still permit a macroscopic contact angle to obtain that can be analyzed using Young's equation.

INTRODUCTION

The contact angle of a liquid on a solid is a key determinant for many technological processes (e.g., brazing or infiltration) and for microstructure formation in multiphase materials that involve wetting by liquids.[1,2] Moreover, contact angles can be used in combination with measured liquid surface energies to calculate works of adhesion which are a measure of chemical bonding across interfaces. They can leverage the interfacial fracture resistance, absent unaccounted adsorption effects, and, in concert with wetting characteristics which dictate flaw severity, can be among the factors that control mechanical strength.[2-7]

Contact angles have been extensively studied in metal/ceramic systems over several decades.[8-14] There have been steady improvements in the quality of data plus studies of wetting and liquid surface energies for ever more systems.[15-29] Both temperature and composition dependencies have been addressed but a coherent picture is still lacking. This paper focuses on the latter as being fundamental and discusses the inherent, sometimes unappreciated influences of $p(O_2)$ on contact angles and works of adhesion in metal/oxide systems. It also outlines some nearly unavoidable complicating issues that can shift macroscopic contact angles from their equilibrium values. After giving a conceptual picture, it assesses behavior for stable liquid metals on Al_2O_3 as a model system.

BACKGROUND

The equilibrium value of the contact angle, θ, at the junction of a liquid and a rigid surface for a chosen system is determined by a balance of energies of the interfaces between the liquid, solid and vapor phases, Fig. 1, as expressed by:[30-32]

$$\gamma_{lv} \cos \theta = \gamma_{sv} - \gamma_{sl} \tag{1}$$

The temperature and composition dependencies of θ can be measured directly, and γ_{lv} can also be determined based on the deviation from sphericity in sessile drop experiments or by other direct methods.[32] Having this combination permits computation of the work of adhesion, W_{ad}, and of $(\gamma_{sv} - \gamma_{sl})$,

$$W_{ad} = \gamma_{lv} + \gamma_{sv} - \gamma_{sl} \tag{2}$$

$$W_{ad} = \gamma_{lv}(1 + \cos \theta) \tag{3}$$

Interpretation has several problems, including: the need for additional interfacial energy data suitable for the temperature and composition; concern whether effects of composition, including impurities, can be understood in a context of equilibrium adsorption; and the question about how the force balance at the triple junction can be locally satisfied in a manner that does not invalidate application of the Young-Dupré relation.

Specifically, attaining complete equilibrium at a junction requires motion of the triple line both horizontally and vertically, which leads to two independent relations based on:[32]

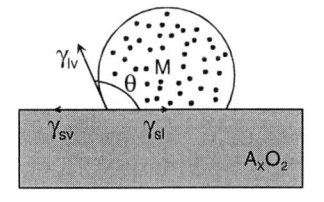

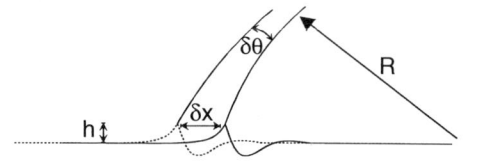

$$\gamma_{lv}/\sin \phi_{lv} = \gamma_{sv}/\sin \phi_{sv} = \gamma_{sl}/\sin \phi_{sl} \tag{4}$$

Figure 1. The contact angle, θ, depends on the force balance among the surface energies of the liquid, γ_{lv}, solid, γ_{sv}, and solid-liquid interface, γ_{sl}, that is attained by horizontal motion of the junction. The diagram in (b) shows an expected shape of the ridge and associated undercutting that would form by any local diffusion near the triple line, in response to forces in horizontal and vertical directions at the junction. If this perturbation can move readily with the triple junction, the classical Young-Dupré relation pertains.

where ϕ_i is the dihedral angle opposite the ith interface. Eqs. (1-3) effectively assume the vertical motion is negligible or reversible. At low temperatures, the vertical component of surface tension is resisted by elastic distortions of the solid, which may be exhibited on solids at low fractions of their melting points (perhaps $< T_m/4$). If any local diffusion can occur, a small ridge will ensue, Fig. 1b. This is the counterpart to the grooves commonly found where grain boundaries intersect free surfaces.[33] Several consequences have emerged from an ongoing analysis of these ridges.[34] Motion of the triple line can be limited by drag from the ridge, leading to spreading rates which are orders of magnitude slower than for viscous flow controlled motion.[35,36] If a ridge is very small ($h/R \ll 0.1$, Fig. 1b), it can be carried with the triple junction and experience little growth, thereby, allowing the macroscopic angle θ to approach that given by Eq. (1), while simultaneously satisfying Eq. (4) locally. If the actual macroscopic angle is sufficiently far from the equilibrium value, the liquid front will detach from the ridge and move rapidly toward a position giving the proper angle. When θ is relatively near the equilibrium value, the ridge grows in height, h, and will exert ever greater drag forces.

Oxide/metal systems are inherently ternary. The oxide and metal phases can coexist in a specific range of $p(O_2)$, within which compositions of the bulk phases and interfaces may vary, even if by small amounts (except for a metal with its own oxide). Thus, variations of γ's with temperature reflect vibrational entropy and composition changes of the phases, which often cause θ to diminish with heating, as well as adsorption effects. Identifying and outlining the inherent adsorption regimes will be emphasized here.

For equilibrium, all interfaces must exhibit equilibrium levels of adsorption requiring diffusion among the phases. The effects on equilibrium interfacial energies are given by the Gibbs adsorption isotherm, which is:[37,32]

$$d\gamma = -kT \sum \Gamma_i \, d \ln a_i \tag{5}$$

The Γ_i refer to the excess amounts of each species per unit area of interface, which are present solely due to the interface being present. The equilibrium levels are dictated by the activities, a_i, of the components in the environment and so appear or disappear reversibly with creation or loss of interfacial area.

A further understanding requires knowing one more of the interfacial energies, which are harder to measure. For Al_2O_3, interpolating between an average from atomistic computations[38,39] for T = 0 and measured values for the liquid at the melting point[40] gives γ_{sv} = [2.0 - 0.058T/K] J/m² for the pure, stoichiometric surface. This temperature dependence matches those from experiments using cleaner liquid metals.[18] The magnitude is consistent with cleavage energies, being lower than most which include irreversible components, but approaching some from subcritical crack growth at intermediate temperatures.[41] This is also similar to values deduced from combinations of sessile drop and grain boundary grooving angles, without considering effects from impurities.[1,20]

FORMULATION

The stability range for an oxide/metal system with two metals, is dictated by conditions to form the stable oxide of A, taken as the more active metal, and the first stable oxide of M, the more noble one, i.e.,

$$xA + O_{2\{v\}} \rightarrow A_xO_2 \qquad \Delta G_{ox}^A \tag{6a}$$
$$xM + O_{2\{v\}} \rightarrow M_xO_2 \qquad \Delta G_{ox}^M \tag{6b}$$

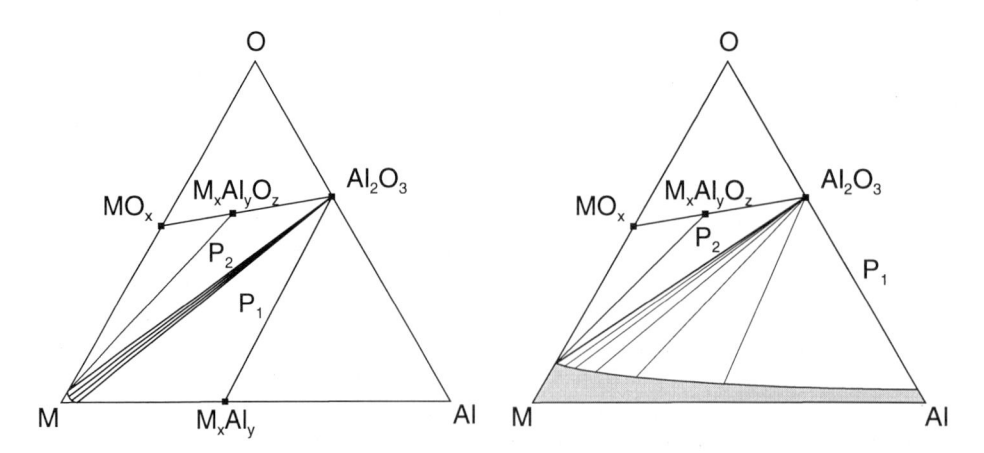

Figure 2. Typical form of phase diagrams involving a mutually stable metal/oxide pair. The two phases are compatible over a range of $p(O_2)$ that is delineated by the values of $p(O_2)$ at which the base metal becomes oxidized, P_2, and where the oxide becomes reduced, P_1. At lower temperatures, all the condensed phases are solid (a); whereas, for the hotter situation depicted (b) both pure metals and all intermediate compositions are molten, as typified by many metal/Al_2O_3 couples.

The coexistance range for the system A_xO_2 and M is, thus,

$$p_1(O_2)^{eq} \equiv P_1^{\ddagger} \leq p(O_2) \leq P_2^{\ddagger} \equiv p_2(O_2)^{eq} \tag{7}$$

and within this range where A_xO_2 is present, activities are related as

$$a_A{}^x p(O_2) = C \exp[(\Delta G_{ox}{}^A)/kT] \tag{8}$$

The free energies refer to those under standard conditions (which are often denoted $\Delta G°$). The equilibrium oxidation conditions for A and M, $p_1(O_2)^{eq}$ and $p_2(O_2)^{eq}$, may be modified by formation of compounds, Fig. 2. This can cause adjustments to $P_1^{\ddagger}$ or $P_2^{\ddagger}$ compared to values for the binary metal-oxygen systems, $P_i°$, which derive from Eqs. (6). That is, reduction of A_xO_2 is expedited by any formation of an AM_z intermetallic compound or dissolution of A into M; similarly, oxidation of M is made slightly easier by formation of a compound oxide, MA_uO_v.[42-44] At higher temperatures, the metals can both be molten, and often all intermetallic compounds are also molten giving a fully miscible liquid metal phase. Then, the composition of liquid in equilibrium with A_xO_2 varies, being pure A at $P_1°$ and having a rapidly increasing M content with rising $p(O_2)$; it becomes nearly pure M after a few orders of magnitude increase in $p(O_2)$, based on Eq. (8).

Adsorption prospects for the oxide/metal system, in principle, involve any of the three species at each of the three interfaces. *The following gives plausible reactions that yield multicomponent adsorption at interfaces and makes comparisons to similar reactions in the bulk in order to fix ideas and to permit quantification.* It must be appreciated that the reactions to form interfacial species explicitly refer to reversible adsorption, not reactions to make thin layers of what are essentially bulk phases on an energetic basis, e.g., of an oxide on a metal. It is fruitful to anticipate high, low and intermediate $p(O_2)$ regimes, all within the range in Eq. (7).

First consider the higher $p(O_2)$ regime. Oxygen may adsorb onto the metal, which can be expressed assuming dissociative adsorption as:

$$O_{2\{v\}} \rightarrow 2O_{\{IF\}} \qquad \Delta G_{ad}{}^i \tag{9}$$

The adsorption energies, $\Delta G_{ad}^{①}$ for O onto A and $\Delta G_{ad}^{②}$ on M, can be compared to those for oxidation to form the lowest valence, stable oxide Eq. (6).

The Langmuir adsorption isotherm for dissociative adsorption, as in Eq. (9), is:[32,45]

$$\Gamma = \Gamma^* \{p(O_2)^{1/2}/[p(O_2)^{1/2} + p^{*1/2}]\}[(1-X_l)] \qquad (10)$$

where Γ^* is the areal density of surface sites, and the quantity in { } describes the fraction that are occupied. The last term adjusts this to yield the excess quantity when the atomic fraction of surface active species dissolved in the liquid, X_l, is not trivial compared to the concentration near the interface. This model provides a simple description of adsorption statistics and permits straightforward integration of Eq. (5). When X_l is small, so the distinction between adsorbate and excess vanishes, this yields

$$\gamma = \gamma^\circ - \Gamma^* kT \ln \{1 + [p(O_2)/p^*]^{1/2}\} \qquad (11)$$

where γ° is for pure material. A characteristic pressure correlates with the activity for 1/2 coverage of available surface sites and scales as[32,45] $p^* \propto \exp[(\Delta G_{ad})/kT]/2$. This permits estimates of $p(O_2)^{eq}/p^*$ as:

$$P_i^\circ/p^{i*} = B \exp[(\Delta G_{ox}^i - \Delta G_{ad}^i)/kT] \qquad (12)$$

where B is a constant and $p^{①*}$ is for A and $p^{②*}$, for M.

Experimentally determined amounts of adsorbed oxygen, Γ_O, on various metals, largely deduced from surface energy measurements via Eqs. (5) and (11), are compared in Table I with two idealized bases for monolayer coverage, ML. One is the average site density for a metal surface and the other is that for oxygen in the related oxide. Despite variability among studies, trends are clear. Typically oxygen is adsorbed on a metal near some p^* and $p(O_2)^{eq}/p^*$ may be 3-8 orders of magnitude, Table I. The saturation levels tend to yield Γ_O of 1/3 to 1+ ML. So a useful guideline is that *adsorption is a frequent, but not inevitable, precursor to bulk reaction* - taken in the sense of occurring at lower activity (lower free energy). This is directly applicable for the M surface at the high part of the

Table I. Adsorption Characteristics

Metal	Al	Au (liq)	Ag (liq)	Ag (s)	Cu (liq)	Cu - S (liq)*	Fe (liq)	Fe - S (liq)*	Ni (liq)
Γ_O (nm^{-2})	17	0?	3-8	11	2-4	7-9+	9-14+	8-9	12
$V_{MO2/x}^{-2/3}$ (nm^{-2})	17		7.0	7.0	8.6	7.6	12	10	14
$V_M^{-2/3}$ (nm^{-2})	14.4	14.9	14.3	14.3	18.3	18.3	18.3	18.3	19.3
$[p(O_2)^{eq}/p^*]^{1/2}$			100	10^3	10^4		20	3000	
Refs.	25	21	22-24	46	12-15	24	26,28	26-28	29

* Values for S rather than O coverage.
+ Some of the data imply values as much as a factor of two higher near the phase boundary.
The last row uses the quantity $[p(O_2)^{eq}/p^*]^{1/2} \approx X_l^{Oeq}/X_l^{O*}$ in the liquid, to allow comparison to sulfide cases where the ratio of concentrations in the liquid is known.
$V_{MO2/x}^{-2/3} = (MW/\rho N_A)^{-2/3}$, O ion density per plane of the first oxide.
$V_M^{-2/3} = (MW/\rho N_A)^{-2/3}$, atom density per plane of the liquid metal.

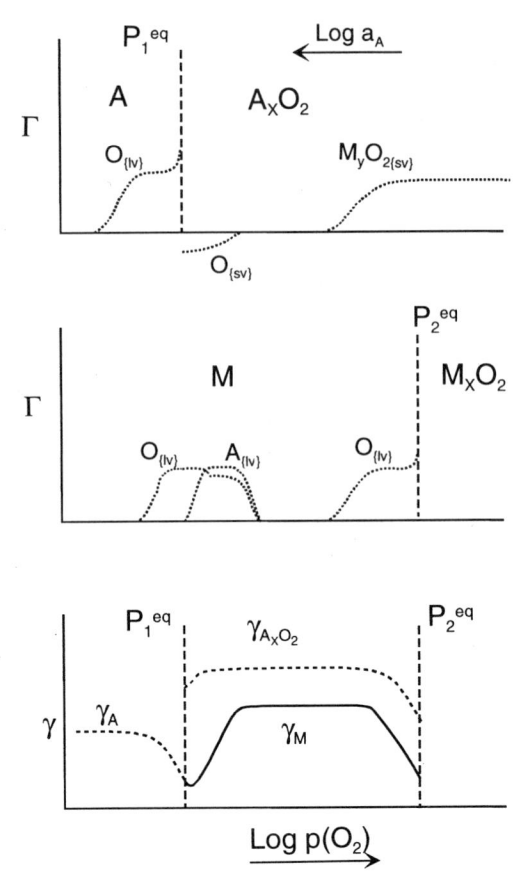

Figure 3. Diagrams illustrating the expected adsorption on the surfaces of the metal and oxide phases containing A in (a) and for M in (b), over a range of $p(O_2)$ emphasizing the coexistence range of the metal/oxide pair, and then showing the requisite behavior of the surface energies in (c). The features typify those expected where the metal is liquid with complete miscibility for A-M liquids.

stable $p(O_2)$ range and has implications for the low end of the coexistence field and for other interfaces where a generalized form of this idea will repeatedly apply.

So typically γ_{lv} is discernibly reduced from $\gamma°$ at p^* and with further increasing $p(O_2)$ is much lower at $p(O_2)^{eq}$, as shown schematically in Fig. 3. As follows from Eqs. (11) and (12), if $p(O_2)^{eq}/p^* \gg 1$, then the surface tension at the phase boundary, $\gamma^{\ddagger}$, compared with that for the pure material scales as:

$$\gamma° - \gamma^{\ddagger} \propto \Gamma^* (\Delta G_{ox}{}^i - \Delta G_{ad}{}^j) \qquad (13)$$

For the oxide surface, the situation could be simple dissociative adsorption of O_2 onto it, as in Eq. (9), but often the energy for this, $\Delta G_{ad}{}^{③}$, is expected to be small. Such adsorption may be important for transition metal oxides, especially lower valence ones, e.g., FeO and TiO.[47] For refractory oxides, e.g., Al_2O_3, this is less likely to apply at $p(O_2)$ ranges of interest.[47,48] More likely to occur is reactive adsorption of oxygen plus some M onto the surface of A_xO_2. This can be expressed as:

$$yM_{\{l\}} + O_{2[v]} \rightarrow M_yO_{2\{v\}} \qquad \Delta G_{ox\text{-}1}{}^M \qquad (14)$$

$$yM_{\{l\}} + O_{2[v]} \to M_yO_{2\{sv\}} \to yM_{\{sv\}} + 2O_{\{sv\}} \qquad \Delta G_{ox\text{-}1}{}^M + \Delta G_{ad}{}^{④} \qquad (15)$$

where y *is distinguished from x*, Eq. (6b), to anticipate the adsorbing species possibly being metastable in bulk relative to M_xO_2. The fully dissociated equation is more appropriate at partial coverage, but it is unclear that it is much better near saturation; so no distinction is considered for the present. This formulation permits ready comparison with the activity for the bulk oxide formation in Eq. (6b). If for simplicity, the adsorbing oxide in (14) is taken as the lowest valence, stable oxide of M, then $y = x$, and,

$$P_2°/p^{④*} \propto 2\exp[-\Delta G_{ad}{}^{④}/kT] \qquad (16)$$

Thus, in the presence of M liquid with unit activity, adsorption of the noble oxide, M_xO_2, onto the surface of the substrate oxide would occur at lower $p(O_2)$ than the bulk oxidation point for M if $\Delta G_{ad}{}^{④} < 0$. Of course, the adsorbing oxide could be another species; for example, SnO, which is metastable in bulk versus SnO_2, could adsorb preferentially.

Based upon experiments[49-52] and atomistic computations[45,53,54] for oxides adsorbing onto Al_2O_3 or MgO, e.g., for MgO, CaO or Y_2O_3 on Al_2O_3, expected adsorption energies are $\Delta G_{ad}{}^{④} \approx 1\text{-}3$ eV. Also some observations reveal that TiO_x adsorbs to varying degrees on Al_2O_3 surfaces in both oxidizing and reducing atmospheres.[55] Thus, similar values of $\Delta G_{ad}{}^{④}$ for NiO adsorbing onto Al_2O_3 seem certain to pertain for Ni-Al_2O_3. If the idea can be generalized to other metal/Al_2O_3 systems, $P_2°/p^{④*} \approx 10^4 \text{-} 10^{11}$ would be typical. As depicted in Fig. 3, this would reduce the oxide surface energy, γ_{sv}, in the high $p(O_2)$ part of the coexistence range for the A_xO_2-M pair.

For the oxide/metal interface, the adsorption is usefully visualized similarly,

$$yM_{\{l\}} + O_{2\{v\}} \to M_yO_{2\{sl\}} \qquad \Delta G_{ox\text{-}1}{}^M + \Delta G_{ad}{}^{⑤} \qquad (17)$$
$$P_2°/p^{⑤*} \propto 2\exp[(\Delta G_{ox}{}^M - \Delta G_{ox\text{-}1}{}^M - \Delta G_{ad}{}^{⑤})/kT] \qquad (18)$$

This produces excess oxygen, Γ_O, at the interface and an associated reduction in interfacial energy, γ_{sl}. Such reactions have been proposed,[11] but direct computational or experimental evidence regarding the magnitude of $\Delta G_{ad}{}^{⑤}$ is sparse. Prior assessments of the $p(O_2)$ sensitivity of θ have recognized that γ_{sl} must diminish at higher oxygen activity,[11-13] but these often made unwarranted assumptions that negate the actual values deduced, i.e., assuming nothing changes on the oxide surface. However, some bonding to the M liquid on one side of the interface and to the oxide on the other should justify this oxide adsorbate being more stable than on the A_xO_2 surface, $\Delta G_{ad}{}^{⑤} < \Delta G_{ad}{}^{④}$, or than on the liquid. This seems especially likely if the M at the interface is only partially ionized, $y > x$.

At low $p(O_2)$, analogous although less simple behavior can be anticipated. For the oxide surface, some local reduction with preferential loss of oxygen leaving a metal rich species adsorbed on the surface can be expected. Thus,

$$A_xO_{2\{s\}} \to (x/y)A_yO_{2\{v\}} + (1 - x/y)O_{2\{v\}} \qquad -\Delta G_{ox\text{-}2}{}^A \qquad (19)$$
$$A_xO_{2\{s\}} \to (x/y)A_yO_{2\{sv\}} + (1 - x/y)O_{2\{v\}} \qquad -\Delta G_{ox\text{-}2}{}^A + (x/y)\Delta G_{ad}{}^{⑥} \qquad (20)$$

where $y > x$. This may occur in the A-O binary. It would become increasingly likely as the activity of A rises at low $p(O_2)$, Eq. (8), as depicted in Fig. 3.

However, as the liquid M is present at high activity, some may adsorb onto the oxide. This could follow a simple $p(O_2)$ independent adsorption on the A_xO_2, but if interactions between the two metals are favorable, adsorption should be preferred when the A_xO_2 surface is partially reduced. This, can be represented utilizing Eq. (20) as:

$$(zx)M_{\{l\}} + A_xO_{2\{s\}} \rightarrow (x/y)A_yO_2M_{zy\{sv\}} + (1-x/y)O_{2\{v\}} \qquad (21)$$
$$-\Delta G_{ox\text{-}2}{}^A + (x/y)(\Delta G_{cm\text{-}1}{}^M + \Delta G_{ad}{}^{\circledast})$$

where the first two energies describe the formation of a compound having the composition of the adsorbate which may be metastable and substoichiometric with respect to A_xO_2. The surface excess can be contemplated in terms of a loss of oxygen, i.e., $\Gamma_O < 0$, or equivalently as excess A, in addition to the excess of M. The coefficient z strictly reflects the ratio of the excess quantities, Γ_M/Γ_A, on the oxide surface. The importance of such a reaction can be rationalized based on near-neighbor bonding: adsorbing an M would replace a dangling A bond on the partially reduced A_xO_2 surface with an A-M bond plus a dangling M bond. This dictates that such adsorption would be increasingly favorable when the A-M interaction is stronger and, also likely, if γ_{Mv}/γ_{Av} is smaller.

Note, adding Eq. (6) to Eqs. (19-21) yields equivalent relations which emphasize the adsorbate being an excess of A rather than a deficiency of oxygen. Also, doing so makes it clearer that if $\Delta G_{ad}{}^{\circledast} < 0$, the surface will reduce at a higher $p(O_2)$ than does the bulk, analogous to behavior described for oxidizing a metal, Eq. (12). Then, for the reaction in Eq. (20), simplifying to $y \rightarrow \infty$ gives:

$$p^{\circledast*}/P_1{}^\circ \propto \exp[-\Delta G_{ad}{}^{\circledast}/kT]/2 \qquad (22)$$

For the metal surface, two different considerations are insightful. First note that for pure A, normally, O will adsorb at $p(O_2) < P_1{}^\circ$, Eq. (12), so that a metal surface would be saturated with O at $p(O_2) = P_1{}^\circ$. Also the liquid M typically dissolves some of metal A, Fig. 2, and the solubility varies with $p(O_2)$, being essentially $X_l^A \rightarrow 1$ at $P_1{}^\circ$ and vanishing at higher $p(O_2)$, as Eq. (8) pertains. However, any A at the liquid alloy surface would strongly attract oxygen adsorbate at these $p(O_2)$ levels. A simple model is as follows. At the surface, A atoms would adsorb O with the same probability as occurs at $P_1{}^\circ$ (where $\Gamma_O{}^*$ applies), but as the concentration of A in the melt diminishes, Γ_O would diminish in proportion. The simplest consistent relations are that $\Gamma_A = 0$ and accordingly that

$$\Gamma_O = \Gamma_O{}^*(P_1{}^\circ/p(O_2))^{1/x} \qquad (23)$$

This should be a lower bound, as the A-O attraction would enhance the surface population of A, inducing an excess of both Γ_O and importantly of Γ_A on the liquid metal.

An approach that allows this would be attractive especially if in the metallic phase, A were surface active on M. This is likely with Al_2O_3, as Al has a lower surface energy than do several metals, e.g., Cu, Pt, Ni;[16,17] this is particularly so comparing γ_{lv} for an oxygen covered Al surface and for the clean noble metals. Once more, it is pertinent to invoke that the adsorption energy could induce early partial reduction of the A_xO_2 and adsorption of the more metallic species, now onto the liquid. Such an adsorbed A_yO_2 molecule would be bonded to the metallic liquid on one side of the interface and equilibrated with O_2 from the vapor on the other. Then, two parallel reactions must be compared; first,

$$A_xO_{2\{s\}} \rightarrow (x/y)A_yO_{2\{lv\}} + (1-x/y)O_{2\{v\}} \qquad -\Delta G_{ox\text{-}2}{}^A + (x/y)\Delta G_{ad}{}^{\circledast} \qquad (24)$$

once again with $y > x$. However, if the major species dissolving within the liquid is fully metallic, the controlling reaction is essentially Eq. (6a). So, as $p(O_2)$ rises above $P_1{}^\circ$, the level of A decreases much faster in the bulk liquid than on the surface, Eq. (24). Thus, the concentration of A_yO_2 on the surface would far exceed that of either A or O in the liquid for $p(O_2) \gg P_1{}^\circ$ and so would essentially constitute surface excess.

The considerations expressed in Eqs. (23) and (24) anticipate that on the metal $\Gamma_O > 0$. Critically, the latter also requires that $\Gamma_A > 0$. To appreciate the impact, note that owing to the constraint in Eq. (8), the Gibbs adsorption relation is:

$$d\gamma = -(kT/2)\{\Gamma_O - (2/x)\Gamma_A + (2d\ln a_M/d\ln p(O_2))\Gamma_M\}\, d\ln p(O_2) \tag{25}$$

Thus with $\Gamma_M = 0$, the ratio of Γ_O/Γ_A dictates whether γ rises or falls with changing $p(O_2)$. In particular, in the situation depicted in Eq. (24), the coefficient in brackets in Eq. (25) would be negative somewhat above $P_1°$, and so the *surface energy would increase* with $p(O_2)$. However, for the reaction in Eq. (24), the amount of adsorbate must be decreasing with increasing $p(O_2)$. Thus, the γ_{lv} would be at a minimum at some $p(O_2) > P_1°$ and would increase, likely to the value for pure M at higher $p(O_2)$, Fig. 3c.

This behavior, illustrated in Fig. 3b, can actually be easily understood by considering the situation of an essentially pure M phase in equilibrium with A_xO_2 and a declining $p(O_2)$. Upon approaching some $p^{⑧*}$, partial reduction of A_xO_2 and associated adsorption of A_yO_2 onto the M would initiate yielding an excess of this species on the liquid, if $\Delta G_{ad}^{⑧}$ is sufficiently negative. With further decrease in $p(O_2)$ (increase in a_A), the surface would saturate, and still later, dissolution of A into the bulk would increase, growing large near $P_1°$. Thus, upon approaching the phase boundary, Γ_A on the liquid would vanish, owing to the large amount in solution, but Γ_O would approach the saturation limit for pure A. An estimate of the onset condition for this adsorption on the liquid is:

$$p^{⑨*}/P_1° \propto \exp\{[(y/(y-x))(\Delta G_{ox-2}{}^A - (x/y)\Delta G_{ad}^{⑧}) - \Delta G_{ox}{}^A]/kT]\} \tag{26}$$

For the interface, the expected reductive adsorption is also usefully expressed as,

$$(zx)M_{\{1\}} + A_xO_{2\{s\}} \to (x/y)A_yO_2M_{zy\{sl\}} + (1-x/y)O_{2\{v\}} \tag{27}$$
$$-\Delta G_{ox-2}{}^A + (x/y)[\Delta G_{cm-1}{}^M + \Delta G_{ad}^{⑨}]$$

Again the relevant excess quantities can be expressed as an oxygen loss, $\Gamma_O < 0$, plus some excess M deriving from an adjusted local density near $p^{⑨*}$. Note that $(\Delta G_{cm-1}{}^M + \Delta G_{ad}^{⑨}) < \Delta G_{ad}^{⑥}$ as well as $(\Delta G_{cm-1}{}^M + \Delta G_{ad}^{⑦}) < \Delta G_{ad}^{⑥}$ if an A-M interaction is strong. In fact, at the metal/oxide interface, an A-M interaction is unavoidable, as bulk M is necessarily present; whereas on the oxide surface, M atoms will only be present if the interaction induces them to adsorb from the bulk M. Thus, $\Delta G_{ad}^{⑨} < \Delta G_{ad}^{⑦}$ should be frequent.

However, modifications ensue as the liquid becomes A rich at very low $p(O_2)$ and a_M decreases. First, if the adsorption reaction follows Eq. (27), then some M adsorbate will occupy the interface as the bulk M content nearly vanishes. In the context of Eq. (25), the Γ_M term would become predominate relative to the negative Γ_O. The resulting reversal in sign of the coefficient means that γ_{sl} versus $p(O_2)$ (or $X_l{}^A$) would have a minimum, as for γ_{lv}, but likely being deeper and occurring at smaller M content, i.e., very near $P_1°$. Even without segregation of M at the A/A_xO_2 interface, i.e., if $z = 0$ in Eq. (27), some reduction in Γ_A may occur upon approaching $P_1°$. In the limit, Γ_A must approach that for the binary interface. It is not obvious whether this would be as great as for the hypostoichiometric A_xO_2 surface; if it were intermediate between that value and the $\Gamma_O > 0$ expected on the A surface, it would be small. This level of Γ_A would produce an inflection in $\gamma_{sl} - p(O_2)$ but not a minimum as with M adsorption. An insight about this limiting value emerges from realizing that $-\Gamma_A$ is equivalent to the excess (relaxation) volume that would arise from bonding stoichiometric surfaces of A and A_xO_2 together; from this, it is perhaps plausible that Γ_A is > 0 but is less than for the oxide surface. For the oxide surface, a modification may also pertain; namely as the activity of M decreases, any tendency for M adsorption onto the surface via Eq. (21) in preference to Eq. (20) may be diminished.

Finally, a parallel exists at high $p(O_2)$; A_xO_2 could adsorb onto the metal as a complex with M_yO_2. This could be represented by adjusting Eq. (15), analogous to the addition to Eq. (20) giving Eq. (21), instead of using Eq. (9).

Finally, these considerations provide viable arguments for adsorption being relevant on all three interfaces at both extremes of the $p(O_2)$ range of stability. If $P_2°/P_1°$ is large enough, an intervening regime should exist in which neither net oxidation nor reduction type adsorption reactions pertain. This would be the intrinsic, or *plateau regime* wherein the interfaces would be representative of the hypothetical binary system of A_xO_2-M. In principle, M could be adsorbed onto the oxide surface, and A_xO_2 could be adsorbed onto the metal, both reactions being independent of $p(O_2)$. However, the energies for these are probably not favorable if the range of $\ln p(O_2)$ is very large; then, these surfaces should be pure and stoichiometric, i.e, $\Gamma_i = 0$. Similarly, absent an oxidation/reduction reaction occurring at the interface, the adsorption can be described in terms of the excess volume associated with the interface, again independent of $p(O_2)$.

Less information exists about behavior at intermediate and low regimes of $p(O_2)$ than for the high $p(O_2)$ end. The objective is to know explicitly how surface stoichiometry or adsorption depends on temperature and $p(O_2)$, for oxide surfaces and the interfaces as well as the metals. A plethora of information derives from surface spectroscopy studies in UHV (ultrahigh vacuum). Unfortunately, relationships to equilibrium conditions must be inferred as the surfaces formed by cleaning or deposition are typically only metastable owing to low gas arrival rates, and activities are not meaningfully defined in cold wall chambers wherein gas phase equilibration with the surfaces does not obtain. Thus, annealing effects reveal trends but not adsorption energies for statistical models.

For sapphire, LEED evidence reveals that the (0001) surface changes reversibly at ~1000°C between oxidizing and reducing conditions, as obtained in UHV. It was proposed that a surface transition involves shifting between stoichiometric and partially reduced states;[56] further surface science studies have justified a refined scheme for the latter with $\Gamma_{Al} = 14$ nm^{-2}.[57] However, from Eq. (22), it can be realized that if the surface can be reduced in UHV, then p^* for 1/2 coverage of excess Al must be orders of magnitude greater than $P_1°$, which at 1000°C is ~10^{-35} atm and inaccessible in a UHV system (e.g., the bulk sapphire is not reduced). Thus, from Eq. (22), $-\Delta G_{ad}^{⑥}$ may be several eV for basal planes, but the behavior of other planes is unknown.

In addition, many M/Al_2O_3 combinations exhibit stable intermetallics; this indicates Al-M interactions are strong and that mixed adsorption such as contemplated in Eq. (21) or (27) is likely. Some XPS and Auger spectroscopy studies of Al_2O_3 with a few atomic layers of deposited metal have revealed bonding features found for intermetallics, e.g., with Ni, for which annealing at higher $p(O_2)$ can eliminate such attributes.[58] This reinforces the idea that reductive adsorption at interfaces pertains at very low $p(O_2)$ and could be favored relative to the surface nonstoichiometry, i.e., $(\Delta G_{cm-1}^M + \Delta G_{ad}^{⑨}) < \Delta G_{ad}^{⑥} < 0$.

Interpreting such spectroscopic information requires distinguishing among transitional bonding at an interface having a stoichiometric oxygen level plus substoichiometric A balanced with partially ionized M versus reactions involving excess Al or O. Recent evaluation of spatially resolved EELS peaks from a STEM have revealed evidence of Cu^{+1} being present at a Cu/Al_2O_3 interface, consistent with concepts, Eq. (17), for oxygen adsorption at high $p(O_2)$.[59] Studies using high resolution TEM[60] or atom probe[61] have sought to deduce whether the last atomic layer of oxide is oxygen or metal at such interfaces, despite imprecisions regarding exact populations of transition planes. One study of Cu/MgO interfaces found excess oxygen.[61] These interfaces were plausibly equilibrated at $p^{④*} < p(O_2) < P_2°$, but the role of $p(O_2)$ requires affirmation. In addition, Pd/Al_2O_3 or

Pd/MgO interfaces were found to act as strong traps for hydrogen after being annealed in air but not after annealing in reducing conditions, revealing a major change occurs in the interfacial species which was interpreted as excess oxygen serving as the H traps.[62] Lastly, both computations[63,64] and TEM[60b] studies of Nb/Al$_2$O$_3$ (0001) have indicated that Nb ions preferably replace the last row of Al; clearly this situation is not stoichiometric, but corresponds to having $\Gamma_O > 0$ and so $p(O_2)$ dependent interfacial energy.

Several experimental problems, in addition to ridging, can yield results which deviate from equilibrium behavior for a metal/oxide couple. Impurities are a continual concern. There is a trend that measured surface energies for metals have risen over the years[11,17,19] which owes, in part, to improved system purity. With highly refractory liquid metals, large errors in θ can occur as the liquid tends to be oxidized. Typically, systems must be heated well above the melting point to eliminate the oxide skin on the liquid.[9,10,65-67] For some metals, e.g., Sn, the furnace atmosphere can become reducing relative to $p(O_2)^{eq}$ at higher temperatures.[66,67] However, with Al containing metals, this situation is not easily attained. Instead, evaporation of Al, Al$_2$O, etc. is sufficiently fast that a passive oxide layer cannot develop;[65,68] in essence, such systems are in a regime of active oxidation. Finally, equilibrating oxygen among the atmosphere, liquid and interfaces may occur slower than expected, even when ridge drag limited spreading provides extra time to accomplish this. However, enhanced wetting, termed "dry spreading," can obtain in situations where adsorbate would be on the surface at equilibrium, but is absent while the liquid actually spreads.[35,69,70] This can markedly promote spreading in aqueous or organic systems, but effects in those of interest here are not yet appreciated.

EQUILIBRIUM WETTING

Wetting studies for many liquid metals on Al$_2$O$_3$ have been reported in the literature. The reported behavior for less reactive systems coupled with recent observations will be assessed in terms of models and concerns for systematic errors just described.

Sessile drop experiments for several pure metals (Sn, Au, Cu, Ni, Pd, Pt) and Al on sapphire have been conducted *in vacuo* or in Ar, using furnaces with Ta or C heating elements.[71] For metals with easily reduced oxides, experiments were conducted 50-70°C above the melting point. The θ was monitored *in situ* in a horizontal tube furnace for metals with melting points below 1100°C. For experiments in a Ta element, top-loaded furnace, used for higher melting point metals or with closed crucibles, θ can only be assessed on the cold, solidified drop. Experience with Cu and Au revealed that these angles were within ±3° of those observed while hot. The $p(O_2)$ was evaluated by analyzing the oxidation of various metal foils (Ni, W, Mo, Cr, Ti ..). Finally, wetting of Ni and Cu at the phase boundary for M/M$_x$O$_2$ has been determined by using a closed alumina crucible and emplacing the metal/sapphire assembly in an inner crucible surrounded by a powder mixture of the metal, its oxide and sometimes alumina.

Drops of Cu on sapphire having both receding and advancing fronts take some minutes to approach the *same equilibrium value*, Fig. 4, which is deemed to satisfy Eq. (1). Ongoing studies affirm that the slow approach is controlled by formation and motion of submicron ridges, Fig. 1b, as seen by AFM.[34] Similar behavior was achieved in a tubular furnace with Ta or C heaters *in vacuo* or gettered Ar, in the top-loaded Ta furnace in Ar, and in a closed sapphire cell inside the tubular furnace *in vacuo*. The $\theta \approx 114°$ at intermediate $p(O_2)$ in all furnaces suggesting that impurity effects have been minimized. At the phase boundary, θ is just above or below 90° depending whether at $P_2^{\ddagger}$ or P_2°, respectively.

With Al/sapphire, it is well known that near the melting point the strong oxide skin on the liquid impedes fluid flow.[9] With increasing temperature, this skin ruptures and then disappears; a shiny metallic surface and nearly spherical droplet shape indicate that the contact angle is dictated by capillary forces. Our experiments were at 1150°C, where the droplet becomes "clean" under conditions where active oxidation predominates.[65,68] In these circumstances, mass loss from the droplet is not trivial.[10] A consequence is that the contact angle slowly decreases with time, then the contact front abruptly moves inward with an attendant increase in θ, and the cycle repeats, Fig. 4. Evidently the liquid front is pinned by a growing triple line ridge, while evaporation reduces the droplet volume so fast that eventually θ departs from equilibrium enough for the junction to break-away and rapidly re-approach the equilibrium value. Hence, the actual equilibrium θ is the extreme not an average over this cycle, Fig. 4. Analogous departures from the equilibrium angle can be seen with Cu when held *in vacuo* for long periods.[34]

There are extensive data for Au, Cu and Ag on Al_2O_3. With Au, little change in γ_{lv} or θ with $p(O_2)$ is exhibited over the range 10^{-22} - 1 atm.[21,71] Evidently O_2 does not adsorb onto Au nor adsorb to form inherently hyperstoichiometric Al_2O_3 surfaces below 1 atm $p(O_2)$ near 1100°C. The behavior with Ag is similar to that for Cu, which will be assessed to yield a hypothetical set of γ vs. $p(O_2)$ functions that describe the wetting.

First note, the binary Al/Al_2O_3 system is unique in that the two phases only coexist at $P_1°$; the equilibrium θ for this is ~80°, Fig. 4. Studies have given γ_{lv} = 1.05 J/m² for clean, liquid Al (based on Auger spectroscopy) and shown some reduction in γ_{lv} with O adsorption, which can exceed a ML.[25] Unfortunately, γ_{lv} data were not measured to the phase boundary. Older data are as low as 0.5 J/m² perhaps reflecting O adsorption. However, if $p*/p(O_2)^{eq}$ is typical of other metals, then γ_{lv} should be low compared to the pure situation. Thus, we adopt $\gamma_{lv}^{\ddagger}/\gamma_{lv}°$ = 0.4 at $P_1°$, which is arbitrary but typical for γ_{lv} when Al is equilibrated with Al_2O_3.

The γ_{lv} for Cu exhibit expected trends for plateau and high $p(O_2)$ behavior,[12-15,19] Fig. 5. Some differences among reports are attributed to impurities which lower γ_{lv}. Contact angles for Cu/sapphire, Fig. 6a, exhibit more variability, with lower θ levels found more recently,[12,13,15,71-73] but no sensitivity to orientation has emerged.[13,15] Based on concepts above, a set of three interfacial energies, shown in Fig. 6c, are proposed as rendering a "best fit" to the θ and γ_{lv} data and satisfying other considerations. The associated θ and W_{ad} from these, Eqs. (1) and (3), are compared to originally determined quantities in Figs. 6a and 6b, respectively. The higher γ_{lv} and recent, lower θ values are deemed most representative of

Figure 4. Evolution of contact angles showing two different effects of ridging. The behavior for Cu/Al₂O₃ in dry Ar shows the convergence for two types of drops with initially different shapes so that one front advances and the other recedes toward the equilibrium θ. The slow evolution is dictated by drag from the triple line ridge. The set of data for Al/Al₂O₃ *in vacuo*, show combined effects of rapid evaporation and successive formation of ridges that pin the droplet periphery. Evaporation of liquid Al causes θ to change until a breakaway condition is reached; whereupon, the droplet front breaks away from the ridge, and quickly approaches a new equilibrium. The dotted lines show the equilibrium values of macroscopic contact angle for each case.

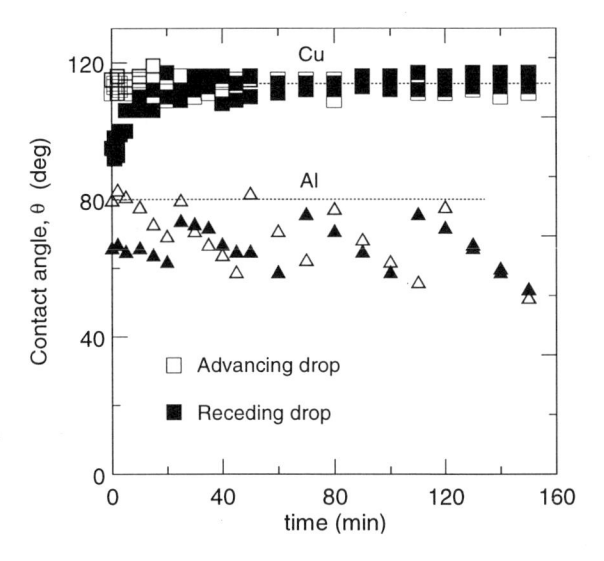

Figure 5. Measured values of liquid metal surface energy, γ_{lv}, versus $p(O_2)$ for Cu liquid on an Al_2O_3 substrate as reported by various investigators. Also shown is the expected value for pure Al/Al_2O_3 at $p(O_2)^{eq}$; this is the limiting value for the binary Cu-Al liquid equilibrated with Al_2O_3. The heavy line is also that in Fig. 6-c, the "best-fit" value.

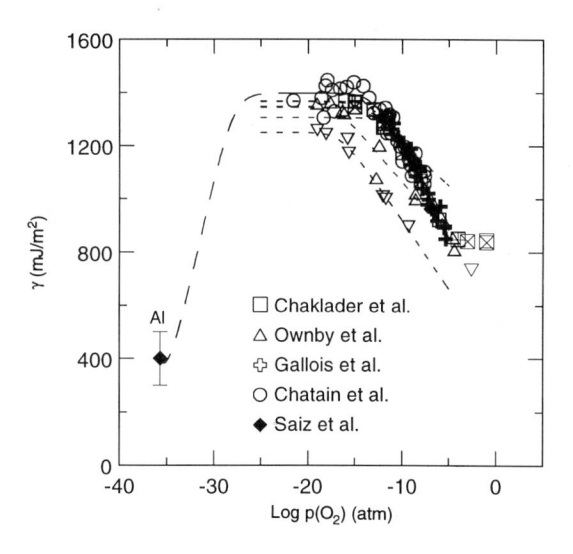

the pure system. At high $p(O_2)$, θ drops to < 90° which implies that γ_{sl} decreases below γ_{sv}; but γ_{sv} could also decrease, both owing to Cu-O adsorption, Eqs. (15) and (17). The choice among having more adsorbate on the interface, initiating adsorption at a lower $p(O_2)^*$ compared to that for the Al_2O_3 surface, or a combination is rather arbitrary. Details of the choices yield fine structure in the plots of θ and W_{ad} vs. $p(O_2)$.

At temperatures where Cu is molten, Al and all intermetallics are also molten. Thus, at low $p(O_2)$, behavior for Cu-Al alloys equilibrated with Al_2O_3 should approach that for pure Al, as depicted in Figs. 2 and 3; the Al content in the liquid predominates at $p(O_2) = P_1°$ and declines at higher $p(O_2)$. Comparing experiments[14,15,19,22] reveals little effect on γ_{lv} from exposure to pure Al_2O_3 for Cu (or Ag) at intermediate $p(O_2)$. So with rising $p(O_2)$, γ_{lv} must adjust from that for Al/Al_2O_3 to the higher one for pure Cu obeying a relationship deriving from Eqs. (24) with (10), shown in Fig. 5. Also, we assume the Al_2O_3 surface is simply somewhat hypostoichiometric (Al rich) at low $p(O_2)$, and that reductive adsorption also occurs at the interface but with requisite higher values of p^* and coverage so that θ declines from the plateau level for Cu/Al_2O_3 to ~ 80° at the phase boundary (the Al/Al_2O_3 limit, Fig. 4). The construction in Fig. 6 yields plausible variations of the γ's to accomplish this (assuming that Γ_{Al} for the interface decreases near $P_1°$). A set of measured angles for Cu-Al alloys,[73] Fig. 6b, is qualitatively consistent, but seem too high, likely owing to issues from Al_2O_3 skin formation on the metal.

With Ni/Al_2O_3, contact angles for the apparent plateau and high $p(O_2)$ limit are similar to those for Cu,[71] and an expected effect of high $p(O_2)$ on γ_{lv} has been reported.[29] Also, θ for Ni-Al alloys has been found to decline with Al content[74] qualitatively as depicted for Cu at low $p(O_2)$, Fig. 6a. For Pt and Pd, the θ's are rather similar but effects of $p(O_2)$ are unknown. However, in nongettered Ar, Pt droplets in an open system migrate and exhibit a slightly lower θ than found in a closed sapphire cell, 110° versus 114°; this, plus enhanced faceting suggests that an unidentified adsorbate can affect the sapphire surface and induce dry spreading.[71] The latter θ is regarded as nearer the plateau level, but more investigation is needed. In addition, the low $p(O_2)$ regime should be more accessible for high melting metals, but it has not been identified with Pt or Pd. The $P_1°$'s increase steadily with temperature. Moreover, from thermodynamic data, $P_1^{\ddagger}/P_1°$ is $10^3 - 10^7$ for Al to form Ni_3Al with Ni or to dissolve 50% in Pt.[43,42] Finally, favorable reduction and adsorption at interfaces, Eqs. (21) and (27) would make $p^*/P_1°$ even greater. Thus, the plateau region for Ni/Al_2O_3 may be much narrower than for Cu/Al_2O_3, by half in terms of $\ln p(O_2)$.

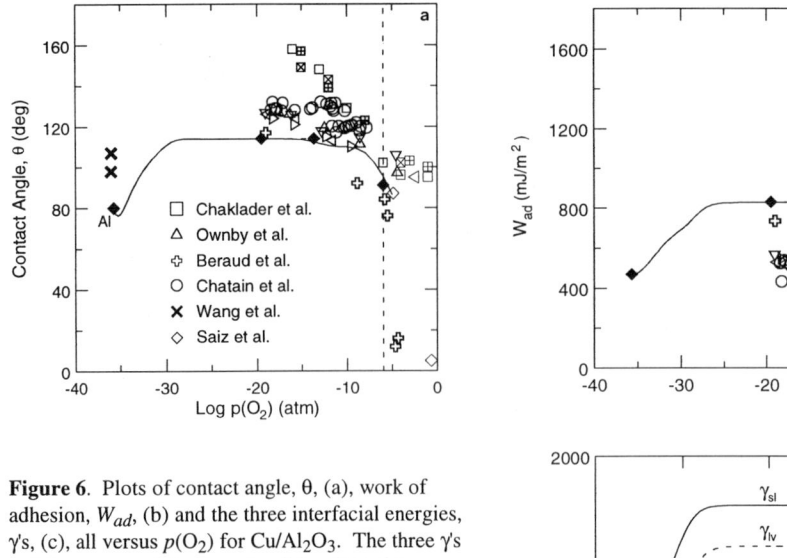

Figure 6. Plots of contact angle, θ, (a), work of adhesion, W_{ad}, (b) and the three interfacial energies, γ's, (c), all versus $p(O_2)$ for Cu/Al$_2$O$_3$. The three γ's in (c) are computed using plausible adsorption isotherms as "best-fit" values that describe the measured levels of θ and $γ_{lv}$ and meet other constraints. The resulting computations of θ and W_{ad} are shown as heavy lines in (a) and (b). This calculated W_{ad} is compared with values implied directly from various experimental measurements of θ and $γ_{Cu-v}$.

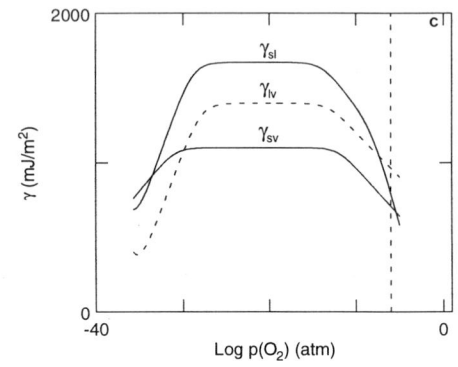

With Sn/Al$_2$O$_3$, the oxide skin disrupts during heating at ~850°C. By 1000°C in dry Ar, the furnace atmosphere is below $p(O_2)^{eq}$, the droplet is well behaved, and evaporation is slow. Upon heating beyond the point of skin rupture, θ reportedly rises until reaching a higher level that is maintained with further heating.[66,67] In the context of model behavior, the angle attained just after cracking the skin is near the minimum expected from inherent adsorption when $p(O_2) \approx P_2°$. With further heating, the atmosphere is increasingly below $P_2°$ and θ rises, approaching the plateau value when the ambient is less than $p^{②*}$. Also, Sn-Al alloys at high temperature reveal a low $p(O_2)$ regime; it has been reported that θ drops with Al content, even below that for Al.[67] This seems consistent with $γ_{lv}$ being less for oxygen covered Al than for pure Sn. Tin may typify other low melting point metals in that surfaces are impure unless heated well beyond melting.

DISCUSSION

A conceptual model for the mutual adsorption in a metal/metal-oxide system has been developed based on generalizing well known behavior for oxygen interactions with metals. Typically oxygen is adsorbed onto a metal at a much lower $p(O_2)$ than is needed to oxidize it. Thus, adsorption reactions are proposed for all the interfaces present, at both ends of the stable range of $p(O_2)$. These will prevail under conditions where no third phase is present. However, the adsorption reactions may anticipate the bulk reactions that actually terminate the coexistence field; those reactants, or metastable variants, are stabilized by attractions to the interfaces. That is M_xO_2 may be adsorbed at a lower activity than needed to form it in bulk. Similarly, adsorption can involve reduction, often modified by A-M interactions, at higher $p(O_2)$ than is required for bulk phase formation.

This motivates the assumption invoked repeatedly that the multicomponent adsorbate has a fixed stoichiometry, e.g., Eq. (15), which, in turn, justifies using Langmuir type adsorption models and simple estimates of adsorption energies that expedite assessing relevance, e.g., Eq. (16). Models having more realistic adsorbate interactions, that allow mixing of adsorbate and host atoms[24] or incorporate space charge effects[75,76] may be more satisfying. They would yield adsorption energies that depend upon coverage, and coverage-activity-temperature relations would differ in detail. Nonetheless, the key aspect of substantial coverage occurring at a characteristic $p(O_2)^*$ would remain, and rarely are high temperature data, e.g., O or S on M, adequate to distinguish among models.

Should an intermediate range of $\ell n\, p(O_2)$ exist in which no mutual adsorption occurs involving net oxidation or reduction of interfacial species, then contact angle and work of adhesion values will also exist that can be taken as characteristics for the quasi-binary oxide/metal system; these have been termed "plateau" values. They may describe γ's and W_{ad} for pure, stoichiometric interfaces, which are those tacitly contemplated for most correlations[11,17,77,78] or calculations[79-85] for bonding between pure oxides and metals. In principle, $p(O_2)$ independent adsorption could occur in this regime, although it is deemed unlikely for materials with a large coexistence range of $\ell n\, p(O_2)$, based on behavior of M/Al$_2$O$_3$ systems. However, for materials with smaller $P_2°/P_1°$ ratios, or for which A and M are both multivalent, such intrinsic or plateau behavior is unlikely to exist.

One result is a provisional set of γ vs. $p(O_2)$ functions for Cu/Al$_2$O$_3$. These exhibit notable adsorption effects on the ceramic as well as on the metal and interface at both high and low $p(O_2)$; they are consistent with lower contact angles at either extreme of the coexistence field. Behavior for other metals implies these should be widespread attributes. The model reveals a larger effect of $p(O_2)$ on the γ's than on θ or on W_{ad}. Clearly, predicting effects on W_{ad} is less reliable as it depends upon differences in the strengths of adsorption among the interfaces. Similarly, direct determinations of W_{ad}, based on θ and $γ_{lv}$, disagree markedly among different experiments.

New predictions have emerged regarding behavior at low $p(O_2)$. However, with Al containing systems, this situation is not easily achieved owing to oxidation of the metal in typical furnace atmospheres. Instead, reproducible behavior derives under conditions wherein the outward flux of Al, Al$_2$O, etc. vapor exceeds that of O$_2$ toward the surface, yielding a halo of low $p(O_2)$ around the drop.[65-68] An open question is whether under such

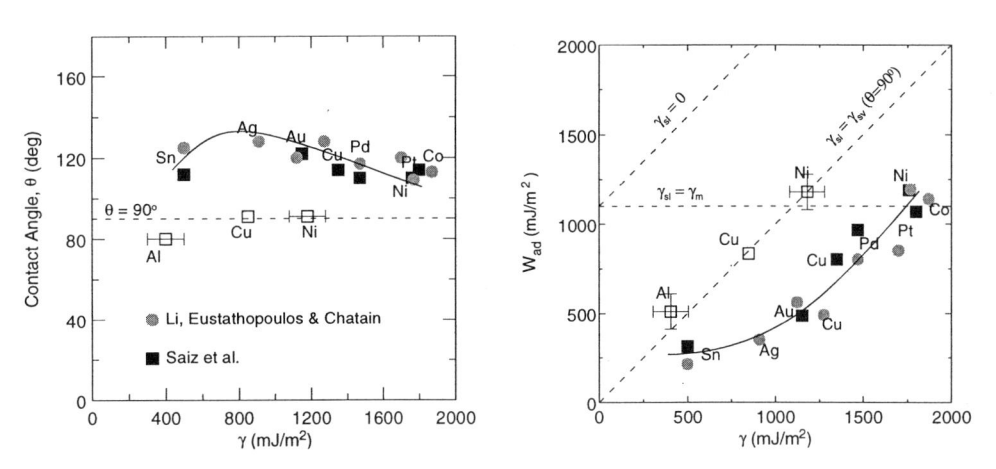

Figure 7. Contact angle, θ, and work of adhesion, W_{ad}, determined for various metals with Al$_2$O$_3$ are plotted against surface energy for the metal. Values appropriate to the plateau condition (stoichiometric interfaces) (full) and the high $p(O_2)$ limit (open) are shown.

conditions of active oxidation, where fluid behavior yields $\theta \sim 80°$, it can be expected that adsorbed O is present on the metal or that any of the interfaces are at equilibrium. It has been tentatively assumed so here. A second uncertainty involves the adsorption at the oxide surfaces and the interfaces near the phase boundary. Noble metal segregation or a negative excess volume at the Al/Al$_2$O$_3$ interface would, respectively, deepen or reduce the minima in θ, Fig. 6a, but independent evidence for either is wanting.

Values of θ and W_{ad} for various metals with Al$_2$O$_3$ are plotted in Fig. 7 versus γ_{lv} for the liquid metals where putative identifications of plateau conditions or high $p(O_2)$ limits are merited.[71] These utilized measured θ with W_{ad} deduced via Eq. (3); the γ_{lv} are taken from the literature.[14-19,21,22,25] Similar results from recent compilations of other work are also presented.[16,17] For comparison, an average $\gamma_{sv} = 1.1$ J/m^2 for stoichiometric Al$_2$O$_3$ is shown. The W_{ad} are systematically higher than in prior correlations.[3]

Plateau levels of W_{ad} strongly increase with γ_{lv}, but do not correlate with ΔG_{ox}^M for these metals, as sometimes appears.[11,77] Apparently, stronger bonding to Al$_2$O$_3$ derives from metals having higher bond energies. This could be envisaged as a local metal-metal interaction[78] or a covalent bonding contribution. Recent quantum mechanical calculations for a few M/MgO systems exhibit a similar trend of W_{ad} but with rather higher levels (1-2 J/m^2);[80-85] these may be too high due to neglected effects of lattice misfit and surface relaxations. A lower limit to W_{ad} for a noninteracting metal/oxide system derives from image and dispersion forces;[48,76,79] an estimate of $W_{ad} \sim \gamma_{oxide}/4$ could be inferred from data for metals with lower γ_{lv}, Fig. 7b. This is below simple estimates ($W_{ad} \sim \gamma_{oxide}$) or recent computations for M/MgO systems (~ 1 J/m^2), perhaps owing to image forces being smaller with larger metal atoms. Adjustments for temperature were not attempted, which could cause some differences relative to calculations of W_{ad} for T = 0. Further work to assure plateau conditions and avoid impurities may improve insights. Finally, a trend that $\theta \sim 90°$, and so $\gamma_{sv} \sim \gamma_{sl}$ at the high $p(O_2)$ boundary is emerging. The associated W_{ad}, if anything, exceed plateau values, but cannot be directly interpreted in terms of bonding across the interfaces as different amounts of work would derive from ideal interfacial fractures that precluded adjustment of adsorption levels for those interfaces.[4]

SUMMARY AND CONCLUSIONS

A liquid metal on an oxide substrate is stable over a range of oxygen activity; thus, adsorption has been considered as dependent on $p(O_2)$, which is a fundamental variable. Models developed exploit the idea that adsorption reactions can often be anticipated to occur as energetic precursors to bulk reactions. In parallel, a variety of data indicate that adsorption effects are significant for all three interfaces, including the oxide surface which has often been neglected owing to a lack of data for γ_{sv}. Adsorption is more likely at the ends of the coexistence field. Thus, where the two limiting $p(O_2)$ values differ greatly, a $p(O_2)$ range may exist within which all interfaces are pure and stoichiometric. The θ and W_{ad} from this plateau condition are the intrinsic values for the metal/ceramic system.

For several metal/Al$_2$O$_3$ pairs examined: the intrinsic θ's are 110-130°; angles at the high $p(O_2)$ limit drop toward or below 90°; and near the minimum $p(O_2)$, θ's below 90° are also indicated. A detailed evaluation yielded provisional interfacial energies for Cu/Al$_2$O$_3$. Realizing the low $p(O_2)$ regime is difficult owing to persistent, thin Al$_2$O$_3$ layers on the metal. At higher temperatures, where Ni, Pd or Pt melt, the low $p(O_2)$ regime may be more important, owing both to easier reduction of Al$_2$O$_3$ and to stronger interactions of Al with these metals. Also, the plateau W_{ad} increases with the metal γ_{lv}, especially once it is above a threshold value. Finally, the actual levels of W_{ad} exceed those typically found earlier. This reflects a combination of experiments with cleaner systems, appreciation of the $p(O_2)$

regimes, and understanding of the potential role of triple line ridging. Analysis of the latter has helped to reveal regimes and conditions wherein the macroscopic contact angle does approach that from the Young equation.

Acknowledgments

This work was supported by the Director, Office of Energy Research, Office of Basic Energy Sciences, Materials Sciences Division of the U. S. Department of Energy under Contract No. DE-AC03-76SF00098. Numerous discussions with Profs. J. A. Pask and M. Rühle and the assistance of S. Foppiano are much appreciated.

References

1. A. P. Tomsia, J. A. Pask & R. E. Loehman, ASM International Engineered Materials Handbook, vol 4 on *"Ceramics and Glasses - Sect. 7, Joining,"* p. 482 (1991).
2. M. G. Nicholas & D. A. Mortimer, *Mat. Sci. & Tech.*, **1** 657 (1985).
3. J. T. Klomp, *Mat. Res. Soc. Symp. Proc.*, **40** 381 (1985).
4. J. P. Hirth & J. R. Rice, *Metall Trans. A*, **11A** 1501 (1980).
5. A. G. Evans & B. J. Dalgleish, *Acta Metall. Mater.*, **40** S295 (1992).
6. B. J. Dalgleish, E. Saiz, A. P. Tomsia, R. M. Cannon & R. O. Ritchie, *Scripta Metall. Mater.*, **31** 1109 (1994).
7. A. P. Tomsia, E. Saiz, B. J. Dalgleish & R. M. Cannon, *Proc. 4th Japan Intl. SAMPE Symposium*, Tokyo, p. 347 (1995).
8. M. Humenik, Jr. & W. D. Kingery, *J. Am. Ceram. Soc.*, **37** 18 (1954); C. R. Kurkjian & W. D. Kingery, *J. Phys. Chem.*, **60** 961 (1956).
9. J. J. Brennan & J. A. Pask, *J. Am. Ceram. Soc.*, **51** 569 (1968).
10. J. A. Champion, B. J. Keene & J. M. Sillwood, *J. Mat. Sci.*, **4** 39 (1969).
11. Ju. V. Naidich, *Prog. Surf. Membr. Sci.*, **14** 353 (1981).
12. T. E. O'Brien & A. C. D. Chaklader, *J. Am. Ceram. Soc.*, **57** 329 (1974); S. P. Mehrotra & A. C. D. Chaklader, *Metall. Trans. B*, **16B** 567 (1985).
13. P. D. Ownby & J. Liu, *J. Adhes. Sci. Technol.*, **2** 255 (1988).
14. B. Gallois & C. H. P. Lupis, *Metall. Trans.*, **12B** 549 (1981).
15. V. Ghetta, J. Fouletier & D. Chatain, *Acta. Mater.*, **44** 1927 (1996).
16. D. Chatain, I. Rivollet & N. Eustathopoulos, *J. Chim. Phys.*, **83** 561 (1986).
17. J. G. Li, *J. Am. Ceram. Soc.*, **75** 3118 (1992).
18. K. Nogi, K. Ogino, A. McLean & W. A. Miller, *Metal. Trans.*, **17B** 163 (1986).
19. K. Nogi, K. Oishi & K. Ogino, *Mater. Trans., JIM*, **30** 137 (1989).
20. P. Nikolopoulos & S. Agathopoulos, *J. Eur. Ceram. Soc.*, **10** 415 (1992).
21. D. Chatain, F. Chabret, V. Ghetta, & J. Fouletier, *J. Am. Ceram. Soc.*, **76** 568 (1993).
22. D. Chatain, F. Chabret, V. Ghetta & J. Fouletier, *J. Am. Ceram. Soc.*, **77** 197 (1994).
23. R. Sangiorgi, M. L. Muolo & A. Passerone, *Acta Metall.*, **30** 1597 (1982).
24. G. Bernard & C. H. P. Lupis, *Metall. Trans.*, **2** 2991 (1971).
25. L. Goumiri & J. C. Joud, *Acta Metall.*, **30** 1397 (1982).
26. F. A. Halden & W. D. Kingery, *J. Phys. Chem.*, **59** 557 (1955).
27. B. J. Keene, K. C. Mills, J. W. Bryant & E. D. Hondros, *Can. Metall. Q.*, **21** 393 (1982).
28. A. Kasama, A. McLean. W. A. Miller, Z. Morita & M. J. Ward, *Can. Metall. Q.*, **22** 9 (1983).
29. K. Ogino, J. Taimatsu & F. Nakatani, *J. Japan Inst. Metals*, **43** 871 (1979).
30. T. Young, *Philos. Trans R. Soc. (Lond.)*, **95** 65 (1805).
31. A. Dupré, *Theorie Mecanique de la Chaleur*, Paris, 368 (1869).
32. A. W. Adamson, *Physical Chemistry of Surfaces*, 4th ed., J. Wiley & Sons, NY (1982).
33. W. W. Mullins, *J. Appl. Phys.*, **28** 333 (1957); *Acta Metall.*, **6** 414 (1958).
34. E. Saiz, A. P. Tomsia & R. M. Cannon, *Acta Mater.*, submitted
35. P.-G. DeGennes, *Rev. Mod. Phys.*, **57** 827 (1985).
36. P. Ehrhard & S. P. Davis, *J. Fluid Mech.*, **229** 365 (1991).
37. J. W. Gibbs, *Collected Works*, vol I, Longmans, Green & Co., NY, pf. 219 (1931).

38. P. W. Tasker, *Adv. In Ceram.*, **10** 176 (1984).
39. I. Manassidis & M. J. Gillan, *J. Am. Ceram. Soc.*, **77** 335 (1994).
40. J. M. Lihrmann & J. S. Haggerty, *J. Am. Ceram. Soc.*, **68** 81 (1985).
41. R. M. Cannon, *Adv. in Ceram.*, **10** 818 (1984).
42. J. T. Klomp, in *Ceramic Microstructures '86: Role of Interfaces*, eds. J. A. Pask & A. G. Evans, Plenum P., N.Y., p. 307 (1986).
43. K. P. Trumble & M. Rühle, *Acta Metall. Mater.*, **39** 1915 (1991).
44. K. P. Trumble, *Acta Metall. Mater.*, **40** S105 (1992).
45. W. C. Mackrodt & P. W. Tasker, *J. Am. Ceram. Soc.*, **72** 1576 (1989).
46. F. H. Buttner, E. R. Funk & H. Udin, *J. Phys. Chem.*, **56** 657 (1952).
47. V. E. Henrich, *Mat. Res. Soc. Symp. Proc.*, **357** 67 (1995).
48. A. M. Stoneham & P. W. Tasker, *Ceramic Microstructures '86: Role of Interfaces*, eds. J. A. Pask & A. G. Evans, Plenum P., N.Y., p. 155 (1987).
49. R. C. McCune & R. C. Ku, *Adv. in Ceram.*, **10** 217 (1984).
50. R. C. McCune, W. T. Donlon & R. C. Ku, *J. Am. Ceram. Soc.*, **69** C196 (1986).
51. S. Baik, D. E. Fowler, J. M. Blakely & R. Raj, *J. Am. Ceram. Soc.*, **68** 281 (1985).
52. S. Baik & C. L. White, *J. Am. Ceram. Soc.*, **70** 682 (1987).
53. E. A. Colbourn & W. C. Mackrodt, *J. Mater. Sci.*, **17** 3021 (1982).
54. M. J. Davies, P. Kenway, P. J. Lawrence, S. C. Parker, W. C. Mackrodt & P. W. Tasker, *J. Chem. Soc., Faraday Trans.* 2, **85** 555 (1989).
55. M. Kitayama & A. M. Glaeser, to be published.
56. T. M. French & G. A. Somorjai, *J. Phys. Chem.*, **74** 2489 (1970).
57. M. Gautier, et al. *J. Am. Ceram. Soc.*, **77** 323 (1994).
58. F. S. Ohuchi & M. Kohyama, *J. Am. Ceram. Soc.*, **74** 1163 (1991).
59. C. Scheu, G. Dehm, H. Müllejans, R. Brydson & M. Rühle, *Microsc. Microanal. Microstruct.*, **6** 19 (1995); M. Rühle, G. Dehm, & C. Scheu, *Ceramic Microstructures '96: Control at the Atomic Level*.
60. G. Necker & W. Mader, *Phil. Mag. Lett.*, **58** 205 (1988); J. Mayer, G. Gutekunst, G. Möbus, J. Dura, C. P. Flynn & M. Rühle, *Acta Metall. Mater.*, **40** S217 (1992).
61. H. Jang, D. N. Seidman & K. L. Merkle, *Interface Sci.*, **1** [1] 61-75 (1993).
62. X. Y. Huang, W. Mader & R. Kirchheim, *Acta Metall. Mater.*, **39** 893 (1991).
63. C. Kruse, M. W. Finnis, V, Y, Milman, M. C. Payne, A. DeVita & M. J. Gillan, J. Am. Ceram. Soc., **77** 431 (1994).
64. D. F. Duffy, J. H. Harding & A. M. Stoneham, *Acta Mater.*, **44** 3293 (1996).
65. V. Laurent, D. Chatain, C. Chatillon & N. Eustathopoulos, *Acta Metall.*, **36** 1797 (1988).
66. I. Rivollet, D. Chatain & N. Eustathopoulos, *Acta Metall.*, **35** 835 (1987).
67. N. Eustathopoulos, D. Chatain & L. Coudurier, *Mat. Sci. Eng.*, **A135** 83 (1991).
68. E. Ricci & A. Passerone, *Mat. Sci. Eng.*, **A161** 31 (1993).
69. I. A. Aksay, C. E. Hoge & J. A. Pask, *J. Phys. Chem.*, **78** 1178 (1974).
70. R. M. Cannon, E. Saiz, A. P. Tomsia & W. C. Carter, *Mat. Res. Soc. Symp. Proc.*, **357** 279 (1995).
71. E. Saiz, S. Foppiano, A. P. Tomsia & R. M. Cannon, to be published.
72. C. Beraud, M. Courbiere, C. Esnouf, D. Juve & D. Treheux, *J. Mater. Sci.*, **24** 4545 (1989).
73. D.-J. Wang & S.-T. Wu, *Acta Metall. Mater.*, **43** 2917 (1995).
74. V. Merlin & N. Eustathopoulos, *J. Mat. Sci.*, **30** 3619 (1995).
75. S. M. Mukhopadhyay & J. M. Blakely, *J. Am. Ceram. Soc.*, **74** 25 (1991).
76. D. F. Duffy, J. H. Harding & A. M. Stoneham, *J. Appl. Phys.*, **76** 2791 (1994); *Acta Metall. Mater.*, **43** 1559 (1995).
77. J. E. McDonald & J. G. Eberhart, *Trans. Met. Soc., AIME*, **233** 512 (1965).
78. P. Hictor, D. Chatain, A. Pasturel & N. Eustathopoulos, *J. Chim. Phys.*, **85** 941 (1988).
79. M. W. Finnis, *Acta Metall. Mater.*, **40** S25 (1992).
80. U Schönberger, O. K. Andersen & M. Methfessel, *Acta Metall. Mater.*, **40** S1 (1992).
81. C. Li, R. Wu, A. J. Freeman and C. L. Fu, *Phys. Rev.*, **B48**, 8317 (1993).
82. T. Hong, J. R. Smith & D. J. Srolovitz, *Acta Metall. Mater.*, **43** 2721 (1995).
83. Y. Li, D. C. Langreth and M. R. Pederson., *Phys. Rev.*, **B52**, 6067 (1995).
84. M. W. Finnis, *J. Phys. Condens. Mater.*, **8** 5811 (1996).
85. R. Benedek, M. Minkoff & L. H. Yang, *Phys. Rev. B (Cond. Mat.)*, **54** 7697 (1996).

INTERFACE MATERIALS FOR OXIDE COMPOSITES

D. B. Marshall and P. E. D. Morgan
Rockwell Science Center
1049 Camino Dos Rios
Thousand Oaks, CA 91360

ABSTRACT

Several interphase systems with the potential to enable damage tolerant behavior in oxide composites are reviewed. The most promising are rare-earth phosphates (monazite), aligned layer structures (β-aluminas/magnetoplumbites), and porous coatings. Observations of the fracture behavior and microstructural stability of $LaPO_4$ coatings in alumina and alumina-zirconia are discussed.

1. INTRODUCTION

There is a need for damage-tolerant ceramic composites in a wide variety of high temperature engineering structures. However, despite the very attractive mechanical properties that have been demonstrated during the past decade in fiber reinforced composites of SiC, Si_3N_4, C, and glass ceramics with C and BN interphases, the number of applications of these materials remains small. This is due in part to their susceptibility to embrittlement in high and intermediate temperature oxidizing environments. The embrittlement is associated with transport of oxygen into matrix cracks and subsequent reaction with the C or BN interphases.[1-5]

To avoid this limitation, several all-oxide composite systems have been explored recently.[6-27] A key requirement for these composites is the identification of interphase materials that will provide a weak link in order to separate cracking in the matrix and fibers (the same role as played by C and BN in the non-oxide composites). The number of choices of oxide-based materials that are stable at high temperature and weakly bonded is not large, since most oxides tend to bond well to each other. Further restrictions are imposed by the need for chemical compatibility of the fibers and matrix, morphological stability at the use temperature, and stability in other environments such as water and CO_2. In the following sections we will first briefly review various approaches that are being taken to develop oxide composites (section 2), then describe recent progress on using rare-earth phosphates as weakly bonded layers.

2. POTENTIAL INTERPHASE SYSTEMS

The choice of potential interphase systems is limited by availability of oxide fibers. Tow fibers are available in polycrystalline alumina and mullite, as well as various lower temperature alumino-silicate, glass-containing compositions. Development of single crystal fibers of sapphire, yttrium aluminate (YAG), and alumina/YAG eutectic is also being investigated in several laboratories. The following potential interphase systems are suitable for composites containing alumina, mullite or YAG.

2.1 Layered Oxide Crystal Structures

The use of layered crystal structures that contain easy cleavage planes has been explored as a means of engineering an interphase that allows debonding. The first attempts involved use of mica (fluorophlogopite).[6] Although adequate debonding could be achieved, high temperature stability was limited by reactions of the fluorophlogopite with alumina and mullite.

Of more interest are the β-alumina/magnetoplumbite family of layer structures, which are compatible with alumina and stable to temperatures up to 1800°C.[7] These structures consist of spinel blocks interleaved with weak layers containing large cations. The spinel layers are much thicker than the alumino silicate layers in mica, thus reducing the fraction of modifying cations and simplifying synthesis, phase relations and compatibility.

Debonding and compatibility have been demonstrated in several systems[7-12] (e.g. Na(Li)-β"-alumina, $BaMg_2Al_{16}O_{27}$, $KMg_2Al_{15}O_{25}$, $LaAl_{11}O_{18}$ and $CaAl_{12}O_{19}$). The main obstacles remaining to the development of a useful fiber coating involve morphology: both of the microstructures of the coating and of the fracture surface once debonding occurs. Because of their anisotropy, the β-alumina/magnetoplumbite grains grow in a plate-like morphology. Therefore, a fiber coating must have either a very small grain size (compared with the coating thickness), or, if the grain size is larger, all grains aligned around the circumference of the fiber. Some success has been achieved in grain alignment,[8-11] although the degree of alignment must be perfect over distances equal to the required pull-out length (isolated grains oriented normal to a fiber have been observed to grow into the fiber and act as effective sites for deflecting a debond crack into the fiber).

The large-grained, aligned microstructure leads to steps on the surface of a debond crack which cause resistance to pull-out of the debonded fiber. The degree of debond roughness that can be tolerated has not been well defined. A very fine grained microstructure may be an advantage in this regard.

2.2 Weak Interfaces

Two types of interphase have been found for which the bond between the interphase and the fibers is intrinsically weak, with toughnesses sufficiently low that debonding occurs and prevents growth of a crack across the interface. One is tin dioxide, which has been shown to provide a diffusion barrier and weak interface between glass or alumina materials and alumina fibers (both polycrystalline and single crystal).[13,14] However, a limitation of this system is that it could never be exposed to an even slightly reducing environment, as the melting point of the lower oxide SnO is low (1080°C) and eutectics are presumably lower.[15]

The other class of interphase is the group of rare earth phosphates that form the monazite and xenotime structures. The interface between these compounds and various refractory oxides is sufficiently weak to allow debonding.[16,17] This, combined with their phase compatibility, high melting points (many over 2000°C), and corrosion resistance, make them appealing as components of refractory oxide composites. Material combinations for which weak bonding and phase compatibility have been demonstrated (at least in closed systems) include: (1) LaPO$_4$/Al$_2$O$_3$, the first monazite-based system for which debonding was observed[16,17] (provided the stoichiometry of the LaPO$_4$ is 1:1, this system is phase compatible and morphologically stable for long periods at temperatures at least as high as 1600°C); (2) LaPO$_4$ with Y-stabilized zirconia;[18] (3) CePO$_4$ with Ce-stabilized zirconia: (4) LaPO$_4$ with mullite;[19] and (5) YPO$_4$ with YAG.[20,21]

2.3 Porous Interphases or Porous Matrix

Several research groups are exploring an approach that relies on debonding within a weak fiber coating rather than at an interface, with the weakness being induced by controlled porosity within the coating.[22-24] In this case, the interphase material itself may bond strongly to the matrix and fiber. Examples include porous interphases of ZrO$_2$ or Al$_2$O$_3$ in composites consisting of Al$_2$O$_3$ fibers and Al$_2$O$_3$ matrix. Although debonding has been demonstrated in such systems, several concerns remain. One is the extent to which fiber strength is degraded by regions of coating that bond strongly to the fiber. Another relates to the roughness of the debonded region and the degree of pullout achievable. Both are in principle controllable to some extent by controlling the size scale of the porosity. However, a limitation of very fine scale porosity is its tendency to coarsen at high temperatures.

In a related approach, some success has been achieved with oxide composites with porous matrices, in which the notion of requiring debonding of individual fibers is abandoned, and instead damage tolerance and notch insensitivity are achieved by splitting within the weak matrix, in regions that are matrix-rich, i.e., between fiber tows in woven composites and in regions with random fluctuations of fiber density in unidirectional composites.[25-27] Non-catastrophic, wood-like fracture has been observed in such composites (examples include composites with Al$_2$O$_3$ fibers and porous mullite, alumina, or AlPO$_4$ matrix). However, one drawback is their low shear strength, which is a consequence of the weak matrix. The most likely solution for improving the shear strength is to use 3-D woven or stitched fiber reinforcements.

3. INTERFACIAL DEBONDING
3.1 Observations of debonding in LaPO$_4$ - containing systems

Examples of debonding associated with LaPO$_4$ in several systems are shown in Fig. 1. The alumina-sapphire system of Fig. 1(a) was fabricated[16] by packing sapphire fibers that had been slurry coated with LaPO$_4$ in alumina powder and hot pressing at 1400°C. The cracks at the bottom of the micrograph were generated by a Vickers indentation located in the matrix beneath the field of view. In Fig. 1(b) a thin layer of LaPO$_4$ sandwiched between polycrystalline alumina shows that debonding occurs even for relatively irregular interfaces that must exist in polycrystalline systems. Figure 1(c) shows a Vickers indentation in a multilayered composite consisting of alternating layers of LaPO$_4$ and two-phase alumina/yttria-partially stabilized zirconia.[18] The composite was fabricated by sequential slip casting, followed by sintering at 1600°C. The LaPO$_4$ layers are evidently able to contain very severe damage due to indentation.

3.2 Mechanics of Debonding

A general observation is that cracks are able to pass from alumina or zirconia-based materials to LaPO$_4$ without causing interfacial debonding. However, cracks passing in the opposite direction cause debonding and are unable to cross the interface. This behavior, which is clearly evident in Figs. 1(a) and (c), is consistent with measured toughnesses and the

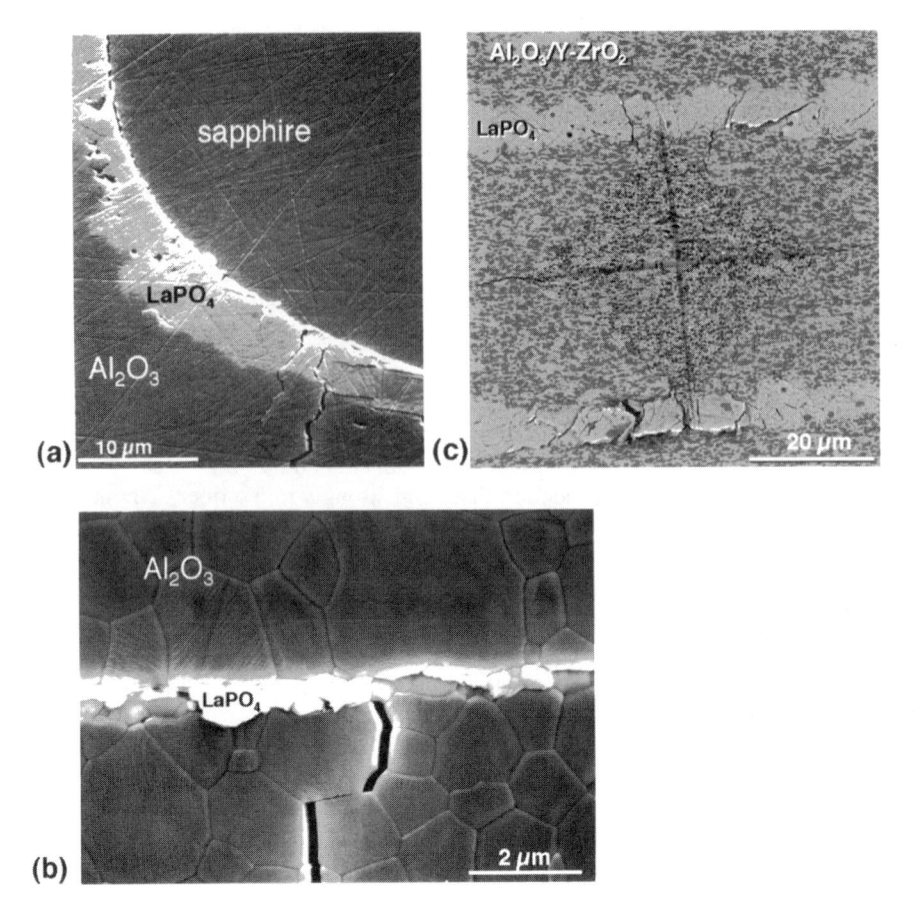

Figure 1. (a) Debonding at LaPO$_4$ - sapphire interface. Coated sapphire fiber in polycrystalline alumina matrix, hot pressed 1400°C for 1 h. Crack produced by Vickers indentation located below the field of view. (b) Debonding at thin layer of LaPO$_4$ in polycrystalline alumina. (c) Cracking and debonding in layer of LaPO$_4$ in polycrystalline, 2 phase Al$_2$O$_3$/Y-ZrO$_2$.

debonding behavior predicted by the analysis of He and Hutchinson.[28] They define a critical value of the ratio, Γ_i/Γ_2, of the toughnesses of the interface and the material into which the crack is about to grow (Fig. 2). If the toughness ratio for a particular system falls above the critical value, the cracks grows through the interface, whereas if the toughness ratio falls below the critical value, debonding occurs. The critical value depends on the elastic mismatch parameter, α, as shown in Fig. 2:

$\alpha = (E'_2 - E'_1)/(E'_2 + E'_1)$

where E'$_1$, and E'$_2$ are the plane-strain Young's moduli for the two materials.

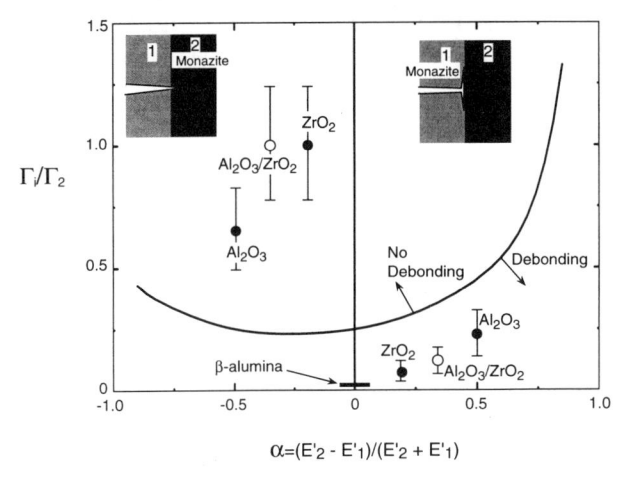

Figure 2. Comparison of measured fracture toughnesses with the debond criterion of He and Hutchinson.[28]

Measured values of the toughness ratio for various combinations of materials (LaPO$_4$ with Al$_2$O$_3$, Y-ZrO$_2$ and Al$_2$O$_3$/Y-ZrO$_2$) are shown in Fig. 2. Values on the right half of this figure (i.e., positive α), which fall below the critical curve, correspond to cracks growing from LaPO$_4$ to the other material, while values to the left (negative α), which fall above the critical curve, correspond to cracks growing from the other material to the LaPO$_4$. The measured values used in these calculations are given in Table 1. The results are in agreement with the above observations.

Table 1. Material Properties

Material	Young's Modulus E (GPa)	Fracture Energy (J/m^2)	Interface Fracture Energy (J/m^2)
LaPO$_4$	133	7	
Al$_2$O$_3$	400	20	4
Y-ZrO$_2$	200	110	7
Al$_2$O$_3$/Y-ZrO$_2$ (1:1)	300	80	7

Also shown for comparison in Fig. 2 is an estimated toughness ratio for a β-alumina coating in alumina. The toughnesses in this case were estimated from measurements on single crystal specimens, using the lengths of indentation cracks growing normal and parallel to the weak layers in polished cross sections normal to the weak layers.[7] The toughness ratio estimated in this manner (0.01) is much lower than for the LaPO$_4$ based systems.

A complete analysis of debonding in these systems should also include analysis of the influence of residual stresses, which have the effect of shifting the critical curve in Fig. 2.[29]

Complete analysis thus requires knowledge of component geometry, and is also dependent on the presence of flaws in the layers and interfaces. Nevertheless, qualitative trends can be easily estimated from known thermal expansion coefficients. Since the thermal expansion coefficient of LaPO$_4$ is almost identical to that of ZrO$_2$ and higher than that of Al$_2$O$_3$, the stresses parallel to a LaPO$_4$ coating in alumina-containing systems are tensile in the LaPO$_4$ and compressive in the other material. Therefore, the tendency for a crack to penetrate the interface when growing towards the LaPO$_4$ layer is enhanced (i.e. the critical curve in Fig. 2 for $\alpha < 0$ is lowered), whereas the tendency for debonding is enhanced when the crack approaches the interface from within the LaPO$_4$ layer (i.e. the critical curve for $\alpha > 0$ is raised). In the LaPO$_4$/ZrO$_2$ system there should be no residual stresses as long as there is no transformation of the ZrO$_2$ from tetragonal to monoclinic phase during cooling from the fabrication temperature. If, on the other hand, transformation takes place during cooling, residual stresses develop with the same sign as for the alumina-containing systems.

4. MICROSTRUCTURAL STABILITY AND COMPATIBILITY

The compatibility of LaPO$_4$ with Al$_2$O$_3$ and Y-ZrO$_2$ at temperatures of 1400°C and 1600°C is illustrated in Fig. 1. An important requirement for compatibility is that the ratio of La to P in the LaPO$_4$ layer be exactly unity. In studies of the LaPO$_4$/Al$_2$O$_3$ system, excess P has been shown to result in formation of AlPO$_4$ at the interface, while excess La causes the formation of La Al$_{11}$O$_{18}$, a magnetoplumbite related structure.[10]

As well as being phase compatible with the fibers and matrix, an interphase material must also be morphologically stable in thin layers. Excessive grain growth, especially in systems with large anisotropy of growth directions can lead to break-up of thin layers. Relative grain growth and interface morphology in a multilayered composite containing large thickness polycrystalline layers of Al$_2$O$_3$, Al$_2$O$_3$/Y-ZrO2, and LaPO$_4$ are shown in Fig. 3. The composite was fabricated by centrifugal casting of slurries followed by hot pressing at 1400°C. The grain size in the LaPO$_4$ layer is approximately the same as in the (undoped) Al$_2$O$_3$, whereas the grain size in the two-phase Al$_2$O$_3$-ZrO2 is much smaller, as observed in many earlier studies. The morphologies of the boundaries between the Al$_2$O$_3$ and Al$_2$O$_3$-ZrO$_2$ layers and between the LaPO$_4$ and Al$_2$O$_3$-ZrO$_2$ layers are similar.

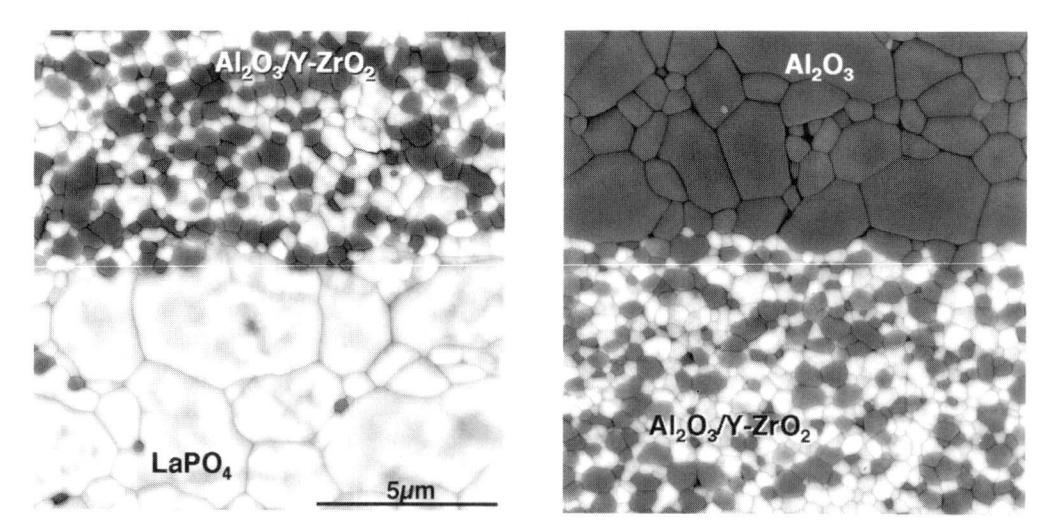

Figure 3. Microstructures of: LaPO$_4$ and Al$_2$O$_3$/Y-ZrO$_2$ layers (left); and (b) Al$_2$O$_3$ and Al$_2$O$_3$/Y-ZrO$_2$ in specimen hot pressed at 1400°C for 1 h (right).

The stability of thin layers of LaPO$_4$ in Al$_2$O$_3$ is shown in Figs. 1(b) and 4. In these cases the grain size in the LaPO$_4$ layers is approximately equal to the layer thickness and considerably smaller than the grain size in the Al$_2$O$_3$ matrix. Therefore, grain growth in the LaPO$_4$ appears to be retarded when the grain size becomes as large as the layer thickness. Moreover, the LaPO$_4$ layers appear to be resistant to breakup by grain growth, even after extended heat treatment (20h) at 1600°C (Fig. 5). This stability is presumably related to the large dihedral angle observed at the junction of a LaPO$_4$ grain boundary with an Al$_2$O$_3$ grain surface.[16]

Thin LaPO$_4$ layers are also effective barriers to grain growth in the Al$_2$O$_3$ matrix, as illustrated in Fig. 4. This micrograph is from a composite that was fabricated by dipping sapphire fibers sequentially in slurries of LaPO$_4$ and Al$_2$O$_3$ to give a multilayered coating, then surrounding the coated fibers with Al$_2$O$_3$ powder and hot pressing at 1400°C for 1 hr. The inner layer of LaPO$_4$ on the fiber in Fig. 4 was not continuous. During fabrication, the sapphire fiber acted as a large seed grain and grew into the polycrystalline Al$_2$O$_3$ matrix in regions where there was no LaPO$_4$ coating. However this exaggerated grain growth was fully confined by the continuous second LaPO$_4$ layer and by the regions where the first layer was present.

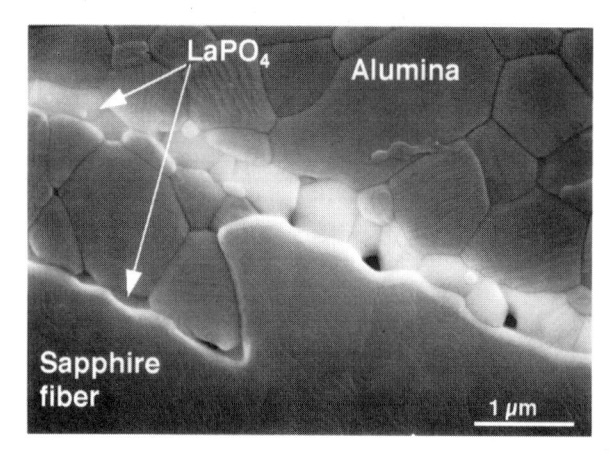

Figure 4. Exaggerated grain growth from sapphire fiber with multiple concentric coatings of LaPO$_4$ and Al$_2$O$_3$: exaggerated grain growth occurred from region where inner coating was missing.

5. INTERFACE FRACTURE MORPHOLOGY

Debonding at LaPO$_4$-coated sapphire fibers occurs cleanly along the LaPO$_4$-sapphire interface, leaving a fracture surface that closely replicates the interface structure (Figs. 5(b) and 7(a)). In this case, cusps are formed where the LaPO$_4$ grain boundaries meet the surface of the sapphire fiber. The dihedral angle between the surface of the cusp is large (~150-170°) and the height of the cusps is small (~ 50 nm), within the range of surface roughness observed on other reinforcing fibers in non-oxide systems. Separate experiments have confirmed that the cusps on the surfaces of the fibers do not degrade their strengths.

The roughness of the fracture surface between layers of polycrystalline Al$_2$O$_3$ and LaPO$_4$ is determined by the grain size and can be substantially larger than for the sapphire fibers. Moreover, in polycrystalline systems, the additional resistance to cracking along a rough interface often leads to secondary cracking within the LaPO$_4$ layers.

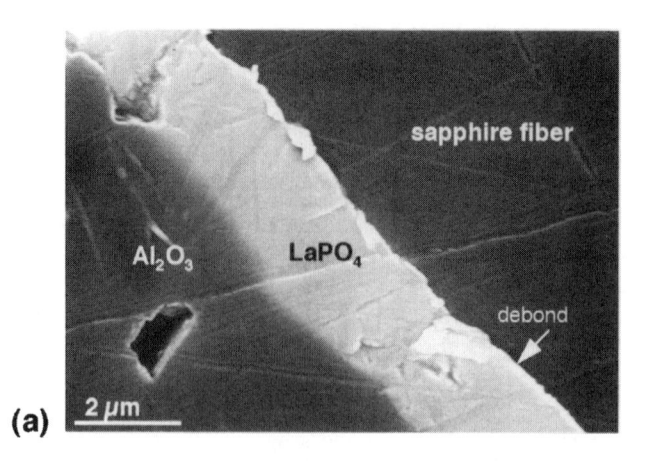

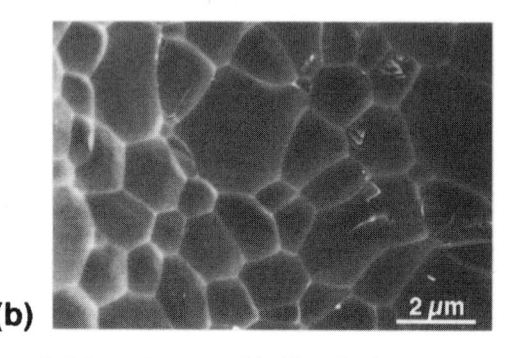

Figure 5. (a) Debonding at LaPO$_4$ coating on sapphire fiber after heat treatment at 1600°C for 20h. (b) Fracture surface from (a).

The LaPO$_4$-sapphire fracture morphology of Fig. 5 may be contrasted in Fig. 6 with the morphology in a system containing an aligned β-alumina coating. The fibers of Figs. 5 and 6 were from the same composite, formed by packing coated sapphire fibers in Al$_2$O$_3$ powder and hot pressing at 1400°C for 1 hr, then heat treating at 1600°C for 20 h. The fiber from Fig. 5 was coated with LaPO$_4$ from a slurry in which the La:P ratio was carefully adjusted to unity, whereas the fiber from Fig. 6 was coated by laser ablation, which resulted in the presence of a small, but measurable excess of La. During the heat treatment at 1600°C, the excess La reacted with the Al$_2$O$_3$ on both sides of the coating to form LaAl$_{11}$O$_{18}$. The reaction product on the inner side adjacent to the fiber was well aligned, with the c-axis normal to the fiber surface (i.e., weak planes parallel to the fiber surface). An indentation crack which approached the fiber in a direction normal to the interface (out of the field of view) of Fig. 6(a) was arrested and caused debonding within the aligned layer of LaAl$_{11}$O$_{18}$. The plate-like morphology of this layer, indicated schematically in Fig. 7(b), is reflected in the morphology of the fracture surface of Fig. 6(b).

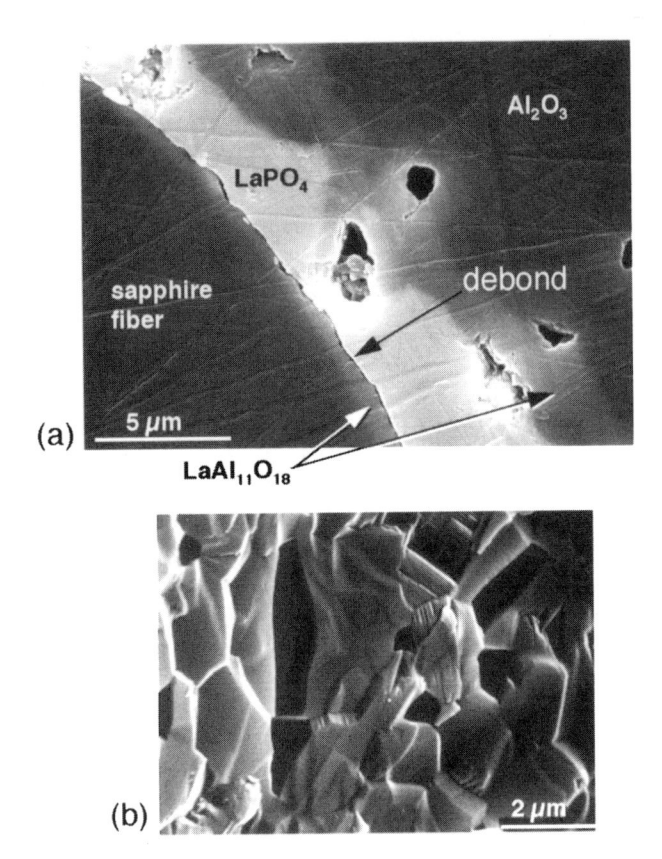

Figure 6. (a) Debonding at layer of aligned $LaAl_{11}O_{18}$ formed by reaction of La-rich $LaPO_4$ coating on sapphire fiber. (b) Fracture surface from (a).

6. CONCLUSIONS

Several potentially viable approaches exist for introducing damage - tolerant behavior in oxide composites; fiber coatings of monazite, fiber coatings of layered oxides, or porous coatings or matrix. All have been shown to allow the necessary fiber debonding. Monazite is especially appealing in view of its long-term, high-temperature microstructural stability. Such composites are at an early stage of development, and advances are needed in fiber coating methods, development of improved oxide fibers (especially creep resistance) and development of methods of matrix fabrication.

ACKNOWLEDGMENTS

Funding for this work was provided by the US Office Naval Research, Contract N00014-95-C-0057, monitored by Dr. S. Fishman.

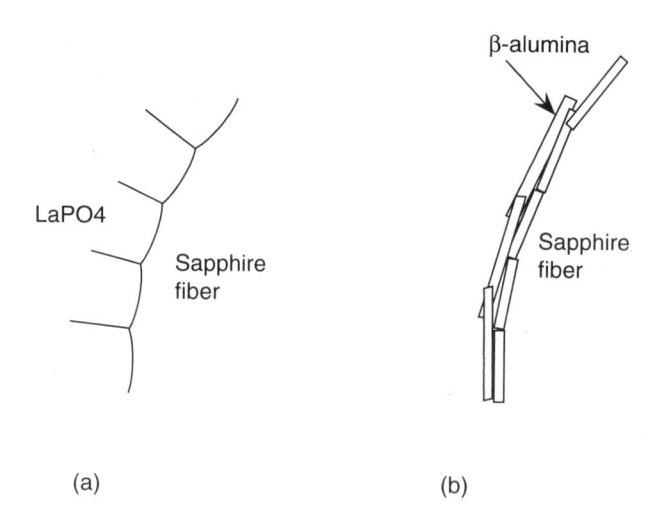

Figure 7 (a) Schematic of microstructure at interface of Fig 5(a). (b) Schematic of microstructure at interface of Fig 6(a)

REFERENCES

1. J. J. Brennan, "Interfacial Characterization of Glass and Glass-Ceramic Matrix/Nicalon SiC Fiber Composites," *Mat. Sci. Res.,* **20** 546-560 (1986).
2. R. F. Cooper and K. Chyung, "Structure and Chemistry of Fiber-Matrix Interfaces in Silicon Carbide-Fiber-Reinforced Glass-Ceramic Composites: an Electron Microscope Study," *J. Mater. Sci.,* **22** 3148-3160 (1987).
3. J. J. Brennan, "Ch. 8"; in Fiber-Reinforced Ceramic Composites. Ed. M. K. D. New York, Noyes, 1990.
4. F. E. Heredia, J. C. McNulty, F. W. Zok and A. G. Evans, "Oxidation Embrittlement Probe for Ceramic-Matrix Composites," *J. Am. Ceram. Soc.,* **78** [8] 2097-2100 (1995).
5. R. H. Jones, C. H. Henager and C. F. J. Windisch, "High Temperature Corrosion and Crack Growth of SiC-SiC at variable Oxygen Partial Pressures," *Mater. Sci. Eng.,* **A198** 103-112 (1995).
6. R. F. Cooper and P. C. Hall, "Reactions Between Synthetic Mica and Simple Oxide Compounds with Application to Oxidation-Resistant Ceramic Composite," *J. Am. Ceram. Soc.,* **76** [5] 1265-1273 (1993).
7. P. E. D. Morgan and D. B. Marshall, "Functional Interfaces in Oxide-Oxide Composites," *J. Mat. Sci. Eng.,* **A162** [1-2] 15-25 (1993).
8. M. K. Cinibulk and R. S. Hay, "Textured Magnetoplumbite Fiber-Matrix Interphase Derived Sol-Gel Fiber Coatings," *J. Am. Ceram. Soc.,* **79** [5] 1233-1246 (1996).
9. M. K. Cinibulk, "Magnetoplumbite Compounds as a Fiber Coating for Oxide-Oxide Composites," *Ceram. Eng. and Sci. Proc.,* **15** [5] 721-728 (1994).
10. D. B. Marshall, P. E. D. Morgan and R. M. Housley, "High temperature Stability of $LaPO_4/Al_2O_3$ composites: Effect of La:P ratio," *J. Am. Ceram. Soc.,* (in preparation).
11. M. K. Cinibulk, "Synthesis and Characterization of Sol-Gel derived Lanthanum hexaluminate powders and films," *J. Mater. Res.,* **10** [1] 71-76 (1995).
12. M. H. Lewis, M. G. Cain, P. Doleman, A. G. Razzell and J. Gent, "Development of Interfaces in Oxide and Silicate-Matrix Composites," *Ceramic Transactions,* **58** 41-52 (1995).

13. A. Maheshwari, K. K. Chawla and T. A. Michalske, "Behavior of Interface in Alumina/Glass Composite," *Mater. Sci. Eng.,* **A107** 269-276 (1989)
14. R. Venkatesh and K. K. Chawla, "Effect of Interfacial Roughness on Fiber Pullout in Alumina/SnO$_2$/Glass Composites," *J. Mat. Sci. Lett.,* **11** 650-652 (1992).
15. P. E. D. Morgan and R. M. Housley, "Some Effects of Eutectic Liquid Under Reducing Conditions in the Alumina-Tin Dioxide-Tin Composite System," *J. Am. Ceram. Soc.,* **78** [1] 263-265 (1995).
16. P. E. D. Morgan and D. B. Marshall, "Ceramic Composites of Monazite and Alumina," *J. Am. Ceram. Soc.,* **78** [6] 553-63 (1995).
17. P. E. D. Morgan, D. B. Marshall and R. M. Housley, "High Temperature Stability of Monazite-Alumina Composites," *J. Mat. Sci. Eng.,* **A195** 215 - 222 (1995).
18. D. B. Marshall, P. E. D. Morgan and R. M. Housley, "Debonding in Multilayered Composites of Zirconia and LaPO$_4$," *J. Am. Ceram. Soc.,* (in press).
19. P. E. D. Morgan, D. B. Marshall and R. M. Housley, "Particulate composite of Mullite-Alumina," *J. Am. Ceram. Soc.,* (in preparation).
20. D.-H. Kuo and W. M. Kriven, "Chemical stability, microstructure and mechanical behavior of LaPO$_4$-containing ceramics," *Mater. Sci. Eng.,* **A210** 123-134 (1996).
21. D.-H. Kuo and W. M. Kriven, "Characterization of Yttrium Phosphate and a Yttrium Phosphate/Yttrium Aluminate Laminate," *J. Am. Ceram. Soc.,* **78** [11] 3121-3124 (1995).
22. J. B. Davis, J. P. A. Lofvander, A. G. Evans, E. Bischoff and M. L. Emiliani, "Fiber Coating Concepts for Brittle Matrix Composites," *J. Am. Ceram. Soc.,* **76** [5] 249-257 (1993).
23. J. B. Davis, J. Yang and A. G. Evans, "Effects of Composite Processing on the Strength of Sapphire Fiber-Reinforced Composites," *Acta Met.,* **43** [1] 259-268 (1995).
24. M. H. Jaskowiak, J. I. Eldridge, J. B. Hurst and J. A. Setlock, "Interfacial Coatings for Sapphire/Al$_2$O$_3$, pp 84, Proc. High Temp Review 1991, NASA Conference Publication 10082, 1991.
25. W. C. Tu, F. F. Lange and A. G. Evans, "Concept for a Damage-Tolerant Ceramic Composite with "Strong" Interfaces," *J. Am. Ceram. Soc.,* **79** [2] 417-424 (1996).
26. F. F. Lange, W.-C. Tu and A. G. Evans, "Processing of Damage-Tolerant, Oxidation-Resistant Ceramic-Matrix Composites," *Mater. Sci. Eng.,* **A195** 145-150 (1995).
27. W. P. Keith and K. T. Kedward, "Shear Damage Mechanisms in a Woven, Nicalon Reinforced Ceramic Matrix Composite," *J. Am. Ceram. Soc.,* (in press).
28. M.-Y. He and J. W. Hutchinson, "Crack Deflection at an Interface Between Dissimilar Materials," *Int. J. Solids Struct.,* **25** 1053-1067 (1989)
29. M.-Y. He, A. G. Evans and J. W. Hutchinson, "Crack Deflection at an Interface Between Dissimilar Elastic Materials: Role of Residual Stresses," *Int. J. Solids Structures,* **31** [24] 3443-3455 (1994)

NEAR ATOMIC SCALE NANOCHEMISTRY AND STRUCTURE: CERAMIC GRAIN BOUNDARIES AND INTERFACES

R.W. Carpenter[1] and W. Braue[2]

[1]Center for Solid State Science and Science and Engineering of Materials Program, Arizona State University, Tempe, AZ 85287-1704
[2]German Aerospace Research Establishment (DLR), Materials Research Institute, Linder Hoehe, D-51147 Cologne, Germany

INTRODUCTION

The local chemistry around microstructural features such as grain boundaries, heterophase interfaces, and inclusions in conventionally processed ceramics is of interest because of its effect on properties, such as the viscosity of thin amorphous phases (Urbain et al., 1981) and its effect on interface bond strength (Evans, 1990). Measurements of local chemistry of interest include qualitative and quantitative composition at these defects and along chemical gradients extending into the adjacent regions, and spectroscopic measurements that provide information about local bonding. Earlier measurements of the structure of these defects by TEM imaging methods by Clarke and Thomas (1977), Clarke (1979) and others showed that they are small, thus chemically sensitive analysis methods with very high spatial resolution are required. In addition these microstructural features are internal in ceramics, and often 2-dimensional (e.g., interfaces), so transmission electron microscopy methods with the feature of interest in edge-on orientation are required. Just as in imaging for defect analysis there is no lower limit on the resolution useful for these chemical analyses. The lower limits attainable are determined by the experimental method and equipment used. At present nanospectroscopy (either electron energy loss or energy dispersive x-ray) or energy selected imaging (using energy loss electrons) are the best choices which produce direct chemical sensitivity. High angle dark field scanning transmission imaging (more often but less descriptively called Z-contrast imaging) is the best indirectly chemically sensitive method for the measurements of interest. To achieve the required spatial resolution field emission sources are required for the small probe methods, which include both nanospectroscopies and Z-contrast imaging.

In this paper we discuss applications of electron energy loss nanospectroscopy, and energy selected and Z-contrast imaging to chemical distributions in silicon nitride or aluminum oxide matrix-silicon carbide whisker composites and sintered α-silicon carbide, and the results from the materials viewpoint.

MATERIALS AND METHODS

The silicon nitride matrix ceramic matrix composites (CMC) were conventionally synthesized by mixing the constituents with 5.5 wt.% yttria and 1.1 wt.% alumina as sintering aids and 20 vol.% β-SiC whiskers, cold pressed and presintered for 1 hour at 1500°C in 0.1 MPa argon. Complete densification was achieved by hot pressing the presintered compact at 1780°C for 1 hour in 190 MPa argon. The aluminum oxide CMC was hot pressed at 1830°C at 60 MPa in vacuum to achieve final densification, without addition of sintering aids or organic precursors. We expected *a priori* that this CMC, densified without sintering aids, would contain smaller amounts of thin amorphous phase in its grain boundaries and interfaces and that turned out to be the case, but the microstructure also contained other interesting features. The α-silicon carbide examined was densified by pressureless sintering at 2000°C in ~1 atm. argon pressure in the furnace. The submicron size silicon carbide powder was doped with 2 wt.% carbon (added as polyphenylene-H-resin) by mixing before green body compaction. The resin contains a trace of sulfur and about 1.5% oxygen as impurities, and converts to amorphous carbon at 800°C. The starting α-silicon carbide powder contained less that 0.2 wt.% metallic impurities, with aluminum, iron and calcium on the particle surfaces being the major constituents.

Microscopes with field emission sources (FEG) were used for the small probe experiments discussed here. A Philips EM400 FEG was used for nanospectroscopy and an HB 501 STEM was used for Z-contrast imaging. The high brightness and forward-peaked electron emission pattern of FEG sources are essential to obtain the spatial resolution required (about 3 nm or smaller) to determine the chemical distributions of interest. The brightness of FEG sources is in the low 10^8 amperes/cm^2•steradian range, much higher than the $\sim 5 \times 10^5$ brightnesses of conventional thermionic sources. This higher brightness permits formation of small focused probes containing sufficient current for high spatial resolution nanospectroscopy (Carpenter, 1989). Methods for calculating focused probe size and current density as functions of the electron optical parameters of microscope illumination systems have been determined (Weiss et al., 1991). Nanospectroscopy and Z-contrast imaging are, in general, count rate limited, meaning that the smallest usable probes must contain enough current to collect nanospectra in times short enough that relative motion between the specimen and focused probe does not appreciably degrade spatial resolution. Relative motion between probes and specimens results from mechanical and electrical instabilities, such as stage drift or current fluctuations in beam steering coils, and is very difficult to evaluate analytically. It can be evaluated experimentally, using edge-on interfaces that are atomically sharp both chemically and structurally (Catalano et al., 1993). If these conditions are satisfied the spatial resolution attainable for FEG nanospectroscopy using thin specimens will be between 3 and 0.5 nm. For the present work electron energy loss (EELS) nanospectrum collection time was typically 100 milliseconds, using a Gatan 601 parallel spectrometer, and our spatial resolution was ~2.5 nm. The Z-contrast results presented here are relatively low resolution, so stability was not limiting. It should be clear from this discussion that simply increasing the current in a small field emission probe to very high values, thus reducing data collection times, is a way to minimize the undesirable degradation of spatial resolution by electromechanical microscope instabilities. Radiation damage to the specimen will usually determine the upper probe current limit when this method is used to achieve the best spatial resolution. For materials with moderate to wide band gaps, which includes most ceramics, the radiation effects will usually occur as so-called ionization damage, resulting in differential mass loss from the small specimen area irradiated by the electron nanoprobe. This type of radiation damage can be most definitively detected by time-resolved nanospectroscopy (Das Chowdhury et al., 1989). To achieve the best possible spatial resolution for nanoprobe experiments, stabilize the microscope used to minimize electromechanical instabilities (which will require significant effort) and then use the highest probe current possible, consistent with radiation damage characteristic of the specimen, to collect data.

Two chemically sensitive imaging methods were used to examine distributions: energy selected imaging which is spectroscopic (i.e. direct chemical sensitivity to particular elements), and Z-contrast STEM imaging which is indirectly sensitive to specimen composition through its contrast dependence on atomic number. Energy selected imaging makes use of inelastically scattered electrons to produce dark field images whose contrast is proportional to specimen composition and thickness, for example electrons that have caused core K-shell excitation of oxygen atoms in the specimen. The energy of the loss electrons used for energy selected imaging is $E_0 - \Delta E$, where E_0 is the unscattered beam energy and ΔE is the characteristic loss ($\Delta E = 531$ eV for the oxygen K-shell). To minimize chromatic aberration induced resolution loss, energy selected images are formed with the loss electrons on the optic axis with a narrow energy pass band in a microscope fitted with an imaging energy filter. We used an omega (Ω) filter on a Zeiss 912 microscope and a 20 eV pass band. Z-contrast images are formed by electrons scattered elastically through large angles, collected by an annular dark field detector. The contrast in these images depends on Z^n, where n is typically 1.5 to 2 and Z is the average atomic number of the specimen, thus specimen regions containing high atomic number elements produce higher contrast (Liu and Cowley, 1993; Pennycook et al., 1995).

EXPERIMENTAL RESULTS

Ceramic Matrix Composites

The $Si_3N_4/SiC(w)$ composites contained a thin amorphous phase in both matrix grain boundaries and whisker matrix interfaces. In both the width (i.e. thickness, viewed edge-on) of the amorphous phase varied over an appreciable range from one matrix grain boundary to another, and from one whisker/matrix interface to another, even in the same TEM specimen. The widths of amorphous phase, which we have called structural width, were measured from HRTEM images. Examples of two different structural widths are shown in figures 1 and 2. In figure 2 b,c we also show profiles of the oxygen and nitrogen distributions across the edge-on grain boundary shown in figure 2a, determined by position resolved electron energy

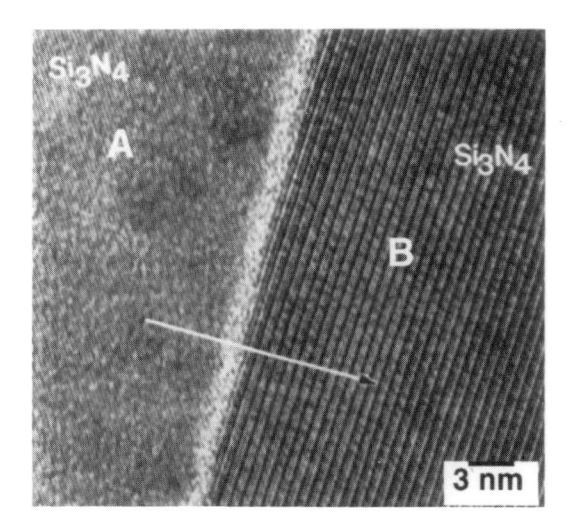

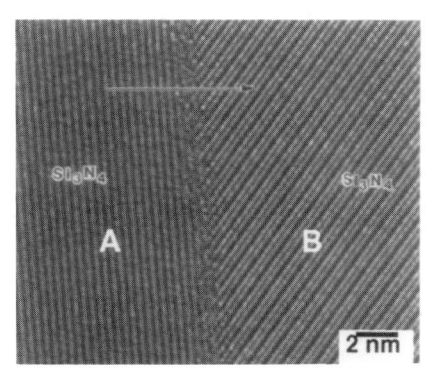

Figure 1. An HRTEM image of a matrix grain boundary containing an easily resolvable thin amorphous film in an $Si_3N_4/SiC(w)$ CMC synthesized using American Matrix whiskers. The boundary is edge-on, with minimum structural width ≈0.8nm. 200KV.

loss nanospectroscopy (PREELS) (Weiss et al., 1992). The structural width of the boundary shown in figure 1 is about 0.8 nm, with some variability along the boundary. The structural width of the grain boundary of figure 2 is about 0.3 nm, about the dimension characteristic of the Si-X tetrahedral units that comprise silicon nitride, oxide and carbide. The full width of the oxygen enrichment and nitrogen depletion regions, which we have called chemical widths, were 41 and 45 nm, respectively. The grain boundary of figure 1 exhibited similar but wider oxygen and nitrogen chemical widths 88 and 87 nm, respectively. There is an apparent correlation between structural and chemical width which we are examining quantitatively. The matrix grain boundaries of all $Si_3N_4/SiC(w)$ composites we examined always exhibited an appreciable chemical width, even when the structural width was a thin disordered region, not identifiable as distinct amorphous phase (Cf. figure 2). These small structural widths do not necessarily imply the presence or absence of chemical impurities without independent spectroscopic evidence.

The interfaces between SiC whiskers and Si_3N_4 matrix grains exhibited laterally discontinuous solute distributions, i.e. regions of zero chemical width, as shown in figures 3 and 4, not continuous distributions as in the case of matrix grain boundaries. The qualitative correlation between structural and chemical widths observed for matrix grain boundaries also

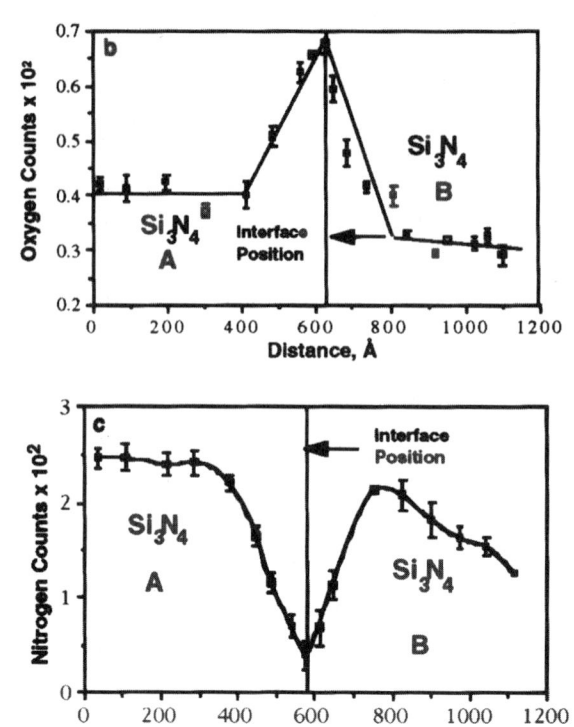

Figure 2. An HRTEM image of an edge-on matrix grain boundary in $Si_3N_4/SiC(w)$ CMC synthesized using Huber whiskers. This boundary has structural width ≈ 0.3nm, approximately the size of the Si-X tetrahedral unit in these covalent ceramics, where X= O, C, or N. (b) Variation in oxygen content along the scan path normal to the boundary plane indicated by the arrow in the image. (c) Variation on nitrogen content across the boundary plane along the same scan path.

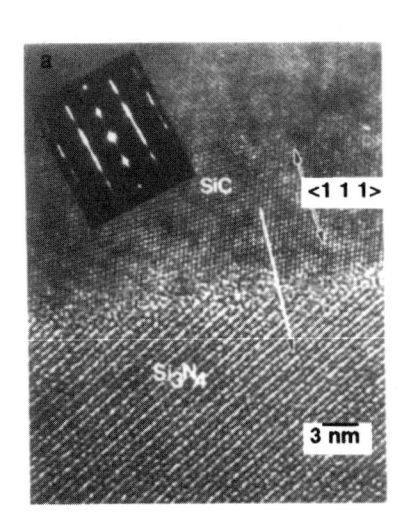

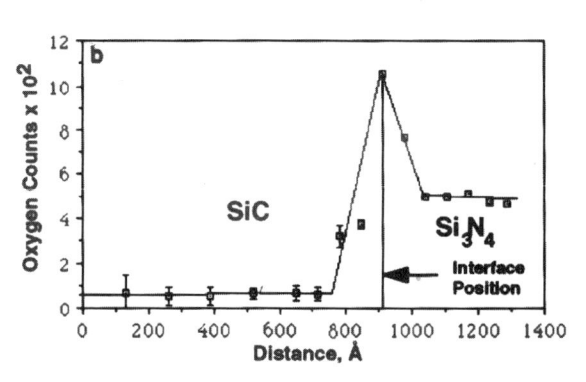

Figure 3. (a) HRTEM image of an edge-on $SiC(w)/Si_3N_4$ interface in a CMC synthesized using American Matrix whiskers. Structural width ≈1.5nm. SiC is <110> zone axis orientation. (b) Variation of oxygen content along the arrowed image scan path. Similar measurements showed nitrogen depletion at the interface.

holds true for whisker/matrix interfaces. However, the PREELS chemical width curves for whisker/matrix interfaces tend to extend farther into the Si_3N_4 matrix grains than into the whiskers. The distributions of yttrium and aluminum in matrix grain boundary and whisker/matrix interface regions were also of interest, because these elements were major constituents of the sintering aid used. Z-contrast STEM imaging was used to detect the yttrium distribution, because nitrogen, carbon and silicon interfere with PREELS detection using the N,M and low loss absorption edges. An amplitude modulated Z-contrast STEM image of an edge-on grain boundary is shown in figure 5a.

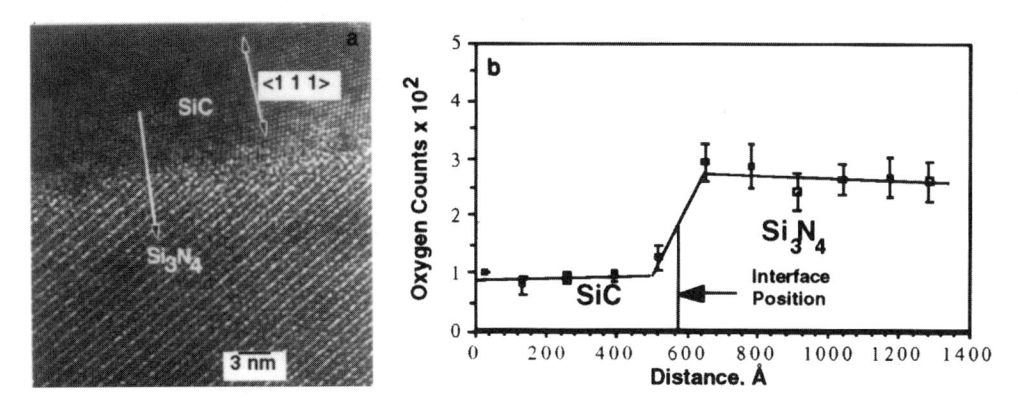

Figure 4. (a) HRTEM image of the Fig.3 interface, but at a position about 20 nm laterally displaced along the interface from the region shown in Fig.3. (b) Variation of oxygen content along the scan path normal to the plane of the interface shown in the image. No sharp oxygen peak was observed at the interface position, indicating that the oxygen distribution was discontinuous along the interface. The background oxygen contents in the whisker and matrix differed.

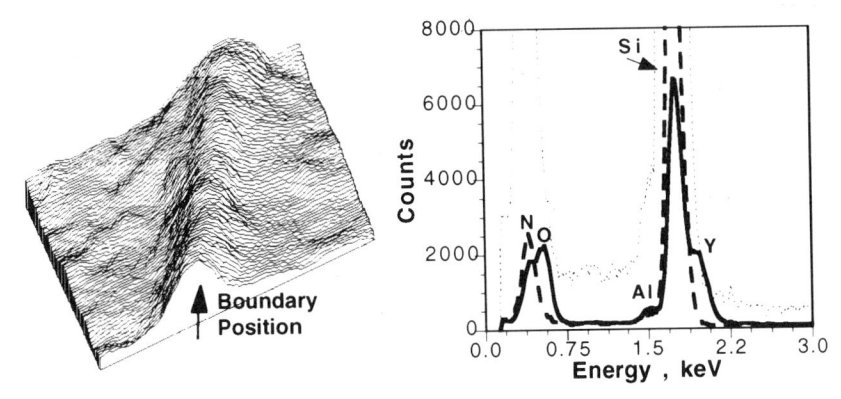

Figure 5. (a) Oblique view of amplitude modulated Z-contrast STEM image of an edge-on matrix grain boundary in an Si_3N_4/SiC(w) CMC synthesized using Huber whiskers. (b) Windowless characteristic x-ray spectra (2 nm probe size) in edge-on boundary plane shown in the image (solid curve), and, at a point 14 nm from the boundary plane, in one of the bounding nitride crystals (dashed curve). The latter curve, when amplified by 10x (narrow dotted curve), exhibits asymmetric shoulders on the Si peak, showing that O, Al, and Y were present in the matrix grain chemical width.

The chemical width of this distribution is ~20 nm and it is symmetrical about the boundary plane. This observation was verified by energy dispersive x-ray nanospectroscopy, which also showed that the aluminum distribution extended into the nitride crystal regions adjacent to the boundary plane (figure 5b). These results show that the yttrium and aluminum distributions follow the oxygen distribution. We performed similar experiments on whisker/matrix interfaces, which showed that where the oxygen distribution was laterally discontinuous, so, too, were the yttrium and aluminum distributions (Liu et al., 1993).

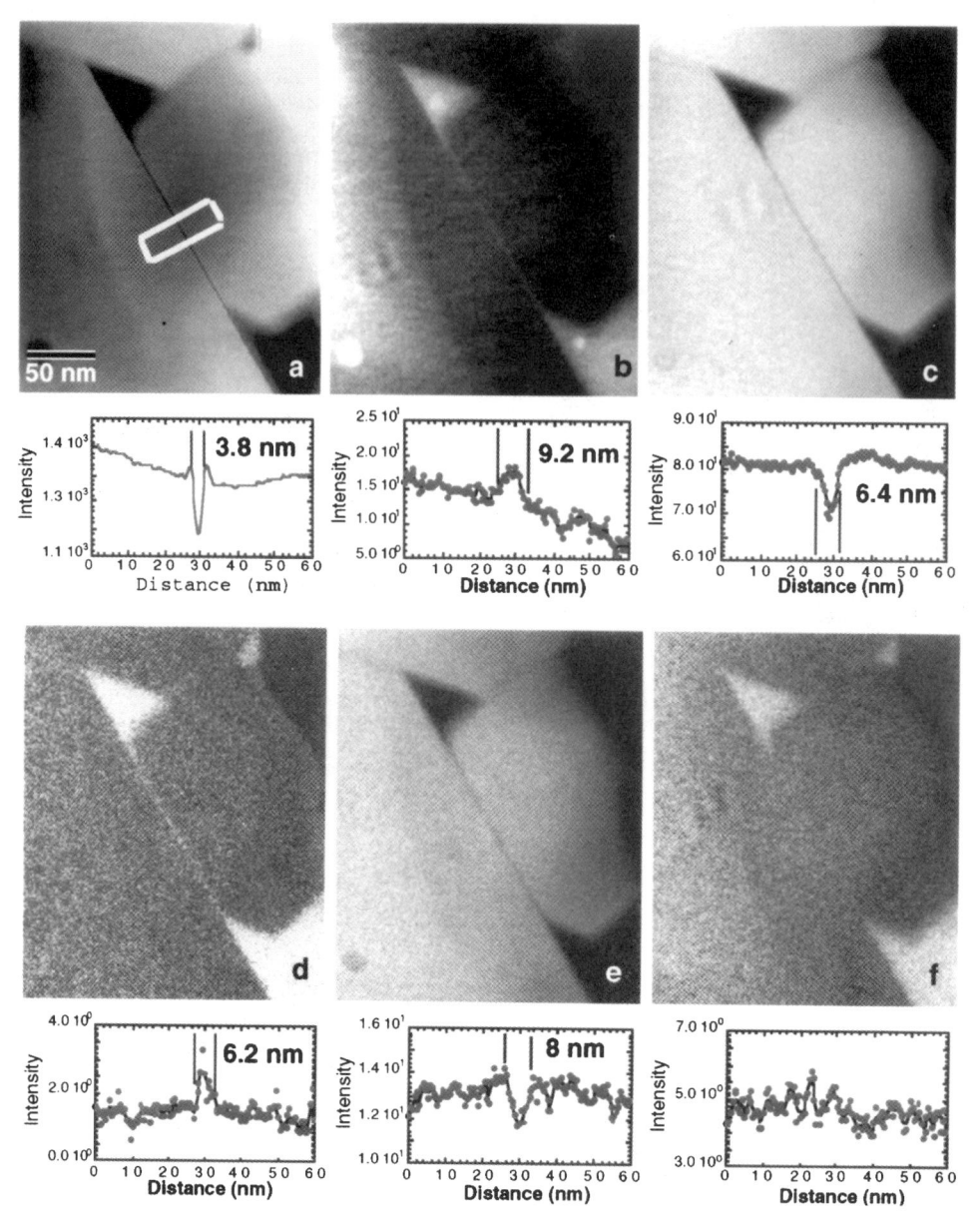

Figure 6. Bright field and energy selected images of an edge-on matrix grain boundary connected to two triple junctions in an Si$_3$N$_4$/SiC CMC. Edge of foil visible at upper right. (a) bright field; (b) Al-L image, 20eV window(W), 10s collection time(CT); (c) Si-L image, 20eV W, 10s CT; (d) O-K image, 20eV W, 30s CT; (e) N-K image, 20eV W, 30s CT; (f) C-K image, 20eV W, 20s CT.

Energy selected images of a short matrix grain boundary and portions of its triple junctions are shown in figure 6. This boundary had an appreciable structural width, i.e. a well defined thin amorphous phase. The energy selected images, figures 6b-6e, show that the boundary region and triple junctions are enriched in aluminum and oxygen and depleted in silicon and nitrogen in excellent qualitative agreement with the PREELS results above. We have attributed the bright contrast in figure 6f to carbon, however the yttrium M core edge, which is broad but not strong, interferes with the carbon K edge and further experiments are required to explain this contrast in detail. The plots under the images show that the chemical widths determined by energy selected imaging are smaller than the corresponding widths determined by PREELS. The difference results from instrumentation and method. For energy selected imaging a large unfocussed probe is used to irradiate the imaged area but for PREELS a small focused high current probe is used to irradiated a nanoarea, thus many more counts are accumulated per unit collection time. The minimum detectable mass fractions (MMF) for the elements of interest is proportional to $(Jt)^{-1/2}$, where J in the incident electron flux and t is the counting time. J is much larger for PREELS, and the corresponding higher detection sensitivity for small amounts of solute makes the method, although slower than energy selected imaging, most useful for determining chemical widths (Carpenter et al., 1995).

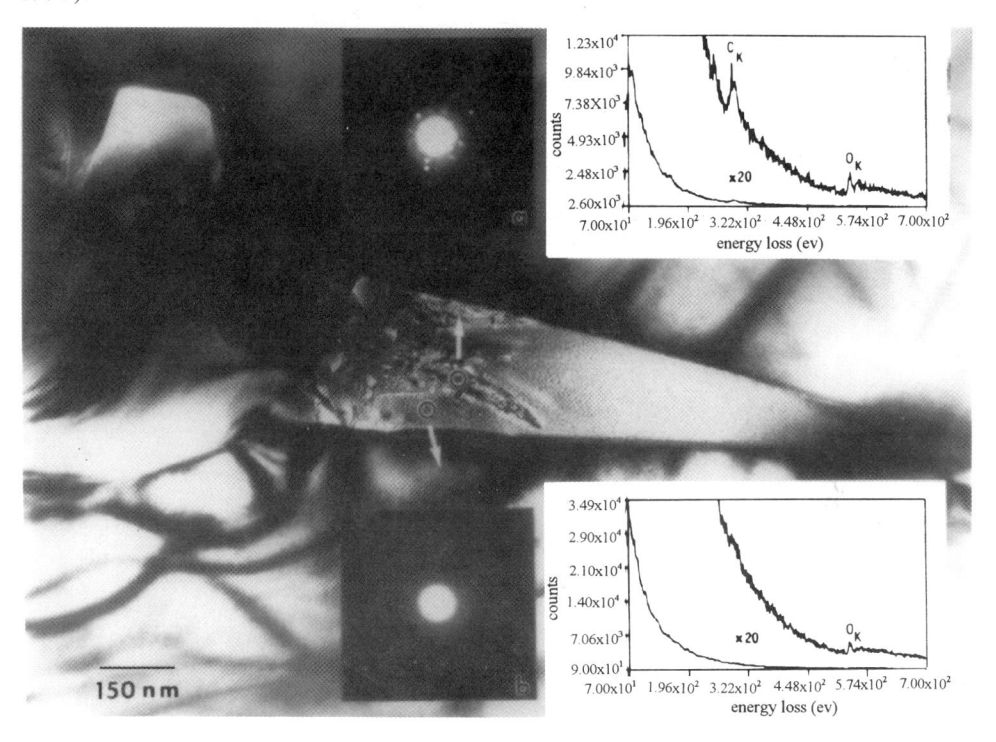

Figure 7. Triple junction intergranular inclusion in an Al_2O_3 matrix/SiC(w) CMC densified without sintering aids. Nanodiffraction showed that the inclusion matrix was amorphous and that it contained turbostratic graphite crystals heterogeneously nucleated at the bounding crystal surfaces. EELS from the regions indicated showed that the matrix was SiO_x, x < 2.0, and confirmed the graphite fibers.

The amount of amorphous phase in grain boundaries and whisker/matrix interfaces depends on the amount of sintering added and surface impurities on the powder starting materials. Similar composites with aluminum oxide matrices containing silicon carbide whiskers but processed without sintering aid additions had smaller amounts of boundary/interface thin amorphous phase (Braue et al., 1990; Wereszczack et al., 1993). High resolution TEM showed that structural widths of boundaries/interfaces in these composites was typically small, ~0.3 nm, and crystal-to-crystal bonding occurred in most boundaries/interfaces. In the alumina-silicon carbide composites we examined, triple junctions contained most of the amorphous phase in an interesting multiphase microstructure, shown in figure 7, was found in some of the triple junctions. The triple junction "pocket" microstructure contained graphite fibers in an amorphous matrix shown by nanospectroscopy to be composed of silicon, aluminum, oxygen and a small amount of calcium. The graphite is believed to have formed either by condensation of a carbon-rich vapor trapped in a glass phase pocket or nucleated from an oxycarbide liquid phase. Note that the graphite fibers were nucleated preferentially near the alumina grain at upper left, where their density is largest. Carbon solubility reaches about 2.5 wt.% at high temperatures in the Al-Si-Mg-O-C system (Homeny et al., 1988), implying that in the latter case heterogeneous nucleation occurred at the alumina-glass interface. In either case the source of carbon was probably excess carbon on the surface of whiskers, although carbon impurity pickup during

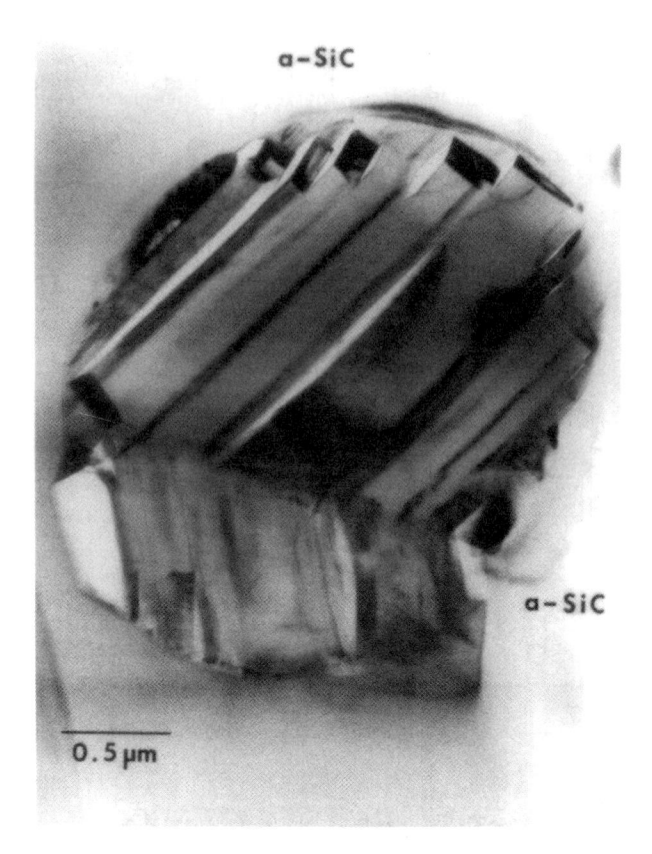

Figure 8. Elliptical intragranular inclusion containing thick aligned graphite laths inside an α-SiC grain, in polycrystalline α-SiC densified with addition of 2 wt.% B plus 2 wt.% C at 2000°C in 1 atm. Argon.

processing may also have contributed. These relatively localized amorphous matrix/graphite crystal regions also occur in some silicon carbides, depending on processing.

Silicon Carbide

Sintered α-SiC containing 2 wt.% boron plus 2 wt.% carbon and reaction bonded SiC (RB-SiC) both contain intragranular and intergranular inclusions with turbostratic graphite inclusions dispersed in an amorphous matrix (Braue and Carpenter, 1990; Braue et al., 1994). In these cases, however, nanospectroscopy showed that the amorphous matrices contain both oxygen (up to ~35 at.%) and carbon. The volume fraction of turbostratic graphite fibers in both intragranular and intergranular inclusions is variable, depending on local composition and processing. Higher carbon content relative to oxygen, determined by nanospectroscopy, produced a larger graphite fraction. The amorphous matrix phase of the inclusions at triple junctions serves as a sink for metallic impurities in solution, such as transition metal impurities, when they are present. The amorphous matrix phase of the inclusions in RB-SiC contains more silicon than the amorphous phase of the other materials examined, due to the excess silicon in this material, and that can result in formation of a small number of silicon and silicon carbide nanocrystals along with graphite crystals in the amorphous phase. The composition of the amorphous phase is dependent on the starting materials and processing for each type of SiC.

Graphite fibers in intragranular inclusions were aligned in parallel directions, as shown in figure 8. Others that we have observed contained a smaller graphite fiber volume fraction, but the fibers were always well aligned, usually in a single direction. In intergranular inclusions at triple junctions, on the other hand, the graphite fiber array was disordered, with fibers in many different nonparallel directions. In both types of inclusions the remaining amorphous matrix was near the center, away from the original interfaces with the bounding crystal or crystals. These microstructures imply that the graphite fibers were nucleated heterogeneously at the interfaces between the amorphous matrix and the bounding crystals. Graphite fibers in intragranular inclusions apparently self-selected preferred nucleation orientations and then grew into aligned fibers. Fibers in intergranular inclusions at triple junctions can nucleate on several crystals all in different orientations, in general, and grow into fibers that are not aligned. The fibers in figure 7 appear to have nucleated mainly on the crystal at upper left, and are relatively well aligned, for a triple junction. Most inclusions at triple junctions that we have observed contained poorly aligned fibers extending from the interfaces of three bounding crystals into the amorphous matrix. It is probable that the intragranular inclusions were formed by trapping intergranular inclusions within grains during grain growth in densification heating cycles. Those inclusions remaining connected to grain boundaries (intergranular) served as traps for impurities on particle surfaces. The inclusion matrices are silicon oxycarbide glasses that reject turbostatic graphite upon cooling (Braue et al., 1994).

The carbon and oxygen distribution in silicon carbides densified by hot pressing with additives was significantly different. Alpha silicon carbide doped with 0.2 wt.% aluminum and densified by hot isostatic pressing exhibited aluminum rich amorphous inclusions that also contained silicon and oxygen at triple junctions, but no carbon was detected and graphite fibers were not observed, suggesting that aluminum silicate liquid was present at temperature (Braue et al, 1988). The grain boundaries in this material exhibited continuous narrow disordered regions that were difficult to image by high resolution TEM, and too narrow to be described as a distinct thin amorphous second phase. Position resolved electron energy loss nanospectroscopy showed that the disordered regions contained a very small amount of oxygen. Alpha silicon carbide mixed with 4.15 wt.% alumina plus 5.85 wt.% aluminum carbide (Al_4C_3) densified by unaxial hot pressing (35 MPa) for 10 m at 1875°C in nitrogen exhibited nanocrystalline reaction zones containing aluminum, silicon, oxygen, carbon and nitrogen discontinuously distributed along grain boundaries and at all triple junctions examined (Carpenter et al., 1991). Graphite fibers were observed in some of the triple junction inclusions but not in the grain boundary reaction zones. The composition of the triple junctions varied over a large range with most containing about 30 at.% aluminum, 50 to 60 at.% oxygen plus carbon, 10 at.% nitrogen, and 5 at.% Si. The oxygen content was typically 30 to 40 at.%, a bit above the carbon content, and nitrogen was only detected in zones where the aluminum content was much higher than the silicon content. This densification process was based on reaction between Al_2O_3 and Al_4C_3 to form Al_2OC, aluminum oxycarbide, which is hexagonal with lattice parameters similar to 2H-SiC and AlN. Aluminum concentration gradients (non zero structural widths) extending into silicon carbide crystals bounding reaction zones were detected by nanospectroscopy. This system did not reach equilibrium during the short densification cycle, but near theoretical density was attained.

These results show that two types of oxygen sinks occur in silicon carbides, and their existence tends to be mutually exclusive. Inter- and intragranular silicon oxycarbide glass inclusions occur in silicon carbides densified with boron and carbon or with excess silicon (RB-SiC). The glass will precipitate graphite fibers if the carbon activity in the glass is high enough. No amorphous grain boundary phases or oxygen segregation to grain boundaries was detected in these materials. If densification was accomplished by addition of a metal (aluminum) that can react with impurity oxygen then aluminosilicate glass at triple junctions and a small amount of oxygen segregation at grain boundaries was observed, but not enough to produce a distinctly identifiable thin amorphous phase. In this case, which appears to be more similar to the microstructure expected for liquid phase densified silicon nitride ceramics, there may also have been a small amount of aluminum at the grain boundaries, but we have not detected it yet, due to the interference between L-shell spectrum lines for silicon and aluminum. When additional densification additives that will raise the activity of oxygen, carbon and aluminum in the liquid phase are used discontinuous nanocrystalline reaction zones were observed at grain boundaries and triple junctions, with graphite fibers in some of the triple junctions. In the latter case, where more additives were used, more and larger reaction zones were observed, however their crystallinity and discontinuities at grain boundaries are departures from normally observed behaviors.

DISCUSSION AND CONCLUSIONS

The applications of recently developed experimental nanoanalytical electron beam methods to the ceramic systems above permit several interesting conclusions and point the way toward further important experiments and experimental methods development.

The chemical width measurements shown with the high resolution image in figure 2 above illustrate the need to use both nanospectroscopy and images for grain boundary/interface analysis. The absence of a well defined thin amorphous phase does not convey much information about the presence or absence of chemical segregation (see also figures 3 and 4 for whisker/matrix surfaces), or about the existence of chemical widths extending into the crystal regions adjacent to interfaces and boundaries. In these silicon carbide whisker/silicon nitride composites the chemical widths are 10 to 120 times larger than corresponding structural widths, corresponding to diffusion constants in the 10^{-15} to 10^{-16} cm^2/sec range.

In silicon carbide there appear to be two primary types of oxygen containing sinks: (i) thin disordered regions at grain boundaries, and (ii) inclusions at triple junctions and inside grains containing a glass matrix of variable composition which, depending on local carbon activity, may contain graphite fibers apparently formed by an exsolution or precipitation reaction during cooling. The glass phase usually retained metals intentionally added or present as impurities. The inclusion type sinks were by far the most common, especially in the specimens where aluminum or its compounds were not used as densification aids. Occasionally inclusions containing only graphite were observed but, interestingly, graphite fibers were not observed along grain boundaries (two grain junctions) in any type of silicon carbide, or in any of the composites we have examined.

The nanochemistry of grain boundaries, triple junctions and heterophase interfaces in the ceramics we have examined so far is more complex than conventional high resolution imaging results of the past indicated, and additional high spatial resolution nanochemical measurements are required before easy generalizations about solute behavior can be made. One important question is the relative importance on the final microstructure of surface impurities on the starting particles compared to small amounts of sintering aids added. From this viewpoint, the only silicon nitride we have examined that did not contain oxygen or other nanospectroscopically detectable impurities was synthesized by high purity chemical vapor deposition (Das Chowdhury et al., 1993). High resolution grain boundary images from this material were qualitatively the same as the matrix grain boundary shown in figure 2 above. There does not yet appear to be any high resolution data available for deoxidized nitride powders sintered without sintering aids, but there is some interesting data for silicon carbide.

Turan and Knowles (1995) recently investigated this question in silicon carbide matrix composites reinforced with 10 or 20 wt.% silicon nitride particles densified without sintering aids, using HRTEM imaging methods. After densification to near theoretical density using the starting particles in the as received condition, some of both carbide and nitride grain boundaries, their triple junctions, and the heterophase boundaries contained small amounts of discontinuously distributed amorphous phase. However, when the particles were deoxidized prior to densification, to remove native oxide from the particle surfaces, attainable density was lower and HRTEM images showed that most grain boundaries/interfaces did not contain thin amorphous phase. Their limited nanospectroscopy data showed that the glass in large triple junctions contained carbon; in particular the silicon-L near edge fine structure of their EELS data (their fig. 13a) showed that the glass contained less oxygen than SiO_2 and was probably a silicon oxycarbide glass (Skiff et al., 1987; Kim and Carpenter, 1990). These observations and our data on CVD silicon nitride show that ceramic particle surface chemistry, as well as sintering aids, have significant effects on the amount of thin amorphous boundary/interface phase, its thickness and distribution, and, presumably, its composition in both silicon carbide and silicon nitride matrices. Given the large chemical widths we have so far observed in most of these materials, and the variable thin amorphous film configurations, it is unlikely that any of them reached equilibrium. We expect that the composition of grain boundary/interface regions will depend on initial surface chemistry, sintering aids and processing cycle.

It is important to note that thin amorphous interfacial phases with associated chemical widths have been observed for the first time recently in naturally occurring ceramic compounds in ultramafic rocks. Wirth (1996) used HRTEM imaging and electron nanoprobe energy dispersive x-ray spectroscopy to show that thin (1 to 2 nm) discontinuous amorphous films of variable width (structural width) occurred in grain boundaries in olivine from ultramafic mantle xenoliths in a basinite from San Carlos, Arizona. The amorphous phase was mainly Al_2O_3 (~35 wt.%), SiO_2 (~38 wt.%), and FeO (~10 wt.%) in regions wide enough for quantitative measurements, which was quite different from the olivine matrix (~50 wt.% MgO, 40 wt.% SiO_2, 9 wt.% FeO). The major impurities in the olivine matrix, Al_2O_3, CaO, TiO_2 and Cr_2O_3 (~1.5 wt.% total), were concentrated in the amorphous phase along with about 4 wt.% MgO. The chemical widths in the crystalline matrix grains adjacent to the amorphous grain boundary phase, which was formed by grain boundary melting, were significantly wider than the amorphous phase. In regions where the structural and chemical widths were large enough for quantitative measurement using the thermionic electron source available, the chemical widths were 200 to 500 nm wide. It now appears that chemical widths associated with thin amorphous boundary/interface phases of even vanishingly small dimensions may be a general phenomenon in most synthetically or naturally processed ceramics when a liquid phase occurs during processing.

The data presented indicate that chemical interfaces at grain and heterophase interfaces in many ceramics investigated are diffuse and imply that this is a general effect, probably present in other ceramics. Because diffusion is slow is these materials, it is unlikely that any of them reached equilibrium, but local metastability can be assumed. High spatial resolution composition measurements for the thin amorphous phase extending into the chemical widths in the bounding crystals are planned to investigate the diffusion behavior and equilibrium state in several ceramic systems. Small probe position resolved electron nanospectroscpy will be used for most of the measurements. The experimental data profiles are convolutions of the actual composition profiles with the current distribution in the focused probe and the thickness profile of the specimen along the scan path, and can recovered by deconvolution. Methods for calculation and measurement of the probe current distribution function are known (Weiss et al., 1991). Methods for measuring and controlling specimen thickness profiles developed during specimen preparation are being developed, to complete the data necessary for deconvolution.

Acknowledgment

We are pleased to acknowledge support for this research by the U.S. Department of Energy, Basic Energy Sciences, Division of Materials Sciences, under grant DE-FG03-94ER45510 (Dr. Otto Buck) and by the German Aerospace Research Establishment, Cologne.

REFERENCES

Braue, W. and Carpenter, R.W., 1990, Analytical electron microscopy of graphite-rich inclusions in sintered α-silicon carbide, J. Mat. Sci. **25**:2943.

Braue, W. Goering, J. and Ziegler, G., 1988, Proc. 3rd Int. Symp. on Ceramic materials and components for engines, Las Vegas, NV., Nov. 27-30, V.J. Tennery (ed.), American Ceramic Society:818.

Braue, W., Carpenter, R.W. and Smith, D.J., 1990, High-resolution interface analysis of SiC-whisker-reinforced Si_3N_4 and Al_2O_3 ceramic matrix composites, J. Mat. Sci. **25**:2949.

Braue, W., Das Chowdhury, K. and Carpenter, R.W., 1994, Oxygen sinks in SiC-based ceramics, Mat. Res. Soc. Symp. Proc. Vol. **327**:275.

Carpenter, R.W. 1989, High resolution interface analysis, Matl. Sci. and Engr. A107:207.

Carpenter, R.W., Bow, J.S., Kim, M.J., Das Chowdhury, K., and Braue, W., 1995, Chemical widths at composite interfaces: relationships to structural widths and methods for measurement, Mat. Res. Soc. Symp. Proc. Vol. **357**:271.

Carpenter, R.W., Braue, W. and Cutler, R.A., 1991, Transmission electron microscopy of liquid phase densified SiC, J. Mater. Res. **6**:1937.

Catalano, M. Kim, M.J., Carpenter, R.W., Das Chowdhury, K. and Wong, Joe, 1993, The composition and structure of SIPOS: a high spatial resolution electron microscopy study, Jour. Mater. Res. **8**:2893.

Clarke, D.R. and Thomas, G., 1977, Grain boundary phases in hot-pressed MgO fluxed silicon nitride, J. Amer. Ceram. Soc. **60**:491.

Clarke, D.R., 1979, On the detection of thin intergranular films by electron microscopy, Ultramicros. **4**:33.

Das Chowdhury, K. Carpenter, R.W. and Weiss, J.K., 1989, Radiation damage characteristics of silicon oxynitride ceramics, Proc. 47th Ann. Mtg. Elec. Micros. Soc. America, San Francisco Press, Inc.:428.

Das Chowdhury, K., Carpenter, R.W. and Braue, W., 1993, Grain boundaries in silicon nitride, Proc. 51st Ann. Mtg. Micros. Soc. Amer., San Francisco Press Inc.:920.

Das Chowdhury, K., Carpenter, R.W., Braue, W., Liu, J. and Ma, H., 1995, Chemical and structural widths of interfaces and grain boundaries in silicon-nitride-silicon carbide whisker composites, Jour. Amer. Ceram. Soc., **78**:2579.

Evans, A.G., 1990, Perspective on development of high-toughness ceramics, J. Amer. Ceram. Soc. **73**:187.

Homeny, J., Nelson, G.G. and Risbud, S.H., 1988, Oxycarbide glasses in the Mg-Al-Si-O-C sytem, Jour. Amer. Ceram. Soc. **71**:386.

Kim, M.J. and Carpenter, R.W., 1990, Composition and structure of native oxide on silicon by high resolution analytical electron microscopy, J. Mater. Res., **5**:347.

Kohl, H. and Rose, H., 1985, Theory of image formation by inelastically scattered electrons in the electron microscope, Adv. in Electronics and Electron Phys. **65**:173.

Liu, J. and Cowley, J.M., 1993, High-resolution scanning transmission electron microscopy, Ultramicros. **52**:335.

Liu, J. Das Chowdhury, K., Carpenter, R.W. and Braue, W., 1993, Elemental analysis of matrix grain boundaries in SiC whisker reinforced Si_3N_4 based composites, Mat. Res. Soc. Symp. Proc. Vol. **287**:329.

Pennycook, S.J., Jesson, D.E., Chisholm, M.F. Browning, N.D., McGibbon, A.J. and McGibbon, M.M., 1995, Z-contrast imaging in the scanning transmission electron microscope, Journ. Micros. Soc. Amer. **1**:231.

Skiff, W.M., Carpenter, R.W. and Lin, S.H., 1987, Near-edge fine-structure analysis of core-shell electronic absorption edges in silicon and its refractory compounds with the use of electron-energy-loss microspectroscopy, J. Appl. Phys., **62**:2839.

Turan, S. and Knowles, K.M., 1995, A comparison of the microstructure of silicon nitride-silicon carbide composites made with and without deoxidized starting material, Jour. of Microscopy, **177**:287.

Urbain, G. Cambier, F., Deletter, M. and Anseau, 1981, Viscosity of silicate melts, Trans. J. Brit. Ceram. Soc. **80**:139.

Weiss, J.K., Carpenter, R.W. and Higgs, A.A., 1991, A study of small electron probe formation in a field emission gun TEM/STEM, Ultramicros., **36**:319.

Weiss, J.K., Rez, P. and Higgs, A.A., 1992, A computer system for imaging and spectroscopy in analytical electron microscopy, Ultramicros. **41**:291.

Wirth, R., 1996, Thin amorphous films (1-2 nm) at olivine grain boundaries in mantle xenoliths from San Carlos, Arizona, Contrib. Mineral Petrol, **124**:44.

ANION SEGREGATION AT Si₃N₄ INTERFACES STUDIED BY HIGH-RESOLUTION TRANSMISSION ELECTRON MICROSCOPY AND INTERNAL FRICTION MEASUREMENTS: A MODEL SYSTEM

H.-J. Kleebe [*] and G. Pezzotti [#]

[*]University of Bayreuth, Materials Research Institute (IMA I)
D-95440 Bayreuth, Germany
[#]Department of Materials, Kyoto Institute of Technology, Matsugasaki, Kyoto 606, Japan

ABSTRACT

Structural characterization of internal interfaces in a Si₃N₄ model system was performed by high-resolution transmission electron microscopy (HRTEM). The materials required hot-isostatic pressing (HIPing) for complete densification since they contained no external sintering aids except a small fraction of anions, *i.e.*, fluorine or chlorine. These materials are considered a model system, because the non-crystalline secondary phase between Si₃N₄ grains is composed of pure silica glass with only F or Cl segregated at the glass pockets and interfaces. Internal friction curves were obtained at temperatures up to 2000°C and were generally composed of an exponential background superimposed by a well-defined grain-boundary relaxation peak. The incorporation of anions into the glass structure simultaneously changes (i) the intergranular film thickness and (ii) the effective glass viscosity, which is lowered owing to the formation of non-bridging bonds. Hence, the interface structure/chemistry dominates the macroscopic properties of the bulk material at both room temperature and at high service temperatures.

INTRODUCTION

Ceramic materials have gained wide interest owing to their potential for structural applications, in particular at elevated temperatures. Densification of Si₃N₄-based ceramics is commonly performed by liquid-phase assisted sintering in order to overcome the low volume diffusion of Si₃N₄.[1] Sintering additives used to achieve complete densification react with SiO₂, present on the Si₃N₄-particle surface owing to oxidation from exposure to atmosphere, and form a silica-rich liquid at elevated temperatures.[2] Upon cooling, remains of this liquid are present at triple-grain junctions and along grain boundaries as a secondary glass phase. The high-temperature performance of Si₃N₄ primarily depends on the volume fraction and composition of such glass residues, owing to low glass-softening temperatures.[3]

Since Si_3N_4 ceramics are comprised of highly refractory matrix grains surrounded by a less refractory vitreous phase, the modification or elimination of low-melting secondary phases is required to overcome the aforementioned obstacles hampering potential application. One approach to improve high-temperature performance of Si_3N_4 ceramics is to simply reduce the glass-volume fraction. A number of studies were reported on Si_3N_4 materials HIPed without the addition of sintering aids.[4,5] Substitution of lanthanide oxides for commonly used metal oxides e.g., the addition of Yb_2O_3 as sintering aid, has been explored to form a more refractory intergranular phase.[6,8] Post-densification heat treatment is also a widely accepted technique to minimize the amount of glass residue by forming highly refractory crystalline phases, however, a complete crystallization of the glass pockets cannot be achieved during annealing.[7,8] Hence, crystallization reduces the volume fraction of residual glass, but can simultaneously lead to an enrichment of impurities within the residual amorphous phase (in particular, when the impurities cannot be dissolved in the newly formed secondary crystal lattice[6]) which greatly limits high-temperature applicability.

The interface structure, i.e., the intergranular film width, is a fingerprint of the interface chemistry. Segregation of cations at the boundary results in changes in interphase thickness.[6,9] Recent studies have shown that, apart from cation impurities, the incorporation of anions (such as fluorine) can cause a significant decrease in subcritical crack-growth resistance and creep resistance.[10,11] Moreover, internal friction measurements performed at a maximum temperature of 1800 °C also revealed a clear grain-boundary relaxation peak, attributed to the presence of fluorine at the interfaces.[12] Thus, in addition to ist volume fraction, the chemistry and respective structure of the amorphous secondary phase, are important with respect to the high-temperature performance of the material. Therefore, further details about structural and chemical characteristics of ceramic residual glasses, as well as their influence on micromechanical processes and the macroscopic performance are required.

In this paper we report on the microstructure/property correlation of a F- or Cl-doped Si_3N_4 with no further external sintering aids. Emphasis is placed on the structural and chemical characterization of the vitreous secondary phase performed by HRTEM and analytical electron microscopy (AEM), respectively. Moreover, correlations between the atomic structure of the amorphous intergranular phase and the bulk mechanical properties of the model system are discussed in light of the internal friction data.

EXPERIMENTAL PROCEDURES

High purity α-Si_3N_4 starting powder (E-10, Ube Ind. Ltd., Ube, Japan) was doped with fluorine or chlorine during processing. Fluorine was introduced into the powder by pulverized polytetrafluoroethylene (Teflon, E.I. du Pont de Nemours, Wilmington, DE/USA) while chlorine was added by the use of hexachloroethane. The powder blends were preheated in vacuum at 1200 °C in order to decompose the dopant and to eliminate excess carbon via CO gas. Incorporation of fluorine or chlorine into the SiO_2-glass structure is thought to be achieved at temperatures above 1000 °C. Pre-heated specimens were encapsulated in an evacuated boron-silicate glass tube to enable complete densification as well as to avoid the evaporation of dopant. Isostatically pressed powder compacts were densified by hot-isostatic pressing (HIPing) at 1900 °C for 2 hours at 180 MPa Ar-gas overpressure. Full density (>99.5% theoretical density) was achieved under the applied processing conditions. As a reference material, an undoped Si_3N_4 material was fabricated under nearly identical HIPing parameters. In order to adjust for a difference in grain size, observed when heat treated under identical processing conditions, a 50 °C higher HIPing temperature was applied for the undoped compared to the F- or Cl-containing materials.

Interface structure and interface chemistry of the Si_3N_4 materials, with fluorine or chlorine (and silica) as the only sintering aids, were studied by transmission electron microscopy (TEM). TEM-foil preparation was performed by standard techniques, which involve diamond-blade cutting, mechanical grinding, dimpling, and Ar-ion beam thinning to perforation, followed by coating with a thin carbon layer to minimize electrostatic charging under the electron beam. TEM and AEM, were performed using a Philips CM20FEG (field emission gun) operating at 200 kV with a point resolution of 0.24 nm. The microscope was fitted with an energy-dispersive X-ray Ge detector (Tracor Voyager 2100) and an electron energy-loss spectrometer with parallel detection (PEELS, Gatan 666). To accurately determine the thickness of intergranular films at two-grain junctions, HRTEM studies were performed using a JEOL 4000EX (top entry) microscope operating at 400 kV with a point resolution of 0.18 nm. The {10-10} lattice spacings of the ß-Si_3N_4 (0.658 nm), as depicted in the HRTEM micrographs, were used as internal reference to quantitatively measure grain-boundary film width.

Internal friction data were monitored by a torsional pedulum apparatus, specially designed for measurements at very high temperatures.[13] Internal friction was measured at a frequency of 10 Hz by the free-decay method, while the grain-boundary viscosity values, as a function of F- or Cl-content, were calculated from the grain-boundary peak.

RESULTS AND DISCUSSION

Transmission Electron Microscopy of F- and Cl-Doped Si_3N_4

The overall microstructures of the F- and Cl-doped Si_3N_4 materials, observed on a more-macroscopic scale, was undistinguishable from each other and even from the undoped (anion-free) reference material. Figure 1 shows a low magnification TEM image of the Cl-doped Si_3N_4 material. The microstructures of both anion-containing materials were composed of fine grained, equiaxed ß-Si_3N_4 grains with only a small fraction of slightly larger and partly elongated grains. The ß-Si_3N_4 grains were typically about 0.2-1 µm in diameter. No crystalline phases other than the matrix grains were observed. The formation of Si_2N_2O during HIPing was not expected, since the maximum processing temperature was higher than the decomposition temperature of Si_2N_2O of about 1830 °C [14] and due to the low intrinsic silica content of the material.

Apart from crystalline matrix phases, a small amount of residual glass was observed within the F- and Cl-doped materials. As typical for Si_3N_4 ceramics, the amorphous residue was present at triple pockets and along grain boundaries. The size of triple-grain junctions was on the order of 10-200 nm in diameter. A careful TEM inspection of such triple pockets, involving goniometer tilting during observation, revealed no secondary crystalline phase. Thus, devitrification of the glass did not occur upon cooling. The multi-grain glass pockets, which contain most of the amorphous residue, are interconnected by a three-dimensional network of thin grain-boundary films. As will be discussed in the following chapters in more detail, these intergranular films revealed a characteristic thickness, which is closely related to the interface chemistry, i.e., the anion segregation. The Si_3N_4 microstructure observed in these materials differs in two aspects from common Si_3N_4 grades: (i) a rather fine grained Si_3N_4 matrix and (ii) a highly pure secondary glass phase with no other crystalline phases present. This defines a model system comprised of globular Si_3N_4 grains, about 1 µm in diameter, covered by a high-purity glass which, at first approximation, can be seen as pure fused silica. Excess glass is bound to small triple junctions, typically being 10-200 nm in size. These boundary conditions allow, in principle, to propose a simplified model of the glass structure, a pure SiO_4^{4-}-tetrahedra network. Moreover, the incorporation of fluorine or chlorine can also be treated in this manner, since the presence of anions will only alter the glass structure, *i.e.*, the respective network structure (no solid solution of either fluorine or chlorine with Si_3N_4 has been detected).

Fig. 1: TEM micrograph showing the fine-grain microstructure of the Cl-doped Si_3N_4 material. Note that this microstructure is indistinguishable from the overall microstructure observed in the F-doped Si_3N_4 as well as the undoped reference material.

Densification of such high purity materials via HIPing is a prerequisite to fabricate model systems that allow a systematic change of the interface structure by altering the glass composition and, therefore, enable a correlation between the monitored mechanical response and atomic changes at the interface.

Interface Structure - High-Resolution Transmission Electron Microscopy

HRTEM was employed to study the interface structure of both the F- and Cl-doped material. Owing to the liquid phase formation at the applied high HIPing temperature, although only a small volume fraction of glass is thought to be formed, a continuous amorphous intergranular film was present at grain boundaries. Different TEM techniques allow for the detection of intergranular films and the quantitative evaluation of their thickness. High-resolution lattice imaging is a method capable of the resolution necessary to obtain detailed information of the grain boundary and the intergranular phase itself. It has been shown to be applicable to quantitatively evaluate the intergranular film thickness in Si_3N_4 materials with an accuracy of ±0.1 nm.[15] The presence of such amorphous interfacial glass profoundly affects the mechanical properties of Si_3N_4-based ceramics, in particular at high service temperatures. It is important to note that such interfacial glass films reveal, for a given interface chemistry, a constant thickness, which seemingly suggests that the film thickness can be seen as a fingerprint of the local grain-boundary chemistry.

Figure 2 shows a characteristic intergranular film observed at a grain boundary in the F-doped composite. Quantitative evaluation of the grain-boundary film thickness of 10 boundaries resulted in an average equilibrium width of 1.1 nm. The undoped reference composite revealed a film thickness of 1.0 nm. Si_3N_4 materials doped with SiO_2 as the only sintering aid always showed a constant film thickness of 1.0 nm, independent of processing technique applied and volume fraction of silica added (excess silica is present at triple pockets). Thus, a small increase in grain-boundary film thickness of Δ=0.1 nm (10% wider) was determined, which can be rationalized by the change in interfacial glass composition.

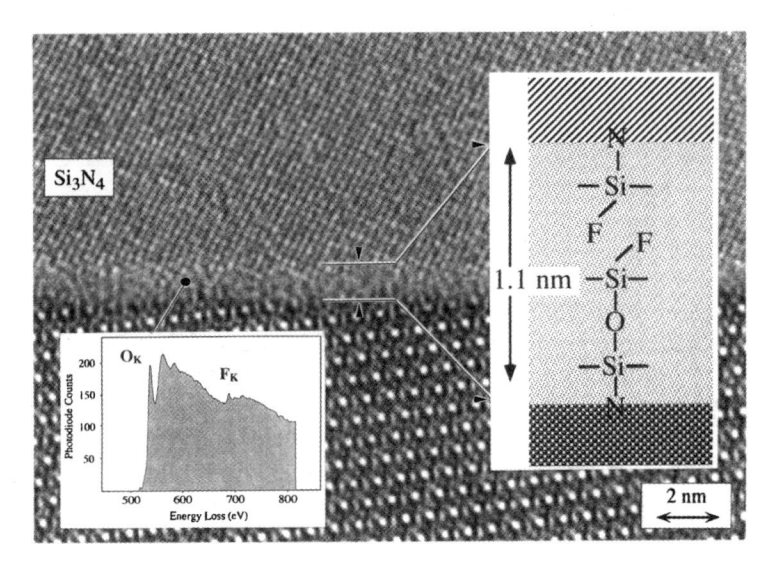

Fig. 2: HRTEM image of an intergranular grain-boundary film observed in the F-doped Si$_3$N$_4$ material. In contrast to the undoped (F-, Cl-free) reference material, which showed a film width of 1.0 nm, an increase in film thickness of about 10 % was monitored.

Electron energy-loss spectroscopy revealed the presence of fluorine both at triple-grain junctions and grain boundaries. Determining the F/O-ratio for the spectra obtained from triple-grain pockets and Si$_3$N$_4$ interface regions, no major compositional changes were detected. The F-content in both cases was evaluated to be about 25 at% fluorine (O/F-ratio within the silica glass). This result seemingly suggests a homogeneous secondary phase composition at interfaces within the F-doped Si$_3$N$_4$. Assuming a homogeneous F-distribution within the SiO$_2$-glass structure, a constant and characteristic film thickness was expected. Therefore, the HREM observations are consistent with the obtained EELS data, which both indicate an altered interfacial glass structure/composition, as compared to the undoped composite.

HREM and AEM studies showed that the variation in grain-boundary film thickness is related to changes in F-content of the intergranular films. Here, compared with the pure SiO$_2$-containing material (with h = 1.0 nm), a small increase in film thickness was expected by the incorporation of anions. This is due to the fact that the replacement of one bridging oxygen ion versus fluorine, requires two anions to allow for charge balance. Since the coupled and equally charged F$^-$-ions (both placed at the former oxygen-bridge site) will repell each other, a widening of the film thickness is observed. The anion-distribution is thought to be homogeneous within the F-doped glass.

In contrast to the results of the F-containing material, the presence of Cl could not be detected within the residual glass of the Cl-doped sample utilizing either technique EDX or EELS. Although a small fraction of Cl was still present in the HIPed materials, as determined by ICP analysis (compare Table 1), it is concluded that their content is below the detection limit of the applied analytical TEM-techniques. That a small fraction of Cl was indeed incorporated into the residual glass structure could indirectly be demonstrated by internal friction measurements, $i.e.$, by the greatly lowered peak-top temperature of the Cl-doped sample, as discussed in more detail in the following section.

Tab. 1: Anion concentration within the HIPed Si_3N_4 materials determined by ICP analysis. The F- or Cl-content within the residual glass was estimated based on the oxygen content of 1.4 wt% within the bulk.

Amount of Dopant	Chemical Analysis	Estimated Fraction in SiO_2 Glass	Internal Friction Peak Temperature
undoped	-	-	1710 °C
low F-content	0.05	2.1 wt%	1654 °C
high F-content	0.14	5.8 wt%	1586 °C
low Cl-content	<0.001	<0.05 wt%	-
high Cl-content	0.016	0.69 wt%	1239 °C

Internal Fraction Measurements - Micromechanical Analysis

Internal friction measurements of the F- and Cl-doped materials as well as the undoped reference sample were performed at very high temperatures up to 2000 °C. The internal friction curves were generally composed of an exponential background superimposed by a well-defined grain-boundary relaxation peak in the temperature regime between 1250 °C and 1750 °C. With increasing anion concentration at the interface, the grain-boundary relaxation peak is shifted to lower temperatures (see also Table 1, F-content), which indicates a reduction in the effective glass viscosity. The peak-top temperature, evaluated by the internal friction measurement, decreased by 56 °C and 124 °C for the low and high F-addition, respectively. Following the proposed structure modification, a continuous incorporation of F into the glass will also weaken the glass network due to the formation of non-bridging sites. This lowers the effective viscosity at the interface and hence leads to the monitored lowering of the peak-top temperature during the internal friction measurements.

A lowered cohesive interface strength also rationalizes the observation that crack propagation shows a transition between transgranular to mainly intergranular fracture, when F or Cl is added to the system. This latter effect is pronounced in the F-containing system, since the effective F-content is higher compared to the respective Cl-content. Compared to the undoped, high-purity composite, anion-containing samples also reveal markedly higher creep rates [12], owing to a decrease in the aparent grain-boundary viscosity.

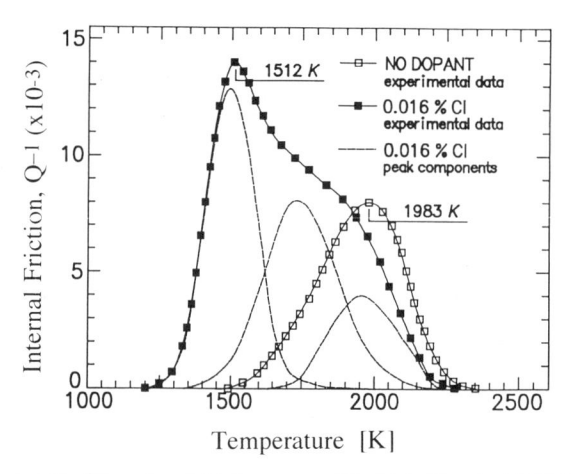

Fig. 3: Peak components of the internal friction curve of the high Cl-material (Table 1). The broad peak of the Cl-doped sample is deconvoluted into three complementary components (broken lines).

Internal friction curves of the Cl-contaning and undoped Si_3N_4, as a function of temperature also show a well defined relaxation peak. It has recently been established that such peaks arise from the sliding process along grain boundaries within the polycrystal due to the applied external stress. From the present data it is recognized that Cl-additions produce a remarkable shift of the internal friction curve towards lower temperatures, much more pronounced compared to the F-containing sample, even though the anion concentration is lower in the Cl-doped Si_3N_4 (see Table 1). The temperature corresponding to the internal friction peak temperature decreases for several hundred degrees upon Cl-addition and, moreover, the peak shape markedly broadens, as depicted in Figure 3. The broadened experimental peak of the Cl-doped material can be devided into three complementary peak components. It is noteworthy that the position of the high-temperature peak component of the Cl-containing Si_3N_4 coincides with the peak of the undoped reference material. This surprising result strongly suggests that, despite the Cl-addition, a certain fraction of the grain boundaries in the doped polycrystal still slide at a rate dictated by the viscosity of the undoped SiO_2. An inhomogeneous dispersion of the dopant within the specimen may cause this internal friction peak broadening. However, such a non-uniform distribution of Cl at the interface would also result in a variation in grain-boundary film thickness. HRTEM studies of the boundaries within the high Cl-doped material indeed revealed a difference in film thickness, as shown in Figure 4.

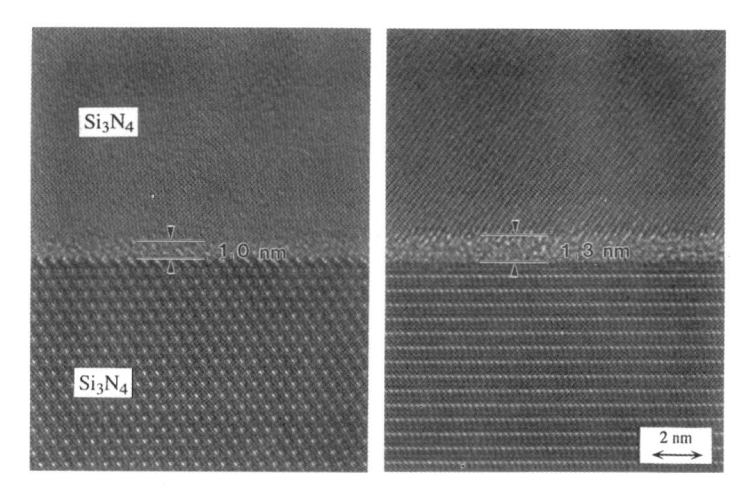

Fig. 4: HRTEM images of intergranular grain-boundary films observed in the Cl-doped Si_3N_4 material. In contrast to the F-doped material, a variation in film width (bimodal) was observed, which is thought to be a result of an non-uniform dispersion of anions at interfaces.

It is thought that the variation in film width results from local inhomogeneities of the Cl-distribution at interfaces. Due to the rather low Cl content within the material, it is assumed that a uniform dispersion of the Cl-anions could not be achieved during densification. A careful chemical analysis of grain-boundary films has not been performed yet. At triple pockets, however, no Cl could be detected by EDX or EELS (detection limit). Since the experimental HRTEM and internal friction results suggest an inhomogeneous distribution of Cl at the interface, it might be possible to detect Cl at boundaries where the anion is enriched. Further work employing the high-resolution SIMS technique is in progress to address this open question.

CONCLUSIONS

Variations in interface chemistry owing to the incorporation of anions such as fluorine or chlorine, and thus changes of the atomic structure at the interface, can strongly affect the mechanical response of the bulk material by lowering the cohesive strength of internal interfaces. At room temperature, this phenomenon leads to an increase in intergranular fracture during crack propagation. At high service temperatures, a shift of the damping temperature curves to lower temperatures is monitored, when compared to the undoped Si_3N_4 material, owing to a lowered effective glass viscosity.

The given results seemingly suggest that the interconnected network of the residual glass dominates the performance of such liquid-phase sintered Si_3N_4 materials. Hence, bulk material performance is also affected by the anion concentration, *i.e.*, the atomic glass structure.

Acknowledgements: The authors would like to thank Prof. T. Nishida and Prof. G. Ziegler for their support throughout this work. Prof. K. Urabe is greatly acknowledged since the HRTEM studies of the Cl-doped material were performed at Ryukoku University.

References

1. K. Kijima and S. Shirasaki, "Nitrogen Self-Diffusion in Silicon Nitride," J. Chem. Phys., 65, 2668-2671 (1976).
2. F.F. Lange, "Fabrication and Properties of Dense Polyphase Silicon Nitride," Am. Ceram. Soc. Bull., 62 [12] 1369-74 (1983).
3. W.A. Sanders and D.M. Mieskowski, "Strength and Microstructure of Si_3N_4 Sintered with ZrO_2 Additions," Adv. Ceram. Mater., 1, [2] 166-73 (1986).
4. K. Homma, H. Okada, T. Fujikawa, and T. Tatuno, "HIP Sintering of Silicon Nitride without Additives," Yogyo Kyokaishi, 95 [2] 229-34 (1987).
5. G. Pezzotti, "Si_3N_4/SiC-Platelet Composite Without Sintering Aids: A Candidate for Gas Turbine Engines," J. Am. Ceram. Soc., 76 [5] 1313-20 (1993).
6. H.-J. Kleebe, J. Bruley, and M. Rühle, "HREM and AEM Studies of Yb_2O_3-Fluxed Silicon Nitride Ceramics With and Without CaO-Addition," J. Eur. Ceram. Soc., 14, 1-11 (1994).
7. D.A. Bonnell, T.Y. Tien, and M. Rühle, "Controlled Crystallization of the Amorphous Phase in Silicon Nitride Ceramics," J. Am. Ceram. Soc., 70 [7] 460-5 (1987).
8. M.K. Cinibulk, G. Thomas, and S.M. Johnson, "Fabrication and Secondary-Phase Crystallization of Rare-Earth Disilicate - Silicon Nitride Ceramics," J. Am. Ceram. Soc., 75 [8] 2037-43 (1992).
9. H.-J. Kleebe, M.J. Hoffmann, and M. Rühle, "Influence of Secondary Phase Chemistry on Grain-Boundary Film Thickness in Silicon Nitride," Z. Metallkd., 83 [8] 610-7 (1992).
10. D.R. Clarke, "Densification of Silicon Nitride: Effect of Clorine Impurities," J. Am. Ceram. Soc., 65 [2] C-21-C-23 (1982).
11. I. Tanaka, K. Igashira, H.-J. Kleebe, and M. Rühle, "High-Temperature Strength of Fluorine-Doped Silicon Nitride," J. Am. Ceram. Soc., 77 [1] 275-7 (1994).
12. G. Pezzotti, K. Matsuchita, H.-J. Kleebe, Y. Okamoto, and T. Nishida, "Viscous Behavior of Grain and Phase Boundaries in Fluorine Doped Si_3N_4-SiC Composites," Acta Metall. Mater., 43 [12] 4357-70 (1995).
13. G. Pezzotti, H.-J. Kleebe, K. Matsuchita, T. Nishida, M. Sakai, "Cohesive Energy of Interfaces and Toughness of Fluorine Doped Si_3N_4/SiC Composites," J. Ceram. Soc. Jpn., 104 [1] 17-22 (1996).
14. Z.K. Huang, P. Greil, and G. Petzow, "Formation of Siliconoxinitride from Si_3N_4 and SiO_2 in the Presence of Al_2O_3," Ceramics International, 10, 14-7 (1984).
15. M.K. Cinibulk, H.-J. Kleebe, and M. Rühle, "Quantitative Comparison of TEM Techniques for Determining Amorphous Intergranular Film Thickness," J. Am. Ceram. Soc., 76 [2] 426-32 (1993).

TESTING INDUCED NANOSCALE INSTABILITIES OF SECONDARY PHASES IN Si_3N_4, THE TEM APPROACH TO HIGH-TEMPERATURE MICROSTRUCTURES

W. Braue [#] and H.-J. Kleebe [*]

[#] German Aerospace Research Establishment (DLR), Materials Research Institute, D-51147 Cologne, Germany
[*] University of Bayreuth, Materials Research Institute (IMA I), D-95440 Bayreuth, Germany

ABSTRACT

The significance of dynamic microstructures in liquid-phase sintered Si_3N_4 due to the constraints of HT-testing is addressed by focusing on three Si_3N_4 materials which have been tested on a different time scale extending from several minutes to well in excess of 1000 hours. As shown by high-resolution and analytical transmission electron microscopy, HT-testing can introduce nanoscale morphological and compositional instabilities of both the amorphous and crystalline secondary phases, providing a surprisingly sensitive tool to monitor the response of Si_3N_4 materials at high temperatures, which is otherwise insufficiently described by the as-sintered microstructure.

INTRODUCTION

Despite the maturity of liquid-phase sintered Si_3N_4 after decades research focused on this class of structural ceramics, the significance of long-term environmental stability and resistance to high-temperature deformation is still valid and has recently been readdressed.[1] In general, the high-temperature performance of liquid-phase sintered Si_3N_4 materials are limited by time-dependent failure in stress-temperature-time space due to viscous deformation of the non-crystalline oxinitride secondary phase.[2,3] It is this intrinsic microstructural feature in dense Si_3N_4 which leaves a narrow margin for the substantial improvement of high-temperature properties. During the last decade, a strong research effort has been directed towards the structure and composition of internal interfaces applying todays high-resolution TEM techniques for imaging, diffraction and spectroscopy on a truly atomic level.[4,5] In the common rationale for the microstructure/property correlation of Si_3N_4, which is widely accepted in the materials community, a homogeneous static microstructure with negligible compositional gradients among the secondary phases is assumed, although the limitations of this approach has been pointed out in previous research.[6,7]

However, during high-temperature (HT) testing of liquid-phase sintered Si_3N_4, the combined effects of high temperature and localized stresses can introduce dynamic nanoscale processes involving the decomposition of secondary phases, although the testing conditions are considered well below the stability limits of the refractory additive system employed.

Ceramic Microstructure: Control at the Atomic Level
Edited by A. P. Tomsia and A. Glaeser, Plenum Press, New York, 1998

Recent TEM investigations have shown that diffusion-controlled instabilities of the crystalline secondary phase and corresponding compositional fluctuations of the amorphous phase define a surprisingly sensitive tool to monitor the nanoscale microstructural response during HT-deformation of Si_3N_4.[10] This effect is particularly appreciated since the dislocation activity in the matrix grains is limited, due to the high Peierls stresses required for covalent compounds such as Si_3N_4. Hence it is difficult to correlate testing conditions with a single unequivocal deformation event.

We apply the TEM approach to three different Si_3N_4 materials which have been exposed to high-temperature testing on a different time scale extending from several minutes to well in excess of 1000 hrs. The microstructural response to the different testing conditions at triple-grain junctions and grain boundaries is monitored by high-resolution imaging and analytical transmission electron microscopy emphasizing the dynamic nature of liquid-phase sintered Si_3N_4 microstructures during HT-testing.

EXPERIMENTAL PROCEDURES

Three different liquid-phase sintered Si_3N_4 materials were employed in this study: (i) a HIP-RBSN material doped with 4 wt% Y_2O_3 + 2 wt% Al_2O_3 (courtesy of G.A. Schneider, Technical University Hamburg-Harburg, Germany) which has been employed in a rapid-quench experiment from high temperatures, as described elsewhere.[8] Here, crack propagation induced through focused radiant heating of a thin prenotched disk-shaped specimen was investigated at 835°C, 925°C and 1010°C. (ii) a RBSN-based sintered silicon nitride material (SRBSN), doped with 5 wt% unstabilized ZrO_2. Densification and microstructural development are discussed in detail elsewhere.[9] High-temperature deformation characteristics of SRBSN in air were qualitatively evaluated in loading/unloading experiments in four point bending with successive temperature increments of 50°C between room temperature and 1200°C.[10] (iii) a MgO-doped hot-pressed Si_3N_4 (Norton Co., Worchester, MA/USA, NC-132 grade). The specimen investigated (courtesy of G.D. Quinn, NIST, Gaithersburg/MA, USA) has been derived from a stress-rupture experiment in air.[2]

The nanoscale microstructure of the tested specimens was investigated by conventional and analytical transmission electron microscopy (CBED, EDS, PEELS) employing a Philips CM 20 FEG and a Philips EM 430 transmission electron microscope operating at 200 kV and 300 kV, respectively.

RESULTS AND DISCUSSIONS

Creep pore formation/devitrification of the amorphous secondary phase after short-term HT-testing (30 min.)

The as-received HIP-RBSN material displayed a noticeable homogeneous microstructure with an amorphous Y-Al-Si-O-N phase at grain boundaries and triple-grain junctions lacking any crystalline secondary phases. Focused radiant heating of the prenotched HIP-RBSN disk resulted in a lateral temperature gradient of the thin specimen thus imposing a stress gradient of similar shape. This stress gradient generated a single macroscopic intergranular crack due to tensile stresses close to the crack tip. After testing at 835°C and 925°C, the microstructure at the vicinity of the crack wake was virtually indistinguishable from that of the as-received HIP-RBSN after cooling to room temperature. While the bulk microstructure randomly inspected in the microscope remained unchanged after testing at 1010°C, the microstructure along the crack path changed completely as a function of distance to the crack according to the effective stress gradient generated in the experiment. Single creep pores were formed in those amorphous triple-grain junctions located at the outer zone of the wake. These are displaced by several microns from the crack path where the stress intensity has already decayed. In the immediate vicinity of the crack, however, the pore density in the triple-grain junctions was significantly higher (Fig. 1a).

Moreover, the residual amorphous phase displays partial devitrification forming δ-$Y_2Si_2O_7$ aggregates of remarkable uniform crystal orientation of the Si_3N_4 grains. It is believed that the generation of two virgin internal surfaces due to the crack propagation is mandatory for the subsequent devitrification of the amorphous phase.

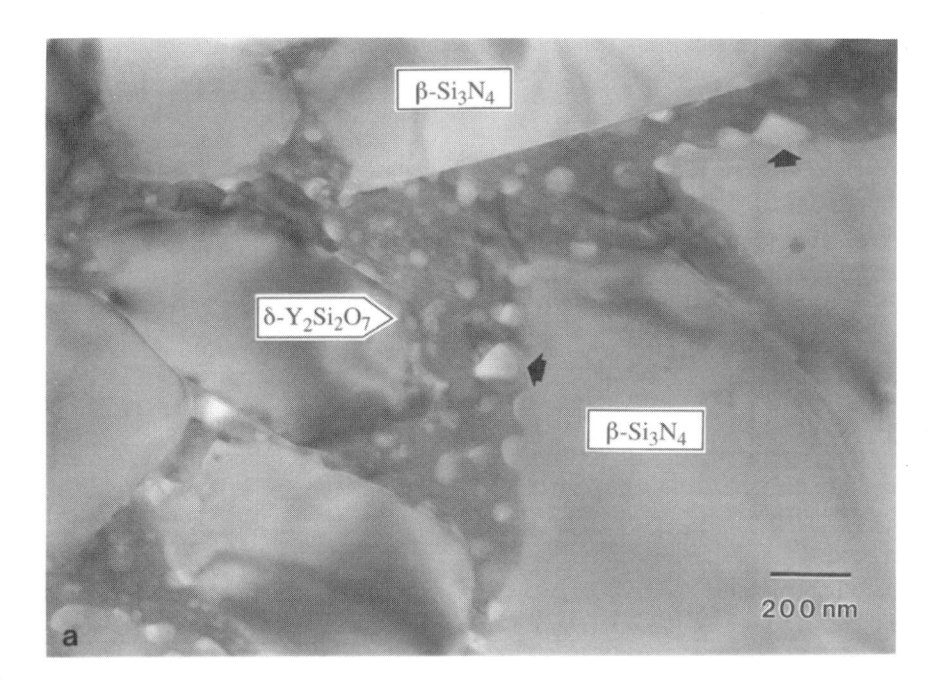

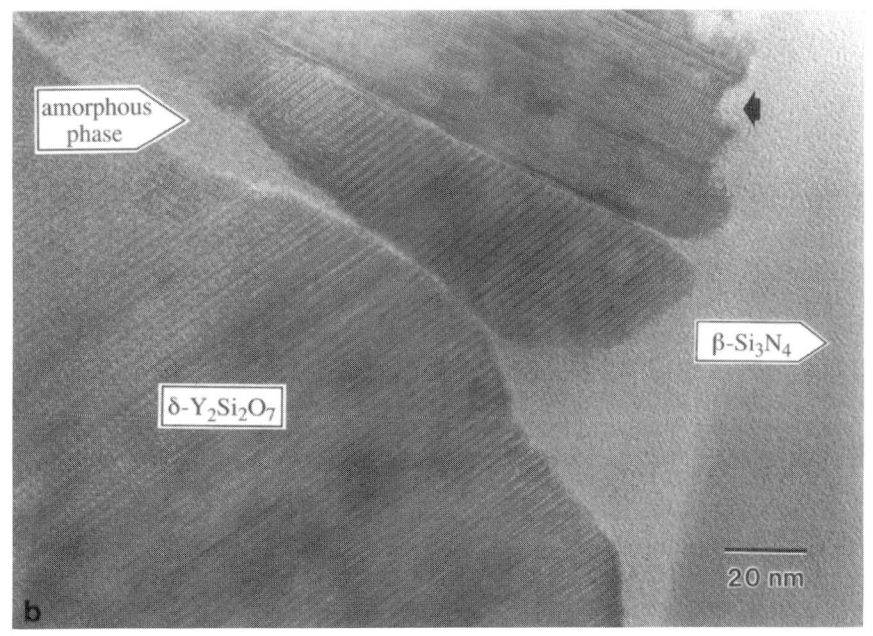

Fig. 1: TEM micrographs of (a) the culmination of pore formation in an amorphous triple-grain junction within a $Y_2O_3+Al_2O_3$-doped HIP-RBSN followed by partial devitrification (δ-$Y_2Si_2O_7$) as located in the immediate vicinity of the crack wake, (b) HREM image of the δ-$Y_2Si_2O_7$ phase showing a heavily faulted internal structure.

Following Stoke's law, oxygen diffusion along the crack wake into the amorphous phase is enhanced at $T > T_g$ and the bulk of the glass is shifted towards a more silica-rich composition thus allowing nucleation and growth of δ-$Y_2Si_2O_7$ polycrystals even during this short-term testing. Typically, δ-$Y_2Si_2O_7$ crystals are heavily faulted on {100} planes and form a single crystal aggregate upon coalescence (Fig. 1b). As the nucleation of creep pores yields relaxed triple-grain junctions, the growth conditions for $Y_2Si_2O_7$ are different from constrained crystallization employing undeformed glassy pockets. Although the δ-phase is the high-temperature polymorph of $Y_2Si_2O_7$ above 1535°C [11], it's dominance during devitrification of Si_3N_4 at rather moderate annealing/testing temperatures has been explained by kinetic constraints.[12]

It is important to emphasize that this characteristic microstructure along the crack wake was not observed after testing at 925°C. It took the small temperature increment imposed during the 1010°C experiment to exceed the glass transition temperature of the non-crystalline phase which is typically between 950°C and 1000°C for Y-Al-Si-O-N oxinitride glasses with low nitrogen solubility.[13] Due to the reduced viscosity at $T \geq T_g$, the formation of creep pores within the amorphous triple-grain junction is enhanced along the crack wake. In this context, devitrification of the glassy phase may actually provide a tool to impinge individual creep pores thus preventing the generation of a macroscopic defect in the microstructure via pore coalescence.

Solution/reprecipitation of refractory crystalline secondary phase (m-ZrO_2) after mid-term HT-testing (1-2 h)

The as-sintered secondary phase assemblage of zirconia-doped SRBSN consists of irregularly shaped m-ZrO_2 particles at triple-grain junctions and a silica-rich noncrystalline phase exhibiting a film thickness nearly three times less than that in other sintering additive systems.[9] Besides the different thermal expansion coefficients of the constituents, the volume increase of the martensitic phase transformation upon cooling gives rise to residual stresses in SRBSN which are monitored by the strong strain contours along β-Si_3N_4/ZrO_2 phase boundaries. Because of the refractoriness of the SiO_2-ZrO_2 additive system, one would expect the nanoscale microstructure of SRBSN to undergo little if any change during HT-testing. However, TEM studies provide evidence that the high-temperature deformation of ZrO_2-doped Si_3N_4 introduces a variety of dynamic processes at triple-grain junctions and internal interfaces which are closely related to the partial dissolution of the crystalline secondary phase and have recently been correlated with the different HT-deformation regimes.[10]

Starting at testing temperatures as low as 950°C, dissolution of m-ZrO_2 in the preexisting glass proceeds via a dissolution/reprecipitation process which is enhanced by the local stress distribution given by the combination of residual stresses in the microstructure and the external load applied. The inital stage of the dissolution process is accompanied by a noticeable globularisation and surface reconstruction of the m-ZrO_2 particles due to the formation of {001}, {010} and {011} macroscopic facets. Although this mechanism does not alter the total amount of crystalline and amorphous secondary phase, partial dissolution of m-ZrO_2 locally increases the amount of amorphous phase (Fig. 2a), thus raising the zirconium content compared to the preexisting glass, as shown by small-probe microanalysis.[9] Cool-down from testing temperatures results in the reprecipitation of a high amount of submicron lath-shaped crystals at multi-grain junctions, which have also been identified as m-ZrO_2 via CBED and HREM. Two different nucleation processes are involved at this stage, (i) homogeneous nucleation from bulk glass and (ii) heterogeneous nucleation supported by low-energy facets of the β-Si_3N_4 grains (Fig. 2b). The fact that triple-grain pockets containing dissolved /reprecipitated m-ZrO_2 particles are adjacent to areas exhibiting a non-modified secondary phase indicates an inhomogeneous stress-state within the SRBSN microstructure.

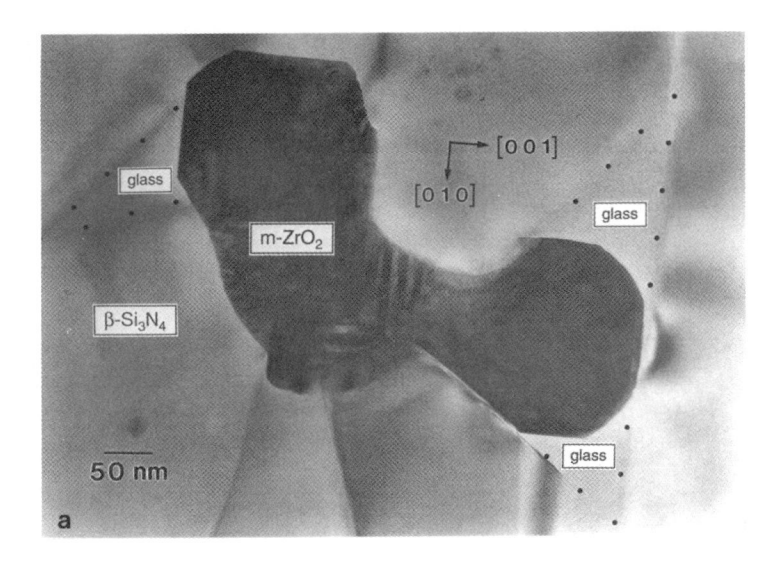

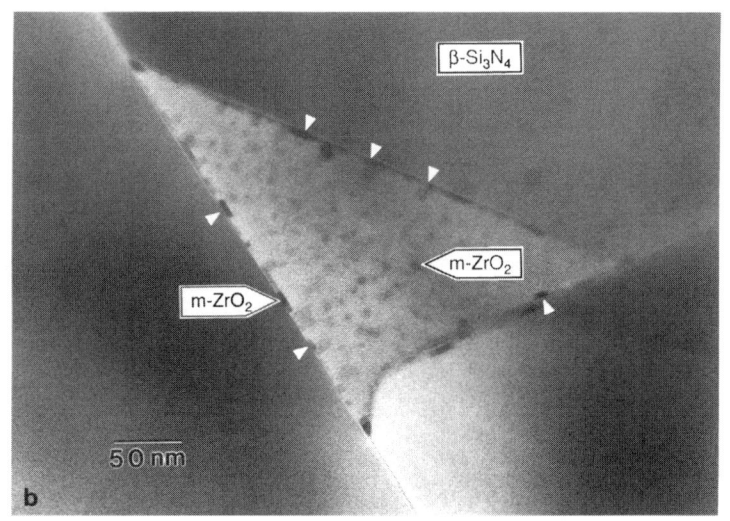

Fig. 2: TEM micrographs of (a) partial dissolution of a m-ZrO$_2$ secondary phase particle in SRBSN as characterized by faceting on low-index planes and local formation of additional glassy phase pockets adjacent to the facets (as indicated by dot symbols), (b) nucleation and growth of submicron m-ZrO$_2$ precipitates within a triple-grain junction surrounded by β-Si$_3$N$_4$ crystal facets.

Formation of a new grain-boundary structure after long term HT-testing (≥1000 h)

A commercial MgO-doped HPSN (NC-132) material employed in a series of stress rupture experiments in order to generate a fracture mechanism map has been selected to highlight the constraints of dynamic Si$_3$N$_4$ microstructures during long-term HT-testing. The family of NC-132-type ceramics defines a reference Si$_3$N$_4$ material dating from the late

1970's which has been intensively studied in previous research (for review see refs. 2,6). The specimen investigated has been tested at 266 MPa and 1100°C in air until it failed after 14941 (!) hours due to creep fracture.[2] Because of the sheer duration of exposure to high temperatures and stresses this specimen is rated unique and it offers the rare chance to the microscopist to study a ceramic microstructure which may have truly approached thermodynamic equilibrium in terms of environmental stability.

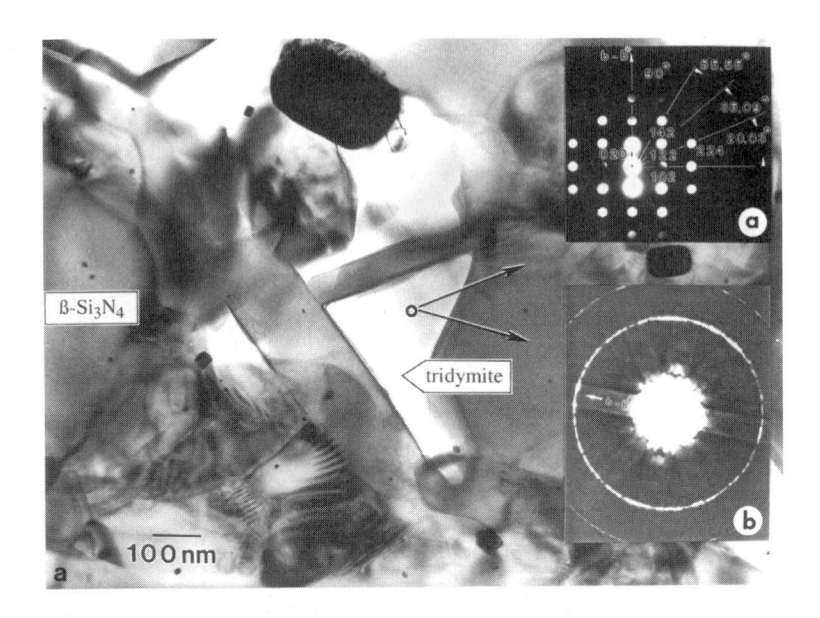

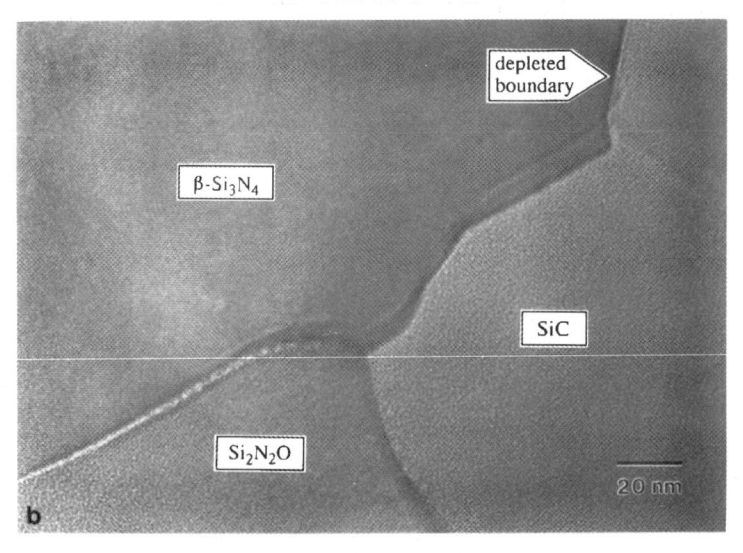

Fig. 3: TEM micrographs of (a) formation of tridymite at a triple-grain pocket after long-term exposure of NC-132 (MgO-doped HPSN) to high temperature, (b) depleted interface along the SiC/β-Si$_3$N$_4$ phase boundary revealing a well faceted interface with no residual amorphous phase present.

At the 1100°C testing temperature, bulk oxidation of NC-132 is gradually becoming effective.[14] Nevertheless, the transition temperature T_g of the low-viscosity Mg-Si-O-N glassy is clearly exceeded [15] emphasizing that grain- and phase boundaries are already acting as fast diffusion paths for the outward diffusion of the sintering additive and impurities during passive oxidation. Compared to the as-received NC-132, noticeable microstructural changes were detected in the matrix grains and the grain-boundary network. The α-Si$_3$N$_4$ grains incorporate dislocation loops which have not been previously observed by the presenting authors in the as-sintered Si$_3$N$_4$ bulk materials. These structural defects have recently been reported from commercial α-Si$_3$N$_4$ powder particles.[16] Our finding suggests that the formation of intrinsic vacancies due to the partial replacement of nitrogen by oxygen in the α-Si$_3$N$_4$ structure may be a likely mechanism favored by the extreme duration of this stress rupture experiment. On a gross scale, the onset of creep deformation in an oxidizing enviroment disconnects the percolation of the previously continuous grain-boundary network and creates two interrelated microstructural characteristics: (i) the formation of creep-induced cavities and (ii) depleted grain boundaries. The creep pore formation at multi-grain junctions is frequently associated with tungsten-bearing particles (WC). Newly-formed, highly-faulted SiC grains are often enriched in the immediate vicinity of the cavity. This is probably due to the oxidation of WC, thus locally reducing the effective oxygen partial pressure.[17-19] At some locations, obviously uneffected by nucleation of creep pores, former glassy triple grain pockets far below the surface scale have been completey transformed to tridymite (Fig. 3a). The long-range outward diffusion of grain-boundary segregants such as magnesium and impurities into the oxide surface scale causes a reduction of the metal/silicon ratio in the grain-boundary composition which decreases the film thickness of the amorphous interlayer, as observed in other oxidized Si$_3$N$_4$ materials.[20] The depletion mechanism observed during the long-term HT-testing of NC-132 gives rise to a grain boundary structure unusual for as-sintered Si$_3$N$_4$ microstructures (Fig. 3b). It is characterized by slightly curved, yet microfaceted boundaries, which lack any detectable glassy phase. They frequently contain triangle-shaped epitaxial Si$_3$N$_4$ deposits [21] at the intersection of a matrix grain boundary with the surface of a Si$_3$N$_4$ grain which stem from solution/reprecipitation of Si and N enriched during cation depletion (Mg, impurities) and collapse of former triple-grain junction. The benefits of grain boundary "de-segregation" in terms for reduced grain-boundary diffusion rates via an oxidizing heat-treatment prior to HT-deformation has previously been noted.[22,23] The strain generated during depletion of the grain-boundary network gives rise to severe microcrack formation and a high dislocation density in the β-Si$_3$N$_4$ grains.

CONCLUSIONS

The case studies presented have provided ample evidence that room-temperature microstructures of liquid-phase sintered Si$_3$N$_4$ materials yield rather limited information with respect to the response of individual microstructural elements during high-temperature testing at temperatures above T_g. Furthermore, the results emphasize the fact that HT-deformation of Si$_3$N$_4$ is directly controlled by the stability of nanoscale microstructure of both the amorphous and crystalline secondary phases. Hence, in order to study ceramic microstructures in close correlation to their mechanical response at elevated temperatures, it is mandatory to address the effects of dynamic microstructures and appreciate the fact that polyphase Si$_3$N$_4$ microstructures may indeed change simultaneously with even short-term HT-testing. Moreover, the results emphasize the role of high-resolution and analytical transmission electron microscopy as complementary techniques to the macroscopic approach employed in materials characterization. The microscopist may identify a critical microstructural feature before it may actually be monitored by a macroscopic property or testing procedure. Despite the structural integrity of Si$_3$N$_4$ during testing on a macroscopic scale, the stability of Si$_3$N$_4$ microstructures is established as a sensitive dynamic parameter on a nanometer scale, which suggests that the true long-term stability of liquid-phase sintered Si$_3$N$_4$ materials is restricted to service temperatures below the transition temperature of the non-crystalline secondary phase.

Acknowledgements: The authors are indebted to G.A. Schneider, Technical University Hamburg-Harburg, Germany as well as G.D. Quinn and S.M. Wiederhorn, NIST, Gaithersburg, MA/USA for stimulating discussions and providing sample materials.

References

1. R. Raj, Fundamental research in structural ceramics for service near 2000°C, *J. Am. Ceram. Soc.* 76: 2147 (1993).
2. G.D. Quinn, Fracture mechanism maps for advanced structural ceramics, part 1: methodology and hot pressed silicon nitride results, *J. Mater. Sci.* 25: 4361 (1990).
3. G.D. Quinn and W. Braue, Fracture mechanism maps for advanced structural ceramics, part 2: sintered silicon nitride, *J. Mater. Sci.* 25: 4377 (1990).
4. H.-J. Kleebe, M.J. Hoffmann and M. Rühle, Influence of secondary phase chemistry on grain boundary film thickness in silicon nitride, *Z. Metallkd.* 83: 8 (1992).
5. H.-J. Kleebe, M.K. Cinibulk, I. Tanaka, J. Bruley, J.S. Ventrano and M. Rühle, High-resolution electron microscopy studies on silicon nitride ceramics, in: Tailoring of Mechanical Properties of Si_3N_4 Ceramics, M. J. Hoffmann and G. Petzow, eds., Kluwer Academic Press, 1994.
6. W. Braue, G. Wötting and G. Ziegler, The impact of compositional variations and processing conditions on secondary phase characteristics in sintered silicon nitride materials, 883, in: Ceramic Microstructures 1986-The Role of Interfaces, J. A. Pask and A.G. Evans, eds., Plenum Press, 1986.
7. D.R. Clarke, High-temperature microstructure of a hot-pressed silicon nitride, *J. Am. Ceram. Soc.* 72: 1604 (1989).
8. M.K. Cinibulk, H.-J. Kleebe, G.A. Schneider and M. Rühle, Amorphous intergranular films in silicon nitride ceramics quenched from high temperatures, *J. Am. Ceram. Soc.* 76: 2801 (1993).
9. H.-J. Kleebe, W. Braue and W. Luxem, Densification studies of SRBSN with unstabilized zirconia by means of dilatometry and electron microscopy, *J. Mater. Sci.* 29: 1265 (1994).
10. W. Braue, H.-J. Kleebe and J. Göring, Secondary phase instability in ZrO_2-doped silicon nitride during high-temperature testing, in: 4th Euro Ceramics, Vol. 3, S. Meriani and V. Sergo, eds., Gruppo Editoriale Faenza Editrice, Italy, 1996.
11. K. Liddell and D.P. Thompson, X-ray diffraction data for yttrium silicates, *Br. Ceram. Trans.* 85: 17 (1986).
12. W. Braue and G.D. Quinn, Partial devitrification of sintered silicon nitride during static fatigue testing, *Mat. Res. Soc. Symp. Proc.* 287: 347 (1993).
13. T. Rouxel, M. Huger and J.L. Besson, Rheological properties of Y-Si-Al-O-N glasses-elastic moduli, viscosity and creep, *J. Mater. Sci.* 27: 279 (1992).
14. S.C. Singhal, Thermodynamics and kinetics of oxidation of hot-pressed silicon nitride, *J. Mat. Sci.* 11: 500 (1976).
15. G. Leng-Ward and M.H. Lewis, Oxynitride glasses and their glass-ceramic derivatives, 106, in: Glasses and Glass Ceramics, M. H. Lewis, ed., Chapman & Hall, 1986.
16. C.M. Wang, Fine structural features in α-Si_3N_4 powder particles and their implication, *J. Am. Ceram. Soc.* 78: 3393 (1995).
17. S.M. Wiederhorn and N.J. Tighe, Proof testing of hot-pressed silicon nitride, *J. Mater. Sci.* 13: 1781 (1978).
18. M. Backhaus-Ricoult and Y.G. Gogotsi, Identification of oxidation mechanisms in silicon nitride ceramics by transmission electron microscopy studies of oxide scales, *J. Mater. Res.* 10: 2306 (1995).
19. D.R. Clarke, A comparison of reducing and oxidizing heat treatments of hot-pressed silicon nitride, *J. Am. Ceram. Soc.* 66: 92 (1983).
20. M.K. Cinibulk and H.-J. Kleebe, Effects of oxidation on intergranular phases in silicon nitride ceramics, *J. Mater. Sci.* 28: 5775 (1993).
21. J.S. Ventrano, H.-J. Kleebe, E. Hampp, M.J. Hoffmann and R.M. Cannon, Epitaxial deposition of silicon nitride during post-sintering heat treatment, *J. Mater. Sci. Letters* 11: 1249 (1992).
22. B.S.B. Karunaratne and M.H. Lewis, Grain-boundary de-segregation and intergranular cohesion in Si-Al-O-N ceramics, *J. Mater. Sci.* 15: 1781 (1980).
23. F.F. Lange, B.I. Davis and D.R. Clarke, Compressive creep of Si_3N_4-MgO-alloys: III, *J. Mater. Sci.* 15: 616 (1980).

GRAIN-BOUNDARY FILMS IN A SILICON NITRIDE CERAMIC AT HIGH TEMPERATURES

M. K. Cinibulk[1] and H.-J. Kleebe[2]

[1]Wright Laboratory Materials Directorate
Wright–Patterson Air Force Base, OH 45433

[2]Institut für Materialforschung, Lehrstuhl Keramik und Verbundwerkstoffe
Universität Bayreuth, 95440 Bayreuth, Germany

ABSTRACT

The high-temperature microstructure of an MgO-sintered Si_3N_4 (NC-132) was investigated. Thin samples were heated to temperatures between 1350°C and 1650°C for various times and then quenched to "freeze-in" the high-temperature microstructure. The grain-boundary film thickness was found to depend on temperature and residence time prior to quenching. Rapid heating to temperatures just above the eutectic temperature, followed shortly by quenching, resulted in large increases in intergranular film thickness due to solution of Si_3N_4 in the glass; the large variation in film widths observed at different grain boundaries indicated a condition of nonequilibrium. For higher temperatures and/or longer times at temperatures, the increased amorphous phase at the grain-boundaries could be redistributed to the multiple-grain junctions by either viscous flow or diffusion of Si_3N_4 due to a chemical potential gradient in the amorphous phase. Redistribution of glass resulted in film thicknesses slightly greater than those found at room temperature, due to small compositional changes of the glass. Equilibrium film thicknesses were obtained when liquid phase redistribution was not kinetically limited.

INTRODUCTION

The behavior of most liquid-phase sintered ceramics depends primarily on amorphous grain-boundary phases at high temperatures. While much work has been devoted to studying the behavior of these materials at elevated temperatures (1000°–1500°C), little research has focused on their actual microstructures at temperature. It is generally assumed that what is observed at room temperature is representative of the microstructure at elevated temperatures; in a multiphase ceramic consisting of a refractory phase and other less refractory phases, e.g., silicon nitride containing an amorphous grain-boundary phase, this may not be the case.

The first evidence indicating that the high-temperature microstructure in such a multiphase ceramic may be different from that observed at room temperature was found in an MgO-hot-pressed Si_3N_4 (NC-132) quenched from ~1450°C[1]. While the microstructure at ~1400°C was macroscopically similar to that at room temperature, the Si_3N_4 grain morphology was reported to be more rounded with a noticeable increase in the volume fraction of the intergranular amorphous phase. Even more intriguing was that the grain-boundary film thickness was reported to be much greater (2–8 nm) than that observed at room temperature (~1 nm). Furthermore, unlike the intergranular films observed in all slowly cooled Si_3N_4 materials, the thickness was not constant throughout the material.

A more systematic approach was taken by Cinibulk et al.[2] to initially confirm these results. While we too observed an intergranular-film thickness increase at these temperatures, we also found that the thickness again assumed a constant width at higher temperatures. The establishment of a constant film thickness at all grain boundaries indicated the existence of an equilibrium film thickness.[3] The observation that the films are not constant at all grain boundaries in Si_3N_4 at the lower temperatures implies that they are nonequilibrium films. The focus of the present work was to examine the conditions under which an equilibrium film thickness will or will not be established. Furthermore, the question as to the composition of the amorphous phase at high-temperatures and its influence on the film thickness is addressed.

EXPERIMENTAL

High-temperature microstructures of an MgO-hot-pressed Si_3N_4 (NC-132, Norton Company) were obtained for characterization by transmission electron microscopy (TEM) by quenching thin disc-shaped specimens from elevated temperatures as described in detail previously.[2] In short, thin (100 µm) samples were suspended between two opposed ellipsoidal gold-coated mirrors, with a 150-W tungsten-halogen bulb situated at the focal point of each lamp. The sample was heated to a temperature of 1350°, 1420°, 1470°, 1500°, 1550°, 1600°, or 1650°C, monitored by an infrared pyrometer, and held at temperature for 5 s, 15 s or 300 s, as summarized in Table 1. The sample was then propelled into a water quenching bath at room temperature, which was situated ~3 cm below the center of the specimen. Specimens entered the water within 0.005 s and were cooled to below 1200°C (the minimum temperature at which diffusional processes in the intergranular film were believed to occur) in less than 0.05 s. At these quench rates the cooling times are short enough to negate microstructural changes during quenching and ensure that the microstructure observed at room temperature is representative of that at high temperatures.

Transmission electron microscopy (TEM) specimens were obtained by dimpling the central region of remnants of the quenched discs from both sides to remove any oxidized material. Electron transparency was obtained by ion milling. Conventional and analytical TEM were performed using either a JEOL JEM-2000FX or a Philips CM20FEG, both operating at 200kV. High-resolution transmission electron microscopy (HREM) was performed with a JEOL JEM-4000EX, operating at 400 kV, to quantitatively determine intergranular film thickness.[4,5]

CHARACTERIZATION OF HEAT-TREATED AND QUENCHED MATERIAL

In this study, and in previous work,[2,5] an equilibrium thickness of 0.8 ±0.1 nm at β- Si_3N_4-β- Si_3N_4 grain boundaries was observed in the as-received material (Fig. 1), and has been taken as a base line for comparison with film thicknesses observed in the quenched

Table 1. Intergranular film thicknesses observed at each heat-treatment condition.

Temperature (°C)	Residence time (s)	Film thickness (nm)
As-received	—	0.8
1350	5	0.8
1420	5	2.0–7.0*
1470	300	1.0
1500	5	1.5–3.5*
1550	5	1.0
1550	300	1.1
1600	15	1.2
1650	15	1.3

*Nonequilibrium film thickness, which was not the same at all grain boundaries.

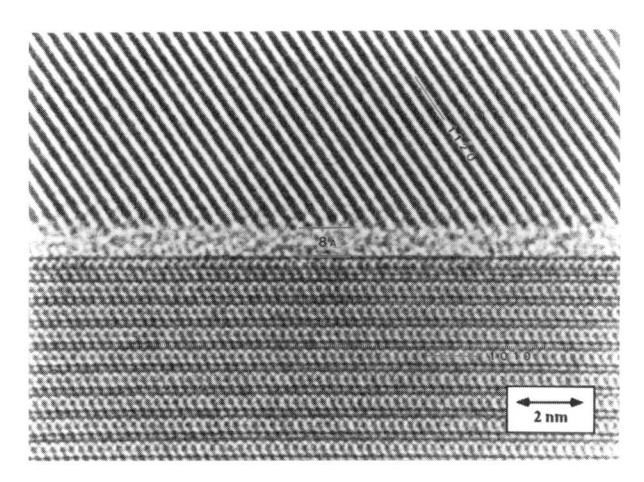

Figure 1. HREM image of a grain boundary in as-received NC-132, with an equilibrium film thickness of 0.8 nm (base line).

specimens. A summary of intergranular film thicknesses observed in specimens at the various temperatures and residence times is given in Table 1.

The lowest temperature from which specimens were quenched was 1350°C; the microstructure and thickness of the intergranular film was identical to that of the as-received, base-line material. At 1420°C/5 s, a substantial increase in the grain-boundary film thickness was observed. Intergranular films ranged from 1.8 nm to 7.0 nm in thickness (Fig. 2(a)). Specimens quenched from 1500°C/5 s also contained a range of film thickness (1.5–3.5 nm, Fig. 2(b)). The increase in film thickness and variation in thickness at different grain boundaries implies a nonequilibrium condition. These thicknesses are of the same magnitude as those reported by Clarke.[1] While, large increases in film thickness were observed there was no notable increase in volume fraction of liquid at multiple-grain junctions.

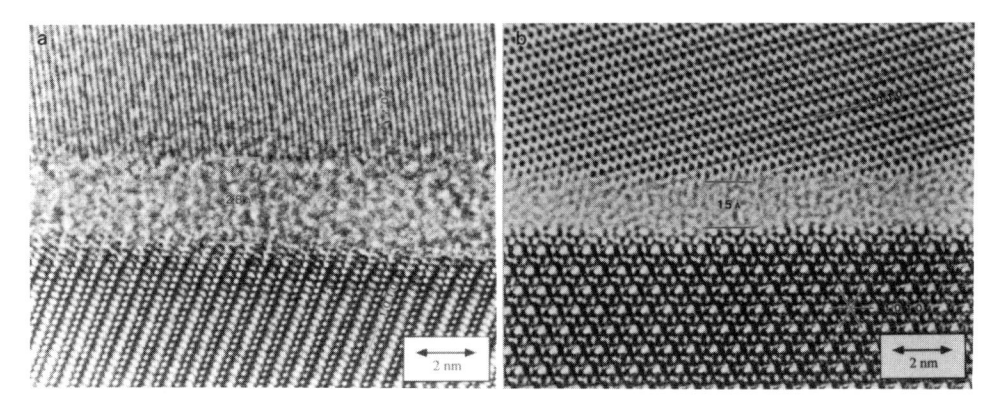

Figure 2. Grain boundaries in materials quenched from (a) 1420°C and (b) 1500°C showing film thicknesses of 2.8 nm and 1.5 nm, respectively. Nonequilibrium films were observed in both materials ranging from 1.5 nm to 7 nm in thickness.

At 1550°C and above, the intergranular films adopted a thickness similar to that observed in the base-line material. Also, all intergranular films had a characteristic, constant thickness at each temperature, as measured by HREM (Fig. 3), indicating that they are equilibrium films. Furthermore, the film thicknesses increased with each increase in temperature. The specimen held at 1470°C for 300 s contained an equilibrium film having a thickness of 1.0 nm (Fig. 4), whereas the specimen held at 1500°C for just 5 s contained nonequilibrium films (Fig. 2(b)).

EQUILIBRIUM AND COMPOSITION OF INTERGRANULAR FILMS

The TEM observations can be explained by considering the probable minimum eutectic temperature of the material system, the diffusional processes occurring in the grain-boundary region, and the forces acting on the intergranular film. The temperature and residence time of the material determines its chemistry and equilibrium state, which dictates the intergranular film thickness observed upon quenching.

The minimum eutectic temperature in the Si_3N_4-Si_2N_2O-Mg_2SiO_4 compatibility triangle has been reported to be 1515°C; however, this can be lowered by impurities, such as Ca, to 1325°C.[6] Evidence for a eutectic temperature below 1500°C in this system has been reported based on the onset of rapid densification near 1300°C[7] and large increases in

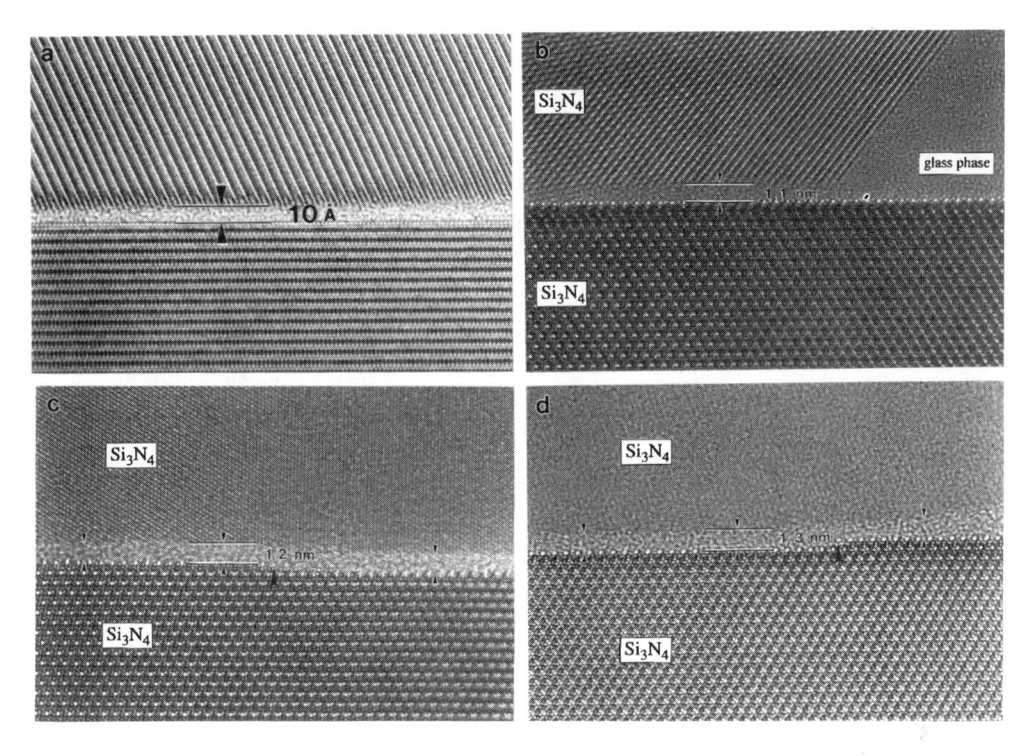

Figure 3. Intergranular films in materials quenched from (a) 1550°C for 5 s, (b) 1550°C for 300 s, (c) 1600°C for 15 s, and (d) 1650°C for 15 s showing equilibrium film thicknesses of 1.0 nm, 1.1 nm, 1.2 nm, and 1.3 nm, respectively.

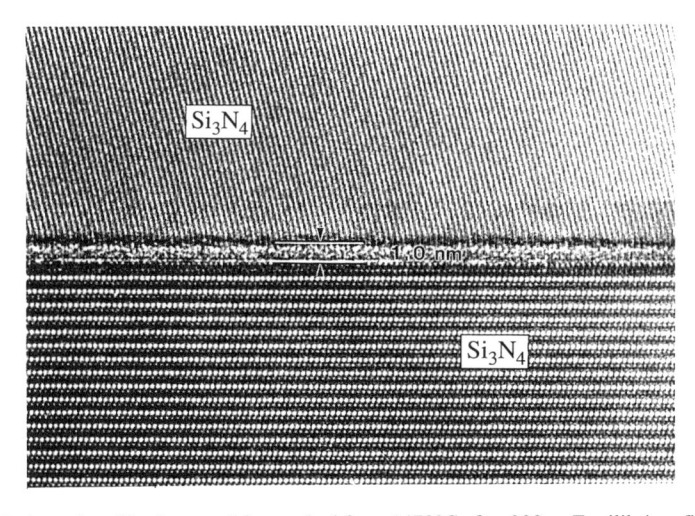

Figure 4. Grain-boundary film in material quenched from 1470°C after 300 s. Equilibrium film thickness is 1.0 nm.

oxidation kinetics above 1450°C.[8] A minimum eutectic temperature closer to 1450°C seems likely for NC-132, as work in the literature[9,10] on high-temperature mechanical and oxidation behavior indicates no change in diffusional processes occurring up to at least 1400°C (usually the upper temperature limit of the investigations). Specimens that were quenched in the present study from 1350°C did not show any change in film thickness or overall microstructure from the base-line material. Since the temperature is probably below the eutectic temperature, the intergranular phase would not be expected to change. Similar observations have been made of a Si_3N_4 ceramic containing a crystalline secondary phase where the melting of the secondary phase was used to determine when the eutectic temperature had been exceeded.[2]

Specimens heated to between 1420°C and 1500°C for just 5 s exhibited significant changes in intergranular film thickness. At these temperatures above the eutectic, an increased solubility of Si_3N_4 into the glass phase would be expected. The solution of Si_3N_4 into the grain-boundary liquid increases with increasing temperature. This increased solubility of Si_3N_4 in the film would initially increase its thickness. The widening of an intergranular film has also been shown to occur in a Yb_2O_3-sintered Si_3N_4 when quenched from a temperature above its eutectic temperature,[2] and in polycrystalline olivine (($Mg,Fe)SiO_4$) just above the bulk melting point of the intergranular silicate glass, during *in situ* heating in a scanning electron microscope.[11]

Forces, as proposed by Clarke et al.[3,12] continue to act upon the intergranular film in an attempt to re-establish an equilibrium film thickness by relocation of liquid, the amount of which is increased due to finite solution of Si_3N_4 in the intergranular phase, from the grain boundaries to the multiple-grain junctions. This could occur either by viscous flow of the liquid or by diffusion due to a chemical potential gradient that is established by solution of Si_3N_4 in the film, analogous to mechanisms operating during creep.[13-16] Reprecipitation of Si_3N_4 at triple-grain junctions was not observed; however, if these regions are under relative tension compared to the grain boundaries under compression, this could occur. At intermediate temperatures and short residence times, the nonequilibrium state of the films (characterized by the varying film thickness in a given specimen) is likely due to kinetically limited matter transport in the highly viscous amorphous phase. The rate of thinning of a liquid film between two sintering particles has been shown to be inversely proportional to its viscosity.[17,18] This is supported by the observations of equilibrium films in the specimen quenched from the same intermediate temperature range, but held at temperature for times long enough for diffusional processes to occur. The films have a thickness that is constant (1.0 nm), but slightly greater than what is observed in the base-line material. These observations are summarized schematically in Fig. 5

For the specimen quenched from 1470°C/300 s, as well as those specimens quenched from 1550°C and above, the equilibrium film thicknesses increased with temperature. The thermal expansion of the glass is orders of magnitude too small to account for the increased film thicknesses alone. While the steric repulsive force is not expected to change with temperature, the dispersion force is proportional to temperature,[3] but, the logarithmic film-thickness dependence on temperature is too weak to give the increases in film thicknesses observed at 1650°C vs. 1350°C.

It has been shown that the equilibrium film thickness is strongly dependent on the composition of the film and the bounding grains,[3,5,12,19-22] but not on the volume fraction of the glass.[23,24] The increased solubility of Si_3N_4 with increasing temperature must alter the composition of the intergranular glass. Preliminary efforts to determine compositional changes in the glass in the specimens quenched from above 1500°C indeed indicate an increased N content, determined by electron-energy loss spectroscopy. Increased N content in bulk oxynitride glasses has been shown to result in a significant increase in both the dielectric constant and the index of refraction.[25] The magnitude of the van der Waals

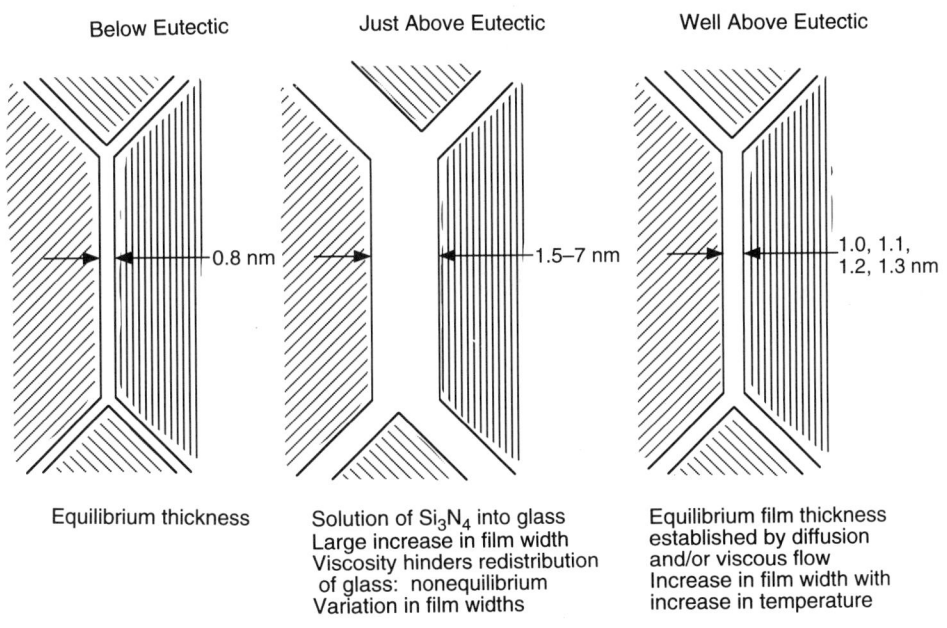

Figure 5. Schematic of features of amorphous films observed after heating to various temperatures.

Below Eutectic — 0.8 nm — Equilibrium thickness

Just Above Eutectic — 1.5–7 nm — Solution of Si_3N_4 into glass. Large increase in film width. Viscosity hinders redistribution of glass: nonequilibrium Variation in film widths

Well Above Eutectic — 1.0, 1.1, 1.2, 1.3 nm — Equilibrium film thickness established by diffusion and/or viscous flow. Increase in film width with increase in temperature

dispersion force depends on the dielectric constants and indices of refraction of both the grains and the glass.[3] The trend in increasing film thickness with increasing temperature suggests that the dispersion force is weakened by the increase in the dielectric constant and refractive index of the glass, due to enhanced solution of nitrogen in the film at higher temperatures. Alternatively, the steric repulsive force, which depends on the structural correlation length and the degree of epitaxy of the glass to the Si_3N_4 grains, could be increased by increased nitrogen content in the glass. The electrical double layer contribution to the overall repulsive force could also be affected by the increase in the nitrogen concentration in the glass films.

CONCLUSION

Our observations reveal that two separate events occur upon heating a liquid-phase sintered ceramic to high temperatures. The first is that there is a distinct increase in the thickness of the grain-boundary films and that there is no common film thickness at all boundaries. This occurs at temperatures close to the eutectic temperature of the system where solution of grains in the glass occurs, but attainment of equilibrium by redistribution of intergranular glass is kinetically limited by its viscosity. At significantly higher temperatures and/or residence times, diffusional processes occur more rapidly, ensuring that an equilibrium film thickness is attained.

The second event that occurs at higher temperatures is the increase in equilibrium film thickness with increasing temperature. The chemistry of the glass film is expected to be different at high temperatures, due to increased solubility of Si_3N_4. The change in glass composition can alter the balance of forces responsible for establishing an equilibrium film thickness. The increase in Si_3N_4 content in the amorphous MgSiON intergranular phase probably leads to a net increase in the repulsive force normal to the grain boundary, resulting in a widening of the intergranular films with increasing temperatures.

Little difference in the morphology of the Si$_3$N$_4$ grains was observed in samples quenched from as high as 1650°C. Furthermore, no observable change in intergranular-phase volume fraction was noted, indicating that the increased solubility of Si$_3$N$_4$ in the liquid phase was apparently enough to alter the properties of the film but not enough to noticeably increase the volume fraction of the amorphous phase.

ACKNOWLEDGMENTS

This work was initiated during our stay at the Max-Planck-Institut für Metallforschung, Stuttgart, Germany; we thank M. Rühle for supporting the initial stages of this work. We are indebted to G.A. Schneider and R. Mager for their assistance with the heat treatment and quenching experiments.

REFERENCES

1. D.R. Clarke, High-temperature microstructure of a hot-pressed silicon nitride, *J. Am. Ceram. Soc.*, **72**:1604 (1989).
2. M.K. Cinibulk, H.-J. Kleebe, G.A. Schneider, and M. Rühle, Amorphous intergranular films in silicon nitride ceramics quenched from high temperatures, *J. Am. Ceram. Soc.*, **76**:2801 (1993).
3. D.R. Clarke, On the equilibrium thickness of intergranular glass phases, *J. Am. Ceram. Soc.*, **70**:15 (1987).
4. M.K. Cinibulk, H.-J. Kleebe, and M. Rühle, Quantitative comparison of TEM techniques for determining amorphous intergranular film thickness, *J. Am. Ceram. Soc.*, **76**:426 (1993).
5. H.-J. Kleebe, M.K. Cinibulk, R.M. Cannon, and M. Rühle, Statistical analysis of the intergranular film thickness in silicon nitride ceramics, *J. Am. Ceram. Soc.*, **76**:1969 (1993).
6. F.F. Lange, Eutectic studies in the system Si$_3$N$_4$-Si$_2$N$_2$O–Mg$_2$SiO$_4$, *J. Am. Ceram. Soc.* **62**:585 (1979).
7. G.R. Terwilliger and F.F. Lange, Pressureless sintering of Si$_3$N$_4$, *J. Mater. Sci.*, **10**:1169 (1975).
8. W.C. Tripp and H.C. Graham, Oxidation of Si$_3$N$_4$ in the range 1300° to 1500°C, *J. Am. Ceram. Soc.*, **59**:399 (1976).
9. D. Cubicciotti and K.H. Lau, Kinetics of oxidation of hot-pressed silicon nitride containing magnesia, *J. Am. Ceram. Soc.*, **61**:512 (1978).
10. R.L. Tsai and R. Raj, The role of grain-boundary sliding in the fracture of hot pressed Si$_3$N$_4$ at high temperatures, *J. Am. Ceram. Soc.*, **63**:513 (1980).
11. S. McKernan, Environmental scanning electron microscopy of ceramics at high temperature, in: *Microbeam Analysis*, J. Friel, ed., VCH Publishers, New York (1994).
12. D.R. Clarke, T.M. Shaw, A.P. Philipse, and R.G. Horn, On a possible electrical double layer contribution to the equilibrium film thickness of intergranular glass films in polycrystalline ceramics, *J. Am. Ceram. Soc.*, **76**:1201 (1993).
13. R. Raj, Creep in polycrystalline aggregates by matter transport through a liquid phase, *J. Geophys. Res.*, **87**:4731 (1982).
14. G.M. Pharr and M.F. Ashby, On creep enhanced by a liquid phase, *Acta Metall.* **31**:129 (1983).
15. J.E. Marion, A.G. Evans, M.D. Drory, and D.R. Clarke, High temperature failure initiation in liquid phase sintered materials, *Acta Metall.*, **31**:1445 (1983).
16. J.R. Dryden, D. Kucerovsky, D.S. Wilkinson, and D.F. Watt, Creep deformation due to a viscous grain boundary phase, *Acta Metall.*, **37**:2007 (1989).
17. F.F. Lange, Liquid-phase sintering: are liquids squeezed out from between compressed particles?, *J. Am.Ceram. Soc.*, **65**:C-23 (1982).
18. O.-H. Kwon and G.L. Messing, A theoretical analysis of solution–precipitation controlled densification during liquid-phase sintering, *Acta Metall. Mater.*, **39**:2059 (1991).
19. H.-J. Kleebe, M.J. Hoffmann, and M. Rühle, Influence of the secondary phase chemistry on grain boundary film thickness in silicon nitride, *Z. Metallkd.*, **83**:610 (1992).
20. I. Tanaka, H.-J. Kleebe, M.K. Cinibulk, J. Bruley, D.R. Clarke, and M. Rühle, Calcium concentration dependence of the intergranular film thickness in silicon nitride, *J. Am. Ceram. Soc.*, **77**:911 (1994).
21. Y.-M. Chiang, L.A. Silverman, R.H. French, and R.M. Cannon, Thin glass film between ultrafine conductor particles in thick-film resistors, *J. Am. Ceram. Soc.*, **77**:1143 (1994).
22. C.-M. Wang, X. Pang, M.J. Hoffmann, R.M. Cannon, and M. Rühle, Grain boundary films in rare-earth-glass-based silicon nitride, *J. Am. Ceram. Soc.*, **79**:788 (1996).
23. P. Greil and J. Weiss, Evaluation of the microstructure of β-SiAlON solid solution materials containing different amounts of amorphous grain boundary phase, *J. Mater. Sci.*, **17**:1571 (1982).
24. I. Tanaka, H.-J. Kleebe, M.K. Cinibulk, J. Bruley, and M. Rühle, Amorphous grain-boundary films in SiO$_2$-containing Si$_3$N$_4$ ceramics, unpublished work.
25. R.A.L. Drew, S. Hampshire, and K.H. Jack, Nitrogen glasses, *Proc. Br. Ceram. Soc.*, **31**:119 (1981).

MICROSTRUCTURE AND INTERGRANULAR PHASE DISTRIBUTION IN Bi_2O_3-DOPED ZnO

Yet-Ming Chiang, Jonq-Ren Lee, and Haifeng Wang

Department of Materials Science and Engineering
Massachusetts Institute of Technology
Cambridge, MA 02139

ABSTRACT

The atomic-level distribution of Bi solute in polycrystalline ZnO has been investigated through model experiments utilizing HREM and STEM as analytical tools. In solid-liquid equilibrium, the capillary (sphering) pressure is systematically varied by controlling grain size and liquid fraction. Intergranular phase thicknesses and dihedral angles are found to vary with capillary pressure, a phenomenon which can be understood from the effect of applied pressure on the interparticle force balance and interfacial energy. In the solid state, Bi-saturated boundaries are, surprisingly, also found to contain an amorphous film of nanometer thickness. It is shown that this is the equilibrium configuration of the doped grain boundary, and that in contrast to conclusions from previous studies, virtually all boundaries in Bi-doped ZnO contain a thin intergranular film.

INTRODUCTION

The processing, microstructure, and properties of Bi_2O_3-doped ZnO, as in many ceramics, are directly related to the atomic-level structure and chemistry of interfaces. While this system is best known as an exemplar of electrically active (varistor) grain boundaries,[1-3] it has also been widely studied as a model liquid phase sintered system,[4-6] as well as one exhibiting "activated" sintering in the solid state.[7,8] The wetting behavior of grain boundaries, and the existence and stability of intergranular films, have been of much interest.[4,9-13] Controversy remains over the role of intergranular films in processing and electrical behavior. While early studies presumed the existence of a continuous grain boundary phase, a TEM study by Clarke[9] in 1978 concluded that most grain boundaries in commercial varistors were devoid of intergranular phase. In view of this result, subsequent researchers (including present authors) attributed Bi enrichment observed at grain boundaries by scanning transmission electron microscopy (STEM)[10,14-15] and Auger spectroscopy[16,17] to enrichment within a crystalline solid solution, or to adsorption layers of atomic thickness. In contrast, Olsson and Dunlop[12,13] more recently reported that many, but not all, grain boundaries and heterophase interfaces in commercial varistors do contain some form of intergranular phase. In this work we aim to resolve the conditions under which Bi-doped ZnO boundaries do or do not contain intergranular films.

The concept that intergranular films can take on a equilibrium thickness of nanometer nanometer dimensions due to a balance between attractive and repulsive colloidal forces[18] is now reasonably well-accepted. Several papers in these proceedings address this subject for systems such as Si_3N_4 and TiO_2, in which the intergranular phase is siliceous. ZnO-Bi_2O_3 is an interesting counterpart in that it represents a system with comparatively weaker interfacial forces. The Hamaker constant, which scales the strength of the van der Waals attraction, has been determined by the full-spectral method[19] using vacuum ultraviolet spectroscopy of ZnO crystals and ZnO-Bi_2O_3 quenched eutectic liquid.[20] For ZnO grains separated by the eutectic composition (or vice versa), the Hamaker constant was determined to be 4 zJ, which is but a fraction of the value found for other ceramics[19,21] with siliceous intergranular films, such as silicon nitride,[22,23] titania,[24] alumina,[25] and ruthenate thick-film resistors.[26] As we show, the short range repulsive interactions also appear to be weak enough that a modest superimposed capillary pressure (<10 MPa) noticeably affects the film thickness. Thus ZnO-Bi_2O_3 is a "colloidally soft" system in which the effect of pressure can be used to test concepts of equilibrium-thickness films.

This paper summarizes systematic experiments which we have conducted to investigate the phenomenology and stability of intergranular films in ZnO-Bi_2O_3, and their role in the processing and electrical properties of polycrystalline ZnO varistors. In liquid-solid equilibrium, representing the liquid-phase sintered condition, the secondary phase morphology and composition of grain boundaries have been observed as a function of capillary pressure. In the solid-state, which represents the post-sintering evolution of varistor grain boundaries, the equilibrium morphology of Bi-doped ZnO boundaries has been studied, and high pressure experiments have been conducted to investigate the nature of the intergranular phase. The latter subjects are also discussed in greater detail in recent publications.[27,28]

SAMPLE PREPARATION

All studies were conducted on a series of polycrystalline samples prepared from ZnO powders homogeneously doped with Bi_2O_3, and in some cases, CoO to improve varistor properties. The powders were synthesized by dispersing $ZnCl_2$ and $4BiNO_3(OH)_2BiO(OH)$ in aqueous solution and precipitating zinc hydroxide through the addition of a solution of 7.4M reagent grade ammonium hydroxide.[29] The precipitate / water mixture was washed using deionized water, the suspension was atomized into liquid nitrogen, and the mixture was freeze-dried. The dried hydroxide powders were calcined in air for two hours at 500°C. The resulting powder compositions were determined using inductively-coupled plasma emission spectroscopy (ICP); all compositions referred to in this paper are the actual analysed compositions. These powders were hot-pressed in air, using a WC/Co die, at ~650°C and 1 GPa for 2 hours, forming 5.6 mm diameter pellets with a bulk density greater than 95% of the theoretical value for ZnO. A 200μm thickness was ground from each pellet surface to remove a possible die contamination layer. The samples were surrounded by powder of the same composition and heat-treated in platinum capsules to prevent Bi evaporation loss during firing. Where quenching was desired, the capsules were dropped into water for a quench rate estimated to be ~1500°C/sec, following a temperature drop during sample transfer between the hot zone and quench medium of not greater than 20°C.

TEM and STEM was conducted on samples prepared by mechanical grinding, polishing to a thickness of about 10-15 μm, and ion-milling with Ar^+ ions accelerated through 6 kV on a liquid nitrogen cooled stage (Gatan Dual Mill Model 600). The high resolution electron microscopy was carried out using a Topcon EM-002B microscope operating at 200 kV, equipped with a ±10° double-tilt sample holder. The grain boundary segregation was characterized using Fisons/Vacuum Generators HB5 or HB603 STEMs equipped with a Link Systems energy dispersive X-ray analyzer. The integrated Bi excess per unit boundary area (i.e., solute coverage) was determined following the method of Ikeda et al..[30] In order to calibrate the Bi/Zn concentrations, a quenched single-phase glass standard of composition 14.5 mol% ZnO + 29.9 mol% Bi_2O_3 and 55.5 mol% SiO_2 was prepared and analysed under the same STEM operating conditions.

SOLID-LIQUID EQUILIBRIUM

Experiments at Zero Capillary Pressure

To observe the distribution of intergranular liquid in the absence of external capillary pressure, liquid-penetration experiments were performed, after the studies of Flaitz and Pask[31] and Shaw and Duncombe[32] on alumina. A densified polycrystalline sample was immersed in a pre-equilibrated Bi_2O_3-ZnO liquid, prepared by melting a mixture of 30 mol% ZnO + 70 mol% Bi_2O_3 (i.e., ZnO rich with respect to the eutectic liquid). The liquid volume was approximately 10 times the volume of the pellet, and samples were equilibrated at 750°-950°C in air. Since the samples were doped with Bi_2O_3 in excess of the solubility limit, upon heating both the liquid internal to the polycrystals and the immersion liquid quickly reach the equilibrium composition. After firing, the samples were rapidly cooled by withdrawal from the furnace.

At 750° and 850°C, the grains remained "flocced" despite grain coarsening and sphering, accompanied by some liquid penetration into the polycrystal. Figure 1 shows that a pellet equilibrated for 24 h at 750°C expands slightly from the initial diameter, but does not deflocculate to form dispersed ZnO particles. The flocs of coarser grains observed in the liquid near the pellet surface are believed to result from the excess ZnO in the liquid source. A similar microstructure with slightly larger grains is observed after 24 h. at 850°C. At a higher temperature of 950°C, however, deflocculation of grains from the polycrystal became evident, as shown in Figure 2.

In a previous study[15] we reported that immersion in a pure Bi_2O_3 melt causes complete deflocculation of densified samples, in contrast to these results. It is now clear that equilibration between the sample and the pure Bi_2O_3 liquid in that experiment, which requires dissolution of ZnO, was sufficient to cause deflocculation. The present experiments show that in the absence of ZnO dissolution, van der Waals interparticle attraction is strong enough to prevent deflocculation at 750-850°C, similar to observations for ruthenate thick film resistors.[26] The deflocculation at 950°C is likely due to a decrease in van der Waals attraction as the liquid composition becomes enriched in ZnO. The binary phase diagram (Fig. 3) shows that the ZnO concentration of the equilibrium liquid increases from 14 mol% ZnO at the eutectic temperature (740°C) to 25 mol% ZnO at 950°C. A schematic representation of the corresponding change in Hamaker constant is shown in Figure 3. In solid-solid equilibrium, one has a limiting value for ZnO separated by the nearly pure Bi_2O_3 secondary phase. Above the eutectic temperature, the Hamaker constant represents ZnO separated by ZnO-Bi_2O_3 liquid, and should decrease continuously in value with heating as the liquid becomes enriched in ZnO, following the liquidus line. Eventually, the solid and liquid composition may become sufficiently alike that the van der Waals attraction cannot prevent deflocculation.

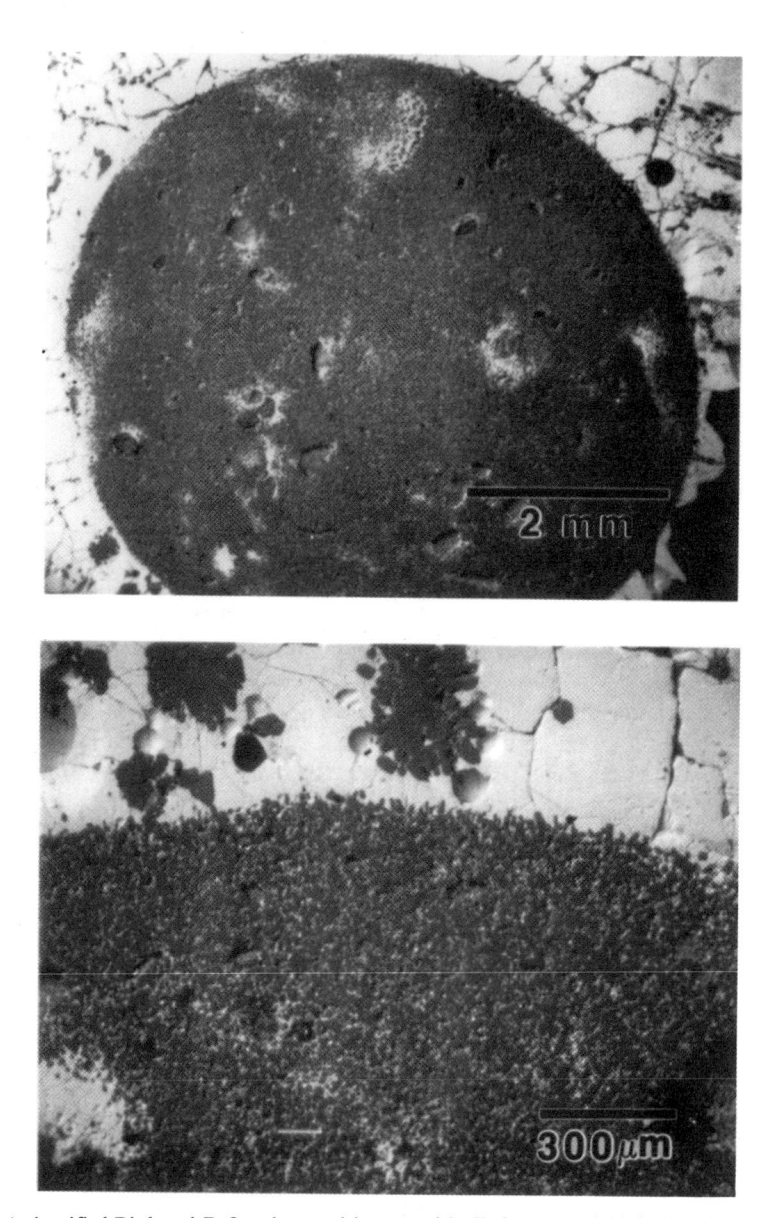

Figure 1. A densified Bi-doped ZnO polycrystal immersed in ZnO-saturated Bi_2O_3 liquid at 750°C for 24 hours does not deflocculate despite liquid penetration. Flocs of coarse grains near the surface originate in the liquid.

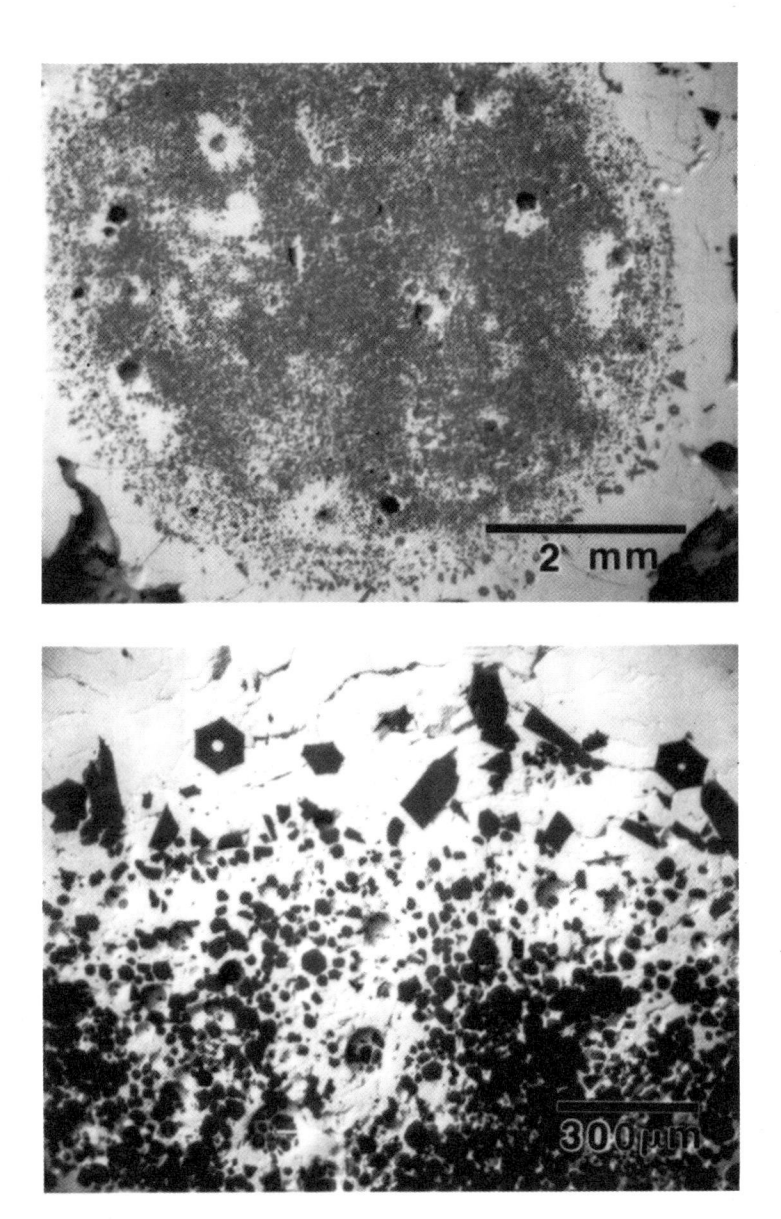

Figure 2. At 950°C for 24 hours, the polycrystal deflocculates significantly, suggesting weaker van der Waals attraction between particles at the higher temperature.

Capillary (Sphering) Pressure Dependence

Isolated crystalline grains in liquid, unconstrained by neighboring grains, tend to evolve towards spherical or faceted Wulff shapes. As discussed by Park et al.[33], in a dense liquid-phase sintered polycrystal the liquid fraction limits the extent of grain sphering, and there exists an internal capillary or "sphering" pressure between grains deformed to fill space. This pressure is balanced by the capillary pressure due to wetting liquid menisci formed where grain boundaries intersect the surface, and the whole is at mechanical equilibrium. Therefore the internal capillary pressure is given by the surface meniscus radius ρ and liquid surface tension γ_L ($P_C = \gamma_L / \rho$), and in principal could be determined by directly observing meniscus radii at the polycrystal surface. It can also be determined from microstructural parameters, however. Park et al.[33] modeled the internal capillary pressure in idealized microstructures consisting of monosized close-packed grains with isotropic interfacial energies, a cross section of which is shown in Figure 4. The capillary pressure is in general a function of the liquid volume, grain size, and interfacial energies (which dictate the grain boundary dihedral angle ϕ and surface liquid contact angle θ). From their results,[33] one observes that when the liquid volume fraction V_L is small (<5%), the capillary pressure is a nearly linear function of liquid fraction and grain size. In this regime, for a zero grain boundary dihedral angle, the capillary pressure is given by the following approximate relationship for all values of surface contact angle:

$$P_C = \frac{\gamma_{SL}}{13 V_L R} \left[\frac{\pi}{3\sqrt{2}(1-V_L)} \right]^{1/3} \qquad (1)$$

in which γ_{SL} is the solid-liquid interfacial energy and R is the grain radius. The capillary pressure can then be estimated from just three parameters: the liquid fraction, which is determined by phase equilibria, the grain size, which is experimentally measurable, and the solid-liquid interfacial energy. The latter has not to our knowledge been measured for ZnO-Bi$_2$O$_3$. We present results below in the form of a normalized capillary pressure, P_C/γ_{SL}, as well as absolute pressures calculated assuming $\gamma_{SL} = 100$ mJ/m^2.

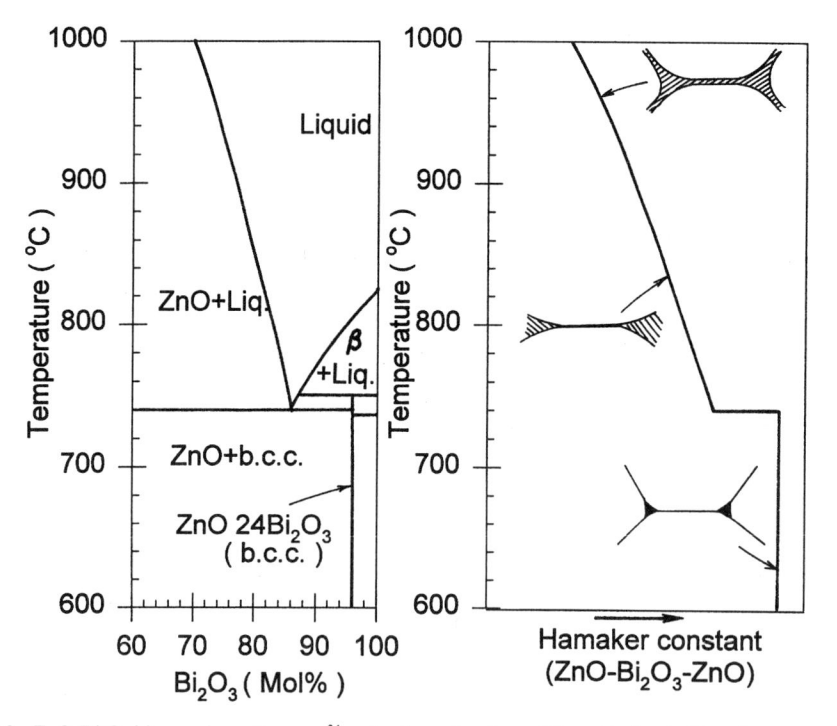

Figure 3. ZnO-Bi$_2$O$_3$ binary phase diagram,[34] and schematic view of the variation in Hamaker constant with temperature, as the secondary phase in equilibrium with ZnO and its composition change.

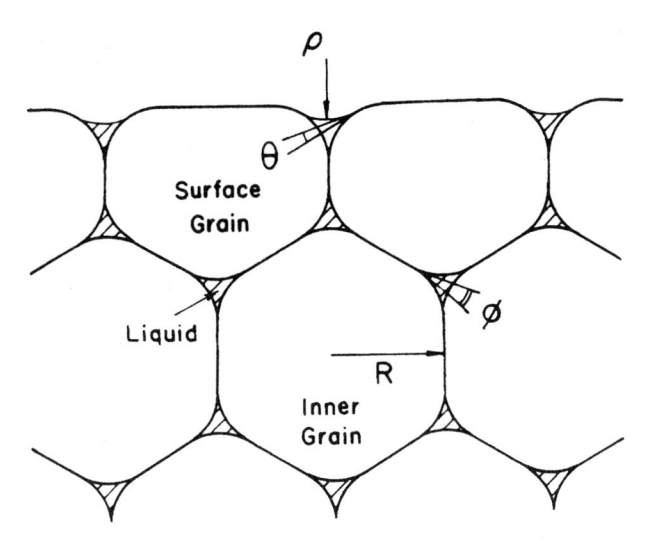

Figure 4. Sphering pressure model for liquid-phase sintered polycrystal consisting of identically sized grains in a close-packed arrangement, from Park et al.[33]

In our experiments the capillary pressure was systematically varied by controlling the grain size and liquid fraction, and the corresponding Bi excess at grain boundaries was measured using STEM. The immersion samples discussed earlier represent the lower limit of zero capillary pressure. In the doped polycrystals, the liquid fraction was calculated from the ICP analysed composition (0.023 mol% to 0.577 mol% Bi_2O_3) using the published phase diagram,[34] with an additional correction[35] for the amount of Bi_2O_3 determined by STEM to be segregated to grain boundaries, and therefore not available as bulk liquid. The corrected liquid fractions ranged from 0.06 to 1.52 vol%. The average grain sizes were determined from polished cross sections using the Heyn intercept method as embodied in ASTM E112[36], and ranged from 1.4 µm to 11.6 µm. The capillary pressure calculated from Eq. (1) using the average grain size ranged up to a maximum of approximately 9 MPa.

As shown in Figure 2, ZnO grains deflocculated from the polycrystal form rounded equiaxed grains at small sizes, but evolve into larger grains of hexagonal cross section which we presume are sections through hexagonal prisms with major and minor axes corresponding to the c and a axes of würtzite respectively (the cause of the internal Bi_2O_3 precipitates formed in a few grains in Fig. 2 is not known). Gambino et al.[11] also observed facetted ZnO grains in samples with substantial liquid. When the liquid fraction is small, however, facetting appears to be suppressed, and cross sections through three-grain-junctions (Figure 5) show nearly circular grain cross sections indicating little anisotropy in the solid-liquid interfacial energy, and apparently zero dihedral angles, which are discussed later. With the exception of a distribution in grain size and aperiodic grain packing, these fully-dense polycrystals closely resemble the morphology assumed in the model (Fig. 4).

The results are shown in Figure 6 as the Bi excess, Γ_{Bi}, plotted agains the calculated capillary pressure for samples equilibrated at 750°C. Each datum represents a single grain boundary measured in a polycrystalline sample, and Γ_{Bi} is given in terms of equivalent monolayers relative to δ-Bi_2O_3, wherein one monolayer is equivalent to 8×10^{14} cm^{-2}. Each

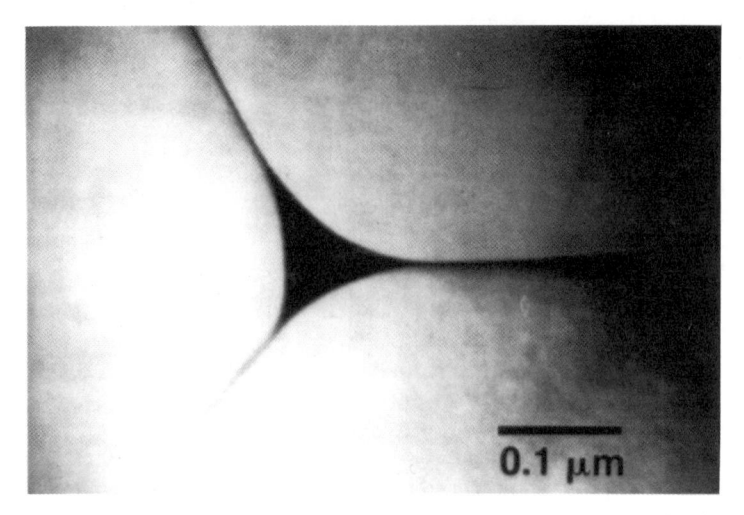

Figure 5. In samples with small liquid fraction, a zero dihedral angle is commonly observed, with no evidence of interfacial facetting.

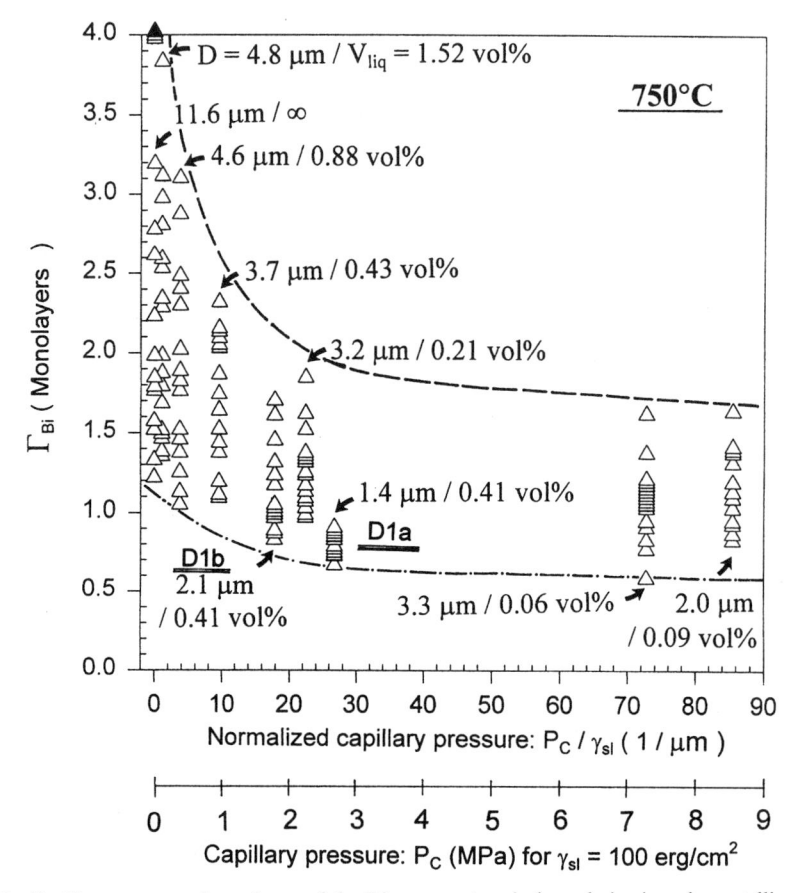

Figure 6. Capillary pressure dependence of the Bi excess at grain boundaries in polycrystalline Bi-doped ZnO fired at 750°C. Each datum represents a single grain boundary. The grain size and liquid fraction are given for each sample.

vertical data set represents a single sample, in which STEM measurements were made at between 10 and 17 randomly-selected grain boundaries. The average grain size D and the liquid fraction are labeled for each sample. The results show two significant features. As the capillary pressure increases, the *range* as well as the *average value* of Γ_{Bi} both decrease. At zero capillary pressure, Γ_{Bi} ranges from a low of 1.2 monolayers to thick liquid films with greater than 4 equivalent monolayers, indicated by the filled symbol in Fig. 6. However, even a small capillary pressure of order 0.1 MPa is sufficient to narrow the range and remove the thick films. With capillary pressures above approximately 1.5 MPa, Γ_{Bi} further narrows to a relatively constant range between 0.7 and 2.0 monolayers. Similar results were observed in samples equilibrated at 850°C, although the average values and range of Γ_{Bi} were slightly larger than at 750°C.

Two samples in Fig. 6, labeled D1a and D1b, showed lower Γ_{Bi} than otherwise comparable samples. These were exceptional in having significant Si impurity levels of ~450 ppm, and showed simultaneous segregation of both Si and Bi to grain boundaries.

Grain Boundary Liquid Films

HREM showed that the above STEM results correspond to composition measurements of intergranular liquid films whose average thickness varies systematically with the sphering pressure. HREM of the sample with zero capillary pressure (Fig. 7a) as well as that with the highest capillary pressure (Fig. 7b) both show continuous intergranular amorphous films, thicker and more varied in thickness in the former. The minimum observed coverage of Γ_{Bi} ~1 monolayer corresponds to a film of ~1 nanometer thickness. These results are consistent with the existence of a short-range repulsive potential resisting closer approach of the two grain surfaces. Even at zero capillary pressure, most films are still thin enough for the short-range interactions conceived in equilibrium-thickness film models[18,20,24] to be applicable. The reasons for the thickness variations are not known in detail, but may result from locally varying capillary pressure due to variations in interfacial energy and grain size, modest anisotropies in the Hamaker constant with crystal orientation,[19] and/or boundary-to-boundary variations in the strength of the short-range repulsions, which are to be expected on the basis of order interactions across a thin amorphous film.[20,24]

We also observed a dependence of the dihedral angle at grain boundaries on the capillary pressure. In the samples with low or zero capillary pressure, finite dihedral angles with or without adjacent facetting could be observed, while at higher capillary pressures, the zero contact angle morphology of Fig. 4 was observed. In the same temperature range (780°C), Kingery[37] reported a range of non-zero dihedral angles, with an average value of 25°, in ZnO containing 0.5% Bi_2O_3 and 0.5% CoO. Those results are consistent with the present ones, since their liquid fraction and grain size are such that the capillary pressure would be very low. We attribute the decrease in dihedral angle upon increasing capillary pressure to the following. In the equilibrium-thickness film model, the energy of a grain boundary containing an equilibrium film is lowered from that of two well-separated solid-liquid interfaces (of energy $2\gamma_{SL}$) by the short-range interactions. The energy of the film is $2\gamma_{SL}-\Delta\gamma$, where $\Delta\gamma$ is the lowering of energy at the equilibrium separation (Fig. 8). The contact angle is necessarily greater than zero, since the boundary (with film) has a lower energy than the two free interfaces. However, an externally applied pressure will thin the film, and increase its energy following the energy-separation curve in

Fig. 8. With sufficient applied pressure, Δγ→0, the energy of the film is no longer lower than that of two free interfaces, and the dihedral angle becomes zero.

Overall, these results show that in liquid-phase sintered polycrystals the morphology, dihedral angle, and atomic-level distribution of liquid phase can be strongly influenced by capillary pressure. Microstructures can in principle be designed for uniform or non-uniform liquid distribution by the appropriate control of sphering pressure. It should be noted that typical commercial varistor formulations, and indeed most liquid-phase sintered

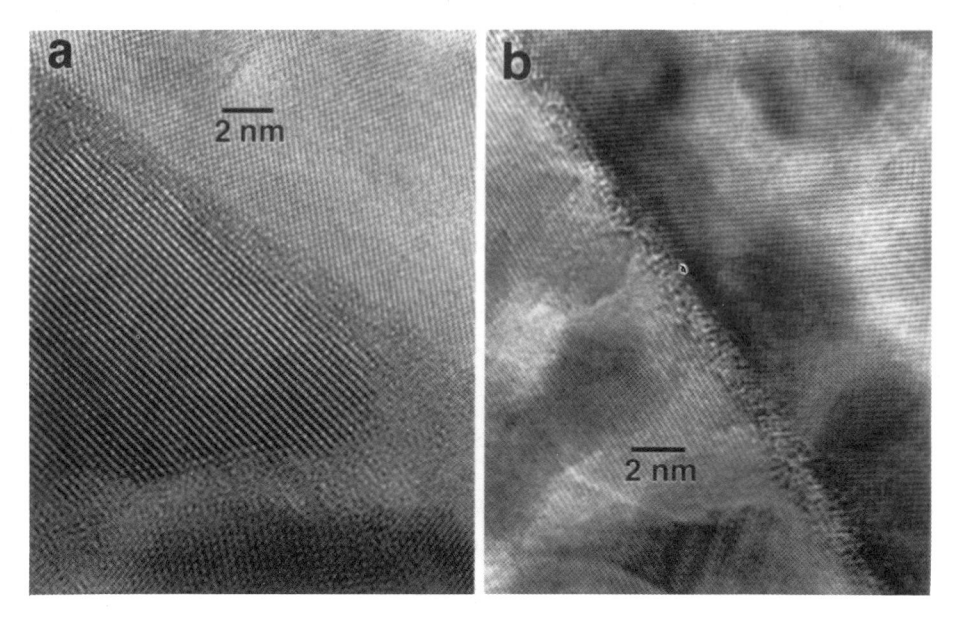

Figure 7. Grain boundary amorphous films (a) between flocced grains in the sample equilibrated at zero capillary pressure, and (b) in sample at highest capillary pressure in Fig. 6.

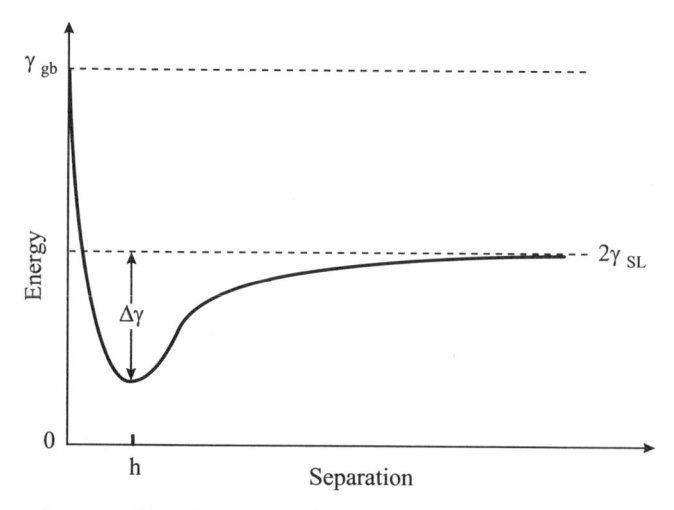

Figure 8. Two well-separated interfaces, each of energy γ_{SL}, may lower their energy due to the balance between van der Waals attraction and short-range repulsion by Δγ at the equilibrium separation h. The energy of the film may be a local minimum, or may be lower than the energy of a crystal-crystal grain boundary, γ_{gb}.

ceramics, contain a sufficiently large grain size and liquid fraction, to be at the low end of our capillary pressure scale. Greater uniformity in phase distribution is expected for materials with a small grain size and a small fraction of uniformly distributed liquid.

SUBSOLIDUS EQUILIBRIUM

Bismuth Solid Solubility and Equilibrium Segregation

Slowly-cooled zinc oxide varistors exhibit segregation of Bi to grain boundaries[10] in quantities previously estimated to be 0.5-1 monolayers.[16,17] Using the present samples, we quantified more precisely by STEM the Γ_{Bi} at grain boundaries equilibrated just below the solidus temperature (725°C), and observed average levels of 8×10^{14} cm^{-2}, with little variation from grain boundary to grain boundary in a single sample.[N] This value is equivalent to 0.9 monolayers with respect to δ-Bi_2O_3 (0.67 monolayers with respect to Zn), and is similar to the liquid-state coverages at high capillary pressures (Fig. 6). We have also determined[27] that the solubility limit of Bi_2O_3 in ZnO near the eutectic temperature is much lower than previously reported.[38] Samples doped with 0.06 mol% Bi_2O_3 exhibited grain junction secondary phases in STEM, while those doped with 0.023 mol% Bi_2O_3 did not, establishing that the solubility lies between these limits. (In contrast, Hwang et al.[38] have reported solubilities as high as 0.21% at 750°C.) This low solid solubility indicates that the ZnO phase in virtually all experimental materials, and certainly in commercial varistors, is Bi-saturated. Grain boundary Bi segregation in all samples may therefore be expected to evolve, upon cooling below the eutectic temperature, towards the equilibrium level.

Equilibrium Intergranular Films in the Solid State

In order to determine whether the equilibrium level of Bi segregation corresponds to a continuous segregation layer or a thin intergranular film, we examined samples in which subsolidus equilibration at 650-700°C was approached along three paths:

1) A sample was fired in solid-liquid equilibrium at 850°C, then annealed at 700°C for 24 h, and cooled at 150°C/hr in air to room temperature. This heat treatment mimics typical varistor processing, whereby grain boundaries are dewetted by liquid during the approach to solid state equilibrium.

2) A sample was densified from doped powder by hot-pressing at 700°C for 2 h, then annealed at the same temperature and 1 atm pressure for 24 h., followed by quenching into water. This sample reaches solid-state equilibrium without ever being exposed to Bi-rich liquid.

3) Grain boundaries were desegregated by applying a high hydrostatic pressure on the order of 1 GPa.[27] The desegregated sample was examined by HREM. Then, a sample in which the equilibrium segregation at 1 atmosphere pressure was restored by annealing at 650°C for 24 h was examined. These samples also never exceeded the eutectic temperature.

The grain boundary segregation observed by STEM in each of these samples is shown in Fig. 9. Each bar in the histograms represents the Γ_{Bi} measured at a single grain boundary in the polycrystals. Comparing Figs. 9a, 9b, and 9d, we see that a similar segregation level is reached upon approaching equilibrium along the three paths. In contrast, Figure 9c shows that pressure-desegregation lowers Γ_{Bi} to virtually undetectable levels.

Upon examining by HREM a total of 40 grain boundaries (~10 in each sample), we found that only the desegregated sample contained atomically abrupt grain boundaries with

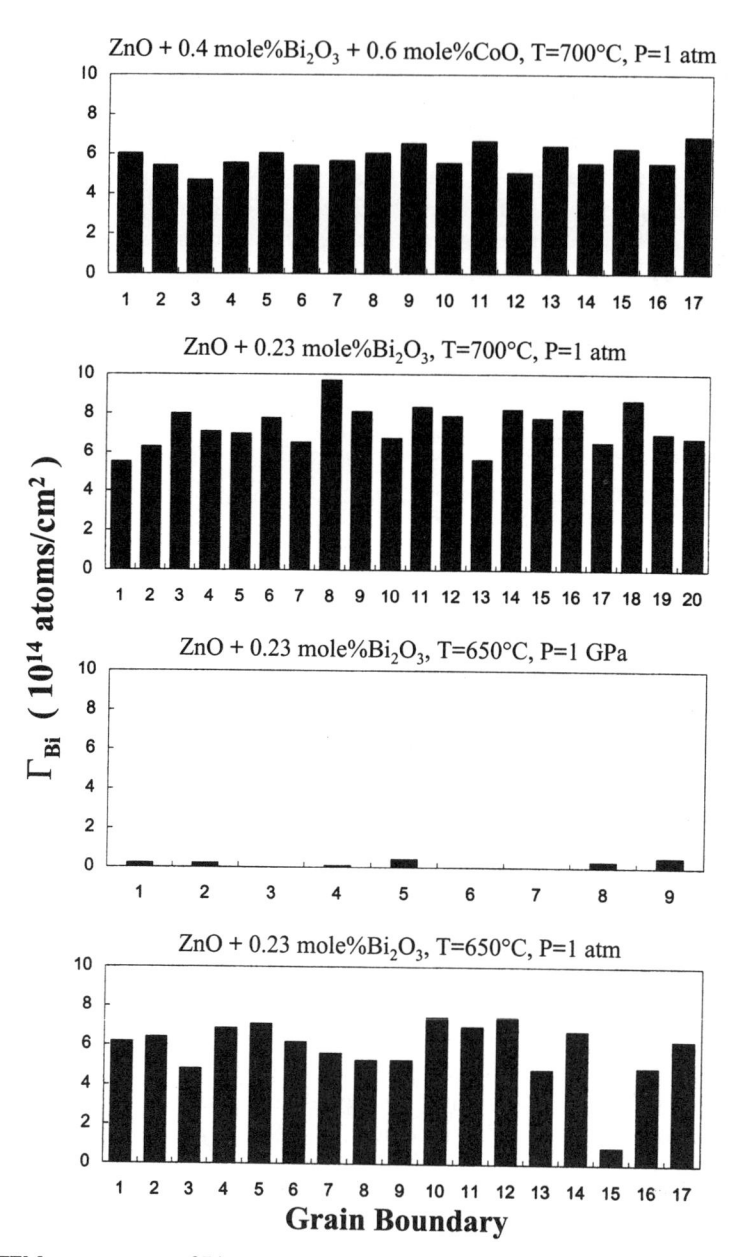

Figure 9. STEM measurements of Bi excess at grain boundaries in samples equilibrated below the eutectic temperature. Each bar represents a single grain boundary in a polycrystalline sample. Note that all except for (c), the high pressure-desegregated sample, exhibit near-saturation levels of segregation.

no intergranular phase. All samples with equilibrium Bi segregation showed an amorphous intergranular phase with a thickness between 1.0 and 1.5 nm, one example of which is shown in Fig. 10. More detailed results appear in reference 28. The average thickness and range of thicknesses of the intergranular films are shown in Table 1. We see that the sample cooled from solid-liquid equilibrium exhibited slightly greater average thickness than did the others. However, it was also different in composition, being co-doped with Bi and Co, for improved varistor characteristics. A very slight segregation of Co in the film was detected. At present, we cannot determine if the slightly greater film thickness in the co-doped samples is due to compositional effects or processing history.

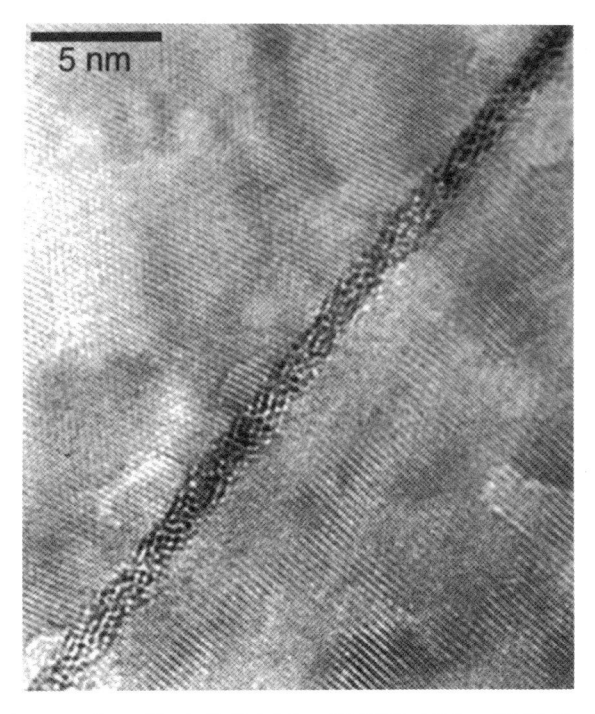

Figure 10. Intergranular amorphous film in ZnO + 0.4 mol% Bi_2O_3 + 0.6 mol% CoO, equilibrated below the eutectic temperature.

Table 1. Intergranular film thicknesses and composition in samples studied.

SAMPLE	FILM THICKNESS (nm) Range	Average	FILM COMPOSITION
ZnO + 0.4 mole%Bi_2O_3+ 0.6 mole%CoO T = 700°C, P = 1 atm	1.0 - 1.5	1.3	4.4 : 1 : 0.05*
ZnO + 0.23 mole%Bi_2O_3 T = 700°C, P = 1 atm	0.9 - 1.1	1.0	2.3 : 1**
ZnO + 0.23 mole%Bi_2O_3 T = 650°C, P = 1 atm	0.7 - 1.3	1.0	2.8 : 1**

*: Ratio of ZnO:Bi_2O_3:CoO; **: Ratio of ZnO:Bi_2O_3.

Also given in Table 1 is the grain boundary film composition, estimated from the average thickness and Γ_{Bi}. The Zn concentration of the film could not be directly measured due to the large background signal. Therefore, to calculate the film composition, a molar volume for the amorphous phase must be assumed. We have assumed a molar volume of 50 cm^3/mole, close to that of Bi_2O_3. (A lower assumed molar volume increases the calculated ZnO concentration in the film.) Note that the compositions of the films are *richer* in ZnO than the eutectic composition. Clearly, they do not correspond to a quenched eutectic liquid. A retained liquid film might furthermore be expected to crystallize some ZnO on the adjacent grains during cooling, which would cause the remaining film to become more enriched in Bi_2O_3. These subsolidus intergranular amorphous films have a equilibrium, ZnO-rich composition that is distinct from any bulk equilibrium phase in the system. In samples quenched from above the eutectic, the similarity in film thickness and Γ_{Bi} suggests that they are also ZnO-rich with respect to the eutectic liquid. That is, even in liquid-solid equilibrium, the intergranular film and the bulk liquid are of distinctly different compositions. As proposed by Cannon,[39] the intergranular amorphous film seems more appropriately viewed as a multilayer adsorbate layer rather than a modified liquid.

The sample which was pressure-desegregated and then restored to equilibrium is particularly interesting, as it shows conclusively that the amorphous intergranular film is energetically favored over a crystal-crystal grain boundary. Film formation from the pressure-desegregated state requires a solid-state amorphization reaction between the crystalline ZnO grains and the crystalline Bi_2O_3 secondary phase. Thin film amorphization reactions have been widely studied,[40] and are driven by a negative free energy of mixing. Typically, the amorphization reaction continues until the films are consumed or until growth is kinetically limited. In contrast, the present reaction to form the amorphous intergranular film stops when the film reaches the equilibrium thickness. The same is observed at the heterointerface between ZnO and Bi_2O_3 at the grain junctions.[28] We believe that film formation is driven by the decrease in total interfacial energy upon forming an equilibrium-thickness intergranular film, and occurs against a volume free energy change which is likely positive, as well as against other barriers such as the strain energy.[28] The results indicate that the amorphous intergranular film not only has an equilibrium thickness (which can represent a local or global energy minimum), but is a minimum energy configuration for Bi-saturated ZnO grain boundaries, being lower in free energy than the crystal-crystal ZnO boundary or two well-separated glass-crystal interfaces, as illustrated in Fig. 8.

We now attempt to reconcile these results with previous TEM observations[9,12,13] which show atomically ordered grain boundaries. Previous STEM studies[10,14,15] have also failed to distinguish the discrete nature of the film. One possible explanation is an important ion-thinning artifact which we have discovered,[28] wherein the amorphous intergranular film is removed at the very edge of thinned foils and "heals" to form a crystal-crystal grain boundary. Examination of very thin sections, as was required at lower accelerating voltages in earlier TEM studies, could have mistakenly interpreted these artifacts for the true nature of the boundary. Without lattice imaging, the grain boundary film appears dark always (in bright field imaging) due to the high atomic number of Bi, and the discreteness of the film is obscured. Alternatively, it may be that commercial varistors so contain a distribution of boundary types markedly different from the simplified varistors examined here. This seems unlikely given the similar amounts of Bi segregation which are reported.[16,17] We do know that low energy, special grain boundaries in ZnO films do not form intergranular films,[41] but these also accept much less Bi segregation. Further study of other varistor types seems necessary for a final resolution of these discrepancies.

Mechanism of Activated Sintering in ZnO-Bi$_2$O$_3$

Bi$_2$O$_3$-doped ZnO is one of several ceramic systems in which accelerated sintering is observed upon heating before the eutectic temperature is reached.[7,8] This phenomenon has been termed "activated sintering," and has not previously been explained on a mechanistic basis. We propose that the solid-state formation of amorphous intergranular films, and possibly of amorphous surface films, is responsible for this effect. Activated sintering is observed in ZnO-Bi$_2$O$_3$ starting at about 650°C, the same temperature at which we see rapid formation of the equilibrium film by solid-state amorphization. There is additional evidence that the amorphous film is a path of rapid atom transport. Desegregation under pressure shows that the amorphous film has greater molar volume than equimolar quantities of crystalline ZnO and Bi$_2$O$_3$. The pressure-desegregation phenomenon has been shown to be completely reversible,[27] with a time scale for desegregation or resegregation of less than 10 min. at 650°C. Since the source and sink for Bi during the process is the crystalline grain junction secondary phase, rapid Bi grain boundary diffusion is implied. SIMS measurements of oxygen diffusion[42] have also shown that the grain boundary diffusion parameter δD_b is increased by a factor of about 10^3 in this temperature range. The enhanced oxygen grain boundary diffusion coefficient is undoubtedly a key function of Bi-doping in creating electrically active grain boundary states in varistors, as proposed by Greuter.[3]

GENERAL REMARKS

With these results, the microstructure development process in liquid-phase sintered ZnO varistors which are slowly cooled to room temperature can be summarized as follows. At typical sintering temperatures of T>1150°C, ZnO grains are constrained from separating only by the limited liquid fraction and resulting capillary forces. Grain boundaries are completely wet by the liquid, and a broad range of intergranular film thicknesses exists. Upon cooling towards the eutectic temperature, there is an increase in van der Waals attraction between grains, accompanied by precipitation of ZnO from solution, which causes the distribution of intergranular film thicknesses to narrow. Most intergranular films will reach a thickness of 1-3 nm. While the dihedral angles at many boundaries may still appear small, "dewetting" of the boundaries will be largely complete if cooling is sufficiently slow. Cooling through the eutectic temperature (740°C) results in crystallization of the Bi$_2$O$_3$ phase, which then meets grain boundaries with a high dihedral angle indicating classical dewetting. However, the grain boundaries retain an amorphous intergranular film of ~1 nm thickness, as this is the equilibrium configuration. Small adjustments in thickness and composition may occur with further subsolidus annealing and oxidation processes which give rise to varistor activity.

Acknowledgements

We thank Yaping Liu for Figure 7a, and H. D. Ackler, R. M. Cannon, G. Ceder, H.L. Tuller, and J. B. Vander Sande for many useful discussions. The support of the U.S.

Department of Energy, Office of Basic Energy Sciences, Grant No. DE-FG02-87ER45307, and NSF Grant No. 9400334-DMR are gratefully acknowledged.

REFERENCES

1. L. M. Levinson and H. R. Philipp, "ZnO varistors for transient protection", *IEEE Transactions on Parts, Hybrids and Packaging*, Vol. PHP-13, 338 (1977).
2. T. K. Gupta, "Application of zinc oxide varistors", *J. Am. Ceram. Soc.*, **73**[7] 1817-40 (1990).
3. F. Greuter, "Electrically active interfaces in ZnO varistors", *Solid State Ionics*, **75**[1] 67-78 (1995).
4. J. Wong, "Nature of intergranular phase in nonohmic ZnO ceramics containing 0.5 Mol% Bi_2O_3", *J. Am. Ceram. Soc.*, **57** (8) 357 (1974), and J. Wong and W. G. Morris, "Microstructure and phases in nonohmic $ZnO-Bi_2O_3$ ceramics", *Ceram. Bull.*, **53**[11] 816 (1974).
5. D. Dey and R. C. Bradt, "Grain Growth of ZnO During Bi_2O_3 Liquid-phase Sintering", *J. Am. Ceram. Soc.*, **75**[9] 2529-34 (1992).
6. A. Peigney, H. Andrianjatovo, R. Legros, A. Rousset, "Influence of chemical composition on sintering of bismuth-titanium-doped zinc oxide," *J. Mater. Sci.*, **27**, 2397-405 (1992).
7. M. Ghirlanda and L. C. De Jonghe, "Sintering of doped ZnO powders: Relationship between grain growth and sintering stress," *Mater. Sci. Forum*, **94-96**, 855-64 (1992).
8. A. Peigney and A. Rousset, "Phase transformations and melting effects during the sintering of bismuth-doped zinc oxide powders," *J. Am. Ceram. Soc.*, **79**[8] 2113-26 (1996).
9. D. R. Clarke, "The microstructural location of the intergranular metal oxide phase in a zinc oxide varistor", *J. Appl. Phys.*, **49**[4], 2407 (1978).
10. W. D. Kingery, J. B. Vander Sande and T. Mitamura, "A scanning transmission electron microscopy investigation of grain boundary segregation in a $ZnO-Bi_2O_3$ varistor", *J. Am. Ceram, Soc.*, **62**[3-4] 221-22 (1979).
11. J. P. Gambino, W. D. Kingery, G. E. Pike and H. R. Phillip, "Effect of heat treatments on the wetting behavior of bismuth rich intergranular phases in ZnO:Bi:Co varistors," *J. Am. Ceram. Soc.*, **72**[4], 642-45 (1989).
12. E. Olsson, L. K. L. Falk and G. L. Dunlop, "The microstructure of a ZnO varistor material", *J. Mater. Sci.*, **20**[11], 4091-98 (1985).
13. E. Olsson and G. L. Dunlop, "Characterization of individual interfacial barriers in a ZnO varistor material", *J. Appl. Phys.*, **66**[8], 3666-75 (1989).
14. Y.-M Chiang and W. D. Kingery, "Compositional changes adjacent to grain boundaries during electrical degradation of a ZnO varistor", *J. Appl. Phys.*, **53**[3], (1982).
15. J.-R. Lee and Y.-M. Chiang, "Bi segregation at ZnO grain boundaries in equilibrium with Bi_2O_3-ZnO liquid," *Solid State Ionics*, **75**, 79-88 (1994).
16. F. Stucki, P. Brüesch, and F. Greuter, "Electron spectroscopic studies of electrically active GBs in ZnO," *Surf. Sci.*, **189/190**, 294-299 (1987).
17. F. Greuter, G. Blatter, M. Rossinelli and F. Stucki, "Conduction mechanism in ZnO-varistors: An overview," *Ceram. Trans.*, **3**, 31-53 (1989).
18. D. R. Clarke, "On the equilibrium thickness of intergranular glass phases in ceramic materials", *J. Am. Ceram. Soc.*, **70**[1] 15-22 (1987).
19. R. H. French, L. K. DeNoyer, R. M. Cannon, and Y.-M. Chiang, "Full spectral calculation of non-retarded Hamaker constants for ceramic systems from interband transition strengths," *Solid State Ionics*, **75**, 13-33 (1994).
20. H. D. Ackler, "Thermodynamic calculations and model experiments on thin intergranular amorphous films in ceramics," Ph. D. thesis, M.I.T., Cambridge, MA, February 1997.
21. H. D. Ackler, R. H. French, Y.-M. Chiang, "Comparisons of Hamaker constants for ceramic systems with intervening vacuum or water: From force laws and physical properties," *J. Colloid Interface Sci.*, **179**, (1996).
22. H.-J. Kleebe, M. J. Hoffman, and M. Rühle, "Influence of secondary phase chemistry on grain boundary Film thickness in silicon nitride," *Z. Metallkd.*, **83**[8], 610-6 (1992).
23. H.-J. Kleebe, M. K. Cinibulk, R. M. Cannon and M. Rühle, "Statistical analysis of the intergranular film thickness in silicon nitride ceramics", *J. Am. Ceram. Soc.*, **76**[8] 1969-77 (1993).
24. H. D. Ackler and Y.-M. Chiang, "Thin intergranular films in ceramics: Thermodynamic calculations and Model experiments in the system titania-silica," these proceedings.
25. S. C. Hansen and D. S. Phillips, "Grain boundary microstructures in a liquid phase sintered alumina," *Phil. Mag. A*, **47**[2], 209-234 (1983).

26. Y.-M. Chiang, L. A. Silverman, R. H. French and R. M. Cannon, "Thin glass film between ultrafine Conductor particles in thick-film resistors", *J. Am. Ceram. Soc.*, **77**[5] 143-52 (1994).
27. J.-R. Lee, Y.-M. Chiang and G. Ceder, "Pressure-thermodynamic study of grain boundaries: Bi segregation in ZnO", *Acta mater.*, in press.
28. H. Wang and Y.-M. Chiang, "Equilibrium configuration of Bi-doped ZnO grain boundaries: Intergranular amorphous films," *MRS Symp. Proc.*, in press, and H. Wang and Y.-M. Chiang, "Thermodynamic stability of intergranular amorphous films in Bi-doped ZnO," submitted to *J. Am. Ceram. Soc.*
29. R. J. Lauf and W. D. Bond, "Fabrication of high field zinc oxide varistors by sol-gel processing," *Ceram. Bull.*, **63**[2] 278-81 (1984).
30. J. A. S. Ikeda, Y.-M. Chiang, A. J. Garratt-Reed and J. B. Vander Sande, "Space charge segregation at grain boundaries in titanium dioxide: II, Model experiments", *J. Am. Ceram. Soc.*, **76** (10) 2447-59 (1993).
31. P. L. Flaitz and J. A. Pask, "Penetration of polycrystalline alumina by glass at high temperatures," *J. Am. Ceram. Soc.*, **70**[7], 449-55 (1987).
32. T. M. Shaw and P. R. Duncombe, "Forces between aluminum oxide grains in a silicate melt and their effect on grain boundary wetting," *J. Am. Ceram. Soc.*, **74**[10] 2495-2505 (1991).
33. H.-H. Park, S.-J. L. Kang, and D. N. Yoon, "An analysis of the surface menisci in a mixture of liquid and deformable grains," *Metall. Trans. A*, **17A**, 325-30 (1986).
34. G. M. Safronov, V. N. Batog, T. V. Stepanyuk and P. M. Fedorov, *Russ. J. Inorg. Chem.*, **16**[3], 460 (1971).
35. J.-R. Lee, "Effect of pressure and thermal history on grain boundary solute coverage in $ZnO-Bi_2O_3$ and relation to varistor properties," Ph. D. thesis, M.I.T., Cambridge, MA, June 1996.
36. ASTM E112-88, "Standard test method for determining average grain size," in Annual Book of ASTM Standards, Vol. 03.01, ASTM, Philadelphia, PA (1994).
37. W. D. Kingery, "Distribution and influence of minor constituents on ceramic formulations," Ceramic Microstructures '86, edited by J. A. Pask and A. G. Evans, *Material Science Research*, Vol. 21, Plenum Press, New York, 1987.
38. J-H. Hwang, T. O. Mason, and V. P. Dravid, "Microanalytical determination of ZnO solidus and liquidus boundaries in the $ZnO-Bi_2O_3$ system," *J. Am. Ceram. Soc.*, **77** [6]1499 (1994)
39. R.M. Cannon, these proceedings.
40. W.L. Johnson, "Amorphization by interfacial reactions," pp. 517-549 in Materials Interfaces, D. Wolf and S. Yip, editors, Chapman and Hall, 1992.
41. I. Majid, Y. Liu, R. W. Balluffi and J. B. Vander Sande, "Grain boundaries of controlled geometry in ZnO Films grown by chemical vapor deposition: Undoped and Bi-doped boundaries", *MRS Symp. Proc., 1996.*
42. J. Claus, H. L. Tuller, J.-R. Lee, and Y.-M. Chiang, unpublished work.

THIN INTERGRANULAR FILMS IN CERAMICS: THERMODYNAMIC CALCULATIONS AND MODEL EXPERIMENTS IN THE SYSTEM TITANIA-SILICA

Harold D. Ackler and Yet-Ming Chiang

Department of Materials Science and Engineering
Massachusetts Institute of Technology
Cambridge MA 02139

INTRODUCTION

Many technical ceramic materials are densified with additives that form a liquid phase at elevated temperatures to provide liquid phase sintering. The exact distribution of the liquid or glassy phase in the final microstructure differs from material to material, depends on processing method, and may have a dramatic effect on final properties. Perhaps the most important feature of this distribution is the thin liquid film along grain boundaries, retained as amorphous or glassy films after cooling. These films are characterized by being nearly constant in thickness along the boundary and on the order of one nanometer in thickness. It is increasingly accepted that they represent an equilibrium separation between grains due to a balance between attractive and repulsive forces (Clarke, 1987). Two examples where these minute microstructural features have a significant influence on final properties include creep behavior in Si_3N_4 (Raj, 1993) and electronic conductivity in ruthenate thick film resistors (Chiang, et al., 1994).

There have been numerous recent experimental studies of boundary films, the major results of which can be summarized as follows: (i) Amorphous films are found at virtually all grain boundaries in Si_3N_4, even between intergrown grains, and also between Si_3N_4 grains and devitrified glass pockets (for example see Vetrano, et al., 1993 or Kleebe, et al., 1993). (ii) The chemistry of the liquid affects the film thickness (Tanaka, et al., 1994; Chiang, et al., 1994). (iii) Film thicknesses at different homophase grain boundaries in a single Si_3N_4 sample are generally observed to be within a few angstroms of the mean (Kleebe, et al., 1993). However, in ruthenate thick film resistors (Chiang, et al., 1994), fine grained Si_3N_4 (Pan, et al., 1996), and TiO_2 with intergranular SiO_2 (Ackler and Chiang, 1996), wider distributions are observed. (iv) Films do not crystallize during long anneals well below the crystallization temperature for the bulk glass (Vetrano, et al., 1993).

While a film of constant thickness may exist at thermodynamic equilibrium, processing causes many systems to approach the equilibrium configuration from different extremes. The limiting cases consist of grains initially in contact, and those initially wells eparated by liquid. In liquid phase sintered Si_3N_4 and RuO_2-based thick film resistors, grains are initially separated by glass due to the native oxide layer on Si_3N_4 particles, and the low melting and surface-wetting glass used in thick film resistors. On the other hand, in Al_2O_3 grain boundaries may form at particle contacts before the sintering liquid forms, in which case grain boundary penetration is required to form a film. The resultant microstructures may be path dependent.

In this paper we first present an improved model for the interaction energy across a thin intergranular film which is based on the Allen-Cahn interface concept, wherein the degree of order varies away from a crystal-liquid interface while composition remains constant. Then, model experiments on the influence of initial particle separation on final microstructures in the system TiO_2-SiO_2 are discussed. The system exhibits mutually insoluble solid phases, a eutectic liquid that forms at 1550°C with a low TiO_2 content, and a large difference in refractive indices (2.56 for TiO_2 and 1.5 for SiO_2) which provides a large van der Waals attraction between particles (French, et al., 1995). Experimental observations show that TiO_2 grains initially in contact and those initially separated by silicate liquid come to different final microstructures in this ceramic-glass system. The thermodynamic calculations support the existence of an energy barrier between grain boundaries free of glass and those containing a film, consistent with the path-dependence of film stability.

THERMODYNAMIC MODELS

Clarke Model

This model, which has motivated much of the recent work on grain boundary films, considers a balance of attractive and repulsive interactions between the faces of two grains separated by a liquid film (Clarke, 1987). In the absence of externally applied forces, the attraction is solely attributed to van der Waals forces, which may be balanced by steric forces (due to structure in the liquid) and perhaps electrostatic forces (due to charged species on the grain faces). The overall interaction energy is minimized at an "equilibrium" separation between grain faces. The interaction energy for van der Waals forces acting against steric repulsion is given by

$$E(h) = \frac{-A_{121}}{12\pi h^2} + \frac{4a\eta_0^2 \xi}{\tanh\left(\frac{h}{2\xi}\right)} \qquad (1)$$

where A_{121} is the Hamaker constant for bodies of material 1 separated by material 2 (equivalent to A_{212} in the nonretarded limit), h is the separation between the faces, $a\eta_0^2$ is an ordering force (steric force) considered to be proportional to the heat of fusion of the liquid, and ξ is a correlation length of molecular dimensions. Equation (1), plotted in Figure 1, predicts a minimum in energy at a small separation of ~2 nm, using parameters appropriate for Si_3N_4. However, the energy rises without bound as the separation approaches zero. If at zero separation the energy must reach that of a grain boundary, then an energy barrier may exist between the two energy minima corresponding to a grain boundary and a grain boundary film (Straw and Duncombe, 1991). This barrier is shown schematically in Figure 1. For a sufficiently large barrier, clearly the initial state, *i.e.* separated vs. contacting grains, may determine the final structure.

Contribution of Liquid Structure to Film Thermodynamics

The steric repulsion term in the Clarke model was derived from diffuse interface theory, and the resulting terms $a\eta_0^2$ and ξ depend on thermodynamic parameters in a way which makes the assignment of specific numerical values somewhat difficult. Notice that this term describes only repulsive interactions; i.e. all terms are positive in value.

An improvement of the treatment of liquid structure in thin films has recently been developed (Ackler, et al., 1996) from diffuse interface theory (Cahn and Hilliard, 1958; Allen and Cahn, 1979), which predicts that liquid structure alone can result in attractive as well as repulsive interactions. The results of this model are summarized here. Assuming that the crystal surface is able to impose structure or order in the immediately adjacent liquid, the degree of order in the liquid is described by the order parameter ϕ. An order parameter of $\phi=0$ corresponds to the amorphous or liquid state, while $\phi=1$ corresponds to the fully ordered or crystalline state. The composition is at this point considered uniform across the film and only the result of changes in structure are taken into account. We assume that both structural and compositional gradients exist and are likely correlated. A complete treatment must include both effects.

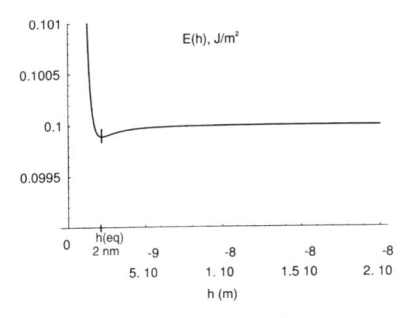

Figure 1. Interaction energy $E(h)$ using $A_{121} = 25$ zJ, $a\eta_0^2 = 100$ MPa, $\xi = 2.5$ Å. Minimum in energy at ~2 nm corresponds to predicted equilibrium film thickness. If the energy must reaches that of a grain boundary at $h = 0$, an energy barrier is implied.

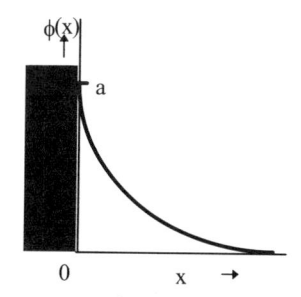

Figure 2. Schematic profile of order in liquid at crystal surface. Crystal imposes order $\phi = a$ at the surface ($x = 0$) which decays to zero as $x \to \infty$.

A schematic profile of the order imposed in a liquid near a crystal surface is shown in Figure 2. At the interface, $x = 0$, ϕ has the value $0 < a < 1$, and decays to 0 as $x \to \infty$. (For the remainder of this paper, the variable "a" is the value of the order parameter, and not the variable "a" in the Clarke model, Equation (1).) A grain boundary film consists of two opposing crystal/liquid interfaces. The excess free energy of the liquid in the film over that of bulk liquid is composed of volumetric and gradient terms. The volumetric energy difference Δf due to ordered structure is approximated[1] after the work of Landau (1967), by

---

[1] It should be pointed out that this energy expression is zero at the "melting" point for arbitrary ϕ, which will lead to no volumetric energy contribution. A more rigorous volumetric energy is a fourth order, "double well" function of ϕ with minima at fully amorphous and crystalline states which would provide a nonzero volumetric energy at that temperature. The use of such a function makes the problem intractable analytically, and introduces no additional physics over the quadratic expression used here for temperatures away from the "melting" point (see Ackler, et al. 1996, for details).

$\Delta f(\phi(x)) = \Delta f_{c-1}\phi(x)^2 = v\phi(x)^2$, where $\phi(x)$ is the spatially varying order parameter, Δf_{c-1} is the difference in free energy between fully crystalline and fully amorphous states, and $v = \sqrt{|\Delta f_{c-1}|}$ is defined as the volumetric energy coefficient. The gradient energy (see Cahn and Hilliard, 1958) is given by $\varepsilon(\nabla\phi(x))^2$ where ε the gradient energy coefficient and ∇ is the gradient operator. (Our ε is equal to the term $\sqrt{(2K)}$, where K is the gradient energy coefficient of Cahn and Hilliard.)

The interfacial energy per unit area is the integral of volumetric and gradient energies over the film thickness. For a film of thickness h centered at $x=0$, the integration is from $-h/2$ to $+h/2$:

$$\sigma(h) = \int_{-h/2}^{h/2}\left[v^2\phi(x)^2 + \varepsilon^2(\nabla\phi(x))^2\right]dx \qquad (2)$$

The minimum interfacial energy is obtained for the profile in order, $\phi(x)$, that satisfies the Euler equation

$$\nabla^2\phi(x) - \frac{v^2}{\varepsilon^2}\phi(x) = 0 \qquad (3)$$

subject to the boundary conditions $\phi(-h/2) = a$ and $\phi(+h/2) = b$, where b is the surface order parameter at the second crystal surface. For a film, we obtain

$$\phi(x) = \frac{1}{2}\left\{\cosh\left(\frac{v}{\varepsilon}x\right)\left[(a+b)\operatorname{sech}\left(\frac{hv}{2\varepsilon}\right)\right] - \sinh\left(\frac{v}{\varepsilon}x\right)\left[(a-b)\operatorname{csch}\left(\frac{hv}{2\varepsilon}\right)\right]\right\} \qquad (5)$$

which, when substituted in Equation (2) gives

$$\sigma(h) = \varepsilon v\left[-2ab + a^2\cosh\left(\frac{hv}{\varepsilon}\right) + b^2\cosh\left(\frac{hv}{\varepsilon}\right)\right]\operatorname{csch}\left(\frac{hv}{\varepsilon}\right) \qquad (6)$$

where the interfacial energy is a function of thickness h.

On the other hand, for a single interface like that in Figure 2, the integration is from $x=0$ to $x=\infty$, the boundary conditions are $\phi(0)=a$ and $\nabla\phi(\infty)=0$, and the order profile is given by

$$\phi(h) = a\left(\sinh\left(\frac{v}{\varepsilon}x\right) - \cosh\left(\frac{v}{\varepsilon}x\right)\right) \qquad (4)$$

which, when substituted into Equation (2), gives $\sigma = a^2\varepsilon v$.

To evaluate these results, numerical values for the volumetric and gradient energy coefficients may be obtained from the energy of a single crystal/liquid interface. For the interface between a crystal and its own liquid, we take $a = 1$. From published crystallization data for SiO_2 at 1800°C (Scherer, et al., 1970) we obtain $\Delta f_{c-1} = 8 \times 10^5$ J/m^3, which gives $v = \sqrt{|\Delta f_{c-1}|} = 900$ J/m^3. To evaluate ε, we use the result that $\sigma = a^2\varepsilon v$. Turnbull (1950a, 1950b) has shown that there exists an empirical relationship between a material's heat of fusion and its crystal/liquid interfacial energy:

$$\sigma = n \frac{\Delta h_{C-L}}{N_A^{1/3} \Omega^{2/3}} \tag{7}$$

where Δh_{C-L} is the heat of fusion (574 J/mol for SiO_2, Scherer, et al., 1970), Ω is the molar volume (2.6×10^{-5} m^3/mol), N_A is Avagadro's number, and n is a numerical coefficient typically between 0.3 and 0.5. Lower values of n are appropriate for materials with lower entropies of fusion (Thomas and Stavely, 1952). Given that SiO_2 has one of the lowest entropies of fusion known, we shall use a coefficient of 0.25. This gives $\sigma = 33$ mJ/m^2 for pure SiO_2. Thus, from published glass crystallization data, we are able to self consistently determine values for the parameters ε and ν. For the TiO_2-SiO_2 system at 1600°C, the values $\varepsilon = 3 \times 10^{-5}$ J$^{1/2}$/m$^{1/2}$, $\nu = 1100$ J$^{1/2}$/m$^{1/2}$ will be used. These values are approximations based on extrapolating the properties of pure SiO_2 to describe those of a two component liquid (Ackler et al., 1996).

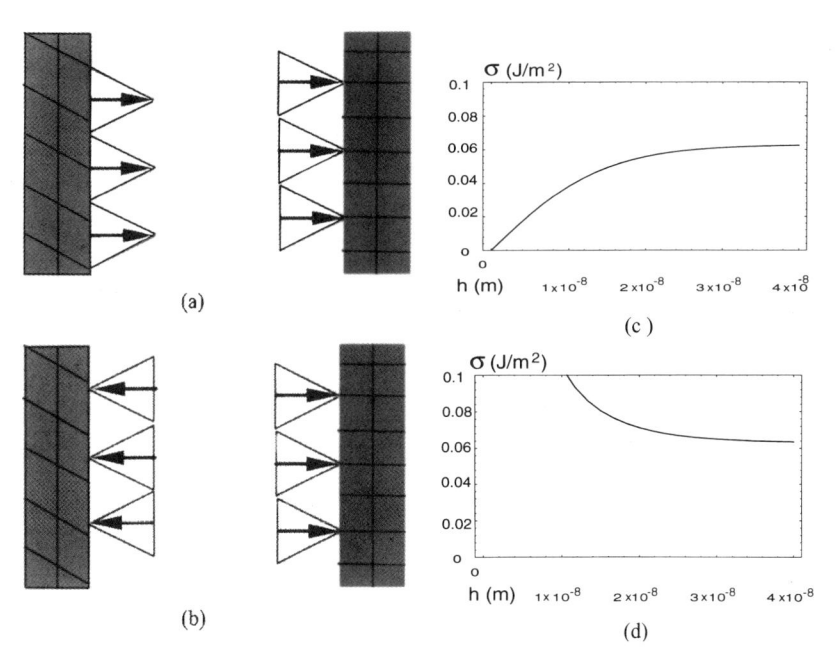

Figure 3. Schematic representation of ordering of glass network molecules at crystal surfaces bounding thin film. (a) $a=b=1$, corresponding energy vs. separation in (c). (b) $-a=b=1$, corresponding energy vs. separation in (d)

We next take the order parameter a and b at each crystal face bounding a film to be a vector quantity. The physical justification for this is illustrated in Figure 3. A two dimensional sheet of molecular units (*i.e.* a sheet of silica tetrahedra joined at three of their four vertices) can have both a direction and a magnitude of order. By this definition, $-1<\phi<0$ and $0<\phi<1$ corresponds to molecules that 'point" in the $-x$ and $+x$ directions, respectively. In any system it is reasonable to expect these parameters to take on a range of values, bounded by two extreme cases of complete order (Figure 3). In Figure 3a all molecules are pointed in the $+x$ direction, so a and b are both $+1$. In contrast, in Figure 3b the molecules on the opposing faces are pointed in opposite directions, so *a* and *b* are +1 and -1, respectively. The corresponding energy given by Equation (6) is plotted in Figures 3c and 3d, respectively. For case 3a, the energy is monotonically attractive, showing that liquid structure alone can result in attractive interactions across a film. No equilibrium separation will occur. For case 3b, the energy is monotonically repulsive, and strong enough that a typical van der Waals attraction will not result in an energetic minimum (Ackler, et al., 1996), again resulting in no equilibrium separation.

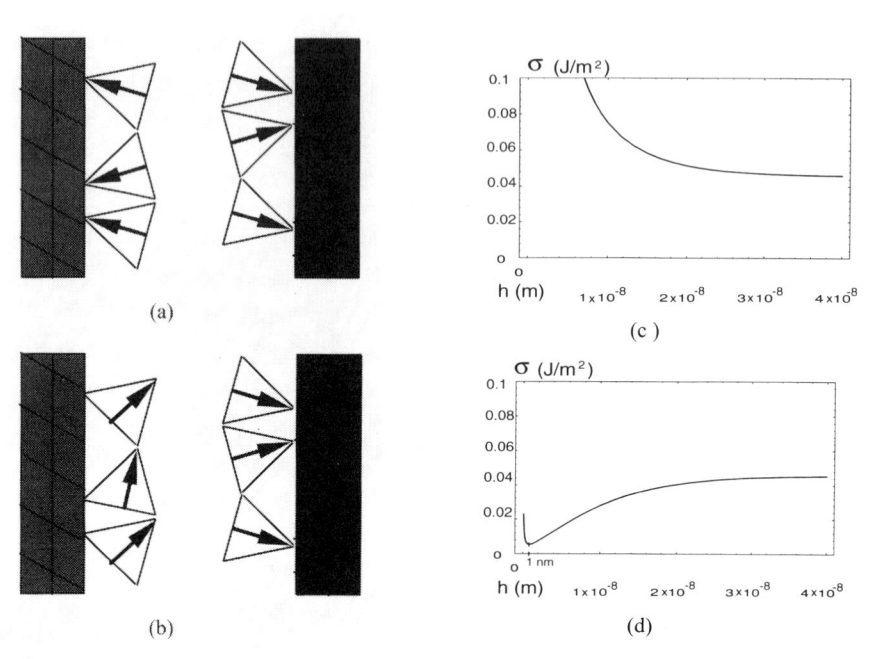

Figure 4. Schematic representation of ordering of glass network molecules at crystal surfaces bounding thin film. (a) $1 < a/b < 0$, corresponding energy vs. separation in (c). (b) $0 < a/b < 1$, corresponding energy vs. separation in (d).

However, in general the liquid will not be completely ordered at each crystal surface, and the order parameter will differ in both magnitude and sign between the two surfaces. Figure 4 shows two examples of such cases. In Figure 4a, *a* and *b* are of different magnitude and opposite sign. The interaction energy for this configuration is shown in Figure 4c for a=-0.8 and b=0.9, and is again monotonically repulsive and large compared to a typical van der Waals attraction (Ackler, et al., 1996). In Figure 4b, the order parameters are now of the same sign. Now the energy is attractive at large separations, but goes through a minimum at a nanometer separation before it rises steeply as the film thickness goes to zero, as shown in Figure 4d. This predicts an equilibrium separation on the order of a nanometer, due solely to the effects of imposed liquid order.

If we now add the van der Waals attraction to these results, the total film energy is

$$\sigma(h) = \frac{-A_{121}}{12\pi h^2} + \varepsilon v \left[-2ab + a^2 \cosh\left(\frac{h v}{\varepsilon}\right) + b^2 \cosh\left(\frac{h v}{\varepsilon}\right) \right] \csc h\left(\frac{h v}{\varepsilon}\right) \qquad (8)$$

Figure 5 gives the interaction energy for a Hamaker constant of 25 zJ, $\varepsilon = 3 \times 10^{-5}$ J$^{1/2}$/m$^{1/2}$, $v = 1.1 \times 10^3$ J$^{1/2}$/m$^{1/2}$, $b = 0.9$, and three values of a. The Hamaker constant was calculated using the full spectral method described by French, et al. (1995) for the TiO$_2$-SiO$_2$ system (Ackler, et al., 1995). These values are appropriate for experiments discussed later. The plots illustrate three different types of behavior. For small differences in order the film energy decreases monotonically with thickness, shown by the curve for $a = 0.85$, indicating a preference for glass free boundaries. For $a = 0.72$, the film energy decreases with thickness to a minimum before rising steeply without bound. For the intermediate case where $a = 0.78$, the energy exhibits a minimum at finite separation, but the van der Waals attraction is sufficient to overcome the short range steric repulsion, and another energy minimum results as the thickness goes to zero. This results in two minimum energy states at finite and zero thickness, separated by an energy barrier.

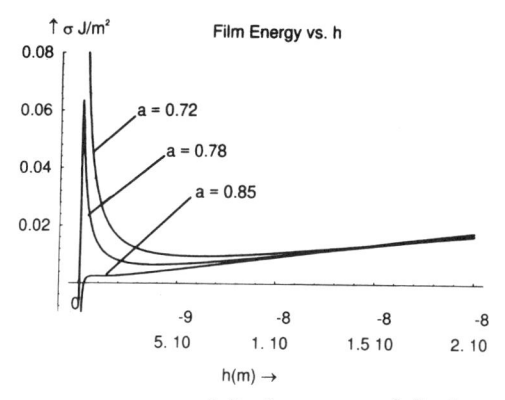

Figure 5. $\sigma(h)$ for $A_{121} = 25$ zJ, $\varepsilon = 3 \times 10^{-5}$ J$^{1/2}$/m$^{1/2}$, $v = 1.1 \times 10^3$ J$^{1/2}$/m$^{1/2}$, $b=0.9$, and these values of a.

In any system, a range of values for a and b is expected. If ϕ can vary from 0 to 1 at every interface, according to the present model a variety of observed interfacial structures should result, including glass-free grain boundaries (when a and b are nearly equal) or grain boundaries and films separated by an energy barrier. This would seem to contradict observations of uniform films. However, in many systems the crystal may impose a more limited range of order on the liquid. Furthermore, as discussed in detail elsewhere (Ackler, et al., 1996), the ratio v/ε also has strong influence on the range of predicted film thicknesses. For a given range over which a and b can independently vary, the range of film thicknesses narrows as v/ε increases. Also, since v/ε increases as the temperature increases, the observed film thicknesses should decrease, and the distribution narrow (Ackler, et al., 1996).

Qualitatively, some of the results of this model are similar to those of the Clarke model. The primary improvements are that the energy coefficients v and ε are explicitly defined and readily obtained from published data, and that the use of surface order parameters provides an explicit relation between crystal-liquid ordering and film behavior. This improved model also demonstrates the possibility of an attractive force due to liquid structure alone, not predicted by the Clarke model, which is conceptually similar to the "hydration" forces between surfaces in aqueous media (Israelachvili, 1992). Existing results showing that different systems exhibit different ranges of film thickness uniformity, and the experiments below showing a broad range of film thicknesses in the TiO$_2$-SiO$_2$ system as well as an energy barrier between grain boundaries and films, are in agreement with these theoretical results.

EXPERIMENTS IN TiO$_2$-SiO$_2$

The effects of approaching the equilibrium separation from different initial separations can be investigated through controlled materials processing. Dense polycrystals have all grains initially in contact and can be immersed in a glass melt that has been equilibrated with the polycrystal material. This permits the interparticle interactions beginning from zero separation to be investigated by glass penetration (Flaitz and Pask, 1987, Shaw and Duncombe, 1991). We immersed dense TiO$_2$ polycrystals in a high SiO$_2$ melt that had been equilibrated with TiO$_2$. To achieve an initial state of well-separated particles, we used a TiO$_2$ powder coated with SiO$_2$, following a processing method similar to that used by Sacks, et al., (1991).

Experimental Procedures

For the penetrated polycrystal experiments, high purity TiO$_2$ powder[2] was hot pressed in a 5 mm inside diameter tungsten carbide die for 4 h at 750°C in air with an applied pressure of approximately 500 MPa. The exterior surfaces of the pellets were ground with 600 grit SiC paper to remove any contamination from the die, and they were sliced into several pieces ($\cong 2 \times 2 \times 1$ mm) with a diamond saw and cleaned. The glass was made by mixing ~10% TiO$_2$ powder with colloidal SiO$_2$[3] to produce a melt with the equilibrium composition at 1600°C (Phase Diagrams for Ceramists, 1964). The mixture was dried and heated in air in a 50 ml platinum crucible to about 500°C for about 5 h, to 1000°C for 1 h, then to 1675°C for 1 h and water quenched. Pieces of the polycrystals were packed in crushed of glass in 5 mm die × 2 mm deep platinum crucibles and annealed at 1600°C for 1 h in air. The samples were prepared for TEM by slicing, grinding to ~100 μm thick, dimpling, and ion thinning. They were inspected using scanning electron microscopy[4] (SEM) to determine the distribution of glass throughout the sample, and high resolution electron microscopy[5] (HREM) to determine grain boundary morphologies (direct crystal-crystal contacts vs. intergranular films) at atomic level resolution.

For the coated powder experiments, high purity TiO$_2$ powder[6] of approximately 0.5 μm particle size was coated with SiO$_2$ by the hydrolysis of tetraethylorthosilicate (TEDS) (Sacks, et al., 1991). The van der Waals interaction drops off by approximately 10 nm (Israelachvili, 1992), so the glass coating should not be much thicker than this to provide adequate attraction to bring particles together. A coating several nm thick on each particle will keep them apart while providing only about 10-15 vol% liquid at temperatures.[7] The coating process, described elsewhere in detail (Ackler and Chiang, 1996), produced a cake of coated powder about 0.5 mm thick. Pieces of the cake were mixed in distilled water and shaken to disperse the particles. Carbon coated TEM grids were dipped into the dilute particle suspensions and dried for TEM inspection of the as-coated powder. Pieces of the cake were put in a covered platinum pan and vacuum annealed at 500°-600°C for 4 h at $< 5 \times 10^{-5}$ torr to completely dry/degas the cake, then fired in air at 500°C for 30 min, then at 1600°C for 5 or 10 min. The pieces were thinned and polished from one side, dimpled, and ion milled for inspection of the grain boundaries in HREM and scanning transmission electron microscopy (STEM).

---

[2] ultra pure TiO$_2$, ~34 nm particle size, Nanophase Technologies, Darien IL USA
[3] Nyacol colloidal SiO$_2$ sol, Nyacol Products, Inc. Ashland MA USA
[4] JEOL 6320 FEG SEM, JEOL Ltd. Tokyo Japan; Link Analytical EDXS, Oxford Instruments, England
[5] Topcon EM-002B TEM, Topcon Technologies, Paramus NJ USA
[6] Cerac 99.9% TiO$_2$ powder, ~ 0.5 μm particle size, Cerac, Inc. Milwaukee WI USA
[7] Higher volume fractions of glass result in significant differential thinning during ion milling, preventing the fabrication of samples suitable for good HREM

Results and Discussion

The glass was observed to penetrate through the polycrystal along triple grain junctions only, leaving grain boundaries in contact, as shown in Figure 6a. EDS data in Figure 6b, from the pocket marked "A" in Figure 6a, shows it contains silicon from the glass. The TEM image in Figure 6c shows glass pockets in the middle of the sample. The glass did not separate the grain boundaries, in contrast to the results of Flaitz and Pask (1987) and Shaw and Duncombe (1991) for pure Al_2O_3 penetrated by an anorthitic glass melt. (Flaitz and Pask (1987) did observe a dramatic reduction in penetration of MgO-doped Al_2O_3 as compared to pure Al_2O_3, indicating the penetration of grain boundaries by liquid glasses is system specific.) In the present experiments, out of over fifty grain boundaries inspected, only two were found with glass films, 0.6 nm and 0.7 nm thick, respectively.

The powder coating process produced a ~5 nm layer of SiO_2 on the TiO_2 particles, shown in Figure 7a. Roughly half of the particles were initially in contact, permitting the formation of grain boundaries, or were multicrystalline with grain boundaries present before

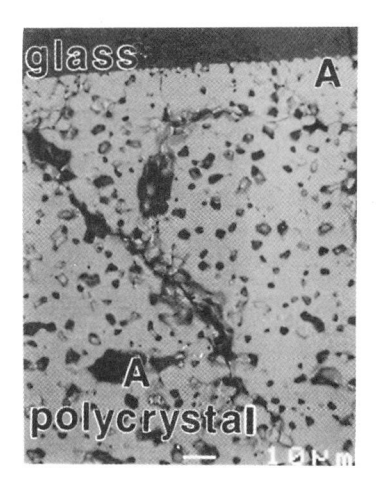

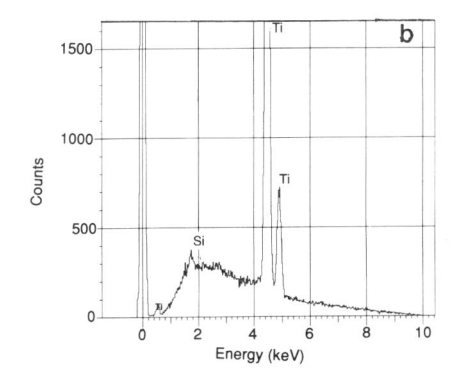

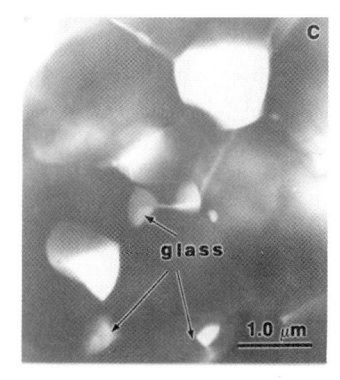

Figure 6. (a) Back scattered electron micrograph of glass penetrated TiO_2 polycrystal. Pockets contain SiO_2, as indicated by EDS spectrum from pocket "A" in (b), showing Si from glass. (c) TEM micrograph showing glass pockets in sample interior.

coating as shown in Figure 7b. Some agglomerates have void spaces in their interiors which were filled with SiO$_2$ during the coating process, indicated by the arrow in Figure 7b. The coated powder had an initial particle size of approximately 0.5 μm. After 5 and 10 min at 1 600°C, the TiO$_2$ grains in these SiO$_2$-coated powder samples had coarsened to average grain diameters of 5-10 μm.

In the sample annealed for 5 min, 12 grain boundaries and 17 glass films were observed out of the 29 boundaries resolved at high magnification. Figure 8 shows one of these glass films. The thicknesses of the 17 films varied between 0.6 and 5 nm, with a mean value of 1.4 ± 0.5 nm. In the sample annealed for 10 min, 37 grain boundaries and 20 glass films were observed out of 66 clearly resolved boundaries. Film thicknesses in these samples ranged from 0.6 to 7 nm, with a mean of 1.8 ± 1.5 nm. An additional 9 boundaries showed thicker films around 10 nm, which were not included in the calculation of the mean.

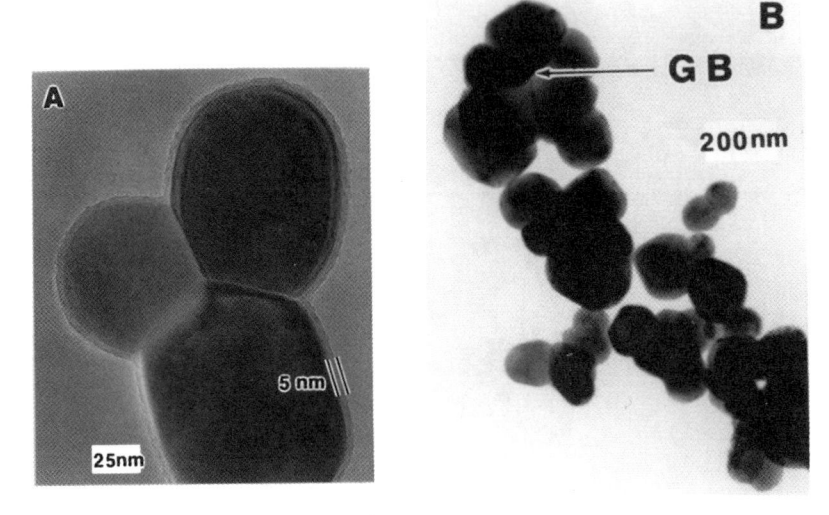

Figure 7. SiO$_2$ coated particles. (a) 5 nm SiO$_2$ coating outlines particles. (b) Note severe agglomeration and grain boundaries between subgrains, and small SiO$_2$ filled void spaces at arrow.

Nonuniform thickness along the latter boundaries suggests they are not "equilibrium" films. Thus, of the total of 95 clearly resolved boundaries in coated powder samples, 49 grain boundaries and 37 films (plus 9 thick films) were observed, and the cumulative thicknesses ranged from 0.6 to 7 nm with a mean of 1.7 ± 1.4 nm.

There is clearly a statistically significant difference in the fraction of grain boundaries with thin glass films between the two experiments, with only 4% of the boundaries in the glass melt penetrated polycrystal contained glass films, compared to 43% of the boundaries in the SiO$_2$-coated powder material. The processing differs only in the duration of anneal and the initial particle separation. The failure of the glass melt to penetrate initially sintered grain boundaries while easily penetrating three-grain junctions demonstrates that the grain boundaries represent at least a local thermodynamic minimum. The stability of glass films throughout microstructural evolution in the coated-particle experiments shows they also represent at least a local thermodynamic minimum. The coexistence of both boundary morphologies is clear evidence of the coexistence of two states, stable or metastable, separated by an energy barrier, each largely inaccessible to the other. The magnitude of the energy barrier, the depth of the minima, and the relative energies of the two minima are not known in detail. However it is clear that in at least this system, and possibly many others, the final grain boundary structures may be determined by the initial interparticle separation. In Si_3N_4 materials in which the powder particles are separated by a native oxide, glass-free grain boundaries may be inaccessible using the liquid phase sintering methods currently employed, and samples with initial zero separation have yet to be studied.

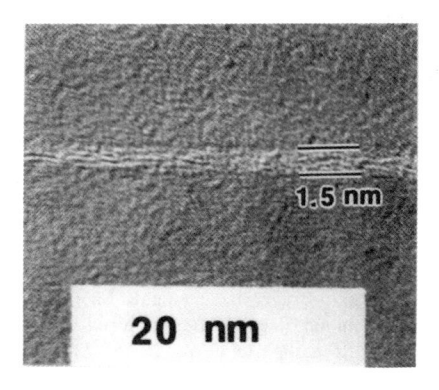

Figure 8. 1.5 nm thick grain boundary film in SiO_2-coated sample annealed 5 min at 1600°C in air.

CONCLUSIONS

An improved model of the interaction energy between grains separated by a thin liquid film is summarized. This model predicts the existence of an energy barrier between glass-free grain boundaries and boundaries with thin films. In addition, this model provides detailed insight to the important parameters governing the behavior of intergranular liquid films, and indicates the ranges of their values in which various types of behavior may be anticipated. Model studies of initially contacting vs. initially well-separated grains in TiO_2-SiO_2 provide experimental evidence for the existence of an energy barrier. This energy barrier implies that the initial separation between grains (grains in contact vs. separated by liquid) may determine the final grain boundary structure.

ACKNOWLEDGEMENTS

We thank Rowland Cannon, Craig Carter, and Roger French for useful discussions. This work was supported by the U.S. Department of Energy, Basic Energy Sciences Division, Materials Sciences Division under contract No. DE-FG02-87-ER45307, and some experimental facilities by the MRSEC Program of the National Science Foundation under award number DMR 94-00334.

REFERENCES

Ackler, H. D. and Chiang, Y-M., 1996, "Thin Intergranular Amorphous Films in the Model Binary Ceramic System TiO$_2$-SiO$_2$", to be submitted to *J. Am. Ceram. Soc.*

Ackler, H. D., Carter, W. C., and Chiang, Y-M., 1996, "Contribution of Liquid Structure to Thermodynamics of Thin Liquid Films Constrained between Crystal Faces", to be submitted to *J. Chem. Phys.*

Ackler, H. D., French, R. H., Chiang, Y-M., 1995, unpublished research.

Allen, S. M., and Cahn, J. W., 1979, "A Microscopic Theory for Antiphase Boundary Motion and its Application to Antiphase Domain Coarsening", *Acta Met.*, 27 1085-1095.

Cahn, J. W., and Hilliard, J. E., 1958, "Free Energy of Nonuniform System. I. Interfacial Free Energy", *J. Chem. Phys.*, 28 [2] 258-267.

Chiang, Y-M., Silverman, L. A., French, R. H., Cannon, R. M., 1994, "The Thin Glass Film between Nanocrystalline Conductor Particles in Thick Film Resistors", *J. Am. Ceram. Soc.*, 77 [5] 1143-1152.

Cinibulk, M. K., Thomas, G., and Johnson, S. M., 1990, "Grain-Boundary-Phase Crystallization and Strength of Silicon Nitride Sintered with a YSiAlON Glass", *J. Am. Ceram. Soc.*, 73 [6] 1606-1612.

Clarke, D. R., 1987, "On the Equilibrium Thickness of Intergranular Glass Phases in Ceramic Materials", *J. Am. Ceram. Soc.*, 70 [1] 15-22.

Clarke, D. R., Shaw, T. M., Philipse, A. P., and Horn, R. G., 1993, "Possible Electrical Double-Layer Contribution to the Equilibrium Thickness of Intergranular Glass Films in Polycrystalline Ceramics", *J. Am. Ceram. Soc.*, 76 [5] 1201-1204.

French, R.H., Cannon, R.M., DeNoyer, L.K., and Chiang, Y-M., 1995, "Full Spectral Calculation of Non-Retarded Hamaker Constants for Ceramic Systems from Interband Transition Strengths", *Sol. St. Ionics*, 75 13-33.

Flaitz, P. L. and Pask, J. A., 1987, "Penetration of Polycrystalline Alumina by Glass at High Temperatures", *J. Am. Ceram. Soc.*, 70 [7] 449-455.

German, R. M., 1985, <u>Liquid Phase Sintering,</u> Plenum Press.

Israelachvili, J. N., 1992, <u>Intermolecular and Surface Forces, Second Edition</u>, Academic Press, Ltd. London.

Kleebe, H.-J., Cinibulk, M. K., Tanaka, I., Bruley, J., Cannon, R. M., Clarke, D. R., Hoffman, M. J., and Rühle, M., 1993, "High-Resolution Electron Microscopy Observations of Grain-Boundary Films in Silicon Nitride Ceramics", MRS Symposium Proc. 287 65-78.

Landau, L. D., 1967, "On the Theory of Phase Transitions", Section 29 in <u>Collected papers of L. D. Landau,</u> D. T. Haar, editor, Gordon and Breach, Science Publishers.

Lee, J. R. and Chiang, Y-M., 1995, "Bi segregation at ZnO grain boundaries in equilibrium with Bi2O3 ZnO liquid", *Sol. St. Ionics*, 75 79-88.

Pan, X., Mayer, J., Rühle, M., and Niihara, K., 1996, "Silicon Nitride Based Ceramic Nanocomposites", *J. Am. Ceram. Soc.*, 79 [3] 585-590.

<u>Phase Diagrams for Ceramists</u>, 1964, TiO$_2$-SiO$_2$ phase diagram from R. C. DeVries, R. Roy, and E. F. Osborn, *Trans. Brit. Ceram. Soc.*, 53 [9] 531 (1954) American Ceramic Society, Figure 113.

Pompe, W. and Clarke, D. R., 1996, unpublished research

Raj, R., 1993, "Fundamental Research in Structural Ceramics for Service Near 2000°C", *J. Am. Ceram. Soc.*, 76 [9] 2147-2174.

Sacks, M. D., Bozkurt, N., and Schieffele, G. W., 1991, "Fabrication of Mullite and Mullite-Matrix Composites by Transient Viscous Sintering of Composite Powders", *J. Am. Ceram. Soc.*, 74 [10] 2428-2437.

Scherer, G., Vergano, P., and Uhlman, D.R., 1970, "A Study of Quartz Melting", *Phys. Chem. Glasses*, 11 [3].

Shaw, T. M. and Duncombe, P. R., 1991, "Forces between Aluminum Oxide Grains in a Silicate Melt and Their Effect on Grain Boundary Wetting", *J. Am. Ceram. Soc.*, 74 [10] 2495-2505.

Tanaka, I., Kleebe, H.-J., Cinibulk, M. K., Bruley, J., Clarke, D. R., and Rühle, M., 1994, "Calcium Concentration Dependence of the Intergranular Film Thickness in Silicon Nitride", *J. Am. Ceram. Soc.*, 77 [4] 911-914.

Thomas, D. G. and Staveley, L. A. K., 1952, "A Study of the Supercooling of Drops of Some Molecular Liquids", *J. Chem. Soc.*, 4569-4577.

Turnbull, D., 1950a, "The Supercooling of Aggregates of Small Metal Particles", *Trans. AIME, J. Metals*, 188 1144-1148.

Turnbull, D., 1950b, "Formation of Crystal Nuclei in Liquid Metals", *J. App. Phys.*, 21 1022-1028.

Vetrano, S., Kleebe, H.-J., Hampp, E., Hoffmann, M. J., and Rühle, M., 1993, "Yb$_2$O$_3$-fluxed sintered silicon nitride". Part 1. Microstructure characterization", *J. Mat. Sci.*, 28 3529-3538.

Wagstaff, F. E. and Richards, K. J., 1966, "Kinetics of Crystallization of Stoichiometric SiO$_2$ Glass in H$_2$O Atmospheres", *J. Am. Ceram. Soc.*, 49 [3] 118-121.

Wang, C-M., Pan, X., Hoffmann, M. J., Cannon, R. M., and Rühle, M., 1996, "Grain Boundary Films in Rare-Earth-Glass-Based Silicon Nitride", *J. Am. Ceram. Soc.*, 79 [3] 788-792.

ATOMIC STRUCTURE OF THE Σ5 (210)/[001] SYMMETRIC TILT GRAIN BOUNDARY IN YTTRIUM ALUMINUM GARNET

Geoffrey H. Campbell and Wayne E. King

University of California
Lawrence Livermore National Laboratory
Chemistry and Materials Science Directorate
Livermore, CA 94551

INTRODUCTION

Yttrium aluminum garnet (YAG) is a promising high temperature structural material. Grain boundaries in YAG have been shown to play a critical role in its high temperature behavior. The creep rate of polycrystalline YAG[1] is many orders of magnitude higher than the creep rate of single crystal YAG,[2] indicating that grain boundaries provide a high-diffusivity path for the cations, which facilitates dislocation climb. The contributions of grain boundary sliding and cavitation to the creep rate have not been explicitly investigated and further understanding at the atomic level is required if the high temperature properties are to be controlled.

Atomistic simulations in ceramic materials have been limited compared to those in metals. Commonly, ceramics have comparatively complex crystal structures. For example, the unit cell of YAG contains 160 atoms, even though it is one of the more simple structures with cubic symmetry. Hence, the size of atomistic simulation ensembles are comparatively large, leading to large calculations. Other difficulties are associated with the long range nature of the Coulomb interaction, which causes a slow convergence in summing the contributions to the energy by all the charged ions in the crystal. Whereas in metals the interaction calculation can be cut off after a few nearest neighbors, the ionic interactions must be summed to infinity. Nevertheless, point defect properties of a wide range of ceramics,[3] including YAG,[4] have been calculated. Planar defects have also been simulated in simple oxides and the alkali halides, including surfaces,[5] stacking faults,[6] and grain boundaries.[7,8]

A limited number of experimental studies of grain boundary atomic structure in ceramics have been performed by high - resolution transmission electron microscopy (HREM), notably in NiO,[9] TiO_2,[10] and Al_2O_3.[11,12] Comparisons of model predictions to experimental data on the atomic structure of grain boundaries in metals have been very successful in clarifying the limits of some models while providing confirmation of the accuracy of new models.[13] This study will provide relevant experimental data which can readily be compared to the predictions of developmental atomistic models of ceramics. Atomistic simulations with experimentally verified accuracy can be used to develop fundamental understanding of observed material behaviors such as diffusion,[4] which further the understanding of the influence of grain boundaries on the high temperature behavior of YAG.

EXPERIMENTAL METHODS

Materials

Single crystals of pure (undoped) YAG were grown by the Czochralski method and specimens were cut from the lower - stress regions of the boule (Union Carbide Crystal Products, San Diego, CA). Specimens were oriented by Laue x-ray backscatter diffraction and cut into cylinders of 19 mm (0.75 in.) diameter and 12.7 mm (0.5 in.) in length. The faces of the cylinders were cut and polished parallel to the (210) plane to within 0.1°. The faces were polished flat to $\lambda/10$ (55 nm) (Valpey - Fischer Corp., Hopkinton, MA). After the relative rotation relationship was established, a reference flat common to both crystals was ground and polished on the sides of the cylinders in order to re-establish the orientation immediately prior to bonding.[14]

Bonding

The grain boundaries were prepared by ultra - high vacuum (UHV) diffusion bonding.[15] Auger electron spectroscopy (AES) of the as-introduced surface showed carbon as the main surface impurity. Sputtering the surface with 1 keV Xe^+ at a 15° grazing incidence removed the surface contamination to below the detection limit of AES. The rough sides and bottom of the crystals are presumed to remain contaminated after sputtering and thus the crystals were heat treated in UHV at 1200°C for 2h to degas any volatile contamination. Subsequent Auger spectroscopy of the surfaces to be bonded showed no contaminants after annealing.

Immediately prior to bonding, the cleaned surfaces were placed in contact and the crystals were aligned in the twist orientation by reflecting a laser from the reference flat on the sides of the crystals. The alignment was within a reflected spot diameter, which corresponds to approximately 0.1°.

The crystals were bonded with an applied load of 1430 N corresponding to a pressure of 5 MPa and the bonding temperature was 1550°C. The samples were held for six hours at the bonding temperature. During the entire bonding cycle, the total vacuum level never exceeded 1×10^{-5} Pa (7×10^{-8} torr). Residual gas analysis showed H_2 and CO to be the most common residual gases while at temperature.

Characterization by HREM

Specimens for examination in the transmission electron microscope (TEM) were prepared by diamond cutting, polishing, and dimpling. Thinning to final electron transparency was performed by ion milling with 6 keV Ar^+ at 13° grazing incidence. Specimens were lightly coated with carbon to prevent static charging in the microscope.

Two types of specimens where prepared: (i) with the common [001] for both crystals as the specimen normal and (ii) with the common [1 $\bar{2}$ 0] as the specimen normal. Both orientations have the grain boundary viewed edge - on. The first is equivalent to viewing the boundary parallel to the tilt axis and the second is viewing the boundary perpendicular to the tilt axis.

Specimens were observed in a high resolution TEM (JEM-4000EX, JEOL, Tokyo, Japan) operating at 400 keV with an objective lens spherical aberration coefficient of 1.0 mm and a spread of focus of 8 nm. Micrographs were acquired on film at an electron - optical magnification of 800 kX with an illumination semi-angle of convergence of 0.91 mrad and an objective aperture diameter equivalent to 13.1 nm^{-1} in reciprocal space.

High Resolution Image Simulation

High resolution image simulation was performed with the EMS suite of computer programs.[16] Crystal structure information was obtained from Wyckoff.[17] Model grain boundary structures were created by simple geometric manipulation of the perfect crystal. This method follows the Coincident Site Lattice model[18] for forming grain boundary structures.

Images of the perfect crystal were calculated by the Bloch wave method with 500 waves included in the calculation. Images of the grain boundary were calculated by the multi-slice

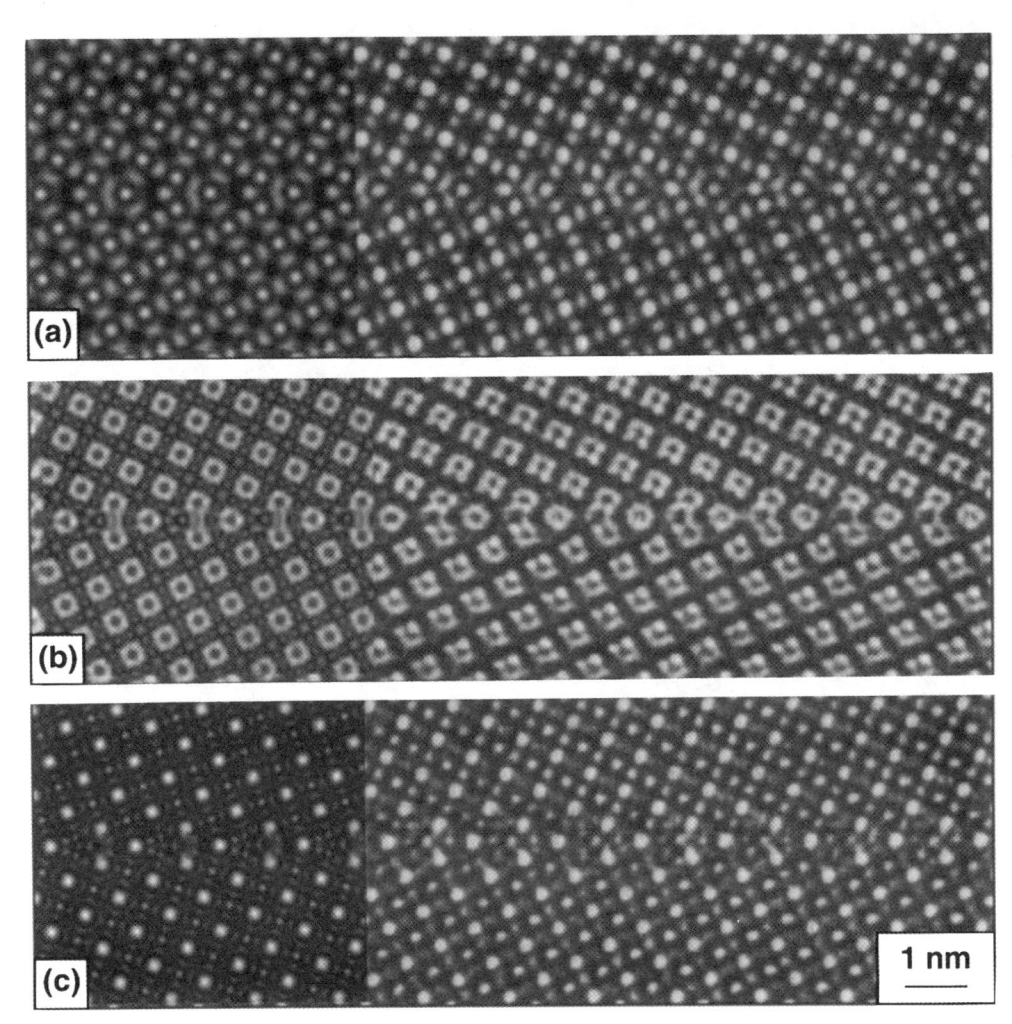

Figure 1. A selection of three micrographs taken of the identical region of a $\Sigma 5$ (210)/[001] symmetric tilt grain boundary in YAG. The images differ only in the microscope focus used to acquire them. The focus deviation values from Gaussian were determined to be (a) 25 nm, (b) -25 nm, and (c) -88 nm. The images are taken along the common [001] direction of the adjacent crystals (parallel to the tilt axis of the boundary). The simulated high resolution images derived from the model structure are shown as inserts on the left.

technique. The entire repeat unit of the boundary structure was used for each calculation. In the [001] direction the projected potential of four slices of the boundary repeat unit were calculated. In the [1$\bar{2}$0] direction, eight slices were used. The calculation matrix used 190 samples/nm (1024 × 512 matrix dimensions in [001] and 1024 × 256 in [1$\bar{2}$0]).

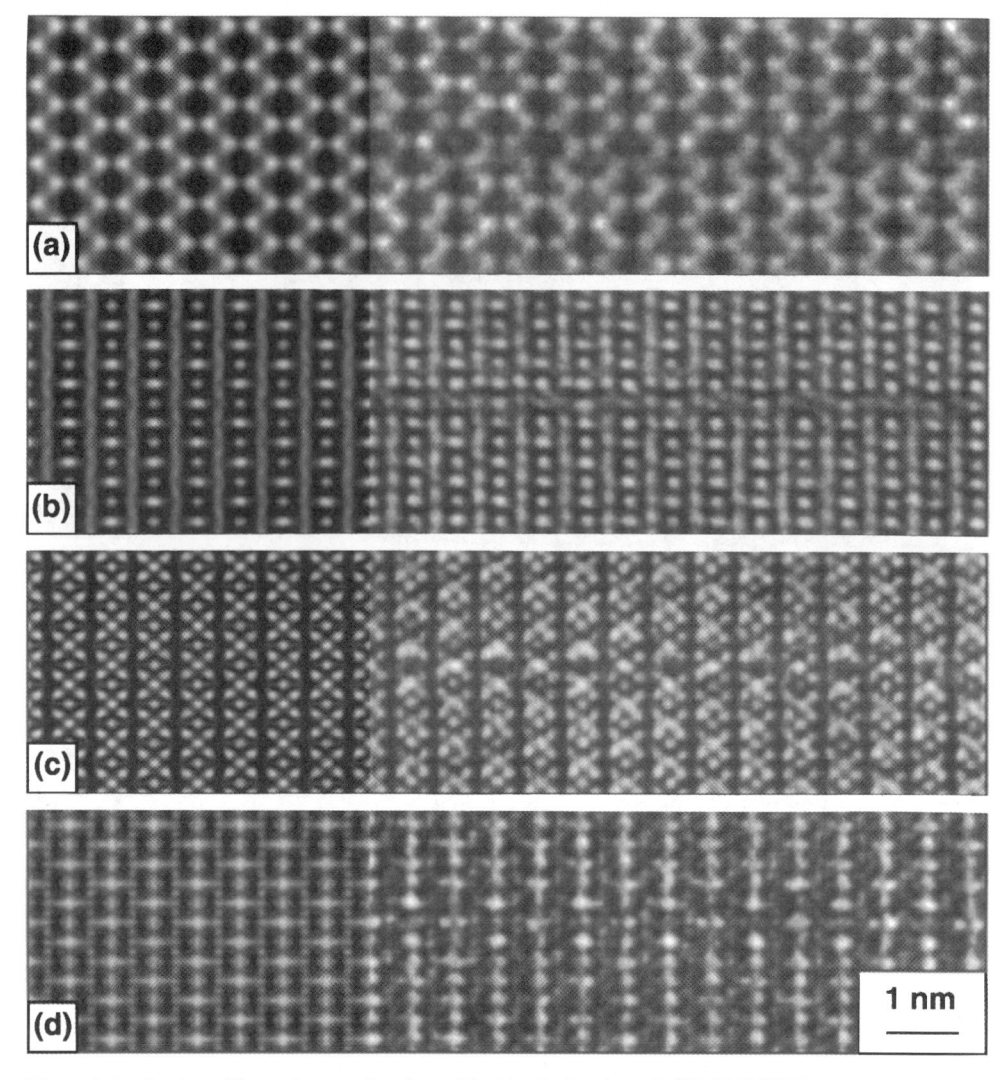

Figure 2. A selection of four micrographs taken of the identical region of a $\Sigma 5$ (210)/[001] symmetric tilt grain boundary in YAG. The images differ only in the microscope focus used to acquire them. The focus deviation values from Guassian were determined to be (a) 65 nm, (b) -5 nm, (c) -60 nm, and (d) -100 nm. The images are taken along the common [1 $\bar{2}$ 0] direction of the adjacent crystals (perpendicular to the tilt axis of the boundary). The simulated high resolution images derived from the model structure are shown as inserts on the left.

RESULTS

Bonding

The diffusion bonded interface produced between the crystals was found to have a small fraction (<1%) of its area decorated with small (~10 μm) residual voids. These voids appeared to be uniformly distributed on the grain boundary. Except for these small voids, bonding was complete across the surface between the crystals even at the edges of the cylinders.

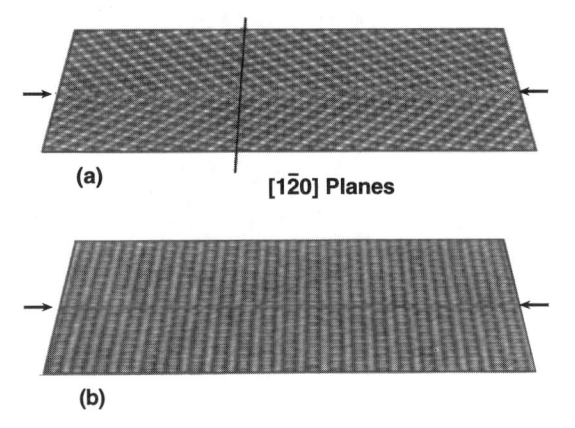

Figure 3. Glancing angle perspective views across the grain boundary using the images seen in (a) Figure 1(a) and (b) Figure 2(b). Due to the distortion, the magnification is not constant across the image, therefore no scale is indicated. The boundary runs horizontally in both cases. It can be seen that the contrast features along the planes perpendicular to the boundary are not displaced when crossing the boundary. A line is drawn in (a) to aid the eye. These images indicate that mirror symmetry of the atomic structure in three dimensions is present at the boundary.

High Resolution Electron Microscopy

Images of the grain boundary were acquired at several values of the microscope focus and at several locations. High resolution images were acquired in both the direction parallel to the tilt axis and perpendicular to the tilt axis. These images are shown in Figures 1 and 2 along with simulated high resolution images as inserts that are discussed in a later section.

The high resolution images can be inspected directly for rigid body translations of the adjacent crystals with respect to one another. The micrographs viewing the boundary parallel to the tilt axis (Figure 3a) show that the crystals are not displaced from a mirror symmetric relation of the crystals to within approximately 0.1 Å. The same holds true for viewing in the direction perpendicular to the tilt axis (Figure 3b). Inspection of the boundary for dilation is more difficult. A first inspection was performed using the technique of Merkle.[19] This technique shows the boundary to be free of dilation to an accuracy of about 0.2 Å.

Atomic Structure and Image Simulation

The garnet crystal is a cubic structure belonging to space group Ia$\bar{3}$d (space-group number 230). It is composed of a body-centered sublattice of Al (octahedrally coordinated sites) with Al-Y pairs on face centers (tetrahedrally coordinated Al and dodecahedrally coordinated Y). The unit cell is composed of 8 formula units, $Y_3Al_5O_{12}$, with a lattice parameter of 1.201 nm. The projection of the unit cell along [001] is shown in Figure 4.

On the basis of the observations of the rigid body displacements, the candidate atomic models can be restricted to only those showing mirror symmetry on the atomic scale about the boundary plane. Inspection of the unit cell of YAG suggests two positions at which the mirror plane can be placed. These planes are indicated in Figure 4. However, simulated HREM images have revealed that only the plane labelled B gives results comparable to the experimental

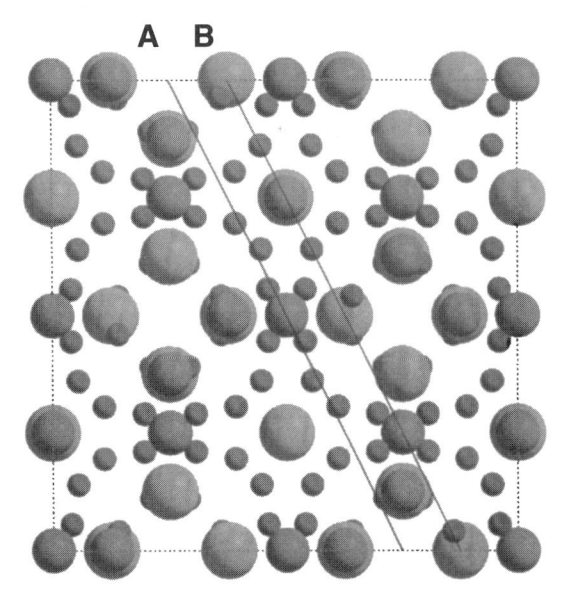

Figure 4. The unit cell of YAG as viewed along [001]. The small spheres are oxygen. The middle-sized spheres are aluminum. The large spheres are yttrium. Two candidate planes for forming the plane of mirror reflection are indicated and labeled A and B.

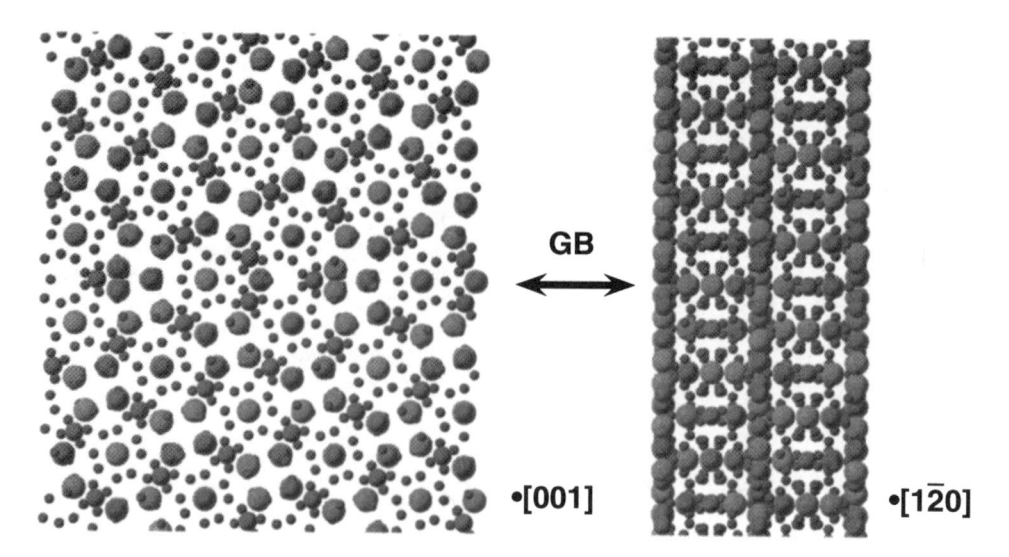

Figure 5. Models of the atomic structure of the Σ5 (210)/[001] symmetric tilt grain boundary in YAG formed by performing a mirror reflection operation at (a) plane A and (b) plane B in Figure 3. One structural unit of the boundary is shown. The plane of the grain boundary is marked by an arrow for each model. The top illustration of the models is a view parallel to the tilt axis along [001] and the bottom illustration is a view perpendicular to the tilt axis along [1 $\bar{2}$ 0].

images.

The resultant grain boundary structure from performing a twinning operation at plane B is shown in Figure 5. The structural unit of the boundary has a length of 5.370 nm in [1 $\bar{2}$ 0] and 1.201 nm in [001].

Image simulations for the atomic model are compared with all the experimental images in Figure 1 for viewing along the tilt axis and in Figure 2 for viewing perpendicular to the tilt axis. It is easily seen from this comparison that the grain boundary model agrees reasonably well with the data.

DISCUSSION

The high resolution images show this grain boundary to be atomically flat and straight for extended distances, up to several hundred nanometers. This behavior suggests that the boundary is faceting to a low energy configuration or at least low energy with respect to small deviations from the common (210) crystal plane.

The other remarkable aspect to this boundary is the mirror symmetry exhibited at the atomic scale. Despite examining the boundary at many widely spaced locations and on different TEM specimens, the relative translation state was identical. In all cases, mirror symmetry of atoms on either side of the boundary was found. This behavior is in contrast to that found in NiO,[9] TiO$_2$,[10] or the fcc metals,[20] where different translational states are observed.

The view of the boundary as projected along the tilt axis clearly shows the twin plane. The structure formed in Figure 5 provides simulated images which closely match the experimental images. The structure of the boundary as projected along [001] is well approximated by the model.

At variance with the view along the tilt axis, the view perpendicular to the tilt axis (Fig. 2) shows a distinct mismatch between the simulated and experimental images of the boundary. In the proposed atomic model, the boundary would be invisible as viewed along [1 $\bar{2}$ 0]. The boundary is visible in the experimental micrographs. Two possibilities exist for creating the contrast change at the boundary: a change in composition and/or a change in structure. A

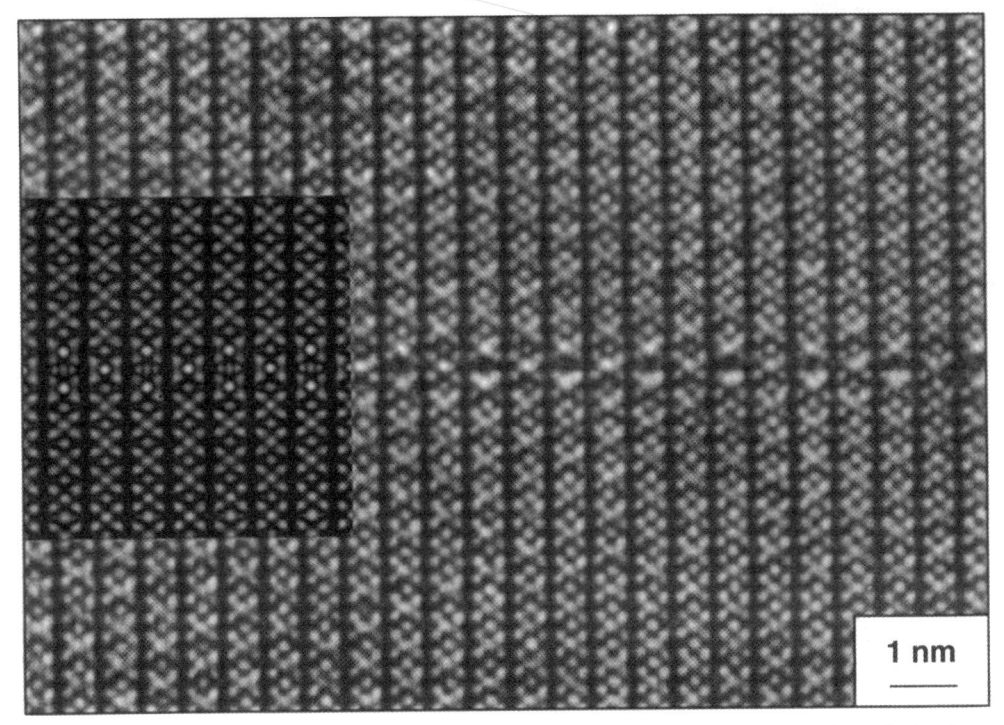

Figure 6. High resolution image simulation using an atomic model whose cation site compositions in the vicinity of the boundary have been altered to bring the simulation into better agreement with the experimental image. The experimental image is the same as that appearing in Figure 1(c).

change in structure would be highly restricted to displacements of atoms in only the [001] direction in order to maintain the image contrast when viewed in projection along [001]. Changing the composition of the boundary could be achieved by replacing Y atoms for Al atoms or vice versa. By changing the cations near the boundary, the contrast from the resulting simulated image can be brought into better agreement with the experimental image (Figure 6). But these changes are made without a physical basis beyond matching the contrast in the simulated and experimental images on a qualitative basis and are thus open to some doubt. In order to achieve a more thorough understanding of the boundary structure, atomistic simulations and more quantitative analysis of the high resolution images[21] would need to be performed to put proposed grain boundary structures on a sound physical basis.

CONCLUSIONS

The $\Sigma 5$ (210)/[001] symmetric tilt grain boundary in YAG was produced by UHV diffusion bonding precisely oriented single crystals. The boundary has been characterized by HREM along two different directions, parallel and perpendicular to the tilt axis. Models of the atomic structure of the boundary were formed following the Coincident Site Lattice scheme. The resulting models are equivalent to twins formed at the atomic scale. The high resolution images show no rigid crystal translations away from the perfect mirror reflection relation. Comparison of the simulated images using the atomic model as input with the experimental images identifies the plane of mirror symmetry. The atomic model is shown to be in good agreement with the experimental images when viewed parallel to the tilt axis, but disagrees with the images perpendicular to the tilt axis. The agreement between the simulated images and the experimental images can be improved by changing the composition of the grain boundary with respect to the bulk. To reach a more certain conclusion on the structure of the grain boundary will require the additional support of theoretical calculations.

ACKNOWLEDGMENTS

We would like to thank John Petrovic for supplying the YAG single crystals. We would also like to thank Ivar Reimanis for stimulating and helpful discussions. We would like to thank Doug Medlin for the use of the 4000EX at Sandia National Laboratories, Livermore. Finally, we would like to thank Walt Wein for his careful work on specimen preparation. This work performed under the auspices of the United States Department of Energy, Office of Basic Energy Sciences and Lawrence Livermore National Laboratory under Contract No. W-7405-Eng-48.

REFERENCES

1. T. A. Parthasarathy, T.-I. Mah and K. Keller, Creep Mechanism of Polycrystalline Yttrium Aluminum Garnet, *J. Am. Ceram. Soc.* 75:1756 (1992).
2. G. S. Corman, Creep of Yttrium Aluminium Garnet Single Crystals, *J. Mater. Sci. Lett.* 12:379 (1993).
3. C. R. A. Catlow and W. C. Mackrodt. *Computer Simulation in Solids*, Springer Verlag, Berlin (1982).
4. L. Schuh, R. Metsalaar and C. R. A. Catlow, Computer Modelling Studies of Defect Structures and Migration Mechanisms in Yttrium Aluminium Garnet, *J. Eur. Ceram. Soc.* 7:67 (1991).
5. P. W. Tasker, The Surface Energies, Surface Tensions and Surface Structure of the Alkali Halide Crystals, *Philos. Mag. A* 39:119 (1979).
6. P. W. Tasker and T. J. Bullough, An Atomistic Calculation of Extended Planar Defects in Ionic Crystals, Application to Stacking Faults in the Alkali Halides, *Philos. Mag. A* 43:313 (1981).
7. D. M. Duffy and P. W. Tasker, Computer Simulation of <001> Tilt Grain Boundaries in Nickel Oxide, *Philos. Mag. A* 47:817 (1983).
8. D. M. Duffy and P. W. Tasker, Computer Simulation of <011> Tilt Grain Boundaries in Nickel Oxide, *Philos. Mag. A* 48:155 (1983).
9. K. L. Merkle and D. J. Smith, Atomic Structure of Symmetric Tilt Grain Boundaries in NiO, *Phys. Rev. Lett.* 59:2887 (1987).
10. U. Dahmen, S. Paciornik, I. G. Solorzano and J. B. Vandersande, HREM Analysis of Structure and Defects in a $\Sigma 5$ (210) Grain Boundary in Rutile, *Interface Sci.* 2:125 (1994).
11. T. Höche, P. R. Kenway, H.-J. Kleebe, M. Rühle and P. A. Morris, High - Resolution Transmission Electrom Microscopy Studies of a Near $\Sigma 11$ Grain Boundary in α - Alumina, *J. Am. Ceram. Soc.* 77:339 (1994).
12. F.-R. Chen, C.-C. Chu, J.-Y. Wang and L. Chang, Atomic Structure of $\Sigma 7(0112)$ Symmetrical Tilt Grain Boundaries in α - Al_2O_3, *Philos. Mag. A* 72:529 (1995).
13. G. H. Campbell, S. M. Foiles, P. Gumbsch, M. Rühle and W. E. King, Atomic Structure of the (310) Twin in Niobium: Experimental Determination and Comparison to Theoretical Predictions, *Phys. Rev. Lett.* 70:449 (1993).
14. W. L. Wien, G. H. Campbell and W. E. King, Preparation of Specimens for Use in Fabricating Bicrystals by UHV Diffusion Bonding, in: *Microstructural Science*, D. W. Stevens, E. A. Clark, D. C. Zipperian and E. D. Albrecht, ed., ASM International, Materials Park, OH (1996).
15. W. E. King, G. H. Campbell, A. W. Coombs, G. W. Johnson, B. E. Kelly, T. C. Reitz, S. L. Stoner, W. L. Wien and D. M. Wilson, Interface Science of Controlled Metal/Metal and Metal/Ceramic Interfaces Prepared Using Ultrahigh Vacuum Diffusion Bonding, in: *Joining and Adhesion of Advanced Inorganic Materials*, A. H. Carim, D. S. Schwartz and R. S. Silberglitt, ed., Materials Research Society, Pittsburgh, PA (1993).
16. P. A. Stadelmann, EMS - A Software Package for Electron Diffraction Analysis and HREM Image Simulation in Materials Science, *Ultramicroscopy* 21:131 (1987).
17. R. W. G. Wyckoff. *Inorganic Compounds $R_x(MX_4)_y$, $R_x(M_nX_p)_y$, Hydrates and Ammoniates*, Interscience Publishers, New York (1965).
18. W. Bollmann. *Crystal Defects and Crystalline Interfaces*, Springer - Verlag, Berlin (1970).
19. K. L. Merkle, Quantification of Atomic - Scale Grain Boundary Parameters by High - Resolution Electron Microscopy, *Ultramicroscopy* 40:281 (1992).
20. D. L. Medlin, M. J. Mills, W. M. Stobbs, M. S. Daw and F. Cosandey, HRTEM Observations of a $\Sigma = 3$ {112} Bicrystal Boundary in Aluminum, in: *Atomic - Scale Imaging of Surfaces and Interfaces*, D. K. Biegelsen, D. J. Smith and S. Y. Tong, ed., Materials Research Society, Pittsburgh, PA (1993).
21. W. E. King and G. H. Campbell, Quantitative HREM Using Non-linear Least-squares Methods, *Ultramicroscopy* 56:46 (1994).

SYNTHESIS AND MICROSTRUCTURE OF MULLITE FIBERS GROWN FROM DEEPLY UNDERCOOLED MELTS

W.M. Kriven[1], M.H. Jilavi[1], D. Zhu[1], J.K.R. Weber[2], B. Cho[2], J. Felten[2] and P.C. Nordine[2]

[1]Department of Materials Science and Engineering
University of Illinois at Urbana-Champaign
105 S. Goodwin Ave., Urbana, IL 61801

[2]Containerless Research Inc.
906 University Place
Evanston, IL 60201

INTRODUCTION

Structural fibers that exhibit high strength, low creep, and oxidation resistance at high temperatures are a key requirement in the development of advanced ceramic-ceramic composite materials.[1-3] Oxide materials offer considerable potential to meet this need; many oxides have the required high intrinsic properties. However, the synthesis of oxide fibers with sufficiently high practical properties and acceptably low production cost has not yet been achieved.

There is considerable interest in the use of interface-weakening and interphase-weakened coatings for fibers to toughen ceramic-ceramic composites.[4,5] The coatings provide a weak connection between the fiber and matrix material at which crack propagation is stopped by separation of the crack tip at the fiber matrix interface or by failure in the weak interphase coating. Each coating material of interest places further requirements on the fiber (and matrix) materials, since the fiber and matrix must exhibit thermochemical stability with the coating.

We have recently shown that oxide glass fibers can be formed from many new materials by pulling the fibers from deeply undercooled melts in which melt viscosities are sufficient for fiber pulling operations. This paper presents our initial investigations on the recrystallization and structure of mullite glass fibers, in work to develop the fibers for high temperature structural applications. We also present results on the synthesis of additional glass fibers from several oxide materials chosen for their capability to provide high temperature strength, high modulus, chemical stability in their oxidizing environments, creep resistance, and/or compatibility with matrix and interface-modifying coatings.

EXPERIMENTAL PROCEDURES

Fiber Synthesis

Bulk crystalline oxide materials were first formed into dense specimens of *ca.* 0.3 cm diameter by melting in a laser hearth melter.[6] Mullite of stoichiometric $3Al_2O_3 \cdot 2SiO_2$ composition was obtained from Kyoritsu (KM mullite 101, Kyoritsu, Nagoya, Japan). The specimens were then levitated and melted with a CO_2 laser beam. Initial experiments demonstrated that deep undercooling of the melt was possible when the heating laser intensity was reduced under the containerless conditions.[7] Melt temperatures equal to 0.8 of the absolute melting point were typically obtained, and some of the materials formed bulk glass specimens upon cooling to ambient temperatures.[8,9] Fibers were pulled from the levitated drops of undercooled liquid by inserting a 100 µm diameter tungsten stinger into the molten drop and quickly withdrawing the stinger with a spring or stepper motor driven actuator. Fiber pulling velocities ranged from 5 to 120 cm/s.

The materials were chosen for their capability to provide high temperature strength, high modulus, chemical stability in oxidizing environments, creep resistance, and/or compatibility with matrix and interface-modifying coatings.

Fiber Characterization

Scanning electron microscopy (SEM) and microchemical analysis (EDS) were used to evaluate the surface morphology, the diameter of fibers along their lengths, and the chemical composition of the fibers. Transmission electron microscopy (TEM) and selected area electron diffraction (SAD) techniques in a Philips CM12 TEM were used to characterize fiber microstructure. Fibers prepared by ion milling, were examined both lengthwise and in cross-section.

Crystallization and Annealing

Glass fibers of the mullite composition were annealed in air at temperatures of 1100°, 1200°, 1300°, and 1500°C for one hour. Lengths of fiber were placed on a platinum foil holder and heated at a rate of 5°C/min to the annealing temperature.

Mullite fibers *ca.* 20 µm in diameter were coated with yttrium phosphate (YPO$_4$), by a laser ablation technique[10] and embedded in a mullite matrix. The composite was then sintered at 1500°C in air for 2 hours. The resulting compact was sectioned and examined by SEM.

Mechanical Evaluation

Tensile strengths of glass and recrystallized fibers were measured using an Instron Material Testing Machine. The fiber strength was measured using a method developed at UIUC for high-modulus single-filament materials. The fiber diameters were 20-30 µm, the gauge length was fixed at 23 mm, approximately 1000 times the fiber diameter. Test fibers were bonded into a specially designed frame with "M-Bond" epoxy adhesive. The measurement technique and interpretation was consistent with ASTM standard D3379-75. Tests were performed in an Instron 1205 tensile machine operated under computer control. Fibers were pulled at a constant strain rate = 5 µm/s. A high sensitivity load cell was used to measure load. A computer was used to record the load and displacement as a function of time. Broken fibers were examined by optical microscopy to determine the cross section so that stress could be determined.

RESULTS

Glass Fiber Synthesis

Table 1 presents results of fiber synthesis experiments. In some cases, the composition is given in terms of the mole percentage of its component oxides. Typical glass fiber diameters are indicated in column 2. Tensile strengths were measured for a few glass and recrystallized mullite fibers as given in the third column. The YAG-composition glass fibers were pulled to investigate fiber pulling for a material with exceptional creep properties. The forsterite composition glass fibers were investigated because these materials are chemically compatible with enstatite, MgO•SiO$_2$, an interphase-weakening material.[4]

The fibers were pulled at rates of 5 to 120 cm/s. The fiber diameter could be controlled via the melt viscosity (degree of undercooling) and the pulling rate. The preferred temperature range for drawing fibers was typically 70-140°C below the melting point for Al$_2$O$_3$-SiO$_2$ and the YAG compositions, and was not measured for forsterite compositions. At higher temperatures, rather short fibers were obtained, or the stinger pulled from the melt without drawing a fiber. At lower temperatures, the fiber drawing force tended to pull the entire sample from its levitated position. In preliminary experiments the fibers were pulled at rates which increased continuously, leading to tapered fibers. Later experiments, using the stepper-motor driven puller which rapidly accelerated to a constant pulling rate, led to uniform fiber diameters.

Table 1. Fiber materials, diameters, and tensile strengths

Material	Fiber Diameters, μm	Tensile Strength, GPa
60%Al_2O_3-40%SiO_2 Mullite (glass)	1-50	5.7 ± 0.6
Mullite recrystallized at 1100°-1200°C	8-19	0.7 - 1.0
50-72%Al_2O_3 and 28-50%SiO_2 (glass)	1-50	not tested
62.5%Al_2O_3 and 37.5%Y_2O_3 YAG (glass)	8-43	2.4 ± 0.1
50%Al_2O_3-50%CaO (glass)	10-50	not tested
66.6%MgO-33.4%SiO_2 forsterite (glass)	10-50	not tested

Mechanical Evalulation

Several mullite fibers were tested at ambient temperature. A representative stress-strain plot for a mullite fiber is presented in Figure 1. The mean tensile strength of five fibers was 5.61 ± 0.71 GPa. Fibers were typically amorphous and contained few defects and flaws. The small number of flaws was advantageous since it provided more consistent mechanical properties.

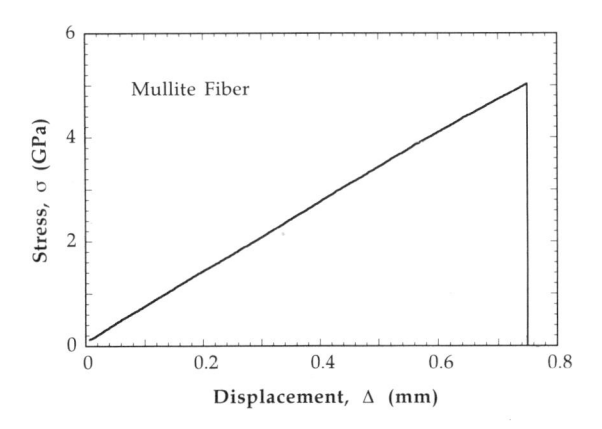

Figure 1. Load-displacement graph for a $3Al_2O_3 \cdot 2SiO_2$ mullite fiber pulled in air at 1400°C. It was tensile tested at a strain rate of 5 μm/s at ambient temperature.

Microstructure Characterization

Fibers pulled at constant velocity using the stepper motor-driven puller had uniform diameters of 10-40 μm. These fibers were pulled in lengths up to 50 cm. They were easy to handle and could be bent to a radius of 2-3 mm. Figures 2a and b are scanning electron micrographs of a mullite fiber surface and polished cross section. The fiber surface was smooth, homogeneous and free from imperfections. SEM examinations showed that the fibers exhibited brittle fracture to produce smooth and dense fracture surfaces.

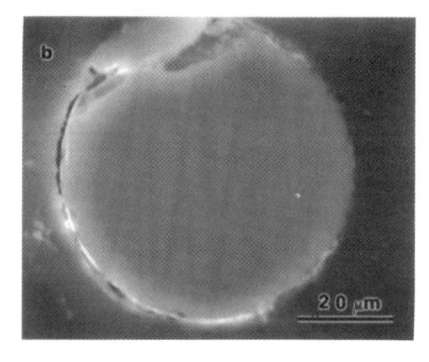

Figure 2. Scanning electron micrograph of a 3:2 mullite fiber. The fiber surface was smooth, homogeneous and free from imperfections. The fiber was pulled in oxygen from a drop of liquid undercooled to *ca.* 1675 K; (a) fiber surface (b) fiber cross section.

Figure 3. SEM micrograph showing regions along the length of a fiber pulled at a rate that increased from 5 to 30 cm/s as the fiber was pulled. The diameter of the fiber decreased from: (a) approximately 20 μm where it touched the stinger to (b) 1 μm at the tip.

In general fibers grown from boules of $3Al_2O_3 \cdot 2SiO_2$ composition retained their stoichiometry. However, chemically inhomogeneous regions were occasionally found near the ends of some fibers or in the fibers of 70 mol% composition that were pulled during the initial experiments. Figure 4 illustrates an occassional inhomogeneous area in a fiber. These effects were observed in fibers pulled at relatively high temperatures and of the 70% Al_2O_3 composition. Fibers obtained from the 60% mol% (3:2 mullite) composition and at somewhat lower drawing temperatures did not exhibit these effects.

Some fibers 20 μm in diameter, which were coated with yttrium phosphate by a laser ablation technique[10] and embedded in a mullite matrix by hot pressing at 1500°C, retained their shape during high temperature processing. Figure 5 shows a cross section of such a fiber, which had high chemical and dimensional stability.

Fibers made from stoichiometric mullite ($3Al_2O_3 \cdot 2SiO_2$) were examined by TEM. Figure 6 is a transmission electron micrograph of a thinned section of a mullite fiber. The corresponding diffraction rings obtained by selected area diffraction (SAD) at several points on the fiber were diffuse indicating that the fiber was amorphous. The absence of extinction contours and grain boundaries confirmed this observation.

Recrystallization Studies

Results of the annealing experiments are summarized in Table 2. The fibers were annealed in air. Fibers annealed at 1500°C broke into several pieces. Fibers annealed at 1300°C retained their shape but were too delicate to handle. Fibers annealed at 1100° and 1200°C could be handled and tensile tests were performed on several of these specimens.

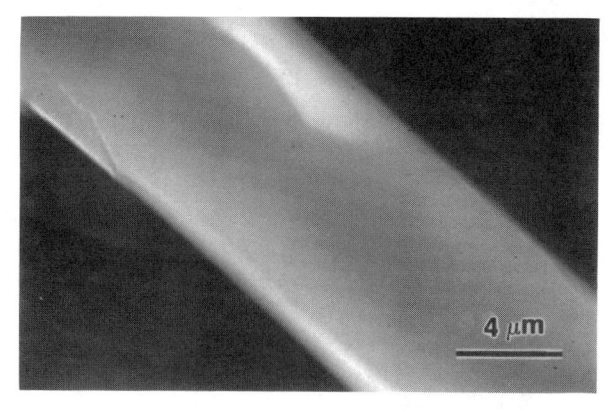

Figure 4. SEM micrograph of a fiber showing apparent phase separation. This was pulled in air from a melt of 70 mol% Al_2O_3 at 1500°C. The melt was heated extensively prior to fiber pulling, and may have experienced a composition change due to vaporization of its components.

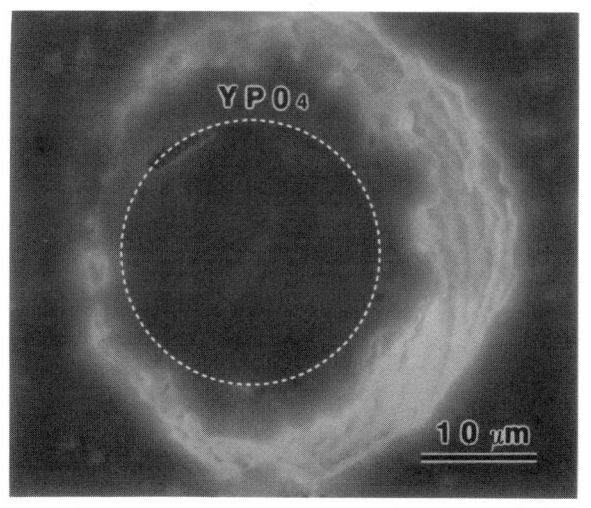

Figure 5. SEM micrograph of YPO_4-coated mullite fiber embedded in a mullite matrix and sintered at 1500°C for 2 hours. The micrograph shows the high chemical and dimensional stability of the coated mullite fiber composite. The circle of dots indicates the circumference of the fiber. Bright electrical charging artifacts surround the coating.

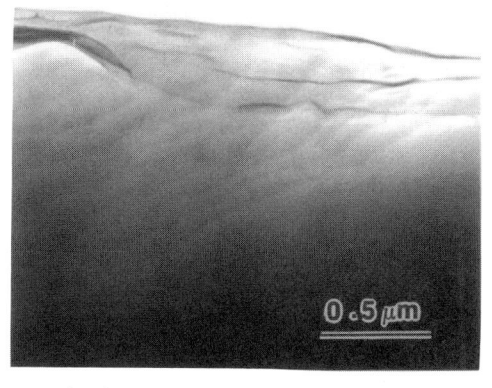

Figure 6. TEM micrograph of an amorphous mullite fiber. The specimen surface ripples are artifacts from the ion milling operation used to prepare the specimen. The corresponding diffuse diffraction pattern indicated that the material was amorphous.

Table 2. Procedures and results for annealing amorphous mullite fibers

Temperature (°C)	Ramp Rate (°C/min)	Result
1100	5	0.7 GPa tensile strength
1200	5	1.0 GPa tensile strength
1300	5	Broken when handled
1500	5	Broken during anneal

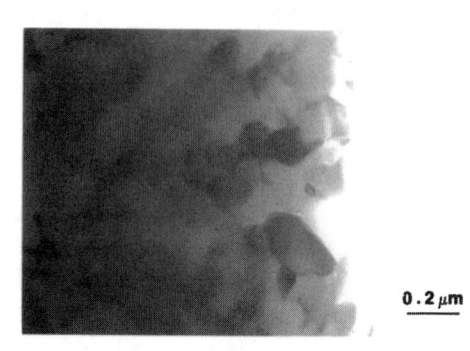

Figure 7. Bright field TEM micrograph of polycrystalline $3Al_2O_3 \cdot 2SiO_2$ mullite fiber annealed at 1200°C for one hour in air.

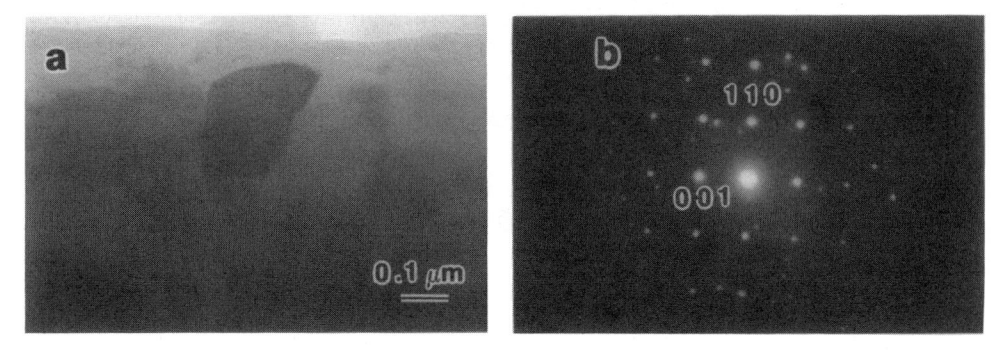

Figure 8. (a) Bright field TEM micrograph of an individual mullite grain and (b) corresponding selected area diffraction pattern indexed as $3Al_2O_3 \cdot 2SiO_2$ mullite in the $[1\bar{1}0]$ zone axis.

Fibers annealed at 1200°C were examined by TEM. Figure 7 shows that the microstructure was polycrystalline with a sub-micron grain size. Bright field (BF) and selected area electron diffraction (SAD) of a typical individual grain is given in Figure 8 which was indexed as stable $3Al_2O_3 \cdot 2SiO_2$ mullite. This was confirmed by microchemical analysis using energy dispersion X-ray spectroscopy (EDS) in the TEM.

DISCUSSION

Fibers are pulled from undercooled melts and at high pulling rates, which results in cooling rates in excess of 1000°C/s in the region just outside the liquid source. Crystallization is suppressed and glass fibers are formed. In order to pull fibers, it is necessary to pull at a rate greater than the crystallization velocity of the undercooled liquid from which the fibers are being pulled. If this condition is not met, the undercooled liquid

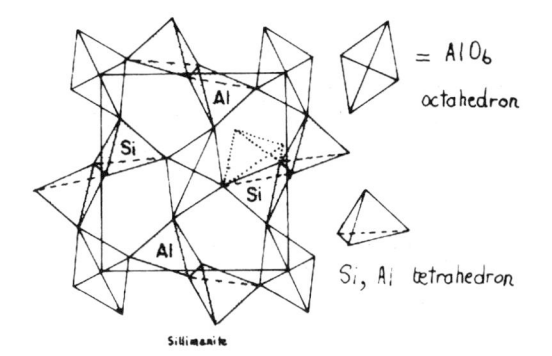

Figure 9. Schematic crystal structure of sillimanite (Al$_2$O$_3$•SiO$_2$) in (001) projection, showing chains of edge-shared AlO$_6$ octahedra and alternating chains of AlO$_4$/SiO$_4$ tetrahedra

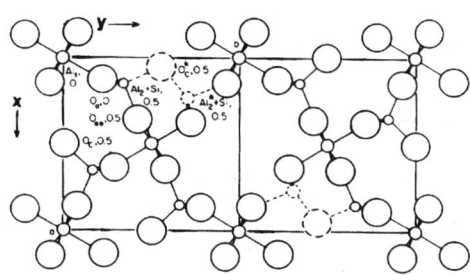

Figure 10. Crystal structure of mullite (3Al$_2$O$_3$•2SiO$_2$) in the (001) orthorhombic projection illustrating statistical removal of one corner-shared AlO$_4$/SiO$_4$ chain. After Burnham.[13,14]

will crystallize. The need for fast pulling to avoid crystallization of the drop makes the process inherently fast with pulling rates greater than 1 m/s for 20 μm diameter fibers. The production rate compares favorably with the pedestal growth[11] and edge-defined, film-fed growth[12] techniques which provide *ca.* 100 μm diameter crystalline fibers grown at much slower rates.

Fibers coated with yttrium phosphate and embedded in a mullite matrix retained their shape during the 1500°C sintering process. This result shows that these materials have good potential for use in high temperature applications.

Microstrutural analysis shows that the fibers have a smooth and homogeneous surface. The tensile strength of the amorphous 3Al$_2$O$_3$•2SiO$_2$ fibers is *ca.* 6 GPa which significantly exceeds the *ca.* 2.1 GPa strength of commercially available fine grained alumina fibers (Nextel 720®). Alumino-silicate fibers of the 3Al$_2$O$_3$•2SiO$_2$ mullite composition were crystallized at 1200°C into random orientations and sub-micron grain size. Their tensile strength was of the order of 1 GPa. This method of crystallization by heating the whole fiber at once, at the annealing temperature, yielded a microstructure which suggested that nucleation occurred at multiple sites and in an uncontrolled way. Future work is planned in which the fibers will be directionally annealed to achieve control of nucleation and growth and perhaps obtain crystallographic alignment within the fiber.

The crystal structures of mullite (3Al$_2$O$_3$•2SiO$_2$) and the high pressure sillimanite (Al$_2$O$_3$•SiO$_2$) polymorph on which it is based [13-17] are schematicaly illustrated in Figures 9 and 10. They are basically chain structures, consisting of chains of edge-shared AlO$_6$ octahedra arranged parallel to the orthorhombic [c] axis, and placed at the corners and center of each unit cell. In sillimanite, there are a further four chains of alternating AlO$_4$ and SiO$_4$ corner-shared tetrahedra which are arranged symmetrically within the unit cell. In mullite, only three chains of AlO$_4$/SiO$_4$ tetrahedra are arranged within each unit cell so as to be statistically randomized to X-ray diffraction, but give rise to the complex incommensurate modulations observed by TEM.[18-20] Thus extremely high strength and creep resistance would be expected from single crystal fibers having the [c] axis aligned parallel to the fiber length. One may speculate that even a polycrystalline, textured fiber with aligned [c] axes[21] should yield significantly improved tensile strengths, as compared to randomly crystallized polycrystalline fibers.

CONCLUSION

Oxide glass fibers with fracture strengths up to 6 GPa can reproducibly be made by pulling from undercooled molten oxide precursors. Glass fibers (e.g. mullite, YAG) can be recrystallized to form equilibrium phases corresponding to the composition of the starting material. Alumino-silicates of the 3Al$_2$O$_3$•2SiO$_2$ composition crystallize into the equilibrium 3:2 mullite form.

ACKNOWLEDGEMENT

This work was supported by a Small Business Innovation Research (SBIR) Grant number F49620-95-C-0067 through the US Air Force Office of Scientific Research, under the supervision of Dr. Alexander Pechenick.

REFERENCES

1. R.A. Lowden and M.A. Karnitz, "A Survey of the Status of Ceramic Reinforcement Technology and Its Relationship to CFCCs for Industrial Applications," Oak Ridge National Laboratory, Oak Ridge, TN (1996).
2. A.G. Evans, "Perspective on the Development of High Toughness Ceramics," J. Am. Ceram. Soc., **73** [2] 187-206 (1990).
3. R.J. Kerans, R.S. Hay, N.J. Pagano and T.A. Parthasarathy, "The Role of Fiber-Matrix Interfaces in Ceramic Composites," Am. Ceram. Soc. Bull., **68** [2] 429-442 (1993).
4. P. E. D. Morgan and D. B. Marshall, "Ceramic Composites of Monazite and Alumina," J. Am. Ceram. Soc., **78** [6] 1553-1563 (1995).
5. W.M. Kriven, "Displacive Transformations and Their Applications in Structural Ceramics," J. De Physique IV, **5**, C8, 101-110 (1995).
6. J.K.R. Weber, J.J. Felten and P.C. Nordine, "New Method for High Purity Ceramic Synthesis," Rev. Sci. Instrum., **67**, 552-24 (1996).
7. J.K.R. Weber, D.S. Hampton, D.R. Merkley, C.A. Rey, M.M. Zatarski and P.C. Nordine, "Aero-acoustic Levitation - A Method for Containerless Liquid Phase Processing at High Temperatures," Rev. Sci. Instrum., **65**, 456-465 (1994).
8. J.K.R. Weber, D.R. Merkley, C.D. Anderson, P.C. Nordine, C.S. Ray and D.E. Day, "Enhancement of Calcia-Gallia Glass Formation by Containerless Processing," J. Am. Ceram. Soc., **76**, 2139 (1993).
9. J.K.R. Weber, C.D. Anderson, S. Krishnan and P.C. Nordine, "Structure of Aluminum Oxide Formed from Undercooled Melts," J. Am. Ceram. Soc., **78**, 577-82 (1995).
10. C.M. Huang, Y. Xu, F. Xiong, A. Zangvil and W.M. Kriven, "Laser Ablation Coatings on Ceramic Fibers for Ceramic Matrix Composites," J. Mat. Sci. and Eng., **A191**, 249-56 (1995).
11. A. Sayir and L.E. Matson, "Growth and Characterization of Directionally Solidified $Al_2O_3/Y_3Al_5O_{12}$ (YAG) Eutectic Fibers," NASA Conf. Publications 10082 High Temp. Review, Oct 29th - 30th (1991), pp. 83.1-83.13.
12. J.M. Collins, H.E. Bates and J.J. Fitzgibbon, "Growth and Characterization of Single Crystal YAG Fibers," Materials Directorate, Wright Laboratory, Air Force Material Command, Wright Patterson AFB, OH Report WL-TR-94-4085, June 1994.
13. C. W. Burnham, "The Crystal Structure of Mullite," Yearb. Carnegie Inst., **62**, 158-165 (1963).
14. C. W. Burnham, "The Crystal Structure of Mulllite," Yearb. Carnegie Inst., **63**, 223-228 (1964).
15. S. Durovic, "Refinement of the Crystal Structure of Mullite," Chem. Zvesti, **23**, 113-128 (1969).
16. R. Sadanaga, M. Tokonami and Y. Tackeuchi, "The Structure of Mullite, $2Al_2O_3 \cdot SiO_2$ and Relationship with the Structures of Sillimanite and Andalusite," Acta Crystallogr. **15**, 65-68 (1962).
17. S. O. Agrell and J. V. Smith, "Cell Dimensions, Solid Solutions, Polymorphism and Identification of Mullite and Sillimanite," J. Am. Ceram. Soc., **43** [2] 69-76 (1960).
18. W. E. Cameron, "Mullite: A Substituted Alumina," Am. Mineral. **62**, 747-755 (1977).
19. W. E. Cameron, "Composition and Cell Dimensions of Mullite," Am. Ceram. Soc. Bull., **56** [11] (1977).
20. W. M. Kriven and J. A. Pask, "Solid Solution Range and Microstructures of Melt-Grown Mullite," J. Am. Ceram. Soc., **66** [9] 649-654 (1983).
21. W. M. Kriven, R. A. Gronsky and J. A. Pask, "Dislocations and Low-angle Grain Boundaries in Mullite," this volume.

THE CUBIC-TO-HEXAGONAL TRANSFORMATION TO TOUGHEN SiC

Warren J. MoberlyChan, J. J. Cao, C. J. Gilbert,
R. O. Ritchie, and L. C. De Jonghe

Materials Sciences Division
Lawrence Berkeley National Laboratory
University of California, Berkeley
Berkeley, CA 94720

ABSTRACT

Silicon carbide is a desirable high temperature structural material, however, its poor fracture toughness at room temperature has limited its practical application. Recent processing developments have toughened the microstructure with interlocking, plate-like grains and an ~1 nm thick amorphous grain boundary. Intergranular fracture around these elongated grains leads to crack deflection and elastic bridging behind an advancing crack tip, thereby giving rise to the fracture resistance. ($K_c > 9$ MPa m$^{1/2}$.) Furthermore, the increase in interlocking that develops with the high aspect ratio results in an increase in room temperature strength even as grains grow. The plate-like microstructure arises from the cubic-to-hexagonal transformation in SiC, which must be judiciously controlled in order to insure full densification, to obtain a high aspect ratio, and to optimize the microstructure of the sintering additives. Thus this work evaluates this transformation using XRD, SEM and high resolution TEM. Although a multitude of literature discusses a transformation invoked by the motion of stacking faults and partial dislocations, this study has characterized a growth-induced transformation, with the initial beta grain acting as a seed on which the alpha grows. Further growth of dual-phase grains develops asymmetric faccting on the plate-like surfaces, which, coupled with crystalline triple junction phases, complicate the intergranular crack path and increase the effect of interlocking.

INTRODUCTION

In "the art of the possible" ceramic microstructures of '86, R. J. Brook [1] separated "processing and microstructural design as two sectors" that could only be connected by judicious thoughts and actions of mankind. However, he recognized that the "quest for uniform microstructures with their associated reproducible properties must be coupled with designs of more deliberately complex microstructures." The present development of SiC with a "record toughness" [2, 3], although historically based on an attempt to design a deliberate composite microstructure [4, 5, 6], has directly coupled processing [7, 8] to a "naturally" designed complex microstructure through a critical characterization of the evolution of the SiC ceramic microstructure [9-13].

Polycrystalline SiC continues to offer a promise of superior high temperature structural properties; however, difficulty in uniform processing over the past three decades [14-17] has only exaggerated its inherent brittleness at room temperature. And even with the most uniform SiC microstructure produced from these decades (the commercially available Hexoloy SA [18]), the room temperature toughness of 2 (or 3) MPa m$^{1/2}$ [2, 7, 19] leaves SiC well behind other structural materials, including other nonoxide ceramics such as Si_3N_4 [20]. However, processing developments in the last half decade have improved the fracture toughness of research-grade SiC to a practical 10 MPa m$^{1/2}$ [2, 7, 21-25]. Control of a phase transformation and development of a plate-like microstructure during processing are responsible for toughening these SiC ceramics. These *in situ* toughened SiC incorporate 3 to >20 vol. % secondary phases, which not only enhance densification (and transformation, and microstructural development) but also provide a weak medium for crack deflection around the plate-like grains. It is this tortuous fracture around elongated grains that promotes the high toughness. Unfortunately, the majority of these tough SiC have >5% secondary phases which leads to a reduction in strength, especially at high temperature. This work characterizes the controlled processing of SiC with <5% additives to produce a complex, yet uniform, microstructure with optimal fracture toughness and even superior strength.

The development of an *in situ* toughened microstructure in SiC requires an understanding of the cubic-to-hexagonal phase transformation. SiC can have the cubic zincblende structure or many hexagonal polytypic structures based on different stacking of the basic wurtzite structure. The observations of over 200 different polytypic structures have led to nearly as many published reports of phase transformations in SiC. (For review articles on polytypic structures and transformations in SiC, see references [26-28, 10].) The commonly accepted transformation mechanism is characterized in a series of papers by Heuer, et al [29-32]. The basis of their transformation involves the motion of partial dislocations, as occurs for FCC-to-HCP in metals [33], to change the stacking of planes from ABCABC.... to a hexagonal polytype, typically the α-6H with an ABCACB.... stacking. Recent high resolution lattice imaging has documented this transformation mechanism in single crystal SiC [34]. However, when this dislocation-induced transformation is applied to polycrystalline materials, it should result in a refinement of grain (or variant) size. During the densification of ceramics, grain growth is nearly inevitable, and always the rule for the processed SiC microstructures presented in the literature. Thus it is more common that a "seed" hexagonal grain preferentially grows at the expense of nonequilibrium cubic grains, thereby having the overall transformation dominated by seeding [25, 28, 32, 10]. The difficulty in manufacturing reproducible microstructures through seed processing is exemplified by three recent reports that claimed the SiC phase transformation required >10% seed [35], <2% seed to densify [36], and no seeds [37], all for the same starting SiC powders and additives, and similar processing conditions. The present work characterizes the SiC transformation as a consequence of only grain growth, where the new planes in a growing grain take on the stacking of the thermodynamically preferred α-4H polytype [38], without a need for intentional seeding nor dislocation motion [10].

The SiC phase transformation occurring during grain growth can be controlled to develop a nonequiaxed microstructure. Anisotropic growth of the hexagonal structure can be enhanced at lower temperatures to develop either whiskers (especially for the α-2H polytype [39]) or plates. Historically, the growth of plate-like grains has been a bane of mechanical properties in SiC. The impingement of plate-like grains would prevent further densification, and the nonuniform distribution of large plates provided critical flaws that lowered the toughness and strength of SiC [28]. Thus Hexoloy SA, with its uniform, equiaxed α-6H microstructure, minimal (<5%) secondary phases, and minimal (~2-5%) porosity [7] has represented a technological achievement [19], even though its fracture toughness is poor [2]. In the present work, the use of hot pressing and judicious additives enables full densification to be incurred prior to the majority of the phase transformation. Controlling the subsequent transformation in a fully dense medium enables plate-like grains to develop, which produces a uniformly interlocked microstructure with improved strength as well as toughness.

EXPERIMENTAL PROCEDURES

Two primary requirements for processing a good SiC ceramic are uniform distribution of microstructure and full densification. Two additional desires are a low volume fraction of (usually mechanically deleterious) secondary phases resulting from additives and a possible microstructural development which actually enhances the intrinsic mechanical properties (strength and toughness) of the ceramic. However, the former two essential requirements often promote undesirable results for the later two. As the densification of SiC has been improved in the past decades, the typical microstructures exhibit a substantial growth in grain size [16] and/or a major volume fraction of secondary phases [21-23, 35, 37], both of which lower the strength at ambient and high temperature.

The present ABC-SiC is named after its minimal processing additives (3% Al, 0.6% B, and 2% C). The starting submicron (average 0.2 microns) β-SiC powders [40] and additives are ultrasonically mixed in toluene, wet and dry sieved, and hot-pressed at 50 MPa in a graphite die and furnace. Hot pressing temperatures range from 1600°C to 1950°C with durations of 15 min. to 4 hr. Further details of the ceramic processing are provided in references [7, 8].

The characterization of these ABC-SiC ceramics includes X-Ray Diffraction (XRD); Scanning Electron Microscopy (SEM) of polished surfaces, fracture surfaces, and surfaces etched in molten salt; Transmission Electron Microscopy (TEM); high resolution lattice imaging (HR-TEM); Energy Dispersive X-ray Spectroscopy (EDS); Electron Energy Loss (EELS) Filtered Imaging; and Auger Electron Spectroscopy (AES) of polished and fractured surfaces. Mechanical testing consists of 4-pt. bend strength tests [7] and toughness measurements on compact tension samples [2]. Additional details of the characterization and the mechanical testing of related ABC-SiC microstructures are presented in references [2-13].

Research of the phase transformation in SiC is commonly presented in terms of XRD data; however, the complexity of the SiC microstructure can mislead the interpretation of polytypic phases. A primary concern is the shapes of the grains, as this can lead to a nonrandom distribution of orientations in a powder sample. A typical method of sampling for XRD is to grind the densified material into powder and then press into a pellet for the XRD analysis. However, these plate-like SiC microstructures intergranularly fracture, resulting in plate-like powders. Subsequent pressing of these powders rotates them such that an artificially high fraction have their basal planes parallel to the top of the pressed XRD pellet. This preferred orientation provides a relatively high intensity for diffraction from the basal planes, which has the same d-spacing in all hexagonal and cubic polytypes. Resulting XRD data may be misinterpreted as having a high volume fraction of cubic phase, when none may exist. For this research, where processing produces no preferential orientations of microstructure, it has been determined that a polished surface of polycrystalline SiC actually provides a more representative XRD pattern. Since plate-like surfaces and internal defects are always parallel to the basal plane in these ABC-SiC ceramics, however, the intensities of XRD peaks other than that of the basal plane are diminished. Furthermore, the presence of mixed polytypic phases within each grain (to be discussed in the analysis of the microstructure) also produces destructive interference effects, which again reduce the intensities of nonbasal diffraction peaks. A more in-depth discussion of the limitations of XRD analysis for polytypic structures is presented in reference [10]. Suffice to summarize that XRD analysis alone is insufficient without a microstructural assessment of internal defect structure, especially as mechanical properties correlate more with grain shapes than with crystal structures [7, 10].

EXPERIMENTAL RESULTS

Although quantitative XRD has limitations when applied to the complex microstructures (plate-like shapes, high density of stacking faults, multiple-phase grains, anisotropic distribution of faults and phases, fine grain size, etc.) of SiC, it may still provide a statistically qualitative assessment of the phase transformation as a function of processing conditions. Individual XRD datum are presented in earlier publications [7, 8, 10], with Figure 1 plotting the volume fraction of SiC transformed as a function of processing temperature. All points on the line represent ceramics hot pressed for 1 hour with 3% Al additive. However, the transformation is also a function of time and concentration of additives. Whereas, 1 hr at 1900°C and 3% Al produces ~75% transformation, 4 hr. provides complete transformation and 1% Al causes negligible transformation to occur at the same temperature. The limits of XRD analysis are further revealed by TEM observations of small regions of hexagonal phase after processing with 1% Al and residual β-3C portions which persist in all grains after processing for 4 hr. Yet the XRD analyses are sufficiently reproducible to enable its utility for monitoring processing.

SEM imaging of the grain growth and of the development of a plate-like microstructure in ABC-SiC has been previously presented [7, 10]. The original β-cubic powders are submicron and equiaxed, whereas the α-hexagonal grains after hot pressing at 1900°C are >10 microns in diameter and have aspect ratios upwards of 10. However, the SEM images do not establish the phase(s) present nor elucidate the transformation mechanism(s). Furthermore, the molten-salt etching technique for preparation of SEM sample surfaces preferentially removes smaller grains and secondary phases, thereby deceptively enhancing the appearance of large, flat plate-like surfaces.

TEM imaging measures a statistically smaller portion of material, yet it provides an easier means (than SEM and/or XRD) by which to establish the transformation mechanism, as well as to understand how the grain shapes develop and influence mechanical properties. Figure 2 is a Bright Field (BF) TEM image of the microstructure after hot pressing at 1700°C for 1 hr. Although electron diffraction corroborates the XRD analysis of ~100% beta phase still being present, the grain growth during densification results in the larger grains developing parallel, planar boundaries and an aspect ratio (<2) akin to plates. BF imaging exhibits the internal defect structure of the SiC grains, which are confirmed by electron diffraction and HR-TEM to be twins and stacking faults. A cubic structure allows for a multiplicity of 4 sets of {111} planes; however, each grain develops twins and stacking faults on only one set of parallel (111) planes. The anisotropic nature of these faults brings about anisotropic growth of the cubic grains, with growth being slower perpendicular to the planar surfaces.

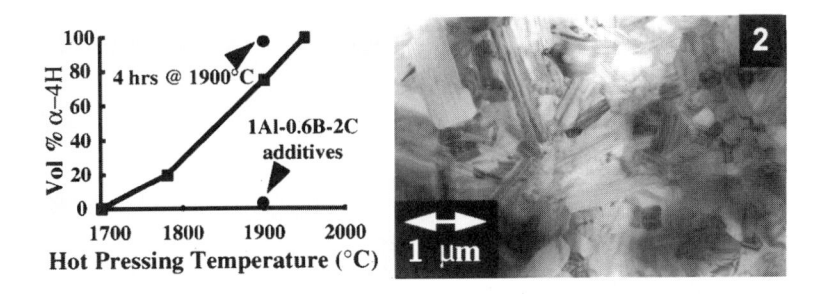

Figure 1. Plot of transformation as a function of temperature, based on XRD. Data on line represents 1 hr. at temperature and additives concentrations of 3 at% Al, 0.6% B, and 2% C. Transformation is also a function of time and concentration, with less additives causing no transformation and 4 hr. causing full transformation, both at 1900°C. **Figure 2**. BF-TEM image of β-SiC hot pressed 1 hr. at 1700°C. As grains grow, they elongate parallel to the stacking faults and microtwins, which occur on only one {111} plane in each grain.

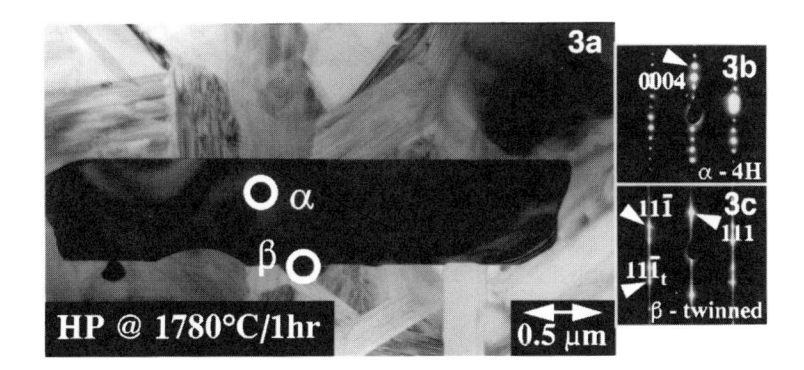

Figure 3. (a) BF-TEM image of ABC-SiC hot pressed at 1780°C. As grains grow, new planes take on the stacking of the thermodynamically favored α-4H phase. Anisotropic growth of the dual-phase grain develops a plate-like shape. (b) and (c) Electron diffraction patterns from the α and β regions of the grain.

Figure 3a is a BF-TEM image of the SiC microstructure after hot pressing 1 hr. at 1780°C. Although XRD implies minimal phase transformation has occurred, TEM depicts large elongated grains surrounded by smaller, more-equiaxed grains. Selected Area Electron Diffraction (SAD) patterns in Figures 3b and 3c, from the "top" half and the "bottom" half of the plate-like grains, respectively exhibit the α-4H hexagonal and β-3C cubic structures. The surrounding smaller grains are identified by electron diffraction as primarily β-3C phase. Electron diffraction and HR-TEM of the middle of the larger plate-like grains often depict a region of mixed phases and higher density of faults. Stacking faults are common throughout all layers of the plate-like grain, although the density of such faults is less in regions of grains grown at higher temperatures [10]. (The microtwins exist only in the lower beta portions of the grains, as mirror twinning on the (0001) basal plane is not possible in the hexagonal polytypic structures.) As well as elucidating internal defect structure, BF-TEM depicts subtle differences in the shapes of the top and bottom of grains, which reveal aspects of the transformation mechanism, as well as the toughening mechanism. The tops of plate-like grains have atomically flat (0001) planes terminating them, whereas the bottoms are rough and terminated by $\{\bar{1}11\}$ facets of the β-3C phase. The differences between the (0001) and (000$\bar{1}$) surfaces of plate-like grains remain evident after processing above 1900°C.

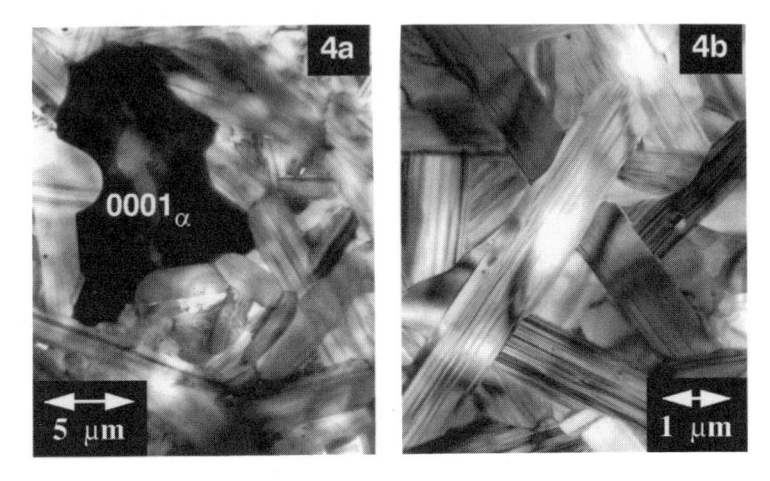

Figure 4. BF-TEM images: (a) Acquired on [0001] zone-axis, depicting irregular diameter of plate-like grain in ABC-SiC. (b) Cross-sectioned image shows grains grow in an interlocked fashion with the same stacking faults growing on opposite sides of an interlock.

The large grain in Figure 3a represents one of the first to grow and transform; however, all grains eventually grow (via Ostwald ripening) into plate-like, dual-phase grains. Since the microstructure has been densified prior to this growth/transformation, the grains cannot develop uniformly round diameters. Radial elongation of a grain is interrupted by other elongating grains, as is depicted in the BF image of Figure 4a, acquired at the [0001] normal to a plate-like grain. The development of a completely dense and plate-like microstructure requires the growth of interlocking plates, as depicted in the cross sectioned grains of Figure 4b. The two grains in Figure 4b have been slightly tilted from $\{\bar{2}110\}$ orientations, to enable the imaging of internal faults. Although one plate-like grain is interrupted by a second, the internal faults appear continuous across the second grain (see arrows). This establishes that the first grain grew around the second [10]. Once a fault nucleates, elongated growth extends it to the radial limits of a plate-like grain. The lack of different stacking faults on each plane, which would produce vertical faults such as Double Position Boundaries within the grain, establishes the difficulty in nucleating each new basal plane and the rationale for growing elongated plate-like grains.

Secondary phases do persist even though the processing of ABC-SiC included minimal additives. Again their structure(s) and location(s) within the overall microstructure reveal their influence on both transformation and toughening mechanisms. Occasional large (>1 micron) pockets of secondary phases persist; however methodologies for homogenization prior to densification have reduced these pockets to a level of minor influence. Prior characterization [4-9, 11] has determined two ternary phases, $Al_8B_4C_7$ and Al_4O_4C, form as a reaction between additives and native oxide on the surfaces of the original SiC powders. Of more interest is the portion of additive (<1%), which as a liquid phase wets amongst the initial SiC powders and persists along grain boundaries and small isolated triple junction channels between SiC grains after hot pressing. Figure 5 is an HR-TEM image sectioning through a triple junction channel, where the additive has crystallized to form the Al_4O_4C phase. Such triple junction channels, although only a few nanometers in width, extend many microns along the flat surface of a plate-like grain. Liquid phase additives are often difficult to crystallize in nonoxide ceramics and have been reported to require special nucleating agents and special crystallization processing steps. Even when they nucleate, they often grow as a single large grain amongst hundreds of

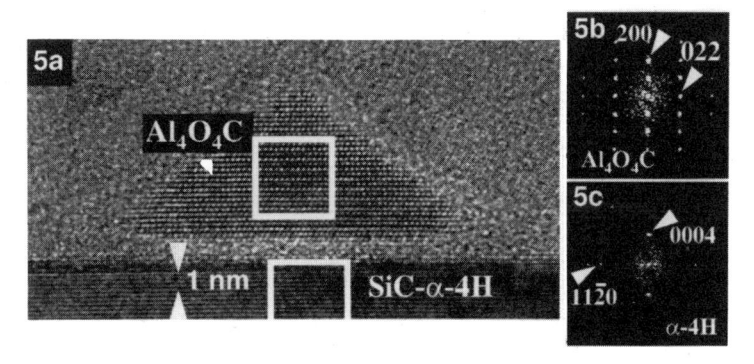

Figure 5. HR-TEM image of crystalline triple junction. Computer-generated diffractograms determine the crystal structure (in this case Al_4O_4C), and an epitaxial orientation relationship with the matrix SiC grain.

matrix grains [41]. No additional processing is necessary to crystallize the additives in ABC-SiC. The volume fraction of additives in ABC-SiC is sufficiently reduced to isolate the triple junction channels; however, their limited nucleation results in crystalline grains extending many microns along a triple junction channel. When sectioning this crystalline triple junction for HR-TEM, it is statistically rare to section through the point of nucleation. However, the epitaxial orientation relationship between matrix grain and crystallizing secondary phase is evident in Figure 5. The thin amorphous layer between the lower matrix grain and secondary phase is believed to arise due to the presence of steps on the surface of the matrix grain boundary. When the crystallizing secondary phase grows past such a step, it offers the opportunity to detach from the matrix grain and continue growing without having to accommodate an interfacial mismatch stress. The amorphous heterophase boundary layer has been reported to be thicker than the amorphous homophase boundary between matrix grains [41], and similar observations have been made for the layers above this crystalline triple junction channel in ABC-SiC [42]. However, the amorphous layer residing between the epitaxial secondary phase and the lower matrix grain is typically thinner than the ~1 nm thick amorphous layer between two SiC grains [9, 10, 12, 42]. Although EELS imaging establishes a uniform chemistry for all amorphous layers between SiC grains [12], the different chemistries at heterophase boundaries have not yet been quantified for ABC-SiC nor other ceramics [41]. A chemical difference may not only influence thickness, but also cause cracks to deflect "up and over" the crystalline triple junctions rather than the apparently easier path beneath the triple junctions [2, 7, 8, 12]. An earlier comparison between a 1 nm thick amorphous layer in ABC-SiC and a 2-5 nm layer in Hexoloy SA has shown thickness of amorphous layer does not correlate to the ease of intergranular fracture [9].

Strength, based on 4-pt bend tests of polished beams, and toughness, based on R-curve tests on compact tension specimens, have been acquired with testing methods described in earlier reports [2, 7, 8]. Figures 6a and 6b plot the change in microstructure and change in mechanical properties as a function of hot pressing temperature (all for 1 hr.). Whereas the SiC grain size continues to increase with temperature, the aspect ratio and properties increase and then decay when a sufficiently high temperature causes "oversintering" [43]. It is interesting to note that an increase in grain size from <0.2 µm to >10 µm is associated with a doubling of the strength. Yet it is more apparent that both strength and fracture toughness track the aspect ratio, as the increasing aspect ratio causes an increase in interlocking of grains. The development of a continuous network of interlocks not only enhances strength, but also improves the Weibull Modulus [8].

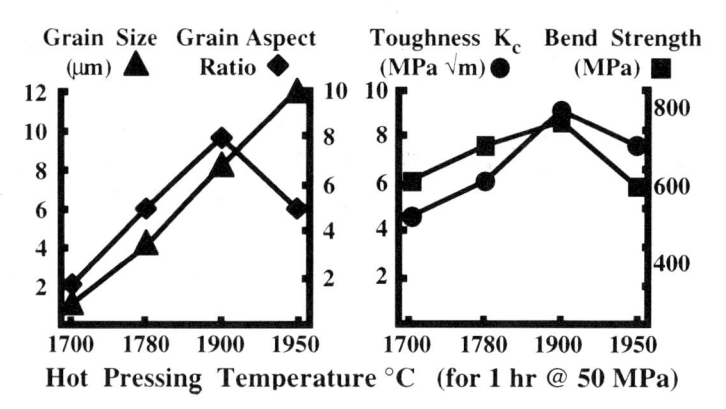

Figure 6. (a) Plot of grains size and aspect ratio as a function of processing temperature. (b) Plot of strength and toughness as a function of processing temperature. Mechanical properties track aspect ratio, with strength increasing even as grains grow >50-fold, due to increased interlocking.

DISCUSSION

The SiC Phase Transformation

The processing of ABC-SiC satisfies the two essential criteria of a structural ceramic: uniform microstructure and full densification. In this ceramic, the Al additive enables liquid-phase sintering that, assisted by minimal pressure, enhances densification. Coble [44] attributed the lack of shrinkage during sintering of SiC to the dominance of "surface diffusion and/or evaporation/condensation over lattice or grain boundary diffusion." Although the early stages of densification may incur some "evaporation", it is apparent that liquid phase sintering dominates at most temperatures, and especially during the cubic-to-hexagonal transformation. The Al, enhanced by the B and C additives, reacts with the native oxide on SiC powders, thereby allowing grain boundary (or liquid phase) diffusion to occur at temperatures well below the typically >2000°C used for sintering SiC.

Since diffusion of SiC through the liquid phase is easier than bulk diffusion at the processing temperatures incurred in ABC-SiC, this liquid phase provides an easier mechanism for the transformation than exists in a single crystal [34]. In order for partial dislocations to induce full transformation, there must exist either an (internal grain) source of dislocations on each basal plane and/or a cross-slipping mechanism [34]. Since such cross-slipping is thermally activated, it is expected to be active only at a temperature where bulk diffusion is prominent. However, it must be remembered that stacking faults are easily formed during the growth of SiC. Thus the metastable β-3C grain can act as a seed for the growth of α-4H. Once a stacking fault has been nucleated onto the β-3C grain, in essence the α-4H phase has begun to grow. The thermodynamically favored α-4H regions will then Ostwald ripen at the expense of metastable, smaller β-3C grains. The anisotropic nature of the α-4H phase leads to enhanced elongation of plate-like shapes. However, in this elongation of basal planes, the β-3C planes at the bottom of the plate are treated as stacking faults and become elongated too, except for regions where their expansion would necessitate the consumption of other impinging α-4H grains. Thus the continued elongation of β-3C planes is typically discontinuous (see Figure 8a), and the bottom of these dual-phase grains remains irregular and faceted during growth.

The sintering, phase transformation, and development of a mechanically improved microstructure in SiC can be summarized in Figure 7, with all dependent on growth via diffusion through the liquid phase at the grain boundaries. The densification (Figure 7b) need not precede the beginning of the transformation (Figure 7c); and the decrease in aspect ratio at higher temperatures implies that interlocked, plate-like grains (Figure 7d) are not the most equilibrium microstructure. However, the transformation to an elongated microstructure prior to densification inhibits subsequent densification [28]; and an equiaxed microstructure [7, 9, 19] does not provide toughness nor strength comparable to that of ABC-SiC. The hot pressing enables use of minimal additives, enhances densification in the early stages of grain growth, and allows lower temperature processing for controlled microstructural development.

Microstructural Toughening Mechanisms

Two aspects of the microstructure developed in ABC-SiC become mutual requirements for its high toughness. The microstructure has an elongated morphology and

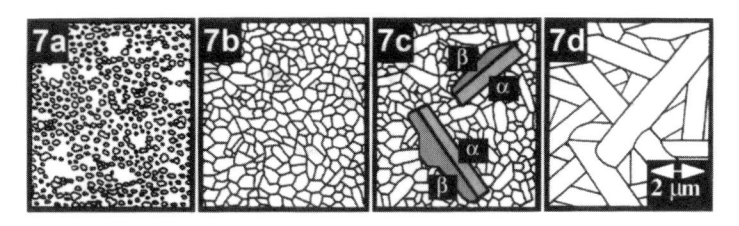

Figure 7. Schematics of sintering / growth / transformation of SiC. (**a**) Initial powders. (**b**) Densification due to liquid-phase sintering. (**c**) During growth, the stacking converts to α-4H, creating dual-phase grains. Continued growth is anisotropic. (**d**) Plate-like grains in dense medium form an interlocking microstructure.

a chemically weak interface at grain boundaries, both of which are intentional elements designed into toughened composite structures. Other than these two toughening features, the "uniform" microstructure, with its full density and minimal deleterious phases, provide higher strength and Weibull Modulus than commercial polycrystalline SiC [8]. Commercial Hexoloy SA has equiaxed grains of similar size to ABC-SiC and a transgranular fracture mode, but these two features couple to provide poor toughness. Processing of ABC-SiC at temperatures below 1700°C produces a relatively equiaxed microstructure with intergranular fracture between the submicron β-3C grains [4, 6]. However, the intergranular fracture alone produces negligible toughening and actually diminishes strength [6, 7]. Similarly, SiC ceramics with elongated microstructures but a transgranular fracture mode exhibit poor toughness [28, 45]. In order to assess the fracture mechanics, the intergranular fracture in the amorphous layer and the influence of aspect ratio must be characterized.

Assessment of the mechanical influence of a ~1 nm thick amorphous layer at grain boundaries is strongly dependent on the ability to analyze (or even detect) it. Reports on SiC, as well as other nonoxide ceramics, provide numerous discussions of (or lack thereof) amorphous grain boundary layers. A commonly accepted technique for analysis of such grain boundary layers is HR-TEM, where the disordered nature of the amorphous layer can be viewed as distinct from the atomic periodicity in the crystalline grains. However, the application of HR-TEM is not full-proof. First, a micrograph exhibiting an amorphous layer requires ideal imaging capabilities of the microscope and a "special" (but not too special) grain boundary orientation in the sample. Grain boundaries between two ceramic grains with easily-imaged, low-index-zone-axis orientations represent boundaries so special that it can become energetically favorable for them not to have an amorphous layer [41, 46]. However, their analysis can be statistically irrelevant in a polycrystalline matrix. On the other hand, imaging of boundaries between two high-index-zone-axis orientations may only provide sufficient lattice fringe resolution to observe the higher order periodicity of the polytype. This leaves the resolution limit for detecting a grain boundary layer no better than the repeat unit of the polytype; being >1 nm for α-4H and >1.5 nm for α-6H [15]. Of course, the dominant difficulty in HR-TEM is having a sufficiently large area of grain boundary being atomically flat and oriented parallel to the electron beam. In ABC-SiC, the majority, as well as flattest and most mechanically meaningful, of grain boundaries are those which comprise the basal facet of one plate-like grain (as in the lower SiC grain of Figure 5) and a random facet of the second (upper) grain. Secondly, the appearance of an amorphous layer in an ideally oriented grain boundary does not preclude its formation due to an artifact of sample preparation or having unresolvable imaging conditions for a "crystalline" grain boundary layer. For example, etching to thin a sample for TEM analysis often results in preferential etching at grain boundaries. Since the SiC samples are prepared by ion etching, the grain boundaries may be susceptible to being amorphized at a rate faster than for the grains on either side of the grain boundary. Likewise, if a crystalline grain boundary layer does not have an exact lattice matching to one of the grains on either side of the grain boundary, the optimal imaging conditions may differ in the microscope. Thus any difference in thickness, chemistry, periodicity, and/or orientation associated with a crystalline grain boundary layer makes its imaging differ from the SiC grains. The appearance of an amorphous layer in a HR-TEM image becomes "more" conclusive when additional techniques are used to chemically analyze the layer. In ABC-SiC ceramics EDS and EELS imaging establish an enrichment of Al and O in the amorphous boundary layers [10, 12, 42]. In addition, AES of fracture surfaces determine the Al-O layer is ~1 nm thick, and has a minor concentration of S [9] segregated to the interfaces. Basically, the observation of an amorphous layer by HR-TEM, especially with corroboration by another technique(s), is fairly conclusive of the presence of the layer; but the lack of observation is only inconclusive evidence.

Figure 8. (a) BF-TEM image of cross-sectioned plate-like grain in ABC-SiC hot pressed at 1900°C. The plate-like surfaces of grains are not flat; irregular β-{$\bar{1}$ 11} facets are on bottoms and asperities of crystalline triple junction phases on tops (see Figure 5). (b) Schematic of fracture path, which is confined to the ~1 nm thick amorphous phase surrounding the plate-like grain. The asperities resist sliding and hinder pullout.

The tracking of toughness with aspect ratio (Figure 6) suggests crack bridging as the toughening mechanism. Furthermore, energetic calculations [2] indicate the dramatic toughening evident in an R-curve test of ABC-SiC (~4 times the intrinsic toughness of SiC) surpasses the toughening attributed to only elastic bridges behind a crack tip. However, the fracture mechanics of bridging and pullout commonly consider simple geometries such as symmetric fibers and straight crack fronts. The complexity of intergranular fracture around interlocked plate-like shapes produces a tortuous crack path, and applied math [47, 48] for even simple perturbations from a planar crack front predicts toughening. The interlocked microstructure of plate-like grains in ABC-SiC does not allow for a simple pullout such as that predicted for a fiber. Furthermore, the surfaces of the plate-like grains are not truly flat, having irregular {$\bar{1}$ 11} facets on the β-3C surface and even small crystalline triple junction phases attached to the top α-4H surface. Figure 8a is a BF image of a cross-sectioned, plate-like grain with diffraction contrast indicating the asperities on both sides. SEM imaging of fracture surfaces [2, 8, 10, 12] show the fracture path is above these crystalline triple junctions and not straight through the amorphous phase beneath them (Figure 5). Intergranular fracture is confined to the ~1 nm thick amorphous layer which follows a tortuous path around the asperities (Figure 8b). The asperities hinder sliding of the plate and frictional stress is increased. However, frictional toughening requires energy dissipation via the actual sliding of the plate, which the locked microstructure hinders. No wear of the crystalline asperities on a plate-like surface is observed for monotonic fracture. Although the crack propagates past the elastic bridge, it becomes difficult for it to open more than the thickness of the amorphous layer.

CONCLUSIONS

Improvement in strength and toughness of polycrystalline SiC is attained through control of microstructural development. Characterization of the cubic-to-hexagonal phase transformation and of the role of secondary phases is critical to understand both how the microstructure is reproducibly attained and how the microstructure enhances the mechanical properties. The phase transformation is determined to be growth induced, without needing intentional seeding nor dislocation motion. The additives provide a liquid phase during processing, and hot pressing enables densification prior to transformation. With both densification and transformation occurring at relatively low temperatures, the transformation is controlled to produce a uniform, interlocked, dual-phase, plate-like microstructure with a strength higher than that attained for a comparable equiaxed microstructure. After processing the additives persist as a ~1 nm thick amorphous grain boundary phase that invokes intergranular fracture. However, the tortuous crack path around the plate-like grains, as well as around crystalline triple junctions and irregular β-$\{\bar{1}\,11\}$ facets on the bottom of the plate-like grain, enhances toughening due to bridging of the interlocked grains behind the crack tip.

ACKNOWLEDGMENTS

The authors wish to acknowledge M. Gopal, Y. He, M. Sixta, and Professor G. Thomas for helpful discussions. The staff members of the National Center for Electron Microscopy at LBL are thanked for the utilization of microscopes and accessory equipment. This work was funded by the Director, Office of Energy Research, Office of Basic Energy Sciences, Materials Science Division, with the U. S. Dept. of Energy, under Contract # DE-AC03-76SF00098.

REFERENCES

1. R. J. Brook, "Ceramic Microstructures: The Art of the Possible," in Ceramic Microstructures '86, ed. J. A. Pask and A. G. Evans, 15-24 (1986).
2. C. J. Gilbert, J. J. Cao, W. J. MoberlyChan, L. C. De Jonghe, and R. O. Ritchie, "Cyclic Fatigue and Resistance-Curve Behavior of an *In Situ* Toughened Silicon Carbide with Al-B-C Additions, " *Acta Metall. Mater.*, **44** [8] 3199-3210 (1996).
3. R. O. Ritchie, C. J. Gilbert, J. J. Cao, W. J. MoberlyChan, J. M. McNaney, and L. C. De Jonghe, "On the Mechanisms of Subcritical Crack Growth under Monotonic and Cyclic Loads in a Toughened Silicon Carbide at Ambient and Elevated Temperatures," in Ceramic Microstructures '96, ed. A. P. Tomsia, (in these proceedings).
4. T. D. Mitchell, L. C. DeJonghe, W. J. MoberlyChan, & R. O. Ritchie, "Silicon Carbide Platelet / Silicon Carbide Composites", *J. Am. Ceram. Soc.*, **78** [1], 97-103 (1995).
5. W. J. MoberlyChan, J. J. Cao, M. Y. Niu, and L. C. De Jonghe , "SiC Composites with Alumina-Coated α-SiC Platelets in β-SiC Matrix", Microbeam Analysis (EMSA 1994), edited by J. Friel, VCH Publishers, NY, 49-50 (1994).
6. W. J. MoberlyChan, J. J. Cao, M. Y. Niu, L. C. De Jonghe, and A. F. Schwartzman, "Toughened β-SiC Composites with Alumina-Coated α-SiC Platelets", in High Performance Composites, edited by K. K. Chawla, TMS, 219-229 (1994).
7. J. J. Cao, W. J. MoberlyChan, L. C. De Jonghe, C. J. Gilbert, and R. O. Ritchie, "*In Situ* Toughened Silicon Carbide with Al-B-C Additions," *J. Amer. Ceramic Soc.*, **79** [2] 461-469 (1996).
8. J. J. Cao, "*In Situ* Toughened SiC Ceramics with Al-B-C Additions and Oxide-Coated SiC Platelet / SiC Composites," Ph. D. Thesis, University of California, Berkeley (1996).
9. W. J. MoberlyChan, J. J. Cao, and L. C. De Jonghe, "On the Role of Amorphous Grain Boundaries and the β - α Transformation to Toughen SiC," (in submission to *Acta Metall. Mater.*).
10. W. J. MoberlyChan, J. J. Cao, and L. C. De Jonghe, "The Cubic-to-Hexagonal Phase Transformation and Microstructural Development in Polycrystalline Silicon Carbide," (in submission to *J. Amer. Ceramic Soc.*)
11. W. J. MoberlyChan, J. J. Cao, & L.C. De Jonghe , "The β-3C to α-4H Transformation in SiC with Al, B, and C Additions: Kinetics Dominated by Growth Akin to Ostwald Ripening", MSA Proc., (G. W. Bailey), p.658 (1996).
12. W. J. MoberlyChan, J. J. Cao, & L.C. De Jonghe , "On the Crystallography of Triple Points in Nonoxide Ceramics: Epitaxial Relationships are a Function of Statistics and Geometry", MSA Proc. (G. W. Bailey), p.648 (1996).

13. W. J. MoberlyChan, J. J. Cao, & L.C. De Jonghe, "Controlling Microstructural Evolution to *in situ* Toughen and Strengthen Silicon Carbide", <u>Microscopy Proceedings, (MSA)</u> (edited by G. W. Bailey), p.692 (1996).
14. K. K. Kappmeyer, "The Importance of Microstructural Considerations in the Performance of Steel Plant Refractories," in <u>Ceramic Microstructures '66</u>, ed. R. M. Fulrath and J. A. Pask, 522-558 (1966).
15. L. U. Ogbuji, "Grain Boundaries in Carbon- and Boron-Densified SiC: Examination by High Resolution Transmission Electron Microscopy," in <u>Ceramic Microstructures '66</u>, ed. J. A. Pask and A. G. Evans, 713-723 (1976).
16. C. Greskovich and S. Prochazka, "Selected Sintering Conditions for SiC and Si_3N_4 Ceramics," in <u>Ceramic Microstructures '86</u>, ed. J. A. Pask and A. G. Evans, 601-610 (1986).
17. K. Maeda, Y. Takeda, A. Soeta, K. Usami, and S. Shinozaki, "Electrical Barriers at Grain Boundaries in Silicon carbide materials with BeO Addition," in <u>Ceramic Microstructures '86</u>, ed. J. A. Pask and A. G. Evans, 757-765 (1986).
18. Hexoloy SA is commercially available from Carborundum, Inc., Niagara Falls, NY.
19. K. Y. Chia and S. K. Lau, "High Toughness Silicon Carbide," *Ceram. Eng. Sci. Proc.* **12** [9-10] 1845-1861 (1991).
20. W. Braue, G. Wötting, and G. Ziegler, "The Impact of Compositional Variations and Processing Conditions on Secondary Phase Characteristics in Sintered Silicon Nitride Materials," in <u>Ceramic Microstructures '86</u>, ed. J. A. Pask and A. G. Evans, 883-896 (1986).
21. D. H. Kim and C. H. Kim, "Toughening Behavior of Silicon Carbide with Additions of Yttria and Alumina," *J. Am. Ceram. Soc.*, **73** [5] 1431-1434 (1990).
22. M. A. Mulla and V. D. Krstic, "Mechanical Properties of β–SiC Pressureless Sintered with Al_2O_3 Additions," *Acta Metall. Mater.*, **42** [1] 303-308 (1994).
23. N. P. Padture and B. R. Lawn, "Toughness Properties of a Silicon Carbide with an *in situ* Induced Heterogeneous Grain Structure," *J. Amer. Ceramic Soc.*, **77** [10] 2518-2522 (1994).
24. S. S. Shinozaki, "Unique Microstructural Development in SiC Materials with High Fracture Toughness," *Matls. Res. Soc. Bull.*, Feb. 42-45 (1995).
25. V. D. Krstic, "Optimization of Mechanical Properties in SiC by Control of the Microstructure," *Matls. Res. Soc. Bull.*, Feb. 46-48 (1995).
26. A. R. Verma and P. Krishna, in <u>Polymorphism and Polytypism in Crystals</u>, J. Wiley & Sons, 113 (1966).
27. D. Pandey and P. Krishna, "Mechanism of Solid State Transformations in Silicon Carbide, in <u>Silicon Carbide</u>, ed. R. C. Marshall, J. W. Faust, and C. E. Ryan, U. South Carolina Press, 198-206 (1973).
28. N. W. Jepps and T. F. Page "Polytypic Transformations in Silicon Carbide," in <u>Prog. Crystal Growth and Characterization</u>, ed. P. Krishna, Pergamon Press, **V7**, 259-307 (1983).
29. A. H. Heuer, G. A. Fryburg, L. U. Ogbuji, and T. E. Mitchell, "The β to α Transformation in Polycrystalline SiC: I, Microstructural Aspects," *J. Am. Ceram. Soc.*, **61** [9] 406-412 (1978).
30. T. E. Mitchell, L. U. Ogbuji, and A. H. Heuer, "The β to α Transformation in Polycrystalline SiC: II, Interfacial Energies," *J. Am. Ceram. Soc.*, **61** [9] 412-413 (1978).
31. L. U. Ogbuji, T. E. Mitchell, and A. H. Heuer, "The β to α Transformation in Polycrystalline SiC: III, The Thickening of a Plates," *J. Am. Ceram. Soc.*, **64** [2] 91-99 (1981).
32. L. U. Ogbuji, T. E. Mitchell, A. H. Heuer, and S. Shinozaki, "The β to α Transformation in Polycrystalline SiC: IV, A Comparison of Conventionally Sintered, Hot-Pressed, Reaction-Sintered, and Chemically Vapor-Deposited Samples," *J. Am. Ceram. Soc.*, **64** [2] 91-99 (1981).
33. C. R. Houska, B. L. Averback, and M. Cohen, "The Cobalt Transformation," *Acta Metall.*, **8**, 81-87 (1960).
34. P. Pirouz and J. W. Yang, "Polytypic Transformations in SiC: the Role of TEM," *Ultramicroscopy*, **51**, 189-214 (1993).
35. E. J. Winn and W. J. Clegg, "The Processing of In-Situ Toughened Ceramics," oral presentation at Amer. Ceramic Soc., New Orleans Nov., (1995).
36. M. J. Hoffman, "In-Situ Toughening of Non Oxide Ceramics," oral presentation at Amer. Ceramic Soc., New Orleans Nov., (1995).
37. S. H. Robinson, D. J. Shanefield, and D. E. Niesz, "Microstructural Control of Silicon Carbide Via Liquid Phase Sintering," oral presentation at Amer. Ceramic Soc., New Orleans Nov., (1995).

38. C. Cheng, V. Heine, and I. L. Jones, "Silicon Carbide Polytypes as Equilibrium Structures," *J. Phys. Condensed Matter*, , **2** 5097-5113 (1990).

39. G. A. Bootsma, W. F. Knippenberg, and G. Verspui, "Phase Transformations, Habit Changes and Crystal Growth in SiC," *J. Crystal Growth*, **8**, 341-353 (1971).

40. Starting β-SiC powders are BSC-21 from Ferro, Inc., Cleveland, OH.

41. J. S. Ventrano, H. -J. Kleebe, E. Hampp, M. J. Hoffmann, M. Rühle, R. M. Cannon, "Yb_2O_3 - fluxed sintered silicon nitride: Microstructural characterization," *J. Matls. Sci.*, **28**, 3529-3538 (1993).

42. R. M. Cannon and W. J. MoberlyChan, "Crystalline and Amorphous Triple Points in SiC," work in progress.

43. W. J. MoberlyChan, J. J. Cao, and L. C. De Jonghe, "Oversintering of SiC and SiC/Al_2O_3-Coated SiC-Platelet Composites", (to be submitted to *Comm. Amer. Ceramic Soc.*)

44. R. L. Coble, "Development of Microstructure in Ceramic Systems," in Ceramic Microstructures '66, ed. R. M. Fulrath and J. A. Pask, 658-680 (1966).

45. S. Dutta, "Sinterability, Strength and Oxidation of Alpha Silicon Carbide Powders," *J. Matls. Sci.*, **19**, 1307-1313 (1984).

46. H. Ichinose, Y. Inomata, and Y. Ishida, "HRTEM Analysis of Ordered Grain Boundaries in High Purity Alpha-SiC," in Ceramic Microstructures '86, ed. J. A. Pask and A. G. Evans, 255-262 (1986).

47. H. Gao and J. R. Rice, *J. Appl. Mech.*, **56**, 828 (1990).

48. A. F. Bower and M. Ortiz, "A Three-Dimensional Analysis of Crack Trapping and Bridging by Tough Particles," *J. Mech. Phys. Solids*, **39** [6] 815-818 (1991).

49. C. J. Gilbert and R. O. Ritchie, "On the Quantification of Bridging Tractions During Subcritical Crack Growth Under Monotonic and Cyclic Fatigue Loading in a Grain-Bridging Silicon Carbide Ceramic", (submitted to *Acta mater.*)

MORPHOLOGY AND MICROSTRUCTURE OF AlN SINGLE CRYSTALS ON Si (111): A COMBINATION OF SURFACE ELECTRON SPECTROSCOPIES AND TRANSMISSION ELECTRON MICROSCOPIES

F. Malengreau
Centre d'Etudes de Chimie Métallurgique - CNRS
15, rue Georges Urbain, F-94407 Vitry Sur Seine Cedex (France)
Laboratoire Interdisciplinaire de Spectroscopie Electronique
FUNDP, Rue de Bruxelles, 61, B-5000 Namur (Belgium)

S. Hagège
Centre d'Etudes de Chimie Métallurgique - CNRS
15, rue Georges Urbain, F-94407 Vitry Sur Seine Cedex (France)

R. Sporken, M. Vermeersch, and R. Caudano
Laboratoire Interdisciplinaire de Spectroscopie Electronique
FUNDP, Rue de Bruxelles, 61, B-5000 Namur (Belgium)

D. Imhoff
LPS-CNRS, F-91405 Orsay (France)

INTRODUCTION

In the last 10 years, the III-nitride semiconductors (AlN, GaN, InN and ternary compounds) have revealed promising physical and optical properties. Potential applications include blue light emitting diodes and ultraviolet (UV) detectors, as well as high power and high temperature electronic devices. Aluminum nitride has a wide band gap (6.2 eV). Because of its piezoelectric properties, it can also be used in field effect transistors as a gate dielectric film and in thin film microwave acoustic resonators.[1] A high thermal conductivity (320 W/mK for ideally pure single crystals) and a thermal expansion coefficient close to that of Si (4-5 x 10^{-6}/K for AlN and 3 x 10^{-6}/K for Si) make the AlN/Si system a good candidate for microelectronics and for use in metal-insulator-semiconductor (MIS) devices. These applications require high quality crystals and low contamination in order to obtain near-ideal material characteristics.

High quality aluminum nitride layers have already been obtained by MOCVD,[2] or reactive MBE (molecular beam epitaxy),[3] but up to now, reactive sputtering has only resulted in highly oriented films[4] or in epitaxial films containing a high density of dislocations. In a study of the AlN/Si interface, Meng et al.[5] showed evidence of the growth of wurtzitic AlN and of an epitaxial orientation relationship for films grown at temperatures ranging from 600 to 1000°C. They also showed that the large misfit (> 20%) between AlN and Si may be accommodated by a coincidence ratio of 5/4. Finally, real-time measurements of substrate bending during the deposition showed the presence of variation of intrinsic stresses resulting from the island growth mode.

Due to the refractory nature of the material, a minimum growth temperature ($T/T_m \geq 0.3$) is usually needed to obtain epitaxial growth and high quality single crystals. However, when the lattice misfit is too high, even energy produced by high temperature is not sufficient to overcome the strains and reorganize the film, and the result is a columnar growth.[6] It is possible to overcome the problems that give rise to lower quality films by depositing a buffer layer of AlN[7] or GaN[8] at low temperature. With the presence of this buffer layer, dislocation density decreases rapidly as thickness increases.

EXPERIMENTAL PROCEDURE

The substrates were Si(111) that had been resistively heated at a temperature of 1050°C during deposition. The films were grown by reactive rf sputtering of a pure Al target in a mixed atmosphere of N and Ar with a base pressure of 1×10^{-7} Pa and a total deposition pressure of 5.5×10^{-1} Pa. Precise control of the partial pressure of each gas in the chamber allowed us to obtain higher plasma stability at very low deposition rates (500 Å/hour at 1050°C). Ex-situ chemical preparation of the substrate is based on the RCA method. The growth procedure, described in detail in another paper,[6] was based on the deposition of a first layer, several tens of angstroms thick, at a lower temperature of 700°C. On top of this "buffer" layer, the thick film was grown at 1050°C.

Electron spectroscopy measurements have been performed in-situ, and the growth was performed by deposition/analysis cycles in order to obtain information from the interface. Surface diffraction patterns were obtained on four-grid LEED (low energy electron diffraction) optics. Auger measurements were performed with a cylindrical mirror analyzer (CMA) and an axial electron gun ($0 < E_p < 10$ keV). The HREELS (high resolution electron energy loss spectroscopy) spectrometer was a double hemispherical system (SEDRA-ISA Riber). The spectra presented here were measured in specular geometry with a 45° incidence angle. The primary energy of the electron beam was 7 eV. The x-ray photoelectron spectroscopy (XPS) spectrometer was a Scienta ESCA 300. The x-ray source is a rotating anode. The electron beam is produced by a two-stage gun and directed onto an Al band on the rim of the anode, producing Al Kα radiation (1486.7 eV) after monochromatization. Photoelectrons are collected by a multi-element lens and focused onto a slit/aperture pair at the entrance plane of the hemispherical electron energy analyzer.

Transmission Electron Microscopy (TEM) was carried out in cross section. Samples were mechanically thinned to 30 µm and then ion milled with Ar$^+$. TEM, both conventional and high resolution, was performed on a Topcon 0002-B operating at 200 kV and reaching a resolution of 1.8 Å. Energy loss measurements were performed in a VG STEM microscope equipped with a field emission gun (FEG) and a PEELS Gatan spectrometer. The resolution of the spectrometer was about 0.5 eV in the elastic peak, and the size of the probe varied between 8 and 18 Å.

RESULTS

Electron spectroscopies provide chemical information on the reactions occurring at the interface. HREELS spectra show a strong evolution during the first stages of growth (Figure 1(a)). During the deposition of the buffer layer, two vibrations appear at 85 and 124 meV, and the angular dispersion of the dipole lobe broadens (~ 20°), suggesting high surface roughness and a three-dimensional growth of the layer. When the temperature is increased, a main vibration quickly appears at an energy of 110 meV, and the roughness diminishes, demonstrated by a decrease of the angular dispersion to a very low value for the thick film (~ 6°). In a previous paper,[6] we have shown that it is possible to calculate the spectra by means of dielectric theory. These simulations suggested that the first layer deposited at low temperature is highly disordered and rough and that this buffer layer is strongly modified by the transition to the final growth temperature. We have also shown that in order to obtain good agreement between theory and experiment, it was necessary to include a silicon nitride layer, modeled by a Si-centered tetrahedron forming a SiN_4-type molecule. Further, we had to assume a layer thickness greater than 100 Å for this "nitride".

XPS measurements confirm the presence of a Si-N binding at the beginning of the growth (Figure 1(b)), but with a lower concentration than HREELS simulations suggested. The concentration is much lower for this growth method than for the one-step growth

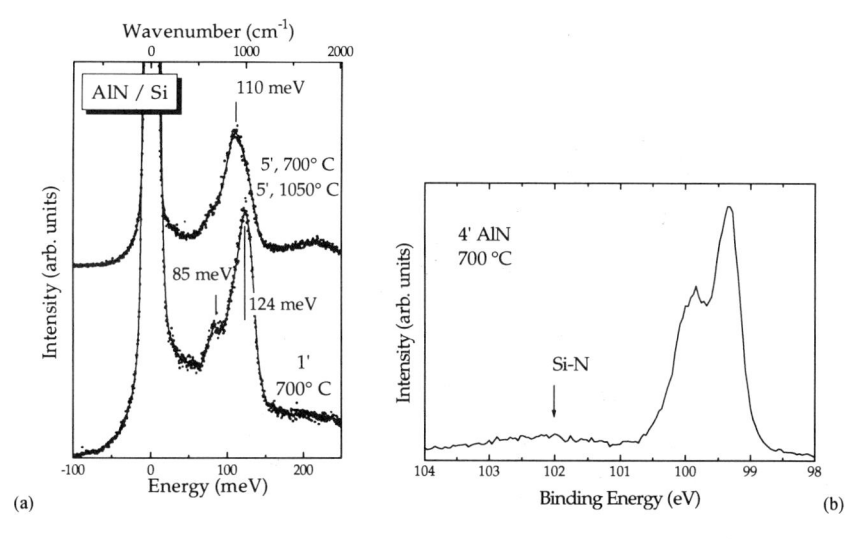

Figure 1. (a) HREELS spectra of the AlN/Si interface formed by the two-step growth method at different stages of deposition. (b) Corresponding XPS measurements (Si 2p line) for the very first stage of growth.

performed at high temperature for the whole deposition. No oxygen was detected by XPS for this sample at the interface. A general overview of the thick film in AES (Auger electron spectroscopy) found only residual oxygen at concentrations lower than 1 at.%. No carbon contamination was detected in the sensitivity range of AES.

Typical LEED and TEM 0.35-µm-thick diffraction patterns are shown in Figure 2. These patterns are characteristic of hexagonal AlN, with lattice parameters very close to the bulk values.

PEELS measurements have been focused on the Si-L, N-K, and Al-L absorption edges. A typical line scan spectrum recorded at the interface is shown in Figure 3. A first examination of this spectrum shows that the transition between the edges is abrupt, indicating a sharp interface, but the evolution of the fine structure of the edges at the interface does not present any strong change.

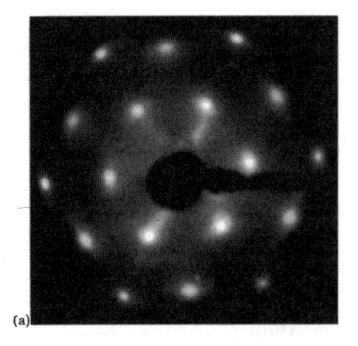

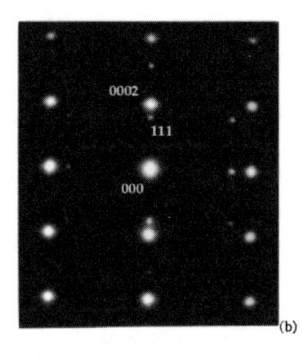

Figure 2. (a) LEED pattern of the AlN(001) grown by the two-step growth method (Ep = 81 eV). (b) Cross section diffraction pattern showing superimposition of AlN (1010) with Si (211) zone axis patterns.

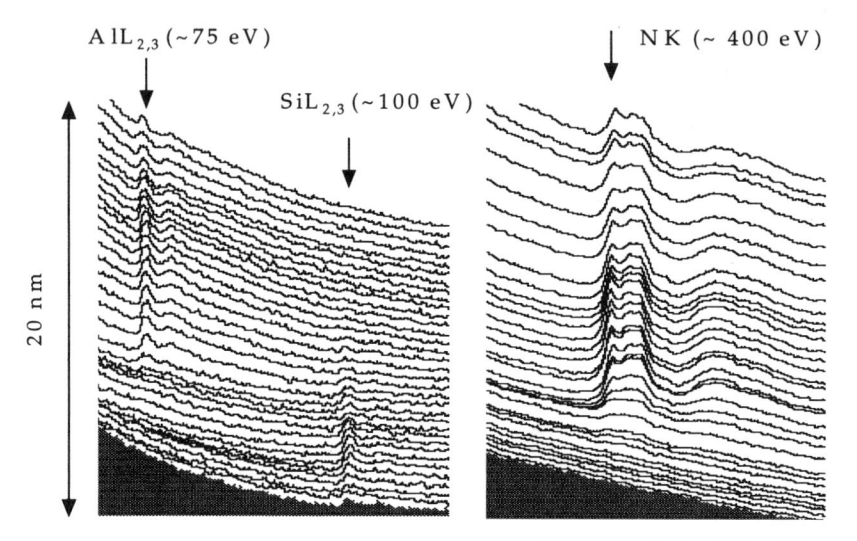

Figure 3. Typical PEELS line scan spectrum measured in our samples for AlN and Si.

The Si and N reference edges are shown in Figure 4 for the different possible compounds constituting the system. The Al edge is similar to the N edge with three peaks in its fine structure. The references have been recorded on our own samples for AlN and Si. From this figure, we can first note that the AlN film is very pure. Indeed, we have been able to confirm that the presence of oxygen contamination results in a sharp increase of the peak at lowest energy.[9] This means that the oxygen concentration detected by Auger was mainly concentrated at the surface. This higher surface concentration is probably due to the annealing of the sample in the deposition chamber in N_2 atmosphere.

We can also note, comparing the nitrogen edge for AlN and Si_3N_4, that a shift occurs between AlN and β-Si_3N_4, but not between AlN and α-Si_3N_4. However, the unambiguous separation of these compounds will only be possible in the most favorable cases. The same comparison of the Si-L edge shows a significant shift between pure Si and Si_3N_4 (~ 5 eV). We will focus later on the fine structure of the edges presented in Figure 4 together with the corresponding references to extract the information needed.

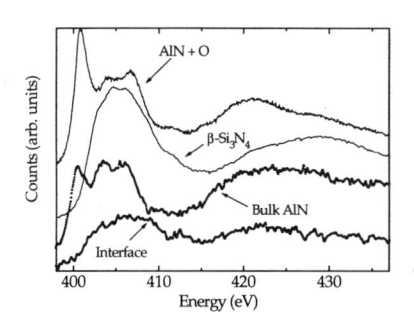

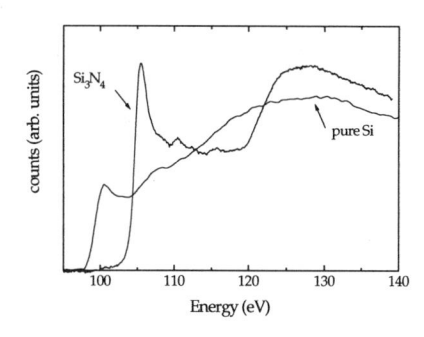

Figure 4. Detail, in the interface region, of the line scan of N-K edges shown in Figure 3 together with the different references for these edges.

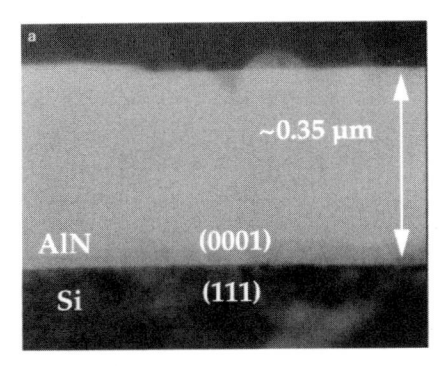

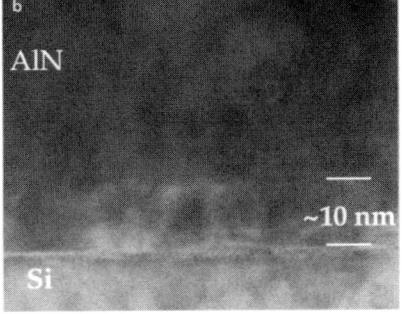

Figure 5. TEM images of a 0.35-μm-thick AlN film formed by the two-step growth method. (a) Low resolution image. (b) High resolution image.

TEM images of 0.35-μm-thick film are shown in Figure 5. The quality of the film is confirmed; at low resolution, we can see that the layer is homogeneous and the interface abrupt (Figure 5(a)). No major defects were observed, and the film does not present a columnar growth as in the case of a one-step deposition at high temperature.[6] The homogeneity of image contrasts suggests that the misorientation of the crystallites constituting the film remains very low (~ 0.1 degree). At higher resolution (Figure 5(b)), we can distinguish an interfacial layer of AlN, about 100 Å thick; this layer can obviously be correlated to the layer deposited during the first step of the growth. We can see stronger differences in the contrasts of this first layer than in the thick film, which indicates that it is constituted of very small domains that are slightly more misoriented (Figure 5(b)).

In high resolution imaging, we can see that, while the buffer layer seems to be less organized, the interface between the AlN and the substrate is structurally sharp. We can easily recognize the interfacial arrangement of atomic columns and precisely differentiate the upmost layer of the substrate from the first AlN atomic layer (Figure 6).

DISCUSSION

TEM imaging has shown the presence of a first layer in the AlN film, caused by the deposition of the buffer layer at lower temperature. This layer is constituted of small crystallites whose misorientation is of the order of 1 degree. It has been shown in the HREELS section that the surface presents extreme roughness during the growth of this buffer layer, suggesting the formation of islands which finally coalesce. The presence of

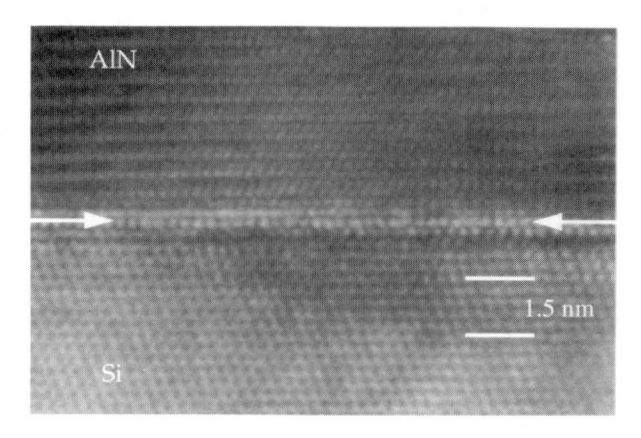

Figure 6. HRTEM image of the interface between the buffer layer and the substrate.

defects observed in the structure of the film confirms the initial three-dimensional growth. However, as the interface between the interfacial layer and the main film is rather sharp, it is possible that the increase in temperature results in a reorganization of the buffer layer from the top to the substrate, in which case the thickness of the observed layer is probably smaller than that of the original buffer layer. It is indeed probable that a reorganization front moves towards the substrate until the strains in the layer are too high to be overcome by the energy of heating. The thick film deposited on top of the buffer layer is similarly constituted, but with a much larger size (100 nm instead of 10 nm) and a smaller misorientation (~ 0.1 degree instead of 1 degree). This shows that the final layer deposited by the two-step growth method results in a well-organized film, as was suggested by electron spectroscopy.

Conventional electron microscopy and diffraction do not show any evidence of symmetry, suggesting the presence of a silicon nitride at the interface. Even high resolution imaging did not show, in the limits of the resolution, any structure that would correspond to such a compound. There is, therefore, a contradiction with the results from electron spectroscopy, even though it is probable that the concentration of the Si-N binding is lower than the one estimated from HREELS simulations, as already suggested by XPS.

A fine analysis of the line scan of energy losses through the surface yields the following conclusions. First, we saw that the interface is chemically sharp; the nitrogen and Al are not detected in the substrate and Si is not included in the AlN film. A zoom of the interface region (Figure 4) for the N-K edge shows a fine structure that is different from the one of the aluminum nitride and close to the structure of silicon nitride. However, we were not able to observe a strong change in the Si-L edge, suggesting that, even if there is a change in the interface region, this modification does not unambiguously correspond to the formation of a silicon nitride, and the existence of a possible nitride is more localized at the interface, in contradiction to what we first estimated.

Overestimation of the N concentration in HREELS simulations is attributed to the assumption of a SiN_4 type molecule, based on what is probably an oversimplified model. Another possible origin for this contradiction is the fact that XPS and HREELS are in-situ measurements, whereas HREM is an ex-situ measurement. The fact that electron spectroscopy measurements are performed during the deposition while TEM and PEELS results correspond to the study of the final layer after the total deposition time and annealing of the film, must be noted here. We may have to assume that a chemical reorganization of the film is occurring during growth and annealing.

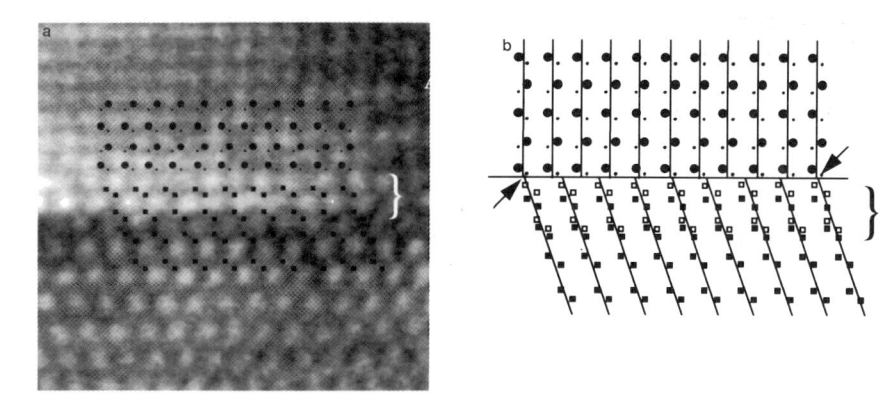

Figure 7. (a) Schematic representation of the 5/4 near coincidence orientation at the AlN/Si interface. (b) Superimposition on the HREM image, showing the displacement of the last three Si layers.

At high resolution, the interface between the substrate and the buffer layer is sharp. We did not detect any important relaxation of the lattice parallel to the surface. We may assume that during the growth of small crystallites in the buffer layer, the epitaxial orientation with the substrate allows less accumulation of global strains at the interface. This reduced accumulation may be due to the near 5/4 coincidence ratio at the interface, despite the large misfit, which can be seen from the superimposition in Figure 7(b). For this orientation, the areas of good fit are densely distributed within the interface. The structural relaxation of poorly matched areas would have a negative energetic impact by affecting the coherency of well-matched areas. Therefore, during the growth of crystallites, the amount of accumulated strain may be relaxed at the low angle boundary between them. The very low deposition rate and the relatively high temperature used during our experiment were critical for this mechanism to operate. Time and energy are necessary to allow the film to relax at different stages of its growth. We can also conclude from such an abrupt interface that the chemical interaction at the interface probably remains small.

Conversely, perpendicular to the interface, we observed a compression of the last two atomic layers of Si to about 30% of the bulk interdistance for both layers. Figure 7 shows the modification observed. For this figure, image simulations have determined that the white dots correspond to the tunnels of the structure. In Figure 7(b), the open squares show the normal positions of the atoms in the bulk, while the black squares show the positions estimated from the experimental image, showing the transformation from rigid superimposition of the lattices and the observed compression of the Si layers. This compression is not yet explained and can hardly be correlated to the formation of the silicon nitride layer, the lattice parameters being quite different than those of Si_3N_4. However, it is clear that this contraction is not only due to the surface; it is indeed probable that even if the free surface of silicon undergoes a rearrangement, the presence of aluminum nitride will compensate for this effect. We have seen that PEELS measurements do not currently provide enough information to determine the nature of this layer, due to the small size of the region under investigation and the intrinsic nature of the edges. A finer analysis of this region is in progress.

SUMMARY

To summarize, we have studied the growth of AlN single crystal on silicon at high temperature and by the two-step growth method. A comparison of the information obtained from electron spectroscopy and transmission electron microscopy has shown the complementary nature of the technique, but revealed a possible contradiction between ex-situ and in-situ investigations.

ACKNOWLEDGMENTS

R. Sporken acknowledges financial support from the National Fund for Scientific Research (NFSR, Belgium). This work was supported by the EEC Human Capital and Mobility program, a research program on interfacial materials (Région Wallonne, Belgium), the NFSR (Belgium), and by the Belgian Prime Minister's Services-Science Policy Programming within the framework of the "Interuniversity Attraction Pole in Interface Science" and the "Impulse Program on High Tc Superconductors."

REFERENCES

1. K. S. Stevens, M. Kinniburgh, A. F. Schwartzman, A. Ohtani and R. Beresford, Demonstration of a silicon field-effect transistor using AlN as a gate dielectric, *Appl. Phys. Lett.* **66**:3179 (1995).
2. A. Saxler, P. Kung, C. J. Sun, E. Bigan and M. Razeghi, High quality aluminum nitride epitaxial layers grown on sapphire substrates, *Appl. Phys. Lett.* **64**:339 (1994).
3. L. B. Rowland, R. S. Kern, S. Tanaka and R. F. Davis, Epitaxial growth of AlN by plasma-assisted, gas-source molecular beam epitax, *J. Mater. Res.*, **8**:2310 (1993).
4. W. J. Meng, J. Heremans and Y. T. Cheng, Epitaxial growth of aluminium nitride on Si(111) by reactive sputtering, *Appl. Phys. Lett.*, **59**:2097 (1991).
5. W. J. Meng, J. A. Sell, T. A. Perry, L. E. Rehn and P. M. Baldo, Growth of aluminum nitride thin films on Si(111) and Si(001): structural characteristics and development of intrinsic stresses, *J. Appl. Phys.*, **75**:3446 (1994).
6. F. Malengreau, M. Vermeersch, S. Hagège, R. Sporken, M. D. Lange and R. Caudano, Epitaxial growth of aluminum nitride layers on Si(111) at high temperature and for different thicknesses, in Journal of Materials Research, in press.
7. T. Warren Weeks, Jr., M. D. Bremser, K. Shawn Ailey, E. Carlson, W. G. Perry and R. F. Davis, GaN thin films deposited via organometallic vapor phase epitaxy on a (6H)-SiC(0001) using high-temperature monocrystalline AlN buffer layer, *Appl. Phys. Lett.*, **67**:401 (1995).
8. J. N. Kuznia, M. Asif Khan, D. T. Olson, R. Kaplan and J. Freitas, Influence of buffer layers on the deposition of high quality single crystal GaN over sapphire substrates, *J. Appl. Phys.*, **73**:4700 (1993).
9. see also: C. Colliex, R. Brydson, V. Serin et S. Matar, to be published.

NANOMETER LEVEL CHARACTERIZATION OF RAPIDLY DENSIFIED CERAMICS AND GLASS-SEMICONDUCTOR COMPOSITES

Subhash H. Risbud and Valerie J. Leppert

Division of Materials Science and Engineering
Department of Chemical Engineering and Materials Science
University of California, Davis, CA 95616

ABSTRACT

The nanostructure of rapidly densified ceramic powders was characterized to determine the grain boundary structure and chemistry at the near-atomic level. Clean grain boundaries and excellent grain-to-grain contact were observed in the alumina powders densified to almost full density, and very little grain growth was seen in sintered samples. High resolution TEM studies of a number of composite samples containing CdS or ZnSe nanocrystals dispersed in a glass matrix revealed the size and distribution of the nanocrystallites. Data showing the effect of the size of the semiconductor nanoparticles on the optical absorption or luminescence is discussed.

INTRODUCTION

Structural characterization of features controlling the properties of ceramics and composite materials with glass and ceramic matrices has steadily progressed toward smaller and smaller dimensions since the first Ceramic Microstructures Conference in 1966. In current day materials research, the micrometer level optical and scanning electron microscopy techniques typical of the mid sixties are being consistently supplemented by the powerful insights provided by elegant and detailed structural and chemical information down to the atomic level. Thus, it has become common for most modern materials research activities to include nanostructural and nanochemical characterization with, for example, high resolution electron microscopy (HREM), electron energy loss spectroscopy (EELS), scanning tunneling microscopy (STM), atomic force microscopy (AFM) and related techniques. In this paper we summarize some important aspects of the nanostructure of rapidly densified ceramic powders, and melt cast and reheated semiconductor-glass composites under investigation in our laboratory. The interest in the structure of the rapidly densified ceramics stems from the unusual plasma activated sintering (PAS) method that we have used to achieve near-theoretical density in a variety of ceramic powders in 10-15 minutes of sintering[1-5]. The characterization of the dispersed semiconductor nanocrystal-glass matrix composites[6-11] is significant because the size of the semiconductor particles has a strong effect on the optical properties (e.g. blue-shifted absorbance and luminescence related to the altered band gap of the semiconductor) relevant to applications in optical communication, flat panel display and light emitting device technologies.

EXPERIMENTAL PROCEDURES, RESULTS AND DISCUSSION

PAS of Ceramic Powders

A schematic sketch of the PAS process in shown in Figure 1. PAS is the sintering technique that generates a plasma in the particles of material to be sintered with application of an adequate pressure to the material and utilizing the activation of the particle surfaces caused by the plasma. The plasma is generated by instantaneous electric pulsed-power application control with high current and adequate voltage settings. The entire process of densification using the PAS technique consists of four major control stages. These are pressing, pulsed-power application, continuous-power application, and cool down as shown in Figure 2. In the PAS process, pressure and temperature are the most important variables to achieve successful sintering. The time envelope of these variables can be controlled by dynamic adjustment of the piston motion, pulsed electric power (duty, cycles of voltage and current), continuous electric power (voltage and current), and flow rate of the coolant.

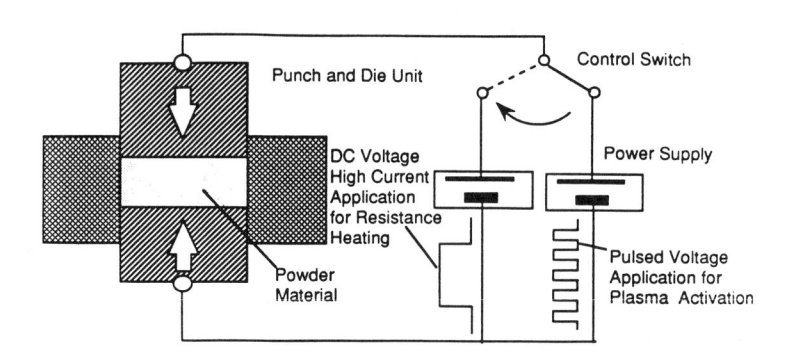

Figure 1. Schematic diagram of PAS system.

In current usage of PAS equipment, multiple trial-and-error operations are used, understandably leading to poor productivity in processing consolidated materials. For best results, optimization of the voltage level and the current envelope (vs. time) is needed, and our present research is addressing these issues. To build a scientific framework for this potential technology, the mechanism of the PAS process must be evaluated based on experimental data regarding the quality of consolidated materials (as evaluated by microstructure and X-ray diffraction) as a function of manipulations of the process variables in the PAS system. The specific process control parameters in the PAS process include: temperature (desire high gradient); time (short times); pressure (needs to be controlled as sintering progresses); voltage (discharge action); current (magnetic field action); atmosphere (selective gas and pressure including vacuum condition). The current technology for successful sintering of a particular material is heavily dependent on experience-based knowledge, and although much science has evolved, rapid consolidation by PAS or other plasma techniques is still in its infancy. A summary of the PAS processing parameters and some ceramics processed in our laboratory is given in Tables 1 and 2 below, followed by details of $\propto$-Al_2O_3 PAS and nanostructure characterization.

The as received $\propto$-alumina powders were loaded into graphite dies (2 cm in diameter) without any prior cold compaction. Chemical analysis of the powders indicates (oxides of each element in ppm): 457 Mg, 101 Fe, 83 Si, 38 Ga, 38 Zn, 33 Ti, 17 Na, 16 Mn, 5 Zr, less than 4 Mo and Cr, 3 Ca, and less than 1 Li and Ni. Plasma Activated Sintering was then

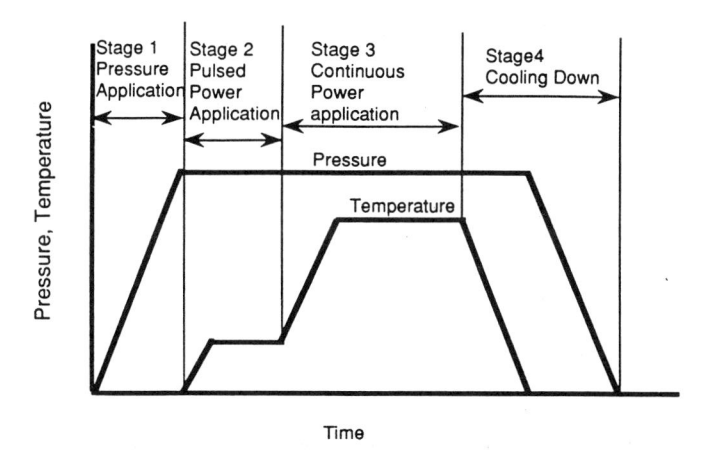

Figure 2. Plasma-assisted sintering pressure and temperature profile.

performed using a typical activation step consisting of 30 to 60 ms On/Off electrical discharge pulses lasting a total duration of one minute. Following this activation step the powders in the die were heated resistively using a 1500 A DC current to about 1150 °C. Six different experimental variations were attempted to achieve the maximum density in the solid samples after PAS at below 1150 °C, thus avoiding exaggerated grain growth and retaining the alumina particle size in the final nanostructure. Highest densities were obtained with the following empirically optimized PAS processing schedule: 90 seconds of plasma activation pulses (each pulse of 60 millisecond duration) at room temperature followed by heating to 200 °C, repeating the 90 seconds of 60 ms pulses, heating to 1150 °C, holding for 10 minutes and returning the sample to room temperature. An average of three measurements on the same sample gave densities of about 99.2 % of the theoretical value (3.986 g/cm^3).

The microstructure and grain boundary chemistry of the rapidly sintered bulk disks were examined by high resolution electron microscopy and electron energy loss spectroscopy. Planar TEM specimens were prepared by mechanical thinning and dimpling. Final thinning to electron transparency was done by argon ion-beam milling. High resolution electron

Table 1. Typical PAS processing parameters.

Parameter	Value
Applied pressure	15 MPa
Plasma voltage	25 V
Plasma current	750 A
Plasma on-state duration	80 msec
Plasma off-state duration	80 msec
Plasma duration	30 sec
Resistance heating voltage	70 V
Resistance heating current	2000 A
Total sintering time	< 10 minutes

Table 2. Summary of PAS processed ceramics.

Material and Application	PAS Temperature (°C)	PAS Time	Percent Theoretical Density
AlN (electronic packaging substrate)	1650-1800	5-10 min	> 99%
Al_2O_3 (cutting tool)	1400-1500	1-5 min	>98%
NZP (M-Zr-phosphate) (low thermal expansion ceramic)	1200-1300	5-12 min	>99%
Bi-Pb-Sr-Ca-Cu-O (superconducting ceramic oxide)	740-800	30 sec - 8 min	>98%
WC-Co (cutting tool)	1300-1400	6 min	>99%

microscopy (HREM) experiments were performed in a Topcon 002B microscope at 200 kV, with an interpretable resolution limit of 0.18 nm. The chemistry at and near grain boundaries was analyzed by position-resolved electron energy loss spectroscopy (EELS) using a Philips EM400ST field emission gun microscope coupled to a Gatan parallel-detection electron energy loss spectrometer at 100 kV. A 3 nm electron probe with about 1 nA of current was used for this purpose. All EELS spectra taken during this work were from specimen regions thin enough for quantitative analysis so that probe broadening did not yield significant loss of spatial resolution. A liquid nitrogen cooled specimen holder was used for all analytical experiments, to minimize specimen contamination and local specimen heating.

Figure 3 shows a bright-field TEM image of the PAS processed Al_2O_3 sample. The image reveals a structure with an average grain size of about 650 nm, with the distribution in grain sizes ranging between approximately 250 nm and 1050 nm. A very small number of voids were also observed within a few grains. The grain boundaries were examined by HREM imaging to determine whether or not thin amorphous regions existed between boundary surfaces. Figure 4 shows a typical HREM image of the grain boundary in PAS consolidated Al_2O_3. The observed grain boundary is in edge-on orientation with respect to the incident electron beam direction. The upper grain is in $[2\bar{2}01]$ zone axis orientation and the bottom one is in $[1\bar{1}01]$ zone axis orientation. The structural width of the boundary in Figure 4 is about 0.4-0.5 nm. The boundary appears to be very clean even at this fine scale of resolution, similar to the grain boundaries we had earlier observed in PAS processed AlN[1]. There was no evidence of an amorphous or impurity phase layer at any of the grain boundary regions observed and the entire sample showed direct bonding between adjacent Al_2O_3 grains. The chemical width corresponding to oxygen distributions in grain boundary regions was examined using nanospectroscopy. Electron energy loss spectroscopy (EELS) from the grain boundary showed aluminum and oxygen as the major elements and no other detectable elements were observed at and near grain boundaries. The normalized oxygen-K core loss edge signal across the grain boundary showed a small decrease in oxygen signal (~5%) at the boundary relative to the grain matrix. Slight differences in the edge threshold and near edge fine structure of O-K edge between grain matrix and grain boundary were also evident.

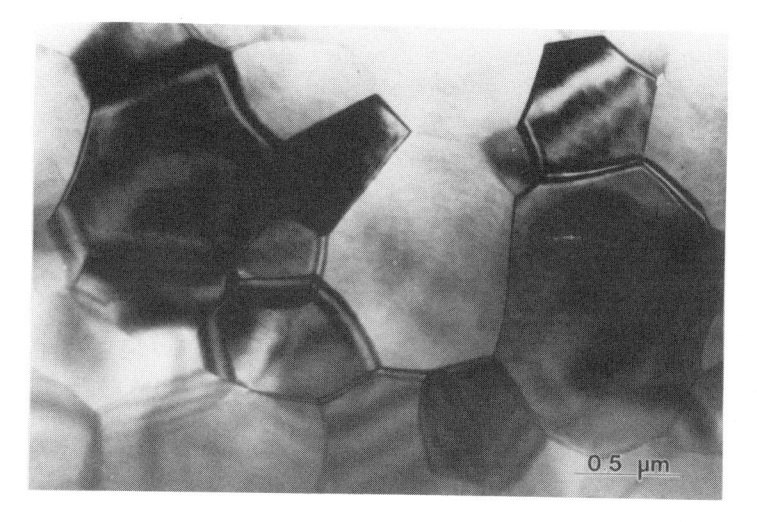

Figure 3. Bright-field TEM image of PAS processed Al_2O_3. No evidence of an amorphous or impurity phase layer was found at any of the grain boundary regions observed. A very small number of voids were observed within a few grains.

Figure 4. HREM image of a typical grain boundary in PAS consolidated Al_2O_3. The boundary appears clean structurally and there is direct bonding between grains

SEMICONDUCTOR NANOCRYSTAL-GLASS COMPOSITES

Nanocrystals were formed by the process of precipitation in silicate glasses originally developed to make color filter glasses. Glass melting, fast quenching, and careful heat treatment were the three major materials preparation steps; and they involve dissolving the semiconductor in the glass melt during high-temperature melting, forming supersaturated solutions during fast quenching, then nucleating, growing, and coarsening quantum dots in glass by heat treatment. The formation of nanometer size quantum dots through these heat treatments constitute the major advantage of this fabrication technique, while the low solubility of semiconductors in silicate glasses and broad quantum-dot size distribution[6-11] are among the disadvantages. The coloration caused by nanocrystal precipitation in glasses is vividly seen in some of the commercial cut-off filter glasses made by the Corning and Schott Glass companies. These glasses exhibit two or three absorption peaks other than the simple absorption edge shifts corresponding to different stages of heat treatment. Combining the microstructural data and the absorption spectra, the absorption edge shifts can be attributed to the size-dependent energy gap of semiconductor nanocrystals. The absorption peaks on the shoulder of the absorption edge can be explained as quantum-confined exciton-transition peaks. The easy availability of these commercial cut-off color glasses led to early studies of quantum confinement effects but commercial glasses were not originally designed for the study of these new physical phenomena. In fact, it is impossible to avoid mixed nanocrystallites in these materials and thus the best samples for research are carefully designed base-glass compositions with deliberate semiconductors additions.

Samples containing CdS or ZnSe semiconductor were prepared by co-melting 1-3 wt. % of the semiconductor in an experimental silicate glass containing SiO_2 (56%), K_2O (24%), BaO (9%), B_2O_3 (8%) and CaO (3%) in wt. % of each oxide. Excess semiconductor was added in order to compensate for vaporization losses, with the retained amount of semiconductor estimated to be not less that 10-15% of the batched amount. Twenty-five grams of the pre-mixed semiconductor and glass powders were melted in alumina crucibles at 1400 °C. Melts were rapidly quenched upon removal from the oven between two brass plates. Clear glass sheets were obtained by quenching the CdS melts, and these samples underwent an additional heat treatment step ("striking") in order to promote nucleation and growth of semiconductor nanocrystals. An optimal heat treatment range of 577 °C to 735 °C was determined using differential thermal analysis (DTA). Variation of time and temperature during striking produced samples ranging in color from light greenish yellow to bright yellow. Quenched ZnSe melts produced glass sheets reddish yellow in appearance and these samples were not subjected to striking. However, additional ZnSe samples were prepared by allowing the melt to cool in air ("as cast"). These samples were reddish-orange in color.

Samples were prepared for transmission electron microscopy (TEM) analysis by successive mechanical polishing, dimple grinding and Ar ion-beam milling. CdS TEM samples were examined with a 120 keV Phillips EM-400 TEM, with high resolution electron microscopy performed on a 400 keV JEOL 400EX HRTEM and a 200 keV Topcon 002B HRTEM. Nanospectroscopy and nanodiffraction data experiments were performed at 100 keV in a Phillips EM400ST/FEG microscope coupled to a Gatan parallel-detection electron-energy-loss spectrometer. CdS samples were also examined with a KRATOS 1.5 MeV HVTEM (high voltage transmission electron microscope). ZnSe TEM samples were examined with a 100 keV Hitachi 600 STEM, with high-resolution transmission electron microscope (HRTEM) images obtained with a 200 keV Topcon 002B HRTEM.

Microanalytical and microdiffraction data were collected with a JEOL JEM 200CX AEM (155 eV resolution for MnKα radiation, 72° take-off angle).

The CdS particles (Figure 5) observed in the base glass were randomly distributed in the matrix and exhibited ellipsoidal morphology, with semi-axes of 25.5 and 11 nm. All these CdS particles were found to be polycrystalline, while the average size of nanocrystals in each individual particle was about 3.3 nm, with a standard deviation of about 0.6 nm. Note that these CdS polycrystalline particles are nearly an order of magnitude larger than the predicted Bohr radius of exciton for CdS. Individual nanocrystals are easily visible around the periphery of the polycrystalline particle. Each polycrystalline particle contained about 500 nanocrystals with a unimodal particle-size distribution. No evidence of very small single-crystal CdS regions appeared in the matrix. We determined that the structure of these nanocrystals was the wurtzite-type hexagonal, with lattice constants corresponding to CdS, from microdiffraction experiments.

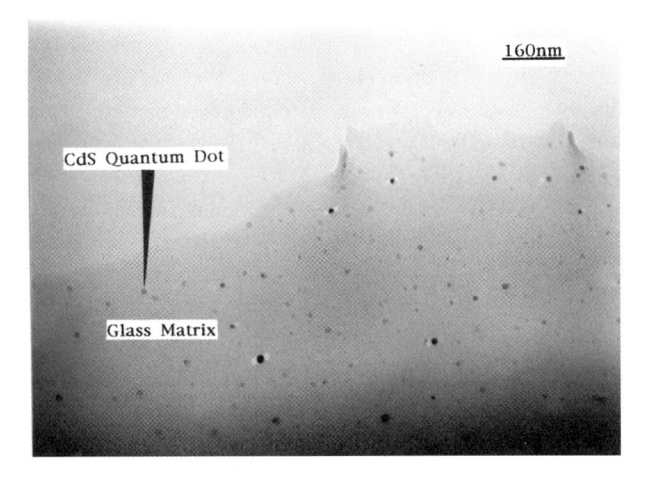

Figure 5. HVTEM bright-field image of CdS quantum dots in a glass matrix heat treated on a hot-stage in HVTEM at 600 °C for 103 minutes.

It is well known that atomic arrangements on grain boundaries differ from bulk crystals and cause perturbations of the lattice potential. The magnitude of these perturbations is not known quantitatively in most cases, but it is not generally large. When impurity segregation occurs to grain boundaries, however, the perturbation can be much larger if the segregant differs greatly in electronic structure from the atomic constituents of the crystal of interest. In the present case, oxygen may have segregated to the CdS grain boundaries.

Bright-field TEM images of ZnSe quantum dots grown in a borosilicate glass matrix reveal particles to be randomly distributed throughout the sample and to have a uniform size distribution (Figure 6). The average particle diameter obtained for the as cast sample is 5.5 ± 1.7 nm, while quenching of the sample further reduces the average particle diameter to 3.7

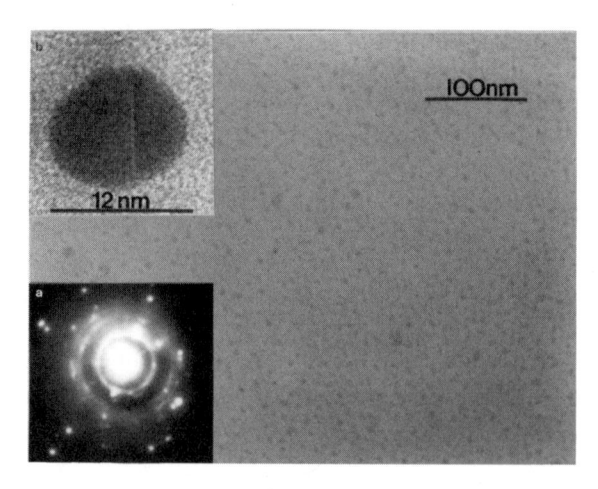

Figure 6. Bright-field TEM of ZnSe in glass (as cast). Inset (a) is a HRTEM of an individual ZnSe quantum dot and inset (b) depicts an SAED pattern for the sample corresponding to cubic ZnSe.

± 1.1 nm by halting crystal growth at an earlier stage. Particle size for both samples is less than the exciton diameter of ZnSe (11.4 nm). The quenched sample had an absorption edge at 3.1 eV, blue-shifted from the bulk value of 2.65 eV, consistent with quantum confinement effects. Band-edge as well as visible-infrared peaks were present in the photoluminescence spectra of these samples. A HRTEM image of a ZnSe nanoparticle in glass is shown in the top inset of Figure 6. Most nanocrystals were single crystal, with a few larger internally twinned structures such as this one, observed in the cast sample. An interfacial region between the crystal and the glass matrix was not observed, indicating that surface oxidation did not occur. SAED data (bottom inset of Figure 6) shows diffraction ring spacings matching those of the cubic phase of ZnSe.

EDXS was used to pole several nanocrystals for their composition. Most particles were found to be composed of zinc and selenium. An observed variation in zinc to selenium peaks between different nanocrystals may be indicative of a range in stoichiometry or may be due to overlap of the zinc and copper grid peaks. A few particles were observed to consist entirely of selenium or selenium oxide. The presence of oxygen in the glass matrix prohibits distinguishing between the element or its oxide. Strong convergence of the electron beam on these selenium based particles was found to result in rapid dissolution of the particles into the glass matrix (time for dissolution was on the order of 1 sec)., confounding attempts to further characterize these samples through microdiffraction. Recent results on these samples using Raman spectroscopy indicate that Se may also be redissolved through laser irradiation[12].

ACKNOWLEDGMENTS

This work was supported by the NSF Electronic Materials Program through Grant DMR94-11179. HREM, EDXS and ARM data were obtained at the National Center for Electron Microscopy at the Lawrence Berkeley National Laboratory under U.S. Department of Energy Contract # DE-AC-03-76SF00098. We acknowledge collaborations with Dr. Moon J. Kim of Arizona State University - Tempe.

REFERENCES

1. Risbud, S.H., Groza, J.R., and Kim, M.J., Clean grain boundaries in aluminium nitride ceramics densified without additives by a plasma-activated sintering process, *Phil. Mag. B*, 69:525 (1994).
2. Risbud, S.H., and Shan, C.H., Resistivity drops at >240 K and diamagnetic AC susceptibility up to 300 K in rapidly consolidated YBCO, *Mater. Lett.*, 20:149 (1994).
3. Shan, C.H., Risbud, S.H., Yamazaki, K., and Shoda, K., Rapid consolidation of Bi-Pb-Sr-Ca-Cu-O powders by a plasma activated sintering process, *Mater. Sci. Eng. B*, 26:55 (1994).
4. Groza, J.R., Risbud, S.H., and Yamazaki, Plasma activated sintering of additive-free AlN powders to near-theoretical density in five minutes, *J. Mater. Res.*, 7:2643 (1992).
5. Risbud, S.H., Shan, S.H., Kim, M.J., and Mukherjee, A.K., Retention of nanostructure in aluminum oxide by very rapid sintering at 1150 degrees C, *J. Mater. Res.*, 10:237 (1995).
6. Borelli, N.F., Hall, D.W., Holland, H.J., and Smith, D.W., Quantum confinement effects of semiconducting microcrystallites in glass, *J. Appl. Phys.*, 61:5399 (1987).
7. Potter, B.G., and Simmons, J.H., Quantum size effects in optical properties of CdS-glass composites, *Phys. Rev. B*, 37:10838 (1988).
8. Liu, L.C. and Risbud, S.H., Quantum-dot size-distribution analysis and precipitation stages in semiconductor doped glasses, *J. Appl. Phys.*, 68:28 (1990).
9. Risbud, S.H., Liu, L.C., and Shackelford, J.F., Synthesis and luminescence of silicon remnants formed by truncated glassmelt-particle reaction, *Appl. Phys. Lett.* 63:1648 (1993).
10. Liu, L.C., and Risbud, S.H., Real-time hot-stage high-voltage transmission electron microscopy precipitation of CdS nanocrystals in glasses: experiment and theoretical analysis, *J. Appl. Phys.*, 76:4576 (1994).
11. Risbud, S.H., and Underwood, H.B., Synthesis and optical properties of GaAs nanoclusters sequestered in amorphous matrices, *J. Mater. Syn. Proc.*, 1:225 (1993).
12. Su, Z., Rodriques, P.A.M., Yu, P.Y., and Risbud, S.H., Selenium molecules and their possible role in deep emission from glasses doped with selenide nanocrystals., *J. Appl. Phys.*, 80:1 (1996).

ACCOMMODATION OF VOLUME CHANGES DURING DIFFUSION-CONTROLLED METAL PRECIPITATION INSIDE AN OXIDE MATRIX

Monika Backhaus-Ricoult

Centre d'Etudes de Chimie Métallurgique - CNRS
15 Rue G.Urbain, 94 407 Vitry, France

ABSTRACT

Under certain conditions, partial reduction of mixed oxides leads to diffusion-controlled formation of metal precipitates inside the oxide matrix. Local volume changes related to this precipitation process are important. Depending on the mechanism of metal formation, it can be either negative, as in the case of substitutional mechanisms, or positive, as in the case of interstitial mechanisms. In both cases it induces large stresses during precipitate growth. At high temperatures these stresses may be relaxed by plastic deformation of the matrix. In the present work, internal reduction of doped magnesia and alumina single crystals is investigated. For magnesia, metal precipitation is always accompanied by formation of free matrix dislocations and stresses due to volume changes are fully compensated by dislocation climb. For doped alumina, no formation of matrix dislocations is observed in the temperature range investigated; stresses due to precipitation and precipitate growth are relaxed by diffusional creep. Only when the metal phase precipitates out of the supersaturated mixed oxide during cooling pore formation at the precipitate interfaces is observed.

INTRODUCTION

Many heterogeneous solid state reactions involve a volume change. Important local volume changes occur during precipitate formation within a metal or oxide matrix upon internal oxidation and reduction.[1-4] In general, precipitates and host matrix have different lattice parameters. The surrounding matrix suffers then the volume changes due to nucleation and precipitate growth and must adapt to it. As a result, the whole material deforms elastically and, above a certain deformation, plastically. Since oxide matrices (case of internal reduction) are more rigid than metal matrices (case of internal oxidation) and their deformation is more difficult, growth of metal precipitates in oxides may be more easily affected by the reaction-related deformation and result in characteristic features for the morphology of such precipitates and of the reduction scale microstructure.

For this reason, we studied the internal reduction of various mixed oxides. During exposure to low oxygen activity, metal precipitation occurs inside the oxide matrix. Depending on the point defect chemistry of the starting mixed oxide, different metal formation mechanisms may occur. In general, substitutional mechanisms with negative reaction volume and interstitial mechanisms with positive reaction volume can be distinguished. The former ones create tensile stresses at the growing precipitate interface, while the latter ones cause compressive stresses. Since the volume change between oxide and metal is significant, large stresses build up around the precipitates, which must be relaxed. Stress relaxation in a rigid oxide matrix can occur through different mechanisms.

In the present work, we have selected nickel-doped magnesia and chromium-doped alumina as model materials for internal reduction stu.'ies at different reaction temperatures and for various doping levels. Details of the microstructure of the reduced scale, obtained by transmission electron microscopy, are interpreted in terms of metal formation mechanism, stress generation at the growing metal precipitate interface and stress relaxation.

EXPERIMENTAL PROCEDURES

The materials used were:
- nickel-doped MgO single crystals and polycrystals with doping levels between 2 and 10%,
- chromium-doped Al_2O_3 single crystals with doping levels of 1 and 2% and polycrystals with doping levels of 3, 6, 10 and 20%.

They were cut into 5mm x 5mm x 3mm samples and surface polished down to 1 µm finish. Doped MgO crystals were reduced in a chamber, in the presence of a C/CO buffer or under gas flow of different CO/CO_2 gas mixtures.[2] Reaction temperatures of 1400°C, 1200°C and 900°C were chosen. Reaction times varied between 1 and 500 h. Chromium-doped alumina crystals were reduced in a closed reaction chamber at oxygen activity 10^{-20} (Al/Al_2O_3 buffer), at temperatures between 1400 and 1700°C for 1 to 200 h. Details of the experimental set up are described in reference [4].

Polished cross-sections of the reduced samples were observed by optical microscopy, analytical scanning electron microscopy and conventional, analytical, and high resolution transmission electron microscopy (JEOL 2000 FX; TOPCON 2b with a point-to-point resolution of 0.18 nm). For TEM studies, cross sections were either mechanically polished with diamond lapping films to a thickness of about 50 µm and further thinned by argon bombardment in a cold stage at 5 kV, 0.6 mA, or mechanically thinned on a tripode with a tilt of 2.5°, down to vanishing thickness at the thin edge. Carbon coating was necessary to avoid charging of the thin specimens under the electron beam.

EXPERIMENTAL RESULTS

Under the above described reaction conditions, internal reduction has occurred in all mixed oxides, as evidenced by the presence of metal precipitates inside the host oxide matrix. The reduction front is found to move with time from the outer surface towards the bulk, according to a classical parabolic rate law, indicative of a diffusion controlled reaction.

For each oxide system, only a very restricted number of metal precipitate morphologies and of relative orientation relationships between precipitates and host matrix is found. Both, morphology and orientation relationship may eventually vary with the following reaction parameters: reaction temperature, driving force for reduction and doping level of the mixed oxide.

TEM observations of the reduction scales reveal the detailed morphology of the precipitates and of their surrounding matrix, as summarized below.

MgO-Ni

Pure metallic nickel (fcc) is observed in cube-on-cube orientation, as polyhedra ranging in size from 20 nm to 1 µm. The larger polyhedra mainly show {100} facets with {111} truncation of the corners of the cubes. Differently, for smaller precipitates the relative size of {111} facets increases. The metal and the oxide matrix are in intimate contact at the interfaces, no pores or cracks being observed at reaction temperatures above 1200°C. HREM revealed atomically flat (100)//(100) interfaces. The lattice misfit across these interfaces is large, 15%, giving rise to interfacial misfit dislocations every seventh lattice plane. However, misfit strains are located at the interface and do not extend over more than a few atomic planes.

All metal precipitates are linked to free matrix dislocations, which form a very dense network all over the reduced layer. Dislocations present in the MgO matrix have Burgers vector a/2 [110]MgO. Considerable dislocation densities are observed in oxides initially containing about 3% nickel oxide. A typical reduction scale is presented in Figure 1.

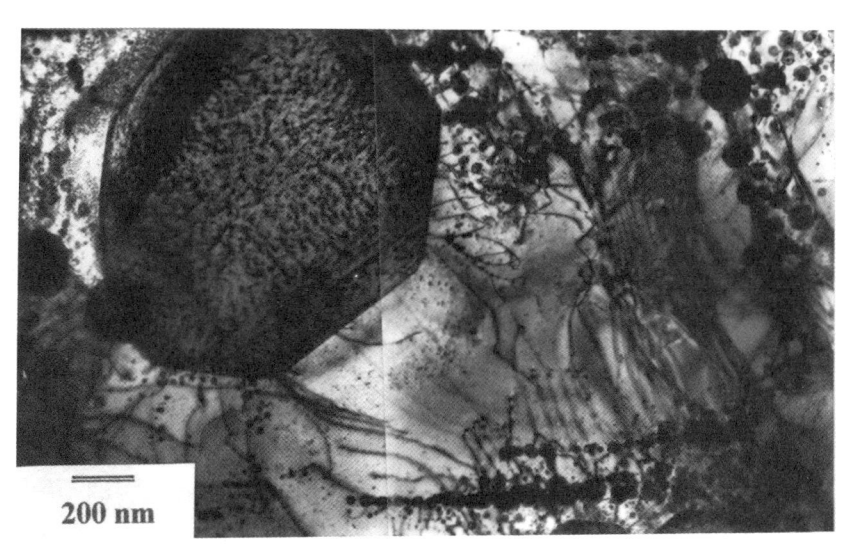

Figure 1. Typical reduction scale of $(Mg_{0.97}.Ni_{0.03})O$ reduced at 1400°C in C/CO for 50 h, showing nickel precipitates interconnected by a/2 [110]$_{MgO}$ dislocations (bright field TEM image close to [100] MgO zone axis)

Al$_2$O$_3$ - Cr

As in the previous case, metal precipitates form within the reduction scale. During diffusion-controlled precipitation, bcc chromium adopts the following orientation relationship with the R3R (quasi hexagonal) alumina matrix:

$(0001)_{Al2O3}$ // $(111)_{Cr}$ and $[11\bar{2}0]_{Al2O3}$ // $[011]_{Cr}$

The lattice parameter measured (at room temperature) in the chromium precipitates is systematically greater than that of pure chromium by about 3%. This increase in lattice parameter can be related to the difference in thermal expansion of alumina and chromium. Extrapolation of the lattice parameter measured for the precipitates at room temperature to the reaction temperature, by use of thermal expansion data, yields the theoretical pure chromium lattice parameter. This suggests that the chromium lattice is elastically stretched during cooling inside the rigid alumina matrix. The absence of pores between precipitate and oxide matrix points out to a very strong interface.

At low temperature and low doping levels in the starting mixed oxide, when precipitate growth remains slow and is controlled by long-range diffusion, mainly chromium rods with hexagonal cross section, long prismatic facets and rounded tips are observed, which are elongated parallel to the c-axis of alumina, see Figure 2a. High resolution electron microscopy and low-angle convergent beam diffraction of the interface in samples (cooled at room temperature) showed that the precipitate rods are surrounded only by an extremely weak strain field around their cross section and by a larger strain field around the tips of the rods.[5] No dislocations are observed around the rods, only in very rare cases single dislocations are detected at the rod tips of chromium precipitates.

At high reaction temperatures and high chromium oxide contents in the starting mixed oxide, thick chromium platelets with large basal facets $(0001)_{Al2O3}$ // $(111)_{Cr}$ and smaller prismatic facets $\{11\bar{2}0\}_{Al2O3}$ // $\{011\}_{Cr}$ form during diffusion-controlled internal reduction, see Figure 2b. Precipitates and matrix are once more in intimate contact; no pores or cracks are observed for this type of precipitates. HREM and LACBED reveal no strain field around those precipitates at all. Local lattice misfit at the interfaces is accommodated by misfit dislocations at the interfaces, but the corresponding strain fields remain restricted to the first two to three atom layers. No free matrix dislocations are observed in alumina over all the reduction scale!

Other small, almost spherical, precipitates are observed, shown in Figure 2c. Their origin is different, since they form from the supersaturated mixed oxide during cooling. They show a different orientation relationship compared to the rods and platelets[4,6] and are faceted along dense packed matrix planes. These precipitates, in contrast to those formed by diffusion-controlled reduction, are always attached to a large interfacial pore!

DISCUSSION

As described in earlier models, internal reduction of mixed oxides does not occur by simple oxygen transport through the reaction scale, as in the case of internal oxidation of alloys. On the contrary, this reaction instead involves a number of different point defect fluxes, depending on the defect chemistry of the mixed oxide. These defects react at the reaction front, where partial reduction of the more noble transition metal takes place and leads to metal formation. While the global macroscopic reaction is only the reduction of the more noble metal oxide (in solid solution) to its metallic state, the local reaction may differ for the different oxides. Metal may form by different mechanisms, interstitial and substitutional ones, depending on the type of oxide and its defect chemistry. Each possible reaction mechanism yields metal formation with a reaction rate depending on concentration and mobility of the implied defects, and a characteristic reaction volume. The reaction volume is defined as being the volume change resulting from metal precipitate growth at the reaction front. At this point, we can already understand that each mechanism corresponds to a certain precipitate growth rate and to a time-dependent stress field resulting from the reaction volume increase.

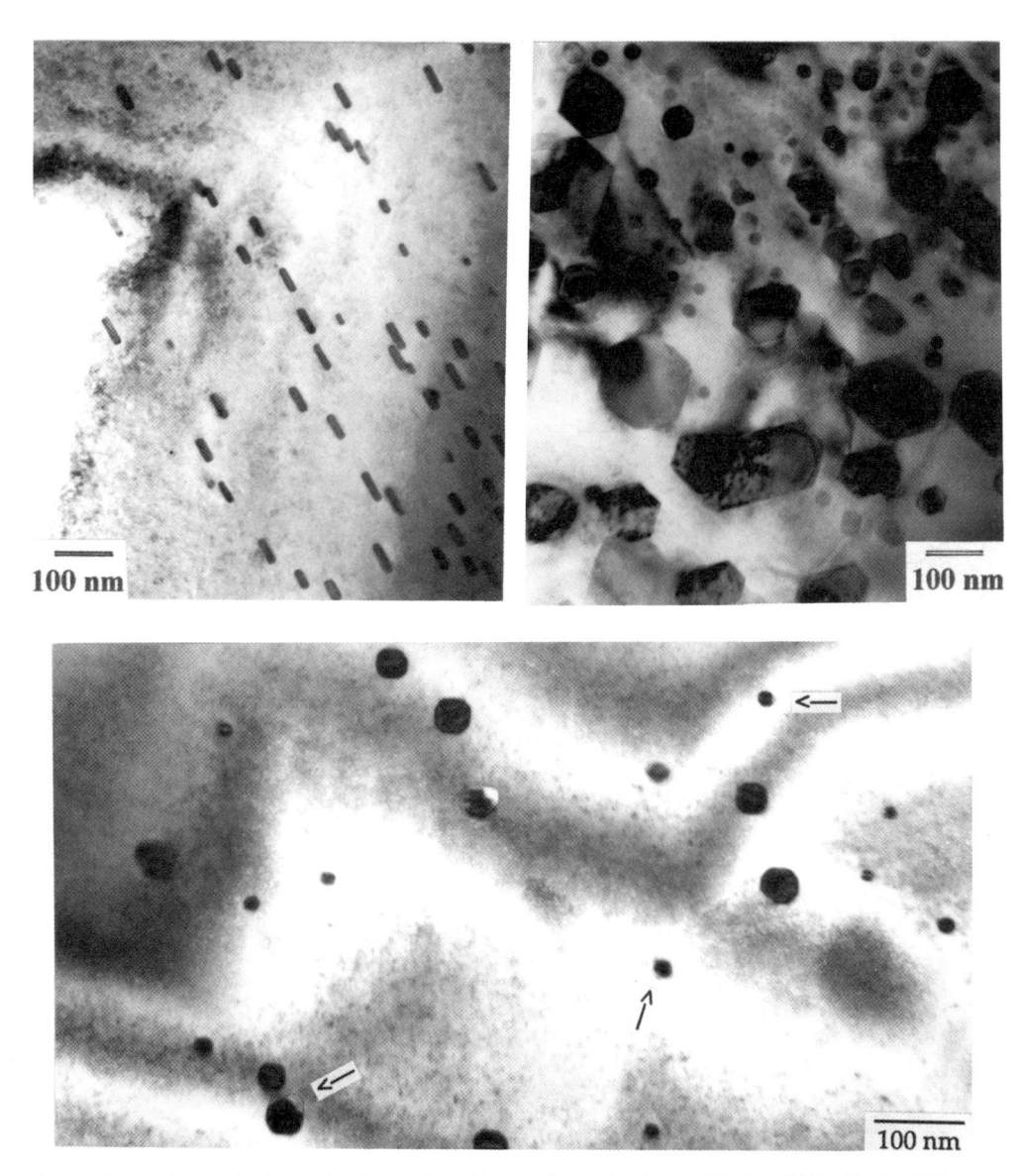

Figure 2. Typical reduction scale of a) $(Al_{0.97}Cr_{0.03})_2O_3$ reduced at 1500°C in Al/Al_2O_3 for 50 hours, showing rod-shaped chromium precipitates, b) $(Al_{0.8}Cr_{0.2})_2O_3$ reduced at 1600°C in Al/Al_2O_3 for 50 hours, showing plate-shaped chromium precipitates, c) $(Al_{0.97}Cr_{0.03})_2O_3$ reduced at 1700°C in Al/Al_2O_3 for 50 h, showing spheroidal chromium precipitates with small interfacial pores formed upon cooling.

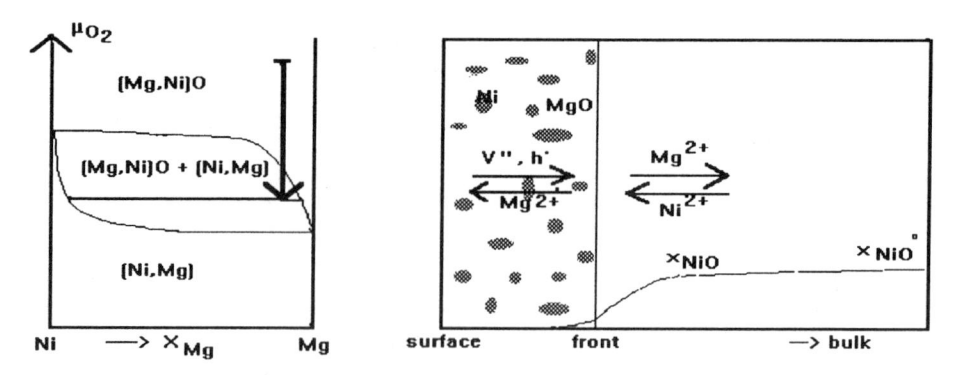

Figure 3. Internal reduction scheme for (Mg,Ni)O with the corresponding phase diagram

For initial precipitate nucleation and growth, elastic deformation of the metal precipitate and of the more rigid oxide matrix occurs, but very rapidly the elastic limit of the matrix is reached. Increase of stress at the precipitate interface slightly slows down further precipitate growth. However, stresses continue to build up at a rate faster than the decrease of precipitate growth rate. Therefore, very rapidly either fracture or plastic deformation of the oxide matrix takes place in order to release the increasing stress.

In the following, we interpret our experimental observations for partially reduced doped magnesia and doped alumina in these terms and explain the observed microstructures by the difference in stress relaxation in the two materials.

Reduction of (Mg,Ni)O

Differently charged cation vacancies, $V^{n'}{}_{Mg}$, and transition metal (M) cations on magnesium cation sites, $M_{Mg}{}^{x}$ together with electron holes, $h^{·}$, constitute the majority point defects in transition metal doped magnesia [7,8]. Upon exposure to low oxygen activity of doped magnesium oxide samples previously equilibrated at high oxygen activity, point defect fluxes establish in the oxygen potential gradient: cation vacancies diffuse from the inside of the crystal to the outer surface, cation counterfluxes occur in the opposite direction and charge carrier transport across the scale provides electroneutrality, Fig. 3. By a local decrease of the oxygen activity, the mixed oxide solid solution eventually becomes thermodynamically unstable, see schematic phase diagram of Fig. 3, and (allowing a certain supersaturation for the nucleation), metal precipitation takes place inside the oxide matrix. Models for the global reaction kinetics are presented in reference [1], however, local metal formation mechanism and precipitate growth kinetics are more complex.

The local chemical reaction at the metal-oxide interface, which yields further growth of the precipitate, can be formulated as

$$Ni^{x}{}_{Mg} = Ni + V''{}_{Mg} + 2h^{·}$$

Together with metallic nickel, Ni, cation vacancies $V''{}_{Mg}$ are formed, which diffuse in the oxygen potential gradient, across the reduction scale to the outer surface. The above metal formation reaction obeys the laws of conservation of electroneutrality, mass, anion and cation sites in the oxide matrix, but the volume is not maintained because new metal lattice sites are created. Following the above reaction equation, nickel metal forms on an interstitial site of the oxide lattice. Formation of a single atom or a small nickel cluster could still be envisioned. However, growth of 10 nm- or even 1 μm-sized precipitates is impossible because of the enormous compressive stresses which build up in the precipitate and in the surrounding matrix during the growth process. The local build-up of stress can be estimated from the reaction volume, which in the case of an interstitial metal formation corresponds to the metal volume, $+ V^{prec}_{Ni}$.

Elastic deformation of magnesia being very limited, even at high temperature, the matrix readily undergoes plastic deformation. Deformation maps of magnesia[9] indicate plastic deformation by dislocations for the considered temperature-stress range. Dislocations easily form at the precipitate interfaces, where stresses are highest. Dislocations in MgO are typically a/2 [110], as free matrix dislocations observed in the reduction scale. By dislocation climb, local stresses are relaxed.

In the case where dislocation climb is a fast mechanism compared to precipitate growth, the chemical reduction reaction at the interface in vicinity of a dislocation can be written in the following way, where the dislocations behaves as an internal surface, where MgO lattice units are destroyed by climb of cation-anion vacancy pairs $\{V^x_{Mg}\text{-} V^x_O\}$:

$$Ni^x_{Mg} + O^x_O = Ni + \{V^x_{Mg} - V^x_O\} + 1/2\, O_2$$

If dislocation climb is not fast enough to ensure local stress relaxation, a stress increase takes place during precipitate growth and leads to further formation of dislocations. The observed dislocation density should therefore be directly correlated to the stress relaxation rate and vary in different systems or for different temperatures with the climb rate of such dislocations.

If plastic deformation of the matrix allows complete relaxation of the stresses produced during precipitate growth, the precipitates can grow under thermodynamical local equilibrium conditions to low energy morphologies in low energy orientations (minimizing interfacial energy and residual strain energy at the interface), as discussed in reference [6].

If stress relaxation becomes insufficient (because reaction temperature is too low to allow nucleation of dislocations and/or too low to activate dislocation climb or glide and/or too low to activate diffusional creep), local stresses increase and then slow down and even eventually stop the chemical reaction. This was obser ed for nickel doped magnesia at low reaction temperatures, 900°C.[2] Since stress increase is much faster than simultaneous decrease of precipitate growth rate, physical destruction of the interface by crack formation (then seen as a pore) may become possible. Such cracks, forming interconnected brances and being filled with nickel metal were observed at 900°C [2], where plastic deformation of the matrix remains rather slow.

Observations as reported for the (Mg,Ni)O were made in many other systems during internal reaction.[1,3] The above results are relevant to precipitation in systems, where plastic deformation follows a power law creep, with large exponents characteristic of dislocation creep.

Reduction of doped alumina

Our experimental observations concerning the microstructure of doped alumina differ from those made for doped magnesia. No free matrix dislocations are observed and stresses around precipitates are small or non-existant, even after cooling.

We can ascribe this behaviour to a different defect chemistry and to the specific plastic deformation behaviour of alumina.

Under high oxygen partial pressure, aluminum vacancies are considered the main defects in pure alumina.[9] However, at very low oxygen partial pressure and in the presence of doping elements partly in their 2^+ valence state, oxygen vacancies and aluminum interstitials may become majority defects in doped alumina. We used the data of references 11-13 to calculate defect concentrations as a function of the oxygen partial pressure (Figure 4). Even if these data are based on simplifying assumptions which affect the precision of defect concentrations, we see that at low oxygen activity $V_O^{\cdot\cdot}$, $Al_i^{\cdot\cdot\cdot}$ and e' are the dominant point defects.

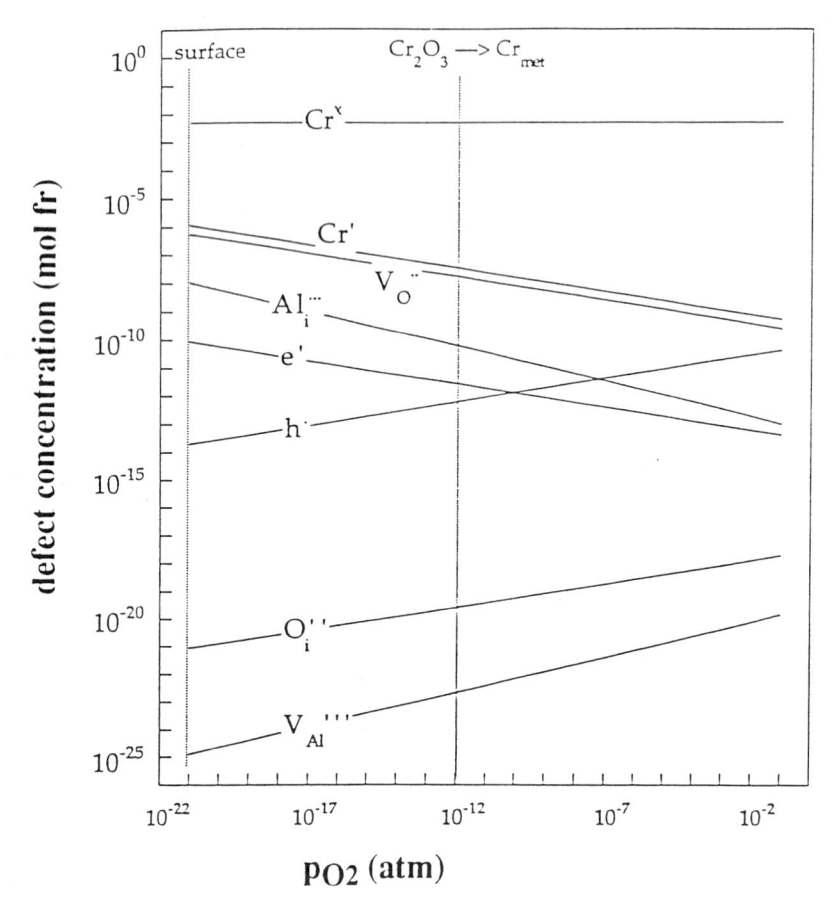

Figure 4. Presentation of the defect concentrations in $(Al_{0.97}Cr_{0.03})_2O_3$ as a function of oxygen partial pressure

With the above result concerning the types of point defects present in chromium-doped alumina at low oxygen partial pressures, we propose a mechanism for the internal reduction of chromium-doped alumina which is based on an oxygen vacancy and an oxygen counterflux, as well as an aluminum interstitial flux and electron transport across the reduced layer, see Fig. 5.

The defects in presence allow two different mechanisms for the formation of chromium metal:
- via aluminum interstitials by an interstitial mechanism, where the newly-formed metal (Cr) grows on interstitital sites inside the oxide

$$Al_i^{\cdot\cdot\cdot} + Cr^x{}_{Al} + 3\,e' = Cr + Al^x{}_{Al}$$

- via oxygen vacancies by a substitutional mechanism, where together with the metal *(Cr)* an unoccupied structural unit [2 V_{Al}'''+ 3 $V_O^{..}$] forms in the lattice, which provides inside the oxide matrix the space necessary for the newly-formed metal to grow:

$$3\ V_O^{..} + 2\ Cr^x_{Al} + 6\ e' = 2\ Cr + [2\ V_{Al}''' + 3\ V_O^{..}]$$

Metal formation by interstitial mechanism is accompanied by a positive reaction volume and yields large compressive stresses, while metal formation by substitutional mechanism results in tensile stresses.

Since both mechanisms contribute to the metal formation, a global reaction equation for metal formation can be formulated by adding the contributions, each scaled by the factor of the product of defect concentration and defect mobility. The corresponding stress production at the interface due to the global reaction volume follows then from the volume change with time.

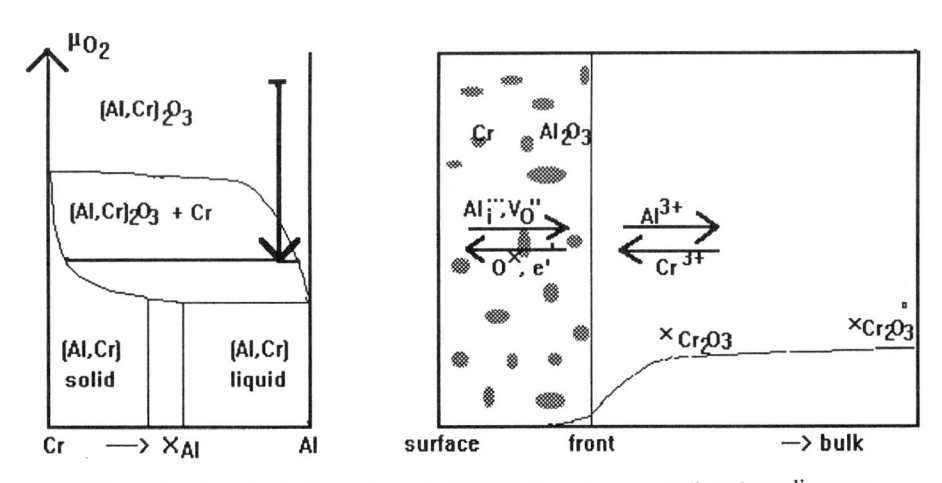

$$dV^{reaction}/dt = A\ \lambda\ \{+ V_{Cr}^{mol}\ D_{V_O^{..}}\ \nabla c_{V_O^{..}}$$
$$- (1/2\ V_{Al2O3}^{mol} - V_{Cr}^{mol})\ D_{Al_i^{...}}\ \nabla c_{Al_i^{...}}\}$$

Figure 5. Internal reduction scheme for (Al,Cr)$_2$O$_3$ and corresponding phase diagram

(A is the considered reaction cross section, λ the volume fraction of chromium metal formed at the reaction front)

The global stress production at the reaction front can be calculated by considering the total sample cross section and the volume fraction of chromium metal, which depending on the starting oxide composition and the oxygen partial pressure (complete or partial depletion of the mixed oxide by chromium) can vary between 1 and 20%.

More interesting than the global stress production at the reaction front is the local stress production at the growing precipitate interface. Using a very simplified model of rod-shaped precipitates with the rod tip at the reaction front being the (only) growing interface, we can write the stress production as

$$d\sigma^{interface})/dt = Y \int_{\xi interface}^{+\infty} \varepsilon\ (\xi)\ d\xi\ ,$$

217

Y being the Young modulus, $\sigma(\varepsilon^{\text{interface}})$ the unrelaxed interfacial stress and $\varepsilon(\xi)$ the deformation. Anisotropy of the elastic behaviour has to be taken into account. $\varepsilon(\xi)$ is given by the characteristic elastic wave propagation. Details of this approach using different simplifications will be published elsewhere.

Stress at the interface, due to precipitate growth, locally modifies the point defect concentrations. In the simplest approach, we obtain an exponential dependence of the corresponding defects on stress, σ, and molar volume, V^{molar}:

$$c_{V_O^{\cdot\cdot}}(\sigma) = c_{V_O^{\cdot\cdot}}(0) \, \exp(-\sigma \, V^{\text{molar}}/RT)$$

$$c_{Al_i^{\cdots}}(\sigma) = c_{Al_i^{\cdots}}(0) \, \exp(\sigma \, V^{\text{molar}}/RT)$$

(with $c_{V_O^{\cdot\cdot}}(\sigma)$ and $c_{V_O^{\cdot\cdot}}(0)$ being the oxygen vacancy concentration at σ and in the stress-free state, respectively; $c_{Al_i^{\cdots}}(\sigma)$ and $c_{Al_i^{\cdots}}(0)$ being the aluminium interstitial concentration at stress σ and in the stress-free state, respectively). Because the point defect concentrations are stress-dependent, the boundary conditions at the reaction front in the diffusion problem for the internal reduction are altered.

Any stress build-up at the interface accelerates the mechanism which allows this stress to relax. If the interstitial mechanism is dominant, compressive stresses build up, which in turn reduce the aluminum interstitial concentrations at the interface and increase the oxygen vacancy concentration, thereby promoting substitutional metal formation, which decreases the compressive stresses. If oxygen vacancies are the majority defects, tensile stresses will result from substitutional metal formation, thereby reducing the substitutional reaction rate and increasing the interstitial reaction rate.

Following the example of other oxides (with mobilities of interstitial atoms by a factor 100 to 1000 higher than that of vacancies[14,15]) the defect fluxes of aluminum interstitials and oxygen vacancies can be assumed to be of the same order of magnitude, which keeps the stress production small. In addition, the stress dependence of the boundary defect concentrations minimises the stress production by the stress-flux coupling of the two defects.

As a consequence, metallic chromium precipitates can grow during internal reduction in alumina, without any stress build-up. This explains our experimental observation of perfect chromium precipitates forming inside alumina, without any pores or free matrix dislocations.

However, our experimental observations on precipitates which formed during cooling from the supersaturated mixed oxide solid solution show the consequences of missing strain relaxation in the alumina matrix. Those precipitates are always attached to a pore. The formation of a pore is interpreted as a rupture of the alumina-chromium interface which is a consequence of high build up of stresses at the metal-ceramic interface and insufficient relaxation during cooling (lower diffusion rates and short cooling times).

SUMMARY

The influence of local volume changes at the metal/oxide front of internal reduction on diffusion-controlled metal precipitate growth has been studied. Metal precipitation during internal reduction can, depending on the considered mixed oxide and its defect chemistry, either be produced by a substitutional mechanism which is accompanied by a negative reaction volume and a tensile stress built-up in the precipitate surroundings, or by an interstitial mechanism, which goes hand in hand with the build-up of compressive stresses at the interface.

The interfacial stresses alter the equilibrium defect concentrations at the interface and slow down the precipitate growth rate. However, the increase of stress occurs much faster than the decrease of the precipitate growth rate. Therefore, the material's elastic limit is easily reached and plastic deformation (or fracture) of the matrix occurs.

In the present work we discussed two examples, metal precipitation in doped magnesia and in doped alumina:

Deformation maps indicate that plastic deformation of magnesia occurs by dislocations in the considered reaction condition range. Dislocations form at the metal precipitate-matrix interface and climb, resulting in stress relaxation. The dislocation density depends on the stress concentration rate during precipitate growth and on the stress relaxation rate related to dislocation climb. Experimental observations of reduction scales of nickel-doped magnesia show high densities of free matrix dislocation which interconnect the precipitates. Dislocation loops and helices indicate strong dislocation climb activity.

In chromium doped alumina substitutional and interstitial metal formation mechanisms are simultaneously activated and give rise to precipitate growth rates of the same order of magnitude. As a consequence, interfacial stresses building up during the growth of the precipitate couple the two partial growth rates and yield stress-free precipitate growth. Experimental observations of reduction scales of chromium-doped alumina confirm this statement. No free matrix dislocations, pore formation, crack formation or elastic deformation was observed around the chromium metal precipitates which formed by diffusion-controlled growth. In contrast, pore formation at precipitate interfaces was observed for metal precipitation from the supersaturated solid solution upon cooling, when diffusive transport becomes too slow to relax the present stresses.

REFERENCES

1. H. Schmalzried and M. Backhaus-Ricoult, « Internal solid state reactions », *Progress in Solid State Chem.* 22: 1 (1993)
2. M. Backhaus-Ricoult and S. Hagege, « Internal reduction of (Mg,Cu)O », *Phil. Mag.* A 67: 1471 (1993)
3. M. Backhaus-Ricoult and D. Ricoult; « Electron microscopy of internally reduced (Mg,Ni)O », *J. Mat. Sci.* 23: 1309 (1988)
4. M. Backhaus-Ricoult, A. Peyrot, P. Moreau and S. Hagège., « internal reduction of Cr-doped alumina », *J. Am. Ceram. Soc.* 77:423 (1994)
5. S. Hagège, M. Backhaus-Ricoult, C. Deininger, « LACBED of residual stresses around metal precipitates » to be published
6. M. Backhaus-Ricoult, « Growth and equilibrium morphology of metal precipitates formed inside an oxide matrix », *J.Interface Science 4 : 285 (1997)*
7. B. Wuensch, S.C.Semken, F.Uchikoba, H.Yoo, « The mechanism for self-diffusion in magnesium oxide » in *Ceramic Transactions: Point defects and related properties*, ed J.L.Routbort, T.O.Mason, p.79 (1992)
8. B. Wuensch, W.C. Steele, T. Vasilos. , « Cation self-diffusion in single-crystal MgO », *J. Chem. Phys.* 58: 5258 (1972)
9. J.P. Poirier; *Creep of crystals*, ed. A.H.Cook, W.B.Harland, N.F.Hughes Cambridge University Press, p.233 (1985)
10. B. Lesage, A.M. Huntz, « Diffusion du chrome et du fer dans l'alumine monocristalline », *Scripta Met.* 14: 1143 (1980)
11. B.V. Dutt, J.P. Hurrel, F.A. Kroeger, « High temperature defect structure of cobalt-doped alumina », *J. Am. Ceram. Soc.* 58: 420 (1975)
12. B.V. Dutt, F.A. Kroeger, « High temperature defect structure of iron-doped alumina », *J. Am. Ceram. Soc.* 58: 474 (1975)
13. A. Peyrot, PhD thesis, « Etude physicochimique et microstructurale de la précipitation du chrome dans l'alumine », University Orsay-Paris XI (1995)
14. M. Martin, Habilitationsschrift, »Ueber das Verhalten von Uebergangsmetalloxiden im Sauerstoffpotentialgradienten », Universitaet Hannover (1993)
15. M.K Loudjani, R.Cortes, « X-ray absorption spectroscopy study of the local structure and the chemical state of yttrium in polycristalline alumina », *J. Eur. Ceram. Soc.* 14: 67 (1994)

CONTROL OF THE MICROSTRUCTURE OF POLYCRYSTALLINE DIAMOND AND RELATED MATERIALS VIA AN ENHANCED CVD PROCESS

Melanie L. Carasso, Sherry S. Staehle, Paul A. Demkowicz,
Donald R. Gilbert, Rajiv K. Singh, and James H. Adair

Department of Materials Science and Engineering
University of Florida
Gainesville, FL 32611-6400

INTRODUCTION

High quality diamond films have potential applications in high speed, high temperature electronic devices and hard, wear resistant coatings for cutting tools. Diamond is an ideal material for substrates or thin films in integrated circuits because it has the combined properties of high thermal conductivity, high electrical resistivity, and low dielectric constant.[1-5] In addition to passive roles as heat sinks, diamond is also being evaluated as a device material for microwave frequency and semiconductor applications. Diamond is desirable for electronics operating at microwave frequencies because of its low susceptibility to X-ray, ultraviolet, and gamma radiation damage.[3,5] The interest in diamond as a semiconductor stems from its ability to operate at temperatures between 100°C and 500°C, beyond the range of most smaller band gap semiconductors.[6]

The transparency of diamond to visible and infrared radiation, combined with its resistance to abrasion, chemical attack, and radiation damage, make it a potential candidate for windows and lenses or as a coating for these materials[3]. These products are suitable for use in aerospace vehicles and instruments and in other harsh chemical environments[4].

The first successful synthesis of diamond was achieved in 1955 by researchers at General Electric under pressure and temperature conditions of 55,000 atm and 2000°C[7]. Synthetic diamonds produced in this way are commonly used for cutting, grinding and polishing, but cannot be formed into thin films or coatings. Renewed interest in diamond synthesis has developed recently, focusing in particular on the possibility of synthesizing diamond from the vapor phase at much lower pressures and temperatures. Some early success was achieved independently by Derjaguin[7], Eversole[8], and Angus[9], but the growth rates for these processes were typically very slow (Ångströms per hour) and co-deposition of graphite was a consistent problem, compromising the quality of the diamond produced. These techniques were based largely on the thermal decomposition of a carbon-containing gas and subsequent deposition of carbon atoms onto a substrate, a process known as chemical vapor deposition (CVD).

Further innovations in the growth of diamond films from the vapor phase were introduced in the 1970s. Techniques involving gas activation via plasmas or hot filaments increased growth rates considerably and also reduced the amount of simultaneous graphite deposition[3]. The use of atomic hydrogen was found to assist the elimination of graphitic carbon by selectively etching deposited graphite and other undesirable sp^2-bonded phases from the growth surfaces[3,7,10]. The deposition of high quality diamond films at cost-effective rates is the primary goal of current diamond film research.

The growth of diamond films can be enhanced by seeding the substrate to create nucleation sites prior to CVD. Seeding techniques usually involve ultrasonication of the substrate in diamond suspensions[11,12] or scratching the substrate with diamond paste[2,13,14]. These methods usually result in damage to the substrate. Valdes et al.[15] developed an electrophoretic deposition technique which was non-damaging to the substrate, however the process is limited to substrates which are both conductive and resistant to (usually non-aqueous) electrolyte solutions.

In this paper we report the development of a nondestructive technique for seeding substrates with diamond. The procedure involves electrostatic deposition of diamond particles onto the substrate via an adsorbed layer of charged polymer[16]. Using an electron cyclotron resonance (ECR) CVD procedure, we have been able to grow continuous diamond coatings on these seeded substrates over relatively large areas.

CHARACTERIZATION OF DIAMOND

The surface chemistry of diamond in aqueous suspension has not yet been clearly resolved in the literature. Using electrokinetic and contact angle measurements on diamond exposed to different surface treatments, Shergold and Hartley[17,18] concluded that the groups on the surface of diamond could include carboxylic acids, phenols, quinones, carbonyls, and lactones. In contrast, Hansen et al.[19] deduced from contact angle and X-ray photoelectron spectroscopy experiments that the diamond surface was comprised of hydroxyl groups and no carboxylic or benzoic sites. To characterize the surface chemistry of diamond in aqueous suspension, we measured the zeta potential[20] of diamond particles as a function of pH.

Diamond powder[a] was washed thoroughly with deionized water to a constant specific conductivity (< 5 µS/cm) then suspended in potassium chloride[b] solutions of varying concentration. Samples of diamond suspension were taken and the pH was adjusted by the addition of potassium hydroxide[b] or hydrochloric acid[b]. The zeta potential of the samples was measured using a Brookhaven ZetaPlus device[c]. The ZetaPlus detects the movement of the particles by a frequency shift in the light scattered by the particles (Doppler shift). Student's T test was used to calculate the 95% confidence interval for the measured zeta potentials.

The zeta potential of water-washed diamond as a function of pH is shown in Figure 1. The diamond particles were negatively charged over the pH range 2 to 12, with zeta potentials between -20 mV and -40 mV above pH 3. Increasing the concentration of KCl

---

[a] General Electric Series 300, 0-0.5 µm diameter, General Electric Superabrasives, Worthington, Ohio.
[b] Fisher Scientific, Fair Lawn, New Jersey.
[c] Brookhaven Instruments Corporation, Holtsville, New York.

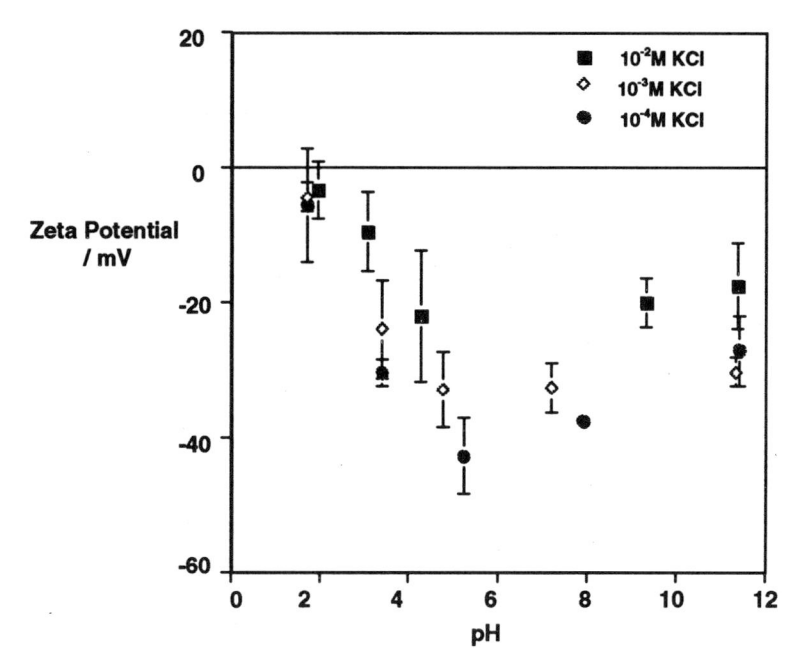

Figure 1. Zeta potential of diamond as a function of pH for various KCl concentrations.

resulted in a decrease in zeta potential, a phenomenon known as double layer compression.[20]

Further experiments showed that washing the diamond powder with base (0.1 M sodium hydroxide[b]) instead of deionized water resulted in more negative zeta potentials over the pH range. Washing with acid (0.1 M nitric acid[b]) led to less negative zeta potentials. This indicates that there are oxidizable groups on the surface of diamond, leading to ionizable protons which control the surface chemistry in aqueous suspensions.

ELECTROSTATIC DEPOSITION

The electrostatic deposition method for diamond particles on arbitrary surfaces is based on consideration of the surface charges which develop on particles and substrates when immersed in water. Silicon wafers are commonly used as substrates in diamond deposition studies. The native surface layer on these wafers is silicon dioxide, a material which has been extensively characterized and carries a negative charge in aqueous environments above pH 2[21]. Hence the electrostatic forces operating between diamond particles and the native surface of silicon are repulsive. These repulsive forces must be overcome in order to attach diamond particles to silicon substrates. In the electrostatic deposition technique, this is accomplished via a surface pretreatment of positively charged polymer to create an electrostatically attractive 'bridge' between the diamond and silicon surfaces.

The cationic polymer used in our experiments was poly(ethyleneimine) (PEI)[d], a highly branched polymer which exhibits a positive charge over most of the pH range[22]. We have found that this polymer adsorbs strongly to silicon, providing a firm anchor for the diamond seeds. Silicon wafers[e] were treated with a 1 - 2% weight percent aqueous

---

[d] Poly(ethyleneimine), 50% in water, Eastman Kodak Company, Rochester, New York.
[e] Recticon Enterprises, Inc., Pottstown, Pennsylvania.

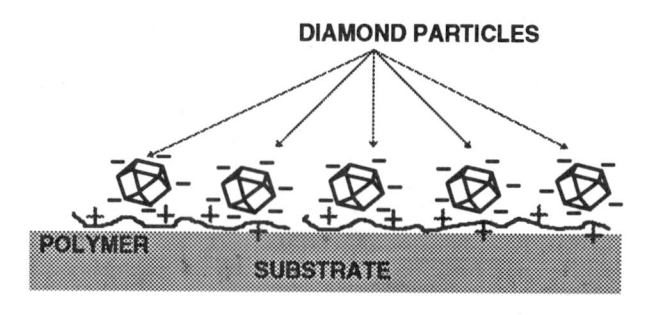

Figure 2. PEI and diamond adhered to substrate.

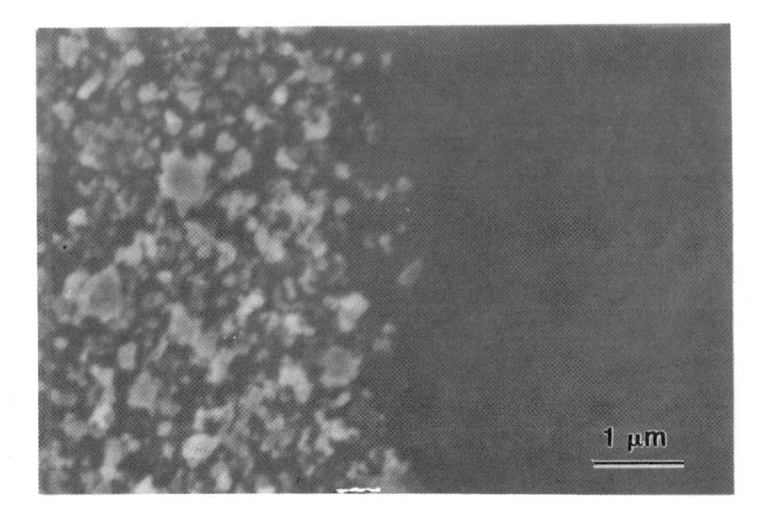

Figure 3. Scanning electron micrograph of a PMMA substrate. The left half was treated with PEI, then the whole surface was exposed to colloidal diamond.

solution of PEI for 2 minutes to impart a positive charge to the substrate. The substrates were vigorously rinsed with deionized water and suspended in aqueous colloidal diamond suspension for 2 minutes, then rinsed again with deionized water and allowed to air dry. Figure 2 shows a schematic representation of the resulting seed layer.

The effectiveness of the polymer treatment in the seeding process is illustrated in the scanning electron micrograph[f] shown in Figure 3. The substrate used in this experiment was poly(methyl methacrylate) (PMMA)[g], which carries a negative surface charge above pH 4[23]. Only the left half of the substrate was exposed to PEI solution, then the entire sample was treated with colloidal diamond. Diamond seed particles adhered exclusively to the polymer-coated region of the surface.

The concentration of the diamond suspension used during the electrostatic deposition affects the number density of seeds on the substrate. Figures 4a and 4b are scanning electron micrographs of silicon wafers treated with 1.0% v/v diamond and 0.1% v/v diamond respectively, after the PEI coating. Exposure to the more concentrated diamond suspension clearly produced a more dense coverage of the substrate.

---

[f] JEOL JSM-6400, JEOL USA, Peabody, Massachusetts.
[g] Courtesy of Dr. Roger Pryor, Wayne State University, Detroit, Michigan.

(a) **(b)**

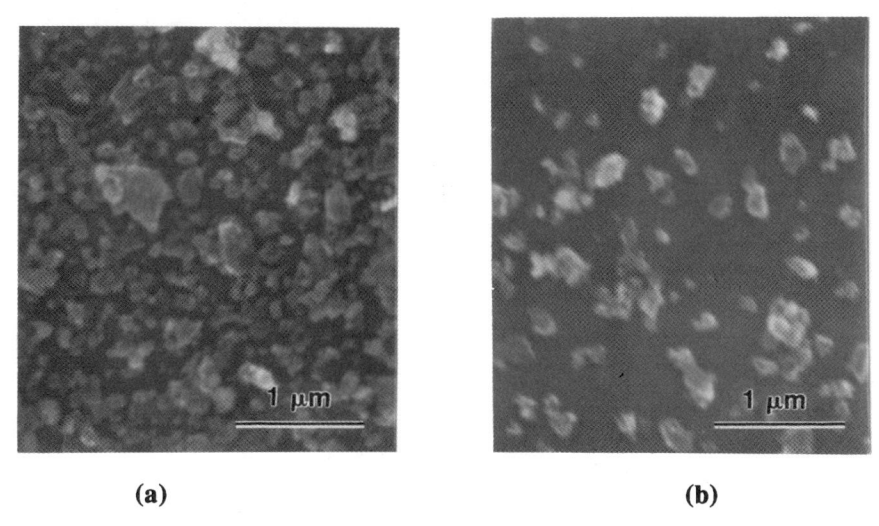

Figure 4. Scanning electron micrographs of silicon substrates after electrostatic deposition in (a) 1.0 weight % diamond, (b) 0.1 weight % diamond.

ECR CVD

Diamond thin films were grown on substrates which had been seeded using the electrostatic deposition procedure. Film growth was achieved using an electron cyclotron resonance (ECR) CVD technique, which has been described previously[24,25]. The ECR system is shown schematically in Figure 5. Rings of permanent magnets on the outside of the ECR module generate a magnetic field of strength 875 Gauss inside the vacuum chamber. Microwave energy of frequency 2.45 GHz and power 1000 W is directed through the quartz window at the top of the ECR module. The neutral source gas mixture of methanol and hydrogen (typically 1% CH_3OH:99% H_2, total pressure 1 Torr) is injected into the ECR chamber, where it is activated by the ECR plasma to produce the active species necessary for diamond growth. The overall reaction which occurs is[3]

$$CH_4(g) \rightarrow C(diamond) + 2H_2(g)$$

Diamond deposition occurs on the seeded substrate, which is placed on a heated plate (~700°C) in the vicinity of or downstream from the ECR zone. The co-deposition of graphite is controlled by atomic hydrogen in the gas mixture.

Although the ECR method does not produce particularly rapid growth rates, it does offer several advantages over other CVD techniques. The coupling of the plasma with the magnetic field generates a uniform distribution of the activated species, which enables large substrate areas to be coated. The low pressure and temperature of the plasma does not interfere with the heating of the sample plate, allowing good control over the substrate temperature.

Figure 6 shows an SEM image of a polycrystalline diamond film grown under ECR conditions for 8.5 hours on an electrostatically seeded silicon wafer. The film was dense and continuous over the entire substrate area, with a relatively uniform crystal size of about 1 µm. Figure 7 is an AFM image[h] of a similar sample, also demonstrating the continuity and density of the film. A sample grown under ECR for 2 hours was used for AFM step height analysis[i], giving a film thickness of 0.5-0.8 µm.

---

[h] Digital Instruments Nanoscope III, contact mode, Si_3N_4 tip.
[i] Topometrix, contact mode, Universal stage, Si_3N_4 tip.

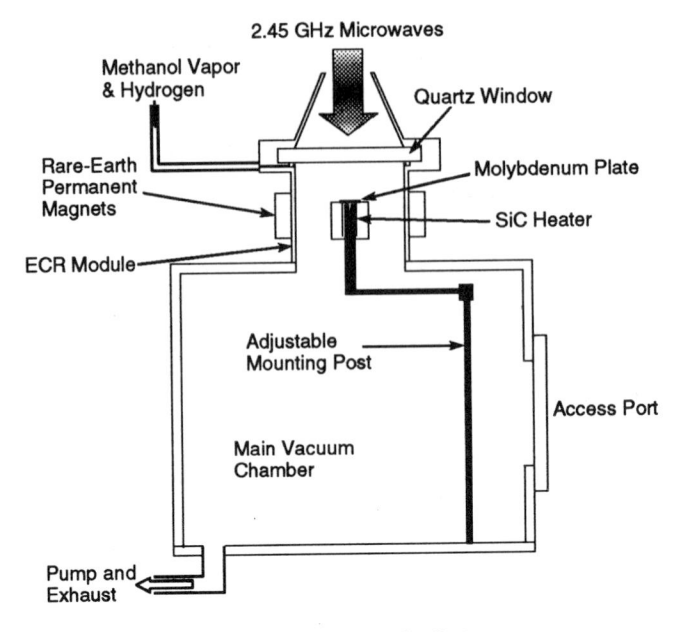

Figure 5. The microwave ECR CVD system.

Figure 6. Scanning electron micrograph of diamond film after electrostatic deposition and 8.5 hours ECR.

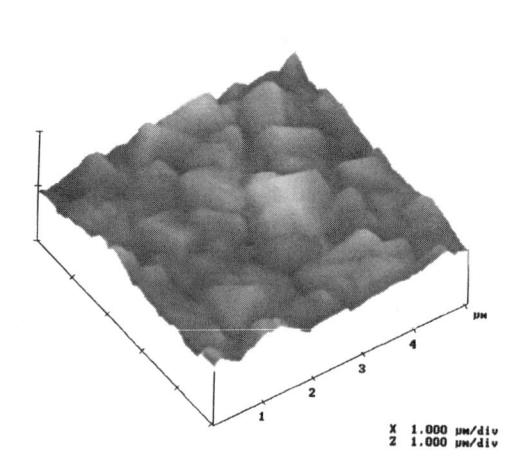

Figure 7. Atomic force microscopy image of diamond film.

SUMMARY

A nondestructive electrostatic deposition technique for seeding substrates with colloidal diamond has been described. An adsorbed layer of charged polymer is used to attach the diamond particles to the underlying substrate. The growth of continuous diamond films over areas up to 80 cm^2 (12.5 sq. in.) has been achieved via ECR CVD.

REFERENCES

1. Hoover, D.S., *Solid State Tech.*, Feb., 89 (1991).
2. Ramesham, R., Roppel, T., Ellis, C., and Rose, M.F., *J. Electrochem. Soc.*, **138**, 1706 (1991).
3. Spear, K.E., *J. Am. Ceram. Soc.*, **72**, 171 (1989).
4. Field, J.E., ed., *The Properties of Diamond*, Academic Press, London (1979).
5. Shenai, K., Scott, R.S., and Baliga, J., *IEEE Transactions on Electron Devices*, **36**, 1811 (1989).
6. Geis, M.W. and Angus, J.C., *Scientific American*, **267**, 84 (1992).
7. Derjaguin, B.V. and Fedoseev, D.B., *Scientific American*, **233**, 102 (1975).
8. Eversole, W.G., U.S. Patent No. 3,003,188 (1962).
9. Angus, J.C., Will, H.A., and Stanko, W.S., *J. Appl. Phys.* **39**, 2915 (1968).
10. Angus, J.C., Buck, F.A., Sunkara, M., Groth, T.F., Hayman, C.C., and Gat, R., *MRS Bulletin*, **14**, 38 (1989).
11. Ramesham, R. and Roppel, T., *J. Materials Research*, **7**, 2785 (1992).
12. Masood, A., Aslam, M., Tamor, M.A., and Potter, T. J., *J. Electrochem. Soc.*, **138**, L67 (1991).
13. Davidson, J.L., Ramesham, R., and Ellis, C., *J. Electrochem. Soc.*, **137**, 3206 (1990).
14. Narayan, J. and Chen, X., *J. Appl. Phys.*, **71**, 3795 (1992).
15. Valdes, J.L., Mitchel, J.W., Mucha, J.A., Seibles, L., and Huggins, H., *J. Electrochem. Soc.*, **138**, 635 (1991).
16. Adair, J.H. and Singh, R.K., U.S. Patent No. 5,485,804 (1996).
17. Shergold, H.L. and Hartley, C.J., *Int. J. Mineral Processing*, **9**, 219, (1982).
18. Hartley, C.J. and Shergold, H.L., *Chemistry and Industry*, **6**, 244 (1980).
19. Hansen, J.O., Copperthwaite, R.G., Derry, T.E., and Pratt, J.M., *J. Colloid Interface Sci.*, **130**, 347 (1989).
20. Hunter, R.J., *Foundations of Colloid Science, Volume I*, Oxford University Press, Oxford, (1986).
21. Iler, R.K., *The Chemistry of Silica*, Wiley-Interscience, New York, (1979).
22. Horn, D., in: *Polymeric Amines ands Ammonium Salts*, E.J. Goethals, ed., Pergamon Press, Oxford (1980).
23. Hunter, R.J., *Introduction to Modern Colloid Science*, Oxford University Press, Oxford, (1993).
24. Singh, R.K., Gilbert, D., Tellshow, R., Holloway, P.H., Ochoa, R., Simmons, J.H., Koba, R., *Appl. Phys. Lett.*, **61**, 2863 (1992).
25. Gilbert, D.R. and Singh, R., in: *Advances in Coatings Technologies for Corrosion and Wear Resistant Coatings*, Srivatsa, A.R., Clayton, C.R., and Hirvonen, J.K., eds., The Minerals, Metals & Materials Society, (1995).

MICRODESIGNED INTERFACES:
NEW OPPORTUNITIES FOR STUDIES OF SURFACES AND GRAIN BOUNDARIES

Mikito Kitayama, James D. Powers,
Lawrence Kulinsky and Andreas M. Glaeser

University of California, Berkeley, and Center for Advanced Materials,
Lawrence Berkeley National Laboratory
Berkeley, CA 94720

ABSTRACT

The development of methods for producing highly controlled internal voids in ceramics provided a new class of experimental methods for studying the properties of surfaces and interfaces in ceramics. Recent work, based on refined capabilities for producing microdesigned internal defect structures has broadened the range of problems that can be addressed. This paper reviews and previews recent research focussing on improving our understanding of surface diffusion in ceramics, providing experimentally determined values of surface energies in doped and undoped sapphire, and on developing new approaches to generating graded microstructures and single crystals by solid-state routes.

INTRODUCTION

In the mid-1980's, a method for fabricating highly controlled internal defect structures in ceramics was developed.[1] This method relies on a combination of photolithographic processing, ion beam etching, and solid-state diffusion bonding. This technique has been adapted and applied to the study of a wide variety of phenomena in single-crystal sapphire and polycrystalline alumina. In sapphire, it enables studies of high temperature crack healing in materials of known chemistry and crystallographic orientation. The morphological (Rayleigh) instability of controlled-radius pore channels of known crystallographic orientation can also be studied in doped and undoped material, allowing the interplay between crystallographic and chemical effects to be probed. When applied to single crystal/polycrystal assemblies, the method permits quantitative measurements of pore-boundary separation in alumina, and provides a highly accurate method for studying grain boundary migration. Studies of all these behaviors have been performed, and extensively reported in the literature.[2-7]

This paper focuses on the more recent developments and extensions of these processing techniques, and the experiments which these developments have made possible. The use of lithography to produce preperturbed pore channels has provided a powerful new method for investigating surface stability and surface diffusion in sapphire. New insights on the role of surface energy anisotropy are emerging. By combining the lithography-based methods with atomic force microscopy, a new experimental approach for characterizing the Wulff shape of undoped and doped sapphire has emerged. The effects of temperature and impurities on relative surface energies are now being examined, and links between morphological stability and dopant-induced surface stabilization are indicated. In tandem with these studies of solid-vapor interfaces, solid-solid interfaces are also being investigated, with particular emphasis on the role that low level dopant additions and dopant valence gradients can play on microstructural evolution. Doping stimulated anisotropies in grain boundary properties may allow the formation of novel microstructures exhibiting desirable properties or property gradients. Progress in these areas is reviewed here.

I. MORPHOLOGICAL EVOLUTION OF PRE-PERTURBED PORE CHANNELS IN SAPPHIRE

One of the first problems to be addressed using controlled-morphology pore arrays was the development of Rayleigh instabilities in long pore channels.[2,5,6] In theory, channels should develop perturbations of a characteristic wavelength, which increase in amplitude and eventually cause the channel to break up into pores with a spacing determined by the surface energy anisotropy of the material (and the controlling transport mechanism, assumed in these experiments to be surface diffusion). By measuring the eventual pore spacings, it should be possible to identify the key parameters of breakup, which in turn provides information about the governing surface energy anisotropy. In practice, the perturbation growth rate varies weakly enough with wavelength that the eventual pore spacing distribution is quite broad, and it is difficult to identify the characteristic wavelength with any confidence.[5,6] For this reason, an experimental method utilizing pre-perturbed pore channels was developed.

The introduction of an artificial, controlled-wavelength perturbation of sufficient amplitude into a pore channel strongly biases the channel to break up with the specific wavelength of the initial perturbation. This minimizes the competition and interference between naturally-developing perturbations, and so the kinetics of breakup are controlled by a single perturbation and represent its growth (or shrinkage) rate. By introducing channels with imposed perturbations of several different wavelengths, differences in evolution rate can easily be observed, and the results can easily be fit to models for channel breakup. This technique also allows determination of the other characteristic parameter of Rayleigh breakup, the minimum wavelength perturbation which will grow, which is itself a function of the anisotropy.

Examples of channels in sapphire containing well-defined sinusoidal perturbations are shown in Figure 1. As anticipated, when samples containing such channels were annealed at high temperature (1650-1700°C), a broad range of (imposed) wavelengths emerged within

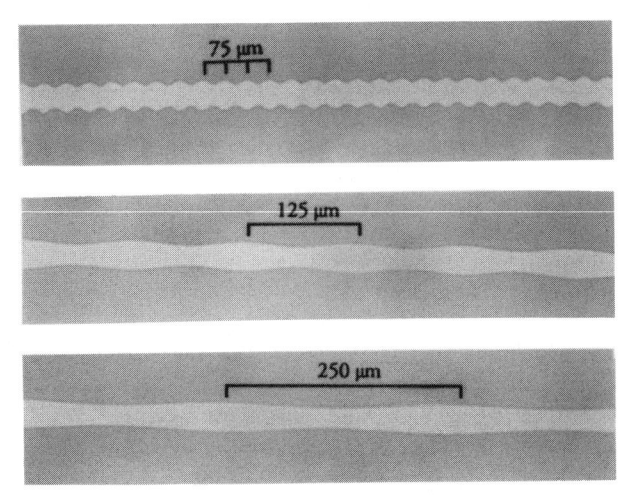

Figure 1.   Pre-perturbed pore channels in sapphire prior to evolution. Note the well-defined sinusoidal perturbations, and that the imposed wavelength can be varied in a controlled manner.

which the channels eventually broke up into isolated pores with an interpore spacing identical to the initial imposed wavelength (see Figure 2). This is an important result, as it indicates that the imposed perturbation dominates the evolution, and therefore, measured parameters such as the breakup rate can be considered as functions of the imposed wavelength only.

By measuring the time required for each channel to break up into pores, it is possible to determine both key parameters of breakup: the thermodynamic minimum wavelength (λ_{min}), below which perturbations will not grow, but rather damp out, and the kinetic maximum wavelength (λ_{max}), the wavelength which evolves most quickly. The relevant data and our best estimates for the values of these parameters are indicated in Figure 3 and Table 1.

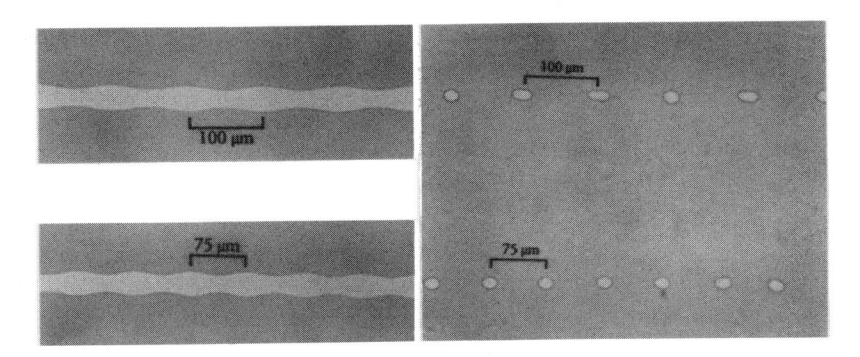

Figure 2. Channels before and after breakup. Note that the spacing of isolated pores (at right) is identical to the original wavelength imposed on the channel.

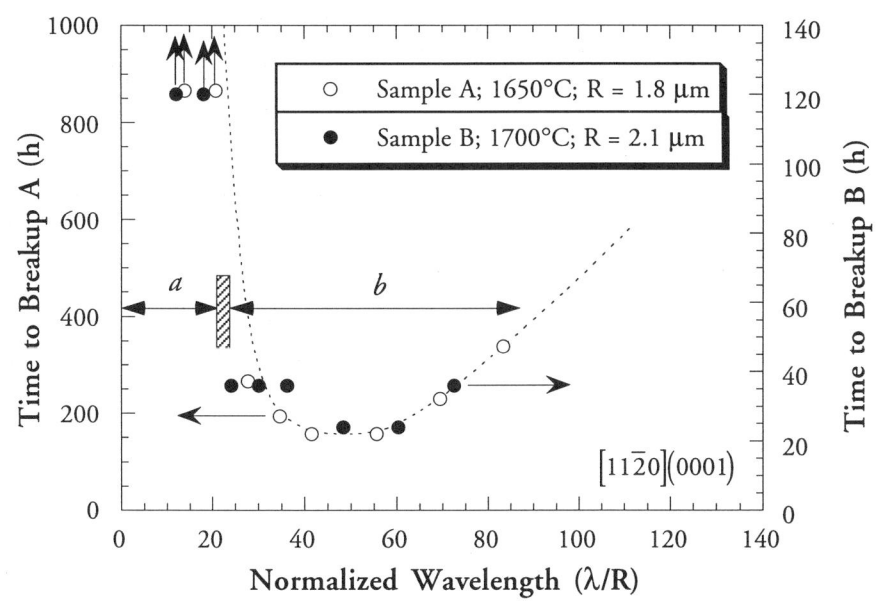

Figure 3. Effect of wavelength and temperature on the breakup kinetics of pore channels of equivalent circular radius R, oriented parallel to [11$\bar{2}$0] in the basal plane of undoped sapphire. *a* denotes the region where $\lambda < \lambda_{min}$ and *b* the region in which the imposed wavelength controls the behavior. The wavelength yielding minimum breakup time defines λ_{max}.

Table 1 Tabulation of λ_{min} and λ_{max} for two undoped samples.

	Isotropic	Sample A, 1650°C	Sample B, 1700°C	
Orientation	All	$[11\bar{2}0](0001)$	$[11\bar{2}0](0001)$	$[1\bar{1}00](0001)$
λ_{min}	$2\pi R \approx 6.28R$	$21R < \lambda_{min} < 28R$	$18R < \lambda_{min} < 24R$	$12R < \lambda_{min} < 18R$
λ_{max} (surface diffusion)	$\sqrt{2} \cdot \lambda_{min} \approx 8.89R$	$42R < \lambda_{max} < 56R$	$48R < \lambda_{max} < 60R$	$24R < \lambda_{max} < 30R$

A more detailed analysis of the behavior of these channels indicates that although the anisotropy effect will shift the values of λ_{min} and λ_{max}, the behaviors, when re-expressed in terms of appropriate normalized wavelength and normalized time parameters, can all be reduced to a single curve. The success of this normalization implies that the underlying modelling provides a useful first-order correction for the effects of anisotropy. Thus, parameters quantifying the surface energy anisotropy can be extracted from the data, and the effects of anisotropy can be accounted for when the surface diffusivity is estimated. A more complete discussion of the data and the theoretical underpinnings has been published elsewhere.[7]

The ability to introduce such controlled-geometry pore structures as the pre-perturbed channels used in these experiments clearly gives us the capability of designing samples with a degree of experimental control unattainable in previous work. This is one of the most attractive features of the method of microdesigned interfaces. Another is its flexibility: it can be applied to a number of different problems, as can be seen in the next section.

II. TEMPERATURE DEPENDENCE OF THE WULFF SHAPE IN TI-DOPED ALUMINA

The Wulff shape graphically displays the surface energy anisotropy of a material, and can be used to assess the relative surface energies of different crystallographic planes.[8,9] Studies of the Wulff shapes of low vapor pressure metals have been performed utilizing ultra-high-vacuum furnace environments to maintain clean surfaces.[10,11] In contrast, little experimental work has been performed on ionic crystals, in large part because of the relatively greater experimental difficulties. Recently, lattice simulations have provided estimates of the surface energies and heats of segregation for various low-index (and presumed stable) planes in alumina.[12,13] Dopants, particularly those that segregate anisotropically, can radically alter the Wulff shape. This work focused on developing a new technique for characterizing the Wulff shape of alumina and assessing the effect of Ti doping on the Wulff shape.

Pores, individually 16 × 16 × 0.5 µm with a 16 µm edge to edge spacing, were laid out in 250 × 250 pore arrays, resulting in a total of 62,500 lithographically introduced internal pores.[14] Such arrays were generated on surfaces of various crystallographic orientations. For Ti-doped sapphire, the originally flat cavities are oriented with their large faces parallel to the $(11\bar{2}0)$ face of sapphire. A second substrate of identical orientation was placed upon the etched surface and aligned to produce, at worst, a very low angle misorientation twist boundary. The two sapphire substrates were bonded to one another by hot pressing at 1300°C with an applied pressure of ≈10 MPa. Very little adjustment of the pore shape occurred during the bonding process. Subsequent anneals were conducted at 1600°C and 1800°C. Due to the transparency of the sample, it was possible to observe the shape changes using optical microscopy, and to continue the annealing until no further shape changes were evident.

To expose the equilibrated pores, the annealed specimens were polished, with the bond plane containing the pores at a slight incline (1-2°) to the plane of polish. This made it possible to expose a strip of the pore array, while maintaining the other pores fully enclosed by sapphire. The pores exposed in this manner were examined using a scanning electron microscope (SEM) and an atomic force microscope (AFM) to identify the facet structures.

Figures 4a and 4b are SEM micrographs of exposed pores in Ti-doped alumina annealed for 160 h at 1600°C and for 48 h at 1800°C, respectively. The substrate has an a($11\bar{2}0$) orientation. The c(0001) and m($10\bar{1}0$) orientations are mutually orthogonal and the directions are as indicated in the figures. Facets that intersected the pore perimeter and were normal to the substrate face were identified as c and r($10\bar{1}2$) planes from their intersection angles and knowledge of the crystallography of alumina. Facets within the pore were identified using AFM line scan analysis. A line scan was performed along the c-direction intersecting the middle of a pore annealed at 1600°C. The facet was found to be inclined by 27.7° with respect to the a-plane for this trace direction, which indicates these facets are p($11\bar{2}3$) planes (28.8° theoretical). Two distinct facets observed inside a pore annealed at 1800°C were also identified using an AFM line scan, and measured inclination angles of 42.4° indicate that these facets are r-planes (43.0° theoretical). All facets identified using these methods are indicated in the figures. While p-facets are clearly visible in the specimen annealed at 1600°C, they are a much less prominent feature in the specimen annealed at 1800°C.

By alternating between the two anneal temperatures, and monitoring the shape changes, it was possible to establish that the shape changes were reversible, strongly suggesting that the shapes observed after prolonged anneals at 1600°C and 1800°C are the Wulff shapes of Ti-doped alumina at these temperatures. The fact that the ($11\bar{2}0$) plane originally bounding the etched-in cavities decomposed into other planes indicates that the a-plane is not a low energy plane in Ti-doped alumina. By combining the results of AFM measurements, and SEM examination of the facet structure, it was possible to calculate the surface energies relative to that of the (0001) plane. In these calculations, the surface energy of the basal plane is set equal to one, i.e., $\gamma_{0001} = 1$. The results are summarized in Table 2, and compared to the findings for undoped alumina at 1600°C derived from the study of Choi et al.[15] If Ti would segregate to all faces and lower the surface energy by the same fractional amount, then the Wulff shape would be unaffected by doping. The comparison shows that the introduction of Ti leads to the loss of the ($11\bar{2}0$) plane, and as observed in undoped sapphire, the energy of the m($10\bar{1}0$) plane is high, and thus this plane is not part of the Wulff shape. This suggests anisotropic segregation. Limited experiments suggest that the surface tensions of *liquids* decrease with increasing temperature. If the relative energies of different facets in solids change with changes in temperature, then the Wulff shape will be temperature dependent as observed. When dopants are present, and the segregation behavior is facet specific and temperature dependent, changes in the temperature may induce changes in the relative degrees of segregation, and additional contributions to the temperature dependence of the Wulff shape may arise.

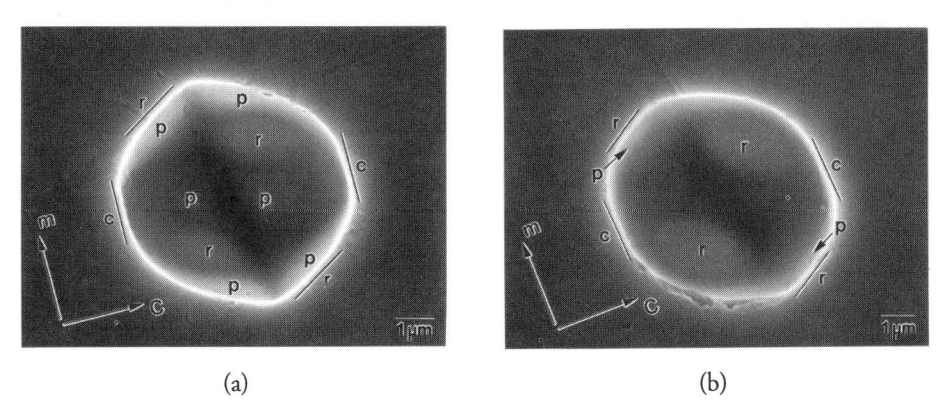

(a)   (b)

Figure 4.   SEM micrograph of the pore shape of Ti-doped alumina (left) annealed at 1600°C for 160 h, and (right) annealed at 1800°C for 48 h.

Table 2. Relative surface energies of Ti-doped alumina[*1]

Plane	NIST[*2] Undoped @ 1600°C	This work Ti-doped @ 1600°C	This work Ti-doped @ 1800°C
c(0001)	1	1	1
m($10\bar{1}0$)	>1.16[*3]	>1.33[*3]	>1.46[*3]
s($10\bar{1}1$)	1.07±.02	>1.30[*3]	>1.35[*3]
r($10\bar{1}2$)	1.05±.02	1.06±.01	1.11±.01
a($11\bar{2}0$)	1.22±.05	>1.22[*3]	>1.37[*3]
p($11\bar{2}3$)	1.06±.02	≈1.06	≈1.2

[*1] setting the surface energy of the basal plane as unity; [*2] reference 15; [*3] not observed, lowest possible values.

III. TITANIUM EFFECTS ON SINTERING AND BOUNDARY MOBILITIES IN ALUMINA

In the work described above, dopant and anisotropy effects on the evolution of pore structures in sapphire have been studied. From this work (and other previous work in the field), it becomes apparent that small amounts of impurity can cause significant changes in surface energies, evolution rates, and anisotropy. These results have implications for microstructure development in polycrystalline systems as well. By studying impurity effects in polycrystalline systems and relating the results to what is observed in simpler systems, it should be possible to develop a more complete understanding of the different ways that impurities alter properties and behavior. Currently, the work is focusing on Ti-doped alumina, a system which exhibits some very interesting and unusual properties. Descriptions of two of the experiments being performed, some of the results from each, and some implications for materials design which arise from these results are discussed here.

When attempting to study the effects of a specific dopant on processing behavior, it is important to limit the extent to which other impurities are present in the material. In this work, the effects of only 500 ppma Ti were being studied, which mandates an extremely low background impurity level. For this reason, ultra-pure alumina (<40 ppma total cation impurity) was used in this research. Processing steps were designed to minimize contamination, and indeed, chemical analysis (by spark source spectroscopy) performed after all processing and firing steps indicates no increase in background impurity level. Details on the processing are given elsewhere.[16]

In the first set of experiments, the effect of Ti on microstructure development during sintering was studied. During this study, it became apparent that firing atmosphere plays a significant role. Titanium is a multivalent impurity in alumina; Ti^{3+} and Ti^{4+} can both be present. In this work, Ti dopant was incorporated as TiO_2 (Ti^{4+}); however, it was necessary to sinter in vacuum to avoid contamination from the environment, which would favor the reduction of Ti from 4+ to 3+. There is qualitative evidence that such a reduction occurs during sintering of these compacts. The outer shell (≈1mm thick) of each compact acquires a pink color, indicative of the presence of Ti^{3+}; the inside remains white, however, suggesting no significant reduction at the compact center. Analysis of the absorption spectrum across these samples confirms the change in valence.

The difference in valence between the center and exterior of these samples induces a corresponding and significant difference in microstructural evolution. During sintering at high temperature (1550°C), the interior, Ti^{4+}-dominated region, develops an anisotropic, faceted microstructure, while the edge regions develop a much more equiaxed microstructure (see Figure 5). In undoped samples, there was no significant difference between sample edges and centers; the development of these gradient microstructures appears to arise from the interplay between the multivalent impurity and the reducing nature of the atmosphere.

This difference in microstructure between the sample center and edge, driven by impurity valence differences, has interesting implications for materials design. Note that there are no gradients involved in either the green compact or during subsequent processing, and yet a significant microstructural gradient exists in the finished ceramic. This suggests the possibility of creating ceramics with graded microstructures from uniform starting materials, using simple processing techniques. Since the major cost involved in making functionally graded materials is usually in the preparation of graded precursors and/or the control of temperature or other gradients during processing, such a simple technique, utilizing properties of the impurities themselves rather than the processing techniques, seems promising.

In the second set of experiments to be discussed here, the effects of Ti impurity of grain boundary mobility in alumina were determined. The anisotropic, faceted grain growth observed in the sintering work suggests the existence of anisotropy in boundary mobility (velocity per unit driving force); this was one of the reasons for pursuing this data.

Some previous experimental data on boundary mobilities in alumina[13,17,18] are plotted in Figure 6. A theoretically determined estimate of the intrinsic boundary mobility is also indicated in this figure.[18] The difference between the intrinsic and experimental values, a difference of several orders of magnitude, is generally attributed to the presence of impurities, which through their interaction with the grain boundary, exert a drag force that inhibits migration. Closing this gap by developing aluminas with significantly higher mobilities could have important applications for materials design. Alumina precursors, either bulk parts or continuous fibers, could be formed and then quickly converted into single crystal sapphire via growth from a sapphire seed. The crystallographic orientation of the final product could also

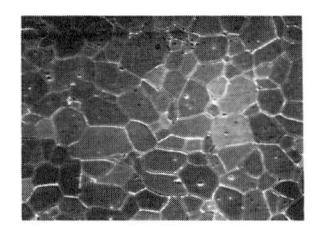

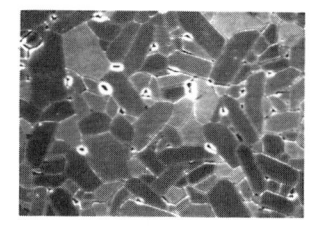

Figure 5. Microstructures of Ti-doped alumina sintered in vacuum for 2 h at 1550°C. Note microstructural differences between (a) sample edge and (b) sample center

be fixed by controlling the orientation of the seed, or by exploiting a highly anisotropic boundary mobility. In this work, as discussed below, we find that such an increase in boundary mobility is brought about by titanium additions.

An oriented sapphire wafer, diffusion bonded to a polycrystalline alumina substrate, will act as a seed for abnormal grain growth at high temperature. By measuring the migration rate of the sapphire-polycrystal boundary as it grows into the polycrystal and combining this data with information about the average grain size, it is possible to determine the boundary mobility M_b. By using sapphire seeds of different, known orientations, it is possible to determine the degree of anisotropy in M_b. In this work, two sapphire orientations were used: c-plane, or (0001), and a-plane, or (11$\bar{2}$0). This allows direct comparison with previous work in undoped alumina, where sapphire seeds of the same orientation were used, but alumina compacts of lower overall purity were studied.[3]

Doping with 500 ppm Ti significantly *increased* the boundary mobility in samples annealed at 1600°C with both c-plane and a-plane oriented sapphire seeds (see Figure 7 and Table 3). Such an enhancement is surprising, as impurities can usually be expected to reduce M_b by a solute or precipitate drag mechanism. The direction of these results is consistent with previous measurements by Horn and Messing[19], however, the degree of increase is much greater in this work. Both sets of data are indicated in Figure 6 as well, for comparison with prior work and with intrinsic values.

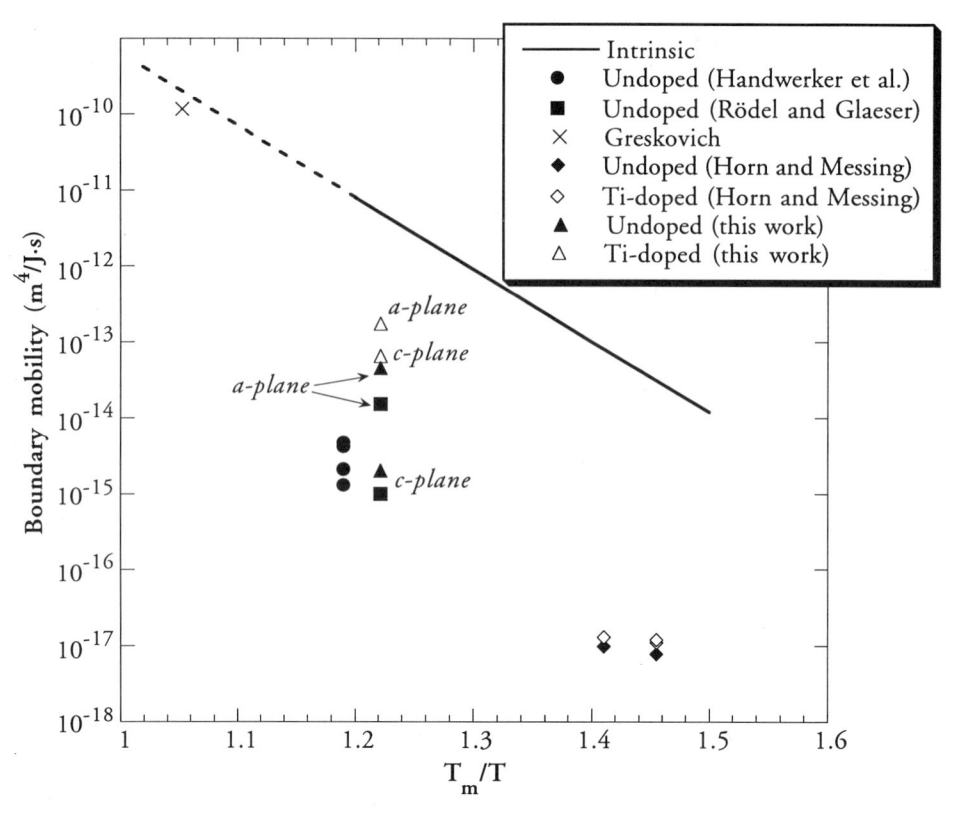

Figure 6. Comparison of experimentally determined boundary mobilities in alumina (data taken from references 3, 17, 18, 19) and a calculated estimate of the intrinsic mobility (from reference 18). Data from this work is also indicated.

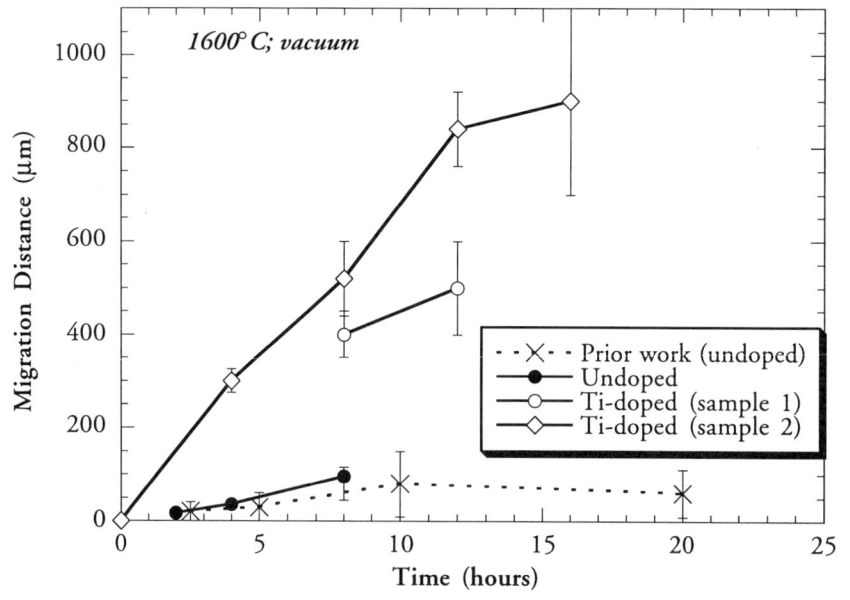

Figure 7. Grain boundary migration in Ti-doped and undoped alumina, using an a-plane oriented sapphire seed. The dotted lines represent previous work on undoped alumina. Results for c-plane sapphire were similar although mobilities are lower.

Table 3: Boundary mobilities in undoped and Ti-doped alumina ($M_b \times 10^{15}$ m^4/J·s)

Sapphire seed orientation	c-plane (0001)	a-plane (11$\bar{2}$0)
Undoped alumina (prior work; ref. 3)[*]	1	10-20
Undoped alumina (this study)	2	20-70
Ti-doped alumina (this study)	30-100	150-200

These results are consistent with the fact that grain growth in Ti-doped alumina is faster than in undoped material, as determined in these experiments. We note that since the polycrystal being consumed is dense, this variation of the grain size does not include potential effects of Ti on the sintering trajectory; the effect of Ti on grain boundary migration is being isolated. One possible explanation for the enhancement is a compensation effect: Ti^{4+} could interact with the divalent impurities (primarily Mg^{2+} and Ca^{2+}) and monovalent impurities to create stoichiometric complexes which reduce the degree of boundary segregation, and thus reduce the impurity drag. Note that, in these experiments, all firing was performed in vacuum, which favors a high Ti^{3+}/Ti^{4+} ratio. Therefore, the concentration of Ti^{4+} is much lower than the overall Ti concentration, and may in fact be comparable to the total concentration of divalent impurity, which is well below 100 ppm in these samples. If the mobility is in fact being altered by a compensation mechanism, then the mobility should be strongly dependent on both the overall Ti concentration and on the processing environment, which will also change the Ti^{4+} concentration. In ongoing work, the Ti content and firing atmosphere are being varied in order to observe the dependence of M_b on these parameters.

A significant anisotropy in boundary migration was evident in both undoped and doped samples, as well. It is not obvious from the current data that Ti-doping would promote anisotropic grain growth over undoped material, but future experiments with different orientations and firing atmospheres may shed some more light on this behavior. In any case, this technique for determining M_b is proving very useful for probing how impurities affect boundary behavior during processing.

SUMMARY

Microdesigned interfaces enable experiments which give us simple and direct access to information about surface and interface properties (energies, diffusivities, etc.). This and related techniques are continually being adapted, improved and extended in order to design model experiments which exert progressively more and more control over the systems being studied. Currently, the method is being used with great success to probe the thermodynamics and kinetics of surfaces and interfaces in alumina.

Studies of the high-temperature evolution of internal preperturbed pore channels have yielded quantitative measurements of the key parameters of breakup, λ_{min} and λ_{max}. These parameters provide information on the nature and degree of surface energy anisotropy in undoped and doped material, and allow us to correct for variations in surface energy anisotropy when determining the surface diffusivity.

A powerful new method for determining Wulff shapes has been developed, and applied to Ti-doped alumina. In continuing work, the Wulff shapes of undoped, Mg-doped, and Ca-doped alumina are being determined. The Wulff shape has been shown to vary with temperature, and the demonstration that the shape changes are reversible supports the view that true equilibrium pore shapes are being assessed.

Microstructure development in Ti-doped alumina has been shown to depend on the valence state of the dopant. This result raises the possibility of using multivalent impurities to develop graded microstructures with simple processing methods. Doping with 500 ppm Ti significantly increases the grain boundary mobility in alumina, and may suggest new methods of producing single crystals by solid-state methods. Work to determine the cause of this increase is ongoing.

ACKNOWLEDGEMENTS

The research described in this paper was primarily supported by the National Science Foundation under Grant No. DMR-9222644. Portions of the work performed by M. Kitayama relating to pore shape measurements, and some of the prior research referenced in this paper were supported by the Director, the Office of Energy Research, Office of Basic Energy Sciences, Materials Sciences Division of the U.S. Department of Energy under Contract No. DE-AC03-76SF00098. We also acknowledge an NSF Equipment Grant No. DMR-9119460 which allowed the acquisition of hot pressing equipment critical to this work. Throughout the period of research we have benefitted from many helpful discussions with our colleagues. We are particularly grateful for many stimulating conversations with Rowland Cannon Jr.. Finally, this work would not have been possible without the kind and continuing assistance of the staff of the Microfabrication Laboratory at the University of California..

REFERENCES

1. J. Rödel and A. M. Glaeser, Microdesigned interfaces: new opportunities for materials science, *J. Ceram. Soc. Japan*, **99**:251 (1991).
2. J. Rödel and A. M. Glaeser, High temperature healing of lithographically introduced cracks in sapphire, *J. Am. Ceram. Soc.*, **73**:592 (1990).
3. J. Rödel and A. M. Glaeser, Anisotropy of grain growth in alumina, *J. Am. Ceram. Soc.*, **73**:3293 (1990).
4. J. Rödel and A. M. Glaeser, Pore drag and pore-boundary separation in alumina, *J. Am. Ceram. Soc.*, **73**:3302 (1990).
5. J. D. Powers and A. M. Glaeser, High-temperature healing of cracklike flaws in Mg and Ca ion-implanted sapphire, *J. Am. Ceram. Soc.*, **75**:2547 (1992).
6. J. D. Powers and A. M. Glaeser, High-temperature healing of cracklike flaws in titanium ion-implanted sapphire, *J. Am. Ceram. Soc.*, **76**:2225 (1993).
7. L. Kulinsky, J. D. Powers and A. M. Glaeser, Morphological evolution of pre-perturbed pore channels in sapphire, *Acta Materialia*, **44**:4115 (1996).
8. G. Wulff, XXV. Zur Frage der Geschwindigkeit des Wachsthums und der Auflösung der Krystallflächen, *Z. Kristallgr.*, **34**:449 (1901).
9. C. Herring, Some theorems on the free energies of crystal surfaces, *Phys. Rev.*, **82**:87 (1951).
10. M. McLean, Determination of the surface energy of copper as a function of crystallographic orientation and temperature, *Acta Metall.*, **19**:387 (1971).
11. J. C. Heyraud and J. J. Metois, Equilibrium shape and temperature; lead on graphite, *Surf. Sci.*, **128**:334 (1983).
12. W. C. Mackrodt, Atomistic simulation of the surfaces of oxides, *J. Chem. Soc. Faraday Trans. 2*, **85**:541 (1989).
13. W. C. Mackrodt and P. W. Tasker, Segregation isotherms at the surfaces of oxides, *J. Am. Ceram. Soc.*, **72**:1576 (1989).
14. J. Rödel and A. M. Glaeser, A technique for investigating the elimination and coarsening of model pore arrays, *Materials Lett.*, **6**:351 (1988).
15. J. Choi, D. Kim, B. J. Hockey, S. M. Wiederhorn, C. A. Handwerker, J. E. Blendell, W. C. Carter and A. R. Roosen, The equilibrium shape of internal cavities in sapphire, *J. Am. Ceram. Soc.*, in press (1996).
16. M. Kitayama and J. A. Pask, Formation and control of agglomerates in alumina powder, *J. Am. Ceram. Soc.*, **79**:2003 (1996).
17. C. A. Handwerker, P. A. Morris, and R. L. Coble, Effects of chemical inhomogeneities on grain growth and microstructure in Al_2O_3, *J. Am. Ceram. Soc.*, **72**:30 (1989).
18. M. F. Yan, R. M. Cannon, and H. K. Bowen, Grain boundary migration in ceramics, in *Ceramic Microstructures '76*, Ed. R. M. Fulrath and J. A. Pask, Westview Press, Boulder, 1977.
19. D. S. Horn and G. L. Messing, Anisotropic grain growth in TiO_2-doped alumina, *Mat. Sci. and Eng. A*, **A195**:169 (1995).

EFFECT OF PACKING STRUCTURE OF POWDER PARTICLES IN A CERAMIC GREEN BODY ON MICROSTRUCTURE DEVELOPMENT DURING DENSIFICATION

Keizo Uematsu and Nozomu Uchida

Nagaoka University of Technology
Department of Chemistry
Kamitomioka, Nagaoka, Niigata, Japan 940-21

INTRODUCTION

Properties of virtually all ceramics are severely degraded by the presence of defects, which develop from powder packing irregularities in green bodies during the densification process.[1-3] To produce good ceramics, it is crucial to characterize these irregularities in green bodies and to establish their relation to defects in the resultant ceramics. A recent study, using a new characterization tool,[4-6] showed that all green bodies examined contained extremely large particles,[5-8] aggregates[5,6] and pores.[9-11] Large pores in a green body were found to grow[12,13] in accord with theory,[14] forming large voids in the microstructure during compact densification. The behavior of large aggregates and particles during the densification process has not been well studied, and examination of their behavior appears to be interesting.

This paper examines the behavior of large particles and aggregates during densification. A new characterization tool, developed by the authors several years ago,[4-6] was again used to observe the morphology of these features in green and sintered bodies. To critically examine the current level of ceramic engineering, we chose two systems.

One system is an advanced alumina powder of granulated form, representing the best of today's ceramics. The other is of a typical industrial grade low-soda alumina. The former compact consists of very fine particles of the average size ≈0.1 µm. It also contains a few aggregates, as has been reported in our previous papers.[15] The size of some aggregates is over 20 µm. In the aggregates, fine alumina particles are densely packed and share a common crystallographic orientation. The unique highly textured structure of the aggregate endows it with optical properties similar to those of an alumina single crystal, and causes a brightness change with rotation under a crossed polarized light microscope. Vigorous grinding might have broken these aggregate into smaller aggregates, or perhaps even to primary particles. Clearly, the manufacturer of the granules had left the regrinding to the user. Our use of the as-purchased granules for this study might not have been expected by the manufacturer, and the results may therefore be unfair to them. Nevertheless, these aggregates are preserved in the powder compacts, again generating the change of brightness with rotation during microscopic examination under crossed polarized light. The applied force of compaction is clearly insufficient to break them into fragments.

The subject of interest in the second system is large particles. They are found in all commercial powders which have been examined. They are believed to initiate abnormal grain growth during densification. In addition to the large aggregates and particles, both compacts contain other structural irregularities which are common in a green body made with powder granules. One is the low density regions of network structure at the boundaries of granules, the origin of which has been ascribed to the binder segregated at the surface of granules.[16] The other is the particle orientation in the boundary region of granules. The significance of this in ceramic processing has not been well understood. It may cause shrinkage anisotropy and residual internal stress in ceramics.

The structures of the specimens before firing is characterized by a tool called the immersion liquid technique.[4-6] With this method, the granules and green body are made transparent by immersion liquids which suppress the reflection of light at the particle-liquid interface. Detailed structures of granules and green bodies are observed by the subsequent examination with a transmission optical microscope and/or a polarized light microscope. For examining the structure of sintered bodies, the optical microscopic methods are again applied. The immersion liquid is not needed because a thin sintered ceramic specimen is transparent. No other common analytical tool can match the high sensitivity of this method in characterizing detrimental features. The bulk examination of the entire specimen provides us with detailed information of rare features, characteristic of the features of interest in ceramics, and features whose detection and characterization is very difficult using conventional tools. Comparison of structures before and after sintering may provide us with detailed information on microstructure development during densification.

EXPERIMENT

Two types of specimens were used for this study. One is made from commercial high purity alumina powder of very fine particles size (0.1 μm) and of granulated form (TM-DS, Taimeikagaku, Nagano, Japan), and the other is made from low soda alumina powder (160SG-1, Showadenko, Tokyo, Japan) of average particles size 0.6 μm. The first were formed into compacts by uniaxial pressing (10 MPa) followed by isostatic pressing(100 MPa). For densification, the compact was sintered at 1573 or 1873K for 3600 s in air. The bodies sintered at the former and the latter temperatures were then hot isostatically pressed at 1673 and 1823K, using a pressure of 100 MPa applied for 3600 s, respectively. The second powder of low-soda alumina was made into granules by spray drying with PVA (PVA-105, Kuraray, Tokyo, Japan) as a binder. A compact was prepared by uniaxial pressing at 50 MPa, followed by isostatic pressing at 100 MPa. To allow structural examination, these green bodies were heated to 773K and held for 1800 s in air to remove the binder.

A small piece was cut from the body and then thinned with grinding paper to ≈0.1 mm thickness by hand polishing. Bromonaphthalene (refractive index, n = 1.69) or methyleneiodide (n = 1.74) were used to make the sample transparent for microstructural examination in transmission mode using normal and crossed polarized light microscopes, respectively. For the examination in the former mode, the liquid must have a moderate refractive index mismatch with alumina. This is required since optical contrast is generated by the scattering of light at solid-liquid interfaces. For the latter mode of examination, the optical contrast is generated by the retardation of light, and good matching of refractive indices of the relevant materials is favorable to obtain high transparency and high optical contrast. For easy displacement of air with the liquid, a drop of liquid is placed at the edge of porous specimen and allowed to be absorbed by it. No evacuation is required provided occasional entrapment of air bubbles does not significantly interfere with the observation. A cover glass must be placed on the specimen immersed in the liquid to protect the lens; these liquids are corrosive to optical glasses.

After examination of the structure, the thin specimen made from the low soda alumina was heated for densification and for a subsequent examination. Structural change during sintering was followed by repeating the observation and heating. The same region within the specimen was observed after sintering at successively higher temperatures up to 1825K. The density of the specimen was determined using another pellet of the same material, which was subjected to the same thermal history. For examination of HIPed alumina, a thin slice was cut and the both faces were polished with a diamond powder (0.5 µm) to a thickness of approximately 50 µm for similar optical microscopic observations. No immersion liquid is applied for this observation.

RESULTS AND DISCUSSION

Figure 1 shows optical micrographs taken using normal transmission mode of alumina ceramics HIPed at 1573 and 1823K. The dark features in the micrograph are pores. Separate examinations show that these materials have full density and average grain sizes 2.8 and 5.9 µm, respectively.[17,18] In contrast to the naive expectation of a pore-free structure, both of them clearly contain many pores in the volume. The shape of large pores appears to be irregular, whereas, smaller pores are more equiaxed. The size and number of pores appear similar at both sintering/HIPing temperatures in accord with our former study.[17,18] The maximum pore diameter has been reported to exceed 150 µm for these ceramics.[18] In contrast to the first impression obtained from the micrographs, the total volume of these pores is actually low. The apparently high concentration of pores is due to the unusually high depth capability of the present examination tool. All pores located at various depth are observed simultaneously. This feature is a clear contrast to conventional methods, with which only the surface defects, such as pores exposed on a polished or fracture surface, can be observed. The structure shown at the center of the micrograph (a) is clearly formed from large crack-like structures at the contact region or boundary between granules. In a separate paper, we have reported a clearer micrograph of the pore network[15] formed from these irregularities. They arose from the network of low density region at the boundaries of granules. The lower temperature of HIPing (1627K) in the study left the pore network after the densification.

Figure 2 shows the crossed polarized light micrographs for alumina ceramics HIPed at 1673 and 1823K. In this observation mode,[5] an optically uniaxial material, such as

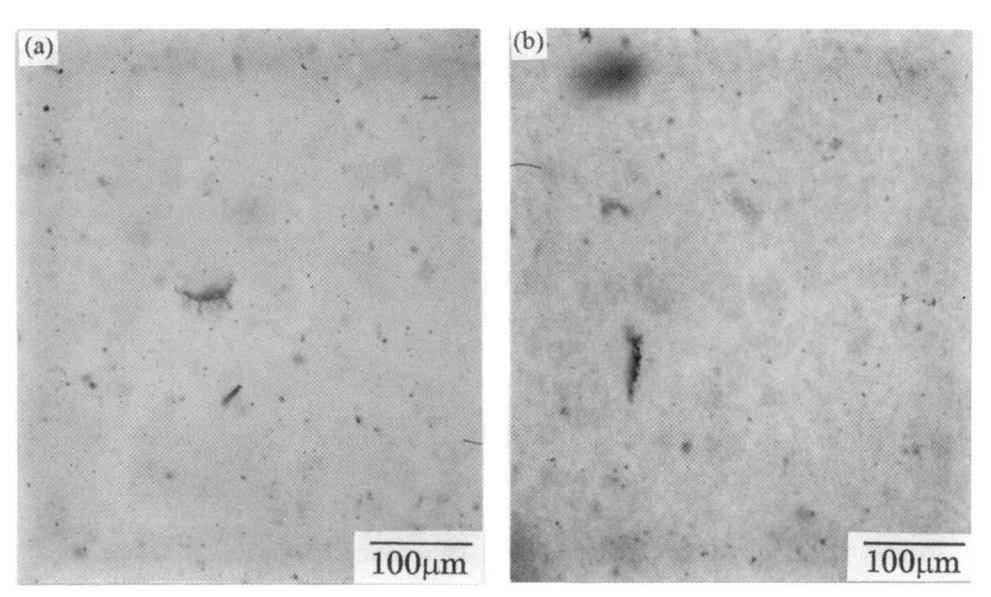

Figure 1 Optical micrographs of specimens examined at normal transmission (a) Sintered at 1573K and HIPed at 1673K, and (b) sintered at 1873K and HIPed at 1823K.

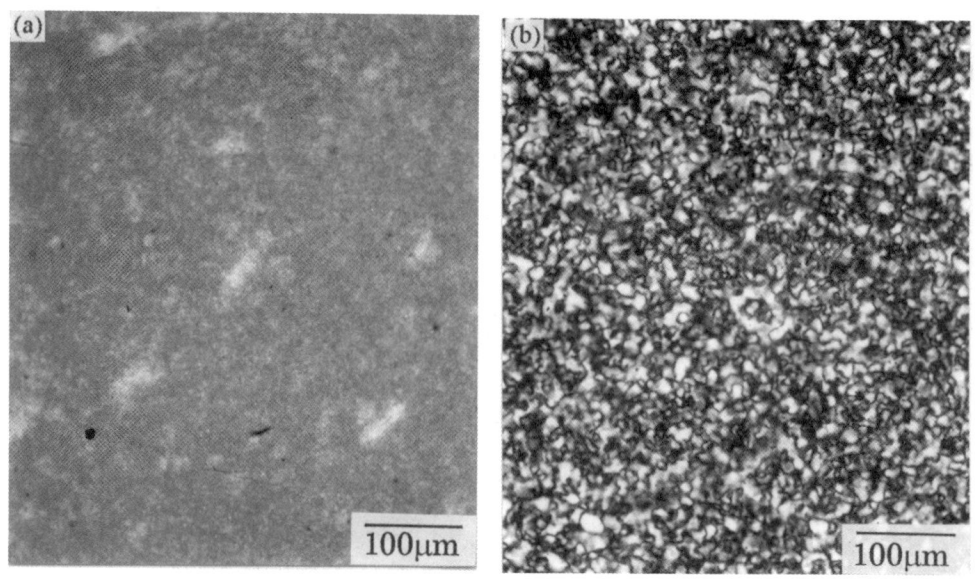

Figure 2   Crossed polarized light micrographs of alumina ceramics (a) sintered at 1573K and HIPed at 1673K, and (b) sintered at 1873K and HIPed at 1823K.

alumina, appears bright and changes its brightness with the rotation every 45°. In the specimen sintered and HIPed at low temperature, two feature are noted in the micrograph. One is fairly well defined; the features are elongated and flat in shape, and appear bright. The elongated features tend to orient in a specific direction. Their origin will be discussed below. The other type of feature is less-well defined; of many features of lesser brightness are distributed throughout the matrix. Change of brightness with rotation was noted for both features, showing that both of them are due to an optically anisotropic structure in the ceramics. In the specimen treated at a higher temperature, extensive growth of matrix particles is noted, severely interfering with the observation of structures which might possibly have developed from aggregated particles. Maximum efforts were devoted to detailed examination of the structure. However, even with this effort, we could not find any region which suggests the presence of large particles of optically uniaxial properties. Again, extensive grain growth appears to be absent in this specimen.

    Figure 3 shows an SEM micrograph of a representative microstructure of a specimen HIPed at 1823K which was then polished and thermally etched. Including the region of this micrograph, the microstructure was uniform for all areas examined. Abnormally large grains with size exceeding 20 µm were not found in any region. This result indicates that abnormal grain growth is absent under the given processing condition, sintering at 1873K and HIPing at 1823K.

    To analyze the above results further, Figure 4 shows an optical micrograph of the green body taken in the crossed polarized light mode; this work has already been presented elsewhere.[15] The comparison of Figs. 2(a) and 4 shows that the origin of the well-defined features in Figure 2(a) can be attributed to structures developed from the large aggregates in the green compacts. Due to the somewhat elongated shape, they tend to become aligned with their longest axis normal to the direction of uniaxial compaction during sample preparation. The similar sizes of the features in these figures shows that the aggregate did not grow significantly during densification. This behavior is in clear contrast to that of matrix particles. They grew to more than 20 times the size of the original particles. The absence of a large grain in Figure 2 (b) again suggests that aggregates do not grow significantly during densification, while matrix particles grew markedly. Recall that the particles of 0.1 µm had grown to 5.8 µm, after sintering at 1873K and HIPing at 1823K.

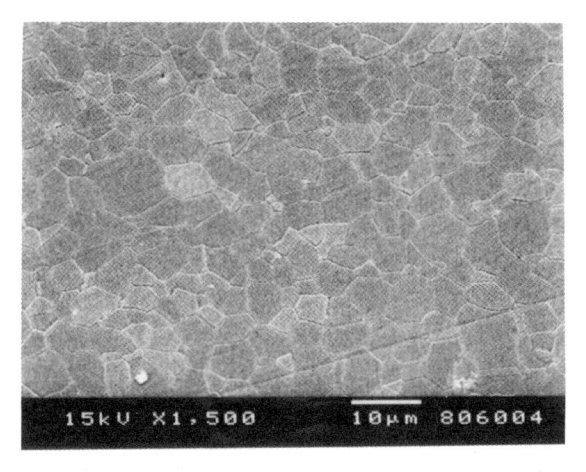

Figure 3   SEM micrograph of alumina ceramic made with advanced alumina powder by sintering at 1873K and HIPing at 1823K

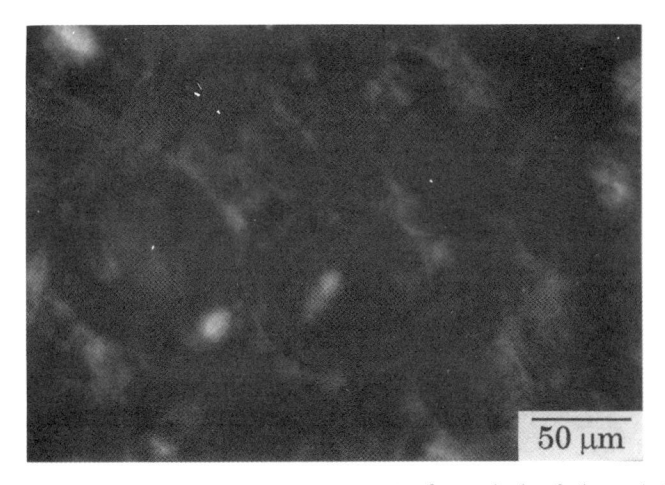

Figure 4   Crossed polarized light micrograph of green body of advanced alumina

Figure 5 shows the growth of a large particle and change of particle packing structure with densification for commercial low-soda alumina specimen sintered in a stepwise manner at successively higher temperatures. In the specimen sintered at 1273K and with the relative density 54%, the same as that of the green body, a large alumina particle is noted at the center of the micrograph in Figure 5 (a). In addition, the matrix shows a mosaic pattern of brightness. In our former paper, the origin of this pattern was attributed to the oriented matrix particles[6] in granules, which were plastically deformed during uniaxial compaction. Both the large particles and the matrix change their brightness with rotation at approximately the same angle. This optical response with rotation shows that the overall direction of orientation is the same for matrix and large particle, although the matrix particles are too small to be individually inspected. Similar large alumna particles have been commonly noted in all of our green bodies made from commercial grade alumina powders of various sources. The presence and orientation of large alumina particles has also been reported in an injection molded green body.[7,8]

After the heat treatment with step-wise increment of temperature, no drastic change was noted in the micrograph of specimen up to 1723K, where the relative density reached 87%. Minor changes were found, however. They include the shortened distance between bright features and the further reduction of clearness for less-well defined mosaic structure of the matrix. The former occurs due to the linear shrinkage of the body by densification. The linear shrinkage is estimated for the change of relative density from 54% to 87%, and should be 15%, which is in good agreement with the change of distance between bright features, 15%. The significance of the latter feature is not clear at present, although it must have been caused by a certain structural change of the matrix at the micro scale. All pores must be open at this low relative density. The transparency of the specimen was largely lost after sintering at 1823K, where the relative density and grain size reached 97% and 1.5 µm, respectively. The scattering of light at the pore surface is significant under this condition, where all pores have been closed, and the liquid is unable to penetrate into them. It completely obscures the change of brightness which might happen with the rotation in the matrix region. Nevertheless, the large alumina particles are clearly visible in the micrograph. Their size and shape appear to remain almost the same even after the significant densification at this temperature.

A new feature appeared at this temperature. Slightly elongated round features were found in the matrix, possibly the macrostructure developed from the corresponding structure of granules in the green body. The dark line of network is tentatively attributed to the boundaries of granules. This dark contrast at the granule boundary may suggest that grains are more randomly oriented in this region than in granules. For more detailed examination of the structure in this region, better optical information is needed. Further efforts are needed to obtain better micrograph with emphasis on improvement of sample preparation. Except for the development of this unique structure, no other feature, such as a new large particle, developed in the structure. Clearly, no abnormal grain growth started at this temperature. The temperature must have been too low for its initiation.

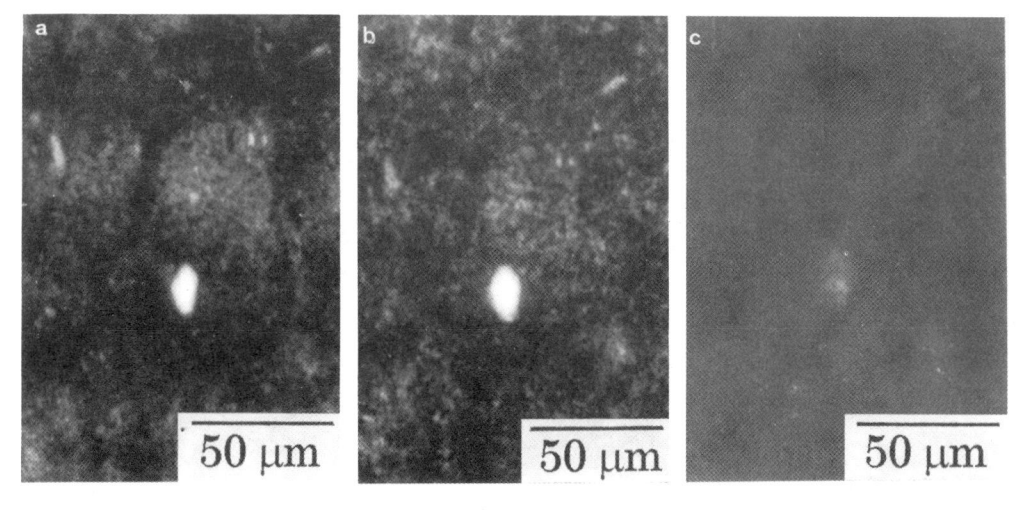

Figure 5 Crossed polarized light micrographs at the same position in alumina ceramic sintered at various temperature. (a) 1273K, (b) 1723K, (c) 1823K

CONCLUSIONS

The change of pore size and morphology during densification was examined in alumina green bodies. The following conclusions were reached.

1. Many pores were left in high purity alumina ceramics even after HIPing to nominally full density. The size and concentration of pores increased with increasing sintering/HIPing temperature.

2. Large aggregates in high purity alumina neither grew nor changed their shape significantly even in sintering and HIPing treatment.

3. Large particles in commercial grade alumina ceramic did not grow and did not initiate abnormal grain growth at temperatures up to 1823K.

4. The packing structure of powder granules in green body forms apparently the similar structure in the specimen sintered to near full density. The particle orientation at granule boundaries was possibly responsible to the development of this unique structure.

REFERENCES

1. W. D. Kingery, Firing - The proof test for ceramic processing, in: Ceramic Processing before Firing, pp. 291, Ed. J. Y. Onoda and L. L. Hench, Wiley, New York, (1978).
2. H. Takahashi, N. Shinohara, K. Uematsu, J. Tsubaki, Influence of granule character and compaction on the mechanical properties of sintered silicon nitride, *J. Am. Ceram. Soc.*, 79:843 (1996).
3. Y. Iwamoto, H. Nomura, I. Sugiura, J. Tsubaki, H. Takahashi, K. Ishikawa, N. Shinohara, M. Okumiya, T. Yamada, H. Kamiya and K. Uematsu, Microstructure evolution and mechanical strength of silicon nitride ceramics, *J. Mater. Res.*, 9:1208 (1994).
4. K. Uematsu, J.-Y. Kim, Z. Kato, N. Uchida and K. Saito, Direct observation method for internal structure of ceramic green body - Alumina green body as an example, *Nippon seramikkusukyokai Gakujutsuronbunshi*, 98:515 (1990).
5. K. Uematsu, Immersion microscopy for detailed characterization of defects in ceramic powders and green bodies, *Powder Technology*, in press
6. K. Uematsu, H. Ito, Y. Zhang and N. Uchida, Novel characterization method for the processing of ceramics by polarized light microscope with liquid immersion technique, *Ceram. Trans.*, 54:83 (1995).
7. K. Uematsu, H. Ito, S. Ohsaka, H. Takahashi, N. Shinohara and M. Okumiya, Characterization of particle packing in an injected molded green body, *J. Am. Ceram. Soc.*, 78:3107 (1995).
8. K. Uematsu, S. Ohsaka, H. Takahashi, N. Shinohara, M. Okumiya, Y. Yokota, K. Tamiya, S. Takahashi and T. Ohira, Characterization of micro- and macrostructure of injection molded green body by liquid immersion method, *J. Europ. Ceram. Soc.*, in press.
9. K. Uematsu, J.-Y. Kim, M. Miyashita, N. Uchida and K. Saito, Direct observation of internal structure in spray-dried alumina granules, *J. Am. Ceram. Soc.*, 73:2555 (1990).
10. K. Uematsu, M. Miyashita, J.-Y. Kim, Z. Kato and N. Uchida, Effect of forming pressure on the internal structure of alumina green bodies examined with immersion liquid technique, *J. Am. Ceram. Soc.*, 74:2170 (1991).
11. J.-Y. Kim, M. Miyashita, M. Inoue, N. Uchida, K. Saito and K. Uematsu, Characterization of internal structure in Y-TZP powder compacts, *J. Mater. Sci.*, 27:587 (1992).
12. K. Uematsu, M. Miyashita, J.-Y. Kim and N. Uchida, Direct study of the behavior of flaw-forming defect in sintering, *J. Am. Ceram. Soc.*, 75:1016 (1992).
13. J.-Y. Kim, M. Miyashita, N. Uchida and K. Uematsu, Change of internal/pore structure in alumina green body during initial sintering, *J. Mater. Sci.*, 27:6609 (1992).
14. W. D. Kingery and B. Francois, The sintering of crystalline oxides, I. Interaction between grain boundaries and pores, in: Sintering and Related Phenomena, Ed. G. C. Kuczynski, N. A. Hooton and C. F. Gibbon, Gordon and Breach, New York (1976).

15. K. Uematsu, Y. Zhang and N. Uchida, Characterization and formation mechanisms of defects in alumina powder compacts, *Ceram. Trans.*, **51**:263 (1995).
16. K. Uematsu and Y. Zhang, Systematic understanding of ceramic processing and related interfacial phenomena, in:Materials Science Monographs, 81, Science of Ceramic Interfaces II, Ed. J. Nowotny, Elsevier Science, Amsterdam, pp. 399 (1994).
17. M. Miyashita, J.-Y. Kim, N. Uchida and K. Uematsu, Effect of applied pressure of HIP on the removal of processing void in alumina ceramics and the resultant improvement of mechanical property, *MRS Symp. Proc.*, **251**:165 (1992).
18. K. Uematsu, M. Sekiguchi, J.-Y. Kim, K. Saito, Y. Mutoh, M. Inoue and Y. Fujino, Effect of processing conditions on the characteristics of pores in hot isostatically pressed alumina, *J. Mater. Sci.*, **28**:1788 (1993).

INFLUENCES OF INTERPARTICLE STRESSES ON INTERMEDIATE-STAGE DENSIFICATION AND GRAIN GROWTH

Alan W. Searcy, Jeffrey W. Bullard and Dario Beruto

Division of Materials Research,
Berkeley National Laboratory
and Department of Materials Science and Mineral Engineering,
University of California, Berkeley, CA 94720

Department of Materials Science and Engineering,
University of Illinois at Urbana-Champaign, Urbana, IL 61801

Istituto di Ingegneria e Scienza dei Materiali
Facolta di Ingegneria, Universita degli Studi di Genova, Genoa, Italy

INTRODUCTION

Sintering models have assumed that the only necessary step of densification is diffusion of matter from between pairs of particles to their surfaces.[1-5] But in a particle compact a second step is also necessary - movement of all particles of the compact toward its center. And that centripetal particle movement may become the rate-limiting step of densification because of constraints imposed by interparticle bonding.[6,7]

The suggested constraints to centripetal movement would act as tensional or shear stresses[6,7] that oppose the driving force for densification. Therefore, we looked for evidence that intermediate-stage densification is opposed by such constraining stresses. The present paper reports evidence that stresses do indeed play a central role in intermediate stage sintering. We use simple three-dimensional models to demonstrate that the extensive grain growth that occurs during intermediate stage sintering of typical compacts can only occur if stresses are developed that are strong enough to rupture the solid network. We suggest that rupture reflects the destabilizing of some interparticle necks as a result of differential densification, and we suggest that particle movement in response to tensional stresses can be the rate-limiting step of both grain growth and of densification during intermediate stage sintering.

MODELS FOR INTERMEDIATE STAGE SINTERING

As background to presentation of our analysis, we will briefly identify those findings from earlier theoretical studies that have particularly influenced our thinking. Kuczynski's classical two-sphere model[4] for sintering was designed to describe only the initial few percent of densification and, therefore, could properly neglect the influence of grain boundary energy on the driving force for densification. But in 1977, Joe Pask, who we honor at this conference, and his then student Carl Hoge pointed out that pairs of identical crystalline particles and arrays of crystalline particles should approach metastable equilibria when further increases in grain boundary area would make the increase in grain boundary energy greater than the concurrent decrease in surface energy of the particle array.[8]

Lange and Kellett[9,10] modeled densification as the decrease in center-to-center distance between isotropic particles arranged in a variety of geometries and also modeled coarsening and particle rearrangement. They concurred in Hoge and Pask's conclusion that in arrays of relatively low coordination numbers and/or high grain boundary energies metastable equilibria could be reached before densification is complete. They also concluded that during later intermediate stage sintering (that is, while most pores still remain open in a continuous interconnected network) coarsening would facilitate densification by restoring its driving force.

One can see, however, from quantitative calculations of Cannon and Carter,[11,12] who used typical values for surface and grain boundary energies to evaluate the approach to metastable equilibrium by linear arrays of identical particles, that those metastable equilibrium configurations would not be reached untill linear shrinkages of 30-70% (see Fig. 5 in ref. [11]. The isotropic volumetric shrinkage in a powder compact corresponding to such changes in linear dimensions would be 66-97%. Thus, it would appear that typical powder compacts with initial porosity of 40-60% would not reach equilibrium during intermediate-stage sintering. We will reach the conclusion below that in compacts formed of nonaggregated particles of a near identical initial size, most coarsening during intermediate stage sintering results from network rupture, which is always detrimental to densification.

Carter and Cannon evaluated the sintering of lines of initially identical isotropic particles by alternate mechanisms that either completely suppress line shortening (densification in the model) or that required shortening to occur by the most thermodynamically favorable path.[12] Their model incorporates coarsening that could result from transfer of matter from one linear particle array to another or from some multiparticle segments of an array to other multiparticle segments. Among their valuable findings, one of particular relevance to the present paper is illustrated in Fig. 1. The figure shows a linear particle array that is acted on by a tensional force F. Cannon and Carter remark that line shortening cannot occur if the applied tensional stress, F/A, where A is the grain boundary area, is equal to σ, the mechanical equivalent of the chemical potential that drives line shortening. Lesser applied forces reduce the net driving force for shortening, and greater tensile forces could "pull the system apart".[11] Cannon and Carter identify differential densification (that is the densification of different adjacent subregions of a compact at different rates) as one source of tensile stresses in those regions that are densifying at relatively low rates, and they identify conditions that could destabilize a linear array enough to produce interparticle rupture.

EVIDENCE FOR NETWORK RUPTURE

Neither the analyses of Lange and Kellett,[9,10] of Cannon and Carter,[11,12] nor of any other investigators of whom we are aware has been used to explain the fact that in real compacts *average pore cross sections, as well as particle cross sections, commonly increase from the initiation of sintering until a substantial fraction of any possible densification has occurred.*[13] Growth in average pore cross sections from the initiation of sintering in monodispersed and nonaggregated compacts implies that interparticle rupture begins soon after the initiation of sintering. This conclusion follows from consideration of the geometry of interpenetrating networks:

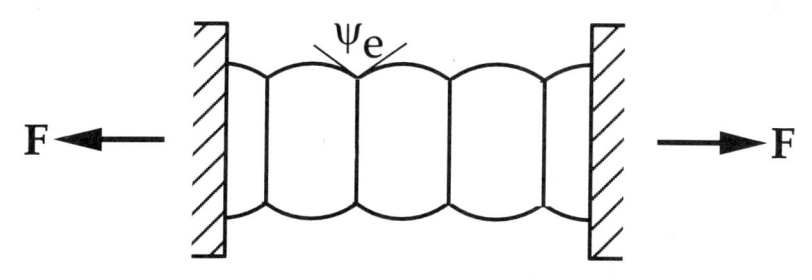

Figure 1. Linear array of truncated spherical particles at provisional equilibrium state with fixed dihedral angle, subjected to a tensile force F.

By definition, intermediate stage sintering is the regime in which the pores of a compact are all linked in one continuous three dimensional network that interpenetrates a continuous solid network. Fig. 2 shows a small volume element of a model for such an assembly. The model particle network is composed of orthogonal arrays of lines of identical isotropic particles like those used in one-dimensional models of sintering.[9-12] For simplicity in describing the features of present interest, each line of particles is replaced by a rod that has the same diameter as the particles. If we assume that each rod is rigidly bonded to the orthogonal rods with which it is in contact, then densification, which shortens each rod and increases its diameter, would necessarily decrease the volume of the compact and decrease the average pore cross section. There is no evident way that average pore cross sections in an interpenetrating pore-particle network can increase substantially during intermediate stage sintering - as they are commonly found to do[13] - unless the particle network is separated into segments that are then incorporated into thickened adjacent grains of the remaining network as densification proceeds. And in fact, Prochazka has pointed out that two interpenetrating networks cannot retain the same relative sizes during a change in scale unless a network *is* broken and reformed.[14]

Further evidence that tensional stresses during intermediate stage sintering reach high enough levels to cause interparticle rupture can be deduced from the geometrical constraints placed on grain growth in an interpenetrating pore-particle network. The diameter of the rods of Fig. 2 can be taken as a rough measure of the average grain size. It is evident from simple geometry that if the green density of the model compact is one-half the theoretical density, an increase of less than a factor of two in the rod diameters would fill the available pore volume. But typically grains grow by much more than a factor of two during intermediate stage sintering.[15] For example, Coble found that in Al_2O_3 compacts of relative green density $\rho_o=0.47$, the average grain size increased by about a factor of ten, from 0.2 μm to ~ 20 μm, before the continuous pore network broke up when the relative density reached $\rho \sim 0.9$.[16]

Fig. 3 shows the cross-sections of changes in the rod model that are implied by observations like Coble's. Such dramatic increases in grain size appear to require that tensional forces be strong enough to cause repeated interparticle rupture.

HOW TENSILE STRESSES CAN CAUSE EITHER DENSIFICATION OR RUPTURE AND GRAIN GROWTH

Under especially favorable circumstances, it is possible to produce green compacts that undergo no significant grain growth during intermediate stage sintering. For example, Wang and Raj were able to achieve negligible intermediate stage grain growth with densely packed, colloidally prepared Al_2O_3.[17,18] The reason for this favorable result is not well understood.[17] The behavior of Coble's Al_2O_3 compacts,[16] in which the average grain size

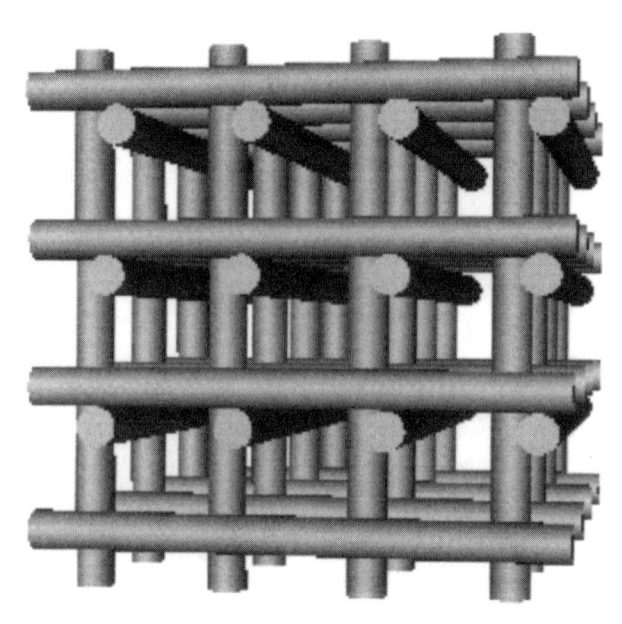

Figure 2. Model of an intermediate-stage sintering three-dimensional particle network as arrays of mutually orthogonal cylinders in tangential contact. Each cylinder represents a multiparticle chain similar to that shown in Fig. 1.

increased by a factor of ten during intermediate stage sintering, is more representative of typically-prepared metal oxide and metal compacts.[15]

We have shown above that this more typical intermediate stage grain growth is only possible in monodispersed powders if interparticle rupture occurs repeatedly. We will describe next how interparticle rupture can result from variations in the local driving force for densification and variations in the local impedance to particle movements.

Suppose that a compact is formed in an arrangement like that of Fig. 2. With each line composed initially of touching spherical particles of equal size, suppose coarsening is suppressed, and suppose that molecules diffuse from between the centers of each pair of particles as conceived in the classic two-sphere model for densification. If the molecular fluxes were exactly identical for each grain boundary of the compact, there might be no impedance either to the movement of all particfles of each volume element toward the center of that volume element or to the movement of all volume elements of the compact toward the compact center. This kind of densification can be called unconstrained.

But it is unrealistic to expect all of the molecular fluxes to be identical in real compacts. Hoge and Pask drew attention to the fact that the behavior of real compacts is complicated by variations in particle size, particle shapes, particle packing, and surface and grain boundary energies. In a system in which particle movement is unconstrained, variations in the rate of line shortening (densification) would simply reflect the sum of the shortening of distances between particle pairs without regard to whether or not the shortening between different pairs occurs at markedly different rates.

In a system formed into a rigid, noncoarsening particle network, however, constraints imposed by the network on particle movement would markedly change the kinetics. Movement of a pair of particles that would be exceptionally fast in an unconstrained system would in a network like that of Fig. 2 be opposed by largely tensional back stresses imposed by the cross-links between particle rows. If the network were rigidly pinned at each cross

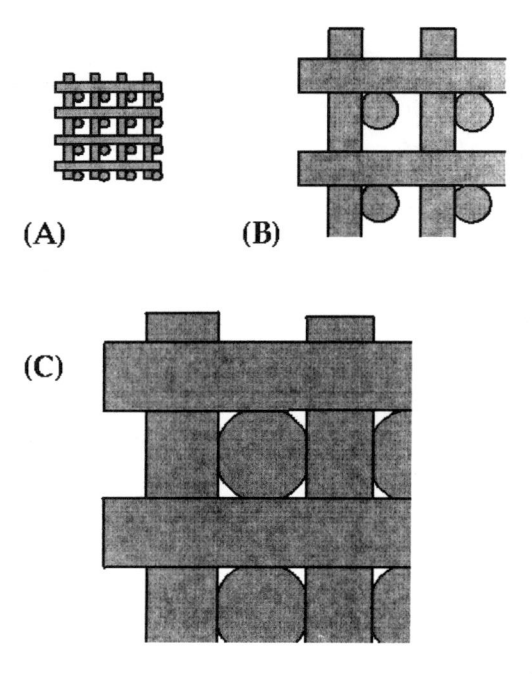

Figure 3. Schematic representation of the kinds of microstructural changes in the cylinder model implied by the data on sintering of Al$_2$O$_3$ reported by Coble.

link, it might densify at a much lower rate than the unconstrained network of the same structure by growth of each grain boundary area at a rate reduced by network-imposed back stresses like those modeled by Cannon and Carter.[11,12] This rigid network behavior can be called fully constrained.

But an important undated movie made about 1976 by R. M. Fulrath of sintering at constant heating rates in a high-temperature scanning electron microscope shows that real compacts are not well described by either an unconstrained or a fully constrained model. Film sequences of compacts formed of monodispersed particles of ceramics, of a glass, or of a metal show particles in subregions within the field of view to undergo sudden bursts of movement that initially are not correlated in any obvious way with movements - or inactivity - of particles in other subregions of the surface. Among observable events are particle rearrangements,[19] and interparticle rupture. We think that the sudden bursts of movement occur when interparticle rupture removes, or at least reduces, the local constraints to particle movement.

Although Carter and Cannon predicted network rupture,[12] they did not discuss the influence rupture would have on grain growth or on subsequent densification. As a basis for doing so here, it is useful to first distinguish between local densification and global densification.

A volume element of a compact like that shown schematically in Fig. 2 can be said to undergo local densification if the distances between centers of particles of that volume element are shortened. The compact undergoes global densification if its total volume decreases.

Global densification of a compact requires movement of its particles over distances that increase directly with their distances from the center of mass. Local densification in a particular volume element contributes to global densification only to the extent that the particles in that volume element, as they move toward each other, pull toward the compact center all the volume elements that lie beyond that particular element on a vector drawn

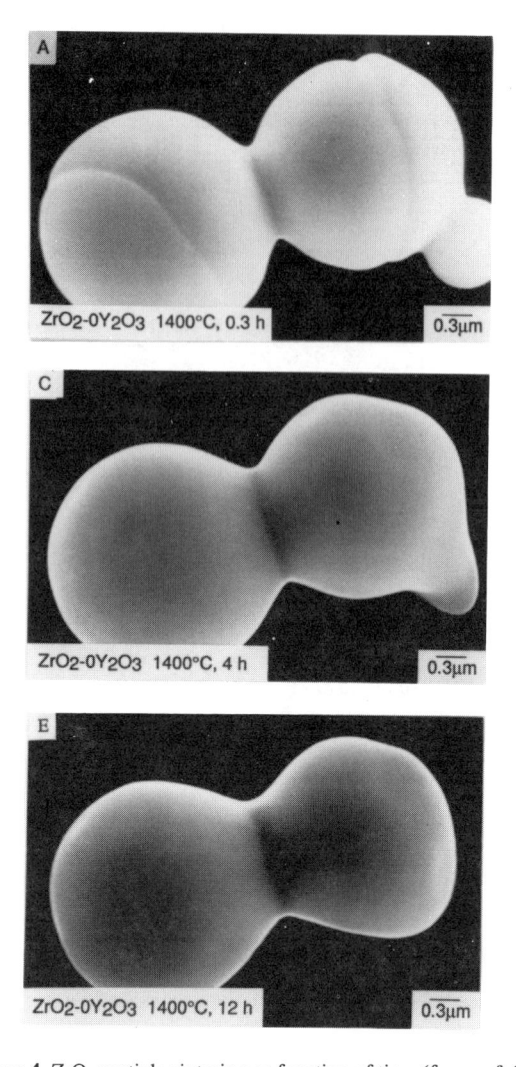

Figure 4. ZrO$_2$ particle sintering as function of time (from ref. 21).

outward from the compact center. Absent the concurrent movement of other volume elements, even complete densification within a particular volume element would simply enlarge its surrounding pores without any change in the global density.

At any given time there will be variations among volume elements, not only in the local driving force for densification, but also in the tensional stresses that act to restrain the local densification. If, for example, those variations cause the particles within two adjacent volume elements all to move at the same rate toward other particles within each volume element, but if the centers of the two volume elements are held in place, then those necks that join particles across the shared boundary between the two volume elements would lengthen. But lengthening an initially stable neck rapidly makes it unstable to thinning and rupture.[12,20] Because a thermodynamic instability is responsible for the rupture, it can occur, not only in brittle solids like Al$_2$O$_3$, but also in ductile solids like Cu.[15]

Interparticle rupture in a line of particles abruptly removes the constraints to movement of the two particles that had shared the ruptured neck. Fig. 4, from a paper by Slamovich and Lange[21], can be used to illustrate the influences that we expect the elimination of constraint on the particle movement to have on local rates of grain growth and densification.

Figs. 4(A), 4(C), and 4(E) show the shape evolution of ZrO_2 particles after sintering on sapphire surfaces for times of 0.3 h, 4 h, and 12 h at 1400°C. In Fig. 4(A) four necks can be discerned. Of these, all but the relatively thin neck in the center have dihedral angles (and ratios of grain boundary area to surface area) qualitatively like those predicted to be present when unconstrained metastable equilibrium (with respect to further shrinkage) is reached between typical crystalline ceramic particles (compare Fig. 4(A) to Fig. 1).[9-12] The relatively thin neck in the center differs appreciably from those predictions in that the surface near the neck has a saddle shape, rather than being strictly convex, and in that the ratio of neck area to particle surface area is relatively low. The fact that, at that neck, metastable equilibrium with respect to continued shrinkage has not been reached after either 0.3 h or 4 h is confirmed by the fact that the cross-sectional area of the neck has increased in each successive frame.

It is possible that, if the two remaining particles in Fig. 4(E) happen to be in registry, then the neck between them may persist simply because there is no grain boundary at the neck along which relatively fast diffusion can occur. But we think the relatively slow elimination of that neck may illustrate the influence of a cross linkage on particle movement: perhaps three of the four necks in Fig. 4(A) disappear rapidly by a combination of the densifying and/or coarsening paths modeled by Lange and Kellett,[10] but bonding to the alumina substrate probably substantially reduces the rate at which the two remaining particles can approach their shared neck.

CONCLUSIONS

Global densification requires particle movement toward the center of a compact. That movement is driven by tensile stresses that are transmitted through the continuous particle network and which reflect some kind of average of the distribution of driving forces for local densification. But differential rates of local densification during intermediate stage sintering of typical compacts create variations in local tensile and shear stresses that can cause rupture at the weakest interparticle necks. Rupture is followed by rapid local densification and coarsening by those particles adjacent to the point of rupture.

Thus tensile and shear stresses are necessary to global densification, which is usually the objective of sintering, but local differences in these stresses cause grain growth, which is usually not desired. We hope to identify means of increasing the average sintering stress while suppressing local differences.

References

1. B.Y. Pines, Mechanism of sintering, J. Tech. Phys., **16**:737 (1946).
2. V.A. Ivensen, Investigation of process of density increase in the sintering of single-phase metal powder compact, J. Tech. Phys., **17**:1301 (1947).
3. A. J. Shaler and J. Wulff, Mechanism of sintering, Ind. and Eng. Chem., **40**:838 (1948).
4. G. C. Kuczynski, Self-diffusion in sintering of metallic particles, Trans. AIME **185**:169 (1949).
5. W. D. Kingery and M. Berg, Study of the initial stages of sintering solids by viscous flow, evaporation-condensation, and self-diffusion, J. Appl. Phys., **26**:1205 (1955).
6. A. W. Searcy, D. Beruto, and M. Capurro, Effective driving force for densification." Oral presentation delivered at the Annual Meeting of the American Ceramic Society, Cincinnati, OH, April 30, 1991.
7. D. Beruto and M. Capurro, Autostresses induced by point defects in sintering phenomena: effect on mass transport and sintering stress. J. Mater. Sci., **28**:4693 (1993).
8. C. E. Hoge and J. A. Pask, Thermodynamics and geometrical considerations of solid state sintering, Ceramurgia Int., **3**:95 (1977).
9. B. J. Kellett and F. F. Lange, Thermodynamics of densification: I, sintering of simple particle arrays, equilibrium configurations, pore stability, and shrinkage, J. Am. Ceram. Soc., **72**:725 (1989).

10. F. F. Lange and B. J. Kellett, Thermodynamics of densification: II, grain growth in porous compacts and relation to densification, J. Am. Ceram. Soc., **72**:735 (1989).
11. R. M. Cannon and W. C. Carter, Interplay of sintering microstructures, driving forces, and mass transport mechanisms, J. Am. Ceram. Soc., **72**:1550 (1989).
12. W. C. Carter and R. M. Cannon, Sintering microstructures: instabilities and interdependence of mass transport mechanisms, p. 137 in: Ceramic Transactions, Vol. 7, Sintering of Advanced Ceramics, C. E. Handwerker, J. E. Blendell, and W. Kaysser, eds., American Ceramic Society, Westerville, OH (1990).
13. O. J. Whittemore and J. A. Varela, p. 777 in: Sintering: Key Papers, S. Somiya and Y. Moriyoshi, eds., Elsevier Applied Sciences, New York (1990).
14. S. Prochazka, Surface area, average mean curvature and chemical potential in porous bodies, Vol. 7 of Ceramic Transactions, p. 127. C. E. Handwerker, J. E. Blendell, and W. Kaysser, eds., American Ceramic Society, Westerville, OH (1990).
15. T. K. Gupta, Possible correlation between density and grain size during sintering, J. Am. Ceram. Soc., **55**:276 (1972).
16. R. L. Coble, Sintering crystalline solids: II, J. Appl. Phys., **32**:793 (1961).
17. J. Wang and R. Raj, Estimate of the activation energies for boundary diffusion from rate-controlled sintering of pure alumina, and alumina doped with zirconia or titania, J. Am. Ceram. Soc., **73**:1172 (1990).
18. J. Wang and R. Raj, Activation energy for the sintering of two-phase alumina/zirconia ceramics, J. Am. Ceram. Soc., **74**:1959 (1991).
19. H. E. Exner and E. Arzt, Sintering processes, p. 1885 in: Physical Metallurgy, 3rd Ed., R. W. Cahn and P. Haasan, eds., Elsevier Science Publishers BV, Amsterdam (1983).
20. A. W. Searcy, J. W. Bullard, and W. C. Carter, Possible explanations of transient neck formation between pairs of {100} faceted particles, J. Am. Ceram. Soc., **79**:2443 (1996).
21. E. B. Slamovich and F. F. Lange, Densification Behavior of Single-Crystal and Polycrystalline Spherical Particles of Zirconia, J. Am. Ceram. Soc., **73**:3368 (1990).

THE Al$_2$O$_3$-SiO$_2$ SYSTEM: LOGICAL ANALYSIS OF PHENOMENOLOGICAL EXPERIMENTAL DATA

Joseph A. Pask

Department of Materials Science & Mineral Engineering
University of California, Berkeley, CA 94720

ABSTRACT

Presence or absence of αAl$_2$O$_3$ as a source of Al$_2$O$_3$ molecules in the starting materials used for determining the phase equilibria in the SiO$_2$/Al$_2$O$_3$ system is shown to be responsible for reported differences in phase diagrams. Equilibrium amounts of Al$_2$O$_3$ molecules dissolved in alumino-silicate liquids from αAl$_2$O$_3$ as a source are determined by the bond strength of the source. Under a balance, or saturation of the liquid with αAl$_2$O$_3$, the liquid molecular structure does not become saturated with clusters of Al$_2$O$_3$ molecules. Monophasic starting materials, however, can provide sufficient less strongly held Al$_2$O$_3$ molecules to possibly saturate the alumino-silicate liquid molecular structure. The significance of this condition is that a cooling alumino-silicate liquid precipitates αAl$_2$O$_3$ spontaneously only when its atomic structure becomes saturated with clusters of Al$_2$O$_3$ molecules. However, when the free energy of the liquid relative to Al$_2$O$_3$ becomes equal to the free energy necessary to dissolve αAl$_2$O$_3$, precipitation of αAl$_2$O$_3$ can not occur spontaneously but precipitation can occur on introduction of "seeds" or nucleating agents of αAl$_2$O$_3$. Discussions of these phenomena relative to thermodynamic stability are presented in the text.

INTRODUCTION

Published diagrams of experimentally determined phase equilibria in the SiO$_2$/Al$_2$O$_3$ system indicate major disagreements in the high alumina compositions. These are evident in Fig. 1 which shows two of the most frequently quoted representative diagrams and indication of a third. Differences are evident in the position of the αAl$_2$O$_3$ liquidus and in the melting behavior of mullite as well as its solid solution range versus temperature.[1-6] These differences have been reported many times and **can not be** literally attributed to experimental errors. Discussions of experimental evidence have been published.[7-10] An application of further thermodynamic reasoning to the mechanisms and conditions of the solid state chemical reactions that take place provides an understanding, explanation and acceptance of the reported experimental results, either as stable or metastable equilibria.

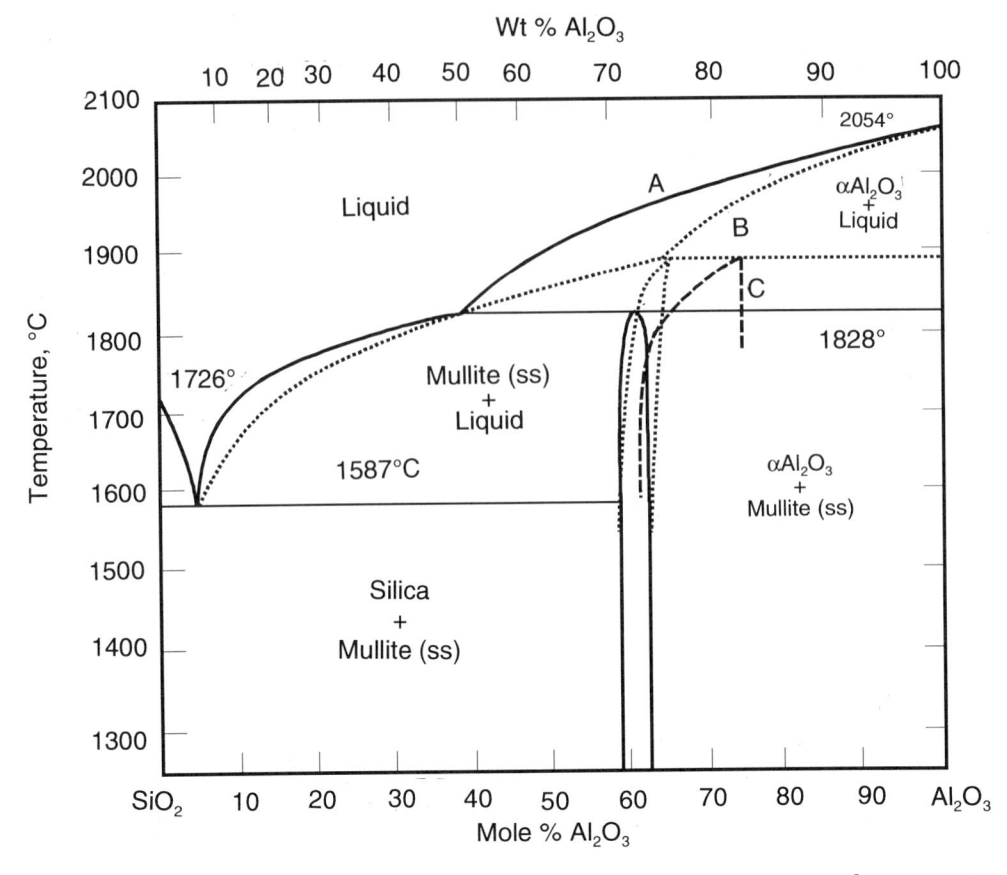

Figure 1. Superimposed phase equilibria diagrams: A - (solid line) Aksay and Pask,[5] B - (dotted line) Klug, Prochazka and Doremus,[6b] C - (dash line) Aksay and Pask.[5]

Experimental determinations of Al_2O_3-SiO_2 equilibria have been based primarily on the use of two different types of starting materials: 1) Mixtures of αAl_2O_3 and SiO_2 powders are heated up to a set temperature and held until steady state phase compositions are reached, then quenched and analyzed; and 2) Materials with fixed chemical compositions of Al_2O_3 and SiO_2 are held and homogenized above the liquidi temperatures to form a single phase liquid, lowered to test temperatures of interest, annealed to reach the steady state phase composition, quenched and analyzed. In order to realize the significance of these approaches in determining the phase equilibria it is necessary to understand the mechanisms and kinetics of the solid state reactions involved and their relationship to the possible reaction paths.

Two solid state chemistry principles play major roles in determining and explaining the differences in the experimental reaction results: a) mechanisms and controlling factors of isothermal chemical reactions; and b) factors that control precipitation of crystalline phases from cooling homogenized alumino-silicate melts. Logical deductive reasoning applied to the experimental phenomenological data develops a fundamental understanding of the reaction mechanisms that lead to the reported phase equilibria.

SOLID STATE CHEMICAL REACTIONS

It is logical that in solid state chemical reactions the spatial distribution of the atoms or molecules in the starting mixtures is a fundamental factor in determining the mechanisms and kinetics of the overall reactions. Starting mixtures of Al, Si, and O atoms equivalent to proportions in mullite will be used as examples. Heat treatment of a homogeneous starting mixture, such as a monophasic gel, results in rapid direct exothermic conversion to mullite The kinetics are relatively fast since at some temperature that provides sufficient activation energy the atoms and electron bonds readjust spontaneously with a minimum amount of mass transport and with a strong exothermic effect ($-\Delta H$) to realize the most negative free energy phase composition ($-\Delta G$).

On the other hand, if the same overall mixture of Al, Si, and O atoms is present as a homogenized powder of αAl_2O_3 and fused SiO_2 particles, the reaction mechanisms and resulting kinetics are modified and become relatively complicated and difficult to analyze. Reactions are initiated at contact interfaces of unlike particles and continue by diffusion mechanisms which normally are slow processes. Another complication frequently arising is the presence of intermediate reactions that lead to a final overall reaction resulting in the same final product. On completion, the final equilibrium phase composition is the same in both approaches, but the microstructures and kinetics are different.

Analyses of diffusion couples of αAl_2O_3 (sapphire) and fused SiO_2 at a number of temperatures provide a convenient and useful method for determining and understanding the nature of the reaction mechanisms that occur at the interfaces.[5,11]. A mullite layer forms along the interfaces at temperatures below about 1828°C which is the incongruent melting point. A kinetic analysis of the diffusion data as determined by Davis in his thesis dissertation,[11] however, revealed that mullite did not grow to an observable thickness below 1634°C. It was suggested that solution of Al_2O_3 molecules in the fused SiO_2 was faster than the growth rate of mullite below this temperature. The driving force for solution of Al_2O_3 becomes the driving force for releasing the Al_2O_3 molecules from αAl_2O_3.

Since the atomic bonding in αAl_2O_3 is strong, it is postulated that an initial partial reduction at the interface occurs to make alumina molecules available. This mechanism appears to be associated with the partial initial reduction of interfacial Al_2O_3 to transitory layers which are dissolved relatively fast as an overall net endothermic reaction ($+\Delta H$). It is postulated that this reaction is facilitated by a corresponding relatively large increase in entropy ($+\Delta S$) which results in a favorable ($-\Delta G$) according to ($\Delta G = \Delta H - T\Delta S$). On saturation of the alumino-silicate liquid with Al_2O_3 molecules further reduction of alumina layers does not occur without the unavailability of the driving force for solution of alumina. Mullite then begins to grow at the liquid/αAl_2O_3 interface whose rate is controlled by a slow interdiffusion process through the growing mullite layer. As a result, this exothermic reaction takes place over a relatively longer period of time and temperature. It thus does not appear as a noticeable exothermic reaction peak that would form for a more spontaneous reaction.

It is evident that, in a simplified form, this overall NET chemical reaction can be treated as a two-step process. Step I is concerned with the energy balance of the silicate liquid with Al_2O_3 molecules. Since the bonding strength in αAl_2O_3 is high, the net thermal energy associated with release of Al_2O_3 molecules as described is endothermic ($+\Delta H$). It is postulated that the formation of the intermediate layers of transitory aluminas contributes to an increase in entropy for the reaction ($+\Delta S$). The result is a negative free energy ($-\Delta G$) for Step I. In the listing of equations for Step I, the intermediate substeps have not been determined. Step II is concerned with the exothermic reaction between alumina-saturated silicate liquid and αAl₂O₃ to form mullite. The OVERALL NET reaction based on the summation of the two steps indicates the formation of mullite (-ΔG).

Step I
Ia $2SiO_2(l) + Al_2O_3 \rightarrow 2SiO_2 \cdot Al_2O_3(l)$ $-\Delta G_I$
Ib $\alpha Al_2O_3 \rightarrow Al_2O_3$ $+\Delta G_{II}$
Net I $2SiO_2(l) + \alpha Al_2O_3 \rightarrow 2SiO_2 \cdot Al_2O_3(l)$ ΔG_{soln}

Step II
IIa $2SiO_2 \cdot Al_2O_3(l) + 2Al_2O_3 \rightarrow 3Al_2O_3 \cdot 2SiO_2$ $-\Delta G$
IIb $2(\alpha Al_2O_3) \rightarrow 2Al_2O_3$ $+\Delta G$
Net II $2SiO_2 \cdot Al_2O_3(l) + 2(\alpha Al_2O_3) \rightarrow 3Al_2O_3 \cdot 2SiO_2$ $-\Delta G_{mul}$
NET $2SiO_2(l) + 3(\alpha Al_2O_3) \rightarrow 3Al_2O_3 \cdot 2SiO_2$ $-\Delta G_{react}$

Precipitation from Cooling Alumino-Silicate Melts

This procedure is extensively used for determination of phase equilibria in the SiO₂/Al₂O₃ system[4,6]: a mixture with a fixed chemical composition of Al₂O₃ and SiO₂ is melted and homogenized at temperatures above the αAl₂O₃ liquidus. This liquid serves as the starting material for the equilibria studies. It is lowered to the temperature of interest, annealed to reach steady state phase compositions, quenched and analyzed.

The understanding of this approach is based on the fact that the atomic structure of a liquid is temperature dependent. Above the liquidi temperatures, the thermal energy is large enough to keep the atoms and molecules of any selected composition (in our case, compositions having more than approximately 50 wt% Al₂O₃) sufficiently dispersed so that they do not cluster to form nuclei or crystalline precipitates. The liquid thus has a negative free energy relative to solution of additional Al₂O₃ molecules. On cooling, the free energy of solution decreases until it becomes equal to the free energy necessary to release molecules of Al₂O₃ from αAl₂O₃ as a source ($-\Delta G_{soln} = -\Delta G_{molecule}^{\alpha Al_2O_3}$). Technically, the liquid is now saturated with αAl₂O₃, but $\Delta G_{soln} \neq 0$, i.e., the liquid structure itself is not saturated with Al₂O₃ molecules to permit formation of clusters and nuclei. With continued cooling, the liquid increases in density and its structure rearranges so that clusters of Al₂O₃ molecules develop and the free energy of the liquid relative to solution of Al₂O₃ becomes zero ($\Delta G_{soln} = 0$), i.e., the liquid structure becomes saturated, technically identified as supersaturation with Al₂O₃ molecules. The significance of this sequence is that spontaneous precipitation of the clustering species as the most thermodynamically stable crystalline compound occurs only on development of supersaturation. Supersaturation exists at the lower temperature αAl₂O₃ liquidus in Fig. 1B.

If nuclei or "seeds" of the precipitating phase are introduced into the liquid at temperatures below the upper αAl₂O₃ liquidus (Fig. 1A) and above the lower αAl₂O₃ liquidus (Fig. 1B), crystallization of αAl₂O₃ can take place up to the equilibrium composition without previous supersaturation of the liquid structure which is required for spontaneous precipitation. If the upper liquidus of Fig. 1A were not existent, these seeds or nuclei would dissolve in the liquid.

SPECIFIC EXAMPLES OF REACTIONS

Two alumino-silicate chemical compositions (65wt% and 83wt% Al₂O₃) will now be used to further illustrate the critical effect of processing procedures on crystallization results. Differences in the phase equilibria development will be followed by referring to the phase equilibria diagrams shown in Fig. 1: Diagram A derived with Al_2O_3 present as αAl_2O_3 in starting materials,[2,5,11] and Diagram B derived with αAl_2O_3 present as molecules in a cooling monophasic phase,[4,6] and Diagram C indicating the formation of $3Al_2O_3 \cdot 2SiO_2$ type of mullite.[5]

65 wt% Al_2O_3 Composition

A mixture of 65wt.% Al_2O_3 and fused SiO_2 particles homogenized above 1900°C form an alumino-silicate liquid of that composition. On cooling this liquid as a starting material and utilizing Fig. 1A, the free energy negativity of the liquid relative to solution of Al_2O_3 decreases progressively. At about 1900°C, the free energy of solution relative to Al_2O_3 molecules becomes equal to the free energy that would be necessary to remove an Al_2O_3 molecule from αAl_2O_3 ($-\Delta G_{soln}^{melt} = +\Delta G_{molecule}^{\alpha Al_2O_3}$). Spontaneous precipitation of αAl_2O_3, however, does not occur because the liquid's atomic structure does not become saturated with Al_2O_3 molecules and thus $-\Delta G_{soln}^{melt} \neq 0$. With continued cooling to about 1850°C (Fig. 1B), the liquid structure continues to be unsaturated with αAl_2O_3 molecules. But, the liquid atomic structure becomes saturated with mullite molecules ($-\Delta G_{soln}^{melt} = 0$ for solution of mullite), causing mullite crystals to precipitate spontaneously at 1850°C. With continued cooling, the liquid and mullite adjust their equilibrium compositions until 1587°C and then react to form cristobalite and mullite.

If the 65wt% Al_2O_3 starting material consists of a mixture of αAl_2O_3 and fused SiO_2 particles that is heated up to a temperature between 1828°C and 1900°C, it forms an equilibrium mixture of αAl_2O_3 and alumino-silicate liquid according to the Fig. 1A diagram. On cooling, the ratio of this mixture equilibrates, and at 1828°C it reacts to form mullite and alumino-silicate liquid. At 1587°C an equilibrium composition of cristobalite and mullite forms. The presence of αAl_2O_3 from 1828 to 1900°C prevents the saturation of the liquid with molecules of mullite at 1850°C.

On the other hand, an addition of "seeds" or nuclei of αAl_2O_3 to the alumino-silicate liquid (that was formed above 1900°C) between 1900°C and 1850°C results in an equilibrium amount of αAl_2O_3 and liquid represented by the upper liquidus curve (Fig. 1A). If this higher temperature αAl_2O_3 liquidus did not exist, the "seeds" would be dissolved by the liquid. The presence of "seeds" and resulting αAl_2O_3 crystallization also affects the atomic rearrangement of the liquid structure so that the mullite saturation and precipitation at 1850°C does not occur. At 1828°C the liquid and αAl_2O_3 react to form mullite and excess liquid.

83 wt.% αAl_2O_3 Composition

Let us similarly analyze a mixture containing 83wt% Al_2O_3 which is homogenized above about 2000°C (Fig. 1B). On cooling, the negativity of the free energy of solution relative to Al_2O_3 decreases until at about 2000°C it becomes equal to the free energy necessary to remove a molecule of Al_2O_3 from αAl_2O_3. Spontaneous precipitation of αAl_2O_3, however, does not occur because the atomic structure of the liquid does not become saturated with stable clusters of Al_2O_3 molecules. With continued cooling, the liquid structure changes continually to about 1980°C at which point the liquid atomic structure becomes saturated with stable Al_2O_3 clusters and the free energy of the liquid relative to the solution of Al_2O_3 becomes zero ($\Delta G_{soln}^{melt} = 0$). Then, αAl_2O_3 precipitates spontaneously. Precipitation of αAl_2O_3 continues to 1890°C while the liquid's atomic structure decreases in Al_2O_3 but remains continually saturated with Al_2O_3 molecules. When the liquid composition changes to about 76wt% Al_2O_3, it crystallizes to form the $2Al_2O_3 \cdot SiO_2$ mullite. With continued decrease of temperature the mullite continues to equilibrate with αAl_2O_3 (Fig. 1B).

If the same starting composition of 83 wt% Al_2O_3 is prepared as a powder mixture of particles of αAl_2O_3 and fused SiO_2, the following reaction sequence occurs when the mixture is first heated just below 2000°C and cooled to 1828°C according to Fig. 1A. At 1828°C a reaction occurs to form the $3Al_2O_3 \cdot 2SiO_2$ type of mullite with an excess of αAl_2O_3.

Mullite Compositions

Three composition ranges have been reported for mullite: nominally identified as $3Al_2O_3 \cdot 2SiO_2$, $2Al_2O_3 \cdot SiO_2$ and $3Al_2O_3 \cdot SiO_2$. The range observed is determined by the conditions for their formation.

The first type of mullite identified was formed by heating kaolinitic clay. Because the mullite that formed interfaced completely with alumino-silicate melts in the absence of αAl_2O_3, the equilibrium mullite solid solution composition was richer in SiO_2, nominally $3Al_2O_3 \cdot 2SiO_2$ (70-72 wt% Al_2O_3). On the other hand, mullite grown at the interface of an αAl_2O_3/fused SiO_2 diffusion couple has an equilibrium solid solution composition of about 74-75 wt% Al_2O_3. It does not form above 1828°C indicating that mullite on heating in the presence of αAl_2O_3 melts incongruently at this temperature to form αAl_2O_3 and alumino-silicate solution. In the absence of αAl_2O_3 as a nucleating agent this type of mullite melts congruently at about 1890°C.[4]

The second type of mullite is represented by single crystals of mullite grown in the absence of αAl_2O_3 from fused homogeneous melts by the Czochralski technique at about 1890°C with a composition of about $2Al_2O_3 \cdot SiO_2$ (77 wt% Al_2O_3). This cooling melt reaches free energy equilibrium equivalent αAl_2O_3 at about 1960°C but αAl_2O_3 does not precipitate spontaneously because the molten liquid structure is not saturated with Al_2O_3 molecules. With continued cooling to 1890°C, the liquid structure still does not become saturated with Al_2O_3 molecules but it becomes structurally saturated with $2Al_2O_3 \cdot SiO_2$ molecules which results in spontaneous growth of the single crystal (about 77 wt% Al_2O_3). The absence of αAl_2O_3 and lack of saturation of liquid with Al_2O_3 molecules permitted the liquid to cool to about 1890°C with the formation of mullite of the 2:1 type. Actually it is difficult to determine experimentally whether the mullite melts congruently or incongruently. A shift of the αAl_2O_3 liquidus (Fig. 1B) only by about 0.5 wt% at the 1890°C melting temperature of mullite with about 78 wt% αAl_2O_3 can determine whether melting is technically congruent[6] or incongruent.[6b]

The third type of mullite forms from starting mixtures containing compositions of about 84 wt% Al_2O_3 (nominally $3Al_2O_3 \cdot SiO_2$) that are homogenized to form liquids at temperatures above the liquidus and then quenched. It is theorized that the rapid cooling results in retention of some of the high temperature atomic structure of the liquid leading to the precipitation of the high Al_2O_3 mullite structure.[3] The phase equilibria involving this type of mullite have not been determined.

SUMMARY

It has been shown that in Al_2O_3/SiO_2 phase equilibria studies the presence or absence of Al_2O_3 in the form of αAl_2O_3 in the starting materials affects the reaction sequence in reaching phase equilibria, particularly in the higher temperatures and larger Al_2O_3 contents. An understanding of the results is based on the understanding of the mechanisms involved in two types of solid state chemical reactions: (1) precipitation of a phase from a cooling liquid, and (2) an isothermal solid state reaction.

The solution of molecules into its atomic structure by a melt is associated with the release of some of its free energy ($-\Delta G_{soln}^{melt}$ relative to solution of the given molecules) which should be more negative than the free energy necessary to release the molecule from its host crystal structure ($-\Delta G_{melt} > +\Delta G_{cryst}$). With continued solution $-\Delta G_{soln}^{melt}$ becomes less negative and $+\Delta G_{cryst}$ remains constant. When ($-\Delta G_{melt} = +\Delta G_{cryst}$), the solubility limit is reached based on availability of the molecule energetically, i.e., the liquid is energetically saturated with the source compound of the molecule but of insufficient amount to cluster and nucleate. However, with another source from which the molecules are more easily available energetically the liquid $-\Delta G_{soln}^{melt}$ becomes zero. This occurrence indicates that the liquid structure has become saturated with the molecules and it can not accommodate additional molecules which then form clusters of the molecules which nucleate to form the most energetically favorable compound containing this molecule by spontaneous precipitation. In this case, the liquid is defined as being supersaturated relative to this compound.

This logical fundamentally-based analysis of experimental data indicates the criticality of whether the Al_2O_3 in the starting compositions is present as αAl_2O_3 or incorporated into a homogenized melt which serves as the starting material. The derived phase equilibrium diagrams thus could be labeled accordingly to reflect this difference: either as the $\alpha Al_2O_3/SiO_2$ system or the Al_2O_3/SiO_2 system. The diagram representing the $\alpha Al_2O_3/SiO_2$, system (indicated as Fig. 1A) can be considered to be thermodynamically stable.

ACKNOWLEDGEMENT

Helpful discussions with Prof. Alan W. Searcy are gratefully acknowledged.

REFERENCES

1. H. Schneider, K. Okada and J. A. Pask, MULLITE AND MULLITE CERAMICS, John Wiley & Sons, West Sussex, England, 251 p. (1994).
2. N. L. Bowen and J. W. Greig, "The System Al_2O_3-SiO_2", *J. Am. Ceram. Soc.*, 7. 238-254 (1924).
3. N. A. Toropov and F. Y. Galakhov, "New Data for the System $Al_2O_3SiO_2$", *Dokl. Akad. Nauk SSSR*, 78, 299 (1951).
4. S. Aramaki and R. Roy, "Revised Phase Diagram for the System Al_2O_3-SiO_2," *J. Am. Ceram Soc.*, 45, 229-242 (1962).
5. I. A. Aksay and J. A. Pask, "Stable and Metastable Equilibria in the System Al_2O_3-SiO_2," *J. Am. Ceram. Soc.*, 58, 507-512 (1975).
6b. F. J. Klug, S. Prochazka and R. H. Doremus, "Alumina-Silica Phase Diagram in the Mullite Region," *J. Am. Ceram. Soc.*, 70, 750-759 (1987).
6a. S. Prochazka and F. J. Klug, "Infrared-Transparent Mullite Ceramics," *J. Am. Ceram. Soc.*, 66, 874-880 (1983).
7. J. A. Pask, "Phase Equilibria in the Al_2O_3-SiO2 System with Emphasis on Mullite," in CERAMIC DEVELOPMENTS: PAST, PRESENT AND FUTURE, edited by C. C. Sorrell and B. Ben-Nissan, 34-36, 1-8 (1988).
8. J. A. Pask, "Phase Equilibria of Mullite," in MULLITE AND MULLITE CERAMICS, edited by H. Schneider, K. Okada and J. A. Pask, John Wiley & Sons, 83-104 (1994).
9. J. A. Pask, "Critical Review of Phase Equilibria in the $3Al_2O_3·2SiO_2$ System," in MULLITE AND MULLITE MATRIX COMPOSITES edited by S. Somiya, R. F.

Davis and J. A. Pask, Ceramic Transactions, v. 6, Am. Ceram. Soc., Inc., 1-14 (1990).
10. J. A. Pask, "Importance of Starting Materials on Reactions and Phase Equilibria in the Al$_2$O$_3$-SiO$_2$ System," *J. European Ceram. Soc.*, 16, 101-108 (1996).
11. R.F. Davis and J.A. Pask, "Diffusion and Reaction Studies in the System Al$_2$O$_3$-SiO$_2$," *J. Am. Ceram. Soc.*, 55, 525-531 (1972).
12. S. Risbud, V. Draper and J.A. Pask, "Dependence of Phase Composition on Nuclei Available in SiO$_2$-Al$_2$O$_3$ Mixtures," *J. Am. Ceram. Soc.*, 61, 471-472 (1978).
13. W.M. Kriven and J.A. Pask, "Solid Solution Range and Microstructures Of Melt-Grown Mullite," *J. Am. Ceram. Soc.*, 66, 649-654 (1983).

SINTERING BEHAVIOR, MICROSTRUCTURE AND PROPERTIES OF FINE SILICON NITRIDE CERAMICS

Alida Bellosi, Frederic Monteverde, Cesare Melandri, and Stefano Guicciardi

CNR - IRTEC, Research Institute for Ceramics Technology
Via Granarolo 64 - 48018 Faenza (Italy)

INTRODUCTION

Potential engineering applications of structural ceramics are greatly limited by their intrinsic brittleness, i.e. their low fracture toughness. With respect to Si_3N_4-based materials, the improvement of strength and toughness based on the development of "in situ" reinforced microstructures has been recently investigated,[1-5] where elongated grains are grown in a fine-grained matrix: the resulting bimodal grain size distribution can improve fracture resistance and toughness. However, microstructural features such as grain size and aspect ratio, type and distribution of secondary phases, flaw population influence properties and performances. Thus the tailoring of microstructures, the optimisation of processing parameters and their correlations to resulting properties are of great interest. In Si_3N_4-SiC composites and nanocomposites, an increase in hardness, strength and toughness can be obtained because of the effects of nanosize particles which pin and pile-up the dislocations creating subgrain boundaries within the matrix.[6-10] Recent studies suggested that the incorporation of SiC particles into Si_3N_4 matrices influence densification, microstructure and properties.[7-10] Furthermore Si_3N_4-SiC composites with microstructural refinement and improved sinterability were obtained with the addition of ultrafine Si/C/N powders to submicronic commercial Si_3N_4 powders.[11-13] These materials, with homogeneous and fine microstructures, exhibited a plastic deformation, thereby predicting a useful behaviour for superplastic forming.[9]

This present contribution deals with the production and characterization of: i) fine grained "in situ" reinforced dense monolithic Si_3N_4 starting from commercial powders, ii) nanostructured Si_3N_4-SiC composites from the addition of 30 wt% Si/C/N ultrafine powders to the same silicon nitride powder. On the basis of previous results, the system $Y_2O_3+La_2O_3$ was selected as a sintering aid, because it has revealed its effectiveness in improving high temperature mechanical properties.[4] Two different amounts of sintering aids (4 wt% and 6 wt% maintaining a unitary ratio Y_2O_3/La_2O_3) are compared and the effects of two different mixing methods (ultrasonification and attritor mixing) are evaluated, as the homogeneity of the starting components is crucial, particularly when nanosize powders are used. All the mixtures were densified by hot pressing. Densification behaviour, microstructure and mechanical properties of the dense bodies are discussed.

EXPERIMENTAL PROCEDURES

The powders used in these experiments are:
- commercial Si_3N_4 (UBE SNE10, Japan), equiaxed particles of mean grain size 0.16 µm, s.s.a. (B.E.T.) 11.5 m^2/g, mean aggregate size 0.48 µm, 97 wt% α- and 3 wt% β-Si_3N_4, impurities O 1.3wt%, Cl <100 ppm, Fe<100 ppm, Ca<50 ppm, Al<50 ppm;
- ultrafine Si/C/N (ENEA, Italy) synthesized by CO_2 laser irradiation of SiH_4, NH_3 and hydrocarbons mixture[14,15], atomic composition corresponding to 23 wt% SiC and 77 wt% Si_3N_4 (as determined by XRD spectra), s.s.a. (BET) 68.7 m^2/g, mean spherical diameter 30 nm.

The following compositions were chosen:

Powder Compositions	Method of Preparation
Si_3N_4+3 wt% La_2O_3+3 wt% Y_2O_3	ultrasonically mixed, labelled as LY33u
Si_3N_4+30 wt% Si/C/N+3 wt% La_2O_3+3 wt% Y_2O_3	ultrasonically mixed, labelled as LY33Fu
Si_3N_4+2 wt% La_2O_3+2 wt% Y_2O_3	ultrasonically mixed, labelled as LY22u
Si_3N_4+30 wt% Si/C/N+2 wt% La_2O_3+2 wt% Y_2O_3	ultrasonically mixed, labelled as LY22Fu
Si_3N_4+2 wt% La_2O_3+2 wt% Y_2O_3	attritor mixed, labelled as LY22a
Si_3N_4+30 wt% Si/C/N+2 wt% La_2O_3+2 wt% Y_2O_3	attritor mixed, labelled as LY22Fa

The sintering aids were incorporated as follows:

i) *Ultrasonic mixing of LY33u and LY22u.* Dispersion of additive powders in isobuthyl alcohol by ultrasonic stirring, then addition of silicon nitride powder under continuous mixing and subsequent ultrasonic stirring by pulsed cycles. Drying under magnetic stirring at 90°C and sieving.

ii) *Ultrasonic mixing of LY33Fu and LY22Fu.* Due to difficulties in dispersing ultrafine powders and in order to avoid the increase of the oxygen content in these highly reactive powders, the blends were processed under more accurate procedures. Before powder addition, a solution of water free n-hexane, dispersant on the base of phosphate ester (Emphos PS-21A, Witco) and binder (Triolein) were prepared. In sequence the sintering aids, ultrafine Si/C/N and Si_3N_4 powders were one by one gently added and stirred by ultrasonic pulses: the slurries were continuously mixed with a magnetic stirrer, dried heating up to 80°C, softly crushed and sieved. All the operations were carried out under a continuous stream of pure nitrogen.

iii) *Attritor mixing of LY22a.* Addition of Si_3N_4 and additive powders in isobuthyl alcohol mixing the slurry within a lined Teflon mill and using Si_3N_4 milling balls for 4 hours at 400 rpm. Drying under magnetic mixing and sieving.

iv) *Attritor mixing of LY22Fa.* A lined Teflon mill, Si_3N_4 milling balls and n-hexane as liquid medium were used. At first sintering aids and silicon nitride powder were dispersed by attritor, then Si/C/N powder was added and attritor milled for 4 hours. The slurry was dried by a rotavapor under flowing nitrogen and then sieved.

The distributions of the additives and crystalline phases of the mixtures were estimated by SEM and XRD; oxygen and nitrogen content was measured by LECO (Table 1).

All the mixtures were hot pressed under vacuum in an induction-heated graphite die using a pressure of 30 MPa and a temperature of 1850°C. Continuous shrinkage of the samples was recorded during hot pressing (Table 2).

The microstructure of the dense materials was examined on polished, plasma etched and fracture surfaces (parallel and perpendicular to the applied pressure) by an SEM equipped with a backscattered electron detector and energy dispersive microanalyzer. Mean grain size (*d*), thickness (*t*), length (*l*) and aspect ratio (*a*) of elongated β-Si_3N_4 were evaluated (Table 3).

Crystalline phases of dense materials, as well as the orientation of the β-Si$_3$N$_4$ grains, were determined by XRD. In order to evaluate the recrystallization of the secondary phases, annealing tests were performed at 1400°C for 6 h in a flowing nitrogen atmosphere.

The thermal expansion coefficient (λ) was evaluated up to 1400°C, using heating rate of 5°C/min. The Young's modulus (E) was measured on a 0.8x8.0x28 mm sample, using the frequency resonance method.[16] The flexural strength (σ) was measured on 2.0x2.5x25 mm bars, in a 4-pt bending fixture, with 20 mm as the outer span and 10 mm as the inner span, and a crosshead displacement of 0.5 mm/min. The fracture toughness (K$_{IC}$) was evaluated using the Direct Crack Measurement (DCM) method, with a load of 98.1 N in a Zwick hardness tester using the formula proposed by Anstis et al[17]. The Vickers microhardness (HV1.0) was measured on polished surfaces with a load of 9.81 N. The values of these properties are reported in Table 4.

Crack propagation studies were performed by SEM and correlated to composition and process of these materials.

Table 1. Specific surface area (BET), oxygen and nitrogen content (wt%) of the powder mixtures.

	LY33u	LY33Fu	LY22u	LY22Fu	LY22a	LY22Fa
s.s.a.(m^2/g)	11.9	27.0	11.8	27.9	10.9	26.6
O	3.4	7.0	2.8	3.5	2.9	3.7
N	38.2	37.5	39.3	41.0	39.0	36.0

Table 2. Values of the relative density (starting ρ$_o$, at 1850°C ρ$_{1850}$), final density ρ$_f$, temperature of shrinkage onset T$_1$, temperature at which second densification stage starts T$_2$.

Sample	ρ$_o$ (%)	ρ$_{1850}$ (%)	ρ$_f$(gr/cm^3)	T$_1$ (°C)	T$_2$ (°C)
LY33u	48	79	3.27	1640	1740
LY33Fu	48	87	3.23	1630	1720
LY22u	50	71	3.24	1660	1750
LY22Fu	48	84	3.21	1590	1730
LY22a	47	80	3.24	1580	1720
LY22Fa	47	94	3.24	1490	1670

Table 3. Some microstructural parameters of the hot pressed samples: mean grain size (d), length (l), thickness (t), aspect ratio (a) of the elongated β-Si$_3$N$_4$ grains, preferred orientation of the β-Si$_3$N$_4$ grains I$_{200}$/I$_{002}$ on a surface parallel to the applied pressure (random value 6.6), crystalline phases.

Sample	β-Si$_3$N$_4$ grains (μm)			Aspect ratio	I$_{200}$/I$_{002}$	α-Si$_3$N$_4$	SiC	Si$_2$N$_2$O
	d	l	t	a				
LY33u	.38±.60	2.3±1.2	.28±.18	7.8±3.1	2.3	9	-	-
LY33Fu	.22±.40	.38±.27	.13±.09	6.9±4.2	3.3	4		8
LY22u	.41±.10	2.5±1.2	.23±.13	7.8±1.9	2.5	traces	-	
LY22Fu	.20±.03	.33±.27	.19±.16	5.7±2.9	2.8	6		9
LY22a	.45±.08	1.2±.6	.31±.15	6.3±2.4	1.7	-	-	
LY22Fa	.24±.03	-	-	-	2.1	3		8

Table 4. Properties of the materials tested: thermal expansion coefficient (20-1200°C) λ, Young's modulus E, Vickers microhardness HV1.0 and fracture toughness K_{IC} (on surfaces parallel and perpendicular to the direction of the applied pressure), strength σ.

Sample	$\lambda(10^{-6}C^{-1})$	E(GPa)	HV1.0 (GPa) perp	parall	K_{IC} (MPa√m) perp	parall	σ (MPa) RT	1000°C	1200°C	1400°C
LY33u	3.81	325	18.9±.3	17.6±.6	5.7±.1	5.2±.2	941±67	826±29	974±71	771±56
LY33Fu	3.98	312	19.1±.3	18.4±.3	4.6±.1	4.4±.3	753±80	736±52	668±111	595±91
LY22u	3.74	325	17.8±.3	16.6±.5	5.6±.2	5.0±.2	1137±65	889±119	789±32	678±57
LY22Fu	3.82	318	19.2±.4	18.3±.4	4.8±.2	4.2±.3	722±18	672±56	616±35	614±73
LY22a	3.83	324	17.1±.2	16.3±.3	5.3±.2	4.9±.3	875±62	796±81	698±19	615±18
LY22Fa	3.88	323	18.6±.2	17.8±.3	4.5±.2	4.3±.1	915±163	812±140	690±67	595±41

RESULTS AND DISCUSSION

Powder mixtures

The specific surface area of the powder mixtures are, as expected, strongly increased by the presence of the ultrafine powder (Table 1). As crystalline phases, peaks due to additive phases (La_2O_3, Y_2O_3 and some $La(OH)_3$) were detected in the mixed powders. SEM analyses indicated the presence of randomly distributed aggregates of additives (an example is shown in Fig. 1a,b), which however have low dimensions (<3 µm). The oxygen content in the mixtures containing Si/C/N is higher, owing to its initial composition and partly to the powder treatment, which, notwithstanding the control of the process, involves oxidation due to the reactivity of these ultrafine powders.

Sintering behaviour

The sintering behaviours during hot pressing (T=1850°C, P=30 MPa) of the powder mixtures are compared in Fig. 2, where the relative density is reported as a function of temperature from commencement of densification up to 1850°C, and as a function of time over the soaking period. All the mixtures reach nearly full density (99-100% of relative density), however the densification rate is higher for the compositions containing ultrafine Si/C/N, as shown by the densification trends in Fig. 2. During heating up to 1850°C samples LY33Fu LY22Fu and L22Fa reached respectively the relative density of 87%, 84%, and 94%; on the contrary samples LY33u LY22u and LY22a reached respectively 79%, 71%, and 80%.

The temperature (as registered during hot press runs) at which the shrinkage starts (T_1, Table 2), corresponds to the first densification stage (particles rearrangement and fragmentation) and ranges from 1490°C to 1660°C. The temperature, at which the second densification stage starts (solution- diffusion-repreciptation), corresponds approximately to the liquidus temperature (T_2, Table 2) and ranges from 1670°C to 1750°C. Both these temperatures are dependent on: -powder processing method (the attritor milling lowers these temperatures due to the presence of contaminants and to a modification of the particle surface chemistry); -the presence of ultrafine Si/C/N powder, which allows lower T_1 and T_2 (due to the oxygen content and to the higher sinterability related to the extremely fine grain size); -the additive amount: at constant compositions and powder treatments, a reduced amount of additives increases T_1 and T_2.

Microstructure

The microstructure of hot pressed materials is shown in Figures: 3a,b (sample LY22u), 3c,d (sample LY22Fu), 4a,b (sample LY22a), 4c,d (sample LY22Fa). Some microstructural characteristics are reported in Table 3. The materials are fully dense, with the presence of β-Si$_3$N$_4$ grains, i.e. so called "in situ" composite or self reinforced materials. The aspect ratios of the elongated grains vary in the range 6-8 for monolithic Si$_3$N$_4$ and are lower in the composites.

Equiaxed grains are also present and prevalent in the composites from the Si$_3$N$_4$-Si/C/N mixtures (sample LY22Fa has mostly equiaxed grains). Moreover these samples evidence a finer microstructure (mean grain size 0.20-0.24 μm) resulting from the homogeneous nucleation, while silicon nitride materials have mean grain sizes in the range 0.38-0.45 μm.

Residual α-Si$_3$N$_4$ phase was detected in various amounts up to ~9%. In the composites crystalline SiC (8-9%) and Si$_2$N$_2$O (2-5%) were also observed.

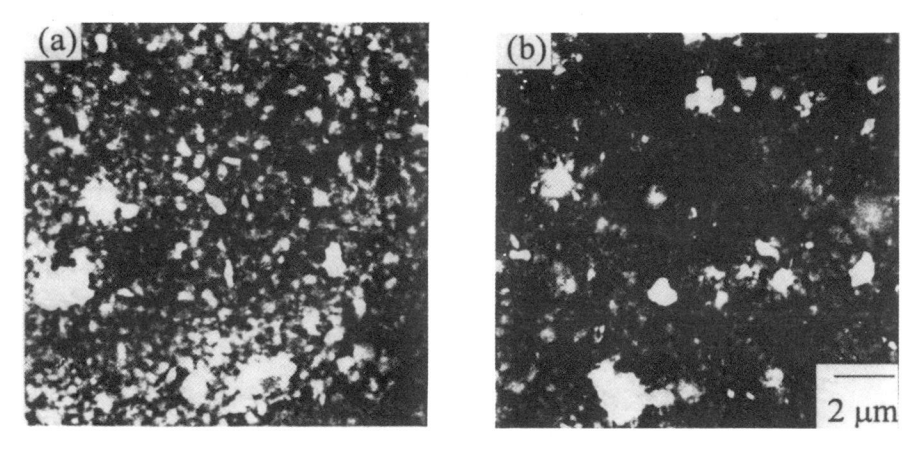

Figure 1. BSE micrographs from cold pressed powders mixed by attritor (a) and ultrasonically mixed (b).

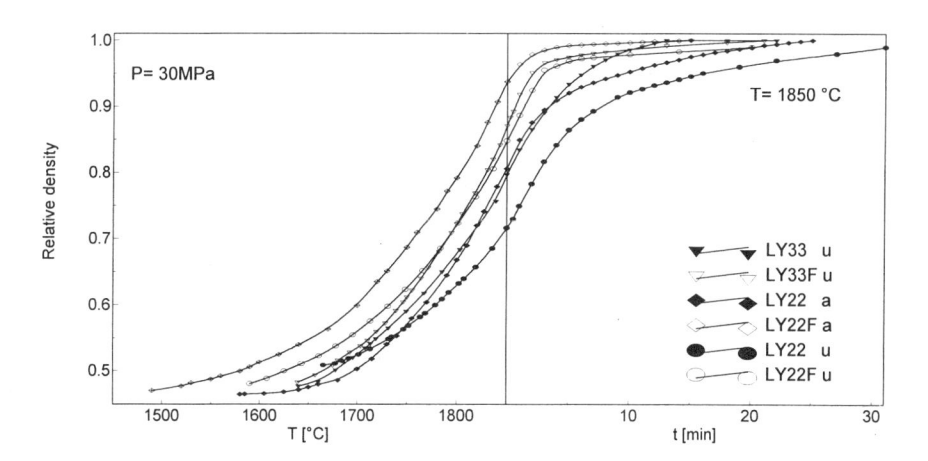

Figure 2. Densification curves of the hot pressed samples.

A very strong preferred orientation of the β-Si$_3$N$_4$ grains, indicated by the value of the ratio I$_{(200)}$/I$_{(002)}$ was measured by X-ray diffraction (Table 2) with alignment of the basal planes of the polycrystalline grains in direction parallel to that of the applied pressure. The preferred orientation is reduced in the composites. These data are confirmed by the microstructural characteristics on the surfaces parallel and perpendicular to hot pressing direction (Figures 3-4).

From the reported results it is determined that densification, grain growth and morphology are related to the starting compositions and to the processing conditions. During the α→β transformation of the α-rich silicon nitride powder, β-nuclei are formed with an orientation that initiates the growth of primary rod-like grains. The amount, type and distribution of the liquid phase at the operating temperature (which also depend on powder treatments and on the presence of ultrafine Si/C/N particles) are parameters that influence the number and size of the nuclei and their growth rate. In fact the microstructures exhibit clear differences in several features as the ratio of columnar/equiaxed grains, grain size shape and aspect ratio, relevant anisotropies in surfaces parallel or perpendicular to the applied pressure, type, amount and distribution of the secondary phases.

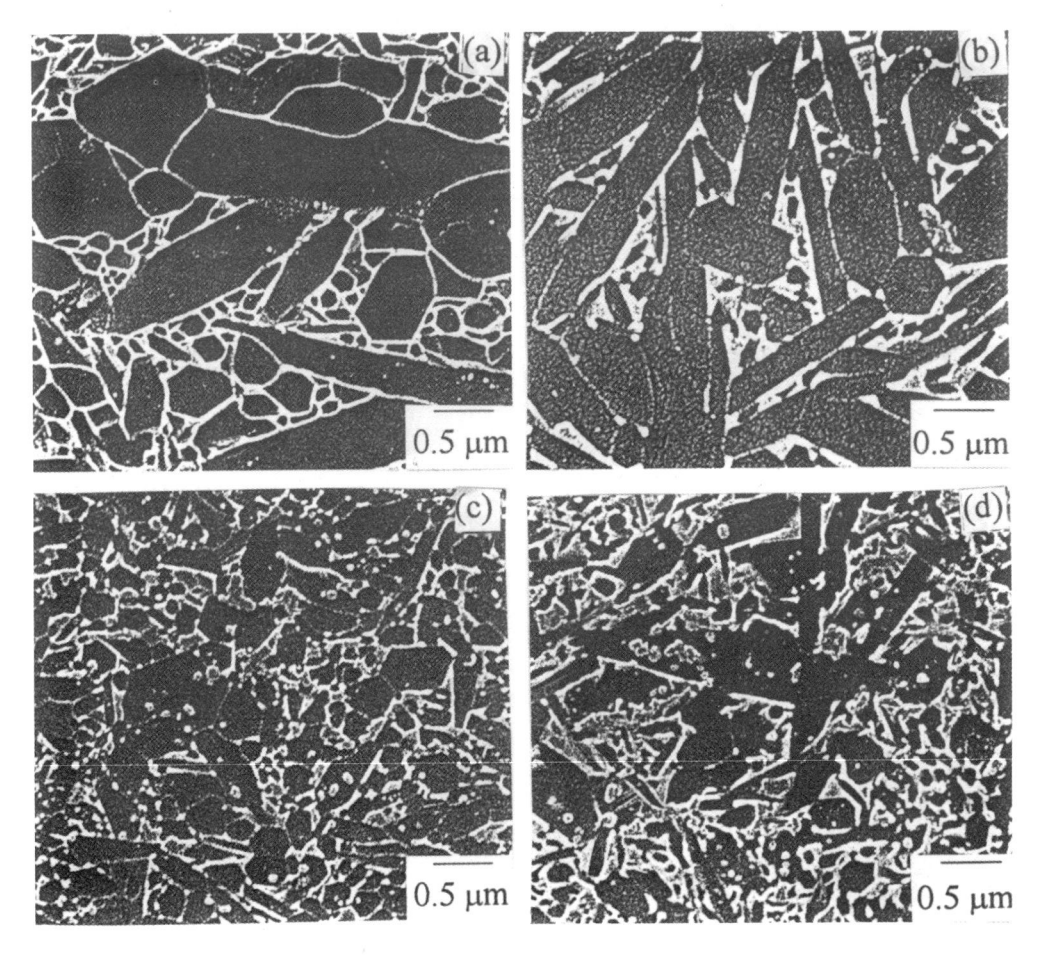

Figure 3. SEM micrographs from plasma etched surface of samples LY22u (a,b) and LY22Fu (c,d), parallel (leftside) and perpendicular (rightside) to the direction of the applied pressure.

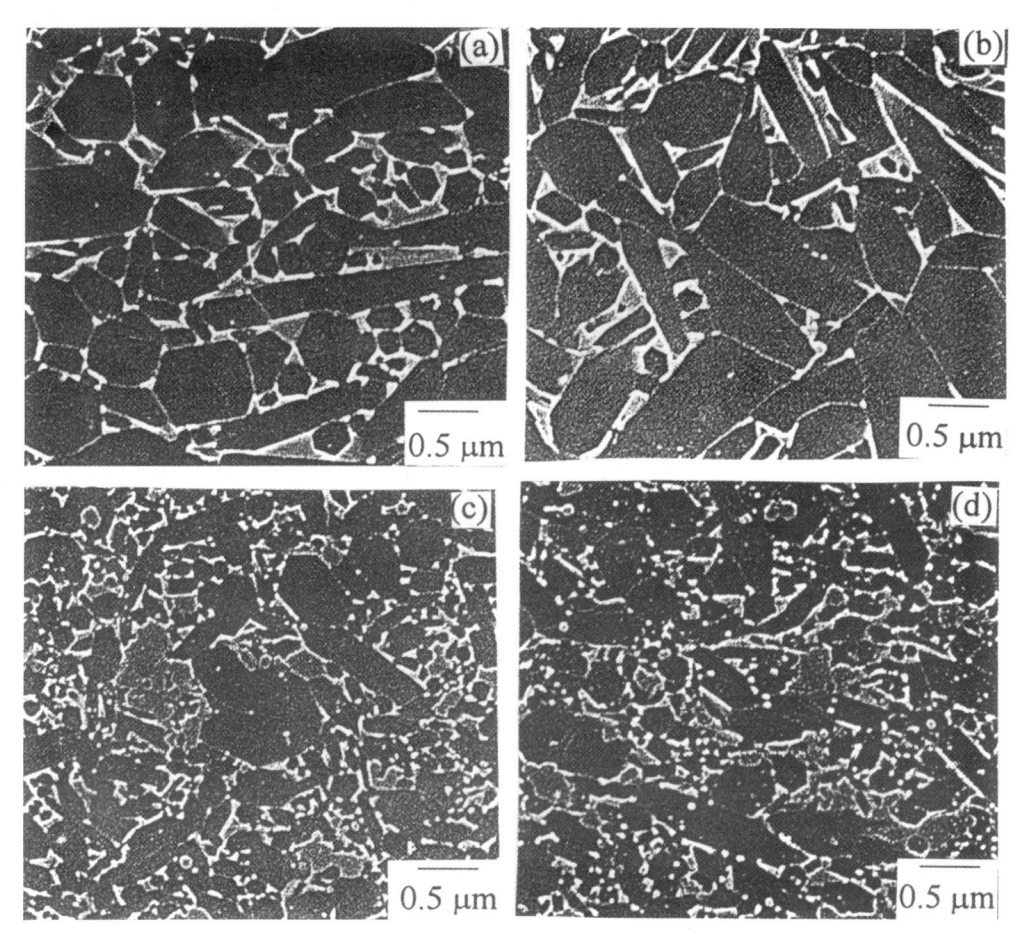

Figure 4. SEM micrographs from plasma etched surface of samples LY22a (a,b) and LY22Fa (c,d), parallel (leftside) and perpendicular (rightside) to the direction of the applied pressure.

The liquidus temperatures T_2 are in agreement with the La_2O_3-Y_2O_3[18] and La_2O_3-Y_2O_3-SiO_2[19] phase diagrams, however deviations among the different samples have to be ascribed to the oxygen content and to the presence of other contaminants occasionally introduced during processing or to the composition and particle surface chemistry of the ultrafine Si/C/N powder.

As the viscosity of the liquid phase at the working temperature is higher for compositions with high liquidus temperatures, the diffusion-controlled processes, densification and grain growth, are peculiar for each sample because the diffusion coefficient is inversely related to the viscosity of the liquid phase[1]. Generally a more viscous liquid phase limits mass transport, causing a retarded grain growth and the development of a fibrous microstructure, as observed mainly in samples LY33u and LY22u. On the other side, sample LY22a reveals a coarser microstructure with a lower aspect ratio, obviously related to a lower viscosity of the liquidus, as inferred by its lower liquidus temperature.

Regarding the nanocomposites produced with the addition of ultrafine Si/C/N, the number of initial nuclei and the reduced viscosity of the liquidus, due to the presence of more oxygen, play an important role in establishing the grain growth and the final microstructural characteristics.

Comparing the monolithics and the composites, it is evident that the latter system has the advantage to associate an excellent sinterability to the possibility to produce fully dense nanosize Si_3N_4-SiC composites, but the presence of elongated grains is greatly decreased.

In the materials tested, generally only traces of crystalline grain boundary phases were detected after hot pressing. The softening of these glassy phases, verified during thermal expansion tests, occurred within two temperature ranges: the former at 1050-1200°C and the latter at 1250-1400°C. During annealing heat treatments (1400°C for 6 hours), several grain boundary phases crystallised, in amounts of 2-4%: Lanthanum- and Yttrium-silicates, (mainly in the composites) and Lanthanum- and Yttrium-Silicon Oxide Nitrides (mainly in monolithic samples).

Thermal and mechanical properties

Thermal expansion. The values of thermal expansion coefficients (Table 4) show a non linear behaviour with temperature owing to the presence of intergranular glassy phases. The composites produced with the addition of Si/C/N powder have higher thermal expansion coefficients due to the presence of SiC and of Si_2N_2O. For the same system LY22, the thermal expansion coefficient is higher in samples from attritor milled powders than from ultrasonificated ones, inferring that the attritor mill introduces oxidation or contamination which increases the amount of the intergranular phases.

Young's modulus. The modulii for monolithic materials (LY33u, LY22u, LU22a) are similar or higher than the values of nanocomposites. These higher values are due to the presence of oxynitride glasses, which are the main constituents of the intergranular phases. They are stiffer than the oxide glasses mainly found in samples produced with the addition of Si/C/N. Moreover in this case the presence of Si_2N_2O is supposed to balance the effect of SiC.

Hardness. The different values measured on surfaces parallel or perpendicular to the applied pressure during hot pressing account for microstructural anisotropies (i.e. the grain alignment that occurs during sintering) and, in agreement with previous results[20], show a trend to increase with the amount of residual α-Si_3N_4.

Toughness and crack propagation. Higher toughnesses are measured in monolithic Si_3N_4 samples than in composites Si_3N_4-SiC, owing to a toughening effect resulting from elongated microstructure (Table 3, Figures 3-4). The fracture toughness in the surface perpendicular to hot pressing is higher than the toughness parallel to pressing.

Concerning the surface parallel to the hot pressing direction, the grain alignment in the direction perpendicular to hot pressing axis has to be considered. When an indentation is made on this surface, the crack lengths are longer perpendicularly to the hot pressing axis than in the parallel direction (Fig.5a). Therefore, when the crack propagates parallel to the hot pressing axis, the crack tip is hindered by unfavourably oriented grain boundaries and is deflected (Fig 5b). When the crack propagates in the direction perpendicular to the hot pressing axis, it encounters a lower number of unfavourably oriented grain boundaries and the whole crack path is longer (Fig. 5c).

Generally, in the monolithic samples (LY33u, LY22u, LY22a) the self reinforced microstructure allows crack propagation which partly follows the grain boundaries, but for greater extents runs through the grains (especially elongated grains). The interference of these grains with the crack path causes deflection, debonding, bridging, and branching. Figure 6a highlights areas where reinforcing mechanisms act against crack propagation and lower the stress concentration factor on the crack tip, contributing to the toughening of the

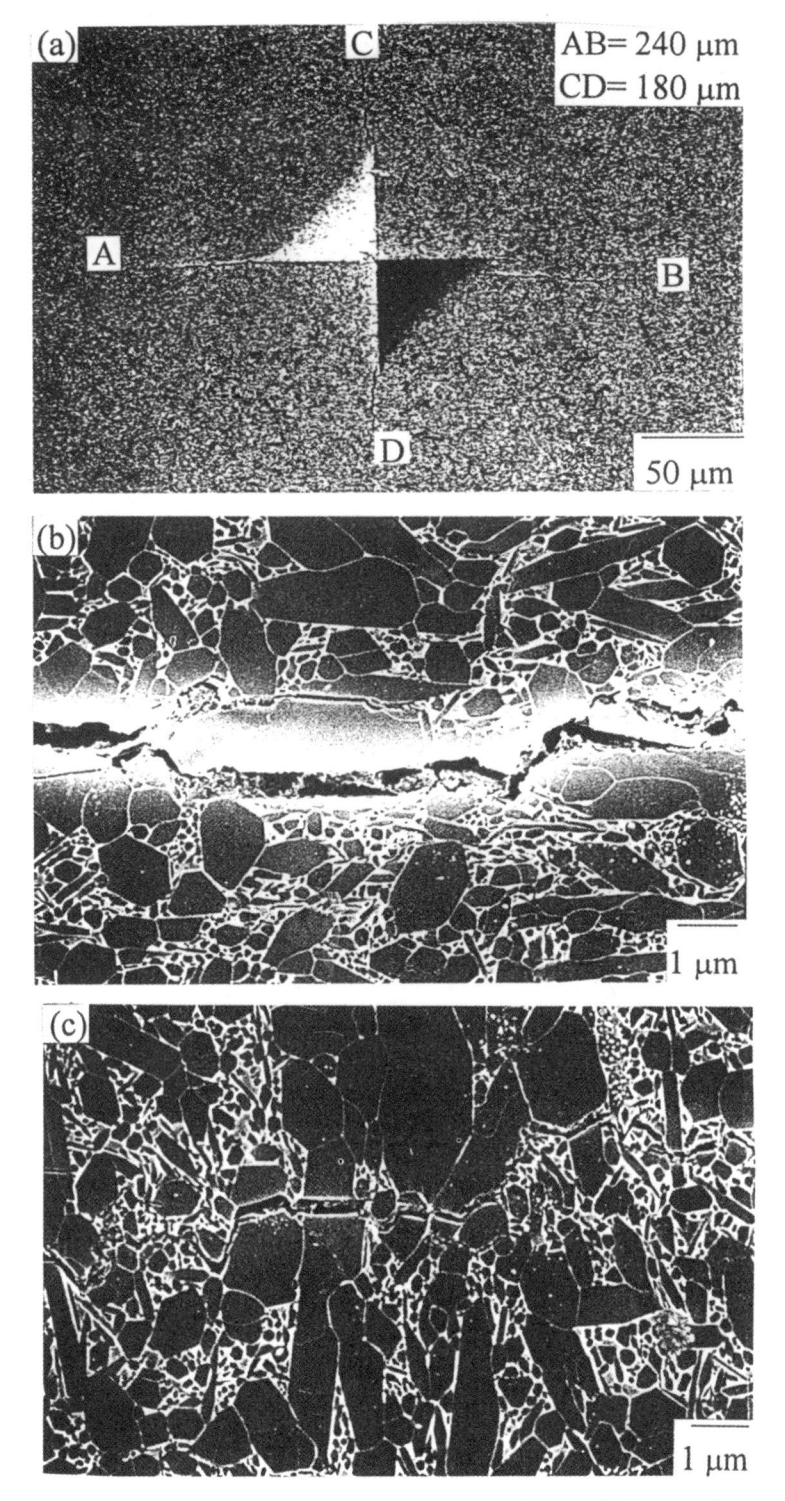

Figure 5. SEM micrographs showing an indentation and the relative cracks on a surface parallel to the applied pressure (a), a detail of crack path AB (b) and CD (c).

"in situ" composites. Crack bridging does not often consist of individual grains but of several grains connected by the intergranular phase. Also modulus load transfer from large and elongated grains to the surrounding matrix and crack branching disperses the constraint level on multiple crack tips. Crack deflection is more frequently observed in samples with high aspect ratios (>7) and is more pronounced in materials with the largest grain size. In the nanocomposites (LY33Fu, LY22Fu, LY22Fa) cracks propagate in a more straight path (Fig. 6b).

Room temperature flexural strength and analysis of critical defects. Quite striking differences in strength were measured (Table 4), although all values were generally high. For samples produced from ultrasonificated mixtures, the composites Si_3N_4-SiC show lower strengths than silicon nitride materials, while in attritor milled samples the values are almost the same. Therefore, the defect population, acting as critical flaws, was influenced by the powder processing methods and by the addition of ultrafine powders to submicronic powders.

The fracture generally initiated in areas with a high concentration of additives (probably hard aggregates in the raw materials) which favoured anomalous grain growth and the segregation of Y- and La-based compounds (Fig. 7a).

In the nanosize composites, the defects are porous areas associated with a poor homogeneity of the mixed Si_3N_4-Si/C/N powders (Fig. 7b).

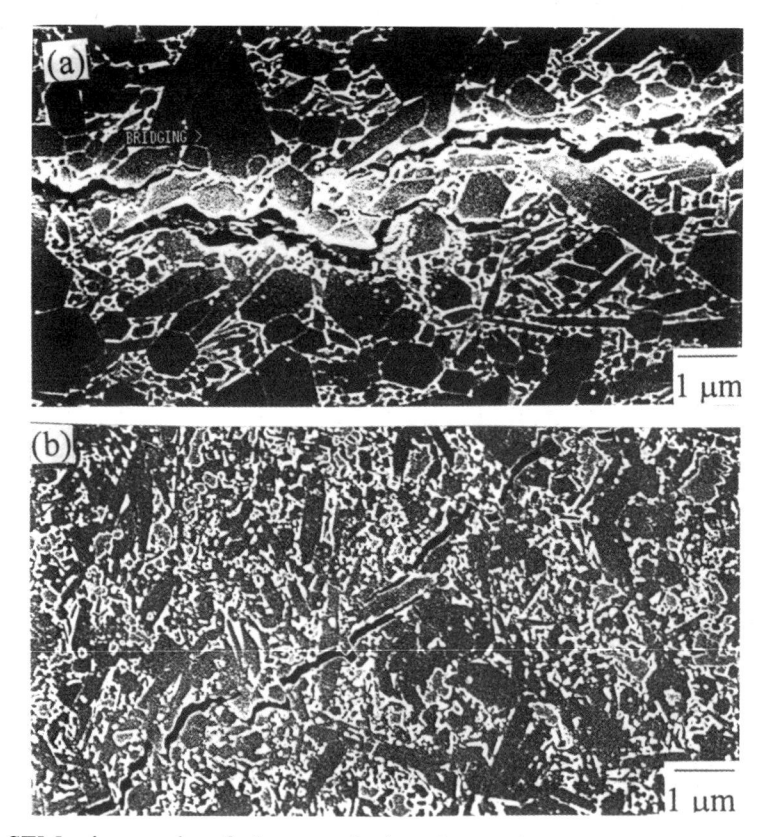

Figure 6. SEM micrographs of plasma etched surfaces, showing a crack path and typical reinforcing mechanisms (crack deflection and bridging) in monolithic samples (a) and nanocomposites (b).

High temperature flexural strength. The strengths measured at high temperature are influenced by the relaxation of the residual stresses due to machining, the oxidation attack during testing and the characteristics of the intergranular phases which can promote crack blunting but, at the same time, also subcritical crack growth. Some comments can be presented based on statistical comparison, considering however that the large spread of the fracture data could obscure significant differences.

At 1000°C the relaxation of the machining stresses and the softening of the grain boundary phase take place. As a result, the fracture strengths are usually lower than the corresponding ones at R.T. In LY33u and LY22u materials this is quite evident, while in other samples the phenomenon is not so pronounced.

At 1200°C, strengths are relatively unchanged with respect to 1000°C and the load-displacement curves of the specimens are still elastically linear up to the fracture strength for all the materials. Only two materials (LY22a and LY22Fa) have lower strength at 1200°C than at 1000°C. Both these samples were produced by attritor mixed powders, which, as previously shown, induces a lower liquidus temperature, therefore presumably less refractory grain boundary phases. As the observed flaw population remains almost the same for all the materials up to 1200°C, this strength degradation is due to a lower fracture toughness at 1200°C, obviously related to the characteristics of the grain boundary phases.

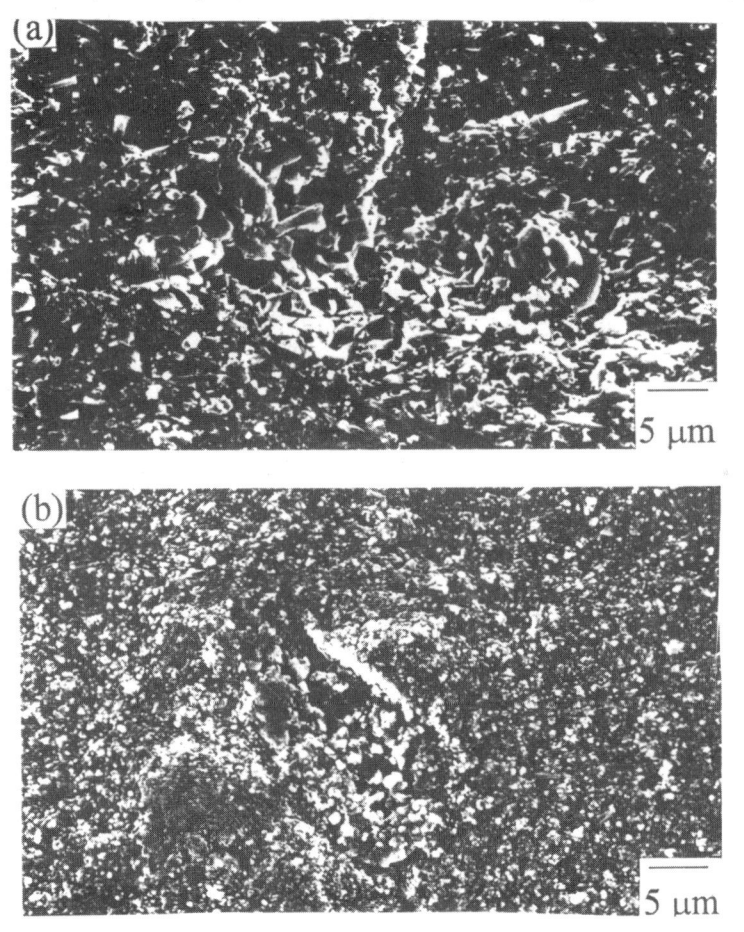

Figure 7. SEM micrographs from a fracture surface of monolithic (a) and composite (b) material showing typical defects originating failure in the tested samples.

At 1400°C, fracture originates in all cases mainly in the oxidized surface layer and a small deviation from linearity of the load-displacement curve is also noted. Moreover at this temperature, the glassy phase at the grain boundary becomes quite viscous and subcritical crack propagation can occur. A strength decrease is found for LY22u, LY22a, LY22Fa, as temperature increases from 1200°C to 1400°C.

CONCLUSIONS

Fine "in situ" reinforced Si_3N_4 materials with improved thermal and mechanical properties were produced by hot pressing with the addition of $Y_2O_3+La_2O_3$ as sintering aids. Si_3N_4-SiC nanocomposites were obtained by the addition of 30 wt% of ultrafine Si/C/N powders to silicon nitride powder.

The influence of the amount of sintering aids and of the powder processing methods was evaluated. Densification behaviour, microstructure, thermal and mechanical properties were related to the starting composition and to the microstructural characteristics.

Fibrous microstructures of the Si_3N_4 grains (aspect ratios in the range 6-8) were obtained in monolithic samples; the Si_3N_4-SiC composites have mainly equiaxed, very fine (~200 nm) grains, and the few elongated grains have lower aspect ratios. The sinterability is strongly improved by the addition of the ultrafine Si/C/N powder, and the powder treatment has a clear effect: attritor mixing results in mixtures with higher sinterability than the ultrasonificated ones. These phenomena have been explained in terms of variations in amount, type, and viscosity of the liquid phase at the hot pressing temperatures.

Regarding the mechanical properties the best results were registered in monolithic silicon nitride (σ>1100MPa at R.T. and up to ~800MPa at 1400°C). In addition different effects of the powder treatments were observed: ultrasonic mixing gives the best results for monolithic samples, while attritor milling improves the homogeneity when ultrafine Si/C/N powder is added, limiting therefore the critical defects in the Si_3N_4-SiC nanocomposites. High temperature performances are strongly dependent on flaw population and on the amount and viscosity of grain boundary phases, which are related to starting composition and to powder processing methods.

Acknowledgements

The authors are grateful to A. Borsella, S. Martelli and S. Botti of ENEA, Div. Inn. Fis, C.R. Frascati for the production and characterization of the ultrafine Si/C/N powders, to D. Dalle Fabbriche of IRTEC for the technical assistance.

REFERENCES

1. G. Wotting, and G. Ziegler, Influence of powder properties and processing conditions on microstructure and mechanical properties of sintered Si_3N_4, *Ceramics Int.*, **10**:468 (1984).
2. F. Cambier, A. Leriche and V. Vandeneede, Powder characterization and optimisation of fabrication and processing for sintered Si_3N_4, part 1, in: *Ceramic Materials and Components for Engines*, W: Bunk and N: Hausner, eds., Verlag Deutsche Keramische Ges., Cologne - Germany (1986).
3. F. F. Lange, Fabrication and properties of dense polyphase silicon nitride, *Am. Cer. Soc. Bull.*, **62**:1369 (1983).
4. A. Bellosi and G. N. Babini, Microstructural design of Si_3N_4 with tailored properties, in:

Key Engineering Materials vol (89-91), Silicon Nitride '93, M. J. Hoffman, P. F. Becher and G. Petzow, eds., Trans Tech Publications Ltd, Aedermannsdorf - Switzerland (1994).
5. A. Bellosi, F. Monteverde and G. N. Babini, Influence of powder treatment methods on sintering, microstructure and properties of Si_3N_4-based materials, in print in the Proceedings: NATO ARW "Engineering Ceramics '96", Bratislava-Slovakia (1996).
6. K. Niihara, K. Izaki and T. Kawakahi, Hot pressed Si_3N_4-32% SiC nanocomposites from amorphous Si-C-N powder with improved strength above 1200°C, *J. Mat. Sci Lett.*, **10**:112 (1990).
7. B. A. Bender, R. P. Ingels, W. J. McDonough and J. R. Spawn, Novel ceramic microstructure and nanostructure for advanced processing, *Adv. Ceram. Mat.*, **1**:137 (1989).
8. A. Sawaguchi, K. Toda and K. Niihara, Mechanical and electrical properties of Si_3N_4-SiC nanocomposite material, *J. Am. Ceram. Soc.*, **74**:1142 (1991).
9. M. Mayne, D. Bahloul-Hourlier, P. Goursat and J-L. Besson, Microstructure and mechanical properties of Si_3N_4-SiC nanocomposites, *Fourth Euroceramics*, vol.4, A. Bellosi, ed.,Gruppo Editoriale Faenza Editrice, Faenza - Italy (1995).
10. H. Drost, M. Friedrich, R. Mohr, R. Mach and E. Gey, Property modification of nanoscaled Si-C-N composites by variation of the synthesis conditions, ibid.
11. J. Dusza, P. Sajgalik and M. J. Reece, ibid.
12. P. Sajgalik, and J. Dusza, Preparation and properties of fine Si_3N_4 ceramics, in: *Key Engineering Materials*, vol **89-91**, *Silicon Nitride 93*, M. J. Hoffman, P. F. Becher and G. Petzow, eds., Trans Tech Publications Ltd, Aedermannsdorf-Switzerland (1994).
13. A. Bellosi, Si_3N_4-SiC ceramic nanocomposites, in: *Nanophase Materials - Material Science Forum* vol **195**, E. Bonetti and D. Fiorani, eds., Trans Tech Publications Ltd, Aedermannsdorf-Switzerland (1995).
14. R. Giorgi, S. Turtù, G. Zappa, E. Borsella, S. Botti, M. C. Cesile and S. Martelli, Microstructural properties of laser synthesized Si/C/N nanoparticles, *Appl. Surf. Science*, **93**:101 (1996).
15. E. Borsella, S. Botti, R. Martelli, R. Alexandrescu, M. C. Cesile, A. Nesterenko, R. Giorgi, S. Turtù and G. Zappa, Laser synthesis of ceramic nanocomposite powders, to be published in Silicates Industriels.
16. ASTM C 848-78 (Reapproved 1983).
17. G. R. Anstis, P. Chantikul, B. R. Lawn and D. B. Marshall, A critical evaluation of indentation techniques for measuring fracture toughness: I, direct crack measurement, *J. Am. Ceram. Soc.*, **64**:533 (1981).
18. E. M. Levin, and H. F. McMurdi, in: *Phase Diagram for Ceramists*, M. K. Reser, ed., The American Ceramic Society, Inc. Columbus, Ohio (1975).
19. E. M. Levin, C. R. Robbins, and H. F. McMurdie. in: *Phase Diagrams for Ceramists*, M. K. Reser, ed., The American Ceramic Society, Inc. Columbus, Ohio, (1969).
20. Z. Lences, S. Guicciardi, C. Melandri, and A. Bellosi, Preparation, microstructure and mechanical property relationship in Si_3N_4 based ceramics, *British Ceramic Transactions*, **94**:138 (1994).

SIMULATION OF ANISOTROPIC GRAIN GROWTH BY OSTWALD RIPENING

Veena Tikare[1] and James D. Cawley[2]

[1]Sandia National Laboratories
Albuquerque, NM 87110-1411
[2]Case Western Reserve University
Cleveland, OH 44106

ABSTRACT

In this paper, a two-dimensional Potts model that can simulate anisotropic grain growth by Ostwald ripening will be presented. The model defines a Wulff shape for the grains and allows each individual grain to grow in its local environment which is controlled by the solid/liquid interfacial energy, the spatial distribution of neighboring grains, area fraction of grains, wetting by and distribution of the liquid matrix, and the concentration gradients in the liquid. The results of this simulation technique will be presented with emphasis on the kinetics and grain shape evolution and will be compared to those of isotropic grain growth. Finally, the limitations of the Potts model in such microstructural evolution processes will be discussed.

INTRODUCTION

In fully wetting systems, after densification, grain growth occurs by the solution - reprecipitation mechanism and is therefore appropriately considered in the context of Ostwald ripening. Ostwald ripening is the increase in average grain size driven by the reduction in interfacial energy between the grains and surrounding matrix by diffusion through the matrix phase. Modeling anisotropic grain growth by Ostwald ripening is the focus of this investigation.

Predictive modeling of microstructural evolution presents unique challenges because of the large number of thermodynamic, mechanistic and spatial variables that must be considered simultaneously. Analytic models make many simplifying assumptions to make the problem tractable. In spite of the many assumptions, analytic models have increased the understanding of microstructural evolution processes; however, they have failed to predict the details of the microstructure or to incorporate complexities such as anisotropy[1,2,3,4]. Motivated by the desire to eliminate these assumptions, many simulation techniques have been developed[5,6,7]. Among these is the Potts model, a statistical-mechanical simulation technique, which has proven to be robust for modeling a diverse set of microstructural evolution problems, including normal and abnormal grain growth[8,9] particle-inhibited grain growth[10], grain growth in multiphase systems[11], Ostwald ripening[12] and sintering[13]. In this investigation, the Potts model was modified to simulate isotropic and anisotropic grain growth in liquid-phase-sintered ceramics by Ostwald ripening.

MODEL

The microstructure was represented digitally by discretization on a two-dimensional, square lattice with periodic boundary conditions in both the X- and Y-directions. This lattice was populated with a canonical ensemble of two types of entities, designated A and B. These entities may be thought of as meso-scale domains of a particular phase. Entity A was the primary constituent of the solid phase, and entity B of the liquid phase. The A-entities were allowed to assume one of Q degenerate states, $q_A \in [1,2,...,Q]$ where $Q = 100$ for most simulations. Each state, q_A, is an arbitrary label for the purposes of tracking individual grains, and can be considered to be a particular crystallographic orientation. The B-entities could assume only one state, $q_B = -1$. The negative number has no significance; it simply makes computation easier. The grains are features consisting of contiguous sites of the same state, q_A.

The Potts model used in this investigation incorporated interfacial energy by defining interaction energies between neighboring sites. The Hamiltonian, H, for the simulation and was calculated as the sum of the interaction energy between each site, i, and each of its first and second nearest neighbors, j,

$$H = \frac{1}{2} \sum_{i=1}^{N} \sum_{j=1}^{8} E(q_i q_j) \quad (1)$$

where N was the total number of sites, E was the bond energy between neighboring sites in arbitrary units of energy and was a function of the state of the two neighboring sites, and q_i was the phase and orientation of the entity occupying site i. The Hamiltonian had units of energy, the same as that for E, the bond energy. The bond energy, E, was defined so that only neighbors of different states have a bond energy.

The solid-liquid (A-B) interfacial energy, γ_{sl}, and the grain boundary energy between two solid grains (A$_i$-A$_j$) of different orientation, γ_{ss}, were defined by their respective bond energies. A fully miscible liquid was considered in the simulation; thus, liquid-liquid interfaces were unnecessary in the simulation. Assigning only one state to the B-sites ensured that no liquid-liquid interfaces could occur. The interfacial energies, γ_{sl} and γ_{ss}, can be calculated as follows:

$$\gamma_{qiqj} = \frac{E_{qiqj} N_{qiqj}}{L_{qiqj}} \quad (2)$$

where E_{qiqj}, the bond energy, is a function of neighboring phases and states, q_i and q_j, $Nqiqj$ is the number of the unlike bonds at the interface, and $Lqiqj$ is the length of the interface. The units of interfacial energy are in energy per unit length rather than energy per unit area because two-dimensional space is being used for this simulations. A consequence of eq. (2) is that the ratio of γ_{sl} to γ_{ss} can be varied by varying the ratio of E_{AB} to E_{AA}. The use of a square lattice to digitize the microstructure introduced some anisotropy to the interfacial energies. For isotropic grain growth simulations, both first and second nearest-neighbor interactions were considered to minimize lattice anisotropy effects as shown by Holm et al[14]. For anisotropic grain growth, anisotropy was introduced by making the bond energies directionally dependent. Half of the A-entities, even numbered q's, were given high bond energies in the Y-direction and low bond energies in the X-direction; and vice versa for A-entities with odd numbered q's. Furthermore, only first nearest neighbors were considered in eq. (1).

Transport was simulated by exchanging neighboring sites with each other. Conserved dynamics were used to maintain constant composition, i.e. fixed number of A- and B-sites. The step-by-step simulation technique used is reviewed below:

1. The two-dimensional, square lattice was populated with the desired composition (fraction of A-sites and B-sites). The lattice was initialized by placing the A- and B-sites at randomly selected lattice sites. The A-sites were assigned random q-values, between $q_A = 1$ and Q, and all B-sites were assigned $q_B = -1$
2. A lattice site was chosen at random, then one of its neighboring sites was chosen at random from its eight first- and second-nearest neighbors.
3. If one site was an A-site and the other a B-site, the simulation continued to step 4; if not, then to step 7. If both were A-sites, the exchange was not permitted because solid state diffusion was not allowed. If both sites were B-sites, the exchange does not change the microstructure; therefore consideration of this exchange was unnecessary.
4. A new orientation for the A-site was chosen at random from Q (=100) possible orientations.
5. The new configuration, A- and B-sites exchanging places with the A-site assuming the new orientation, was considered and the change in energy for this configuration was calculated using the Hamiltonian given in eq. (2).
6. The classical Metropolis algorithm[15] was used to determine if the new configuration was accepted or rejected. The Metropolis algorithm uses Boltzmann statistics to determine the exchange probability, ω, as follows:

$$\omega = \exp\left(-\frac{\Delta H}{k_B T}\right) \quad \text{for} \quad \Delta H > 0$$

and (3)

$$\omega = 1 \quad \text{for} \quad \Delta H \leq 0$$

where k_B is the Boltzmann constant, and T is absolute temperature and has units of bond energy divided by the Boltzmann constant. As is conventional for Monte Carlo simulations, all temperatures are given as $k_B T$ with units of energy rather than just temperature.
7. The number of iterations is incremented by 1/N, where N is the total number of sites and the program is returned to step 2. The number of iterations is given in units of Monte

Carlo step, MCS. At 1 MCS, the number of attempted exchanges was equal to the total number of sites in the simulation space. It has been shown that MCS scales linearly with time by some constant, τ, for such simulations[16,12] thus MCS may be considered a measure of time in arbitrary units.

RESULTS AND DISCUSSION

The microstructures of isotropic and anisotropic grain growth simulations are shown in figures 1 and 2. The grain morphologies of the two are very different. In the isotropic simulations, the grains appear regular and equiaxed. In the anisotropic simulations, half the grains are elongated in one direction and rest in the perpendicular direction, since the variation in bond energies is restricted to these two directions. These highly elongated grains are faceted on the edges parallel to the high bond energy direction. The elongated and faceted grains persist throughout the grain growth simulations.

Figure 1. Isotropic grain growth simulation microstructures at time, t = 4, 35 and 100 KMCS.

Figure 2. Anisotropic grain growth simulation microstructures at time, t = 29, 52 and 100 KMCS.

The data from these simulations are plotted as average grain size versus time on logarithmic scales in figures 3 and 4, where grain size is the average grain radius. The anisotropic grains size is also given as the average radius of circular grains of equivalent area. Although, this seems a non-physical measure, it was chosen as the best one to allow comparison to both isotropic grain growth simulations and predictions by analytical models. The grain growth curves for the anisotropic simulations exhibit two distinct regions; the first is a non-linear region and the second is a linear region indicating power law behavior. Examination of the simulated microstructures revealed that during the initial, non-linear region, grains are nucleated and solid precipitates onto nuclei from a supersaturated liquid. During precipitation all grain grow, but during the second stage, grains grow by Ostwald ripening, some grains grow at the expense of others. It is this Ostwald ripening regime that is of interest in this investigation.

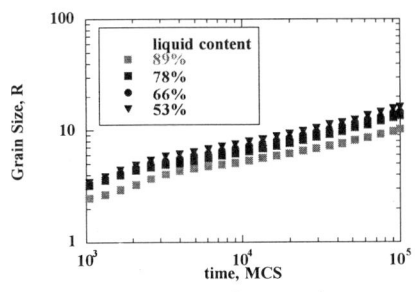

Figure 3. Isotropic grain growth curves.

Figure 4. Anisotropic grain growth curves.

Ostwald ripening theories predict power law behavior as given by the following equation:

$$R^n - R_o^n = Kt \tag{4}$$

where R is the average grain radius at time $t > 0$, R_o is the initial grain size at time $t = 0$, K is the grain growth constant and n is the grain growth exponent. The data do indicate power law behavior with the inverse slope in the Ostwald ripening regime equal to n. The grain growth exponent, n, increased with increasing solid fraction as seen in figure 4. This behavior was in sharp contrast with the isotropic grain growth behavior, which has constant n as shown in figure 3, and with Ostwald ripening theories which also predict constant n, when the grain growth mechanism remains the same.

To understand the origin of varying n, we considered how the grain lengths and widths grew during Ostwald ripening. The grain length was measured as the second moment around the axis going through the center of mass in the direction of low bond energy and grain width was the same except in the direction of high bond energy, as defined by the following relationship:

$$I_L = \frac{\sum_{i=1}^{a}(X_i - X_c)^2 + \frac{1}{12}}{a} \qquad I_W = \frac{\sum_{i=1}^{a}(Y_i - Y_c)^2 + \frac{1}{12}}{a} \tag{5}$$

where I_L and I_W are the second moments around the low bond energy direction and high bond energy direction, respectively, and are used as measures of grain length and width, respectively. X_C is the location of the axis of low energy through the center of mass, X_i - X_C is the distance of the site i from that axis, and a is the total number of sites in the grain. Similarly, Y_c is the location of the axis of high energy through the center of mass and Y_i - Y_c is the distance from that axis.

The average length moment and width moment in the Ostwald ripening regime are plotted in figure 5 as a function of time on logarithmic scales for different solid contents. Again, power law behavior is observed for both moments at all three solid contents as indicated by the linear relationship observed in these plots. However, the exponent of the power law was different for

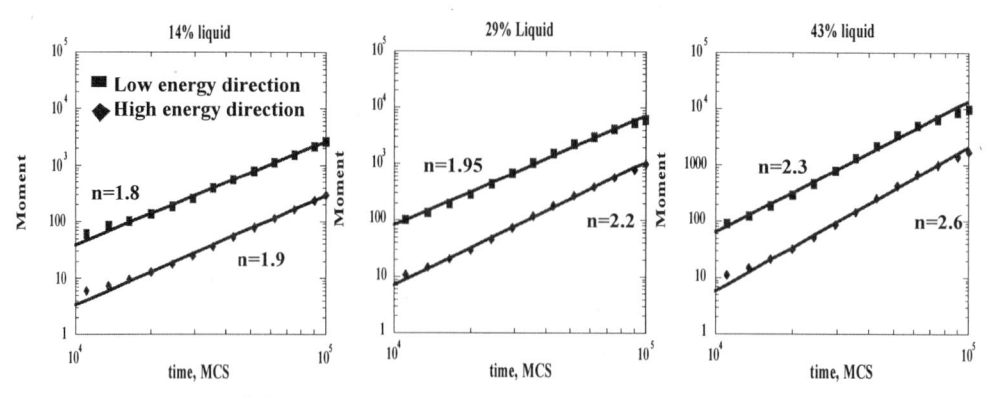

Figure 5. Moments of inertia as a function of time during anisotropic grain growth at different liquid contents.

each moment at the different solid fractions. All three simulations exhibited higher powers for the width moment than the length moment suggesting that the grains grew wider faster than they grew longer. Furthermore, the rate of grain widening increased with decreasing solid fraction. These observations suggested that the shape as characterized by the anisotropy of the grains was changing both with time and solid fraction.

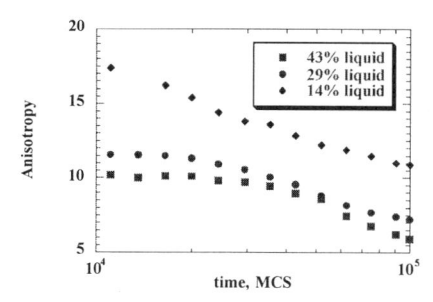

Figure 6. Anisotropy of grains during grain growth.

This was confirmed by determining the anisotropy of the grains; anisotropy, A, was quantified as the ratio of the two moments as given by:

$$A = \frac{I_L}{I_W} \qquad (6)$$

Anisotropy of grains is plotted as a function of logarithmic time in figure 6. Anisotropy decreased with increasing time and decreasing solid fraction. This is consistent with the microstructures shown in figure 2 which show that initially long grains form. These grains are constrained in the long direction by grains elongated in the perpendicular direction, but almost never by grains elongated in the parallel direction. These initially long grains, which are constrained along their length, widen but remain constrained along their length, thus anisotropy decreases.

The change in grain growth exponent, n, with solid fraction may be due the differing constraints in the two different directions for the various solid contents. In the isotropic grain growth case, the relative ease of growth in any direction is constant, but not so for the anisotropic case. The constraint is changing which in turn changes the rate of growth in the different directions. At high solid fractions the constraint along the width of the grains is high thus growth in this direction lags behind growth in the length direction as seen in figure 7.

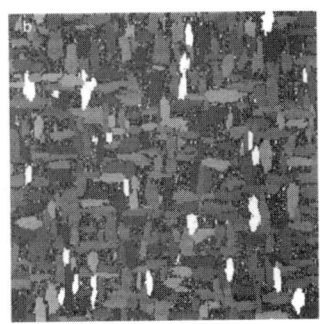

Figure 7. Anisotropic grain growth microstructure at time t = 100 KMCS for liquid content = 14, 29 and 43%.

The decrease in anisotropy with decreasing grain fraction was unexpected. In silicon nitride ceramics which have anistropies similar to the one used in this study it was found that at higher liquid contents grains were more elongated then at lower liquid contents. This increase in anisotropy with decreasing grain fraction was attributed to less constraint at lower grain fractions. The contrary behavior in the simulations is a consequence of the limitation of the grain anisotropy to two directions and to two dimensions.

The restriction of anisotropy to two directions is similar to rafting in metals where elastic stress field constrains growth of percipitates in direction aligned with the applied stress. It is also similar to precipitation of a second phase within grains such that the precipitation is aligned in particular crystallographic direction of the parent crystal. These are observed in many martensitic systems such as MgO partially stabilized zirconia.

Conclusions

Anisotropic grain growth by Ostwald ripening can be simulated by the Potts model by assigning different bond energies between neighboring sites along the different lattice directions. The grain growth exhibited power law behavior, but the grain growth exponent was not constant; it increased with decreasing liquid content. Grain anisotropy decreased with increasing grain size and decreasing liquid.

Acknowledgment:

This work was supported by the U.S. Department of Energy under contract number DE-AC04-95AL85000.

---

[1] I.M. Lifshitz and V.V. Slyozov, "The Kinetics of Precipitation From Supersaturated Solid Solutions," J.Phys. Chem. Solids, 19 [1-2] 35-50 (1961).
[2] C. Wagner, "Theory of the Ageing of Precipitates by Redissolution," Zeitschrift fur

Elektrochemie, 65 [7-8] 581-591 (1961).

[3] P.W. Voorhees and M.E. Glicksman, "Solution to the Multi-Particle Diffusion Problem with Application to Ostwald Ripening - II. Computer Simulations," Acta Metall. 32 [11] 2013-2030 (1984).

[4] R.T. DeHoff, "A Geometrically General Theory of Diffusion Controlled Coarsening," Acta Metall. 39 [10] 2349-2360 (1991).

[5] K.W. Mahin, K. Hanson and J.W. Morrid Jr., "Comparative Analysis of the Cellular and Johnson-Mehl Microstructures through Computer Simulations," Acta Metall. 28 443-453 (1980).

[6] E.A. Ceppi and O.B. Nasello, "Computer Simulation of Bidimensional Grain Growth," Scripta Metall. 18 1221-1225 (1984).

[7] A. Soares, A.C. Ferro and M.A. Fortes, "Computer Simulation of Grain Growth in a Bidimensional Polycrystal," Scripta Met. 19 1491-1496 (1985).

[8] D.J. Srolovitz, M.P. Anderson, P.S. Sahni and G.S. Grest, "Computer Simulation of Grain Growth - I. Kinetics," Acta Metall., 32 793-802 (1984).

[9] D.J. Srolovitz, G.S. Grest, and M.P. Anderson, "Computer Simulation of Grain Growth - V. Abnormal Grain Growth," Acta Metall., 33 [3] 2233-2247 (1985).

[10] E.A. Holm and D.J. Srolovitz, "Effects of Inert Particle Morphology on Grain Growth in Composites," Proc. of Morris E. Fine Symp., edited by P.K. Liaw, J.R. Weertman, H.L. Marcus and J.S. Santner, Minerals, Metals and Materials Soc. 187-192 (1991).

[11] E.A. Holm, D.J. Srolovitz, J.W. Cahn, "Microstructural Evolution in Two-Dimensional Two-Phase Polycrystals," Acta Metall., 41 [4] 1119-1136 (1993).

[12] V. Tikare, Numerical Simulation of Grain Growth in Liquid Phase Sintered Materials, Case Western Reserve Univ., Cleveland, OH (1994).

[13] G.N. Hassold, I-W Chen, and D.J. Srolovitz, "Computer Simulation of Final-Stage Sintering: I. Model, Kinetics, and Microstructure," J. Am. Ceram. Soc. 73 [10] 2857-2864 (1990).

[14] E.A. Holm, J.A. Glazier, D.J. Srolovitz and G.S. Grest, "Effects of Lattice Anisotropy and Temperature on Domain Growth in the Two-Dimensional Potts Model," Phys. Rev., A 43 2662-2668 (1991).

[15] N. Metropolis, A.W. Rosenbluth, M.N. Rosenbluth, A.N. Teller and E. Teller, "Equation of State Calculations by Fast Computiong Machines," J. Chem. Phys., 21 1087-92 (1953).

[16] Y. Limoge and J.L. Bocquet, "Monte Carlo Simulation in Diffusion Studies: Time Scale Problem," Acta Metall., 36 [7] 1717-1722 (1988).

MULLITIZATION BEHAVIOR OF ALPHA ALUMINA/SILICA MICROCOMPOSITE POWDERS

Michael D. Sacks, Keyun Wang, Gary W. Scheiffele, and Nazim Bozkurt*

Department of Materials Science and Engineering
University of Florida
Gainesville, FL 32611

ABSTRACT

The mechanism of mullite formation was investigated using submicrometer composite particles which consisted of alpha alumina cores and amorphous silica coatings. Mullitization behavior was monitored using X-ray diffraction analysis, differential thermal analysis, and scanning electron microscopy. The transformation occurred with an incubation period which was followed by stages of rapid mullite growth (up to ~70% conversion) and slower mullite growth. The first growth stage occurred primarily by nucleation and growth within the siliceous matrix. Available evidence indicates that the growth rate was controlled by dissolution of alumina in the siliceous phase. The second stage of mullitization occurred primarily by interdiffusion of alumina and silica through the mullite grains formed during the first stage. The transformation rate in the second stage was increased significantly by using mullite seed particles which produced smaller grain sizes during the first stage (and, thereby, decreased the interdiffusion distances needed to complete the reaction). This seeding approach allowed fabrication of bulk mullite samples with nearly 100% relative density and fine grain size (~0.4 μm) after heat treatment at only 1400°C (2 h).

INTRODUCTION

The mechanism of mullite formation (and the temperature range over which mullitization occurs) depend on the scale of chemical homogeneity of the aluminum silicate precursors. When the reaction is carried out using alumina and silica in the

---

* Permanent address: Department of Metallurgical Engineering, Istanbul Technical University, Maslak, 80626 Istanbul, Turkey.

form of bulk samples or relatively coarse powder mixtures, mullite forms at the interface between the reactants.[1,2] Growth of the mullite requires chemical interdiffusion of aluminum and silicon ions through the interfacial layer. Relatively high temperatures (>1500°C) are required for the reaction to proceed to an appreciable extent.

Mullitization temperatures can be lowered considerably when the reaction is carried out using chemically-synthesized precursors with fine-scale (<100 nm) mixing of the aluminum- and silicon-containing components. The sol-gel methods have been investigated extensively because of their potential to produce high purity materials with relatively low processing temperatures for both densification and mullitization. Sol-gel precursors to mullite are usually divided into two general categories -- "single phase" and "diphasic" -- based on the precursor chemical homogeneity and the resultant crystallization behavior. In amorphous single-phase systems (which have a molecular or near-molecular scale of mixing of aluminum and silicon ions), mullite crystallizes rapidly by a strong exothermic reaction at or below ~1000°C.[3-18] The reaction is apparently nucleation-controlled with an activation energy of approximately 300 kJ/mol.[7] Mullite formation from diphasic mixtures (in which alumina-rich and silica-rich precursors are segregated on a scale of approximately 1-100 nm) occurs more slowly by a weakly exothermic reaction at temperatures typically in the range of ~1150-1350°C.[8-23] Studies of the microstructure development and reaction kinetics indicate that mullitization of diphasic gels occurs by nucleation and growth with activation energies in the range of ~900-1100 kJ/mol.[12,16,19,20]

Wei and Halloran[19,20] studied the phase transformation behavior in gels prepared from colloidal boehmite (as the alumina source) and tetraethylorthosilicate, TEOS, (as the silica source). They observed a temperature-dependent incubation period prior to mullite formation. The incubation period was followed by a rapid nucleation event and subsequent growth of mullite from a constant density of nuclei. Microstructure observations indicated that growth proceeded preferentially through the siliceous "matrix" phase since alumina particles were often engulfed by the growing mullite grains. The activation energies for the incubation and growth stages were similar (i.e., 987 ± 63 and 1070 ± 200 kJ/mol, respectively) suggesting that both were controlled by the same rate-limiting process. Based on microstructure observations and reaction kinetic data, Wei and Halloran concluded that the growth rate was controlled by either short-range diffusion near the mullite/alumina/silica interface or by the reaction at the interfacial region. Li and Thomson[12] also studied the mullitization behavior of gels prepared from colloidal boehmite and TEOS and obtained results which were generally in agreement with those reported by Wei and Halloran. (For example, they observed an incubation period prior to mullitization and reported an activation energy of 1028 ± 37 kJ/mol for the growth stage.) Li and Thomson claimed that the mullite growth rate was diffusion-controlled based on the reaction kinetic data and on microstructure observations. Huling and Messing[16] investigated the mullitization kinetics for a variety of tailored diphasic aluminum silicate gels. One of the samples was a "colloidal gel" prepared by mixing fine particles of amorphous silica and boehmite. The activation energy for mullitization in this sample was very similar (1034 ± 124 kJ/mol) to the values obtained by Wei and Halloran and Li and Thomson despite the significant difference in the spatial distribution of the silica in the gel. This led Huling and Messing to conclude that the phase transformation was controlled by the alumina component in the gel. They suggested that the rate-limiting step in mullite growth was the "release" of alumina into the siliceous phase. Sundaresan and Aksay[24] reanalyzed data in the literature and concluded that mullite nucleation and growth occurred within the siliceous matrix. Incubation was interpreted as the period during which alumina dissolves in the silica until the amount exceeds the critical concentration required for nucleation. Growth of mullite would require continued dissolution of alumina in the silica. Sundaresan and Aksay concluded that the similarity in the activation energy values for incubation and growth indicated that alumina dissolution was the rate-limiting step in both stages.

The source of alumina during the mullitization reactions in heat-treated diphasic gels is either a transitional alumina phase or aluminum silicate "spinel." There are few detailed studies of mullite formation using diphasic powders in which the alumina source is the alpha phase. Recently, Sacks et al.[25-29] studied the densification, transformation behavior, and microstructure development during heat treatment of submicrometer composite particles which consisted of inner cores of α-Al_2O_3 and outer coatings of amorphous silica. The α-Al_2O_3 particles were approximately an order of magnitude larger (~0.2 μm average diameter) than the typical alumina sources used in sol-gel mixtures. Nevertheless, powder compacts prepared from the microcomposite particles could be viscously sintered to almost 100% relative density (i.e., to a nearly pore-free state) at relatively low temperature (~1300°C). These compacts were subsequently converted to dense mullite after reaction between the alumina and silica constituents at higher temperature.

In this paper, we present evidence concerning the mechanism of mullitization in samples prepared with silica-coated α-Al_2O_3 microcomposite particles. As discussed below, investigations of the reaction kinetics and microstructure development indicate that mullite formation occurs in two stages. The first stage of mullitization occurs primarily by nucleation and growth within the siliceous matrix. The second stage occurs primarily by interdiffusion of alumina and silica through the mullite grains formed during the first stage.

EXPERIMENTAL

The procedure for synthesis of the microcomposite particles ("coated particles") has been described previously.[25] The method involves precipitating a (hydrous) silica coating on α-Al_2O_3 core particles (fractionated AKP-50, Sumitomo Chemical America, New York). The silica precipitation was carried out by base-catalyzed hydrolysis/condensation of TEOS (Fisher Scientific, Fair Lawn, NJ). Microcomposite powders with different thicknesses of the silica coating (and different overall alumina/silica ratios) were prepared by varying the alumina/TEOS ratio in the initial suspensions used for precipitation.[25] The powders used in this study had overall alumina/silica weight ratios of ~66/34, ~74/26, and ~83/17 after heat treatment at 1600°C. These chemical compositions were determined on sintered powder compacts using an electron microprobe analyzer (Superprobe 733 Scanning Electron Microprobe, JEOL, Tokyo, Japan) equipped with X-ray wavelength dispersive spectrometers. At 1600°C, equilibrium Al_2O_3-SiO_2 phase diagrams[2,30-32] indicate that "66/34" samples are in the mullite + silica-rich liquid phase field, "74/26" samples are near the boundary between the mullite (solid solution) and mullite + alumina phase fields, and "83/17" samples are in the mullite + alumina phase field.

As noted earlier, some studies indicate that mullite can form by nucleation and growth within the siliceous phase. Hence, it was of interest to determine the chemical composition of the siliceous phase in heat treated microcomposite powders. This was accomplished by first dissolving the siliceous phase by aging the microcomposite powder in a concentrated (~49 wt%) hydrofluoric acid (HF) solution for 100 h. (Silica dissolves readily in concentrated HF, while α-Al_2O_3 has very low solubility under the same conditions.) The HF solution/powder mixture was subsequently filtered to separate the solution containing the dissolved components from the undissolved α-Al$_2$O$_3$ core particles. The solution was then analyzed for the Si and Al content using inductively coupled plasma (ICP) spectroscopy (Model Plasma II, Perkin-Elmer Co., Norwalk, CT). Due to the low solubility of α-Al$_2$O$_3$ in HF, it was expected that the measured Al content would arise mostly from alumina that had dissolved in the siliceous phase during heat treatment of the microcomposite powder.*

The mullitization reaction in powders was followed using differential thermal analysis, DTA, (Model STA-409, Netzsch Co., Exton, PA) and X-ray diffraction, XRD, (Model APD 3720, Philips Electronics Instrument Co., Mt. Vernon, NY). The DTA measurements were carried out at 5°C/min using a flowing air (~60 cm^3/min) atmosphere. XRD was carried out using Ni-filtered CuKα radiation. Quantitative analyses of the phase content of heat treated powders utilized CaF$_2$ as an internal standard. XRD was also carried out on polished surfaces of bulk samples. The latter samples were prepared from the microcomposite powders by slip casting using procedures described in detail previously.[25] Dried powder compacts were heat treated in air at temperatures up to 1650°C.

Scanning electron microscopy, SEM, (Model JSM-6400, JEOL, Tokyo, Japan) was used for microstructure observations on polished bulk samples. Samples which contained significant amounts of siliceous phase were chemically etched using dilute (~0.5-2.0 wt%) hydrofluoric acid solutions. Samples which were substantially converted to mullite were thermally etched at ~1450°C in air. Samples were sputtered coated (Desk II, Denton Vacuum, Inc., Moorestown, NJ) with a thin gold/palladium layer prior to the SEM observations. Grain size measurements were made by the lineal intercept method.

RESULTS AND DISCUSSION

The mullitization kinetics for a "74/26" microcomposite powder were monitored by quantitative XRD. Figure 1 shows the extent of reaction (percentage converted to mullite) vs. heat treatment time at temperatures in the range of 1375-1450°C. The plots show an initial incubation period, a period of rapid mullitization (up to ~70% conversion), and a final period of slower mullitization.

The observation of an incubation period is consistent with the behavior reported for mullitization of diphasic aluminum silicate gels.[12,16,20] The activation energy obtained for incubation from the data in Fig. 1 was 1094 ± 47 kJ/mol which is similar to the value of reported by Wei and Halloran.[20] As noted earlier, incubation was interpreted by Sundaresan and Aksay[24] as the period in which alumina dissolves into the silica until the critical concentration for nucleation of mullite is exceeded. However, prior studies have not provided any direct evidence for the presence of dissolved alumina in the siliceous phase. In this study, the alumina contents in the siliceous coatings of heat treated microcomposite powders were assessed using the HF dissolution/ICP analysis method described earlier. Samples were analyzed after heat treatments of 8 h at 1350°C or 1375°C. The latter condition was chosen since it produced a sample which was in the late stage of the incubation period, but which did not yet contain any mullite (i.e., as determined by XRD). The measured alumina content increased from ~5.9 wt% to ~13.7 wt% for the 1350°C and 1375°C samples,

---

* Unfortunately, the surfaces of the as-prepared alumina core particles were highly hydroxylated. Aluminum hydroxides have much greater solubility in HF than the alpha form of alumina. Thus, it was expected that at least some of the measured Al content determined for HF-treated microcomposite powders came from the core particles.

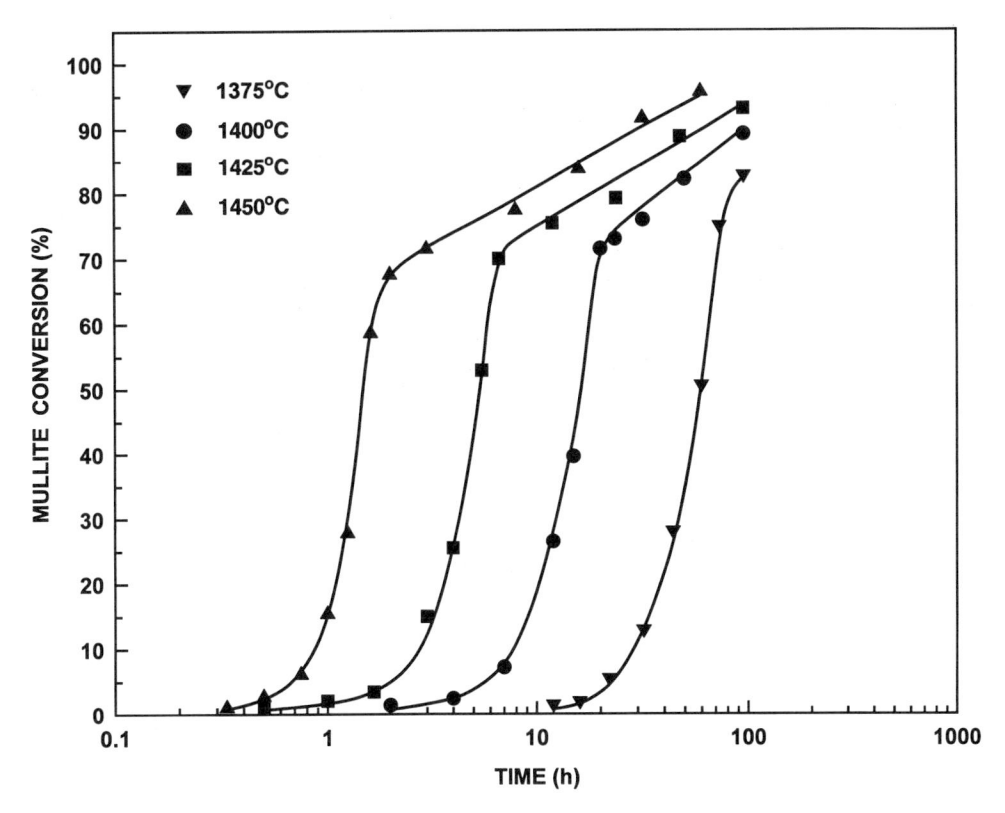

Figure 1. Plots of the extent of reaction (conversion to mullite) vs. heat treatment time at the indicated temperatures.

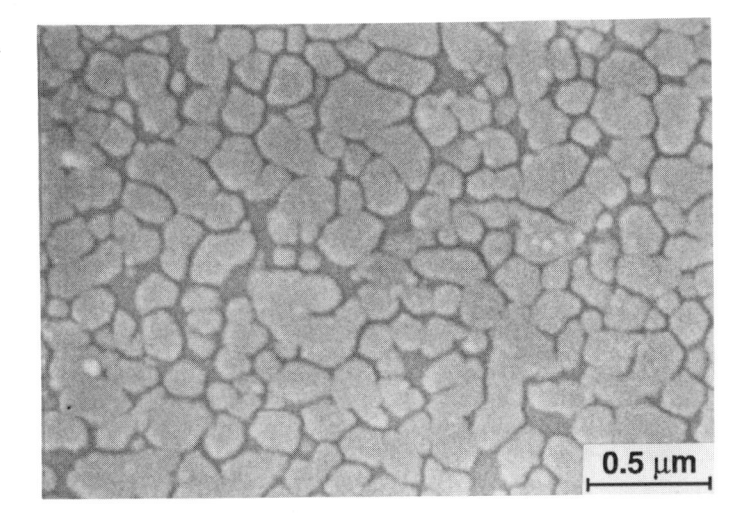

Figure 2. Scanning electron micrograph of a sample with composition of ~74 wt% Al$_2$O$_3$/26 wt% SiO$_2$ which was heated treated at 1400°C for 2 h.

respectively. These results are consistent with the interpretation of incubation as the period in which alumina dissolves in silica up to the critical nucleation concentration.

The data from the period of rapid mullite formation in Fig. 1 showed a good fit to the Avrami model for nucleation and growth processes. The Avrami time exponents for the various reaction temperatures indicated in Fig. 1 were in the range of ~2.4-2.8. If we assume that the reaction occurs by growth of a constant number of nuclei which are formed immediately after the incubation period, the time exponent indicates that mullitization occurs by a combination of two- and three-dimensional growth. This behavior can be rationalized by considering the microstructure in the early stages of mullite formation. Figure 2 shows a bulk sample prepared by sintering a microcomposite powder compact at 1400°C for 2 h. The microstructure shows mostly alumina particles dispersed in a dense siliceous "matrix" phase. (XRD showed that just a trace amount of mullite was present in this sample.) During the densification process, the alumina core particles move closer together (on average) and much of the siliceous coatings is redistributed to "fill" the void space formed by the initial packing of the microcomposite particles. Thus, Fig. 2 shows that the siliceous phase is present as both larger, three-dimensional "pockets" and smaller, nearly two-dimensional films formed between alumina particles.* Therefore, both three-dimensional and two-dimensional growth of mullite would be expected if mullite nucleates and grows within the siliceous phase.

An activation energy of 1120 ± 16 kJ/mole was obtained from the data in Fig. 1 for the rapid mullitization stage (i.e., up to ~70 wt% conversion). A similar value (1042 ± 32 kJ/mol) was obtained from constant-heating-rate DTA experiments with the "74/26" microcomposite powder.[28] These activation energies are in good agreement with values reported in several investigations of mullitization of diphasic gels.[12,16,20] In addition, the results of this study and the study by Wei and Halloran[20] have shown that essentially the same activation energies are obtained for the stages of incubation and rapid mullite growth. This observation suggests that both processes are controlled by the same rate-limiting step. For diphasic gels, Huling and Messing[16] and Sundaresan and Aksay[24] concluded that the reaction rate is controlled by the dissolution of alumina in the siliceous phase. The same rate-controlling step is apparently operative during mullite formation from silica-coated α-Al_2O_3 composite particles. This was indicated from a study of the effect of the alumina/silica ratio on the reaction temperature during the first stage of mullitization. The alumina/silica ratio was varied by preparing microcomposite particles with varying thickness of the silica coatings on the same alumina core particles. Figure 3 shows TEM micrographs of microcomposite powders having overall alumina/silica weight ratios of ~66/34, ~74/26, and ~83/17. DTA plots for these powders are shown in Fig. 4A. The temperature for the first stage of the mullitization reaction decreased as the alumina/silica ratio increased. The DTA peak temperatures were 1518°C, 1499°C, and 1489°C for the "66/34", "74/26", and "83/17" microcomposite powders, respectively.† The effect of alumina/silica ratio on the mullitization temperature was also confirmed using XRD analysis.[29] Bulk compacts were prepared from the "66/34", "74/26", and

---

* Figure 2 shows many instances in which thin siliceous films are present between flattened alumina particles. As noted earlier, substantial dissolution of alumina in the siliceous phase occurs just before the initial formation of mullite. The flattening of alumina particles presumably occurs due to preferential dissolution at the particle regions with the highest curvatures.

† The DTA results show that the mullitization reaction for microcomposite particles is endothermic. In contrast, the mullitization reactions in single-phase and diphasic sol-gel samples are exothermic. The reason for this difference is discussed in detail elsewhere.[27]

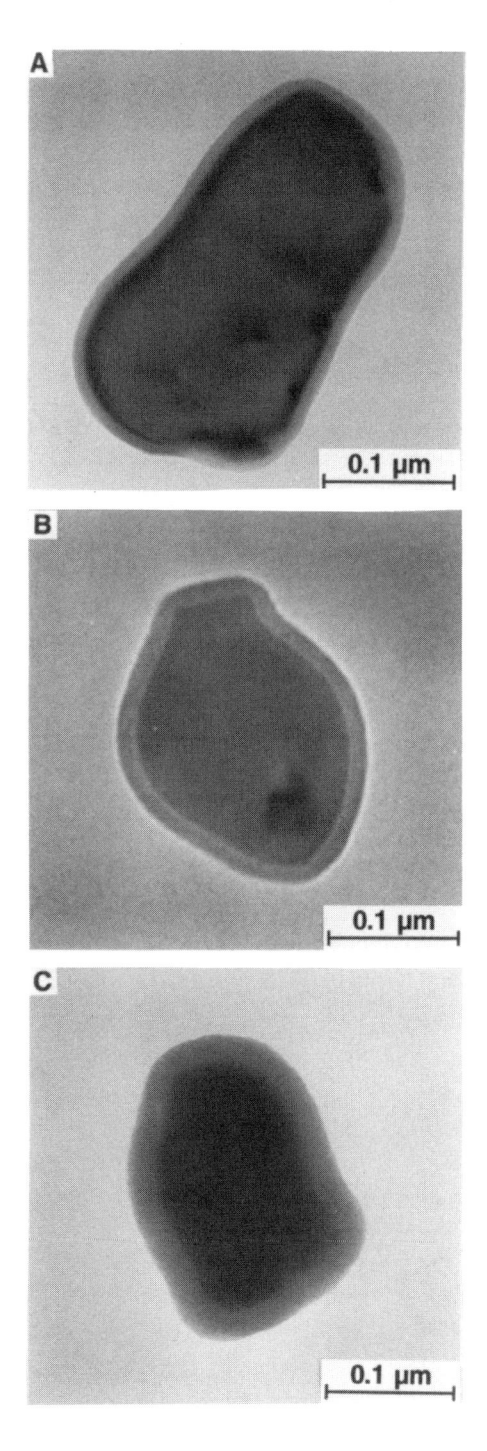

Figure 3. Transmission electron micrographs of microcomposite particles with varying thickness of the amorphous SiO$_2$ layer. Overall compositions are (A) ~83 wt% Al$_2$O$_3$/17 wt% SiO$_2$, (B) ~74 wt% Al$_2$O$_3$/26 wt% SiO$_2$, and (C) ~66 wt% Al$_2$O$_3$/34 wt% SiO$_2$. The same Al$_2$O$_3$ core particles were used to prepare each powder.

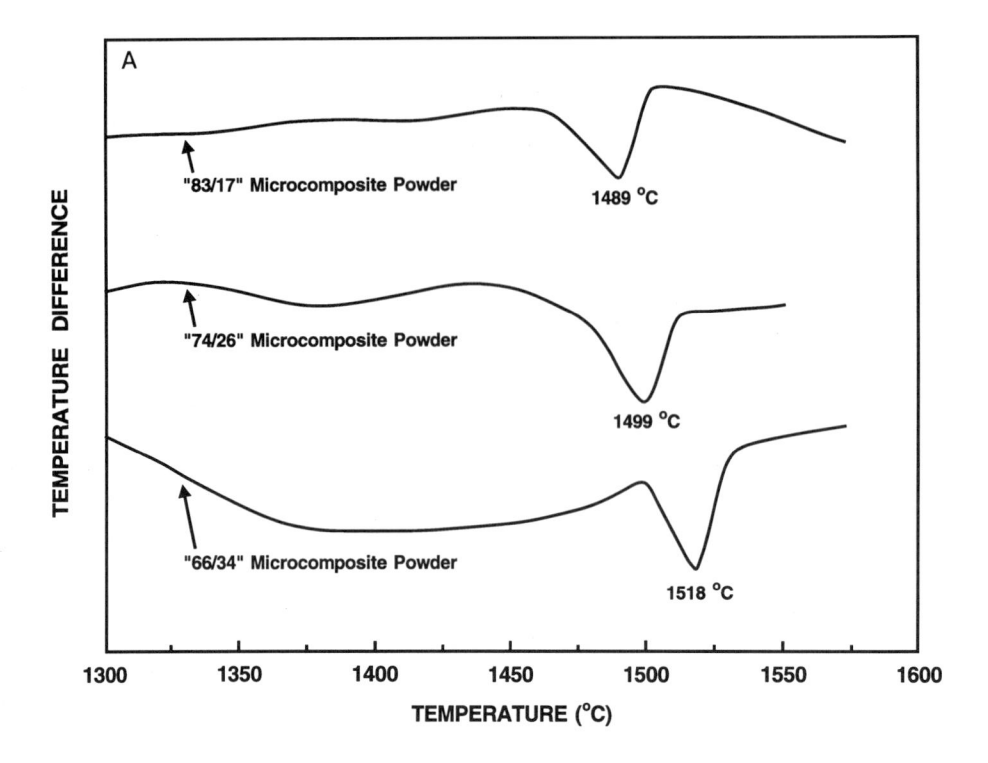

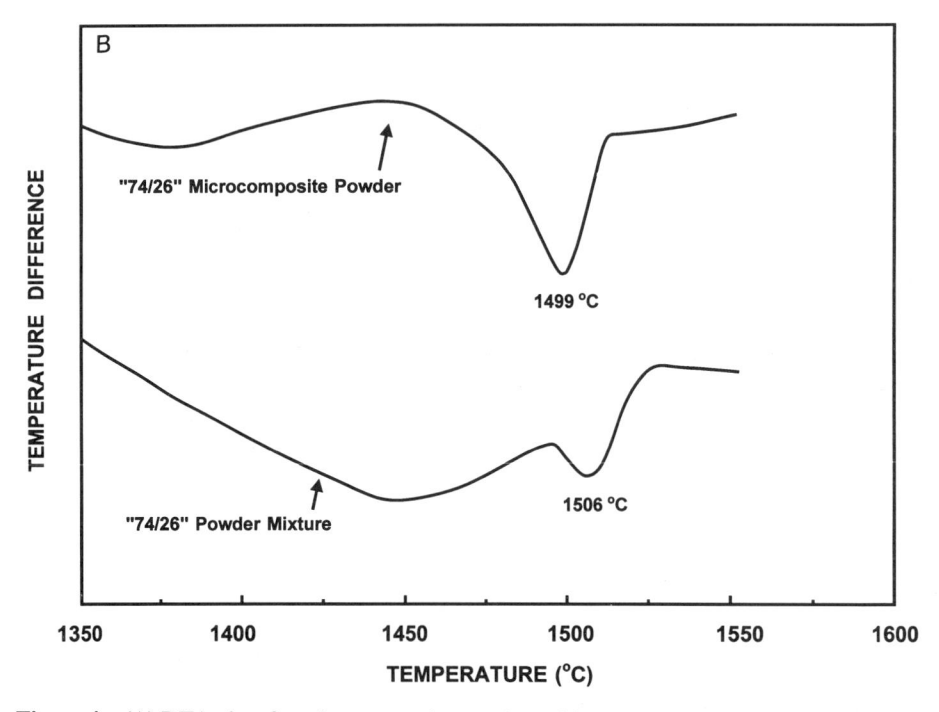

Figure 4. (A) DTA plots for microcomposite powders with compositions of ~66 wt% Al_2O_3/34 wt% SiO_2, ~74 wt% Al_2O_3/26 wt% SiO_2, and ~83 wt% Al_2O_3/17 wt% SiO_2. (B) DTA plots for a microcomposite powder and a powder mixture of Al_2O_3 and SiO_2 particles. Both powders had an overall composition of ~74 wt% Al_2O_3/26 wt% SiO_2.

"83/17" microcomposite powders and subsequently were heat treated for 2 h at eight different temperatures in the range of 1300 - 1650°C. The XRD results were consistent with the DTA results in Fig. 2 in that the temperature for the onset of mullitization increased as the alumina/silica ratio decreased.[29] The dependence of the initial mullitization temperature on the alumina/silica ratio is consistent with dissolution of alumina as the rate-controlling step during the rapid mullitization stage. In samples with higher alumina/silica ratio (e.g., "83/17"), a smaller amount of the alumina core particles needs to be dissolved in the silica to reach the critical nucleation concentration and to maintain the necessary concentration for growth of mullite. Conversely, a larger amount of alumina needs to be dissolved in the higher silica content "66/34" samples in order to reach the same solute concentration as in the "83/17" samples. Thus, if the reaction rate is controlled by alumina dissolution, the onset of mullitization would be inhibited in samples with lower alumina/silica ratio (and enhanced in samples with higher alumina/silica ratio). This is consistent with the DTA (Fig. 4A) and XRD results.[29]

It can be argued that diffusion of dissolved alumina through the siliceous phase might also be the rate-controlling reaction step during the rapid mullitization stage. The average diffusion distance for the dissolved alumina increases as the thickness of the silica layer of the microcomposite particles increases. Therefore, if the reaction rate is controlled by alumina diffusion, mullitization would be inhibited by lower alumina/silica ratios (i.e., thicker silica layers). However, other evidence suggests that diffusion does not control the reaction rate. An experiment was carried out with two different powders having same overall composition of ~74 wt% alumina/~26 wt% silica: (i) microcomposite particles and (ii) a mechanical mixture of α-Al_2O_3 and amorphous silica particles. The same alumina powders were used for each type of powder and the silica in each case was prepared by base-catalyzed precipitation of TEOS. Thus, the spatial distribution of the alumina and silica phases was the only significant difference between the powders. In the microcomposite particles, the silica was initially present as a relatively uniform coating (thickness ≈ 15-20 nm) around each alumina core particle. In contrast, the distribution of the silica was initially less uniform in the powder mixture; the silica particles had a median diameter of ~160 nm. Hence, longer average diffusion distances would be required in the powder mixture in order to achieve uniform dissolution of alumina in the siliceous phase.* Despite this difference, DTA plots comparing the two powders (Fig. 4B) show that the mullite transformation temperature was only slightly higher for the powder mixture. The peak temperatures differed by only ~7°C, i.e., ~1499°C and ~1506°C for the microcomposite powder and the powder mixture, respectively. In contrast, the microcomposite powder samples with "83/17" and "66/34" compositions showed a much larger difference in DTA peak temperatures (i.e., ~29°C according to Fig. 4A) despite a relatively small difference in the alumina diffusion distances. (The thickness of the silica layer increases only ~10-15 nm as the overall alumina/silica weight ratio decreases from 83/17 to 66/34.) Therefore, it is less likely that diffusion of alumina in the siliceous phase controls the onset of the mullitization and, instead, dissolution of alumina is apparently the rate-controlling reaction step.

---

* It should be noted that the silica particles flowed extensively during heat treatment of the mixed powder sample. Transmission electron microscopy (TEM) observations on a sample heat treated at 1225 °C (48 h) revealed that many alumina particles had at least portions of their surfaces coated with relatively thin silica layers (i.e., comparable to the layer thicknesses in coated powders). However, TEM and SEM also revealed silica-rich sintered regions extending at least several micrometers, as well as many sintered clusters of alumina particles. Hence, it is apparent that average diffusion distances were larger in the mixed powder sample.

Microstructure observations were consistent with the kinetic data in Fig. 1 in that two different growth stages were indicated. Bulk powder compacts prepared with "74/26" microcomposite powders can be sintered to essentially 100% relative density after heat treatment at 1300°C (2 h). The microstructure of these samples consists of alumina core particles distributed in a dense, continuous siliceous matrix. Except for some dissolution of the alumina in the siliceous phase, there is relatively little change in microstructure when samples are further heat treated to the point where the incubation stage is just completed. This is illustrated in Fig. 2 which shows a sample after heat treatment at 1400°C. (XRD showed that this sample contains only a trace of mullite.) With further heat treatment at 1500°C, the first stage of mullitization is completed and the microstructure undergoes considerable coarsening. Figure 5A shows that the mullite grains were approximately an order of magnitude larger than the starting size of the microcomposite particles. This grain development was consistent with nucleation of mullite at a relatively low number of sites followed by growth throughout the interconnected siliceous phase. Each grain was able to grow until impinged upon by other growing grains. With a low number of nucleation sites, grains grew to relatively large size. This reaction mechanism leads to the entrapment of alumina particles within the growing mullite grains (Fig. 5B). Subsequent elimination of the entrapped alumina particles (via reaction with residual silica) requires chemical interdiffusion through the mullite grains. Since mullite diffusion coefficients are relatively low and diffusion distances are relatively long (i.e., grains are in the micrometer range), the rate of mullitization decreased significantly during the second stage, as indicated in Figure 1. XRD studies on bulk samples indicated that essentially complete reaction was not achieved until heat treatment at ~1600°C (2 h).[26]

Kinetic data (partially shown in Fig. 1) for the second stage of mullitization showed a good fit over the range of ~70-90 wt% mullite conversion to the interdiffusion equation by Carter.[33] An activation energy of 562 ± 131 kJ/mol was obtained for data collected in the temperature range of 1400-1500°C. This value is somewhat lower than activation energies reported from diffusion and creep experiments in mullite.[34-37] (The latter studies give values in the range of ~750 ± 70 kJ/mol.) This may reflect the fact that mass transport during the second stage of the reaction will not occur entirely through the mullite lattice and/or along mullite grain boundaries (i.e., some alumina transport through silica-rich regions is still expected).

Long range chemical interdiffusion during the second stage of the mullitization reaction can also account for the development of the small pores which are evident (see Fig. 5A) within the large mullite grains of samples heat treated at 1500°C. (As shown in Fig. 2, bulk samples are essentially pore-free after sintering and before any significant mullitization occurs. The pores shown in Fig. 5A develop during the late stages of the reaction.) Several observations were made regarding the porosity in the 1500°C samples: (i) Almost all pores were within the mullite grains. (ii) Many of the pores had circular or rounded cross-sections, i.e., similar to the morphology of the intragranular alumina particles. (iii) Although quantitative measurements were not made, it was noted that most of the pores were smaller in size than the intragranular alumina particles. (iv) Regions with a large number of intragranular particles tended to have relatively few nearby intragranular pores (Fig. 5B). Conversely, regions with a relatively large number of intragranular pores tended to have relatively few nearby intragranular particles (Fig. 5A). Based on these observations, it can be concluded that the elimination of the intragranular alumina particles (as mullitization proceeded) was directly associated with the formation of the intragranular porosity. This suggests the possibility that the porosity was created by a Kirkendall effect, i.e., due to differences in diffusion rates for the reacting species within the mullite grains. (If this interpretation is correct, it indicates that alumina diffusion in mullite is much more

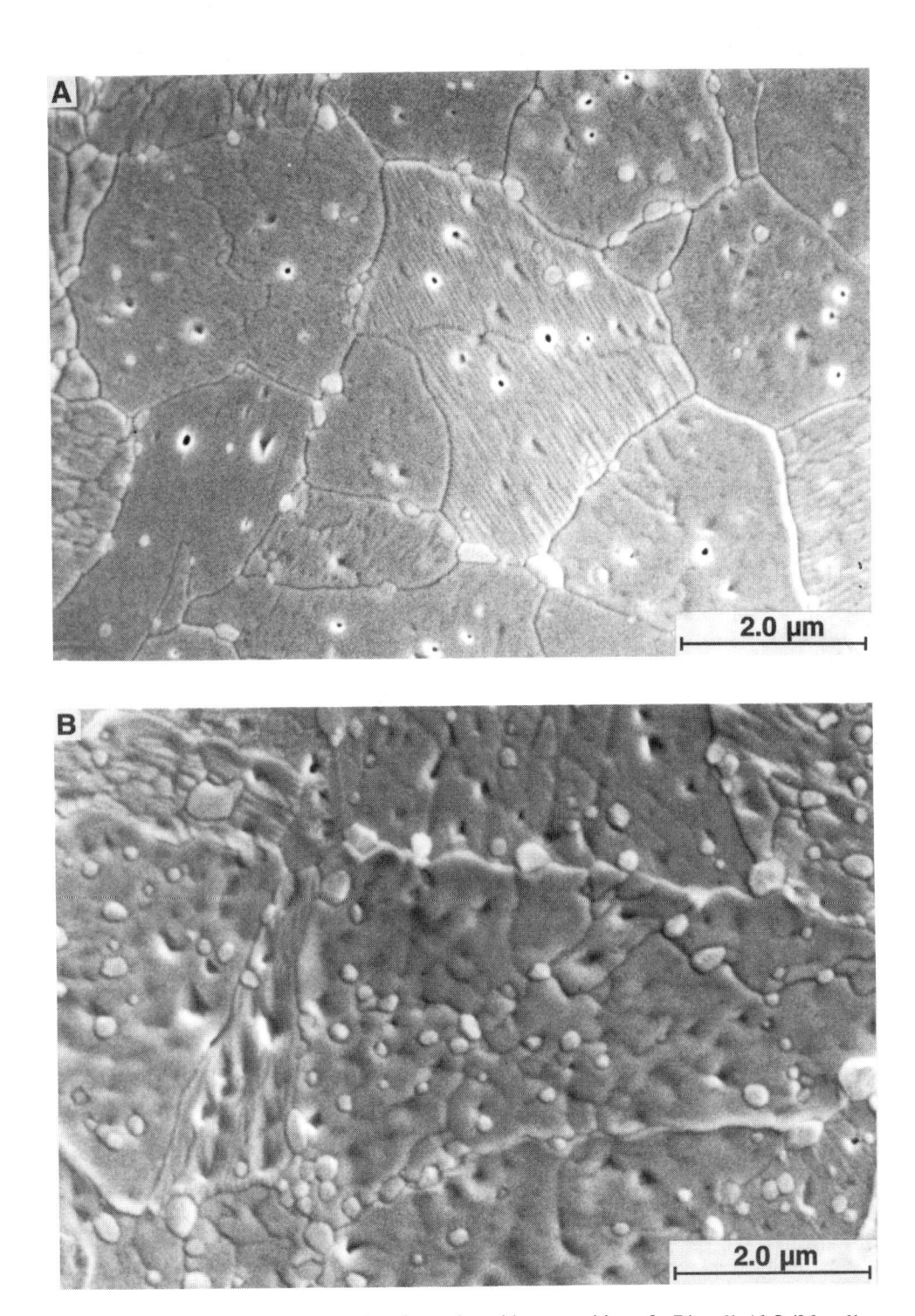

Figure 5. Scanning electron micrographs of samples with composition of ~74 wt% Al$_2$O$_3$/26 wt% SiO$_2$ which were heated treated at 1500°C for (A) 8 h and (B) 2 h.

rapid than silica diffusion.)

The mullitization behavior and microstructural evolution were altered dramatically by incorporating 2 wt% of fine (~0.1 μm) mullite seed particles in the microcomposite powder compacts.[26] The seed particles had relatively little effect on powder compact densification, as both seeded and unseeded samples reached essentially full density (nearly zero porosity) after sintering at 1300°C for 2 h. However, mullite formation was significantly enhanced in the seeded compacts. Heat treatment at 1400°C for 2 h resulted in nearly complete mullitization, as indicated by XRD (Fig. 6) and microstructure observations (Fig. 7). In addition, the average grain size was only ~0.4 μm, i.e., approximately an order of magnitude smaller than in unseeded samples after complete mullitization. These observations were consistent with mullite formation by nucleation and growth within the siliceous phase during the first stage of the reaction. The seeds provided a high concentration of (extrinsic) growth sites so that the typical mullite grain underwent less growth before impinging upon other growing grains. Hence, a smaller grain size was observed. In turn, this reduced the distances required for chemical interdiffusion during the second stage of mullitization, so the reaction was completed at much lower temperature. The shorter interdiffusion distances also minimized porosity development via the Kirkendall effect. The small, rounded, intragranular pores that were so common in the unseeded samples were rarely observed in seeded samples.

The importance of chemical interdiffusion through the grains during the second stage of mullitization was also demonstrated in experiments carried out with samples having varying alumina/silica ratio. As noted earlier, unseeded bulk samples with the "74/26" composition required heat treatment at 1600°C for 2 h to achieve essentially complete mullitization.[26] In contrast, the reaction was completed (as determined by XRD) in the unseeded "83/17" and "66/34" samples after heat treatments at only 1450°C (2 h) and 1550°C (2 h), respectively.[29] One factor that could contribute to the lower temperatures for complete reaction is that the amount of mullite formed in "66/34" and "83/17" samples is less than in "74/26" samples. (The former compositions are well into two-phase regions of the equilibrium Al_2O_3-SiO_2 phase diagram, while the latter composition is in or near the single-phase mullite solid solution range.) However, this is not considered the primary reason for the difference in reaction temperatures. This is recognized by recalling that "74/26" samples were almost fully reacted to mullite at only 1400°C in compacts which contained ~2% mullite seed particles.[26] The low reaction temperature in that case was attributed to much smaller chemical interdiffusion distances during the second stage of mullitization (i.e., because seeding resulted in the development of much smaller mullite grain sizes during the first stage of the reaction). The lower reaction temperatures in the (unseeded) "83/17" and "66/34" samples in this study are also attributed to enhanced chemical interdiffusion during the second stage of the reaction, although not because of smaller mullite grain sizes. Figure 8 shows polished and etched microstructures for "83/17", "74/26", and "66/34" samples which were heat treated at 1500°C for 8 h. The grain sizes of the samples are quite similar.* However, the "83/17" and "66/34" samples show high

---

* Quantitative microscopy measurements showed that the average mullite grain sizes were 3.0, 3.5, and 3.2 μm for the "66/36", "74/26", and "83/17" samples, respectively. (Standard deviations were 1.7, 1.8, and 1.9 μm, respectively.) In unseeded samples, the grain size is controlled by the intrinsic nucleation concentration that develops during the first stage of the mullitization reaction. This concentration is apparently relatively insensitive to the alumina/silica ratio, i.e., at least for the composition range used in this study.

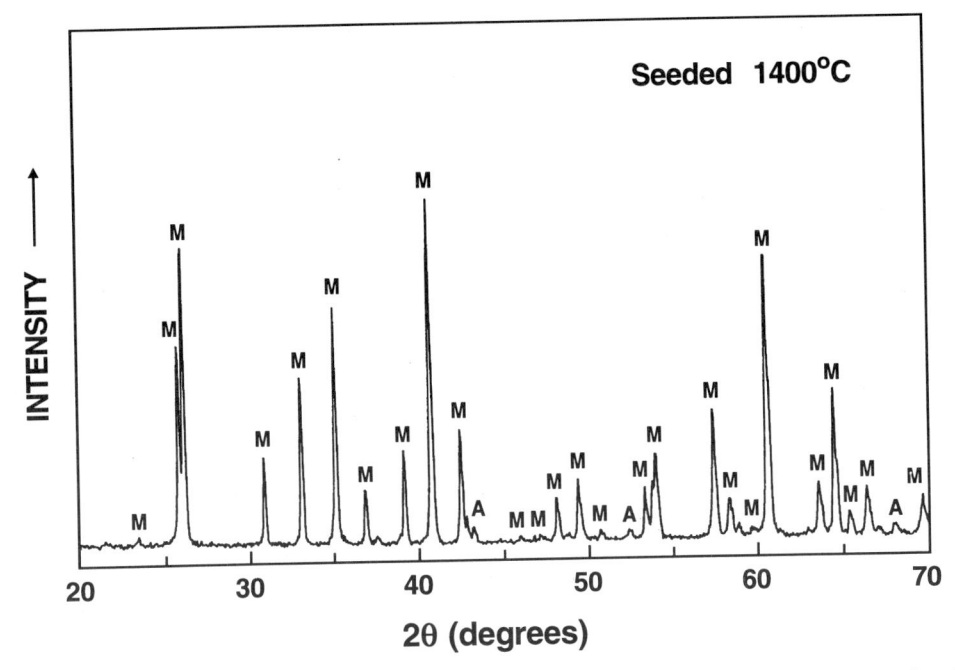

Figure 6. X-ray diffraction pattern for a seeded sample with composition of ~74 wt% Al_2O_3/26 wt% SiO_2 which was heat treated at 1400°C for 2 h. (M is mullite and A is α-Al_2O_3.)

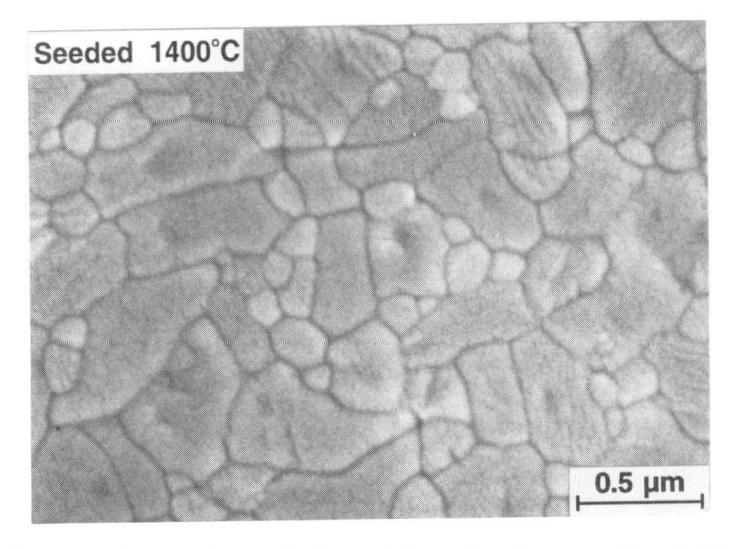

Figure 7. Scanning electron micrograph of a seeded sample with composition of ~74 wt% Al_2O_3/26 wt% SiO_2 which was heat treated at 1400°C for 2 h.

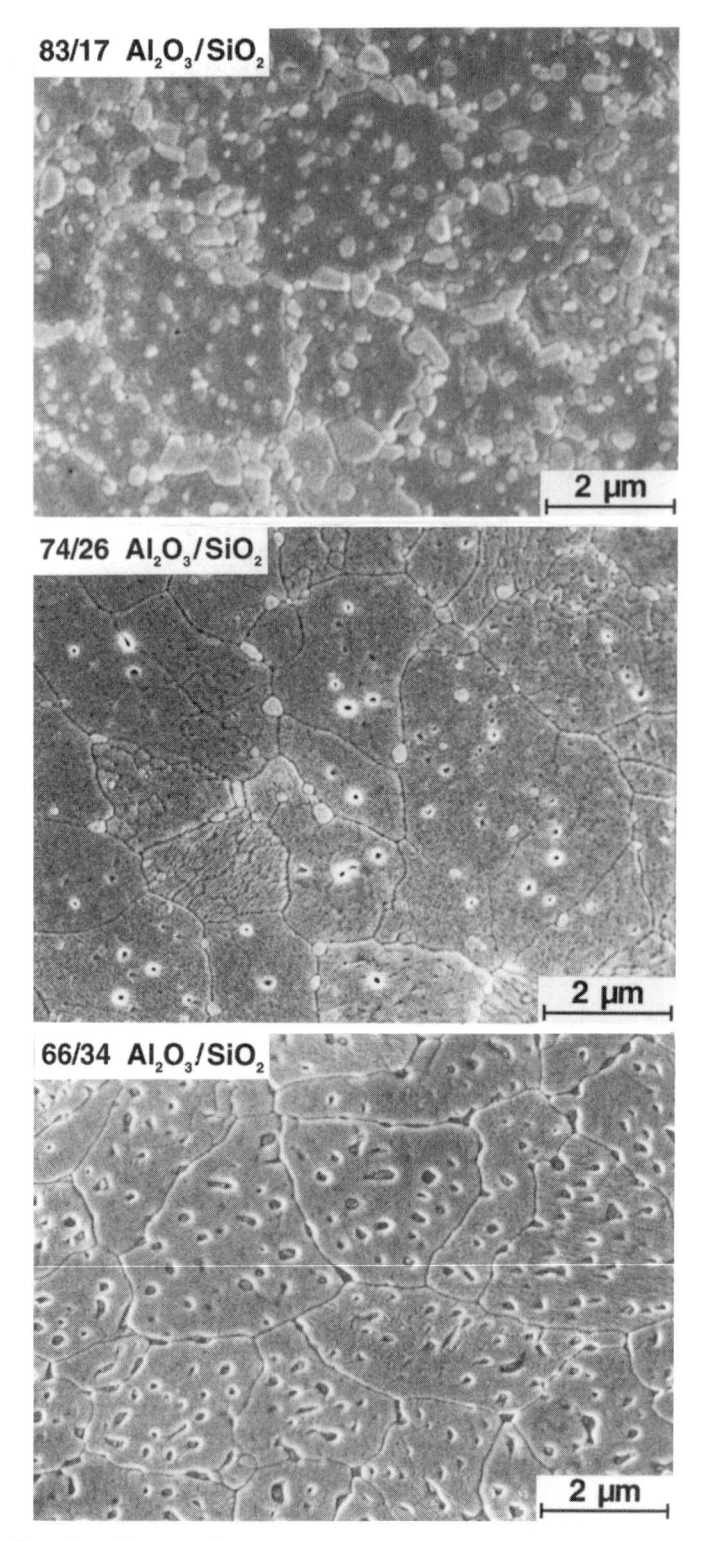

Figure 8. Scanning electron micrographs of unseeded samples with the indicated compositions (Al$_2$O$_3$/ SiO$_2$ weight ratios) which were heat treated at 1500°C for 8 h.

concentrations of fine (submicrometer), second-phase regions at intragranular and intergranular locations. (In the "83/17 sample, the reaction has already been completed so the second-phase particles are all alumina. The "66/34" sample contains mostly silica-rich "particles," although some alumina particles must still be present because XRD showed that the reaction was not quite completed for this heat treatment condition.) It is evident from Fig. 8 that one of the reactants in the "83/17" and "66/34" samples (i.e., alumina and silica, respectively) is present at high concentrations throughout the second stage of mullitization. (In contrast, the "74/26" sample develops relatively low concentrations of both reactants within the grains as the reaction proceeds during the second stage, i.e., because the sample is converting to single-phase mullite.) As a consequence, higher chemical interdiffusion fluxes will be maintained within the mullite grains of the "83/17" and "66/34" samples throughout the second stage. In addition, diffusion distances are shorter in these samples since the average distance of separation between alumina-rich and silica-rich regions is smaller. Furthermore, the interdiffusion distances will remain smaller, compared to the "74/26" sample, because one of the reactants will remain in excess throughout the second stage of mullitization.

Figure 8 also shows that the "74/26" sample contains fine, intragranular pores. In contrast, pores are not observed in the "83/17" and "66/34" samples. As noted earlier, pore formation in unseeded "74/26" samples is probably due to the Kirkendall effect, i.e., due to the difference in diffusion rates for alumina and silica within the mullite grains and the relatively long diffusion distances required to complete mullitization. The absence of pores in the "83/17" and "66/34" samples is attributed to the smaller chemical interdiffusion distances during the second stage of the reaction and is consistent with the lower temperatures required to complete mullitization in these samples. The question does arise, however, as to why mullitization is completed at significantly lower temperature for the "83/17" sample ($\leq 1450°C$, 2h) compared to the "66/34" sample ($\leq 1550°C$, 2h). One possible reason is that mass fluxes may be higher and chemical interdiffusion distances may be shorter in the "83/17" sample. The "83/17" sample in Fig. 8 has a higher second phase content (i.e., ~29 vol% compared to ~10 vol% for the "66/34" sample in Fig. 8) and a smaller average interparticle separation distance (i.e., ~0.5 μm vs. ~1.0 μm for the "66/34" sample in Fig. 8). Another possible reason for the difference in reaction temperatures is a difference in the diffusion rates for alumina and silica within the mullite grains. As noted earlier, the development of intragranular pores in the unseeded "74/26" samples (Figs. 5 and 8) suggests that alumina diffusion within mullite is considerably faster. This could account for the lower temperature needed to complete the second stage of mullitization in the "83/17" sample.

SUMMARY

The mechanism of mullitization was investigated in samples prepared with silica-coated α-Al_2O_3 microcomposite particles. The phase transformation was characterized by an incubation stage and two growth stages. Chemical analysis of heat-treated microcomposite powders provided direct evidence that alumina dissolved in the siliceous glass during the incubation period. This was followed by nucleation and growth of mullite within the siliceous phase. Available evidence indicated that dissolution of alumina was the rate-limiting step during the incubation and initial growth stages. Microstructure observations indicated that mullite grains were able to grow continuously throughout the interconnected siliceous matrix until impinged upon by other growing grains. Due to a low intrinsic nucleation concentration, this resulted in an average grain size which was approximately an order of magnitude larger than the size of the original microcomposite particles. The growth path during the first stage of mullitization led to the entrapment of many alumina particles within the growing mullite grains. Hence, continued growth of mullite during the second stage required interdiffusion of alumina and silica through the grains. The rate of mullitization decreased during this stage because of the relatively long diffusion distances and the low diffusion coefficients for mullite. However, the reaction rate was increased significantly by incorporating a small amount (~2%) of fine (~0.1 μm) mullite seed particles in the microcomposite powders. The seed particles acted as extrinsic sites for nucleation and growth which led to much smaller grain sizes after the first stage of mullitization. This decreased the chemical interdiffusion distances required for eliminating entrapped alumina particles during the second stage of mullitization. Hence, it was possible to produce bulk mullite samples with nearly 100% relative density (i.e., nearly pore-free) and fine grain size (~0.4 μm) after heat treatment at only 1400°C for 2 h.

ACKNOWLEDGEMENTS

Experimental assistance from Y.-J. Lin is gratefully acknowledged. This work was supported by the: Defense Advanced Research Projects Agency and Office of Naval Research (contract nos. MDA972-88-J-1006 and N00014-91-J-4075); National Science Foundation, Division of Materials Research, Ceramics and Electronics Materials Program (grant no. DMR-8451916); Florida High Technology and Industry Council.

REFERENCES

1. W.L. de Keyser, Reactions at the point of contact between SiO_2 and Al_2O_3, pp. 243-257 in *Science of Ceramics*, Vol. 2, G.H. Stewart, ed., Academic Press, London, U.K. (1965).
2. R.F. Davis and J.A. Pask, Diffusion and reaction studies in the system Al_2O_3-SiO_2, *J. Am. Ceram. Soc.*, 55 [10]: 525-531 (1972).
3. J. Ossaka, Tetragonal mullite-like phase from co-precipitated gels, *Nature (London)*, 19 [4792]: 1000-1001 (1961).
4. S. Kanzaki, H. Tabata, T. Kumazawa, and S. Ohta, Sintering and mechanical properties of stoichiometric mullite," *J. Am. Ceram. Soc.*, 68 [1]: C-6 - C-7 (1985).
5. B.E. Yoldas and D.P. Partlow, Formation of mullite and other alumina-based ceramics via hydrolytic polycondensation of alkoxides and resultant ultra- and micro-structural effects, *J. Mater. Sci.*, 23 [5]: 1895-1900 (1988).
6. J.C. Huling and G.L. Messing, Chemistry-crystallization relations in molecular mullite gels, *J. Non-Crystalline Solids*, 147/148: 213-221 (1992).
7. D.X. Li and W.J. Thomson, Mullite formation kinetics of a single-phase gel, *J. Am. Ceram. Soc.*, 73 [4]: 964-969 (1990).
8. D.W. Hoffman, R. Roy, and S. Komarneni, Diphasic xerogels, a new class of materials: phases in the system Al_2O_3-SiO_2, *J. Am. Ceram. Soc.*, 67 [7]: 468-71 (1984).
9. R. Roy, S. Komarneni, and D.M. Roy, Multi-phasic ceramic composites made by sol-gel technique, pp. 347-359 in *Better Ceramics Through Chemistry*, Mat. Res. Soc. Symp. Proc., Vol. 32, C.J. Brinker, D.E. Clark, and D.R. Ulrich, eds., Elsevier Publishing Co., New York (1984).
10. K. Okada and N. Otsuka, Characterization of the spinel phase from SiO_2·Al_2O_3 xerogels and the formation process of mullite, *J. Am. Ceram. Soc.*, 69 [9]: 652-656 (1986).
11. M.J. Hyatt and N.P. Bansal, Phase transformations in xerogels of mullite composition, *J. Mater. Sci.*, 25: 2815-2821 (1990).
12. D.X. Li and W.J. Thomson, Kinetic mechanisms for the mullite formation from sol-gel precursors, *J. Mater. Res.*, 5 [9]: 1963-1969 (1990).
13. B.E. Yoldas, Effect of ultrastructure on crystallization of mullite, *J. Mater. Sci.*, 27: 6667-6672 (1992).
14. J.C. Huling and G.L. Messing, Hybrid gels for homoepitactic nucleation of mullite, *J. Am. Ceram. Soc.*, 72 [9]: 1725-1729 (1989).

15. J.C. Huling and G.L. Messing, Hybrid gels designed for mullite nucleation and crystallization control; pp. 515-26 in *Better Ceramics Through Chemistry IV*, Mat. Res. Soc. Symp. Proc., Vol. 180, B.J.J. Zelinski, C.J. Brinker, D.E. Clark and D.R. Ulrich, eds., Materials Research Society, Pittsburgh, PA (1990).
16. J.C. Huling and G.L. Messing, Epitactic nucleation of spinel in aluminum silicate gels and effect on mullite crystallization, *J. Am. Ceram. Soc.*, 74 [10]: 2374-2381 (1991).
17. C. Gerardin, S. Sundaresan, J. Benzinger, and A. Navrotsky, Structural investigation and energetics of mullite formation from sol-gel precursors, *Chem. Mater.*, 6: 160-170 (1994).
18. D.X. Li and W.J. Thomson, Tetragonal to orthorhombic transformation during mullite formation, *J. Mater. Res.*, 6 [4]: 819-824 (1991).
19. W.-C. Wei and J.W. Halloran, Phase transformation of diphasic aluminosilicate gels, *J. Am. Ceram. Soc.*, 71 [3]: 166-172 (1988).
20. W.-C. Wei and J.W. Halloran, Transformation kinetics of diphasic aluminum silicate gels, *J. Am. Ceram. Soc.*, 71 [7]: 581-587 (1988).
21. C.-S. Hsi, H.-Y. Lu, and F.-S. Yen, Thermal behavior of alumina-silica xerogels during calcination, *J. Am. Ceram. Soc.*, 72 [11]: 2208-2210 (1989).
22. S. Rajendran, H.J. Rossell, and J.V. Sanders, Crystallization of a coprecipitated mullite precursor during heat treatment, *J. Mater. Sci.*, 25: 4462-4471 (1990).
23. G. Klaussen, G.S. Fischman, and J.L. Laughner, Microstructural evolution of sol-gel mullite, *Ceram. Eng. Sci. Proc.*, 11 [7-8]: 1087-93 (1990).
24. S. Sundaresan and I.A. Aksay, Mullitization of diphasic aluminosilicate gels, *J. Am. Ceram. Soc.*, 74 [10]: 2388-2392 (1991).
25. M.D. Sacks, N. Bozkurt, and G.W. Scheiffele, Fabrication of mullite and mullite-matrix composites by transient viscous sintering of composite powders, *J. Am. Ceram. Soc.*, 74 [10]: 2428-2437 (1991).
26. M.D. Sacks, Y.-J. Lin, G.W. Scheiffele, K. Wang, and N. Bozkurt, Effect of seeding on phase development, densification behavior, and microstructure evolution in mullite fabricated from microcomposite powders, *J. Am. Ceram. Soc.*, 78 [11]: 2897-2906 (1995).
27. K. Wang and M.D. Sacks, Mullite formation by endothermic reaction of alpha alumina/silica microcomposite particles, *J. Am. Ceram. Soc.*, 79 [1]: 12-16 (1996).
28. M.D. Sacks, K. Wang, G.W. Scheiffele, and N. Bozkurt, Activation energy for mullitization of alpha alumina/silica microcomposite particles, *J. Am. Ceram. Soc.*, 79 [2]: 571-573 (1996).
29. M.D. Sacks, K. Wang, G.W. Scheiffele, and N. Bozkurt, Effect of composition on mullitization behavior of alpha alumina/silica microcomposite powders, to be published in *J. Am. Ceram. Soc.*
30. S. Aramaki and R. Roy, Revised phase diagram for the system Al_2O_3-SiO_2, *J. Am. Ceram. Soc.*, 45 [5]: 229-242 (1962).
31. I.A. Aksay and J.A. Pask, Stable and metastable equilibria in the system SiO_2-Al_2O_3, *J. Am. Ceram. Soc.*, 58 [11-12]: 507-512 (1975).
32. F.J. Klug, S. Prochazka, and R.H. Doremus, Al_2O_3-SiO_2 system in the mullite region, *J. Am. Ceram. Soc.*, 70 [10]: 750-759 (1987).
33. R.E. Carter, Kinetic model for solid-state reactions, *J. Chem. Phys.*, 34 [6]: 2010-2015 (1961).
34. P.A. Lessing, R.S. Gordon, and K.S. Mazdiyasni, Creep of polycrystalline mullite, *J. Am. Ceram. Soc.*, 58 [3-4]: 149 (1975).
35. P.C. Dokko, J.A. Pask, and K.S. Mazdiyasni, High temperature mechanical properties of mullite under compression, *J. Am. Ceram. Soc.*, 60 [3-4]: 150-155 (1977).
36. I.A. Aksay, *Diffusion and Phase Relationship Studies in the Alumina-Silica System*, Ph.D. Thesis, University of California, Berkeley, CA (1973).
37. A.P. Hynes and R.H. Doremus, High-temperature compressive creep of polycrystalline mullite, *J. Am. Ceram. Soc.*, 74 [10]: 2469-2475 (1991).

PROCESSING OF TEXTURED CERAMICS BY TEMPLATED GRAIN GROWTH

M. M. Seabaugh, S.H. Hong, and G. L. Messing

Department of Materials Science and Engineering
The Pennsylvania State University, University Park, PA 16802

INTRODUCTION

Over the last twenty years, the processing community has concentrated on preparing fine-grained, equiaxed ceramic microstructures. For many systems, it is now routine to accomplish this objective and the related microstructure-property goals. More recently, it is apparent that there is a potentially large number of interesting microstructure-property relationships that remain to be explored. These microstructures, like Si_3N_4, are based on anisotropic grain growth which results in a self-reinforced ceramic.

Alternatively, fibers, whiskers, and platelets have been added to fine-grained ceramic composites to improve the strength and fracture toughness by crack deflection[1]. In reinforced composites, when the inclusion phase becomes interconnected, it forms a rigid framework which can inhibit the densification of the matrix[2] and require expensive processing methods to achieve high densities. The fraction of reinforcing phase is therefore limited to the percolation limit for the inclusion shape. This fraction can be raised by orienting the inclusion phase, often by tape casting[3], extrusion, or hot forming processes[4].

Some ceramic systems develop microstructures similar to ceramic composites through anisotropic grain growth. In anisotropic materials, the large grains often demonstrate whisker or platelet morphologies, and such microstructure development can be controlled by the addition of proper dopants or liquid phase formers and by the addition of template seed particles which serve as favorable growth sites for anisotropic grains. Like ceramic composites, densification of anisotropic materials can be inhibited if a rigid framework of large grains forms too early in the process.

The regulation of anisotropic grain growth opens many new opportunities. Textured ceramics hold promise for improved and unique electrical[5], piezoelectrical[6], mechanical[7] and other properties[8]. Textured microstructures can be achieved by applying mechanical[7], electrical, magnetic[9], and temperature[10] gradients during processing. Common mechanical methods for inducing orientation include hot-forging[5], hot pressing[6], tape casting[7], slip casting[11], and extrusion. Although anisotropic grains can be oriented by hot forming processes, these processes are too expensive. A lower cost option is to control texture development by regulating growth *in situ* during the processing of the initial green body.

One way to control anisotropic microstructure orientation and overcome the poor densification of reinforced ceramics is seeded or templated grain growth. Templated grain growth implies the use of aligned particles (i.e. seeds) to obtain oriented grain growth in the direction of the seed particles. Thus, the template particles must be anisometric or have anisotropic properties (e.g., magnetic) for proper orientation control. Textured Si_3N_4[7] and SiC[11], and Al_2O_3[12] produced by templated grain growth resulted in high strength and fracture toughness.

The development of a self-reinforced, textured microstructure can be produced *in situ* by combining seeding, anisotropic grain growth, and shear forming processes. Understanding how the processing variables affect the green and fired microstructures of textured ceramics will allow a more complete understanding of the range of achievable microstructures and improve the control of the microstructure development in such materials. In this paper, we discuss the processing parameters which affect microstructure development with particular emphasis on the effects of template orientation, template/matrix grain size ratio, template concentration and liquid phase content on the development of a textured microstructure. We have chosen to demonstrate the process of templated grain growth with TiO_2-doped mullite[13] and the CaO-SiO_2-alumina systems[14,15] because of the demonstrated propensity for anisotropic grain growth and the availability of alumina and mullite template particles.

EXPERIMENTAl PROCEDURES

Two systems were used to study the effect of green processing parameters on the development of textured microstructures by templated grain growth (TGG); a calcium aluminosilicate liquid phase sintered alumina, and a titania-doped mullite system. The alumina system was prepared by dispersing high purity alpha alumina in water at a concentration of approximately 60 wt% using an organic polyelectrolyte dispersant. Calcium carbonate powder and colloidal silica were added to the slurry at approximately 5 wt% of the solids. Approximately 5 wt% alumina platelets (solids basis), 5-10 μm and 20-25 μm in major axis direction with an aspect ratio of 5 to 10 were added. The slurry was sonicated for 15 minutes, stirred for 24 hours, then sonicated again for 15 minutes. Polyvinyl alcohol (PVA)) and glycerol were added as 4.4 and 4.6 wt% of the total solids, respectively. The slurry was then stirred for 24 hours.

The mullite system was prepared from a diphasic mixture consisting of boehmite and colloidal silica. Mullite whisker seeds were prepared by dissolving mullite pellets consisting of anisotropic grains. Whiskers were added in concentrations of up to 10 wt%. Titania was added as a fine powder at a concentration of 3 wt%. PVA and glycerol were also used as binders and plasticizers in these samples, following a similar processing method to that of the alumina.

The slurries were tape cast at speeds of 2, 4 and 8 cm/sec at blade heights between 500 and 1500 μm, at a viscosity of 150 mPa•sec. Tapes were dried under a controlled humidity of approximately 80% relative humidity for two days, then cut into 18 by 24 mm layers and laminated at 50 MPa for 3 minutes. Mullite samples were calcined at 500°C for 2 hours, while alumina laminates were calcined at 1000°C for 3 hours. Samples were fired at temperatures between 1450 and 1650°C for times between 1 and 300 minutes.

RESULTS AND DISCUSSION

Orientation Effects

In order to obtain a textured final microstructure via templated growth, template particles must first be oriented in the green body. An example of nonuniform orientation in a sintered mullite microstructure is shown in Figure 1a. Due to the low viscosity of the slurry, and the use of a single blade during casting, the initial orientation of the whisker particles in this sample was not uniformly oriented through the tape thickness. The mullite sample shown has developed strong texture in the upper two-thirds of the sample, but the bottom one third of the sample shows almost no orientation parallel to the casting direction.

Given this result in the mullite system, the relationship between template orientation and tape casting conditions in the alumina system was quantified by the fluid flow model and equations for couette and pressure were examined[16]. Over a wide range of casting conditions, little change in template orientation could be observed by SEM. Two samples with 1 wt% and 25 wt% platelets are shown in Figure 1b and 1c. Both samples were cast at 8 cm/sec with a blade height of 500 μm and calcined at 1000°C for 3 hours. Both samples show strong orientation through the entire thickness of the layer, indicating that for these casting conditions, the velocity gradient through the layer is sufficient to orient platelets, and this orientation is unaffected by the concentration of platelets. This was also true for all the slurry viscosities and casting speeds tested in the alumina system.

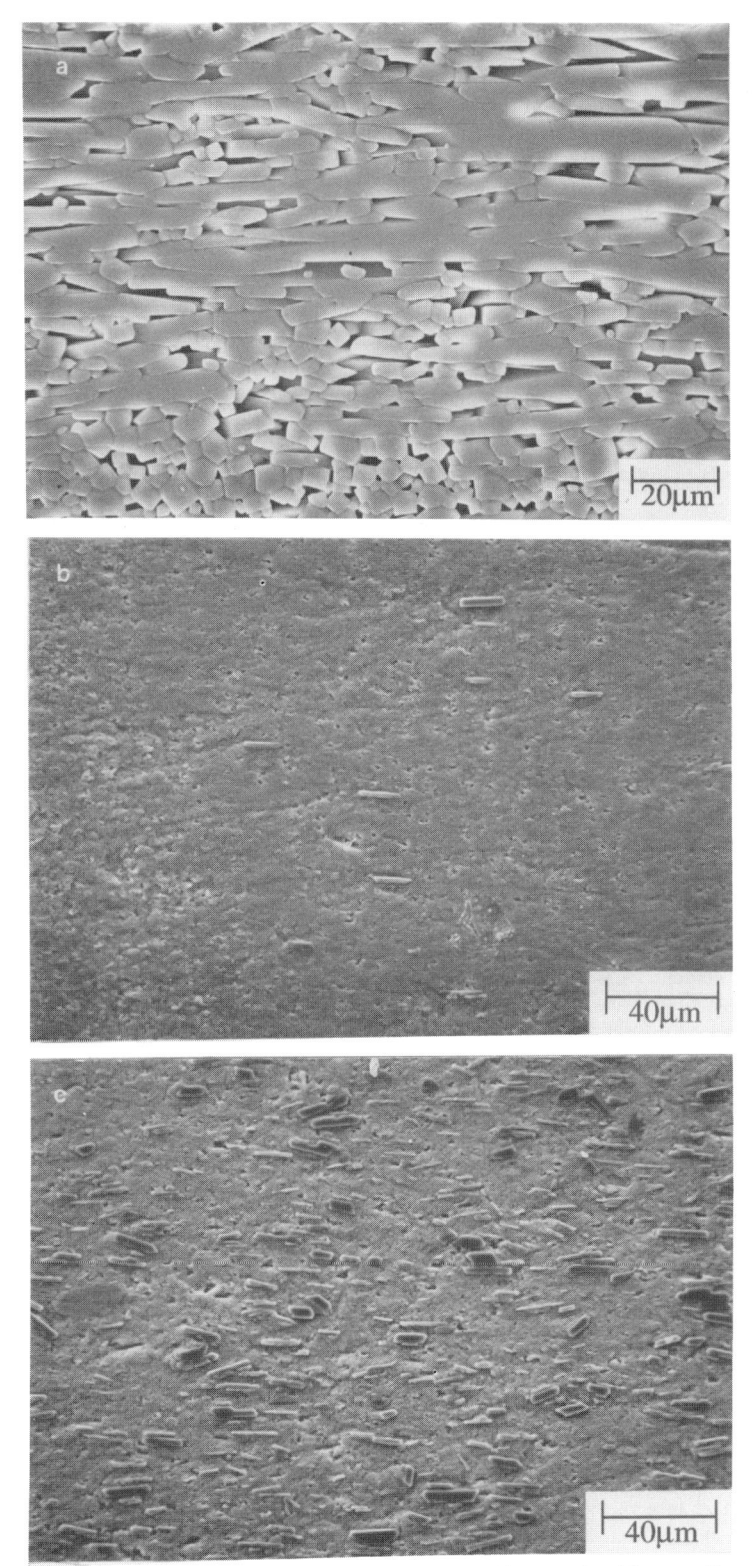

Figure 1. Orientation effects during tape casting. a) Sintered mullite sample with nonuniform through-thickness orientation. b) Tape cast alumina sample with 1 wt% templates. c) Tape cast alumina sample with 25 wt% templates. Note similar orientation in b) and c).

Template Concentration

While initial template concentration has little effect on the initial template orientation in the tape cast samples, the number of template particles play a significant role in the determination of the final microstructure. With an increasing concentration of template particles, the number of matrix grains available for consumption by the templates decreases and the distance between anisometric grains also decreases, creating an environment which favors the development of smaller anisometric grains with lower aspect ratios.

The low temperature sintered microstructure (1450°C, 5 h) of 2 wt% templated mullite (Figure 3a) consists of fine matrix grains and the original template particles. Upon sintering at 1600°C for 5 hours, the template grains grow much faster than the matrix grains, and some of the template grains grow to lengths more than 100 μm. In contrast, a 10 wt% templated sample sintered under the same high temperature conditions consists almost exclusively of anisometric grains with similar sizes. The decrease in length and the aspect ratio can be linked to the increased likelihood of impingement and the decreased amount of matrix grains available as a source of material for anisotropic grain growth.

Template/Matrix Grain Size Ratio

The template grain size is important in determining the final microstructure. This follows from arguments first presented by Hillert[17] on conditions favoring abnormal grain growth. Two alumina microstructures comparing the effect of increasing template size are presented in Figure 3. Figures 3a and 3b show microstructures resulting from alumina samples sintered at 1650°C for 10 minutes with 5 wt% templates of two different initial diameters, 5-10 and 20-25 μm and thicknesses of 2-4 μm. The sintered microstructures are markedly different. With the small platelets, the elongated grains have an average length of 23 μm, an average thickness of 8.5 μm and an average aspect ratio of 2.8. However, in the large platelet case, the grains have an average aspect ratios of 5.8, an average length of 47 μm and an average thickness of 8.1 μm.

Examination of these two samples indicates that while both samples begin with well oriented template particles in a fine matrix, the smaller platelets are more easily impinged in the longitudinal direction by matrix grains during grain growth, which have grown to approximately 5-7 μm in diameter. Once the grains impinge they coarsen by growing thicker, yielding lower aspect grains. Another reason the grains in the sample with small templates have a low aspect ratio is that with equivalent weight percents of templates, there are roughly ten times as many seed particles in the small platelet case compared to the large platelet case. This increased number decreases the platelet to platelet separation distance and thus increases template to template impingement earlier in the growth process.

Two theories may also help explain why the larger templates grow more easily. First, the Hillert criteria for abnormal grain growth[17] cites the presence of a grain twice the size of the mean grain size of the dense matrix as a condition for abnormal growth. Second, from Ostwald ripening theory, the larger the disparity between matrix and template grains, the faster the growth rate.

Liquid Phase Content

The amount of liquid present during sintering can be altered by changing the amount of calcia and silica, the $CaO:SiO_2$ ratio, or the sintering temperature. In these two samples, the amount of silica and calcia was changed, while the $CaO:SiO_2$ ratio and sintering conditions remained constant. Figure 4 shows samples with 4 vol% template particles and $CaO+SiO_2$ contents of 0.5, and 5 wt% fired at 1650°C for 2 hours; the equilibrium liquid phase contents in the samples at 1650°C were 1.72 and 17.2 vol%, respectively.

In 4a, the sample completely densified with 1.72 vol% liquid. Anisometric grains developed from template particles and a textured microstructure resulted. However, a few matrix grains and some randomly oriented anisometric grains remain. The anisotropic grains have an average diameter of 15 μm and an average thickness of 3.5 μm. The average aspect ratio is low, at 4.3. The low liquid content limited the growth of the anisotropic grains, and the grains are not strongly faceted.

In Figure 4b, the sample densified in the presence of 17.2 vol% liquid. The grains became highly faceted and the microstructure became well aligned during sintering. The average grain diameter remained nearly identical to figure 4b with an average diameter of 21 μm, but the

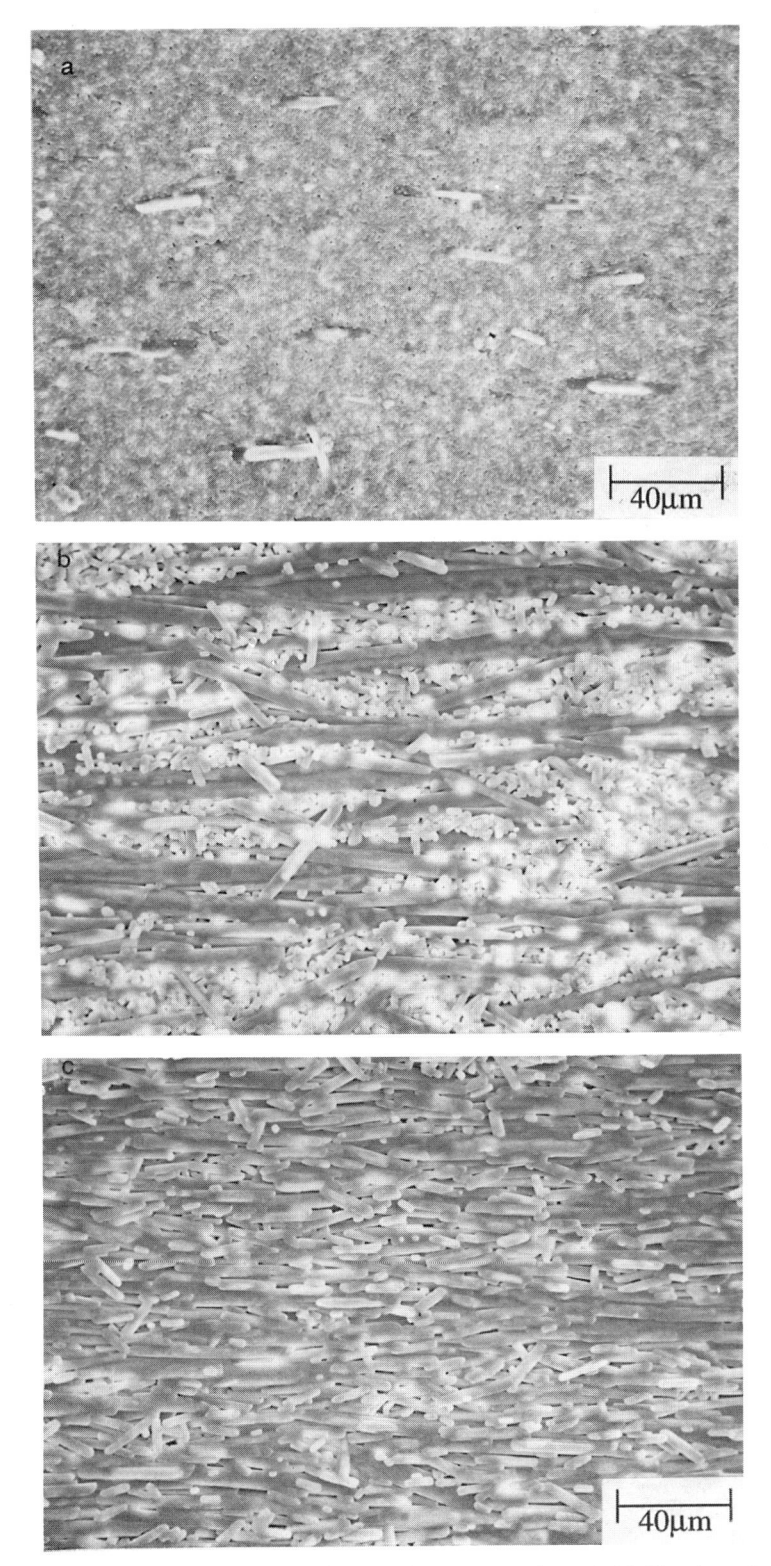

Figure 2. Microstructure evolution with varying template concentration. a) Mullite sintered at 1450°C, 5 h, 2% template particles. b) Mullite sintered at 1600°C, 5 h, 2% template particles. c) Mullite sample sintered at 1600°C, 5 h, 10% template particles. Note the change in aspect ratio between b) and c).

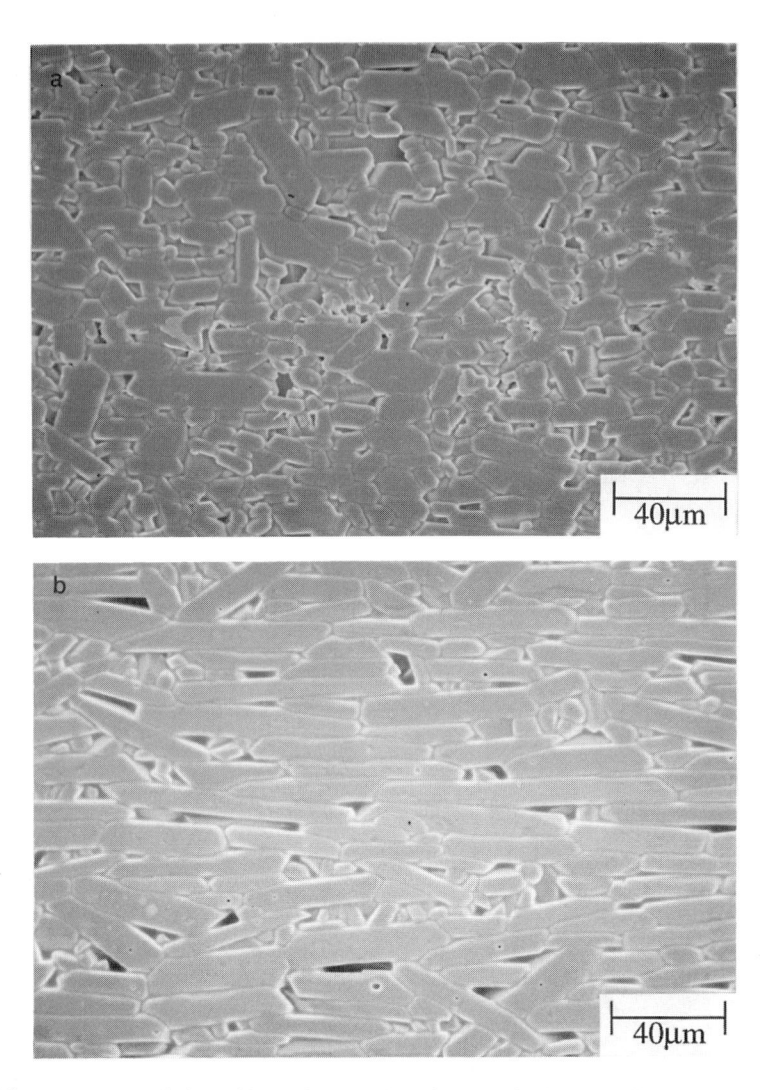

Figure 3. Microstructure evolution with varying template size. a) Liquid phase sintered alumina sintered at 1650°C, 10 min, 5% templates 5-10 μm initial diameter. b) Liquid phase sintered alumina sintered at 1650°C, 10 min, 5% templates 20-25 μm initial diameter.

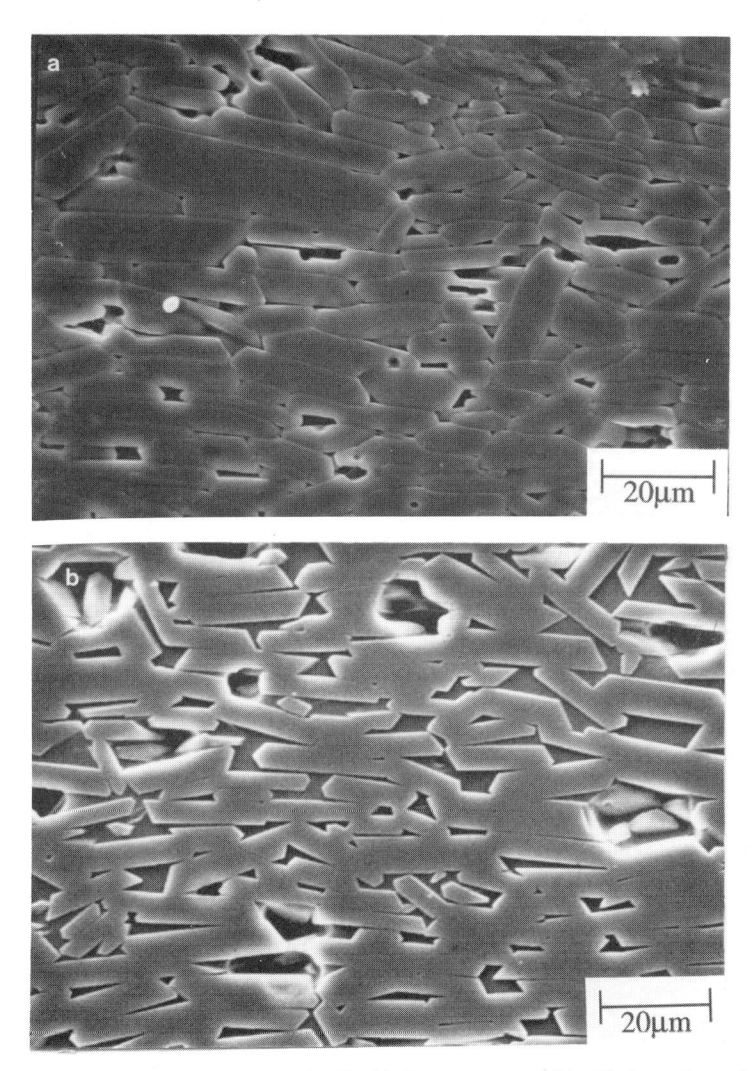

Figure 4. Microstructure evolution with varying liquid phase content. a) Liquid phase sintered alumina sintered at 1650°C, 2 h, 1.72 vol% liquid phase. b) Liquid phase sintered alumina sintered at 1650°C, 2 h., 17.2 vol% liquid phase.

grains are thicker at 5 µm, resulting in a lower aspect ratio of 4.1. Where grains contact, no grain boundaries are visible. Interestingly, the grains appear to have rotated to align themselves with other anisometric grains. This suggests that TGG allows the development of stronger texture than possible by growth of the template particles only.

The increased faceting and the alignment of anisometric grains with increased liquid content indicates the liquid provides a fast diffusion path for template growth and a means for grain rearrangement during anisotropic growth. In systems with large amounts of liquid, off-axis grains may be susceptible to consumption. After the matrix is consumed, off-axis grains dissolve, and are redistributed on large template particles, minimizing the energy of the system.

SUMMARY

Templated grain growth is a unique approach to regulate microstructure development and to obtain a range of ceramic microstructures differing in the size and amount of anisotropic grains. Important criteria for templated grain growth include anisotropic crystal structure, anisotropic growth rates for the material, and the presence of template grains larger than the matrix grain size. The development of texture in these microstructures depends critically on the initial microstructure, including the orientation and the size and concentration of template particles. Growth of anisometric grains is dependent upon the matrix grain size, the liquid content, and the relative size of the matrix and template particles.

Acknowledgements: The authors gratefully acknowledge the support of the ONR Grant N00014-94-0007. M. Seabaugh gratefully acknowledges the receipt of a National Science Foundation Graduate Research Fellowship.

REFERENCES
1. P. F. Becher, "Microstructural design of toughened ceramics," *J. Am. Ceram. Soc.*, **74**[2] 255 (1991).
2. E. A. Holm and M. J. Cima, "Two-dimensional whisker percolation in ceramic matrix-ceramic whisker composites," *J. Am. Ceram. Soc.*, **72**[2] 302 (1989).
3. M. Wu and G.L. Messing, "Fabrication of oriented SiC-whisker-reinforced mullite matrix composites by tape casting," *J. Am. Ceram. Soc.*, **77**[10] 2586 (1994)
4. I. Balberg and N. Binenbaum, "Computer study of percolation threshold in a two-dimensional anisotropic system of conducting sticks," *Phys. Rev. B: Condens. Matter*, **28**[7] 3799 (1983).
5. T. Takenaka and K. Sakata, "Grain orientation and electrical properties of hot-forged $Bi_4Ti_3O_{12}$," *Jap. J. Appl. Phys.*, **19**[1] 31 (1980).
6. H. Igarashi, K. Matsunaga, T. Taniai, and K. Okazaki, "Dielectric and piezoelectric properties of grain-oriented $PbBi_2Nb_2O_9$ ceramics," *Am. Ceram. Soc. Bull.*, **57**[9] 815 (1978).
7. K. Hirao, M. Ohashi, M. E. Brito, and S. Kanzaki, "Processing strategy for producing highly anisotropic silicon nitride," *J. Am. Ceram. Soc.*, **78**[6] 1687 (1995).
8. G. E. Youngblood and R. S. Gordon, "Texture-conductivity relationship in polycrystalline lithia-stabilized β"-alumina," *Ceram. Internal.*, **4**[3] 93 (1978).
9. A. Holloway, R. W. McCallum, and S. R. Arrasmith, "Texture development due to preferential grain growth of Ho-Ba-Cu-O in 1.6-T magnetic field," *J. Mater. Res.*, **8**[4] 727 (1993).
10. A. Halliyal, A. S. Bhalla, and R. E. Newnham, "Polar glass ceramics-a new family of electroceramic materials: tailoring the piezoelectric and pyroelectric properties," *Mat. Res. Bull.*, **18**[8], 1007 (1983).
11. M. D. Sacks, G. W. Scheiffele, and G. A. Staab, "Fabrication of textured silicon carbide via seeded anisotropic grain growth," *J. Am. Ceram. Soc.* **79** [6] 1611 (1996).
12. S.-H. Hong and G. L. Messing, "Densification and anisotropic grain growth in titania-doped diphasic mullite gels," accepted to *J. Am. Ceram. Soc.*.
13. T. Carisey, I.Levin, and D. G. Brandon, "Microstructure and mechanical properties of textured Al_2O_3," *Materials Science and Engineering A* **195** 189 (1995)
14. M. Seabaugh, D. Horn, I. Kerscht, S.-H. Hong, and G. L. Messing, Anisotropic grain growth in alumina ceramics, in *Sintering Technology*, R. M. German, G. L. Messing, and R. A. Cornwall, eds., Marcel Dekker, New York, 1996.
15. M. M. Seabaugh, I. H. Kerscht, and G. L.Messing, "Texture development by templated grain growth in liquid phase sintered α-alumina," to be published in *J. Am. Ceram. Soc.*.
16. Y. T. Chou, Y. T. Ko, and M. F. Yan, "Fluid flow model for ceramic tape casting," *J. Am. Ceram. Soc.*, **70**[10] C-280-C-282 (1987).
17. M. Hillert, "On the theory of normal and abnormal grain growth," *Acta Metall.*, **13**, 227-38 (1965).

CHARACTERIZATION OF SECOND PHASES IN TRANSLUCENT ALUMINA BY ANALYTICAL TRANSMISSION ELECTRON MICROSCOPY

George C. Wei[1], Seung-Joon Jeon,[2] Changmo Sung,[2] and William H. Rhodes[1]

[1]Osram Sylvania Incorporated, Beverly, MA 01915
[2]Center for Advanced Materials, University of Massachusetts, Lowell, MA 01854

ABSTRACT

Microstructures of magnesium aluminate spinel and yttrium aluminate garnet second-phases in translucent alumina were characterized by transmission electron microscopy and convergent beam electron diffraction techniques. High order Laue zone line measurement was employed to examine chemical composition of spinel through lattice parameter determination. Phase equilibrium, the formation of the second phases, and their roles during sintering, are discussed.

INTRODUCTION

Translucent alumina is typically made by hydrogen sintering of compacts of high-purity, finely divided alumina powders doped with magnesia.[1] The MgO sintering aid prevents abnormal grain growth to allow annihilation of pores pinned at grain boundaries. The roles of MgO were summarized as an inhibitor for grain-boundary motion by a grain-boundary segregation mechanism,[2,3] and as an anisotropy homogenizer.[4,5] The microstructure of MgO-doped alumina is typically composed of equiaxed grains and magnesium aluminate ($MgO \cdot Al_2O_3$) spinel second phase. Often, Y_2O_3 and MgO are used as a double dopant system for translucent alumina, taking advantage of a lower sintering temperature brought about by the presence of a liquid phase during sintering. The MgO dopant remains an effective grain growth inhibitor and porosity can be nearly eliminated in the presence of the liquid phase.[6] The microstructure of translucent alumina doped with MgO and Y_2O_3 typically consists of grains of a bimodal size distribution along with spinel and yttrium aluminate ($3Y_2O_3 \cdot 5Al_2O_3$) garnet (YAG) second phases.

Although many studies[7-16] in the literature were devoted to monolithic spinel, there was a lack of an in-depth study of the structure and composition of the spinel second-phase particles typically present within translucent alumina. In one paper[17] spinel particles inside translucent alumina were charaterized by electron microprobe analysis as 77 wt % Al_2O_3 ($MgO \cdot 1.32 Al_2O_3$) corresponding to a lattice parameter of 0.8049 nm. Because of evaporation of the MgO dopant during sintering, the distribution of spinel in a thin-walled translucent alumina was typically non-uniform with spinel particles more populous in the center vs. the near-surface region of the wall.[17] Transmitted-light optical microscopy of thin sections, was commonly used in observing spinel particles in translucent alumina.[18]

The primary purpose of this study is to characterize the spinel second-phase particles in MgO and Y_2O_3-doped alumina using transmission electron microscopy (TEM) and convergent beam electron diffraction (CBED) techniques. It is of importance to determine the precise chemical composition of the spinel phase (that can display a significant range of non-stoichiometry) which is formed during sintering of translucent PCA, in order to understand the microstructural evolution and the role of the dopants in the system. The distribution, composition, and structure of spinel particles were investigated as a function of the dopant level and sintering cycle. High order Laue zone (HOLZ) lines obtained from the CBED technique were used extensively to examine the chemical composition of the spinel second-phase through lattice parameter determination. The present study also included characterization of the YAG second phase.

Formation of $MgAl_2O_4$ spinel from MgO and Al_2O_3 has been studied.[7-10] The mechanism for spinel formation was suggested as the interdiffusion of Al^{3+} and Mg^{2+} ions through the oxygen lattice. Navias[8] heated sapphire either in direct contact with or in the neighborhood of MgO blocks in hydrogen. He reported spinel was formed at temperatures as low as 1500°C. The physical properties including lattice constant, Vickers hardness, and refractive index, varied continuously in the reacted zone, indicating a gradual change in the chemical composition of the spinel.

The MgO dopant in an alumina green compact is expected to either dissolve in alumina and/or fully react with alumina to form spinel during air prefiring and during the heating portion of the sintering cycle. In about the same temperature range, YAG was also reported to form in Y_2O_3-doped alumina.[17] The Al_2O_3-MgO-Y_2O_3 ternary system has been studied.[6,21] A ternary eutectic point was found[6] (equivalent oxide composition: 30.3 wt. % alumina, 65.5 wt. % YAG, 4.2 wt. % spinel; temperature: 1761°C).

As is evident from the known phase diagrams of the MgO-Al_2O_3 system,[16,17,19,20] alumina has a large solid solubility in spinel and supersaturated single-phase spinel can form in the range from $MgO \cdot Al_2O_3$ to $MgO \cdot 5Al_2O_3$, whereas MgO has a limited solid solubility in the spinel. The lattice parameter decreases with increasing Al^{3+} substitution since three larger (0.66Å) Mg^{2+} ions are replaced by two smaller (0.51Å) Al^{3+} ions and one vacancy in the alumina-rich spinel region. It has been reported that formation of spinel from MgO and Al_2O_3 at a molar ratio of n, proceeds in accordance with:[11]

$$MgO + n\, Al_2O_3 \rightarrow (1+3n)/4\, Mg_{4/(1+3n)}\, Al_{8n/(1+3n)}\, O_4 \quad (1)$$

The lattice parameter (a) of spinel varies with composition (on the alumina-rich side of the spinel phase field in the MgO-Al_2O_3 phase diagram) according to:[11,12,15]

$$a\,(Å) = (8.6109 + 23.7195\, n) / (1 + 3n) \quad (2)$$

Thus, the theoretical lattice parameter of spinel can be calculated as a function of composition based on Equation (2). Also, supersaturated solid solutions of alumina-rich nonstoichiometric spinel can decompose into Al_2O_3 and $MgAl_2O_4$ below the solvus line during slow cooling. Equation (2) is expected to be applicable to spinel in translucent alumina doped with MgO and Y_2O_3, since the solubility of Y_2O_3 in spinel is believed to be very low, because of the relatively large size of Y^{3+} ions. The solubility of MgO in sapphire was recently measured (e.g., only 55 ppm at 1700 °C [22]). Similarly, the solubility of Y_2O_3 in Al_2O_3 was also low (<20 ppm at 1850°C).[23]

Alper et al.[16] reported that the lattice parameter was constant in spinel compositions ranging from the stoichiometric spinel to MgO-rich spinel containing up to about 37.5 wt. % equivalent MgO level (n = 0.65). This was confirmed by Chiang and Kingery,[12] in MgO-rich spinel ranging up to n = 0.883 (31 wt. % MgO). It also agreed with the findings[24] that, at MgO levels such as 32 and 34 wt. %, MgO crystalline phase and spinel were present in MgO-Al_2O_3 mixtures heated at 1750°C, and air quenched. Cooling rate was known[25] to affect the distribution of cations in tetrahedral and octahedral sites, and therefore, the lattice parameter of spinel.

EXPERIMENTAL PROCEDURES

The plane-view specimens for TEM study were prepared by cutting the translucent alumina samples into sections from the area of interest. The TEM specimens were prepared in the usual manner, first grinding the sample to a thickness of about 200 μm with a 20 µm diamond wheel followed by diamond core drilling to produce 3 mm disks. These disks were dimple-ground, and argon ion milling was used to complete the thin foil preparation. These samples were observed by TEM/EDXS. Four zone axes including [111], [113], [114] and [116] were employed to identify the changes or shifts of high order Laue zone (HOLZ) lines due to lattice parameter variations in the spinel phase. The CBED patterns for [113] zone axis were used extensively in deriving the lattice parameters. Computer simulations of HOLZ lines in several zone axes were carried out using DFTools computer programs (public software made at Lehigh University, Bethlehem, PA).

The method of detecting the changes of spinel composition involved studying CBED patterns from thin foil specimens. CBED patterns are formed when a convergent beam of electrons is diffracted by a thin single-crystal region of the sample. The electron energy has a significant effect on the positions of the HOLZ lines. The value of the accelerating voltage in the TEM must be calibrated with a known crystal. CBED patterns at a high camera length (1600 mm) of pure Si were indexed as zone axes [114] and [113], which were used to accurately determine the accelerating voltage as 119.11 kV (for instance, nominal accelerating voltage is 120 kV). The lines of the CBED patterns, known as high order Laue zone (HOLZ) lines, are the elastic analog of Kikuchi lines. HOLZ lines arise from elastic scatter of the part of the convergent beam of electrons which is at the exact Bragg angle. The positions of the HOLZ lines are a sensitive function of lattice parameter and therefore, using suitable standards and/or computer simulations, it is possible to obtain composition information of spinel phase as a function of dopant level or sintering temperature since lattice parameters vary with chemical composition.

This HOLZ line technique can be used to measure lattice parameter to an accuracy of 2 parts in 10,000. In order to use the HOLZ lines for absolute lattice parameter determination, it is necessary to calibrate with specimens of known lattice parameters. In this study, three spinel standards were used: (1) $1MgO \cdot 0.9Al_2O_3$, (2) $1MgO \cdot 1Al_2O_3$, and (3) $1MgO \cdot 2Al_2O_3$, Table 1. The standard spinel samples were prepared by mixing predetermined amounts of MgO and alumina, heating at 1650 °C for 2 h in air, and air quenching. The compositions of the standards were determined by wet chemical analysis. HOLZ lines of several zone axes were obtained from these samples for a comparison of lattice parameter and spinel composition. A series of CBED patterns and their computer simulated patterns at the [113] zone axis was obtained to check the sensitivity of the HOLZ line patterns to the spinel composition. Different lattice parameter values give different simulated HOLZ line patterns. Matching computer simulations of HOLZ line positions (calculated at exactly 119.11 kV accelerating voltage) with the experimental CBED-HOLZ line patterns involved comparison of the ratio of certain HOLZ lines that intersected into triangles. In this way, the lattice parameter for the spinel standards was derived, and spinel second-phase particles in translucent alumina were analyzed. The lattice parameters of the three spinel standards were measured as a = 0.8100 nm ($1MgO \cdot 0.9Al_2O_3$), a = 0.8081 nm ($1MgO \cdot 1Al_2O_3$), a = 0.8021 nm ($1MgO \cdot 2Al_2O_3$), all in agreement with the literature.[11,12,16] The calculated lattice parameters for spinel phase of the three compositions were a = 0.8097 nm, a = 0.8083 nm and a = 0.8007 nm, respectively. The CBED patterns for other zone axes such as [111], [114], and [116], were employed to confirm the consistency of the results of lattice parameter measurements. Standards 1 and 2 showed excellent agreement between the measured and calculated lattice parameter, Table 1. The agreement in spinel standard 3 is not as good; it could be due to a non-uniformity in that particular standard sample.

Table 1. Spinel standards.

Standard	Composition	Measured lattice parameter (nm)	Calculated lattice parameter (nm)
1	$MgO \cdot 0.9Al_2O_3$	0.8100	0.8097
2	$MgO \cdot Al_2O_3$	0.8081	0.8083
3	$MgO2 \cdot Al_2O_3$	0.8021	0.8007

RESULTS AND DISCUSSION

The above showed the sensitivity of the HOLZ line patterns to the lattice parameter of spinel. The CBED patterns from many spinel particles were examined in translucent alumina as a function of the dopant level, sintering temperature, and cooling rate of the alumina samples. The dopant levels in the samples used in the present study were in the range[18] (100-2500 ppm) typically used in translucent alumina. The dopant levels were catagorized as high (H), medium (M), and low (L). The sintering temperatures (1800-1950 C) of the translucent alumina samples are listed in Table 2. The results of lattice parameter measurements are also summarized in Table 2. In any of the given translucent alumina samples, typically both intergranular and intragranular spinel phases were observed. Their sizes ranged from about 0.2 to about 5 µm. The relative population of the spinel second phase located inside the grains vs. at grain boundaries was difficult to assess, since no statistical and quantitative evaluation was made. The spinel particles listed in Table 2 were selected for their suitability (e.g., TEM sample thickness uniformity, and the size of spinel) to generate sharp CBED patterns; it does not mean that only one type of (intragranular or intergranular) spinel is present in that sample. However, YAG second phase was universally located at grain boundaries. The sizes of the YAG particles (~0.5 to ~20 µm) are generally much larger than spinel.

The consistency of the lattice parameters within one sample was thought to be good, considering the variation of spinel distribution from the center to the near-surface region of the samples. Such a gradient might cause a variation of the MgO level in the spinel particles; spinel in the center region could be more MgO-rich than the ones near the surface.

The first three samples were doped with increasing MgO level at a constant Y_2O_3 level, and all were sintered at a relatively low temperature and cooled down slowly. HOLZ line patterns for the [113] zone axis along with simulated patterns are in Fig. 1-3. Table 2 showed the ranges of the lattice parameters and corresponding n values for samples 1 to 3, were close to each other; a = 0.8097-0.8066 nm, n = 0.89-1.14 (low MgO level); a = 0.8097-0.8038 nm, n = 0.90-1.45 (medium MgO level); and a = 0.8042-0.8078 nm, n = 1.04-1.40 (high MgO level). The ranges covered stoichiometric spinel, suggesting that considerable decomposition of the nominally alumina-rich spinel that could prevail at the sintering temperature, occurred during the slow cooldown.

MgO-rich spinel existed in translucent alumina, as two spinel particles (n = 0.89, 0.90), which were within the range shown in the Al_2O_3-MgO phase diagram,[16] were identified, Table 2. However, the binary system indicated that only alumina-rich spinel should exist in translucent alumina containing MgO sintering aid. The presence of the MgO-rich spinel in the translucent alumina doped with MgO and Y_2O_3 was intriguing. One possibility was that the spinel portion of the binary system was altered considerably by the presence of the ternary eutectic,[6,21] which made it possible to simultaneously form alumina-rich and MgO-rich spinel in the equilibrium Al_2O_3-MgO-Y_2O_3 system.

A schematic Al_2O_3-MgO-Y_2O_3 phase diagram along with the Al_2O_3-MgO binary phase diagram are shown in Fig. 4. During the heating portion of the sintering cycle, a ternary eutectic[6,21] of composition "C_e" (in Fig. 4) forms at 1761°C. Compositions normally employed in translucent alumina fall within the triangle, Al_2O_3-spinel-eutectic, so the amount of the eutectic liquid depends on the Y_2O_3 concentration. As temperature rises above the ternary eutectic point, the liquid changes in composition along the T_e-T_{se} line connecting the spinel-Al_2O_3 eutectic and the ternary eutectic. In the temperature range of 1761 to 1900°C, the phases in equilibrium are solid alumina, solid spinel, and a liquid of a composition determined by the intersection of the isothermal plane with the boundary line T_e-T_{se}, with an overall composition of "A" in Fig. 4. If temperatures above the Al_2O_3-spinel eutectic point (1900°C) are used, the phase assemblage consists of one solid phase (corundum) and a liquid. However, for normal sintering temperatures (1800-1900°C) liquid phase sintering enhances densification in the presence of solid spinel and presumably segregated MgO at the grain boundaries. It is somewhat surprising that MgO remains effective in preventing pore entrapment in the presence of this strong driving force for densification and grain growth. It has been demonstrated that MgO retards grain growth

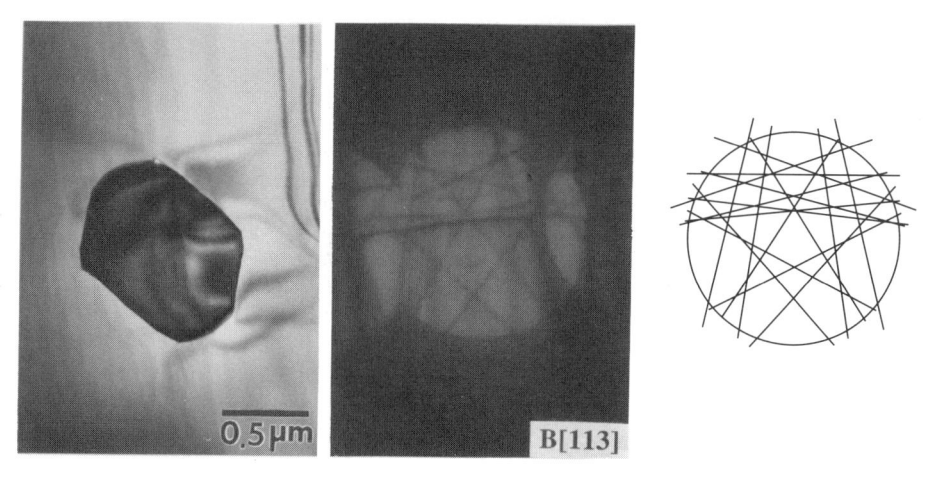

Fig. 1. Bright-field TEM image and CBED patterns for a spinel particle in sample 1 (a = 0.8097 nm).

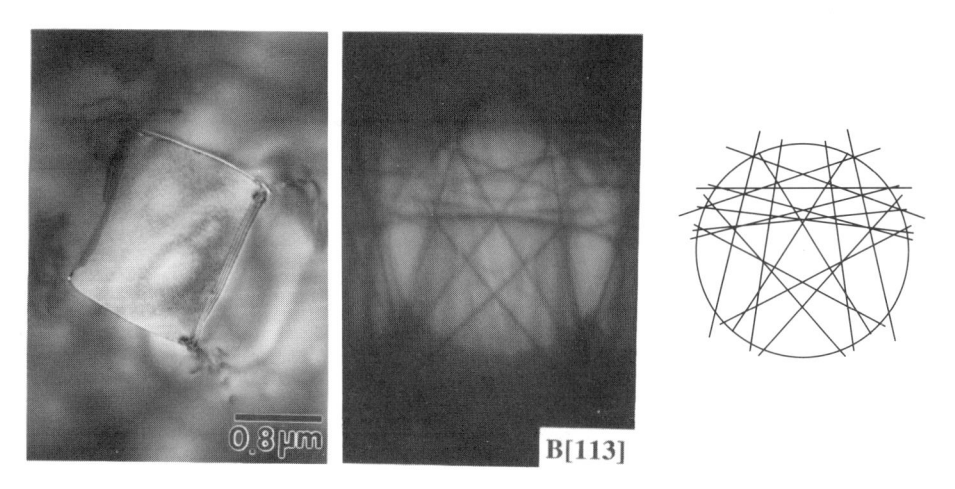

Fig. 2. Bright-field TEM image and CBED patterns for a spinel particle in sample 2 (a = 0.8038 nm).

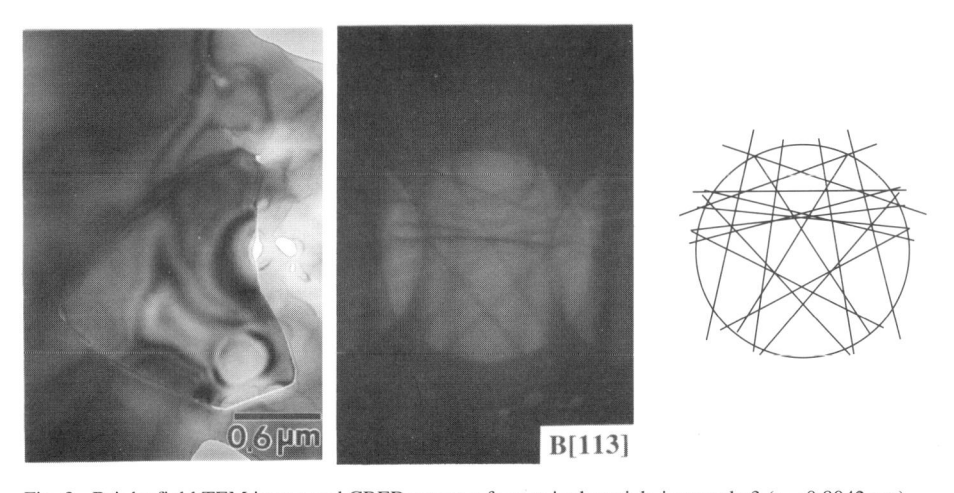

Fig. 3. Bright-field TEM image and CBED patterns for a spinel particle in sample 3 (a = 0.8042 nm).

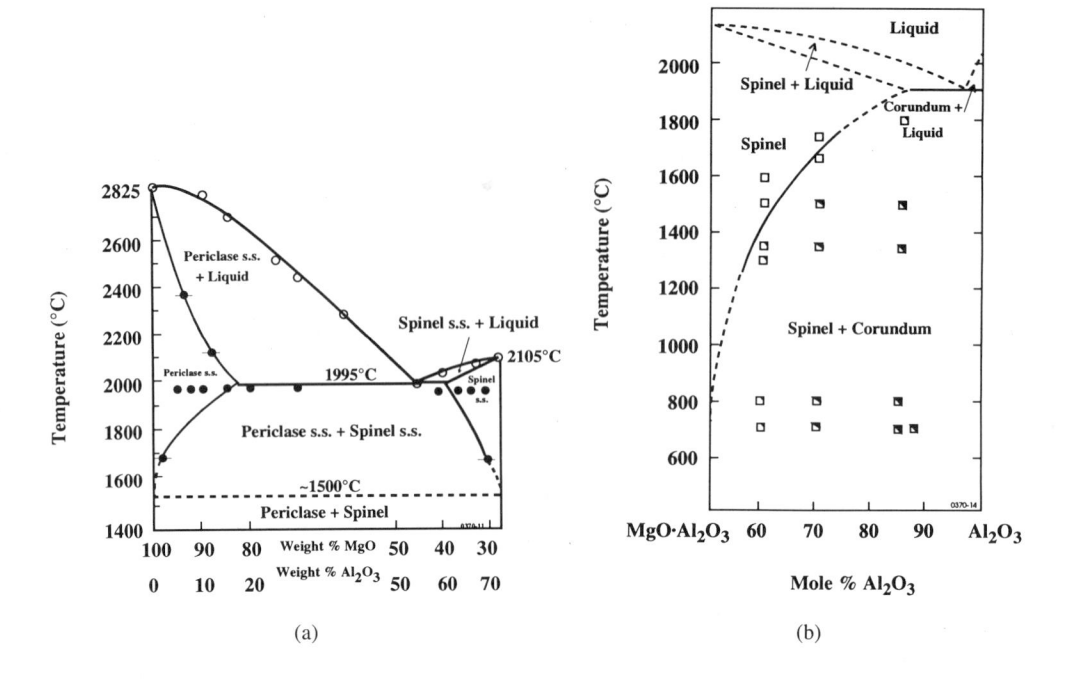

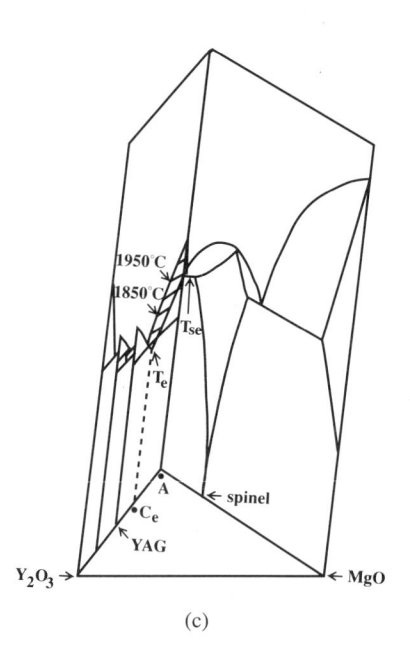

Fig. 4. Phase diagram of the Al$_2$O$_3$- MgO binary system[16,19] (a) and (b), and a schematic of the Al$_2$O$_3$- MgO- Y$_2$O$_3$ ternary system[21], (c). The composistion of the spinel particles in the present study are marked in (a) and (b).

Table 2. Results of TEM and CBED ([113] zone axis) analyses of spinel second phase in translucent alumina.

sample	sinter temp[*]	cooling rate[†]	MgO level[*]	Y2O3 level[*]	spinel location	lattice parameter ±0.0005 nm	composition, MgO·nAl2O3 n =	spinel shape
1	L	S	L	M	inside grains	0.8097	0.89	angular
1	L	S	L	M	inside grains	0.8069	1.11	round
1	L	S	L	M	inside grains	0.8066	1.14	angular
2	L	S	M	M	inside grains	0.8043	1.38	angular
2	L	S	M	M	inside grains	0.8097	0.90	angular
2	L	S	M	M	inside grains	0.8038	1.45	angular
3	L	S	H	M	inside grains	0.8078	1.04	angular
3	L	S	H	M	inside grains	0.8066	1.14	angular
3	L	S	H	M	inside grains	0.8042	1.4	angular
4	H	F	H	M	inside grains	0.8006	2.03	round
4	H	F	H	M	inside grains	0.7990	2.48	round
4	H	F	H	M	at grain boundary	0.7971	3.31	lenticular
4	H	F	H	M	inside grains	0.7997	2.26	round
5	M	S	M	none	at grain boundary	0.8067	1.13	angular
5	M	S	M	none	at grain boundary	0.8054	1.26	angular
5	M	S	M	none	at grain boundary	0.8049	1.31	angular

[*] L: low, M: medium, H: high.
[†] S: slow, F: fast.

by a factor of 50 in high purity Al2O3 but only by a factor of 5 in Al2O3 containing some liquid phase.[26] As the sample is cooled through the eutectic point, additional spinel, Al2O3, and all the YAG phase precipitates out. The spinel that is found within the grains probably became entrapped by grain growth accompanying the sintering phase, and this morphology is found for Al2O3 sintered with and without Y2O3 and its resultant liquid phase. Some spinel is found at grain boundaries (again for Al2O3 doped with or without Y2O3) and YAG, on the other hand, is always found at grain boundaries, presumably due to its precipitation from the liquid grain boundary phase.

Within one sample, there appeared to be a difference in the lattice parameter of intergranular vs intergranular spinel, although the difference was small. Both types of spinel were examined in sample 4, and the difference in lattice parameter was 0.7990 vs. 0.7971 nm. The CBED patterns for sample 4, are in Figs. 5 and 6. The corresponding n value was higher for spinel at grain boundaries (n = 3.31) relative to the ones inside the grains (n = 2.26-2.48). This was understandable, because the spinel at grain boundaries

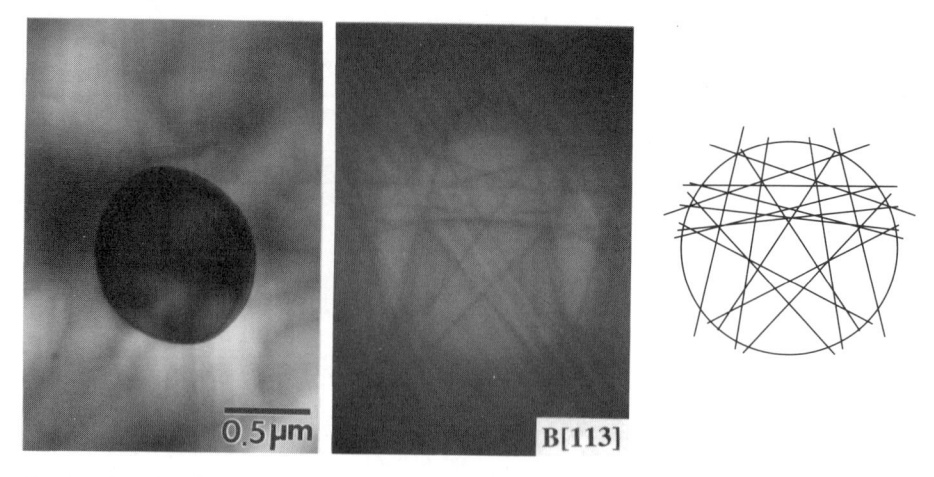

Fig. 5. Bright-field TEM image and CBED patterns for a spinel particle in sample 4 (a = 0.7990 nm).

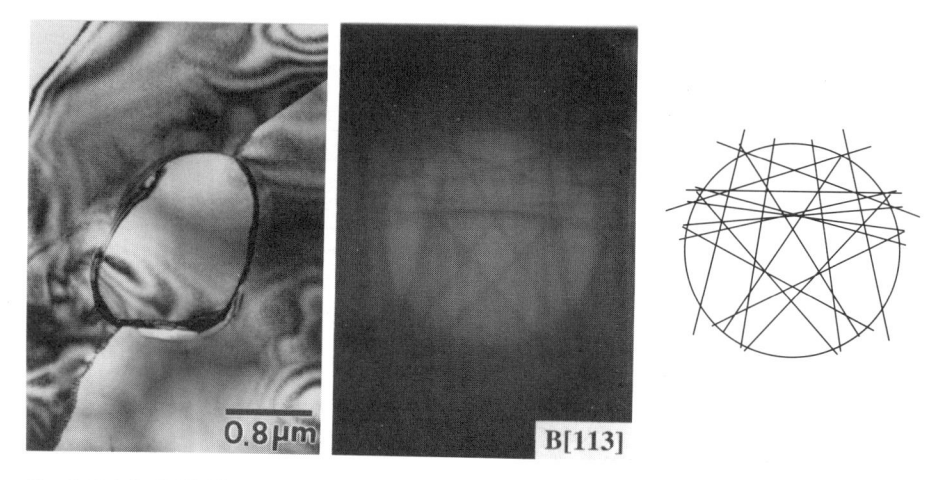

Fig. 6. Bright-field TEM image and CBED patterns for a spinel particle in sample 4 (a = 0.7971 nm).

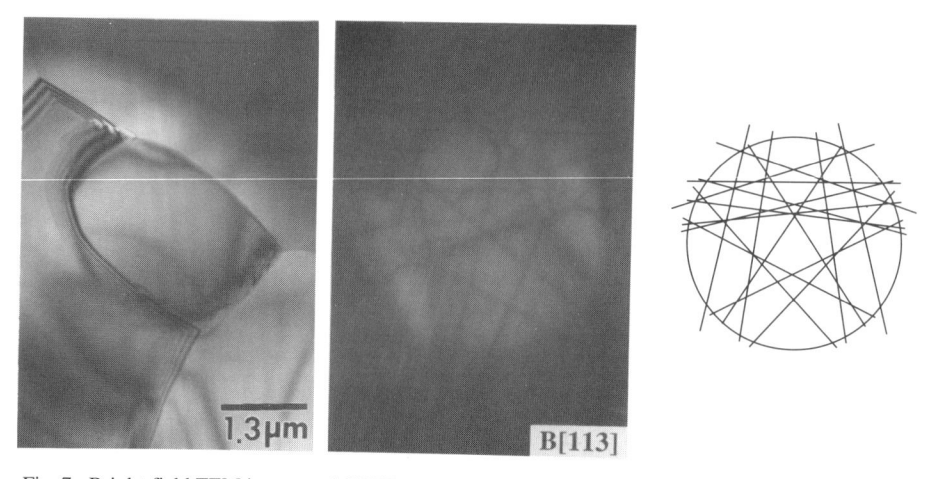

Fig. 7. Bright-field TEM image and CBED patterns for a spinel particle in sample 5 (a = 0.8054 nm).

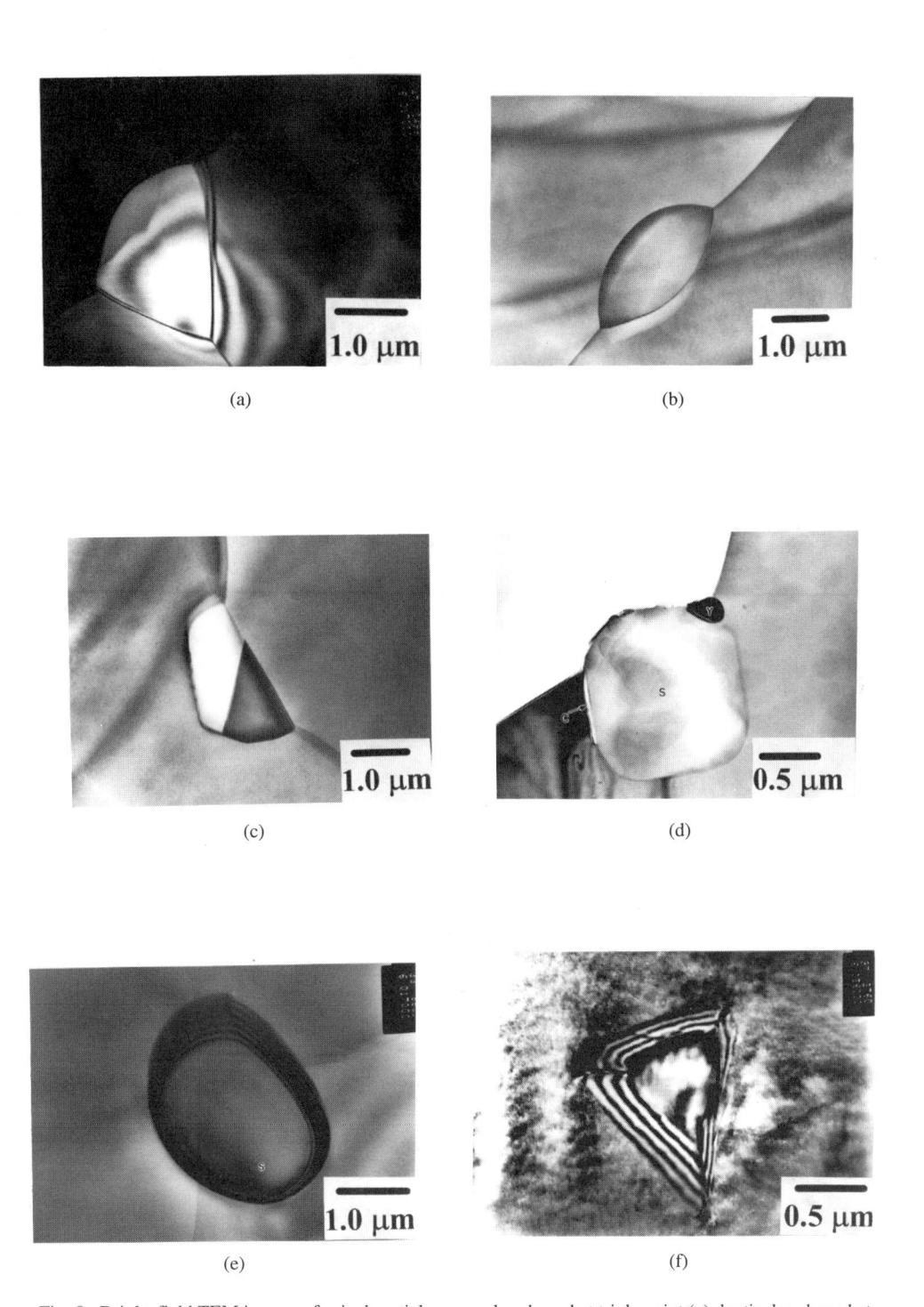

Fig. 8. Bright-field TEM images of spinel particles: angular-shaped at triple point (a), lenticular-shaped at grain boundary (b), two spinel crystals at grain boundary (c), spinel (S) and YAG (Y) at boundary along with a gap (C), between spinel and alumina matrix (d), round-shaped spinel (S) inside-grain (e), and angular-shaped inside grain (f).

could readily lose MgO along the fast transport paths (i.e., grain boundaries) during sintering and cooldown, whereas the spinel embedded inside the grains could not. This explanation was consistent with the finding that grain boundaries in pure, polycrystalline spinel were typically more alumina-rich than the bulk of spinel grains.[13]

Higher sintering temperature and fast cooldown produced spinel that was more alumina-rich. For example, both samples 3 and 4 contained the same dopants, but spinel in sample 4 (sintered at a higher temperature) has n values of 2.26-3.31, in contrast with the smaller n values (n = 1.04-1.4) in sample 3 (sintered at a lower temperature). This is in agreement with the phase diagram;[19] during slow cooling, the spinel composition should follow the solvus line, and fast cooling would not allow the high-alumina spinel phase to track the equilibrium line.

In alumina doped solely with MgO (sample 5), the spinel particles were alumina-rich. The CBED patterns are shown in Fig. 7. No MgO-rich spinel was found. The n values (n = 1.13-1.31) of the spinel in sample 5 were reasonably close to those of the doubly-doped samples 2 and 3 (lower sintering temperature and slow cooling), but lower than sample 4 (higher sintering temperature and fast cooling). The difference between spinel in sample 5 vs. sample 4 could be due to the double dopant vs. single dopant system. It could also be rationalized by the higher sintering temperature and faster cooling rate for sample 4.

The morphology of spinel second phase inside grains varied from angular to round. Both shapes were observed for intragranular spinel, regardless of the sintering temperature or dopant system. A gap or micro-void often appeared to be present in a portion of the entire spinel-alumina interface for angular, intragranular spinel, Fig. 8. This might be due to the volume shrinkage associated with decomposition of alumina-rich spinel into alumina and spinel during cooldown. Sometimes, dislocations were observed in the alumina matrix near the corners of the intragranular spinel phase, Fig. 8, indicating internal strains. In extreme cases, microcracks formed probably as a result of the expansion mismatch between spinel and alumina during cooldown. In contrast, round-shaped spinel inside grains appeared to be strain-free, and coherent with the alumina matrix, Fig. 8. Comparison of the intragranular spinel composition in sample 4 vs. in sample 3 (Table 2), suggested that the round-shaped spinel was more alumina-rich than the angular ones.

The intergranular spinel phase was typically angular and located at triple points. Lenticular spinel was observed at the boundary between two alumina grains, Fig. 8. Intergranular spinel composed of two crystals, was also observed. Triple points in the doubly-doped alumina, sometimes, consisted of mixtures of spinel, YAG, and corundum, indicating precipitation during cooldown.

The features of strain field and microcracking were not observed with YAG second phase in translucent alumina. YAG second phase particles were typically defect-free, Fig. 9. One additional unique feature of YAG second phase compared with spinel, was that it seemed to be able to dissolve Ca or Si impurities commonly present in alumina powders, Fig. 9.

SUMMARY AND CONCLUSIONS

- TEM and CBED-HOLZ line technique, plus calibration with spinel standards of known compositions, was successfully used in measuring the lattice parameter and determining the composition of spinel in translucent alumina.
- Intergranular and intragranular spinel particles with an angular or round morphology were present in alumina doubly doped with MgO and yttria, or singly doped with MgO.
- Higher sintering temperatures appeared to produce spinel second phase that was more alumina-rich.
- MgO-rich spinel found in some doubly-doped samples could be a result of modification of the spinel phase field in the MgO-Al_2O_3 system caused by the addition of yttria.
- YAG second phase was always distributed at grain boundaries, and it could absorb impurities such as CaO or SiO_2 in alumina.
- The existence of mixtures of spinel, YAG, and corundum at triple points, was an evidence of liquid phase sintering operative in alumina doped with MgO and yttria.

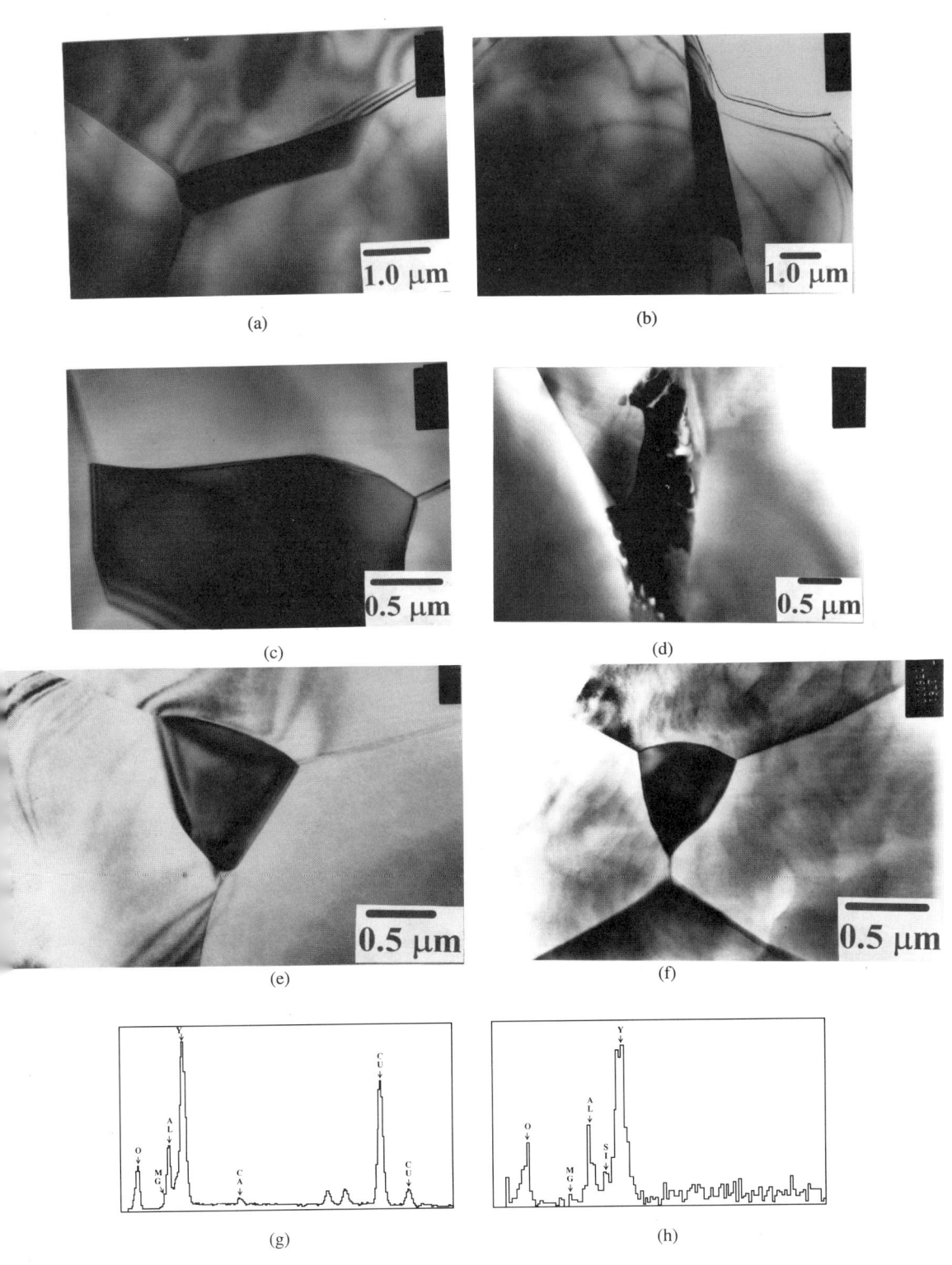

Fig. 9. Bright-field TEM images of YAG particles: angular-shaped at grain boundaries (a) and (b), lenticular-shaped at grain boundary (c), mixture of YAG, spinel, and corundum at triple point (d), Ca-containing YAG at triple point (e), and Si-containing YAG at triple point (f), along with EDXA spectra: (g), (h), of YAG particle in (e), and (f), respectively.

REFERENCES

1. R.L. Coble, Sintering alumina: effect of atmosphere, *J. Am. Ceram. Soc.* 45 [3] 123-27 (1962).
2. K.K. Soni, A.M. Thompson, M.P. Harmer, D.B. Williams, J.M. Chabala, and R. Levi-Setti, Solute segregation to grain boundaries in MgO-doped alumina, *Appl. Phys. Lett.* 66 [21] 2795-97 (1995).
3. S.J. Bennison and M.P. Harmer, A history of the role of MgO in the sintering of α-Al_2O_3, *Ceram. Trans.* 7: 13-49 (1990).
4. S. Baik and J.H. Moon, Effects of magnesium oxide on grain-boundary segregation of calcium during sintering of alumina, *J. Am. Ceram. Soc.* 4 [4] 819-822 (1991).
5. C.A. Handwerker, J.M. Dynys, R.M. Cannon, and R.L. Coble, Dihedral angles in MgO and Al_2O_3: distributions from surface thermal grooves, *J. Am. Ceram. Soc.* 73 [5] 1371-77 (1990).
6. W.H. Rhodes, Phase chemistry in the development of transparent polycrystalline oxides, pp. 1-42, in *Phase Diagrams for Advanced Ceramics*, ed. A. Alper, Academic Press, New York, NY (1995).
7. R.E. Carter, Mechanism of solid state reaction between magnesium oxide and alumina and between magnesium oxide and ferric oxide, *J. Am. Ceram. Soc.*, 44 [3] 116-120 (1961).
8. L. Navias, Preparation and properties of spinel made by vapor transport and diffusion in the system MgO-Al_2O_3, *J. Am. Ceram. Soc.*, 44 [9] 434-446 (1961).
9. R.C. Rossi and R.M. Fulrath, Epitaxial growth of spinel by reaction in the solid state, *J. Am. Ceram. Soc.*, 46 [3] 145-149 (1963).
10. W.P. Whitney and V.S.Stubican, Interdiffusion studies in the system MgO-Al_2O_3, *J. Phys. Chem. Solids*, 32: 305-312 (1971).
11. H.U. Viertel and F. Seifert, Physical properties of defect spinel in the system MgO·Al_2O_3-Al_2O_3, *Neues. Jahrb. Mineral., Abh.*, 134 [2] 167-182 (1979).
12. Y. Chiang and W.D. Kingery, Grain-boundary migration in nonstoichiometric solid solutions of magnesium aluminate spinel: I, grain growth studies, *J. Am. Ceram. Soc.* 72 [7] 271-77 (1989).
13. Y. Chiang and W.D. Kingery, Grain-boundary migration in nonstoichiometric solid solutions of magnesium aluminate spinel: I, effects of grain-boundary nonstoichiometry, *J. Am. Ceram. Soc.* 73 [5] 1153-58 (1990).
14. M. Matsui, T. Takahashi, and I. Oda, Influence of MgO vaporization on the final-stage sintering of MgO-Al_2O_3 spinel, pp. 562-73 in *Advances in Ceramics, Vol. 10, Structure and Properties of MgO and Al_2O_3 ceramics*, ed. W.D. Kingery, Am. Ceram. Soc., Columbus, OH 1984.
15. A. Navrotsky, B.A. Wechsler, K. Geisinger, and F. Seifert, Thermochemistry of $MgAl_2O_4$-$Al_{8/3}O_4$ defect spinels, *J. Am. Ceram. Soc.* 69 [5] 418-22 (1986).
16. A.M. Alper, R.N. McNalley, P.G. Ribbe, and R.C. Doman, The system MgO-$MgAl_2O_4$, *J. Am. Ceram. Soc.* 45 [6] 263-268 (1962).
17. Z. Nakagawa, Effect of additives on solid state sintering of alumina, pp.74-88, *Trans. Mater. Res. Soc. of Jpn*, No.11, ed. S. Somiya, M. Doyama, and Y. Agata (1992).
18. R. J. Charles, S. Prochazka, C. E. Scott, Alumina ceramic, US patent No. 4285732 (1981).
19. D.M. Roy, R. Roy, and E.F. Osborn, The System MgO-Al_2O_3-SiO_2 and Influence of Carbonate and Nitrate Ions on the Phase Equilibria, *Am. J. Sci.* 251: 337-361 (1953).
20. F. Colin, Contribution to the study of phases obtained during the reduction of some nAl_2O_3·MO oxides, *Rev. Hautes Temp. Refract.* 5 [4] 267-83 (1968).
21. P.A. Bosomworth, M.P.Harmer, H.M. Chan, and W.H. Rhodes, *Bull. Am. Ceram. Soc.* 65: 499 (1986).
22. K. Ando and M. Momoda, Solubility of MgO in single-crystal Al_2O_3, *J. Jpn. Ceram. Soc.* 95 [4] 381 (1987).
23. J.C. Cawley and J.W. Halloran, Dopant distribution in nominally yttrium-doped sapphire. *J. Am. Ceram. Soc.*, 69 [8] C195-96 (1986).
24. R.E. Shannon, D.L. Johnson, and M.E. Fine, Precipitation hardening of spinel with excess MgO, *J. Am. Ceram. Soc.* 57 [6] 269 (1974).
25. E. Stoll, P. Fischer, W. Halg, and G. Maier, Redetermination of the cation distribution of spinel ($MgAl_2O_4$) by means of neutron diffraction, *J. Phys. (Orsay, Fr.)*, 25: 447-48 (1964).
26. S.J. Bennison and M.P. Harmer, Grain growth kinetics for alumina in the absence of a liquid phase, *J Am. Ceram. Soc.* 68 [1] C22-24 (1985).

ORIGIN AND CONTROL OF ABNORMAL GRAIN GROWTH IN ALUMINA

Jong-Chul Nam, Il-Joon Bae, and Sunggi Baik*

Department of Materials Science and Engineering
Pohang University of Science and Technology (POSTECH)
Pohang, 790-784, Korea
(* Corresponding author)

ABSTRACT

A number of models have been presented to explain the role of MgO on sintering and grain growth of alumina. In view of recent experimental results showing that abnormal grain growth (AGG) is related with a small amount of glass-forming impurities, the beneficial role of MgO has to be re-evaluated in conjunction with the microstructural evolution of alumina in the presence of such impurities, particularly Ca and Si. With a fixed amount of Si, and thus for a known critical grain size for AGG, varying amounts of MgO were introduced to determine how MgO influences the critical grain size required for AGG. The mobilities of dry and wet boundaries were also determined as a function of MgO concentration with the fixed Si concentration. The experimental results revealed that the primary role of MgO is to reduce the migration rate of wet boundaries. Little effect on the migration rate of dry grain boundaries has been observed.

INTRODUCTION

The primary role of MgO for achieving fully dense and translucent alumina by sintering is to prevent abnormal grain growth (AGG) in the final stage of densification.[1] However, its mechanistic understanding is still lacking. The reason is that the mechanism of AGG has not been fully explained. Without understanding why AGG occurs in alumina during sintering, it has not been possible to explain clearly how the small addition of MgO prevents the AGG.

Recent studies[2,3] revealed that AGG is not an intrinsic property of alumina but an extrinsic property due to the presence of glass-forming impurities such as Ca, Si, K, and Na in the powder or those introduced during powder processing or sintering, in particular. In the presence of small amounts of glass-forming impurities, the portions beyond their solubility limits would accumulate at grain boundaries as grains grow continuously during the final stage of densification, form thin intergranular liquid films, and induce a sudden increase in the rate of grain boundary migration, thereby triggering AGG. Therefore, for a given concentration of impurities, AGG occurs at a critical grain size when the impurity content of grain boundaries reaches a critical level sufficient to form intergranular glassy phases of thermodynamically stable thickness. For instance, Fig. 1 shows the average grain sizes above which AGG was observed as a function of excess Si concentration the amount above its solubility limit, for ultrapure (>99.999%) alumina sintered in a 'clean condition'.[4] In spite of varying powder processing as well as sintering temperature and time conditions, an inverse relation between the average grain size and the excess Si concentration is uniquely depicted, which could be used to estimate the minimum thickness of stable intergranular aluminosilicate films for inducing AGG to be about 3.7 nm. It is particularly interesting to note that very large migration rates are realized as the grain boundaries are wetted by such thin liquid films. In the case of alumina, the migration rate of wet boundaries were estimated from the growth rates of abnormal grains and found to be 10^2 - 10^3 times faster than dry boundaries with an equivalent impurity level.

A number of models have been presented that propose mechanisms by which a small amount of MgO controls AGG in alumina.[1,5] In view of the results showing that AGG is related with a small amount of glass-forming impurities, the role of MgO has to be re-evaluated in conjunction with the microstructural evolution of alumina in the presence of such impurities, particularly Ca and Si. It is now firmly established that (i) the solubility of Mg in α-alumina is much lower than that determined by Roy and Coble[6] using polycrystalline alumina in 1968; a recent study by Ando[7] showed that the solubility of Mg in single crystal sapphire is 120 ppm at 1800°C, and 70 ppm at 1700°C in cationic atomic concentration, and (ii) Mg tends to segregate strongly to all interfaces including free surfaces[8,9], pore surfaces[10], and grain boundaries[11,12] of alumina. Therefore, if a certain amount of MgO is added usually above its solubility limit in alumina, dynamic redistribution of Mg ions occurs in the course of densification and grain growth as a solute within alumina grains, and as a segregant to powder surfaces, to pore surfaces, or to grain boundaries. Some may be consumed as spinel

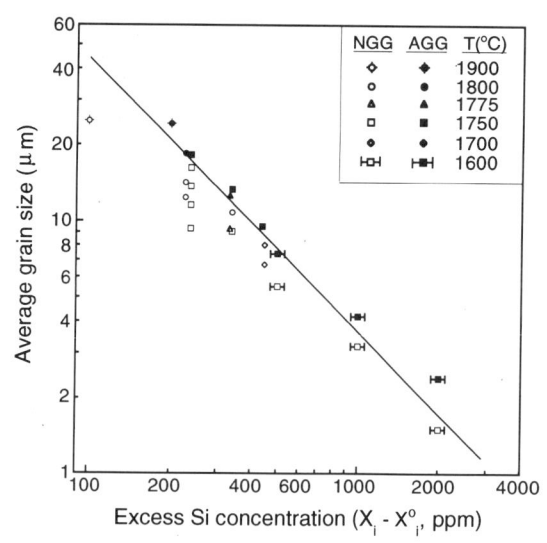

Figure 1. Average grain sizes at the on-set of abnormal grain growth determined as a function of Si concentration above its solubility at various sintering temperatures. The inverse relationship between average grain size and excess Si concentration implies that the average Si concentration at grain boundaries is constant at the moment of abnormal grain growth in alumina.

precipitates. The central question is how a small amount of MgO, typically of the order of a few hundred ppm or a thousand ppm at most, negates the unfavorable influence of glass-forming impurities such Si and Ca present in commercial-purity alumina or introduced during subsequent processing.

In view of our new understanding of AGG in alumina[2-4], the following four possibilities are worth being considered seriously for further studies. First, MgO may increase the critical grain size for AGG determined by the concentration of glass-forming impurities by reducing the amount of impurities being accumulated to grain boundaries. Handwerker et al.[13] proposed such a possibility of reduction in silicate glass by increasing the solubility of Si^{4+} in Al^{3+} matrix via charge and lattice-strain compensation with Mg^{2+} solutes. Second, MgO may enhance the homogeneous distribution of glass-forming impurites and induce concurrent appearance of an intergranular glass phase. Then, AGG would be prevented even though the grain growth rate may still increase at the moment of glass formation. Indirect evidence of this possibility can be found in the studies demonstrating that anisotropy in grain boundary segregation of Ca[11] or in grain boundary migration[14] was clearly reduced by an addition of MgO. Third, MgO may reduce the migration rate of *wet* boundaries either by reducing solid-liquid interfacial energies or by changing their anisotropy. Song and Coble[15] found in fact further addition of MgO to alumina doped intentionally with SiO_2 + CaO induced a change in grain morphology from platelike to equiaxial shape. Fourth, MgO may simply reduce the mobility of *dry* boundaries and change the average grain size - density trajectory favorably to avoid the critical grain size condition for AGG before reaching a full density. Experimental evidence to support this notion has been published.[16,17] However, it has been unclear whether the observed reduction in grain growth rate was in fact due to the reduction in the migration rate of dry grain boundaries or due to the reduction in the migration rate of wet boundaries. The experimental method presented in this study provided us with a unique opportunity to discern among four possible mechanisms which one is the primary function of MgO in alumina.

EXPERIMENTAL PROCEDURE

Samples are prepared from ultrapure α-alumina powder (AKP-5N, Sumitomo Chemicals, Tokyo, Japan) of purity exceeding 99.999%. Varying amounts of MgO up to 400 ppm were added to the powder in addition to 400 ppm of SiO_2. Magnesium acetate tetrahydrate, $(CH_3CO)_2Mg\cdot 4H_2O$ of 99.999% purity and tetraethyl orthosilicate (TEOS) of 99.999+% purity were used for doping MgO and SiO_2, respectively. The doping concentrations were determined on the basis of cationic mole ratio of M/Al (M = Mg, Si). The dopants were dissolved in ethyl alcohol dispersed with the powder, hydrolyzed by adding a proper amount of pure water, and dried in a warm ultrasonic bath. The doped powders were calcined at 1000°C for 2 h in a sapphire tube loaded in an alumina tube furnace. Sintering was performed at 1750°C also in a sapphire tube loaded in a tungsten-mesh furnace under flowing Ar gas. Extreme care was taken to prevent introduction of any other impurities except the dopants throughout the entire processing. The processing steps taken in this study have been shown to be contamination-free.[2,11]

The sintered samples were ground and thermally etched at 1400°C for 20 min in the sapphire tube loaded in the tungsten furnace. Average grain sizes, G, were measured by the linear-intercept method using the relationship, G = 1.5L, where L is the average intercept length. Maximum grain sizes, G_{max}, were determined by averaging the largest and smallest dimensions of the largest grain observed in the cross-sectional micrographs. We define AGG if $G_{max} \geq 10G$, normal grain growth (NGG) if $G_{max} \leq 3G$, and bimodal grain growth (BGG) if $3G < G_{max} < 10G$. Densities were measured by the water immersion method.

RESULTS AND DISCUSSIONS

Figure 2 shows the average and maximum grain sizes versus sintering time with different doping concentrations of Mg. The average grain sizes were determined in the region that contains no abnormal or bimodal grains. Figure 3 illustrates micrographs showing NGG, BGG, and AGG in the samples doped with 60 ppm of Mg and sintered for different time periods. The average grain sizes follow the normal cubic-grain-growth kinetic equation, $G^3 = Kt$, where K is the grain growth rate constant and t is the sintering time. Interestingly, the effect of MgO on the average grain size is insignificant. Whereas, G_{max} decreases drastically with increasing MgO addition, and falls below 10G showing BGG up to 100 ppm of Mg and below 3G above 200 ppm of Mg showing NGG behavior. A similar result has been observed in the ultrapure alumina doped with CaO as a liquid former.[5]

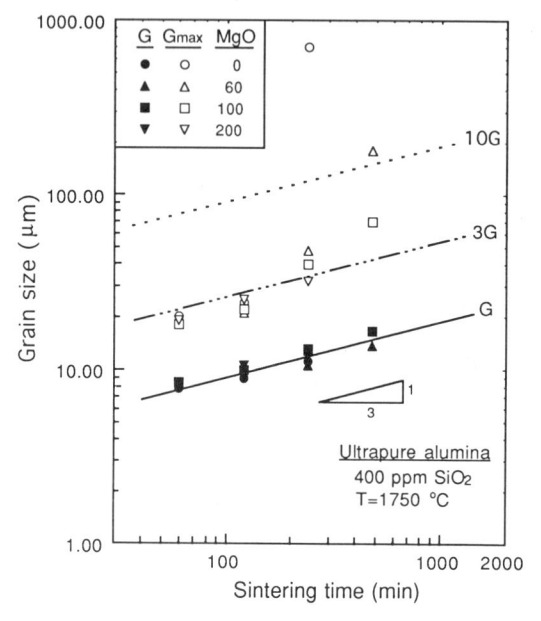

Figure 2. Average and maximum grain sizes versus sintering time at 1750°C for ultrapure alumina co-doped with 400 ppm of Si and varying amounts of Mg. MgO has little effect on the average grain sizes.

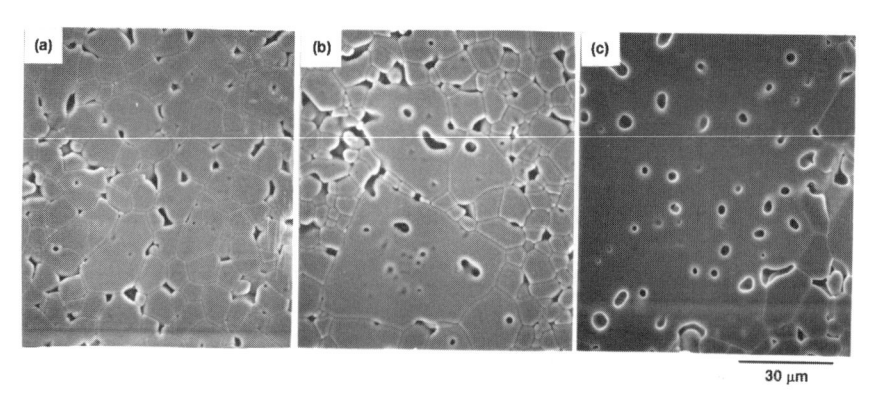

Figure 3. Scanning electron micrographs representing (a) a normal grain growth behavior after 120 min, (b) a bimodal grain growth behavior after 240 min, and (c) an abnormal grain growth behavior after 480 min of sintering at 1750°C for ultrapure alumina co-doped with 400 ppm of Si and 60 ppm of Mg.

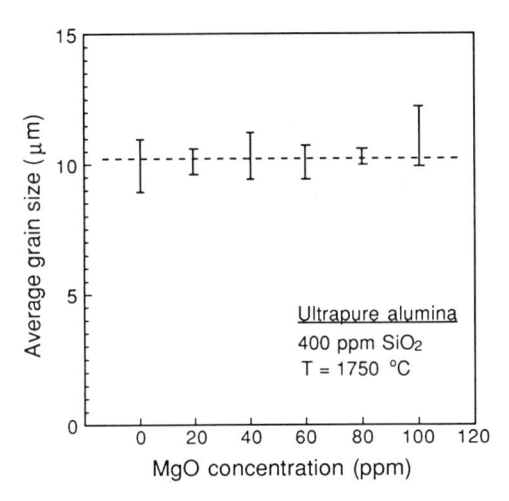

Figure 4. Average grain size ranges within which AGG or BGG initiates as a function of doping concentration of Mg.

Figure 4 shows the range of average grain sizes within which AGG or BGG starts appearing as a function of Mg concentration. We expect quite different behaviors for the change in critical grain sizes for AGG (or BGG) as a function of Mg concentration for the four different proposed mechanisms of MgO. With increasing MgO concentration, larger grain sizes are expected for the first mechanism, a reduction in the amount of glassy phase by increasing solubility of glass-forming impurities, or for the second mechanism, a promotion of homogeneous distribution of glass-forming impurities. In both cases, with increasing Mg concentration, larger grain sizes are required for achieving a sufficient level of Si at grain boundaries to form liquid phases. On the other hand, for the cases of third and fourth mechanisms which propose reductions in grain boundary mobilities of wet and dry boundaries, respectively, it is expected that an addition of MgO will have little influence on the critical grain sizes. The result shown in Figure 4 clearly favors the third or fourth mechanism. Addition of MgO has little effect on the critical grain size.

Now, the remaining question is whether the primary effect of MgO is on the mobility of dry or wet boundaries. Figure 5 compares the average velocity of grain boundary migration calculated by the average grain sizes, dG/dt, with that estimated by the maximum grain sizes, dG_{max}/dt, in the region (or the sintering time period) showing NGG-AGG or NGG-BGG transition. A drastic decrease in dG_{max}/dt, which represents presumably the average velocity of wet boundaries, is clearly depicted with increasing amount of Mg, while the change in dG/dt is only marginal. It clearly demonstrates that the major role of MgO for suppressing AGG is to reduce the migration rate of wet boundary formed above the critical grain size for a given concentration of glass-forming impurities. Apparently, MgO hardly affects the mobility of dry boundaries. As shown in Figure 6, its influence on the grain size - density trajectory is also insignificant in comparison to the drastic effect on the mobility of wet grain boundaries as shown in Figure 5.

The migration rate of a grain boundary is a product of its mobility and driving force. It can be argued that very large migration rates of wet boundaries might originate from a large increase in the driving force for atomic transfer across the intergranular liquid layer as a result of anisotropic solid-liquid interfacial energies, or from a large increase in the rate of atomic transfer through the liquid layer. As such, it seems reasonable to argue that the mechanism by which a small amount of MgO reduces the migration rate of wet boundary is either by

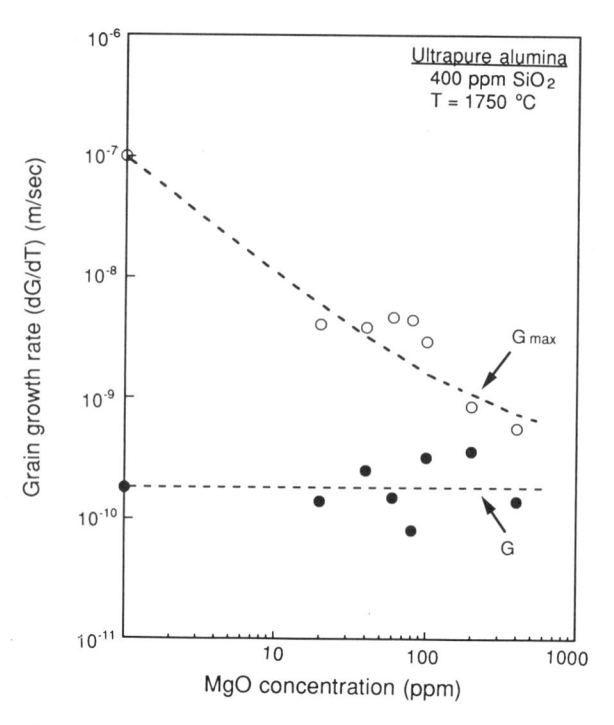

Figure 5. Grain boundary migration rates of dry and wet boundaries estimated by the growth rate of average grains and maximum grains, respectively, plotted as a function of doping concentration of Mg.

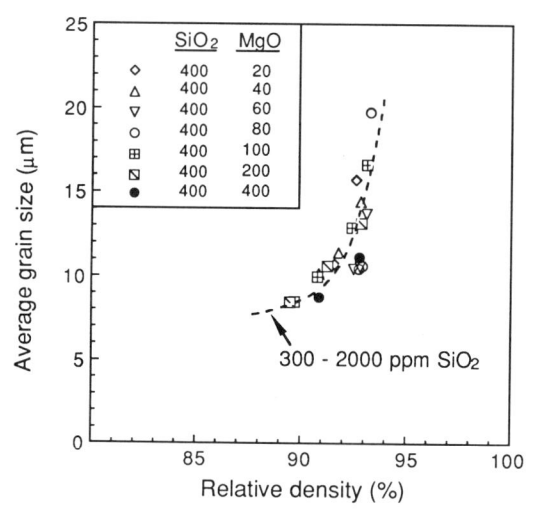

Figure 6. Average grain size - relative density trajectory obtained with ultrapure alumina co-doped with 400 ppm of Si and varying amounts of Mg up to 400 ppm after sintering at 1750°C.

changing the solid-liquid interfacial energies reducing its driving force or by reducing the boundary diffusivity which determines its mobility. Anisotropic solid-liquid interfacial energies of α-alumina in contact with anorthite glass have been measured.[18] Further work is necessary to confirm if additions of MgO indeed modify the degree of anisotropy in the solid-liquid interfacial energies or the diffusivity through the intergranular liquid film.

CONCLUSIONS

An experiment has been designed and performed to identify the primary role of MgO during sintering of alumina containing small amounts of liquid-forming impurities. The critical grain size required for abnormal grain growth, which is determined initially by doping concentration of Si, is found to be insensitive to the additional doping concentration of Mg. The migration rate of wet boundaries which appeared when the critical grain size was exceeded is found to decrease continuously with increasing Mg concentration. Hence, we conclude that the primary role of MgO in suppressing abnormal grain growth in alumina is to reduce the mobility of intergranular liquid films of a few nanometers in thickness.

REFERENCES

1. S. J. Bennison and M. P. Harmer, A history of the role of MgO in the sintering of α-Al_2O_3, p. 13 in Ceram. Trans., Vol. 7: *Sintering of Advanced Ceramics*, C. A. Handwerker, J. E. Blendell and W. A. Kaysser, ed., American Ceramic Society, Columbus, OH (1990).
2. S. I. Bae and S. Baik, Sintering and grain growth of ultrapure alumina, *J. Mater. Sci.*, 28:4197 (1993).
3. S. I. Bae and S. Baik, Determination of critical concentrations of silica and/or calcia for abnormal grain growth in alumina, *J. Am. Ceram. Soc.*, 76:1065 (1993).
4. I. J. Bae and S. Baik, Abnormal grain growth of alumina, to appear in *J. Am. Ceram. Soc.* (1997).
5. S. I. Bae and S. Baik, Critical concentration of MgO for the prevention of abnormal grain growth in alumina, *J. Am. Ceram. Soc.*, 77:2499 (1994).
6. S. K. Roy and R. L. Coble, Solubilities of magnesia, titania, and magnesium titanate in aluminum oxide", *J. Am. Ceram. Soc.*, 51:1 (1968).
7. K. Ando, Impurity diffusion and the effect of the dissolution of impurities of diffusivities of constituent ions in α-Al_2O_3, p. 297 in: *Sintering '87*, Vol. 1., S. Somiya et al., ed, Elsevier Publ. Ltd., London (1988).
8. S. Baik, D. E. Fowler, J. M. Blakely, and R. Raj, Segregation of Mg to the (0001) surface of doped sapphire, *J. Am. Ceram. Soc.*, 68:281 (1985).
9. S. Baik, Segregation of Mg to the (0001) surface of single-crystal alumina: quantification of AES results, *J. Am. Ceram. Soc.*, 69:C-101 (1986).
10. C. Sung, G. C. Wei, J. Ostreicher, and W. H. Rhodes, Segregation of magnesium to the internal surface of residual pores in translucent polycrystalline alumina, *J. Am. Ceram. Soc.*, 75:1796 (1992).
11. S. Baik and J. H. Moon, Effect of magnesium oxide on grain-boundary segregation of calcium during sintering of alumina, *J. Am. Ceram. Soc.*, 74:819 (1991).
12. K. K. Soni, A. M. Thompson, M. P. Harmer, D. B. Williams, J. B. Chabala, and R. Levi-Setti, Solute segregation to grain boundaries in MgO-doped alumina, *Appl. Phys. Lett.*, 66:2795 (1995).
13. C. A. Handwerker, P. A. Morris and R. L. Coble, Effect of chemical inhomogeneities on grain growth and microstructure in Al_2O_3, *J. Am. Ceram. Soc.*, 73:130 (1989).
14. J. Rödel and A. M. Glaeser, Anisotropy of grain growth in alumina, *J. Am. Ceram. Soc.*, 73:3292 (1990).
15. H. Song and R. L. Coble, Origin and growth kinetics of platelike abnormal grains in liquid-phase-sintered alumina, *J. Am. Ceram. Soc.*, 73:2077 (1990).
16. M. P. Harmer, Use of solid-solution additives in ceramic processing", pp. 679-96 in Advances in Ceramics, Vol. 10; *Structure and Properties of MgO and Al_2O_3 Ceramics.*, W. D. Kingery, ed., American Ceramic Society, Columbus, OH (1984).
17. S. J. Bennison and M. P. Harmer, Effect of magnesia solute on surface diffusion in sapphire and the role of magnesia in the sintering of alumina, *J. Am. Ceram. Soc.*, 73:833 (1990).
18. D.-Y. Kim, S. M. Wiederhorn, B. J. Hockey, C. A. Handwerker, and J. E. Blendell, Stability and surface energies of wetted grain boundaries in aluminum oxide, *J. Am. Ceram. Soc.*, 77:444 (1994).

THE ORIGIN AND GROWTH KINETICS OF PLATE-LIKE ABNORMAL GRAINS IN LIQUID PHASE SINTERED BARIUM TITANATE

D. Kolar, A. Rečnik* and M. Čeh

Jožef Stefan Institute, University of Ljubljana, Jamova 39, 1001 Ljubljana, Slovenia

ABSTRACT

During sintering in an air atmosphere, barium titanate ceramics frequently exhibit exaggerated grain growth. Discontinuous grain growth is particularly rapid in the presence of the Ti-rich eutectic and results in coarse, equiaxed grains with broad size distribution. During sintering under reducing conditions, above the $BaTiO_3$-$Ba_6Ti_{17}O_{40}$ eutectic temperature, barium titanate ceramics exhibit characteristically exaggerated anisotropically grown grains, which were identified as the hexagonal polymorph, known to be formed above 1460°C in air, and ~1330°C in hydrogen. The hexagonal grains are platelets with high aspect ratio.

Change in morphology strongly depends on sintering temperature, impurity/dopants concentration, and amount of liquid phase. Kinetics studies confirmed that the liquid eutectic promotes the cubic-hexagonal transformations. The growth mechanism of hexagonal grains is dissolution of small cubic grains and precipitation on the hexagonal grains which preferentially grow in the direction of the prism planes. The initially high aspect ratio >50 decreases after grain impingement. Increase of the amount of liquid phase and increase of temperature increase the number of nucleation sites which results in smaller grain size after short sintering time.

INTRODUCTION

The abnormal[1] or discontinuous grain growth, sometimes referred to as secondary recrystallization, is a frequently observed phenomenon during sintering of ceramics. It is characterized by the rapid growth of a limited number of grains ('nuclei") to sizes much larger than that of the average grain population. In some systems the abnormal grains exhibit preferential growth in a particular direction, giving rise to rod-like or plate-like grain morphology.

In the manufacturing of ceramics, coarse-grained microstructures with anisotropic

---

*visiting scientist at the Max-Planck-Institut, Institut für Werkstoffwissenschaft, Seestrasse 92, 70174 Stuttgart, Germany

grains are usually undesirable since they adversely effect the properties, and reproducibility of the manufacturing process. For some applications, however, the grain growth anisotropy may be beneficial, as in the case of magnetically hard ferrites[2]. Recently, exaggerated growth of rod-like or plate-like grains in silicon nitride and silicon carbide engineering ceramics, raised considerable interest, making possible in-situ toughening of ceramics[3-9].

In spite of considerable interest, the phenomenon of anisotropic abnormal grain growth is not well understood. Possible reasons include[10]: (I) local density and pore size heterogeneity, (II) concentration heterogeneity of dissolved atoms, (III) partial wetting by liquid phase, (IV) anisotropic grain boundary energy, and (V) heterogeneous liquid phase distribution.

$BaTiO_3$ is one of the first ceramic materials for which anomalous grain growth was recorded[11]. Sintering of $BaTiO_3$ based materials is normally carried out with a small excess of TiO_2 as a sintering aid. The TiO_2 excess reacts with $BaTiO_3$ to produce $Ba_6Ti_{17}O_{40}$ which forms with $BaTiO_3$ an eutectic melt at about 1320°C. The liquid phase triggers discontinuous grain growth which results in coarse-grained structure.

The accepted mechanism for microstructure development is nucleation and growth from liquid phase, similar to the Ostwald ripening process[12-15]. The grains grown in air-sintered compacts are isotropic. Recently, it was reported that in a hydrogen atmosphere grains grow anisotropically with plate-like morphology[16].

The aim of this investigation was to analyze the mechanism of anisotropic exaggerated grain growth in $BaTiO_3$ in more detail. The system may well serve as a model system for anisotropic exaggerated grain growth in general.

EXPERIMENTAL

The starting material was high purity commercial grade $BaTiO_3$ (Transelco Code 219/9). The supplier's specification lists max. impurities Al_2O_3 50 ppm, CaO, SrO and SiO_2 100 ppm each, $BaO+SrO/TiO_2$ 0.995-1.000 and Coulter average grain size 1.3 -1.5 µm. Various amounts of $Ba_6Ti_{17}O_{40}$, prepared by calcining $BaTiO_3$ and TiO_2 at 1250°C and finely ground were added to the starting material and homogenized. The powder was pressed into pellets (ϕ = 6 mm, h = 3 mm) without binder and sintered in Ar-8 v/o H_2 gas mixture, moistened by passing through H_2O at 25°C. After 1 hour soaking at 1300°C (to ensure densification) the samples were sintered at 1340-1370°C for various times and rapidly cooled. The samples were characterized by x-ray diffraction (XRD), optical and electron microscopy. Digital processing and analysis was carried out to evaluate the microstructural features quantitatively.

X-ray diffraction patterns were recorded on a Philips 1710 PW diffractometer. The relative proportions of hexagonal $BaTiO_3$ in the samples were determined by a matrix-flushing method proposed by Chung[17].

RESULTS AND DISCUSSION

The abnormal anisotropic grains observed in the $BaTiO_3$ ceramics sintered under reducing conditions are a hexagonal polymorph of $BaTiO_3$[16]. Microstructure development depends on the nucleation of hexagonal grains and their growth. Microstructural characteristics such as grain size, anisotropy (aspect ratio) and grain size distribution are strongly influenced by the preparation conditions of the ceramics. The important parameters are (1) reducing atmosphere, (2) purity of starting $BaTiO_3$ powder, (3) amount of liquid phase at sintering temperature, (4) sintering temperature, and (5) time.

Microstructures of BaTiO$_3$ Ceramics Sintered in a Reducing Atmosphere

Figures 1a-1f show microstructures (at various magnifications) of pure BaTiO$_3$ ceramics, and BaTiO$_3$ ceramics with 2 m/o Ba$_6$Ti$_{17}$O$_{40}$, after sintering at 1340°C in a reducing atmosphere for various times. The microstructures of ceramics processed with shorter sintering times show bimodal size distributions and demonstrate anomalous anisotropic growth of a limited number of grains in a fine-grained matrix. The size of the cubic matrix grains is around 5 µm. Larger magnification (Figure 1e) reveals the presence of a solidified liquid phase among the grains and a relatively high quantity of smaller globular inclusions in the coarse anisotropic grains. By EDX and TEM analysis it was confirmed that the inclusions are either cubic BaTiO$_3$ crystallites or drops of solidified liquid phase with a composition close to Ba$_6$Ti$_{17}$O$_{40}$. Inclusions indicate a relatively high growth rate of anisotropic grains. The fine pores in the large grains are believed to be due to escaping O$_2$ from BaTiO$_3$. Prolonged sintering results in a coarse-grained microstructure with a less pronounced bimodal size distribution (Figure 1f). It was established previously[16] that the large anomalously grown anisotropic grains are the hexagonal polymorph of BaTiO$_3$, whereas small isotropic matrix grains are the tetragonal polymorph (cubic at sintering temperature). Elongated grains on 2-dimensional microphotographs are, in reality, plates, as demonstrated in the SEM picture (Figure 2). The grains grow preferentially in the direction of the prism planes.

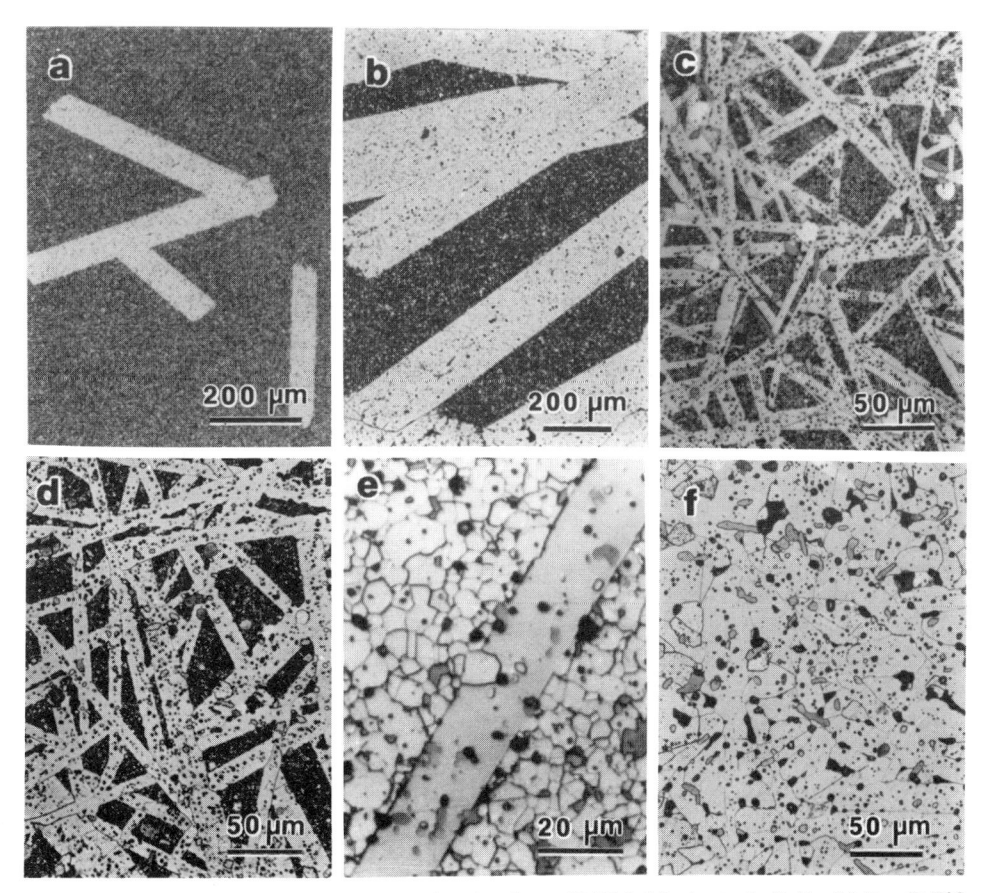

Figure 1. Microstructure of BaTiO$_3$ ceramics after sintering at 1340°C, 30 minutes in H$_2$/Ar. (a) Pure BaTiO$_3$ 30 minutes, (b) pure BaTiO$_3$ 110 minutes, (c) BaTiO$_3$ - 2 m/o Ba$_6$Ti$_{17}$O$_{40}$ 15 min, (d) and (e) 30 minutes, (f) 120 minutes. Note different magnifications.

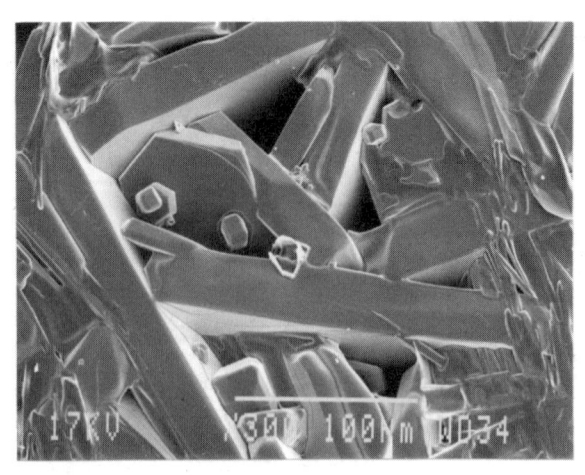

Figure 2. SEM micrograph of surface of $BaTiO_3$ + 2 m/o $Ba_6Ti_{17}O_{40}$ ceramics after sintering at 1400°C, 2 hours in H_2/Ar.

Role of the Reducing Atmosphere and BaTiO₃ Purity

Pure $BaTiO_3$ transforms from the cubic to hexagonal polymorph in an air atmosphere above 1430°C[18] or 1460°C[19]. In hydrogen, the transformation temperature is lowered to 1330°C[20]. This lowering of transformation temperature is regarded as essential for anisotropic growth of $BaTiO_3$ grains in a reducing atmosphere. Microstructure development based on a dissolution-precipitation mechanism demands the presence of hexagonal nuclei at the temperature of liquid phase formation. Without hexagonal nuclei the abnormal growth results in isotropic grains of the cubic polymorph, as observed in air-sintered samples. The possible appearance of hexagonal nuclei at higher temperature is less effective in inducing exaggerated growth due to an already coarsened isotropic matrix.

The temperature of cubic-hexagonal transformation critically depends on the purity of $BaTiO_3$. Conversion is inhibited by a number of impurities, especially Si, Fe and Sr such as are present in barium titanate of commercial purity[20]. In such powders anisotropic growth may not occur. In our experiments, the addition of 0.1 w/o SiO_2 to $BaTiO_3$ was sufficient to change the microstructure of $BaTiO_3$ sintered at 1340-1350°C from anisotropic to isotropic.

Kinetics of the Cubic-Hexagonal Transformation and Microstructure Development of BaTiO₃ in a Reducing Atmosphere

Kinetics of the C-H transformation at 1340°C was measured by quantitative XRD and by quantitative evaluation of microstructures. Figure 3 shows the degree of transformation vs. time for pure $BaTiO_3$ and $BaTiO_3$ with 2 m/o $Ba_6Ti_{17}O_{40}$. Both kinetic curves exhibit S-shape, typical for a phase transformation governed by nucleation and growth[21]. Pure $BaTiO_3$ displays an extended induction period. The transformation starts after 30 minutes and is completed after 120 minutes. The addition of $Ba_6Ti_{17}O_{40}$ to $BaTiO_3$ considerably reduces the observed induction period noticed for pure $BaTiO_3$ and enhances the process of transformation. A conversion of 50% to the hexagonal polymorph is noticed after only 20 minutes. Independent measurements of recrystallised areas in the sintered microstructures are in good agreement with the results of the quantitative XRD analysis (Figure 3) which confirms the hexagonal nature of the recrystallised grains. Quantitative evaluation of parallel sets of microstructures of $BaTiO_3$ + 2 m/o $Ba_6Ti_{17}O_{40}$ ceramics sintered isothermally at 1450°C revealed that the area density of recrystallized grains (nuclei) remains constant with time at approximately 52 ± 10/nuclei/mm². In the first

30 minutes, the average length increased approximately linearly with time with the rate of 10 μm/min, and then started to level off due to impingement of grains. For the same reason, the average aspect ratio (length/diameter) of grains which was ~14 after a short (10 minutes) sintering time, started to decrease and reached 8 after 30 minutes.

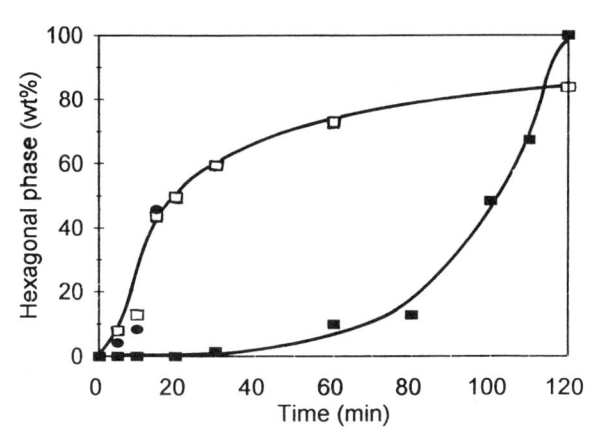

Figure 3. Percentage conversion to the hexagonal BaTiO$_3$ as a function of time at 1340°C for BaTiO$_3$ by XRD (■), BaTiO$_3$ + 2 m/o Ba$_6$Ti$_{17}$O$_{40}$ by XRD (□), and by area measurement (●).

The Influence of Liquid Phase on Microstructure Development

Beside the presence of hexagonal nuclei, liquid phase is essential for the exaggerated anisotropic grain growth in reduced BaTiO$_3$. Figure 4 shows the microstructure of pure BaTiO$_3$ ceramic sintered with a thin layer of Ba$_6$Ti$_{17}$O$_{40}$ powder on the top of the pellet at 1340°C. Upon formation of the eutectic liquid, this liquid penetrated the pellet to some

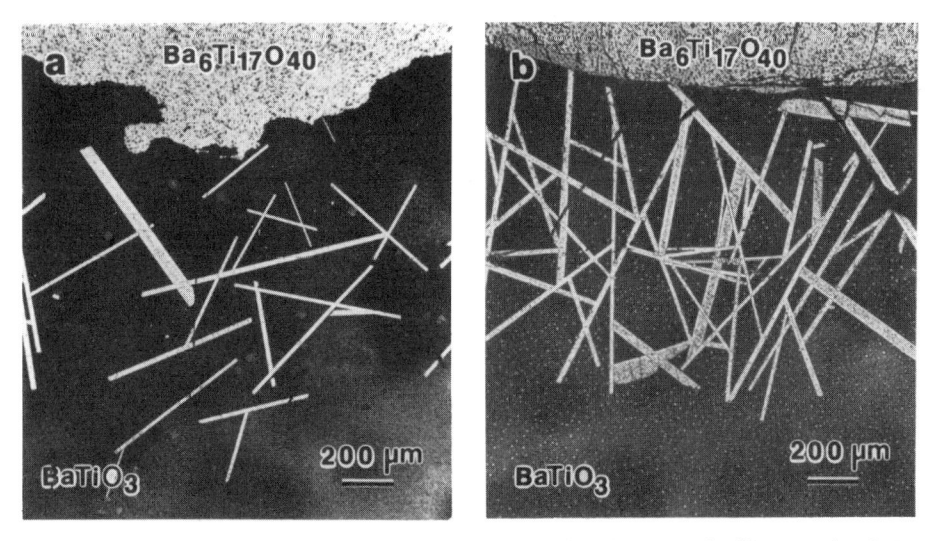

Figure 4. The influence of Ba$_6$Ti$_{17}$O$_{40}$ layer on recrystallization of BaTiO$_3$ (growth of hexagonal grains). (a) 15 minutes and (b) 30 minutes at 1340°C.

extent, thereby increasing the number of nucleation sites and accelerating the anisotropic growth. Maximum growth rate in preferential directions, estimated from pictures, was found to exceed 30 µm/min. The maximum aspect ratio exceeds 50.

Figures 5a-5c demonstrate the influence of increasing amount of liquid phase on microstructure development of $BaTiO_3$ at 1340°C, sintered for 30 minutes. Exaggerated anisotropic growth above the temperature of eutectic formation is also observed in "stoichiometric" $BaTiO_3$. It is believed that local heterogeneity may lead to occasional formation of an eutectic melt, which creates a few nucleation sites and starts the growth of a limited number of anisotropic grains. Those grains may grow unhindered to very large dimensions (Figures 1a, b). Rapid growth of a few exaggerated grains in stoichiometric $BaTiO_3$ was also reported[15], whereas such phenomenon was not observed in BaO-excess $BaTiO_3$.

Diagrams 6a,b show the dependence of density of discontinuously growing grains and equivalent area diameter as a function of the amount of $Ba_6Ti_{17}O_{40}$ after 30 minutes sintering at 1340, 1350 and 1360°C. The expected trends may be distinguished. The number of nucleation sites increases with the amount of liquid phase and temperature. This may be explained as a consequence of better distribution of the liquid phase in the structure. With increasing number of grains growing the average diameter decreases due to the grain impingement.

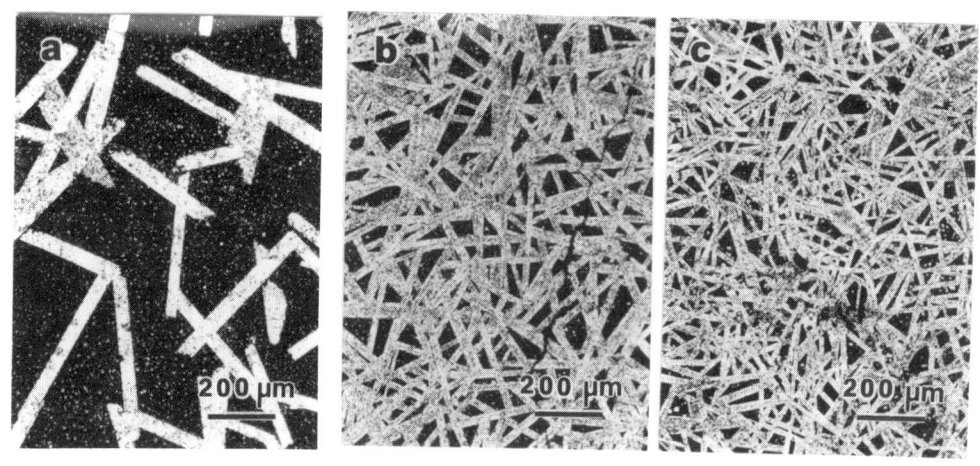

Figure 5. The influence of the amount of $Ba_6Ti_{17}O_{40}$ forming eutectic with $BaTiO_3$ on the microstructure after 30 minutes sintering at 1340°C: (a) 0.5 m/o, (b) 2 m/o, (c) 4 m/o $Ba_6Ti_{17}O_{40}$.

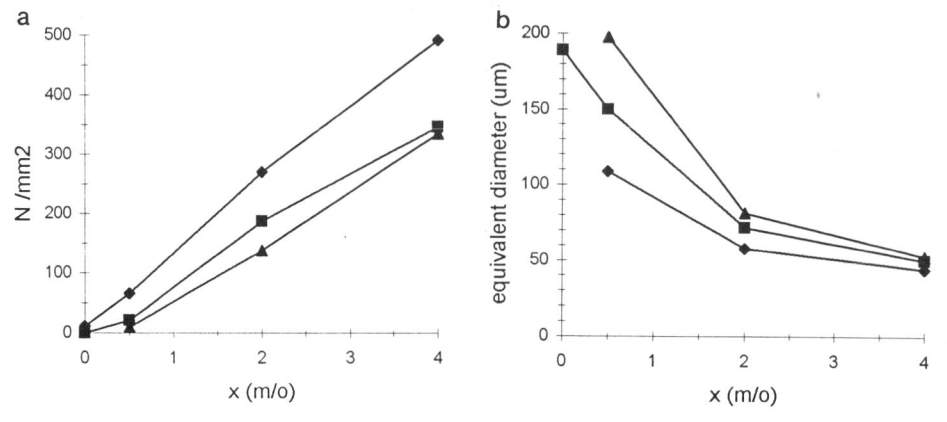

Figure 6: (a) Area nuclei density and (b) equivalent nuclei diameter in $(1-x)BaTiO_3-xBa_6Ti_{17}O_{40}$ ceramics as a function of x, sintered 30 minutes at 1340 (▲), 1350 (■) and 1360°C (♦).

Mechanisms of Exaggerated Grain Growth

Exaggerated growth of isotropic $BaTiO_3$ grains with a cubic crystal structure in an oxidizing atmosphere has been ascribed to the presence of parallel (111) twins[22]. According to the proposed mechanisms, the fast growth of twins is pronounced along the layers of face shearing octahedra at so called reentrant angles. Similarly, a discontinuous growth along the layers of Ti_2O_9 groups may occur in hexagonal $BaTiO_3$[16], though at a much higher rate since there are infinitely more reentrant angles in a hexagonal structure than in a singular case of parallel (111) twins in the cubic $BaTiO_3$. The formation of face sharing octahedra in the form of Ti_2O_9 groups is supposed to be initiated by Ti^{3+} ions, created in reducing atmosphere. The origin of hexagonal nuclei is not clear at present. A limited number of hexagonal nuclei may exist already in starting $BaTiO_3$ powder as a consequence of the calcination step, or they may originate from instability of the cubic $BaTiO_3$ in a reducing atmosphere.

SUMMARY

Barium titanate ceramic sintered under reducing conditions, above a $Ba_6Ti_{17}O_{40}$ eutectic temperature, exhibits exaggerated anisotropically grown grains. The recrystallization critically depends on impurities, which increase the temperature of the cubic-hexagonal transition. The anisotropic grains are an hexagonal polymorph and the growth mechanism is dissolution of small cubic grains and precipitation on to the hexagonal grains which preferentially grow in the direction of prism planes. The initially high aspect ratio (maximum observed >50) decreases after grain impingement. With increasing amount of liquid phase and temperature in the region of 1340-1360°C, the frequency of nucleation sites increases resulting in smaller grain sizes at short sintering times. Ti^{3+} ions, induced by reducing atmosphere, play an important role in the formation of hexagonal nuclei.

REFERENCES

1. M. Hillert, On the theory of normal and abnormal grain growth, *Acta Metall.*, 13(3):227-38 (1965).
2. A.L. Stuijts, Sintering of ceramic permanent magnetic material, *Trans. Brit. Ceram. Soc.*, 55: 57 (1956).
3. E. Tani, S. Umebayasni, K. Kishi, K. Kobayashi and M. Nishijima, Gas-pressure sintering of Si_3N_4 with concurrent addition of Al_2O_3 and 5 wt. % rare-earth oxide: high fracture toughness Si_3N_4 with fibre-like structure, *Am Ceram. Soc. Bull.*, 65(9):1311-15 (1986).
4. C.J. Hwang and T.Y. Tien, Microstructural development in silicon nitride ceramics, *Materials Science Forum* 47:84-109 (1989).
5. F. Wu, H. Zhuong, L. Ma and X. Fu, Self-reinforced silicon nitride by gas-pressure sintering, *Ceram. Eng. Sci. Proc.*, 14(1-2):321-32 (1993).
6. W. Dressler, H.J. Kleebe, M.J. Hoffmann, M. Rühle and G. Petzow, Model experiments concerning abnormal grain growth in silicon nitride, *J. Eur. Ceram. Soc.*, 16:3-14 (1996).
7. N.P. Padture, In situ toughened silicon carbide, *J. Amer. Ceram. Soc.*, 77(2): 519-23 (1994).
8. M.A. Mulla and V.D. Krstic, Pressureless sintering of β-SiC with Al_2O_3 additions, *J. Mater. Sci.*, 29: 934-38 (1994).
9. Y.W. Kim, M. Mitomo and H. Hirotsuru, Grain growth and fracture toughness of fine-grained silicon carbide ceramics, *J. Amer. Ceram. Soc.*, 78(11):3145-48 (1995).
10. P. Greil, pp. 68-70 in: *Keramik,* H. Schaumburg, ed., Teubner, Stuttgart (1994).
11. J.E. Buke, Grain growth in ceramics, pp. 109-116 in: *Kinetics of High Temperature Processes,* W.D. Kingery, ed., Wiley, New York (1959).

12. D.G.K. Hennings, R. Jansen and P.J.L. Reynen, Control of liquid-phase enhanced discontinuous grain growth in barium titanate, *J. Amer. Ceram. Soc.*, **70**(1):23-27 (1987).
13. D. Kolar, Discontinuous grain growth in multiphase ceramics, pp. 529-545 in: *Ceramic Transactions*, Vol. 7, C.A. Handwerker, J.E. Blendell, W.A. Kayser, eds., Am. Ceram. Soc. (1990).
14. J. Chen, W. Jin and Y. Yao, Study of the anomalous grain growth of $BaTiO_3$ ceramics, Ferroelectrics **142**: 153-159 (1993).
15. T. Yamamoto and T. Sakuma, Influence of small cation nonstoichiometry on the grain growth of $BaTiO_3$, *Eur. J. of Solid State Inorg. Chem.*, **32**:731-740 (1995).
16. A. Rečnik and D. Kolar, Exaggerated growth of hexagonal barium titanate under reducing sintering conditions, J. Am. Ceram. Soc., **79**(4):1015-18 (1996).
17. F.H. Chung, Quantitative Interpretation of X-ray Diffraction Patterns of Mixtures. I. Matrix-Flushing Method for Quantitative Multicomponent Analysis, J. Appl. Cryst., **7**:519-525 (1974).
18. K.W. Kirby and B.A. Wechsler, Phase relations in the barium titanate-titanium oxide system, *J. Am. Ceram. Soc.*, **74**:1841-47 (1991).
19. D.E. Rase and R. Roy, Phase equilibria in the system $BaO-TiO_2$, *J. Am. Ceram. Soc.*, **38**(3):102-13 (1959).
20. R.M. Gleister and H.F. Kay, An investigation of the cubic-hexagonal transition in barium titanate, *Proc. Phys. Soc.*, **76**:763-771 (1960).
21. C.N.R. Rao and K.J. Rao, Phase transformations in solids, pp. 131-185 in: *Progress in Solid State Chemistry*, 4, H. Reiss, ed., Pergamon, Oxford (1967).
22. H. Schmelz and H. Thomann, Twinning in $BaTiO_3$ ceramics, *Ceram. Forum Int.*, **61**:199-205 (1984).

MICROSTRUCTURAL CONTROL OF ZINC OXIDE VARISTOR CERAMICS

Richard C. Bradt* and Susan L. Burkett^

Metallurgical and Materials Enginering*
Electrical and Computer Engineering^
The University of Alabama
Tuscaloosa, AL 35487 USA

INTRODUCTION

It has been nearly a quarter of a century since Matsuoka (1) reported the non-ohmic properties of ZnO ceramics. During that period ZnO-based ceramics with their highly non-linear current-voltage characteristics have accrued more than half of the varistor market in numbers and nearly 80% of the market in value as reported by Okinaka and Hata (2). These electronic devices, often referred to as MOV's, or metal oxide varistors, find applications as surge absorbers, surge arrestors and in micromotors. Although these ZnO-based devices probably have their single largest volume usage in high-voltage power distribution systems, it is becoming increasingly common to protect household applicances and personal computers with varistor devices. It is likely that most readers of this article utilize such a device to protect their personal computers.

The non-ohmic characteristics of ZnO ceramics are directly related to the electronic states at the grain boundaries of these polycrystalline ceramic devices. ZnO varistors are classical grain-boundary-active electronic ceramic devices. Although perhaps minor controversy exists relative to some of the issues related to the device physics, it has been intensively studied and is generally agreed to have many characteristics in common with Schottky barriers formed at metal-semiconductor junctions (3). Even though Seitz, et al. (4) report that different grain boundaries in these MOV's function differently, the breakdown voltage of each individual grain boundary in a polycrystalline ZnO varistor is generally agreed to be about 3 V. If one considers a polycrystalline ZnO varistor as the series sum of the individual grain boundary barriers, then it is evident that the breakdown voltage of the device depends on its thickness and the grain, or crystal size of the ZnO microstructure. As Chiang et al. (5) point out, a 1 mm thick varistor with a 10 µm average grain size will exhibit a breakdown voltage of about 300 V. Obviously, the grain growth control of the ZnO component of varistor formulations during the manufacturing process is an absolute necessity.

Unfortunately, commercial ZnO varistor formulations are extremely complex from a chemistry perspective. This has been noted by both Gupta (6) in his comprehensive review and also by Olsson (7) in her thesis. What is evident, however, is that the numerous additive oxides do segregate to the ZnO grain boundaries and assume critical roles in the development of the electrical properties of ZnO varistors. Empirical varistor recipes have developed as a consequence of trial and error types of cookbook studies, often without any fundamental knowledge of the specific electronic character of the segregated oxide additions at the ZnO polycrystalline grain boundaries, nor even the details of the actual segregation process. This grain boundary segregation is expected to affect the grain boundary mobility and thus directly determine the grain growth characteristics of the ZnO in the varistor formulations. It is evident that there exists a strong interdependency between the effects of the compositional formulations of ZnO-based varistors and both the electrical characteristics and the microstructure of the ceramic during the processing of these electronic ceramics.

It is appropriate to briefly review the three general categories of chemical oxide additives that are utilized in most commercial ZnO varistor formulations. These are the varistor-action oxides, specific current-voltage modifiers and the spinel forming grain growth inhibitors. In the first category are BaO, Bi_2O_3, Pr_6O_{11} and La_2O_3. It is the grain boundary segregation of these rather large cation oxides that is directly responsible for the varistor action of the ZnO ceramics, as has been discussed by Cordaro, et al. (8) and Alim (9). Generally, the cations from these oxides are much too large to exhibit any significant solubility in the ZnO wurtzite structure, thus it is to be expected that they will segregate to the grain boundaries. As evident in the studies of Hwang, et al. (10) for Bi_2O_3 additions and Alles and Burdick (11) for additions of Pr_6O_{11}, these oxides also assume critical roles in the liquid phase sintering phenomenon during the ZnO varistor manufacturing process. As reviewed by German (12), microstructural coarsening, or grain growth is a normally expected process during the latter stages of liquid phase sintering. It might be expected that in addition to their varistor action effects, the above oxides will most certainly also directly affect the grain growth characteristics of the ZnO.

The second group of additive oxides that are frequently contained in commercial ZnO varistor formulations are specific current-voltage modifiers. The 3-d transition metal oxides are believed to directly affect the leakage current and also contribute to the current-voltage characteristics. These oxides may be expected to exist in some form of distribution between segregation at the grain boundaries and solution within the wurtzite structure of the ZnO grains. The distribution coefficients are not known with any certainty. The group of oxides including Al_2O_3, Ga_2O_3 and In_2O_3 all have cations that are soluble in the ZnO wurtzite structure and function as donors to increase the conductivity of the ZnO grains. Alkali oxides are also known to be present (13,14). Unfortunately, detailed defect chemistry studies of most of the aforementioned oxide additions to ZnO have never been completed, thus the precise roles of these oxide additions are not fundamentally understood.

The grain growth inhibitor additions to ZnO are those oxides which generally react with the ZnO to form the spinel phase at the grain boundaries, thus creating a grain boundary drag mechanism which reduces the rate of grain growth (15). The general reaction for the formation of spinels may be written as follows:

$$AO + B_2O_3 = AB_2O_4.$$

However, in the case of numerous oxide additions to ZnO in the varistor formulations, the resulting spinels may be defect spinels with vacancies compensating for aliovalent cations. Whereas the normal variety of spinel may be formed with any Al_2O_3 which is present, the addition of Sb_2O_3 yields the $Zn_7Sb_2O_{12}$ spinel, which is clearly of the defect variety. It is not unreasonable to expect that with the multicomponent formulations of several different spinel forming oxide additives in many commercial ZnO varistors, solid solutions of several different cations within the spinels may exist within the microstructure of a commercial ZnO varistor.

The aforementioned complexity presents a difficult and multifaceted problem when addressing the grain growth of ZnO in varistor ceramics, for the numerous components are certain to exhibit interdependencies as well as individual effects. An obvious route to assess some of those effects is to first consider the grain growth of "pure" ZnO, then apply some of the same considerations to commercial ZnO varistor formulations where data is available, just to establish the appropriateness of the analysis, and finally to systematically address the effects of the many additives on an individual basis. Admittedly, the latter is a challenging undertaking, but fortunately there exist some systematic trends that are enlightening.

GRAIN GROWTH IN POLYCRYSTALLINE CERAMICS

The phenomenon of grain growth in polycrystalline ceramics during the sintering process of powders is an extremely important one, for many different properties of ceramics depend on the final grain size of the ceramic body. Two different approaches have been applied to study this phenomenon and both have merit. The most fundamental approach is to address the mobility of individual grain boundaries within the microstructure. This has an advantage for ceramic bodies with large grain sizes where an individual grain boundary can actually be directly monitored. However, in the instance of polycrystalline ZnO varistor ceramics, it is usual to begin with very fine powders, typically 0.10 µm in diameter, or less. These may grow to sizes of 10 µm or larger during processing. Thus a single grain in the final ceramic device may consist of 1,000,000 or more original particles and may have contained a similar number of different grain boundaries. For this reason the average grain size approach seems to be the more appropriate instead of attempting to track any single grain boundary within the ZnO microstructure.

Grain growth in a polycrystalline material can be described by the kinetic grain growth equation which consists of several descriptive parameters (16). It is usually written as follows:

$$G^N - G_0^N = K t \exp(-Q/RT).$$

where G_0 is the initial grain size and G is the grain size after growth for the time t at the absolute temperature T. The term K is a geometrical preexponential constant. The two parameters, N and Q, are the grain growth exponent and the activation energy for grain growth, respectively. These can be determined through activation analysis of the grain growth process and are considered characteristic of the mechanism of the grain growth phenomenon.

In actual practice, an isothermal grain growth experiment is conducted by maintaining the specimen at a given temperature and measuring the grain size as a function of time. If the initial grain size is small relative to the final grain sizes for the experimental times, then the term containing G_0^N can be ignored and the above equation can be rewritten in the form:

$$\log G = (1/N) \log t + (1/N)\log K \exp(-Q/RT),$$

from which the grain growth exponent can be determined from the slope of the (log G) versus (log t) plot, as N is the reciprocal of the slope. Once the N value has been determined for the system of interest, then the activation energy can be determined from an Arrhenius plot of $\log(G^N/t)$ versus $(1/T)$ as described by rearranging the terms of the above equation as follows:

$$\log (G^N/t) = \log (K) - (Q/RT).$$

The activation energy, Q, can be easily determined from the slope. The geometrical constant,

K, is readily determined from the intercept. This type of grain growth activation analysis has been previously applied to numerous systems from pure metals to compex multiphase minerals with equally successful results (17,18). Usually the activation energy for grain growth is equivalent to, or nearly equivalent to that for diffusion of one of the atomic or ionic species, the rate controlling one for the grain growth process.

GRAIN GROWTH IN PURE ZINC OXIDE

Fortunately, there have been a number of fundamental studies published on the grain growth of "pure" zinc oxide, where the term pure is used simply because it is relativey easy to obtain very high ZnO purity levels through the oxidation of multiply distilled zinc metal. The 1960's publications of Gupta and Coble (19,20) established the baseline to which all future studies have been referenced. This body of studies has been reviewed most recently by Senda and Bradt (21). It is the latter data which is summarized in the figures below.

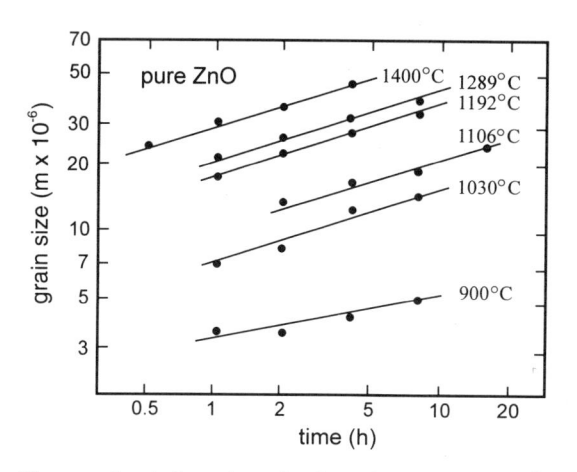

Figure 1. The (log G) versus (log t) plot to determine the grain growth exponent for "pure"ZnO. (21)

In the left graph, the grain sizes for isothermal sintering are plotted for different firing times at a series of different sintering temperatures to determine the grain growth exponent. For all temperatures between 1030°C and 1400°C, the slope is 1/3, yielding a grain growth exponent of 3. The results for 900°C do yield a slope that is less than 1/3 and suggest an exponent larger than 3, indicative of a slower rate of grain growth. It is not understood exactly why the lower temperature yields a larger exponent indicative of a slower rate, but perhaps it is because of the lack of densification and an increased drag on the individual grain boundaries. It is significant that other researchers, including Nicholson (22), Gupta and Coble (19,20), Dutta and Spriggs (23) and Readey et al. (24), also observe a grain growth exponent of 3 even though their ZnO powders were different purities and different particle sizes. As 3 is a grain growth exponent that can be classically derived from fundamental principles as the exponent for grain growth of a single phase material, it is an acceptable experimental result for pure ZnO.

The Arrhenius plot in Figure 2 is constructed by incorporating the grain size exponent of 3 into the data and replotting as log (G^3/t) versus $(1/T)$ to obtain the activation energy for the ZnO grain growth process. Interestingly, when the results obtained at 900°C are incorporated into this form of presentation, they are in agreement with the extension of the higher temperature data, even though the original 900°C results did not appear to ascribe to the exponent of 3. The resulting activation energy is about 224 kJ/mol in general agreement with

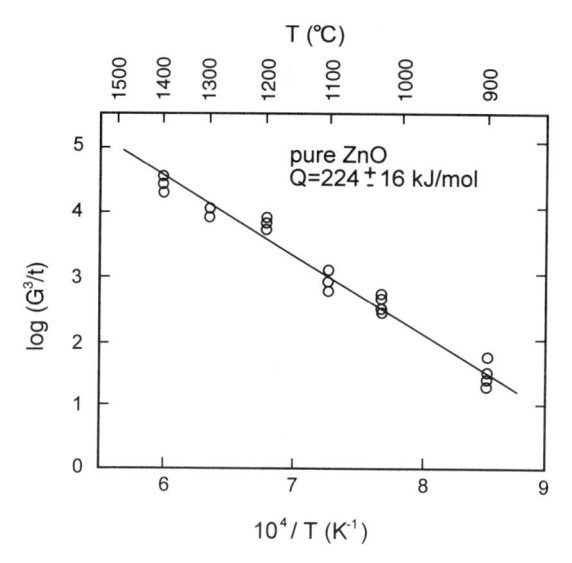

Figure 2. An Arrhenius plot of the data in Figure 1 as (log G³/t) versus (1/T) to determine the activation energy for grain growth in "pure" ZnO. (21)

previous investigators and supportive of the mechanism proposed by Gupta and Coble of interstitial zinc diffusion as the rate controlling mechanism for ZnO grain growth.

With the activation parameters of N = 3 and Q = 224 kJ/mol established for "pure" ZnO grain growth, it is now possible to examine both commercial-like ZnO varistor formulations and specifically designed binary and ternary oxide systems based on ZnO to assess their grain growth characteristics and activation parameters for a basis of direct comparison.

GRAIN GROWTH IN COMMERCIAL-LIKE VARISTOR FORMULATIONS

It is not suprising that commerical producers of ZnO-based varistors have not published the ZnO grain growth characteristics of their products. However, activation analysis of the grain growth of ZnO as outlined above is little more than an academic exercise if it cannot be related to actual commercial ZnO varistor formulations. Chen, Shen and Wu (25) have applied the above activation analysis approach to simulated varistor compositions which include Bi_2O_3, Sb_2O_3, Cr_2O_3, CoO and MnO additions. Of course the valences and the phase assemblages of the compositions most certainly changed from the starting mixture during the firing of the ZnO ceramic. However, these general compositions are certainly sufficiently complicated enough to resemble actual ZnO varistor formulations.

Applying a similar activation analysis as that reported above for the "pure" ZnO, Chen, Shen and Wu (25) reported several significant differences, while generally confirming the technical validity of the experimental approach. First they observed that the grain growth exponent was 6 for each of the four different compositions that they studied. An N value of 6 is indicative of a significantly reduced rate of grain growth when compared to the various "pure" ZnO ceramics which have been studied and all of which exhibit an N value of 3. The various oxide additives obviously produce reduced rates of grain growth compared with the "pure" ZnO materials. Also significant in the study of Chen, Shen and Wu is the activation energy for these complex varistor formulations. They report an activation energy for ZnO grain growth of 301 kJ/mol for the multicomponent ceramics, significantly larger than the 224 kJ/mol value for the "pure" ZnO reported by Senda and Bradt (21).

From this study of Chen, Shen and Wu (25) it must be concluded that the fundamental activation analysis approach that identifies with pure materials is also generally applicable to the compex formulations of ZnO varistor ceramics. Furthermore, when applied to these commercial varistor-like compositions, the grain growth exponent, or N value is increased from 3 to 6 and the activation energy for grain growth also increases substantially. These results suggest that the rate controlling mechanism for ZnO grain growth is not the same in these complex varistor-like multicomponent ZnO formulations as it is for "pure" ZnO.

GRAIN GROWTH IN BINARY ZnO - Bi_2O_3 COMPOSITIONS

Although several different oxide additives may impart varistor action to ZnO through segregation at the grain boundaries, the use of Bi_2O_3 is practically dominant in the commercial sector for its dual role in aiding densification through the promotion of liquid phase sintering. (26,27) The electron microscopy studies of Olsson (7), Clarke (28) and Kingery, et al. (29) all illustrate the development of a continous network of a Bi_2O_3-rich phase at the ZnO grain boundaries in sintered ZnO varistor ceramics. The role that this grain boundary liquid layer phase assumes in the grain growth process of the ZnO during liquid phase sintering is thus of primary importance, for most commercial ZnO varistor formulations are processed under these liquid phase sintering conditions.

The liquid phase sintering process and subsequent coarsening has been clearly detailed by Lay (30). He illustrates that two processes can be rate controlling for the coarsening process as smaller grains are dissolved in the grain boundary liquid phase and then reprecipitated on the larger particles, thus creating general microstructural coarsening. The two possible rate controlling processes are: (i) the phase boundary reaction between the particles and the grain boundary liquid phase, and (ii) the diffusion of the solute through the liquid phase. The former shows no dependence of the rate of grain growth on the grain boundary liquid phase content, whereas the latter shows a grain growth rate dependence that is inverse to the grain boundary liquid phase thickness, or content. The wider the grain boundary liquid phase, the slower the rate of grain growth.

Two studies specifically address the role of the Bi_2O_3 content on the ZnO grain growth using an activation analysis approach. Senda and Bradt (21) and Dey and Bradt (31) investigated overlapping compositional ranges, the former 1 - 4 weight % and the latter 3 - 12 weight %. In the 1 - 4 weight % range, the ZnO grain growth exponent was observed to be 5, similar to the N value of 6 observed for the commercial varistor compositions sintered in the presence of a Bi_2O_3-rich liquid grain boundary phase by Chen, Shen and Wu (25). The activation energy for grain growth was consistently about 150 kJ/mol, significantly less than the 224 kJ/mol value for "pure" ZnO. In this range, the grain size increased with the initial addition of the Bi_2O_3, but then remained nearly constant. Over the 3 - 12 weight % compositions the ZnO grain size decreased with increasing Bi_2O_3 content, which is equivalent to a grain size decrease with an increase in the width of the Bi_2O_3-rich grain boundary liquid phase. The grain growth exponent was still 5, but the activation energy for ZnO grain growth increased to about 270 kJ/mol. It is evident that there is a rate controlling mechanism change for the ZnO grain growth from the lower to the higher Bi_2O_3 contents in this binary system and that the mechanism in both regions is different from that in "pure" ZnO.

The microstructural coarsening mechanism, or ZnO grain growth rate controlling mechanism for the higher liquid phase contents in this system is most certainly one of diffusion through the Bi_2O_3-rich grain boundary liquid phase. That is consistent with the decreasing ZnO grain size with increasing liquid phase content during sintering. Unfortunately, there is no fundamental data published on the diffusion of ZnO, or any species thereof, through a Bi_2O_3-rich liquid phase similar to that which exists at the grain boundaries, thus the activation energy cannot be directly compared. As the activation energy does not vary, nor does the ZnO grain size very much at the lower Bi_2O_3 levels, it is logical to conclude that the phase boundary reaction between the solid ZnO grains and the Bi_2O_3 - rich grain boundary liquid phase is the rate controlling mechanism. Again the N value of 5 is fully consistent with ZnO grain growth in Bi_2O_3-rich grain boundary liquid phase containing commercial ZnO varistor formulations.

If these results are compared directly with those of the varistor-like compositions previously studied by Chen, Shen and Wu (25), then it is evident that the rate controlling process for ZnO grain growth in commercial ZnO varistor formulations containing a Bi_2O_3-rich liquid phase must be different from that in solid state sintered ZnO. The grain growth exponent is larger, 5 or 6 as opposed to 3 and the activation energy is different, too, much larger than the value for "pure" ZnO as summarized for different investigators (21).

ZnO GRAIN GROWTH IN ADDITIVE OXIDE SPINEL FORMING SYSTEMS

The initial systematic study of the effect of a spinel forming additive on the grain growth of ZnO was a complete solid state binary system investigation of the addition of Sb_2O_3 to ZnO (15). The result was a significant reduction of the ZnO grain size during firing and the formation of the $Zn_7Sb_2O_{12}$ spinel in the form of small crystals located at the ZnO grain boundaries. It is apparent that the presence of the spinel particles on the grain boundaries results in a Zener-drag effect reducing the grain boundary mobility and thus the rate of ZnO grain growth (32). Similar to the observations for the commercial ZnO varistor compositions, the grain growth exponent, or N value was 6 and the activation energy increased well beyond that for pure ZnO to nearly 600 kJ/mol. The presence of this spinel is certainly a grain growth inhibitor for ZnO as both the grain growth exponent and the activation energy for grain growth increase substantially with the antimony oxide addition.

Another factor develops within the ZnO microstructure with the addition of Sb_2O_3. It is the formation of a single inversion twin within each of the ZnO grains. This structural feature has been explained as several planes of cubic spinel stacking sequence within the hexagonal wurtzite structure (33). The similarity of the (111) plane atomic arrangement of the spinel structure and the (0001) of the wurtzite structure allows the twins to develop when the appropriate spinel forming oxide is added to ZnO. The crystallographic restrictions of the wurtzite inversion twin allows just head to head configurations, thus only a single twin can be accomodated within each grain. Although the effects of these twins on the electrical properties of varistors has received some speculation, their role remains unexplained.

Two ternary systems of spinel forming additions to the binary $ZnO-Bi_2O_3$ system have been systematically investigated (34,35). Those are the separate additions of Al_2O_3 and TiO_2 which yield the normal $ZnAl_2O_4$ spinel and the Zn_2TiO_4 spinel, respectively. Although the latter system is complicated by the initial formation of the $Bi_4Ti_3O_{12}$ compound, it decomposes and the zinc titanate spinel is formed above 1050°C. The formation of the spinel phase in both of these systems has a similar effect of grain growth inhibition as the zinc antimonate spinel in the solid state study previously described, even though liquid phase sintering occurs. The grain growth exponents are 4 and 6, respectively, both greater than the value of 3 for the solid state grain growth of "pure" ZnO and similar to the value for the complex commercial varistor formulations. The activation energies for grain growth of the "pure" ZnO in the two ternary systems are also significantly greater than the values for "pure" ZnO. The alumina system exhibits an activation energy for ZnO grain growth of about 400 kJ/mol while the titania containing system is about 360 kJ/mol, compared with the value of 224 kJ/mol for the ZnO. One must conclude that the addition of spinel forming additives, whether in the solid state, or in the presence of a Bi_2O_3-rich grain boudary liquid phase during firing, has a strong inhibition effect on the grain growth of ZnO. The spinel increases the grain growth exponent and also increases the activation energy for the grain growth of ZnO well above the experimental values for "pure" ZnO. Furthermore, the activation energies for the inhibition of ZnO grain growth in the spinel forming systems is specific to the individual spinel which forms in each particular system.

The activation energy for ZnO grain growth is spinel specific in the different spinel-containing multicomponent systems, including those containing a Bi_2O_3-rich grain boundary liquid phase. This fact suggests that the mechanism of ZnO grain growth inhibition in those systems must be specifically related to some aspect of the spinel phase itself. It must be related to the Zener-drag process involving the spinel particles at the ZnO grain boundaries, grain boundaries which are coated with a Bi_2O_3-rich liquid phase during the sintering and grain growth process at the ceramic firing temperatures (32). Figure 3 illustrates some of the potential diffusion paths for the motion of a ZnO grain boundary with a Bi_2O_3-rich grain boundary liquid phase and a spinel particle located within the liquid grain boundary phase, and perhaps even spanning that grain boundary phase.

Five possible processes are suggested for controlling the motion of the grain boundary with the spinel particle attached from position (i) to (ii). These include: (1) diffusion within the ZnO lattice, (2) the phase boundary reaction at the particle-liquid grain boundary interface, (3) diffusion through the Bi_2O_3-rich grain boundary liquid phase, (4) surface, interface or grain boundary diffusion, and (5) diffusion within the spinel particle lattice itself. Nunes and Bradt (34) address each of the possibilities for the alumina containing system, concluding that it is the diffusion within the $ZnAl_2O_4$ spinel particles located at the grain boundaries that controls the grain growth of the ZnO in that system. Although the activation analyses for diffusion through most of the suggested paths have never been completed, so that there are no experimental activation energies for most of the suggested possible paths, there exists data for processes (1) and (2) and data for diffusion in the zinc aluminate spinel. It is the collective consideration of those results that leads to the spinel particle diffusion conclusion. The activation energies for the other two spinel containing systems supports that conclusion, for both are different and both exceed the reported activaton energy for the grain growth of "pure" ZnO.

CONCLUSIONS

The microstructural control of ZnO in the manufacturing process is a critical aspect for the use of ceramic systems based on ZnO as varistors. From a materials engineering

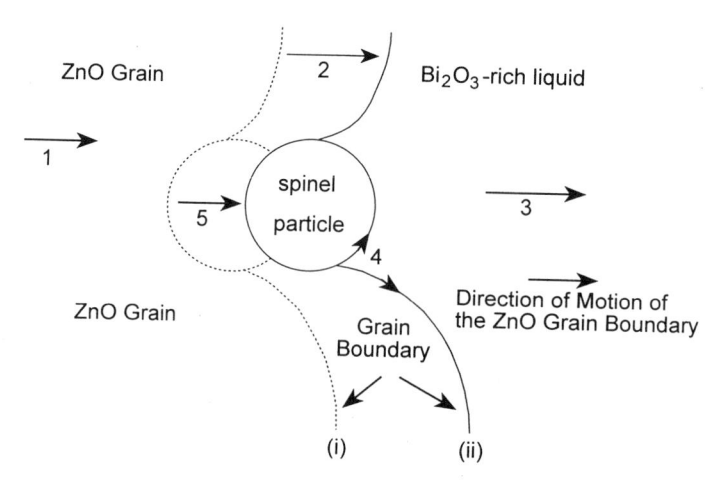

Figure 3. A schematic of a migrating ZnO grain boundary with a spinel particle attached.

perspective, the grain growth of ZnO is well understood in "pure" ZnO and in some simple varistor ceramic systems. For example, the procedures to limit the grain growth of ZnO with chemical additions of oxides that form spinel phases at the grain boundaries is basically understood. The above understanding has enabled the technical community to design functional, reliable, ZnO-based ceramic varistor formulations. It can be stated with confidence that the ceramic/materials engineering aspects of these varistors is with significant understanding. However, when the situation is taken a step futher and questions are asked regarding the detailed chemical aspects of the formulations and the electronic states of the individual cation constituents, then the understanding is at a considerably less comfortable level.

ACKNOWLEDGEMENTS

The authors acknowledge the contributions to this manuscript from many different individuals, primarily in the form of technical discussions regarding the processing and grain growth of commercial ZnO varistors. These include M. Alim, D. Dey, M. Green, T. Gupta, M. Helfand, L. Levinson, S. Nunes, T. Senda, H. Suzuki and T. Yamaguchi. We apologize if we inadvertently omitted anyone.

REFERENCES

1. M. Matsuoka, "Nonohmic properties of zinc oxide ceramics", Jpn. J. Appl. Phys. 10 (6),736 (1971).
2. H. Okinaka and T. Hata, "Varistor, thermistor manufacturing in Japan", Amer. Cer. Soc. Bull. 74 (2), 62 (1995).
3. L. M. Levinson and H. R. Philipp, "Zinc oxide varistors - a review", Amer. Cer. Soc. Bull. 65 (4), 639 (1986).
4. M. A. Seitz, A. K. Verma and R. W. Hirthe, "AC electrical behavior of individual MOV grain boundaries", Cer. Trans.:Adv. in Var. Tech. 3, 135 (1989).
5. Y-M. Chiang, D. P. Burnie III and W. D. Kingery, 225 in "Physical Ceramics", J. Wiley, N.Y., N.Y. (1997)
6. T.K. Gupta, "Influence of microstructure and chemistry on the electrical characteristics of ZnO varistors", 493 in "Tailoring Multiphase and Composite Ceramics", Plenum Pub. Corp., N.Y., N.Y. (1986).
7. E. Olsson, "Interfacial microstructure in ZnO varistor materials", Ph.D. Thesis, Dept. of Physics, Chalmers University of Technology, Goteborg, Sweden (1988).
8. J. F. Cordaro, Y. Shim and J. E. May, "Bulk electron traps in ZnO varistors", J. App. Phys. 60 (2), 4186 (1986).
9. M. Alim, "Influence of intrinsic trapping on the performance characteristics of ZnO-Bi$_2$O$_3$ based varistors", Active and Passive Elec. Comp. 17, 99 (1994).
10. J-H Hwang, T. O. Mason and V. P. Dravid , "Microanalytical determination of ZnO solidus and liquidus boundaries in the ZnO-Bi$_2$O$_3$ system", J. Amer. Cer. Soc. 77 (6),1499 (1994).
11. A. B. Alles and V. L. Burdick, "The effect of liquid phase sintering on the properties of Pr$_6$O$_{11}$-based ZnO varistors", J. App. Phys. 70 (11), 6883 (1991).
12. R. M. German, 127 in "Liquid Phase Sintering", Plenum Pub. Corp., N.Y., N.Y. (1985).
13. T. K. Gupta, "Effect of K$_2$O on the grain growth of ZnO", J. Amer. Cer. Soc. 54 (6), 413 (1971).
14. T. Watari and R. C. Bradt, "Grain growth of sintered ZnO with alkali oxide additions", J. Cer. Soc. Jap. 101 (10), 1085 (1993).
15. T. Senda and R. C. Bradt, "Grain growth of zinc oxide during the sintering of zinc oxide - antimony oxide ceramics", J. Amer. Cer. Soc. 74 (6), 1296 (1991).
16. M. N. Rahaman, 445 in "Ceramic Processing and Sintering", M. Dekker, Inc., N. Y., N. Y. (1996).
17. M. Hillert, "On the theory of normal and abnormal grain growth", Acta Metall. 13 (4), 227 (1965).
18. J. B. Baldo and R. C. Bradt, "Grain growth of lime and periclase phases in a synthetic doloma", J. Amer. Cer. Soc. 71 (9), 720(1988).
19. T. K. Gupta and R. L. Coble, "Sintering of ZnO I: densification and grain growth", J. Amer. Cer. Soc. 51 (9), 521 (1968).
20. T. K. Gupta and R. L. Coble, "Sintering of ZnO II: density decrease and pore growth during the final stage of the process", J. Amer. Cer. Soc. 51 (9), 525 (1968).
21. T. Senda and R. C. Bradt, "Grain growth in sintered ZnO and ZnO-Bi$_2$O$_3$ ceramics", J. Amer. Cer. Soc. 73 (1), 106 (1990).
22. G. C. Nicholson, "Grain growth in zinc oxide", J. Amer. Cer. Soc. 48 (4), 214 (1965).
23. S. K. Dutta and R. M. Spriggs, "Grain growth in fully dense ZnO", J. Amer. Cer. Soc. 53 (1), 61 (1970).

24. D. W. Readey, T. Quadir, and J. H. Lee, "Effects of vapor transport on microstructure development", 485 in "Ceramic Microstructures '86", Plenum Pub. Corp., N. Y., N. Y. (1986).
25. Y-C. Chen, C-Y. Shen and L. Wu, "Grain growth processes in ZnO varistors with various valence states of manganese and cobalt", J. App. Phys. 69 (12), 8363 (1991).
26. R. Einzinger, "Metal oxide varistors", Ann. Rev. Mat. Sc. 17, 299 (1987).
27. T. K. Gupta, "Application of ZnO varistors", J. Amer. Cer. Soc. 73 (3), 1817 (1990).
28. D. R. Clarke, "The microstructural location of the intergranular metal oxide phase in a zinc oxide varistor", J. App. Phys. 49 (4), 2407 (1978).
29. W. D. Kingery, J. B. Van der Sande and T. Mitamura, "A scanning transmission microscopy investigation of grain boundary segregaton in a ZnO - Bi_2O_3 varistor", J. Amer. Cer. Soc. 62 (3-4), 221 (1979).
30. K. W. Lay, "Grain growth in UO_2 - Al_2O_3 in the presence of a liquid phase", J. Amer. Cer. Soc. 51 (7), 373 (1968).
31. D. Dey and R. C. Bradt, "Grain growth of ZnO during Bi_2O_3 liquid-phase sintering", J. Amer. Cer. Soc. 75 (9), 2529 (1992).
32. R. C. Bradt, S.I. Nunes, T. Senda, H. Suzuki, and S, L. Burkett, "Grain Growth Control via in-situ particle formation and Zener drag (pinning) during sintering of ZnO", 389 in "Sintering Technology", M. Dekker, Inc., N.Y., N.Y. (1996).
33. T. Senda and R. C. Bradt, "Twinning in ZnO ceramics with Sb_2O_3 additions", J. Jap. Cer. Soc. 99 (9), 727 (1991).
34. S. I. Nunes and R. C. Bradt, "Grain growth of ZnO in ZnO-Bi_2O_3 ceramics with Al_2O_3 additions", J. Amer. Cer. Soc. 78 (9), 2469 (1995).
35. H. Suzuki and R. C. Bradt, "Grain growth of ZnO in ZnO-Bi_2O_3 ceramics with TiO_2 additions", J. Amer. Cer. Soc. 78 (5), 1354 (1995).

METAL/OXIDE INTERFACES:
CHEMISTRY, WETTING, ADHESION, AND OXYGEN ACTIVITY

Dominique Chatain[1], Véronique Ghetta[2] and Jacques Fouletier[3]

[1]Centre de Recherche sur les Mécanismes de la Croissance Cristalline*
CNRS, Campus de Luminy, Case 913, 13288 Marseille Cedex 9, France
[2]Laboratoire de Thermodynamique et Physico-Chimie Métallurgiques,
ENSEEG-INPG, BP75, 38402 Saint Martin d'Hères Cedex, France
[3]Laboratoire d'Ionique et d'Electrochimie du Solide,
ENSEEG-INPG, BP75, 38402 Saint Martin d'Hères Cedex, France

INTRODUCTION

This paper is a review of recent work on the effect of oxygen on liquid metal/oxide interfaces and metal surfaces. The system considered consists of a two-phase ternary at fixed temperature. The metallic phase and the oxide are in equilibrium when the oxygen activity ranges between a lower limit, where the oxide is reduced, and an upper limit, where the metal is oxidized. In between, small concentrations of oxygen are in solution in the metallic phase, and a large amount of oxygen may segregate at the surface and at the interface of the metallic phase, thus inducing drastic changes in the wetting and the adhesion in metal/oxide systems.

Experimental work reported in the literature gives clear evidence of the segregation of oxygen at the free surface of metals, thus their surface free energy decreases when the oxygen activity increases above a critical value.[1-6] But the behavior of oxygen at the metal/oxide interface is unclear. As no direct measurement of the liquid metal/oxide interface free energy can be performed, it must be deduced from simultaneous measurements of the wetting angle and the surface free energy of the metallic liquid (through Young's equation), after demonstrating that the surface free energy of the oxide remains unchanged with increase in oxygen activity. Some work in the literature shows that the contact angle of metals on oxides decreases with the oxygen activity.[2,4,7-10] Consequently, taking into account Young's equation and the fact that the contact angle of pure metals on oxides is generally greater than 90°,[11] the metal-oxide interfacial free energy should decrease more rapidly than the metal surface energy when the oxygen activity increases.

---

* Laboratoire associe aux Universités d'Aix-Marseille 2 et 3

Recently, we have shown that the contact angle is not a monotonically decreasing function of oxygen activity.[10,12] The existing model, established to explain the effect of oxygen at metal/oxide interfaces, cannot explain this new experimental feature.[13]

This model assumes the same behavior for oxygen adsorption at the metal surface[3,14] and at the metal-oxide interface, leading to a continuous decrease of interfacial free energy as oxygen activity increases. The results of a set of wetting experiments on Au, Ag, Cu and Au-Cu alloys/sapphire systems are presented to review and clarify previous conclusions, and in particular, to reassess semi-empirical rules relating to the improvements of wetting and adhesion of metals on oxides when oxygen is added to the metal.

EXPERIMENTAL PROCEDURE

Our experimental sessile drop set-up has already been described in detail.[15] For the experiments presented here, oxygen activity is controlled by buffered gas mixtures at a total pressure of 10^5 Pa, using an electrochemical pump-sensor device. Table 1 summarizes the gas mixtures used and the corresponding oxygen activity ranges. In-situ changes of the oxygen activity are achieved step by step from 10^{-10} up to 1. The stages are long enough for chemical equilibrium to be reached between the liquid, the solid, and the gas. These stages typically last from 20 to 300 minutes, depending on the buffer capacity of the gas mixtures.

Table 1. Gas mixtures and corresponding oxygen activity ranges at 1370 K.

Gas mixture	Oxygen activity (a_O)
He + H_2/H_2O	$10^{-10} - 10^{-6}$
He + CO/CO_2	$10^{-7} - 10^{-4}$
He + O_2 (electrochemical pump)	$10^{-3} - 10^{-1}$
He + O_2 (gas mixture)	$10^{-1} - 1$

The alumina substrates are non-oriented sapphire platelets (10 to 15 mm in diameter and 0.5 to 1 mm thick), 99.993 wt.% pure, purchased from RSA.[†] Before the wetting experiments, the platelets are ultrasonically rinsed with acetone and then annealed at 1473 K in high vacuum (10^{-5} Pa) for two hours in order to eliminate OH groups adsorbed on their surface. Surface roughness, R_a, is 10 nm or less as measured by optical interferometry. The liquid drops are prepared from 99.9995% pure copper and silver, and 99.9999% pure gold. Their weights vary from 0.6 to 1.4 g. For pure metals, the working temperature is 10 K above the melting point of the metal. For the Cu-Au alloys, it is 10 K above the melting point of copper.

From the drop shadow profile, contact angles, θ, are measured and the liquid surface free energy, γ_{LV}, determined for the largest drops using the computational method developed by Rotenberg et al.[16] Contact angles are either measured by drawing a tangent to the profile of the drop at the triple point, or calculated from the drop profile. Either method produces consistent results. The accuracy of the γ_{LV} and θ determinations depends mainly on the symmetry of the drop.

From the measured contact angle and the liquid surface free energy, the difference between the solid-liquid interfacial free energy and the solid surface free energy, $(\gamma_{SL} - \gamma_{SV}^*)$, and the work of adhesion can be calculated using Young and Dupré equations:

[†] Rubis Synthétique des Alpes, BP 16,38 560 Jarrie, France.

$$\cos\theta = \frac{\gamma_{SV} - \gamma_{SL}}{\gamma_{LV}} \tag{1a}$$

$$W = \gamma_{SV} + \gamma_{LV} - \gamma_{SL} \tag{1b}$$

As long as no reaction occurs at the interface, the experimental results can be interpreted using these equations.

Thermodynamics of the M-O/Al$_2$O$_3$ Systems

The thermodynamics of the M-O/Al$_2$O$_3$ systems have been reported in previous papers.[10,12] At the lowest experimental oxygen activity of 10^{-10} the metallic phases studied do not reduce the alumina. Liquid Au does not undergo oxidation, but MeAlO$_2$ compounds form at high oxygen activity for Me = Cu or Ag. This occurs at an oxygen activity of about 10^{-3} for Cu, and above 1 (the upper experimental limit of oxygen activity) for Ag.[17]

RESULTS AND DISCUSSION

Oxygen Adsorption at the Surface of Liquid Metals: First Order Transitions by Means of "Oxide Molecules"

Figure 1 displays the schematic behavior of the surface free energy of the metallic liquids as a function of oxygen activity. A plateau in the low oxygen activity range is followed by a linear decrease of the surface free energy as the oxygen activity increases. These two ranges correspond to low and high oxygen adsorption domains, respectively. The two lines of Figure 1 can connect either smoothly or sharply. If they connect at a singular point in a_O^*, it is the result of a first order transition in the oxygen adsorption. While the existence of such a transition can be strongly suggested by the shape of the surface free energy lines, as a function of oxygen activity, its proof requires direct measurement of oxygen adsorption. From the slope of the second line, and using the Gibbs adsorption equation, the absolute oxygen adsorption at the surface, Γ_O^{LV}, can be deduced[10]:

$$\Gamma_O^{LV} = -\frac{d\gamma_{LV}}{d\ln a_O} \tag{2}$$

where R is the gas constant, T the temperature, and a_O the oxygen activity.

Table 2 summarizes the characteristics of oxygen adsorption at the surface of the metallic liquids studied.

The surface free energy of gold is constant over the range of oxygen activity from 10^{-10} to 1.[14] This corresponds to the low oxygen adsorption domain depicted in Figure 1.

The surface free energies of the other metallic phases studied (silver, copper, and copper-gold alloys) display the two oxygen adsorption domains of Figure 1. The experimental results[10,12] suggest that the connection between the lines corresponding to the two oxygen adsorption domains could be sharp, which indicates the presence of a first order transition in the oxygen adsorption at a_O^*. Table 2 shows that a_O^* increases as the oxygen activity corresponding to the limit of solubility increases. This is the case for Ag compared to Cu, and the case for the Cu-Au alloys when their Au content increases.

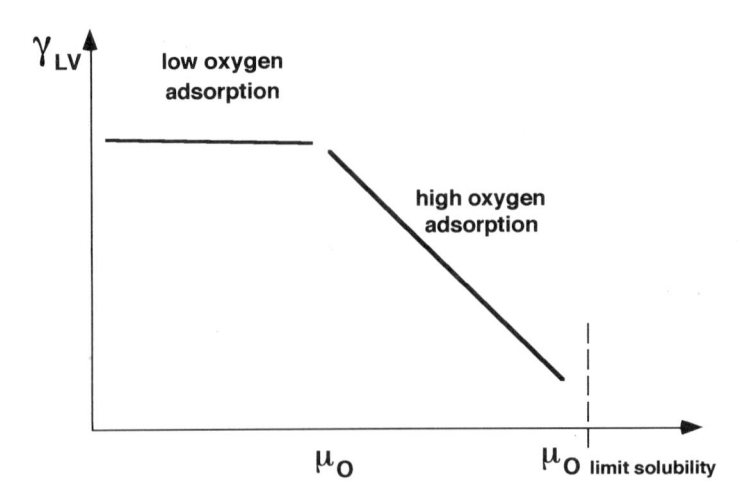

Figure 1. Schematic behavior of the surface free energy as a function of oxygen chemical potential, i.e., oxygen activity (μ_O) = RT ln(a_O)).

If a monolayer model is considered, the surface area per adsorbed oxygen mole would be 220±20×10³ m²/mol. But in a closed-packed layer, the area available for each metal atom is about 50×10³ m²/mol, and that for each O^{2-} ion would be about 40×10³ m²/mol. As proposed by Kozakevitch[18] and then by Bernard and Lupis,[3] we could interpret the saturation of the surface by considering the adsorption of an "oxygen plus metal molecule" having a stoichiometric ratio and a maximum of metal-oxygen bonds in the surface plane.

Table 2. Oxygen adsorption at the surface of the studied metallic phases.

metal or alloy	T (K)	a_O^*	Γ_O^{LV} (mol/m²)	ref.
Au	1363	/	/	15
Ag	1253	1.5 x 10⁻³	4 x 10⁻⁶	10
Cu	1365	1.0 x 10⁻⁶	5 x 10⁻⁶	12
Cu-0.12 Au	1365	1.0 x 10⁻⁶	"	12
Cu-0.15 Au	1365	1.6 x 10⁻⁶	"	12
Cu-0.20 Au	1365	5.0 x 10⁻⁶	"	12
Cu-0.60 Au	1365	1.0 x 10⁻⁵	"	12
Cu-0.85 Au	1365	6.3 x 10⁻⁴	"	12
Cu-0.90 Au	1365	>1.6 x 10⁻²	"	12

The molar area of the (111) planes of Cu_2O and Ag_2O fits the area inferred from the experiments well.[1,3] Thus, such oxygen adsorption can be understood as the formation of a 2D-oxide.

The segregation at metallic surfaces of an elementary "oxide molecule" was also suggested by Naidich[13] and is consistent with two features:

- If the two components are considered to be Me and "Me-O" instead of Me and O, a repulsive interaction between the former two components, instead of an attractive one, would result. This is very consistent with "strong oxygen" segregation.

- The existence of perfect wetting of Cu by Cu_2O at 1500 K,[19] and a first order transition in the oxygen adsorption are consistent with the prewetting and wetting models of Cahn[20] and Widom.[21]

These models predict that when a three-dimensional phase transition is approached, an adsorption layer transition can occur at an interface before the chemical potential for the three-dimensional transition is reached. This transition can be viewed as the formation of a two-dimensional layer precursor for the three-dimensional phase.

Oxygen Adsorption at the Metal/Alumina Interfaces: Other First Order Transitions

Figure 2 shows the schematic dependence of the interfacial free energy between a metallic liquid and an oxide-on-oxygen activity. This curve displays a plateau at low oxygen activity, corresponding to a domain of low oxygen adsorption. It then shows a downward step in interfacial free energy, which corresponds to a first stage of high oxygen adsorption at the metal/oxide interface. Beyond a second plateau, in the high oxygen activity range, a second decrease of the interfacial free energy is shown, which corresponds to a second high oxygen adsorption domain. Taking alumina for the oxide and using the Gibbs adsorption equation, the slope of the γ_{SL} - a_O curve allows the relative adsorption of oxygen with respect to aluminum at the interface to be calculated:

$$\Gamma_O^{SL} - \frac{3}{2}\Gamma_{Al}^{SL} = -\frac{1}{RT}\frac{d\gamma_{SL}}{d\ln a_O} \qquad (3)$$

The order of magnitude of the slopes of the two adsorption domains suggests that "oxide molecules" also adsorb at the interface. The original behavior of this curve is a nonmonotonically decreasing slope when the oxygen activity increases. Equation 3 shows that the plateaus of the γ_{SL} - a_O curve correspond to oxygen activity ranges where the oxygen adsorption with respect to the aluminum adsorption is zero. Taking into account a larger aluminum activity in the low oxygen activity range, the first plateau would correspond to an Al_2O_3 interface. The second plateau could be understood as a reconstruction of the interface occurring when the ratio of Al and (O-Me) goes to the value 2/3. It follows the adsorption of an "oxide molecule" which could contain both Al and Me. The Al-Me-O oxide molecule that segregates in this oxygen activity range could be seen as the two-dimensional layer precursor of the three-dimensional mixed oxide of the two metals that forms at higher oxygen activity.

In the systems studied, the second high oxygen adsorption domain prevails in an oxygen activity range that corresponds to that of a similar domain at the metallic surface. As the magnitudes of the adsorptions at the surface and at the interface given by Equations 2 and 3 are close to each other, and as the aluminum activity is very low, the Me-O "oxide molecule" could thus also be the segregating component at the interface. Table 3 gives the characteristics of the oxygen adsorption at the metal/oxide interface.

For gold, no oxygen adsorption is detected in the oxygen activity range explored.

For pure silver and copper, the interfacial free energies display the two high oxygen adsorption domains. In the case of silver, they are well-separated, but in the case of copper, there is no plateau between the two domains. That means that the reconstruction of the oxide at the interface is readily followed by the adsorption of the "Me-O" molecule.

In the case of Cu-Au alloys, increasing additions of gold displace the two high oxygen adsorption domains towards higher oxygen activities. The second domain is displaced more than the first, in such a way that it disappears first as the gold concentration of the Cu-Au alloys increases. This can be explained by considering that the interfacial activity of the

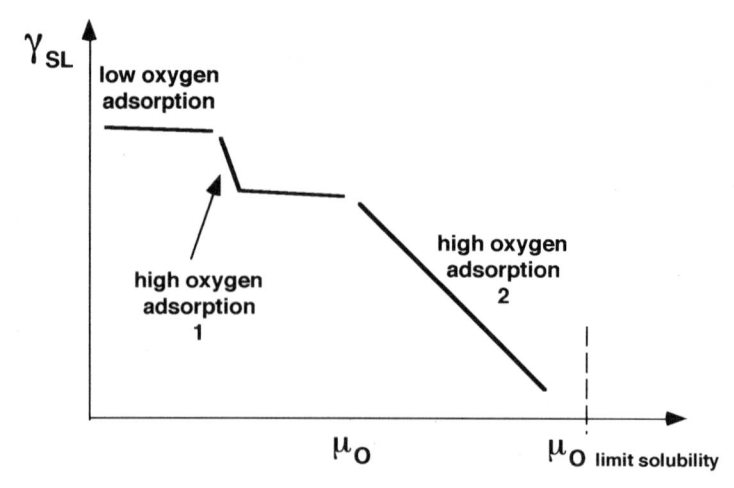

Figure 2. Schematic behavior of the interface free energy as a function of oxygen chemical potential, i.e., oxygen activity ($\mu_O = RT \ln(a_O)$).

segregating "oxide molecule" depends on the activity of its metallic components. Since the activity of copper decreases as gold is added to it, the activity of oxygen must increase to maintain a constant interfacial activity of the "Cu-O" molecule. In the case of the "Cu-Al-O" molecule, addition of gold also decreases the activity of Al. Thus, the copper-aluminum oxide molecule tends to become less and less interface-active compared to the copper oxide molecule as gold is added to copper.

Table 3. Oxygen adsorption at the interface between the studied metallic phase and alumina.

Metal or alloy	T (K)	a_O^{*1}	a_O^{*2}	$(\Gamma_O^{SL} - \frac{3}{2}\Gamma_{Al}^{SL})_{"2"}$ (mol/m²)	ref.
Au	1363	/	/	/	15
Ag	1253	3.0 x 10⁻⁷	1.5 x 10⁻³	6.0 x 10⁻⁶	10
Cu	1365	1.0 x 10⁻⁶	1.0 x 10⁻⁶	5.0 x 10⁻⁶	12
Cu-0.12 Au	1365	1.0 x 10⁻⁶	1.0 x 10⁻⁶	"	12
Cu-0.15 Au	1365	1.6 x 10⁻⁶	1.6 x 10⁻⁶	"	12
Cu-0.20 Au	1365	1.0 x 10⁻⁴	5.0 x 10⁻⁶	"	12
Cu-0.60 Au	1365	/	1.0 x 10⁻⁵	/	12
Cu-0.85 Au	1365	/	/	/	12
Cu-0.90 Au	1365	/	/	/	12

Wetting and Adhesion

The wetting behaviors of the different metallic liquids on alumina as a function of oxygen activity will now be described. The results represent the original measurements from which the information on oxygen segregation at metal/oxide interfaces were deduced, and are presented at the end of this paper for the sake of clarity of the interpretation.

Figure 3 shows the wetting curves obtained for Ag, Cu, and the Cu-Au alloys on alumina as a function of oxygen activity. From Young's equation, and taking into account

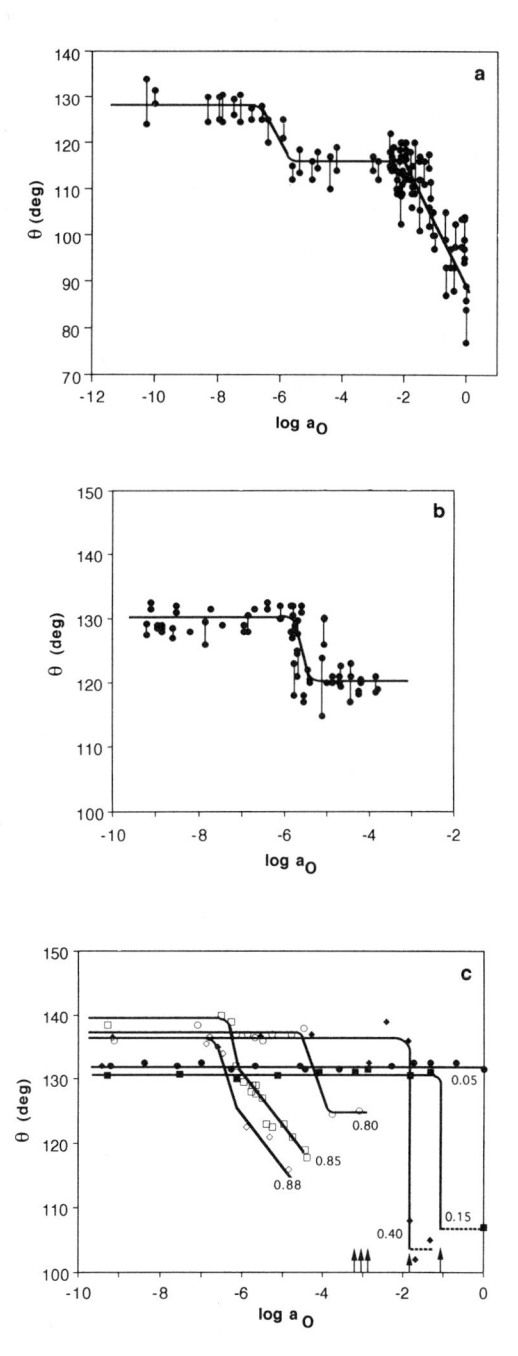

Figure 3. Contact angles as a function of oxygen activity. (a) Ag contact angles. (b) Cu contact angles. (c) Au-Cu contact angles. (Note that the atomic fraction of copper in the alloy is indicated on each curve, and arrows at the bottom of the frame show the oxygen activity corresponding to the limit of oxygen solubility in the alloys.)

that the surface free energy of the oxide is constant,[15] it can be seen that in the case of a contact angle greater than 90°, the surface and interface free energies act inversely in the improvement of wetting.

For silver, the initial oxygen adsorption at the interface induces a decrease of the contact angle. In the high oxygen activity range, the second oxygen adsorption at the interface is energetically more efficient than that at the surface, and produces a further decrease of the contact angle of silver on alumina.

For copper, the initial oxygen adsorption at the interface also induces an improvement of wetting, but in the high oxygen activity range the energetic effects of oxygen adsorption at the surface and at the interface are balanced in such a way that the contact angle remains constant.

In the case of the Cu-Au alloys, two types of wetting behavior can be distinguished. For high Cu content, the contact angle decreases when oxygen activity increases, due to the dominant energetic effect of oxygen adsorption at the interface. For the high gold content alloys, the wetting does not change until the oxygen activity corresponding to the limit of solubility of oxygen in the metallic liquid is reached. In this case, in the low oxygen activity range, there is no adsorption of the Cu-Al-O "oxide molecule" at the interface, and in the high oxygen activity range, the energetical effects of the adsorption of the Cu-O "oxide molecule" at the surface and at the interface are balanced.

In the case of pure Cu and high gold content Cu-Au alloys, some experiments were performed for oxygen activity exceeding the value corresponding to the formation of $AlCuO_2$. In this case, the contact angle gradually decreases, reaching stable values of about 90° and 100°, respectively, after several hours. This behavior is consistent with the formation of the $AlCuO_2$ phase at the interface, as previously observed at higher temperatures,[19] or for the solid Cu/Al_2O_3 interface.[22,23] The metallic liquid spreads on the solid as the $AlCuO_2$ compound forms, and the spreading stops when the contact angle reaches its equilibrium value. This is consistent with the analysis of reactive wetting performed by Cannon et al.[24]

At the present time, the occurrence of a contact angle "step" in the low oxygen activity range has been demonstrated experimentally for different metallic liquids on alumina.[10,12] In the case of Cu/Al_2O_3, most of the θ values measured in the non-reactive range by other authors[2,25] are generally consistent with our experimental curve. The previously proposed shape of the wetting curve is a plateau followed by a monotonic decrease of θ as a_O increases. It should be pointed out that this curve was fitted to scattered experimental θ values, some of them measured in the reactive range where $AlCuO_2$ forms at the interface. In our opinion, this proposed shape is erroneous, and has been influenced by theoretical arguments which assume a systematically stronger energetic effect of oxygen adsorption at the metal/oxide interface than at the surface.[13]

The work of adhesion, W, of metallic liquids on Al_2O_3 as a function of a_O is calculated using the Young-Dupré equations (Equation 1) from the fitted curves of γ_{SL} and θ. Due to the different oxygen adsorption modes at the surface and the interface, the W-a_O curve cannot show a systematic shape. This is illustrated in Figure 4, which displays the adhesion curves for the Cu-Au alloys on alumina. These latter results indicate that the "conventional wisdom," on the basis of which an improvement of adhesion with increased oxygen content in the liquid metal is expected, should be revised in light of the experimental results for Cu and Ag in contact with Al_2O_3.

CONCLUSIONS

From experimental results on Ag/Al_2O_3, Cu/Al_2O_3, and $Au-Cu$ alloys/Al_2O_3, we have shown the adsorption modes of oxygen at the surface of a metallic liquid and at the interface

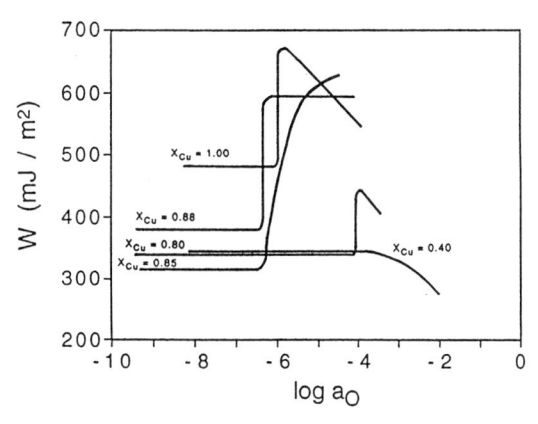

Figure 4. Work of adhesion of Au-Cu alloys as a function of oxygen activity.

between these liquids and an oxide. The results also suggest the occurrence of first order transitions in oxygen adsorption isotherms at metal surfaces and metal/oxide interfaces. Two separate adsorption transitions of oxygen can be detected at the metal/oxide interface. These transitions, which can be understood as the formation of 2D-interfacial oxides, correspond to so-called two-dimensional phase transitions. They are related to the well-studied phenomenon of surface premelting of solids.

A second important point resulting from these features concerns the wetting and the adhesion behavior of a metallic liquid on an oxide when oxygen is dissolved in the metallic liquid. The experimental results show that the formation of a 2D-interfacial oxide is the best means of increasing the work of adhesion and of improving wetting. But when oxygen adsorption at the liquid metallic surface and at its interface with the oxide occur in the same oxygen activity range, oxygen cannot produce a systematic improvement of wetting and adhesion.

Thermodynamic modeling of these results should now be performed, and tested with further specific experiments. Experiments on solid metal/oxide interfaces should also be of interest to understand the roles of the chemistry and the structure of the two contacting phases. Particularly, experiments on oriented single-crystals are needed to understand the so-called "2D-oxide formation" in comparison with premelting of particular crystallographic surface orientations of pure metals.

REFERENCES

1. B. Gallois and C. H. P. Lupis, Effect of oxygen on the surface tension of liquid copper, *Metall. Trans. B*, **12**:549 (1981).
2. P. D. Ownby and J. Liu, Surface energy of liquid copper and single crystal sapphire and the wetting behavior of copper on sapphire, *J. Adhes. Sci. Technol.*, **2**:255 (1988).
3. G. Bernard and C. H. P. Lupis, The surface tension of liquid silver alloys: part II Ag-alloys, *Metall. Trans.*, **2**:2991 (1971).
4. K. Ogino, H. Taimatsu and F. Nakatani, Effect of oxygen on surface tension of liquid Co and Co-Fe alloys and the wettability of Al_2O_3, *J. Japan Inst. Metals*, **46**:957 (1982).
5. R. Sangiorgi, M. L. Muolo and A. Passerone, Surface tension and adsorption in liquid silver-oxygen alloys, *Acta Metall.*, **30**:1597 (1982).
6. K. Ogino and H. Taimatsu, Effect of oxygen on the surface tension of liquid nickel and the wettability of alumina by liquid nickel, *J. Japan Inst. Metals*, **43**:871 (1979).
7. T. E. O'Brien and A. C. D. Chaklader, Effect of oxygen on the reaction between copper and sapphire, *J. Am. Ceram. Soc.*, **57**:329 (1974).
8. S. P. Mehrotra and A. C. D. Chaklader, Interfacial phenomena between molten metals and sapphire, *Metall. Trans.*, B**16**:567 (1985).

9. H. Taimatsu, K. Ogino and F. Nakatani, Effect of oxygen on the wettability of magnesium oxide by liquid nickel, *J. Japan Inst. Metals*, **50**:176 (1986).
10. D. Chatain, F. Chabert, V. Ghetta and J. Fouletier, New experimental set-up for wettability characterization under monitored oxygen activity: II, wettability of sapphire by silver-oxygen melts, *J. Am. Ceram. Soc.*, **77**:197 (1994).
11. D. Chatain, L. Coudurier and N. Eustathopoulos, Wetting and interfacial bonding in ionocovalent oxide liquid metal systems, *Revue Phys. App I*, **23**:1055 (1988).
12. V. Ghetta, J. Fouletier and D. Chatain, Oxygen adsorption isotherms at the surfaces of liquid Cu and Cu-Au alloys and their interfaces with Al_2O_3 detected by wetting experiments, *Acta Mater.*, **44**:1927 (1996).
13. J. V. Naidich, The wettability of solids by liquid metals, in: *Progress in Surface and Membrane Science*, D. A. Cadenhead and J. F. Danielli eds, Academic Press, London, UK, 353 (1981).
14. E. Ricci, A. Passerone and J. C. Joud, Thermodynamic study of adsorption in liquid metal-oxygen systems, *Surface Science*, **206**:533 (1988).
15. D. Chatain, F. Chabert, V. Ghetta and J. Fouletier, New experimental set-up for wettability characterization under monitored oxygen activity: I, role of oxidation state and defect concentration on oxide wettability by gold, *J. Am. Ceram. Soc.*, **76**:1568 (1993).
16. Y. Rotenberg, L. Boruvka and A. W. Neumann, Determination of surface tension and contact angle from the shapes of axisymmetric fluid interfaces, *J. Colloid Interface Sci.*, **93**:169 (1983); P. Cheng and A. W. Neumann, Computational evaluation of axisymmetric drop shape analysis profile, *Colloid Surf.*, **62**:297 (1992).
17. *Technisch Physische Dienst, Delft, Netherlands* (1968), cited in the JCPDS - International Center for Diffraction Data file.
18. P. Kozakevitch, Surface activity in liquid metal solutions, S.C.I. Monograph, **28**:223 (1966).
19. M. D. Baldwin, P. R. Chidambaram and G. R. Edward, Spreading and interlayer formation at the copper-copper oxide/polycrystalline alumina interface, Metall. *Mater. Trans.* A, **25**:2497 (1994).
20. J. W. Cahn, Critical point wetting, *J. Chem. Phys.*, **66**:3667 (1977).
21. B. Widom, Structure of the α/γ interface, *J. Chem. Phys.*, **68**:3878 (1978).
22. K. A. Rogers, K. P. Trumble, B. J. Dalgleish and I. E. Reimanis, Role of oxygen in microstructure development at solid-state diffusion-bonded Cu/Al_2O_3 interfaces, *J. Am. Ceram. Soc.*, **77**:2036 (1994).
23. C. Beraud and C. Esnouf, Liaison cuivre-alumine: évolution et caractérisation microscopique des composés interfaciaux, *Microsc. Microanal. Microstructure,* **1**:69 (1990).
24. R. M. Cannon, E. Saiz, A. P. Tomsia, and W. C. Carter, Taxonomy of reactive wetting, in *Structure and Properties of Interfaces in Ceramics*, MRS symposium proceedings (Fall 1994).
25. D. Chatain, M. L. Muolo and R. Sangiorgi, Wettability of sapphire by Ag, Cu, and Ag-5Cu under different oxygen pressures, in: *Designing Ceramic Interfaces II:* S. D. Peteves, ed., CEC Publ., Luxembourg, 359 (1993).

CERAMIC/METAL REACTIONS and MICROSTRUCTURES in CERAMIC JOINTS

Stathis D. Peteves, Mervi Paulasto, Giacomo Ceccone and
Michael G. Nicholas

Institute for Advanced Materials, JRC, European Commission,
P.O. Box 2, 1755 ZG, Petten, The Netherlands

INTRODUCTION

Ceramics and in particular combinations of ceramics and metals appear to offer many technical and economic advantages to design engineers, permitting products to be made with better wear resistance, higher service temperatures and so on, but if these aspirations are to be met it is usually necessary for the ceramic and the metal to form high integrity interfaces. The lattice binding characteristics of ceramics and metals are usually markedly different, and hence it can be difficult to fabricate technologically adequate interfaces. Nevertheless, nature contains many examples of high integrity ceramic-metal interfaces, such as those between Al or stainless steels and their tenacious protective surface oxides or between metal matrices and strengthening arrays of carbide precipitates. Notable characteristics of these natural interfaces are that they are not only strong, but are in localised chemical equilibrium and there are strong arguments that localised equilibrium is needed if high integrity synthetic interfaces are to be formed.[1]

Achieving local chemical equilibration of interfaces between disparate ceramics and metals requires chemical reactions to occur, but experience shows that only certain reactions are beneficial and that their progress must be controlled if useful interfaces are to be fabricated. A prime example of this is the development and optimisation of active metal brazes in recent years that has permitted the easier fabrication of structurally sound joints with a wide range of ceramics.[2] In fact, the progress of chemical reactions can affect every stage of the formation of a synthetic interface from the initial achievement of contact, through the development of a stable chemistry and microstructure, to the optimisation of mechanical properties. This paper provides an overview of our efforts to understand and optimise such effects in order to promote our ability to fabricate ceramic/metal interfaces, drawing on our experience of the brazing and diffusion bonding of Al_2O_3, Si_3N_4 and ZrO_2.

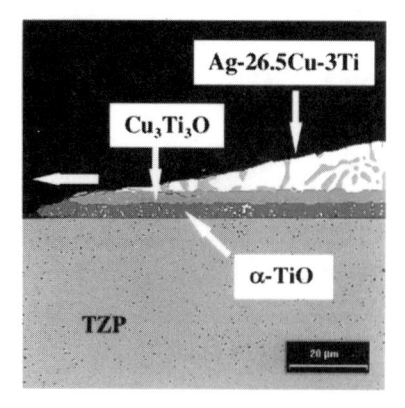

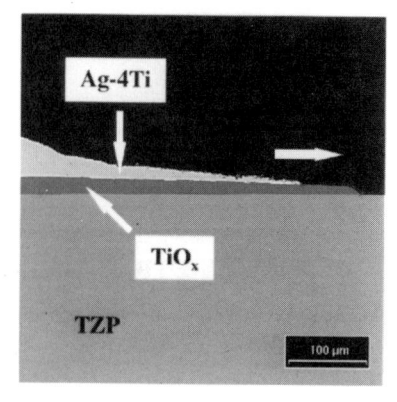

Figure 1. SEM images of cross-sectioned solidified Ti-active metal filler sessile drops on ZrO_2 (3% Y_2O_3). Left: Ag-26.5%Cu-3%Ti (900°C, 10 min, $p=5 \times 10^{-7}$ mbar), Right: Ag-4%Ti (1000°C, 10 min, $p=5 \times 10^{-7}$ mbar). The reaction products between the filler metal and the ceramic extend across the contact region.

INITIAL INTERFACE FORMATION

The first and paramount requirement for reactive bonding is that the joining process allows an intimate and true contact between the workpieces to be realized. Achieving contact when diffusion bonding ceramics to metals is not easy. Usually the process relies on the use of soft metal interlayers, smooth mating surfaces and most important by high temperatures and pressures to accommodate surface irregularities, and thus, formation of an intimate contact by localized deformation of the metal component.

Matters are easier if brazing is the joining route, provided that the filler metal/alloy wets the ceramic. This is not a problem nowadays when using active metal brazes, which have been shown to wet well ceramics in moderate vacuum/inert environments as soon as they melt.[3] Hence, if the braze wets and spreads, an intimate interface can be easily realised by the liquid penetrating and filling all irregularities at the solid surface.

Having during the joining process achieved contact at the atomic level, interfacial reactions may occur. However, while reactions result from contact, these can also cause it if the reaction product layer(s), RPL, spreads laterally. Experimental data do point out that lateral growth does occur and affects both brazing and diffusion bonding. Diffusion bonded interfaces also presumably grow by lateral extension of RPLs, although there is limited evidence regarding it.[4] This extension will certainly be slow, but not important because diffusion bonding is invariably slow, and it is only necessary to close micro-voids.

The evidence of RPL lateral growth and interface formation is strong in the case of brazing, as illustrated in Fig. 1, which shows the cross-sections of solidified active filler metal sessile drops on ZrO_2. It should be noted that these images may indicate that the liquid front follows the RPL raft, rather than spilling over onto the unreacted ceramic surface. Needless to say, this may not be what actually happens at temperature since thermal contraction of the bulk liquid while cooling and solidifying, cross-sectioning and polishing may have altered the true situation. Regardless, it is clear that the liquid front does not lead or lag behind the raft edge by much.

Reactive sessile drops spread on ceramics, but do so slowly.[5,6] Linear rates of typically 1 µm/s or less have been measured for a Ag-39%Cu-3.7%Ti* drop spreading on Al_2O_3 at

---

* Contents are in wt% unless noted otherwise.

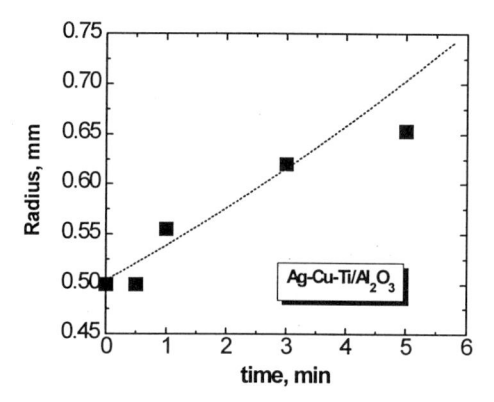

Figure 2. The radius of a sessile drop of Ag-39%Cu-3.7%Ti spreading on Al₂O₃ at 950°C in vacuum.

950°C in vacuum, Fig. 2. Even slower linear spreading, was observed in the Au-15%Ni-1.8%V / Si₃N₄ system, Fig. 3, with a measured rate in the order of 0.02 µm/s in the wetting regime. This system is also reactive, yet not as strongly when compared with the Ti-active brazes, yielding the formation of VN$_x$ as the interfacial reaction product, Fig. 4. Recently, several factors have been identified contributing to reactive wetting, although no generally accepted theory capable of describing it is yet available.[5,7] Comments on these factors is outside the scope of this paper, however a link between the slow spreading rate and reactivity of the filler metal with the Si₃N₄ may be noteworthy. Based on thermodynamic calculations[8], at the triple point, i.e., at the wetting front, V-nitrides are likely unstable for the given experimental conditions of T=1050°C and $pN_2 \leq 10^{-9}$ bar. Hence, the difficulties with forming and stabilizing VN could cause the observed very slow spreading rates.

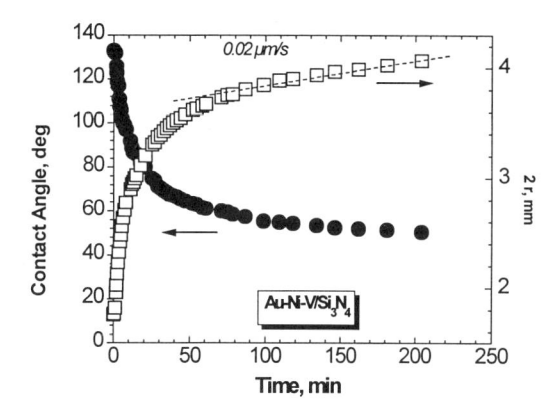

Figure 3. Contact angle and radius of a drop of Au-15%Ni-1.8%V spreading on Si₃N₄ versus time at 1050°C in vacuum ($p = 10^{-6}$ mbar).

EQUILIBRATION

Provided that a true contact is being established at the ceramic/metal interface at high temperatures, if they are not in equilibrium, chemical reactions will take place. While overall equilibrium is never reached, localised equilibrium can be attained at the various

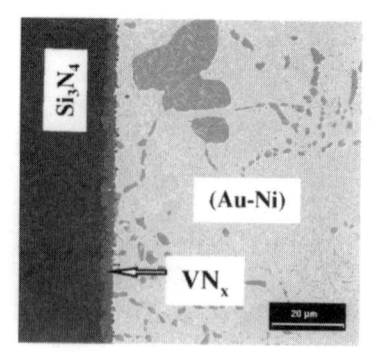

Figure 4. SEM micrograph of the AuNiV/Si₃N₄ interface brazed at 1050°C for 30 min.

interfaces and the progress of reactive bonding is then dictated by the equilibrium thermodynamics and kinetics of the system. Thermodynamics indicates which are the possible phases that can be stable, whereas kinetics dictate how fast or how much of a new phase can be formed.

If the thermodynamic and kinetic data, e.g., phase diagrams, activity and diffusion coefficients is complete, one could predict the sequence of reactions, the interphases that form, and even their thicknesses and morphologies.[9] However, for most ceramic/metal systems of practical interest such knowledge is incomplete, thus restricting the predictions that can be made.[9-11] In principle, the following set of data are required: (I) interdiffusion coefficients, (II) isothermal sections of the pertinent phase diagram that along with mass balance considerations, relative diffusivities and the rule that intrinsic diffusion only proceeds in the direction of decreasing activity of the components[12], can rule out which reaction paths are not possible, and (III) dependence of the isothermal phase diagram section on the joining environment, e.g., on the partial pressure of nitrogen in the case of nitride bonding. Some of these issues will be illustrated below for the reactions between Si₃N₄ and Ti-alloys.

Ti- Ag-Cu / Si₃N₄ System

Reactions between Ti-containing filler alloys and Si₃N₄ have been studied, and the formation of titanium-nitride and -silicides have been reported.[13] Based on the microstructural observations, the following reaction chemistry is often proposed:

$$Ti + 1/9\ Si_3N_4 \Rightarrow 4/9\ TiN + 1/9\ Ti_5Si_3 \quad (1)$$

In these considerations it is generally assumed that the activity of Ti is equal to one. These reaction products are indeed found to form when a thin layer of pure Ti reacts with Si₃N₄, with the complete reaction path at 1100°C found as Si₃N₄/TiN$_x$/Ti₅Si₃[N] + α-Ti/α-Ti/β-Ti.[14] Figure 5 shows the isothermal section of the Ti-Si-N system calculated at 1100°C, where the experimentally found reaction sequence is highlighted.[14] This sequence is in accord with the mass balance criterion, and calculations of the activities of Ti, Si and N as a function of the atom-ratio of the elements also show that the sequence obeys the rule for intrinsic diffusion down an activity gradient.

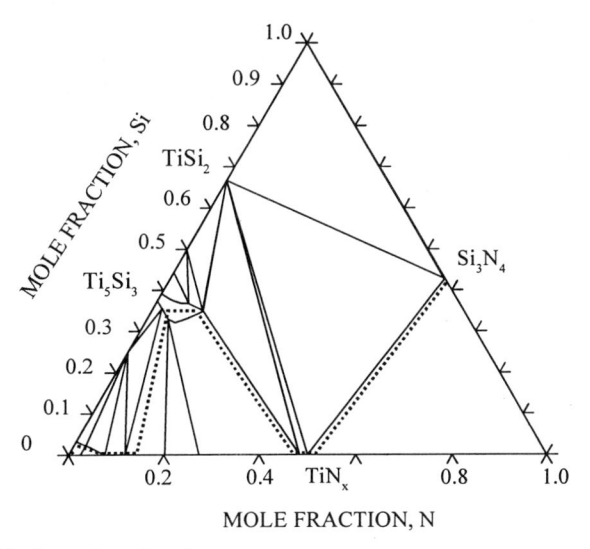

Figure 5. Calculated isothermal section of the Ti-Si-N system at 1100°C. Dotted line indicates the diffusion path between Ti and Si$_3$N$_4$.

However, in brazes the activity of Ti is strongly affected by the solvent and other solutes, and therefore various reaction products can form depending on the specific composition of different brazes. Naturally, the activities of silicon and nitrogen in the solution phase also affect the evolution of various reaction products. Figure 6 presents the activity of Ti calculated in a liquid phase containing Ag, Cu, Ti, Si and N.[15] The activity of titanium in a AgCu liquid containing 2 at-% Ti is shown with a dashed line. When this liquid begins to dissolve the Si$_3$N$_4$ in a Si/N ratio of 3 to 4, the activity of Ti decreases (line A). This provides a driving force for diffusion of titanium to the interface leading to an enhanced dissolution of Si$_3$N$_4$ and to a further decrease of the Ti-activity (line B and C).

In an equilibrium state, nearly all titanium has gathered at the interface, and the formation of Ti-silicides or other compounds is dependent on the equilibrium activity of Ti and the corresponding Si-activity in the Ti-rich layer. Although the calculated activities of

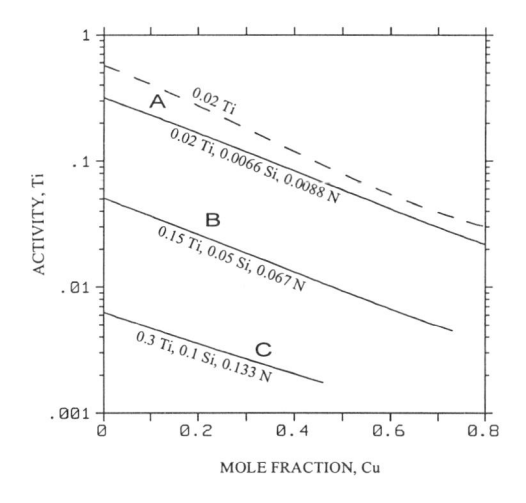

Figure 6. Activity of Ti in the AgCu-2 at-% Ti liquid and in the AgCuTiSiN liquid (Solid lines) at 1000°C.

Ti and Si in the liquid do not exclude the possibility of Ti$_5$Si$_3$[N] formation, what forms likely adjacent to TiN is a Ti-Si-Cu-N compound.[14-17] Furthermore, according to Fig. 6, the activities of both titanium and silicon decrease strongly with increasing Cu-content in the liquid, which leads to lower equilibrium activities of both Ti and Si in Cu-rich brazes than in Ag-rich brazes. Therefore, it is expected that the reaction products in the Ag-rich and Cu-rich brazes are different.

MICROSTRUCTURAL DEVELOPMENT

For technically satisfactory joints, of Si$_3$N$_4$ ceramics at least[#], the evidence points to the need for a separate phase with a finite thickness at the interface. It appears that attainment of equilibrium is not possible with mere atomic clusters or that appropriate surface energies can not be achieved. After nucleation of the reaction products at the interface, the RPL usually thickens with a rate depending on the efficiency of supply of the relevant atoms across this reaction layer, and hence, on the diffusivity of the relevant species within it.

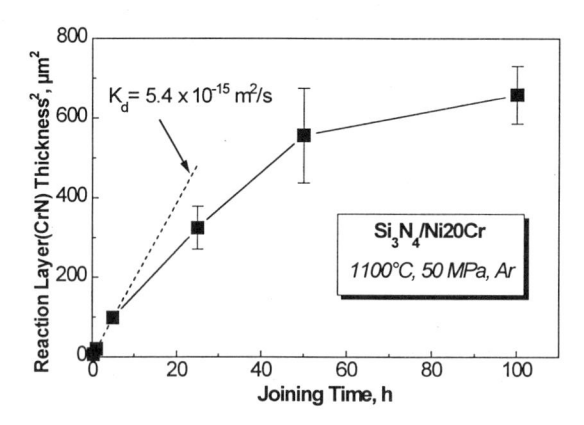

Figure 7. The square of the interfacial reaction layer (CrN) thickness versus time for Si$_3$N$_4$/Ni20Cr/Si$_3$N$_4$ joints bonded in Ar at 1100°C.

Under the aforementioned conditions, the thickness of the RPL is parabolically dependent on time, as shown in Fig. 7, for diffusion bonded Si$_3$N$_4$ with a Ni-20%Cr interlayer. For the given experimental conditions, the interfacial reaction product in this system is CrN.[20] The deduced activation energy for the growth of this reaction layer likely represents that for diffusion of nitrogen within the CrN, or that of Cr diffusion in the Ni20Cr alloy. What is worth noting is that the RPL growth constants found in this system are in good agreement with those of other Metal/Si$_3$N$_4$ systems, Fig. 8, in which Cr also is the active interfacial element.[21-23]

Obviously, when the reactions are fast and/or the concentration of the active species is low, the reactant will be exhausted early and the RPL will grow no thicker. This is often observed when brazing ceramics with low Ti-content activated filler metals, Figure 9.[24]

[#] It has been demonstrated that, for a few metal/oxide ceramics systems, e.g. Nb/sapphire, it is possible to have a strong bond without per se an interfacial reaction layer.[18,19]

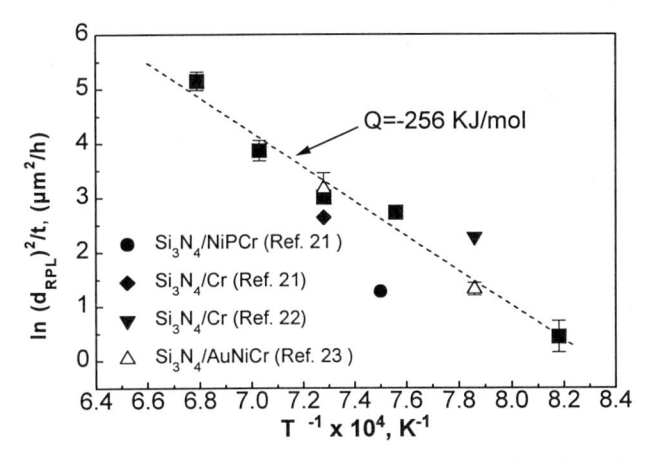

Figure 8. Arrhenious plot for the growth rate constant of the CrN reaction layer formed at the interface of Si$_3$N$_4$/Ni-20Cr joints (1 h, Ar).

Similarly, the interfacial reaction product will cease to thicken when the concentration of the active element in the bulk drops below the necessary concentration limit for the reaction to continue. This is indeed the case for the Si$_3$N$_4$/Ni20Cr system, in which for the given experimental conditions CrN stops forming when the Cr concentration in the interlayer falls below 10%.[20]

On the other hand if the reactant is always available, i.e., as the metal workpiece in the case of ZrO$_2$ bonded to Ti via a Au interlayer, the overall reaction front continues growing with bonding time whereas individual reaction layers may stop growing or even disappear after a given bonding time.[24]

MECHANICAL PROPERTIES

There is strong evidence that temperature, time and reactant concentration affect the strength of ceramic joints, since these fabrication parameters control the extent of the reactions, and the thickness and morphology of the RPL. Experimental evidence from both

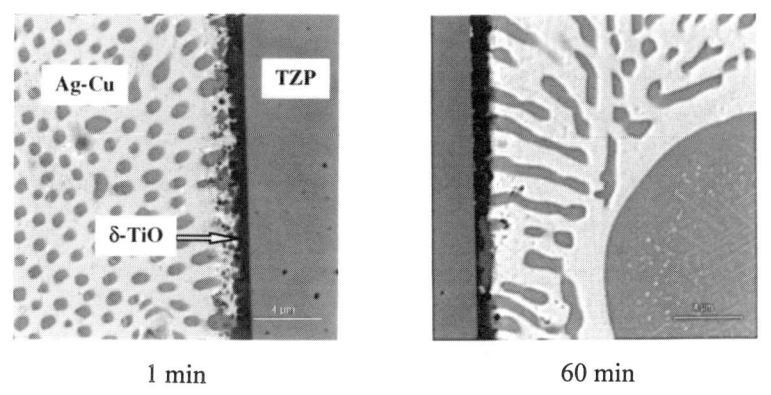

Figure 9. SEM images of ZrO$_2$ brazed interfaces with Ag-34.7%-1.65%Ti (50 µm thick) at 900°C for different times.

diffusion bonding and brazing of Si_3N_4, as well as of other ceramics, indicates that thin RPLs are generally beneficial, as illustrated in Figs. 10 and 11. Although in the case of brazed joints the strength is not given as a function of the reaction layer thickness (e.g. TiN_x + Ti-Cu-Si-N) it is known that this varies proportionally with the amount of active filler metal used. For both systems it is evidenced that optimum strength is achieved for certain yet low thickness reaction layers.

There exists no clear evidence that the chemistry of the RPL affects the mechanical properties. However, experimental data for diffusion bonded ceramic joints are in contrast with the logical expectation that solid solutions may be more appropriate that new phase interphases. Thus, stronger joints have been produced when bonding Si_3N_4 with Fe-Cr and Ni-Cr alloys, which under certain experimental conditions permit the formation of Cr-

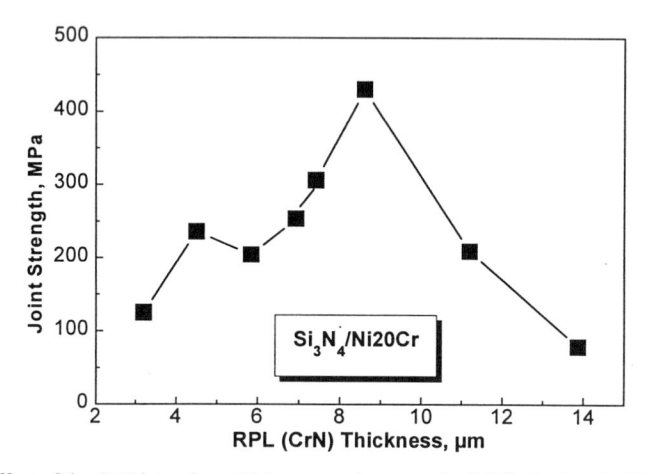

Figure 10. Effect of the CrN-interphase thickness on the strength of diffusion bonded Si_3N_4 joints with Ni20Cr interlayers.

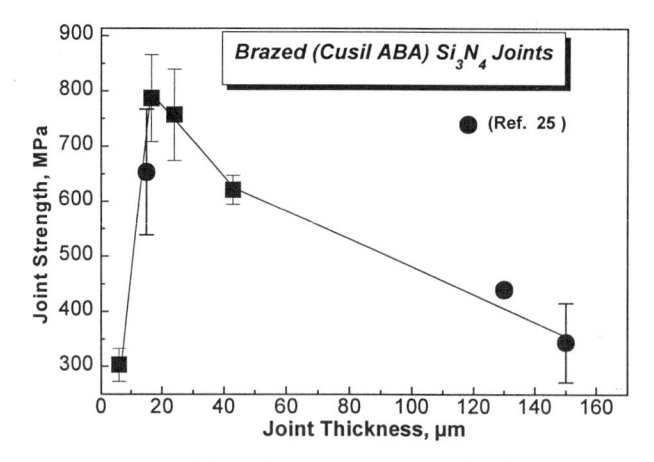

Figure 11. Strength of brazed Si_3N_4/Si_3N_4 joints as a function of the filler metal (Ag-35%-1.65%Ti) thickness.

nitrides at the interface, than with Ni and Fe interlayers that lead to formation of solid solutions, Ni[Si] and Fe[Si,N].[26]

Fracture paths vary from system to system and do not allow for conclusive remarks to be made regarding the relative strengths of the ceramic/RPL and bulk metal interlayer/RPL interfaces. However, it is generally observed that for Si_3N_4 brazed joints the fracture path is at the ceramic-metal nitride reaction layer, Fig. 12, whereas for diffusion bonded joints most often fracture coincides with the Kirkendall plane of the interdiffusion process during bonding, Fig. 13.

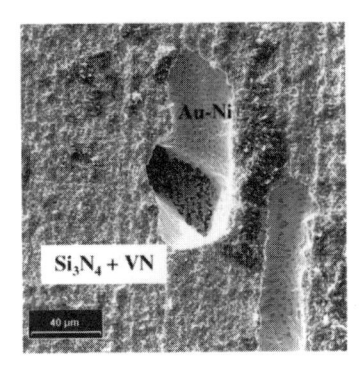

Figure 12. Fracture surface (metal side) of a Au-15%Ni-1.8%V brazed Si_3N_4.

Figure 13. Fracture surface (metal side) of diffusion bonded Si_3N_4 with Fe interlayer.

CONCLUSION

Studies of ceramic-metal systems suggest that all major aspects of joint formation and performance are profoundly affected by the nature, extent or rapidity of chemical reactions between the workpieces and the joining materials. The need to direct and control these reactions, therefore, provides a framework for the optimised selection of joining materials and process parameters. The framework has similarities to that developed for other systems, e.g., glass-metal systems, but it is as yet speculative in places and more detailed quantitative work is needed to provide greater confidence. In particular, more needs to be known about the role of reaction kinetics in the growth of diffusion bonded interfaces, and about the influence of reaction product chemistry on the joint mechanical properties.

ACKNOWLEDGEMENTS

The Institute work described here was conducted under the European Commission research and development programme.

REFERENCES

1. J.A. Pask and A. Tomsia, Wetting, spreading and reactions at liquid/solid interfaces, in: *Surfaces and Interfaces in Ceramic and Ceramic-Metal Systems,* J. Pask and A. Evans, eds., Plenum Press, New York, NY (1981).
2. M.G. Nicholas, *Joining of Ceramics*, Chapman Hall, London, UK (1990).
3. H. Mizuhara, E. Huebel and T. Oyama, High-reliability joining of ceramic to metal, *Ceram. Bull.,* 68:159(1989).

4. P. Moretto, M. Moulaert, P. Glaude, P. Frampton, G. Ceccone and S.D. Peteves, Interfacial reactions and kinetics between Si_3N_4 and Ni-Cr alloys, in: *Designing Ceramic Interfaces II,* S.D. Peteves, ed., Commission of the European Communities, Luxembourg (1993).
5. K. Landry and N. Eustathopoulos, Dynamics of wetting in reactive metal/ceramic systems: linear spreading, *Acta Metall. Materialia,* in print.
6. M.G. Nicholas and S.D. Peteves, The kinetics of liquid braze spreading, in: *High Temperature Capillarity,* N. Eustathopoulos, ed., Reproprint, Bratislava, Slovakia, (1994).
7. R.M. Cannon, E. Saiz, A.P. Tomsia, W.C. Carter, Reactive wetting taxonomy, *Mater. Res. Soc. Proc.,* Vol. 357, MRS, Pittsburgh (1995).
8. M. Paulasto, G. Ceccone, S.D. Peteves, R. Voitovich and N. Eustathopoulos, Brazing of Si_3N_4 with Au-Ni-V-Mo filler alloy, *Ceram. Trans.,* in print.
9. F.J.J. van Loo, J.H. Gulpen and A. Kodentsov, The role of diffusion and thermodynamics in metal-ceramic interactions, in: *Joining Ceramics, Glass and Metal,* H. Krappitz and H.A. Scaeffer, DGG, Frankfurt, Germany (1993).
10. G.P. Kelkar, K.E. Spear and A.H. Carim, Thermodynamic evaluation of reaction products and layering in brazed alumina joints, *J. Mater. Res.,* 9:2244 (1994).
11. M. Backhaus-Ricoult, Physicochemical processes at metal-ceramic interfaces, in: *Metal-Ceramic Interfaces,* M. Ruhle, A.G. Evans, M.F. Ashby and J.P. Hirth, eds., Pergamon Press, Oxford, UK (1990).
12. F.J.J. van Loo, Multiphase diffusion in binary and ternary solid-state systems, *Prog. Solid St. Chem.,* 20:47 (1990).
13. S.D. Peteves, G. Ceccone, M. Paulasto, V. Stamos and P. Yvon, Joining Silicon Nitride to itself and to metals, *JOM,* 48:48 (1996)
14. M. Paulasto, Activation mechanism and interfacial reactions in brazing of Al_2O_3 and Si_3N_4 ceramics with Ag-Cu-Ti filler alloys, *Ph.D. Thesis,* Helsinki Univ. of Technology, Espoo, Finland (1995).
15. M. Paulasto and J.K. Kivilahti, Formation of interfacial microstructure in brazing of Si_3N_4 with Ti-activated Ag-Cu filler alloys, *Scripta Metall. Et Materialia,* 32:1209 (1995).
16. A.H. Carim, Transitional phases at ceramic-metal interfaces: orthorhombic, cubic and hexagonal Ti-Si-Cu-N compounds, *J. Am. Ceram. Soc.,* 73:2764 (1990).
17. C. Iwamoto and S.-I. Tanaka, Interface nanostructure of brazed Silicon Nitride, in: *Proc. Ceramic Microstructure '96: Control at the Atomic Level,* A.P. Tomsia and A.M. Glaeser, eds., Plenum Press, in print
18. M. Ruhle and W. Mader, Structure and Chemistry of Metal/Ceramic Interfaces, in: *Designing Interfaces for Technological Applications: Ceramic-Ceramic, Ceramic-Metal Joining,* S.D. Peteves, ed. Elsevier Appl. Sci., London, U.K (1989).
19. B.J. Dalgleish, E. Saiz, A.P. Tomsia, R.M. Canon and R.O. Ritchie, Interface Formation and Strength in Ceramic-Metal Systems, *Scripta Metall. et Mater.,* 31:1109(1994).
20. S.D. Peteves, M. Moulaert and M.G. Nicholas, Interface Microchemistry of Silicon Nitride/Ni-Cr alloy joints, *Metall. Trans.,* 23A:1773 (1992).
21. T. Takashima, T. Yamamoto and T. Narita, Reaction-layer of Silicon Nitride ceramics to metal joints using Ni-base solder, *J. Ceram. Soc. Japan,* 100:913 (1992)
22. I.I. Gab, V.S. Zhuravlev, D.I. Kurkova and Y.V. Naidich, Adhesive, capillary phenomena and formation of diffusion bonded and brazed joints of nonmetallic materials using aluminium, in: *High Temperature Capillarity,* N. Eustathopoulos, ed., Reproprint, Bratislava, Slovakia, (1994).
23. G. Ceccone, M.G. Nicholas, S.D. Peteves, A.P. Tomsia, B.J. Dalgleish and A.M. Glaeser, An evaluation of the partial transient liquid phase bonding of Si_3N_4 using Au coated Ni-22Cr foils, *Acta Mater.,* 44:657 (1996)
24. J.V. Emiliano, R.N. Correia, P. Moretto and S.D. Peteves, Zirconia-Titanium joint interfaces, *Materials Sci. Forum,* 207-209:145 (1996).
25. R.E. Loehman, A.P. Tomsia, J.A. Pask and S.M. Johnson, Bonding mechanisms in silicon nitride brazing, *J. Am. Ceram. Soc.,* 73:552 (1990).
26. S.D. Peteves, Joining nitride ceramics, in: *Ceramics: Charting the Future,* P. Vincenzini, ed., Techna Srl., Faenza, Italy (1995).

JOINING TECHNOLOGY FOR CERAMIC/METAL COMPOSITE STRUCTURES

Katsuaki Suganuma

Institute of Scientific and Industrial Research
Osaka University
Mihogaoka 8-1, Ibaraki
Osaka 567, Japan

INTRODUCTION

Joining of ceramics to metals has a variety of applications both in the structural and the electronics fields. One of the great benefits of the adoption of joining into the structural applications is to provide reliability to the ceramic components by backing up with metal components. For the electronics applications, the connection between electrodes and a ceramic substrate is one of the main requirements for joining them. Recently, the author has developed a new process for joining ceramics and metals by using pressure casting of metals and this method is called the SQ process[1]. In this process, ceramic components are heated up to 500°C - 800°C in a mold. Liquid metal in the temperature range between 700°C and 800°C is poured into the mold and pressure is applied immediately. Liquid metal solidifies rapidly under a pressure and tight binding at the interface is accomplished within a short period. This process has many benefits for the production of ceramic/metal joint structures. Several applications in the electronics fields have been under examination. In the present paper, the basic idea of the SQ process and the nature of the interfaces produced by this process will be introduced.

BASIC CONCEPT OF SQ PROCESS

Squeeze casting is one of the casting methods that have been widely used for the industrial production of metallic components. Among various casting methods, squeeze casting has the great advantage of producing strong and reliable products having fine microstructures. This feature is attributed to high cooling rate and void elimination under a high pressure. The basic concept of the SQ process for the fabrication of ceramic/metal composite structures is schematically shown in Fig. 1. A ceramic to be joined is fixed in a mold and they are pre-heated to a given temperature. Liquid metal liquid is poured into the mold.

Immediately, a pressure is applied on the liquid surface. Under a pressure, liquid metal solidifies very quickly and a void-less interface is formed between the ceramic and the metal. As the metal to be joined to ceramics, one can use any kind of metal when the mold and the casting atmosphere are appropriately selected. However, from a practical standpoint, metals with low melting temperatures are recommended primarily because for economic reasons. Aluminum and its alloys, magnesium, zinc and tin are the typical useful metals and, if inert atmosphere is available for casting, one can use higher melting temperature metals such as copper and silver. Some of the benefits of the SQ process are summarized as follows:

1) Excellent economy - Mass producibility
2) Suppression of severe reaction
3) Little oxidation even in air
4) Large/complex shape joining
5) Easy formation of interlayer
6) Defect-free perfect contact
7) Freedom from wettability concerns

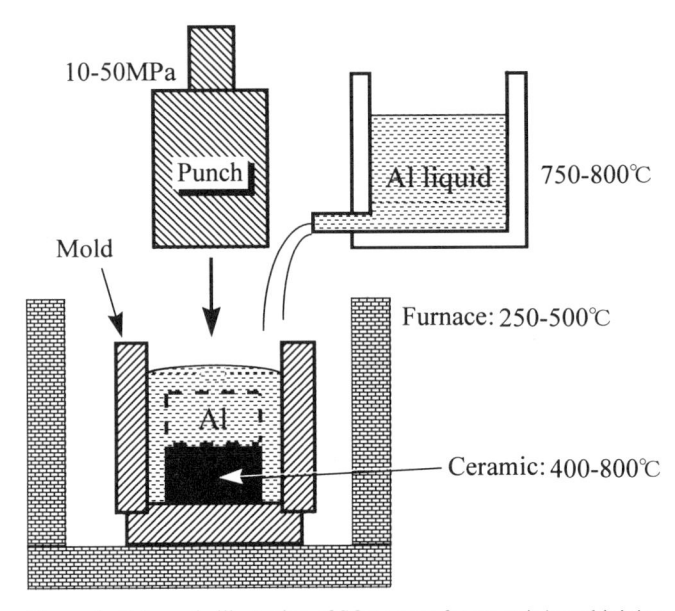

Figure 1. Schematic illustration of SQ process for ceramic/metal joining.

Several variations for joining are shown in Fig. 2. The simplest type (A) is joining a ceramic part to a metal that comes as a liquid. The second one (B) is forming the interlayer structure, in which a liquid metal comes to form an interlayer between a ceramic and a metal. A simple calculation of a metal flow showed that a quite narrow gap between a ceramic and a metal, i.e., a gap of a few μm width, can be filled with a metal liquid under a pressure of 10 to 100 MPa [1]. This process forms interfaces under a dynamic movement of metal flow. Clean and perfect contact at the interface can be achieved even in an air atmosphere. In the following sections, several examples of interfaces produced by the SQ process are shown.

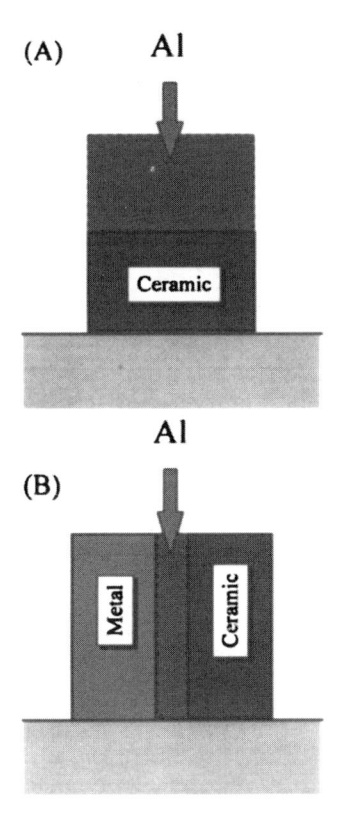

Figure 2. Two types of joining geometries.

JOINING ALUMINUM TO ALUMINA

Because aluminum is one of the quite reactive metallic elements, it reacts with most ceramics severely. The severe interface reaction causes the reduction of strength and reliability of a joint structure. Commercial alumina usually contains much amount of sintering additives such as silica and these react with liquid aluminum severely. It is, therefore, hard to obtain a strong interface in the composite structure. Fig. 3 shows typical interface strengths between alumina-silica and aluminum. The two blocks of alumina were brazed together with aluminum at 800°C for 10 min both in air and in an inert atmosphere. By the conventional process, no joint could be obtained in an air atmosphere, while a strength of only 50 MPa, a value which seems too low for most practical applications, was achieved by inert atmosphere brazing. In contrast, the interface strength formed by the SQ process almost reaches the strength of the parent alumina. When the aluminum layer becomes thick, the joint strength decreases to some extent. However, in the latter case, no interface fracture was observed and a crack always propagated in the aluminum layer. Such a great difference between the conventional method and the SQ process is attributed to the difference in reaction duration in both processes. In the conventional method, one need at least a few minutes to achieve interface formation and this duration sometimes causes severe reaction at interfaces. In the SQ process, on the other hand, typical reaction duration is a few seconds, and even within such a short duration, a tight interface can be formed under a pressure. At the interface, neither voids nor severe reaction were observed, as shown in Fig. 4.

The reliability of the interface formed by the SQ process is also high enough for practical uses. For instances, the Weibull modulus of the alumina joint obtained by the SQ process is beyond 10, that is, the same level of that of alumina used in the experiment.

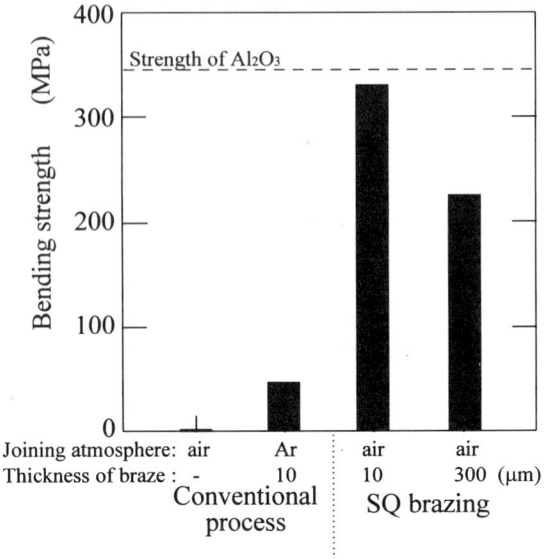

Figure 3. Alumina joint strength brazed with aluminum by conventional brazing and the SQ method.

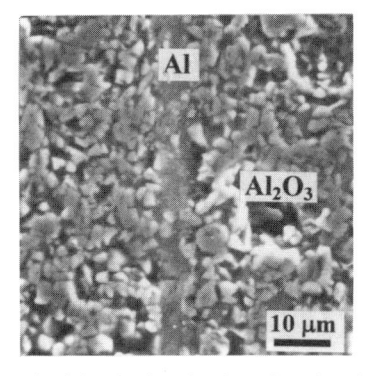

Figure 4. SEM photograph of the alumina/aluminum interface formed by the SQ process.

JOINING ALUMINUM TO SILICON NITRIDE

Silicon nitride is one of the non-reactive ceramics against aluminum liquid but a strong interface can be formed by the conventional method[2]. By the SQ process, a strong silicon nitride/aluminum interface is also formed. Fig. 5 compares joint strengths obtained by conventional brazing, by the SQ process and by solid-state joining. By the conventional method, high strength up to 450 MPa was obtained when joined at 800 °C for 10 min in an inert atmosphere. A slightly lower strength was obtained by the SQ process in air. However, non-wetting areas were frequently observed in the joint obtained by the conventional process[3], while neither non-wetting area nor interface de-bonding were observed in the joint prepared by the present process. By the conventional brazing, non-wetting areas seem to form easily in a non-wetting system such as silicon nitride/aluminum. The solid-state bonded joint, which was obtained by the processing at 650 °C for 1 h under a pressure of 10 MPa, exhibited extremely low strength. This sometimes happens because aluminum has a strong oxide skin on it, and this oxide prevents the formation of an intimate contact in such a low temperature range. One of the great advantages of the SQ process is that it can achieve perfect contact during joining regardless of the wettability of a given joining system.

The interface microstructure of silicon nitride and aluminum formed by the SQ process is basically the same one as that obtained by the conventional process. Fig. 6 shows the typical TEM microstructure of the interface. The reaction layer is approximately 100 nm thick consisting of two layers[4]. One next to the aluminum layer is amorphous containing aluminum, silicon and oxygen. This layer frequently contains particles from the oxide sintering aides in the silicon nitride. The other layer is fine crystalline β'-sialon, containing silicon, aluminum oxygen and nitrogen as the elements. This layer has a nano-structure as shown in Fig. 7.

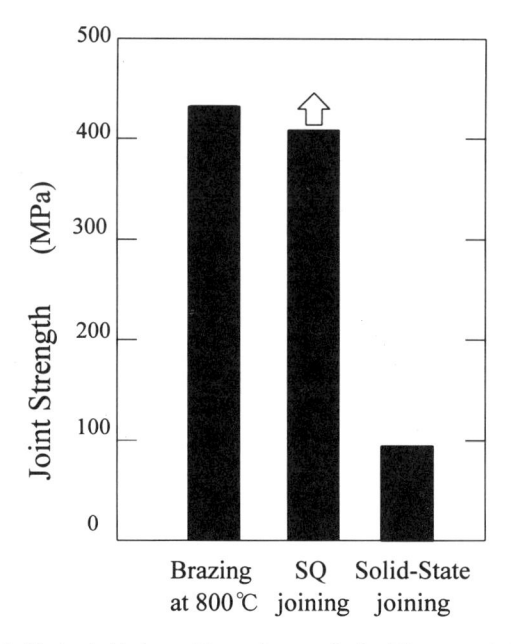

Figure 5. Joint strength of silicon nitride brazed by various methods. The arrow indicates no interface fracture was observed but fracture occurred in the braze layer.

APPLICATION TO THROUGH HOLE CONNECTION

One of the advantages of the SQ process is that there is no need to be concerned about the interface shape. Usually, the joint structure is limited by an interface shape. A complex interface shape cannot be filled with a braze for two reasons. Large metal shrinkage occurs during solidification of a braze and limited wettability between a ceramic and a braze metal prevents liquid metal from spreading over the interface. A plane-to-plane joint structure has been adopted in the practical cases of joining from these reasons. The SQ process enables one to make any kind of complex shaped interface because the liquid metal solidifies under

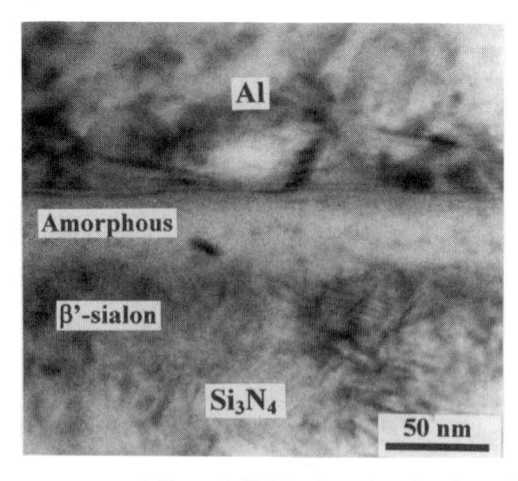

Figure 6. TEM microstructure of silicon nitride/aluminum interface formed by the SQ process.

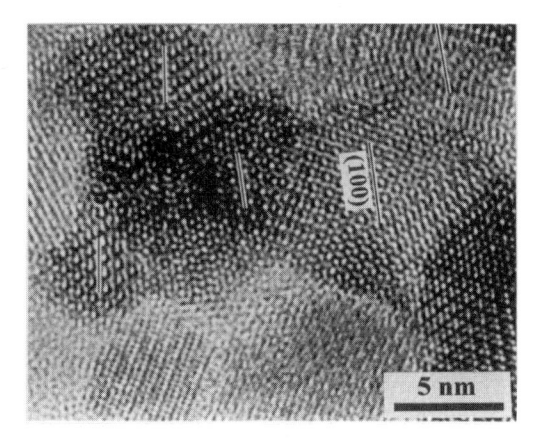

Figure 7. Nano-structure of β'-sialon layer formed at the on silicon nitride/aluminum interface by SQ process (TEM).

pressure. These characteristics can expand the application of joining technologies. Several applications are now under consideration and one of the typical cases is introduced in the present paper.

Recent advances in electronic packaging technology enable us to have highly integrated small devices for extremely downsized new products such as laptop/palmtop personal computers and mobile handy phones. Multilayered printed circuits and ceramic packages are one of the key technologies to fabricate them. Multilayer ceramic substrates are sintered by the cofiring process, in which a stack of printed green ceramic sheets with internal metal electrodes are sintered simultaneously. One of the serious problems in this process is achieving a tight bond between the electrodes and the substrate. Fig. 8 shows the typical defect structure observed in an actual device. Differential shrinkage and thermal expansion mismatch between the substrate and electrode produce poor bonding at the interface. Frequently, no bonding is achieved in a ceramic packages as shown in Fig. 8. Making reproducible good bond at the interface is required for making reliable products. The SQ process can be applied to producing the through-hole connection in multilayer ceramic devices. Fig. 9 illustrates the simplified idea of infiltration by the SQ process. First, a ceramic substrate is sintered with through-holes without any electrode. Under a pressure, an electrode such as aluminum or copper is infiltrated in a liquid-state through complex through-holes in a substrate and solidifies immediately. No defective structure is observed in an experimental sample as shown in Fig. 10. EPMA analysis revealed that no macroscopic interdiffusion occurs across the interface.

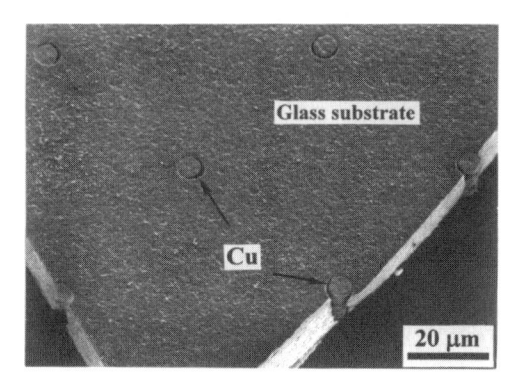

Figure 8. Defect structure observed in a multilayer ceramic package fabricated by the cofiring method (SEM).

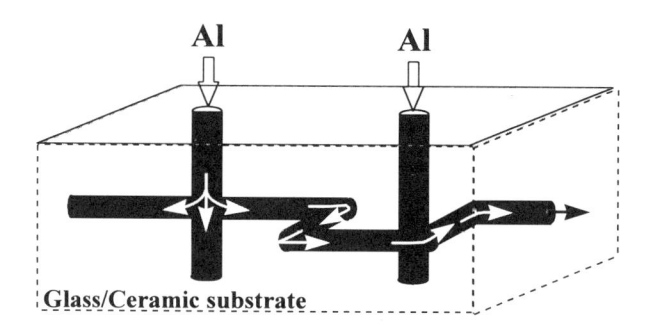

Figure 9. SQ process for through-hole connection of multi-layer ceramic packages.

SUMMARY

This paper presented the outline of the new SQ process for joining ceramics/metals and showed the potential application for electronic packaging technology. The process itself has been widely used in the production of metallic parts and it is well known that the products from this process possess excellent reliability. In the application to the fabrication of ceramic/metal composite structures, even though the metal used in the process must have a low melting point, there are also great advantages such as excellent economy (mass-producibility), suppression of severe reaction, little oxidation even in air, large/complex shape joining, easy formation of the interlayer, defect-free perfect contact and freedom from

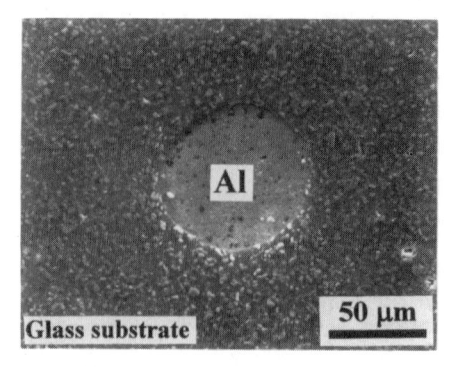

Figure 10. SEM photograph of through-hole pattern fabricated by the SQ process.

wettability constraints. The caution using this process is not to thermal shock the ceramic substrate when metal liquid is poured into a mold. Usually the preheating temperature is set to be the same temperature as that of the liquid metal. Several questions remain on the joining parameters and the details of the interface formation mechanism in the SQ process. For example, how does the roughness of ceramics influence the strength and reliability of the whole structure and what is the true mechanism for eliminating surface contamination of a liquid metal and a ceramic substrate during casting? With such information, one can control the process parameters in a straightforward manner.

REFERENCES

1. K. Suganuma, New process for brazing ceramics utilizing squeeze casting, *J.Mater.Sci.*, **26**: 6144 (1991).
2. X. S.Ning, K. Suganuma, T. Okamoto, A. Koreeda and Y. Miyamoto, Interfacial strength and chemistry of additive-free silicon nitride ceramics brazed with aluminum, *J.Mater.Sci.*, **24**:2879 (1989).
3. M. Morita, K. Suganuma, T. Okamoto, Effect of pre-heat-treatment on silicon nitride joining with aluminum braze, *J.Mater.Sci.Letters,* **6**:474 (1987).
4. K. Suganuma, Strength and microstructure of silicon nitride/aluminum interface fabricated by SQ brazing, *J. Europ. Ceramic Society,* **11**:43 (1993).

MICROANALYSIS OF BURIED METAL/CERAMIC INTERFACES USING NEUTRON REFLECTION

P. Xiao[1], B. Derby[1], J. Webster[2] and J. Penfold[2]

[1]Department of Materials, University of Oxford
 Oxford OX1 3PH, U.K.
[2]ISIS Facility, Rutherford Appleton Laboratory, Chilton, U.K.

INTRODUCTION

A better understanding of the interfacial phenomena between liquid metals and solid ceramics is very important in developing many technologies of materials processing. The interfacial energy, a key physical factor of the interface, is controlled by local chemistry. The segregation of small quantities of alloying elements to interfaces can have a dramatic influence on the energy as quantified by the Gibbs' adsorption isotherm. Interfaces between solid ceramics and liquid metals tend to be of high energy and are normally sufficiently high so as to prevent wetting. It is well known that the addition of small quantities of an active element, notably Ti, to many liquid metals substantially reduces the contact angle on ceramic surfaces. This is accompanied by the segregation of Ti to the metal/ceramic interface and the formation of an intermediate phase believed to be TiO.[1] The presence of TiO has been confirmed by X-ray diffraction of the exposed interface[2] after solidification. It has not been possible to directly characterise the composition of the metal/ceramic interface during wetting experiments because of the presence of the substrate and the wetting droplet which bury the interface. In this article we present results using the technique of specular neutron reflection to characterise such buried interfaces. The results show the capability of neutron reflection for detecting the presence of a thin interfacial layer at a solid ceramic/solid metal interface and the segregation of Ti at a solid ceramic/liquid metal interface.

The specular reflection of neutrons is now established as a technique for the study of a wide variety of problems involving surfaces and interfaces.[3] In a neutron reflection experiment, the specular reflection is measured as a function of the wave vector transfer, Q, perpendicular to the reflecting surface.

$$Q = 2k \sin\theta = \frac{4\pi \sin\theta}{\lambda} \qquad (1)$$

Ceramic Microstructure: Control at the Atomic Level
Edited by A. P. Tomsia and A. Glaeser, Plenum Press, New York, 1998

where k, θ and λ are the incident wave vector, wavelength and angle of incidence of the neutrons respectively. The wave vector transfer is related to the neutron refractive index profile normal to the interface and is often simply related to the neutron scattering length density profile $\rho(z)$, which is given by

$$\rho(z) = \sum_i b_i N_i(z) \qquad (2)$$

where b_i and N_i are the neutron scattering lengths and number densities of component i. The reflectivity is given approximately by,[3]

$$R(Q) = \frac{16\pi}{Q^4} |\rho'(Q)|^2 \qquad (3)$$

where $\rho'(Q)$ is the Fourier transform of the gradient of the scattering length density profile normal to the interface. Thus data from specular neutron reflection can be used to determine composition and density gradients of interfaces.

It has been shown[4] that the intensity of reflected and transmitted neutrons follow the same laws as electromagnetic radiation with the electric vector perpendicular to the plane of incidence and hence most standard optical formalisms can be used with only minor modification.[5] The refractive index of a condensed phase to unpolarised thermal neutrons, n, is commonly written

$$n = 1 - \lambda^2 A \qquad (4a)$$

with

$$A = \frac{b^2 N}{2\pi} \qquad (4b)$$

where λ is the neutron wavelength, b_c is the mean coherent neutron scattering and N is the density of atom sites. Neutron scattering lengths vary from element to element and from isotope to isotope in an almost random manner across the periodic table. Mean scattering lengths per atom and atomic densities at room temperature of a range of elements and compounds, of standard terrestrial isotope mix, used in this study are given in Table 1. The negative scattering length of Ti ensures that the presence of Ti segregated to interfaces will provide a large change in local scattering length density. Hence neutron reflection will be particularly sensitive to the presence of Ti.

Table 1. Neutron scattering lenghts[6] and atom site details at room temperature

Material	Mean Atomic Weight (gmol^{-1})	Density ($\times 10^3$ kgm^{-3})	Coherent Neutron Scattering Length ($\times 10^{-15}$ m)	Scattering Length Density (Nbc) ($\times 10^{-6}$ m^{-2})
Sn	118.69	7.30	6.225	2.39
Ti	47.90	4.54	-3.438	-1.39
Cu	63.55	8.96	7.718	6.55
Ag	107.87	10.50	5.922	3.47
Al$_2$O$_3$	20.39	3.97	4.862	5.71

Standard optical techniques can hence be used to model specular neutron reflection exactly, and have been described in detail elsewhere.[7] In summary the methods depend on classical optical theory by which the specular reflection at an interface is described by Fresnel's law with the reflectivity, R equal to one for incidence angles, θ, smaller than the critical angle and for θ greater than the critical angle by

$$R = \left[\frac{n_1 \sin\theta_1 - n_2 \sin\theta_2}{n_1 \sin\theta_1 + n_2 \sin\theta_2}\right] \quad (5)$$

where θ_1 and θ_2 are the glancing angle of incidence and the angle of refraction and n_1 and n_2 are the refractive indices of the two media respectively. For the case of a thin film present at an interface between two media the reflectivity can be written exactly[8]

$$R = \left[\frac{r_{01} + r_{12} e^{-2i\beta}}{1 + r_{01} r_{12} e^{-2i\beta}}\right] \quad (6)$$

where r_{ij} is the Fresnel coefficient at the ij interface such that

$$r_{ij} = \left[\frac{n_i \sin\theta_i - n_j \sin\theta_j}{n_i \sin\theta_i + n_j \sin\theta_j}\right] \quad (7a)$$

$$\beta = \frac{2\pi n_1 t \sin\theta}{\lambda} \quad (7b)$$

here t is the film thickness and the subscripts 0, 1, 2 refer to the first media, the interfacial layer and the second media respectively. The above method can be extended to more complex interfaces by dividing them into a series of discrete parallel layers. This is then solved using a matrix method of Abeles.[9] This treatment so far assumes ideal smooth interfaces whereas in reality there will be some diffuse scattering because of the interfacial roughness. This effect can be accommodated by modifying the Fresnel coefficients in equation 7.[10]

Thus it is possible to compare the measured specular reflection with models of interface structure and this is the approach used in this study. Because of the number of variables in the model, interface composition, interlayer thickness and interfacial roughness, it is often not possible to use a parametric variation of the model coupled with statistical "goodness of fit" tests to uniquely determine the interfacial structure. However, knowledge of local microchemistry can be used to impose constraints on the model parameters and it is still possible to extract useful information about many interfaces from analysis of reflection data.

EXPERIMENTAL PROCEDURE

In order to measure the neutron reflectivity as a function of wave vector transfer it is necessary to vary either the angle of incidence of the neutron beam or the neutron wavelength (equation 1). The neutron reflectivity profiles were measured using the CRISP instrument on the ISIS neutron source at the Rutherford Appleton Laboratory.[11] ISIS is a pulsed neutron source which provides a white beam of neutrons whose wavelengths can be distinguished by time of flight (TOF). This allows a fixed sample geometry and hence constant illumination area and Q resolution.

To access the buried interface, a single crystal solid substrate, in this case sapphire, is used aligned away from any Bragg reflection angle, if a polycrystalline substrate is used multiple Bragg scattering results in unacceptable attenuation of the neutron beam reflected from the buried interface. The neutron beam enters the sapphire block through the end face and is reflected from the ceramic/metal interface at a glancing angle. The reflected beam emerges through the opposite end face. The faces through which the beam enters and exits the block are almost perpendicular to the beam, avoiding further reflection. Very low glancing angles are used to maximise the reflected signal and a large interfacial area is illuminated by the neutron beam. This is shown schematically in Fig. 1

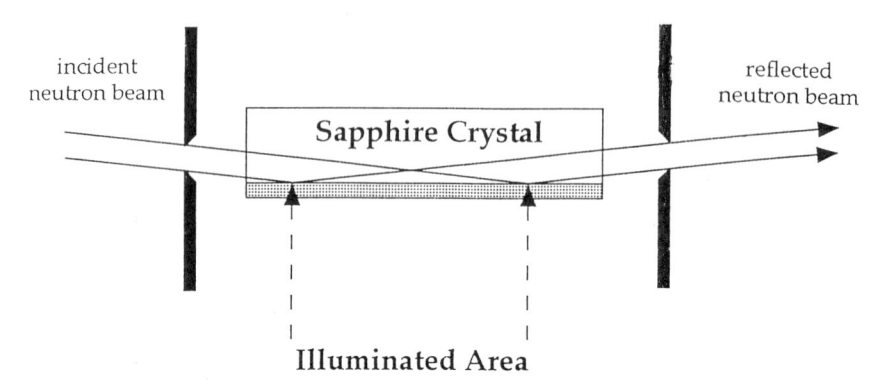

Figure 1. Schematic illustration of the sample orientation showing the collimated neutron beam entering the sapphire crystal and reflecting from the interface under study

Three specimens containing metal/ceramic interfaces were prepared for neutron reflection experiments. The Ag-Cu-Ti/sapphire specimen was prepared by applying a liquid Ag 70%-Cu 28%-Ti 2% alloy (Cusil ABA active brazing alloy: Wesgo, Belmont, CA, USA) on the large polished surface of a sapphire block (10 cm × 5 cm × 1 cm) at 850°C for 10 minutes also in a vacuum of 10^{-4} mbar. Neutron reflection was carried out at the Ag-Cu-Ti/sapphire interface at room temperature. The Sn-Ti/sapphire (Al_2O_3) specimen was prepared by the evaporation of a thin (5 nm) Ti film on a polished surface of a sapphire block, then applying a liquid Sn-0.2% Ti alloy on the coated surface in Ar at 750°C. Neutron reflection at the liquid Sn-Ti/sapphire interface was carried out at 310°C in air. The buried interface was not oxidised in this case due to relatively low temperature and a large area of the interface (5 cm × 7 cm).

RESULTS AND DISCUSSION

Neutron reflectivity profiles from the Ag-Cu-Ti/sapphire interface were measured at room temperature with a range of wave vector transfer of 0.0135 to 0.1755 $Å^{-1}$. Figure 2 shows the neutron reflectivity data (dotted line) with the range of wave vector transfer of 0.0135 to 0.06 $Å^{-1}$ while the reflected signal is not distinguishable from the background when the wave vector transfer is over 0.06 $Å^{-1}$. Fringes in the reflectivity profiles indicate the presence of more than one interface between two media. However, interfacial roughness damps these fringes. Normally the presence of Ti at a metal/ceramic interface results in a redox exchange reaction forming a Ti oxide. The interactions between Ti and O are complex and a range of compounds are possible each showing a considerable range of stoichiometry with anion:cation ratios in the range of 0 - 2. The scattering length densities of these materials are shown in Table 2 and thus any interfacial layer is likely to have a scattering length density in the range $-1.39 \times 10^{-6} < Nb_c < 2.66 \times 10^{-6}$ m^{-2}.

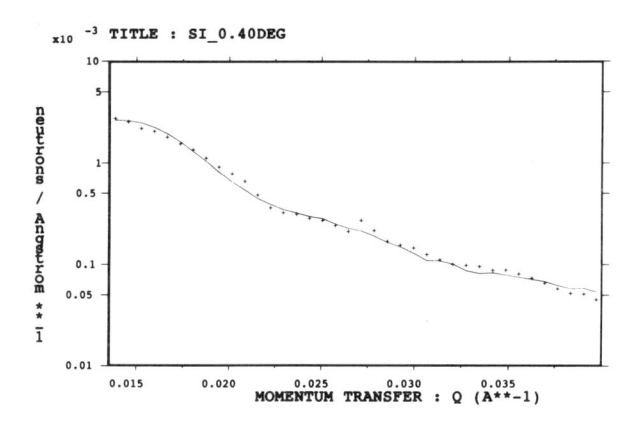

[Figure: Reflectivity profile plot, SI_0.40DEG, neutrons/Angstrom*10^{-3} vs MOMENTUM TRANSFER Q (Å^{-1})]

Figure 2. Reflectivity profile measured from the sapphire/Cu$_3$Ti$_3$O interface (+). The solid line superimposed on the data are fits to the reflectivity assuming a thin (670 Å) interfacial layer of scattering lengthdensity 2.5×10^{-7} m^{-2} present at the interface between sapphire and Cu$_3$Ti$_3$O.

Table 2. Coherent scattering length densities (Nb_c) for a range of Ti containing compounds

Material	Nb_c ($\times 10^{-6}$ m^{-2})
Ti	-1.39
Ti$_2$O	-0.25
TiO	1.33
Ti$_2$O$_3$	2.07
TiO$_2$	2.66
Cu$_2$Ti$_4$O	0.83
Cu$_3$Ti$_3$O	2.66

We examined a polished cross-section of the solidified sample in a scanning electron microscope equipped with a microanalysis facility to determine the chemical composition of the interface on a coarser scale. The cross-section of the sapphire/Ag-Cu-Ti specimen (Figure 3) shows the presence of a thick interfacial layer between Al$_2$O$_3$ and the Ag-Cu-Ti brazing alloy. This was analysed using wavelength dispersive spectroscopy which found a Ti:Cu ratio of 1:1 and the presence of oxygen. The possible reaction products between Al$_2$O$_3$ and Ag-Cu-Ti brazing alloys identified in the past[13-16] are given in Table 2, accompanied by their neutron scattering length densities. The most likely compound is Cu$_3$Ti$_3$O. Such a compound has a cubic structure with lattice parameter as 11.265 Å. It is believed to be almost stoichiometric at 925°C when it is in equilibrium with Cu$_2$Ti$_4$O and Ti$_2$O phases at 925°C, from the Cu-Ti-O phase diagram.[14] The fringes evident in the reflectivity profile of Figure 3 indicate the presence of an interfacial layer between Cu$_3$Ti$_3$O and Al$_2$O$_3$, which cannot be observed in Figure 3. The Cu$_3$Ti$_3$O layer is too thick to generate observable fringes in the range of wave vector transfer investigated. The two well defined sharp peaks at 0.27 Å^{-1} and 0.41 Å^{-1} are probably caused by a Bragg reflection condition being satisfied with one of the phases present and do not affect the interpretation of the data. Assuming the presence of a single layer at the interface between the Cu₃Ti₃O and sapphire, and we were able to obtain a reasonable match between our model and the experimental results (Figure 2). This was achieved using a single intermediate layer of scattering length density 2.5×10^{-7} m^{-2} with thickness 67 nm and roughness 11.5 nm. From the data of Table 2, the possible products at the interface with scattering length densities closest to this are Cu₂Ti₄O and Ti₂O. The thickness of the reaction layer we have identified is similar to those identified in the literature by transmission electron microscopy.[17]

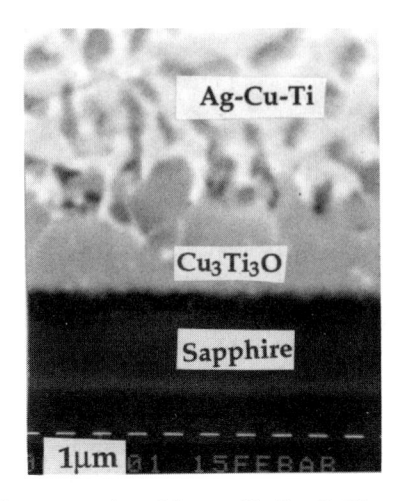

Figure 3. SEM micrograph of the cross-section of the sapphire/Ag-Cu-Ti specimen showing a thick interfacial reaction layer.

The solid Sn/sapphire interface has been previously characterised at room temperature using neutron reflection.[18] Here neutron reflectivity profiles from the Sn-0.2%Ti/sapphire interface were measured both at room and high temperatures with a range of wave vector transfer of 0.0135 to 1 Å^{-1}. It is reasonable to assume that the Ti film at the interface dissolved into the liquid alloy during the preparation of the specimen, according to the Sn-Ti phase diagram.[12] Figure 4 shows the reflectivity data from two measurements with the neutron beam reflected at a) the sapphire/liquid Sn-Ti interface at 310°C, b) the same specimen at room temperature after the liquid metal has solidified. Two reflectivity profiles show their difference at a range of low Q. The difference could be from a) the different densities of solid and liquid phases, b) different interfacial structures of the solid/solid specimen and the solid/liquid specimen. The segregation of Ti at the liquid Sn-Ti/sapphire interface should be within a region of a few atomic layers from the interface, and its thickness should be much smaller than the 670 Å thick layer at the Ag-Cu-Ti/sapphire interface. To determine a very thin interfacial layer of a few Å, neutron reflectivity should be measured and modelled to a high value of wave vector transfer Q for the interpretation of the fine interfacial structure. Figure 5 shows the modelled reflectivity profiles from the liquid alloy/sapphire interface with different values of the interface roughness. The best fit is the model of the metal/ceramic interface with no roughness, which cannot be true since the polishing technique used cannot achieve this level of surface smoothness. Assuming the interface roughness is 8 Å, the best fits are predicted with the presence of an interfacial layer with the scattering length density of 1e^{-6} which is between the scattering density of Ti (-1.39e^{-9}), and that of liquid Sn (2.2e^{-6}) (Fig. 6). This suggests that the interfacial layer is a mixture of Sn and Ti. Fig. 7 shows the modelled profile from the solid Sn-Ti/sapphire interface with an interfacial layer which is the same as that predicted in Fig. 6. The calculated reflectivity profile shows a different shape from the experimental results. The density change of Sn due to solidification has little effect on the reflectivity profile. Therefore, the difference between the reflectivity profiles of the solid/solid and solid/liquid specimens is due to the change of interfacial structures after the solidification of the liquid Sn-0.2%Ti alloy. The better fits are achieved by assuming a 200 Å thick interfacial layer containing Ti at the liquid/solid interface (Fig. 8). A thicker layer would be expected in this case because of solute rejection of Ti during the solidification of the Sn.

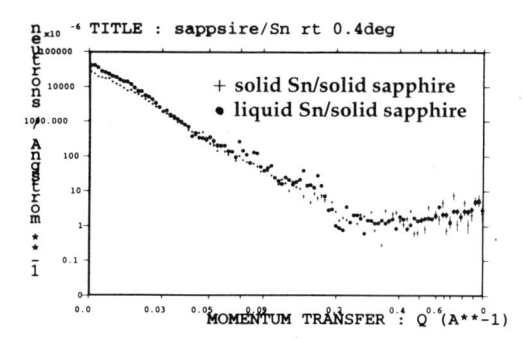

Figure 4. Reflectivity profiles measured for a: solid sapphire/liquid Sn-0.2%Ti alloy (+), b) solid sapphire Sn-0.2%Ti alloy interface (o).

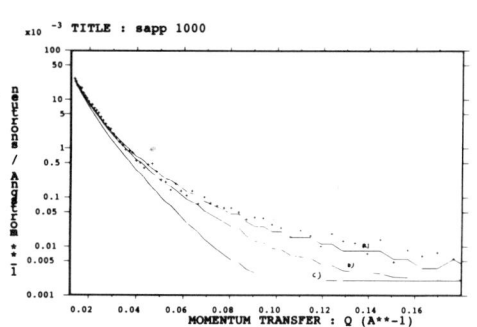

Figure 5. Reflectivity profile measured from the sapphire/liquid alloy interface (+) compared with model predictions (solid line) with no interfacial layer but with different interface roughness a) 0, b) 10 Å, c) 20 Å.

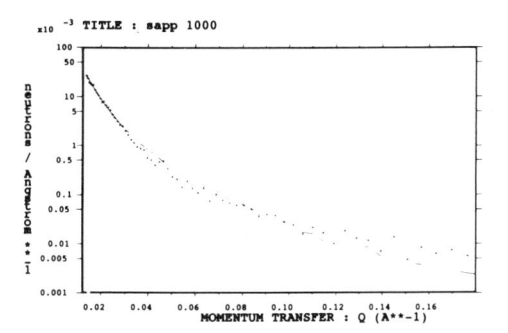

Figure 6. Neutron reflectivity profile predicted from the sapphire/liquid metal interface compared with experimental data (+) assuming a 2 nm thick interfacial layer containing Ti with an interface roughness of 0.8 nm.

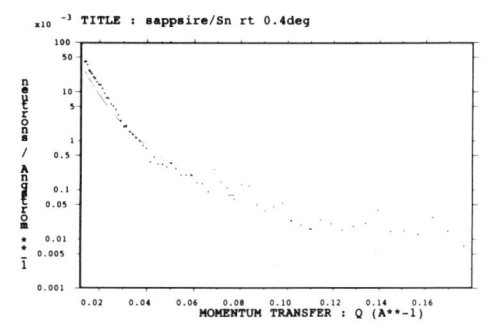

Figure 7. Neutron reflectivity profile measured from the sapphire/solid metal interface (solid line) compared with predicted profile (+) using the interface structure of Fig. 6, but with the metal as a solid phase.

CONCLUSIONS

We have determined the local structure of three metal/ceramic interfaces using specular neutron reflection. The technique have been found to be sensitive to the presence of very thin layers of Ti either introduced deliberately or segregated to metal/ceramic interface because of the negative scattering length of Ti. The thickness and chemical compositions determined for the Ti containing interfacial phases are consistent with those measured by

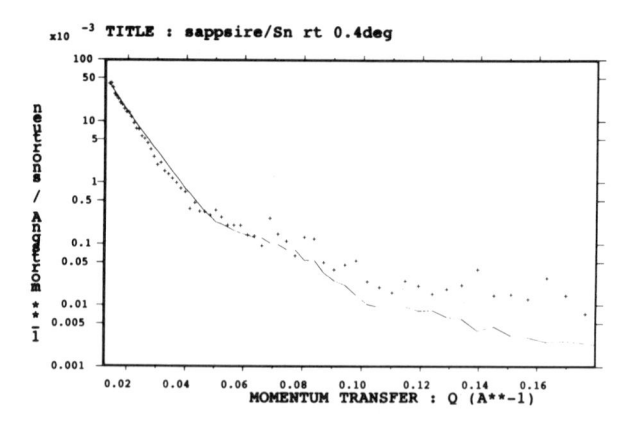

Figure 8. In comparison with Figure 7, better fits are obtained for the neutron reflectivity profile from the sapphire/solid metal interface by assuming a 12 nm thick interfacial layer containing a high concentration of Ti.

other analytical techniques and reported in the literature. The buried liquid metal/solid ceramic interface has been characterized using neutron reflection. An interfacial layer due to the segregation of Ti at the liquid/solid interface was detected. The structure of the liquid/solid interface changed significantly after solidification of the liquid phase in the metal/ceramic specimen. In conclusion we have demonstrated that specular neutron reflection can be used to determine the local structure and chemistry of metal/ceramic interfaces.

REFERENCES

1. M. G. Nicholas, in "Joining Ceramics", Ed. M. G. Nicholas, 73 (1990) Chapman and Hall.
2. Y. Naidich and V. S. Zhuravlev, *Refractories* (USSR), **15**:55 (1974).
3. J. Penfold & R. K. Thomas, *J. Phys. Condens. Matter*, **2**:1369 (1990).
4. M. L. Goldberger and F. Seitz, *Phys. Rev.*, **71**:294 (1947).
5. J. Leckner, "Theory of Reflection", (1987) Martinus Nijhoff.
6. V. F. Sears, Neutron Scattering Lengths and Cross-Sections, *Neutron News*, **3**:26 (1992).
7. J. Penfold, in "Neutron, X-Ray and Light Scattering", Eds. P. Linder and T. Zemb, 223 (1991) Elsevier.
8. M. Born and E. Wolf, "Principles of Optics" (1970) Pergamon.
9. O. S. Heavens, "Optical Properties of Thin Films" (1955) Butterworth.
10. J. Penfold, *J. Physique.*, C7 suppl., **10**:50, 99 (1989).
11. J. Penfold, R. C. Ward and W. G. Williams. *J. Phys. E: Sci Inst.*, **20**:1411 (1987).
12. E.A. Brandes and G. B. Brook, "Smithells Metals Reference Book", 7th Edn, (Butterworths, London, 1992)
13. G. P. Kelkar, K. E. Spear and A. H. Carim, *J. Mater. Res.*, **9**:2244 (1994).
14. G. P. Kelkar and A. H. Carim, *J. Amer. Ceram. Soc.*, **76**:1815 (1993).
15. W. Byun, H. Kim and M. Yun, *Script. Metall. Mater.*, **31**:1143 (1994).
16. H. Hao, Y. Wang, Z. fin and X. Wang, *J. Mater. Sci.*, **30**:1233 (1995).
17. M. L. Santela, J. A. Horton and J. J. Pak, J. Amer. Ceram. Soc., **73**:1785 (1990).
18. P. Xiao, B. Derby, J. Webster and J. Penfold, *Acta Mater.*, (in press)

RECENT PROGRESS IN SURFACE ACTIVATED BONDING

Tadatomo Suga

The University of Tokyo, RCAST
Meguro-ku, Tokyo 153, Japan

INTRODUCTION

Solid bodies only join if their surfaces are brought into contact and a bond is formed. The phenomenon itself seems quite natural and simple. However, various bonding methods exist today, indicating that the joining phenomenon is far from simple.

The majority of solids form stable chemical compounds at room temperature. In particular, metals bond to almost all elements due to the many free electrons that contribute to the bonds. While the surface energy of metals is in the range of 0.2 to 2.7 J/m^2, the surface energy of solids and their surface entropy (the temperature coefficient of the surface energy) is in almost all cases a multiple of this range.[1] Furthermore, for many substances the work of adhesion W_A, namely the difference of the surface energies γ_1, γ_2 and the interface energy γ_{12} ($W_A = \gamma_1 + \gamma_2 - \gamma_{12}$), can be described to be proportional to the square root of the product of the surface energies:[2]

$$W_A = 2\phi\sqrt{\gamma_1\gamma_2} \tag{1}$$

As a matter of fact, the measurement of the interface energies between Al_2O_3 and Au, Ag, Cu, Ni, γ-Fe results in values for ϕ ranging between 0.20 and 0.29, with ϕ defined as in Equation 1. This means that those material combinations that fulfill Equation 1 with the given ϕ are promising candidates for stable bonds.

The question, then, is why are there so many different bonding methods? Why, for example, is bonding between ceramics and metals so difficult to achieve?

The reason is that the previously stated condition for bonding (the surfaces have to be in contact) is in reality not very often fulfilled. The real surface is passivated by adsorption of various gases and organic substances. As the surfaces of metals are in general oxidized, contact with ceramics eventually means contact between the oxide layer on the metal and the ceramics. This does not necessarily mean that bonding is not possible. If the bond between the two materials is not based on ion bonds but rather covalent bonds, the surfaces are in an activated state due to the open or "dangling" bonds. For this activation of the surfaces, in general, a high temperature process is needed in order to open the bonds. This is the reason why the process of joining metals and ceramics usually takes place at elevated temperatures. However, at these high temperatures other reactions also take place that lead to the formation of other reaction products in the interface region. These reaction products very often have a brittle character, systematically degrading the strength and stability of the bond.

Accordingly, to join metals to ceramics, a method is needed to remove the oxide layer and subsequently to activate the surfaces in order create dangling bonds. Furthermore, the method should avoid high temperatures.

The surface activated bonding (SAB) method, as described in the present paper, is a room temperature bonding method based on the controlled removal of the surface oxide layers by low energy atom irradiation or cold hydrogen plasma exposure.

SURFACE ACTIVATED BONDING (SAB)[4-7]

Surface activated bonding (SAB) is one of the most promising methods of room-temperature materials interconnection. The activation can be achieved either by ion irradiation, fast atom beam (FAB) irradiation, or hydrogen radical beam irradiation. The bonding machine is shown in Figure 1.

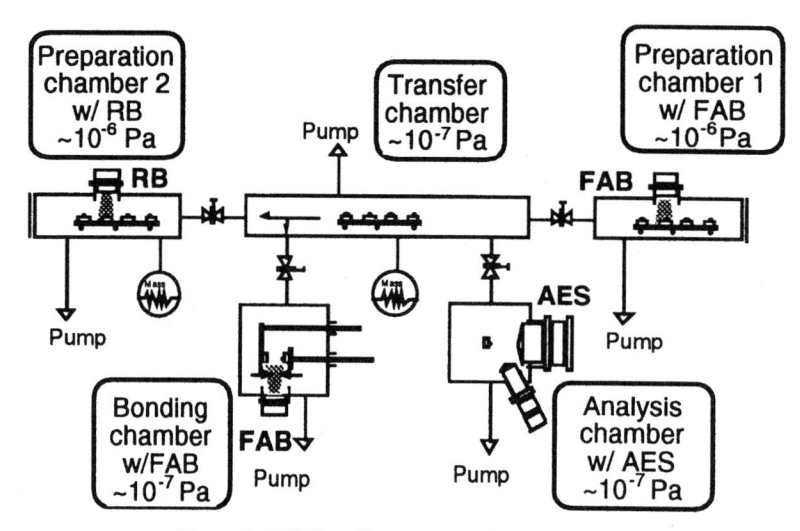

Figure 1. UHV bonding apparatus for SAB process

Until now metals such as Al, In, Sn, Pb, Cu, Au, Ag, Ni, Ti, Fe, semiconductors Si and diamond, and ceramics like Al_2O_3, ZrO_2, SiC, Si_3N_4 and AlN have been successfully bonded. The bonding strength for metal-metal joints will almost reach the strength of the metal—in many cases the joint fails not at the interface but within one of the materials. In the case of Si_3N_4 - Al joints, for instance, tensile strengths of about 100 MPa are achieved.

The method has a number of advantages:

- Since bonding is carried out at room temperature, materials with different coefficients of thermal expansion can be joined without the problem of residual stresses.

- The bonding at room temperature does not lead to a reaction layer. In particular, no solder material is needed.

- With bonding at room temperature, devices that cannot be exposed to heat treatment can be connected.

Figure 2 shows the cross section of an Al-Al joint that was bonded in ultra high vacuum (pressure better than 10^{-8} Pa) at room temperature: direct bonds are realized on an atomic level. The tensile strength reached 120 MPa. The structure of this interface is modified in vacuum by reaction with residual gases, in particular water. Figure 3 shows an interface with an intermediate layer of a hydroxide that was formed in a vacuum of 10^{-5} Pa. The bond with this type of interface still shows a strength of 50 MPa. The interface between Al and Si_3N_4 shows in Figure 4 as an amorphous layer on the Si_3N_4 side that might result from the ion radiation.

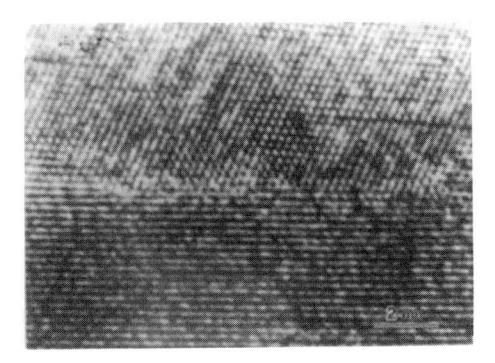

Figure 2. High resolution TEM image of an Al-Al interface fabricated by the SAB method at room temperature.

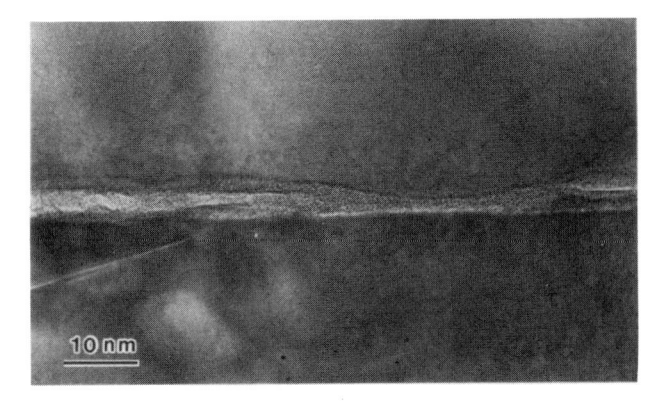

Figure 3. High resolution TEM image of an Al-Al interface with an amorphous intermediate layer, which might be formed during the surface activation process in a vacuum containing water.[6]

Furthermore,
- Al - Al joints can be formed by this method even with an oxygen atom beam.
- Al - Pb joints (a combination of materials that do not form solid solutions) can be formed only after irradiation times much longer than the time needed to remove the oxide layers on the surfaces.

Taking these observations into consideration, the joining mechanism should be understood not simply as a contact phenomenon of clean metal surfaces, but as a more dynamic process initiated by ion bombardment.

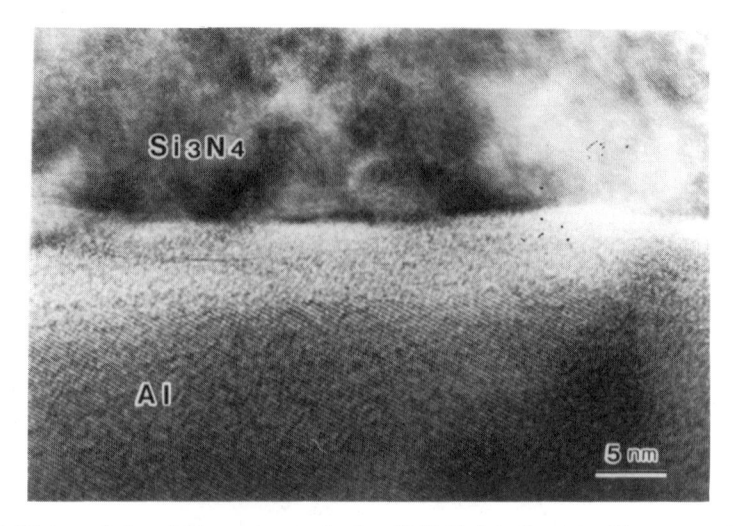

Figure 4. High resolution electron micrograph of an Al-Si$_3$N$_4$ interface joined in a conventional high vacuum. A glassy phase formed on the Si$_3$N$_4$ surface during argon irradiation.[6]

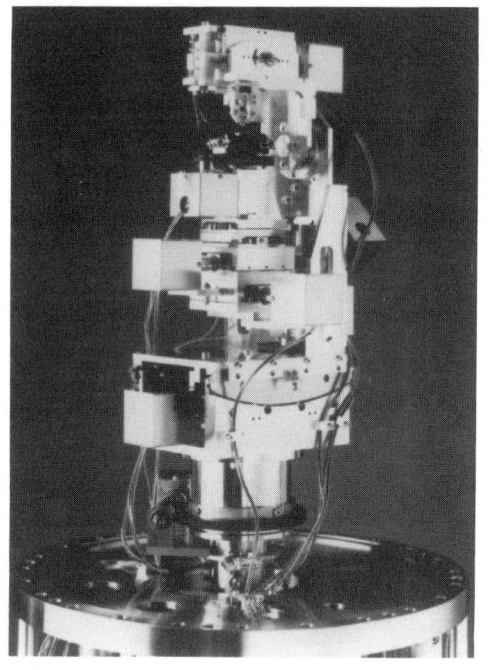

Figure 5. Micromanipulator for macroassembly using the SAB concept.

APPLICATION OF THE SURFACE ACTIVATED BONDING METHOD

SAB is, in other words, aimed at lowering the energy level needed for bonding by utilizing the surface activation process. Accordingly, a variety of other applications besides the room temperature bonding have been made.

- The oxide layer on the surface of solder materials can be removed by ion bombardment. Even if the surfaces are exposed to air for a certain time (resulting in the formation of a thin oxide film) "fluxless soldering" is still possible.[8]

- For the electrodes that are used in the wire bonding process, activation of the surfaces leads to a cleaning effect—although exposed to the atmosphere—and eventually to facilitated bonding conditions.[9]
- The bonding temperature can be lowered by the surface activation process even for conventional diffusion bonding.[10] SAB promises to bond devices that will not stand elevated process temperatures, such as opto-electronic devices.
- SAB allows fabrication of high-functional clad materials.[1]
- Direct semiconductor wafer bonding becomes possible when the adsorbed water on the surfaces is removed by ion bombardment instead of by the conventional annealing method.[12]

Furthermore, the technology can be considered to play a role in the assembly of micro machines. Figure 5 shows an apparatus that was constructed for the purpose of UHV micro assembly. Tiny parts can be manipulated on a nanometer scale under UHV SEM observation and can eventually be bonded at room temperature using the SAB method.[13]

CONCLUSION

The surface activated bonding method is based on a very simple principle, so a variety of applications in different fields are possible.

REFERENCES

1. L.E. Murr, *Interface Phenomena in Metals and Alloys*, Addison-Wesley Pub. Co. (1975).
2. R.J. Good, L.A. Girifalco, G. Kraus, *J. Phys. Chem*, **62**:1418 (1958).
3. R.M. Pillar, J. Nutting, *Phil. Mag.*, **16**:181 (1967).
4. T. Suga, Y. Takahashi, H. Takagi, Ceramic Transactions, **35**:323 (1993).
5. T. Suga, K. Miyazawa, *Acta Scripta Metall, Proc. Ser.* **4**:189 (1990).
6. T. Suga, Y. Takahashi, H. Takagi, et al., *Acta Metall. Mater.*, **40**:s133 (1992).
7. T. Suga, *Proc. 2nd Int. Symp. on Semiconductor Wafer Bonding*, Hawaii, 314 (1993).
8. T. Nishikawa, et al., *Ann. Mtg. Japan Welding Soc.*, **53**:198 (1993).
9. R. Kajiwara, et al., *Ann. Mtg. Japan Welding Soc.*, **55**:110 (1994).
10. O. Ohashi, *Ann. Mtg. Japan Inst. Mater.*, 169 (1993).
11. K. Saijoh. *Ann. Mtg. Japan Inst. Mater.*, 462 (1990).
12. H. Takagi, K. Kikuchi, R. Maeda, T.R. Chung, T. Suga, *Phys. Lett.*, **68**:2222 (1996).
13. T. Suga, N. Hosoda, *Proc. MEMS'95* (1995).

CERAMIC-METAL INTERFACES IN ELECTRONIC CERAMICS
---INTERFACE BETWEEN AlN CERAMICS AND CONDUCTORS

Koichi Niwa, Koji Omote, Yasushi Goto and Nobuo Kamehara

Fujitsu Laboratories LTD.,
10-1, Morinosato-Wakamiya, Atsugi 243-01, Japan

INTRODUCTION

AlN has a high thermal conductivity and thermal expansion coefficient similar to silicon, and as a result it is a good packaging material for Si semiconductor chips. To increase the thermal conductivity, typically $CaCO_3$ and Y_2O_3 are used as a firing additives, these additives may lead to the formation of $AlCa_2O_4$ and $Y_3Al_5O_{12}$ liquid phases during firing, and to the elimination of oxygen from AlN ceramics.[1,2]

For AlN use as a substrate in circuit boards, metallization is a very important step. W is typically used in the metallization of AlN ceramics.[3,4] There are some reports of W/AlN co-firing in the literature.[5-7] Several research papers suggested that there is no interdiffusion of either W or Al at the interface. However, Kurokawa et al. reported that a 200-nm-thick W interdiffusion layer formed at the interface of the AlN ceramics, and an intricate interlocking AlN-W grain structure developed.[6] However, both the reaction at the interface and the effect of adding AlN to W conductors as a function of firing temperature are not yet well understood.

We investigated the effect of sintering additives on the interface of AlN ceramics and W conductors and the role of the liquid phase which occurs during firing. The effect of cooling rate after firing on the interface between the AlN ceramics and W conductors was also investigated.

EXPERIMENT

Two types of AlN green sheets (AlN-2 wt% $CaCO_3$, AlN-5 wt% Y_2O_3) were produced. AlN green sheets were composed of 99.9% purity AlN powders and 99.99% purity $CaCO_3$, and 99.9% Y_2O_3. These green sheets were manufactured by the doctor blade method. The W conductor was formed by the screen printing method. Two types of W pastes (W-0%AlN, W-0.5%AlN) were produced. W-0%AlN pastes were composed of W powder and organic vehicle. W-0.5%AlN pastes were composed of 0.5 wt% AlN, 99.5 wt% W powders, and organic vehicle. The average particle size of the W powders was 1.1 μm. The purity of the W powder was 99.9%.

AlN with $CaCO_3$ was dewaxed at 650°C for 30 min in a nitrogen atmosphere. The firing conditions ranged from 10 minutes at 1400°C to 9 hours at 1800°C in a W element furnace.

AlN with Y_2O_3 was dewaxed at 450°C in air, and fired at 1000 to 1800°C in a graphite element furnace.

Quenched samples were fabricated from AlN with $CaCO_3$ green sheets and W-0.5% AlN paste. The dewaxed body was then set into a quartz ampoule. It was cooled from 1400°C to room temperature in 20 sec.

The adhesion strength of AlN ceramics and W conductors was measured by performing tensile tests in Instron.

The microstructure of the interface between AlN ceramics and W conductors was observed by SEM, TEM, and the atomic element distribution was measured by EPMA using Energy-Dispersive (EDS) and Wavelength-Dispersive (WDS) X-ray spectrometer.

RESULTS

The SEM microstructure of the fractured interface between the AlN ceramics with $CaCO_3$ and W conductor fired for 9 hours at 1700°C is shown in Figure 1. The polished cross section of the interface is shown in Figure 2. Both pictures show that the interface of AlN ceramics with $CaCO_3$ and the W-0% AlN conductor is clearly defined and sharp. On the other hand, the interface between AlN ceramics with $CaCO_3$ and W-0.5% AlN conductors is rough, uneven, showing an interlocked structure. Table 1 shows the adhesion strength between the AlN ceramic with $CaCO_3$ and W conductor fired for 9 hours at 1700°C. The adhesion strength of the AlN ceramic/W-0.5%AlN conductor interface was two times higher than that between the AlN ceramic and the W-0%AlN conductor. The high strength results from the interface interlocking between the AlN ceramic with $CaCO_3$ and the W-5% AlN conductor.[6]

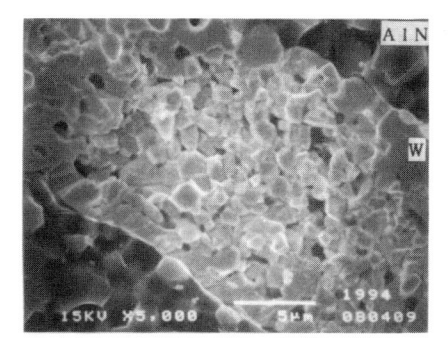

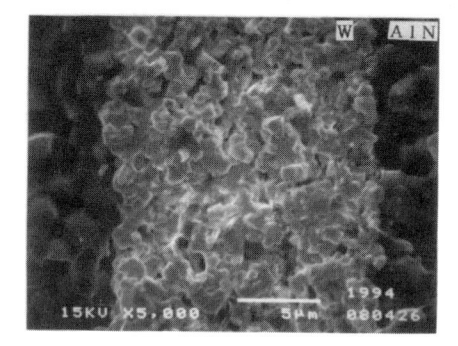

AlN/W 1700°C, 9 h   AlN/W-0.5% AlN, 1700°C, 9 h

Figure 1. Fractured cross-section of AlN ceramics with CaCO3 and W conductors fired for 9 hours at 1700°C

Table 1. Adhesion strength between the AlN ceramics with $CaCO_3$ and W conductors fired for 9 hours at 1700°C.

	Adhesion Strength (MPa)
W	9.0
W-0.5%AlN	>20

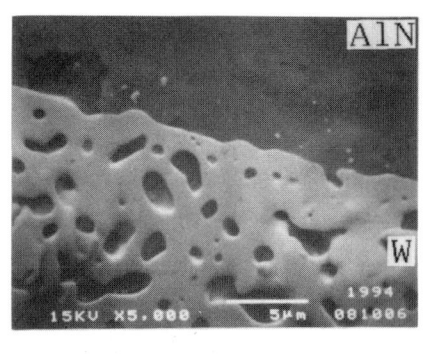

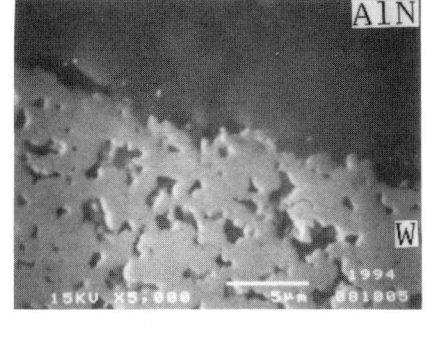

AlN/W 1700°C, 9 h   AlN/W-0.5% AlN, 1700°C, 9 h

Figure 2. Polished cross-section of AlN ceramics with CaCO3 and W conductors fired at 1700°C for 9 hours.

An EDX analysis of the interface in the W-0%AlN specimen indicates that a substantial amount of Ca exists in the W conductor layer. TEM photographs at various firing temperatures of the AlN ceramic with CaCO₃ and W conductors are shown in Figures 3 and 4. At temperatures less than 650°C, W was not observed in the AlN ceramics. At temperatures over 1400°C, W was observed in the AlN ceramic. The W particles at 1400°C have a small size, however, for firing temperatures over 1600°C they have a large size. The penetration depth of W particles into AlN ceramics is 2-3 μm.

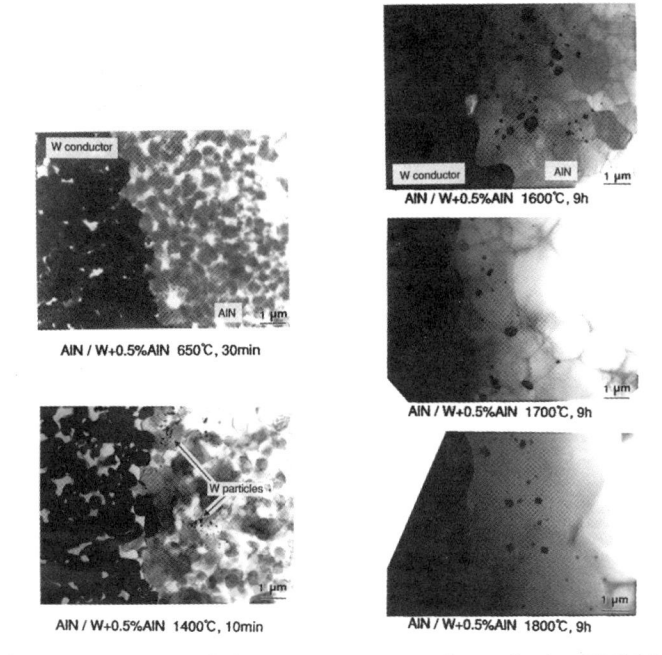

AlN / W+0.5%AlN 650°C, 30min

AlN / W+0.5%AlN 1600°C, 9h

AlN / W+0.5%AlN 1400°C, 10min

AlN / W+0.5%AlN 1700°C, 9h

AlN / W+0.5%AlN 1800°C, 9h

Figure 3. TEM micrographs of AlN ceramics and W conductors fired at 1700°C for 9 hours.

The SEM and EDX image of the AlN fracture surface with Y₂O₃ fired at 1850°C for 5 hours is shown in Figure 5. W particles are observed in the AlN ceramics. Figure 6a shows the XRD patterns of W plate peeled off from an AlN ceramic with Y₂O₃ after firing. When the firing temperature was less than 1000°C, only a W peak is observed, but at 1100-1200°C, W₂C and WC are observed. Above 1400°C, the WC and W₂C peaks disappear. The tungsten carbide seems to influence the formation of W particles in the AlN ceramics. Figure 6b shows the XRD pattern of the AlN surface after removal of the W plate which was fired at over 1600°C, Y-Al-O and W peaks are observed. Probably secondary or ternary liquid phases composed of WC, W_2C and sintering additives such as Y-Al-O exist above 1400°C.

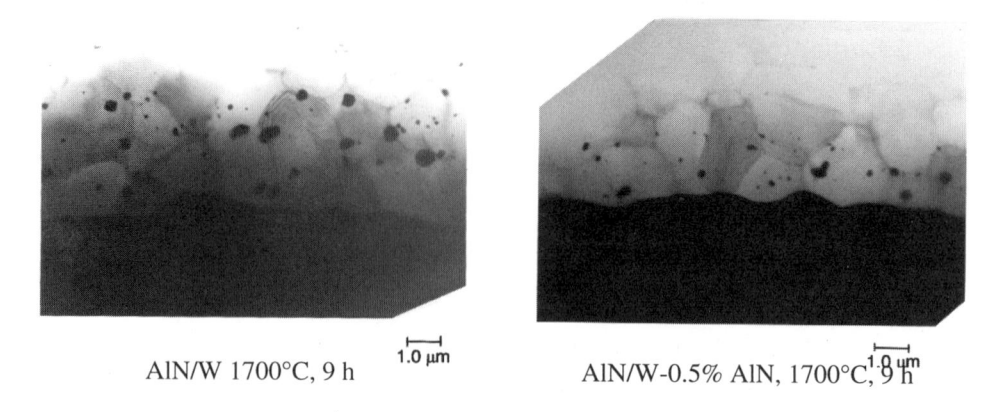

Figure 4. TEM micrographs of AlN ceramics and W conductors fired at 1700°C for 9 hours.

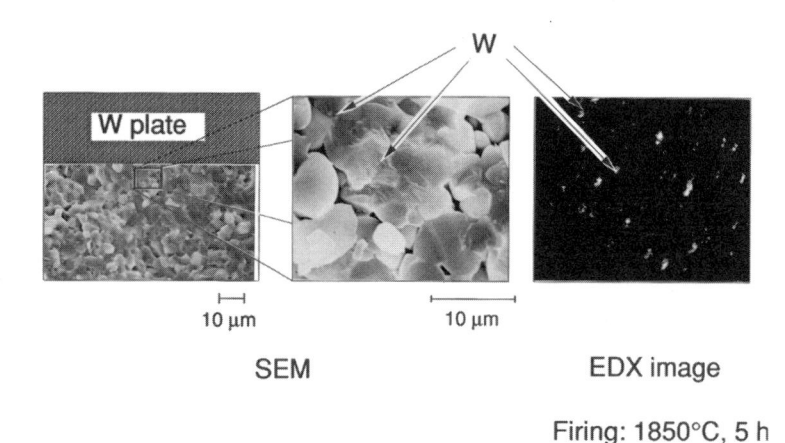

Figure 5. Fractured surface of SEM and EDX image of AlN ceramics with Y_2O_3 fired at 1850°C for 5 hours.

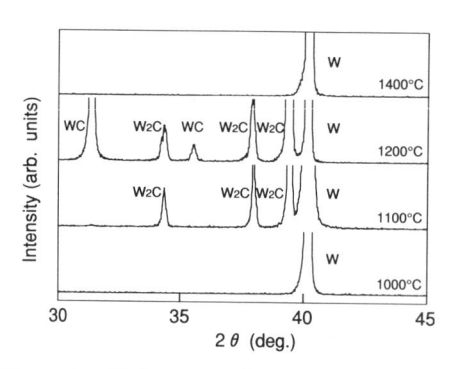

Figure 6a. XRD pattern of peeled off W plate from AlN ceramics with Y_2O_3 after firing.

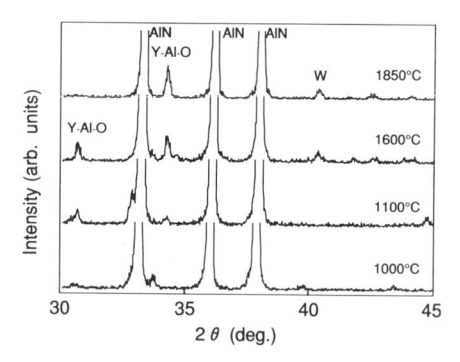

Figure 6b. XRD pattern of peeled off AlN ceramics with Y_2O_3 from W plate after firing.

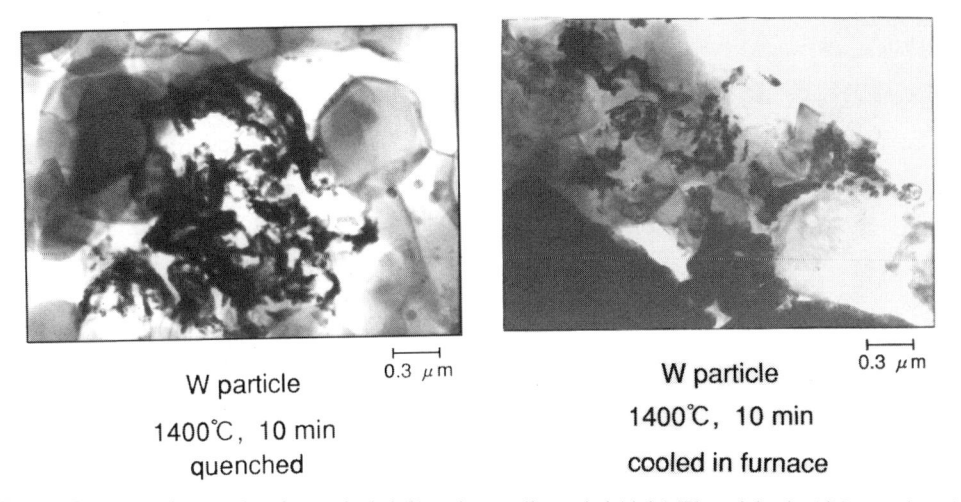

Figure 7. TEM Micrographs of quenched (left) and naturally cooled (right) W particles in AlN ceramics with CaCO$_3$.

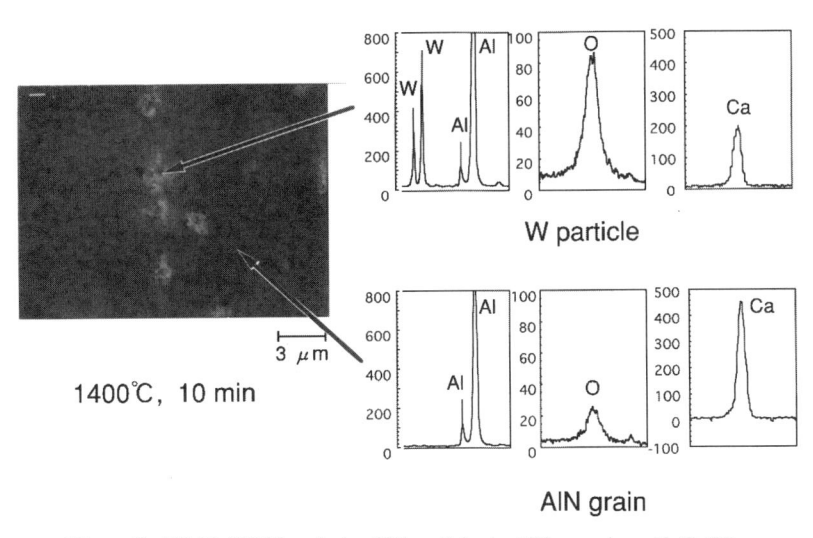

Figure 8. EPMA-WDX analysis of W particles in AlN ceramics with CaCO$_3$.

A TEM image of quenched and slowly cooled samples showing W particles in AlN ceramic with CaCO$_3$ is shown in Figure 7. In the quenched sample, the W particles are agglomerated within the glass phase in the AlN ceramics. On the other hand, in the slowly cooled sample, small W particles are distributed within the AlN grains and on the grain boundaries. Figure 8 shows EPMA-WDX analysis of W particles in the AlN ceramics with CaCO$_3$. It is clear that a significant amount of oxygen exists in the W particles although little oxygen is measured in the AlN ceramics. This suggests that the W-oxide reacts with Al$_2$O$_3$ and Ca-Al-O, and then forms a liquid phase.

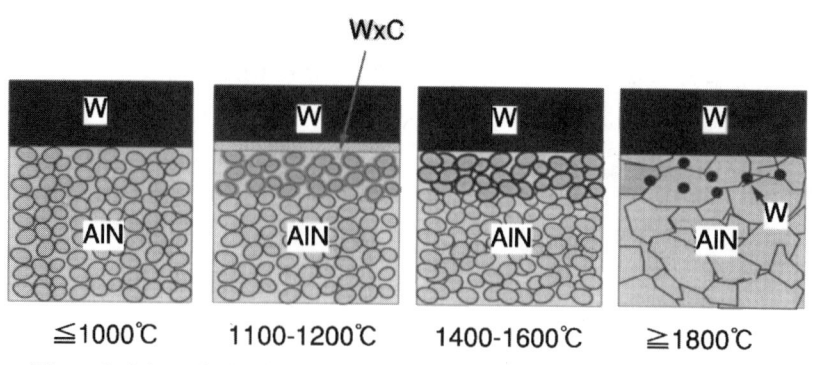

Figure 9. Schematic drawing of reaction between AlN ceramics with Y_2O_3 and W plate.

DISCUSSION

AlN powders typically contain a surface oxide layer. This surface oxide layer plays a key role in the formation of liquid phases during firing. According to the phase diagram, the Al_2O_3 and CaO system has a eutectic at as low as 1390°C.[8] As shown in Figures 4 and 8, in the W-0% AlN system a $CaAl_2O_4$ liquid phase is believed to form in the AlN ceramic, which penetrates into the W conductor layer until its concentration equilibrates with the AlN ceramic and W conductor. Subsequently, the $CaAl_2O_4$ liquid evaporates leaving residual Ca in the W conductor. However, in the W-0.5%AlN system, Al_2O_3 is already present in the W conductor on the imbedded AlN grains. Liquid phases are thought to form in both the conductor and ceramic during firing and then evaporate or diffuse to the AlN/W interface leaving little Ca residue in the W conductor. The liquid phase formation in the W conductor and interdiffusion to the AlN ceramic interface produces the interlocking structure.

Similar results are shown in Figure 6; secondary or ternary liquid phases composed of WC, W_2C and sintering additives such as Y-Al-O are formed at temperatures above 1400°C. The liquid phases penetrate into the AlN ceramic, and then selective elements in the liquid phase evaporate while W particles remain in the AlN ceramic. A schematic drawing of the reaction between AlN-Y_2O_3 ceramics and W plate is shown in Figure 9. In AlN ceramics with Y_2O_3, W_xC forms on the W plate at temperatures higher than 1100-1200°C.

Previously, researchers have suggested[6] that there is no interdiffusion of either Al or W in the W conductor at the AlN/W interface. However, as shown in Figures 3 and 7, W particles were observed in the AlN ceramics with $CaCO_3$ additives after firing. Ternary or higher component W compounds such as carbides or oxides form glasses upon cooling at relatively low temperatures compared to the firing temperature. The glass phase is most probably a remnant of prior liquid phases. Therefore, W because of its high melting temperature does not directly diffuse into the AlN ceramic as cited previously, but it is most likely carried by the liquid phases during firing.

CONCLUSIONS

In this study we investigated the interface between AlN ceramics and W conductors. CaO in the AlN ceramic forms a liquid phase with Al_2O_3, and the liquid phase moves towards the W conductor and reacts with Al_2O_3 in the W conductor. As a result an interlocking interface formed which contributes to strong adhesion between the AlN ceramic and W conductor.

W particles observed in the AlN ceramic seem to be carried by the liquid phase of Ca-Al-O at temperatures over 1400°C. W particles are also observed in the AlN ceramics with Y_2O_3 additives. In this case, W_2C reacts with Al_2O_3; the liquid phase increases the W mobility into the AlN ceramic.

Quenched samples (Figure 7) suggest that W may react as an oxide such as WO or WO_3 with Aluminum Oxide and/or Ca-Al-O compounds. The ternary system of WO_3, Al_2O_3-CaO will have a low eutectic point below 1000°C. This forms a liquid phase in AlN/W interface during sintering. Then the liquid penetrates into AlN. Finally, W particles precipitate. Therefore, W in the AlN ceramic is probably not caused by diffusion of W metal, but by the formation of a ternary or binary liquid phases consisting of W oxide or carbide.

REFERENCES

1. K. Komeya, A. Tsuge, H. Inose and H. Ohta, Effect of $CaCO_3$ addition on the sintering of AlN, *J. Mat. Sci. Let.*, **1**:325 (1982).
2. K. Komeya and H. Inose, The influence of fibrous aluminum nitride on the strength of sintered AlN-Y_2O_3, *Trans. J. Brit. Ceram. Soc.*, **70**:107 (1971).
3. M. Tsukada, E. Udagawa, N. Kamehara and K. Niwa, Cofiring of an aluminum nitride and tungsten system, Proc. of 1st Jap. Inter. Symp., **1**:408 (1989).
4. H. Makihara and N. Kamehara, Additive sintering process of aluminum nitride ceramics, *Mat. Res. Soc. Symp. Proc.*, **249**:437 (1991).
5. E. Udagawa, Y. Imanaka, N. Kamehara and K.Niwa, Influence of firing-gas pressure on the microstructure and thermal conductivity of AlN ceramics, *J. Mat. Sci. Let.*, **9**:116 (1989).
6. N. Kurokawa and H.Taniguti, Interfaces between conductor and AlN ceramics, *J. Mat. Sci. Lett.*, **3**:471 (1984).
7. K. Niwa, K. Omote and Y. Goto, Abst. 97th Annual Meeting of American Ceramic Society, Cincinnati, 97, 170 (1995).
8. A. K. Chatterjee and G. I. Zhmoidin, The phase equilibrium diagram of the system $CaO-Al_2O_3-CaF_2$, *J. Mater. Sci.*, 7:93 (1972).

DENSIFICATION OF TUNGSTEN CONDUCTORS IN COFIRED ALUMINUM NITRIDE MULTILAYER SUBSTRATES BY THE ADDITION OF MANGANESE OXIDE

Jun Monma, Takaaki Yasumoto, Takashi Takahashi, and Nobuo Iwase

Toshiba Corporation
Materials and Devices Development Center
Yokohama, Kanagawa, Japan

ABSTRACT

The sintering behavior of tungsten (W) conductors embedded in aluminum nitride (AlN) multilayer substrates was improved by adding manganese dioxide (MnO_2) to AlN green sheets. To obtain AlN suspensions, MnO_2 powder (purity: 99.99%) was mixed with AlN powder, yttrium oxide (Y_2O_3) powder (as a sintering additive), casting organic additives, and organic solvents by ball milling. Each suspension was cast into AlN green sheets and tungsten conductors were then printed on them. To obtain AlN multilayer substrates, the green sheets were laminated and then sintered at 1780, 1820, and 1850°C for 4 hours in a nitrogen atmosphere. The microstructure of the fracture surface of the AlN multilayer substrate with added MnO_2 indicated that the densified W conductor layers were without pores. Segregation of manganese (Mn) was observed among W grains by EPMA in the AlN multilayer substrate heated up to 1500°C. Moreover, rapid W grain growth was observed in the specimen, and Mn wetted these W grains well. These results indicated that molten Mn penetrated into W grain boundaries from the AlN matrix and accelerated W grain growth by liquid phase sintering during AlN multilayer substrate cofiring. Therefore, the AlN multilayer substrates, including the densified W conductors, were prepared by the addition of MnO_2 into AlN green sheets.

INTRODUCTION

Aluminum nitride (AlN), with its high thermal conductivity,[1] has received considerable attention recently for use in semiconductor devices. As these devices increase in complexity and acquire denser circuits, they require better means of dissipating heat. Moreover, AlN has a thermal expansion coefficient that approximately matches that of silicon and high electrical resistivity.[2,3] These excellent properties give rise to the use of AlN ceramics as cofired multilayer substrates.[1,4–8] Tungsten (W) metal is generally employed as the refractory metal for conductors embedded in cofired AlN multilayer substrates because of its high melting point.[6–8] However, pure W tends to leave many pores in the conductor layers after pressureless sintering. Luh et al. employed a hot press process to enhance densification of pure W embedded in AlN substrates.[6] Densification of W affects the requirements of AlN multilayer substrates, including hermeticity and surface flatness.

The possibility of lowering the sintering temperature of W by adding various activating agents has been attracting the attention of investigators.[9–15] In this study, we used manganese dioxide (MnO_2) as a sintering additive for W. Any sintering additive for W should be segregated from the AlN matrix to W conductor layers during cofire sintering at approximately 1800°C because AlN and W should be densified concurrently without W melting before densification of AlN. Thermodynamic studies indicate that MnO_2 should be reduced in a nitrogen atmosphere during the firing.[6] Kasori et al. reported that MnO_2 was eliminated from AlN ceramics during sintering at 1800°C,[16] and Petrunin et al. reported that molten Mn wetted W well at 1500°C and penetrated to a great depth along the W grain boundaries without forming any intermediate phases.[11] Therefore, we fabricated AlN multilayer substrates from AlN green sheets with added MnO_2 to densify W conductors by liquid phase sintering. After sintering at 1820°C for 4 hours in a nitrogen atmosphere, the AlN multilayer substrates with 5 wt.% MnO_2 added were densified and the tungsten conductors in them were also densified without pores.

EXPERIMENTAL PROCEDURE

The AlN powder commercially available (Tokuyama F-type) is produced by carbothermal reduction of aluminum oxide (Al_2O_3). As a sintering additive, yttrium oxide (Y_2O_3) was used. The Y_2O_3 powder manufactured by Mitsubishi Chemical Industries is 99.99% pure, according to the supplier. To bring out a black appearance of AlN multilayer substrates, tungsten oxide (WO_3) was added. The WO_3 used was a 99.9% pure powder manufactured by Soekawa Chemicals. As the sintering additive for W conductors in AlN multilayer substrates, manganese dioxide (MnO_2) was used. The MnO_2 powder manufactured by Kojundo Chemical Laboratory was 99.99% pure. The W paste is commercially available paste manufactured by DuPont. The average W particle size in the paste is 0.8 µm. The metallic impurities in the paste detected by chemical analysis are listed in Table 1. Manganese was not detected as an impurity.

Table 1. Tungsten powder characteristics in tungsten paste.

Vendor of Paste	DuPont (No. 8748)
Average Particle Size	0.8 µm
Metallic Impurities (ppm)	Ca(45), Fe(40), Ni(7), Cr(5), Mg(3), Cu(<0.5)

Table 2 shows the starting compositions of inorganic materials. The composition without MnO_2 was also prepared to reveal the effects of addition of MnO_2. These inorganic powders were mixed with organic solvents, dispersant, acrylic copolymer binder and plasticizer by ball milling to obtain AlN suspensions. Aluminum nitride green sheets were prepared from the suspensions by tape casting. The green sheet thickness was about 0.5 mm. These green sheets were laminated to about 4 mm thickness by uniaxial pressing under 7.4 MPa, and then cut into 16-mm-square specimens to prepare AlN ceramics. Owing to the preparation of AlN multilayer substrates, tungsten paste was printed on the AlN green sheets by screen printing to make electric circuits. Seven printed green sheets were stacked and laminated by uniaxial pressing under 7.4 MPa. These laminated sheets were cut into the 40-mm-square specimens as multilayer substrates before sintering. Binder burnout for these specimens was carried out in a nitrogen atmosphere with wet hydrogen (a mixture of hydrogen and water vapor) by heating to 975°C. After binder burnout, both specimens were heated up to 1200, 1400, 1500, 1600, and 1820°C in a nitrogen atmosphere in a furnace with carbon heating elements to investigate the sintering process of the W conductors. The heating rate was 10°C/min up to 1200°C and 2.5°C/min thereafter. Upon reaching the desired temperatures, the heating elements were cut off to cool the specimens to room temperature in the furnace. The other specimens were sintered at 1850, 1820, and 1780°C for 4 hours in a nitrogen atmosphere after binder burnout.

Table 2. Starting Compositions of inorganic materials.

	AlN	Y_2O_3	WO_3	MnO_2
Composition 1	90.62 wt.%	4.00 wt.%	0.38 wt.%	5.00 wt.%
Composition 2	94.12 wt.%	4.00 wt.%	0.38 wt.%	1.50 wt.%
Composition 3	95.12 wt.%	4.00 wt.%	0.38 wt.%	0.50 wt.%
Composition 4	95.62 wt.%	4.00 wt.%	0.38 wt.%	0 wt.%

Scanning electron microscopy (SEM), x-ray diffraction, and electron probe microanalysis (EPMA) were the principal characterization tools employed. To evaluate the amounts of Mn remaining in these samples, quantitative chemical analysis was also performed on AlN ceramics heated to various temperatures (between 1200 and 1820°C). The W grain size of the surface W conductors on the AlN multilayer substrates heated up to various temperatures was evaluated by SEM. The microstructures of W conductors were also observed with SEM at the fracture surface of the AlN multilayer substrates. X-ray diffraction analysis was used to identify the second phases present in the surface W conductors on the AlN multilayer substrate heated up to 1400°C. The X-ray diffraction pattern was obtained using CuKα radiation converged to 100 μm in diameter. Distributions of Mn in the AlN multilayer substrates heated up to various temperatures were evaluated by EPMA. The samples for EPMA were cut and then the cross-sectional surfaces were polished to facilitate analysis of the interior of the samples.

RESULTS

Figure 1 shows the amounts of MnO_2 calculated based on the amount of Mn remaining in the AlN ceramics that had 5 wt.% MnO_2 added after heating up to different temperatures in a nitrogen atmosphere. With increasing temperatures, the amounts of residual Mn in AlN ceramics decreased. Manganese dioxide disappears from AlN ceramics during the sintering process, as reported by Kasori.[16] Figure 1 shows that the elimination of Mn has already begun to occur at 1400°C, and that Mn is largely eliminated at about 1600°C. In each specimen sintered for 4 hours at 1850, 1820, and 1780°C, respectively, the Mn has also been eliminated, as listed in Table 3.

Figure 2 shows the W grain size determinations obtained in the surface conductors on AlN multilayer substrates sintered at 1820°C for 4 hours. By increasing the amount of MnO_2 addition, the average sizes of W grains in the specimen surface conductors increased. On average, W grains on the specimen with 5 wt.% MnO_2 added increased by about five

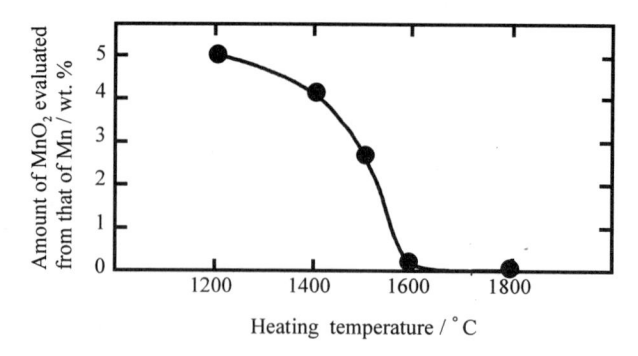

Figure 1. Amounts of MnO$_2$ evaluated based on Mn remaining in AlN ceramics after heating to various temperatures. The AlN ceramics were fabricated from AlN green sheets containing 5 wt.% MnO$_2$.

Table 3. Amount of MnO$_2$ evaluated from that of Mn remaining in AlN ceramics.

Sintering Condition		Amount of MnO$_2$ evaluated from that of Mn (wt.%)	
Temperature	Time	Before sintering	After sintering
1850°C	4 hours	5.00	0.01
		1.50	0.01
1820°C	4 hours	5.00	0.03
		1.50	0.03
		0.50	0.01
1780°C	4 hours	5.00	0.05
		1.50	0.04
		0.50	0.03

times that of W grains on the specimen without MnO$_2$. Figure 3 shows the surface W grain growth results on the specimens with 5 wt.% MnO$_2$ added after heating up to 1200, 1400, 1500, 1600, and 1820°C, respectively. As shown in Figure 3, when 5 wt.% MnO$_2$ is added, the grain growth of surface W conductors on the specimens proceeds rapidly above 1200°C and approaches saturation at 1600°C. Thus, W grain growth should be accelerated until Mn has been eliminated from AlN ceramics.

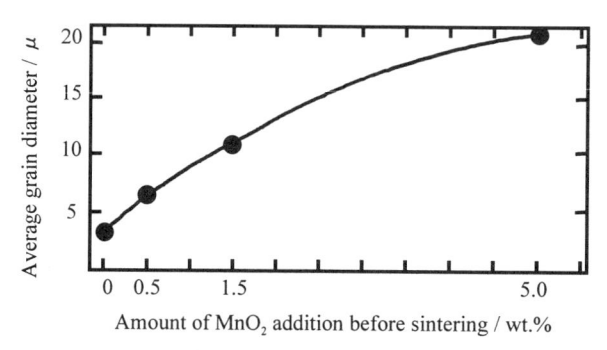

Figure 2. Average tungsten grain diameters in the surface conductors on AlN multilayer substrates sintered at 1820°C for 4 hours.

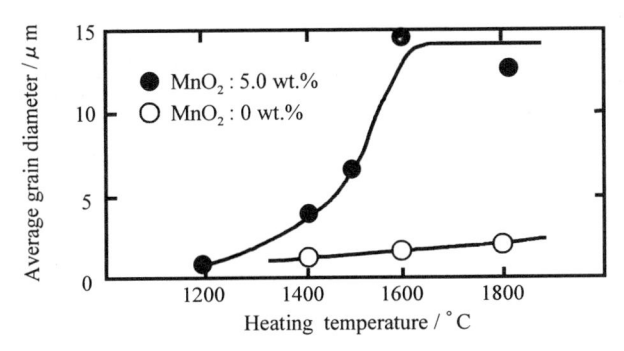

Figure 3. Average tungsten grain diameters in the surface conductors on AlN multilayer substrates after heating up to various temperatures. AlN multilayer substrates were fabricated from AlN green sheets containing 5 wt.% MnO₂ and 0 wt.% MnO₂.

SEM micrographs of the fracture surfaces of the AlN multilayer substrates are shown in Figure 4. The microstructure of the specimen without MnO₂ shows pores in the W conductor layer, whereas the microstructure of the specimen with 1.5 wt.% MnO₂ added shows no pores in the densified W conductor layer.

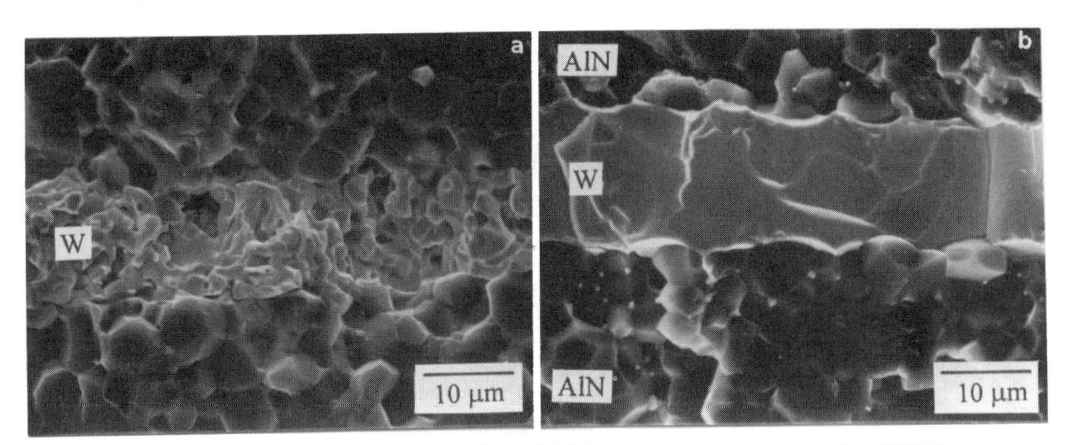

Figure 4. SEM micrographs of the fracture surfaces of AlN multilayer substrates sintered at 1820°C for 4 hours. (a) Microstructure resulting from the specimen without MnO₂. (b) Microstructure resulting from the specimen with 1.5 wt.% MnO₂ added.

Distribution of elemental Mn around W conductors in AlN multilayer substrates was investigated by elemental analysis with EPMA. Figure 5 shows the segregation of elemental Mn around the W conductor layer in the AlN multilayer substrates by the addition of 5 wt.% MnO₂ after heating up to 1500°C. Figure 6 shows the microstructure of the polished cross section of the specimen analyzed with EPMA at the same field of view as shown in Figure 5. The Mn penetrates the W conductor layer and wets the rounded W grains, as shown in Figures 5 and 6. Figure 6 shows, moreover, that a solid neck between two W grains has formed. Formation of a solid neck between two rounded W grains is also typical of liquid-phase sintering. Thus, the Mn segregated into the tungsten conductor layers from the AlN ceramic matrix, and the W grain growth was presumably accelerated by liquid-phase sintering up to 1500°C. The concentration of Mn around W conductors analyzed with EPMA decreased abruptly in the specimen heated up to 1600°C. Elemental Mn was almost entirely eliminated from the W conductor layers in the specimens heated above 1600°C, and similarly from the AlN ceramics, as shown in Figure 1.

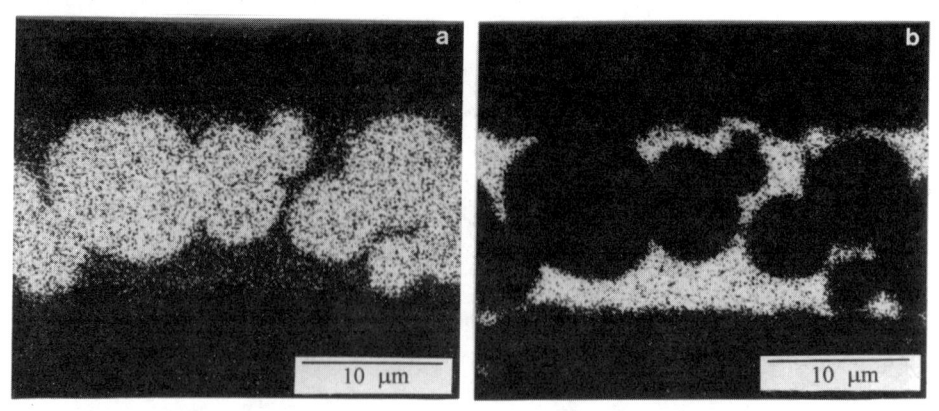

Figure 5. X-ray images of the polished cross section of the AlN multilayer substrates with 5 wt.% MnO$_2$ added after heating up to 1500°C. (a) X-ray image of W element. (b) X-ray image of Mn element.

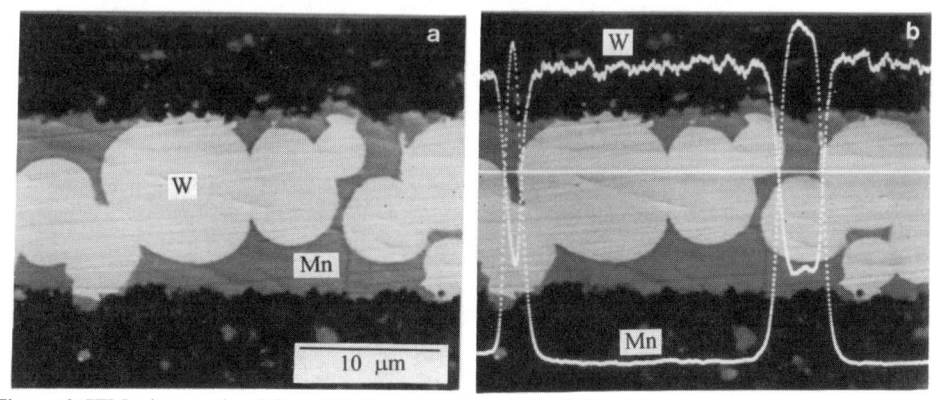

Figure 6. SEM micrographs of the polished cross section of the AlN multilayer substrate with 5 wt.% MnO$_2$ added after heating up to 1500°C. (a) Microstructure of the tungsten conductor layer embedded in the AlN multilayer substrate. (b) EPMA line profile of Mn and W for the polished cross section of the AlN multilayer substrate.

The x-ray diffraction profile for the surface tungsten conductor on the specimen heated to 1400°C exhibited a pattern corresponding to γ-Mn: manganese metal. Figure 7 shows the profile which indicates patterns of W, AlN, Y$_3$Al$_5$O$_{12}$, Al$_5$Y$_3$O$_{12}$, Al$_2$Y$_4$O$_9$, WO$_3$ and γ-Mn. The identification of γ-Mn in the W conductor layers indicates that the MnO$_2$ has been reduced to manganese metal. The manganese liquid phase should be present in activated sintering of W above the melting point of Mn:1245°C.

DISCUSSION

Tungsten conductors in the cofired AlN multilayer substrates were densified by the addition of MnO$_2$ into AlN green sheets. Grain growth of W was presumably accelerated by the segregated Mn liquid phase. Small wetting angle of the liquid phase,[15] high solubility of the solid to the liquid phase, and extremely small solubility of the liquid to the solid phase[12,15] are important factors in obtaining the densified structure after the liquid-phase sintering.

In the W-Mn system, molten Mn wets W well at 1500°C and penetrates into the W grain boundaries.[11] No intermediate phase was found in soldering experiments at 1500°C.[11] The calculated enthalpy of solution of 1 g-atom liquid W in liquid Mn at infinite dilution is ΔH=28 kJ, and that for solution of 1 g-atom liquid Mn in liquid W is ΔH=23 kJ.[17]

(Enthalpy values are calculated by treating atoms as macroscopic pieces of metal and connecting energy effects with contact interactions.[17]) Since the calculated enthalpy values are positive, stable intermetallic compounds should be hard to form in the W-Mn system. In this study, the intermetallic compounds in the W-Mn system are not found (as shown in Figure 7). Thus, W should be removed from the bulk and dispersed as colloidal particles in the molten Mn rather than truly dissolved, as described by Petrunin and Grzhimal'skii.[11] The W starting material particles are of colloidal size: 0.8 μm. These colloidal W particles should diffuse into the molten Mn and precipitate on the surface of larger W particles above 1245°C. The enthalpy of adsorption of Mn on an average W crystal plane was theoretically estimated, moreover, as $\Delta H = -315$ kJ/mol.[18] Thus, the W particles are wetted (adsorbed) by the molten Mn, thus reducing the free surface energy at the boundary between the W particle and the molten Mn. This reduced interphase energy should accelerate precipitation of colloidal W particles and W grain growth in the W conductor layers.

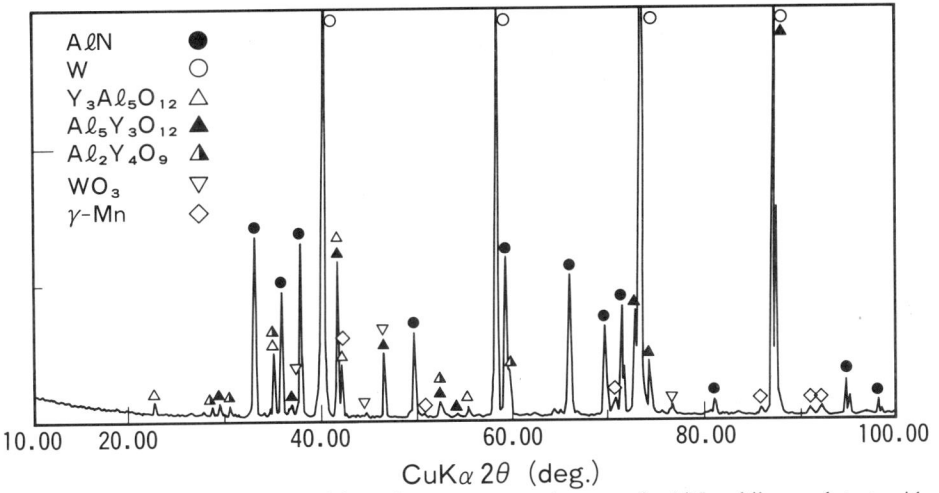

Figure 7. X-ray diffraction pattern of the surface tungsten conductor on the AlN multilayer substrate with 5 wt.% MnO_2 added after heating up to 1400°C, with using Cu Kα radiation converged to 100 μm in diameter.

These results led to the conclusion that the W conductors were densified by the liquid-phase sintering in molten Mn that segregated from the AlN matrix and penetrated the W grain boundaries during the cofiring of AlN multilayer substrates fabricated from AlN green sheets with added MnO_2.

SUMMARY

Enhanced sintering of W conductors embedded in AlN multilayer substrates fabricated from AlN green sheets with MnO_2 added as a sintering additive for W was experimentally investigated. After sintering the AlN multilayer substrate at 1820°C, tungsten conductors were densified without pores by adding MnO_2 into AlN green sheets.

Molten Mn segregated from the AlN matrix into tungsten conductor layers and penetrated the W grain boundaries, wetting the grains well. The tungsten should be dispersed into molten Mn and reprecipitated during the AlN multilayer substrate sintering. Thus, since W grain growth was accelerated by liquid phase sintering, the W conductor layers were densified without pores. The densified W conductors should satisfy such ceramic multilayer substrate requirements as hermeticity and surface flatness.

REFERENCES

1. K. Seitz, H.M. Guther and A. Roosen, Large scale production of AlN substrates: A challenge in ceramic processing, in: *Third Euro-Ceramics Vol. 1*, P. Duran and J. F. Fernandez, eds., Faenza Editrice Iberica S. L., Spain (1993).
2. A.V. Virkar, T.B. Jackson and R.A. Cutler, Thermodynamic and kinetic effects of oxygen removal on the thermal conductivity of aluminum nitride, *J. Amer. Cer. Soc.* **72**:2031 (1989).
3. L.M. Sheppard, Aluminum Nitride: A versatile but challenging material, *Amer. Cer. Soc. Bull.* **69**:1801 (1990)
4. Ellice Y. Luh and John W. Lau, Current processing capabilities for multilayer aluminum nitride, *The International Journal of Microcircuits and Electronic Packaging* **16**:89 (1993).
5. L.E. Dolhert, J.W. Lau, J.H. Enloe, E.Y. Luh, A.L. Kovacs, and J. Stephan, Performance and reliability of metallized aluminum nitride for multichip module applications, *The International Journal for Hybrid Microelectronics* **14**:113 (1991).
6. Ellice Y. Luh, Jack H. Enloe, Leonard E. Dolhert, John W. Lau, Alan L. Kovacs, and Michael R. Ehlert, Design of metallizations and components for aluminum nitride packages for VLSIC, *IEEE Transactions on Components, Hybrids and Manufacturing Technology* **14**:538 (1991).
7. Akinobu Shibuya, Yasuhiro Kurokawa, and Yuzo Shimada, Highly thermal conductive AlN PGA package, *IMC 1992 Proceedings* (Yokohama, Japan, June), 285 (1992).
8. Kiyoshi Hirano, Hisashi Sakuramoto, Shinya Mizuno, Syusei Kuratani, and Koichi Uno, Development of aluminum nitride multilayer pin grid array packages, IMC 1990 Proceedings (Tokyo, May), 249 (1990).
9. I. Amato, On the Mechanism of Activated Sintering of Tungsten Powders, Mater. Sci. and Engr., **10**:15 (1972).
10. W.J. Huppmann and H. Riegger, Liquid phase sintering of the model system W-Ni, *The International Journal of Powder Metallurgy and Powder Technology*, •4:243 (1977).
11. I.E. Petrunin and L.L. Grzhimal'skii, Interaction of tungsten with copper, manganese, silver and tin. Metal Science Heat Treatment Metals [USSR], **24** (1969).
12. R.M. German and B.H. Rabin, Enhanced Sintering through Second Phase Additions, *Powder Metallurgy*, **28**:7 (1985).
13. C.J. Li and R.M. German, Enhanced sintering of tungsten-phase equilibria effects on properties, *The International Journal of Powder Metallurgy and Powder Technology*, **20**:149 (1984).
14. J.L. Johnson and R.M. German, Phase equilibria effects on the enhanced liquid phase sintering of tungsten-copper, *Met. Trans. A*, **24**:2369 (1993).
15. Seng-Ki Joo, Seok-Woon Lee and Tae-Hyoung Ihn, Effect of cobalt addition on the liquid-phase sintering of w-cu prepared by the fluidized bed reduction method, *Metallurgical and Materials Transactions A*, **25**:1575 (1994).
16. Mitsuo Kasori, Fumio Ueno and Akihiko Tsuge, Effects of transition-metal additions on the properties of AlN, *J. Amer. Cer. Soc.*, **77**:1991 (1994).
17. A.R. Miedema, F.R. de Boer, R. Boom and J.W.F. Dorleijn, Model predictions for the ethalpy of formation of transition metal alloys, *CALPHAD*, **1**:341 (1977).
18. A.R. Miedema and J.W.F. Dorleijn, Quantitative predictions of the heat of adsorption of metals on metallic substrates, *Surf. Sci.*, **95**:447 (1980).

WETTABILITY OF Cu-BASED ALLOYS ON ALUMINA AND JOINING OF ALUMINA WITH MICRODESIGNED NICKEL-CHROMIUM ALLOY INTERLAYER

Kunihiko Nakashima [1], Katsumi Mori [1] and A. M. Glaeser [2]

[1]Department of Materials Science & Engineering, Faculty of Engineering, Kyushu University, Fukuoka 812, Japan
[2]Center for Advanced Materials, Lawrence Berkeley National Laboratory, and Department of Materials Science & Engineering, University of California, Berkeley CA 94720, USA

INTRODUCTION

Monolithic ceramic materials are being widely developed for a variety of applications ranging from use as structural components to high-performance electronic substrates. Recently, however, major efforts have been made to use ceramics in structural applications under severe operating conditions. Joining of ceramics to ceramics and to metals is considered one of the key technologies that will either expand or restrict the potential use of ceramic materials in a wide range of applications, such as heat engines, heat exchangers and recuperators.

Recent work by some of the authors[1-3] and Y. Iino[4] has explored the PTLP (Partial Transient Liquid-Phase) method of ceramic-ceramic joining. PTLP bonding is a joining method that seeks to combine the most attractive features of diffusion bonding and brazing. The technique employs a multilayer, initially inhomogeneous interlayer. These microdesigned multilayer interlayer structures form a thin layer of a transient liquid phase (TLP) at relatively low temperatures to facilitate joining with refractory metal-based interlayers. At the temperature of PTLP bonding, diffusion of the less refractory cladding metal into the refractory core metal, reaction of the less refractory cladding metal with the refractory core metal, or a combination of these processes can lead to the disappearance of the liquid. This approach allows the formation of bonds at relatively low temperature, and interdiffusion and resulting homogenization during bonding forms a more refractory interlayer that offers the potential for high temperature use.

In the case of such multilayer interlayers in PTLP bonding, the transient liquid phase is sandwiched between a ceramic and a refractory core metal. The contact angles of a liquid metal on a ceramic are high, and those on a core metal are typically quite low. A necessary condition for favorable redistribution of the liquid along the interface is that the *sum* of these two contact angles be less than 180°. Modifications to the transient liquid phase chemistry that reduce the contact angle on a ceramic are an important factor when modifying the process to improve strength characteristics. Such modifications can arise due to dissolution of the refractory core metal into the transient liquid film during PTLP bonding, in an effort to establish local equilibrium between the liquid phase and the core layer. An understanding of the effect of the associated changes in liquid film chemistry on its wetting behavior is crucial to successful ceramic-metal joining.

The present paper describes the effect of additions (Cr, Ni and 80Ni·20Cr) on the wetting characteristics of liquid Cu on Al_2O_3, and the study of PTLP (Partial Transient Liquid-Phase) bonding of Al_2O_3 using microdesigned multilayer Cu/80Ni·20Cr/Cu interlayers.

EXPERIMENTAL PROCEDURES

Materials

Two kinds of polycrystalline alumina, 99.5% and 99.9% pure (SSA-S and SSA-999W, ≈ 98% dense; Nikkato Co. Ltd.), were used in this study. The 99.5% pure Al_2O_3 contains a small amount of glassy phase. The material was in the form of $20.0 \times 20.0 \times 23.0$ mm^3 blocks for joining experiments and in the form of $10.0 \times 10.0 \times 2.0$ mm^3 substrates for wetting experiments. The bonding and substrate surfaces were machine-polished to a 1 μm (grit size) finish, and subsequently cleaned in isopropyl alcohol, and blown dry. Cleaned alumina blocks and substrates were degreased and preannealed in air at 1000°C for 2 h to remove organic contaminations from the surface.

The 80Ni·20Cr and Cu foils used for joining experiments were respectively 125 μm and 5 μm thick, and reportedly ≤99% and 99.99+% pure. The 80Ni·20Cr foil contains Si as its major impurity. Just prior to use, the 80Ni·20Cr alloy and copper foil pieces (20.0 mm × 23.0 mm) were degreased, subjected to standard etching techniques, and thoroughly rinsed in nano-pure distilled water, and dried.

The rods of pure Cu for experiments examining the wetting characteristics were a high purity (99.99+%), oxygen-free high conductivity copper and served as the base material for alloying. The same cleaning procedure used for Cu foils was employed for the rods. Powders of Cr (≥99%), Ni (≥99.9%) and 80Ni·20Cr (≥99%) were used as additions for the wetting experiments. The additions were placed into holes drilled into the copper rods, and alloys were formed *in situ*.

Sessile Drop Experiment

Wetting experiments were conducted at 1150°C in a carbon element furnace[5]. Rods of pure Cu or of Cu with powders of Ni, Cr or 80Ni·20Cr added were placed onto polished and cleaned alumina substrates. The sample was heated at a rate of 10°C/min to 1150°C under a vacuum of 2.7 to 5.3×10^{-3} Pa, and then maintained at the temperature for 360 min to match the time of the joining experiments. During the anneal, the shape of the molten droplet was photographed at set time intervals. The contact angle and the surface tension were calculated by using Bashforth and Adams' table[6] or by the drop profile curve fitting method[7].

PTLP Bonding and Flexure Testing

Block/foil/block assemblies[5] were bonded in a graphite die in a vacuum hot press with an applied load of 5.1 MPa. After an adequate vacuum was achieved, the temperature was raised to 1150°C at 4°C/min, maintained at 1150°C for 360 min, and then lowered to room temperature at 2°C/min. During this cycle, the vacuum was in the range of 2.7 to 5.3×10^{-3} Pa.

Following bonding, assemblies and as-received Al_2O_3 material were sectioned into beams (≈3.4 mm × 3.4 mm × 40 mm) to compare the strength of bonded assemblies and the inherent strength of the Al_2O_3. Tensile surfaces of all beams were fine-ground and finished by diamond polishing. The final abrasive used was 1 μm diamond. Tensile face edges were beveled to remove machining flaws that could initiate failure. Beams were tested at room temperature using four-point bending with an inner span of 8 mm, outer span of 25 mm and a displacement rate of 0.05 mm/min. Strengths were calculated from the load at failure using standard relationships derived for monolithic elastic materials.

Microstructural and Microchemical Characterization

After completion of the wetting experiments, the cooled samples with solidified droplets were cut perpendicular to the interface. The cross sections of these samples and of the as-bonded beams were examined using a scanning electron microscopy (SEM) and an electron probe micro analyzer (EPMA).

RESULTS AND DISCUSSION

Sessile Drop Experiments

Figure 1 presents the time change of the contact angle of pure liquid Cu and pure liquid Cu with powders of Ni, Cr and 80Ni·20Cr added on a 99.5% Al_2O_3 substrate, as well as of pure liquid Cu on a Ni plate and on an 80Ni·20Cr plate at 1150°C.

The contact angle formed by pure Cu and Cu-rich Cu-Ni alloys seemed to oscillate during the first 100 min of the measurement period, possibly in response to dissolution-induced and therefore time-dependent changes in the liquid-vapor and liquid-substrate interfacial energies. After 100 min the contact angle assumed an essentially constant value that ranged from 115° to 125°, but showed no simple dependence upon the Ni content[5]. In contrast, the contact angle of pure Cu on a Ni plate varied between 10° to 20°, and seemed to reach a constant value after ≈50 min.

The contact angles formed by droplets of Cu with Cr additions decreased with time. As the Cr content was increased, equivalent contact angles were achieved after shorter anneal times. For a Cr level of 6 wt%, contact angles of ≤70° were obtained during the last 30 min of the measurement, and the contact angles were still decreasing when the observations were halted. Similar contact angle trends are observed in reactive metal brazing systems such as Cu-Ag-Ti, where a reaction product is formed on the ceramic surface that is more easily wetted by the residual braze alloy[8]. Kritsalis et al.[9] have examined the effect of adding Cr to Cu on the contact angle formed at 1150°C by the molten alloy on monocrystalline alumina. Their results show a decrease in contact angle with increasing Cr content, however, no reaction product

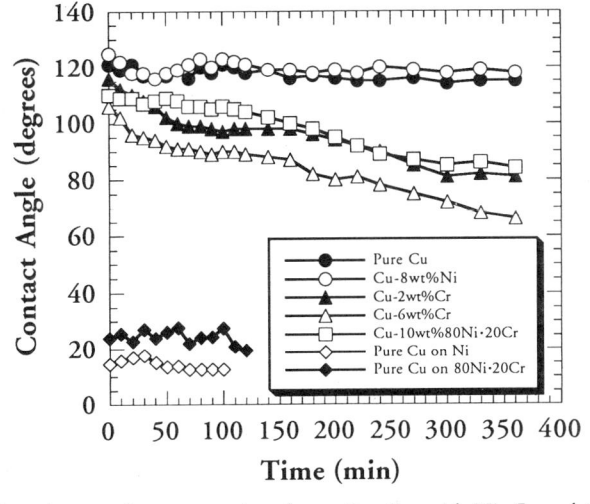

Figure 1. Time change of contact angles of pure Cu, Cu with Ni, Cr and 80Ni·20Cr added on 99.5% Al_2O_3, and of pure Cu on a Ni and an 80Ni·20Cr plate at 1150°C.

forms at the interface. The beneficial effect of Cr additions is related to adsorption effects on the solid-liquid interfacial energy. The present results are only sufficient to show an enhancement of the Cr concentration near the solid-liquid interface[5]. Therefore, it is possible that dissolution of the added Cr is proceeding, and that the time dependence of the contact angle reflects the time dependence of the droplet chemistry and the liquid-solid interfacial energy.

The contact angle formed by droplets of Cu alloyed with 10 wt% 80Ni·20Cr also decreased gradually with time, and contact angles of ≤90° were obtained during the last 90 min of the measurement as shown in Fig. 1. The contact angle of pure Cu on an 80Ni·20Cr plate was in the range of 20° to 30°, and was almost constant during measurement.

The relationship between the contact angles measured at 360 min and the additive content is shown in Figure 2. The contact angles of Cu with Ni additions did not change substantially or systematically with increasing Ni content. The contact angles formed by droplets of Cu with Cr additions could be lowered substantially, and decreased systematically with increasing additive level. The contact angles of Cu with 80Ni·20Cr added decreased more gradually with increasing additive level. However, if the contact angles are replotted in terms of the Cr content in 80Ni·20Cr alloys, the results are identical to those obtained with pure Cr additions. The results indicate that the contact angle of Cu-Cr alloys is not influenced by Ni additions, and suggest that the Cr level in the droplet is the dominant influence on the contact angle. Clearly Cr additions have a beneficial effect on wetting of Cu-rich droplet, reducing the contact angle.

The wettability of liquid metals against ceramics is often evaluated in terms of the work of adhesion, W_{AD}. W_{AD} is usually determined from a sessile drop experiment in which the surface tension, γ_{LV}, and the contact angle, θ, are measured, and is defined by Young-Dupre equation;

$$W_{AD} = \gamma_{LV}(1 + \cos\theta)$$

Figure 3 shows the variation of W_{AD} with addition content in Cu-based alloys measured at 360 min. The W_{AD} of Cu with 80Ni·20Cr added was also plotted against the Cr content in 80Ni·20Cr alloys. In the calculation of the W_{AD}, the values of γ_{LV} and θ obtained in a previous study[5] were used. The W_{AD} of Cu with Ni added was unchanged, however, that of Cu with Cr added increased significantly with additive content. The W_{AD} of Cu with 80Ni·20Cr added gradually increased, and these values are rearranged by the Cr content in 80Ni·20Cr alloys as shown in Fig. 3. These variations of the W_{AD} reflect the changes in the contact angle with addition content shown in Fig. 2.

The findings of the present study have some features in common with those of Kritsalis et al.[9]. Both studies show that the contact angle decreases with increasing Cr content, and that Cr additions do not cause a decrease in the liquid-vapor interfacial energy. This implies that the cause of the decreasing contact angle stems from a Cr addition related effect on the liquid-solid interfacial energy.

The differences appear to be that in the present case, the contact angle decreases with time, and after longer anneal periods than examined by Kritsalis et al., much more substantial decreases are observed. Moreover, the contact angle appears to depend upon the Cr content, with two different sources of Cr with differing Cr:O ratios yielding equivalent results. Due to the relatively higher oxygen content in our Cu, and relatively lower oxygen content in the Cr-containing materials, the overall oxygen levels in our study are higher than those in the work of Kritsalis et al., and the oxygen content is not as strongly affected by additions as in the study of Kritsalis et al.. We note, however, that the final oxygen contents measured in the droplets are generally above those expected based upon the oxygen contents of the Cu and Cr or Ni-Cr alloy used, do not vary with additive content in the manner expected, and are generally higher than those examined by Kritsalis et al.. An additional source of oxygen is suggested.

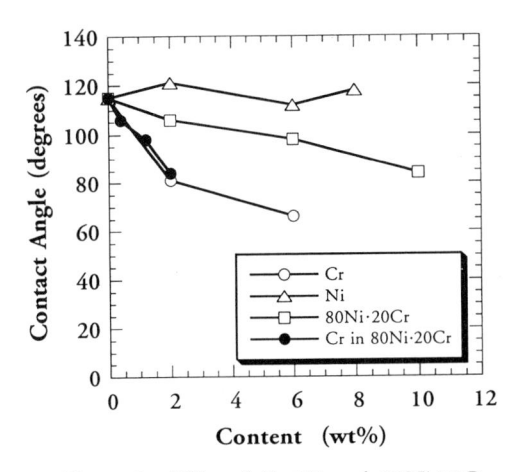

Figure 2. Effect of Cr, Ni and 80Ni·20Cr content on contact angle of liquid Cu on 99.5% Al$_2$O$_3$ after 360 min at 1150°C.

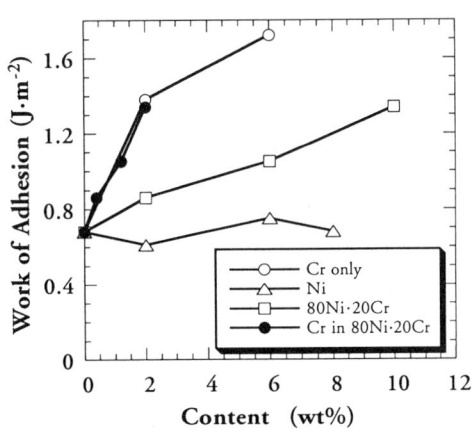

Figure 3. Effect of Cr, Ni and 80Ni·20Cr content on work of adhesion between liquid Cu and 99.5% Al$_2$O$_3$ after 360 min at 1150°C.

We speculate that in the present case, a slow interfacial reaction between the Cu-Cr alloy and the glassy phase in the 99.5% Al$_2$O$_3$ may be occurring. This could provide additional oxygen to the droplet, and, if the interfacial reaction between the droplet and glassy phase is sluggish at 1150°C, cause the time dependent contact angle. After short anneal periods (≤30 min) where reaction would be limited, the contact angles would be expected to agree with the values reported by Kritsalis *et al.*. There is unfortunately only limited overlap between the compositions examined in this study and that of Kritsalis *et al.*. The Cu-2 mass % Cr alloy in our study should initially have $X_O \approx 3.85 \times 10^{-4}$ (mole fraction of oxygen) and the contact angle varies between 116° and 108° during the first 30 min of testing. This compares with ≈110° reported by Kritsalis *et al.* for a similar oxygen (and Cr) level. For this particular droplet, the oxygen content after 360 min was within the range expected based upon oxygen levels in Cu and Cr, however, the contact angle decreased to ≈81°. Recent work by some of the authors[10] indicated that the contact angles formed by droplets of Cu alloyed with 2wt% Cr on a 99.9% Al$_2$O$_3$ substrate (high purity alumina) at 1150°C also decreased with time, however, after 60 min the contact angle assumed an essentially constant value that ranged from 114° to 118°.

Clearly, further studies will be required to identify the experimental and material differences that contribute to the observed differences in wetting and adhesion behavior. What is clear from the present results, is that the addition of Cr (or the interaction of Cr and O in the molten alloy) can significantly improve the wettability of liquid Cu on alumina. In addition, similar benefits result even when Ni is introduced into the liquid. Thus, the combination of Cu and Cr or Ni-Cr could provide a transient liquid phase that would facilitate the formation of reliably strong joints.

Joining Experiments

The indication that Cr beneficially impacted wetting led to PTLP bonding efforts in which Al$_2$O$_3$ was joined using a Cu/80Ni·20Cr/Cu interlayer. Bonding was performed at 1150°C using a 360 min anneal at temperature. The results of flexure tests of as-bonded beams are plotted in Figure 4. Figure 5 also provides optical micrographs of "reconstructed" beams with a relatively low failure strength (401MPa) and the highest flexure strength (616MPa) achieved in present experiments. The 99.5% and 99.9% pure Al$_2$O$_3$ reference beams have an average failure strength of ≈280 ± 25 MPa and of ≈567 ± 44 MPa, respectively.

The average strength and the standard deviation for 10 specimens were 259 MPa

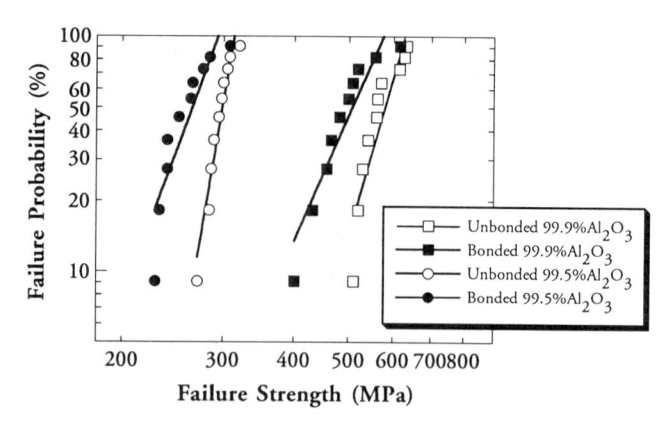

Figure 4. Plot of failure probability versus room temperature failure strength for unbonded and for Cu/80Ni·20Cr/Cu interlayer bonded 99.9% and 99.5% Al_2O_3.

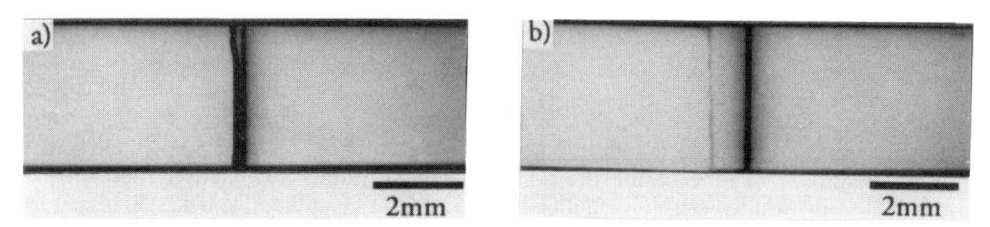

Figure 5. Reassembled bonded 99.9% Al_2O_3 flexure specimens with strengths of a) 401 MPa and b) 616 MPa.

and ±25 MPa for 99.5% Al_2O_3, and 494 MPa and ± 61 MPa for 99.9% Al_2O_3, respectively. In both cases, failure occurred either entirely within the ceramic, or primarily in the ceramic with limited crack propagation along the alumina /interlayer interfaces as shown in Fig. 5. The standard deviations in strength are also comparable to that characterizing the unbonded alumina. It is tempting to attribute the shift of the distribution relative to that of unbonded alumina to the effect of residual stress; the 80Ni-20Cr alloy should have a higher yield stress than pure Ni and thus support a larger thermal expansion mismatch stress without plastic flow. However, such an explanation is not entirely satisfactory since in several of the specimens, failure occurred at more than several interlayer thickness from the interface, where the residual stress should be minimal.

Figure 6 shows the chemical profiles of Cu, Ni and Cr at the interlayer in an as-bonded sample determined by EPMA. The resulting Cu/80Ni·20Cr /Cu thickness ratio leads to average interlayer composition of ≈75 at % Ni, ≈21 at % Cr, and ≈4 at % Cu. Of greater importance is the observation that interdiffusion and homogenization is substantial but not complete. The solidus temperature is not established, however, the solidus temperature of this interlayer is expected to be well above the original processing temperature (1150°C), and the liquidus temperature is about 1400°C[11].

It is useful to compare the bonding conditions and the strengths of Ni/Al_2O_3 diffusion bonding and of Cu/Al_2O_3 eutectic bonding, summarized in Table 1, together with the present work and the previous work of PTLP bonding[1]. For diffusion bonding, the results of Klomp[12] show a trend towards higher strength with increasing bonding stress. When bonding stress are too low, weak bonds are formed even at elevated bonding temperature[13]. The results of

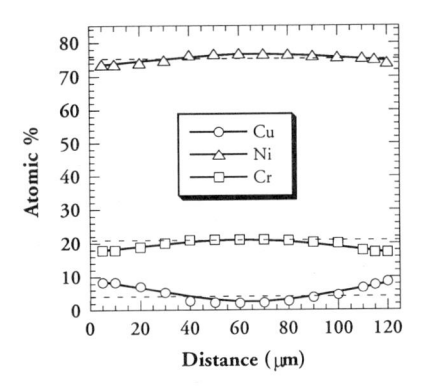

Figure 6. Chemical profiles of Cu, Ni and Cr at the interlayer in an as-bonded sample determined by EPMA.

Table 1. Comparison of bonding method, conditions and failure strengths.

Method & Interlayer	Pressure	Bonding Time	Temperature	Strength
Diffusion bonding[12] Ni/Al2O3	1.5, 9.8, 98 MPa	20 min	1350°C	47, 64, 196 MPa
Eutectic bonding[15] Cu/ Al2O3	–	10 min - 24 h	1075°C	35 - 250 MPa
PTLP[1] Ni/Cu/ Al2O3	5.1 MPa	360 min	1150°C	Average 160 MPa
PTLP(Present Work)* 80Ni·20Cr/Cu/L-Al2O3	5.1 MPa	360 min	1150°C	Average 259 MPa
PTLP(Present Work)** 80Ni·20Cr/Cu/H- Al2O3	5.1 MPa	360 min	1150°C	Average 494 MPa

* L- Al2O3 ; 99.5% pure Al2O3 ** H- Al2O3; 99.9% pure Al2O3

Calow and Porter[14] show that weak bonds result if the bonding temperature is too low. The result of Cu/Al$_2$O$_3$ eutectic bonding[15] shows that the average failure strength of 250 MPa was obtained after bonding at 1075°C for 20 h.

Comparatively, PTLP bonding allows the formation of ceramic-ceramic joints with high melting point metals at temperatures that are generally several hundred degrees lower than those required for the diffusion bonding or brazing. As shown in Table 1, useful strengths can be achieved even with low bonding pressures. We believe that PTLP bonding provides a joining method that has the potential to combine some of the most desirable features of solid-state and liquid based joining methods, and will, together with other transient liquid phase joining methods play an increasingly important role in the joining of advanced ceramics, and composites.

SUMMARY AND CONCLUSIONS

The effect of additions (Cr, Ni and 80Ni·20Cr) on the wetting characteristics of liquid Cu on Al$_2$O$_3$ has been studied using a sessile drop method at 1150°C. Through the use of microdesigned multilayer Cu/80Ni·20Cr/Cu interlayers, PTLP(Partial Transient-Liquid Phase) bonding of alumina has been achieved at 1150°C. The key conclusions are as follows;

(1) Ni additions have no effect on reducing the contact angle of liquid Cu on an Al$_2$O$_3$ substrate. Additions of 80Ni·20Cr or Cr alone reduce the contact angle of liquid Cu on an Al$_2$O$_3$. The contact angle reduction depends upon the Cr content of the sessile drop.

(2) The average strength and the standard deviation were 259 MPa and ±25 MPa for 99.5%

Al$_2$O$_3$, and 494 MPa and ±61 MPa for 99.9% Al$_2$O$_3$, respectively. In both cases, failure occurred either entirely within the ceramic, or primarily in the ceramic with limited crack propagation along the alumina/interlayer interface.
(3) At the bonding temperature, some dissolution of 80Ni·20Cr core layer into liquid Cu film and the diffusion of liquid Cu into 80Ni·20Cr core layer occurred, and homogenization was substantial but not complete. However, the solidus temperature of this interlayer is expected to be well above the original processing temperature.
(4) The results demonstrate that at the bonding temperature, the dissolution of Cr into liquid Cu film improved the wettability of the liquid Cu film on Al$_2$O$_3$, and facilitate the formation of strong joints in PTLP Cu/80Ni·20Cr/Cu interlayer bonded Al$_2$O$_3$.

Acknowledgment

This work was partly supported by the Grant-in-Aid for Developmental Scientific Research (B) (2) from the Ministry of Education, Science and Culture, Japan (No.07555524) which is acknowledged. Scanning electron microscopy (SEM) and electron probe micro analyzer (EPMA) of IMA laboratory, and four-point bending machine of the Onodera laboratory in the Department of Materials Science and Engineering at Kyushu University were employed. The assistance of these laboratories is greatly appreciated.

REFERENCES

1. M. L. Shalz, B. J. Dalgleish, A. P. Tomsia and A. M. Glaeser, Ceramic joining II, Partial transient liquid phase bonding of alumina *via* Cu/Ni/Cu multilayer interlayers, *J. Mater. Sci.*, 29: 3200 (1994).
2. M. L. Shalz, B. J. Dalgleish, A. P. Tomsia, R. M. Cannon and A. M. Glaeser, Ceramic joining III, Bonding of alumina via microdesigned Cu-Nb interlayers, *J. Mater. Sci.*, 29: 3678 (1994).
3. B. J. Dalgleish, A. P. Tomsia, K. Nakashima, M. Locatelli and A. M. Glaeser, Low temperature routes to joining ceramics for high-temperature applications, *Scripta Metall. et Mater.*, 31: 1043 (1994).
4. Y. Iino, Partial transient liquid-phase metals layer technique of ceramic-metal bonding, *J. Mater. Sci Letters.*, 10: 104 (1994).
5. H. Mastumoto, M. R. Locatelli, K. Nakashima, A. M. Glaeser and K. Mori, Wettability of Al$_2$O$_3$ by liquid Cu as influenced by additives and partial transient liquid-phase bonding of Al$_2$O$_3$, *Materials Transactions, JIM*, 36: 555 (1995).
6. F. Bashforth and A. C. Adams, *An Attempt to Test the Theory of Capillarity*, Cambridge University Press, London, U. K., (1883).
7. K. Harada and Y. Muramatsu: Measurement of the surface tension and contact angle of molten metals by contour curve fitting method, *J. Japan Inst. Met.*, 48: 43 (1988).
8. R. E. Loeman and A. P. Tomsia: Joining of ceramics, *J. Am. Ceram. Soc. Bull.*, 67: 37 (1988).
9. P. Kritsalas, L. Curdurier and N. Eustathopoulos, Role of clusters on the wettability and work of adhesion of the Cu-Cr/ Al$_2$O$_3$ system, *J. Mater. Sci. Lett.*, 9: 1332 (1990).
10. A. Kuwaki: *Bachelor Thesis of Kyushu University*, unpublished, (1995).
11. K. P. Gupta, *Phase Diagrams of Ternary Nickel Alloys*, Indian Institute of Metals, Calcutta, 153 (1990).
12. J. Klomp, Solid-state bonding of metals to ceramics, in: *Science of Ceramics, Vol. 5*, C. Brosset and E. Knopp, ed., The Swedish Institute for Silicate Research, (1970).
13. M. C. Gee, Surface reaction between alumina and nickel at high temperatures, *Brit. Ceram. Proc.*, 34: 261 (1984).
14. C. A. Calow and I. T. Porter, The solid-state bonding of nickel to alumina, *J. Mater. Sci.*, 6: 1566 (1971).
15. Sung Tae Kim and Chong Hee Kim: Interfacial reaction product and its effect on the strength of copper to alumina eutectic bonding, *J. Mater. Sci.*, 27: 2061 (1992)

INTERFACE NANOSTRUCTURE OF BRAZED SILICON NITRIDE

Chihiro Iwamoto and Shun-ichiro Tanaka

Tanaka Solid Junction Project, ERATO,
Research Development Corporation of Japan
1-1-1, Fukuura, Kanazawa-ku, Yokohama 236, Japan

ABSTRACT

Nanomorphologies and migration around brazed silicon nitride interfaces were analyzed in order to clarify the brazing mechanism by high resolution transmission electron microscopy techniques. When Si_3N_4 was brazed using Ag-Cu-Ti alloy, TiN nanoparticles formed adjacent to the ceramics as reaction products were commonly accompanied by C-phase, Ti_5Si_3 and Ag-Cu eutectic phases. TiN nanoparticles were arrayed along the [0001] axis of Si_3N_4, and the structure of the Si_3N_4/TiN interface was wavy in one direction. The main orientation relations between TiN and β-Si_3N_4 were determined from the observed lattice images to be $(0001)Si_3N_4//(220)TiN$ and $(10\bar{1}0)Si_3N_4//(111)TiN$. The nature of the crystallographic relationships is also discussed from the polar plane matching viewpoints. It was suggested that the origin of the wavy structure was related to the dissolution process of Si_3N_4.

INTRODUCTION

Because of its productivity, reliability, and high mechanical strength, brazing has been widely used to join ceramics to ceramics or ceramics to metal. These structures have been widely studied on a micron scale. It has been found that various phases of reaction layers are formed adjacent to ceramics during brazing, and the bonding mechanism has been proposed and gradually clarified by many researchers.[1-5]

However, the nanostructure and the formation mechanism of brazed interfaces was not fully understood on an atomic scale. For example, in the SiC/Ag-Cu-Ti system,[6] an excellent lattice matching has been found at the joined interface. On the other hand, in the AlN/Ag-Cu-Ti system,[7] orientation relationships at the AlN/TiN interface were found to differ from those predicted on the basis of lattice matching alone.

This paper describes the detailed nanostructure of the Ag-Cu-Ti brazed β-Si_3N_4 interface observed by high resolution transmission electron microscope (HREM). We also investigate the migration of the interface between β-Si_3N_4 and TiN directly by HREM hot-stage. The origin and the mechanism of brazed interface morphology and epitaxial relationship are also discussed.

EXPERIMENTAL PROCEDURE

Polycrystalline Si_3N_4 was manufactured by pressureless sintering with Al_2O_3 and Y_2O_3 as sintering aids (Toshiba, Japan). The Si_3N_4 specimen was cut into a plate (5 mm × 5 mm × 1 mm), and the surface was polished with diamond pastes and rinsed with an acetone. The composition of thin foil brazing alloy (50 mm thickness) was Ag-27.4 wt.%Cu-4.9 wt.%Ti (GTE Wesgo, USA). Ag-Cu-Ti metallic foils were placed between two Si_3N_4 plates. The specimen was heated up to 1173 K under a vacuum of 4×10^{-5} Pa, and then kept 0 sec. After brazing, the specimen was cooled in the furnace. The bonded specimen was sliced perpendicular to the interface, followed by mechanical polishing and thinning using an Ar^+ beam. The thinned specimen was observed by a high resolution electron microscope (JEOL JEM-2010) at an accelerating voltage of 200 kV.

A double tilt heating holder (Gatan) was employed for heating the sample in order to investigate the interface migration process directly. The Si_3N_4 thin substrate on which brazing alloy was put was heated up to 1123 K, cooled down to 973 K, and heated up to 1123 K again. During the heating cycle, images were recorded with a fiber-optically coupled TV system.

RESULTS AND DISCUSSION

Nanomorphology of the Brazed Si_3N_4 Interface

The reaction phase of the brazed Si_3N_4 consisted of five regions: β-Si_3N_4, TiN, C-phase, Ti_5Si_3 and Ag-Cu eutectic phase. The TiN layer consisted of TiN nanoparticles with an average diameter of about 10 nm, as reported in previous studies.[8,9]

Figure 1 shows an individual TiN nanoparticle with a rod-like precipitate arrayed approximately parallel to the $(10\bar{1}0)$ plane of Si_3N_4. Figure 2 shows a high-resolution image with the incident beam parallel to the [0001] direction of Si_3N_4. The interface between β-Si_3N_4 and TiN is found to be wavy in the (0001) plane.

Crystallographic Orientation Relationships between β-Si_3N_4 and TiN

In Figure 2, the incident beam axis of TiN crystallites is <110>, parallel to the [0001] direction of β-Si_3N_4, as lattice image analysis has shown. The orientation relation of the observed plane is given as $(0001)Si_3N_4//(110)TiN$. TiN and β-Si_3N_4 had several orientation relations because this relation had freedom of rotation in the plane.

The number of rotated TiN grains with the angle between $(200)TiN$ and $(10\bar{1}0)Si_3N_4$ was counted and is shown in Figure 3. The greatest number of orientation relations between two planes was observed at a rotation angle of about five degrees. The relation at this angle is equivalent to $(10\bar{1}0)Si_3N_4//(111)TiN$.

From the viewpoint of polar plane matching, the $(10\bar{1}0)Si_3N_4//(111)TiN$ is a special relationship. The {111} lattice planes of TiN are polar planes formed by either N or Ti atoms, and similarly the {$10\bar{1}0$} lattice planes of Si_3N_4 are also polar planes formed by either N or Si atoms. In the relation $(10\bar{1}0)Si_3N_4//(111)TiN$, polar planes are parallel, fulfilling the requirement that electrostatic potential be minimized.

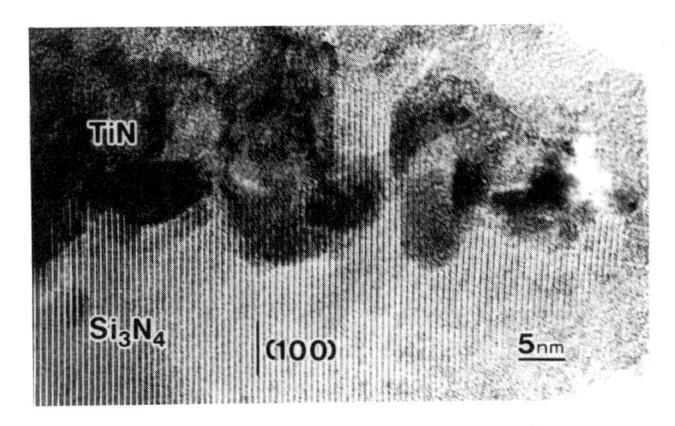

Figure 1. TiN nanoparticles arrayed approximately parallel to Si_3N_4 ($10\bar{1}0$) plane and also parallel to macroscopic interface between Si_3N_4 and TiN.

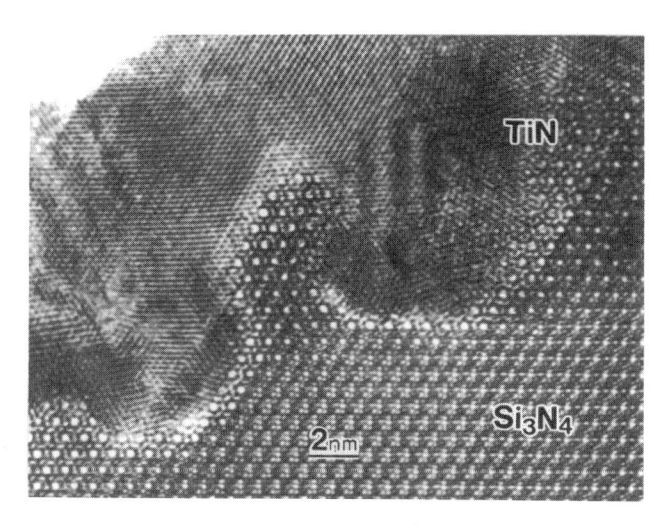

Figure 2. High resolution image of edge-on interface between Si_3N_4 and TiN taken with the incident beam parallel to Si_3N_4 [0001] direction. TiN crystallite direction is always <110>, being parallel to the incident beam.

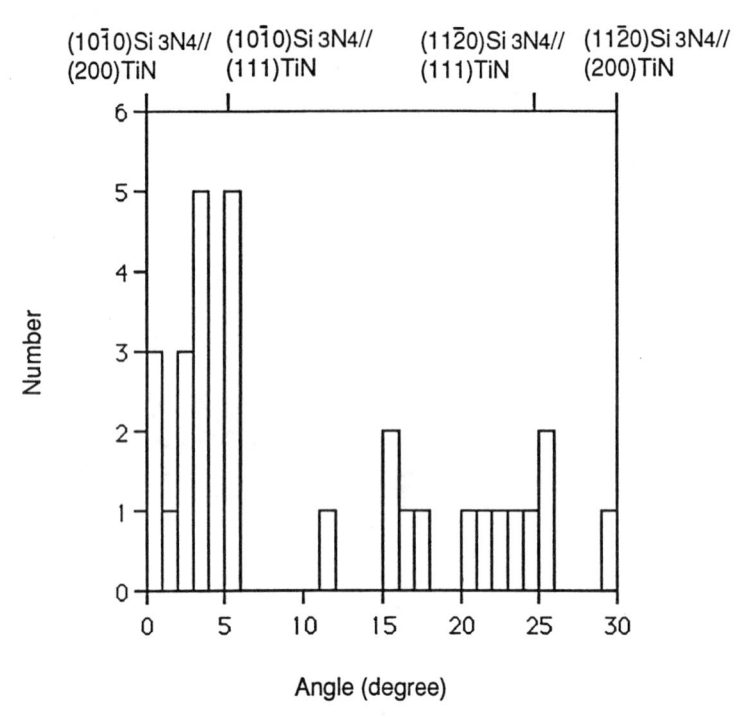

Figure 3. Histogram of the observed interface rotated around the axis of [0001]Si$_3$N$_4$//[110]TiN. Angle 0 degree is equivalent to (10$\bar{1}$0)Si$_3$N$_4$//(200)TiN. The greatest number is observed at five degrees, corresponding to the orientation relation of (10$\bar{1}$0)Si$_3$N$_4$//(111)TiN.

Brazing of Si$_3$N$_4$ in HREM *in situ*

We investigated the high temperature behavior of the interface in order to clarify the mechanism of interface formation using a HREM heating holder.

When the specimen was heated up to 1123 K, Ag-Cu-Ti alloy was molten and wetted the surface of the Si$_3$N$_4$ smoothly. The speed of the molten Ag-Cu-Ti alloy edge depended on the specimen temperature. Figure 4 is a high-resolution micrograph of Ag-Cu-Ti phase on the (0001) Si$_3$N$_4$ substrate at a nominal temperature of 973 K after brazing for about 1 minute at 1123K *in situ*. The Ag-Cu-Ti phase edge has a round shape. TiN is already produced in contact with the Ag-Cu-Ti phase and Si$_3$N$_4$ substrate. The morphology of the interface between TiN and Si$_3$N$_4$ seems similar to that in the bulk brazing. When the specimen was reheated to 1123 K, the interface migrated with Si$_3$N$_4$ dissolution. The migration rate of the interface became very low compared to that during the first brazing process. From these observations, it is considered that the interface morphology was produced during the Si$_3$N$_4$ dissolution process. In other words, at 1123 K, the interface does not rearrange without Si$_3$N$_4$ dissolution.

Considering the kinetics of the reaction process, it should begin with the Si$_3$N$_4$ dissolution by Ti followed by TiN formation. According to the calculation by Tanaka *et al.*,[10] an appreciable amount of Ti might be substitutionally dissolved into β-Si$_3$N$_4$ when the formal oxidation number was +4, and Ti should exhibit considerable solution softening because the covalent bond between Ti and N was locally weakened due to the presence of Ti. This report supports our speculation that Ti atoms substituting for Si atoms in the β-Si$_3$N$_4$ lattice weakened the bonds and dissolved the β-Si$_3$N$_4$ at the Ti atomic sites, and that this was followed by the nucleation of TiN at the sites. The orientation relationship

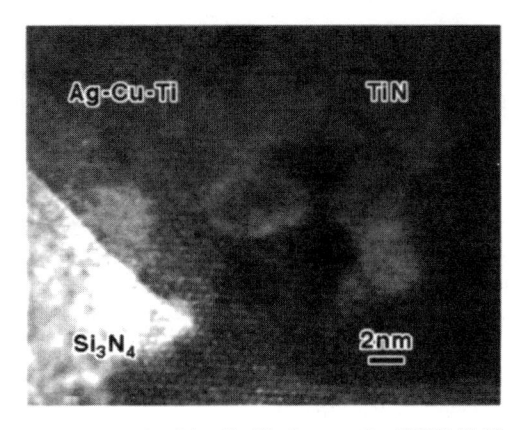

Figure 4. High-resolution micrograph of Ag-Cu-Ti phase on the (0001) Si$_3$N$_4$ substrate at a nominal temperature of 973 K after brazing for about 1 minute at 1123 K *in situ*.

between TiN, which contains Ti atoms substituting on Si sites, and Si$_3$N$_4$ is $(10\bar{1}0)$Si$_3$N$_4$//(111)TiN. This agrees with our experimental result.

Si$_3$N$_4$ {$10\bar{1}0$} planes have three equivalent planes, that is, $(10\bar{1}0)$, $(\bar{1}100)$ and $(0\bar{1}10)$. When Si$_3$N$_4$ dissolves at the {$10\bar{1}0$} plane and TiN nucleates epitaxially on the planes, wavy structure will form in one direction in the brazed Si$_3$N$_4$ interface. One of the reasons for the origin of the wavy structure is thought to be the crystallographic freedom at Si$_3$N$_4$ dissolution.

CONCLUSIONS

Nanomorphologies and migration of the β-Si$_3$N$_4$/TiN interface were analyzed by HREM techniques. TiN nanoparticles were formed adjacent to Si$_3$N$_4$ as reaction products which were accompanied by C-phase, Ti$_5$Si$_3$ and Ag-Cu eutectic phase. The interface of Si$_3$N$_4$/TiN was found to have a one-directional wavy nanomorphology. TiN nanoparticles were arrayed along the [0001] axis of Si$_3$N$_4$. We directly determined the crystallographic relationships between Si$_3$N$_4$ and TiN by HREM image. It was found that main orientation relations, attributed to polar plane matching, minimized the electrostatic potential at the interface. One of the reasons for the origin of the wavy structure was considered to be the crystallographic freedom at the dissolution of Si$_3$N$_4$. As the next step in this work, direct observation of Si$_3$N$_4$ dissolution and the TiN nucleation process is required in order to understand the formation mechanism of the brazed Si$_3$N$_4$ interface.

ACKNOWLEDGMENTS

We would like to thank Prof. Y. Ishida and Dr. H. Ichinose for the stimulating discussions and their interest in this work.

References

1. M. Naka, "Joining mechanism of ceramics to metals using an amorphous titanium-based filler metal," *Mat. Sci. Eng.*, **98**:407 (1988).

2. Y. Nakao, K. Nishimoto and K. Saida, "Reaction layer formation in nitride ceramics (Si_3N_4 and AlN) to metal joints bonded with active filler metals," *ISIJ International*, **30**:1142 (1991).
3. R. E. Loehman, A. P. Tomsia, J. A. Pask and S. M. Johnson, "Bonding mechanisms in silicon nitride brazing," *J. Am. Ceram. Soc.*, **73**:552 (1990).
4. M. G. Nicholas, K. A. Mortimer, L. M. Jones and R. M. Crispin, "Some observations on the wetting and bonding of nitride ceramics," *J. Mater. Sci.*, **25**:2679 (1990).
5. D. H. Kim, S. H. Hwang and S. S. Chun, "The wetting, reaction and bonding of silicon nitride by Cu-Ti alloys," *J. Mater. Sci.*, **26**:3223 (1991).
6. T. Yano, H. Suematsu and T. Iseki, "High-resolution electron microscopy of a SiC/TiC joint brazed by a Ag-Cu-Ti alloy," *J. Mater. Sci.*, **23**:3362 (1988).
7. A. H. Carim and R. E. Loehman, "Microstructure at the interface between AlN and a Ag-Cu-Ti braze alloy," *J. Mater. Res.*, **5**:1520 (1990).
8. Y. Ishida, H. Ichinose and S. Tanaka, "High resolution electron microscopy of silicon nitride-metal bonded interfaces," in: *Ceramic Microstructures '86*, J. A. Pask and A. G. Evans, eds., Plenum Publishing Corporation, 1988.
9. S. Tanaka, "The characterization of ceramic/metal systems joined by an active-metal brazing method," *MRS Int'l. Mtg. on Adv. Mats*, **8**:91 (1989).
10. I. Tanaka and H. Adachi, "Electric structure of 3d transition elements in β-Si_3N_4," *Phil. Mag. B.*, **72**:459 (1995).

THE IMPACT OF INTERFACE NONSTOICHIOMETRY ON GAS/SOLID KINETICS

J. Nowotny

Australian Nuclear Science & Technology Organisation, Materials Division,
Lucas Heights Science and Technology Centre,
Lucas Heights, NSW 2234, Australia

ABSTRACT

The present paper considers defect chemistry for the bulk phase and interface layer of nonstoichiometric compounds, such as NiO or CoO, and their solid solutions with Cr. The effect of segregation-induced concentration gradients and related electric fields in the interface layer on the gas/solid heterogeneous kinetics at elevated temperatures is discussed.

INTRODUCTION

Properties of compounds can be well explained in terms of defect chemistry. However, ideal defect chemistry approximations can only be applied when defect interactions are negligibly low. At higher defect concentrations, interactions between defects become substantial.[1] Hence, defect equilibria should be considered in terms of defect activities rather than concentrations.

Segregation results in a substantial enrichment of the interface layer in lattice defects. Consequently, the extent of interactions between the defects in the interface layer is much higher than in the bulk phase, resulting in the formation of defect complexes and larger defect aggregates. Therefore, the local nonstoichiometry and related defect interactions are a function of the distance from the interface. Accordingly, isomorphic properties within the interface region should be considered within bidimensional layers of the same distance from the interface.

The purpose of the present paper is to consider bulk vs. interface nonstoichiometry in metal oxides as well as the effect of interfaces, specifically the effect of segregation-induced concentration gradients within the interface layer, on gas/solid kinetics for metal oxide/oxygen systems. Before discussing this effect, a short description of basic terms will be given.

POSTULATION OF THE PROBLEM

Nonstoichiometry

Equilibrium concentration of defects and related nonstoichiometry in metal oxides is determined by temperature and oxygen partial pressure. In equilibrium state, the nonstoichiometry in the bulk phase is uniform within the crystal where the local charge neutrality condition must be respected according to which the concentration of all positively charged defects must be equal to the concentration of all negatively charged species at any point of the crystal.

Local charge neutrality is not required at interfaces where, in contrast to the periodic lattice in the bulk phase, it is compensated by a space charge in the vicinity of the interface. In consequence, a gradient of nonstoichiometry and related defect concentration gradients is formed within the interface region.

Since equilibrium requires that electrochemical potential be constant throughout the system for all species involved, an electrical potential gradient must be developed within the interface layer. The effect of this potential gradient on the kinetics of gas/solid heterogeneous processes will be considered below.

Gas/Solid Equilibration

Gas/solid equilibration processes play an important role in processing ceramic materials. When one of the parameters determining equilibrium for a metal oxide/oxygen system, such as temperature or oxygen partial pressure, is changed to a new value, then a new level of nonstoichiometry is established at the gas/solid interface. This new nonstoichiometry is then propagated from the gas/solid interface into the bulk phase until the entire system reaches a new equilibrium state.

Gas/solid equilibration kinetics in nonstoichiometric compounds are rate-controlled by chemical diffusion. Thus chemical diffusion coefficients may be determined from the equilibration kinetic data, and chemical diffusion data may be used to evaluate the time required to reach equilibrium.

The mechanism and kinetics of gas/solid equilibration processes in transition metal oxides have been extensively described by Wagner et al.[2-4]

It has been generally assumed that the equilibration process in compounds is rate-controlled by the mobility of defects in the bulk phase. Accordingly, it has been considered that chemical diffusion data determined from gas/solid equilibration describe the transport kinetics in the bulk phase. This is, however, not the case when segregation results in the formation of a near-surface diffusive resistance.[5]

The purpose of this paper is to consider the effect of segregation-induced defect concentration and related electric fields in the interface layer on the gas/solid equilibration kinetics. This effect will be considered for binary metal oxides such as NiO, CoO, and their solid solutions with Cr.

SEGREGATION-INDUCED EFFECTS

Segregation Equilibrium

Interface layer is not autonomous. Instead, this layer, which is in contact with the bulk phase on one side and the gas phase on the other side, remains in continuous interaction with both phases. Therefore, segregation in solids should be considered in connection with both phases involved. In other words, segregation equilibrium in the interface layer should be considered for the entire gas/solid system involving the bulk phase and the gas phase.

The basic requirement of equilibrium is that the activity of all elements within the system, including the interface layer, must be constant. However, due to activity coefficient variations, the interface layer exhibits substantial changes in concentration and related electrical potential gradients (see Figure 1).

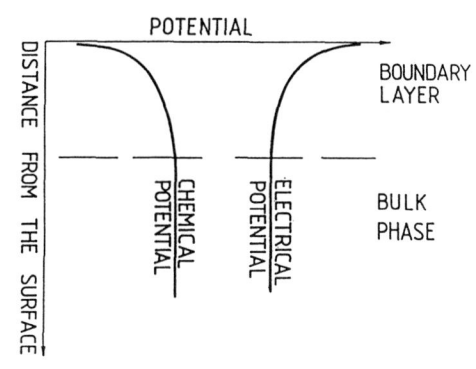

Figure 1. Schematic illustration of segregation-induced chemical and electrical potential gradients within the crystal.

Bulk Defect Chemistry

The predominant defects in NiO and CoO are metal vacancies and electron holes which are formed as a result of oxygen interaction with the crystal surface:

$$1/2\, O_2 \Leftrightarrow V_M^x + O_0^x , \tag{1}$$

$$V_M^x \Leftrightarrow V_M' + h^\cdot , \tag{2}$$

$$V_M' \Leftrightarrow V_M'' + h^\cdot , \tag{3}$$

where V_M^x, V_M' and V_M'' denote neutral, singly ionized, and doubly ionized metal vacancies, respectively, and $h^\cdot$ is an electron hole. Assuming that defect concentrations are equal to their activities the concentration of electron holes may be expressed as:

$$[h^\cdot] = \text{const}\, p(O_2)^{1/n} , \tag{4}$$

where the oxygen exponent, $1/n$, is a parameter dependent on the defect structure. At elevated temperatures, when the majority of defects are charged, the parameter, n, may be determined from measurements of a defect sensitive property such as electrical conductivity.

Ideal defect chemistry assumes that addition of Cr into NiO and CoO results in the formation of donors:

$$Cr_2 O_3 \Leftrightarrow 2Cr_M^\cdot + V_M'' + 3O_0 . \tag{5}$$

Application of the mass action law to Equation (5) results in:

$$[h^\cdot] = \text{const } p(O_2)^{1/4} [Cr_M^\cdot]^{-1/2} . \tag{6}$$

Semiconducting properties may be represented by Fermi energy level, E_F, which may also be considered as a chemical potential of electrons. E_F may be related to the concentration of electronic defects, such as electron holes:

$$E_F = kT \ln N/[h^\cdot] , \tag{7}$$

where N denotes the density of states.

At low nonstoichiometry, when interactions between the defects can be ignored, an ideal approximation is applicable. According to Stoneham,[1] defect interactions become significant above approximately 0.1 at.% resulting in the formation of defect complexes and larger aggregates (Figure 2).

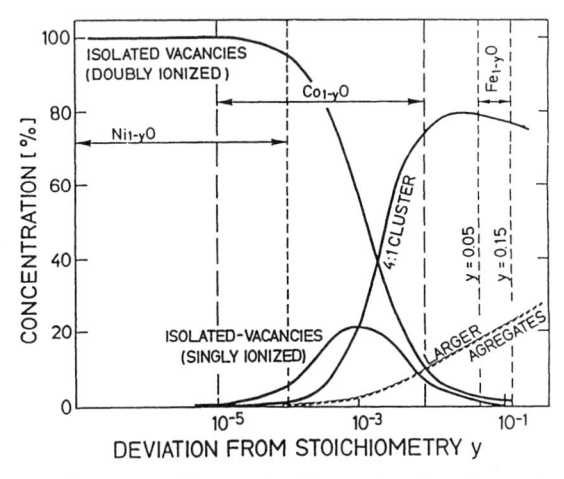

Figure 2. Concentration of defects in transition metal oxides as a function of nonstoichiometry according to Stoneham.[1]

Accordingly, defects in undoped NiO, where nonstoichiometry is at the level of 0.001 at.%, may be considered within the ideal approximation. It has been shown, however, that quenching NiO equilibrated at high temperatures to room temperature results in supersaturation with Ni vacancies and, consequently, leads to strong interactions between the vacancies resulting in the formation of spinel-type structures.[6]

It was shown that the ideal defect model may be applied to undoped CoO only for reduced crystals while at higher $p(O_2)$, when nonstoichiometry reaches about 1 at.%, defect interactions become significant and the ideal approximation is no longer valid.[1,7]

Incorporation of foreign ions, specifically aliovalent ions, results in a substantial increase of interactions between defects. Figure 3 shows the effect of Cr concentration on Fermi energy of Cr-doped CoO determined according to the ideal defect model (Curve 2) and the model derived using activities (Curve 3).[8] As seen, the experimental data are in agreement with the defect model derived using activities rather than that using concentrations.

Interface Defect Chemistry

Nonstoichiometry. An excess of interface energy is the driving force for the segregation of interface defects. Segregation results in a decrease of energy. Duffy and Tasker[9] have reported that the surface of undoped NiO is enriched in Ni vacancies by a factor of about 40. Thus increased nonstoichiometry in the surface layer is still below the 0.1 at.% level or within the ideal approximation.

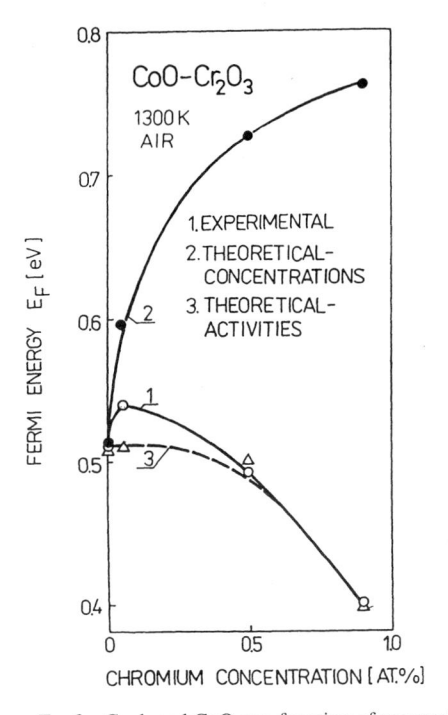

Figure 3. Fermi energy, E_F, for Cr-doped CoO as a function of composition at 1300 K in air.

Nonstoichiometry of CoO is much higher than that of NiO. It appears that segregation-induced enrichment of the CoO surface results in the formation of a low-dimensional overlayer structure involving both Co vacancies and Co interstitials. Interaction between these defects results in the formation of a spinel-like low-dimensional Co_3O_4 structure.[10]

Effect of Aliovalent Ions. The mechanism of incorporating aliovalent ions into the bulk phase is frequently different from that for the interface layer. Figure 4 shows the respective changes of E_F in the bulk phase (determined by Seebeck effect - Curve 2) and at the surface (determined by work function - Curve 3) for Cr-doped NiO.[11] The discrepancy between the two is related to different mechanisms of Cr incorporation into the bulk and the surface layer. While Cr is a donor in the bulk phase of NiO, as described by Equation (5), its incorporation into the surface layer results in a decrease of E_F as if Cr is an acceptor at the surface.[11]

It has been shown that Cr segregates to the surface of NiO and CoO.[12-15] It has also been observed that segregation-induced concentration gradients have an effect on equilibration kinetics.[16-21] It has been documented that the surface layer may be enriched by a factor of 10 and 30.[12,14] Defect structure at this concentration of defects can no longer be considered in terms of ideal approximation.

It is interesting to note that the local solubility of foreign ions within the interface layer may be much higher than that in the bulk phase.[16,17] When segregation-induced enrichment surpasses this solubility limit, the formation of a low-dimensional structure takes place.[18,19] It has been postulated that Cr segregation in both Cr-doped NiO and Cr-doped CoO results in the formation of a spinel-like structure.[11,16-19] Its properties are entirely different from those of the bulk phase.[16-19]

It has been observed that the segregation-induced enrichment of the surface layer in defects depends on the gas phase composition.[13,14] In the case of oxides, the enrichment mainly depends on oxygen activity.

Segregation-Induced Electric Field

Thermodynamic equilibrium requires that electrochemical potential throughout the system is constant:

$$\eta = \mu + e\psi \qquad (8)$$

where η, μ, and ψ denote electrochemical, chemical, and electrical potentials, respectively, and e is elementary charge. Accordingly, the segregation-induced chemical potential gradient results in the formation of electrical potential gradient within the interface layer (Figure 1).

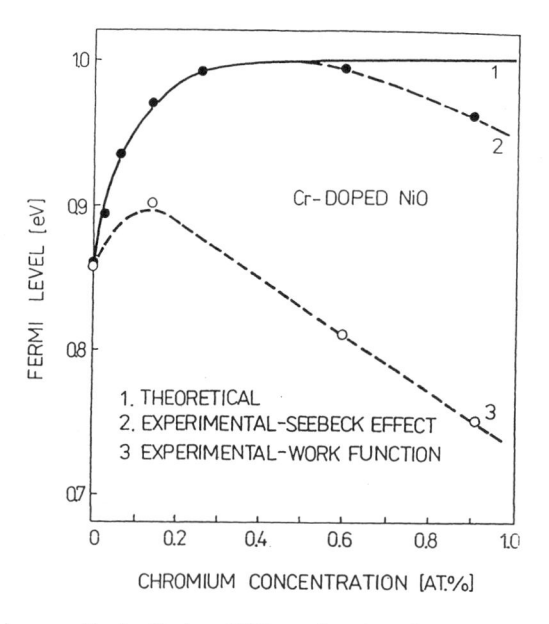

Figure 4. Fermi energy, E_F, for Cr-doped NiO as a function of composition at 1145 K in air.

Effect on the Gas/Solid Kinetics

The segregation-induced electric field has a substantial effect on the transport kinetics of charged defects through this layer.[5,16,18] It has been shown that segregation results in the formation of a near-surface diffusion resistance which may have a retarding effect on equilibration kinetics.[5,18,20] Figure 5 illustrates the effect of the segregation-induced electrical potential barrier on the chemical diffusion coefficient (determined using the solution of Fick's second law, assuming that bulk transport is rate-controlling).

This effect was then confirmed experimentally for undoped and Cr-doped NiO.[19] The effect of the segregation-induced diffusion resistance assumes substantial values and may result in changes of the apparent diffusion coefficient by several orders of magnitude.

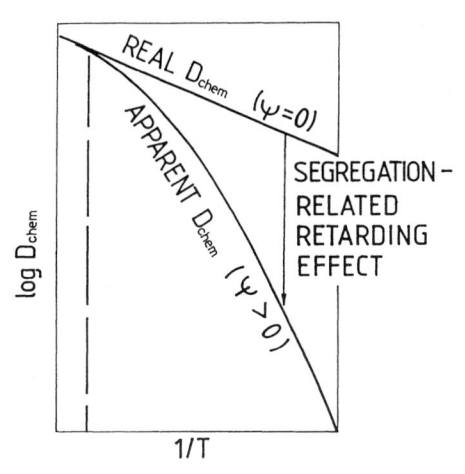

Figure 5. Schematic illustration of the effect of the surface potential barrier on apparent chemical diffusion coefficient.

CONCLUSIONS

Ideal defect chemistry models are applicable only when the concentration of defects is below a certain critical level. Then interactions between defects may be neglected. According to Stoneham,[1] this critical level is approximately 0.1 at.%. This is in agreement with electrical properties of transition metal oxides, such as NiO,[11] CoO,[7,8] and FeO.[21]

When defect concentrations surpass the 0.1 at.% level, substantial interactions between defects lead to the formation of defect complexes and larger defect aggregates. At this point, ideal defect chemistry cannot be applied.

The concentration of defects within the interface layer is substantially higher than in the bulk phase as a result of segregation. This concerns both extrinsic defects, such as impurities, and intrinsic defects, such as metal and oxygen vacancies. Defect structure and defect-related properties within the interface layer should also be considered as a function of distance from the interface. The segregation-induced enrichment results in substantial interactions between the defects within the interface layer. Hence ideal defect models are not applicable.

Defect interactions within the interface layer result in the formation of local structural deformations. When the segregation-induced concentration surpasses a certain critical level, the formation of a bidimensional overlayer structure takes place.[19]

The segregation-induced electric field within the surface layer may have a substantial effect on gas/solid kinetics even at high temperatures. Consequently, the chemical diffusion data determined from the gas/solid kinetics require verification concerning the segregation-induced diffusive component.

REFERENCES

1. M. Stoneham, *Physics Today* **33**:34 (1980).
2. P.E. Childs, J.B. Wagner, Jr., in: *Heterogeneous Kinetics at Elevated Temperatures*, G.R. Belton and W.L. Worrell, eds., Plenum Press, New York, 1970.
3. P.E. Child and J.B. Wagner, Jr,. *Proc. Brit. Ceram. Soc.,* **19**:29 (1971).
4. J.B. Price and J.B. Wagner, Jr., *Z. Physik. Chem., Neue Folge,* **49**:257 (1966).
5. Z. Adamczyk and J. Nowotny, *J. Phys. Chem. Solids,* **47**:11 (1986).
6. W. Carillo-Cabrerra and D.J. Smith, *Solid State Ionics* **32/33**:749 (1989).
7. J. Nowotny and M. Rekas, *J. Am. Ceram. Soc.,* **72**:1 (1989).
8. J. Nowotny and M. Rekas, *J. Electrochem. Soc.*, **131**:94 (1984).
9. D.M. Duffy and P.W. Tasker, *Phil. Mag., Phil. Mag. A,* **50**:155 (1984).
10. J. Nowotny, M. Sloma and W. Weppner, in: *Nonstoichiometric Compounds*, J. Nowotny and W. Weppner, eds., Kluwer Acad., Amsterdam, 1989.
11. J. Nowotny and M. Rekas, *Solid State Ionics* **12**:253 (1984).
12. W. Hirschwald, B. Loechel, J. Nowotny, J. Oblakowski and I. Sikora, *Bull. Polon. Sci., Ser. Sci. Chim.,* **39**:369 (1981).
13. W. Hirschawld, I. Sikora and F. Stolze, *Surf. Interf. Analysis* **3**:157 (1985).
14. I. Sikora, F. Stolze and W. Hirschawld, *Surf. Interf. Analysis* **10**:424 (1987).
15. J. Haber, J. Nowotny, I. Sikora and J. Stoch, *Appl. Surf. Sci.*, **17**:324 (1984).
16. J. Nowotny, I. Sikora and J.B. Wagner, Jr., *J. Am. Ceram. Soc.,* **65**:192 (1982).
17. J. Nowotny, in: *Science of Ceramic Interfaces*, J. Nowotny, ed., Elsevier, Amsterdam, 1991.
18. J. Nowotny and J.B. Wagner Jr., *Oxid. Metals* **15**:169 (1981).
19. J. Nowotny, *Proc. Intern. Conference on Frontier Nanostructured Ceramics*, Fukuoka, October 24-27, 1986.
20. J. Nowotny, in: *Proc. Intern. Symp. on Corrosion of Advanced Materials*, Japan, Dec. 3-7, 1990, Y. Saito, B. Onay and T. Maruyama, eds., Elsevier, 1991.
21. J. Nowotny and M. Rekas, *J. Am. Ceram. Soc.*, **72**:1221 (1989).

REACTIONS AT NANOMETRIC SOLID-LIQUID INTERFACES INFLUENCED BY LOW FREQUENCY ELECTROMAGNETIC FIELDS AND MICROSTRUCTURE DEVELOPMENT OF THE PRECIPITATES

Dario T. Beruto

Istituto di Ingegneria e Scienza dei Materiali
Facoltà di Ingegneria, Università degli Studi di Genova
16129 Genova, Italy

INTRODUCTION

The production of ceramic powders through precipitation processes from solution leads to grains with controlled size and chemical composition only if one can control the formation of the solid particle microstructure at the nanometric and micrometric level.

As often happens in many field of materials science,[1] and also in this research area,[2] it is possible to relate kinetic rates with average solution thermodynamic properties like temperature, pH, pressure, concentrations, etc., but this knowledge, generally, does not provide information about the development of the microstructure of the solid forming grains.

To gain insight on these important steps it is a requirement to understand what happens at the nanometric region of the total volume of the reaction where the solid-liquid transformation is occurring. In these regions, there are nuclei with dimension greater than the critical dimension surrounded by charged Helmholtz layers formed through adsorption of ions from the liquid phase. Thus, if electromagnetic forces are acting throughout the dispersed system, the nature of the solid-liquid interfaces will be affected.

In this paper, it will be shown that, surprisingly, these effects can be produced by low intensity and low frequency electromagnetic applied fields (ELF). The consequences of these changes are such that these techniques might be used to tailor the microstructure of the solid precipitate products. Experimental evidence will be given for calcium phosphates and hydroxyapatite particles obtained from calcium chloride and phosphoric acid solutions.

EXPERIMENTAL TECHNIQUES

In a medium formed by the initial solution and the added precipitating agent, there are cations and anions of different species, hydrated ions, nuclei formed by ions (ions in a cavity and diffused ions[2]), solid grains with colloidal dimensions and agglomerates of particles. All these species can interact with a uniform and/or a nonuniform electrical field, since they are, at least at their solid-liquid interfaces, electrically charged. The electromagnetic forces that are active are mainly the Lorentz forces and the Maxwell stresses.

To evaluate the effects of these forces on the global precipitation process it is necessary to choose, as a standard state, the behaviour of the same solution undergoing the same processing steps but without any exogenous field applied.

From an experimental point of view, this requirement can be satisfied by designing a pair of experiments to be carried-out simultaneously in a symmetrical apparatus where only one part, the active one, is exposed to the electromagnetic field.

If the solution is static (i.e. its barycentric rate is equal to zero) the source of the electromagnetic field can be an alternating current flowing in Helmholtz coils,[3] while if the solution and/or other colloidal systems are flowing in a circuit the source of the electromagnetic field can be a static magnetic field.

The experimental apparatus with Helmholtz-coils sources is illustrated in Figure 1. The details of the apparatus that uses static magnetic fields are covered by patents.[4]

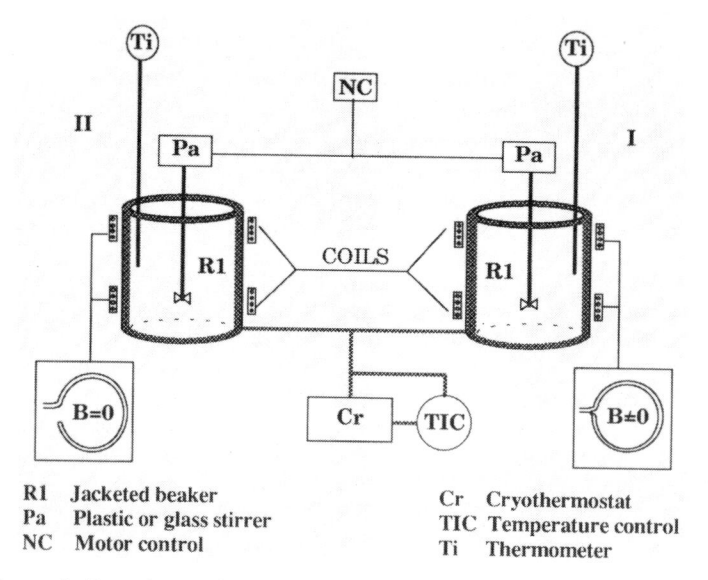

R1 Jacketed beaker
Pa Plastic or glass stirrer
NC Motor control
Cr Cryothermostat
TIC Temperature control
Ti Thermometer

Figure 1. Experimental apparatus with Helmholtz/Dummy-coils source.

A first source of experimental errors that can affect these instruments is extrathermal effects in the systems, due to the electric current flowing into coils. To avoid this inconvenience, the same liquid solutions have been poured into two vessel (I and II in Fig. 1), placed midway between two dummy coils. On the active Helmholtz coils side (I), the alternating current generated by a power amplifier (driven by a sinusoidal signal generator) flows to produce a preselected constant a.c. magnetic field **B** superimposed on the earth's field (in our case, 3 mT). On the passive dummy coils side (II), wires of the same diameter and loop as employed for the active side have been used, but the loop winding has been designed to generate an internal **B** field equal to zero.

Accordingly if any extrathermal effect due to the flow of the electric current is produced this is equal for both experiments, carried out with and without ELF applied field.

Each coil was made up of 22 loops, 14 cm in radius and 18 cm apart. The two coils were fed with alternating current generated by a power amplifier, which was driven by a sinusoidal signal generator. To produce the magnetic field, the two coils were fed in series so as to carry currents in the same direction. The earth's constant magnetic field was not measured. In our experiments the signal-generator frequency was set to 1300 Hz and to 50

Hz, and the magnetic induction was set to an amplitude $B_M = 3$ mT. More precisely, this is the value measured at the symmetry centre of the coil system, that is, on the axis midway between the coils. The amplitude was measured with a 67-loop probe, 26.5 mm² in area; the probe surface was perpendicular to the time-varying magnetic induction, i.e., to the coil axis. B_M was derived from the peak-to peak voltage at the probe terminals, by means of a calibration procedure.

Calcium chloride aqueous solution (0.04 M) and Na_2HPO_4 (0.036 M) were used to precipitate calcium phosphate at 37 °C and at a pH equal to 7.4, following method described in a previous publication.[3]

RESULTS

The detailed mechanism that leads to the phosphates from the calcium chloride aqueous solutions is rather complex,[5] however it is accepted that as soon as nuclei and crystallites of calcium phosphate are formed, a hydrolysis step occurs which transforms gradually the calcium phosphate into hydroxyapatite [HPA, i.e., $Ca_{10}(PO_4)_6(OH)_2$].

The purpose of this paper is not an exploration of the mechanistic details, but instead of exploring the microstructures of a different set of precipitates obtained under the same experimental conditions, with and without the applied ELF. Accordingly at different and preselected precipitation and digestion times, the obtained precipitates were filtered, dried and finally examined with thermogravimetry (TG), X-ray diffractometry (XRD), electronic microscopy (SEM) and N_2-adsorption techniques at 78 K.

The following summarizes the most important results.

SEM observations (Figs. 2 and 3) for the solid precipitates obtained with and without applied ELF, show that they are made up by irregular shaped particles. The products have an average dimension of 10 µm, but the ELF samples have a more rounded shape than those in the samples obtained without ELF. Examinations at higher SEM magnifications give no clear results on the microstructure of these particles.

To obtain information at the nanometric microstructure scale, the N_2 adsorption-desorption measurements at 78 K are very useful.

The specific surface area (BET) of these powders is of the order of 100 m² g⁻¹. These high values cannot be accounted by the high particles dimensions observed in the SEM pictures. Accordingly, the SEM observations are of porous agglomerates. Information on the pore shapes and average dimensions as well as on the internal grain size can be obtained from analysis of the complete N_2 isotherms cycles.

In general, as Figs. 4 and 5 show, the samples are characterized by adsorption isotherms that accordingly to Brunauer's classification[6] are of the IV type, i.e., typical of powders with pores size in the range between 2 nm and 30 nm (mesoporosity range).

Furthermore the hysteresis loops have a shape where adsorption and desorption isotherms are almost parallel into large part of the explored interval P/P_0. According to De Boer,[7] this behaviour reflects capillary cavities with a cylindrical symmetry. From these indications it follows that it is possible to define an average pore radius as :

$$r_{BET} = 1/2 \ (\Delta V_{cap}/S_{BET}) \tag{1}$$

where ΔV_{cap} is the volume of pores, measured at a value of P/P_0 where all pores are filled with liquid nitrogen capillary condensate (here about 0.98), and S_{BET} is the value of the specific surface area evaluated from the isotherms first portion where only physical adsorption phenomena are occurring.

Figure 2. Solid precipitate obtained without applied ELF field.

Figure 3. Solid precipitate obtained with applied ELF field.

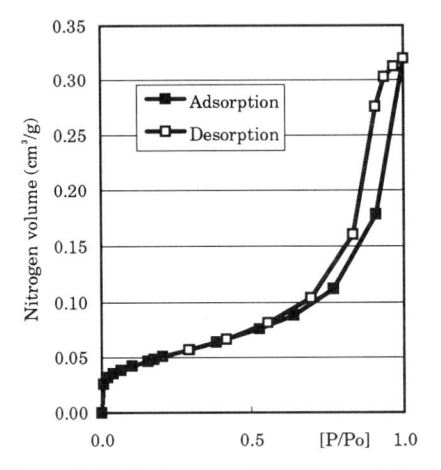

Figure 4. N$_2$-isotherms at 78 K for precipitates obtained without applied ELF field

Figure 5. N$_2$-isotherms at 78 K for precipitates obtained with applied ELF field

From the average dimensions of the pores, assuming that among the internal grains there is no neck formation, the order of magnitude of the grains size inside the agglomerates can be derived.[8] Table 1 summarizes all these microstructural features. From this table it is clear that the effect of the applied exogenous field ELF is that of increasing the average dimension of the nanometric particles by about 35%.

It should be noted that the microstructural range explored by N$_2$-adsorption measurements is larger than that of microporosity. Accordingly, an effect of pore curvature on the adsorption mechanism must be excluded.[9]

A closer examination of the N$_2$-adsorption data allows also to give some information on the chemical nature of the surface sites of the two set of phosphates grains.

It is well known that the value of constant C in the BET equation is related to the adsorbate solid surface interactions,[10] and that this constant is related to the difference between the heat of adsorption and the heat of liquefaction of the adsorbate on the first layer.

Table 1. Microstructural features of calcium phosphates precipitates.

	\multicolumn{8}{c}{ELF electromagnetic field exposure}
	\multicolumn{4}{c}{B = 0 mT}	\multicolumn{4}{c}{B = 3 mT}
Digestione time (min)	S_{BET} (m² g⁻¹)	ΔV (cm³ g⁻¹)	r_{BET} (nm)	C_{BET}	S_{BET} (m² g⁻¹)	ΔV (cm³ g⁻¹)	r_{BET} (nm)	C_{BET}
15	131	0.290	1.10	82.55	108	0.324	1.50	95.97
60	111	0.279	1.26	91.16	72	0.244	1.69	9.91
120	116	0.312	1.34	105.03	74	0.271	1.82	11.4

Hill and Lowell[10] have shown that when sufficient adsorption has occurred to cover the surface with exactly one layer of molecules, the fraction of surface, θ_0 which is not covered by any molecules is dependent on the C_{BET} value through the equation:

$$\theta_0 = (C^{1/2} - 1)/(C - 1) \qquad (2)$$

This implies that if the value of the constant C is high, the uncovered fraction of surface is low, i.e. all the sites of the surface tend to be covered by the physically adsorbed gas in a ratio 1:1. For nominal C values near 100, the fraction of surface unoccupied, when sufficient adsorption has occurred to form a monolayer is 0.091[10] and on the average each occupied site contain about 1.1 molecules. When C becomes lower, this number will increase.

Table 1 shows the experimental C values concerning the N_2-adsorption isotherms on sets of samples obtained with and without the applied ELF field. It is clear that, with the sole exception of the samples obtained in the first period of digestion, the specimens exposed to the applied ELF field, have a significantly lower value of C. Accordingly, at 78 K the surfaces of the ELF ≈1-2 nm grains adsorb less N_2 than is adsorbed by the surfaces of the other set of grains. This difference is certainly related to differences in the chemical nature of the surfaces of these two sets of samples.

N_2-adsorption experiments[11] carried out on $CaCO_3$ and $Ca(OH)_2$ give C values respectively equal to 90 and numbers less than few units. This implies that surface OH⁻ groups have weak interactions with the N_2 adsorbate.

The formation of hydroxyapatite is expected to introduce OH⁻ groups into the phosphate surface made by calcium and phosphates groups. Accordingly, the value of C should decrease with the hydroxyapatite formation. Table 2 suggests that the amount of hydroxyapatite in the solid precipitates is increasing with time of precipitate digestion, and that the ELF samples have a slightly higher HPA content. Such HPA content differences are too small to explain the C_{BET} difference between the ELF samples, and the other samples.

Table 2. Hydroxyapatite content in calcium phosphates after different digestion time.

	\multicolumn{4}{c}{Hydroxyapatite content}
	\multicolumn{2}{c}{Calculated from TG data}	\multicolumn{2}{c}{Calculated from XRD data}
Digestione time (min)	B = 0 mT	B = 3 mT	B = 0 mT	B = 3 mT
15	45 %	50 %	40 %	43 %
60	75 %	78 %	71 %	75 %
120	76 %	80 %	80 %	86 %

The variable that seems to be more significant is not the amount of HPA in the solid phase but the average dimension of its crystallites. Indeed it is interesting to observe that even for the samples treated with the applied ELF the lowering in the C constant occurs only when the average crystallite size is greater than 3 nm.

This observation tends to suggest some relationship between the OH⁻ surface sites distribution and the grain size. It might be possible that, as the hydroxyapatite grain size increases the external surface crystallographic planes have a higher density of OH⁻ sites.

Samples obtained after 120 min of digestion are a solid mixture containing 80% HPA. For these samples it is reasonable to assume that the adsorption properties reflect those of hydroxyapatite surfaces. If so, the hydroxyapatite surfaces, obtained by applying an ELF field and characterized by larger grains, are less reactive towards the N_2 gas, probably because the larger crystallite surfaces are more crowded with OH⁻ groups. Results from others experiments[12] made with **B** equal to 1 mT and frequency equal to 50 Hz, are in agreement with those here illustrated.

DISCUSSION

The experimental evidence above, illustrated and supported also by other cases concerning precipitation of aragonite and calcite from aqueous calcium hydrogencarbonate solutions[13] and from precipitation of barium oxalate[14] and mixed and simple Mg-Al hydroxides[15], are all showing that the application of ELF field to the precipitation processes is enhancing the average grain size dimension of the solid precipitates.

How weak fields of the order of few mT can produces such a change is a very interesting and difficult question. In a previous publication,[3] we have discussed some possible explanations, which serve as the basis of further discussion here.

Ions and hydrated ions, embryos , nuclei of dimensions greater then the critical ones, crystallites and aggregates among the last ones, are the components of the aqueous solutions under study. From the pioneering work by Wrede[3] and by the Princeton and Cornell groups[3] it is known that the effects of non uniform electromagnetic fields are negligible at the molecular level, and progressively increase with the particle size. Kinouchi et al.[16] have shown that to change the average diffusion distance of ions in a static solution, a high-intensity field needs to be used. For instance, to suppress the average diffusion distance of calcium ions in an aqueous solution, the threshold field strength is of the order of 10^6 T. Accordingly ELF fields cannot affect the nature and the structure of hydrated/solvated ions, neither can they affect the probability of collision between cations and anions to form the embryos of the solid products and later the nuclei. Nuclei of dimensions greater than the critical size and agglomerates are therefore the species on which the applied ELF field can act.

It is well known that when a uniform electromagnetic uniform field is applied to a heterogeneous system consisting of a liquid phase and solid particles, the field can concentrate its intensity near the solid/liquid interface. First calculations for this possibility yield a concentration factor of the order of 2-4.[17]

For a nominal system with a geometrical volume equal to the one used in these experiments, with a number of nuclei equal to 10^{15} per cm^3 of solution, an applied field B equal to 3 mT can provide per nucleus an electromagnetic energy of the order of 100 meV.

This energy is greater than kT at 300 K (20 meV), so that it might be used to strain, by a polarization effect, the solid liquid interface around each nucleus if the structure of interfacial solid liquid layer is such to undergo a charges redistribution. As an example to illustrate the model is given by Figure 6.

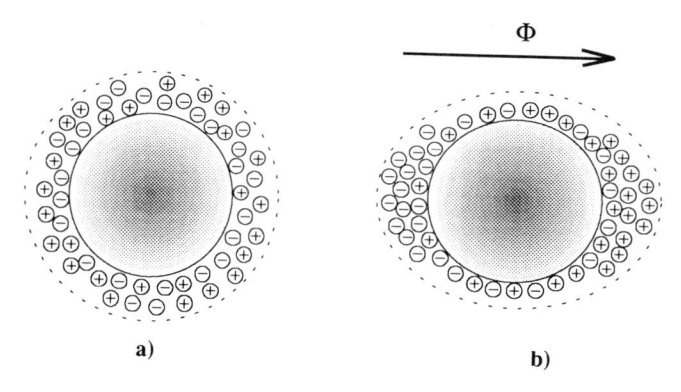

Figure 6. Model to illustrate how ELF field can strain the solid/liquid interface of a nucleus of critical size.

Let assume that a uniform charge density exists on the solid/liquid interface that surrounds the nucleus and/or any colloidal particles (Fig. 6 a). When an external electric field ϕ is applied, the charged density is modified by a polarization effect. If the applied field has a low frequency, the new configuration should be stable over a certain period of time, during which agglomeration and/or other reactive processes can occur.

The total energy of the interfacial solid/ liquid interface would be equal to:

$$\sigma_n = \sigma_{n,u} + \sigma_{n,s} \qquad (3)$$

where σ_n, $\sigma_{n,u}$ and $\sigma_{n,s}$ are respectively, the total interfacial solid/liquid energy per nucleus, the energy per unstrained surface and the strain energy per nucleus.

Evaluation of these quantities made for some ionic solid,[3] assuming that each nucleus of critical size is made up by 100 atoms, leads to $\sigma_{n,u}$ values of the order of 100-300 meV and of $\sigma_{n,s}$ values in the range of 50 meV when the surface polarizable charge is equal to that of about ten electrons. Table 3 summarizes the results. The distortion of charge layer produced by the ELF field yields nuclei which are characterized by an interfacial solid/liquid energy higher than in the unstrained state. The increased interfacial solid/liquid energy ranges between 20% and 40%. For that reason it is possible to explain why the subsequent agglomeration step and/or other reactive surface mechanisms can proceed at an higher rate, giving larger precipitate particles.

Table 3. Interfacial solid-liquid energy for unstrained and strained particles.

Compound	$\sigma_{n,u}$ for nucleus of critical size (meV)	Cluster average dimension (nm)	$\sigma_{n,s}$ strain energy due to polarization (meV)	Increase of $\sigma_{n,u}$ due to $\sigma_{n,s}$ (%)
CaCO$_3$	125	1.66	50	40.00
NaCl	180	2.05	50	27.78
LiF	212	2.05	50	23.58
BaF$_2$	165	2.05	50	30.30
CaF$_2$	300	2.05	50	16.67

It is interesting to observe that the dimension of solid particles on which the applied ELF can act is subject to some restrictions. The first is that the energy of the solid-liquid interface needs to be higher than 20 meV, i.e., the contribution of the thermal fluctuation at the precipitation temperature. The second is that this energy should not be so high as to make the added contribution of the strain energy induced by the ELF applied field negligible. The magnitude of the interfacial total energy is related to the particle size, while the strain energy is somewhat proportional to the number of charges that it is possible to redistribute in the Helmholtz layer. Summing up all these observations it is possible to argue that the particles on which the applied ELF is acting need to have a critical dimension. For the ionic solid discussed here this size seems to be around a few nanometers.

ACKNOWLEDGEMENTS

Discussions with M. Giordani were all very helpful during this work. The author is in debt to all the co-authors of previous publications on this topic carried-out in our Laboratories. This work has been supported by Italian CNR Contribute N. 94.00659.CT11.

REFERENCES

1. C.H. Bamford, C.F.H. Tipper, *Reaction in the Solid State,* Elsevier Publ., Oxford (1980).
2. A.G. Walton, *The formation and Properties of Precipitates*, Intersciences Publ., New York (1967).
3. D.T. Beruto, M. Giordani, Effects of low frequency electromagnetic fields on crystal growth from solutions, in: *Research in Chemical Kinetics*, Vol. 3, R.G. Compton, G. Hancock Eds, Elsevier Science B.V. Publ., Amsterdam (1995).
4. A. Grisoni, *Eur. Pat.* 4004/82-3, 1-7-82, 649754,1982.
5. M.D. Francis, N.C. Web, Hydroxyapatite formation from a hydrated calcium monohydrogen phosphate precursor, *Calc. Tiss. Res.,* 6:335 (1971).
6. S. Brunauer, *The Adsorption of Gases and Vapors*, Oxford University Press, New York (1943).
7. B.C. Lippens, J.H. DeBoer, Studies on pore system in catalysts: V, *J. Catal.*, 4:319 (1964).
8. E.E. Underwood, *Quantitative Stereology*, Addison-Wesley Publ., Reading (1970).
9. S.J. Gregg, K.S.W. Sing, Adsorption. Surface Area and Porosity, Academic Press, London (1982).
10. S. Lowell, *Introduction to Powder Surface Area*, J. Wiley & Sons, New York (1979).
11. D. Nardelli, Ph.D. Thesis, University of Genoa, Italy (1995).
12. D.T. Beruto, A. Chiabrera, M. Giordani, F. Stomeo, S.Verrecchia, "In vitro" effect of low frequency electromagnetic field on calcium phosphates precipitation processes, *Congress EBEA '96*, Nancy, France (1996).
13. D.T. Beruto, M. Giordani, Calcite and aragonite formation from aqueous calcium hydrogencarbonate solutions, *J. Chem. Faraday Trans.*, 89:2457 (1993).
14. R. Berton, D. Beruto, B. Bianco, A. Chiabrera, M. Giordani, Effect of ELF electromagnetic field on precipitation of barium oxalate, *Bioelectrochem. and Bioenerg.*, 30:13 (1993).
15. D.T. Beruto, R. Botter, M. Giordani, G. Gianetti, Influence of non uniform electric field on the coprecipitation of fine particles of Mg-Al hydroxides, in: *Ceramic Powder Processing Science*, H. Hausner, G.L. Messing, S. Hirano Eds., DKG Publ., Köln (1989).
16. Y. Kinouchi, S. Tanimoto, T. Ushita, K. Sato, H. Yamaguchi, H. Miyamoto, Effect of static magnetic fields on diffusion in solutions, *Bioelectromagnetism*, 9:159 (1988).
17. B. Bianco, A. Chiabrera, personal communications.

GROWTH OF INORGANIC CRYSTALS ON THE SURFACES OF TWO-DIMENSIONALLY ASSEMBLED ORGANIC MOLECULES

Kunihito Koumoto[1] and Hong Lin[2]

[1]Department of Applied Chemistry
Graduate School of Engineering
Nagoya University
Furo-cho, Chikusa-ku, Nagoya 464-01, Japan
[2]Institute for Chemical Research
Kyoto University
Uji, Kyoto 611, Japan

INTRODUCTION

The importance of low-temperature materials synthesis should be recognized more widely from the viewpoint of energy saving and environmental protection. In living creatures, in nature inorganic solids with a variety of compositions and structures are synthesized at room temperature. This phenomenon is called biomineralization and has attracted a great number of scientists' interest for many decades.[1-5] Knowledge accumulated in prior studies of biomineralization have been applied to the artificial synthesis and morphology control of materials, and the so-called biomimetic materials chemistry has opened a novel field in materials processing during the past ten years.[4,5]

The basic construction processes of biomineralization consist of (1) supramolecular preorganization, (2) interfacial molecular recognition (templating), and (3) cellular processing. The first stage is the assembly of macromolecules (proteins, polysaccharides, etc.) for the reaction environment where nucleation and growth of inorganic crystals take place. The second, and the final stage is related to various processes with large-scale cellular activity to construct a higher-order architecture.[4] It is the functionalized surface that serves as a template for inorganic nucleation, and the molecular recognition events existing between incipient inorganic nuclei and organic surfaces affect the subsequent crystal growth manner. Molecular recognition is one of the central features of biomineralization, and hence, has a profound significance in biomimetic materials synthesis.

Recent developments in biomimetic materials synthesis involve the use of Langmuir monolayers, Langmuir-Blodgett (LB) films, self-assembled monolayers (SAM), reversed micelles, vesicles, liquid crystals, etc. (Fig. 1) of organic surfactant molecules as templates for oriented crystallization of inorganic substances. Using these molecular assemblies, nanoparticles,[6-10] nanoparticulate films,[11] inorganic-organic nanocomposites,[12-14] and mesophase materials[15-20] have been biomimetically synthesized, and micro-patterning techniques[21-23] for electronic and optoelectronic devices are being developed.

In the present study we have employed Langmuir monolayers for the growth of hydroxyapatite crystals from an aqueous solution[24] (wet process) and LB films for the growth of bismuth crystals from the vapor phase[25] (dry process). It was found that the interfacial molecular recognition somehow assists the construction of nuclei in both wet and dry processes.

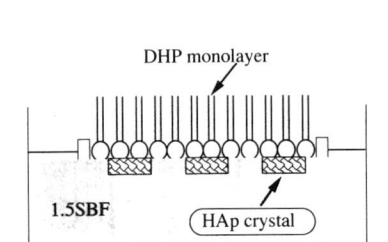

Figure 1. Schematic illustration of HAp crystals growing under a Langmuir monolayer.

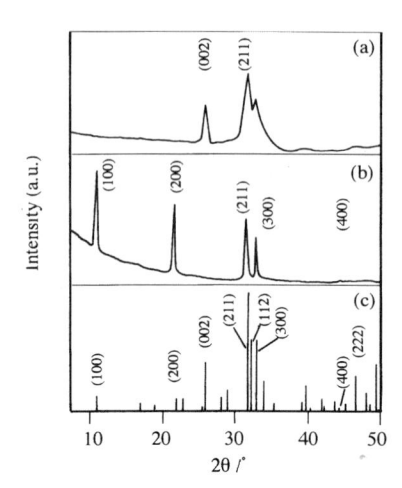

Figure 2. XRD patterns of HAp crystals formed at 36.5°C in 4 d from the 1.5 SBF (a) on a Si substrate placed at the bottom of a reaction vessel, (b) under a compressed Langmuir monolayer of DHP compared with (c) that of the powder listed in a JCPDS card (No. 9-438).

MOLECULAR RECOGNITION IN WET AND DRY PROCESSES
Growth of Hydroxyapatite (HAp) under Langmuir Monolayers (Wet Process)

Simulated body fluid (SBF) with ion concentrations nearly equal to those of human blood plasma was prepared by dissolving NaCl, NaHCO$_3$, KCl, K$_2$HPO$_4$, MgCl$_2$·6H$_2$O, CaCl$_2$, and Na$_2$SO$_4$ into ion-exchanged distilled water. Supersaturated SBF (denoted as 1.5 SBF possessing ion concentrations 1.5 times as high as those of the SBF) was used in the present experiments. These solutions were buffered at pH 7.25 by tri(hydroxymethyl) aminomethane ((CH$_2$OH)$_3$CNH$_2$) and 1M hydrochloric acid (HCl) at 36.5°C, respectively.

Monolayers were prepared in a moving-wall type Langmuir-Blogett (LB) film deposition apparatus (NLE-LB240s-MWA), which has a moving wall width of 50 mm. The prepared 1.5 SBF was used as the subphase for preparing Langmuir monolayers. Chloroform solution (5×10^{-3} mol/l) of the surfactant molecules of dihexadecylphosphate (DHP, (CH$_3$(CH$_2$)$_{15}$)$_2$PO$_4$H) was then spread onto the prepared 1.5 SBF subphase. After the chloroform solvent was evaporated, the surfactant molecules were compressed at the rate of 20 mm/min until the surface pressure reached 25 mN/m to form a solid-phase monolayer. The monolayer was kept at 36.5°C for a certain period of time, and HAp spontaneously grew under the monolayer (Fig. 1).

In the absence of a monolayer, the majority of crystals precipitated were located on the surface of a Si (100) substrate placed at the bottom of a reaction vessel, since a thin silica gel which induced the formation of HAp nuclei was formed on the substrate surface immediately after it was immersed in the 1.5 SBF. Precipitation at the air/solution interface without a monolayer was negligible although small aggregates of crystals were collected occasionally from the surface. These crystals were consistently smaller than collected from the bottom of a reaction vessel. XRD patterns showed the HAp crystals formed on the Si substrate (Fig. 2(a)) were oriented with their (002) and (211) planes parallel to the substrate.

Figure 3. SEM photograph of HAp crystals formed at 36.5°C in 4 d from the 1.5 SBF.

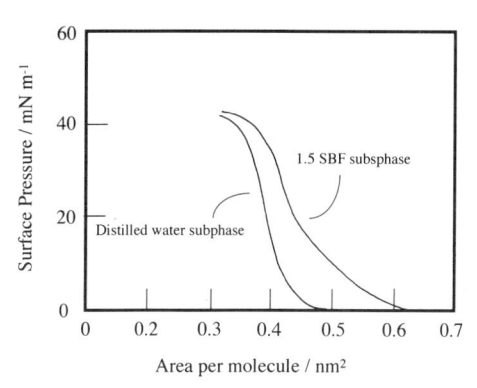

Figure 4. Pressure-area (π-A) isotherms for DHP on the distilled water and the 1.5 SBF at 36.5°C.

Crystallization under the compressed DHP monolayer in the 1.5 SBF at 36.5°C resulted in the growth of oriented HAp crystals with sharp peaks in an XRD pattern (see Fig. 2(b)), indicating their good crystallinity and preferred orientation of (h00) planes though a small diffraction peak of (211) appeared. It can be seen in the SEM photograph (Fig. 3) that the HAp film deposited is ~ 0.3 μm thick and consists of particles of 0.9~1.5 μm in size. Each particle is composed of numerous thread-shaped small crystals.

No precipitation was observed at the bottom of a reaction vessel in the presence of a monolayer for crystallization times less than 5 h. Many crystals precipitated at the bottom after 4 days of growth. However, the precipitate was mostly identified to be NaCl crystals mixed with some unknown phases. This observation clearly suggests that the crystallization of the monolayer surface is kinetically favored over precipitation in a bulk solution.

Figure 4 shows the surface-pressure-area (π-A) curve recorded for DHP molecules spread on the distilled water and the 1.5 SBF subphase at 36.5°C. It is to be noted in the figure that a relatively wider liquid-phase region was observed in the curve when the supersaturated SBF solution was employed as a subphase, indicating that the ions in the subphase had a strong interaction with the DHP headgroups; binding of Ca^{2+} ions to phosphate headgroups must have occurred prior to the nucleation of HAp crystals. The cross sectional area of a molecule read from the π-A curve for the 1.5 SBF subphase at 36.5°C was 0.423 nm^2.

The data obtained so far firmly indicate that a compressed monolayer of DHP promotes crystallizaion of HAp almost exclusively at the monolayer/solution interface with a preferred orientation of (h00) planes parallel to the interface. On the other hand, as shown by Kokubo et al.,[26-28] HAp grown from a 1.5 SBF on substrates of Al_2O_3 and various kinds of organic polymers took preferred orientation of (002) and (211) planes parallel to the substrate. HAp grown on Si (100) also showed a similar orientation as observed in our experiment.

There should be several physical and chemical factors which would determine how an inorganic crystal nucleates and grows on an organic surface. One of the major factors is considered to be the electrostatic interaction associated with the space charge distribution in the vicinity of the interface. The other factor would be the structure and stereochemistry which include the structure of a growing crystal, the manner of arrangement of organic molecules, surface defects of an organic surface, etc.

When a HAp crystal is allowed to grow under a monolayer, cations and anions in the SBF come to the nucleation sites under a monolayer and they would recognize the nature of the monolayer surface and their molecular arrangement. Here, it must be appropriate to regard the interaction at the interface as molecular recognition by the ions to be existing on the faces of incipient nuclei of HAp. Subsequently, they start to grow into large crystals in a controlled manner.

Surfactant monolayers are known to adopt hexagonal or pseudo-hexagonal lattices when compressed,[29-31] and it is assumed that monolayers of DHP should take a similar packing conformation with their hydrophilic head groups lying just under the subphase surface. Then, a comparison of the two-dimensional spacing of phosphate head groups with that of coplanar Ca ions in the (h00) plane of a HAp crystal reveals a fairly good geometric matching along the <001>$_{HAp}$ direction. The interheadgroup spacing of 0.699 nm (calculated from the cross sectional area of a molecule) is close to the coplanar Ca-Ca distance (0.6883 nm) along the <001> direction (Fig. 5), but no such relationship exists for the (001) plane since the Ca-Ca distance on this plane (0.9416 nm) is significantly larger than the corresponding spacing (0.699 nm) of DHP molecules. Consequently, Ca^{2+} ions should have first bonded to phosphate head groups to initiate the formation of HAp nuclei, and this initial arrangement of Ca^{2+} ions must have determined the crystal-axis orientation in the subsequent crystallization process.

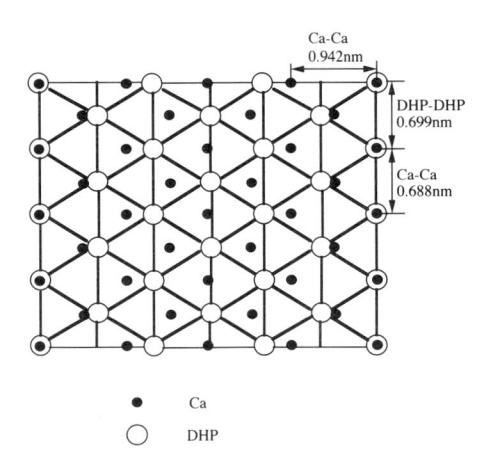

Figure 5. Schematic diagram of geometric matching between DHP molecules and Ca^{2+} ions lying on the (100) plane of a HAp crystal.

Dalas et al.[32] suggested that the phosphorus-containing groups (-P=O) on the polymer surfaces first interact with Ca^{2+} ions forming P-O···Ca bonds that would subsequently act as the active centers for HAp nucleation. They further predicted that a >P(=O)-OH group formed by partial hydrolysis of >P(=O)-OCH$_3$ should have stronger interactions with Ca^{2+} ions than a -P=O group. We have demonstrated that the orientation of HAp crystals grown under Langmuir monolayers depends strongly on the species of functional group,[33] and the most effective functional group for HAp nucleation will be clarified in the near future.

Vapor-Growth of Bismuth on LB Films (Dry Process)

LB films of cadmium stearate (CdSt$_2$) with a hydrophilic surface and with a hydrophobic surface were prepared in the following manner. A chloroform solution (0.14 mg/ml) of stearic acid (SA) was spread onto an aqueous subphase with 0.11 mM CdCl$_2$ adjusted to pH 6.0 with NaHCO$_3$. The monolayer was compressed at a rate of 20 mm/min. A vertical deposition method was used to transfer monolayers to a hydrophilic substrate through the air-water interface. Four-layers of cadmium stearate were deposited to form the hydrophilic surface films, while five-layer films were fabricated to obtain hydrophobic surfaces, respectively. The transfer ratios of the LB films terminated in both hydrophilic and hydrophobic surfaces were almost close to unity during dipping and withdrawing of the substrate. Deposition was carried out under the surface pressure of 25 mN/m at 20.0°C. Films whose last layers terminated in hydrophilic head groups were transferred to a vacuum-chamber immediately to maintain their stable forms.

Bismuth was vacuum-deposited on the glass-supported CdSt$_2$ LB films at room temperature in a high-vacuum evaporation unit (ULVAC, EBX-10D). Care should be taken in using an organic film as a substrate for vacuum deposition. The surfaces of organic compounds are easily affected by the heat radiated from an evaporating source since their thermal conductivity and melting points are both low in general, so that the temperature of the substrate cannot be raised high. For example, the melting point of stearic acid monolayers was measured to be 45-47°C.[34] In this study, the evaporation source, which was a resistively heated alumina crucible containing Bi metal ingots, was positioned about 40 cm below the substrate. Thus, uniform deposition over the 1 cm × 1.5 cm LB film substrate was achieved and the substrate was practically kept at room temperature during vacuum deposition. The pressure inside the vacuum chamber was maintained approximately at 2.7×10^{-4} Pa. The apparent thickness of the deposited Bi was measured and controlled using a quartz monitor.

Bismuth films on both hydrophobic surfaces and hydrophilic surfaces of CdSt$_2$ LB films were polycrystalline and exhibited preferred c-axis orientation as confirmed by the XRD measurement (Fig.6(a)). Within the range of film thicknesses examined, only the diffraction lines of the (003)$_H$, (006)$_H$ and (009)$_H$ planes of hexagonal Bi were observed. This suggests that a high percentage of the basal planes of hexagonal Bi crystals are parallel to the film surface. Figure 6(b) shows the XRD pattern of a 4-layer CdSt$_2$ LB film, which indicates its well-organized layered structure. After bismuth deposition, the layered LB film also exhibited the same XRD pattern as in its original state (Fig. 6(b)), which means that under our deposition conditions the structures of LB films were not damaged by heat or vapor collision.

Deposition of Bi on hydrophilic surfaces of LB films exhibited the most marked effects on crystallization. Firstly, deposition rate and thickness of Bi film are critical to the crystal morphology. Secondly, the lateral crystal growth along the surface was enhanced under some specific conditions, and the size of each crystal was fairly large.

The very low deposition rate, 0.01 nm/s, resulted in a completely different manner of crystallization on the hydrophilic surface. When the thicknesses of Bi films are very thin, say 0.5 nm to 5 nm (averaged value as estimated by the quartz monitor), the Bi crystal exhibited a flat lump morphology, as shown in Fig. 7(a). These lumps, showing a growth along specific directions, were separated but had a striking similarity in shape and orientation. When the Bi films became thicker, the lumps became larger as shown in Fig. 7(b) and (c). Figure 7(c) shows a typical AFM morphology of Bi films with the thickness of 6 nm. A large crystal with very flat surface accompanied by a few fine particles was observed. Similar structures were distributed over the entire 1 cm × 1.5 cm surface.

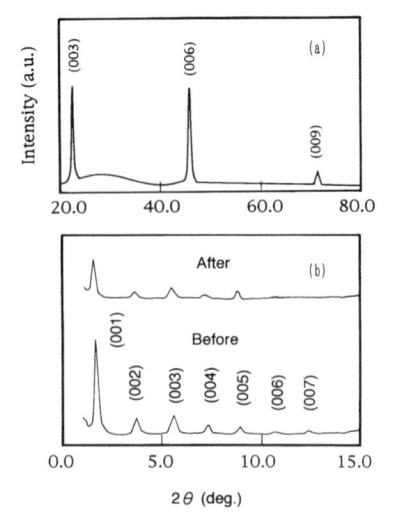

Figure 6. XRD patterns of (a) the Bi film on a LB film and (b) the LB film before and after Bi deposition.

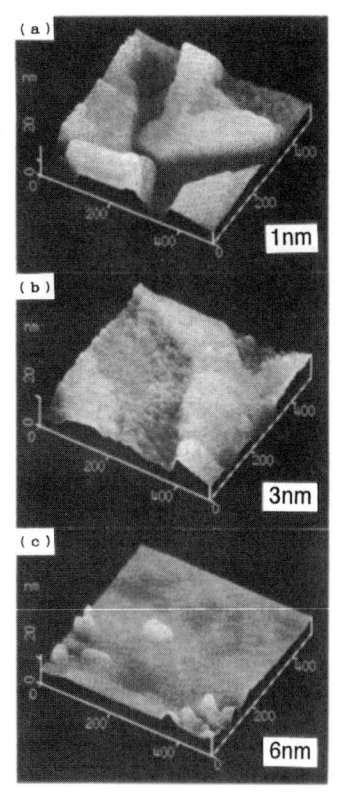

Figure 7. AFM images of Bi deposited on hydrophilic surfaces of LB films; deposition rate is 0.01 nm/s. Apparent thickness of a Bi films is shown in each image.

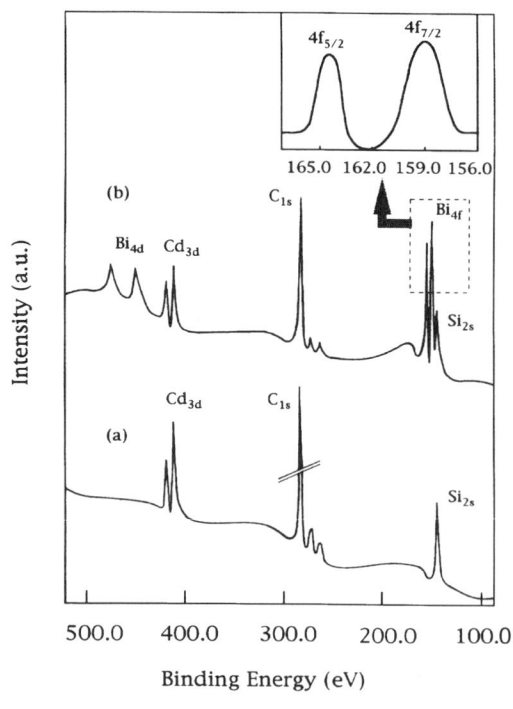

Figure 8. (a) XPS of the CdSt$_2$ monolayer; (b) XPS of the 1 nm-thick Bi film on the CdSt$_2$ monolayer; the insertion shows enlarged peaks of Bi 4f$_{7/2}$ and Bi 4f$_{5/2}$.

XPS analysis was carried out to obtain insight into the interactions between Bi and the LB films (Fig. 8). Broundle et al. have reported that the escape depth of photoelectrons from metal substrates through cadmium arachidate (2.65 nm thick) was 3.6-4.1 nm.[35] Therefore, the area of the metal peak observed in a 1 nm thick Bi film on 2.5 nm thick CdSt$_2$ LB film here can be assumed to be proportional to the amount of metal contained in a monolayer film. Since the C 1s peaks observed in Fig. 8 are almost symmetric and no shoulder peaks appear, an effect of the adsorbed carbon dioxide can be neglected. In Fig. 8, we can see the photoelectron signals for C 1s at 285.0 eV, which corresponds to the ordinary hydrocarbon chain, and for Cd 3d$_{5/2}$ at 405.5 eV, which corresponds to the oxygen-bonded Cd, respectively. In addition, we can see the photoelectron signal for Bi at 159.0 eV which shifted toward the higher energy side (for Bi metal, Bi 4f$_{7/2}$=157.0 eV), indicating a +3 oxidation state for Bi. As shown by Smith et al. by IRS and XPS analysis, vacuum-deposition of copper (Cu) thin films on self-assembled monolayers adsorbed at Au substrate leads to the formation of a carboxylate salt.[36] Consequently, we may determine that Bi is partially bonded to the free carboxyl groups. However we cannot rule out the possibility that partial air-oxidation of Bi had taken place after the sample was taken out of a vacuum chamber.

Bismuth thin films on the substrates of hydrophobic LB films had no anomalous dependence of the structural features on their film thickness, but only exhibited a normal monotonic variation in particle size as shown in Fig. 9. The chemical affinity between Bi and non-charged surfaces of LB films is supposed to be small at room temperature. Thus Bi on hydrophobic surfaces of LB films should be imagined to diffuse more rapidly than on the hydrophilic surfaces, and the diffusivity of Bi adatom in this case is significantly higher on the hydrophobic LB surface than on the surfaces of growing Bi particles. Accordingly, uniform and fine particles were obtained.

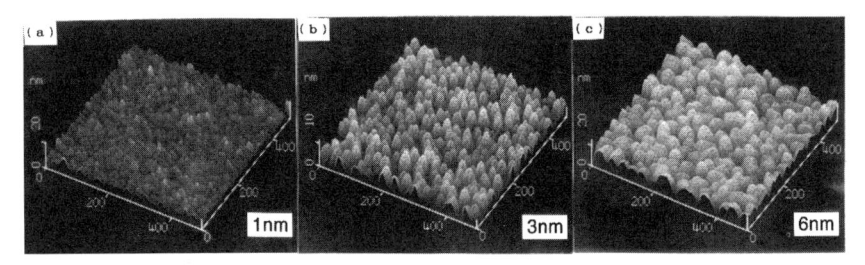

Figure 9. AFM images of Bi deposited on hydrophobic surfaces of LB films; deposition rate is 0.01 nm/s. Apparent thickness of a Bi film is shown in each image. Each protuberance in the images corresponds to a Bi crystalline particle.

A basic difference between hydrophilic surfaces and hydrophobic surfaces of CdSt$_2$ LB films is that there are Cd (II) - bonded carboxylates and free carboxyl groups on the hydrophilic surfaces. According to Kobayashi,[37] Cd ions are attached to about 80% SA molecules at pH 6 of the subphase, and our XPS measurement qualitatively indicated similar Cd bonding situations. The carboxylates and carboxyl groups on the hydrophilic surfaces must have played an important role in crystal nucleation and growth.

Campbell et al.[38] have shown that the organic functional groups bound to the terminal ends of monolayer molecules had pronounced effects on the nucleation and growth of calcium oxalate monohydrate, the more polar surfaces having greater nucleation potential than the non-polar groups. This explanation may be also suitable to the present results. During vacuum deposition, strong interaction between Bi atoms and the carboxylates and carboxyl groups would have resulted in layer-by-layer or layer-plus-island growth, whereas weak interaction resulted only in island growth. Similar results were also obtained for the vapor-growth of metallic zinc.[39]

SUMMARY

Strong interactions between Ca^{2+} ions and phosphate functional groups of a Langmuir monolayer catalyzed the nucleation of HAp crystals from an aqueous solution (wet process). Large crystals of Bi have grown on the hydrophilic surfaces of LB films from the vapor phase, while fine particles have grown on the hydrophobic surfaces (dry process). The results presented here indicate that two-dimensionally oriented molecules have marked effects in terms of interfacial molecular recognition on the inorganic crystallization both in wet and dry processes. It is expected that two-dimensionally assembled organic molecules should be applied more extensively to control the architecture of inorganic crystalline materials.

REFERENCES

1. H. A.Lowenstam, S.Weiner, On Biomineralization, Oxford Univ. Press, New York(1989).
2. K. Simkiss and K. M. Wilbur, Biomineralization: Cell Biology and Mineral Deposition, Academic Press, San Diego (1989).
3. S. Mann, Mineralization in biologicqal systems, Structure and Bonding (Berlin) **54**:125 (1983).
4. S. Mann, Molecluar tectonics in biomineralizaion and biomimetic materials chemistry, *Nature*, **365**:499 (1993).
5. S. Mann, D. D. Archibald, J. M. Didymus, T. Douglous, B. R. Heywood, F. C. Meldrum and N. J. Reeves, Crystallization at inorganic-organic interfaces: biominerals and biomimetic synthesis, *Science*, **261**:1286 (1993).
6. S. Mann, J. P. Hannington and R. J. P. Williams, Phospholipid vesicles as a model system for biomineralization, *Nature*, **324**:565 (1986).
7. S. Bhandarkar and A.Bose, Synthesis of submicrometer crystals of aluminum oxide by aqueous intravesicular precipitation, *J.Colloid Interface Sci.*, **135** (2):531 (1990).
8. B. R. Heywood and E. D. Eanes, An ultrastructural study of the effects of acidic phospholipid substitutions on calcium phosphate precipitation in anionic liposomes, *Calcif. Tissue Int.*, **50**:149 (1992).
9. H. C. Youn, S. Baral and J. H. Fendler, Dihexadecyl phosphate, vesicle-stabilized and in situ generated mixed CdS and ZnS semiconductor particles. Preparation and utilization for photosensitized charge separation and hydrogen generation, *J.Phys.Chem.*, **92**:6320 (1988).
10. F. C. Meldrum, B. R. Heywood and S. Mann, Influence of membrane composition on the intravesicular precipitation of nanophase gold particles, *J.Colloid Interface Sci.*, **161**:16 (1993).
11. J. H. Fendler and F. C. Meldrum, The colloid chemical approach to nanostructured materials, *Adv.Mater.*, **7**(7):607 (1995).
12. X. Peng, Y. Zhang, J. Yang, B. Zuo, L. Xiao and T.Li, Formation of nanoparticulate Fe_2O_3-stearate mutlilayer through the Langmuir-Blodgett method, *J.Phys.Chem.*, **96**:3412 (1992).
13. F. C. Meldrum, N. A. Kotov and J. H. Fendler, Preparation of particulate mono-and multilayers from surfactant-stabilized, nanosized magnetite crystallites, *J.Phys. Chem.*, **98**:4506 (1994).
14. J. Lin, E. Cates and P. A. Bianconi, A synthetic analogue of the biomineralization process: controlled crystallization of an inorganic phase by a polymer matrix, *J. Am.Chem.Soc.*, **116**:4738(1994).
15. C. T. Kresge, M. E. Leonowicz, W. J. Roth, J. C. Vartuli and J. S. Beck, Ordered mesoporous molecular sieves synthesized by a liquid-crystal template mechanism, *Nature*, **359**:710 (1992).
16. J. S. Beck, J. C. Vartuli, W. J. Roth, M. E. Leonowicz, C. T. Kresge, K. D. Schmitt, C. T-W. Chu, D. H. Olson, E. W. Sheppard, S. B. McCullen, J. B. Higgins and J. L. Schlenker, A new family of mesoporous molecular sieves prepared with liquid crystal templates, *J.Am.Chem.Soc.*, **114**:10834 (1992).
17. A. Monnier, F. Schuth, Q. Huo, D. Kumar, D. Margolese, R. S. Maxwell, G. D. Stucky, M. Krishnamurty, P. Petroff, A. Firouzi, M. Janicke and B. F. Chmelka, Cooperative formation of inorganic-organic interfaces in the synthesis of silicate mesostructures, *Science*, **261**:1299 (1993).
18. G. S. Attard, J. C. Glyde and C. G. Golttner, Liquid-crystalline phases as templates for the synthesis of mesoporous silica, *Nature*, **378**:366 (1995).
19. P. T. Tanev and T.J.Pinnavaia, A neutral templating route to mesoporous molecular sieves, *Science*, **267**:865 (1995).
20. D. M. Antonelli and J. Y. Ying, Synthesis and characterization of hexagonally packed mesoporous tantalum oxide molecular sieves, *Chem.Mater.*, **8**:874 (1996).

21. B. C. Bunker, P. C. Rieke, B. J. Tarasevich, A. A. Campbell, G. E. Fryxell, G. L. Graff, L. Song, J. Liu, J. W. Virden and G. L. McVay, Ceramic thin-film formation on functionalized interfaces through biomimetic processing, *Science*, **264**:48 (1994).
22. B. J. Tarasevich, P. C. Rieke, J. Liu, Nucleation and growth of oriented ceramic films onto organic interfaces, *Chem. Mater.*, **8**:292 (1996).
23. N. L. Jeon, P. G. Clem, R. G. Nuzzo and D. A. Payne., Patterning of dielectric oxide thin layers by microcontact printing of self-assembled monolayers, *J. Mater. Res.*, **10** (12):2996 (1995).
24. H. Lin, W. S. Seo, K. Kuwabara, K. Koumoto, Crystallization of hydroxyapatite under Langmuir monolayers, *J. Ceram. Soc. Japan*, **104** (4):291 (1996).
25. H. Lin, W. S. Seo, K. Kuwabara, G. Q. Di, S. Uchiyama, K. Koumoto, Vapor-growth of bismuth on oriented organic films, *J. Surf. Sci. Soc. Jpn.*, **17**(1):31 (1996).
26. M. Tanahashi, T. Yao, T. Kokubo, M. Minoda, T. Miyamoto, T. Nakamura and T. Yamamuro, Apatite coating on organic polymers by a biomimetic process, *J. Am. Ceram. Soc.*, **77**:2805 (1994).
27. M. Tanahashi, T. Yao, T. Kokubo, M. Minoda, T. Miyamoto, T. Nakamura and T. Yamamuro, Apatite formation on organic polymers by biomimetic processing using Na_2O-SiO_2 glasses as mucleating agent, *J. Ceram. Soc. Japan.*, **102** (9):822 (1994).
28. K. Hata, and T. Koukubo, Growth of a bonelike apatite layer on a substrate by a biomimetic process, *J. Am. Ceram. Soc.*, **78**:1049 (1995).
29. T. Inoue, K. Yase, M. Okada, S. Okada, H. Matsuda, H. Nakanishi and M. Kato, Lattice images of Langmuir-Blodgett films of cadmium arachidate arachidate obtained by superconducting cryo-electron microscope, *J. Surf. Sci. Soc. Jpn.*, **11**:29 (1990).
30. E. Meyer, L. Howard, R. M. Overneg, H. Heinzelmann, J. Frommer, H. J. Gruntherodt, T. Wagner, H. Schier and S. Roth, Molecular-resolution images of Langmuir-Blodgett films using atomic force microscopy, *Nature*, **349**:398 (1991).
31. D. K. Schwarts, J. Garnaes, R. Viswanathan and J. A. N. Zasaolzinski, *Science*, **257**:508 (1992).
32. E. Dalas, J. K. Kallitsis and P. G. Koutsoukos, Crystallization of hydroxyapatite on polymers, *Langmuir*, **7**:1822 (1991).
33. H. Lin, T. Yanagi, W. S. Seo, K. Kuwabara and K. Koumoto, Crystal growth of Hydroxyapatite under two-dimensionally oriented organic films, to be published at 9th Int. Symp. on Ceramics in Medicine.
34. T. Kajiyama, Y. Oishi, M. Uchida, Y. Tanimoto and H. Kozuru, Morphological and structural studies of crystalline and amorphous monolayers on the water surface, *Langmuir*, **8**:1563 (1992).
35. C. R. Broundle, H. Hopster and J. D. Swalen, Electron mean-free path lengths through monolayers of cadmium arachidate, *J. Chem. Phys.*, **70**:5190 (1979).
36. E. L. Smith, C. A. Alves, J. W. Anderegg, M. D. Porter and L. M. Siperko, Deposition of metal overlayers at end-group-functionalized thiolate monolayers adsorbed at Au.1. surface and interfacial chemical characterization of deposited Cu overlayers at carboxylic acid-terminated structures, *Langmuir*, **8**:2707 (1992).
37. K. Kobayashi and K. Takaoka, Application of X-ray photoelectron spectroscopy and Fourier transform IR-reflection absorption spectroscopy to studies of the composition of LB films, *Thin Solid Films*, **159**:267 (1988).
38. A. A. Campbell, G. E. Fryxell, G. L. Graff, P. C. Rieke and B. J. Tarasevich, The nucleation and growth of calcium oxalate monohydrate on self-assembled monolayers (SAM), *Scanning Microscopy*, **7**:423 (1993).
39. H. Lin, H. Ando, W. S. Seo, K. Kuwabara and K. Koumoto, Two-dimensionally oriented organic molecules as a substrate for vapor growth of zinc thin films, *Thin Solid Films* in press.

REACTIVE COATINGS ON CERAMIC SUBSTRATES

J. S. Moya[1], J. Requena[1], H. P. Steier[1], A. H. De Aza[2], P. Pena[2] and R. Torrecillas[3]

[1] Instituto de Ciencia de Materiales de Madrid, CSIC, Madrid, Spain
[2] Instituto de Cerámica y Vidrio, CSIC, Madrid, Spain
[3] Instituto Nacional del Carbon, CSIC, Oviedo, Spain

INTRODUCTION

Modification of monolithic substrate surfaces by coating them with a substance different from that of the bulk is currently an important technological issue in structural as well as in functional applications such as: thermal and chemical barriers, wear resistance and tribology, environmental sensors, fibre-reinforced ceramic composites, catalysis, etc.[1] Ceramic coatings can be used to enhance the mechanical, chemical, electrical, optical or thermal properties of substrates, or to impart physical and mechanical properties of the substrates not normally possessed by the substrate itself. Although a wide variety of surface modification techniques are available, e.g. thermal spray processes, sol-gel or slurry methods, chemical vapour deposition (CVD), physical vapour deposition (PVD), etc., it is generally very difficult to modify the surface to any significant depth (i.e. >2 µm) by these routes. This fact limits the adhesion and the possible applications of the obtained coated materials. The solution to this problem requires an increased understanding of chemical and mechanical bonding at the coating-bulk interface.

The main objective of this work is to open a new route to modify the surface characteristics and properties of a ceramic by: (i) coating its free surface with a substance chemically incompatible with the substrate, and subsequently (ii) by controlling the reaction series that could take place at the interface to grow a final textured and mechanically stable coating with a composition and properties different from those of the substrate. A flow-chart of this proccess is given in Fig. 1.

In this sense several examples of reactive coating are possible: a) The formation of a textured layer of calcium hexa-aluminate on an alumina substrate; b) The formation of a stable thick layer of zirconium titanate on a rutile substrate; c) The formation of a textured layer of $BaTiO_3$ on a rutile substrate; and d) The attainment of a mechanically stable zircon-glass-ceramics-zircon multilayer composite.

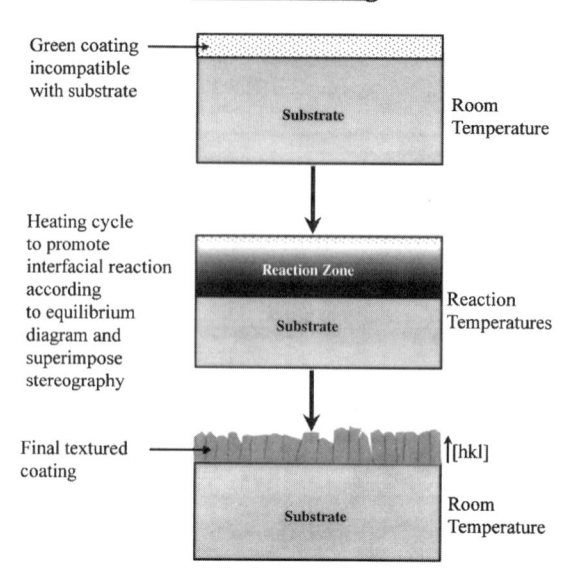

Figure 1. Schematic representation of reactive coating process

REACTIVE COATING OF DOLOMITE ON ALUMINA SUBSTRATE.

Calcium hexaluminate (CaAl$_{12}$O$_{19}$, hibonite)[2] is isostructural with the mineral magnetoplumbite. Both consist of a β-Al$_2$O$_3$ type unit cell built up from two blocks that are isostructural with spinel. These spinel type blocks are separated by a mirror plane, or "conduction plane", with stabilizing cation and oxygen anions, but they differ in the arrangement of the stabilizing cations and oxygen anions in the mirror plane. Hibonite exhibits very promising technological properties, since many cations are available as constituents of the mirror plane. Moreover, some of them can be combined in the same lattice without destroying the hexaluminate structure. Hibonite is phase compatible with alumina up to the peritectic point ≈1800°C, and both have very close average thermal expansion coefficients.[3]

Hibonite presents a wide range of solid solutions with iron oxides and transition cations containing slags,[4] adequate chemical resistance in alkaline environments, high stability in a reducing atmosphere[5] and low solubility in several multicomponent systems.[4-6] Recently hibonite based, materials have found new applications in the field of nuclear waste storage[7] and catalysis.[8] Further applications could be found in the field of solid state ionic conductors due to its β-alumina defect structure.[2]

In the present work we introduce a reactive coating as a feasible route to coat an alumina substrate with a textured layer of hibonite, starting from a reactive coating of dolomite CaMg(CO$_3$)$_2$. The sequence of phases appearing and the mechanism of the interface reaction are explained using the MgO-Al$_2$O$_3$-CaO equilibrium diagram.[9]

Additionally, due to the low cost of dolomite and the simplicity of the overall process a new route to develop high-temperature oxidation-resistant, fiber-reinforced alumina composites or multilayer composites with a weak interface is also proposed.

After polishing and cleaning plates of polycrystaline and dense alumina, a coating was screen printed with a suspension of natural dolomite CaMg(CO$_3$)$_2$ (Prodomasa, Spain) with 30.29 wt% CaO, 22.05 wt% MgO, 0.018 wt% SiO$_2$ and 0.011 wt% Al$_2$O$_3$ as main impurities, and having a specific surface area of 2.5 m²/g; and a particle size <5μm. This suspension was prepared by mixing ethylenglycol (55 wt %) and the dolomite powder (45 wt %) with 1 wt % defloculant (Produkt KV 5088, Zschimmer & Co. Chemische Fabriken, Lanstein/Rhein, Germany) in an agate mortar. Different alumina substrates were screen printed and dried at 80°C for 48 h. The coated substrates were fired at 1550°C and 1650°C for different times.

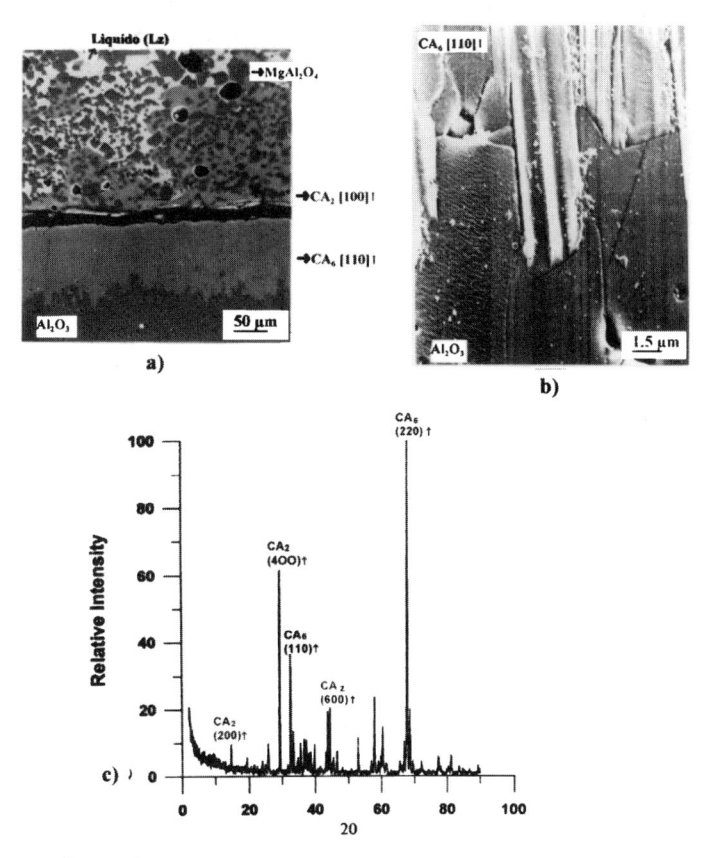

Figure 2. a) SEM micrograph corresponding to the cross section of the polycrystalline alumina substrate coated with dolomite after firing at 1650°C / 4 h. b) Close up SEM micrograph showing the interface between the textured hibonite crystals and the alumina substrate. c) XRD pattern obtained on the hibonite surface showing that is oriented in [110] direction and the remaining CaAl$_4$O$_7$ phase is also oriented along [100] direction.

Fig. 2a shows a SEM view of the cross section of the polycrystalline alumina substrate coated after firing at 1650°C for 4 hours. The upper part of the sample coating is formed by primary spinel cystals and a liquid phase. This part is practically separated from the rest by a longitudinal crack which developed during cooling. At the bottom a textured thick layer of hibonite phase growing from the alumina substrate is also clearly shown. In Fig. 2a close up SEM micrograph of the interface hibonite/Al$_2$O$_3$ is shown where no liquid phase is detected by SEM. Grain boundary microcracks propagate through the hibonite layer along the easy-cleaving basal planes which are perpendicular to the hibonite-alumina interface. The coated surface shows a strongly oriented x-ray diffraction pattern (Fig. 2). This XRD pattern shows that the hibonite is oriented the [110] direction and the remaining CaAl$_4$O$_7$ phase is also oriented along [100] direction. No chipping or interfatial cracks were detected between the textured hibonite layer and the alumina substrates (Fig 2).

The hibonite layer formation and the sequence of phase appearance at the interface reaction can be explained considering the isothermal section at 1650°C of the MgO-Al_2O_3-CaO ternary system (Fig. 3).

During heating a liquid phase appears at the interface $CaO \cdot MgO/Al_2O_3$ at temperatures as low as ≈1321°C as indicated by the MgO-Al_2O_3-CaO equilibrium diagram.[9] This liquid corresponds to the lowest eutectic point of the ternary system. The liquid change in composition from point P to Y as the temperature is rising and the reaction between $CaO \cdot MgO/Al_2O_3$ takes place. At this point liquid Y is saturated with spinel, $MgAl_2O_4$, and magnesia, MgO.

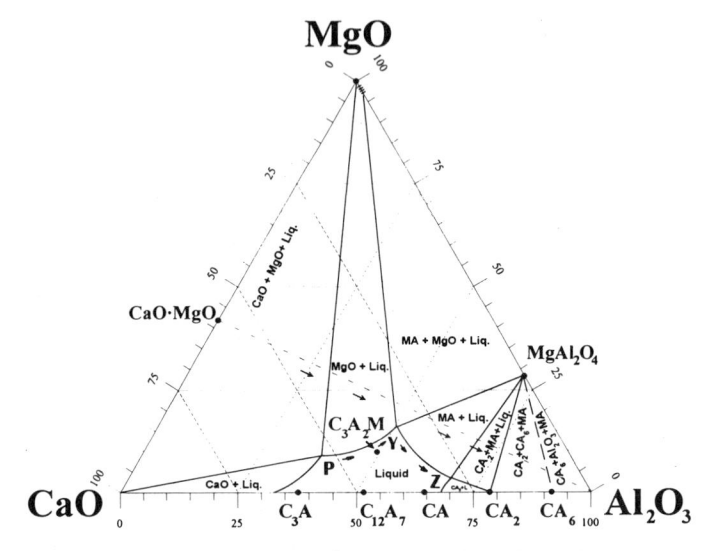

Figure 3. Isothermal section at 1650°C of the MgO-Al_2O_3-CaO ternary system.

As the reaction proceeds magnesia is dissolved and the composition of the liquid in equilibrium with spinel is moving along the line Y-Z, as Al_2O_3 is dissolved. When the liquid composition reaches point Z, then it is saturated in both $MgAl_2O_4$ and $CaAl_4O_7$. Then this last phase nucleates and grows along [100] on the surface of the alumina substrate as shown in Fig. 2. This layer physically separates the spinel plus liquid from the alumina substrate. In a later step the $CaAl_4O_7$ reacts, in the absence of liquid, with the alumina substrate forming hibonite which is compatible with alumina. The hibonite grains also grow oriented, but along the [110] direction. Both growing directions are in close agreement with the data reported by the pioneer study of Kohatsu et al[10] using a CaO/Al_2O_3 couple diffusion system. The crystal orientation of the textured layer is not dependent upon the crystal orientation of the alumina substrate.

Hibonite exhibits large thermal expansion coefficient anisotropy[11], $\alpha_{p(20-800)} = 11.8 \cdot 10^{-6}$ °C^{-1} and $\alpha_{b(20-800)} = 7.5 \cdot 10^{-6}$ °C^{-1}, where the subscripts b and p indicate basal planes and perpendicular to the basal planes, respectively. Hitchcock el al.[12] have reported that single crystals of these materials show a large fracture toughness anisotropy with K_b/K_p ~0.1. In the present work the hibonite layer is oriented with the (110) planes parallel to the interface. Consequently the basal planes are perpendicular to the interface. During cooling the hibonite layer is therefore subjected to a tensile stress ($\sigma_R = 1/2 \, [E \cdot \Delta\alpha_{max} \cdot \Delta T]$ where $\Delta\alpha =$

$\alpha_{alumina}-\alpha_{p(hibonite)} = -3.8\cdot10^{-6}$ °C^{-1}). Because of the hibonite's low k_b value any defect located on the top layer surface is easily propagated as a crack perpendicular to the substrate (Fig. 2). This fact releases the residual stresses developed at the coating interface and avoids the possible development of parallel cracks and chipping.

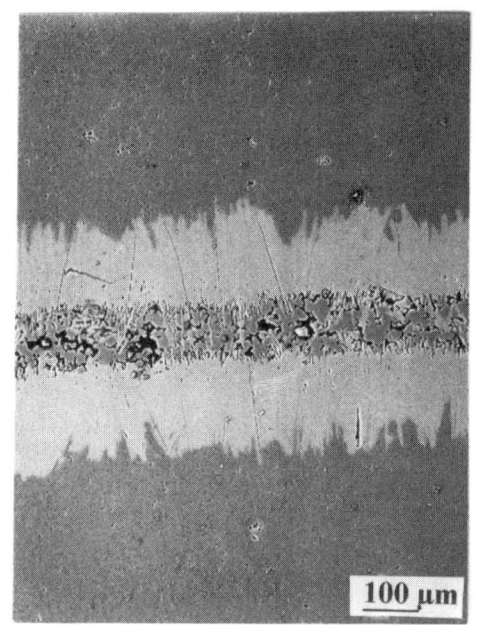

Figure 4. SEM micrograph of the cross section corresponding to Al_2O_3/$MgCa(CO_3)_2$/Al_2O_3 sample heat treated at 1650°C/4h.

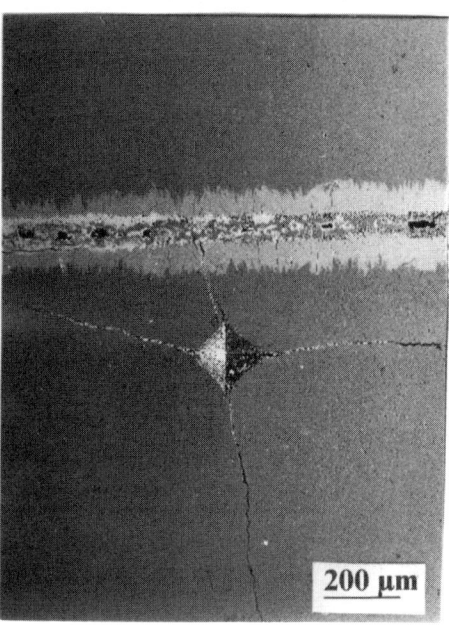

Figure 5. SEM micrograph showing the cracks generated by a Vickers indenter in the "sandwich" type specimen

A similar experiment can be made with a "Sandwich" type specimens where the dolomite coating is the intermediate layer. After firing at 1650°C 4h., because of the symmetrical arrangement of the layer, microcracks develop instead of a long single crack at the interface (Fig. 4). The mechanical stability of this joining has been qualitatively tested by Vickers indentation (Fig. 5). The indentation crack easily penetrates the hibonite layer and due to the high density of microcracks it is trapped in the intermediate glassy rich layer. This type of coating can be used as a weak interface in alumina fiber reinforced alumina matrix composites. In this kind of application the interface between fibres and matrix must be sufficiently weak to allow for crack deflection and debonding at the fiber-matrix interfaces.[13] Morgan et al[14] have shown a similar effect in sapphire fiber reinforced polycrystalline alumina matrix using monazite as a weak interface.

REACTIVE COATING OF ZIRCON ON RUTILE SUBSTRATES

$ZrTiO_4$ based ceramics exhibit several unique properties, which make them important in a broad range of functional applications. In the electronics industry and wireless communication systems[15] these materials are used -mostly in solid solution with SnO_2 - because of their high dielectric constant ε_r at microwave frequencies (up to 40), high Q value (up to 13000) and small temperature coefficient τ_t of resonance frequency (nearly 0 ppm/°C), which are all particularly required to be applied to the dielectric resonator. Thus it is possible to miniaturize integrated resonance circuits without degrading the Q value in order to realize devices like microwave filters, oscillators and antenna.[16]

Further opportunities for specific applications result from the high temperature stability [17] and the possibility to select the temperature coefficient of their dielectric constant and resonant frequency by the quantity of dopant.[18] The literature also reports applications of ZrTiO$_4$ based ceramics as acid-base bifunctional catalyst[19] and as a host lattice for inorganic mixed phase pigments because of its high diffraction index n.[20]

The placement of ZrTiO$_4$ thick films on a ceramic substrate may have both scientific and technological interest. The aim of the present work is to coat a rutile substrate by a mechanical unstable thick layer of ZrTiO$_4$ using reactive coating technique[21,22] on the base of high-temperature relationships in the ZrO$_2$-SiO$_2$-TiO$_2$ system.

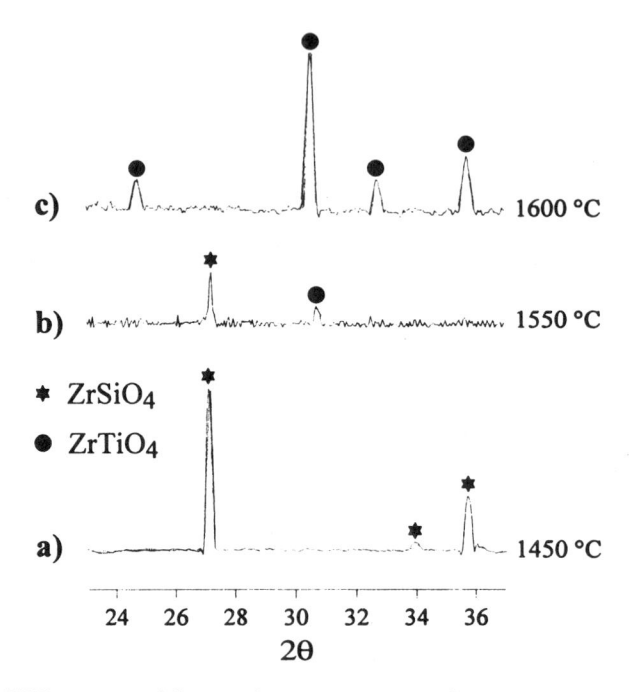

Figure 6. XRD-patterns of the samples coated at a) 1450°C, b) 1550°C and d) 1600°C.

Polished plates of polycrystaline rutile were coated by screen printing using a suspension of zircon powder with an average particle size of 1.3µm. The suspension was prepared by mixing the zircon powder (50 wt%) with ethylenglycol (50wt%) and 1 wt% defloculant. The screen printed substrates were dried at 100°C for 24 h and heat treated for 2 h at 1450°C, 1550°C and 1600°C respectively. Heating and cooling rates were always 5°C/min.

After heat treatment the phases present in the different samples were analyzed by XRD-analysis. The polished cross sections were observed by reflected-light optical microscopy (RLOM), scanning electron microscopy with energy dispersive x-ray analysis (SEM-EDS). Fig. 6 shows the XRD patterns obtained on the coated layers corresponding to the samples fired at 1450, 1550 and 1600°C. As observed, at 1450°C only the zircon starting material is detected. At 1550°C ZrTiO$_4$ appears as an additional phase, and at 1600°C no zircon is detected and only ZrTiO$_4$ appears as a crystalline phase.

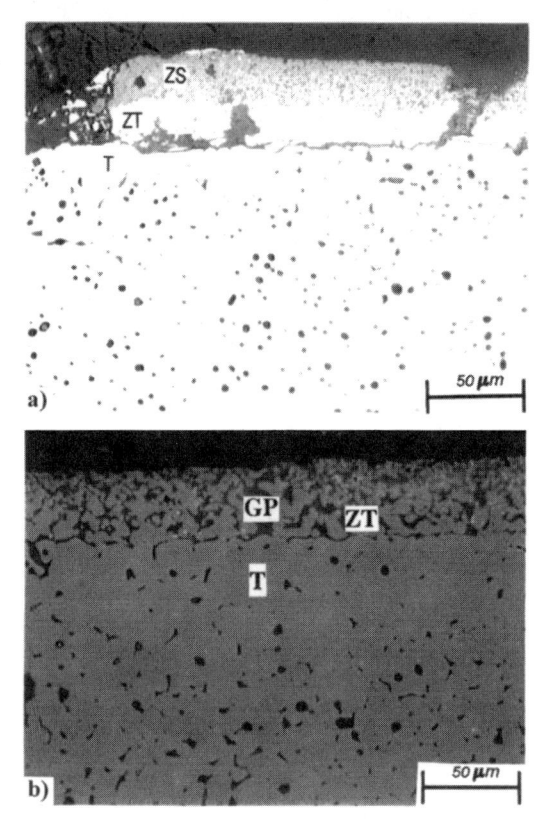

Figure 7. Reflected light optical micrographs of the cross sections of the samples coated at a) 1550°C and b) 1600°C (T:=TiO$_2$, ZS:=ZrSiO$_4$, ZT:=ZrTiO$_4$, GP:=Glassy Phase).

Fig. 7 shows reflected light optical micrographs of the polished cross-section of samples heated at different temperatures. At 1450°C (Fig. 6) no reaction between the rutile-substrate and the ZrSiO$_4$-coating has taken place. The zircon powder has been slightly densified through sintering without forming a stable layer upon the substrate. At 1550°C (Fig. 7a) the reaction:

$$ZrSiO_4 + TiO_2 \text{---------}> ZrTiO_4 + liq \qquad (1)$$

partially takes place and an intermediate layer of ZrTiO$_4$ is formed. However cracking and chipping were observed as a consequence of residual stresses developed during the reaction process due to the volume change ($\Delta V \approx 6\%$) involved.

No cracks are observed at 1600°C (Fig. 7b) and a continuous and mechanically stable thick layer is obtained. This layer consists of ZrTiO$_4$ grains, with an average grain size of $\approx 10\mu m$, embedded in a continuous high silica containing glassy matrix.

The obtained results can be clearly explained considering the ZrO$_2$-SiO$_2$-TiO$_2$ equilibrium phase diagram[23] (Fig. 8). At temperatures below the eutectic point ZrTiO$_4$-TiO$_2$-SiO$_2$ (≈ 1550°C) no liquid phase is present and consequently the reaction at the zircon/rutile interface occurs in the solid state at a slow rate. Above this temperature reaction [1] takes place.

Table 1. EDS analysis and volume fraction of the glassy phase present at the coating layer

Sample	SiO$_2$	Content (wt%)* TiO$_2$	ZrO$_2$	Glassy phase volume fraction Vol%
heated at 1550°C	78.5 ± 2.5	15.5 ± 3.2	5.4 ± 2.1	≈ 20
heated at 1600°C	73.1 ± 2.6	20.6 ± 4.1	6.1 ± 2.5	≈ 35

*average value of six determinations on different areas (5x5 μm^2)

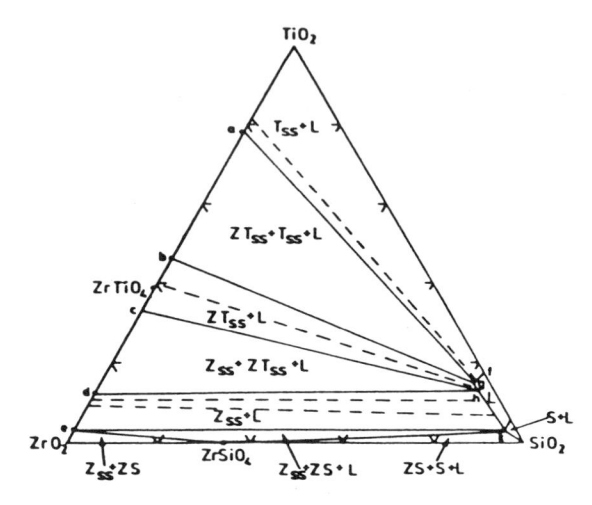

Figure 8. Isothermal section at 1650°C of the TiO$_2$-SiO$_2$-ZrO$_2$ equilibrium phase diagram [23].

Presented in Table 1 is the composition of the glassy phase as well as its volume fraction at both 1550°C and 1600°C. These are in good agreement with the expected values based on the TiO$_2$-SiO$_2$-ZrO$_2$ equilibrium diagram.

In summary a mechanically stable thick layer of ZrTiO$_4$ on a rutile substrate has been obtained from a coating of zircon. After reacting at 1600°C this layer consists of ZrTiO$_4$ grains of about 10 μm average grain size embedded in a silica rich glassy matrix (≈35 vol%).

REACTIVE COATING OF BARIUM TITANATE ON RUTILE SUBSTRATES

BaTiO$_3$-films have well known technological applications. Thick films of BaTiO$_3$ are used as high capacity condensors. Apart from the thickness, the properties of BaTiO$_3$-films depend on such microstructural factors as grain orientation, grain size and internal stresses which affect the distribution of the Curie point (T$_C$), impurity and porosity.[24]

The most interesting of these factors is perhaps the crystal-axis orientation of the grains, since it controls the ferroelectric domain structure.[25] The dielectric constant of a BaTiO$_3$ single crystal is 4100 along the a-axis and 160 along the tetragonal c-axis. The value for coarse-grained material[26] is about 1400.

Brody et al.[27] talk about future applications of BaTiO$_3$ films on a substrate in the fields of nonvolatile random access memory (RAM), optical switches and waveguides, sensors or actuators.

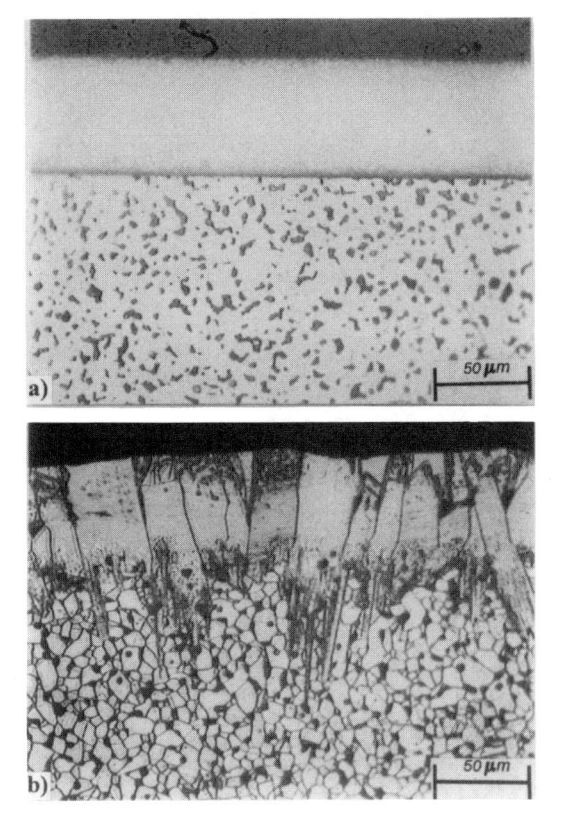

Figure 9. Optical micrographs of the polished cross-sections of BaTiO$_3$ reactive coating on rutile substrate a) before and b) after heat treatment at 1400°C/6 min, 1310°C/1h and subsequent thermal etching.

The main objective of this work is to obtain a textured and mechanically-stable thick layer of BaTiO$_3$ on a titania substrate by reactive coating, taking into account the high temperature relationships in the TiO$_2$-BaO system.

Plates of polycrystaline rutile after polishing were coated by screen printing using a suspension of barium titanate powder with an average particle size of 0.82 µm and BaO/TiO$_2$ ratio of 1.002. The suspension was prepared by mixing the BaTiO$_3$ powder (50 wt%) with ethylenglycol (50 wt%) and 1 wt% deffloculant. The screen printed substrates (Fig. 9a) were dried at 100°C for 24 h and heat treated using the following cycle: heating with 10°C/min to 1400°C → holding 6 min → cooling with 10°C/min to 1310°C → holding 1 h → cooling with 2°C/min to room temperature.

After heat treatment the phases present were analyzed by XRD. Polished cross sections were thermally etched at 1120°C - 30 min and observed by reflected-light optical microscopy (RLOM) and scanning electron microscopy (SEM)

As can be observed in Fig. 9b a stable and textured BaTiO$_3$ coating was obtained after 6 min holding at 1400°C and 1h annealing at 1310°C. This coating is formed by elongated tetragonal BaTiO$_3$ grains with ≈ 50 µm length and ≈ 25 µm width. At the interface long rod-like crystals, 20 - 50 µm in length and 5 - 10 µm in diameter (Figs. 9b, 10a) can be observed.

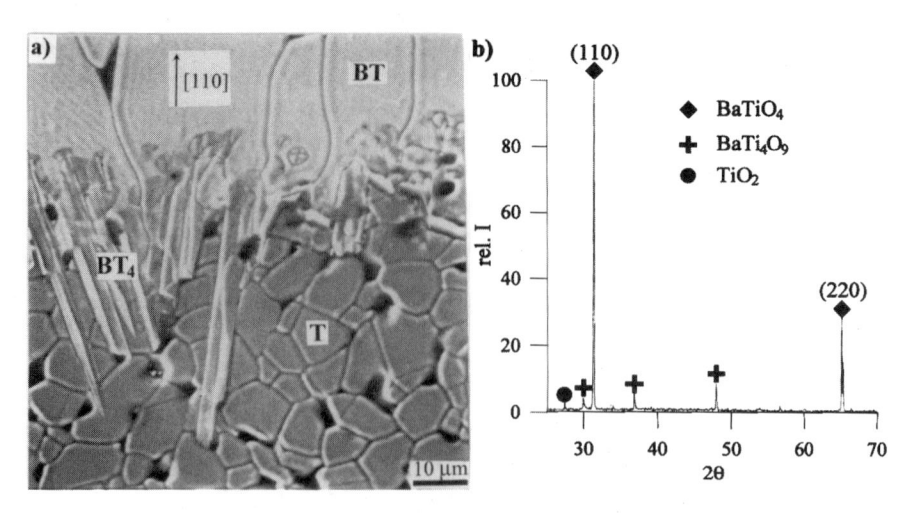

Figure 10. SEM micrograph (a) of the reactive coating interface showing the oriented BaTiO$_3$ (BT) crystals and the rod-like BaTi$_4$O$_9$ (BT$_4$) crystals (T:= TiO$_2$), and (b) the *in situ* XRD pattern of the reactive coating.

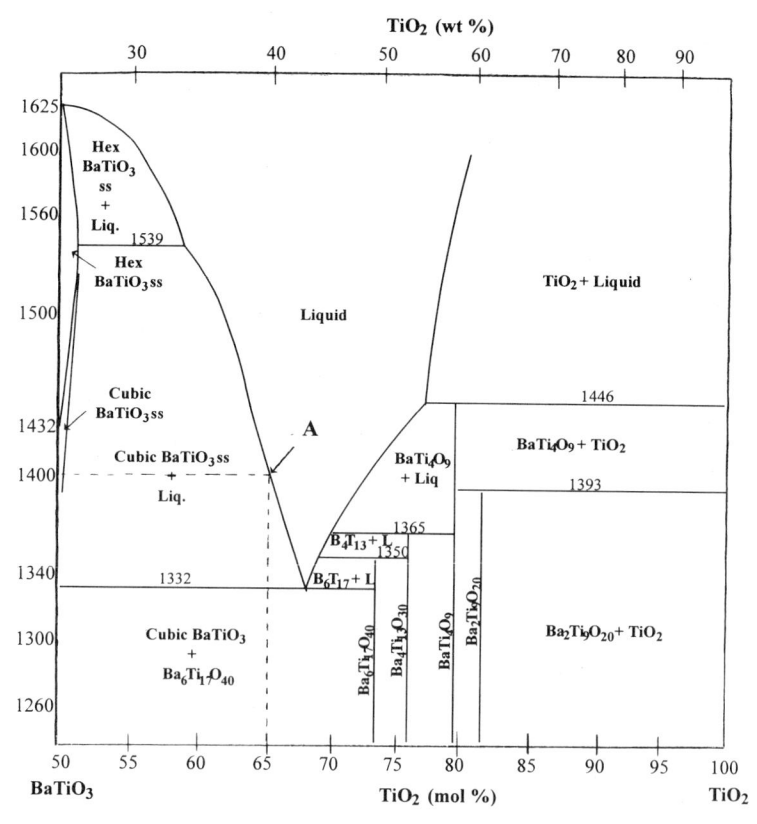

Figure 11. BaTiO$_3$-TiO$_2$ equilibrium phase diagram[12] showing the composition of the liquid phase (A) developed at the coating interface at 1400°C.

The XRD pattern obtained on the top surface of the reactive coated specimen is shown in Fig. 10b. From this pattern it is clear that the BaTiO$_3$-grains are well oriented in the [110] crystallographic direction. The presence of barium tetratitanate (BaTi$_4$O$_9$) compound was also identified.

This particular morphology of the reactive coating on the rutile substrate can be understood considering the high temperature equilibrium relationships in the BaTiO$_3$-TiO$_2$ system as well as the proposed heating cycle.

According to the BaTiO$_3$-TiO$_2$ equilibrium diagram[28] (Fig.11), at 1400°C the composition of the interface between the starting green coating of BaTiO$_3$, and the rutile substrate will move into the BaTiO$_3$ primary crystallization field along the 1400°C isotherm. At the same time a liquid phase, richer in TiO$_2$, with the composition of point "A" shown in Fig. 11 ($\approx$ 66 mol% TiO$_2$) is developed.

During the holding time at this particular temperature the BaTiO$_3$ powder is disolved in this glassy phase and primary BaTiO$_3$ crystals will precipitate at the interface and grow mainly uppward. The liquid is located at the interface between the large BaTiO$_3$ crystals and TiO$_2$ substrate. Because of this this liquid will disolve more TiO$_2$ then its composition will shift to BaTi$_4$O$_9$ primary field favouring the nucleation and growth of the observed primary rod-like crystals of BaTi$_4$O$_9$. This fact also justifies the partial dissolution of the bases of the large BaTiO$_3$ crystals (Fig. 10)

The formation of this particular rod-like compound has also been observed by O´Bryan and Thompson[29] in mixtures of BaTiO$_3$ and TiO$_2$ with $\approx$ 80 mol % TiO$_2$ heated at > 1400°C.

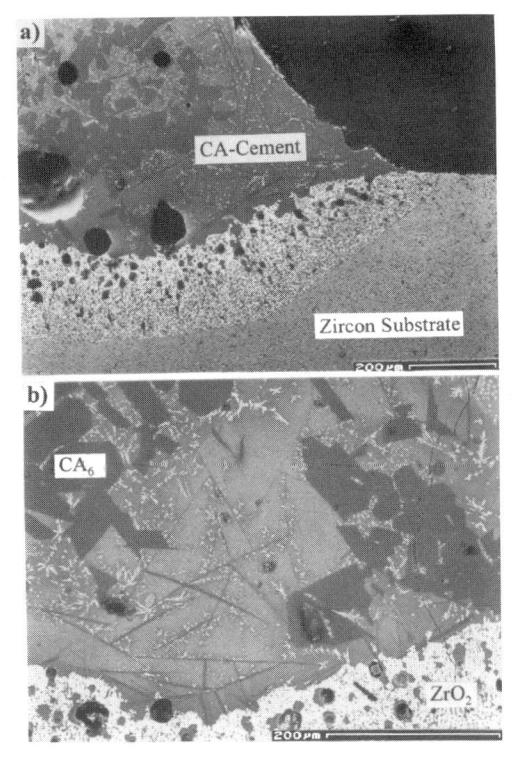

Figure 12. SEM micrograph of the polished cross section of the CA cement zircon substrate after firing at 1615°C at two different magnifications and two different areas.

JOINING ZIRCON SINTERED COMPACTS BY REACTIVE COATING WITH CA CEMENT

Joining of simple ceramic shapes to form complex ceramic parts can be considered as an alternative route to the difficult and expensive methods associated with fabrication of near-net shape ceramic components.[30] The ability to reliably join ceramic parts to form large systems is considered a key technology in enhancing the use of ceramics in high temperature structural applications.[31] However, as has been discussed elsewhere[32], the joining of ceramics is not yet a well-developed art. Most scientific and engineering efforts have been devoted to ceramic metal joining.

Recently zircon is being considered as a matrix in CMC's because of its excellent thermomechanical compatibility with SiC fibres.[33-36]

The main objective of the present work is to develop a low cost method to join a sintered zircon compact by reactive coating with CA-Cement (SECAR 71, Lafarge, France. Al_2O_3, 71.56; SiO_2, 0.23; CaO, 23.58 wt%) containing ink.

The reaction sequence that takes place at the interface and the microstructural evolution have been studied in conjunction with the SiO_2-Al_2O_3-CaO-ZrO_2 quaternary equilibrium diagram.[37]

Dense zircon substrates with polished surfaces were used. An axially pressed cylinder 3 mm in diameter and 5 mm in height of CA-cement powder was placed on the polished surface of the zircon substrate and fired at 1615°C for 1 h. The interfacial reaction was studied in cross sectioned specimens by optical microscope and SEM with EDS microanalysis.

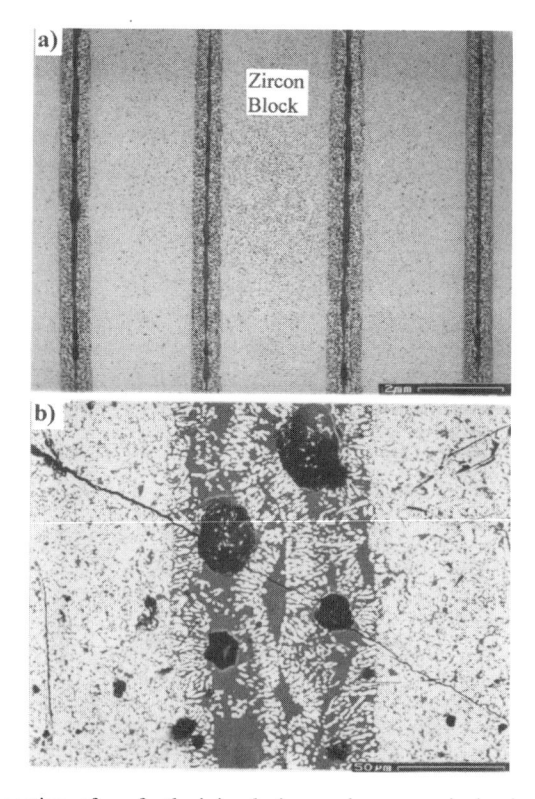

Figure 13. a) Cross section of perfectly joined zircon-glass ceramic laminates and b) indentation print crack passing through the interface without being deflected.

Multilayers of zircon and CA cement were obtained by screen printing the zircon ceramic substrate with a CA-cement ink obtained by mixing the cement powder with isopropilic alcohol in a proportion 2:1. The coated substrates were subsequently stacked subjected to a pressure of about 0.3 MPa. This multilayer composite was fired at 1615°C for 2 h. The polished cross section of the fired multilayer composite was studied by SEM with EDS microanalysis. The residual stress at the interface was studied by Vickers indentation.

Fig. 12 shows the SEM micrographs corresponding to the polished cross section of the CA cement-zircon substrate after firing. The interface appears free of cracks. An interfacial layer of 200 µm (Fig. 12b) is formed by a continuos silica-rich glassy phase with a dispersed and textured zirconia grains of ~ 8 µm. These data were obtained using close up SEM micrographs and EDS which were not shown. At the top of this glassy layer long needle like CA_6 crystals typically formed by precipitation during cooling, surrounded by dendritic precipitates of ZrO_2 grains are observed (Fig.12b).

At 400 µm from the interface equiaxed primary CA_6 crystals (100 µm) are also present surrounded by a CaO-rich glassy phase where small CA crystals have precipitated during cooling (Fig. 12b).

Fig. 13a shows the cross section of zircon-glass ceramic laminates. As can be observed no cracks are present and perfect joining is obtained. The interface consists of by two symmetrical textured layers (~ 300µm) of zirconia grains in a glass-ceramic matrix. The microstructural features of these interfaces can be explained considering the quaternary system ZrO_2 - Al_2O_3 - SiO_2 -CaO proposed by Pena and de Aza[37] and more specifically from the projection through the ZrO_2 corner onto the opposite face of the tetrahedron (Fig. 14). Using this simplified representation it can be stated that the composition of the interface is located on the straight line that joins the SiO_2 (ZS) corner with the point representing the composition of the CA-cement (~ 70 wt% Al_2O_3).

At 1615°C the interface between zircon-CA cement will be liquid. Far from the interface inside the CA-cement bulk (Fig. 12) the composition shifts along the SiO_2 (ZS)-

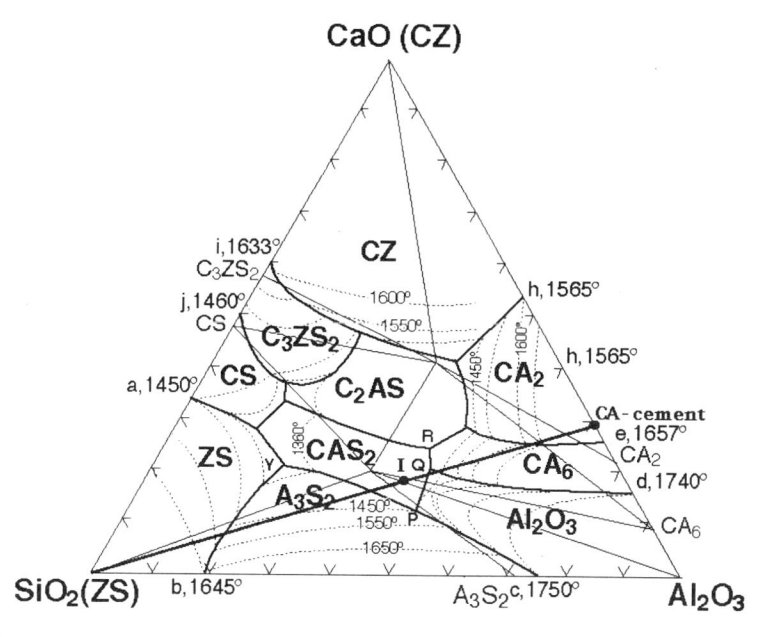

Figure 14. Projection through the ZrO_2 corner, onto the opposite face of the quaternary tetrahedon ZrO_2-Al_2O_3-SiO_2-CaO

70% Al_2O_3 line from point I toward the CA_6 secondary field (Fig. 14). This fact justifies the presence of CA_6 crystals. In the case of the multilayer composite the composition of the interface is richer in silica and then anorthite precipitates from the glassy phase giving an anorthite glass-ceramic joining between the two textured layers of zirconia. These data were obtained using high magnification SEM micrograph and EDS which were not shown.

As a consequence of the high viscosity of the silica-rich glassy phase developed within the original zircon grains, the composition of the zircon-CA cement interface is located in the compatibility tetrahedron ZrO_2-CA_6-CAS_2-Al_2O_3 (point I in Fig. 14). The invariant point of this compatibility tetrahedron is a peritectic located at 1380°C (point Q in Fig. 14).

In Fig. 13b the indentation crack passes through the interface. No deflection of the cracks is produced at the interface. This fact indicates the absence of residual stresses as a consequence of a possible thermal expansion mismatch.

In summary, a reactive coating of CA-cement between two zircon substrates develops stable layer of anorthite-ZrO_2-glassy phase. This layer is mechanically stable and can be used to join zircon. This joint will be stable until the peritectic temperature (1380°C) of the ZrO_2-CA_6-CAS_2-Al_2O_3 compatibility tetrahedron is surpassed.

FINAL REMARKS

The results obtained in the present investigation clearly show that reactive coating is a feasible route to produce textured, mechanically-stable coating on ceramic substrates.

This new route opens the possiblity of new material development for functional applications in electronics, environmetal and optic fields.

It is also possible a) to apply this concept to produce wear resistant coatings on materials, b) to develop a weak interface in fiber reinforced ceramic composites and c) to join ceramic compacts.

ACKNOLEDGEMENTS

This work has been supported by CICYT Spain, through Contract N.° MAT94-0974

REFERENCES

1. R.A. Eppler. *Ceramic and Glasses. Engineered Materials. Handbook Vol. 4.,* ASM international, The Matrials Information Society, pp. 991-3 (1991).
2. J.V. Hoek. *Alkali Metal Corrosion of Alumina, Thermodynamics, Phase Diagrams and Testing,* Ph.D. Thesis, Tech. University Eindhoven, Netherlands, (1990).
3. E. Criado, S. De Aza, and D.A. Estrada, Caracteristicas dilatométricas de los aluminatos de calcio, *Bol. Soc. Esp. Cerám. Vidr.* 14, 3, pp. 271-3 (1975).
4. D.H. Lister, and F.P. Glasser, Phase relations in the system CaO-Al_2O_3-iron oxide, *Trans. Brit. Ceram. Soc.*, 66 pp. 293-305 (1967).
5. E. Criado, and S. De Aza, Calcium hexaluminate as refractory material, *UNITECR'91*, Vol. I, pp. 566-74, Aechen, Germany (1991).
6. P. Pena, and S. De Aza, Compatibility relations in the system ZrO_2-Al_2O_3-SiO_2-CaO, *J. Am. Ceram. Soc.,* 67, C-3-5 (1984).
7. A.B. Harker, and J.F. Flimtoff, Hot isostatically pressed ceramic and glasses forms for immobilizing Hanford high-level wastes, *Adv. Ceram.*, 8, pp. 222-33 (1984).
8. J.D. Hodge, Alkaline earth effects on the reaction of sodium with aluminium oxides, *J. Electrochem. Soc.* 133, 4, pp. 833-6 (1986).
9. A.H. De Aza, P. Pena, and S. De Aza, Research submitted to the *J. Am. Ceram. Soc.*
10. I. Kohatsu, and G.W. Brindley, Solid state reactions between CaO and $\propto$-Al_2O_3, *Zeitschr. Phys. Chem. Neue Folge*, Bd. 60, pp. 79-89 (1968).
11. D. Brooksbank, Thermal expansion of calciumaluminate inclusions and relation to tessellated stresses, *J. of The Iron and Steel Institute*, May, pp. 495-9 (1970).

12. D.C. Hitchcock, and L.C. De Jonghe, Fracture toughness anisotropy of sodium β-alumina, *J. Am. Ceram. Soc.*, 11, C-204-5 (1983).
13. M.K. Cinibulk, Magnetoplumbite compounds as a fiber coating in oxide/oxide composites, *Ceram. Eng. Sci. Proc.*, 15, [5], 721-8 (1994).
14. P.E.D. Morgan, and D.B. Marshall, Ceramic composites of monazite and alumina, *J. Am. Ceram. Soc.*, 78, [6],1553-63 (1995).
15. Y. Zhang, and P.K. Davis, Stabilization of ordered zirconium titanates through the chemical substitution of Ti^{4+} by Al^{3+}/Ta^{5+}, *J.Am.Ceram.Soc.* 77, [3], 743-8 (1994).
16. Y.-c. Heiao, L. Wu, and C.-c. Wei, Microwave dielectric properties of $(ZrSn)TiO_4$ ceramic, Mater.Res.Bull. 23, [12], 1687-92 (1988).
17. F. Azough, and R. Freer, Microstructural development and microvave dielectric properties of zirconium titanate ceramics sintered with Nd_2O_3, *Euro-Ceramics* Vol.2, 2.294-8 (1989).
18. G. Wolfram, and E. Göbel, Existence range, structural and dielectrical properties of $Zr_xTi_ySn_zO_4$ ceramics (x+y+z=2), *Mater.Res.Bull.* 16, [11], 1455-63 (1981).
19. J.A. Navio, M. Macias, and P.J. Sanchez-Soto, Influence of chemical processing in the crystallization behaviour of zirconium titanate materials, *J.Mater.Sci.Lett.* 11, [23], 1570-2 (1992).
20. F.Z. Hund, Mixed-phase pigments based on $ZrTiO_4$, *Anorg.Allg.Chem.* 525, 221-9 (1985) (Ger.).
21. J.S. Moya, A.H. de Aza, H.P. Steier, J. Requena and P. Pena, Reactive coating on alumina substrates. Calcium and barium hexa aluminates, *Script. Met. et Mat.* 31, [8], 1049-54 (1994).
22. M. A. Sainz, R. Torrecillas, and J. S. Moya, Novel Technique for Zirconia-Coated Mullite, *J.Am.Ceram.Soc.* 76, [7], 1869-72 (1993).
23. N.C.H. Lubaba, C. M. Wilson, and N. H. Brett, Phase equilibria and solid solution relationships in the system ZrO_2-SiO_2-TiO_2, *Br.Ceram.Trans.J.* 83, 49-54 (1984).
24. K. Yamashita, T. Hamano, T. Kaga, K. Koumoto and H. Yanagida, The thickness-dependence of dielectric and physical properties of $BaTiO_3$ ceramic thick films, *Jap.J.Appl.Phys.* 22, [4], 580-4 (1983).
25. Y. Ohara, T. Taki, K. Koumoto and H. Yanagida, Crystal-axis oriented ceramics prepared from fibrous barium titanate, *J.Mater.Sci.Let.* 11, 1327-9 (1992).
26. N.C. Sharma, and E.R. McCartney, The dielectric properties of pure barium titanate as a function of grain size, *J.Austral.Ceram.Soc.* 10, [1], 16-20 (1974).
27. P.S. Brody, B.J. Rod, K.W. Bennett, L.P. Cook, P.K. Schenk, M.D. Vaudin, W. Wong-Ng and C.K. Chiang, Preparation, Microstructure and ferroelectric properties of laser-deposited thin $BaTiO_3$ and lead zirconate-titanate films, *Integr.Ferr.* 1, 239-51 (1992).
28. P.P. Phule and S.H. Risbud, Low temperature synthesis and processing of electronic materials in the BaO-TiO_2 system, *J. Mat. Sci.* 25, 1169-83 (1990).
29. H.M. O'Bryan, Jr., and J. Thompson, Jr., Phase equilibria in the TiO_2-rich region of the system BaO-TiO_2, *J. Am. Ceram. Soc.* 57, [12], 522-6 (1974).
30. R. Torrecillas, M.A. Sainz, and J.S. Moya, Alumina-alumina and mullite-mullite joining by reaction sintering process, *Script. Met. et Mat.* 31, 1031-6 (1994).
31. R.D. Watkins, in *ASM Eng. Mat. Handbook Vol. 4, Ceramics and Glasses*, 478-81 (1991).
32. H.P. Kirchner, J.C. Conway jr., and A.E. Segall, Effect of joint thickness and residual stress on the properties of ceramic adhesive joints. I: Finite element analysis of stresses in joints *J. Am. Ceram. Soc.* 70, 104-8 (1987).
33. R.N. Singh, Influence of high-temperature exposure on mechanical properties of zircon-silicon carbide composites, *J. Mat. Sci.*, 26, 1, 117-26 (1991).
34. R.N. Singh, High temperature mechanical properties of a uniaxilly reinforced zircon-silicon carbide composite, *J. Am. Ceram. Soc.*, 73, [8], 2399-406 (1990).
35. J. Llorca, R.N. Singh, Influence of fiber and interfacial properties on fracture behaviour of fiber-reinforced ceramic composites, *J. Am. Ceram. Soc.* 74, [11], 2882-9 (1991).
36. S.K. Reddy, S. Kumar, R.N. Singh, Residual stresses in silicon carbide-zircon composites, *J. Am. Ceram. Soc.* 77, [12], 3221-6 (1994)
37. P. Pena, and S. de Aza, Compatibility relationships of Al_2O_3 and ZrO_2 in the system ZrO_2-Al_2O_3-SiO_2-CaO, *Adv. In Ceram.* Vol. 12, 174-80 (1984).

MULTIPHASE CERAMICS FROM COMBUSTION-DERIVED POWDERS

Roman Pampuch

Department of Advanced Ceramics, AGH,
Krakow, Poland

INTRODUCTION

Ceramic technologies should aim at producing materials with optimum properties for specific functions. In many cases, this means that materials must be well-designed at the microstructural level. However, "a key aspect of microstructures lies in the extent to which they can be realised."[1] To produce components having a designed microstructure with a low degree of random complexity by sintering or hot-pressing powders, high quality powder is required. It is also necessary to realize such a microstructure using cost-effective methods for powder synthesis and production of components. Yet, typical ceramic powder producing methods conform to either the first (high quality powder) or the second (cost effectiveness) requirement only (Table 1). Only combustion synthesis, also called self-propagating high-temperature synthesis (SHS), permits both requirements to be linked most predictably.

In SHS, appropriate choice of particle size and thermal conductivity of the particulate reactant bed allows creation of a situation where the heat of exothermic reactions in a reactive system exceeds heat losses to the environment (Figure l(a)). Successive rise of temperature and increased system reaction brings about an avalanche-like increase of the temperature and of the degree of conversion of reactants to products (Figure 1(b)). These features of combustion synthesis are well-known.

Years of research into solid combustion[2-5] have convinced us that the specific conditions of combustion synthesis permit attainment of high quality powder and, equally important, by a simple process. Using combustion-derived powders to produce β-SiC - B_4C and α/β-Si_3N_4 - β-SiC composites furnishes a good example of these features of combustion synthesis. The first composite is characterized by a high hardness and abrasion resistance,[6] while the latter shows superplastic behaviour and an improved high-temperature strength superior to that of Si_3N_4.[7-9]

Table 1. Features of typical ceramic powder-producing methods and of their products.

Method ⇒ Powder & Method features ⇓	Comminution	Coprecipitation	Hydrothermal	Gas-Phase Reactions	Carbothermic reduction	Solid Combustion
Purity	-	+	+	+/-	+/-	+
Small particle size & narrow size distribution	-	+	+	+	+/-	+
Compositional control	-	+ -	+	+/-	+/-	+/-
Absence of a prolonged milling	-	-	+	+/-	-	+
Low number of main operations	+	-	-/+	-	-/+	+
Low cost of reactants	+	-	-	-	+	-/+
Low cost of installations	-/+	-	-	-	-	+
Facility of scaling up the method	+	-/+	-/+	-	+/-	+

Legend: + = yes; - = no.

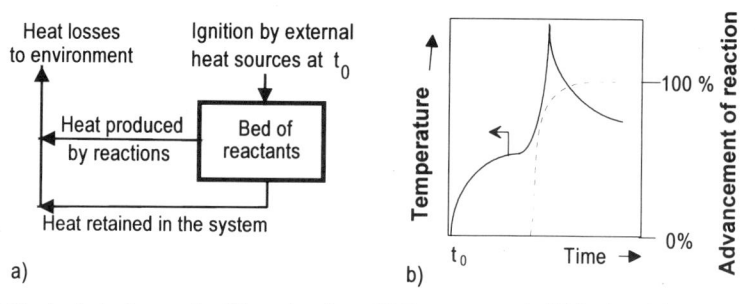

Figure 1. (a) The logical scheme of solid combustion. (b) Temperature (solid line) and degree of conversion to products (dashed line) vs. time at a given point of the reactive mixture.

MICROSTRUCTURE - PROPERTIES RELATIONSHIPS FOR β-SIC - B$_4$C AND α/β-SI$_3$N$_4$ - β-SIC COMPOSITES

The most interesting properties of these composites are observed in the contents of the second component beyond the percolation threshold. Theoretically[10] and experimentally (Figure 2), the percolation threshold should lie at 16 to 19 vol.% of the second component. At such contents, a formation of interpenetrating (duplex) microstructures is possible. Figure 3 shows the bending strength of the β-SiC - B$_4$C composite vs. the boron content. Below the percolation threshold, non-linear changes of strength with composition are observed; the strength initially increases but after reaching a maximum, it decreases again. This is believed to be due to growth of elongated, crack-bridging grains when small amounts of B$_4$C are introduced (see schemas in upper part of Figure 3) and increasing

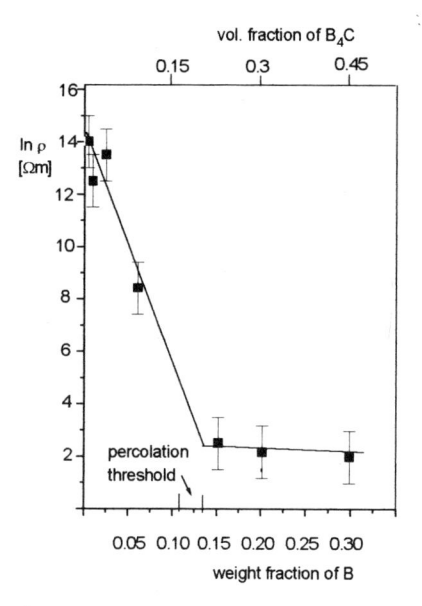

Figure 2. Resistivity of the β-SiC-B₄C composite vs. the boron content, according to Stobierski.[11]

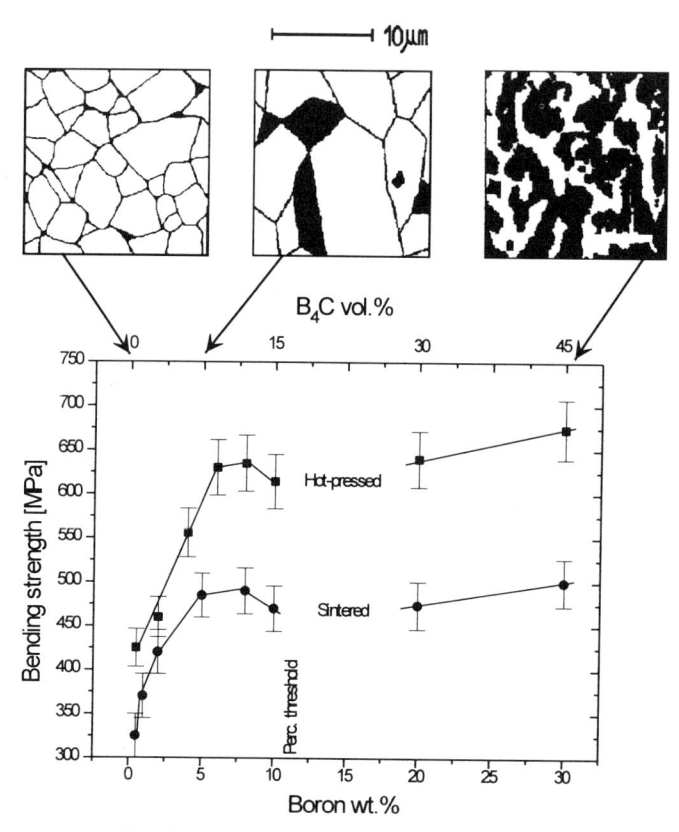

Figure 3. Bending strength of β-SiC-B₄C composites vs. the boron content for samples produced, respectively, by hot-pressing at 1950°C under 25 MPa for 30 minutes and sintering at 2050°C for 30 minutes; the microstructures at various boron contents are shown schematically in upper part of the figure.

inhibition of the grain growth at higher boron (B) or B$_4$C contents. At B contents corresponding to B$_4$C contents superior to the percolation threshold, an interpenetrating (duplex) microstructure is, indeed, observed. Beyond the percolation threshold, increasing amounts of the second phase slightly modify the bending strength in a nearly linear way. Similar changes of bending strength with composition have also been found for the α/β-Si$_3$N$_4$ - β-SiC composite.[8] Namely, non-linear changes occur below the percolation threshold and very small, nearly linear ones above it.

The small, nearly linear changes of mechanical properties with changes of the phase composition are typical for interpenetrating (duplex) microstructures. Because small variations of composition in different pieces of materials cannot be avoided, the insignificant property changes with composition are advantageous because they are virtually equivalent to an elimination of random complexities.

In order to realize homogeneous duplex microstructures, powders should be contiguous to the component phases. In conventional processing, this can be achieved only through very complicated procedures. Solid combustion permits homogeneous composite powders to be produced in a rather simple way, due to the specific mechanism of reactions and transformations under combustion conditions.

MECHANISM OF REACTIONS UNDER COMBUSTION CONDITIONS AND THE FEATURES OF COMBUSTION-DERIVED POWDERS

Let us first consider solid combustion synthesis of composite silicon carbide-boron carbide powders in the Si-B-C system. In silicon systems containing carbon and boron solids, combustion conditions enable attainment of high temperatures in a few minutes' time. Under such conditions, a migrating-thin-reaction-layer mechanism has been identified.[3] It is shown in Figure 4 for the silicon-carbon system. At temperatures of 1950 to 2150°C, the rate of growth of SiC at the C-SiC interface and the rate of dissolution of this primary SiC in liquid silicon at the SiC-Si interface are comparable. Therefore, a thin reaction layer of primary SiC moves into the carbon particles, pulling in its wake a liquid solution of carbon in silicon. On the one hand, this ensures a very high reaction rate. On the other, the mechanism leads to formation of the final product of solid combustion, secondary SiC, by precipitation from the liquid phase. The precipitation occurs in form of successive SiC layers,[12] the ensemble forming metamorphoses of the carbon reactant.[13] If boron is added to the starting reactive mixture, an unstable boron carbide, B$_{13}$C$_2$, can precipitate from the liquid phase together with silicon carbide.

At boron contents of 20 to 30 wt.%, the powder produced by solid combustion consists, indeed, of composite grains formed by alternating SiC and B$_{13}$C$_2$ layers.[6] The latter phase is unstable[14] and when sintered or hot-pressed, it undergoes a spinodal decomposition to B$_4$C and B, the latter reacting with excess carbon to B$_4$C. Such a mechanism favours the formation of interpenetrating (duplex) microstructures. These are observed at high B$_4$C contents (see upper-right part of Figure 3).

Composite β-Si$_3$N$_4$-β-SiC powders can also be produced by combustion, where the grains are an intimate mixture of the component phases. Namely, by combustion of Si + C reactive mixtures in nitrogen under a pressure of 3 MPa,[15] one of the products, Si$_3$N$_4$, gradually decomposes to volatile silicon above 1700°C. Because such temperatures are easily attained on solid combustion, a substantial transport of silicon occurs over the gaseous phase. Both crystallization via the liquid phase (discussed earlier) and mass transport over the gaseous phase should contribute to homogeneous distribution of components in the combustion product. That is, most of the grains of the product, if not all, are constituted both by Si and by C plus N (Figure 5). The combustion-derived powders yield α/β-Si$_3$N$_4$+16 to 20 wt.% β-SiC composites on hot-pressing. These composites are

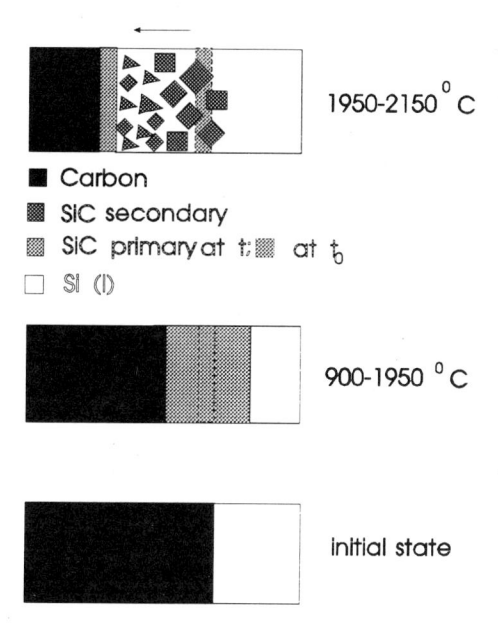

Figure 4. Mechanism of reactions in the Si -C system: growth of the reaction layer thickness at temperatures below 1950°C and the migrating thin-reaction-layer mechanism at higher temperatures, according to Pampuch et al.[3]

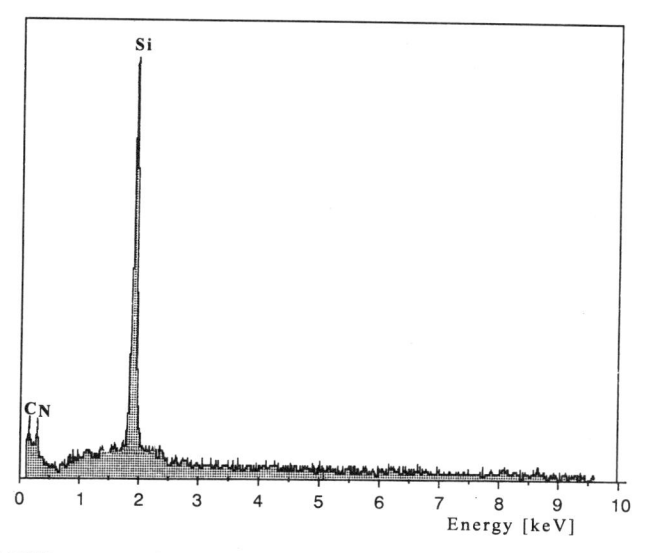

Figure 5. Typical EDX spectrum of a single grain (size about 1 μm) of solid combustion-derived powder. Remark: because of a lower sensitivity of the method to C and N than to Si, the peak heights due to C and N underestimate the C and N content in the grain.[16]

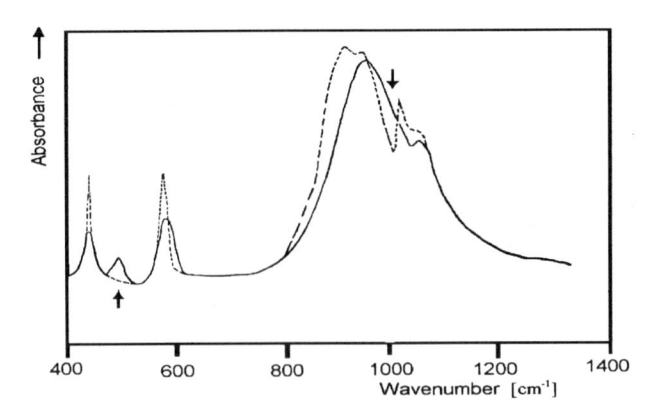

Figure 6. Infrared absorption spectra of silicon nitride powders containing sintering aids (6 wt.% Al_2O_3 + 4 wt.% Y_2O_3). Dashed line: mechanical mixture of Si_3N_4 and of sintering aids. Solid line: combustion-derived powder of the same nominal composition. Arrows indicate the positions of characteristic Si_2N_2O bands.

microstructurally similar to the Si and C plus N composites but are produced by hot-pressing from α/β-Si_3N_4 and β-SiC submicron powders,[7,8] i.e., powders obtained by a more complicated processing.

In order to produce composites having a homogeneous microstructure, uniform distribution not only of the basic powder constituents but also of the sintering aids (if they are necessary) should be the goal. Solid combustion makes this possible by enabling in situ doping, if sintering aids are added to reactants. This is exemplified by β-Si_3N_4 powders produced by combustion of mixtures of Si with 6 wt.% Al_2O_3 and 4 wt.% Y_2O_3 powders in nitrogen at a pressure of 3 MPa. In Figure 6 the dashed line shows the IR absorption spectrum of a mechanical mixture of Si_3N_4 and of the sintering aids. The spectrum of combustion-derived powders of the same nominal composition is shown by a solid line. Besides the dominating bands associated with β-Si_3N_4, two additional bands appear in the latter spectrum, which are the most characteristic ones for Si_2N_2O. This indicates that oxygen (and, by inference, aluminium) was incorporated into the silicon nitride grains during solid combustion.

More homogeneous distribution of basic components and sintering aids in combustion-derived powders is probably the reason for higher homogeneity of the microstructure of β-SiC+B_4C and β-Si_3N_4-β-SiC composites produced from these powders and, consequently, for their higher fracture toughness as compared with similar composites obtained from conventional β-SiC+B_4C and (α/β)-Si_3N_4 + β-SiC powder mixtures (Table 2).

Table 2. Fracture toughness of β-SiC-B_4C and (α/β)-Si_3N_4-β-SiC composites, after Pampuch et al. (1996).[17]

Material (Processing)	K_{ic} [MPam$^{1/2}$]
B_4C (HP)	3.2
β-SiC (HP)	3.0 - 3.5
β-SiC - 20%B_4C (HP, convent. powders)	<3.5
β-SiC - 30%B_4C (HP, SHS-derived powders)	5
α/β-Si_3N_4 (HP)	3.7 - 5.0
α/β-Si_3N_4 - 20%β-SiC (HP, convent. powders)	4.5 - 6.0
β-Si_3N_4 - 20%β-SiC (HP, SHS-derived powders)	7.5

Table 3. Features of combustion-derived powders permitting simplified processing.

Combustion (SHS)-derived powder feature	Effects in comparison with conventional processing
Constitution by intimately mixed basic components	Elimination of less efficient mechanical mixing of the basic particulate components
Incorporation and/or homogeneous distribution of sintering aids	Elimination of less efficient mechanical mixing of sintering aids with the basic powders
A very high degree of conversion of the reactants to products and the relatively high purity of products	Elimination of the steps of removal of not reacted reactants, of side-products, and of impurities
Constitution by soft agglomerates of small composite grains	Substitution of disagglomeration for prolonged grinding

SIMPLIFYING PROCESSING BY USING COMBUSTION-DERIVED POWDERS

The features of combustion-derived powders mentioned above enable simplified processing in comparison with conventional methods, as indicated by the first three items of Table 3 (see also Pampuch et al. (1995)[18] and Lis et al.[19]). The processing simplification advantages of constituting grains by mixing basic components, and of homogeneously incorporating sintering aids are, however, lost if peak temperature on solid combustion is uncontrolled and reaches excessive heights. When peak temperature is greater than sintering temperature, $T_{sintering}$ hard agglomerates are formed and prolonged milling is necessary to produce sinterable powders. As shown in Figure 7 for the case of β-SiC-combustion

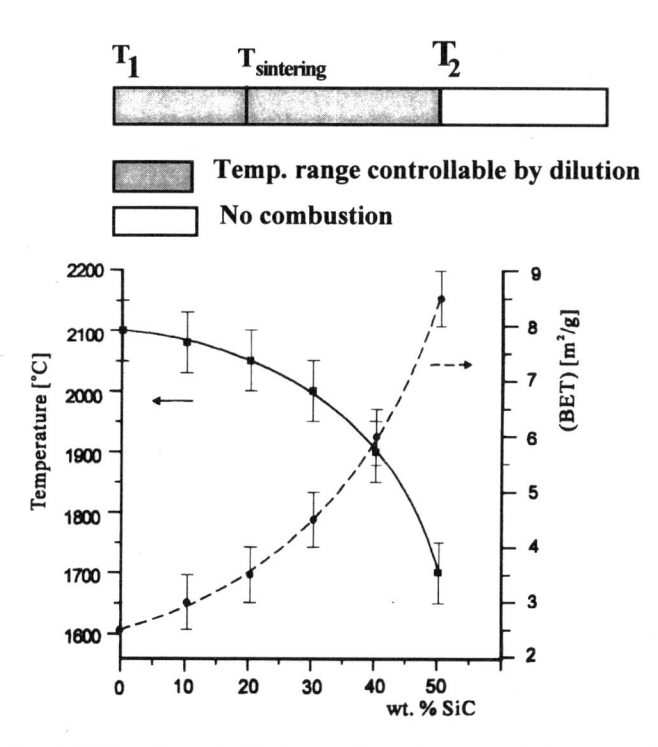

Figure 7. Range of controllable peak combustion temperatures when reactant mixtures are diluted by the inert product (at top) and peak temperatures attained on solid combustion in reactive Si + C mixtures in function of the amount of inert β-SiC added (at bottom, after Stobierski[11]).

synthesis, peak temperature on combustion can be controlled over a broad range by diluting reactants with the inert product. If peak temperature is thus decreased below sintering temperature, soft agglomerates are formed, which can be easily disagglomerated. Similar effects to those observed with silicon carbide have been obtained in the case of the Si_3N_4 combustion synthesis.[19]

AKNOWLEDGEMENTS

The support by the Polish Committee of Research (Grant No. 3P. 470 030 05) is gratefully acknowledged.

REFERENCES

1. R. J. Brook, Ceramic microstructures: the art of the possible, in: *Ceramic Microstructures '86,* J. A. Pask and A.G. Evans, eds., Plenum Press, New York (1987).
2. R. Pampuch, J. Lis, and L. Stobierski, Self-propagating high-temperature synthesis of ceramic powders, in: *Science of Ceramics* vol. 14 D. Taylor, ed., Inst. of Ceramics, Stoke on Trent (1988).
3. R. Pampuch, J. Lis, and L. Stobierski, Heterogeneous reaction mechanisms in the Si-C system under conditions of solid combustion in: *Combustion and Plasma Synthesis of High-Temperature Materials,* Z. A. Munir, J. B. Holt, eds., VCh, New York, Weinheim, Basel Cambridge (1990).
4. R. Pampuch, J. Lis, and L. Stobierski, Synthesis of sinterable β-SiC powders by a solid combustion method, *J. Amer. Ceram. Soc.,* **72**:645 (1989).
5. R. Pampuch, Powders and components produced by using solid combustion, in: *Trends in Emerging Materials and Applications,* A. Bellosi, ed., Faenza Editr., Faenza (1995).
6. L. Stobierski, F. Thevenot, Sintering of boron-containing β-SiC obtained by SHS, in: *Proceed. IUMRS-ICAN'93,* S. Somiya et al., Elsevier Publ. Co., Amsterdam (1994).
7. J. Zeng, I. Tanaka, Y. Miyamoto, O. Yamada, and K. Niihara, *J. Japan Soc. Powders and Powder Metall.* **38**:326 (1991).
8. T. Hirano, K. Niihara, Microstructure and mechanical properties of α/β-Si_3N_4/β-SiC composites, *Materials Letters* **22**:249 (1995).
9. K. Izaki, K. Hakkei, K. Ando, T. Kawakami, and K. Niihara, in: *Ultrastructure Processing of Advanced Ceramics,* J. D. Mackenzie, D. R. Ulrich, eds., John Wiley & Sons, New York (1988).
10. V. V. Skorokhod, Structure-percolation effects in the theory of generalised conductivity of ceramic composites (in Russian), *Papers of the Comm. of Ceramic Sci., Ceramics* **47**:39 (1995).
11. L. Stobierski, Silicon Carbide: constitution, properties, and formation (in Polish), *Papers of the Commision of Ceramic Science, Ceramics* **48**:5 (1996).
12. R. Pampuch, E. Walasek, and J. Bialoskórski, Reactions between carbon and silicon at elevated temperatures, *Ceramics Intern.* **12**:99 (1986).
13. R. Pampuch, L. Stobierski, J. Lis, and M. Raczka, Solid combustion synthesis of SiC, *Mater. Res. Bull.* **22**:1225 (1987).
14. M. Bougoin, F. Thevenot, J. Dubois, and G. Fantozzi, Synthese et proprietes thermomecaniques de ceramiques denses composites carbure de bore-carbure de silicium, *J. of the Less-Common Met.,* **132**:209 (1987).
15. J. Lis, Sinterable powders of covalent compounds produced by self-propagating high-temperature synthesis (in Polish), *Papers of the Comm. of Ceramic Sci., Ceramics* **44**:5 (1994).
16. D. Kata, personal communication to the author.
17. R. Pampuch, L. Stobierski, and J. Lis, Improvement of microstructure and fracture toughness of ceramics by the use of solid combustion-derived powders (in Polish), *Szkło i Ceramika,* **47**:1 (1996)
18. R. Pampuch, L. Stobierski, and J. Lis, SHS powders: the present and future, in: *Ceramic Processing Science and Technology,* H. Hausner, G. L. Messing, and S. Hirano, eds., Amer. Ceram. Soc., Columbus, OH (1995).
19. J. Lis, R. Pampuch, and L. Stobierski, Simplifying processing of sinterable powders by using solid combustion, *Ann de Chimie* **20**:151 (1995).

ABSENCE OF MICROWAVE EFFECT IN CERAMICS: PRECISE TEMPERATURE, THERMAL GRADIENT, AND DENSIFICATION DETERMINATION IN A PROPORTIONAL-POWER MICROWAVE FURNACE

Charles C. Sorrell, Naser Ehsani, Andrew J. Ruys, and Owen C. Standard

Department of Ceramic Engineering
School of Materials Science and Engineering
University of New South Wales
Sydney, NSW 2052
Australia

INTRODUCTION

Many potential advantages have been ascribed to the microwave heating of ceramics.[1-3] From a processing point of view, the most important of these are rapid heating, uniform volumetric heating, and lowered firing temperatures. While the phenomenon of rapid heating is universally accepted, the latter two advantages have not been proved to exist.

Concerning volumetric heating, there have been no reliable measurements of thermal gradients in ceramics owing to the difficulty that most researchers have in placing contact thermocouples inside the microwave cavity while the furnace is operating. Thus, it is most common that noncontact optical means are used to measure the temperature.[4-6] However, in the absence of accurate temperature calibration, noncontact temperature measurements have no demonstrable reliability. Although there now are numerous researchers who use shielded contact thermocouples,[5-9] even as early as 1988,[10,11] none of these has calibrated the temperatures. Most researchers have inferred the temperature and uniformity of heating using secondary techniques, such as phase transformation,[12] microhardness,[13] and grain size.[14] There also has been a significant effort to model the temperature distribution within ceramic samples,[15-17] but no temperature correlations or calibrations have been made.

Concerning lowered firing temperatures, several researchers have claimed the existence of a *microwave effect*, where densification takes place at lowered temperatures owing to enhancement of diffusion,[18-22] but no convincing evidence has been forthcoming. Rothman[18] and Katz et al.[23] have discussed the deficiency in this reasoning in terms of microwaves providing insufficient energy to form Frenkel pairs or enhance lattice vibrations. In short, there have been no confirmed temperature measurements to justify the conclusion of lowered firing temperatures and thus the existence of a *microwave effect*.

EXPERIMENTAL PROCEDURE

Furnaces

The microwave furnace[A] used in this work contains a multimode cavity with several holes for thermocouple access. It operates at a frequency of 2.45 GHz and maximal power of 1.5 kW (all present work was done at 40-75% power). This furnace was commissioned in 1992 when the problems with the available time-slicing microwave furnaces had become very apparent. Thus, this furnace is a true variable-power unit that operates in both the power-control and temperature-control modes. Forward and reflected power are detected and the outputs linearised, so allowing the forward power to be controlled. It is believed that this was the first proportional-power microwave furnace to be marketed commercially.

The resistance furnaces[A] used for comparison between microwave and conventional heating were programmable silicon carbide and molybdenum disilicide muffle furnaces.

Temperature Measurement and Calibration

The temperature in the microwave cavity was measured using commercial sheathed Nicrosil/Nisil (type N) thermocouples up to 1300°C and in-house platinum-sheathed Pt-Pt13Rh (type R) thermocouples for higher temperatures.

The temperature calibration work was done using an experimental setup as shown in Figure 1. The accuracy of the temperature measurement by contact thermocouple was determined by detecting the latent heats of fusion of several metals. The tip of the thermocouple sheath was protected by a dense thin-walled (<0.5 mm) alumina cap placed directly in a hole drilled in a cylinder of spectroscopic-grade (>99.9%) metal. The metals[B] used for melting point determination were tin (232°C), zinc (419°C), aluminum (660°C), silver (961°C), and palladium (1554°C). Each melting point was determined in triplicate using the average of the midpoints of the thermal arrests due to melting and solidification.[24] Heating rates of 300°-1000°C·h^{-1} and cooling rates of 500°-1500°C·h^{-1} were used.

Thermal Gradient Determination

The thermal gradient work with ceramic compacts during microwave heating was done using an experimental setup as shown in Figure 2. The test samples of silicon carbide[C] (strong microwave absorber) and hydroxyapatite[D] (weak microwave absorber) were prepared by thixotropic casting,[25,26] incorporating three axial wells halfway into the 60 mm length of cylinders of 40 mm diameter. The wells were located radially in the center, scored up the surface, and at the midpoint between these two wells. Silicon carbide was heated at 1800°-2000°C·h^{-1} and hydroxyapatite was heated at 900°-1000°C·h^{-1}.

Thermal gradients using a silicon carbide resistance furnace were determined with an identical experimental setup, except the sample was oriented with the axis being horizontal. The maximal heating rate allowed and used in this furnace was 300°C·h^{-1}.

Densification Studies

The densification work during microwave heating was done using an experimental setup as shown in Figure 3. The samples of high-purity 3 mol% Y_2O_3 tetragonal zirconia

---

[A] Ceramic Engineering, Warragamba, NSW, Australia
[B] Johnson Matthey (Australia) Ltd., Thomastown, VIC, Australia
[C] 3.0 μm, Naxos Products, Auburn, NSW, Australia
[D] 5.1 μm (agglomerate size), Plasma Biotal Ltd., Tideswell, Derbyshire, UK

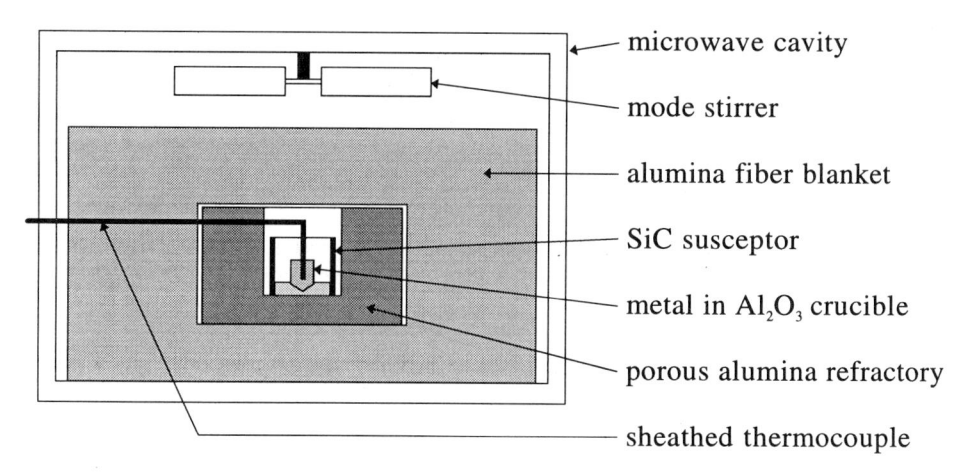

Figure 1. Experimental setup used for temperature measurement and calibration in a microwave furnace.

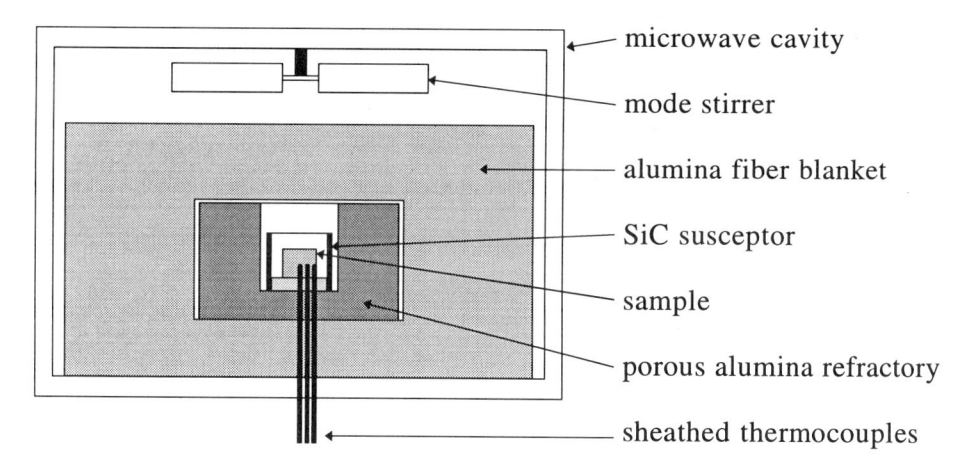

Figure 2. Experimental setup used for thermal gradient determination in a microwave furnace.

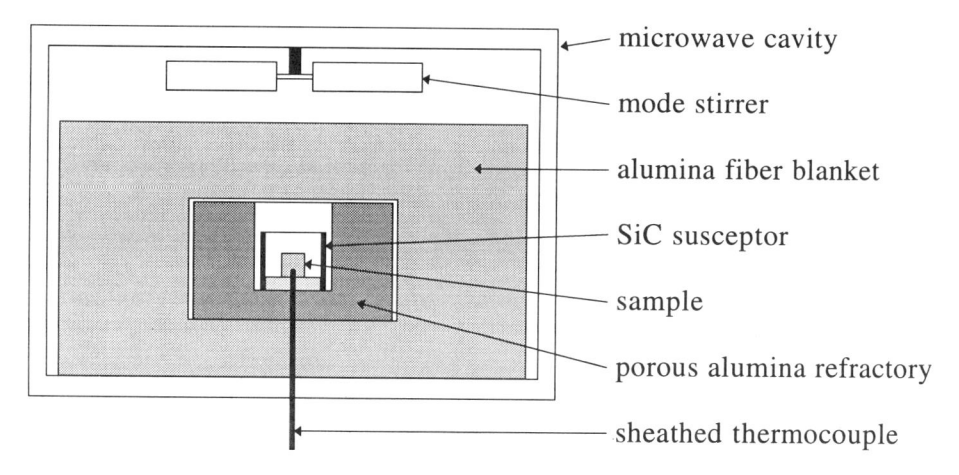

Figure 3. Experimental setup used for densification studies in a microwave furnace.

polycrystal[A] (3Y-TZP; moderate microwave absorber) were of 12 mm height and 15 mm diameter, uniaxially pressed at 10 MPa and cold isostatically pressed (CIP) at 200 MPa. A hole of 4 mm depth was drilled in the central axis of the sample to allow thermocouple insertion in the bottom. A heating rate of 2000°C·h^{-1} to 1000°-1400°C in 100°C increments and soak times of 5, 15, and 30 min were used.

Densification using a molybdenum disilicide resistance furnace was done using a similar setup, where the thermocouple was located in the center of four vertically oriented and tightly packed samples. The maximal permissible heating rate used was 200°C·h^{-1}. Again, 1000°-1400°C in 100°C increments were studied but a soak time of 60 min was used.

Characterization

The bulk densities, open porosities, and closed porosities were determined using the hydrostatic weighing method.[27]

The grain sizes across the diameters of the samples were determined directly from fracture surfaces using a field-emission scanning electron microscope[B] (FESEM) at 20,000X magnification. The line-intercept method[28] (typically 10 lines crossing 200 grains) was used, without the accepted 1.56 correction factor.[29]

The microhardnesses across the internal midpoint diameters of the samples were determined using a Vickers diamond indenter[C] (single indent) with a 500 g load and 15 sec load time. The samples were sectioned longitudinally and the cut surfaces polished by hand using silicon carbide papers, followed by 6, 3, 1, and 0.5 μm diamond pastes. This technique[30] was known to avoid the tetragonal → monoclinic zirconia transformation.

RESULTS AND DISCUSSION

Temperature Calibration

Table 1 summarizes the data obtained for the temperature calibration. It can be seen that the agreement between the measured and theoretical melting points is excellent for all of the metals. Figure 4 shows a typical fusion curve for aluminum. The appearance of this curve is essentially identical to those of the other metals, except for palladium, which showed more temperature fluctuation (small spiking). It is clear that the use of a properly shielded thermocouple provides an accurate measurement of the temperature in a microwave cavity while the furnace is operating provided the following criteria are met:

Table 1. Theoretical and measured melting points of metals in a microwave furnace.

Metal	Theoretical Melting Point (°C)	Measured Melting Point (°C)	Standard Deviation (°C)	Heating Rate (°C·h^{-1})	Cooling Rate (°C·h^{-1})
Sn	232	234.6	0.9	400	500
Zn	419	421.4	0.2	500	1000
Al	660	661.2	6.0	400	1500
Ag	961	955.6	1.8	300	800
Pd	1554	1550.3	5.1	1000	1300

[A] 99.9%, 25 nm, Tosoh Corp., Tokyo, Japan
[B] Model S-4500II, Hitachi Ltd., Tokyo, Japan
[C] Model H2, Leco Corporation, St. Joseph, MI, USA

1) The thermocouple sheath must be a dielectric shield, *e.g.*, a metal.
2) The thermocouple assembly tip must be in direct contact with the sample.
3) The shielding assembly must be thin enough not to act as a thermal insulator.

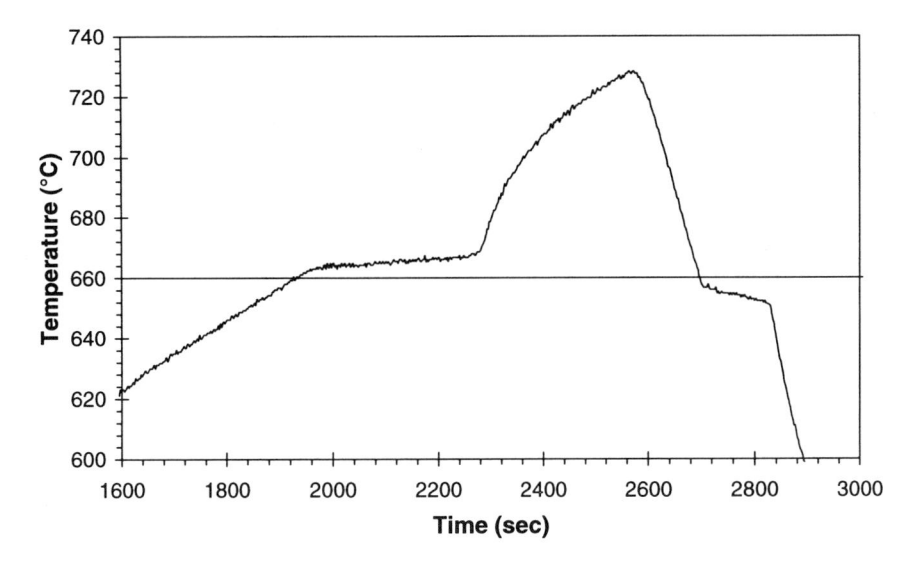

Figure 4. Fusion curve for aluminum heated in a microwave furnace (power on) and then cooled (power off).

Thermal Gradients

Figure 5 compares the thermal gradients measured in hydroxyapatite during microwave resistance and hybrid heating. It can be seen that microwave heating is much faster and reduces the radial thermal gradient. The reversal of the thermal gradient also is apparent. At low temperatures, the surface is hotter owing to the radiant heat from the cylindrical silicon carbide susceptor; at high temperatures, the interior becomes hotter owing to direct microwave absorption.

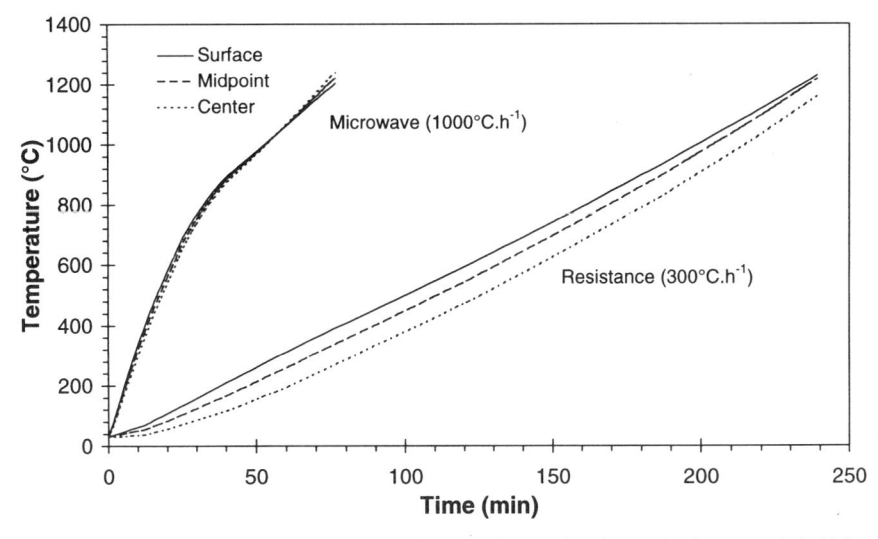

Figure 5. Thermal gradients in hydroxyapatite during resistance heating and microwave hybrid heating.

The reversed thermal gradient can be seen more clearly in Figure 6, which normalizes the surface and midpoint temperatures against the center thermocouple. The onset of microwave absorption for this furnace and insulation assembly probably corresponds to one of two points:

1) At ~280°C, where the curves stop diverging, which is the most likely case
2) At ~660°C, where the curves begin to converge, which probably is due to increasing thermal conductivity resulting from densification

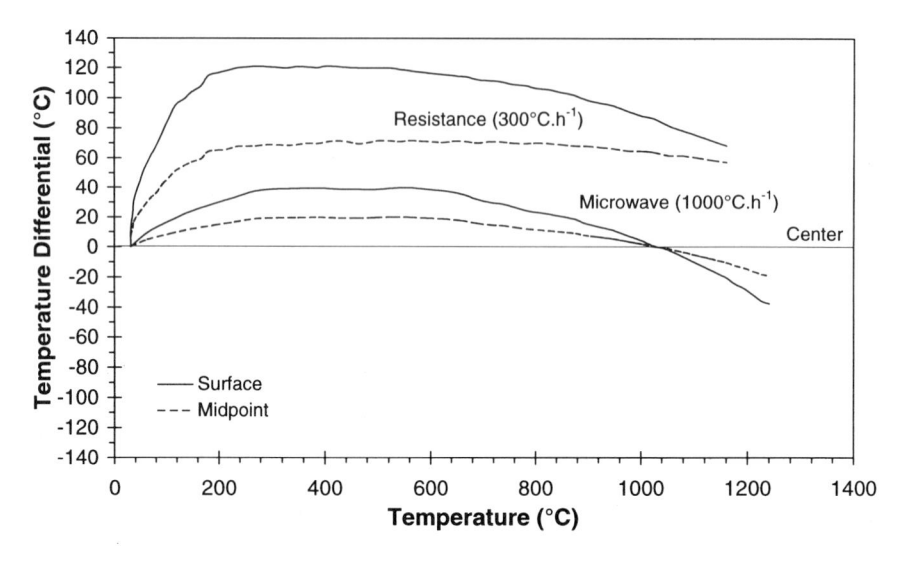

Figure 6. Normalized thermal gradients in hydroxyapatite during resistance heating and microwave hybrid heating.

Figure 7 makes the same comparison for silicon carbide. Although it does not require hybrid heating, this was used. In this case, the reverse thermal gradient is apparent at all temperatures. The (assumed) higher thermal conductivity of silicon carbide is reflected in the lower thermal gradients.

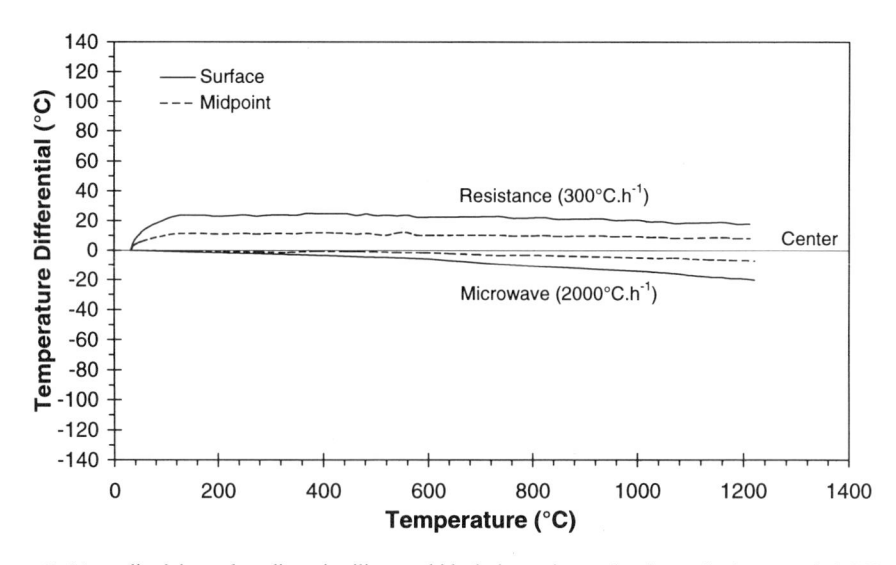

Figure 7. Normalized thermal gradients in silicon carbide during resistance heating and microwave hybrid heating.

The effect of removing the silicon carbide susceptor is shown in Figure 8. In this case, the thermal gradient is considerably increased. Thus, microwave (nonhybrid) heating can be considered to be likely to produce an inferior microstructure when compared to that resulting from resistance heating. These data lead to several important conclusions:

1) The susceptor serves the dual purpose of providing low-temperature heating and balancing the radiant heat loss from the surface of the sample.
2) The shapes of the radiating surfaces of the susceptor(s) should match those of the sample.
3) Temperature measurements that do not consider thermal gradients and radiant heat loss from the sample surface can be misleading.
4) Temperature measurements by thermocouples not in contact with the sample can be misleading; uncalibrated noncontact probes are even more problematic.
5) Susceptors should be used routinely for most ceramics heated by microwaves, irrespective of whether they are weak or strong absorbers.

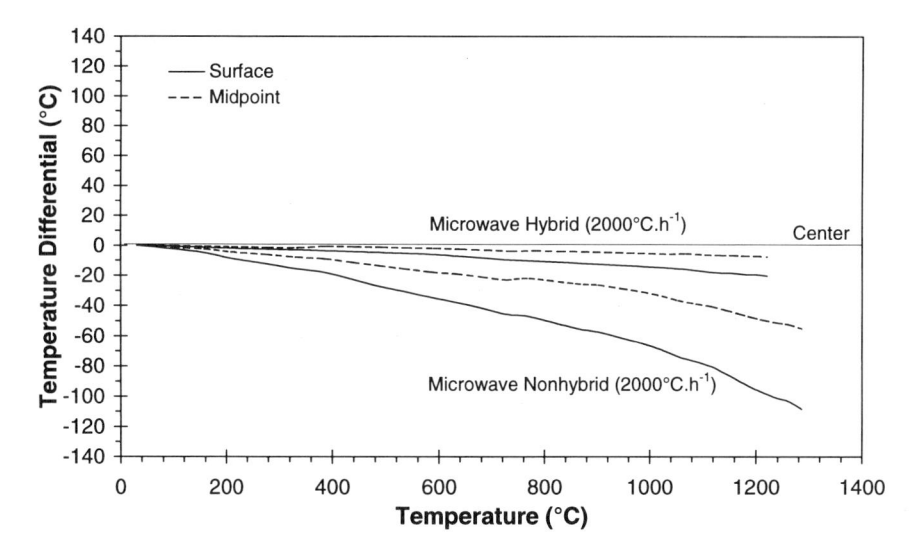

Figure 8. Normalized thermal gradients in silicon carbide during microwave hybrid and nonhybrid heating.

Since dielectric properties change with temperature,[31-33] attempts to minimize thermal gradients should consider temperature effects in both the sample and susceptor. The principle behind this factor is shown in the schematic of the dielectric loss factor versus temperature for zirconia and silicon carbide by Janney et al.,[19] where the absorption of silicon carbide at low temperatures is exceeded by that of zirconia at higher temperatures.

Considerable experimentation in susceptor array design was undertaken in an attempt to minimize the radial thermal gradient. Substantial reduction was achieved through an iterative approach using combinations of susceptor materials and varying the dimensional ratio of the sample:susceptor combination. Figure 9 shows the results of four susceptor array designs for the microwave hybrid heating of green hydroxyapatite of 40 mm diameter. It can be seen that it is possible to reduce the radial thermal gradient to ~5°C throughout the entire heating process and as low as ~1°C (the limit of measurement) during soaking. This is in spite of the fact that the bulk density and thermal conductivity of the sample are changing continuously as are the dielectric properties of the susceptor array. Through this technique, it should be possible to fabricate ceramics of fine and completely uniform grain size with effectively the elimination of thermal and mechanical stresses.

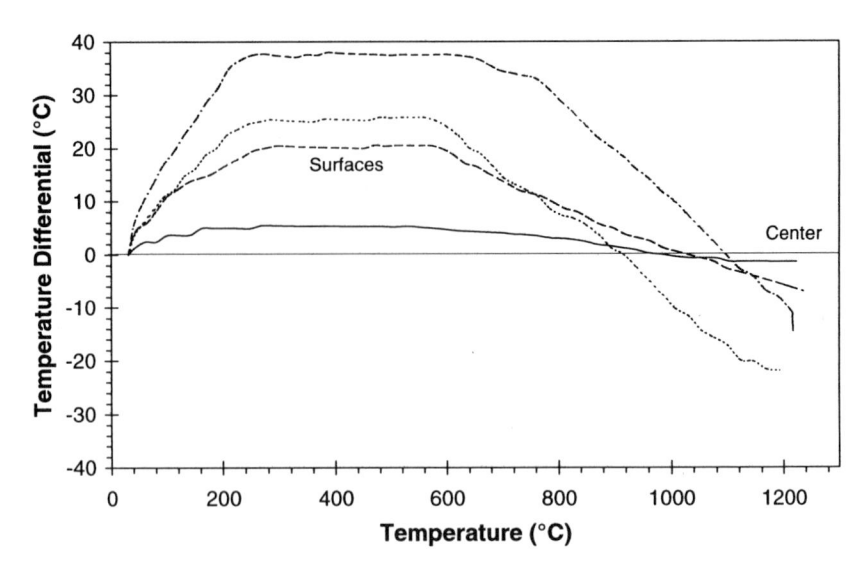

Figure 9. Thermal gradients in green hydroxyapatite (40 mm ∅) measured using four different susceptor arrays.

The data obtained from the experiments on susceptor array design lead to several more significant conclusions:

1) Optimal heating by susceptors requires the use of susceptor combinations that take advantage of changes in dielectric properties with temperature.
2) Appropriate susceptor combinations are obtained by balancing the shapes, size ratio, and weights of the sample and susceptors.
3) The emissivity of the susceptor providing direct radiant heat to the surfaces of the sample should be the same as that of the sample, *viz.*, the same material.

Densification

Three sets of data for the densification of Tosoh Corp. 3Y-TZP have been used for comparative purposes:

1) Resistance heated at 200°C·h^{-1} with a 60 min soak
2) Microwave hybrid heated at 2000°C·h^{-1} with 5, 15, and 30 min soaks
3) Resistance heated at 120°-12,000°C·h^{-1} with a 120 min soak, as reported by Chen and Mayo[34]

Figure 10 illustrates the effect of sintering temperature and soak time on the samples studied in the present work. It can be seen that microwave heating is more effective at ≤1200°C, equivalent at 1300°C, and less effective at 1400°C than resistance heating. In fact, full density was achieved only by resistance heating at 1400°C for 60 min. There is some effect of the time of microwave heating but this decreases as the temperature increases.

Figure 11 shows the bulk density versus sintering temperature. It can be seen that slow resistance heating, rapid microwave heating, and rapid resistance heating show essentially identical trends, with bulk density variations no greater than ~7% at each temperature. It is significant that Chen and Mayo[34] found that the heating rate had no effect on the bulk density because most of the densification occurred during the soak, not during heating. This is in agreement with the observations of Kim and Kim[35] for the same powder. Thus, the main determinant of ultimate bulk density at a given sintering temperature is the soak time.

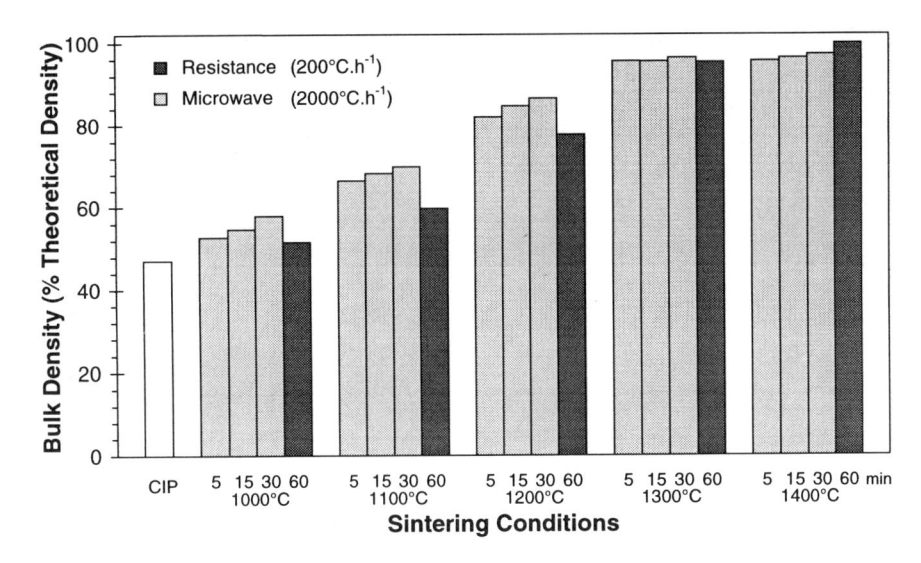

Figure 10. Bulk density as a function of sintering temperature and soak time for slow resistance and rapid microwave hybrid heated 3Y-TZP.

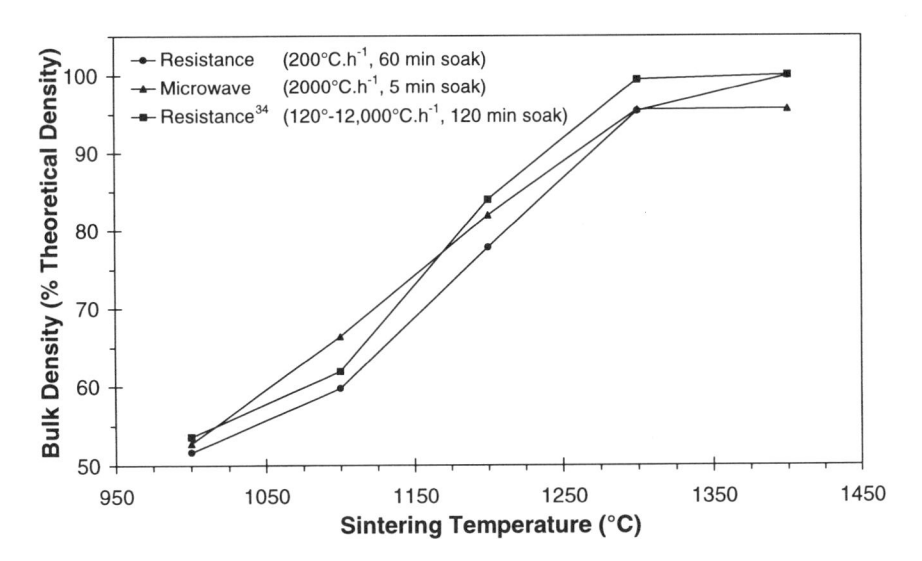

Figure 11. Bulk density as a function of sintering temperature for slow resistance, rapid microwave hybrid, and rapid resistance heated 3Y-TZP.

Figure 12 shows the open and closed porosity versus bulk density. Again, it can be seen that there is essentially no difference between slow resistance and rapid microwave heated samples. These data also show that final-stage sintering and closed porosity begin at ~80% bulk density. As expected, at ≥95% bulk density, only closed pores are present.

Figure 13 shows the grain size and standard deviation versus sintering temperature and time. While it would be logical to assume that rapid microwave heating should result in grain sizes finer than those obtained with resistance heating, some researchers[14,36-38] have observed this and others the converse.[39,40] Dé et al.[41] showed that this effect was dependent on heating temperature and soak time. Chen and Mayo[34] demonstrated convincingly that both the bulk density and grain size were unaffected by the heating rate.

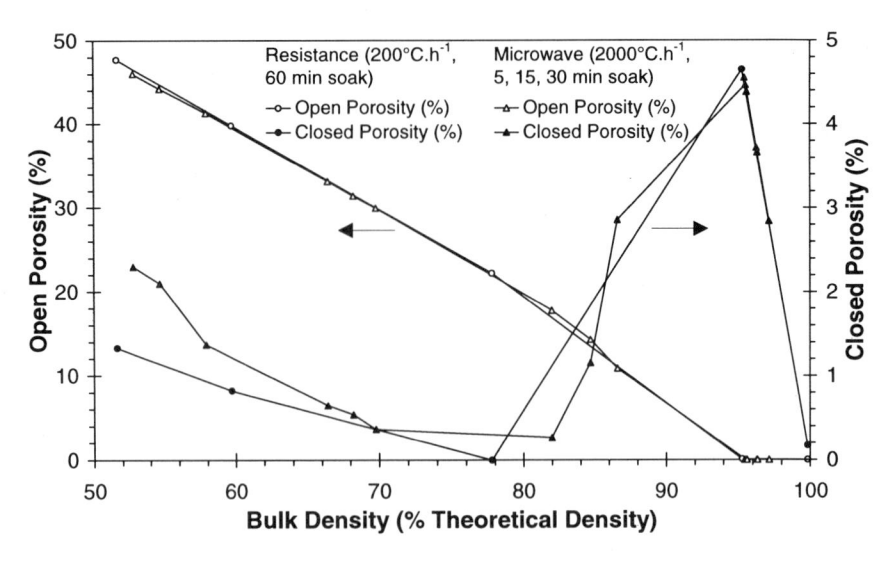

Figure 12. Open and closed porosity as a function of bulk density for slow resistance and rapid microwave hybrid heated 3Y-TZP.

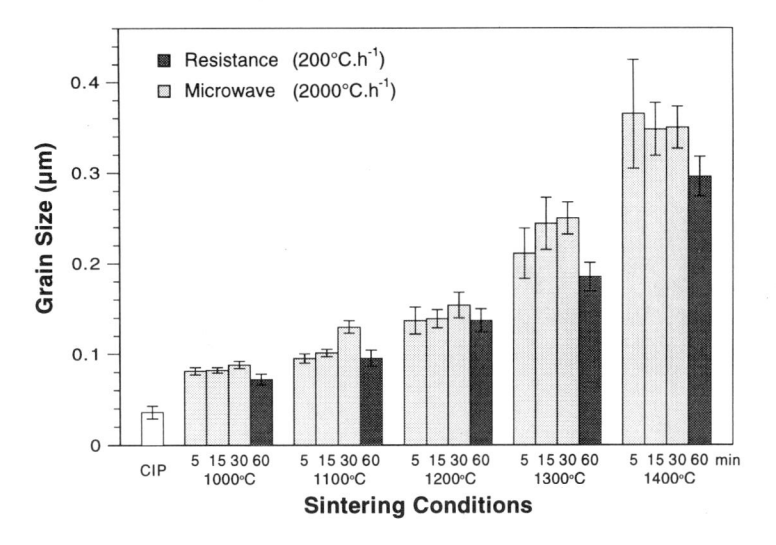

Figure 13. Grain size and standard deviation as a function of sintering temperature and soak time for slow resistance and rapid microwave hybrid heated 3Y-TZP.

The present data indicate that the heating mode has little effect on the grain size at ≤1200°C but, at >1200°C, microwave hybrid heating causes greater grain growth than does resistance heating. Figures 10 and 11 reveal that the microwave hybrid heated samples reached ≥95% bulk density after only 5 min at 1300°C, while the resistance heated samples reached the same density after soaking for 60 and 120 min at the same temperature.

The Chen and Mayo[34] results now allow the present data to be understood. It is well known in ceramics that, once ~95% bulk density is achieved, the grain boundaries can pull away from the pores, causing grain growth but no increase in density. Since ~95% bulk density is achieved very quickly during microwave hybrid heating (5 min at 1300°C), then the grain growth can be attributed to the soak times at the maximal temperature, even though they are of duration ≤30 min. Since ~95% bulk density is achieved slowly during resistance heating (at 1200°C, 81% of the total densification takes place during the 120 min soak[34]), then there is less time for grain growth, despite the longer soak time of 60 min.

Consequently, it is important to note that rapid microwave heating may be disadvantageous microstructurally if careful attention is not paid to kinetics. This observation suggests that considerable flexibility in microstructural evolution can be achieved with suitable variation in microwave heating schedules. In order to achieve the maximal densification while avoiding exaggerated grain growth, careful attention must be paid to the heating rate used to reach ~95% bulk density and the subsequent soak time. Thus, a slower heating rate may be advantageous in terms of both achieving the maximal bulk density and minimizing the thermal gradients, which increase at higher heating rates.

Figure 14 shows the grain size versus sintering temperature. It can be seen that slow resistance heating, rapid microwave heating, and rapid resistance heating show very similar trends and grain sizes, with the exception of the Chen and Mayo[34] data for 1400°C. The reason for this discrepancy is unknown.

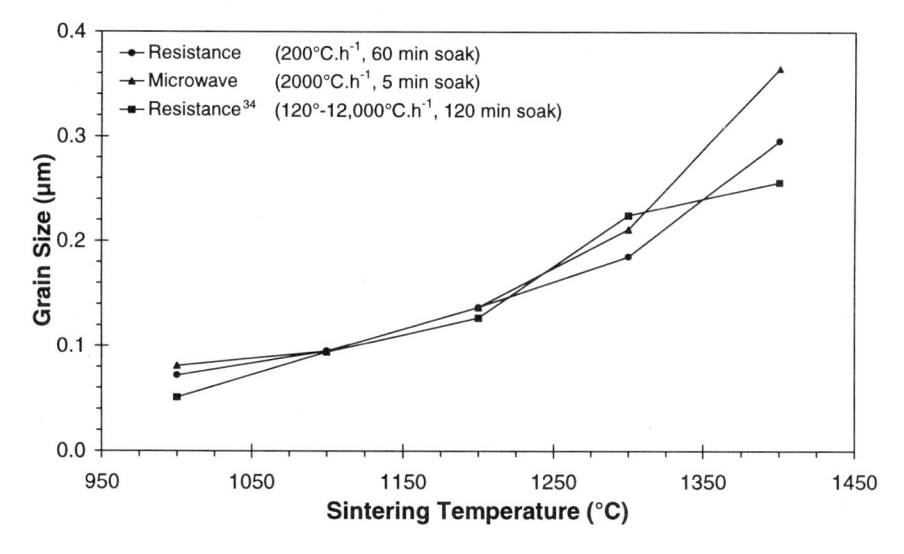

Figure 14. Grain size as a function of sintering temperature for slow resistance, rapid microwave hybrid, and rapid resistance heated 3Y-TZP (1.56 correction factor[29] is removed from the data of Chen and Mayo[34]).

Finally, Figure 15 plots grain size versus bulk density for the three sets of samples. Again, the rapid grain growth at bulk density ≥95% is clear. However, it is plainly apparent that there is no difference between the samples. This is particularly striking in light of the fact that three quite different heating modes were used by two independent research groups and the results are nearly identical. This comparison is critical because it eliminates the problematic uncertainty of the temperature. Since both the trends and experimental values are strikingly similar, this indicates that the heating mode is not significant and that there is no evidence for a *microwave effect*. The only demonstrated effect is that of rapid heating causing rapid densification to ≥95% of theoretical.

Microhardness profiles across samples also were obtained. Figure 16 is revealing in that the Vickers microhardnesses are essentially constant across the samples of lower bulk density (microwave hybrid heated at ≤1200°C), while the microhardness levels themselves are different. However, the samples of ≥95% bulk density (microwave hybrid heated at ≥1300°C) have increased microhardnesses toward the sample surfaces, while the microhardness levels of these two samples show very little difference.

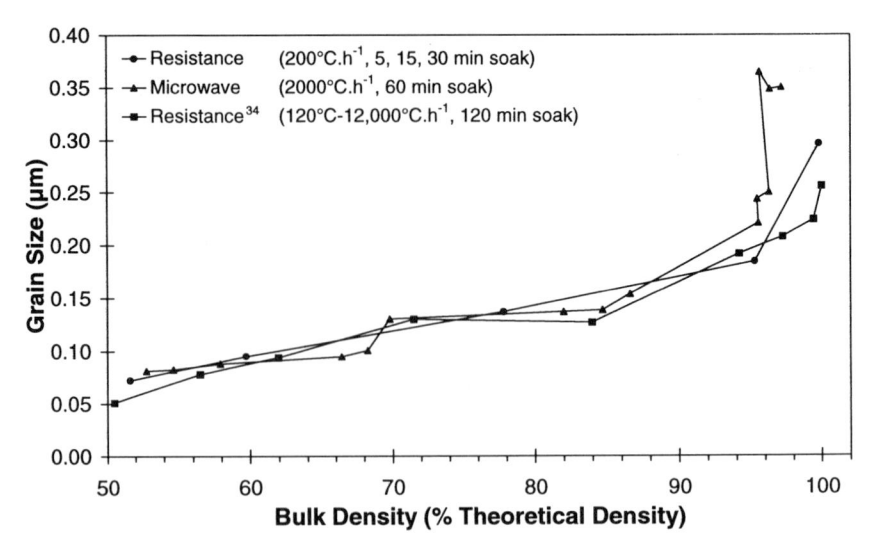

Figure 15. Grain size as a function of bulk density for slow resistance, rapid microwave hybrid, and rapid resistance heated 3Y-TZP (1.56 correction factor[29] is removed from the data of Chen and Mayo[34]).

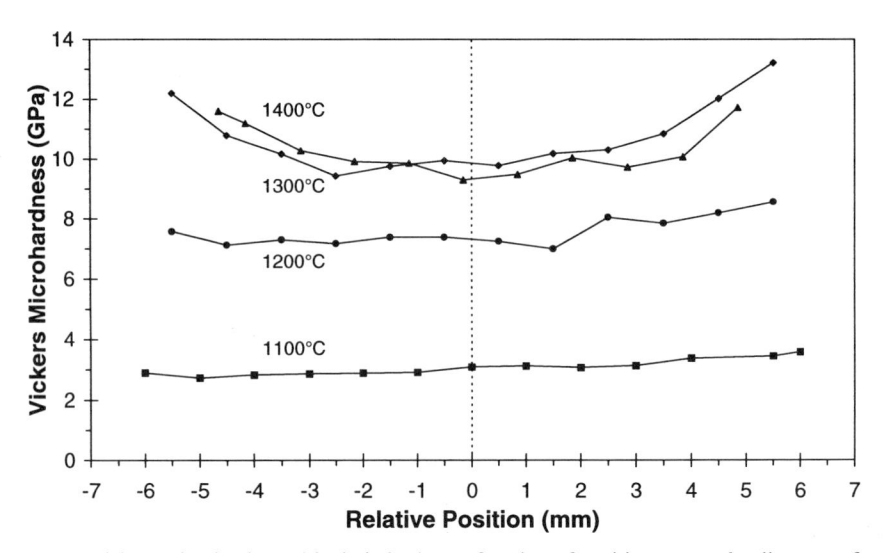

Figure 16. Vickers microhardness (single indent) as a function of position across the diameter of rapid microwave hybrid heated (15 min soak) 3Y-TZP.

It would be logical to assume that the surface hardening is due to surface cooling and smaller grains. However, these samples were heated with a susceptor array that effectively eliminated the radial thermal gradient. Consequently, Figure 17 shows that the grain size is constant across the entire diameter. It should be noted that this effect is observed only in the samples of ≥95% bulk density. The explanation is that this effect is similar to that of toughened glass, where compressive surface stresses develop owing to accelerated cooling of the surface, when the microwave power is off. Thus, the uniform grain size develops during heating with the power on, while the compressive surface stresses arise during cooling with the power off. This effect occurs only during microwave heating, not resistance heating, because the thermal mass of the insulation in the microwave furnace is relatively low, consisting of a thin susceptor array, porous alumina brick, and porous alumina blanket. However, this toughening effect would not be expected to be observed in samples of lower bulk density owing to the lower thermal conductivity and potential stress relief due to the increased sample porosity.

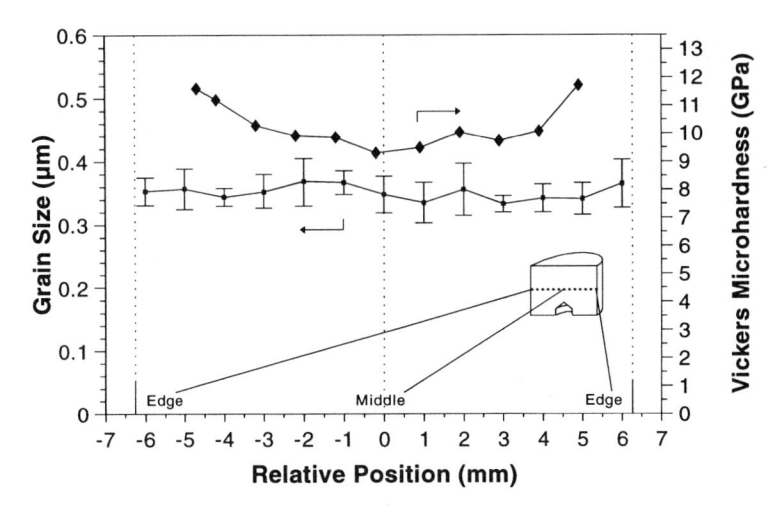

Figure 17. Grain size and Vickers indentation microhardness (single indent) as a function of position across the diameter of rapid microwave hybrid heated (1400°C, 15 min soak) 3Y-TZP.

This interesting result shows that microwave heating (particularly nonhybrid) can be used to toughen dense polycrystalline ceramics without affecting the grain size. It also shows that it there is a hazard in inferring temperature distribution by measurement of microhardness[13] or grain size.[14]

SUMMARY

Purpose of Study

This work was undertaken in order to resolve several issues concerning the microwave and resistance heating of ceramics. Literature data reporting the calibration and measurement of temperature, thermal gradients, and densification during microwave heating are not adequate to answer the question of the possible existence of a *microwave effect*, which depends critically on the method and accuracy of the temperature measurement.

Temperature Measurement

The temperature can be measured accurately using a shielded thermocouple in direct contact with the sample. However, the location of the thermocouple is critical. Figure 8 shows that the difference in temperature between the interior and surface of a sample during microwave heating can be very large (>100°C). The use of an uncalibrated thermocouple located in the vicinity of the sample or an optical probe sighted on the sample surface is inadequate. Large radiant heat losses from sample surfaces necessitate knowledge of the temperatures both inside the sample and on the surfaces. Since these temperature differences are comparable to the enhancement temperatures ascribed to a *microwave effect* (100°-200°C[22,40,42]), this diminishes the credibility of the existence of such an effect.

Susceptors

The use of a susceptor array or combination should be considered a standard procedure because it serves the dual purpose of heating at low temperatures and balancing the radiant heat loss at higher temperatures. Suitable design considerations must be used for both the sample and susceptor characteristics, including dielectric properties, surface shape match, size ratio, weights, and emissivities. The use of a *picket fence*[43] arrangement probably is not optimal for this purpose. Through these considerations, it is possible essentially to eliminate the thermal gradients during heating and soaking, as shown in Figure 9.

Densification

The present work discounts the existence of a *microwave effect* and supports the observation that the discrepancies between microwave and resistance heating are due to incorrect temperature measurement and a rapid heating and densification effect. The latter effect results in the rapid attainment of ≥95% bulk density by microwave heating. These data and those of others[34] show that the heating rate has little or no effect on the ultimate bulk density, grain size, open porosity, and closed porosity, as shown in Figures 10-15. Thus, achievement of microstructural control requires careful monitoring of the kinetics.

Surface Toughening

An interesting effect results from the low thermal mass of the susceptor and insulation with the microwave power off. This causes rapid surface cooling and resultant toughening in dense samples, as shown in Figures 16 and 17.

Microwave Heating of Ceramics

For high-precision densification of ceramics with microwave heating, the interior and surface temperatures of the sample initially must be determined using contact thermocouples. Examination of the heating curves can indicate the temperature of onset of microwave absorption. Subsequent use of a thermocouple not in contact with the sample can be used if it is calibrated against the contact thermocouples. Appropriate susceptor arrays must be devised in order to minimize the radiant heat loss from the sample surfaces and to reduce the thermal and mechanical stresses during heating. Appropriate heating schedules must be used so that exaggerated grain growth at bulk density levels ≥95% can be avoided. The use of these procedures can result in rapid and efficient sintering to full density of ceramics with uniform and fine grain size with no thermal stresses or with surface toughening.

ACKNOWLEDGEMENTS

The authors would like to thank Dr. X. Miao for determining the Vickers microhardnesses of the samples.

REFERENCES

1. W.H. Sutton, Microwave processing of ceramic materials, *Amer. Ceram. Soc. Bull.* 68:376 (1989).
2. W.H. Sutton, Microwave processing of ceramics, p. 3 in: *Microwave Processing of Materials III*, R.L. Beatty, W.H. Sutton, and M.F. Iskander, eds, Materials Research Society, Pittsburgh, PA (1992).

3. W.H. Sutton, Key issues in microwave process technology, p. 3 in: *Microwaves: Theory and Application in Materials Processing II*, D.E. Clark, W.R. Tinga, and J.R. Laia, Jr., eds, American Ceramic Society, Westerville, OH (1993).
4. J. Mershon, Accurate high temperature measurements in microwave environments, p. 641 in: *Microwaves: Theory and Application in Materials Processing*, D.E. Clark, F.D. Gac, and W.H Sutton, eds, American Ceramic Society, Westerville, OH (1991).
5. M.A. Janney, H.D. Kimrey, and J.O. Kiggans, Microwave processing of ceramics: guidelines used at the Oak Ridge National Laboratory, p. 173 in: Ref. 2.
6. D.J. Grellinger and M.A. Janney, Temperature measurement in a 2.45 GHz microwave furnace, p. 529 in: Ref. 3.
7. P.F. Hogan and T. Mori, Development of a method of continuous temperature measurement for microwave denture processing, *Dental Mater. J.* 9:1 (1990).
8. R.H. Plovnick and J.O. Kiggans, Microwave thermal etching of stabilized zirconia, *J. Amer. Ceram. Soc.* 75:3462 (1992).
9. Ph. Boch, N. Lequeux, and P. Piluso, Reaction sintering of ceramic materials by microwave heating, p. 211 in: Ref. 2.
10. I. Ahmad, G.T. Chandler, and D.E. Clark, Processing of superconducting ceramics using microwave energy, p. 239 in: *Microwave Processing of Materials*, W.H. Sutton, M.H. Brooks, and I.J. Chabinsky, eds, Materials Research Society, Pittsburgh, PA (1988).
11. S.L. McGill, J.W. Walkiewicz, and G.A. Smyres, The effects of power level on the microwave heating of selected chemicals and minerals, p. 247 in: Ref. 10.
12. Y. Ikuma, Temperature profile during microwave sintering of ceramics, p. 341 in: *Interfaces of Ceramic Materials: Impact on Properties and Applications*, K. Uematsu, Y. Moriyoshi, Y. Saito, and J. Nowotny, eds, Trans Tech Publications, Zurich (1995).
13. J.J. Thomas, R.J. Christensen, D.L. Johnson, and H.M. Jennings, Nonisothermal microwave processing of reaction-bonded silicon nitride, *J. Amer. Ceram. Soc.* 76:1384 (1993).
14. A. Dé, I. Ahmad, E.D. Whitney, and D.E. Clark, Microwave (hybrid) heating of alumina at 2.45 GHz: I. Microstructural uniformity and homogeneity, p. 319 in: Ref. 4.
15. X.D. Yu, V.V. Varadan, and V.K. Varadan, Modeling microwave heating of ceramics, p. 167 in: Ref. 4.
16. R.K. Singh, J. Viatella, Z. Fathi, and D.E. Clark, Thermal analysis of microwave processing of ceramics, p. 247 in: Ref. 3.
17. J.R. Thomas, Jr., J.D. Katz, and R.D. Blake, Temperature distribution in microwave sintering of alumina cylinders, p. 311 in: *Microwave Processing of Materials IV*, M.F. Iskander, R.J. Lauf, and W.H. Sutton, eds, Materials Research Society, Pittsburgh, PA (1994).
18. M.A. Janney and H.D. Kimrey, Microstructure evolution in microwave sintered alumina, p. 382 in: *Sintering of Advanced Ceramic Materials*, C.A. Handwerker, J.E. Blendell, and W. Kaysser, eds, American Ceramic Society, Westerville, OH (1990).
19. M.A. Janney, C.L. Calhoun, and H.D. Kimrey, Microwave sintering of solid oxide fuel cell materials: I, zirconia-8 mol% yttria, *J. Amer. Ceram. Soc.* 75:341 (1992).
20. S.A. Freeman, J.H. Booske, R.F. Cooper, B. Meng, J. Kieffer, and B.J. Reardon, Studies of microwave field effects on ionic transport in ionic crystalline solids, p. 213 in: Ref. 3.
21. S.J. Rothman, Critical assessment of microwave-enhanced diffusion, p. 9 in: Ref. 17.
22. N.G. Evans, M.G. Hamlyn, and A.L. Bowden, Observation of microwave effects in sintering debased alumina, *Brit. Ceram. Trans.* 95:62 (1996).
23. J.D. Katz, R.D. Blake, and V.M. Kenkre, Microwave enhanced diffusion?, p. 95 in: Ref. 4.

24. W. Hume-Rothery, J.W. Christian, and W.B. Pearson, *Metallurgical Equilibrium Diagrams*, Institute of Physics, London, 1952.
25. A.J. Ruys, S.A. Simpson, and C.C. Sorrell, Thixotropic casting of fibre-reinforced ceramic matrix composites, *J. Mater. Sci. Lett.* 13:1323 (1994).
26. A.J. Ruys, J. Kerdic, and C.C. Sorrell, Thixotropic casting of ceramic-metal functionally gradient materials, *J. Mater. Sci.* 31:4247 (1996).
27. Anonymous, *Australian Standard AS 1774.5–1989. Refractories and Refractory Materials – The Determination of Density, Porosity and Water Absorption*, Standards Australia, Homebush, NSW (1989).
28. Anonymous, *American Society for Testing and Materials ASTM E 112–88. Standard Test Methods for Determining Average Grain Size*, ASTM Committee on Standards, Philadelphia, PA (1988).
29. M.I. Mendelson, Average grain size in polycrystalline ceramics, *J. Amer. Ceram. Soc.* 52:443 (1960).
30. O.C. Standard, *Application of Transformation Toughened Zirconia Ceramics as Bioceramics*, Ph.D. Thesis, University of New South Wales, August 1994.
31. M. Arai, J.G.P. Binner, G.E. Carr, and T.E. Cross, High temperature dielectric property measurements of engineering ceramics, p. 483 in: Ref. 3.
32. N.G. Evans and M.G. Hamlyn, Density dependence of high temperature dielectric properties of debased alumina and the effect of a high loss additive, p. 235 in: Ref. 17.
33. J.G.P. Binner, T.E. Cross, N.R. Greenacre, and M. Naser-Moghadasi, High temperature dielectric property measurements – an insight in microwave loss mechanisms in engineering ceramics, p. 247 in: Ref. 17.
34. D.-J. Chen and M.J. Mayo, Rapid rate sintering of nanocrystalline ZrO_2-3 mol% Y_2O_3, *J. Amer. Ceram. Soc.* 79:906 (1996).
35. D.-H. Kim and C.H. Kim, Entrapped gas effect in the fast firing of yttria-doped zirconia, *J. Amer. Ceram. Soc.* 75:716 (1992).
36. D.K. Agrawal, Y. Fang, D.M. Roy, and R. Roy, Fabrication of hydroxyapatite ceramics by microwave processing, p. 231 in: Ref. 2.
37. M.C.L. Patterson, P.S. Apte, R.M. Kimber, and R. Roy, Mechanical and physical properties of microwave sintered Si_3N_4, p. 310 in: Ref. 2.
38. J. Cai, C.Y. Song, B.S. Li, X.X. Huang, J.K. Guo, and Y.L. Tian, Microwave sintering of zirconia toughened mullite (ZTM), p. 545 in: Ref. 17.
39. C.E. Holcombe and N.L. Dykes, "Ultra" high-temperature microwave sintering, p. 375 in: Ref. 4.
40. A. Dé, I. Ahmad, E.D. Whitney, and D.E. Clark, Effect of green microstructure and processing variables on the microwave sintering of alumina, p. 283 in: *Microwave Processing of Materials II*, W.B. Snyder, Jr., W.H. Sutton, M.F. Iskander, and D.L. Johnson, eds, Materials Research Society, Pittsburgh, PA (1991).
41. A. Dé, I. Ahmad, E.D. Whitney, and D.E. Clark, Microwave (hybrid) heating of alumina at 2.45 GHz: II. Effect of processing variables, heating rates and particle size, p. 329 in: Ref. 4.
42. M.A. Janney and H.D. Kimrey, Diffusion-controlled processes in microwave-fired oxide ceramics, p. 215 in: Ref. 40.
43. M.A. Janney, C.L. Calhoun, and H.D. Kimrey, Microwave sintering of zirconia-8 mol% yttria, p. 311 in: Ref. 4.

STATE AND STRUCTURE OF FREE CARBON IN DENSE SIC DEVELOPED BY SMART PLASMA SINTERING

K. Kijima[1] and Y. Ikuhara[2]

[1]Kyoto Institute of Technology, Kyoto, Japan
[2]University of Tokyo, Tokyo, Japan

ABSTRACT

Plasma sintered SiC-C composites were observed by a transmission electron microscopy in order to investigate the physical state of the carbon, the interface structure between SiC and C, and the grain boundary structure in SiC. The present observation was helpful in discussing the sintering mechanism and in explaining the high hardness and high toughness of plasma sintered SiC-C composites.

INTRODUCTION

SiC containing an excess quantity of carbon is generally hard to sinter by conventional methods, but can be easily sintered by the plasma method to produce an SiC-C composite material.[1,2] The SiC-C composites have a high hardness and high toughness, and are expected to be a material for sliding parts and seals for rotary parts because carbon in the composite acts as a lubricant.

In this study, the microstructures of plasma-sintered SiC-C composites containing B and C were observed by transmission electron microscopy (TEM) to clarify the physical state of the carbon, the interface structures between SiC and C and grain boundary structure in SiC. The results help to clarify the sintering mechanism, and the origin of high hardness and high toughness of plasma-sintered SiC.

EXPERIMENTAL PROCEDURE

α-SiC (average grain size: 0.49 μm), amorphous boron and phenol resin as a carbon source were mixed in ethanol by using a ball mill for 16 h. The mixture, after drying, was formed into a shape in a mold at 30 MPa, and then in an isostatic press at 200 MPa. It was then calcined at 1270 K for 20 min in an Ar atmosphere to carbonize the phenol, producing a bar-like specimen about $5 \times 5 \times 40$ mm^3 in size.

A plasma was produced by a high-frequency generator (frequency: 4 MHz, output: 15 kW). An egg-shaped rf-thermal plasma was produced formed in a water-cooled, double-tube type quartz furnace. The gaseous temperature of the high temperature plasma is about 10000 K in an Ar atmosphere.[2] The specimen was sintered by inserting it into the plasma flame gradually.

Two types of specimens were prepared, A and B containing carbon at levels of 3.0 and 7.0 wt%, respectively. The conditions under which the specimens were prepared are given in Table 1.

TEM observation was performed on ion-thinned specimens using a JEOL TEM 4000 FX.

Table 1. Specimen preparation conditions for sample A and sample B.

Sample	B(wt%)	C(wt.)	Anode Voltage (kV)	Ar gas Flow (SCCM)	Pressure (torr)	Sintering Time (s)	Density (g/cm^3)	Relative Density(%)
A	1.0	3.0	7.0	80	50	300	2.969	92.5
B	2.0	7.0	7.5	40	50	120	3.056	95.2

RESULTS AND DISCUSSION
Sintered SiC Bodies

Figures 1 (a) and (b) show the microstructures of specimens A and B observed by TEM, where the white parts represent pores. As shown, specimen A, has lower density, and was more porous. The average grain size of the specimens A and B was about 1 nm and 1.5 µm, respectively.

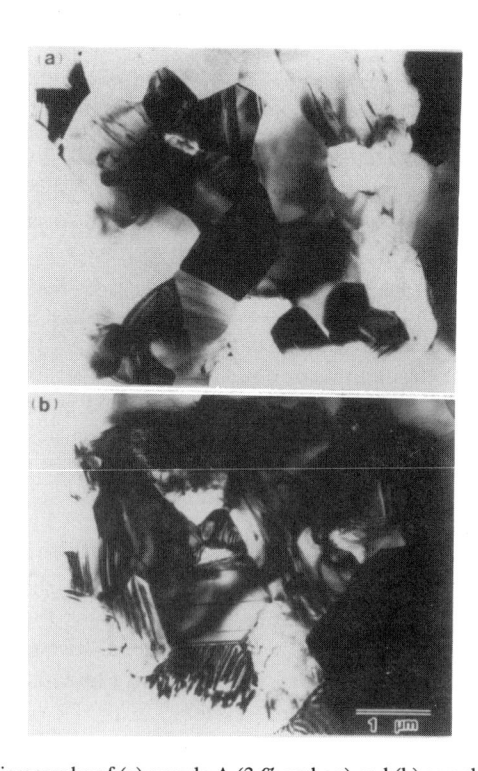

Figure 1. TEM micrographs of (a) sample A (3 % carbon) and (b) sample B (7 % carbon).

Carbon was present around the pores and between the SiC grains in both specimens but was more frequently found between the grains in specimen B. The form and distribution of carbon phases are discussed in detail in the subsequent section.

Sintering density generally decreases as carbon content increases, if specimens are sintered under the same conditions. However, specimen B, containing more carbon, was apparently higher in density than specimen A, conceivably because the former was sintered at a higher anode voltage and lower Ar flow rate, and hence, at a higher temperature.

State of Carbon

As shown in Fig. 2 (c), each pore in specimens A or B was surrounded by fibrous (white arrow) or fine polycrystalline (black arrow) graphite. Figures 2 (a) and (b) show the selected area diffraction patterns of the fibrous and polycrystalline graphite. Each diffraction ring was almost circular, irrespective of the diffraction conditions, indicating that the fibrous and polycrystalline graphite around the pore had no preferred orientation. Since graphite was always found around the pores, as mentioned above, it is possible to postulate that carbon existing at the grain boundaries of SiC diffused to the pores when the grains rearranged during the initial stage of plasma sintering.

Figure 3 shows typical graphite large grains existing at the grain boundary of SiC (white arrow). These graphite grains are single crystals having a relatively large size of 0.4 to 0.7 μm, and are characterized by cracks running parallel to the basal plane.

Thermal expansion coefficients along the a- and c-axis of graphite are 9.5×10^{-7}/K and 2.8×10^{-6}/K,[3] whereas that of SiC is 4.0×10^{-6}/K.[4] Therefore a stress distribution, as illustrated in Fig. 4, arise within the specimens after it is cooled. The Young's moduli along

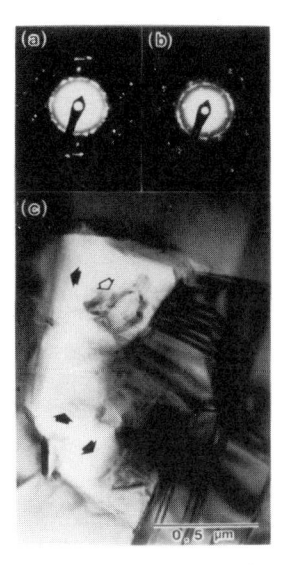

Figure 2. TEM photographs of a region around a pore for sample A (3 % carbon). (a) Diffraction pattern of fibrous graphite and (b) of fine polycrystalline graphite, and (c) TEM image of fibrous graphite (black arrow) and fine polycrystals (white arrows) around a pore.

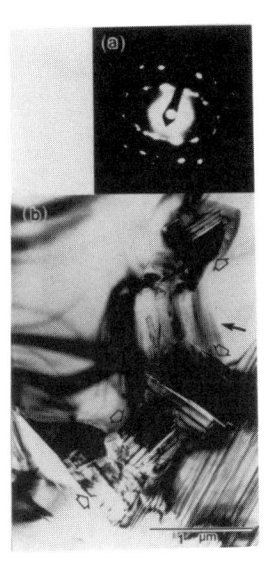

Figure 3. Photographs of a region between SiC grains for sample B (7 % carbon). (a) Diffraction pattern at a region including a particle indicated by the black arrow, and (b) TEM image of single crystals of graphite (white arrows) and fine polycrystal or fibrous crystal of graphite (black arrows).

the a- and c-axis of graphite are 1020 GPa and 36.4 GPa,[5] respectively whereas that of SiC is 400 GPa[4] on average. Presuming that the specimen is heated to 2800 K and cooled to room temperature, a compressive stress of 4070 MPa along the a-axis and a tensile stress of 2730 MPa along the c-axis should be produced in the specimen. These stresses will form the cracks running parallel to the basal plane. Since the minimum stress required to flake the basal plane is about 0.4 MPa,[5] the above mentioned stresses are sufficient to form the cracks. A black contrast, believed to be a stress-induced contrast, appears in the interface between graphite and SiC (Fig. 3), indicating that there is a residual stress after cracking.

Figures 3 (a) and (b) show the diffraction patterns of the region containing a grain, indicated by the black arrow. The pattern is not a circle, but an ellipse. Some spots had high intensity, which correspond to the aforementioned relatively large graphite grains. The other continuous ellipse is due to fine polycrystalline or fibrous graphite, corresponding to the substance indicated by the black arrow. The diffraction ellipse gradually changed to a circle by shortening its longitudinal axis, depending on diffraction conditions, which means that the grains have some preferred orientation.[6] The orientation direction is indicated by the black arrow in the figure. The polycrystalline or fibrous graphite grain has the c-axis oriented basically in the direction indicated by the arrow, rotating another direction around the c-axis. The difference in thermal expansion coefficient between graphite and SiC produced a compressive stress parallel to the basal plane of the graphite, causing the polycrystalline or fibrous graphite grains to be rearranged in such a way that their basal planes overlap.

Figure 4 shows a graphite single crystal sandwiched by the SiC grain. The photograph also shows the cracks running along the basal plane, considered to be produced in similar manner as those described earlier. The black contrast in the SiC grain around the graphite grain represents a stress region.

As discussed above, the graphite present in plasma-sintered SiC-C composite was in the form of (a) a fine polycrystal, (b) a fibrous crystal, or (c) a single crystal. The polycrystalline or fibrous graphite grains were distributed randomly when present around

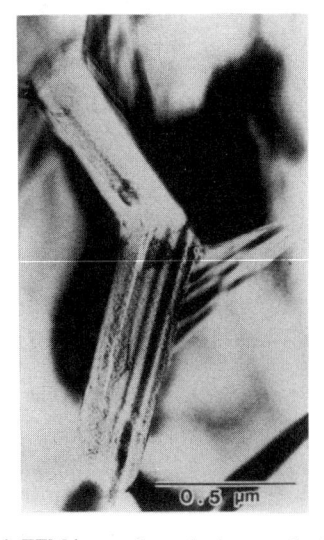

Figure 4. TEM image for a single crystal of graphite in grain boundary of SiC (sample B).

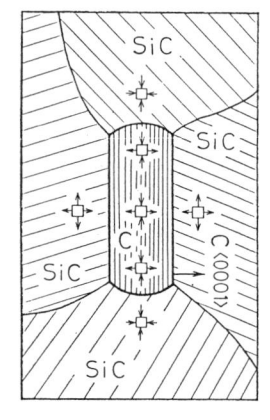

Figure 5. An illustration of stress distribution in SiC-C composite material.

490

the pores, whereas they were oriented when present between the SiC grains. The single crystals were found between the SiC grains, where cracks were formed by compressive and tensile stresses.

Plasma-sintered SiC has higher hardness and toughness than SiC sintered by the conventional method,[2] conceivably resulting from the effects of compressive stress generated by the difference in thermal expansion coefficients between SiC and graphite. A stress distribution, as shown in Fig. 5, is produced around the graphite grains in SiC, where cracks run parallel to the basal plane to cause the tensile stress to be significantly lower than the compressive stress, with the result that the latter stress plays a more important role within the sintered body. Another possible reason for the high toughness is the contribution of cracks in the graphite to inhibit crack propagation through the body.

Interface between SiC and C

Figure 6 is a high-resolution TEM image of the interface between SiC and graphite, showing that the basal plane of graphite is parallel to the interface. The interface energy is not be high comparing with basal plane, because the bonding character of the basal plane is π bonding with low anisotropy. As discussed earlier, a tensile stress is generated on the SiC side and a compressive stress on the graphite side, along the interface, because of the difference of the thermal expansion coefficient of a-axis between graphite and SiC. The figure shows the black contrasts along the interface, reflecting the residual strain. Thus, the interface parallel to the basal plane of graphite was frequently observed in this study.

Figure 7 shows the transformation of SiC from 6H to 3C in the vicinity of the interface between SiC and graphite. Such a phenomenon is sometimes observed in the grain boundaries of SiC sintered by the conventional method.[7] The transformation is considered to have the effect of reducing interface energy. It is also observed in this case that the basal plane of graphite is parallel to the interface between 3C-SiC and graphite. Figure 7 also shows the microcracks on the SiC side, conceivably resulting from tensile stress along the interface, as shown in Fig. 5.

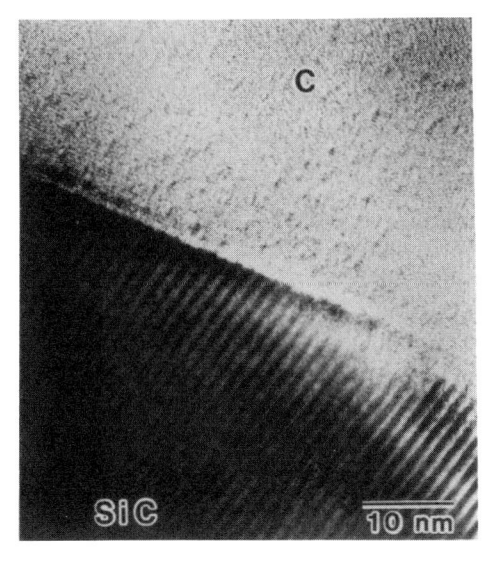

Figure 6. HR-TEM image of the interface between SiC and graphite, indicating that the basal plane of graphite was parallel to the interface (sample A).

Figure 8 shows a case where that the basal plane of graphite is not parallel to the interface between SiC and graphite. In such a case, SiC frequently has a step structure, as seen in the photograph. The height of step is about 1.5 nm, roughly equivalent to the c-axis height of the 6H-SiC unit cell. It was also found that there is about a 3 nm thick layer of low contrast, which represent the amorphous phase, because no lattice image was observed for this phase. It is considered to be the product of the reaction between SiC and graphite. In such a case, the interface will be affected by the σ bond, and will have a high interface energy because of the high anisotropy of σ bonding. The amorphous reaction layer may be formed to reduce the high interface energy.

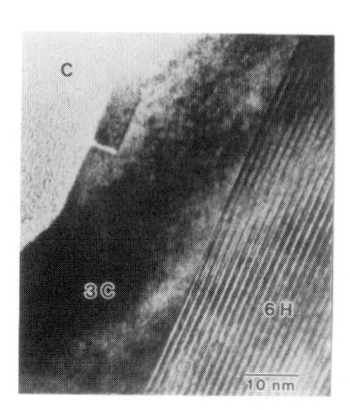

Figure 7. HR-TEM image of the interface between SiC and graphite, indicating that SiC was transformed from 6H to 3C near the interface (sample B).

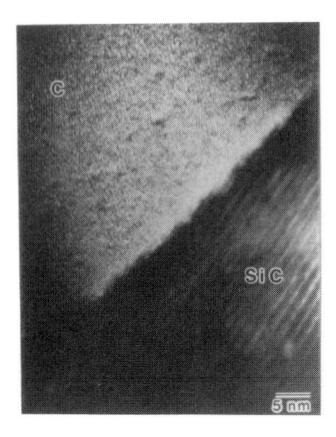

Figure 8. HR-TEM image of the interface between SiC and graphite, indicating that SiC had step structures in an amorphous layer formed by reaction on graphite (sample B).

SiC Grain Boundary

Figure 9 is the high-resolution TEM photograph of the SiC-SiC grain boundaries. This photograph shows no secondary phase. However, structure relaxation might occur within a thickness of 1 nm, judging from the contrast in the grain boundaries.

CONCLUSIONS

The microstructures of plasma-sintered SiC-C composites were investigated by transmission electron microscopy.
1. Fine polycrystals and fibrous crystals of graphite are randomly aggregated around a pore.
2. Graphite between the SiC grains is either in the form of a relatively large, single crystal, or fine polycrystal or fibrous crystal. The single crystals contain cracks parallel to the basal plane, resulting from the difference in thermal expansion coefficients between SiC and graphite, which causes a compressive or tensile stress at the interface. The polycrystalline or fibrous graphite often shows a preferred orientation caused by the compressive stress.
3. The high hardness and high toughness of plasma sintered SiC-C composites conceivably results from residual compressive stress, produced by the difference in thermal expansion coefficients between SiC and graphite. Another possible reason for the high toughness is the contribution of the cracks in graphite to inhibit crack propagation through the body.

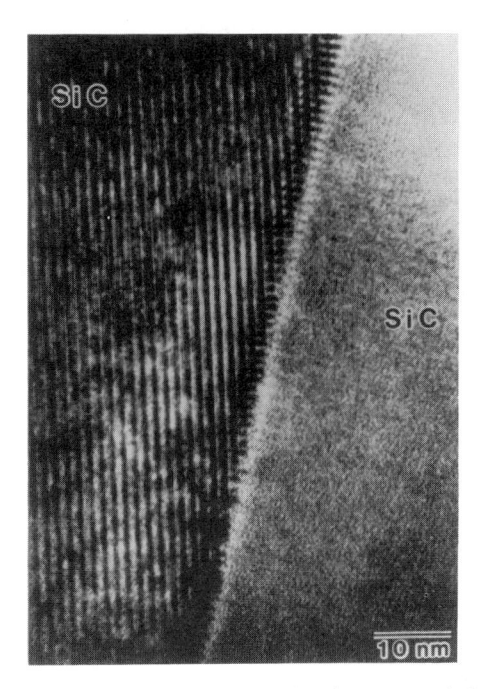

Figure 9. HR-TEM image of grain boundary in SiC, indicating that no second phase existed at the boundary (sample A).

4. At the interface between SiC and graphite, there existed structures in which the basal plane of graphite was parallel to the interface, in which SiC was transformed near the interface, and in which SiC had step structures in an amorphous layer formed by reaction on graphite. These structures might be formed to reduce high energy at the interface.
5. No second phase was observed at grain boundaries in SiC, however, structure relaxation might occur within a thickness of 1 nm.

ACKNOWLEDGEMENT

The authors thank Mr. T. Suzuki of JFCC for his cooperation.

REFERENCES

1. K. Kijima, "Proceedings of The 7th Inter. Symp. on Plasma Chemistry," IUPAC, Eindhoven, **2**:662 (1985).
2. K. Kijima, 6th High-Temperature Structural Material Basic Seminar Proceedings Abstract, 73-78 (1986).
3. Handbook of Chemistry, third edition edited by Chemical Society of Japan, Maruzen II-22 (1984).
4. H. Abe et al., Engineering Ceramics, Giho-do, Tokyo, 16-20 (1984).
5. T. Oku, Introduction to Carbonaceous Materials, revised edition, edited by the Carbonaceous Material Society of Japan, 63-65 (1984).
6. V. A. Drits, Electron Diffraction and High-Resolution Electron Microscopy of Mineral Structures, Springer-Verlag; Berlin 149-154 (1987).
7. K. R. Kinsman and S. Shinozaki, Proc. Int. Conf. Solid Solid Phase Transform. 605 (1002)

COLLOIDAL STABILITY IN COMPLEX FLUIDS

Jennifer A. Lewis

Materials Science and Engineering Department
Beckman Institute for Advanced Science and Technology
University of Illinois at Urbana-Champaign

ABSTRACT

Colloidal stability in complex fluids has been studied with emphasis on depletion effects arising from the presence of nonadsorbed polymeric species. The purpose of this review is to present the various contributions to the interparticle potential in complex colloidal systems, and highlight their importance through experimental observations of structure-property (rheological) relations in such systems.

INTRODUCTION

Colloidal processing offers the potential to reliably produce bulk ceramic films and forms by tailoring suspension "structure" through careful control of interparticle potentials.[1,2] A decade ago, monodisperse colloidal suspensions were considered optimal for achieving the microstructural homogeneity desired in advanced ceramic products.[2] However, emerging methods, such as those based on coagulated networks [3,4] or gel casting [5], rely on colloidal suspensions with increasing complexity. Examples include tailored particle size distributions or fluids containing high salt additions and/or nonadsorbed polymeric species. It is further recognized that several commercial approaches (e.g., tape-cast layers, slip-cast and spray-dried bodies) utilize complex colloidal suspensions. Thus, fundamental knowledge of the relationship between colloidal stability, rheological behavior, and interparticle potentials in complex systems is essential.

This paper focuses on colloidal stability in complex fluids, with specific emphasis on depletion phenomena. The depletion force is a central force in such systems, albeit less well known relative to long-range van der Waals, steric, or electrostatic forces. Depletion phenomena arise between large particles suspended in fluids containing smaller particles (e.g., colloids, nonadsorbed polymer, polyelectrolytes, or micelles).[6-8] Such forces may induce interactions leading to depletion flocculation or restabilization. This paper reviews the various interparticle potentials that contribute to colloidal stability in complex fluids, and then presents experimental observations that highlights their effects on the rheological behavior of weakly flocculated colloidal suspensions in the presence of nonadsorbed polymeric species.

INTERPARTICLE POTENTIALS

Colloidal stability is determined by the total interaction potential energy, V_{tot}, which may be expressed as the sum of the attractive and repulsive potential energies:

$$V_{tot} = V_{vdW} + V_{elect} + V_{steric} + V_{dep} \tag{1}$$

where V_{vdW} is the attractive potential energy due to long range van der Waals interaction forces between particles, V_{elect} is the repulsive potential arising from electrostatic charges on particle surfaces, V_{steric} is the repulsive potential energy arising from the interaction of adsorbed polymer species on approaching particle surfaces, and V_{dep} is the depletion potential energy due the presence of nonadsorbed polymer in solution. As will be discussed below, depletion forces appear to be a limiting case of structural-induced solvation forces.[8] Such forces, however, act over a much larger length scale due to the relative size differences between most depletant and solvent species. Thus, structural contributions arising from solvation (or hydration) forces have been omitted from Eq. (1) due to their short-range nature ($h < 5$ nm).

In the absence of depletion effects, colloidal stability is influenced predominantly by van der Waals, steric, and/or electrostatic interactions. Detailed expressions for these potentials can be found in Refs. 9-11. Colloidal suspensions can be prepared in the dispersed, weakly flocculated, or strongly flocculated states, as shown schematically in Figure 1.

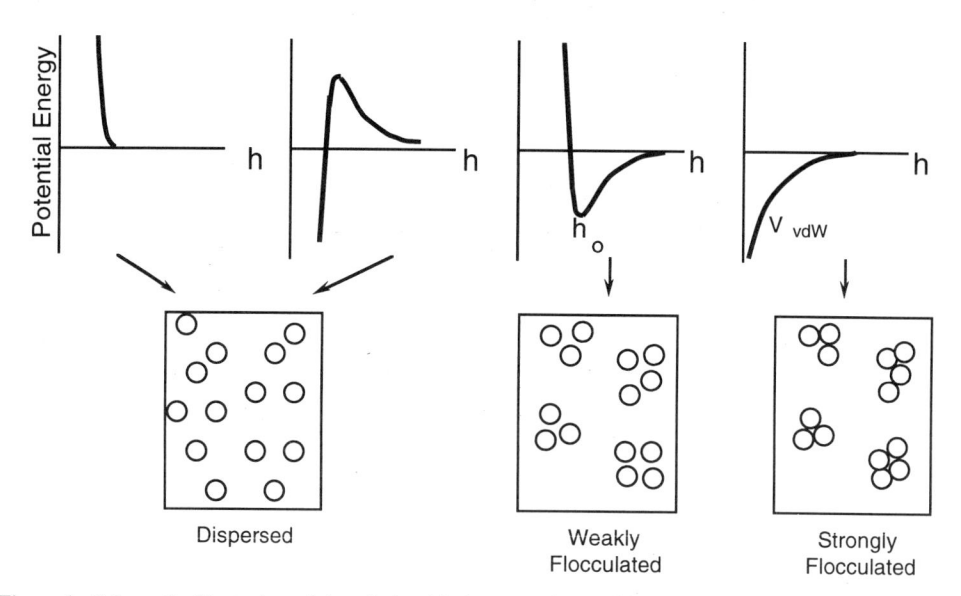

Figure 1. Schematic illustration of the relationship between the total interparticle potential energy and the resulting suspension "structure".

In the dispersed state, discreet particles exist which repel each other upon close approach when the repulsive barrier $\gg kT$. In the weakly flocculated state, particles aggregate in a shallow secondary minimum (well depth $\approx 1 - 20\ kT$) [7] forming clusters (or flocs) of non-touching particle networks, where the minimum equilibrium separation distance between particles is given by h_o. In contrast, particles aggregate into a deep primary minimum in the strongly flocculated state forming clusters of touching particle networks.

Depletion interactions have been associated traditionally with colloidal systems containing nonadsorbing polymeric species.[7] The term, depletion, denotes the existence of a negative concentration gradient near particle surfaces. This distance, known as the depletion layer thickness (Δ), is of the order of the radius of gyration (R_g) of the polymer coils in dilute solution or their correlation length in more concentrated solutions.[7] A schematic representation of such interactions is shown in Figure 2. Upon close approach (h < σ), smaller spheres (e.g., polymer coils) are excluded from the gap between large spheres

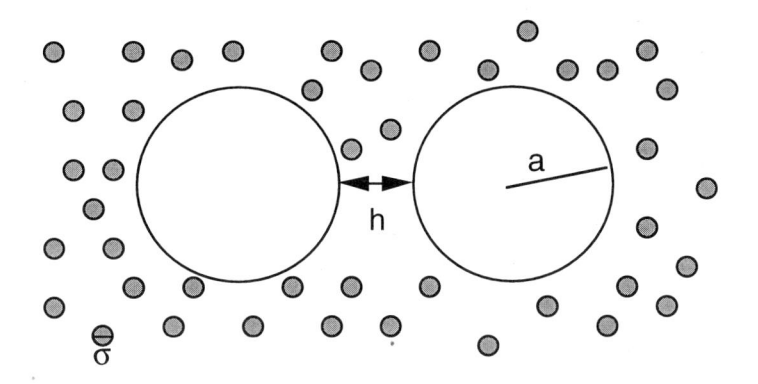

Figure 2. A schematic representation of depletion interactions between large colloidal spheres in a dilute solution of small spherical depletants. [Note: h = interparticle separation distance, a = colloid radius, and σ = depletant diameter.]

giving rise to an osmotic pressure difference which promotes (depletion) flocculation, as first recognized by Asakura and Oosawa.[6] Several researchers have correlated the onset of depletion flocculation to polymer concentrations near the dilute-semidilute transition (ϕ^*), where individual coils intertwine, and the onset of restabilization to concentrations approaching the concentrated regime (ϕ^{**}). However, the mechanism by which nonadsorbed polymer promotes stability is still debated.

Emerging work by Mao et al.[8] and others[13,14] has focused on the interaction between colloidal spheres in dilute solutions of nonadsorbing, mutually avoiding small spheres (of volume fraction, Φ) accounting for the effects of polymer nonideality. In the simple case of hard sphere-hard wall (HSHW) interactions, they recovered the salient features of the Asakura and Oosawa model -- namely that depletion interactions were purely attractive. However, upon modifying the HSHW interaction to account for two-body (and higher order[8]) effects[8,13,14], they found the coexistence of a repulsive barrier and an attractive depletion well at smaller separation distances leading these researchers to suggest that the depletion force is a limiting case of structural-induced solvation forces. The expressions, derived for the depletion interaction potential correct to Φ^2, are given by: [8]

$$V_{dep}(\lambda) = 0 \quad h > 2\sigma \tag{2a}$$

$$V_{dep}(\lambda) = \frac{a\Phi^2 kT}{5\sigma}\left(12 - 45\lambda + 60\lambda^2 - 30\lambda^3 + 3\lambda^5\right), \quad \text{for } 2\sigma > h \geq \sigma \tag{2b}$$

$$V_{dep}(\lambda) = \frac{-3a\Phi kT\lambda^2}{\sigma} + \frac{a\Phi^2 kT}{5\sigma}(12 - 45\lambda - 60\lambda^2), \quad \text{for } h < \sigma \quad (2c)$$

where $\lambda = (h - \sigma)/\sigma$. Using Eq.(2) one calculates the V_{dep} curves shown in Figure 3 for the experimental system discussed below. For these conditions, a repulsive barrier is observed

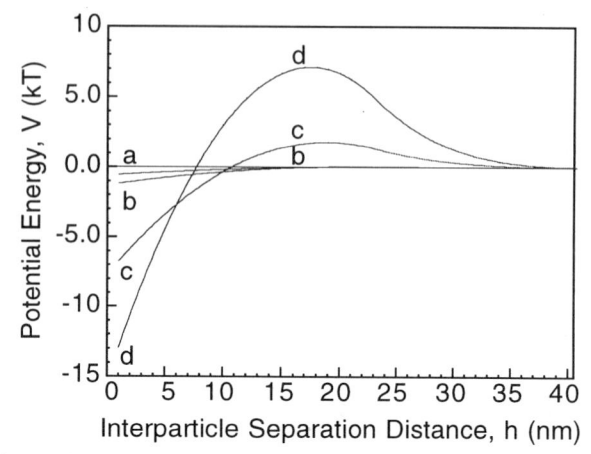

Figure 3. Calculated potential energy of interaction, V ($= V_{dep}$), as a function of interparticle separation distance (h) for spherical colloids ($a = 250$ nm) in nonadsorbed polymer solutions of varying volume fraction (ϕ): (a) 0.0005, (b) 0.0011, (c) 0.0063, and (d) 0.0121. [Note: Polymer diameter ≈ 20 nm, $\Phi = 36 \cdot \phi$ to account for excluded volume effects, and $\phi^* \approx 0.0121$).

at separation distances of the order of coil diameter ($\sigma \approx 2R_g$) which is similar in magnitude to the attractive well depth observed near the interface between particles. As $\phi \longrightarrow \phi^*$, the repulsive barrier was estimated to be a few kT or higher. Such curves display a remarkable resemblance to those obtained for electrostatic stabilization, suggesting that depletion stabilization arises from kinetic (metastable) effects stemming from energy barrier considerations analogous to the electrostatic mechanism.

EXPERIMENTAL OBSERVATIONS

Rheological measurements and optical microscopy were used to characterize the stability of Al_2O_3 - poly(methy methacrylate) - toluene suspensions as a function of nonadsorbed polymer content. The Al_2O_3 powder has an average particle size of 0.5 µm and a distribution between 0.1 - 2.0 µm. Poly(methy methacrylate), PMMA, served as both adsorbed (plateau coverage ≈ 5 mg PMMA/g Al_2O_3) and nonadsorbed species. Full experimental details are given in Ref. 11. In the absence of nonadsorbed PMMA, suspensions exhibited a strong degree of shear thinning behavior as shown in Figure 4. Interparticle potential calculations predict a weakly flocculated system (well depths ≈ 2 - 9 kT), in support of the experimental observations, as shown in Figure 5.

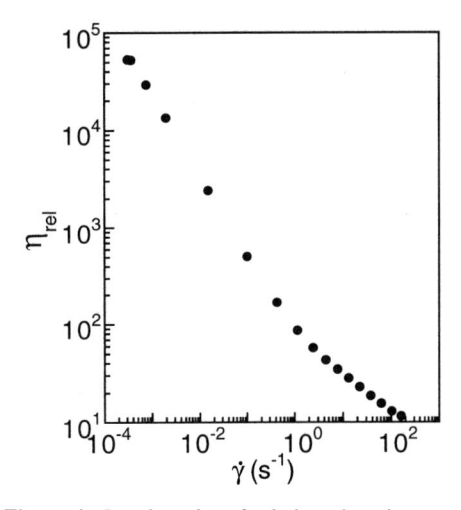

Figure 4. Log-log plot of relative viscosity as a function of shear rate for Al$_2$O$_3$-PMMA-toluene suspensions with $\phi_{Al_2O_3}=0.1$ in the absence of nonadsorbed PMMA.

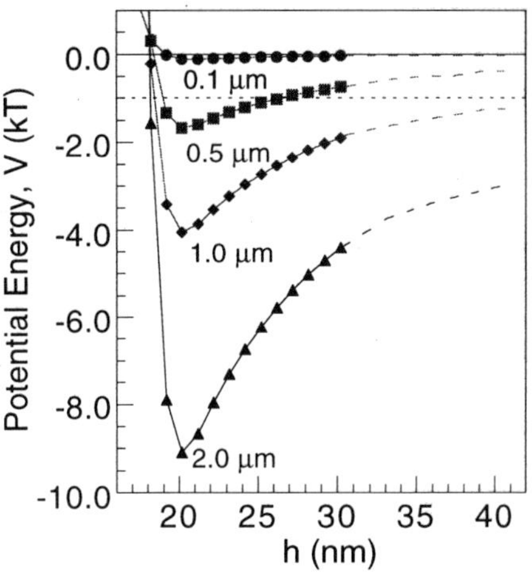

Figure 5. Calculated interparticle potential energy, V (=V_{vdW}+V_{steric}), as a function of separation distance for spherical particles of varying diameter with adsorbed PMMA layers (layer thickness ≈ 11.6 nm).

Figures 6 and 7 depict the dependence of the low shear relative viscosity and characteristic floc size (as observed using optical microscopy) on nonadsorbed PMMA volume fraction, respectively. Clearly, suspension stability improves dramatically with increasing excess PMMA, particularly as the volume fraction of free PMMA, ϕ, $\rightarrow \phi^*$. Such behavior has not been previously predicted over such a wide range of ϕ.

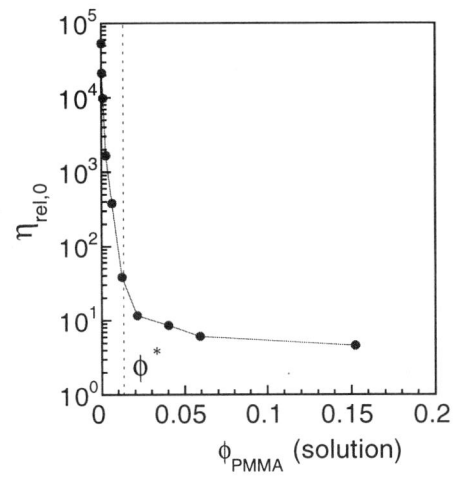

Figure 6. Semi-log plot of the low shear relative viscosity as a function of nonadsorbed PMMA volume fraction.

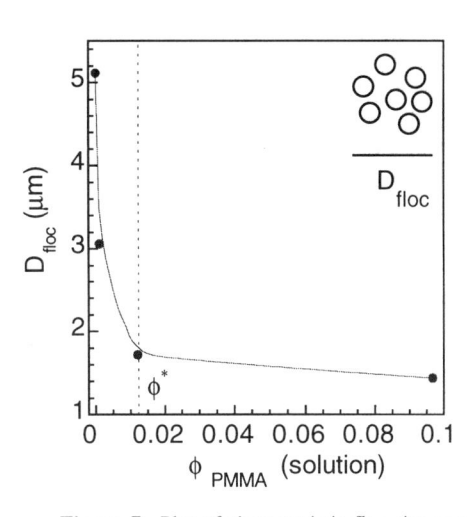

Figure 7. Plot of characteristic floc size as a function of nonadsorbed PMMA volume fraction.

DISCUSSION

One can understand these *unusual* observations by examining the depletion potential calculated for this experimental system (refer to Fig. 3) as well as the initial system stability

499

(refer to Fig. 5). Several approximations are required to model interparticle interactions using Eq. (2), thus V_{dep} is qualitative in nature. V_{dep} was found to exhibit both an attractive minimum and repulsive maximum at larger separation distances, each of the order of kT or higher as $\phi \rightarrow \phi^*$. Upon the addition of excess PMMA, V_{dep} superimposes on the potential energy curves shown in Fig. 5, where the origin ($h = 0$) in Fig. 3 corresponds to an interparticle separation distance of approximately 20 nm in Fig. 5. A driving force for aggregation exists in weakly-flocculated systems in the absence of free polymer, thus the main contribution of V_{dep} is to introduce a repulsive barrier at larger interparticle separation distances. This barrier would deter aggregation provided its of the order of kT or higher leading to improved stability, as observed in Figures 6 and 7. In contrast for stable systems, V_{dep} introduces an attractive minimum that does not otherwise exist. Thus, despite the presence of a modest repulsive barrier which would render some interparticle collisions ineffective, the net response of those systems would be an increase in aggregation kinetics (i.e., depletion flocculation) relative to the initial state, provided an attractive minimum of a few kT or higher exists. [8] In both cases, complete restabilization would be expected when the repulsive barrier exceeds roughly 20 kT.

The above example clearly illustrates that *subtle* contributions to the total interparticle potential, in addition to the well known van der Waals, steric, and/or electrostatic interactions, can have a profound effect on the stability of complex colloidal suspensions. In this work, nonadsorbed polymeric species were observed to enhance suspension stability over a broad range of concentrations as a result of depletion interactions. Related work by Lange and co-workers[3,12], which uncovered the importance of short-range repulsive forces in charge-stabilized systems at high salt concentrations, has similarly highlighted the importance of non-DLVO contributions to the total interparticle potential -- and, thus, to ceramic processing science.

SUMMARY

The reliable production of bulk ceramic films and forms hinges critically on the fundamental understanding of factors influencing colloidal stability in complex fluids. In such systems, *subtle* contributions (e.g., depletion, solvation, or hydration forces) to the total interparticle potential have been shown to dramatically impact suspension stability, and, hence, rheological behavior. Additional research is needed to fully understand how complexities such as small colloidal particles, nonadsorbed polymeric or solvent species, or hydrated ions alter the interactions between larger particles, and, ultimately influence the structural evolution of ceramic suspensions during processing.

ACKNOWLEDGMENTS

I would like to thank A. Ogden and J. Guo for their experimental assistance, and the National Science Foundation (Grant #'s DDM 92-11554 and DMR 94-53446) and the Schlumberger Foundation for their generous support of this research.

REFERENCES

1. F.F. Lange, "Powder Processing Science and Technology for Increased Reliability," *J. Am. Ceram. Soc.* 72:3 (1989).
2. I.A. Aksay, "Microstructure Control Through Colloidal Consolidation"; pp. 94-103 in Advances in Ceramics, Vol. 9 *Forming of Ceramics.* Edited by J.A. Mangels and G.L. Messing. American Ceramic Society, Westerville, OH (1984).

3. F.F. Lange, "New Interparticle Potential Paradigm for Advanced Powder Processing"; pp. 185-201 in Ceramic Transactions, Vol. 22 *Ceramic Powder Science IV*. Edited by S.-I. Hirano, G.L. Messing, and H. Hausner. American Ceramic Society, Westerville, OH (1984).
4. T.J. Graule, F.H. Baader, and L.J. Gauckler, "Shaping of Ceramic Green Compacts Direct from Suspensions by Enzyme Catalyzed Reactions," cfi/Ber.DKG 71:317 (1994).
5. A.C. Young, O.O. Omatete, M.A. Janney, and P.A. Menchofer, "Gelcasting of Alumina," *J. Am. Ceram. Soc.* 74:612 (1991).
6. S. Asakura and F. Oosawa, "Interaction between Particles Suspended in Solutions of Macromolecules," *J. Polym Sci.*, 33:183 (1958).
7. D.H. Napper, *Polymeric Stabilization of Colloidal Dispersions*, Academic Press, London (1983).
8. Y. Mao, M.E. Cates, and H.N.W. Lekkerkerker, "Depletion Force in Colloidal Systems," *Physica A* 222:10 (1995).
9. J.E. Seebergh and J.C. Berg, "Depletion Flocculation of Aqueous, Electrosterically-Stabilized Latex Dispersions," *Langmuir* 10:454 (1994).
10. J.N. Israelachvili, *Intermolecular and Surface Forces*, Academic Press, NY (1992).
11. A.L. Ogden and J.A. Lewis, "Effect of Nonadsorbed Polymer on the Stability of Weakly Flocculated Suspensions," *Langmuir* (in press).
12. B.V. Velamakanni and F.F. Lange, "Effect of Interparticle Potentials and Sedimentation on Particle Packing Density of Bimodal Particle Distributions During Particle Filtration," *J. Am. Ceram. Soc.* 74:166 (1991).
13. J.Y. Walz and A. Sharma, "Effect of Long Range Interactions on the Depletion Force between Colloidal Particles," *J. Colloid Interface Sci.*, 168:485 (1994).
14. P.B. Warren, S.M. Illet, and W.C.K. Poon, "Effect of Polymer Nonideality in a Colloid-Polymer Mixture," *Phys. Rev. E.* 52:5205 (1995).

HIGH TEMPERATURE COLLOIDAL PROCESSING FOR GLASS/METAL AND GLASS/CERAMIC FGM'S

L. Esposito, E. Saiz, A. P. Tomsia and R. M. Cannon

MSD, Lawrence Berkeley National Laboratory
Berkeley, California

ABSTRACT

The paper proposes a basis for understanding the microstructural evolution of functionally graded materials (FGM's) which can provide guidance for tailoring various fabrication processes. Fabrication methods that have been used often rely on a liquid phase being present. However, preliminary experiments and analysis reveal that these specific methods can be considered as subsets of a more general situation that depends upon the colloidal behavior of a solid phase in a liquid. Once this is realized, numerous hybrid methods can be conceived to better fabricate materials including graded ones.

The critical behavior that will occur depends upon the extent to which the liquid wets the solid surfaces as well as the solid-solid interfaces, i.e., grain boundaries. The applicable fabrication methods plus the resulting degrees of interconnectivity of the two phases will, thus, depend on specific combinations of these controlling parameters. The present paper describes a framework in which criteria for forming FGM's are given based upon these considerations: 1) wetting of the solid by the liquid; 2) penetration of grain boundaries of the solid by the liquid; and 3) flocculation or not of the solid particles in the liquid. These issues must be augmented by: 4) dominant sintering mechanism and sintering rate of a network of solid particles immersed in liquid; and 5) materials compatibility.

Various classes of behavior are illustrated based upon interactions of high temperature glasses (mainly within the system Al_2O_3-Y_2O_3-SiO_2) and metals or ceramics.

BACKGROUND

In a functionally graded material (FGM), composition and/or microstructure vary continuously across the material. Such a graded structure can be beneficial for those applications where reducing the residual and applied stress distributions is fundamental for efficiency and life time improvement, as is often needed for bonding of ceramics with metals. Due to their technological interest, the fabrication of metal/ceramic graded materials has been addressed several times in the last decades, but typically lack of basic knowledge about the principles in metal/oxide interactions has limited the reliable fabrication of composites, especially ones suitable for high temperature service.

Table I. The contact angle θ and dihedral angle ψ are proposed as key variables that dictate the expected types of wetting and colloidal interaction between a solid and a liquid phase. Various possible processes for FGM production are listed as functions of the particular class of behavior expected. A few systems are reported as possible FGM candidates or just as examples of a particular type of behavior.

	Colloidal and interfacial (IF) behavior	θ<90° Infiltrate spontaneously		θ>90° Prone to dewetting	
		Processes	Materials	Proc.	Mater.
ψ=0	**Wet interface:** Particles deflocculate Liquids penetrate thru IF's	**Infiltration:** Powder preform disintegrates Dense preform slowly comes apart Kinetics depend on IF chemistry ∴ concentration gradients can develop **Mix powders:** Solid phase disperses in liquid Solid particles are not interconnected Graded architectures by sedimentation possible: No gradient if fine powders Sharp gradient if coarse powders	Al_2O_3/Cu_2O MgO/Si-glass[1] ZnO/Bi_2O_3[2] hot Al_2O_3/Si-glass[3,4] ZrO_2/Si-glass[5]		
ψ>0	**Moist interface:** Particles flocculate giving Stable interparticle film between flocs ∴ flocs shear easily Liquids penetrate thru pure IF's	**Infiltration:** IF penetration changes IF chemistry May drive preform disintegration ? Preform may slump or sag **Mix powders:** Flocculation yields quasi-interconnected particle network Sharp gradient by sedimenting dense flocs	Si_3N_4/Si-glass[6-8] ZnO/Bi_2O_3[2] cool $Pb_2Ru_2O_7/$ Pb-Al-Silicate[9] WC/Co[10,11] SiC/Si-glass[12,13]		
	Moist interface: Particles flocculate giving Stable interparticle film ∴ flocs shear easily Liquids do not penetrate thru pure IF's	**Infiltration:** Powder & presintered compacts are stable during infiltration **Mix powders:** Flocs are deformable	TiO_2/SiO_2[14]		
	Dry Interface: Particles flocculate Flocs are shear resistant Liquids do not penetrate thru IF's	**Infiltration:** Powder & presintered compacts are stable during infiltration **Mix powders:** Sedimentation of open, interconnected flocs ∴ two interconnected phases possible	Refr. met/glass Mo/Si-glass W/Si-glass Trans. met/glass Co/Si-glass Ni/Si-glass	Pressr. Infiltr.	Oxides/ noble met Al_2O_3/Cu Al_2O_3/Pt Si-glasses/ noble met SiO_2/Cu

Viable fabrication methods that utilize a liquid phase having a high melting point include: 1) preparing graded binary powder mixtures followed by liquid phase sintering and 2) liquid infiltration of a graded, sintered, porous preform. However, such liquid containing systems can be profitably viewed as high temperature colloidal suspensions, where a solid phase is suspended in a liquid one, under processing conditions. From this perspective, many phenomena occurring at high temperature can be interpreted in terms of colloidal interactions between dispersed particles in liquids. Obviously, the phenomena governing particular behaviors (mainly wetting, pore elimination, and flocculation) are influenced by numerous variables. Thus, the following fundamentals must be understood well enough to properly control the microstructural development:

1) Wetting of surfaces by liquids: the energetics and rates of spreading plus the extent of reactivity influence wetting, and so control, for example, the infiltration of pore channels.

2) Colloidal interaction of particles in high melting point liquids: the particle architecture that forms depends on the balance between repulsive and attractive forces and on the interparticle shear resistance within flocs.
3) Penetration of grain boundaries by such liquids: again, the influence of this process depends on the kinetics as well as energetics of the system.
4) Character and rate of sintering, including coarsening, for a porous, solid network submerged in a liquid: the strength of a porous preform or flocced network and its susceptibility to being disrupted depend upon the dominant sintering mechanism.
5) Materials compatibility.

The first three of these different aspects can be considered from a unified point of view by focusing on two basic parameters, the contact angle, θ, and the dihedral one, ψ, that control the spatial distributions of liquid and solid. A general framework is proposed (Table I) wherein key interactions between a solid and a liquid phase are described through these two angles. Examples given include systems examined directly by the authors and some from the literature.[1-14] The latter are catalogued, a bit conjecturally, based solely on observed interfacial composition or structure plus general trends in surface wetting[15-20] without showing that all the proposed characteristic behaviors follow.

The contact angle, θ, at the junction of a liquid droplet on a solid usefully describes the equilibrium force balance and the tendency for spreading at short times when a vapor phase is involved (Fig. 1). It is highly desirable to have $\theta < 90°$ to promote pore filling by infiltration or pore removal and so to have methods to attain this. As an example, the $p(O_2)$ in the furnace plays a role in changing the various interfacial energies; there are beneficial effects of being near the edge of the stability range of $p(O_2)$ for oxide/metal pairs.[20] This is an example among others[19-25] to show how it is possible to govern wetting behavior through an understanding of the interfacial forces involved. Adding reactive elements such as Ti or Cr to promote wetting of noble metals on oxides is another.[22-25]

The value of the dihedral angle, ψ, at the junction of a liquid and a grain boundary or interface for a chosen system is dictated by the interfacial energies between the liquid and solid phases (Fig. 2). Depending on the particular balance existing between the γ_{sl} and γ_{gb}, different classes of interfacial wetting behavior can be described and connected with the dihedral angle range (Table I). Again, as the γ's depend on composition, suitable changes in stoichiometry or use of interface-active elements should be useful in adjusting behavior.

If these fundamentals concerning adsorption and wetting of interfaces are conceived in terms of colloidal interactions between grains of solid dispersed within the liquid, several classes of behavior can be envisaged (Fig. 3).[27] Depending alternatively on the predominance of attractive or repulsive forces, flocculation or deflocculation can be expected in a solid/liquid system (flocculation being the attraction of dispersed particles into clusters without particles distorting, and deflocculation being interparticle repulsion giving a stable dispersion). It should be appreciated that a repulsive situation that would lead to particle-particle deflocculation (absent other stronger forces such as gravity) is equivalent to a zero dihedral angle, and conversely.[27] Moreover, flocculating behavior having $\psi > 0$ can entail situations ranging from essentially clean boundaries to ones with increasing amounts of equilibrium adsorbate including seemingly intermediate behaviors

Figure 1. The contact angle θ, assuming that no chemical reaction occurs, depends on the equilibrium force balance among the liquid surface tension, γ_{lv}, the solid surface energy, γ_{sv}, and the energy of the solid liquid interface, γ_{sl}, as expressed by the Young-Dupré equation.[26]

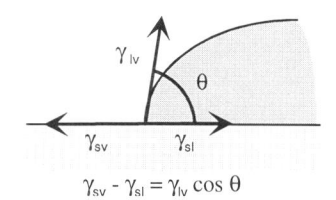

$$\gamma_{sv} - \gamma_{sl} = \gamma_{lv} \cos \theta$$

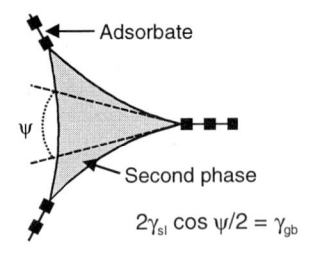

Figure 2. Impurity atoms are partitioned between adsorbate at grain boundaries and the second phase (at triple junctions) in a solid/liquid system (a). The dihedral angle, ψ, indicates the balance between energies for the grain boundary and solid/liquid interface. The boundary energy is strongly affected by any adsorbate as depicted schematically (b) in plots of γ_{gb} showing effects of adding a second component, as solute, up to the solubility limit with several examples based on plausible adsorption isotherms.[27] The equilibrium ψ and equilibrium wetting behavior are dictated by the boundary energy, $\gamma_{gb}*$, for equilibrium adsorption at activities appropriate to the two phase coexistence condition, i.e., at $a_2 = a_2*$.

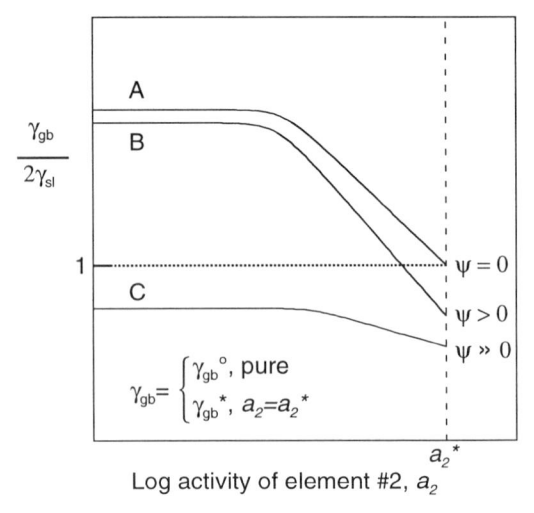

involving multilayers of adsorbed material between solid particles. These latter situations having an equilibrium interparticle spacing (film thickness) are denoted *moist interfaces*, Table I. Changing the liquid chemistry can adjust the thickness of the adsorbate layers, as seen for 0.5-2 nm thick films in Si_3N_4 and $Pb_2Ru_2O_7$, for example,[7-9] or even effect the existence of a film as found for sintered SiC[12] and expected for WC/Co composites.

Depending on the value of ψ, different grain boundary wetting or penetration behaviors can be expected,[28] assuming kinetics are rapid: no penetration of equilibrium liquid, nor deflocculation, when $\psi > 60°$; partial penetration along the triple lines by equilibrated liquid with no deflocculation, when $60° > \psi > 0$; penetration and *interface wetting* by an equilibrated liquid, essentially deflocculation, when $\psi \rightarrow 0$. Then, it may be expected that for a system with a vanishing dihedral angle, a dense sample with grain boundaries would be spontaneously penetrated by liquid. Such has been observed in several circumstances, e.g., silicate glasses have been seen to extensively penetrate dense, polycrystalline Al_2O_3[3,4,29,30] or MgO.[1] Conversely, for a system with a finite dihedral angle, initially dispersed grains would tend to flocculate as has been observed for Si_3N_4 dispersed in oxynitride glasses.[8,31]

However, it is noted that if a material having initially clean grain boundaries is exposed to a liquid containing a species that would adsorb at the boundaries, a large driving force can exist for penetration and deflocculation. This can occur even if the liquid composition is compatible with the bulk solid and if at equilibrium $\psi > 0$, as exemplified by curve B in Fig. 2-b for which $\gamma_{gb}° \gg 2\gamma_{sl} > \gamma_{gb}*$, which may induce transient penetration. (This is analogous to "dry spreading" on free surfaces.[19,32]) Evidence for transient wetting of boundaries in WC/Co and some metallic systems exists,[11,33,34] but this may involve a temperature dependent wetting transition[34] as has also been seen in ZnO/Bi_2O_3 systems.[2,27] Thus, the importance of this remains to be explored along with other metastable or transient adsorption or deflocculation situations.[27] For example, in systems where films would be stable, liquid penetration of clean polycrystalline bodies would be expected, but this does not occur with TiO_2-SiO_2[14] and has not been investigated for SiC-silicates.

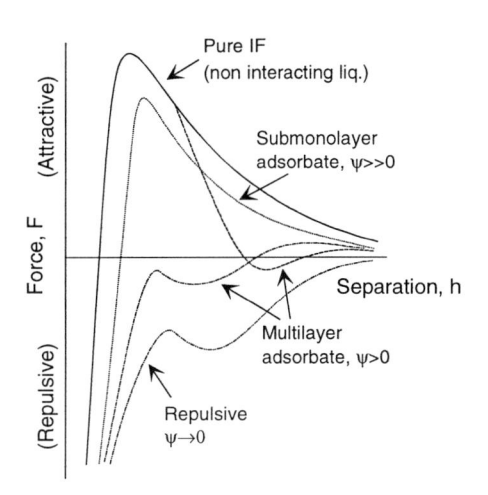

Figure 3. When a solid/liquid system is considered as a colloidal suspension, various classes of behavior are determined by the equilibrium between the attractive and repulsive forces present.[27] This yields: a flocculation situation when the attractive forces are prevalent (i.e., nearly-clean interfaces and a liquid that does not interact with them; $\psi \gg 0$); a deflocculated system when the repulsive forces are predominant at all separations (i.e., liquids penetrate through the interfaces, $\psi \rightarrow 0$); and intermediate situations where a multilayer of adsorbate impurities forms at the interface (i.e., an equilibrium film of fixed thickness, $\psi > 0$).

The several types of materials for which the equilibrium dihedral angle exceeds zero, and complete deflocculation does not obtain, should exhibit important differences in behavior. In one category, *dry interfaces*, which seems to include many simple metal particles in silicate liquids, the grain boundaries are apparently not penetrated by the glass, and are considered to remain free of multilayer segregation of amorphous films, although they may exhibit monolayer levels of adsorption. Alternatively, in systems like Si_3N_4, nearly all of the grain boundaries are nm scale, amorphous, SiO_2 rich films.[6-8] Both types of systems are expected to flocculate at rates that depend upon temperature, viscosity of the glass and particle size; whereas, the shear resistance of the resultant flocs would differ appreciably in the two cases. With dry grain boundaries, it is expected that the flocculated particles would have a high shear resistance and so the flocs would remain open and loose; settling in gravity would not yield well packed regions but instead open, lower density regions (based on the particle phase) within an interconnected metal particle network, analogous to behavior better documented for low temperature suspensions.[35] In contrast, for systems with moist boundaries in which the flocced particles have an intervening amorphous layer, shear within the flocs would be expected to be relatively easier, and would, thereby, permit flocs to rearrange and attain denser packing. This would lead to faster settling in a gravity field, and to a higher ultimate particle packing density. Also, the interparticle network would not be fully interconnected, e.g., the boundary films would disrupt electrical conduction.

Although the type of microstructure and its evolution are primarily dictated by the surface and interfacial wetting behavior, the sintering characteristics and kinetics of the particle network immersed in the liquid are of great importance as well. For instance, if neck growth is comparatively fast, flocs once formed will soon become stronger and more shear resistant. Moreover, if two particles contact and the neck growth is very rapid the two can effectively sinter and coalesce into a single particle before the next particle arrives. In contrast, if the sintering is slow, then a multiparticle network can form in flocculating systems.[36] Even in these cases, it matters whether sintering involves transport of mass from the interparticle grain boundaries (leading to a densification type of behavior), or whether it primarily involves transport within the liquid or at the particle/liquid interface. The first situation causes "shrinkage" of the particle skeleton, perhaps fracturing it into several pieces, and expulsion of the liquid into one or more particle free regions. The latter is essentially parallel to coarsening dominated behavior in conventional sintering,[37] i.e., coarsening of the interparticle network involves rupture of some interparticle contacts, but no net shrinkage of the particle skeleton nor expulsion of the liquid phase.

Thus, applying principles of high temperature colloidal behavior to various methods for producing FGM's via liquid phase mechanisms should be attractive. If the described phenomena are manipulated suitably, it will be possible to control the qualitative aspects of the microstructures and, in combination, by imposing appropriate forces or field gradients during processing to produce desired microstructure and composition gradients. These imposed forces could derive from gravity, temperature gradients or initial composition gradients that would produce reactivity gradients that could be frozen into place.

EXPERIMENTAL

A few exploratory experiments that illustrate some representative behavior were performed. One set of samples was prepared through mixing powders of selected metals, Ni, Co, Mo or W, and the glass with a desired volume ratio. The glass formulation, chosen within the Al_2O_3-Y_2O_3-SiO_2 system (34.8, 26.8, 38.4 cation %, respectively), falls in the ternary peritectic region characterized by a melting point of 1350°C.[38] The softening temperature of the glass determined through a wetting experiment in air is ~1410°C. Pellets of the mixed powders in the chosen metal/glass volume ratio were prepared through linear dry pressing and treated in a tungsten heating element furnace. Flowing, gettered argon (Ti-Zr) was used during these cycles.

Wetting experiments with the same glass on pure metal sheets (Ni, Co, W or Pt) under different $p(O_2)$ conditions (gettered and non-gettered Ar) were performed at 1440°C and contact angles measured after cooling. Wetting and grain boundary penetration behavior of silicate glasses on two dense ceramics, alumina and Si_3N_4 was also investigated. Microstructures were characterized using optical microscopy and SEM-EDS.

HIGH TEMPERATURE COLLOIDAL BEHAVIOR OF METALS IN GLASSES

When a dry Ar atmosphere (having low $p(O_2)$) is used, the $\theta \sim 90°$ or perhaps slightly less for the Ni, Co and W; whereas the angles were 65-75° for these metals in non-gettered Ar. In contrast, the θ with Pt decreased with lower $p(O_2)$, being 45° and 10° in air and dry Ar, respectively.

A colloidal suspension at high temperature can be realized when powders of a high melting point metal and a glass are mixed and treated above the softening point of the glass; its behavior can be interpreted in terms of flocculation/deflocculation and subsequent sintering phenomena. Allowing controlled settling driven by gravity or using graded powder mixes could produce an FGM after cooling the glass.

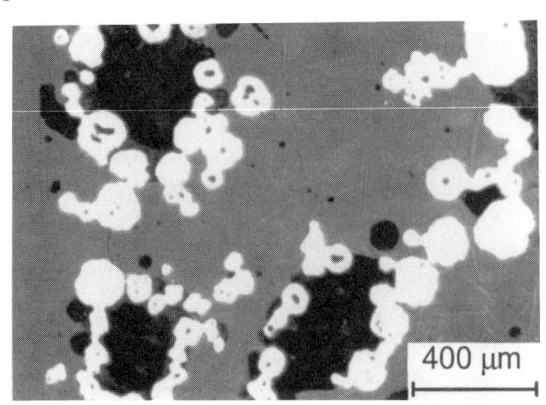

Figure 4. 10 vol % nickel / 90 vol % Al-Y-silicate glass; powder mix treated at 1440°C x 2 hr in dry Ar. Evidently, the marginal wetting of the liquid on the solid particles, $\theta \approx 90°$, leads to configurations in which clusters of the dispersed particles tend to stabilize the porosity in the system. The sintering rate of the metallic particles is high ($T=0.99T_m$).

One undesirable microstructural feature sometimes seen is pores surrounded by flocculating metal particles and containing no glass (Fig. 4); evidently this porosity is stabilized owing, in part, to the high θ (≈90°). It should be noted that once such pores have formed, ridge formation will commence at the triple junctions (analogous to grain boundary grooving) in response to the two dimensional character of the surface force balance there; this will impede further spreading of the liquid.[39,20] As a result, characterization in terms of the so called 1-D contact angle and the Young-Dupré relationship is oversimplified, albeit still useful. The adverse effects of such pinning are most pronounced for systems heated at the highest fraction of the solid melting point, i.e., the Co and Ni.

In these same systems (Ni, Co, Mo and W), ψ » 0° (actually > 90°), the particles invariably flocculated in the powder mixes, and "dry" interfaces tend to form during sintering of the metallic particles following flocculation (Table I, Figs. 5-7). Consistent with the high value of ψ, no penetration of grain boundaries by the silicate liquid was evident in the wetting experiments. (TEM examination of Mo/SiO$_2$ bodies indicates that multilayer segregation is unlikely.[40]) Depending on the metal, the sintering process after flocculation is slow (tungsten and molybdenum, Figs. 5, 6) or fast (cobalt, Fig. 7). The sintering rate in these examples is influenced strongly by the melting temperature of the

Figure 5. 10 vol % tungsten/90 vol % Al-Y-silicate glass; powder mix treated at 1440°C x 2 hr in dry Ar.
Metallic particles flocculate giving stable, open flocs. The sintering of the particle network is comparatively small ($T=0.42T_m$). Sedimentation is slowed down by formation of a continuous skeleton.

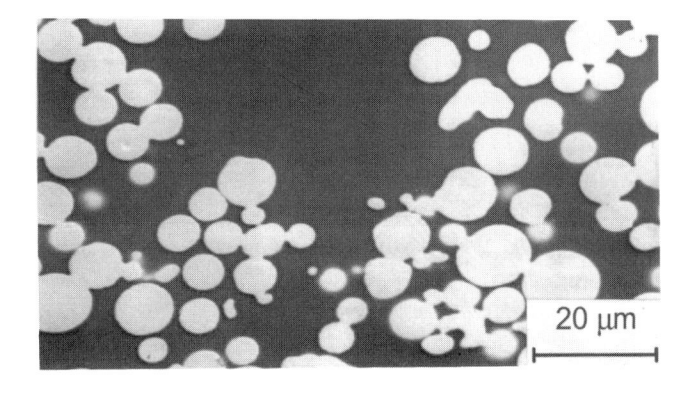

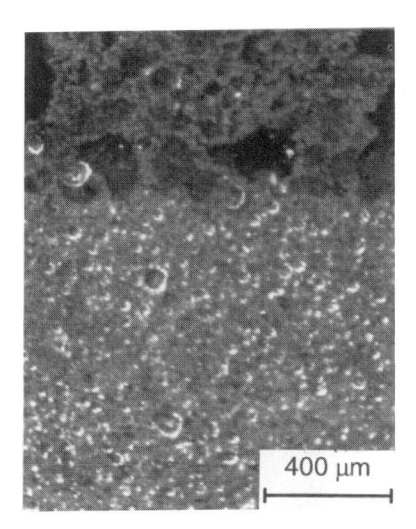

Figure 6. 10, 20, 40 vol % molybdenum/90, 80, 60 % Al-Y-silicate glass; three layer green structure of powder mixes treated at 1440°C x 2 hr in dry Ar.
The flocculation and moderate sintering rate ($T=0.55T_m$) give metal free regions on an intermediate scale. Sedimentation can slowly adjust the starting gradient. Microdesigned graded structures are possible.

metal. In the case of Co, characterized by a melting temperature close to the softening one of the glass ($T = 0.96T_m$), the sintering is fast at the chosen temperature. After only two hours, the network of flocculated metal particles has "densified" noticeably leading to a large metal free region in the glass, but with an undesirably sharp transition. For more refractory metals, as molybdenum ($T = 0.55T_m$) or tungsten ($T = 0.42T_m$), the sintering rate is gradually smaller and after the same heat treatment the metal particles are flocculated but still quite distinct. The Ni ($T = 0.99T_m$) particles, Fig. 4, exhibited less sintering than did the Co particles owing to a much coarser particle size.

Figure 7. 10 vol % cobalt/90 vol % Al-Y-silicate glass; powder mix treated at 1440°C x 2 hr in dry Ar. The sintering rate of the metallic particles is high ($T=0.96T_m$). The glass is expelled on a large scale during sintering of the metal. Nicely graded structures cannot be easily formed via powder mix processes.

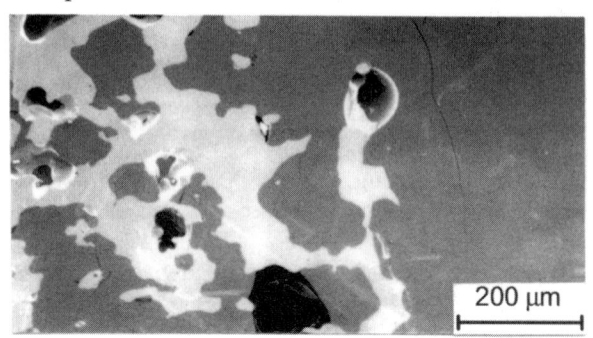

A tendency to form an interconnected skeleton is observed even for a small metal/glass volume ratio (1/9), Fig. 6. Evidently loose, open flocs form and interact such that the settling rate of these is much slower than expected from Stokes settling of isolated particles. Even after heating at 1550°C, little settling had occurred with the ≤ 10 μm W powders at 10% loading. This partly reflects any nonrandom distribution in the initial mixes, but is largely attributed to the shear resistant flocs associated with dry grain boundaries.

GRADED ARCHITECTURES FROM CONTROLLED PENETRATION

A solid/liquid system characterized by $\psi \to 0$ is susceptible to grain boundary penetration in which a dense ceramic body can eventually disintegrate upon exposure to liquid, the energetics and kinetics of the process being governed by the chemistries at the interface and of the liquid. With such systems, it should be possible to build graded structures by contacting a polycrystal to a liquid and controlling the extent of boundary penetration, associated deflocculation and possibly reflocculation.

With dense, sintered alumina in contact with a silicate glass, alumina crystals come apart and float in the glass.[3,4,29,30] Fig. 8 illustrates this for exposure of a 99.5% Al_2O_3 to an Al-Ca-silicate liquid (43, 25, and 32 cation %, respectively). The deflocculation process is reportedly faster for a lower viscosity glass.[30] In addition, there could be an energetic benefit if the silicate liquid contains species that would adsorb at the interface; this may explain that for silicates containing TiO_2, penetration is faster.[29]

Exposing a sintered Si_3N_4 body to the Al-Y-silicate yielded somewhat similar behavior, Fig. 9. Microprobe evaluation revealed that a rather rapid dissolution of Si_3N_4 occurred to saturate the silicate, but penetration and deflocculation continued well beyond this time.[41] To some degree, continued penetration was unexpected, as this is a weakly flocculated system. The explanation may lie in the fact that the glass has a higher Al:Y ratio (1:0.77) than does the sintered body (1:1.20); thus, the initial boundary film in the

Si$_3$N$_4$ -3% Al$_2$O$_3$ -8% Y$_2$O$_3$ is not in equilibrium with the liquid and a driving force exists for penetration as a step toward adjusting the composition of the adsorbate film.

Figure 8. Dense Al$_2$O$_3$ with Al-Ca-silicate glass droplet, treated in air at 1600°C x 2 hr $\psi \approx 0°$; $\theta \approx 20°$.
The liquid silicate totally penetrates most grain boundaries near the surface causing the near surface region to disintegrate. Al$_2$O$_3$ grains largely deflocculate in the glass. Deflocculation kinetics depend on grain boundary and glass chemistry. Graded architectures are thus possible by controlling the grain boundary penetration.

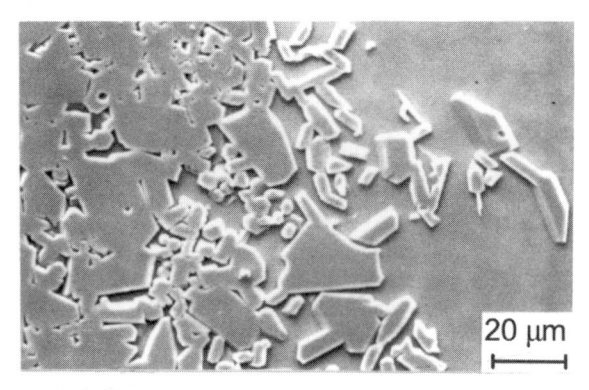

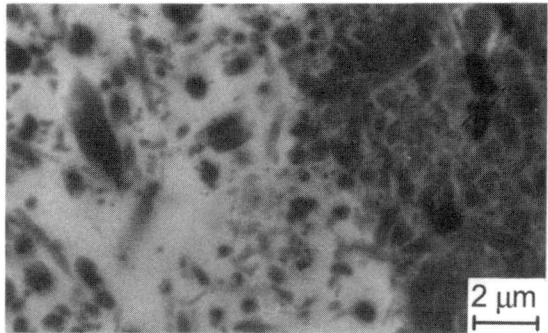

Figure 9. Dense Si$_3$N$_4$ with glass droplet, treated at 1440°C x 6 hr in Ar. $\psi > 0°$; $\theta \to 0$.
The liquid glass has totally penetrated the grain boundaries several grains deep into the material and deflocculated the grains in this region. Again, graded structures can be developed by controlling such penetration by a silicate glass, and any reflocculation of the Si$_3$N$_4$ grains.

SUMMARY

The elements of metal/glass and ceramic/glass interactions are considered in terms of solid/liquid colloidal systems in order to motivate new ways to control microstructures and to make FGM's. Specific fabrication methods that have been used previously, that rely on a liquid phase being present, are: 1) preparation of graded binary powder mixtures and liquid phase sintering; and 2) liquid infiltration of a sintered, porous preform. However, present considerations reveal that these can be considered as subsets of a more general situation that depends upon the colloidal behavior of the solid in the liquid, and so leads to another category: 3) high temperature colloidal processing. Examples for some metal/glass and ceramic/glass model systems that demonstrate different approaches are given and their behavior is interpreted to provide more understanding of the controlling phenomena.

A key issue is that a few underlying, common mechanistic elements control the path of microstructural evolution for several different processes. Thus, by being able to control these, a wider variety of hybrid processes can be utilized, after being tailored as dictated by the processing constraints for the set of materials and type of graded microstructure wanted. For liquid phase based fabrication methods, key phenomena are: wetting, interfacial penetration, colloidal interaction, and network sintering. A categorization of behavior based on these is proposed which is rather speculative as issues of kinetics, metastable or transient behavior, and prediction of interface active species are poorly understood. As few concrete examples exist, much work is needed to correlate interfacial chemistries with actual resultant microstructures, and to address the role of anisotropy.

Acknowledgments

This work was supported by the Laboratory Directed Research and Development Program of Lawrence Berkeley National Laboratory under U. S. Department of Energy under Contract No. DE-AC03-76SF00098.

References

1. C. S. Ramamurthy, M. P. Mallamaci, C. M. Zimmerman, C. B. Carter, P. R. Duncombe & T. M. Shaw, *J. Am. Ceram. Soc.*, submitted.
2. J.-R. Lee & Y.-M. Chiang, *Solid State Ionics*, **75** 79-88 (1994); Y.-M. Chiang, J.-R. Lee & H. Wang, *Proc. Ceramic Microstructures '96: Control at the Atomic Level*.
3. P. L. Flaitz & J. A. Pask, *J. Am. Ceram. Soc.*, **70** 449 (1987).
4. T. M. Shaw & P. R. Duncombe, *J. Am. Ceram. Soc.*, **74** 2495 (1991).
5. M. Rühle, N. Claussen & A. H. Heuer, *Adv. in Ceramics*, **12** 352 (1984).
6. L. K. V. Lou, T. E. Mitchell & A. H. Heuer, *J. Am. Ceram. Soc.*, **61** 392 (1978).
7. I. Tanaka, H.-J. Kleebe, M. K. Cinibulk, J. Bruley, D. R. Clarke & M. Rühle, *J. Am. Ceram. Soc.*, **77** 911 (1994).
8. C.-M. Wang, X. Pan, M. J. Hoffmann, R. M. Cannon & M. Rühle, *J. Am. Ceram. Soc.*, **79** 788 (1996).
9. Y.-M. Chiang, L. A. Silverman, R. H. French & R. M. Cannon, *J. Am. Ceram. Soc.*, **77** 1143 (1994).
10. V. Jayaram & R. Sinclair, *J. Am. Ceram. Soc.*, **66** C137 (1983).
11. R. Warren & M. B. Waldron, *Powder Metal.*, **15** 166 (1972).
12. H.-J. Kleebe, *J. Eur. Ceram. Soc.*, **10** 151 (1992).
13. J. J. Cao, W. J. MoberlyChan, L. C. DeJonghe, C. J. Gilbert & R. O. Ritchie, *J. Am. Ceram. Soc.*, **79** 461 (1996).
14. H. D. Ackler & Y.-M. Chiang, *Proc. Ceramic Microstructures '96: Control at the Atomic Level*.
15. Ju. V. Naidich, *Prog. Surf. Membr. Sci.*, **14** 353 (1981).
16. A. P. Tomsia, A. M. Glaeser & J. S. Moya, *Key Eng. Mater.* (Trans Tech Publ.) **111-112** 191 (1995).
17. K. S. E. Forssberg, *Science of Ceramics*, eds. C. Brosset & E. Knopp, Swedish Inst. for Silicate Research, p. 63 (1970).
18. J. G. Li, *J. Am. Ceram. Soc.*, **75** 3118 (1992).
19. R. M. Cannon, E. Saiz, A. P. Tomsia & W. C. Carter, *Mat. Res. Soc. Symp. Proc.*, **357** 279 (1995).
20. E. Saiz, A. P. Tomsia & R. M. Cannon, *Proc. Ceramic Microstructures '96: Control at the Atomic Level*.
21. F. J. J. Van Loo, J. H. Gülpen & A. Kodenstov, *Joining Ceramics, Glass and Metal*, ed. H. Krappitz, H. A. Schaeffer, Verlag der Deutschen Glastechnischen Gesellschaft e.V., Frankfurt/M, Germany, p. 3 (1993).
22. A. P. Tomsia, J. A. Pask & R. E. Loehman, *ASM Intl. Engineered Materials Handbook*, vol 4 *Ceramics and Glasses* - Sect. 7, "Joining," p. 482 (1991).
23. R. E. Loehman & A. P. Tomsia, *J. Am. Ceram. Soc.*, **77** 271 (1994).
24. C. Kurkjian & W. D. Kingery, *J. Phys. Chem.*, **60** 961 (1956); M. Humenik, Jr & W. D. Kingery, *J. Am. Ceram. Soc.*, **37** 18 (1954).
25. L. Espie, B. Drevet & N. Eustathopoulos, *Metall. and Mater. Trans.*, **25A** 599 (1994).
26. A. W. Adamson, *Physical Chemistry of Surfaces*, 4th ed., J. Wiley & Sons, NY (1982).
27. R. M. Cannon, *Proc. Ceramic Microstructures '96: Control at the Atomic Level*.
28. C. S. Smith, *Trans. AIME*, **175** 15 (1948); *Met. Rev.*, **9** 1 (1964).
29. S. Fujitsu, K. Yamagiwa, E. Saiz & A. P. Tomsia, *Acta Mater.*, submitted.
30. Y. Kuromitsu, H. Yoshida & K. Morinaga, *Proc. 7th Euro Hybrid Microelectronics Conf.*, ISHM, Hamburg, FRG, (1989).
31. M. J. Hoffmann & R. M. Cannon, to be published.
32. P.-G. DeGennes, *Rev. Mod. Phys.*, **57** 827 (1985).
33. R. M. German, *Liquid Phase Sintering*, Plenum Press, NY, Chs. 2-3 (1985).
34. E. I. Rabkin, W. Gust, W. Lojkowski, V. Paidar, *Interface Sci.*, **1** 201 (1993); E. I. Rabkin, L. S. Shvindlerman & B. B. Straumal, *Intl. J. Mod. Phys.*, **B5** 2989 (1991).
35. J. S. Reed, *Principles of Ceramic Processing*, 2nd. ed., John Wiley & Sons, Inc. NY, Ch. 10 (1995).
36. T. H. Courtney, *Metal Trans A.* **8A** 679-84 (1977); *Metal Trans A.* **15A** 1065-74 (1984).
37. C. A. Handwerker, R. M. Cannon & R. L. Coble, *Adv. in Ceram.*, **10** 619 (1984).
38. C. O'Meara, G. L. Dunlop & R. Pompe, *High Tech Ceramics*, P. Vincenzini, Elsevier Science Publ. B.V., Amsterdam, 1987.
39. E. Saiz, A. P. Tomsia & R. M. Cannon, *Acta Mater.*, submitted.
40. W. J. MoberlyChan & A. P. Tomsia, unpublished research.
41. L. Esposito, E. Saiz, A. P. Tomsia, R. M. Cannon & A. Bellosi, *Acta Mater.*, submitted.

MULTICOMPONENT OXIDE COATINGS VIA SOL-GEL PROCESS

Laure Bonhomme-Coury, Muriel Najman, Florence Babonneau, and Philippe Boch

Laboratory of Ceramics and Minerals
Ecole Supérieure de Physique et de Chimie Industrielles
10 rue Vauquelin, 75005 Paris, France

INTRODUCTION

Composites with a glass-ceramic matrix (Li_2O-Al_2O_3-SiO_2 = LAS system) reinforced with continuous SiC fibers exhibit attractive thermal and mechanical properties for aerospace applications. However, because such composites can be corroded by aggressive atmospheres and eroded by dusty environments, they require protective coatings.[1] The present study deals with the preparation (sol-gel route) and characterization of oxide coatings (mullite, cordierite, zirconium titanate, and zirconium silicate), deposited on SiC-LAS composites.

EXPERIMENTAL PROCEDURE

The Composite

The LAS matrix is formed in the glassy state, then devitrified to β-spodumene. Various oxides were added to the base material:[2] ZrO_2 is a nucleating agent, P_2O_5 improves the solubility of ZrO_2 in the glass, MgO behaves as a flux, and Nb_2O_5 modifies the properties of the fiber-matrix interface. The use of Nb_2O_5 is a compromise between mechanical properties, which are improved by the formation of NbC as interphase, and resistance to oxidation, which is decreased by the formation of brittle magnocolombite ($MgNb_2O_6$).

The SiC fibers (*Nippon Carbon*) contain Si, C, and O as elements. β-SiC is in the form of nanocrystalline (< 2 nm) precipitates.[3] In non-oxidizing atmospheres, the fibers exhibit high mechanical properties up to 1000°C,[4,5] but in air the properties are degraded by oxidation, which can begin at temperatures below 800°C.[6]

The composite was a laboratory material,[7] made by infiltrating a woven SiC fabric with a glass-ceramic slurry, then hot-pressing the ply-structure of stacked layers (in a graphite die).

The Protective Oxides

Four criteria were taken into account to select the protective oxides, namely:

- Good refractoriness.
- Low or moderate thermal expansion.
- Intrinsic chemical stability.
- Chemical compatibility with the substrate.

The "chemical compatibility" cannot be ascertained with precision, but implies, as a minimum, that no detrimental interphases form by reaction between the coating and the substrate and, at best, that favorable interphases form to improve the bonding of the coating with the substrate. Four oxides were retained: mullite, cordierite, zirconium titanate, and zirconium silicate, with thermal expansion coefficients at 500°C of 4.6, 1.0–1.5, 7.5 and 6 (in $10^{-6}K^{-1}$), respectively.

The Sol-Gel Routes[8]

Dipping the substrate in a colloidal solution is a simple but efficient route for making a continuous and homogeneous film of colloidal precursor, which is subsequently heat-treated to transform it to an oxide coating. Mixing water with alcoholic solutions of the alkoxide precursors leads to an hydrolysis-condensation process:

Hydrolysis

$M(OR)_x + h\ H_2O \rightarrow M(OR)_{x-h} + h\ ROH$

Condensation

$\equiv M\text{-}OH + RO\text{-}M \rightarrow\ \equiv M\text{-}O\text{-}M\equiv + ROH$

or: $\equiv M\text{-}OH + HO\text{-}M\equiv\ \rightarrow\ \equiv M\text{-}O\text{-}M\equiv + H_2O$

Multicomponent materials need special care to avoid phase separation or preferential segregation. When alkoxides with different hydrolysis-condensation kinetic rates are used (for instance $Si(OR)_4$ and $Zr(OR)_4$), it is necessary to pre-hydrolyze the material that is the less sensitive to water (e.g., $Si(OR)_4$).[9] Moreover, adding water often leads to oxide precipitation, as is the case for Al-, Ti-, and Zr-alkoxides. The chemical modification of the metal alkoxides using chelating agents such as β-diketone or β-keto-ester can be used to control the hydrolysis-condensation process.[10-13] Chelating groups, less labile than alkoxy groups, decrease the growth rate of the oxide network. We used ethylacetoacetate and acetylacetone as chelating agents.

The starting materials were tetraethoxysilane ($Si(OEt)_4$) (*Aldrich*), aluminum sec-butoxide modified with ethylacetoacetate ($Al(OBu^s)_2(etac)$) (*Alfa*), hydrated magnesium acetate ($Mg(OAc)_2 \cdot 4H_2O$) (*Prolabo*), titanium isopropoxide ($Ti(OPr^i)_4$) (*Aldrich*), and zirconium n-propoxide ($Zr(OPr^n)_4$) in n-propanol (*Aldrich*).

After hydrolysis, the mixtures have an oxide content of 2 wt.%. Figure 1 is a flow chart of the sol-gel routes.

The Substrates

In their as-processed state, the substrates exhibit a rough surface. They were degreased (organic solvent, with ultrasonic agitation) and carefully polished before dip-coating.

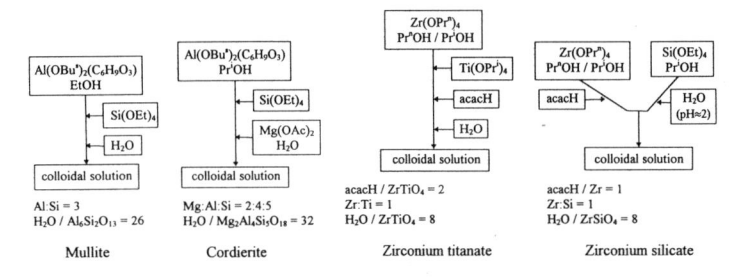

Figure 1. Flow-chart of the sol-gel routes.

The Dip-Coating Process

Three rules must be obeyed to prepare a good coating, namely:

- Using a solution with proper viscosity.
- Using a solution whose stability is well adapted to the process.
- Working with a well-polished substrate.

Solutions with 2 wt.% of equivalent oxide match the first two requirements, and the polishing treatment we used matches the third one. After dipping, the substrate must be carefully extracted from the solution.

Thermal Treatments

First, we studied the powdered materials in air (5°C·min^{-1} up to T_M, then 5 hours at T_M). Second, we studied the coatings, using air for lower temperatures (5°C·min^{-1} up to 800°C), then nitrogen for higher temperatures (5°C·min^{-1} up to 1000°C, then 2 hours at 1000°C).

Characterization

Thermal treatments on xerogels were followed using DTA-TG (5°C·min^{-1}, air flow); crystallization was studied by XRD (Kα-Cu radiation); substrates and coatings were observed by scanning electron microscopy (SEM), and their composition was characterized by EDX. MAS-NMR was used to get more information on poorly crystallized materials. A Bruker MSL 400 spectrometer (79.5 MHz) was used to record ^{29}Si MAS-NMR spectra, and ^{27}Al spectra were recorded using either the same spectrometer (104.3 MHz) or a Bruker MSL 300 MHz (78.2 MHz). Samples were spun at ≈ 5 kHz for ^{29}Si experiments and at 10 to 15 kHz for ^{27}Al experiments. Pulse widths of 2 and 1 µs were used for ^{29}Si and ^{27}Al spectra, respectively. Recycle delays were 60 seconds and 1 second, respectively. Records were made of 500 to 800 scans for ^{29}Si spectra and 600 or 1740 scans for Al spectra. Chemical shifts are referred to tetramethylsilane (TMS) and acidic aqueous solution of aluminum nitrate.

RESULTS AND DISCUSSION

Powdered Materials

Mullite. The gel was dried at 80°C for 24 hours, then finely ground. DTA-TG experiments (Figure 2) show weight loss and endothermal, then exothermal, effects at

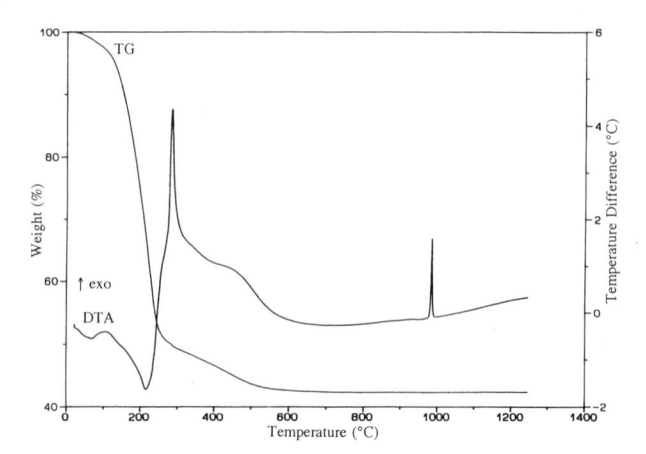

Figure 2. DTA-TG, mullite precursor, air.

temperatures up to 600°C, which are due first to vaporization of alcohol and water, then to decomposition-combustion of organic residues. At 985°C, there is a large exothermic peak, with no weight change, due to the crystallization of a mullitic phase.

The XRD diagrams (Figure 3) show that the powder remains amorphous up to around 900°C; then the mullite crystallizes. At ≈ 1200°C, the main peak (d = 3.40 Å) splits into two peaks, which is characteristic of orthorhombic mullite.

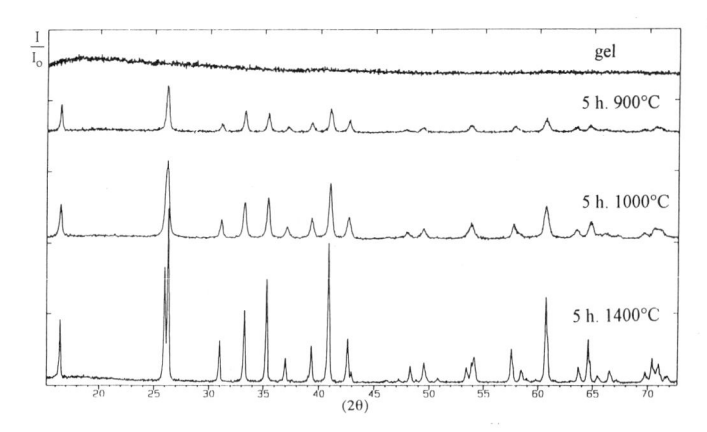

Figure 3. Room-temperature XRD, mullite precursor, treated at temperatures as indicated, air.

The thermal evolution from gel to crystalline product (obtained at 1400°C) was followed by ^{29}Si and ^{27}Al MAS-NMR. The ^{27}Al NMR allows determination of the symmetry of Al sites (sixfold-, fourfold-, or fivefold-coordinated atoms). The ^{27}Al MAS NMR spectra (Figure 4) show a typical evolution for sol-gel synthesis of mullite.[14-17] The spectrum of the gel presents two signals: a small one at about 57 ppm and a large peak whose maximum intensity is at around 3 ppm. These are due to tetrahedral and octahedral environments, respectively. As temperature increases, a new signal appears at around 30 ppm, due to fivefold-coordinated Al. This signal increases with temperature, reaches its maximum at 800°C, then decreases and disappears just when the material crystallizes. The tetrahedral sites take more importance as temperature increases. At 1000°C and 1400°C, two signals subsist, one at around 0 ppm, due to octahedral sites, and another one, wide and complex, located between 45 and 59 ppm, due to tetrahedral sites.[16,18,19]

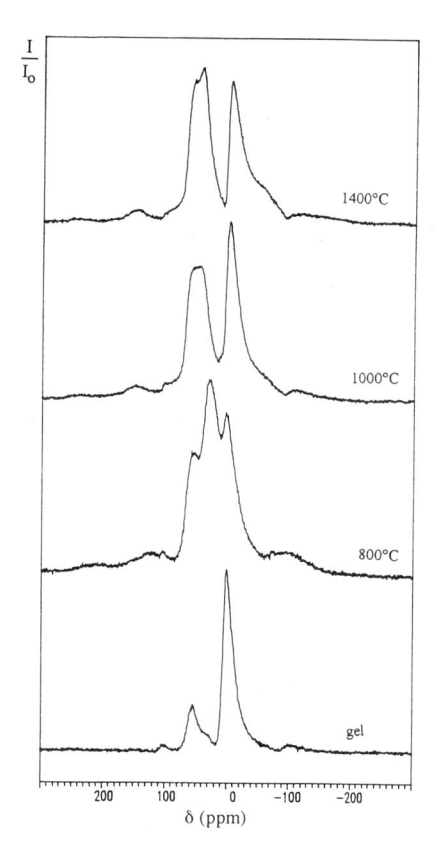

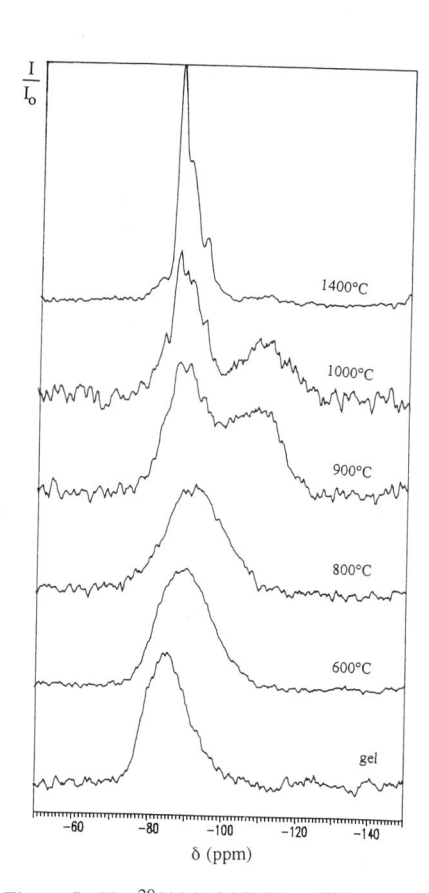

Figure 4. The ^{27}Al MAS NMR, mullite precursor.

Figure 5. The ^{29}Si MAS NMR, mullite precursor.

In this study, ^{29}Si MAS NMR experiments (Figure 5) were useful to investigate non- or poorly-crystallized powders. The spectrum of the gel, with a broad signal centered around −84 ppm, is typical of amorphous material; but it is different from the spectrum of silica gel (usually centered around −100/−110 ppm, due to $Q_{3,0}/Q_{4,0}$ species). This signal can be due to either poorly condensed or more condensed species, but with Si-O-Al bonds ($Q_{3,n}$ and $Q_{4,m}$). In fact, the gel seems to be a mixture of species containing Si-O-Si bonds with species containing Si-O-Si and Si-O-Al bonds. After heating to 900°C and 1000°C, the spectra show a complex signal at around −87 ppm (due to $Q_{4,4}$ species in a mullite-type structure)[18] and a broad signal, at −110 ppm (due to amorphous silica). Once heating reaches 1400°C, the typical spectrum of mullite can be observed, with main peaks located at −86.9, −89.4, and −94.2 ppm. The Al/Si ratio in the crystalline phase was determined with the Winfit[20] decomposition program. For the phase that crystallizes at 900°C, the Al/Si ratio is about 5.8. However, this ratio decreases as temperature increases: it is ≈ 4.7 after treatment at 1000°C and ≈ 3 in the nearly stoichiometric mullite that is obtained at 1400°C.

NMR experiments show that amorphous silica is still present when the 985°C mullitic phase crystallizes. As temperature increases, the silica reacts and the crystalline phase enriches in silicon, therefore decreasing the Al/Si ratio. These results are in agreement with literature data[19,21] concerning the relation between the chemical composition of mullite and the tetragonal-to-orthorhombic transformation.

Cordierite. Results are similar to those described elsewhere.[22] DTA-TG experiments show that, after the usual pyrolysis effects, crystallization occurs at about 980°C. XRD experiments show that a treatment at 850°C for 5 hours leads to metastable μ-cordierite, which transforms to stable, hexagonal α-cordierite at ≈ 1000°C. An exothermic peak at 1080°C is attributed to the crystallization of amorphous residues to α-cordierite.

Zirconium Titanate. DTA-TG curves show the usual effects up to 500 to 600°C, then a complex exothermic peak around 700°C, with weight loss associated with the decomposition of the tightly bonded Zr-hydroxyl groups ($(Zr,Ti)O_{2-x}(OH)_{2x}$), which is followed by the crystallization of $ZrTiO_4$, as mentioned in Munoz-Aguado et al.[23] Chemical analyses carried out on materials heat treated at 500°C show the presence of OH⁻ groups and confirm that the Zr/Ti ratio is equal to unity. XRD experiments show that the xerogel remains amorphous up to 600°C. Above 600°C, orthorhombic $ZrTiO_4$ begins to develop. At higher temperatures, the XRD peaks become narrower; $ZrTiO_4$ is still the only crystalline phase that is observed.

Zirconium Silicate. As for the other compounds, the first events seen by DTA-TG are due to alcohol and water loss and pyrolysis of organic residues (Figure 6). An exothermic peak at 907°C is associated with crystallization of cubic zirconia, as confirmed by XRD. It is known that the amorphous gel can crystallize to cubic or tetragonal zirconia,[24,25] especially when ZrO_2 coexists with amorphous silica.[26] Progressive weight loss due to dehydration continues up to 1200°C, and an exothermic peak is observed at 1208°C on the DTA curve. After heating at 1350°C, zircon ($ZrSiO_4$) coexists with monoclinic zirconia and, at 1450°C, zircon is the main crystalline phase.

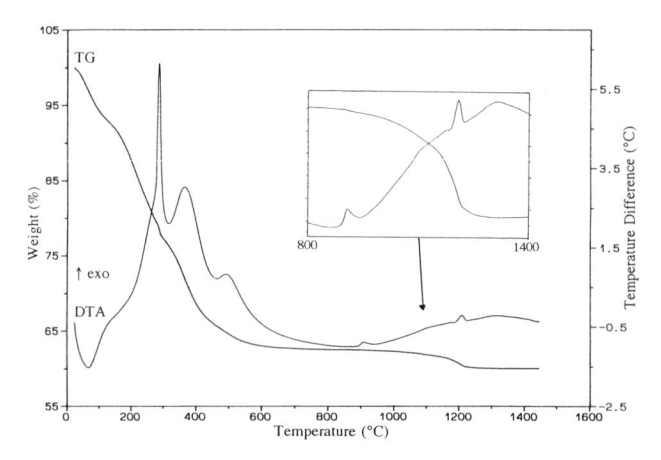

Figure 6. DTA-TG, ZrO_2-SiO_2 precursor, air.

The attribution of the peak at 1208°C (Figure 6) is not clear. XRD experiments carried out on quenched samples, just before and after the peak, do not show any presence of cristobalite, but they show slight changes in the peaks of ZrO$_2$, which could be interpreted as a transition from c-ZrO$_2$ to t-ZrO$_2$. Consequently, three assumptions can explain the exothermic peak:

- c-ZrO$_2$ → t-ZrO$_2$ transformation.
- Crystallization of residual amorphous zirconia to t-ZrO$_2$.
- Microcrystallization of cristobalite.

The first assumption is supported by literature data,[24] which indicate that the c → t transformation can occur in a very narrow temperature range (5°C) and that it depends on particle size. Moreover, the lattice parameters we found agree with literature data.[24] The second assumption is supported by the fact that the end of dehydration frequently coincides with crystallization of non-crystallized residues, in ex-sol-gel materials. Finally, the third assumption is supported by the fact that the presence of ZrO$_2$ was found[27,28] to inhibit the macro-crystallization of cristobalite.

Coatings

For the four oxides, pyrolysis of organic residues in air is completed below 600°C, which means that the treatments at higher temperatures can be conducted in nitrogen, to avoid the oxidation of SiC fibers. It was verified that the change from air to nitrogen has no noticeable influence on the crystallization phenomenon. Heat treatments were carried out in air from room temperature to 800°C (heating rate of 5°C·min^{-1}), with a 30 minute soaking at 800°C, then the heat treatments were continued in nitrogen from 800 to 1000°C. The maximum temperature that the LAS matrix can reasonably endure is 1000°C.

The contraction-dilation changes associated with pyrolysis effects, crystallization, and thermal-expansion mismatch between coating and substrate result in diffuse microcracking within the coatings (Figure 7). However, the aim is to avoid the detachment of the coating

Mullite

Zirconium silicate

Zirconium titanate

Cordierite

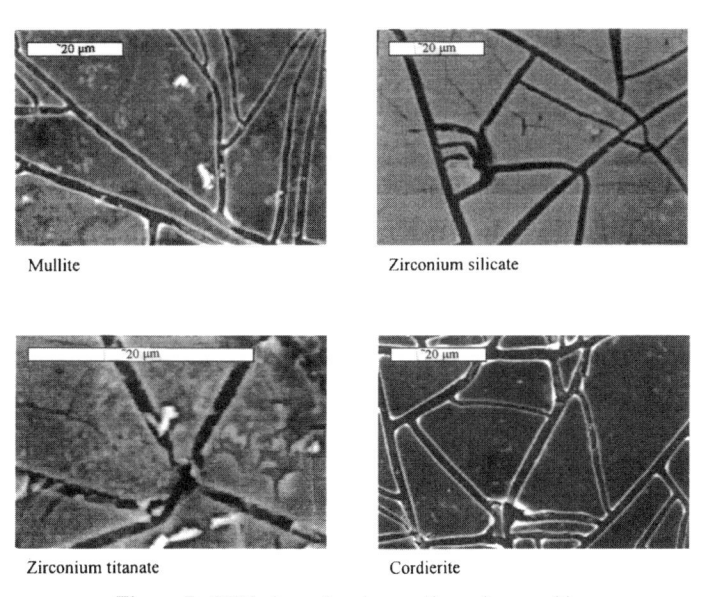

Figure 7. SEM views showing coating microcracking.

from the substrate. Microcracking can be useful for this purpose, by relaxing residual stress. Micrographs of cuts perpendicular to the surface (not presented here) show that the coating thickness varies from 0.2 µm (for cordierite) to 0.9 µm (for ZrO_2-SiO_2 and ZrO_2-TiO_2 materials). Ex-gel coatings are very thin in comparison with plasma-sprayed coatings, but they exhibit much better homogeneity. Moreover, an "n-stage process" can allow the preparation of "n-time thicker" coatings, and ex-gel coatings can be used as underlayer for thicker coatings (e.g. plasma-sprayed coatings).

Oxidation. The composites experience three sorts of oxidation, related to SiC fibers, C, and NbC, depending on temperature, exposition duration, and oxygen partial pressure:

- *SiC fibers* experience[6,29] active oxidation with weight loss, leading to $SiO(g)$ and $CO(g)$, and passive oxidation with weight gain, leading to $CO(g)$ or $CO_2(g)$ and $SiO_2(s)$, which forms a silica coating on the fiber surface. This coating limits the diffusion of oxygen and CO.

- *Carbon*: free carbon is present in the fiber, and graphite is present in the matrix, due to the pollution that is introduced during the preparation stage.

- *NbC*: Nb_2O_5 additives react with C to form NbC.[30] A two-stage degradation of the composite occurs: first as an oxidation of NbC (with weight gain),[31] then as a reaction of Nb_2O_5 with MgO to give magnocolombite ($MgNb_2O_6$). This reaction is associated with a volume expansion, which degrades the composite.

The behavior of non-coated composites and coated composites in oxidizing atmosphere was studied by XRD, SEM, and TG.

For non-coated composites in nitrogen atmosphere, there are no noticeable changes, even at 1000°C. In contrast, samples that are treated in air at 600°C for 30 minutes show evidence of bleaching. At increasing temperatures, the bleaching develops and the sample dilates.

XRD experiments show that the main crystalline phases in the non-oxidized material are β-spodumene (main phase), graphite, NbC, SiC, and MgO. After oxidation, graphite and NbC disappear, and $MgNb_2O_6$ forms. Cristobalite is detected for treatments in air at over 1000°C.

SEM and EDX (not shown here) show that heat treatments in air from RT to 800°C and in nitrogen from 800 to 1000°C degrade the fiber/matrix interface and lead to microcracked zones, with Nb- and Mg-rich granules (mean size ≈ 5 µm). Silica was also detected on the fiber surface.

The TG experiments under dynamic treatment (Figure 8) show a weight loss (due to decarburation), then a weight gain (silica formation) at over 600°C. For isothermal treatments (Figure 9), strong decarburation (weight loss) is observable during the 600°C isotherm. At 800°C, active oxidation is followed by passive oxidation, which leads to a weight gain. At 1000°C, fluctuations between weight loss and weight gain are the mark of coexistence of active oxidation with passive oxidation.

When coated composites are submitted to the same treatments as the non-coated composites, they do not show any bleaching, and XRD experiments cannot detect magnocolombite. However, the sample coated with $ZrTiO_4$ exhibits a partial oxidation of NbC to NbO. SEM and EDX cannot detect Nb- or Mg-rich granules. TG experiments on two-layer cordierite- and mullite-coated samples show that the oxidation rate of cordierite-coated materials is higher than that of mullite-coated ones (Figures 10(a) and 10(b)). While the non-coated samples suffered weight loss ≥ 2.4 wt.% at 600°C and ≥ 1.5 wt.% at 800°C, there is no weight loss for the coated samples. However, there is a weight gain: after 130 hours at 600°C, it is ≈ 1.0 % for cordierite and ≈ 1.1 % for mullite, and after 60 hours at 800°C, it is ≈ 0.9 % for both coatings.

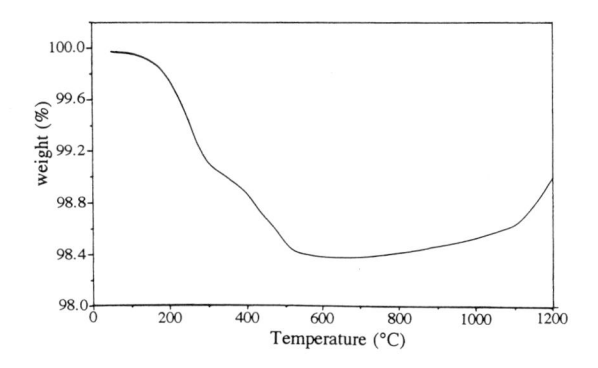

Figure 8. TG curves, non-coated material, dynamic heat treatment, air.

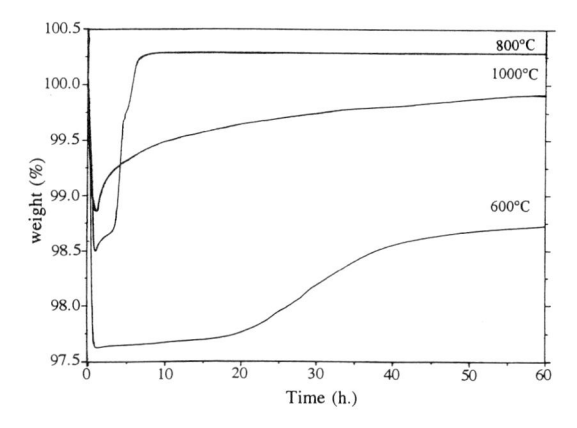

Figure 9. TG curves, non-coated material, isothermal heat treatment, air.

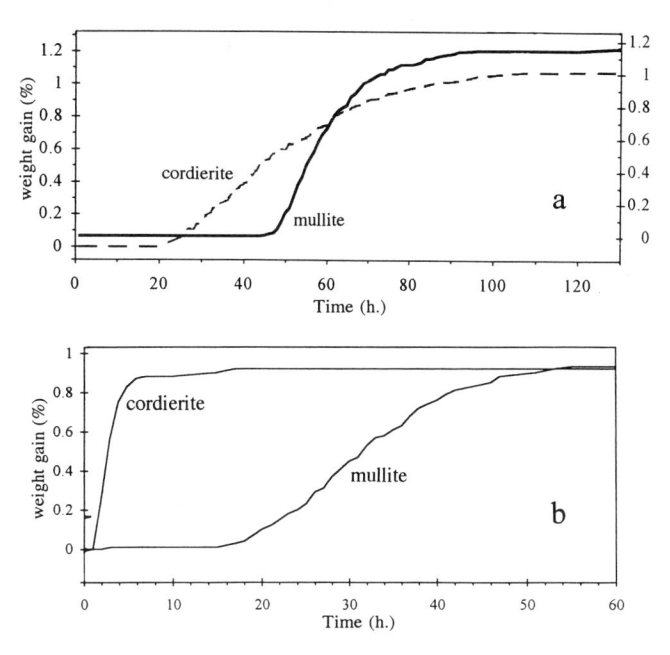

Figure 10. TG curves, cordierite- and mullite-coated materials, isothermal heat treatments, air. (a) T = 600°C (top). (b) T= 800°C (bottom).

The presence of a coating will delay the time when oxidation is detectable from a weight change (Figures 10(a) and 10(b)). At 800°C, this time is close to zero for the non-coated composites, whereas it is about 15 hours for the mullite-coated composites. The cordierite coating was found to be less efficient than the mullite coating.

Abrasion. Exposure to dusty environments (for instance, when a plane flies in a sandstorm) can damage jet-engine parts that are impacted by high-velocity, solid particles.[32,33] Slight abrasion degrades the surface polish of parts, which disturbs the aerodynamic performance of the turbine. Severe abrasion leads to pits and microcracks, which can initiate corrosion and, in the worst cases, failure.

Abrasion tests can be carried out by placing a target in a high-velocity gas stream charged with solid particles. The abrasion resistance is quantified by either the relative weight loss of the target ($\varepsilon = \delta m/m_0$) or the penetration depth (δl). The main parameters of the test are:

- Particle hardness. Alumina and quartz are commonly used, the former being more aggressive and the latter being more representative of real situations.

- Particle shape and size. Angular particles are more aggressive than rounded ones. The size is a critical parameter: fine particles (d < 20 µm) are generally deflected from the target, whereas coarse particles (d >> 20 µm) have a marked influence.

- Particle velocity. For jet engines, the velocities are of hundreds of meters per second.

- Impact angle (α), which can vary from 0 to 90°. For brittle materials, $\varepsilon \propto \sin^2\alpha$.

- Temperature. The material properties generally decrease as temperature increases and, therefore, ε increases with T. However, other effects can play a role, for instance the fact that the ratio between particle velocity and gas velocity varies with temperature. For materials sensitive to oxidation, oxidation interacts with abrasion. This is generally detrimental, but may be beneficial by formation of a coating of silica (in the case of SiC).

The abrasion bench[7] uses a burner with kerosene and air, which allows gas velocity up to Mach 0.99. The test parameters were:

- Quartz (d ≈ 50 µm) as abrasive particles.
- Particle velocity ≈ 170 m·s^{-1}.
- Gas velocity ≈ 500 m·s^{-1}.
- Gas temperature of 450°C or 600°C.
- Impact angle of 30° or 45°.

Table 1 shows the results of tests conducted with 3 g of abrasives (impact angle of 30°, 450°C), for two- and five-layer coated materials.

SEM micrographs (not presented here) show that all the two-layer coatings are perforated, even if cordierite and $ZrTiO_4$ behave better than mullite and ZrO_2-SiO_2. For the five-layer coatings, the two first materials, which are not perforated, are superior to the two others, which exhibit holes, pits, cracking and spalling.

Tests with a greater mass of abrasive (5 or 8 g instead of 3 g) or greater impact angle (45° instead of 30°) show that $ZrTiO_4$ behaves better than cordierite. Tests conducted at 600°C confirm this superiority. Figure 11 shows the improvement provided by a ten-layer $ZrTiO_4$ coating, compared to non-coated composite. The abrasion weight loss of $ZrTiO_4$-coated materials is similar to that of metallic, titanium-based alloys (TA6V).

Table 1: Abrasion weight loss ($\varepsilon = \delta m/m_0$) and penetration depth (δl) determined on non-coated and coated composites (abrasive mass: 3 g, impact angle: 30°, T: 450°C).

Material	Relative weight loss	Penetration depth (µm)
Non-coated composite	$10 \cdot 10^{-3}$	17
2-layer mullite coating	$12 \cdot 10^{-3}$	
5-layer mullite coating	$7 \cdot 10^{-3}$	5
2-layer cordierite coating	$6 \cdot 10^{-3}$	
5-layer cordierite coating	$3 \cdot 10^{-3}$	0.5
2-layer ZrTiO$_4$ coating	$5 \cdot 10^{-3}$	
5-layer ZrTiO$_4$ coating	$7 \cdot 10^{-3}$	0.5
2-layer ZrO$_2$-SiO$_2$ coating	$10 \cdot 10^{-3}$	
5-layer ZrO$_2$-SiO$_2$ coating	$6 \cdot 10^{-3}$	6.5

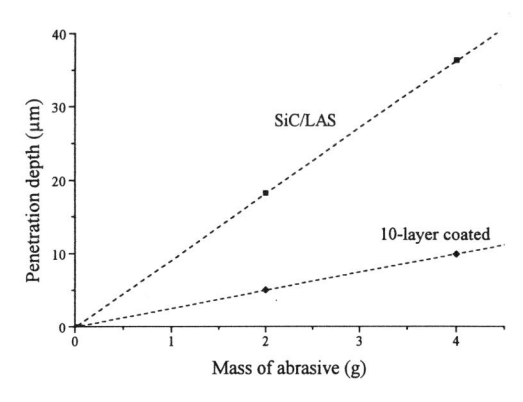

Figure 11. Penetration depth (µm) of 50 µm quartz particles vs. projected-abrasive mass (g). Test at 600°C.

CONCLUSION

The sol-gel route allows oxide coatings to be deposited on SiC-LAS composites. Among the four oxides studied, mullite is the best for oxidation resistance, and zirconium titanate is best for abrasion resistance. In contrast with mullite, which is frequently cited in literature, we have not found any mention of ZrTiO$_4$ as an anti-abrasive coating. The results of the present study show that the general opinion that hard coatings provide better protection than soft ones, all other things being equal, is wrong. Mullite and ZrO$_2$-SiO$_2$ are harder than the composite; but cordierite exhibits similar hardness, and ZrTiO$_4$ is softer. Actually, the dynamics of impact of high-velocity particles on a coated material are complex, involving not only hardness but also other material properties, such as strength, toughness, elastic moduli, damping capacity, acoustic impedance, and interface characteristics. In the absence of a comprehensive theory to link all parameters, the choice of a coating is made primarily by a process of trial and error.

ACKNOWLEDGMENTS

The study was supported by SNECMA (Evry, France). The authors gratefully acknowledge Drs. G. Gauthier and Y. Honnorat for their help.

REFERENCES

1. G. Geiger, Ceramic coatings, *Amer. Ceram. Soc. Bulletin,* **71**:1470 (1992).
2. J. Y. Hsu and R. F. Speyer, Interfacial phenomenology of SiC fibre reinforced $Li_2O \cdot Al_2O_3 \cdot 6SiO_2$ glass-ceramic composites, *J. Mater. Sci.,* **27**:374 (1992).
3. Y. Hasegawa, Synthesis of continuous silicon carbide fibre. Part 6. Pyrolysis process of cured polycarbosilane fibre and structure of SiC fibre, *J. Mater. Sci.,* **24**:1177 (1989).
4. G. Simon and A. R. Bunsell, Mechanical and structural characterization of the Nicalon silicon carbide fibre, *J. Mater. Sci.,* **19**:3649 (1984).
5. G. Simon and A. R. Bunsell, Creep behaviour and structural characterization at high temperature of Nicalon SiC fibres, *J. Mater. Sci.,* **19**:3658 (1984).
6. H. Kim and A. J. Moorhead, Strength of Nicalon silicon carbide fibers exposed to high-temperature gaseous environments, *J. Am. Ceram. Soc.,* **74**:666 (1991).
7. SNECMA, Evry, France.
8. C. J. Brinker and G. W. Scherer, *Sol-Gel Science: the Physics and Chemistry of Sol-Gel Processing,* Academic Press, Boston (1990).
9. B. E. Yoldas, Preparation of glasses and ceramics from metal-organic compounds, *J. Mater. Sci.,* **12**:1203 (1977).
10. C. Sanchez, J. Livage, M. Henry, and F. Babonneau, Chemical modification of alkoxide precursors, *J. Non-Cryst. Solids,* **100**:65 (1988).
11. A. Leaustic, F. Babonneau, and J. Livage, Structural investigation of the hydrolysis-condensation process of titanium alkoxides $Ti(OR)_4$ (OR = OPr^i, OEt) modified by acetylacetone. 1. Study of the alkoxide modification, *Chem. Mat.,* **1**:240(1989).
12. F. Babonneau, L. Coury, and J. Livage, Aluminum sec-butoxide modified with ethylacetoacetate: an attractive precursor for the sol-gel synthesis of ceramics, *J. Non-Cryst. Solids,* **121**:153 (1990).
13. L. Bonhomme-Coury, F. Babonneau, and J. Livage, Investigation of the sol-gel chemistry of ethylacetoacetate modified aluminum sec-butoxide, *J. Sol-Gel Sci. and Technology,* **3**:157 (1994).
14. J. Sanz, I. Sobrados, A. L. Cavalieri, P. Pena, S. de Aza, and J. S. Moya, Structural changes induced on mullite precursors by thermal treatment: a ^{27}Al MAS-NMR investigation, *J. Am. Ceram. Soc.,* **74**:2398 (1991).
15. H. Schneider, L. Merwin, and A. Sebald, Mullite formation from non-crystalline precursors, *J. Mater. Sci.,* **27**:805 (1992).
16. C. Gérardin, S. Sundaresan, J. Benziger, and A. Navrotsky, Structural investigation and energetics of mullite formation from sol-gel precursors, *Chem. Mater.,* **6**:160 (1994).
17. I. Jaymes, A. Douy, D. Massiot, and J. P. Busnel, Synthesis of a mullite precursor from aluminum nitrate and tetraethoxysilane via aqueous homogeneous precipitation: a ^{27}Al and ^{29}Si liquid- and solid-state NMR spectroscopic study, *J. Am. Ceram. Soc.,* **78**:2648 (1995).
18. L. H. Merwin, A. Sebald, H. Rager, and H. Schneider, ^{29}Si and ^{27}Al MAS NMR spectroscopy of mullite, *Phys. Chem., Minerals* **18**:47 (1991).

19. I. Jaymes, A. Douy, P. Florian, D. Massiot, and J. P. Coutures, New synthesis of mullite. Structural evolution study by ^{17}O, ^{27}Al and ^{29}Si MAS NMR spectroscopy, *J. Sol-Gel Sci. and Technology,* **2**:367 (1994).
20. D. Massiot, H. Thiele, and A. Germanus, Programme WinFit, *Bruker Report,* **140**:43 (1994).
21. K. Okada, Y. Hoshi, and N. Otsuka, Formation reaction of mullite from SiO_2-Al_2O_3 xerogels, *J. Mater. Sci. Lett.,* **5**:1315 (1986).
22. L. Bonhomme-Coury, F. Babonneau, and J. Livage, Comparative study of various sol-gel preparations of cordierite using ^{27}Al and ^{29}Si liquid- and solid-state NMR spectroscopy, *Chem. Mater.,* **5**:323 (1993).
23. M.J. Munoz-Aguado, M. Gregorkiewitz, and A. Larbot, Sol-gel synthesis of the binary oxide $(Zr,Ti)O_2$ from the alkoxides and acetic acid in alcoholic medium, *Mat. Res. Bull.,* **27**:87 (1992).
24. K. S. Mazdiyasni, C. T. Lynch, and J. S. Smith, Metastable transitions of zirconium oxide obtained from decomposition of alkoxides, *J. Amer. Ceram. Soc.,* **49**:286 (1966).
25. B. E. Yoldas, Effect of variations in polymerized oxides on sintering and crystalline transformations, *J. Amer. Ceram. Soc.,* **65**:387 (1982).
26. B. E. Yoldas, Zirconium oxides formed by hydrolytic condensation of alkoxides and parameters that affect their morphology, *J. Mater. Sci.,* **21**:1080 (1986).
27. Xiaoming Li and P. F. Johnson, Crystallization and phase transformation of ZrO_2-SiO_2 gels, in: *Mat. Res. Soc. Symp. Proc.,* vol. 180 p. 355, Material Research Society (1990).
28. J. Campaniello, E. M. Rabinovich, P. Berthet, A. Revcolevschi, and N. A. Kopylov, Phase transformations in ZrO_2-SiO_2 gels, in: *Mat. Res. Soc. Symp. Proc.,* vol. 180 p. 541, Material Research Society (1990).
29. W. L. Vaughn, H. G. Maahs, Active-to-passive transition in the oxidation of silicon carbide and silicon nitride in air, *J. Am. Ceram. Soc.,* **73**:1540 (1990).
30. K. M. Prewo, J. J. Brennan, and G. K. Layden, Fiber reinforced glasses and glass-ceramics for high performance applications, *Ceramic Bulletin,* **65**:305 (1986).
31. E. Bischoff, M. Rühle, O. Sbaizero, and A. G. Evans, Microstructural studies of the interfacial zone of a SiC-fiber-reinforced lithium aluminum silicate glass-ceramic, *J. Am. Ceram. Soc.,* **72**:741 (1989).
32. T. Wakeman and W. Tabakoff, Erosion behavior in a simulated jet engine environment, *J. Aircraft,* **16**:828 (1979).
33. G. P. Tilly, Erosion caused by impact of solid particles, in: *Treatise on Material Science and Technology,* vol. 13, Academic Press (1979).

MODELING THE STRUCTURE OF GLASSES AND CERAMIC GRAIN BOUNDARIES (AN INTERSTITIAL APPROACH)

James F. Shackelford

Department of Chemical Engineering and Materials Science
University of California
Davis, CA 95616

ABSTRACT

The characterization of atomic-scale structure has been a central issue in glass science since the pioneering work of Zachariasen and Warren in the 1930's. The understanding of this subject was greatly expanded by the development of amorphous metals in the 1960's. Grain boundary structure in crystalline metals and ceramics is closely related to the structure of corresponding glasses. A useful perspective on structure in these various materials systems is provided by concentrating on their interstitial geometry. The nature of the distribution of interstitial size in silicate glasses has been characterized by experimental methods (such as gas transport measurements) and by computer modeling exercises (such as molecular dynamics). The recent discoveries of medium-range ordering in silicate glasses are best defined in terms of interstitial structure. In order to intelligently design specific glass or interfacial structures at the atomic-scale, it is necessary to fully understand the fundamental structural building blocks which are available. In this regard, the total number of polyhedral "cages" that comprise interstitial structure has been defined. The number available for nonmetallic glasses and grain boundaries (126) is, of course, much larger than that for metals (8). A more compact set can be defined for nonmetals (44 = 28 convex polyhedra + 16 prisms and antiprisms), by considering only "simple" polyhedra which can not be further dissected.

INTRODUCTION

In the second conference in this series, Shackelford and Masaryk[1] introduced the concept of using gas transport as a tool to characterize the interstitial structure of glasses and later used that technique to define the interstitial structure in vitreous silica.[2] In the third conference in this series, Shackelford[3] discussed the use of gas transport as a tool for characterizing the interstitial structure of interfacial phases. Interstitial structure is being used increasingly for the characterization of materials. As will be seen in the next section, the nature of the structure of amorphous metals has traditionally been defined in terms of canonical (or Bernal) holes, a limited set of polyhedra (formed by connecting the centers of adjacent atoms).[4,5] Similarly, the nature of grain boundary structure in metals involves stackings of these relatively simple polyhedra.[6,7] The novel metal structures known as quasicrystals[8,9] are defined in terms of the symmetry of a polyhedron, the icosahedron. The range of fullerene materials includes numerous, molecular cages of carbon atoms with and without interstitial metal atoms. The "simplest" of these structures is the C_{60} molecule with the geometry of the truncated icosahedron.[10] In the field of glass science, neutron diffraction experiments have provided evidence of the medium-range ordering of calcium-centered oxygen octahedra in calcium silicate glass.[11,12]

In order to model the structure of glasses and ceramic grain boundaries in a more quantitative way, it is useful to be able to quantify the geometry of the possible polyhedra which represent the interstitial structure in these systems. The purpose of this paper in the fourth conference in this series is to define the range of interstitial (polyhedral) geometries. Consistent with the theme of this conference, such quantitative information can allow precise modeling and characterization and, therefore, "control at the atomic level."

INTERSTICES IN METALS

For the case of a metallic glass or grain boundary modeled as a stacking of equal-sized spheres, the structural possibilities are quite limited. Non-directional metallic bonding effectively dictates that adjacent, equal spheres will stack together with interstices too small to accommodate another sphere of that size. By joining the centers of atoms clustered around a given interstice, one forms a convex polyhedron characteristic of that interstice. Ashby, et.al.[6] identified the set of eight possible polyhedra as a refinement of the Bernal holes[5] originally used to describe the structure of liquids. In a stacking of spheres with radius, R, polyhedra range in size from the tetrahedron (with an interstice able to accommodate a sphere with radius 0.225 R without distortion) to the icosahedron (able to accommodate a sphere with radius 0.90 R). One should note that the icosahedron has played a central role in the study of quasicrystals, with the 2-, 3-, and especially 5-fold symmetry accounting for their electron diffraction patterns.[8,9]

Metallic Glasses

A good example of the nature of the overall distribution of interstitial size in a metallic glass is provided by the computer simulations of Finney and Wallace.[4] Figure 1 gives an example generated with a hard sphere potential. A log-normal best-fit is shown superimposed on the histogram of Finney and Wallace. As originally argued by Shackelford and Masaryk,[2] the log-normal distribution is appropriate for interstitials in noncrystalline solids, given the analogy to powder particle statistics. Both systems represent the random subdivision of space. It is also interesting to note Figure 2 which is the result of relaxing the structure of Figure 1 by using a soft, rather than hard, potential. In this case, the range of interstitial sizes breaks up into a bimodal distribution. The two peaks indicated by this relaxed structure correspond to the size of the tetrahedal and octahedral sites. The predominance of tetrahedra and octahedra is reminiscent of the fact that these two polyhedra account for all interstitial space in a close-packed metal, with the tetrahedra and octahedra present in a 2:1 ratio. Each of the smaller peaks in the relaxed structure of Figure 2 are also skewed, similar to Figure 1, and each can be fit by a log-normal curve.

It is difficult to know if the bimodal result of the relaxed structure of Figure 2 is indicative of a more realistic interstitial distribution or representative of an approach to crystallization. In either case, the range of sizes represented by each peak is indicative of substantial distortion of the tetrahedral and octahedral polyhedra. In Figure 1, on the other hand, the wide range of interstitial site sizes would suggest that a wide range of polyhedra would be present.

Metallic Grain Boundaries

Ashby, et.al.[6] effectively demonstrated the benefits of representing grain boundary structure in metals in terms of a stacking of polyhedra. Another especially elegant example was presented by Fitzsimmons and Sass[7] in characterizing a $\Sigma 13$ twist boundary in gold. Figure 3 shows an example from Ashby, et.al., viz. a $\Sigma 5$ tilt boundary in an fcc structure. It is important to note that the fcc lattice itself is exclusively a nesting of tetrahedral and octahedral interstices. By contrast, the boundary region is composed of slightly distorted capped trigonal prisms in a regular and repeating arrangement.

INTERSTICES IN NON-METALS

In order to effectively model non-metallic materials as a stacking of polyhedra, it is important to know the possible polyhedra which might appear in such structures. The presence of directional, covalent bonding in traditional noncrystalline solids such as oxide glasses allows for a substantially larger set of interstices than in the relatively close-packed metallically bonded solids.

The following treatment is taken from Shackelford[13] and a more detailed discussion is available there. In order to identify the range of interstitial polyhedra, one needs to identify an appropriate set of geometrical conditions. Analogous to the set of Bernal holes for amorphous metals, a reasonable set for nonmetals would be composed of all constructable polyhedra with n-membered faces. A reasonable range of faces would be 3 $\leq n \leq 10$. A triangular face would, of course, be a lower limit. A 10-membered ring represents a very large structure in an oxide glass[14] and is the largest face in a constructable convex polyhedron with regular faces.[15] In general, limiting the polyhedra to regular faces is not an unreasonable simplification. For example, bond lengths (face edges) will be roughly constant in oxides, and, in general, moderate face distortion would not change the topology of a given polyhedron.

Given the ground rules above, one can systematically catalog the possible polyhedra for non-metallic solids. The results are summarized in Table 1. The general set would include the 5 Platonic (regular) solids. A detailed listing of these five solids along with all other polyhedra is given by Shackelford.[13] The 13 Archimedean (semi-regular) solids would also be included. There are also 8 each prisms and antiprisms for $3 \leq n \leq 10$ rings. Figure 4 illustrates this for the case of n = 7 faces.

As noted in Table 1, there are an additional 92 polyhedra which can be constructed within the initial guidelines of this section. Zalgaller[15] has identified all possible convex polyhedra with regular faces, and Table 1 appropriately labels the additional 92 polyhedra as "Zalgaller solids." Many of these polyhedra, as seen in Shackelford,[13] are "composite" or "compound," i.e., combinations of smaller polyhedra and/or prisms. The "Zalgaller solids" bring the total number of polyhedra to 126.

A further simplification of the canonical hole set is possible by a closer inspection of Zalgaller's comprehensive list of convex polyhedra.[15] By considering only "simple" polyhedra, i.e., those which can not be dissected into two or more other regular-faced polyhedra by a plane(s), the canonical hole set consists of only 28 convex polyhedra,

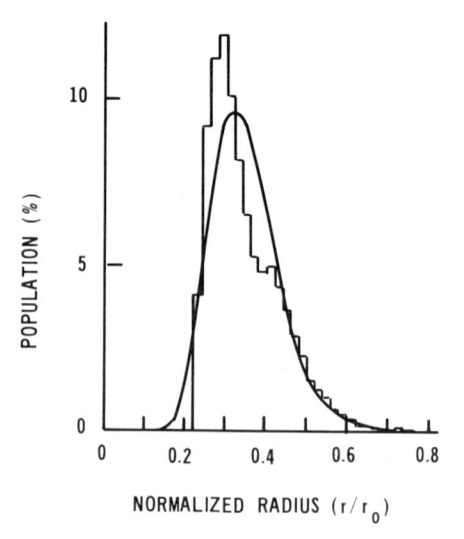

Figure 1. A log-normal best fit curve is superimposed on the histogram of interstitial site sizes in a computer simulation of a metallic glass by Finney and Wallace.[4] In this simulation, a hard sphere potential was used.

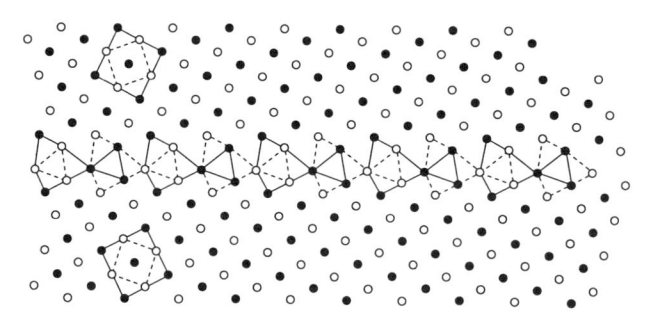

Figure 2. By using a soft sphere potential in the computer simulation of a metallic glass, Finney and Wallace[4] obtained a bimodal distribution of interstitial size, in contrast to Figure 1. The two peaks correspond to the size of tetrahedral and octahedral interstices.

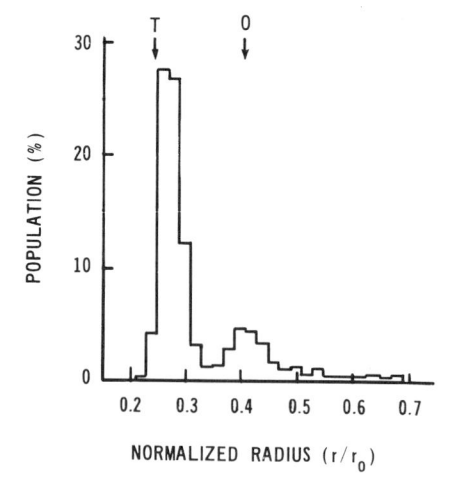

Figure 3. A simulation of a $\Sigma 5$ tilt boundary in an fcc metal by Ashby, et.al.[4] The fcc grain is represented as a stacking of tetrahedral and octahedral polyhedra, while the grain boundary is a regular stacking of capped trigonal prisms.

along with the prisms and antiprisms. The set of 28 simple, convex polyhedra is shown in Figure 5 and includes selected structures from the Platonic, Archimedean, and Zalgaller solids listed by Shackelford.[13] In combination with the 16 prisms and antiprisms, this condensed set represents 28 + 16 = 44 interstitial polyhedra.

Table 1. Canonical Hole Set for Nonmetallic Solids[a]

	5	Platonic solids
	13	Archimedean solids
	8	prisms
	8	antiprisms
	92	"Zalgaller solids"
Total	126	

[a] See text and Shackelford[13] for specific polyhedral descriptions.

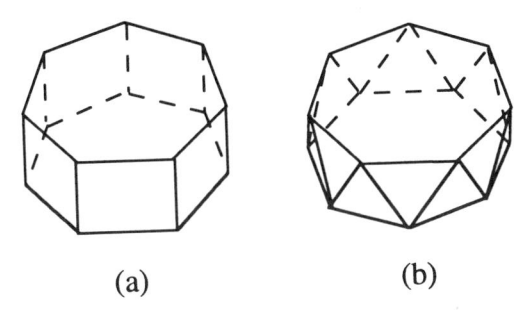

(a)   (b)

Figure 4. A prism and antiprism having a basal face with n = 7 sides. There are 8 each prisms and antiprisms with faces having $3 \leq n \leq 10$ faces. (After Shackelford.[13])

Applications to Glasses and Ceramic Grain Boundaries

Shackelford[13] has characterized the interstitial polyhedra in cristobalite and wollastonite, as the crystalline analogs of vitreous silica and CaO•SiO$_2$, respectively. It would be interesting to do a systematic cataloging of the full range of interstitial polyhedra in computer-generated models of these two glasses. In general, interstitial geometry provides a useful description of the nature of the medium-range structure of noncrystalline solids. One should note that the log-normal distribution of interstitial site size in the amorphous metal of Figure 1 is similar in appearance to that obtained for vitreous silica by both experimental methods[2] and by computer simulation.[16]

Table 1, in general, and Shackelford,[13] in particular, provide a specific and quantitative description of the interstitial structure of noncrystalline solids. As stated at the outset, the goal of this exercise was to provide a realistic set of interstitial geometries representative of the possible structure in glasses and ceramic grain boundaries. With a few simplifying assumptions, the resulting set (126 polyhedra) is clearly much larger than the Bernal holes (8 polyhedra) for metallic glasses and grain boundaries but is nonetheless a tractable number. By not considering compound holes, the number of polyhedra for non-metallic solids is more compact (44 polyhedra). However, one needs to be aware of the likelihood of finding at least some of the compound polyhedra in the larger set introduced in Table 1. Ashby, et.al.[6] illustrated the common occurrence of such compound structures in metallic grain boundaries. Shackelford[13] found examples of compound holes in the wollastonite structure. In fact, the capped trigonal prism which appeared in Figure 3 is a compound hole (a prism with n = 3 plus three M_2 Zalgaller solids)

A general conclusion about characterizing the interstitial structure in this way is that this characterization taken in combination with the description of structural building blocks such as silica tetrahedra and CaO_6 octahedra is a complete structural description of the material. It provides a useful complement to the traditional crystallographic description provided by the unit cell.

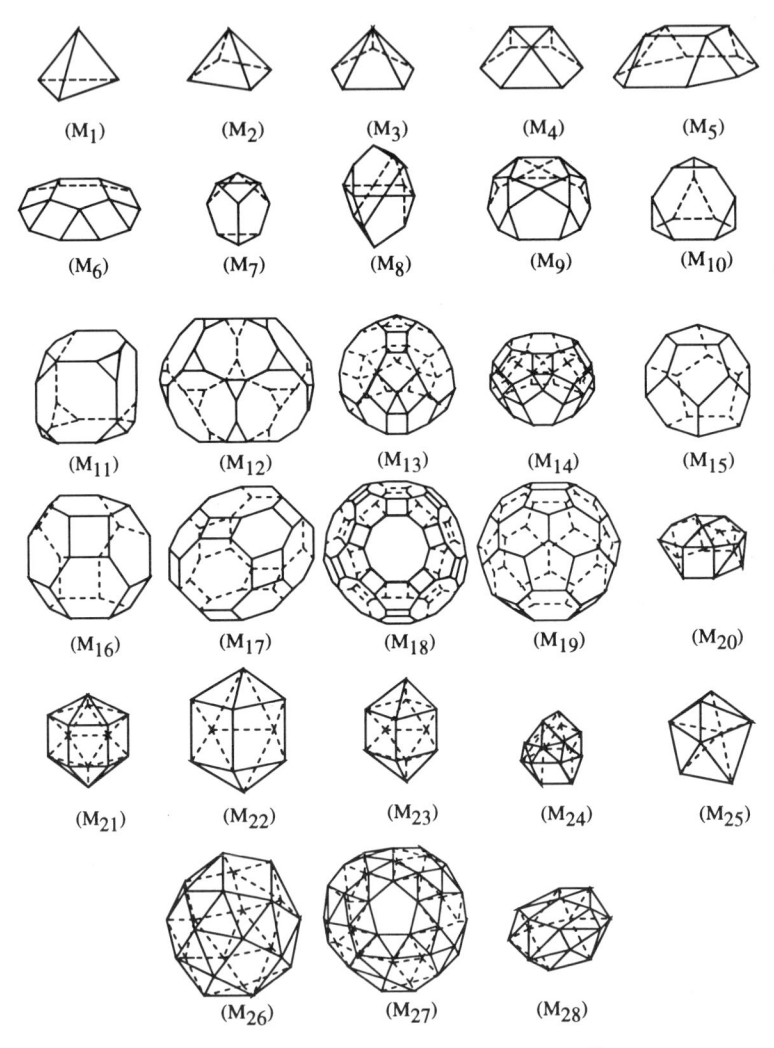

Figure 5. The 28 simple, convex regular polyhedra, as defined by Zalgaller.[15] (After Shackelford.[13]) The notation M_{1-28} is Zalgaller's. In combination with the prisms and antiprisms represented by Figure 4, these 44 solids represent the possible interstitial geometries for nonmetallic glasses and grain boundaries. This is a substantially larger set than the 8 for metals, as described by Ashby, et.al.[6]

Acknowledgments

I am very grateful to Drs. R.L. Loehman, P.F. Green, and R. K. Brow of the Advanced Materials Laboratory (AML), University of New Mexico and the Sandia National Laboratories for hosting a sabbatical leave and facilitating many stimulating discussions in relation to this work.

REFERENCES

[1] J.F. Shackelford and J.S. Masaryk, The gas atom as a microstructural probe for amorphous solids, in: *Ceramic Microstructures '76*, R.M. Fulrath and J.A. Pask, eds., Westview Press, Boulder, CO (1977).

[2] J.F. Shackelford and J.S. Masaryk, The interstitial structure of vitreous silica, *J. Non-Cryst. Solids* 30:127 (1978).

[3] J.F. Shackelford, Gas transport as a tool for structural characterization of interfacial phases, in: *Ceramic Microstructures '86*, J.A. Pask and A.G. Evans, eds., Plenum Press, New York (1987).

[4] J.L. Finney and J. Wallace, Interstice correlation functions; a new sensitive characterization of non-crystalline packed structures, *J. Non-Cryst. Solids* 43:165 (1981).

[5] J.D. Bernal, The structure of liquids, *Proc. R. Soc. London, Ser. A* 280:299 (1964).

[6] M.F. Ashby, F. Spaepen, and D. Williams, The structure of grain boundaries described as a packing of polyhedra, *Acta Metall.* 26:1647 (1978).

[7] M.R. Fitzsimmons and S.L. Sass, The atomic structure of the $\Sigma = 13$ ($\theta = 22.6°$) [001] twist boundary in gold determined using quantitative x-ray diffraction techniques, *Acta Metall.* 37:1009 (1989).

[8] D. Shechtman, et.al., Metallic phase with long-range orientational order and no translational symmetry, *Phys.Rev.Lett.* 53:1951 (1984).

[9] P.J. Steinhardt, Icosahedral solids: a new phase of matter? *Science* 238:1242 (1987).

[10] H.W. Kroto, et.al., New horizons in carbon chemistry and materials science, *MRS Bull.* 19:51 (1994).

[11] P.H. Gaskell, et.al., Medium-range order in the cation distribution of a calcium silicate glass, *Nature* 350:675 (1991).

[12] P.H. Gaskell, Novel aspects of the structure of glasses, in: *Proc. XVI Intl. Cong. on Glass, Vol. 1, Bol.Soc. Esp. Ceram. Vid.* 31-C:25 (1992) 25.

[13] J. F. Shackelford, The interstitial structure of noncrystalline solids, to be published in *J. Non-Cryst. Solids*.

[14] J. F. Shackelford, Triangle rafts - extended Zachariasen schematics for structure modeling, *J.Non-Cryst.Solids* 49:19 (1982).

[15] V.A. Zalgaller, *Convex Polyhedra with Regular Faces*, Consultants Bureau, New York (1969).

[16] S.L. Chan and S. R. Elliott, Theoretical study of the interstice statistics of the oxygen sublattice in vitreous SiO_2, *Phys. Rev. B* 43:4423 (1991).

INTERACTIONS BETWEEN AL$_2$O$_3$ SUBSTRATE AND GLASS MELTS

Kenji Morinaga,[1] Hiromichi Takebe,[1] and Yoshirou Kuromitsu[2]

[1]Department of Materials Science and Technology
Graduate School of Engineering Sciences
Kyushu University
Kasuga, Fukuoka 816, Japan

[2]Central Research Institute
Mitsubishi Materials Corporation
Omiya, Saitama 330, Japan

INTRODUCTION

The quality of bonding between a glass and a ceramic is a factor determining the properties and lifetimes of electronic components such as hybrid ICs, HID lamps, and TV tubes. The bonding is characterized by interfacial reaction and thermal expansion coefficient.

A ceramic component is selected by its appropriate properties for a purpose. On the other hand, as for a glass material, desirable properties can be optimized by the combination of various glass constituents. It is known that there are approximately 50,000 types of glass compositions for present practical uses and over 100,000 types have been studied. This infinite possibility of selective compositions for glass materials is an advantage for the optimization of properties, and a disadvantage for composition selection.

In this study, we summarize the correlation between glass composition and three factors affecting the interfacial morphologies between alumina substrates and glass melts. These three factors are the basicity of glasses, the physical properties of glass melts, and the thermal expansion coefficient of glasses. In the twenty-first century, PbO-free glass frits will be required for the sealing glasses of hybrid ICs; therefore, we have clarified the effect of PbO on the interfacial reactions between alumina substrates and glass melts.

CLASSIFICATION OF BONDING AND OPTIMIZED BONDING MORPHOLOGY

Figure 1 classifies the possible morphologies of bonding between glass melts and ceramics (liquid/solid interfaces). Previous studies[1,2] lead to a conclusion that the best morphology of bonding between a glass melt and an alumina substrate is selective reaction and penetration (SRP). This morphology is used in practice for various glass/ceramic bonding systems. The merit of this morphology is compared with the bonding of two thick papers with adhesive. If one compared the strength achieved when two nondeformable sheets are bonded along the whole plane (as in Fig. 2(a)) or bonded by a process involving selective reaction, as in Fig. 2(b), it is clear that the latter case has the stable interface with a higher strength.

In order to control the interfacial reaction between glass melts and ceramics based on the SRP mechanism, it is necessary to form a compositional inhomogeneity on the ceramic side for selective bonding, to utilize the concept of acid-base for chemical reaction, and to utilize the capillary force with menisci at grain boundaries for penetration. The conditions required of the alumina substrate and glass to realize bonding with the SRP mechanism are discussed in the following chapter.

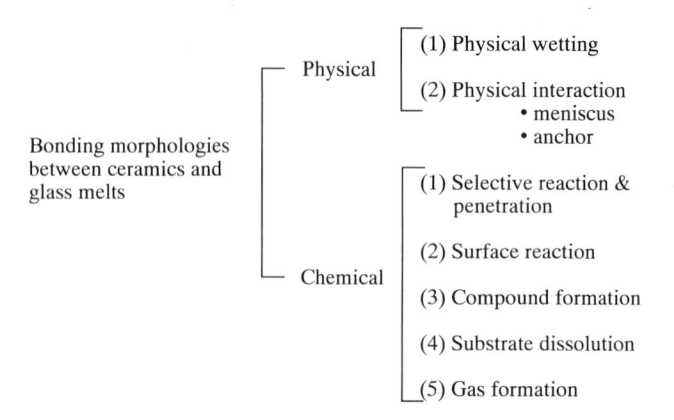

Figure 1. Bonding morphologies between glass melts and ceramics.

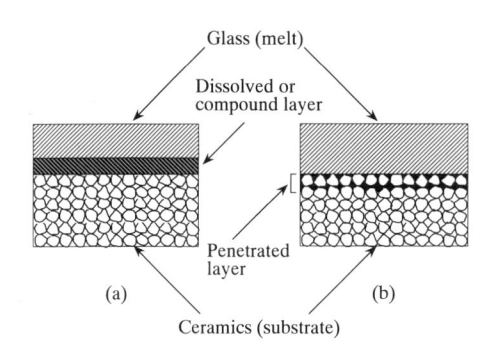

Figure 2. Representative bonding patterns.

SELECTION OF ALUMINA SUBSTRATE AND GLASS MELT COMPOSITION

Selection of Alumina substrate

In order to achieve the bonding morphology shown in Fig. 2 (b), the formation of the inhomogeneity in composition (with microstructure) on the alumina side is required in contrast with homogeneity on the glass melt side. Figure 3 shows the reaction morphologies between a glass melt and two types of alumina substrates: (a) 99%-purity dense alumina and (b) 96%-purity dense alumina with grain boundaries containing SiO_2 and MgO. Figure 3 (a) is a bonding pattern with physical wetting and without any chemical reactions; it delaminates easily at the bonding interface. On the other hand, figure 3 (b) indicates the optimized bonding pattern with the penetrated layer that the SiO_2 constituent at the grain boundaries formed by selectively-reacting with the glass melt.

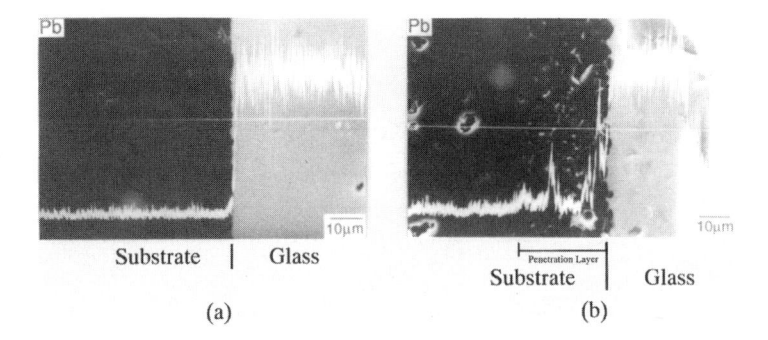

Figure 3. SEM photographs of the interfaces between glass (60PbO-40B$_2$O$_3$(mol%)) and dense alumina substrates ((a) >99%-purity Al$_2$O$_3$, (b) 96%-purity Al$_2$O$_3$)).

This result reveals that the utilization of the 96%-purity alumina substrate with a second-phase at grain boundaries (in this case, SiO$_2$) is desirable for the bonding with glass melt instead of the high-purity dense alumina substrate. That is, a multiphase, inhomogeneous ceramic with an adequate amount of glassy phase leads to the desired bonding morphology between the glass melts and ceramics.

Selection of Glass Composition - The Concept of Basicity -

The properties and chemical reactivities of a glass can be adjusted by its chemical composition. The basicity (acid - base) determines the chemical reactivity at the bonding interface. Oxides are classified into two main types: acidic oxides, e. g., SiO$_2$, B$_2$O$_3$ and basic oxides, e. g., CaO, Na$_2$O, PbO. The driving force for chemical reactions achieving between ceramics and glass melts is due to difference in the basicity of these materials; control of the basicity for each oxide is indispensable for optimizing the interfacial reactions. We proposed a basicity parameter for the glass melt, calculated from composition in terms of coulombic force between the cation and the oxygen ion with the respective normalization of SiO$_2$ and CaO as 0 and 1 as follows.[3]

A parameter A$_i$ expressing the bonding strength between the cation and the oxygen ion based on the coulombic force for glass component i is

$$A_i = \frac{Z_i \times 2}{(r_i + 1.40)^2} \quad (1)$$

where Z$_i$ and r$_i$ are the valency and radius of the cation, and the values 2 and 1.40 are the valency and radius of the oxygen ion, respectively. The reciprocal of A$_i$, B$_i'$, can be a parameter showing oxygen ion activity (basicity) of glass component i

$$B_i' = \frac{1}{A_i} \quad (2)$$

The values of B$_i'$ for SiO$_2$ and CaO are normalized as 0 and 1, respectively, and the parameter B (normalized B$_i'$) of a certain composition for a multicomponent glass is

$$B = \sum_i n_i \cdot B_i \quad (3)$$

where n$_i$ is the cation fraction of glass component i. Table 1 summarizes the values of the basicity parameter B of representative oxides, Al$_2$O$_3$ substrates, and glass melts examined in this study. As an example of the B parameter, BaO has an oxygen ion activity 1.5 times larger than that of CaO; in this way, this approach makes possible the quantitative treatment of basicity.

Table 1. B values of representative oxides, Al₂O₃ substrates, and glass melts examined in this study.

Oxide	B_i	Material	B
K₂O	3.38	Substrate	
Na₂O	2.35	Pure Al₂O₃	0.20
Li₂O	1.72	96% Al₂O₃	
BaO	1.56	Al₂O₃ grain	0.20
PbO	1.31	SiO₂ grain boundary	0.00
SrO	1.27	Glass melt	
CaO	1.00	20PbO · 80B₂O₃	0.17
MgO	0.64	30PbO · 70B₂O₃	0.25
Bi₂O₃	0.51	40PbO · 60B₂O₃	0.35
Al₂O₃	0.20	50PbO · 50B₂O₃	0.45
SnO₂	0.15	60PbO · 40B₂O₃	0.58
GeO₂	0.05	70PbO · 30B₂O₃	0.72
B₂O₃	0.03	50PbO · 50SiO₂	0.65
SiO₂	0.00	60PbO · 40SiO₂	0.78
P₂O₅	-0.10	70PbO · 30SiO₂	0.92
		8K₂O · 92B₂O₃	0.30
		16K₂O · 84B₂O₃	0.56
		24K₂O · 76B₂O₃	0.83
		32K₂O · 68B₂O₃	1.10

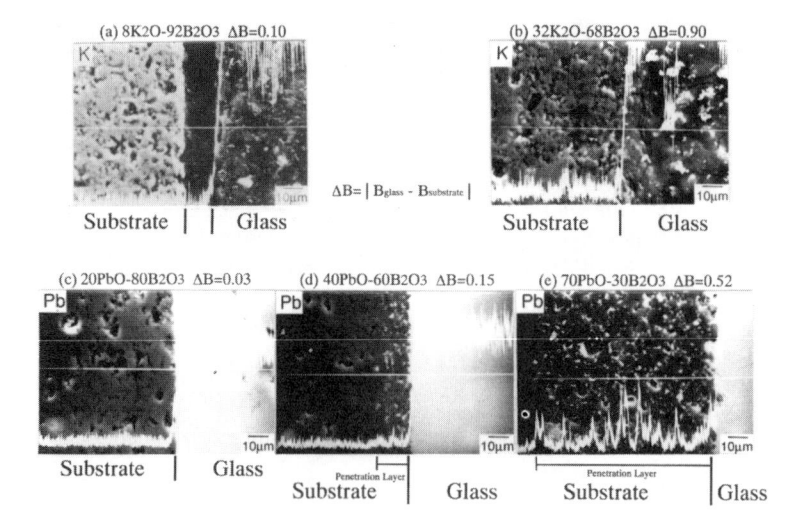

Figure 4. SEM photographs of the interfaces.

The chemical reactions tend to be determined by the B values of the Al$_2$O$_3$ substrate (= 0.20), of the main grain boundary phase SiO$_2$ (= 0.00), and that of multicomponent glass melt, summarized in Table 1. At B < 0.35 of glass melt, the difference of B, ΔB is relatively-small and no chemical reactions are observed; the glass "physically" wets the Al$_2$O$_3$ substrate (Fig. 4 (a), (c)). In the range of 0.35 and 0.60 for B (ΔB = 0.15 - 0.40), the glass melt reacts selectively with a grain boundary phase SiO$_2$, which has a smaller B value than that of Al$_2$O$_3$ (Fig. 4 (d)); at ΔB > 0.40, Al$_2$O$_3$ grains dissolve into the glass melt (Fig. 4 (b), (e)).

There are different interfacial reaction patterns between K$_2$O-B$_2$O$_3$ and PbO-B$_2$O$_3$ glass melt systems. In the system with K$_2$O and the other alkali and alkaline-earth oxides, dissolution of the grain boundaries and of Al$_2$O$_3$ grains into the glass melt is observed. On the contrary, in the system with PbO, the glass melt penetrates into the grain boundaries with chemical reactions; this penetration process is also confirmed in the PbO-SiO$_2$ system.

Factors Affecting Penetration Process

Figure 5 shows the dependence of penetration thickness on reaction time for PbO-B$_2$O$_3$ and PbO-SiO$_2$ glass melts. Two types of alumina substrates: 96% relative density (4% pore) high-purity alumina and 96%-purity dense alumina with second phase grain boundaries were used for experiments. Each glass system has the linear relationship between penetration thickness and reaction time with a slope of 0.5; this indicates that the square of the penetration thickness is proportional to reaction time. In the microstructures of the penetration layers shown in Fig. 4 (d), (e), the glass melt, chemically-reacted with the grain boundary phase SiO$_2$, penetrates into the capillaries formed between the grain boundaries. These results suggest that this penetration phenomenon occurs due to the capillary force at the grain boundaries.

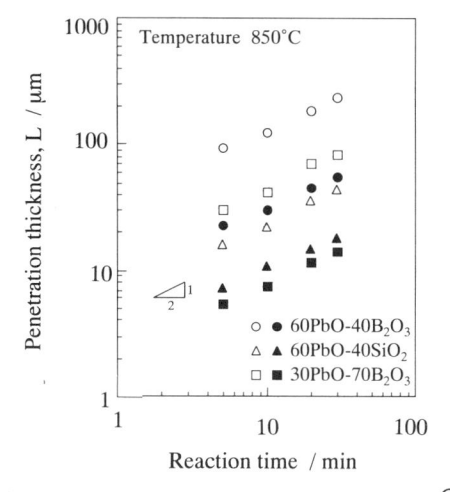

Figure 5. The reaction time dependence of penetration thickness. ○△□ 96% relative density (4% pore) high-purity alumina. ●▲■ 96%-purity dense alumina with the grain boundaries.

The thickness of penetration layer L is generally related to the reaction time t due to the capillary force

$$L^2 = \frac{4K}{r\varepsilon} \times \frac{\gamma \cos\theta}{\eta} \times t \tag{4}$$

where γ and η are the surface tension and viscosity of the melt, θ is the contact angle between melt and substrate, K, r, and ε are the permeability of the porous substrate, the capillary radius, and the pore volume ratio of substrate, respectively. In this experiment, the glass melt spread onto the alumina substrate homogeneously with a contact angle close to zero and each alumina substrate has a microstructure with a constant $A = 4K / r\varepsilon$; therefore, Eq. (4) is simplified as Eq. (5)

$$\frac{L^2}{t} = A \times \frac{\gamma}{\eta} \tag{5}$$

Eq. (5) leads to the conclusion that L^2 is proportional to t; it is consistent with the experimental result of Fig. 5. Table 2 shows the values of γ / η of the glass forming melt. The values of γ / η in the PbO systems indicate one-order larger values, compared with those in the K$_2$O system. Figure 6 shows $(L + d)^2$ versus t for 96% relative density (4% pore) high-purity alumina and L^2 versus t for 96%-purity dense alumina with grain boundaries containing SiO$_2$ and MgO, where L is the penetrated thickness and d is the dissolved thickness of the alumina substrate. As for the 96%-purity alumina substrate, the grain boundaries were selectively-reacted with the glass melts; therefore, no dissolution of alumina grains was observed in this experiment. On the other hand, for the high-purity alumina with 4% pore, alumina grains dissolved into the glass melts. The linear relationship in Fig. 6 suggests that the penetration layer forms due to the mechanism expressed by Eq. (5).

Table 2. Surface tension γ, viscosity η, and γ/η of glass melts.

System	PbO or K$_2$O (mol%)	Surface tension γ (10^{-3} x kg·s^{-2})	Viscosity η (10^{-3} x kg·m^{-1}·s^{-1})	γ/η (10^{-2} x m·s^{-1})
PbO-B$_2$O$_3$	20	82	1023	8
	30	123	468	26
	40	152	214	71
	50	162	79	203
	60	162	39	417
	70	160	25	638
PbO-SiO$_2$	50	208	14791	1
	60	192	501	38
	70	175	83	210
K$_2$O-B$_2$O$_3$	8	98	175	5
	16	128	323	4
	24	155	201	8
	32	166	36	46

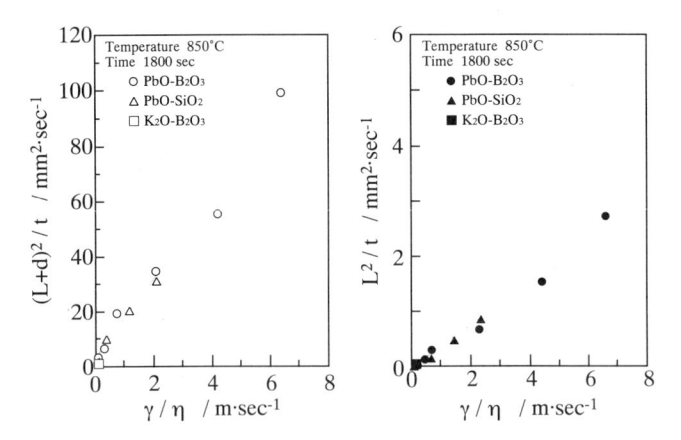

Figure 6. Relationship between $(L + d)^2/t$ and γ/η. ○△□ 96% relative density (4% pore) high-purity alumina. ●▲■ 96%-purity dense alumina with the grain boundaries.

As described above, in order to achieve the bonding morphology based on the optimized SRP mechanism, it is essential that the glass melt with ΔB ($B_{glass} - B_{Al_2O_3}$) = 0.15 - 0.40, which forms a stable glass easily and possesses a large value of γ / η, is selectively-reacted with the grain boundary phase of the alumina substrate.

Estimation of Thermal Expansion Coefficient

The thermal expansion coefficient (TEC) of the glass selected in the previous section should be adjusted close to TEC of Al_2O_3 substrate (70×10^{-7} K^{-1}). We discuss here how the TEC of the glass is predicted from its composition.

The thermal expansion of solid is due to the anharmonic terms of the thermal vibration between atoms (ions). According to classical and statistical thermodynamics, the displacement of the atom from the equilibrium position is expressed as[4]

$$<r - r_{eq}> = \frac{\int_{-\infty}^{+\infty} (r - r_{eq}) \exp(-\frac{\emptyset}{kT}) d(r - r_{eq})}{\int_{-\infty}^{+\infty} \exp(-\frac{\emptyset}{kT}) d(r - r_{eq})} \qquad (6)$$

The potential energy between the cation and the anion in glasses with the ionic bonding is

$$\emptyset_{(r)} = -\frac{Z_+ Z_- e^2}{r} + \frac{B}{r^n} \qquad (7)$$

Coulomb energy (attractive)   Repulsion energy (repulsive)

Eq. (7) is modified using a Taylor series, and then substituted into Eq. (6). Integration yields

$$<r - r_{eq}> = \frac{3kT}{4e^2} \times \frac{1}{\frac{Z_+ Z_-}{r_{eq}^2}(\frac{n-1}{n})} \qquad (8)$$

TEC is expressed from its definition as

$$\alpha \equiv \frac{d}{dT} \times \frac{<r - r_{eq}>}{r_{eq}} = C (\frac{Z_+ Z_-}{r_{eq}})^{-1} \qquad (9)$$

where C is

$$C = \frac{3K}{4e^2} \times \frac{n}{n-1}$$

Eq. (9) is modified for multicomponent glasses

$$\alpha = C \sum_i N_i (\frac{Z_{+i} Z_{-i}}{r_{eq}})^{-1} \qquad (10)$$

where N_i is the cation fraction of constituent i, Z_{+i} and Z_{-i} are the valencies of the cation and oxygen ion, r_{eq} is the mean distance between the cation and the oxygen ion (the sum of the ionic radii).

Figure 7 shows the relationship between the measured α and the calculated value of Eq. (10) from the glass composition of sealing glasses for hybrid IC substrates. α is related linearly to the calculated value. It is possible to predict TEC of a glass from its composition using the compositional parameter of Eq. (10).

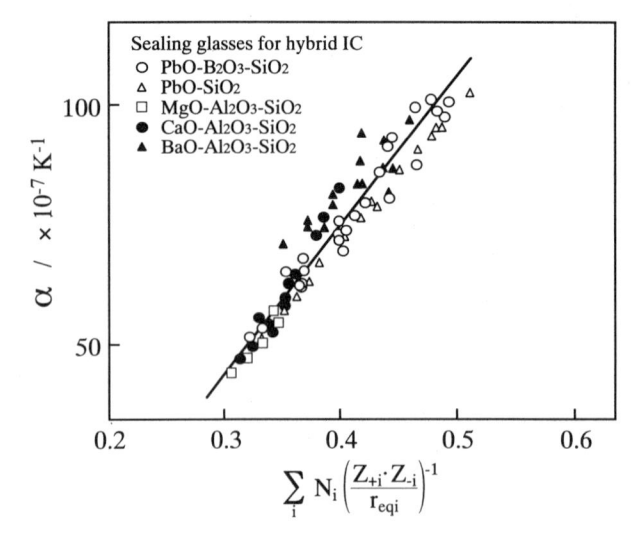

Figure 7. Relationship between the measured thermal expansion coefficient and the calculated compositional parameter of Eq. (10).

Summary

Parameters to control the characteristics of bonded interface between glass melt and ceramics are proposed. In order to achieve the selective-reaction, the 96%-purity dense alumina with the second-phase grain boundaries should be used. A measure of reactivity, that is, the basicity of glass melt can be calculated using the B value. The formation of the penetration layer is determined by the value of γ/η of the glass melt; the penetration thickness is controlled by γ/η and the reaction time. An equation to predict the thermal expansion coefficient from glass composition is proposed; the predicted value is consistent with the measured TEC. PbO affects the chemical reactivity and penetrability of the glass melt, the glass-forming ability and its thermal expansion coefficient. In conclusion, PbO-containing borate, silicate, and borosilicate glasses with excellent glass-forming abilities exhibit high γ/η values due to high surface tension and low viscosity and also have moderate thermal expansion coefficients for alumina substrates; these properties are suitable for sealing glasses for hybrid IC substrates.

REFERENCES

1. K. Harano, K. Yajima, and T. Yamaguchi, Spreading of hybrid IC glass frits on alumina substrates and reaction therewith, Yogyo-Kyokai-Shi. 92: 504 (1984).
2. Y. Kuromitsu, H. Yoshida, and K. Morinaga, Surface treatment of AlN substrate, Procs. of ISHM Europe 89, Hamburg (1989).
3. K. Morinaga, H. Yoshida, and H. Takebe, Compositional dependence of absorption spectra of Ti^{3+} in silicate, borate, and phosphate glasses, J. Am. Ceram. Soc. 77: 3113 (1994).
4. J. Hormadaly, Empirical methods for estimating the linear coefficient of expansion of oxide glasses from their composition, J. Non-Cryst. Solids. 79: 311 (1986).

BIOACTIVE COATINGS ON Ti AND Ti-6Al-4V ALLOYS FOR MEDICAL APPLICATIONS

A. Pazo, E. Saiz, A.P. Tomsia

Lawrence Berkeley National Laboratory
Berkeley, CA, U.S.A.

ABSTRACT

Hydroxyapatite (HA), $Ca_{10}(PO_4)_6(OH)_2$, coatings on metallic medical implants present several advantages: they facilitate joining between the prosthesis and the osseous tissue, and they increase the long term stability and integrity of the implants presently used by medical practitioners and discussed in the literature. The coating techniques provide inadequate adherence of the HA to Ti alloys. More promising coating approaches are based on coating with bioactive glasses that can form HA layers "in vivo".

The present work examines the optimum firing conditions to obtain bioactive silica-based coatings on Ti and Ti-alloys and the mechanism of coating adhesion. Titanium and Ti-6Al-4V alloy were coated with glasses from the system SiO_2-P_2O_5-CaO-MgO-Na_2O-K_2O and fired in different atmospheres. The behavior of glass in body fluids was also studied by "in vitro" tests. Bonding of Ti to silicate glasses in reducing conditions results in poor adhesion due to detrimental interfacial reactions. In air, firing times lower than 1 minute at temperatures between 800 and 850°C yield good bonds accompanied with the controlled formation of a thin interfacial layer of TiO_x. This layer is well adhered to the metal and prevents extensive reaction with the glass. Longer firing times or higher temperatures resulted in dewetting of the molten glass and formation of a thick interfacial oxide layer that adhered poorly to the metal. The addition of small amounts of TiO_2 (3 wt%) broadened the firing range (temperature and time) needed for proper bonding. Similar behavior was observed in a parallel set of experiments performed in N_2, with the result is attributed to the formation of an interfacial TiN_x layer.

INTRODUCTION

The success of surgical and dental implants based on Ti is well documented and proven by considerably clinical experience. This success results, in part, from the phenomenon of osseointegration of coated Ti described by Branemark et al.[1] Coating Ti with hydroxyapatite (HA), $Ca_{10}(PO_4)_6(OH)_2$, facilitates joining between the prosthesis and osseous tissue and increases the long term stability and integrity of implants.

Nevertheless, as noted in a recent review by LeGeros[2], the best method by which to coat metals with apatite remains to be determined.

At present two techniques for HA-Ti coating are used: (i) plasma spraying and (ii) ion sputtering. The first method produces highly textured HA coatings 50 to 200 µm thick. However, it is difficult to control the porosity of a plasma sprayed layer as well as the fraction of HA which decomposes at the high plasma temperature.[3-6] Ion sputtering produces HA layers which are very thin (~1 µm) and dense; however, the lack of open porosity hinders the desired osseointegration.

Both standard coating techniques present common problems: (a) the metal-ceramic bonding is normally a relatively weak physical attachment resulting in frequent spalling, chipping and adherence problems[7-9]; (b) both techniques are "line of sight" methods, therefore preventing the continuous coating of irregular surfaces.

A more promising method involves the use of bioactive (i.e. able to form HA layers in vivo[8]) and biocompatible glasses. These glasses are generally too weak for load-bearing applications, but could be used to improve HA adhesion to Ti by serving as an intermediate layer between the Ti and the HA (or glass/HA mixtures) or even serve alone as HA-producing coatings. However, Hench[8] has indicated that research to develop bioactive glasses that bond well to surgical alloys has been quite complicated for two reasons. The first relates to the way the adherence of glass to metal is conventionally attained. The normal technological procedure of achieving chemical bonding at glass-metals interfaces is to apply the glass to a preoxidized metal.[9] Once hot, the molten glass wets and partially dissolves the oxide. If the glass at the oxide interface immediately becomes saturated because the dissolution rate exceeds its diffusion rate into the bulk glass, then stable chemical bonding is realized.[10] However, this interfacial oxide layer presents some problems. Owing to the open network structure of the bioactive glasses (to enable rapid proton-alkali exchange) the dissolution rate for most oxides in the glass are very fast.[11,12] Moreover, the solution of certain metal oxides in these glasses (including those of Ti and Al) in sufficient concentration during glass/metal bonding can reduce the bioactivity of the glass to zero.[13,14] Thus a careful balance must be struck during processing.

The second problem relates to the high reactivity of Ti, which results in severe chemical reactions at glass-Ti interfaces during processing. These reactions can lead to porous and mechanically weak interfaces, due to the evolution of gaseous reaction products and the formation of brittle, poorly adherent and perhaps highly stressed silicide phases.[15]

The goal of the present work is to examine the optimum processing conditions to obtain bioactive coatings on Ti and Ti-alloys. Key factors for successfully developing such coatings include reducing the reactivity of Ti and the preservation of coating integrity, adherence and bioactivity.

EXPERIMENTAL

Three glasses derived from the SiO_2-Na_2O-K_2O-CaO-MgO-P_2O_5 system were prepared starting from silica sand (99.9%) and reagent grade $NaPO_3$, Na_2CO_3, K_2CO_3, $CaCO_3$ and MgO. The compositions, modified from Bioglass® originally developed by Hench[8], are in Table 1. In order to reduce the coefficient of thermal expansion (CTE) and make it compatible with that of Ti (9.4×10^{-6}/°C), the SiO_2 content was increased and partial substitution of K_2O for Na_2O and MgO for CaO was made. In composition A-6 a small quantity of TiO_2 was added to assess the effects of titania dissolution on bonding and bioactivity.

Table 1. Glass compositions (in wt.%)

Designation	SiO_2	Na_2O	K_2O	CaO	MgO	P_2O_5	TiO_2
Bioglass®	45.0	24.5		24.5		6.0	
A-3	54.5	12.0	4.0	15.0	8.5	6.0	
A-5	56.5	11.0	3.0	15.0	8.5	6.0	
A-6	54.8	10.7	2.9	14.6	8.2	5.8	3.0

The glasses were synthesized by mixing the ingredients in isopropyl alcohol using a high-speed stirrer. After drying (110°C, 48 h), the powder mixture was fired in a Pt crucible for 4 h at 1400°C. The melt was cast into graphite molds to obtain large samples for property evaluation or quenched in a water bath. For coating experiments, the quenched fragments were ground to a particle size of ~10 µm using a planetary mill with silica balls. Thermal and mechanical properties of the glasses were measured by dilatometry and indentation techniques, respectively.

The behaviors of the glasses in body fluid were studied by "in vitro" tests. For this purpose glass plates (10 x 10 x 1 mm) as well as glass powders (~10 µm) were soaked in simulated body fluid[16] (SBF) for different times at a constant temperature of 36.5°C. Samples were then analyzed by X-ray diffraction (XRD) and scanning electron microscopy (SEM). Elemental analysis and X-ray mappings were performed both by energy dispersive (EDS) and wavelength dispersive (WDX) methods. The variation of pH in the SBF was continuously recorded versus time.

Reactivity between HA (99.9% purity obtained by precipitation route[17]) and A-3 glass was studied by XRD and SEM of glass-HA mixtures fired at different temperatures (700-950°C) for times ranging from 5 to 60 minutes.

Ti and Ti-6Al-4V plates (99.9% purity, 10 x 10 x 1 mm) were coated by a glass powder suspension in isopropanol and fired at temperatures between 700-1000°C (heating rate, 40°C/min) in different atmospheres (argon, nitrogen and air). Experiments in air were performed using an Unitek dental. Samples were preheated to 600°C in air; then the oven was evacuated to 100 torr, and simultaneously the sample was heated to the desire peak temperature. The final coating thickness ranged between 25-150 µm. Polished cross sections were examined by optical microscopy and SEM with associated EDS and WDX.

The adherence of each coating was qualitatively evaluated by indentation to reveal the relative crack resistances and weak path fracture. Polished cross-sections of the specimens were subjected to Vickers indentation using loads of 0.5 to 2.2 kg in ambient air.

RESULTS AND DISCUSSION

The bulk properties of the glasses are shown in Table 2. As expected, an increase in SiO_2 content decreases the CTE and increases the softening point of the glass.

"In vitro" tests performed in SBF show that the A-3 glass quickly develops an HA layer on its surface (Fig. 1), while SiO_2-rich compositions (A-5 and A-6) exhibit slower formation of HA. This fact is attributed to the slower pH increase produced when immersing the later in SBF (Fig. 2), since high pH values are required for HA precipitation.[19] Glasses with lower SiO_2 content have a more open network structure which enables a faster proto-alkali exchange and pH increase. Kokubo[20] reported that the compositional and structural characteristics of the HA formed on the surface of a bioactive glass are similar to that of natural HA (human bone). It is thus expected that a strong bond can be formed between the HA grown on the glass surface and the bone.

Table 2. Properties of the glasses

Designation	Glass transition temp. (°C)	Softening point (°C)	CTE (10^{-6}/°C)	Vickers hardness, H_v (GPa)	Young's modulus, E (Gpa)	Toughness, K_{IC} (MPa·m$^{1/2}$)
Bioglass®[18]	---	555	14	---	70	---
A-3	525	575	11.6	3.5-4.0	80-90	0.95-1.07
A-5	558	637	8.5	3.2-3.9	80-90	0.91-1.04
A-6	590	615	---	---	---	---

XRD analyses of HA/A-3 glass mixtures fired at different temperatures do not show new phases after 10 min. firing at T≤800°C. At higher temperatures diopside (CaMgSi$_2$O$_6$) and calcium sodium phosphate (Ca$_6$Na$_3$(PO$_4$)$_5$) form; these are the same phases that result from devitrification of the glass. The results indicate that the glass starts to react with HA at about 850°C with fast reactions occurring above 900°C.

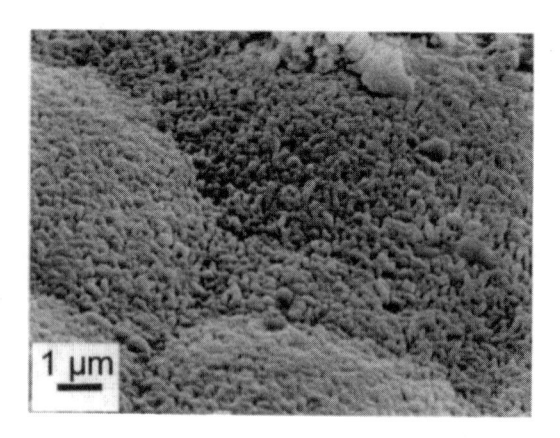

Figure 1. SEM image of A-3 glass plate surface after 21 days in SBF showing HA formation.

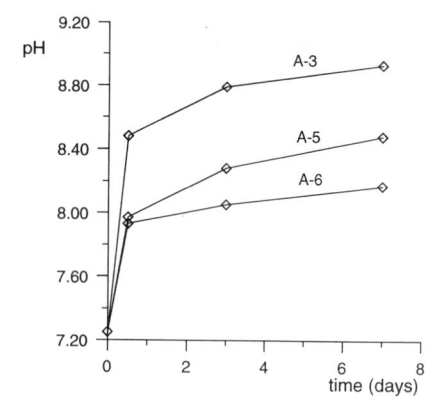

Figure 2. pH evolution after immersion of 3 g of glass powders in 20 ml of SBF.

Figure 3 shows the interface between A-3 glass and a HA grain after 10 minutes at 800°C or 850°C in air. The sample heated at 800°C exhibits no reaction at the interface; conversely, at 850°C the reaction takes place with diopside formation.

Due to the glass dissolution a minimal coating thickness is required to preserve its stability in body fluids. On the other hand, the thicker coatings develop higher thermal stress due to the CTE mismatch. SiO$_2$-rich glasses present a lower CTE (closer to Ti) allowing the fabrication of thicker coatings. However these glasses also present diminished bioactivity. These preliminary studies suggest the existence of an optimum composition range between A-3 and A-5.

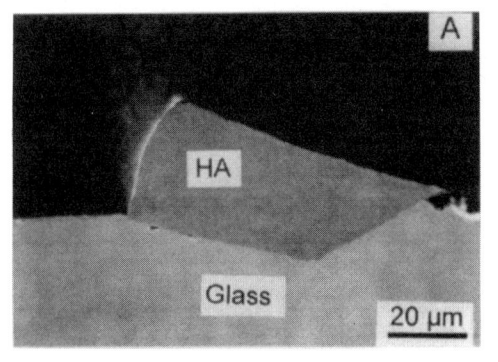

Figure 3. SEM images of the interfaces between A-3 glass and HA fired 10 minutes at 800°C (A) and 850°C (B).

During coating fabrication there is a maximum firing temperature limit determined by the α–β transformation temperature of Ti (880°C) and, in the case of HA-glass mixtures, by the reactivity with HA. Conversely, the glass softening point delimits the minimum processing temperature.

Experiments with glass/Ti assemblies fired in argon produce glass coatings that are fragile and contain cracks (Fig. 4). Bubbles emerging from the liquid glass are observed and indicate that interfacial reactions occur. Ti is capable of reducing SiO_2 and alkaline oxides in the glass, which result in volatilization of oxygen and alkalis and the formation of weakly adherent silicide phases. These reactions include:

$$5Ti + 3SiO_2 = Ti_5Si_3 + 3O_2\uparrow$$
$$8Ti + 3SiO_2 = Ti_5Si_3 + 3TiO_2$$
$$Ti + 2Na_2O = TiO_2 + 4Na°\uparrow$$

The vapor phases resulting from these reactions are responsible for the observed bubbles at the interface. Also the two first reactions form a thin layer of silicide on the Ti surface which is brittle and mechanically incompatible with Ti and glass. These reactions are, therefore, responsible for the mechanically weak seal of silicate glasses to titanium in argon.

Coating experiments in air aim to develop a controlled thin interfacial oxide layer, which adheres to both the metal and glass and prevents extensive reaction between them. Glass coatings fired at temperatures <750°C show poor sintering and adherence. However excellent adhesion is obtained after firings at 800-850°C for times less than 1 minute (Fig. 5). A qualitative assessment of the interfacial strength using indentation cracks showed no interfacial fracture; cracks do not propagate along the interface but rather tend to be driven into the glass (Fig. 6). Coatings with thickness up to 150 µm have been prepared using A-5 and A-6 glasses without any cracking due to thermal expansion mismatch. Glasses with a lower SiO_2 content (e.g., A-3) have higher CTE and consequently require thinner coatings. Future work will determine the optimum thickness for each composition. Higher firing temperatures and times lead to a reaction with associated bubble formation, and dewetting (the glass recedes and forms droplets on the substrate).

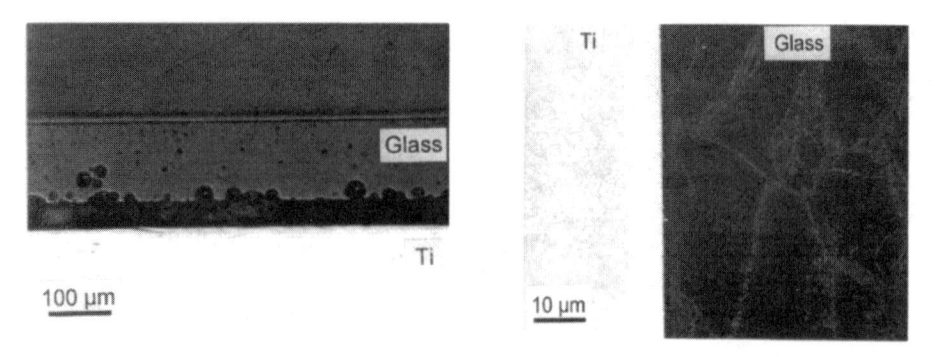

Figure 4. Optical micrograph of a cross-section of A-5 on Ti fired in argon (850°C, 1 min) showing lack of adhesion.

Figure 5. SEM image of a cross-section of A-6 glass on Ti (1min/850°C/air). No reaction products are observed.

Some experiments were conducted entirely in air with slower heating rates and resulted in an excessive oxidation of the titanium before the glass melts. The presence of a thick layer of intergrown TiO_2 (and other Ti-oxides) results in a weak interfacial zone because of physical incompatibility and porosity in the oxide layer (Fig. 7).

Figure 6. Optical micrograph of a cross-section of A-6 glass on Ti (1 min/850°C/air), showing indentation cracks running away from the adherent interface (2.2 kg load, 15 sec., air).

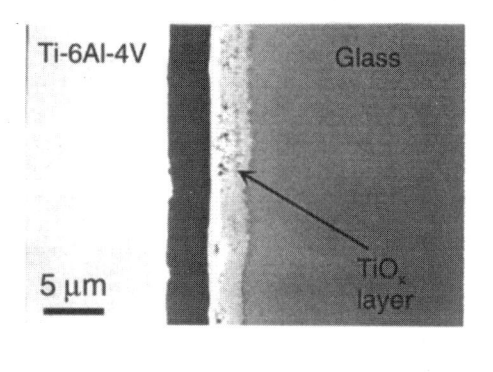

Figure 7. SEM image of an A-3 glass/Ti-6Al-4V interface fired at 960°C for 1 min in air. The TiO_x region is thick and not dense

Figure 8 resumes the effect of temperature and time on the A-5 and A-6 coatings calcined in air. Controlling the substrate oxidation and the dissolution of the interfacial oxide into the glass provides a processing window where coating fabrication is possible. At longer times and temperatures the interfacial oxide layer is completely dissolved and the glass reacts with the metal forming bubbles and silicides. Even if the reaction is prevented the "saturated" glass has a finite contact angle (~35°) on the interfacial layer (Fig. 9) and dewetting takes place. The glass recedes or breaks into droplets (Fig. 10). The addition of small amounts of TiO_2 to the glass reduces the dissolution rate of the interfacial layer, expanding the time-temperature window to enable good bonding. No difference were observed using either pure Ti or Ti alloys as substrates.

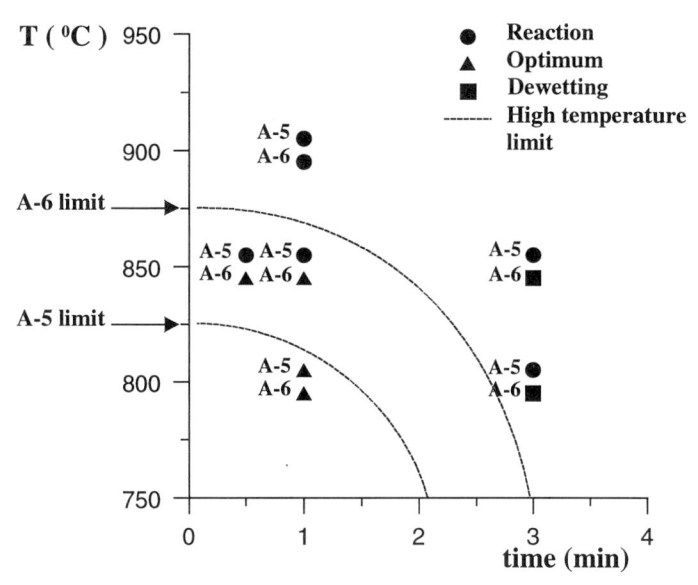

Figure 8. Effect of firing temperature and time in air on coating adherence.

A similar bonding strategy has been attempted by the firing of A-3 glass/Ti assemblies in pure nitrogen.[21] This involves the formation of a physically compatible interfacial Ti_xN layer, which markedly reduces the reactivity and provides good glass-metal adhesion.

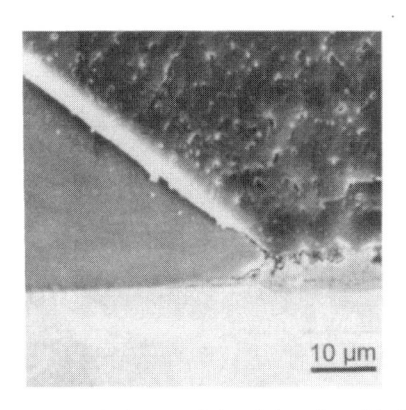

Figure 9. SEM image of the triple point of A-6 on Ti fired at 850°C for 1 min in air.

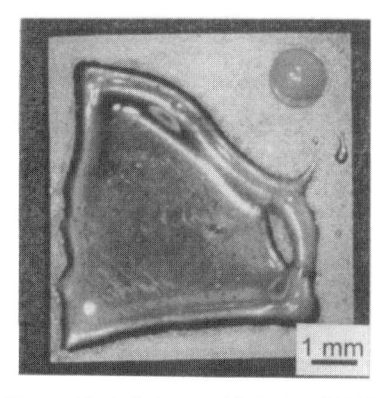

Figure 10. A-6 glass on Ti fired at 800°C for 3 min in air, showing receding of the glass and formation of droplets.

CONCLUSIONS

The results show that by using the SiO_2-Na_2O-K_2O-CaO-MgO-P_2O_5 system, it should be possible to obtain HA-producing glasses with strong adhesion to Ti and also biocompatible glasses that can be used to bond HA layers or to serve as a basis for graded glass/HA composite coatings. At least two workable processes, involving different atmospheres, are clearly applicable for a range of compositions. The first promising glass composition, A-3, has excellent adhesion to Ti, bioactivity and low reactivity with HA under practical thermal conditions. This glass, however, has a higher CTE than desirable and can lead to cracked but not delaminated coatings. Thin coatings neither crack nor delaminate and this composition may still be useful for such coatings. Glasses with higher SiO_2 content have a reduced CTE (A-5) and are adherent, compatible with HA and marginally bioactive. The presence of a small amount of TiO_2 enables a broader range of processing conditions reducing reactivity. These latter compositions, even if not directly bioactive, can be useful for composite or layered glass/HA coatings.

ACKNOWLEDGEMENTS

This work was supported by the supported by NIH/NIDR grant 1R01DE11289

REFERENCES

1. P. I. Branemark, G. Zarb and T. Albrektsson, eds., *Tissue-Integrated Prostheses*, Quintessence Publishing Co., Inc. (1985).
2. R. Z. LeGeros, Calcium phosphate materials in restorative dentistry: a review, *Adv. Dent. Res.,* 2:164 (1988).
3. W. R. Lacefield, E. D. Rigney, L. C. Lucas, J. Ong and J. B. Gantenberg, Sputter deposition of Ca-P coatings onto metallic implants, P. Vincenzini, ed., Elsevier (1991)
4. K. de Groot, Medical applications of calcium phosphate ceramics, *The Cent. Mem. Is. Ceram. Soc. Jap.,* 99:943 (1991).
5. S. Radin and P. Ducheyne, The effect of plasma sprayed induced changes in the characteristics on the in vitro stability of calcium phosphate ceramics, *J. Mat. Sci.,* 3:33 (1992).
6. B. Koch, J. G. C. Wolke and K. de Groot, X-ray diffraction studies on plasma sprayed calcium phosphate-coated implants, *J. Biomed. Mater. Res.,* 24:655 (1990).
7. R. Y. Whitehead, W. R. Lacefield and L. C. Lucas, Structure and integrity of a plasma sprayed hydroxylapatite coating on titanium, *J. Biomed. Mat. Res.,* 27:1501 (1993).
8. L.L. Hench, Bioactive ceramics, in: *Bioceramics: Material Characteristics Versus In Vivo Behaviour*, Annals of the New York Academy of Sciences, Vol.523, P. Ducheyne and J.E. Lemmons, eds., New York, NY (1988).
9. A.I. Andrews. *Porcelain Enamels*, Second Edition, The Garrard Press, Champaign, IL (1961).
10. J. A. Pask, Chemical bonding at glass-to-metal interfaces, in *Technology of Glass, Ceramic, or Glass-Ceramic to Metal Sealing*, MD-Vol.4, W. E. Moddeman, C. W. Merten and D. P. Kramer, eds., The American Society of Mechanical Engineers, New York, N.Y. (1987).
11. L. L. Hench, Glass surfaces, *J. Phys. Colloq.,* **43**:625 (1982).
12. L. L. Hench and D. E. Clark, Physical chemistry of glass-surfaces, *J. Non-Cryst. Solids,* **28**:83 (1978).
13. U. Gross and V. Strunz, The anchoring of glass ceramics of different solubility in the femur of the rat, *J. Med. Mater. Res.,* **14**:607 (1980).
14. U. Gross and V. Strunz, in: *Clinical Applications of Biomaterials,* A. J. C. Lee, T. Albbrektsson and P. Branemark, eds., John Wiley & Sons, New York, N.Y. (1982).
15. F. Feipeng, A. P. Tomsia and J. A. Pask, in: *Proceedings of the 33rd Pacific Coast Regional Meeting of the American Ceramic Society*, San Francisco, CA (1980).
16. J. Gamble, *Chemical Anatomy, Physiology and Pathology of Extracellular Fluid*, Harvard University Press, Cambridge (1967).
17. F. Guitian, R. Conde, C. Santos, A. Pazo and J. Couceiro, Biomateriales ceramicos: síntesis y propiedades de hidroxiapatito y fosfato tricalcico, *Bol. Soc. Esp. Cer. y Vidr.,* **29**:253 (1990).
18. A. Krajewski, A. Ravaglioli, G. De Portu and R. Visani, Physicochemical approach to improving adherence between steel and and bioactive glass, *Am. Ceram. Soc. Bull.,* **64**:679 (1985).
19. P. N. de Aza, F. Guitian, A. Merlos, E. Lora-Tamayo and S. de Aza, Bioceramics-simulated body fuid interfaces: pH and its influence on hydroxyapatite formation, *J. Mat. Sci.: Mat. in Medicine* 7:399 (1996).
20. T. Kokubo, Bonding mechanism of bioactive glass-ceramic A-W to living bone, in: *Handbook of Bioactive Ceramics*, Vol.1, Bioactive Glasses and Glass Ceramics, T. Yamamuro, L. L. Hench and J. Wilson, eds., CRC Press, Boca Raton, FL (1990).
21. A. P. Tomsia, J. S. Moya and F. Guitian, New route for hydroxyapatite coatings on Ti-based human implants, *Scripta Met. Mat.,* **31**:995 (1994).

CHARACTERIZATION OF ALUMINA/SILICON CARBIDE CERAMIC NANOCOMPOSITES

Brian Derby and Martin Sternitzke

Department of Materials
University of Oxford
Parks Road
Oxford, OX1 3PH, UK

INTRODUCTION

Ceramics of an alumina matrix containing SiC reinforcements of approximately 100-300 nm diameter have been reported to have considerably higher strengths than monolithic polycrystalline aluminas. Niihara[1] has reported strength increases from 320 MPa to 1050 MPa and an increase in fracture toughness from 3.1 MPa√m to 4.7 MPa√m. Although strength and possibly toughness increases over those of similar grain size alumina have been reported by a number of workers[2,3,4], maximum strengths of only about 800 MPa have been achieved. In all cases the increase in strength is accompanied by a change in fracture mode from intergranular failure in the monolithic alumina to transgranular failure in the composite.

In brittle materials an increase in strength can normally be attributed either to an increase in fracture toughness of the material or by a decrease in the size and density of fracture initiating flaws. In Niihara's original report[1] the strengthening found in Al_2O_3/SiC nanocomposites was attributed to a reduction in microstructural scale, and hence characteristic flaw dimension, from that of the grain size to a dimension equivalent to the inter-particle spacing. However, using the strength and toughness data reported we find that the defects limiting strength are about 18μm in size, much larger than the SiC particle spacing. Zhao et al. proposed a strengthening mechanism reliant on surface crack blunting[2] to explain the strength increases. This mechanism will be accelerated by high temperature heat treatment and was used to explain the reported increase in strength found with these materials on annealing. Davidge et al.[3] considered that a large increase in strength with an apparently insufficient increase in toughness could be explained by a toughness which increases with crack growth, i.e. R-curve behavior. However, for this to be consistent with published toughness measurements, the R-curve must rise over a very short distance and certainly have reached its plateau by about 100μm crack extension.

Levin et al. have proposed a complex mechanism[5] for strengthening of Al_2O_3/SiC nanocomposites by which crack deflection across the grains is generated by the SiC particles resident on the Al_2O_3 grain boundaries. This change in fracture mode is sufficient to account for an increase in toughness and hence strength. However SiC particles within the grain weaken the composite by inducing a residual tensile stress in the matrix. Hence the strengthening effect of crack deflection is eliminated after reinforcement volume fractions exceed 5%. The model also predicts a particle size effect with the composite toughness strongly dependent on SiC particle size. However, our earlier results[4] have found no such particle size effect, with the toughness measured by indentation remaining constant from SiC particle sizes from 10 nm to 300 nm. Nor have we always found a decrease in strength above 5% SiC[6]. A peak in the fracture strength at some critical volume fraction of SiC is probably an artifact of the processing difficulties encountered with this material as the volume fraction of reinforcement increases.

Many applications for ceramics require good wear resistance and, because the components are often in compressive loading, the strength and toughness of the material are only important in so far as they influence wear resistance. Preliminary studies of the behavior of Al_2O_3/SiC nanocomposites in erosive wear[7] have found a significant increase in wear resistance above that of similar grain size monolithic aluminas. In this study we investigate the wear behavior of ceramic nanocomposites with a range of SiC reinforcement particle sizes and characterize the extent of surface damage after wear and other surface mechanical treatments from measurements of Rayleigh wave velocity made using an acoustic microscope.

EXPERIMENTAL PROCEDURE

The method used for preparing these Al_2O_3/SiC nanocomposites has been explained in detail elsewhere[3,4]. However, in brief, all the composites were fabricated by hot pressing Al_2O_3/SiC powder mixes at 1700°C. AKP53 alumina powder (Sumitomo, Japan), with nominal mean particle size of 200 nm, was used throughout this study. SiC powders were obtained from Lonza (Germany) of grades UF 15 and UF 45; the code number refers to the powder surface area in m^2/g. A finer grade of powder was obtained by sedimenting the UF 45 material in water over a 4 week period. The range of SiC particle sizes achieved are given in Table 1. These powders were mixed by attrition milling in an aqueous media containing a polycarboxylate dispersant (Dispex A40, Allied Colloids, UK). The ultra fine particle SiC was prepared using a novel hybrid process in which Al_2O_3 powders were coated with a polycarbosilane SiC precursor before cross-linking to produce a stable composite powder. These are hot pressed at 1700°C and produce ultra fine SiC particles *in situ*[8]. All nanocomposites were fabricated to contain 5% by volume of SiC. A reference, similar grain size monolithic Al_2O_3, was manufactured using the same Al_2O_3 but now hot pressed at 1500°C to reduce grain growth and obtain a final alumina grain size similar to that found in the nanocomposites. The mean alumina grain size measured in all samples is given in Table 1.

Densities of the materials after processing were measured using Archimedes principle in distilled water. The elastic constants (Young's modulus and Poisson's ratio) were measured using a resonance technique (Grindosonic MK4i, J.W. Lemmens Electronica, Belgium). Four-point bend testing was used to determine the fracture strength of all materials. Beams were machined to size 25 x 2.5 x 2.0 mm (British Standard: BS DD ENV 820 1). The tensile face was polished using a 3μm diamond paste and the tensile edges were also beveled. All tests were carried out on an Instron 1122 load frame at a cross-head displacement speed of 0.5 mm/min using a 4-point bend jig of outer span 20 mm and inner span 10 mm. Hardness was measured using a Vickers diamond pyramid with 5 kg load. Fracture toughness was estimated from the size of surface breaking indentation cracks using the relation of Anstis[9]. Erosive wear was measured using a rotating paddle system schematically illustrated in figure 1. Samples were rotated in an alumina containing slurry and wear was assessed by weight loss; full details of the apparatus are presented in reference 6.

Surface damage in the composites after mechanical treatment was assessed using line focus acoustic microscopy (LFM). In this case the surfaces were prepared using conventional diamond lapping or polishing on a Kent Mark 3 disc polishing machine (Engis Ltd., UK) with four grades of diamond used: 30 μm, 12 μm, 3 μm and 1 μm respectively, to vary the severity of the surface damage. Figure 2 shows a schematic outline of the microscope, the lens produces cylindrical acoustic waves with a frequency of a 225 MHz. These are coupled with the specimen surface using a water droplet and generate surface Rayleigh waves. A full description of the principles of the LFM and the analysis of the theory outlined above can be found in the work of Kushibiki and Cubachi[10]. A complete description of the instrument used in this study can be found in the work of Lawrence[11].

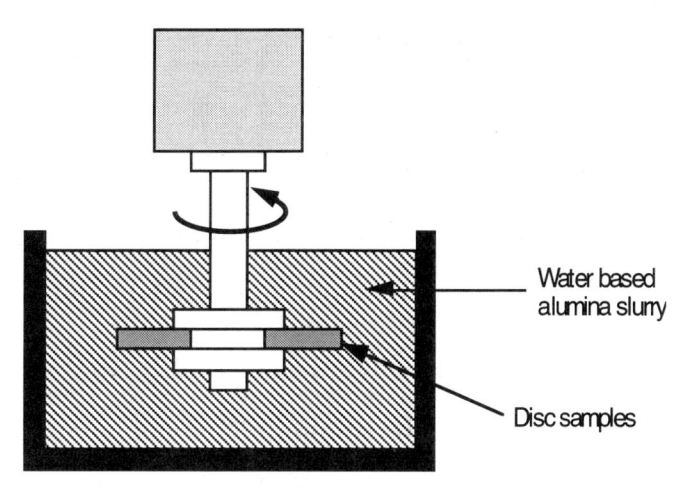

Figure 1. Schematic representation of the wet erosive wear apparatus used in this study.

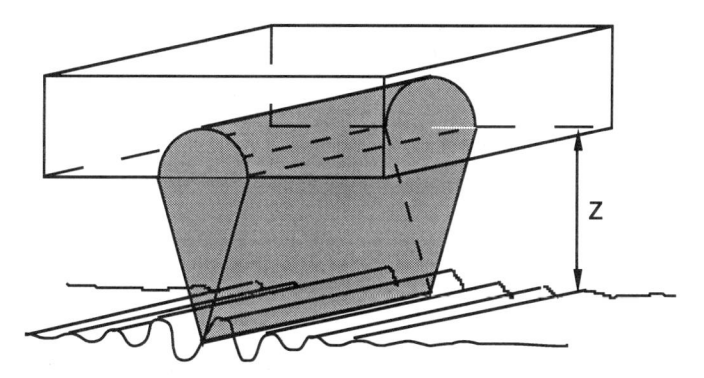

Figure 2. Schematic representation of the line focus acoustic microscope. A cylindrical lens is coupled with the specimen surface and generates Rayleigh waves normal to the cylinder axis.

A plot of the output V of an acoustic lens as a function of the lens/surface separation z is shown in figure 3. This signal $V(z)$ is known as the acoustic material signature. In the samples analyzed $V(z)$ took the form of a maximum at focus ($z = 0$) and gave an approximately exponentially decaying sine wave at increasing negative defocus. The Rayleigh wave velocity (v_R) of the surface being characterized may be determined from the distance between the peaks of the $V(z)$ trace[10]. Rayleigh waves are largely shear in nature and the following semi-empirical relation obtained by Scruby and Drain[12] may be used to relate the Rayleigh wave velocity to the bulk shear wave velocity (v_s).

$$v_s = \left(1.14418 - 0.25771\nu + 0.126617\nu^2\right)v_R \qquad (1)$$

where ν is Poisson's ratio of the material. The shear modulus G can hence be calculated

$$G = v_s^2 \rho \qquad (2)$$

where ρ is the density of the material. These equations assume the material to be elastically

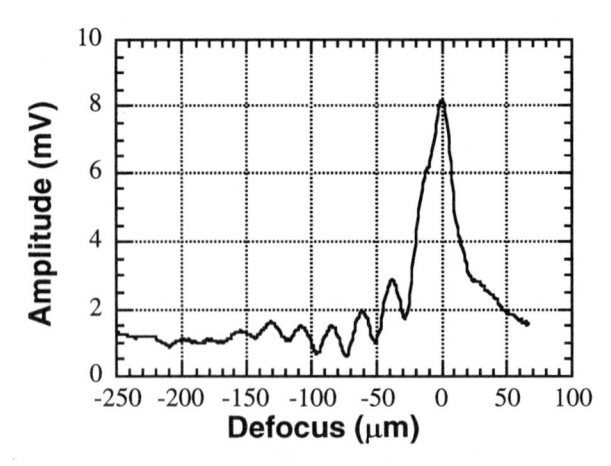

Figure 3. LFM output showing variation in signal amplitude with distance from the surface. Maximum occurs at focus with a monotonically decreasing envelope containing an oscillatory variation.

isotropic which may not be the case for a damaged surface layer but is probably a reasonable approximation.

RESULTS

The mechanical property data are tabulated in Table 1. All the nanocomposites show slightly lower Young's moduli than the monolithic alumina, although minor differences within the composites themselves are within error limits. These probably indicate a small residual porosity in the nanocomposites as we would expect a fully dense nanocomposite to have an elastic modulus about 1% greater than that of alumina because of the higher elastic modulus of SiC. All the nanocomposites are harder than the parent alumina with measured hardness increasing with reinforcement size but these results are probably not significant in view of the error. All nanocomposites were on average stronger than the parent alumina but only the polymer derived material was significantly stronger. Weibull distribution data on these specimens, published elsewhere[4], found that the nanocomposites had very similar Weibull moduli but with their mean strengths displaced to higher values. All the specimens studied had very similar fracture toughness. These results are consistent with previous studies on these material[2,3] except the initial work of Niihara[1] who found much greater strength increases and an increase in toughness.

Figure 4 shows the wear results determined for these materials. The results are broadly in agreement with those reported by Davidge et al.[7] who also found the nanocomposites to have a lower wear rate than measured of monolithic aluminas of equivalent grain size. They only studied nanocomposites of one SiC reinforcement size, comparable to our UF 45 material in that it was made from the same starting powders (but fabricated at another site). The nanocomposites show a lower experimental variation in wear rate with grain size than do the monolithic aluminas and that the size of the SiC reinforcement has little bearing on the results. The worn nanocomposite surfaces showed much reduced grain pull-out over the pure aluminas. This mirrors the well documented change in fracture morphology which is observed in these materials with the nanocomposites failing by transgranular fracture and the monolithic aluminas failing by a mainly intergranular mechanism[1-5].

LFM results from the ceramic surfaces polished using a range of diamond grit sizes are shown in figure 5. This figure shows the Young's modulus measured in the materials, normalized by that measured with the surface polished using a 1 µm diamond grit as the final polish. In all cases there is little reduction in surface elastic modulus as the diamond grit used is increased from 1 µm to 3 µm. With both the 12 µm and 30 µm grit there is a significant reduction in the normalized modulus, with all the nanocomposites showing a

Table 1. Mechanical properties and microstructural data of the nanocomposites and an equivalent grain sized monolithic alumina.

	Alumina	Nanocomposites			
SiC Source		Polymer	Sed. UF 45	UF 45	UF 15
SiC content (wt%)	-	5	5	5	5
SiC particle size (μm)	-	12[a]	55	90	115
Al_2O_3 grain size (μm) (range)	5.0 (0.5-52)	3.5 (0.5-12)	4.0 (0.5-19)	2.1 (0.5-16)	3.5 (0.5-16)
Density (g/cm^3)	3.97	3.86	3.90	3.89	3.92
Young's modulus (GPa)	399	395	394	396	397
Poisson's ratio	0.25	0.24	0.24	0.23	0.25
Hardness (GPa)	17.5±1.6	17.8±1.0	18.3±1.3	18.5±1.5	18.8±1.7
Rupture strength (MPa)	491±63	738±115	549±125	593±95	689±80
Indentation toughness (MPa√m)	3.25±0.27	3.02±0.21	2.76±0.31	3.04±0.39	3.47±0.61

[a] As measured by X-ray line broadening.

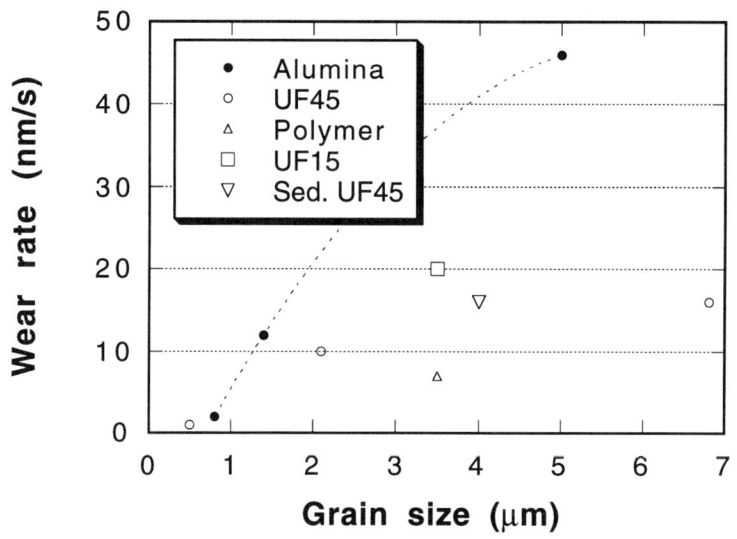

Figure 4. Wear rate as a function of grain size for pure aluminas of different grain size and the ceramic nanocomposites.

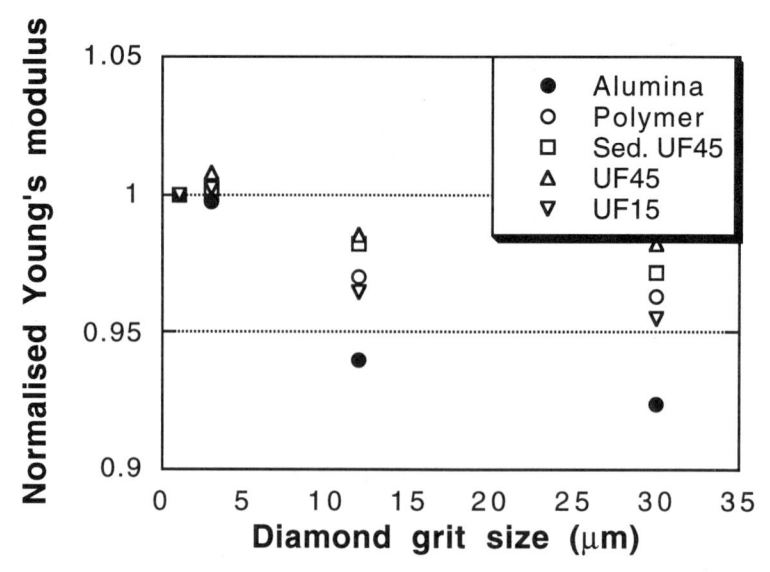

Figure 5. LFM results from polished specimens showing surface damage as a function of final diamond grit size.

smaller reduction than the alumina specimen. We interpret the reduction in elastic modulus with increased diamond grit size as an indication of the extent of surface damage either by grain pull-out or by surface breaking and sub-surface cracks. Thus all the nanocomposites appear to show a significant reduction in the extent of damage induced by polishing when compared with an equivalent grain size alumina. However, there appears to be little variation in the response of the nanocomposites to polishing as a function of SiC particle size with all four materials showing a similar behavior.

DISCUSSION

These results show that there are indeed clear differences in a number of mechanical properties between monolithic alumina and alumina composites containing intragranular SiC particles of 300 nm or smaller. We can confirm, as has been reported by a number of previous studies[1-5], that these materials do show a small yet significant increase in strength above that of monolithic alumina of similar grain size. This is not accompanied by any significant increase in toughness as measured by indentation techniques. Nor is there any measurable R-curve in these materials, at least over a crack length scale of 500 - 5000 μm[13].

In any brittle material the strength of a sample is related to the distribution of critical flaw sizes and the material's intrinsic toughness. Thus any increase in strength must be explained either by an increase in toughness or by a decrease in the size of the maximum flaws present in specimens. In an earlier study[4] we hypothesized that the significant increase in strength was predominantly an effect of a reduced flaw population in these materials. This reduction in flaw population and size range was originally believed to be largely influenced by processing conditions because of an assumed refining effect of the SiC particles on any alumina agglomerates during attrition milling, coupled with their grain pinning effect suppressing large grain growth during the sintering cycle. This can be seen in the grain size range reported in Table 1 with the nanocomposites showing much smaller maximum grain sizes in any sample of grains. A similar conclusion was reached by Zhao *et al.*[2] but they invoked a mechanism by which a surface mechanical treatment was required to produce the increase in strength. Later work by the same group[14] has identified a crack healing mechanism which occurs during intermediate temperature annealing at 1300°C. They proposed that this healing was an important aspect of the increased strength found in these materials and explained the increase in strength found on annealing[1,2].

Our results also suggest that the increase in strength, which is not accompanied by any increase in toughness, is in part caused by a reduction in the surface flaw population because of an increased resistance to contact damage. This increased resistance to damage is clearly seen in the improved wear resistance reported here and elsewhere[7]. It is also seen in the reduction in surface damage after polishing, as measured using the LFM, and in reports elsewhere which comment on the ease at which a good polished finish is achieved with these materials[2]. What is not clear is the mechanism by which this reduction in damage is achieved.

It is reasonable to assume that, because strength is controlled by the largest flaw present in a material, the increase in strength over monolithic alumina of equivalent mean grain size is a result of the suppression of large grain growth in these materials which removes a mechanism for large flaw generation. However it does not explain the reduced wear rate found in these materials. This may be a result of the transition in dominant failure mechanism from intergranular failure in alumina to transgranular failure in the nanocomposites. In the case of polycrystalline alumina the dominant wear mechanism is grain pull-out. Thus the reduction in wear rate with decreasing grain size is a reflection on the reduced unit of wear viz. the volume of a plucked out grain. In the case of the nanocomposites, where grain pull-out is suppressed, wear is probably caused by a complex mechanism of subsurface crack intersection and the wear rate shows neither dependence on grain size nor on SiC inclusion size (figure 4). The wear rate of the nanocomposites is equivalent to that found at alumina grain sizes of $1-2$ µm. We can also hypothesize that this represents the size of wear debris found with nanocomposites and that at alumina grain sizes below this level there will be no additional benefit from the presence of SiC inclusions for wear properties. Unfortunately we do not have LFM results from a range of alumina grain sizes after polishing and so we cannot confirm this hypothesis for this less severe wear environment.

Thus in conclusion we can confirm a number of observations on the mechanical properties of Al_2O_3/SiC nanocomposites. There is a measurable increase in strength above equivalent grain size aluminas. This is possibly caused by a reduction in the flaw population because there is no accompanying increase in material toughness. This may be partly caused by the reduction in size of the largest alumina grains present in the material, presumably because Al_2O_3 grain growth in the material is suppressed. These materials also appear to be more resistant to surface damage as determined by erosive wear and polishing tests and thus may also show reduced surface flaw populations for a given surface treatment. The reason for this resistance is unclear but may be related to the transition in fracture mechanism from intergranular to transgranular on addition of intragranular SiC reinforcement.

ACKNOWLEDGEMENTS

We would like to acknowledge the following for their help and advice throughout this project. R.J. Brook, R.W. Davidge, M.L. Jenkins and S.G. Roberts for many helpful discussions on a range of topics covered in this paper. L. Carroll and B. Su for help in producing specimens, P. Twigg and F. Riley of Leeds University for assistance with the wear measurements, and C.W. Lawrence, E. Dupas and G.A.D. Briggs for assistance with the acoustic microscopy.

This project was supported by the EPSRC through project grant GR/J77542 and the European Union through a BRITE/EURAM Fellowship for M. Strenitzke (Human Capital and Mobility Contract No. ERB BRE2 CT94 3093).

REFERENCES

1. K. Niihara, New design concept of structural ceramics - ceramic nanocomposites, *J. Ceram Soc. Jpn.* 99:974 (1991).
2. J. Zhao, L.C. Stearns, M.P. Harmer, H. Chan and G.A. Miller, Mechanical properties of alumina-silicon carbide nanocomposites, *J. Amer. Ceram. Soc.* 76:503 (1993).

3. R.W. Davidge, R.J. Brook, F. Cambier, M. Poorteman, A. Leriche, D. O'Sullivan, S. Hampshire and T. Kennedy. Fabrication, properties and modelling of engineering ceramics reinforced with nanoparticle of silicon carbide, to appear in *Br. Ceram. Proc.* (1996).
4. L. Carroll, M. Sternitzke and B. Derby, Silicon carbide particle size effects in ceramic nanocomposites, to appear in *Acta Mater* (1996).
5. I. Levin, W.D. Kaplan, D.G. Brandon amnd A.A. Layyous, Effect of SiC submicrometer particle size and content on the fracture toughness of alumina-SiC nanocomposites, *J. Amer. Ceram. Soc.* 78:254 (1995).
6. M. Sternitzke, B. Derby and R.J. Brook, Al_2O_3/SiC nanocomposites by hybrid polymer/powder processing: microstructures and mechanical properties, submitted to *J. Amer. Ceram. Soc.* (1996).
7. R.W. Davidge, P.C. Twigg and F.L. Riley, Effects of silicon carbide nanophase on the wet erosive wear of polycrystalline alumina, *J. Europ. Ceram. Soc.* 16:799 (1996).
8. B. Su and M. Sternitzke, A novel processing route for Al_2O_3/SiC nanocomposites by Si-polymer pyrolysis, *4th Europ. Ceram. Conv.* Ed. A. Bellosi, Vol. 4:109 (1995), Gruppo Ed. Faenza Editrice SpA.
9. G.R. Anstis, P. Chantikul, B.R. Lawn and D.B. Marshall, *J. Amer. Ceram. Soc.* 15:201 (1981).
10. J. Kushibiki and N. Cubachi, Material characterisation by line-focus-beam acoustic microscopy, *IEEE Trans. Sonics. Ultrason.* SU32:189 (1989).
11. C.W. Lawrence, *Acoustic Microscopy of Ceramic Fibre Composites*, D.Phil. thesis, Oxford University, UK (1990).
12. C.B. Scruby and L.E. Drain, Ultrasonic generation by laser, in *Laser Ultrasonic Techniques and Applications,* 221, Adam Hilger, Bristol, UK (1990).
13. M. Hoffman, private communication, TU Darmstadt, Germany (1996).
14. A.M. Thompson, H.M. Chan and M.P. Harmer, Crack healing and stress relaxation in Al_2O_3-SiC "nanocomposites", *J. Amer. Ceram. Soc.* 78:567 (1995).

COMPOSITE POWDER SYNTHESIS

Lutgard C. De Jonghe and YongXiang He

Materials Sciences Division
Lawrence Berkeley National Laboratory
Berkeley CA 94720
and
Department of Materials Science and Mineral Engineering
University of California at Berkeley
Berkeley CA 94720

ABSTRACT

The synthesis of composite powders by a solution-precipitation process is discussed. In this process, thick coatings can be deposited on particles suspended as slurries in solutions. The coatings are produced as a result of controlled heterogeneous precipitation. The process parameters are discussed on the basis of a model combining simple first order reactions. Processing domains can be defined from this description in which the effects of changes in the processing parameters lead to predictable results. Various examples of coated powders prepared by the solution coating method are presented.

THE SOLUTION COATING PROCESS

Heterogeneous precipitation onto solid particles (the core particles) suspended in a solution of salts can be effected by several means. As an example, precipitation can be brought about by a homogenous pH change resulting from the slow decomposition of urea. Such decomposition avoids local supersaturation that is innate in the direct addition of base or acid to the suspension. Heterogeneous precipitation might be represented in a modified LaMer diagram that includes a threshold for heterogeneous nucleation which is lower than that for homogeneous nucleation. Essential features of representing the process by a modified LaMer model require a definition of the rate of change of the pH of the solution, as well as a quantification of both the homogenous and the heterogeneous precipitation thresholds.

In several instances, however, time dependent pH changes are not driving the precipitation reactions, since the pH remains essentially constant during most of the precipitation process. An example of this processing condition is shown in Fig 1, for the case of deposition of an yttria precursor from a salt-urea solution.[1] Observations indicated that during the temperature transient, while the suspension is brought to the process temperature, pH changes occur. However, the coating process only becomes measurable after the suspension has been at the process temperature, where now the pH remains relatively constant. Precipitation is thus not the result of a pH change that would be expected from the decomposition of urea by itself. Rather, the interaction between the urea decomposition and the deposition reaction has the result of keeping the pH approximately level.

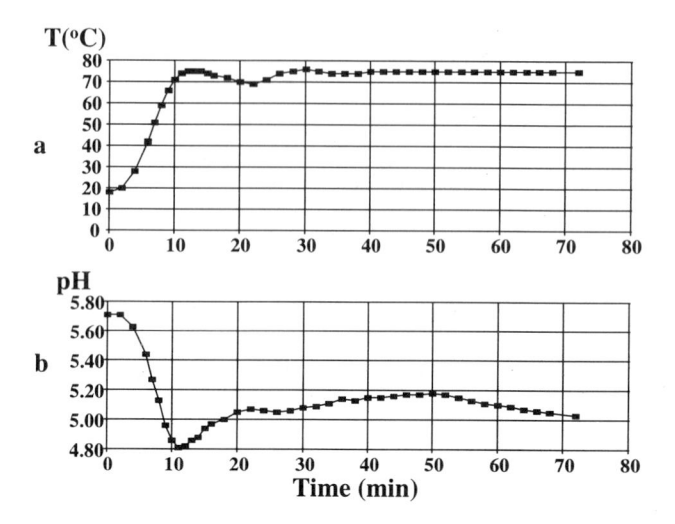

Figure 1. a) Temperature profile for a urea-salt precipitation coating process; b) pH time dependence during a urea-yttria precipitation coating process.[1]

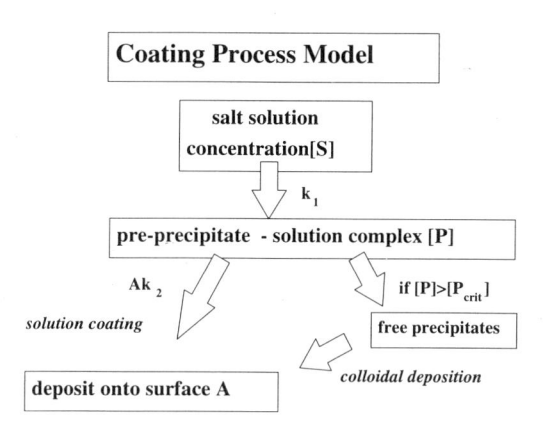

Figure 2. Reaction mechanism for solution coating.

As an alternative to the LaMer description with a changing pH as the key feature, a process mechanism can be proposed as shown in Fig. 2, in which the initial salt solution of concentration $[S_o]$ is converted to a pre-precipitate solution-complex, of concentration $[P]$, that in turn converts to the precipitate on the core particles.

Precipitation in the bulk of the solution is avoided so long as some critical concentration of the solution complex $[P_{crit}]$ is not exceed. If $[P] > [P_{crit}]$, then the coating process will mostly proceed by solution colloid formation in which both solution bulk precipitation and colloidal deposition onto the core particles occurs.

The mechanism is one whereby the coating process can be described by sequential first order reactions.

MODELING OF THE COATING PROCESS

During the transient, *i.e.* before urea decomposition, the cation (M, ignore the charge for simplicity) of the salt is in equilibrium with the precursor solution complex, e.g. (MOH⁻) in the solution:

$$M + H_2O \underset{k_{-1}}{\overset{k_1}{\rightleftharpoons}} MOH^- + H^+ \tag{1a}$$

Let the initial concentration of M be $[S]_0$, then the equilibrium concentration of pre-precipitates, $[P]_e = [MOH^-]_e$, and the proton, $[H^+]_e$, can be determined:

$$[P]_e = [H^+]_e = \frac{-K + \sqrt{K^2 + 4K[S]_0}}{2} \tag{1b}$$

where $K = k_1/k_{-1}$ is the equilibrium constant. Therefore, the concentration of M=[S] at equilibrium, $[S]_e$, is

$$[S]_e = [S]_0 - [P]_e.$$

When the coating bath nears the process temperature, the urea of concentration $[U]_0$ starts to decompose irreversibly, generating OH⁻ as well as CO_2 and NH_4^+,

$$(NH_2)_2CO + 3H_2O \xrightarrow{k_u} CO_2 + 2OH^- + 2NH_4^+ \tag{2a}$$

$$[OH^-] = 2[U]_0[1 - \exp(-k_u t)] \tag{2b}$$

The OH⁻ combines with the H⁺ generated in reaction (1a) and pushes the equilibrium strongly in the forward direction. So, the reverse reaction in (1a) may be neglected:

$$M + H_2O \xrightarrow{k_1} MOH^- + H^+ \tag{3}$$

Then, due to the rise in precursor solution complex the coating reaction becomes significant:

$$MOH^- + [Substrate] \xrightarrow{Ak_2} Coating \tag{4}$$

Since the pH of the solution may be assumed constant during the process, as follows from Fig. 1, the amount of the [OH⁻] generated in reaction (2a) can be considered to consume quantitatively the [H⁺] generated in reaction (3), the instantaneous concentration of M, [S], is given by:

$$[S] = [S]_0 - [P]_e - [OH^-] \tag{5}$$

Let the instantaneous concentration of the precursor solution complex [MOH⁻] be [P]. Now the rate equation can then be written as:

$$\frac{d[P]}{dt} = k_1[S] - Ak_2[P]$$

or

$$\frac{d[P]}{dt} = k_1\{[S]_0 - [P]_e - [OH^-]\} - Ak_2[P] \qquad (6)$$

where A is the total surface area of the substrate and is a constant determined by the amount of core particles initially added to the slurry. Substituting [OH⁻] with equation (2b) and solving (6) give:

$$[P] = e^{-Ak_2 t} \cdot \left\{ [P]_e + \frac{k_1([S]_0 - [P]_e - 2[U]_0)}{Ak_2} \cdot (e^{Ak_2 t} - 1) + \frac{2k_1[U]_0}{Ak_2 - k_u} \cdot \left[e^{(Ak_2 - k_u)t} - 1 \right] \right\} \qquad (7)$$

where $[P]_e$ is given by equation (1b), k_1, k_2, and k_u are rate constants.

Of interest is at what time (t_{max}) the maximum value of [P], $[P_{max}]$ occurs, and how it depends on the experimental parameters. These conditions can be easily found from $d[P]/dt=0$:

$$t_{max} = f([S]_0, [U]_0, k_1, Ak_2, k_u) \qquad (8)$$

and

$$[P_{max}] = f([S]_0, [U]_0, k_1, Ak_2, k_u) \qquad (9)$$

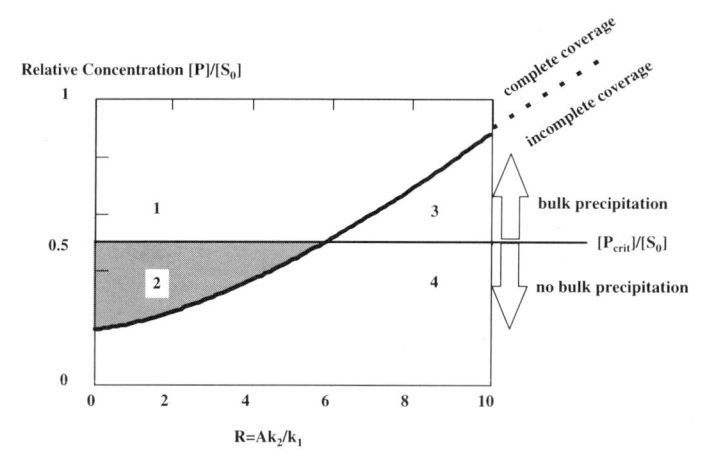

Figure 3. Qualitative dependence of $[P_{max}]$ on surface area A and rate parameters.

A calculated qualitative plot of $[P_{max}]$ vs $R=Ak_2/k_1$ is given in Figure 3 illustrating the dependence of $[P_{max}]$ on the process parameters. A similar plot can be constructed for any one of the experimental variables shown in equation (9).

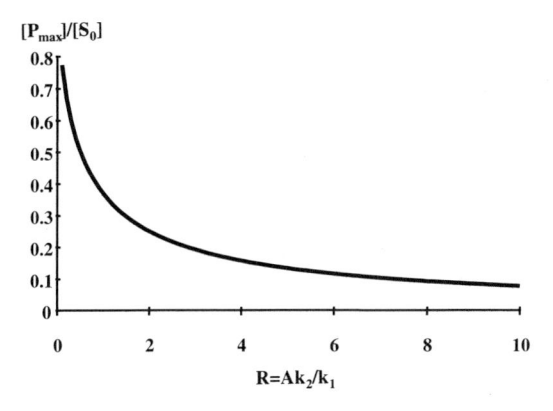

Figure 4. Calculated Processing M, showing the processing domains; 1: complete core particle coverage plus bulk precipitation; 2 : Complete coverage, *i.e.* the desired processing domain; 3: incomplete coverage plus bulk precipitation; 4: incomplete coverage

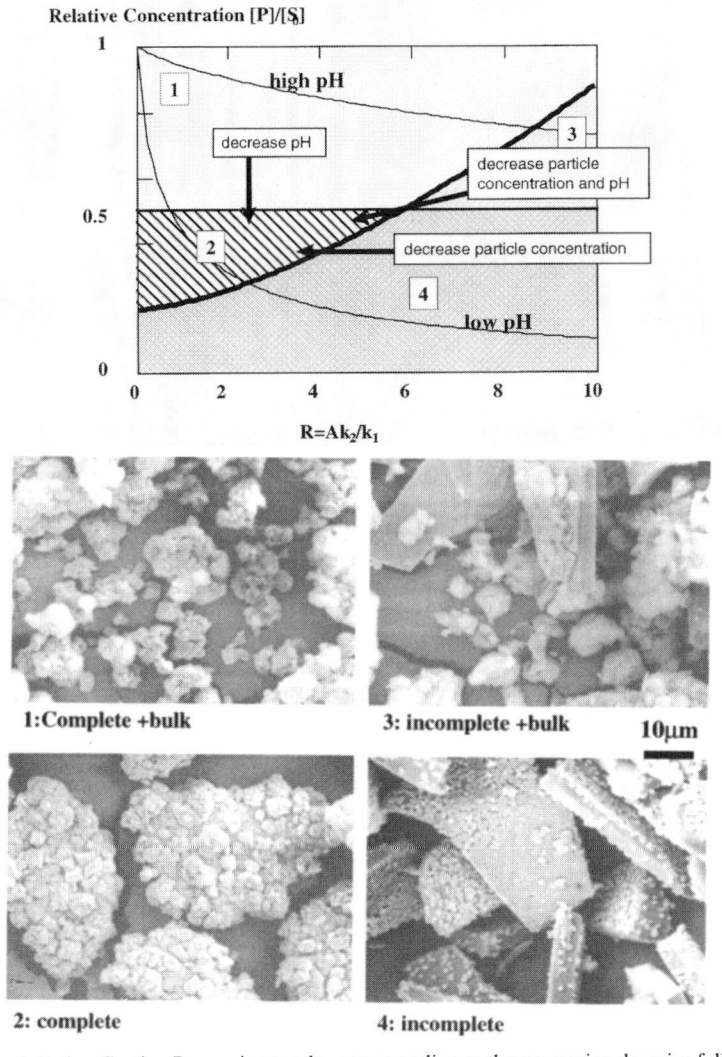

Figure 5. Precipitation Coating Processing results corresponding to the processing domains following the indicated Processing Map directions.

563

Using the description of the coating process developed here, experimental parameters could be varied so that the coating results correspond to the four identified domains in the processing map, Fig.4. The experimental results are summarized in Fig. 5, which demonstrates how processing conditions can be varied to achieve the desired coating result.

With the aid of these expressions, and the assumption that bulk precipitation will occur when [P_{max}] exceeds a critical threshold, [P_{crit}] four domains of coating process results can be identified, as shown in Figure 4. The domains are defined as whether the [P_{max}] is greater or less than the [P_{crit}], and whether the [S_0] is above or below the minimum amount of M corresponding to concentration [S_{lim}], required for complete coverage for given total surface area (A). So, by varying the experimental conditions, one can obtain coatings with: complete coverage and bulk precipitation (1), or incomplete coverage and bulk precipitation (3), or complete coverage and no bulk precipitation (2), or incomplete coverage and no bulk precipitation (4).

COATED PARTICLES

Several examples are shown below of coated powders which have been achieved by the slurry coating process.

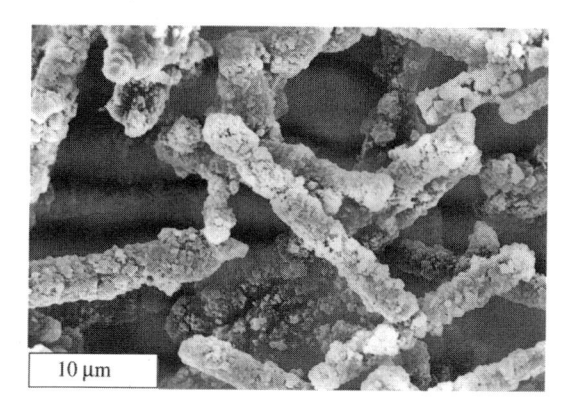

Figure 6. SiC whiskers coated with alumina.
(D. Kapolnek *et al.*)

Figure 7. Alumina-Coated hollow glass spheres
(S. Wu *et al.*)

Figure 8. Uncoated and alumina-coated SiC platelets. (T. Mitchell *et al.*)

Acknowledgments

This work was supported by the Director, Office of Energy Research, Office of Basic Energy Sciences, Materials Sciences Division of the U.S. Department of Energy, under contract No. DE-ACO3-76SF00098, and in part by the U.S. Advanced Research Projects Agency under contract No. AO-8672.

References

1. J. Daniel, *MS Thesis work*, U. California, Berkeley, 1993, unpublished.
2. D. Kapolnek and L.C. De Jonghe, *J. Euro. Ceram. Soc.*, 7:345 (1991)
3. S. J. Wu and L. C. De Jonghe, *J. Mat. Sci.*, 1995
4. T. Mitchell Jr., L. C. De Jonghe, W. MoberlyChan, and R. O. Ritchie, *J. Amer. Ceram. Soc.*, **78**:97 (1995)

SYNTHESIS AND CHARACTERIZATION OF CERAMIC POWDERS BY DUAL IRRADIATION OF CO₂ AND EXCIMER LASERS

Tetsuo Yamada, Yoshizumi Tanaka, Takeshi Suemasu and Yasuhiko Kohtoku

Ube Research Laboratory, UBE Industries, Ltd.,
Ube, Yamaguchi 755, Japan

ABSTRACT

Ultrafine SiC and Si/C/N powders were prepared by photochemical reactions of SiH_4, C_2H_4 and NH_3 under dual irradiation of CO_2 and excimer laser beams. Effects of process variables on the characteristics of these powders were studied. Enhanced crystallization and size reduction of SiC particles were observed by irradiation of excimer laser. Growth of the particles proceeded by CO_2 laser irradiation, whereas refinement of the particles occurred by ablation of the initially formed particles under excimer laser irradiation. Uniformly mixed state of silicon, carbon and nitrogen atoms in the Si/C/N composite particles was confirmed by transition electron microscopy and x-ray photoelectron spectroscopy analyses. Sintering behavior of the Si/C/N powders was examined by microwave sintering and conventional hot-pressing, and the resultant Si_3N_4-SiC composites were consolidated into highly dense bodies with fine grain size and high hardness. It was found that the vapor-phase reaction using dual-type CO_2 and excimer lasers offered a feasible method for controlling composition, crystallinity and particle size in the synthesis of ultrafine sinterable single- and multi-component powders.

INTRODUCTION

Recently, there has been increasing interest in preparation processes and technical applications of ultrafine powders with well-controlled quality. Laser synthesis is a promising technique for the preparation of ultrafine ceramic powders with desired characteristics such as high purity, fine particles, and spherical shape.[1,2] In this technique, solid particles are formed by the reaction of the molecules in a fast-flowing stream of the feedstock reagents. Generally, a continuous-wave CO_2 laser is used in this method, which achieves heating rates of approximately 10^7 K/s and reaction times of approximately 10^{-3} s. Ultrafine powders of Si, SiC, Si_3N_4, Si/C/N, B_4C and Fe/Si/C have been synthesized by a vapor-phase reaction under irradiation of CO_2 laser beam.[3-7] On the other hand, a high energy excimer laser beam greatly impacts the selective excitation and dissociation of chemical bonds of various reagents. There have been few reports on the preparation of ultrafine particles using excimer lasers.[8] In the present study, thermal activation by CO_2 laser radiation and electronic excitation by excimer laser beam were combined in order to synthesize ultrafine SiC and Si/C/N powders with precisely controlled structure and composition. Sintering behavior of the resultant Si/C/N powders was examined by microwave sintering and conventional hot-pressing. Microwave sintering is a new and exciting technique for consolidating ceramic materials, which is facilitating a variety of applications due to rapid and selective heating of constituents and phases.[9] Green compacts of Si/C/N could be heated uniformly by the presence of SiC particulates with high dielectric loss, which couple well with microwave energy, though Si_3N_4 matrixes were relatively low loss in microwave radiation. It is our purpose to report the experimental results on the synthesis of ultrafine SiC and Si/C/N powders by dual irradiation of CO_2 and excimer lasers, with emphasis on the characteristics and sintering behavior of the resultant powders.

EXPERIMENTAL PROCEDURE

A schematic diagram of the apparatus used for preparing ultrafine powders by a laser-induced vapor-phase reaction is shown elsewhere.[10] The apparatus was equipped with a 1.2 kW CO_2 laser emitting at a wavelength of 10.6 μm and a pulsed excimer laser emitting at a wavelength of 193 nm and a maximum pulse energy of 500 mJ by running on an Ar-F gas mixture. Synthesis conditions of SiC and Si/C/N powders are listed in Table 1. Electronic-grade SiH_4, C_2H_4 and NH_3 were employed as the reactant gases. These reactant gases were injected upward through a stainless steel nozzle with an internal diameter of 1.5 or 3.9 mm, and were intersected orthogonally with the CO_2 laser beam having a diameter of 2.5 mm and the excimer laser beam having a 5 × 5 mm cross section. The resultant powders were captured by carrying the mixture of the product and argon gases from the reaction zone into a collection chamber with cylindrical filter.

Chemical composition of the product was analyzed using a LECO TC-136 type N, O Analyzer for the nitrogen and oxygen contents, and a LECO IR-412 type C Analyzer for the carbon content. Characteristics of the resultant powders such as phase composition, morphology, and bonding configuration were examined by powder x-ray diffractometry, transmission electron microscopy (TEM), and x-ray photoelectron spectroscopy (XPS). The single-point BET method was used for measuring specific surface area.

Laser-synthesized Si/C/N powder and a mixture of commercial Si_3N_4 (SN-E10 from UBE Industries, Ltd.) and SiC (T1 from Sumitomo Cement Co.) powders were used as the starting materials of sintering. These powders were mixed with Al_2O_3 (AKP-30 from Sumitomo Chemical Co.) and Y_2O_3 (SU type from Shin-Etsu Chemical Co.) by ball-milling in ethanol, and then dried to obtain the selected composition, 92.0 (Si_3N_4-SiC)-5.0 Y_2O_3-3.0 Al_2O_3 in wt.%. The powder samples were charged into a graphite mold and hot-pressed at 2053 K for 2 hours under 25 MPa. Microwave sintering was accomplished in a multi-mode oven operating at 6.0 GHz. The green compacts pressed isostatically at 150 MPa were placed into a BN crucible, inside of an Al_2O_3 fiberboard box, and sintered at 2053 K for 1 hour in nitrogen atmosphere by microwave heating. The temperature was measured by a molybdenum sheathed W-5%Re/W-26%Re thermocouple touched with the sample. Microstructure of the resultant Si_3N_4-SiC composites was observed by scanning electron microscopy (SEM). Vickers microhardness of the composites was measured under an indentation load of 200 g. Fracture toughness, K_{IC}, of the composites was measured by indentation fracture method according to JIS-R1607.

Table 1. Synthesis conditions of SiC and Si/C/N.

Reactant gas flow rate		
SiH_4	25~100	SCCM
C_2H_4	0~60	SCCM
NH_3	0~200	SCCM
Diameter of feed nozzle	1.5 or 3.9 mm	
Cell pressure	45~95 kPa	
CO_2 laser radiation (CW mode)		
Wave length	10.6	μm
Laser intensity	3.1~6.2	kW/cm^2
Beam diameter	2.5	mm
Excimer laser beam (50-150 Hz)		
Wave length	193	nm (Ar-F)
Pulse energy density	100~200	MJ/cm^2
Cross-section	W5 x H5	mm

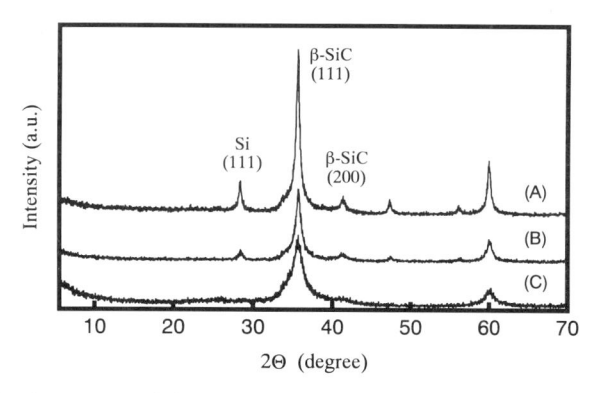

Figure 1. X-ray diffraction patterns of SiC powders prepared by laser irradiation. (a) Excimer laser irradiation. (b) Dual irradiation of CO_2 and excimer lasers. (c) CO_2 laser irradiation.

RESULTS AND DISCUSSION

Controlling Powder Characteristics by Excimer Laser Irradiation

Ultrafine SiC powders were prepared by the reaction of SiH_4 and C_2H_4 under single irradiation of CO_2 or excimer laser and dual irradiation of these lasers. X-ray diffraction patterns of these powders are shown in Figure 1. The distinct diffraction peaks of β-SiC were observed for the powders prepared by excimer laser irradiation and dual irradiation of CO_2 and excimer lasers. Highly crystallized powder was obtained at near-room temperature by irradiation of high energy excimer laser beam. The excimer laser irradiation resulted in good crystal quality. In contrast to this, the powder obtained below 1273 K by CO_2 laser irradiation was poorly crystallized and showed broad diffraction peaks. In a general vapor-phase reaction, both crystallinity and particle size of the products are mainly dependent on crystallization. In our synthesis process, crystallinity can be controlled by a combination of reaction temperature and excimer laser pulse frequency, which enables independent regulation of crystallinity and particle size. Stoichiometric SiC powder, with C/Si atomic ratio of ~1.00, was obtained above 2073 K, though the powder contained free silicon or carbon at lower temperatures.

Since reaction flame temperature depends on both the power density of the CO_2 laser and the gas flow rate of the reactants, changing these two factors regulated the SiC powder reaction flame. Particle size of the resultant SiC powders was evaluated from their specific surface area. The relationships between residence time and particle size prepared by irradiation of CO_2 and excimer lasers are shown in Figures 2 and 3, respectively. The size of particles prepared by CO_2 laser irradiation was increased with increasing temperature and residence time. It was suggested that the growth of SiC particles was governed by either the nucleation and growth of the product induced by supersaturation or the subsequent coagulation of the product caused by Brownian collision.[11]

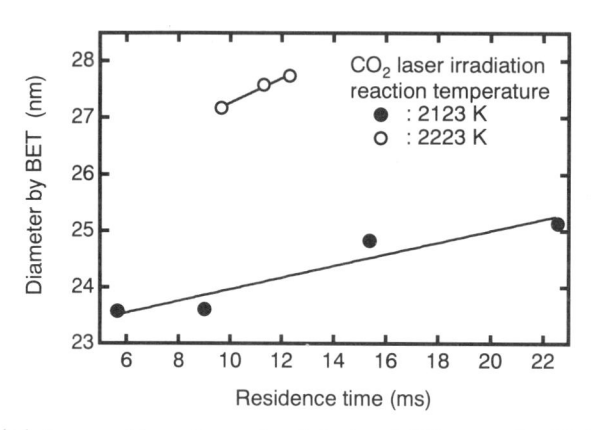

Figure 2. Relationship between residence time and particle size of SiC prepared at 2123 and 2223 K by CO_2 laser irradiation.

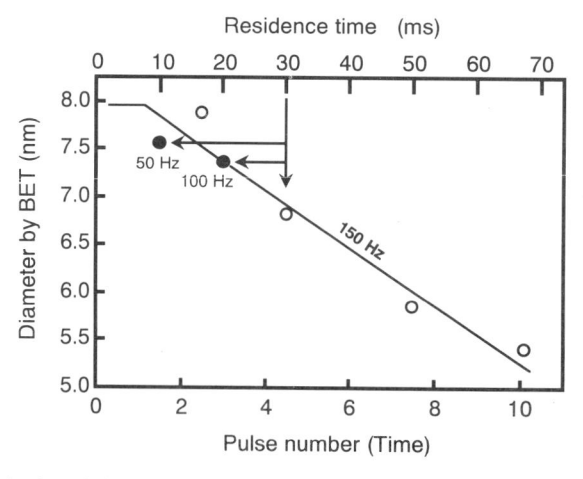

Figure 3. Particle size of SiC powders as a function of pulse number of excimer laser irradiation.

Shielded region
(Without irradiation)

Exposed region
(Laser irradiation)

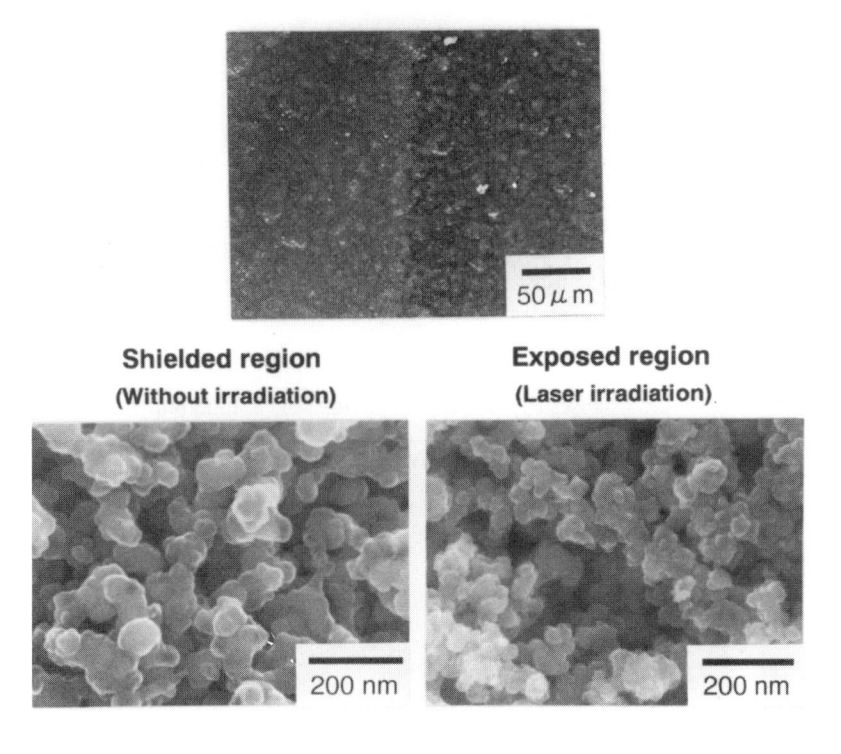

Figure 4. Change of SiC particles irradiated with 10 pulses of Ar-F excimer laser beam.

On the contrary, the size of particles prepared by excimer laser irradiation decreased with increased residence time, or number of pulsed laser irradiations. In order to clarify the reason for this anomalous particle refinement, an excimer laser beam pulse irradiated the unshielded half of the flat surface of SiC powder compacts 10 times. These compacts had been fabricated by deposition of the aerosol particles prepared by CO_2 laser irradiation. Then, the morphology of SiC particles before and after exposure to excimer laser beam was observed by FE-SEM, as shown in Figure 4. It was found that the particle size was drastically decreased by the 10 excimer laser beam irradiation pulses. Laser ablation and condensation caused the shrinkage.[12] Therefore, particle refinement by excimer laser irradiation with increasing residence time was also considered to be a result of laser ablation. Thus it became possible to control the particle size of products by vapor-phase reaction using excimer laser ablation.

Synthesis and Characterization of Si/C/N by Dual Laser Irradiation

Ultrafine Si/C/N powders were prepared by the reaction of SiH_4, C_2H_4 and NH_3 under dual irradiation of CO_2 and excimer lasers. The reaction flame temperature fell with increasing NH_3 gas flow rate, which was caused by the low absorption coefficient of NH_3 for CO_2 laser radiation. A linear relationship was observed between N/(C+N) atomic ratio of the products and the fraction of nitrogen atom in the gas mixture of C_2H_4 and NH_3, $[NH_3]/(2[C_2H_4]+[NH_3])$. Accordingly, the composition of the resultant Si/C/N powder could be extensively controlled by varying the flow rate of these reactant gases. Carbon and nitrogen contents of the powder changed within a wide range from 0 to nearly 55 at.%, respectively. Typical TEM photographs are shown in Figure 5 for the Si/C/N powder with C/N atomic ratio of 0.62. The bright-field image of the product showed aggregated spherical particles with diameters of 20 to 50 nm. The dark-field image of the crystallites with (111) Bragg orientation of β-SiC showed that most of the Si/C/N particles consisted of several randomly oriented crystallites, which were smaller than the particles. Both carbon and nitrogen atoms were detected by the composition analysis of the ultrafine aggregated particles, though resolution limitations made analysis of monodispersed particle difficult. According to the x-ray diffraction measurement of the products, the diffraction peaks of β-SiC were shifted toward higher 2θ position and broadened, which showed the reduction in lattice constant caused by nitrogen dissolution. The diffraction peaks of Si$_3$N$_4$ also showed slight change in the lattice constant caused by carbon dissolution. This mutual solubility of carbon and nitrogen was caused by the rapid heating and quenching in the laser-induced vapor phase reaction.

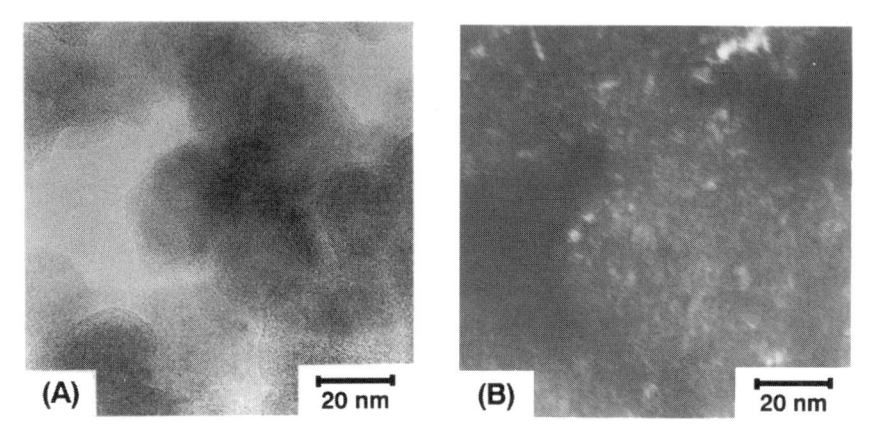

Figure 5. TEM photographs of typical Si/C/N (C/N=0.62) powder prepared by dual-irradiation of CO$_2$ and excimer lasers. (a) Bright-field image of SiC (111) diffraction. (b) Dark-field image of SiC (111) diffraction.

The bonding configuration of the Si atom was evaluated by the chemical shift of XPS spectra. Figure 6 shows the Si 2p spectra of the resultant Si/C/N powders and mechanical mixtures of commercial SiC and Si$_3$N$_4$ powders. Comparison of the products and the mechanically mixed powders was made at nearly the same C/N atomic ratio. The Si 2p spectrum of the mechanical mixture consisted of SiC and Si$_3$N$_4$ peaks whose peak intensities depended on their blending ratio. In contrast to this, the Si 2p spectrum of the Si/C/N powder synthesized by dual laser irradiation was a single peak having a binding energy that depended on its composition. These results showed that the chemical environment around silicon atoms in the Si/C/N powders is quite different from that in the mechanical mixtures. It was suggested that silicon, carbon and nitrogen atoms were homogeneously mixed in the laser synthesized Si/C/N composite powders.

Si/C/N powders doped with Al$_2$O$_3$ and yttrium compound were also prepared in order to obtain highly sinter-active multicomponent powders containing sintering aids. As ultrafine γ-Al$_2$O$_3$ powder could be easily dispersed in Ar carrier gas, Al$_2$O$_3$ was mixed with aerosol particles of Si/C/N by supplying Al$_2$O$_3$ aerosol into a transfer tube between the reaction chamber and the collection chamber. In case of yttrium compound, vapor of

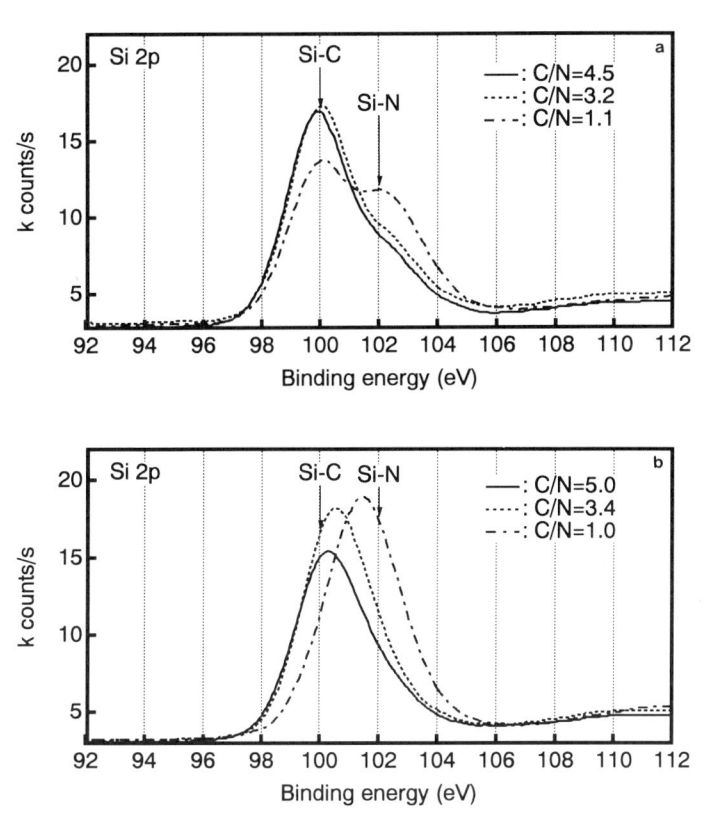

Figure 6. Comparison of Si 2p XPS spectra of laser synthesized Si/C/N powders and mechanical mixtures of SiC and Si$_3$N$_4$ powders. (a) Si/C/N powders by dual irradiation of CO$_2$ and excimer lasers. (b) Mechanical mixtures of β-SiC and α-Si$_3$N$_4$ powders.

Y(DPM)$_3$ complex with boiling point of 546 K was supplied at the reaction zone. As a result of these examinations, Si/C/N powders doped with uniformly dispersed sintering aids were prepared without exposure to air.

Consolidation of Si$_3$N$_4$-SiC Composites by Microwave Sintering and Conventional Hot-Pressing

Sintering behavior of the laser-synthesized Si/C/N powder was compared with that of a mixture of commercial Si$_3$N$_4$ and SiC powders by microwave sintering and conventional hot-pressing. Fully dense Si$_3$N$_4$-SiC composites were fabricated by conventional hot-pressing of both starting materials. Shown in Figure 7 are the comparisons of mechanical properties of hot-pressed Si$_3$N$_4$-SiC composites as a function of SiC content. Vickers microhardness of the Si$_3$N$_4$-SiC composites was raised with increasing SiC content because of the presence of hard SiC particulates, whereas K_{IC} was deteriorated at higher SiC content. This deterioration in K_{IC} was caused by the finer grain size of Si$_3$N$_4$-SiC composites; K_{IC} value degraded by the marked decrease of coarser prismatic Si$_3$N$_4$ grains, which enhanced the amplitude of crack deflection and microcracking. Si$_3$N$_4$-SiC composites obtained from the laser-synthesized Si/C/N powder showed much higher microhardness than those obtained from the conventional mixture of Si$_3$N$_4$ and SiC. This difference in microhardness was related to the grain size of Si$_3$N$_4$-SiC composites: i.e., the microhardness of Si$_3$N$_4$-SiC composites with finer grain size was higher than those of relatively coarser grain size, as confirmed by SEM observation (shown in Figure 8).

Effect of SiC content on densification of Si_3N_4-SiC composites in microwave radiation at 6.0 GHz is shown in Figure 9. Bulk density of Si_3N_4-SiC composites declined with increasing SiC content. Little enhancement of densification was observed for microwave sintering compared with pressureless sintering by conventional heating. The presence of SiC particles affected the amount and viscosity of the molten silicate phase by deoxidating oxide additives at the grain boundaries, resulting in lower densification. Densification of Si_3N_4-SiC composites was also inhibited by a shear stress in the matrix and a tensile hoop stress at the particle-matrix interface generated by a rigid SiC dispersoid particles.[13] The bulk density of Si_3N_4-SiC composites obtained from laser-synthesized Si/C/N powder was higher than that obtained from the conventional mixture of Si_3N_4 and SiC, which verifies the sintering activity of laser-synthesized Si/C/N powder. Laser-synthesized Si/C/N powder has superior properties such as fine particle size, weak agglomeration, and homogeneous distribution of carbon and nitrogen atoms, which leads to higher sinterability than conventional powder mixtures.

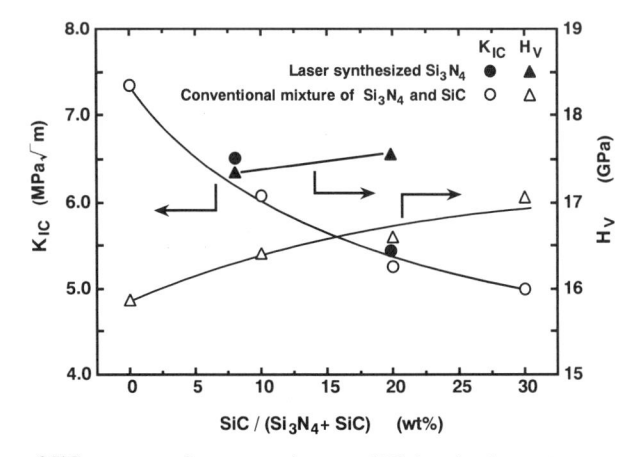

Figure 7. Effects of SiC content on fracture toughness and Vickers hardness of Si_3N_4-SiC composites by hot-pressing.

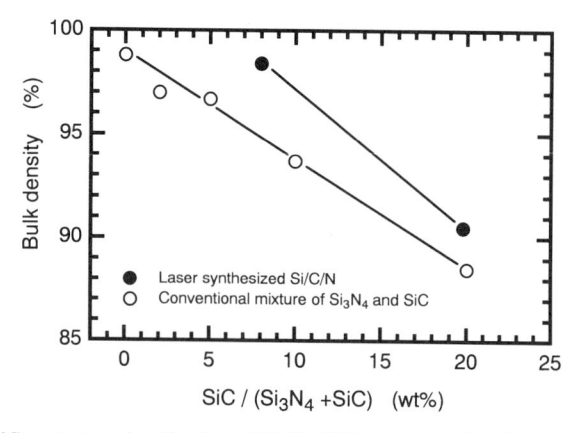

Figure 8. Effect of SiC content on densification of Si_3N_4-SiC composites by microwave sintering (Additive: 3 wt.% Al_2O_3-5 wt.% Y_2O_3).

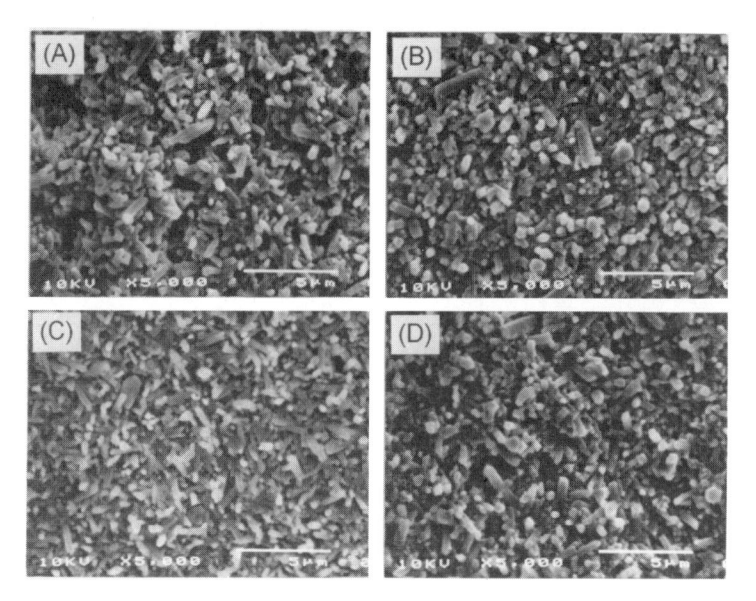

Figure 9. SEM photographs of Si_3N_4-SiC composites. Hot-pressed bodies from (a) laser-synthesized Si/C/N powder and (b) a conventional mixture of Si_3N_4 and SiC powders. Microwave-sintered bodies from (c) laser-synthesized Si/C/N powder and (d) a conventional mixture of Si_3N_4 and SiC powders. Composition: (a), (c) 84.6 Si_3N_4-7.4 SiC-3.0 Al_2O_3-5.0 Y_2O_3 (in wt.%); (b), (d) 82.8 Si_3N_4-9.2 SiC-3.0 Al_2O_3-5.0 Y_2O_3 (in wt.%).

SEM photographs of the microstructure of Si_3N_4-SiC composites obtained from the laser-synthesized Si/C/N powder and the conventional mixture of Si_3N_4 and SiC are shown in Figure 8. Consolidation of the Si_3N_4-SiC composite using the laser-synthesized Si/C/N powder resulted in grain significantly more refined than the conventional powder mixture. Microwave heating led to improved microstructures with the development of fine grain structures similar to hot-pressed bodies.

Production of highly sinter-active Si/C/N powder is very advantageous in consolidating Si_3N_4-SiC composites without applying hot-pressing, which has many restrictions in fabricating complex shapes. Sintering studies of Si/C/N powders doped with uniformly dispersed Y_2O_3 and Al_2O_3 at the reaction zone are now in progress.

SUMMARY AND CONCLUSIONS

Ultrafine SiC and Si/C/N powders with desired composition and characteristics were prepared under dual irradiation of CO_2 and excimer laser beams. Then, sintering behavior of the resultant Si/C/N powders was examined by microwave sintering and conventional hot-pressing. The following results emerged:

- Excimer laser irradiation at near-room temperature was observed to enhance crystallization and size reduction of SiC particles.
- Ablation of the initially formed particles was confirmed to be the dominant mechanism of particle refinement induced by excimer laser irradiation.
- Silicon, carbon, and nitrogen atoms were uniformly mixed in the laser-synthesized Si/C/N composite particles.

- Hot-pressed Si_3N_4-SiC composites obtained from laser-synthesized Si/C/N powder showed higher microhardness than those obtained from conventional mixture of Si_3N_4 and SiC powders.
- Laser-synthesized Si/C/N powder showed higher sintering activity by microwave heating than the conventional mixture of Si_3N_4 and SiC powders.

ACKNOWLEDGMENTS

This work was performed under the Research & Development Program on "Advanced Chemical Processing Technology," consigned to ACTA from NEDO, which is conducted under the Industrial Science and Technology Frontier Program enforced by the Agency of Industrial Science and Technology.

REFERENCES

1. W. R. Cannon, S. C. Danforth, J. H. Flint, J. S. Haggerty and R. A. Marra, *J. Am. Ceram. Soc.*, **65**:324 (1982).
2. W. R. Cannon, S. C. Danforth, J. S. Haggerty and R. A. Marra, *J. Am. Ceram. Soc.*, **65**:330(1982).
3. Y. Suyama. R. A. Marra, J. S. Haggerty and H. K. Bowen, *Am. Ceram. Soc. Bull.*, **64**:1356 (1985).
4. K. Sawano, J. S. Haggerty and H. K. Bowen, *Yogyo Kyokai-shi*, 95:64 (1987).
5. M. Suzuki, Y. Nakata, T. Okutani and A. Kato, *Seramikkusu Ronbun-shi*, **97**:972 (1989).
6. M. Aoki, J. H. Flint and J. S. Haggerty, Ceramic Powder Science II, in: *Ceramic Transactions, Vol.1,* G. L. Messing, E. R. Fuller, Jr., and H. Hausner, eds., The American Ceramic Society (1988).
7. M. Suzuki, Y. Maniette, Y. Nakata and T. Okutani, *J. Am. Ceram. Soc.*, **76**:1195 (1993).
8. J. A. O'Neill, M. Horsburgh, J. Tann, K. J. Grant and G. L. Paul, *J. Am. Ceram. Soc.*, **72**:1130 (1989).
9. T. T. Meek, X. Zhang and M. Rader, Microwaves: Theory and Application in Materials Processing, in: *Ceramic Transactions, Vol. 21*, D. E. Clark, F. D. Gac, and W. H. Sutton, eds., The American Ceramic Society (1991).
10. T. Yamada, in: *Proceedings of ACT Symposium*, ACTA, Tokyo (1994).
11. G. D. Ulrich, in: *Combustion Science and Technology, Vol. 4,* Gordon and Breach Science Publishers Ltd., London (1971).
12. G. P. Johnston, R. E. Muenchausen, D. M. Smith and S. R. Foltyn, *J. Am. Ceram. Soc.*, **75**:3465 (1992).
13. Y. J. Song, D. O'Sullivan, R. Flynn and S. Hampshire, Advanced Ceramics, in: *Key Engineering Materials, Vol. 86-87*, C. Ganguly, S.K. Roy and P.R. Roy, eds. Trans Tech Publications, Switzerland (1993).

IONIC TRANSPORT TO PRODUCE DIVERSE MORPHOLOGIES IN MgAl$_2$O$_4$ FORMATIONS STARTED FROM POWDERS OF VARIED PHYSICAL NATURE

T. Kume, T. Kuwabara, O. Sakurada, M. Hashiba and Y. Nurishi
Department of Applied Chemistry, Faculty of Engineering,
Gifu University, Gifu 501-11, Japan

J. Requena and J. S. Moya
Instituto de Ciencia de Materiales de Madrid, CSIC, Cantoblanco, Madrid, Spain

ABSTRACT

Using the multilayer reaction technique, rapid densification of spinel and rapid grain growth of spinel perpendicular to the reaction layer were observed during its formation in a fine alumina bed near magnesia, leaving growing pores at the D_I/D_M interface in the fine magnesia and fine alumina system. It was observed that rapid densification and an icicle-like growth occurred in the spinel formed in the fine alumina bed in a coarse and dense magnesia and fine alumina system. All these ionic transport processes, including processes sensitive to oxygen content in the atmosphere, were found for the spinel formation started with fine alumina and for the spinel formed near magnesia. The importance of oxygen vacancy for the ionic processes was stressed. Also, the presence of aluminum vacancy was assumed for the cation vacancies, as well as the magnesium and aluminum cations at the interstitial site, and the magnesium at the aluminum site, which have been introduced by other authors. The pore growth at the D_I/D_M interface supported the simultaneous presence of oxygen vacancy and aluminum vacancy.

INTRODUCTION

Several spinel formations and ionic transport processes are sensitive to the oxygen content of the firing atmosphere. Formation of ZnAl$_2$O$_4$ spinel in the fine alumina-zinc oxide system[1-4] was promoted by the firing atmosphere containing oxygen, thereby differing from the system started from the coarse and dense alumina.[3] Promotion of MgAl$_2$O$_4$ formation by the firing atmosphere containing oxygen was also observed in the coarse magnesia and fine alumina system.[5,6] In both cases, the spinel was formed in the fine alumina bed, resulting in a gap between alumina and zinc oxide or magnesia, indicating a transport by a vapor. Succeeding to the spinel formation by the vapor process, an icicle-like

MgAl$_2$O$_4$ was grown in the gap, accompanying densification of the spinel in the alumina bed near magnesia.[5] The growth of the icicle-like spinel was promoted by the atmosphere containing oxygen, while, conversely, the densification of the spinel in the alumina bed was suppressed. Fast grain growth perpendicular to the reaction layer also occurred in the spinel formed in the fine alumina powder bed near magnesia.[8] The emergence of the diverse morphologies for the spinel produced near magnesia suggested that the ionic transport processes are different from the Wagner model[11-14] where the interdiffusion[15] of cations occur and oxygen movement does not.[17] J. Moya, *et al.*, and H. Wohlfrom, *et al.*, found that multilayer obtained by slip casting is suitable for the observation of the trace of the ionic transport in the microstructure.[18,19] Starting with powders of different physical nature changes the mechanism of the ionic transport, as evidenced in the morphologies.

The purpose of this study is to observe spinel formation in a fine magnesia and fine alumina system with slip casted multilayers; to review our preliminary study on spinel formation in coarse and dense magnesia and fine alumina systems; and to discuss the ionic transport on the common bases of the ionic defects in the spinel.

EXPERIMENTAL RESEARCH

Starting materials

Commercially available submicrometer powders were used as starting materials: alumina (Alcoa A-16 SG, Pittsburg, Pa, USA) with an average particle size of 0.5 µm and magnesia (Reagent grade, by Nacalai Tesque, Japan) with an average particle size of 0.3 µm.

Slip Casting Process

Aqueous alumina suspensions with 50% solid contents have been prepared using 1wt% addition of an alkali-free organic polyelectrolyte (Dolapix PC 33, Schimmer Schwartz, Germany). Magnesia suspensions have been prepared in ethyl alcohol media with solid contents ranging from 30 to 50% using 5% dibutylphtalate and 5% formaline sulfonate (Alcoa A-16 SG, Pittsburg, Pa, USA, and E-630, Chukyo Yushi Co., Nagoya, Japan, respectively). The suspensions have been milled in polyethylene balls for 24 hours. After milling bubbles were removed by smooth agitation and the slips were alternatively draincast into plaster mold several times. Then multilayered 40-mm × 40-mm × 5-mm green plates were obtained.

Heating Process

The green plates were placed in Pt boats, inserted in a tube furnace with SiC heating elements, and fired in air at 1300 to 1600°C for 1 to 50 hours. Firing at 1500 to 1700°C for 20 to 80 hours was performed in air in a furnace with super kanthal electric heater. The heating and cooling rates were kept constant (4°C/min. in all the experiments).

Sample Preparation and Microstructural Study

Fired specimens were first embedded in epoxy resin in vacuum, and then cut, ground, and polished down to 2 µm using diamond paste. The observation of the microstructure were carried out with FE-SEM of a TOPCON ABT-150FS.

RESULTS AND DISCUSSION

Spinel Formation in Fine MgO and Fine Al$_2$O$_3$ System

Figure 1 is the cross-section of an early stage spinel layer between multilayered fine MgO and fine Al$_2$O$_3$ after firing at 1300°C. The spinel was produced in the alumina bed to form in two layers with different density in alumina powder bed being the one close to magnesia with higher density.

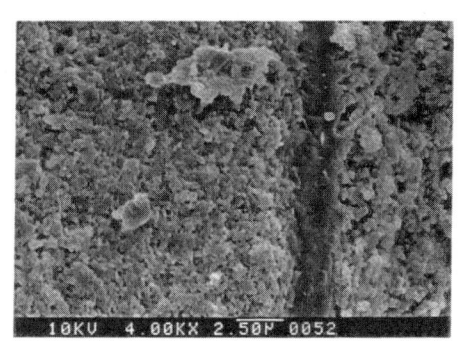

Figure 1. Cross-section of an early stage spinel layer formed in multilayered fine MgO and fine Al$_2$O$_3$ fired at 1300°C for 5 hours in air.

In a later stage, at 1400°C, another densification started from the side near the alumina. At this point, the total spinel layer consisted of three layers. Two layers sintered from both sides near magnesia and alumina by the fast and the delayed sintering, i.e., D$_M$ and D$_A$ layers, sandwiched a porous layer left from the densification, i.e., D$_I$ layer as shown in Figure 2.

Figure 2. Three-layer structure formed between multilayered fine MgO and fine Al$_2$O$_3$ appeared in later stage, after firing at 1400°C for 20 hours in air.

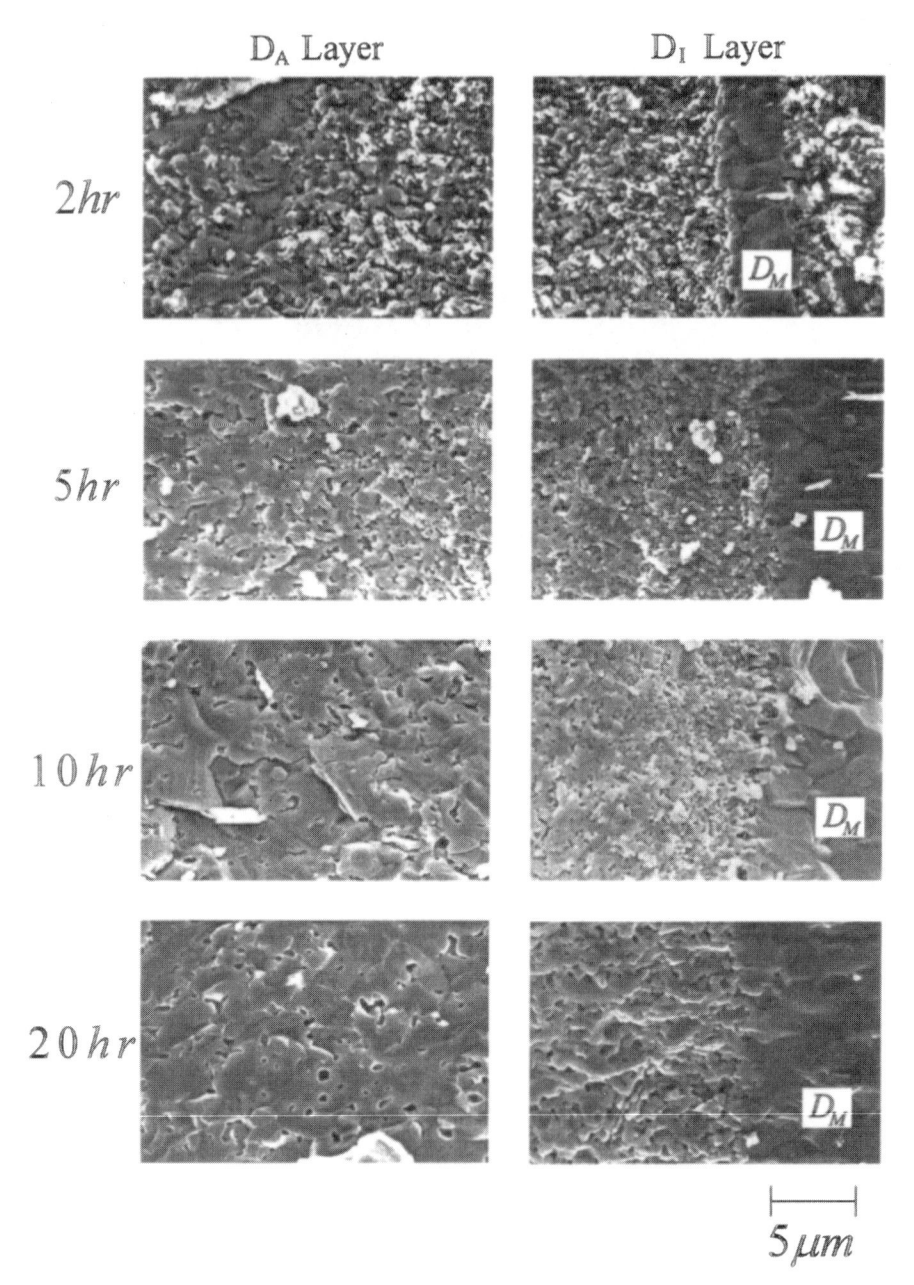

Figure 3. Densification proceeded in D_M, D_I and D_A layer with firing time, respectively

Each microstructure in the D_A, D_I and D_M layers was changed with firing time at 1400°C as shown in Fig. 3. Sintering was proceeded even in the D_I layer by long heating at 1400°C. The degree of sintering, i.e., the difference in the state of pores in each layers differentiated the three layers.

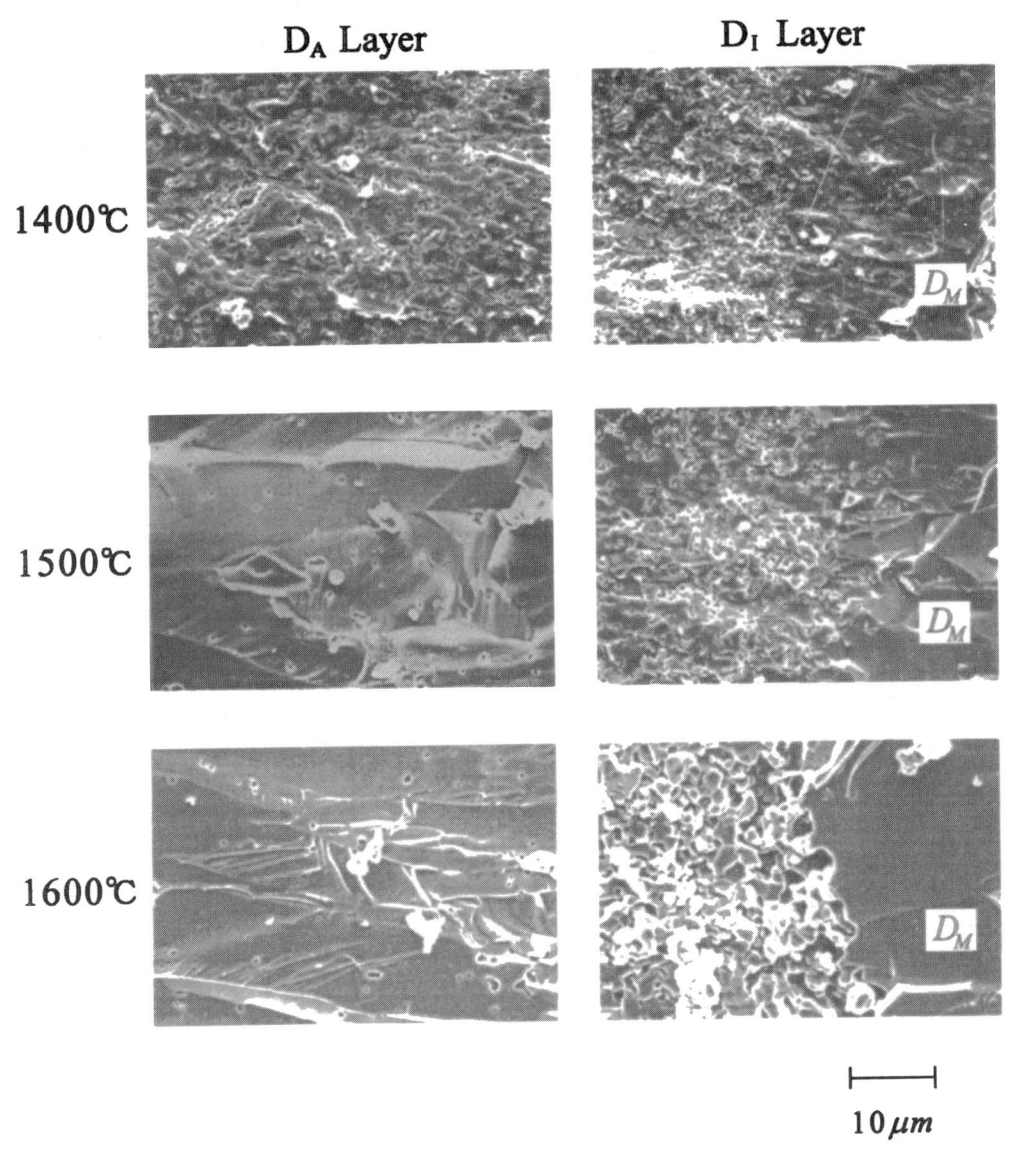

Figure 4. Densification proceeded accelerated by the increase of the heating temperature.

Increasing heating temperature up to 1600°C had a strong effect on the sintering of the spinel layers, more so than the lengthened heating time at 1400°C (see Figure 4). With the increased temperature, the pores in each layer disappeared and the sintering accelerated. At 1600°C, grain growth of densified spinel perpendicular to the reaction interface occurred in the D_M layer, as shown in Figure 5.

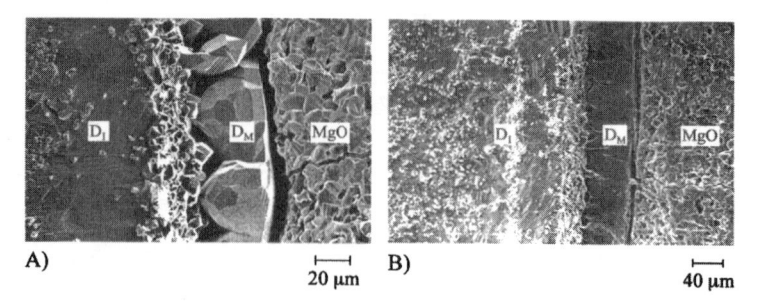

Figure 5. Grain growth occurred perpendicular to the reaction layer. Pore growth occurred at the densification front of the D_M layer. (a) On the cast surface: fired at 1600°C for 50 hours in air. (b) On the fractured surface: fired at 1600°C for 80 hours in air.

Moreover, the grain growth was accompanied by a growth of pores at the D_M/D_I interface in spite of accelerated sintering above 1600°C. Pore growth was observed even on the fractured surface (see Figure 5(b)) of the sample as well as on the surface of the fired sample (Figure 5(a)).

Spinel Formation in Coarse and Dense MgO and Fine Al$_2$O$_3$ Systems

In our preliminary research for coarse and dense magnesia and fine alumina systems, spinel was first formed in the fine alumina bed around the magnesia, leaving a gap between the spinel and magnesia, as shown in Figure 6.[5] It was suggested that a vapor phase caused magnesia to grow the spinel in the fine alumina bed. Spinel formation in an alumina bed due to vapor was promoted in oxygen atmosphere below 1300°C.[6]

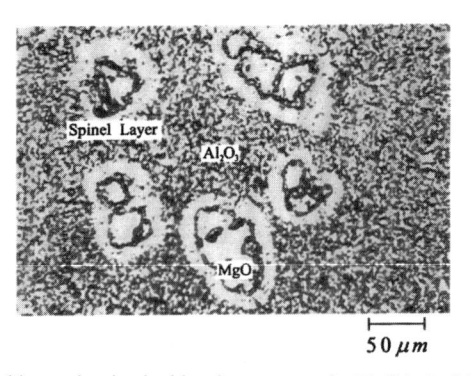

Figure 6. Formation of spinel in an alumina bed leaving a gap at the MgO/spinel interface at an early stage of firing. Coarse and dense MgO and fine Al$_2$O$_3$ fired at 1300°C for 2 hours in air.[*]

---

[*] Figure 6, 7, 8 are quoted from Itoh et al.,[5] with permission, courtesy of North-Holland, New York.

In an intermediate stage at 1300°C, the spinel in alumina bed near magnesia was started to densify and another new spinel was nucleated on the densified spinel in the gap as shown in Fig. 7.

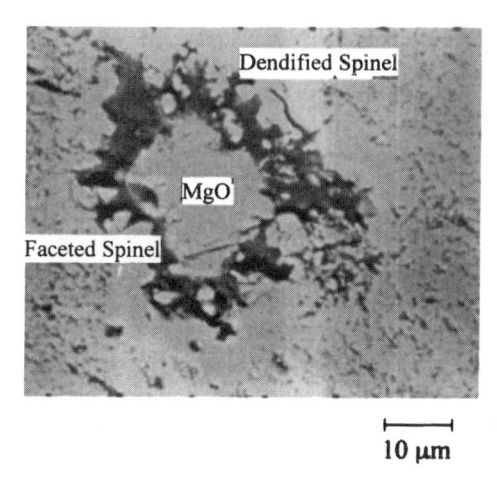

Figure 7. Densification of spinel in an alumina bed and nucleation of another spinel on the densified spinel near magnesia in coarse and dense MgO and Al_2O_3 system occurred at intermediate stage of firing at 1300°C for 10 hr in air.*

The newly nucleated spinel grew like an icicle, exhausting the magnesia in the gap as shown in Figure 8. The growth of the icicle-like spinel was promoted by the increase in the oxygen content in the atmosphere. This growth occurred only in the coarse magnesia system and in the fine magnesia and fine alumina systems.

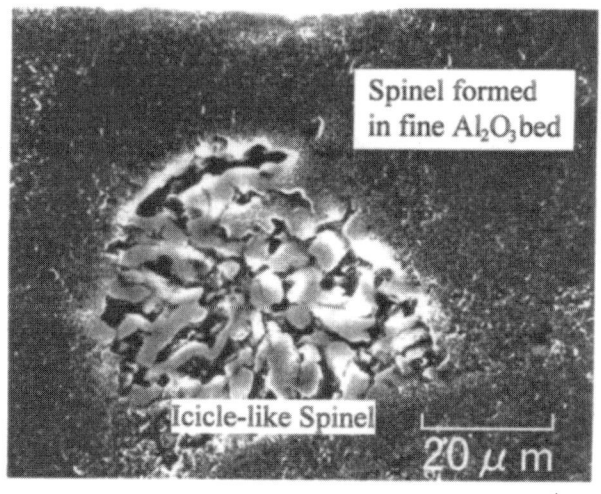

Figure 8. Growth of the new spinel into crystal-like icicles.*

Discussions on the Ionic Transport to Produce Various Morphologies of Spinel

Effect of Firing Atmosphere Retaining Oxygen on the Phenomenological Ionic Transport Processes. When we use fine alumina as a starting powder to form spinel, several atmosphere-sensitive ionic processes occur in the alumina bed near magnesia, as summarized in Table 1.

Table 1. Ionic processes, including atmosphere-sensitive ionic processes, found in the spinel near magnesia.

Processes	O_2	Ar	Conditions
1 Total growth of the spinel layer in alumina bed[5]	í	^	<1300°C
2 The growth of an icicle-like spinel[6]	í		1300°C
3 The fast densification[6,20]	^	í	1300°C
4 The fast grain growth[7–10]			1600°C
5 The fast grain boundary movement[21–24]			1600°C
6 The pore growth at the D_I/D_M interface			1600°C

(Effect of oxygen content on the firing atmosphere)

í: promotive, ^: suppressive

These atmosphere-sensitive phenomenological rate processes should be explained by a mechanism being different from the Wagner model which should have no effect of firing atmosphere due to forbidding of the oxygen movement in the spinel. We have discussed the first process in the table previously, i.e., the parabolic growth of spinel layer, with low activation energy by diffusion of zinc oxide vapor through the fine alumina bed to form $ZnAl_2O_4$, is promoted by the increase in the oxygen contents in the firing atmosphere.[1] The second process, growth of an icicle-like spinel, also occurred in an atmosphere containing oxygen and we have proposed for the growth a cation interdiffusions mechanism partly assisted by the oxygen transport assuming the oxygen vacancy in the solid spinel.[6]

Figure 9 shows the assumed scheme of ionic flow to aid the flow of oxygen brought about by the formation of the oxygen vacancy.[20] Excess magnesia in spinel at the $MgAl_2O_4/MgO$ interface would result oxygen vacancy at the interface, which makes a simultaneous stream of Mg ion and of oxygen vacancy into the spinel in accordance with the assumed scheme. Instead of that, a simultaneous counter-stream of aluminum and oxygen ions have to complement the charge. The sufficient oxygen movement is required for the growth of the icicle-like spinel to balance with the charges of the magnesium ion and oxygen vacancy in the solid state. Promoting effect of the atmosphere containing oxygen on the growth of the icicle-like spinel can be explained by the assumed scheme.[5] Formation of an oxygen vacancy in the spinel at excess magnesia would assist the flow of the magnesium ions. Instead, aluminum and oxygen ions have to move simultaneously in the opposite direction.

The third process in the table, fast densification of the spinel in an alumina bed near magnesia, occurred in an atmosphere without oxygen. Unlike the second process, here the oxygen content in the atmosphere did not promote growth.[5,6] The densification is the reconstruction of the spinel lattice by the movement of oxygen which forms the frame of the lattice and would, therefore, be assisted by the formation of the oxygen vacancy.

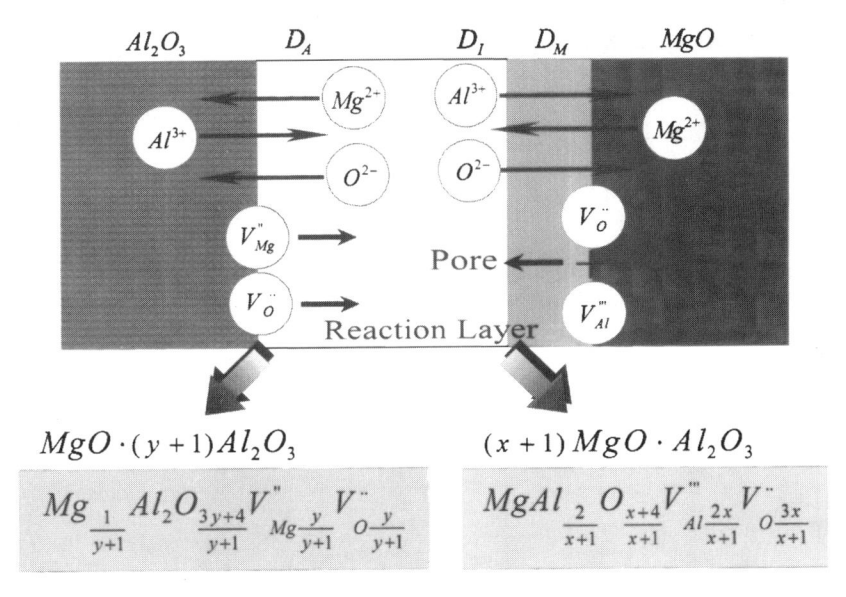

Figure 9. Flow scheme of the ions including the vacancies assumed for the spinel formation reaction and the pore growth.

$MgO \cdot (y+1)Al_2O_3$

$Mg_{\frac{1}{y+1}} Al_2 O_{3\frac{3y+4}{y+1}} V''_{Mg\frac{y}{y+1}} V^{\cdot\cdot}_{O\frac{y}{y+1}}$

$(x+1) MgO \cdot Al_2O_3$

$MgAl_2 O_{\frac{x+4}{x+1}} V'''_{Al\frac{2x}{x+1}} V^{\cdot\cdot}_{O\frac{3x}{x+1}}$

Yamaguchi et al. also found the fourth process, the grain growth of the spinel near magnesia to the direction perpendicular to the reaction layer.[7-10] The grain growth occurred for the spinel in D_M layer in our result. Uematsu et al.[21] and Y. M. Chiang et al.[22-24] also found the rapid grain boundary mobility in $MgAl_2O_4$ spinel at excess magnesia. These fast grain boundary motion would be the driving force for the fast grain growth.

Possibilities of the Vacancies in the Solid Solution of Spinel at Excess Magnesia. Y.M. Chiang et al. proposed two possibilities of ionic transport mechanisms in the fast grain boundary movement brought about in excess magnesia to result fast grain growth.[22-24] The one is the formation of substitutional or interstitial cationic defects, i.e. the magnesium at aluminum site and the magnesium or aluminum at interstitial site, without forming the oxygen vacancy in equation (1) and (2) and the other is the magnesium at aluminum site and the oxygen vacancies keeping electroneutrality in equation (3).

$$4MgO + Al^x_{Al} \rightarrow Mg^x_{Mg} + 4O^x_O + 3Mg'_{Al} + Al^{\cdot\cdot\cdot}_i \quad (1)$$

$$4MgO \rightarrow Mg^x_{Mg} + 4O^x_O + 2Mg'_{Al} + Mg^{\cdot\cdot}_i \quad (2)$$

or

$$3MgO \rightarrow Mg^x_{Mg} + 2Mg'_{Al} + 3O^x_O + V^{\cdot\cdot}_O \quad (3)$$

In phase diagram for magnesia-alumina system, solid solutions were found in excess alumina[25] and also excess magnesia region.[26] In excess alumina to the normal spinel, solid solution without oxygen vacancy is possible because the excess aluminum atom has enough magnesium site to form aluminum at magnesium site keeping balance of charges.[27] In excess magnesia, if oxygen vacancy formation is forbidden, even full substitution of the magnesium at aluminum site is insufficient for the charge balance. Moreover, interstitial magnesium or interstitial aluminum is necessary at extra octahedral site as assumed by Y.M. Chiang in equation (1) and (2).[23] If the oxygen vacancy formation is possible, an equation including aluminum vacancy is possible to balance charges as in the following equation (4) as well as the equation (3).[20,27]

$$4MgO \rightarrow 4Mg_{Mg}^x + 4O_O^x + 3V_O^{\cdot\cdot} + 2V_{Al}^{\prime\prime\prime} \tag{4}$$

Introduction of the Solubility of Vacancies Varying with the Degree of Excess Magnesia in Spinel to Explain Pore Growth. Growth of pores at D_I/D_M interface suggests the simultaneous presence of the cation and anion vacancies.

$$2V_{Al}^{\prime\prime\prime} + 3V_O^{\cdot\cdot} \rightarrow Pore \tag{5}$$

When the equilibrium constant K_d is introduced to the pore formation in Eq. (5), the K_d satisfies Eq. (6). Pores would be formed for the concentration of the vacancies to exceed the limit of solubility product, K_d.[15]

$$K_d = [V_O^{\cdot\cdot}]^3 \, [V_{Al}^{\prime\prime\prime}]^2 \tag{6}$$

The aluminum vacancy should be one of the possible form of the cation vacancies in the solid solution at excess magnesia as well as the magnesium at aluminum site or the magnesium or aluminum at interstitial site assumed by B. Hallstedt.[27] The growth of the icicle-like spinel would be explained by the assistance of the oxygen vacancy and the aluminum vacancy to promote the transport of aluminum ion in the solid spinel. And also the promotion of the densification of spinel in alumina bed in the atmosphere with less oxygen contents can also be explained by the formation of oxygen vacancies in a bulk or at grain boundaries at excess magnesia in spinel at the magnesia side of the spinel. Oxygen vacancy has key role on the characteristic transport phenomena that occurred in magnesia rich spinel like the fast densification.

CONCLUSIONS

In the reaction in the multilayer of fine alumina and fine magnesia, spinel grew in alumina bed. Vapor including magnesium diffused into the tortuous gaps among alumina particles to form spinel in the alumina bed. At low temperature under 1400°C, the fast densification was occurred for the spinel formed in alumina bed near magnesia to form two layer with different density and another densification of spinel started from the side near alumina in delay. At high temperature around 1600°C, the fast grain growth to the direction perpendicular to the reaction layer was occurred in the densified spinel layer near magnesia leaving large grains with thinned end and the grown pores at the D_M/D_I interface. The pore growth would indicate the simultaneous presence of the oxygen vacancy and the aluminum vacancy. The vacancies promote the oxygen and the aluminum transport in solid state respectively during the densification and the grain growth at higher temperature around 1600°C. And also by introducing the solubility product, $K_d = [V_O^{\cdot\cdot}]^3 \, [V_{Al}^{\prime\prime\prime}]^2$, of the oxygen vacancy and the aluminum vacancy, the pore growth would be explained as a result where the vacancy concentrations exceed the decreased K_d with decreasing the excess degree of excess magnesia in the bulk of the solid solution.

REFERENCES

1. H. Okada, H. Kawakami, M. Hashiba, E. Miura, Y. Nurishi and T. Hibino, Effect of physical nature of powders and firing atmosphere on ZnAl$_2$O$_4$ formation, *J. Amer. Ceram. Soc.*, **68**:58 (1985).
2. C. Yokoyama, H. Kawakami, M. Hashiba, E. Miura, Y. Nurishi and T. Hibino, Effect of agglomeration of Al$_2$O$_3$ on the solid state reaction between ZnO and Al$_2$O$_3$, *J. Ceram. Soc. Japan*, **91**:525 (1983).
3. H. Kawakami, H. Okada, M. Hashiba, E. Miura, Y. Nurishi and T. Hibino, "Effect of dispersion of reaction mixtures on the kinetics of formation process of ZnAl$_2$O$_4$, *J. Ceram. Soc. Japan*, **90**:642 (1982).
4. Y. Hukuhara, E. Suzuki, M. Hashiba, E. Miura, Y. Nurishi and T. Hibino, Formation of ZnAl$_2$O$_4$ in ZnO-Al$_2$O$_3$ system accompanied by uneven mixing and composition change, *J. Ceram. Soc. Japan*, **91**:281 (1983).
5. M. Itoh, A. Matsumoto, O. Sakurada, M. Hashiba and Y. Nurishi, Diversity of MgAl$_2$O$_4$ formation associated with changes in physical nature of powders, *Solid State Ionics*, **63-65**:183 (1993).
6. T. Kuwabara, A. Matsumoto, O. Sakurada, M. Hashiba and Y. Nurishi, Detailed microstructure of an icicle-like spinel grown at MgO/MgAl$_2$O$_4$ interface in dense and coarse MgO-Fine Al$_2$O$_3$ powder system, Proceeding of the JFCC International Workshop on Fine Ceramics '96 -Mass and Charge Transport- to appear as the Ceramic Transactions, The American Ceramic Society.
7. I. Yasui, G. Yamaguchi and H. Kurosawa, Interfacial phenomena and the orientation of spinel in MgO-Al$_2$O$_3$ solid state reaction, *J. Ceram. Soc. Japan*, **81**:197 (1973).
8. I. Yasui, G. Yamaguchi and M. Tanaka, Etching studies on the spinel layer formed by solid state reaction between periclase and sapphire, *J. Ceram. Soc. Japan*, **81**:413 (1973).
9. K. Shirasuka and G. Yamaguchi, The orientation of spinel formed by MgO-Al$_2$O$_3$ solid state reaction between 1400 and 1600°C, *J. Ceram. Soc. Japan*, **81**:97 (1973).
10. G. Yamaguchi, K. Shirasuka and M. Munekata, Some aspect of solid state reaction of spinel formation in the system MgO-Al$_2$O$_3$, *J. Ceram. Soc. Japan*, **79**:64 (1971).
11. L. Navias, Preparation and properties of spinel made by vapor transport and diffusion in the system MgO-Al$_2$O$_3$, *J. Amer. Ceram. Soc.*, **44**:434 (1961).
12. R. E Carter, Mechanism of solid-state reaction between magnesium oxide and aluminum oxide and ferric oxide, *J. Amer. Ceram. Soc.*, **44**:116 (1961).
13. R. C. Rossi and R. M. Fulrath, Epitaxial growth of spinel by reaction in the solid state, *J. Amer. Ceram. Soc.*, **46**:145 (1963).
14. A. Yamakawa, M. Hashiba and Y. Nurishi, Growth of zinc aluminate on the surfaces normal to the various crystal axes of an alumina single crystal, *J. Mater. Sci.*, **24**:1491 (1989).
15. V. S. Stubican, C. Greskovich and W. P. Whitney, Interdiffusion studies in some oxide systems, *Mater. Sci. Res.*, **6**:55 (1973).
16. Y. Oishi and K. Ando, Self diffusion of oxygen in polycrystalline MgAl$_2$O$_4$, *J. Chem. Phys.*, **63**:376 (1975).
17. K. P. R. Reddy and A. R. Cooper, Oxygen Diffusion in Magnesium Aluminate Spinel, *J. Amer. Ceram. Soc.*, **64**:368 (1981).
18. J. S. Moya, R. Moreno and J. Requena, Interfacial reaction in zirconia-alumina multilayer composites, *J. Eur. Ceram. Soc.*, **7**:27 (1973).
19. H. Wohlfromm, P. Pena, J. S. Moya and J. Requena, Al$_2$TiO$_5$ formation in alumina/titania multilayer composites, *J. Amer. Ceram. Soc.*, **75**:3473 (1992).
20. T. Kume, T. Kuwabara, O. Sakurada, M. Hashiba, Y. Nurishi, J. Requena and J. S. Moya, Morphologies of spinel and sintering occurred in the MgAl$_2$O$_4$ formation in fine MgO - fine Al$_2$O$_3$ system, Abstract of the 6th Symposium on Reactivity of Solid '95 by Japan Chemical Society (Tokyo), Presentation No. 10, pp. 37-40 (1995).
21. K. Uematsu, R. M. Cannon, R. D. Bagley, M. F. Yan, U. Chowdry and H. K. Bowen, Microstructure evolution controlled by dopants and pores at grain boundaries, pp. 190-205 in Proceeding of an International Symposium on Factors in Densification and Sintering of Oxide and Non-Oxide Ceramics, Tokyo Institute of Technology, Tokyo, Japan, 1979.

22. Y. M. Chiang and W. D. Kingery, Grain-boundary migration in nonstoichiometric solid solution of magnesium aluminate spinel: I, grain growth study, *J. Amer. Ceram. Soc.*, **72**:271 (1989).

23. Y. M. Chiang and W. D. Kingery, Grain-boundary migration in nonstoichiometric solid solution of magnesium aluminate spinel: II, effect of grain-boundary nonstoichiometry, *J. Amer. Ceram. Soc.*, **73**:1153 (1990).

24. Y. M. Chiang and C. J. Peng, Grain-boundary nonstoichiometry in spinels and titanates, Advances in Ceramics, Vol. 23, pp 361-377, Nonstoichiometric Compounds, Edited by C.R.A. Catlow and W.C. Mackrodt, The American Ceramic Society, Westerville OH, (1989).

25. A. M. Alper, R. N. McNally, P. G. Ribbe and R. Doman, The system $MgO-MgAl_2O_4$, *J. Amer. Ceram. Soc.*, **45**:263 (1962).

26. D. M. Roy, R. Roy and E. F. Osborn, The system $MgO-Al_2O_3-H_2O$ and influence of carbonate and nitrate ions on the phase equilibrium, *Am. J. Sci.*, **251**:337 (1953).

27. B. Hallstedt, Thermodynamic assessment of the system $MgO-Al_2O_3$, *J. Amer. Ceram. Soc.*, **75**:1497 (1992).

MICROSTRUCTURE EVOLUTION DURING OXIDATION OF MgO-SiC COMPOSITES

D. J. Aldrich, M. E. F. Camey and D. W. Readey

Colorado Center for Advanced Ceramics
Colorado School of Mines
Golden, CO 80401

INTRODUCTION

Increasing importance has been given to ceramic matrix composites reinforced with SiC particulates, whiskers, and fibers as high temperature structural materials. The oxidation of SiC-reinforced alumina[1-5], mullite[4-6], and alumina-zirconia[7,8] has been studied. In this research, MgO-SiC was chosen as a model material for the main purpose of determining the mechanism of mass transport and the rate-controlling step during oxidation. MgO was chosen as the matrix material for several reasons. The most important is that MgO is an ideal material for studies of mass transport because of the availability of reasonable diffusion data.[9,10] Stability at high temperature in oxidizing environments and the ability to be easily fabricated into a dense composite microstructure also supported the choice of MgO as matrix material. Previous results[11,12] had suggested that magnesium ion diffusion was controlling the rate of oxidation. Therefore, these experiments on both pure and scandium oxide-doped material were performed to confirm this mechanism. Scandium oxide produces cation vacancies in MgO[13,14] according to:

$$Sc_2O_3 \xrightarrow{MgO} 2Sc^{\bullet}_{Mg} + 3O^{X}_{o} + V''_{Mg}.$$

At the temperatures and dopant concentrations used, association of scandium ions with free vacancies to form less mobile defects will occur to some extent[14]:

$$Sc^{\bullet}_{Mg} + V''_{Mg} \rightarrow (Sc^{\bullet}_{Mg}, V''_{Mg})'.$$

Despite defect association, the magnesium vacancy concentration will increase with scandium oxide doping resulting in an increase in Mg^{2+} cation diffusivity which should lead to more rapid oxidation rates in doped samples.

EXPERIMENTAL PROCEDURE

The doped MgO powders were prepared by mixing 5.38 weight percent (w/o) Sc_2O_3 to MgO powder. A saturated solid solution was guaranteed by the presence of a second phase of Sc_2O_3. Scandium was chosen because it has a high solubility in MgO, between 0.005 and 0.044 mole fraction in the temperature interval from 1270 to 1600 °C.[13] The mixed powders were ball milled in polyethylene bottles with Al_2O_3 media for 24 hours. Mixtures were calcined in two steps and ball milled between steps. The first calcination was at 1100 °C for 24 hours and the second at 1650 °C for 20 hours in MgO crucibles. Undoped MgO powder was mixed with 5, 10, and 15 volume percent (v/o) particulate SiC. Doped samples were prepared with only the 10 v/o SiC loading. Samples were hot-pressed in vacuum at 1635 °C for 90 minutes at 40 MPa. The density of the composites was greater than 98.8% of theoretical to ensure the absence of open porosity.

Samples were oxidized in a single-zone tubular furnace in air in an alumina tube. Small slab samples, in the range of 2x2 mm to 5x5 mm, were hung on a platinum wire into the furnace at the desired oxidation temperature for 5, 10, 20, and 30 hours. Temperatures of 1400, 1450 and 1500 °C were used for undoped samples and 1200, 1300 and 1400 °C for doped ones. Samples for SIMS (Secondary Ion Mass Spectroscopy) analysis were oxidized in a mixture of 50% $^{18}O_2$ and 50% $^{16}O_2$.

Samples for TEM were prepared by cutting a section perpendicular to the oxidation front. The sample was polished flat on both sides down to 0.05 µm Al_2O_3 and glued to a 3 mm copper slot grid. It was then dimpled to a thickness of ~10 µm and ion-milled to transparency with 5 kV argon ions at a gun current of 1 mA. Microscopy was performed on both a Philips CM-200 and a CM-30 operating at 200 and 300 kV, respectively. EDS microanalysis (Kevex, Oxford Pentafet) was performed in both nanoprobe and STEM operational modes.

RESULTS AND DISCUSSION

The oxidized region after oxidation consisted of a surface scale that could be macroscopically subdivided in two distinct layers: a white outer layer and a black inner region (dark layer), as shown in Figure 1. Higher magnification of fracture surfaces, Figure 2, shows that the white layer can be further subdivided into a porous and a columnar grain growth layer. The presence of the columnar layer can be seen at the top of Figure 1 as well as a region of greater transparency. X-ray diffraction patterns of unoxidized and oxidized Sc_2O_3-doped samples are shown in Figure 3. The most striking feature is the strong preferred <200> orientation of the oxidized surface. The diffraction pattern represents the strong preferred orientation of the columnar grain surface layer and suggests that the growth direction of the layer was normal to the surface. After removal of the grain growth layer by grinding, only residual Sc_2O_3, MgO, and Mg_2SiO_4, the product of oxidation, are seen in the diffraction pattern. The thicknesses of the dark layer, the white (porous plus columnar) layer, and the columnar large-grained surface layer all increase with time until the white layer has reached about 100 µm. Then both the columnar layer and the dark layer do not increase in thickness but the white layer continues to grow. Scandium doping increases the thicknesses of both the dark layer and the columnar surface layer dramatically as shown in Figure 4, while the porous intermediate layer is thinner than in the undoped material. The porosity in the layer just below the large grains is also apparent in Figure 4. Scandium-doping changes the oxidation behavior considerably which suggests that the rate of magnesium diffusion increases as expected. To provide additional evidence

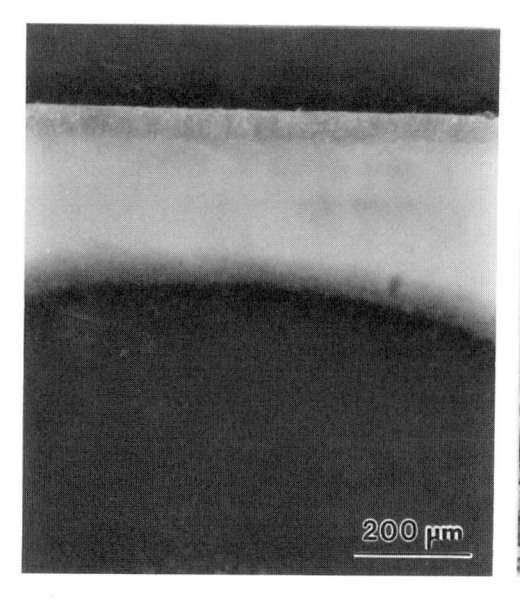

Figure 1. Optical micrograph of a polished sample oxidized at 1400 °C for 30 hours.

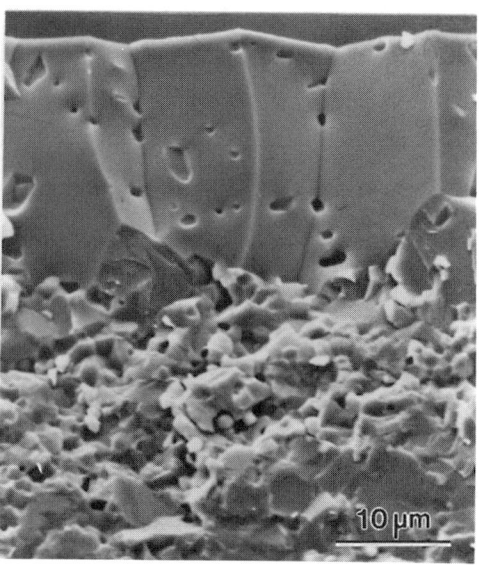

Figure 2. Photomicrograph of an undoped sample containing 10 v/o SiC oxidized at 1400 °C for 20 hours showing the large columnar grain surface layer with porous MgO beneath.

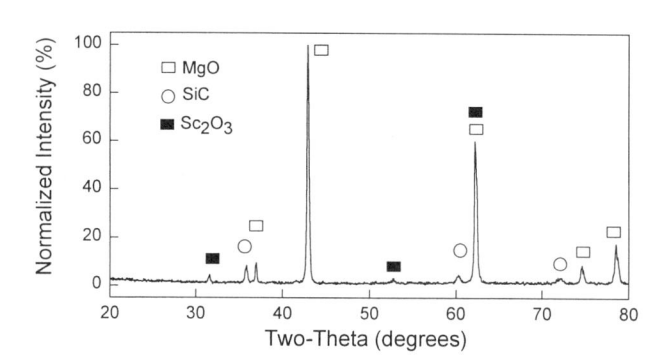

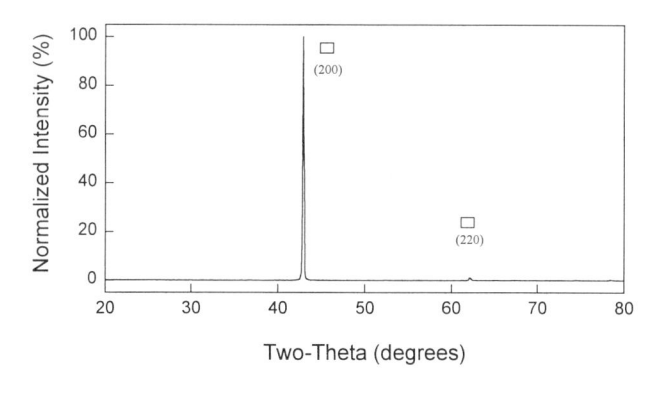

Figure 3. X-ray diffraction patterns for Sc-doped MgO-10 v/o SiC: top, unoxidized; bottom after 30 hours at 1400 °C. Sample surface.

591

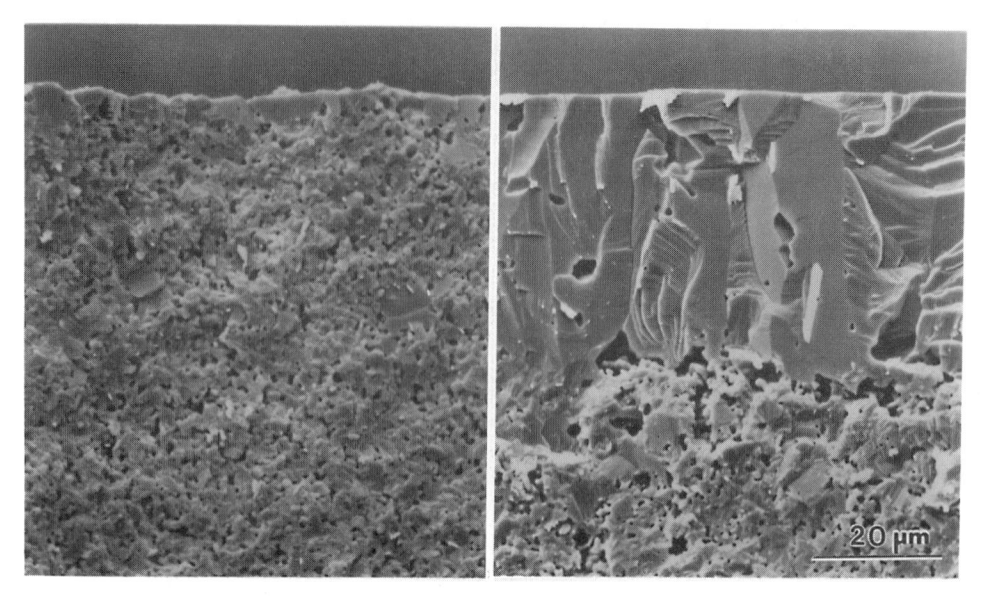

Figure 4. Comparison of the columnar grain growth layers in undoped (left) and Sc$_2$O$_3$-doped (right) samples of 10 v/o SiC oxidized for 30 hours at 1400 °C.

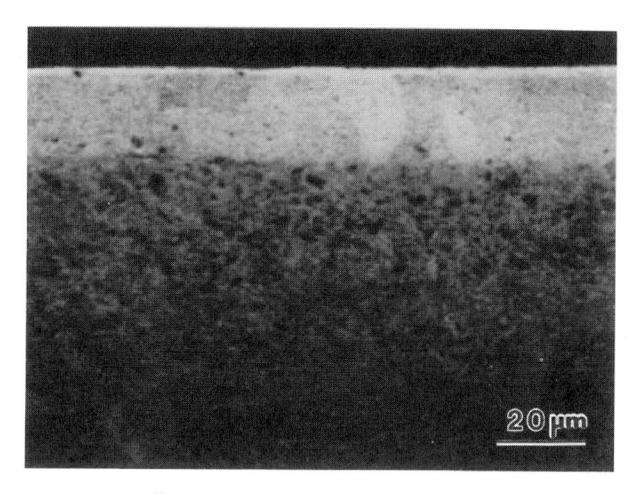

Figure 5. SIMS image showing ^{18}O$_2$ signal for a Sc-doped MgO, 10 v/o SiC oxidized for 30 hours at 1400 °C in a mixture of ^{18}O$_2$ / ^{16}O$_2$.

for magnesium diffusion, oxidation was carried out in an ^{18}O$_2$-enriched atmosphere. If the columnar surface layer and the rate of oxidation are determined by magnesium diffusing out, then only the columnar grain layer should contain ^{18}O after oxidation. The ^{18}O$_2$ signal obtained by imaging SIMS for a Sc-doped MgO, 10 v/o SiC sample oxidized for 30 hours at 1400°C in a mixture of ^{18}O$_2$/^{16}O$_2$ is presented in Figure 5. Bright regions are high in ^{18}O. This figure suggests that the grain growth layer was indeed formed with oxygen from the atmosphere. Some ^{18}O enrichment is also observed in the porous region of the white layer.

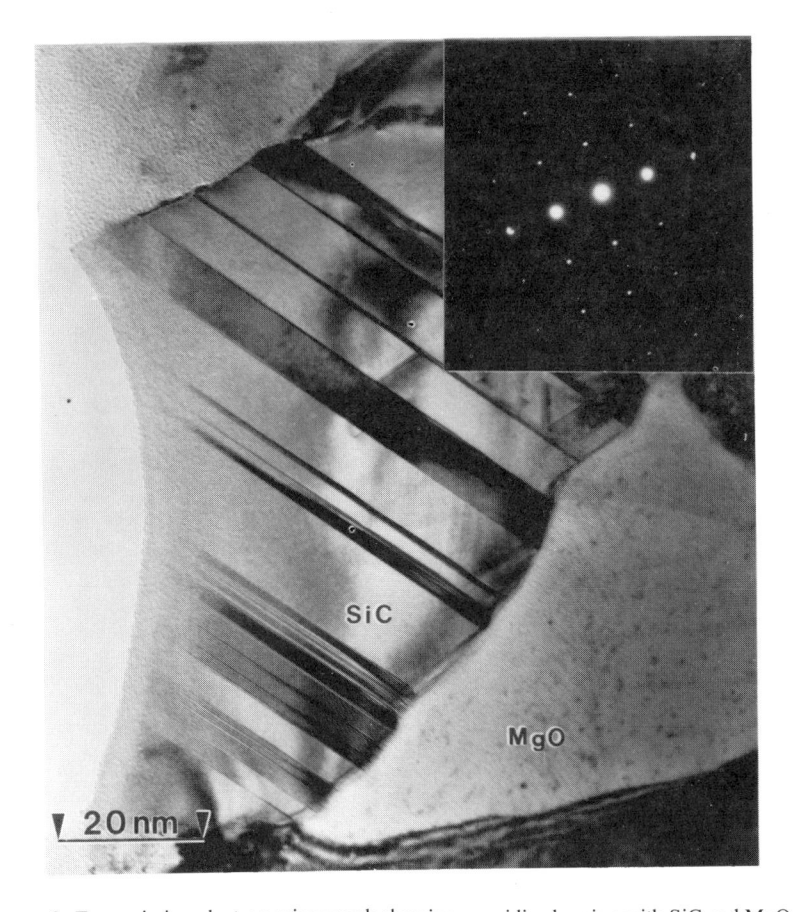

Figure 6. Transmission electron micrograph showing unoxidized region with SiC and MgO.

In the starting material, the SiC and Sc$_2$O$_3$ particles were homogeneously distributed in the MgO matrix. In addition, the microstructure appears to be dense, without any noticeable porosity. Figure 6 is a TEM bright-field image showing SiC in contact with MgO. The interface between these grains appears sharp, indicating that no interdiffusion occurred during hot-pressing. Figure 7 shows a similar SiC grain (with accompanying diffraction pattern) in the partially oxidized region, but the interface with the neighboring particle appears to be rough (indicated by the arrow), unlike the unoxidized material. Examples from other systems, i.e. Al$_2$O$_3$-ZrO$_2$/SiC,[8] would suggest that the oxidation occurs through the following series of reactions:

$$SiC(s) + O_2(g) = SiO_2(s) + C(s)$$
$$C(s) + 1/2\ O_2(g) = CO(g)$$
$$2\ MgO(s) + SiO_2(s) = Mg_2SiO_4(s).$$

The CO generated during oxidation presumably accounts for the pores that are found in the porous white layer. The reactions are consistent with observations that forsterite is present in all of the oxidation scale. However, no distinguishable SiO$_2$ has been located adjacent to the SiC particles in the partially oxidized region. Instead, partially oxidized SiC grains adjacent to Mg$_2$SiO$_4$ were observed, Figure 7. In addition, elemental silicon was observed,

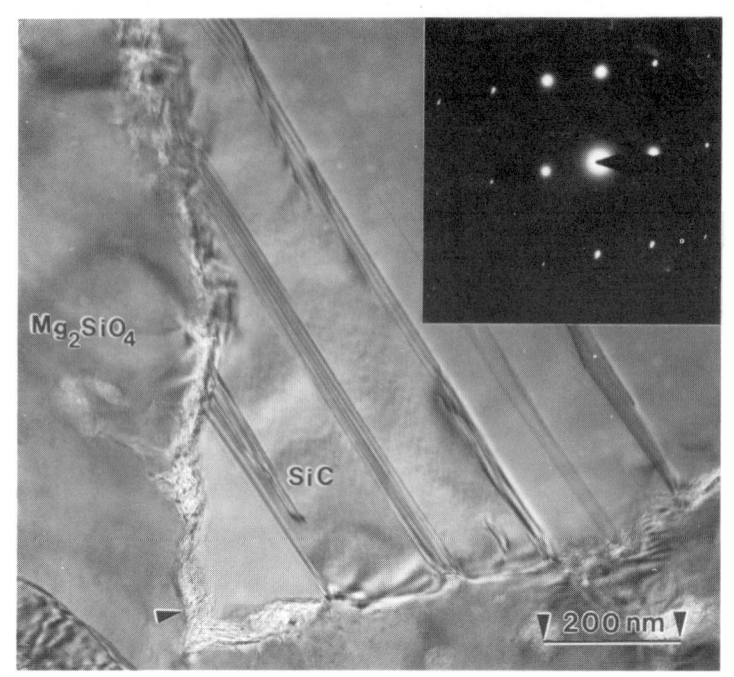

Figure 7. Microstructure in the vicinity of a partially-oxidized SiC grain in the dark region of the oxidized layer.

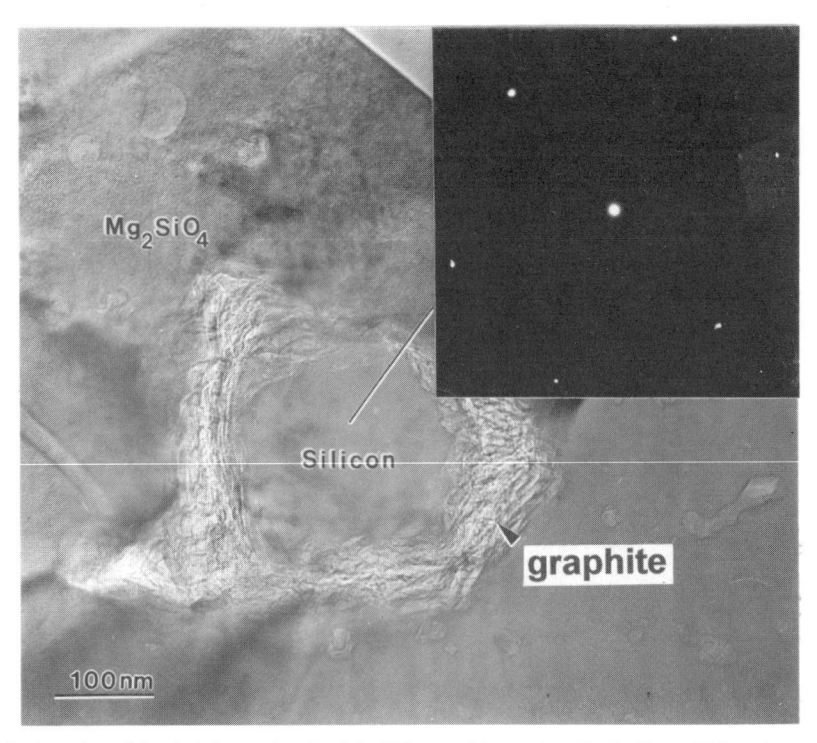

Figure 8. A region of the dark layer showing Mg_2SiO_4, graphite, and residual silicon (diffraction pattern).

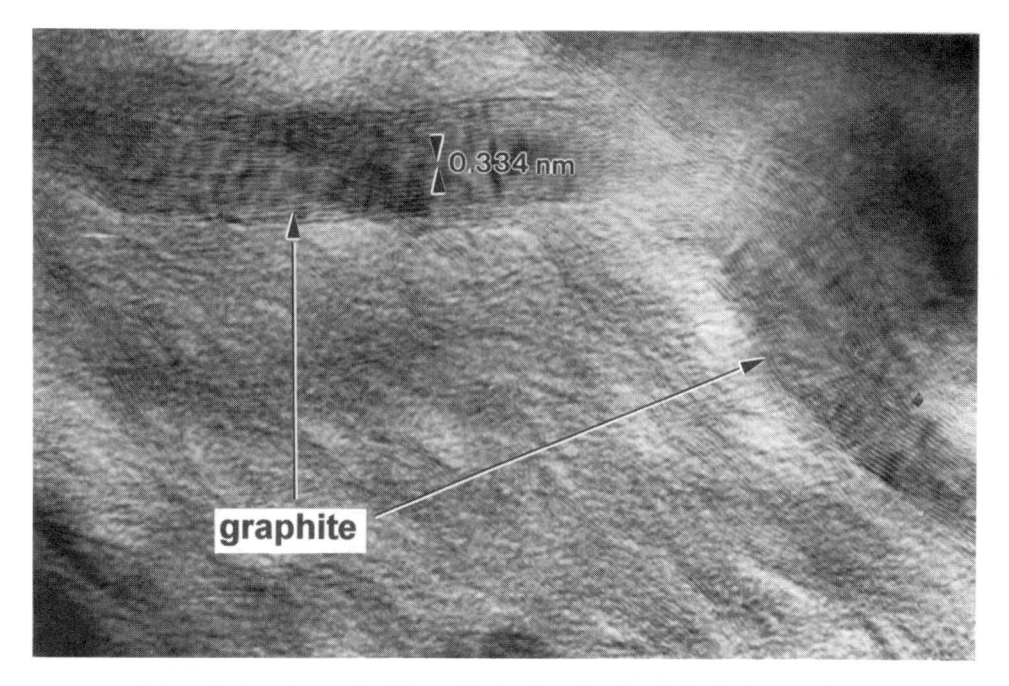

Figure 9. High resolution micrograph of the residual carbon showing the lattice images of graphite planes.

which appears to be the result of a partially-oxidized SiC grain, Figure 8. Finally, HREM imaging of the "reacted" interface in Figure 9 shows that it is graphitic carbon.

It could be speculated that these partially-oxidized regions should contain some amorphous SiO_2 as an intermediate oxidation product, but there is no evidence for SiO_2. The observation of elemental silicon is a surprising result. There are two possibilities to account for it. The first is that Si is being formed as an intermediate oxidation product. A possible non-equilibrium decomposition could be occurring, such as: $SiC(s) + 1/2O_2(g) = Si(s) + CO(g)$. A reaction such as this would explain the lack of SiO_2 in the partially oxidized scale. The second possibility that exists is that residual, unreacted silicon is present in the original starting SiC powder. This is not as likely since the probability of observing this type of relic in the small volume of a TEM sample is low, yet silicon particles are frequently observed.

Conclusions

- Oxidation of SiC-MgO composites leads to three oxidation product scales: a dark layer, a porous white layer, and a columnar grain growth surface layer.
- The columnar growth layer exhibits a strong <200> texture.
- Sc_2O_3 doping enhances the oxidation of MgO-SiC composites and increases the thickness of the surface grain growth layer.
- SIMS data indicate that ^{18}O enrichment occurs only in the grain growth layer.
- TEM investigation of the dark region shows the presence of partially oxidized SiC in contact with graphite and Mg_2SiO_4 which is consistent with models and observations of oxidation in similar composite systems.
- Elemental silicon in contact with large regions of graphite has been observed but is not easily explained by conventional models. It is postulated that the silicon is either an unexpected intermediate oxidation product or a relic of the starting SiC.

Acknowledgments

This research was sponsored in part by the National Science Foundation, grant No. DMR-9110927 and NASA Microgravity Science and Applications Division grant NAG3-1409. Use of the CM-200 at CSM was made possible by NSF grant No. DMR-9204064. The authors would like to thank Mr. Kim Jones at the National Renewable Energy Laboratory for assistance with the TEM, and Dr. Michael Graham at the National Research Council of Canada for the Imaging SIMS data.

References

1. P. F. Becher and T. N. Tiegs, Temperature dependence of strengthening by whiskers reinforcement: SiC whisker-reinforced alumina in air, *Adv. Ceram. Mater.*, 3:148 (1988).
2. F. Lin, T. Marieb, A. A. Morrone, and S. R. Nutt, Thermal oxidation of Al_2O_3-SiC whisker composites: mechanisms and kinetics, in *Mater. Res. Soc. Symp. Proc.*, *Vol. 120* , Materials Research Society, Pittsburgh (1982).
3. J. R. Porter, F. F. Lange, and A.H. Chokshi, Processing and creep performance of SiC-whisker-reinforced Al_2O_3, *Am. Ceram. Soc. Bull.* 66:343 (1987).
4. M. P. Borom, M. K. Brun, and L. E. Szala, Kinetics of oxidation of carbide and silicide dispersed phases in oxide matrices, *Adv. Ceram. Mater.* 3:491 (1988).
5. K. L. Luthra and H. D. Park, "Oxidation of silicon carbide-reinforced oxide-matrix composites at 1375 °C to 1575 °C, *J. Am. Ceram. Soc.* 73:1014 (1990).
6. M. I. Osendi, Oxidation behavior of mullite-SiC composites, *J. Mater. Sci.* 25:3561 (1990).
7. P. Wang, G. Grathwohl, F. Porz, and F. Thummler, Oxidation behaviour of SiC whisker-reinforced Al_2O_3/ZrO_2 composites, *Powder Metallurgy Int.* 23:370 (1991).
8. M. Backhaus-Ricoult, Oxidation behavior of SiC-whisker-reinforced alumina-zirconia composites, *J. Am. Ceram. Soc.* 74:1793 (1991).
9. B. J. Wuensch, W. C. Steele, and T. Vasilos, Cation self-diffusion in single crystal MgO, *J. Chem. Phys.* 58:5258 (1973).
10. B. J. Wuensch, *et al.*, The mechanisms for self-diffusion in magnesium oxide, in: *Point Defects and Related Properties of Ceramics*, Ceram. Trans. Vol. 24, T. O. Mason and J. L. Routbort, eds, American Ceramic Society, Westerville, Ohio (1991).
11. G. W. Hallum, *High Temperature Effects of Oxidation on MgO-SiC Composites*, Ph.D. Thesis, The Ohio State University (1990).
12. M. E. F. Camey, and D. W. Readey, Oxidation kinetics of pure and doped MgO-SiC composites, *Ceram. Eng. and Sci. Proc.* 16:863 (1995).
13. A. F. Henriksen and W. D. Kingery, Solid solubility of Sc_2O_3, Al_2O_3, Cr_2O_3, SiO_2, and ZrO_2 in MgO, *Ceram. Int.* 5:11 (1979).
14. D. R. Sempolinski and W. D. Kingery, Ionic conductivity and magnesium vacancy mobility in magnesium oxide, *J. Am. Ceram. Soc.* 63:664 (1980).

SYNTHESIS, TAILORED MICROSTRUCTURES AND "COLOSSAL" MAGNETORESISTANCE IN OXIDE THIN FILMS

Kannan M. Krishnan[1], A. R. Modak[2], H. Ju[1] and P. Bandaru[1]
[1]Materials Sciences Divison, E.O. Lawrence Berkeley National Laboratory, Berkeley, CA 94720;
[2]present address, Silicon Systems Inc., Santa Cruz, CA

INTRODUCTION

$La_{1-x}M_xMnO_3$ (where M = Sr, Ba or Ca) thin films, exhibiting very high or "colossal" magnetoresistance (CMR), have generated much recent scientific and technological interest[1-4]. In addition to raising fundamental questions on insulator-metal transitions and magneto-transport phenomenon, these materials have potential impact on the future of magnetic field sensing and data storage devices. However, the magnetic and magneto-transport properties of these manganite thin films are believed to be dependent on the optimization of the conditions of growth/annealing, composition, oxidation state, epitaxy and the overall microstructure.

In order to address these issues we have grown $La_{1-x}Sr_xMnO_3$ ($0<x<1$) films (LSMO) using both pulsed laser deposition and a polymeric sol-gel route developed in our laboratory. Pulsed laser deposition provides the ability to deposit ultra-thin films (1-10 nm) and multilayers in addition to being an established, albeit slower, method for the synthesis of thicker (100-1000nm) perovskite films. Polymeric sol-gel synthesis, if suitably optimized, is very versatile and inexpensive. It has additional advantages of overall simplicity, low processing temperature, ease of composition variation, and the ability to produce a wide range of structures. Overall, these two different growth techniques result in different microstructures but in both cases the texture (epitaxy or polycrystallinity) can be effectively controlled by choice of substrates and growth conditions. The crystallography and microstructure of these films were studied using x-ray diffraction and high-resolution transmission electron microscopy. The magnetic/ magnetotransport properties of these LSMO films are then discussed in the context of their growth and microstructural parameters.

THIN FILM GROWTH

LSMO thin films were synthesized[5] by pulsed laser deposition using a 248 nm KrF excimer laser (Lambda Physik, Lextra 200) with a 30ns pulse width and output energy of 400 mJ. The laser was operated at 3 Hz for 15 minutes producing films ~50nm thick. A sintered $La_{0.8}Sr_{0.2}MnO_3$ target was used. The ejected plume of material was deposited simultaneously onto polished $LaAlO_3[100]$ and $Si[100]$ (with native oxide) substrates maintained at 675°C. The base pressure in the chamber was 2x10^{-6} Torr. During the deposition the oxygen pressure in the vacuum chamber was maintained at 100 mTorr by flowing oxygen at 50-70 sccm. After deposition, the films were cooled to room temperature at 5°C/min. in an oxygen pressure between 500-600 Torr.

The polymeric chemical process developed in our laboratory [6] used organometallic precursors of the constituent elements which were synthesized from commercially available compounds. Alkoxides of La, Sr and Mn are possible starting materials due to their ability to complex into atomically well-mixed precursors[7]. However, alkoxides of lanthanum and manganese have limited solubility in organic solvents and hence their respective carboxylates or β-diketonates, were used. Strontium metal, can be readily reacted with different alcohols to form stable alkoxides which are completely soluble in the respective alcohols. After several such considerations and pilot experiments we arrived at lanthanum 2,4-pentanedionate (or Lanthanum acetyl acetonate), manganese(II) acetate and strontium methoxyethoxide as the organometallic precursors. 2-methoxyethanol was used as the base solvent since it has been reported to have advantages over other alcohols in similar sol-gel processes for fabricating perovskite thin films. Individual precursors underwent further processing steps of complexation (in required stoichiometric proportions, i.e. La:Sr:Mn = 0.8:0.2:1) and hydrolysis. To estimate and optimize hydrolysis the precursor complex was hydrolysed, in a pilot experiment, using varying amounts of water from 1 mole H$_2$O/mole manganese to 3 moles H$_2$O/mole manganese. Instead of carrying out gelation tests, each solution was aged for one day and films were deposited. X-ray diffraction was carried out on all the films to determine occurence of extraneous phases resulting from variation in hydrolysis conditions but no difference in the crystal structure was observed for the above range of conditions. This is not unusual because it has been observed before that β-diketonate ligands hydrolyse very slowly, in effect making this process hydrolysis independent.

The complex solutions were spin cast (2000 rpm, 30 sec.) on the required substrate (LaAlO$_3$(100) or Si(100)) accompanied by solvent removal at 300°C. Thermogravimetric analysis (TGA) of the gel dried at a low temperature (80°C) was carried out to estimate film heat treatment temperature. Fig. 1 shows a plot of the weight of the sample versus temperature from the TGA experiment carried out at a heating rate of 10°C/min. The plot reveals an initial gradual weight loss at low temperatures related to solvent removal, followed by a rapid weight loss between 350 and 500°C which is associated with the oxidation of the organics with release of CO$_2$ and H$_2$O. There is further gradual weight loss after 500°C (about 10% of total weight loss) and the weight remains constant after

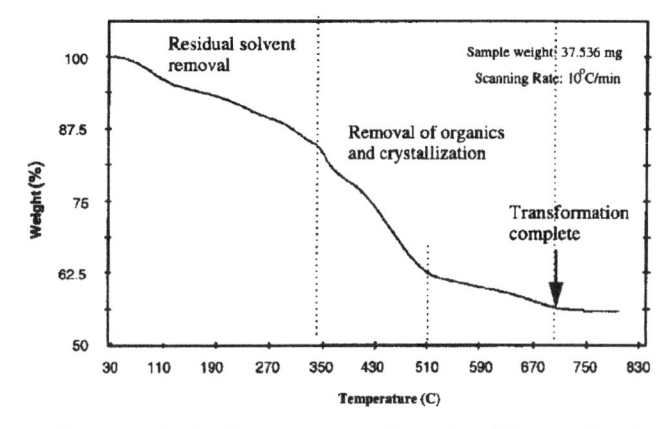

Figure 1. Thermogravimetric analysis of precursor complex gel: weight as a function of temperature measured during heating cycle from 0 - 800 °C at 10°C/min.

about 650°C at which temperature the organometallic-to-oxide conversion was presumed to be complete. The solution was spin cast at 2000 rpm for 30 seconds on the required substrates (either Si(100) or LaAlO$_3$(100)) to form an amorphous condensed film. Amorphous films were heated at 300°C on a hot plate, immediately after spin coating, for trapped solvent removal. Final heat treatment was carried out, based on the above TGA results, at 700°C for 1 hour in air to obtain the oxide thin films. Multiple deposition/drying cycles were used to make thicker films.

CHARACTERIZATION: CRYSTALLOGRAPHY AND MICROSTRUCTURE

PLD grown La$_{0.8}$Sr$_{0.2}$MnO$_3$ Thin Films

Figures 2 a&b show x-ray scans from the LSMO(80/20) films on Si(100) and LaAlO$_3$(100) substrates respectively. A fixed θ (~ 8°), variable 2θ scan (Fig. 2a) of the film grown on Si(100), confirms it to be polycrystalline and randomly oriented. Assuming a cubic structure for the perovskite film, the peaks were indexed as shown and the lattice constant was determined to be a = 3.88 ± 0.05Å. Fig. 2b shows a conventional θ–2θ scan for the film grown on the LaAlO$_3$ substrate (a=3.78Å). The split peaks at each position correspond to the substrate (higher intensity) and the perovskite film. The position of the perovskite peak corresponds to a cubic surface-normal lattice constant, c = 3.92 Å. Measurement of in-plane Bragg reflections, e.g. (111), gave a smaller lattice mismatch and implies that there is some in-plane lattice matching causing a tetragonal distortion in the epitaxial film. The shoulder on the low angle side of the film peak (fig. 2b) is an order of magnitude smaller than the principal peak and may be due to regions of the film that are perfectly lattice matched, in-plane, to the substrate, thus leading to a larger tetragonal distortion.

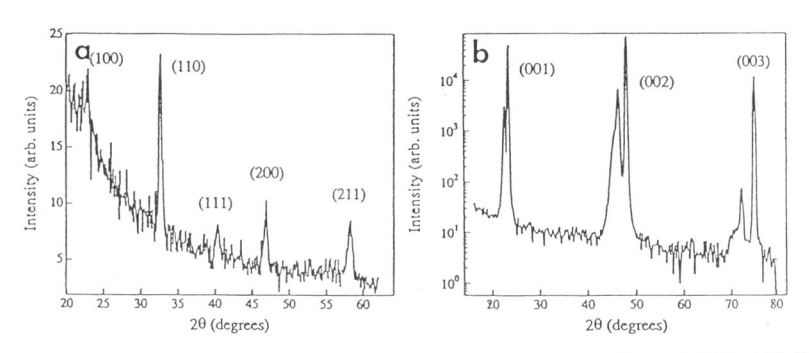

Figure 2. X-ray diffraction data of PLD grown LSMO films on (a) Si(001) and (b) LaAlO$_3$(001)

Transmission electron micrographs of the PLD grown LSMO(80/20) film, on Si(100) substrate, are shown in Fig. 3a &b. In plan view, the film is seen to be polycrystalline with a grain size of 100-200 nm. In cross-section, the films is revealed to be uniform, ~ 450Å thick and having columnar grain morphology. The film/substrate interface was found to be

Sol-gel Derived La$_{1-x}$Sr$_x$MnO$_3$ (0< x< 0.7) Thin Films

The thin films had a smooth reflecting surface and were completely crack-free. The crystal structure was investigated using x-ray diffraction (XRD). Fig. 4(a) and(b) show

free of interdiffusion. The SAD pattern (Fig. 3c) shows a diffraction ring which was indexed to a cubic structure with the same lattice parameters as those observed from XRD.

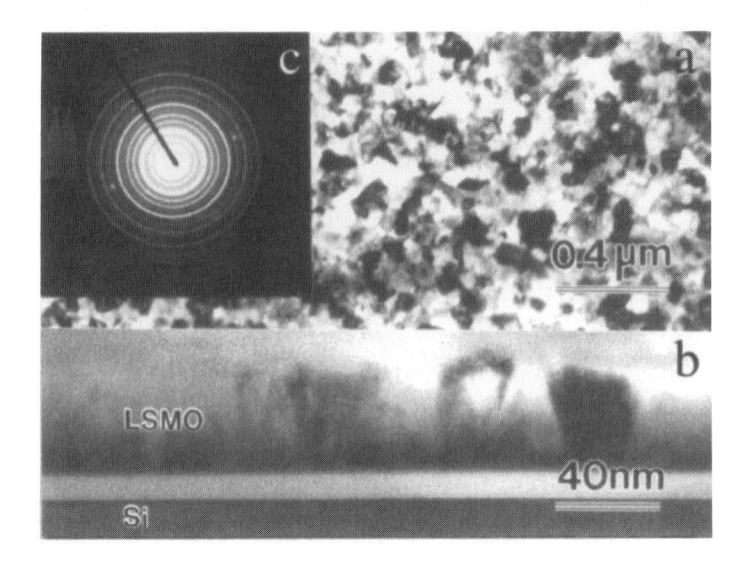

Figure 3. Transmission electron microscopy of LSMO films grown on Si(001) -- (a) plan view, (b) cross-section and (c) selected area diffraction.

normal θ/2θ scans taken from LSMO(80/20) thin films grown on Si(100) and LaAlO$_3$(100) substrates respectively. Single phase, polycrystalline and randomly oriented La$_{0.8}$Sr$_{0.2}$MnO$_3$ thin films were obtained on Si(100) substrates. The XRD pattern from films grown on the LaAlO$_3$ substrate clearly revealed that the films were highly oriented. Diffraction peaks corresponding to only (100) and (200) crystal planes were observed in the normal θ/2θ scans indicating complete (100) normal epitaxy. A plan view, bright field transmission electron micrograph (Fig. 5a) of the LSMO(80/20) thin film grown on Si(100) shows the polycrystalline microstructure with a grain size of 1-2 μm and some inter- and intra-granular porosity (10-20%). In cross-section (Fig. 3b), the film was confirmed to be 0.5μm thick and uniform with some surface roughness (~20 nm). The multilayer structure arising from the multiple coatings (to build film thickness) is distinctly visible. Similar to the PLD films, there was little evidence of interdiffusion at the film/substrate interface. The selected area electron diffraction pattern (Fig. 5c) was indexed to a cubic structure with the same lattice parameters as observed from XRD.

MAGNETIC AND TRANSPORT PROPERTIES

PLD grown La$_{0.8}$Sr$_{0.2}$MnO$_3$ Thin Films

The magnetization M(T, H= 5 kOe) data (Fig. 6a) for the two films shows that their ferromagnetic transition temperatures are different (T$_c$=180K(LaAlO$_3$); T$_c$=230K (Si)). At room temperature, both films show a M(H) behaviour that is linear with applied field consistent with antiferromagnetism (Fig. 6b & c). However, the magnetic moment at 5T for the polycrystalline film (≈0.0035 emu) is higher than that observed for the epitaxial film (≈0.0022 emu). At 10K, both films showed a non-linear hysteresis indicative of their ferromagnetic nature along with a significant high field slope even at applied fields of 5T.

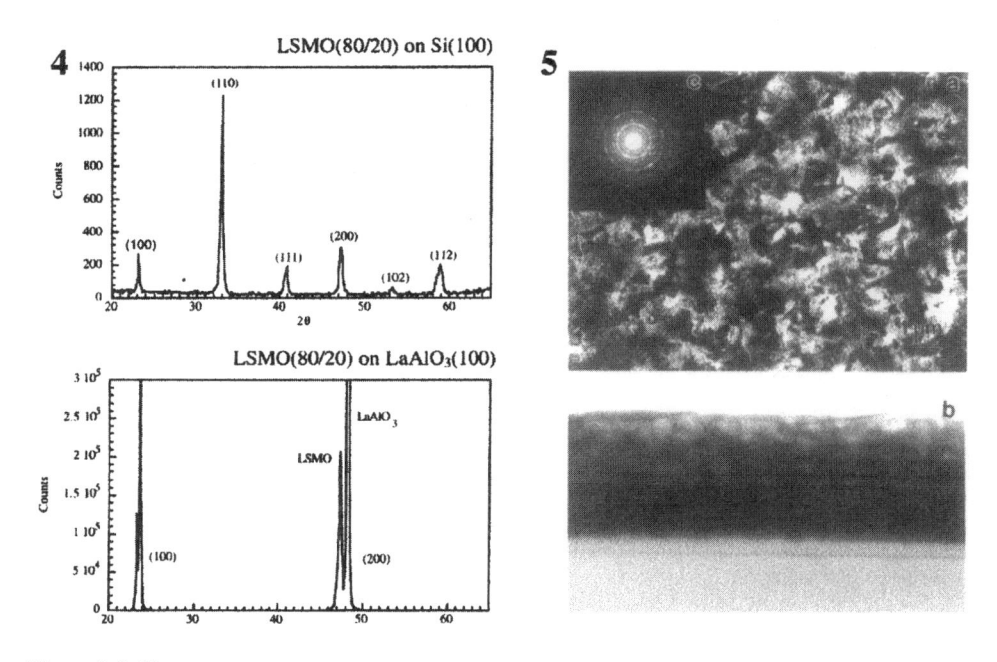

Figure 4. θ–2θ x-ray scans for LSMO films grown on (a) Si(001) and (b) LaAlO$_3$(001)

Figure 5: TEM of polycrystalline LSMO films grown by the sol-gel process on Si(001) -- (a) plan view, (b) cross-section and (c) selected area diffraction.

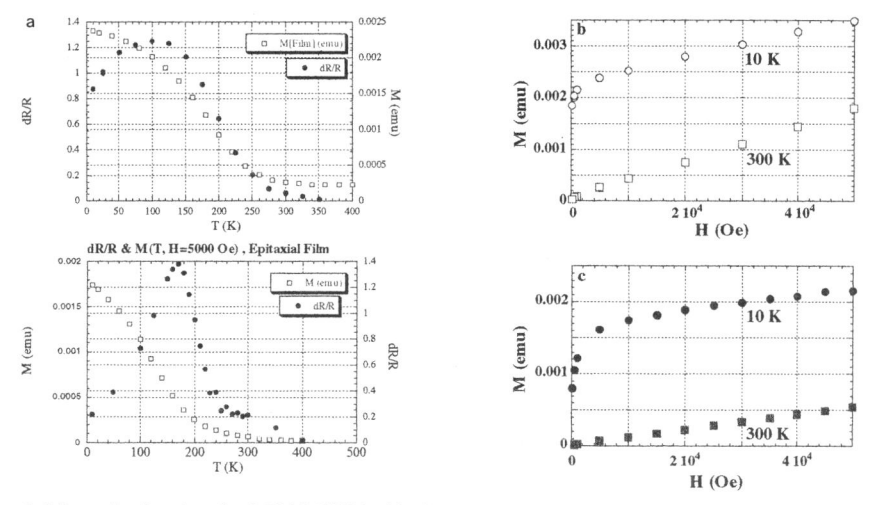

Figure 6: Magnetization data for LSMO(80/20) thin films: (a) M vs T for polycrystallinae and epitaxial films (b) M vs H for polycrystalline film and (c) M vs. H for epitaxial film

The temperature and field dependent resistance values of the films grown on the two different substrates were also different. The overall resistance of the polycrystalline films on Si was 2-3 orders of magnitude greater than that of the single crystal film on LaAlO$_3$. The resistance of the polycrystalline film increased monotonically with decreasing temperature except for a small bump observed around 130K (Fig.7a). At roughly the same temperature, a rather broad peak in the magnetoresistance was observed; the peak magnetoresistance was 125% in a field of 5T. The zero-field resistance of the single crystal film peaked (at 185K) more sharply than the polycrystalline films, and dropped as temperature decreased further indicating a semi-conductor to metal transition (Fig. 7b). In this case, the peak in magnetoresistance correlates reasonably well with the onset of the magnetic transition; the peak magnetoresistance of the single crystal was 140% in a field of 5 T, at a temperature of 170K, which is slightly below the magnetic transition.

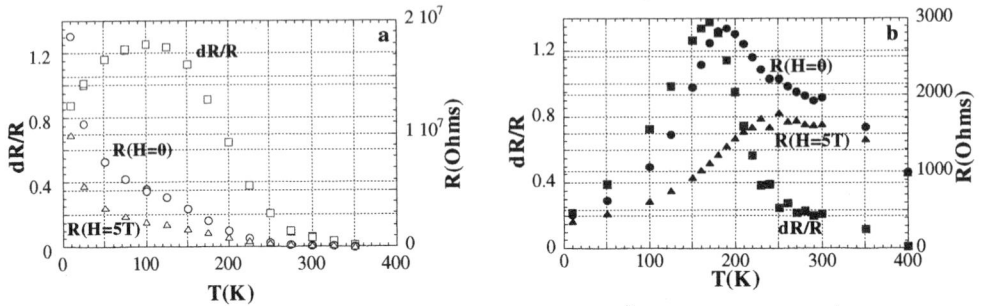

Figure 7. Resistivity data for PLD grown films at H = 0 and 5T: (a) polycrystalline; (b) epitaxial

Sol-gel Derived La$_{1-x}$Sr$_x$MnO$_3$ (0< x< 0.7) Thin Films

The ability to make any composition, rapidly and reproducibly, either epitaxial or polycrystalline, allows us to study the entire composition range with ease. Fig. 8 shows zero-field resistivity data as a function of temperature for a wide range of LSMO compositions grown epitaxially on LaAlO$_3$. Except for an intermediate range of compositions (0.2 ≤ x ≤ 0.4) the films are insulating at all temperatures. For x=0.2, one can clearly observe an insulator to metal transition at ~260 K and for x=0.3, the onset of the transition at 350K is clearly evident.

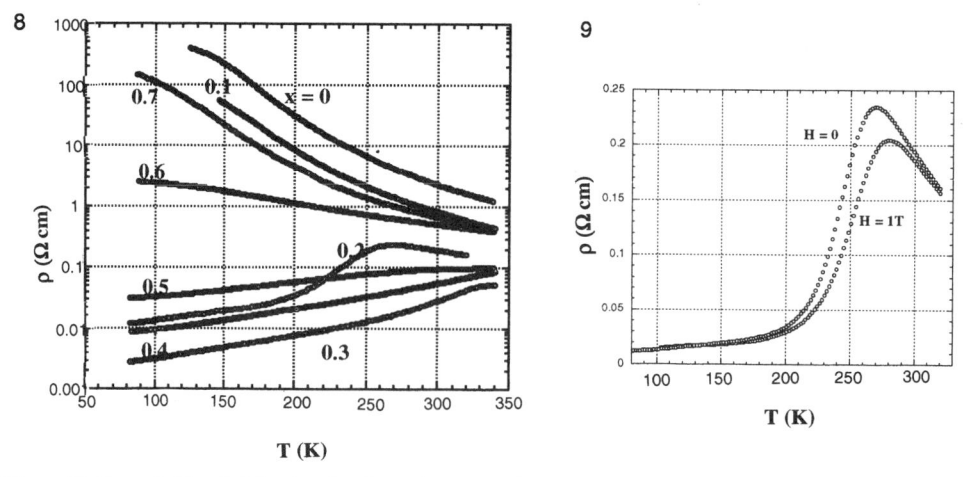

Figure 8: Zero- field resistivity data for the entire range of LSMO compositions grown on LaAlO$_3$
Figure 9: Resistance (at H=0 and 1T) as a function of temperature.

Resistance measurements as a function of temperature (Fig. 9) with H=0 for an optimally annealed LSMO(80/20) thin film grown on LaAlO$_3$(100) show a distinct peak at ~260 K. At H=1T, the peak is supressed substantially and undergoes a shift to higher temperature. In general, the zero field resistance of the polycrystalline films was an order of magnitude higher than the epitaxial films (500 kOhms as opposed to 25 kOhms), similar to the PLD grown films. The epitaxial thin films exhibit a peak MR of 88% at an applied field of 1T.

DISCUSSION

For both growth methods, the choice of substrate determines the orientation of the LSMO film. The Si(100) substrates, with a ~10nm thick oxide layer, imposes no crystallographic constraints on the nucleating phase and leads to polycrystallinity and random orientation. In the case of LaAlO$_3$, the nucleation is strongly influenced by the surface which constrains the films to have an epitaxial relationship. Thus the crystallographic texture is primarily determined by the substrate surface template.

The PLD grown polycrystalline films have a grain size (100-200nm) which is comparable to the film thickness. This is expected since the substrate is held at 675°C providing considerable surface mobility during growth to the incoming atoms. The mechanism, however, is quite different in sol-gel derived thin films -- the grain size in the film is determined not only by the growth kinetics but also by the nature of the polymeric condensed gel[8]. The relatively large grain size ($\approx$ 1-2 µm) observed is a reflection of considerable crosslinking and formation of large colloidal units in the amorphous spun-coat films and their subsequent crystallization into large individual grains. The high porosity (10-20%) is left after removal of the organic mass after which there is very little diffusion-induced densification[8]. Observation of individual layers in the multicoated sol-gel film indicates that preliminary crystallization probably occurs during low temperature solvent removal.

For both PLD and sol-gel derived films, the resistivities of the polycrystalline films are much higher than the corresponding epitaxial films. This can be explained on the basis of the additional scattering from defects and grain boundaries in polycrystalline as opposed to epitaxial films. The differences in transition temperature of LSMO films of identical composition, but grown by these two techniques, could be related to differences in thickness, microstructure as well as composition (oxygen stoichiometry and local compositional inhomogeneties). The differences in the magnitude of magnetoresistance between polycrystalline and epitaxial films is more pronounced for sol-gel films which could be due to their higher thickness (0.5µm as opposed to 450Å for PLD grown fims). However, the cause for the MR enhancement with epitaxy is again related to both morphological and compositional effects and is still not completely understood.

The magnetic moment differences at room temperature between the epitaxial and polycrystalline films can be qualitatively explained. In single crystal LSMO, alternate (001) planes are antiferromagnetically ordered with the moments ferromagnetically ordered along either the [010] or the [100] directions. Hence for an (100) epitaxial film, with the field applied in the plane of the film and assuming that the two spin orientations are equally distributed, the measured susceptibility is: $\chi_{epitaxial} = 1/2\ (\chi_{//} + \chi_\perp)$ where $\chi_{//}$ and $\chi_\perp$ are the susceptibilities parallel and perpendicular to the spin alignment. At the same time, for the polycrystalline film, the measured susceptibility is $\chi_{polycrystalline} = 1/3\ (\chi_{//} + 2\chi_\perp)$ assuming random orientation[9]. Below the Neel temperature, since $\chi_{//} < \chi_\perp$, the polycrystalline film would have a higher measured susceptibility than the (100) epitaxial film. The low temperature magnetic moment differences, however, are rationalized on the basis of possible differences in the amount of Mn^{4+} ions between polycrystalline and epitaxial thin films which in turn affects the moment per Mn atom [10]. The non-equilibrium oxygen content of polcrystalline materials can be expected to be higher than epitaxial films due to enhanced diffusion through grain boundaries, which would force a higher Mn^{4+} to maintain the charge balance. Efforts to quantify the Mn^{4+}/Mn^{3+} ratios in the films utilizing electron energy loss spectroscopy and soft x-ray spectroscopy are currently in progress.

Both the Curie temperature and the ferromagnetic magnetization of the epitaxial film are lower than those of the polycrystal implying that the ferromagnetic moment per Mn atom is smaller in the single crystal than in the polycrystal. In these films the relative atomic fraction of both Sr and O affect the tetravalent Mn content. Further, the change in the ferromagnetic moment per Mn atom with the fraction of tetravalent Mn is particularly large close to 20% Mn^{4+}, which would be expected in the composition corresponding to 20 at. % Sr [2]. Since the total number of La, Sr and Mn atoms in the two films are nominally equal, the difference in the moments could arise mainly from differences in oxygen stoichiometry. The interdiffusion of oxygen would be much faster along grain boundaries than in the bulk and hence, the polycrystal would have a higher equilibrium oxygen content than the single crystal. Future EELS measurements may help to clarify the issue relating to the relative oxygen stoichiometries in the two films.

Overall, the results confirm the important role of the microstructure in determining the properties of LSMO films and suggest the need to make measurements on well defined grain boundaries, e.g. films grown on bicrystal substrates. Such experiments would permit an understanding of the influence of grain boundaries on the double exchange interactions, in addition to magnetoresistance properties.

ACKNOWLEDGEMENTS

This work was supported by the Director, Office of Energy Research, Office of Basic Energy Sciences, Materials Sciences Division of the U.S. Department of Energy under contract no. DE-AC03-76SF00098.

REFERENCES

1. K. Chahara, T. Ohno, M. Kasai and Y. Kozono, Appl. Phys. Lett **63**, 1990 (1993).
2. R. von Helmolt, J. Wecker, B. Holzapfel, L. Schultz and K. Samwer, Phy. Rev. Lett. **71**, 2331 (1993).
3. S. Jin, M. McCormack, T. H. Tiefel and R. Ramesh, J. Appl. Phys. **76**, 6929 (1994).
4. Refer to Proceedings of the 50th MMM Conference, Philadelphia, PA, November 1995.
5. Kannan M. Krishnan, A. R. Modak, C. A. Lucas, R. Michel and H. B. Cherry, J. Appl. Phys., **79**, 5169 (1996)
6. A. R. Modak and Kannan M. Krishnan, J. Amer. Ceram. Soc., 1995, In Press.
7. A. R. Modak and S. K. Dey, Integrated Ferroelectrics **5**, 321 (1994).
8. C. J. Brinker and G. W. Scherer, *Sol-Gel Science: The Physics and Chemistry of Sol-Gel Processing*, (Academic Press, Boston, 1990)
9. B. D. Cullity, *Introduction to Magnetic Materials* (Addison-Wesley, Reading MA, 1972).
10. G. H. Jonker and J. H. van Santen, Physica **16**, 337 (1950)

MAGNETIC AND INTERFACE CHARACTERIZATION OF RF-SPUTTER DEPOSITED MULTILAYERED AND GRANULAR Fe/AlN FILMS

S. Kikkawa, M. Fujiki, M. Takahashi, and F. Kanamaru

ISIR, Osaka University, Osaka 567, JAPAN

INTRODUCTION

Several kinds of iron nitrides with various amounts of nitrogen have been reported. Both $Fe_{16}N_2$ and Fe_4N are interesting magnetic materials because of their large saturation magnetization and small coercivity. The saturation magnetization value gradually decreases with increasing nitrogen content.[1] These nitrides are also superior both in mechanical hardness and chemical stability in comparison to α-Fe, but easily decompose, releasing nitrogen above several hundred °C. Other elements such as Zr, Ti have been added to increase their thermal stability.

It is possible to prepare either γ"- or γ'''-FeN with the rock-salt or zinc-blende type crystal structure by applying rf-sputter deposition because nitrogen is activated to react with iron in rf-plasma.[2] Sputter deposition can also be used to obtain metastable solid solutions such as (Ti,Al)N, (Mo,Zr)N.[3,4] We have been interested in a thermally stable AlN with covalent bond nature as a matrix of a composite film.[3,5] It also serves as a good host for the formation of metastable solid solutions. An interface reaction occurred in multilayered Fe/AlN films despite low preparation temperature.[6] Small amount of nitrogen transferred from the AlN layer and diffused into the α-Fe layer. The magnetic behavior of the interfacial region may be very interesting. Alternatively, ferromagnetic particles may be deposited by a thermal treatment of sputter co-deposited Al-Fe-N thin films. Sputter deposition of iron nitride has not yet been fully investigated. There has been no report of formation of $Al_{1-x}Fe_xN$ solid solutions. Their nitrogen content and thermal behavior may be different with their iron content in the as-deposited film. The nitrogen content in iron nitrides affects their magnetic behavior. Ferromagnetic particle size is also related to their coercivity. Thermal instability of the iron nitrides strongly affects the deposition behavior of ferromagnetic granules in the annealed thin films.

In the present investigation, both multilayered and granular Fe-Al-N films were prepared by applying rf-sputter deposition. Magnetic and interface characterization were performed and related to the microstructural characteristics.

EXPERIMENTAL

Four kinds of iron metal/aluminum nitride multilayered composite film were prepared using two target type rf-sputter deposition equipment (JEOL-430RS):

(A) [Fe(495.7nm)/AlN(434.3nm)]$_1$,
(B) [Fe(178.2nm)/AlN(169.4nm)]$_3$,
(C) [Fe(64.8nm)/AlN(61.6nm)]$_8$,
(D) [Fe(6.5nm)/AlN(6.2nm)]$_{35}$.

Aluminum nitride was formed by reaction sputtering of an Al metal target with nitrogen sputter gas. It was x-ray amorphous because the applied rf-power was only 30W to reduce an intermixing of Fe and AlN at the multilayer interface. Total film thickness was about 1 μm for the former three samples and 0.44 μm for the last sample. The thickness of individual layers was controlled by deposition duration. Experimental details have been described in the previous manuscript.[1]

Al$_{1-x}$Fe$_x$N films were deposited by co-sputtering of a composite target of Al metal chips (5 x 5 mm^2) distributed on an Fe metal disk target of 3 inch diameter. Chemical composition of the film was controlled by varying the Al coverage on the Fe target. The sputtering chamber was evacuated to 5.0 x 10^{-5} Pa prior to film deposition. Nitride films of several μm in thickness were deposited on fused silica substrates by reactive co-sputtering at 1 Pa of an equimolar gas mixture of Ar and N$_2$. Total gas flow rate during deposition was 2.0 sccm. Applied rf-power was 100 W. The products were annealed in a flow of H$_2$ atmosphere between 200 and 500 for 1 h with or without substrates in an alumina boat.

RESULTS AND DISCUSSION
Multilayered film

The multilayered films showed very anisotropic magnetic behavior.[1] They were soft ferromagnets when a magnetic field was applied parallel to the film surface independently of the thickness layers. Their coercivity was less than 5 Oe. On the other hand, they were mostly paramagnetic, with a small degree of ferromagnetic behavior when the magnetic field was perpendicular to the film surface. The coercivity of the latter ferromagnetic component increased with a decrease of the layer thickness in the multilayers. The coercivity increase may be related to a decrease of α-Fe crystallite size with layer thickness observed in x-ray diffraction.[1] Saturation magnetization in the parallel direction showed a very interesting change with the type of multilayer as represented in Fig. 1. The value of magnetization σ was normalized by their iron content. The iron content in the multilayers was analyzed by atomic absorption of their dissolved aqueous solution with HCl. The environment around an iron atom in an Fe layer should be most similar to that in bulk α-Fe (σ = 218 emu/g) in case (A) among the investigated multilayers. The saturation magnetization was ≈ 180 emu/g for film (A). The values were 220 emu/g for film (B), and 210 emu/g for film (C). These values are 22% and 17% larger than the value for film (A). The value decreased to about 160 emu/g for film (D). These changes in σ-value directly correspond to a variation observed on Mössbauer spectra at room temperature.[1,6] A large internal magnetic field component of around 36 T was present with a value of 34 T assigned to α-Fe in spectra for films (B), (C), and (D). Relative amounts of the larger internal magnetic field component were 39%, 19%, and <2% against each total estimated from their spectra, respectively. Components with internal magnetic field smaller than that of α-Fe also appeared in film (C) and the ratio increased in film (D). Even paramagnetic components were observed in the last film. They can be assigned to formation of iron nitride as follows.

The x-ray absorption spectrum of Fe K-edge for film (B) was similar to that for α-Fe as represented in Fig. 2(B). Film (D) showed the presence of a pre-edge at around 7.11 keV assigned to the s→d quadrupole transition of tetrahedrally coordinated iron as shown in Fig. 2(D) as well as in Fig. 2(C). Tetrahedral coordination is present in zinc-blende type γ'''-FeN with paramagnetic behavior.[2] The presence of tetrahedrally coordinated iron was also

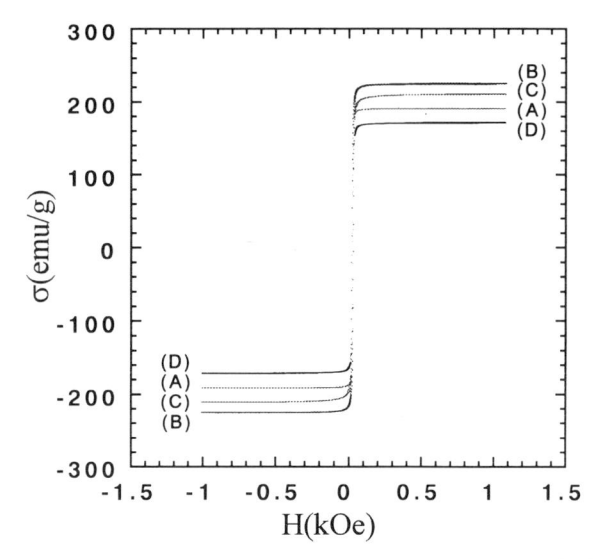

Figure 1. σ-H curves for Fe/AlN multilayered thin films: (A) [Fe(495.7nm)/AlN(434.3nm)]$_1$, (B) [Fe(178.2nm)/AlN(169.4nm)]$_3$, (C) [Fe(64.8nm)/AlN(61.6nm)]$_8$, (D) [Fe(6.5nm)/AlN(6.2nm)]$_{35}$ with an applied magnetic field parallel to the film surface.

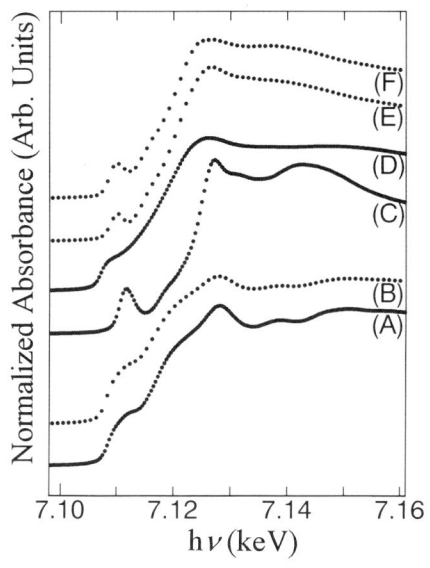

Figure 2 X-Ray absorption spectra of Fe K-edge in He ion yield method for (A): Fe foil, (B): multilayered film (B), (C): γ'''-FeN film, (D): multilayered film (D), (E):as-deposited co-sputtered film (a), and (F): as-deposited co-sputtered film(c).

confirmed by EXAFS calculated from the same x-ray absorption spectrum in Fig. 2(D) with a coexistence of an α-Fe like component. Local structure similar to that in γ'''-FeN might be partly formed in film (D). An XPS depth profile suggested precipitation of metallic aluminum at the Fe/AlN interface.[1,6] Aluminum nitride seems to react with iron metal even during room temperature deposition in nanometer range thickness. The formation of a higher iron nitride at the interface between Fe metal and AlN can also be assumed in sample with a thinner single layer thickness in the multilayers judging from the Mössbauer and XAF spectra.[1,6] Nitrogen atoms taken up into the α-Fe lattice gradually diffused into Fe during the successive sputter deposition. The nitrogen content may be less deeper inside the Fe layer. There may be a lower iron nitride with a very small amount of interstitial nitrogen at the interface between Fe metal and AlN in the multilayers (B) and (C). The giant magnetization may result from an expanded Fe-Fe distance due to the interstitial nitrogen. The situation may be similar to that of the Fe sites having a large internal magnetic field in $Fe_{16}N_2$. Similar interface reaction in several nm thickness has been reported at an interface between a gate insulator Si_3N_4 and an Al metal electrode in a ULSI semiconductor.[7]

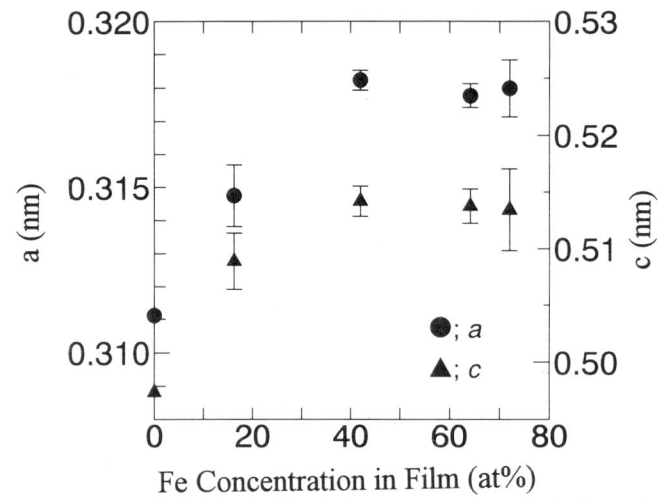

Figure 3 Lattice parameters of $Al_{1-x}Fe_xN$ solid solution vs Fe concentration in the as-deposited film. The values at x = 0 are the data for AlN (JCPDS25-1133).

Granular film

Several kinds of Al-Fe co-sputtered nitride films were deposited with varing iron content. Solid solution of $Al_{1-x}Fe_xN$, isostructural to wurtzite AlN, was observed in the lower iron content region. Rock-salt type γ''-FeN[2] also appeared with an increase of iron content. Appearance of these nitrides is a characteristic of rf-sputter deposition and they are poorly crystallized in these as-deposited films. The hexagonal lattice parameters variations with iron concentration (estimated by EDX) in these films are shown in Fig. 3. Both a- and c-lattice parameters increas with increasing iron content x. They reached a constant value above x=40%. A solid solution limit is present between 20 and 40%. The accuracy is poor because of the large standard deviations in the lattice parameters due to their poor crystallinity. XPS showed that the nitrogen content continuously decreased with increasing iron content x. The nitrogen defect may also contribute to the lattice parameter change with x-value. Fe atoms are tetrahedrally coordinated in the wurtzite type lattice of γ'''-FeN. Some of the Fe atoms are also tetrahedrally coordinated in γ''-FeN because of its nitrogen deficient defect. The presence of Fe tetrahedrally coordinated with 4 nitrogen atoms was confirmed in these as-deposited films by their pre-edge assigned to the s→d quadrupole in XAFS as represented in Figs. 2(E) and (F).

The annealing behavior of the films was a sensitive function of their iron content. Three kinds of films were annealed in flowing H_2 to precipitate ferromagnetic components. Their iron contents in the films were (a) 17 at%, (b) 65 at% and (c) 80 at%, respectively. The wurtzite type solid solution was still present in the film (a) even after an annealing at 500°C. Both the wurtzite type solid solution and γ"-FeN decomposed to crystallize α-Fe particle in the films (b) and (c) above 300°C. TEM micrographs with EDX showed that the Fe particles were homogeneously dispersed in matrices of the annealed films (b) and (c) at 500°C with some size distribution. Their average particle sizes were 100 nm and 200 nm, respectively.

All the as-deposited films were paramagnetic. Films (b) and (c) showed a slight appearance of a ferromagnetic component after annealing at 300°C, and this contribution drastically increased with the formation of -Fe in XRD above 400°C. Magnetic properties are represented in Fig. 4 for the films annealed at 500°C. The film (a) was still mostly paramagnetic with 1 emu/g under an applied magnetic field of 9 kOe even after the annealing. The values are 99 and 172 emu/g for the films (b) and (c), respectively. They correspond to 71 and 97% of the values calculated from a saturation magnetization value for α-Fe (218 emu/g) with their α-Fe fraction analyzed by EDX for the films (b) and (c). Most of the iron has been crystallized to α-Fe in the film (c) after the annealing at 500°C. The magnetization was still not yet saturated at 9 kOe as shown in Fig. 4. Mössbauer spectra were measured at room temperature on these annealed films. Only asymmetrical paramagnetic broad absorption was observed at around zero velocity against the spectrum of α-Fe on the film (a). It might be assigned to the undecomposed $Al_{1-x}Fe_xN$ even after the annealing. Six magnetically coupled absorption lines assigned to α-Fe appeared without the paramagnetic component on the annealed film (c). Magnetic coercivity values estimated from the point at σ-H curves across the σ = 0 axis are 25 and 21 Oe for the annealed films (b) and (c), respectively. The separation between the curves obtained on increasing and decreasing the magnetic field is larger at near saturation levels of than at σ = 0. The change of curve separation with an applied magnetic field may be related to the α-Fe particle size distribution in the films. In any event, the magnetic coercivity decreased with a growth of α-Fe particle size.

The nitrogen content in the as-deposited film was larger in the film with higher Al content as mentioned above. After annealing, there was no N_{1s} XPS peak for films (b) and (c), however film (a) retained its nitrogen. The Fe_{2p} spectra of the annealed film (a) was still broad with a binding energy of E_b ≈ ca.710 eV as observed for the as-deposited film, and depicted in Fig. 5(a). It may be due to the Fe in $Al_{1-x}Fe_xN$. The Fe_{2p} spectra of the films (b) and (c) sharpened after annealing, and exhibited maximum at E_b = 707 eV. This sharpening is due to the loss of the high-energy tail or shoulder in the peak that is caused by the bonding of iron to nitrogen in both γ"-FeN and $Al_{1-x}Fe_xN$; these phases are lost when α-Fe crystallizes during annealing. Al_{2p} spectra showed the presence of Al with E_b ≈ 74 eV which is similar to that of AlN in the annealed film (a). Broad spectra having maxima of E_b ≈ 75eV which can be assigned to Al_2O_3 appeared in films (b) and (c) after annealing. The aluminum was oxidized by the oxygen present in the as-deposited film and also by that in the atmosphere after releasing nitrogen. $Al_{1-x}Fe_xN$ with x=17 at% was thermally stable up to 500°C. The solid solution became unstable with an increase of x as thermally unstable γ"- and γ'"-FeN.[2] Both decompose to precipitate α-Fe in the amorphous Al_2O_3 and release nitrogen.

In summary, both multilayered and granular Fe/AlN films were prepared by rf-sputter deposition technique. X-ray absorption and Mössbauer spectroscopies explicitly showed

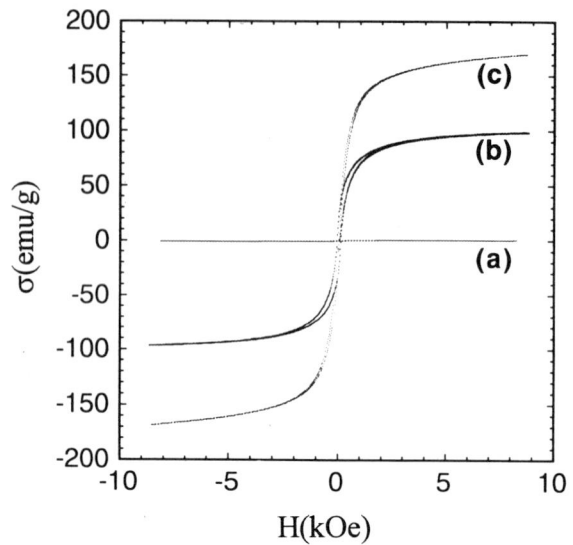

Figure 4 σ-H curves for the co-sputtered Fe-Al-N films with Fe contents of (a) 17 at%, (b) 65 at% and (c) 80 at%, respectively, after annealing at 500°C.

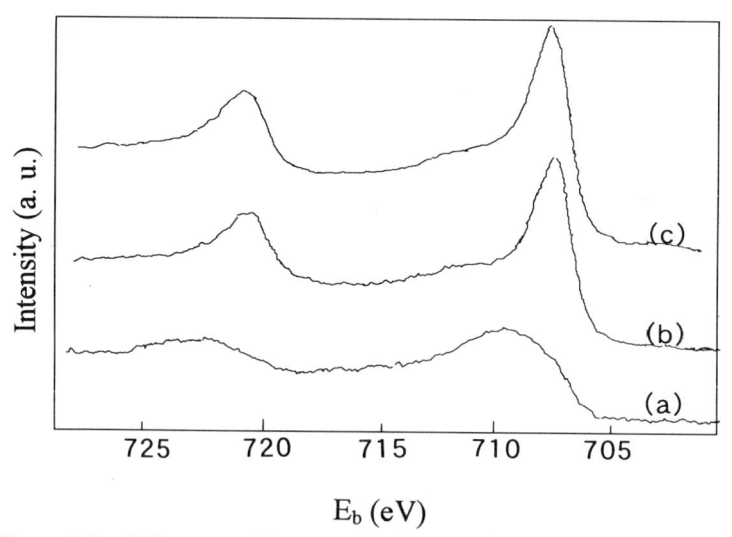

Figure 5 Fe$_{2p}$ XPS spectra of the co-sputtered Fe-Al-N films after annealing at 500°C.

iron nitride at the interface of the multilayered films. Depth profile of XPS suggested precipitation of metallic aluminum there. Aluminum nitride seems to react with iron metal even during room temperature deposition at nanometer thickness scale. Granular magnetic films were also prepared by an annealing of the co-deposited samples. Annealing behavior in flowing H_2 was very sensitive to Fe content of the co-deposited films. An $Al_{1-x}Fe_xN$ solid solution was obtained for x<30 at% and γ''-FeN also coexisted above the solid solution limit in the as-deposited film. The AlN-like matrix could retain nitrogen in the films with less Fe content to form the solid solution. It was oxidized in those with higher Fe contents, releasing nitrogen. Dispersed Fe granules formed during heat treatment in all the co-deposited films.

Acknowledgments

This research was partly supported by Grant-in Aid for Scientific Research on Priority Areas (No.274) from The Ministry of Education, Science and Culture, Physics and Chemistry of Functionally Graded Materials. Part of this work was performed under the approvals of the Photon Factory Advisory Committee (Proposal No.93G160) and also of the Radiation Laboratory in ISIR (7B-14).

REFERENCES

1. S. Kikkawa, Layered nano composites - interface characterization of Fe/AlN multilayered film, *Mater. Trans., JIM*, **37**:420 (1996).
2. M. Takahashi, H. Fujii, H. Nakagawa, S. Nasu, and F. Kanamaru, Structure, electronic, and magnetic properties of sputter-deposited new iron nitrides, Proc. 6th Int. Conf. Ferrites, Tokyo and Kyoto, Japan 1992, 508 (1993).
3. S. Inamura, K. Nobugai, and F. Kanamaru, The preparation of NaCl-type $Ti_{1-x}Al_xN$ solid solution, *J. Solid State Chem.*, **68**:124 (1987).
4. G. Lai, M. Takahashi, K. Nobugai, and F. Kanamaru, Preparation and characterization of superconducting B1-type $Mo_{1-x}Nb_xN$ thin films, *J. Am. Cer. Soc.*, **72**:2310 (1989).
5. S. Kikkawa, M. Takahashi, and F. Kanamaru, Sputter deposition of BN and AlN thin films, *Ceramic Trans.*, **22**:747 (1991).
6. S. Kikkawa, M. Fujiki, M. Takahashi, F. Kanamaru, H. Yoshioka, T. Hinomura, S. Nasu, and I. Watanabe, Interface of iron metal/aluminum nitride multilayered composite films, *Appl. Phys. Lett.*, **68**:2756 (1996).
7. J. Arila and J. L.Sacedón, Reactivity at the Al/Si_3N_4 interface, *Appl. Phys. Lett.*, **66**:757 (1995).

NANO-SIZED OXIDE COMPOSITE POWDERS, BULK MATERIALS AND SUPERPLASTIC BEHAVIOR AT ROOM TEMPERATURE

D. S. Yan, J. L. Shi, Y. S. Zheng
State Key Laboratory on High Performance Ceramics
and Superfine Microstructure
Shanghai Institute of Ceramics, Chinese Academy of Sciences
1295 Ding-Xi Road, Shanghai 200050, China

ABSTRACT

The preparation and characterization of nano-sized zirconia and its composite powders, Y-TZP/Al$_2$O$_3$ by various chemical processes is presented. Different ways to avoid hard-agglomerate formation and the characterization of the powders are discussed. Satisfactory green compact formation and full densification can be achieved at far lower temperatures in comparison with micron sized powder. The incorporation of alumina into Y-TZP nano-powders showed the effect of further lowering the grain size and the enhancement of sinterability as well as the strengthening and toughening of the nano-nano composites. Room temperature tensile fatigue studies on the fully dense nanostructured specimens were pursued. It was first observed that localized superplastic deformation of the grains at and near the fatigue fracture surfaces, was generated by fatigue. Grain boundary diffusion is reasoned to be the major governing micromechanism in operation.

KEYWORDS: Nano-sized composite powders, nanostructured bulk material, tensile fatigue, superplastic deformation, grain-boundary diffusion

I. INTRODUCTION

Nano-sized composite powders and their bulk materials have been a focal point in materials research since they behave so differently from the corresponding bulk materials due to the low-dimensional quantum size effect and the small size effect, etc. As a consequence, unexpected changes of material properties were shown to occur in their electrical, magnetic, optical, thermal properties as well as their strength, hardness, toughness, plasticity and superplasticity and so on. It is, therefore, a fertile field to be explored.

Various methods have been used to process nano-sized powders which can be broadly classified as the chemical and physical routes.[1] For the chemical routes, there are precipitation and coprecipitation methods,[2-6] the hydrolysis of alkoxides[7-8] and the like, the sol-gel method,[9] the micro-emulsion method,[10] the hydrothermal process,[11] various CVD methods and so on. Some of these methods are easily scaled up into application. In parallel, there are also a number of physical methods,[12-15] some of which are specially suitable for the processing of clean metallic nano-sized powders. Of these, some are also easily scaled up. Therefore, for the study of different materials, there is a choice of process or processes that will be best suited to be used.

In this paper, we will concentrate on the processing and characterization of yttria stabilized tetragonal zirconia (Y-TZP) nano-sized powder and its composite with alumina and their compaction and densification behavior. Some preliminary studies on the properties of the nano-nano composite materials will be presented. Well densified nanocrystalline bulk Y-TZP, along with a submicron material of the same composition, were tested at room temperature under cyclic tensile fatigue to failure. Some unexpected new phenomena were observed with the nanocrystalline material which will be briefly discussed.

II. PROCESSING AND CHARACTERIZATION OF NANO-SIZED ZIRCONIA POWDER

Wet chemical methods were chosen to obtain highly pure and well dispersed (at atomic level) nano-sized zirconia and Y-TZP/Al$_2$O$_3$ composite powders, which can also readily be scaled-up. The points of key importance are the control of the nucleation and growth of crystallites and avoiding the generation of hard agglomerates upon drying and calcination due to the formation of bridging oxygen between the crystallites.

A co-precipitation process was used to abtain Y-TZP nano powders. A number of parameters were carefully controlled and optimized which were detailed elsewhere.[16-18] These factors can be summarized in a flow diagram. The crystallite size of the Y-TZP nano-powder can be regulated by varying the calcination temperature (see Fig. 1). It was also shown that there were no hard agglomerates present in the powder which could not be completely broken down under CIP, resulting in a monomodal pore size distribution similar to that of the crystallite size. The bulk material could be densified to >97% TD by adopting a relatively fast firing scheme to 1200 °C; the grain size was ≈ 100 nm.

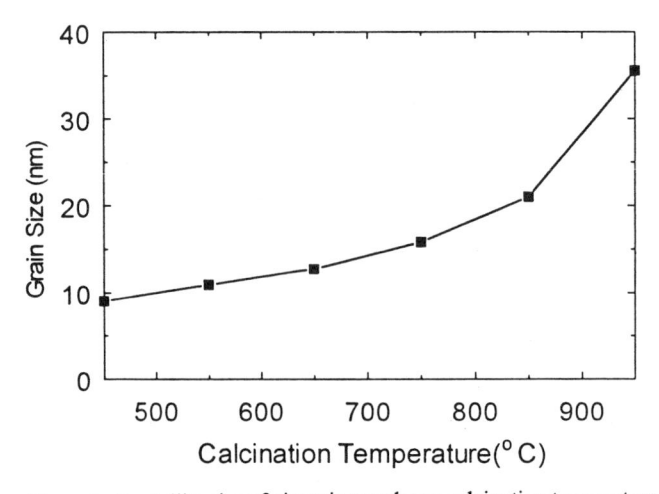

Figure 1 Crystallite size of zirconia powder vs calcination temperature

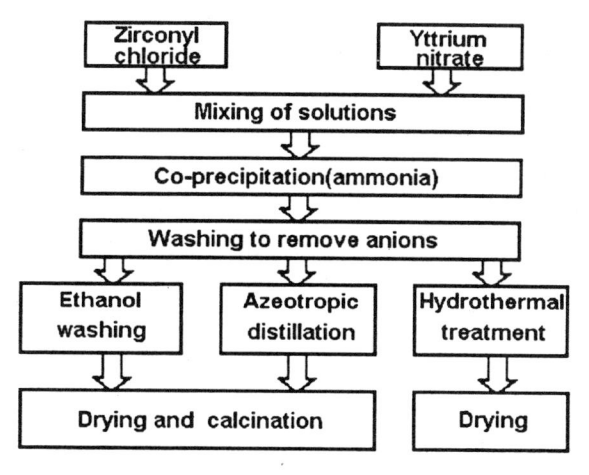

Figure 2 Flow diagram of nanoscale Y-TZP powder preparation by wet-chemical methods

III. PROCESSING, COMPACTION AND DENSIFICATION OF NANO-SIZED Y-TZP/Al₂O₃ COMPOSITE POWDERS

The composite powders were prepared by a similar procedure under controlled conditions.[5] With the incorporation of alumina with yttria stabilized zirconia at the atomic level, the grain size of the primary crystallite formed was significantly reduced (Fig. 3). The (200) lattice plane distance of tetragonal zirconia showed a significant difference between the pure and composite powders (Fig. 4) which was more pronounced at lower calcination temperatures. The effect of lattice distortion also appeared to be more apparent when a smaller amount of alumina, i.e., 5 to 10 mol%, was incorporated into the lattice and concurrently more effectively impeded the grain growth of the Y-TZP crystallites. It was not until 1200 °C or beyond that α-Al₂O₃ began to be excluded out as a separate phase in a vermicular form (Fig. 5-7). The reason for the impedance of the grain growth of the composite powder is believed to be mainly caused by the enhanced segretation of the solute ions near the grain boundaries due to the presence of lattice distortion and the solute drag effects produced therefrom.

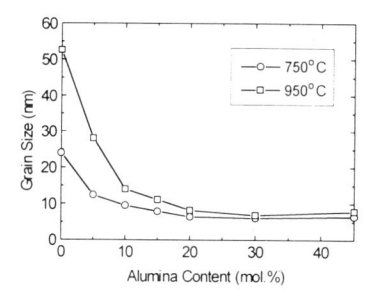

Figure 3 Dependance of the crystallite size on Al₂O₃ content

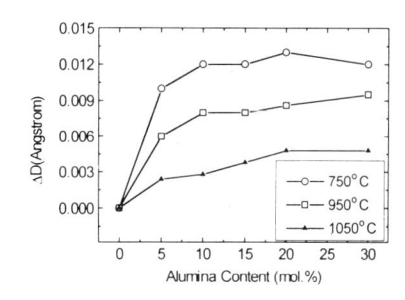

Figure 4 Difference of the (200) lattice plane distance between pure and composite powders

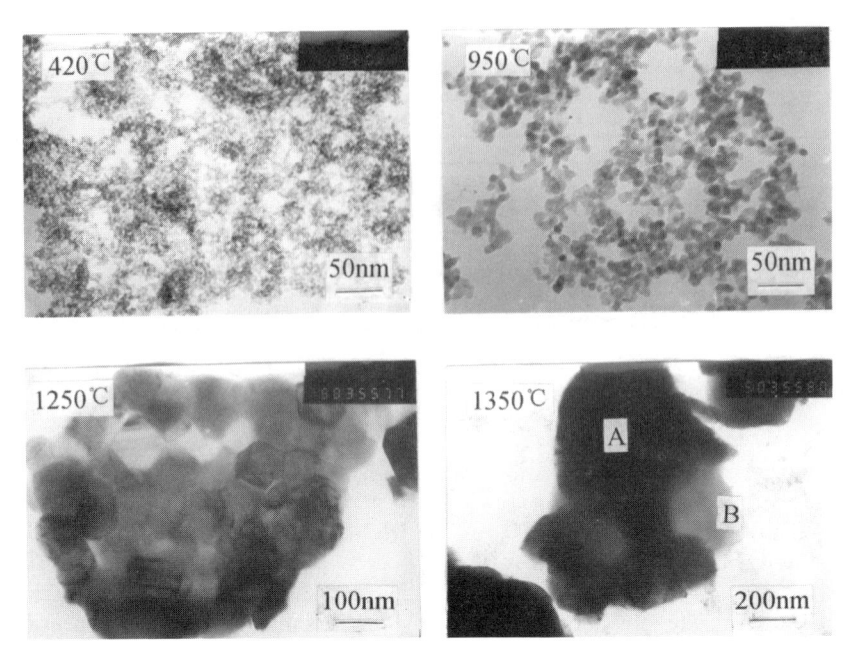

Figure 5 TEM micrographs of Y-TZP/20 mol% Al$_2$O$_3$ composite powders calcined at different temperatures for 120 min.

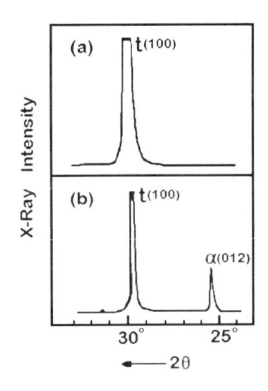

Figure 6 X-ray diffraction patterns of composite calcined at (a) 950 °C and (b) 1250 °C for 120 min.

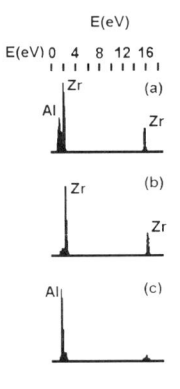

Figure 7 EDS spectra for samples calcined at different temps. (a) 950 °C, (b) 1350 °C, dark phase and (c) 1350 °C, light area

These composite powders could be formed into ideal green compacts by cold isostatic pressing and full densification could be realized at relatively low temperatures when the alumina addition is not more than 20 mol% (Fig. 8).[19] In fact, the densification rate of the composite compacts was actually enhanced with 5 or 10 mol% of alumina incorporation (Fig. 9) due possibly to the much smaller particle size of the powders to begin with. From Fig. 9, it can also be observed that with alumina addition at and beyond 15 mol%, the densification rate showed two peaks with the first one close to 1100 °C and the second peak at higher temperatures, which could be associated with the separating out of the alumina phase and subsequent densification of the diphasic material. Homogeneous nano-nano microstructured composites were obtained with similar grain sizes of the Y-TZP matrix and the second alumina phase. The inhibiting effect of alumina on grain growth of the Y-TZP phase is obvious which can probably be caused by the pinning effect. It can also be seen that with the addition of 45 mol% Al_2O_3, the composite material could not be sintered even when the temperature was raised to 1550 °C.

Our preliminary investigation on the nano-nano bulk composite material showed that the mechanical properties were substantially enhanced with the incorporation of a small amount of alumina in the composite. Some of the results are shown in Table 1.

It is believed that local compressive strain developed at the Al_2O_3/Y-TZP interface tends to absorb a part of the energy at the crack front thus contributing to the enhancement of the strength and the fracture toughness of the material. Further intensive work is now ongoing.

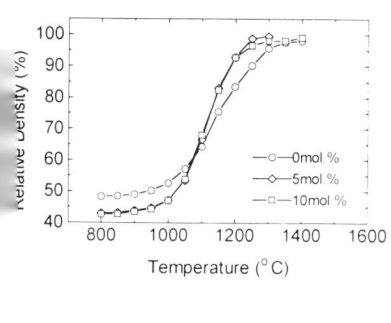

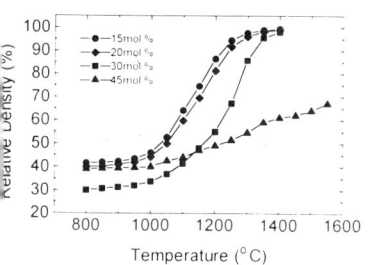

Figure 8 Reletive density vs temperature (5 °C/min)

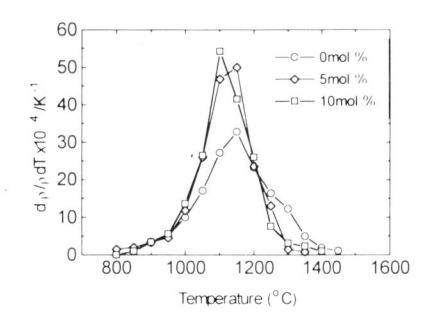

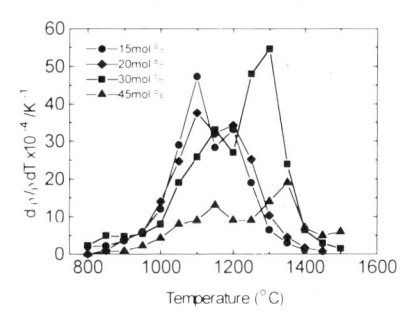

Figure 9 Densification rate vs temperature (5 °C/min)

IV. LOCALIZED SUPERPLASTIC DEFORMATION OF GRAINS AT AND NEAR THE FRACTURE SURFACES BY TENSILE FATIGUE AT ROOM TEMPERATURE

Since the initial report of superplasticity on ceramic material in 1986, a variety of such materials have been shown to exhibit superplastic behavior at high temperature,[20-26] usually close to their original sintering temperatures. Nieh et al.[22] reported that the microstructure of the yttrium stabilized tetragonal zirconia polycrystal samples superplastically deformed at or above 1473 °K remained equiaxed which is consistent to the grain boundary slip mechanism.

Here, we will briefly report on our recent work on fatigue of nanocrystalline bulk Y-TZP material at room temperature. An unexpected phenomenon showing superplastic deformation of the nano-sized grains at and near the fatigue crack propagation region was identified. The possible micromechanism will be discussed.

Large sized bulk samples prepared from Y-TZP nano powders were densified by hot-pressing under a pressure of 25 Mpa. Close to full densification was achieved at 1250°C with an average grain size of 100 ±20 nm. Hot pressed samples were also made at 1400 °C and 1550 °C, giving average grain sizes of 0.35 µm and 1 µm, respectively. Specimens for mechanical property measurements and for tensile fatigue tests were cut from these pieces. Some of the basic properties are listed in Table 2.

For static and cyclic tensile tests, the specimens have a total length of 64 mm and a gauge length of 14 mm. Tensile fatigue tests were carried out on a computer controlled Instron Model 8501 100 KN servo-hydraulic machine at R=0.1 and frequency 0.1 Hz (sinusoidal) at room temperature. The stress was increased at intervals of 10 MPa for 100 nm specimens with 100 cycles at each level. For the 0.35 µm specimens, an interval of 50 MPa stress was used with 100 cycles in between. Failure usually occurred at over 500 cycles for both types of samples.

Table 1. Mechanical Properties of Y-TZP/Al$_2$O$_3$ Nanocomposites

Al$_2$O$_3$, mol%	0	2.5	5	10
Flexural Strength 3-point (MPa)	460±48	640±40	790±74	720±33
K$_{1c}$ (MPa m$^{1/2}$)	8.8±0.83	10.9±1.75	11.3±1.6	9.4±0.7
HV (GPa)	11.3	12.8	13.7	14.5

Table 2 Basic Properties of 3Y-TZP Specimens

Hot-pressing Temp. °C/1h	Grain Size (nm)	Theoretical Density(% of)	Flexural Strength (MPa)	Tensile Strength (MPa)
1250	100±20	98.5	480±45	120±20
1400	350	99.2	1250±43	300±25

Both the fracture surfaces and the side surfaces of the specimens from the fracture edge downwards were carefully observed by AFM. A Nanoscope-III (Digital Instruments Inc., USA) was used. The images were obtained in an ambient environment. Care was taken to avoid sample damage and commercial tips from Digital were used in all these studies.

It was observed a bit beyond our expectation that the nano-sized grain were significantly elongated after tensile fatigue failure (Fig. 10). The ratio of the long and short axes of the grains had values from 3-5 to 8-10. They are generally aligned in parallel. For the 0.35 μm specimens, the microstructure of the surfaces after fatigue failure retained the original equiaxed grain morphology (Fig. 11). Similar grain elongation was also identified on the side faces of the 100 nm specimen within a region a few micrometers from the fractured edge downwards. However, the original nano-sized equiaxed grain morphology remained after the imaging was further tracked down a few more micrometers on the side surface. It may, thus, be concluded that localized superplastic deformation of the grains in Y-TZP specimens with 100 nm grain sizes had occurred at and near the fracture surfaces induced by cyclic tensile fatigue.

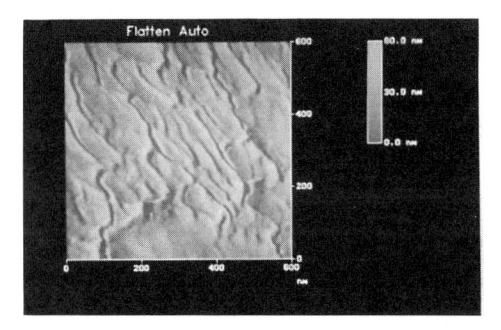

Figure 10 AFM image showing the superplastically deformed grains of the 100 nm 3Y-TZP specimen after cyclic fatigue. Viewed on the fractured surface.

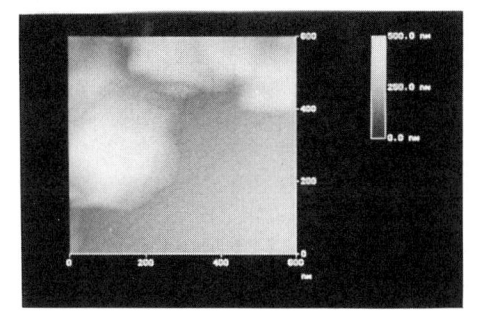

Figure 11 AFM image of 0.35 μm grain-sized 3Y-TZP specimen after cyclic fatigue. The original grain morphology retained. Viewed on the fractured surface.

Two possible mechanisms that could be in operation under the present circumstances are grain-boundary diffusion and dislocation slip. The former is considered to be of major importance. The brief arguments are as follows.

The Coble model of grain-boundary diffusion[27] relating the rate of deformation, $\dot{\varepsilon}$, with grain size, d, may be resorted to as a basis of consideration.

$$\dot{\varepsilon} = B\sigma\Omega\delta D_b / (d^3 KT) \qquad (1)$$

Where σ is the tensile stress, Ω the atomic volume, δ the grain boundary thickness, D_b the grain-boundary diffusivity and KT has the usual meaning. There are two main factors to be considered. The first is the direct effect of grain size alone on the rate of deformation which can be related as $\dot{\varepsilon} \propto d^{-3}$. Thus the 100 nm sized TZP material will exhibit a factor of 40 higher rate of deformation than the 0.35 μm grain-size TZP under similar conditions.

Second, one can estimate the combined contribution of the δD_b terms on the rate of deformation of the 100 nm and the 0.35 μm grain-sized Y-TZP materials by comparing the extent of deformation of these two types of materials after 5000 seconds. For the nanosized grains (Fig. 10), the deformation is ≈ 400%. For the submicron material, the plastic deformation at room temperature is indeed very small, or roughly less than 0.5 %. The difference turns out to be more than 20 times larger for the 100 nm material than that for the 0.35 µm material. Combining the effect of fine grain size together with the difference in δD_b values of these two grades of TZP materials, it may be estimated that the rate of deformation of the 100 nm material could be more than 8×10^2 times greater than that of the 0.35 µm material under similar conditions. This appears to comply with what has actually been observed.

Although the above mentioned arguments favor the grain-boundary diffusion as the controlling mechanism, the contribution by the dislocation mechanism is not to be completely ruled out. As a matter of fact, diffusion motivated dislocation climb and multiplication might be in operation as another or a parallel mechanism to develop slip band-like microfeatures as had also been observed on the side faces down from the fractured edges. However, for an ionic-covalent oxide material, this mechanism would be believed to be of minor importance.

In conclusion, the observations and comparisons are consistent with the view that localized superplastic deformation of the grains in the 100 nm 3Y-TZP material at room temperature driven by cyclic tensile fatigue is essentially governed by grain-boundary diffusion mechanism.

ACKNOWLEDGEMENT

The authors thank Professors C.L. Bai, C.F. Zhu and X.W. Wang of Institute of Chemistry, Chinese Academy of Sciences for performing the AFM studies. Thanks are also due to Professor M.Q. Li and Dr. L. Xu of Institute of Nuclear Research, Chinese Academy of Sciences for investigating the side faces by AFM. Professors Z.Y. Xu of Shanghai Jiaotong University is specially to be thanked for very fruitful and constructive discussions. The authors wish to acknowledge the Grant by the State Science and Technology Commission under the Climb Project on Nano-Materials Science.

REFERENCES

1. C. Suryamarayana, Nanocrystalline materials, Int. Mater. Rev., 40 (1995) 41
2. J.L. Shi and Z.X. Lin, Preparation of zirconia powders by oxalate precipitation, Solid State Ionics, 32/33 (1989) 544
3. J.L. Shi, Z.X. Lin and T.S. Yen, Effect of dopants on the crystallite growth superfine zirconia powders, J. Europ. Ceram. Soc., 8 (1991) 117
4. J.L. Shi, Z.X. Lin, M.L. Yuan and T.S. Yen, Characterization of nanometer zirconia powder, J .Chin. Ceram. Soc., 21 [3] (1993) 2221 (in Chinese)
5. J.L. Shi, B.S. Li, M.L. Yuan and T.S. Yen, Processing of nsno-sized Al_2O_3/Y-TZP composite powders, I. Preparation and characterization of the composite powders, J. Europ. Ceram. Soc., 15 [10] (1995) 959
6. H.B. Qiu, L. Gao, C.D. Feng, J.K. Guo and D.S. Yan (T.S.Yen), Preparation and characterization of nanoscale Y-TZP powder by heterogeneous azeotropic distillation, J. Mater. Sci., 30 (1995) 5508
7. X. Ding, Z.Qi and Y. He, Effect of hydrolysis water on the preparation of nano-crystalline titania powders via a sol-gel process, J. Mater. Sci. Lett., 14 (1995)21
8. L.Gao, H.B. Qiu and Z.R. Huang, Preparation of nanoscale ceramic powders, in "Fabrication and Characterization of Advanced Materials", Eds. S.W. Kinand and S.J. Park, MRS-K, (1995) 251
9. J.I.M. Low, Sol-gel processing of ultrafine ceramic powders, "4th Nisshin Engr. Particle Tech. Intl. Seminar", Dec. 6-8 Osaka, Japan (1995) 1
10. H.B. Qiu, L. Gao, H.C. Qiao, J.K. Guo and D.S. Yan, Nanocrystalline zirconia powder processing through innovative wet-chemical methods, Nano-structured Mater., 6 (1995) 373

11. D.S. Yan, H.B. Qiu, J.L. Shi, Y.S. Zheng and L. Gao, nano-sized powder, bulk materials and their superplastic behavior under fatigue, "4th Nisshin Engr. Particle Tech. Intl. Seminar", Dec 6-8,Osaka, Japan (1995) 36
12. H. Gleiter, Nanocrystalline materials, Progress in Mater. Sci., 33 [4] (1989) 223
13. R. Birringer, Nanocrystalline materials, Mater. Sci. and Engr., A117 (1989) 33
14. H. Hahn, J. A. Eastman and R.W. Siegel, Processing of nanophase ceramics, Ceram. Tran. 1b, Ceram. Powder Sci., (1988)1115
15. D. Vollath and K.E. Sickafus, Synthesis of nanosized ceramics oxide powder by microwave plasma reactions, Nanostructured Mater., 1(1992) 427
16. J.L. Shi, Ph.D Thesis, Shanghai Institute of Ceramics, Chinese Academy of Sciences   (1989)
17. J.L. Shi, Characterization of the pore size distributions of the superfine zirconia powder compacts, J. Solid State Chem., 95[2]. (1992) 412
18. H.B. Qiu, C.D. Feng, L. Gao, J.K. Guo and D.S. Yan, Preparation of monodispersed spherical zirconia particles by homogeneous precipitation, Proceedings 5th Intl. Sym. on Ceram. Mater. and Components for Engine, Eds., D.S. Yan S.X. Shi and X.R. Fu, Shanghai, China, May(1994)618
19. J.L. Shi , J.H. Gao, B.S. Li and T.S. Yan, Precessing of nano-sized Al_2O_3/Y-TZP composite powders, II. Densification and microstructure development, J. Europ. Ceram. Soc., 15 [10] (1995) 967
20. F.Wakai, S. Sakaguchi and Y. matsono, Superplasticity of yttria-stabilized tetragonal ZrO_2 polycrystals, Adv. Ceram. Mater., 1[3] (1986) 259
21. T.G. Nieh, C.M. McNally and J. Wadsworth, Superplastic behavior of 20% Al_2O_3/Y-TZP ceramic composite, Scrip. Metall., 23 (1989) 457
22. T.G.Nieh and J. Wadsworth, Suplastic behavior of a fine grained yttria-stabilized tetragonal zirconia plycrystal (Y-TZP), Acta, Metall., 38 (1990) 1121
23. I-W. Chen and L.A. Xue, Superplastic allumina ceramics with grain growth inhibitors, J. Am. Ceram. Soc., 73[9] (1990) 2585
24. F.Wakai, Y. Kodama, S.Sakaguchi, N. Murayama, K. Izaki and K. Niihara, Asuperplastic covelent crystal composite, Nature, 344 (1990) 421
25. W.J. Kim, J. Wolfenstine and D.D. Sherby, Tensile ductility of superplastic ceramics and metallic alloys, Acta. Metall. & Mater., 39 (1991) 199
26. J. Wittenaner, T.G. Nieh and J. Wadsworth, Superplastic gas-pressure deformation of Y-TZP sheet, J, Am. Ceram. Soc., 76 [7] (1993) 1665
27. R.L. Coble, A model for boundary diffusion controlled creep in polycrystalline materials, J. Appl. Phys., 34 (1963) 1679

ELECTRON BEAM CONTROL OF ALUMINA NANOSTRUCTURE AND Al BONDING

Shun-ichiro Tanaka and BingShe Xu

Tanaka Solid Junction Project, ERATO,
Research Development Corporation of Japan
1-1-1, Fukuura, Kanazawa-ku, Yokohama 236, Japan

ABSTRACT

Oxide-free Al nanoparticles and stable alumina having spherical or faceted shapes were obtained from metastable alumina by electron beam irradiation in transmission electron microscopy (TEM). Al nanodecahedra rotated, revolved, and bonded under successive electron beam irradiation without heating. The driving force for this behavior is momentum transferred from spirally trajected electrons and ions inside the TEM pole piece. A new method of producing onion-like fullerenes and intercalation from amorphous carbon film with the catalytic effect is obtained by means of an electron beam irradiation technique. We propose a possible method of controlling/manipulating nanoscopic phases and their bonding behavior by an electron beam. This approach has the potential to be applied to the fabrication of new materials and new devices.

INTRODUCTION

Various synthesis methods have been developed for metal and ceramic nanoparticles using physical and chemical reactions. However, oxide-free aluminum nanoparticles have not been obtained because of the possibility of oxidation during the process of synthesis. We have succeeded in preparing oxide-free Al nanoparticles with shapes varying from spheres to decahedrons through the process of removing oxygen atoms from metastable aluminium oxide (alumina) by electron irradiation in TEM.[1] Alumina structure and phase have also changed during the process of Al nanoparticle formation accompanied by nanoparticle migration, bonding,[2] and reaction with substrate. This paper describes the morphologies of induced alumina and the bonding behavior of Al nanoparticles obtained by electron irradiation, and also the derivative reaction phenomena with an amorphous carbon substrate.

EXPERIMENTAL PROCEDURE

Our starting material was metastable, 150- to 200-nm-diameter θ-Al$_2$O$_3$ particles irradiated by electrons in a 10^{-5} Pa vacuum on a conventional room-temperature stage of 200 kV high resolution TEM.[1,2] Irradiation intensity of electrons ranged from 0.3 to 3 × 10^{20} e/cm^2 · sec and the diameter of the irradiation area varied from 400 to 120 nm. The nanomorphologies and behaviors of particles were monitored *in situ* by lattice images and electron diffraction patterns. The phase change of alumina was also checked by x-ray diffraction and differential thermal analysis (DTA).

RESULTS AND DISCUSSION

Aluminum Nanodecahedra

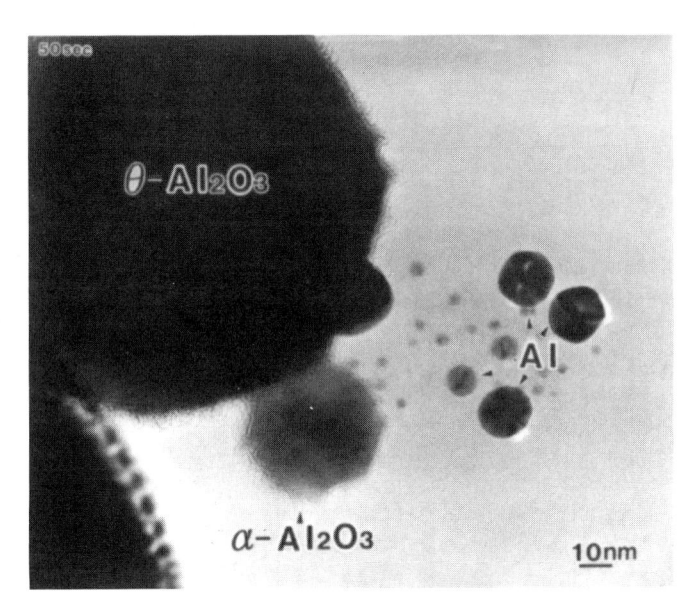

Figure 1. Al and α-alumina nanoparticles formed around θ-alumina by an electron irradiation with the intensity of 10^{20} e/cm^2 · sec.

The results of electron beam irradiation to metastable θ-alumina in TEM are shown in Figure 1. Aluminum and stable nanoparticles were formed around θ-alumina by the electron irradiation. Al nanoparticle has a twinned fivefold structure surrounded by the {111} plane, as proved by the electron diffraction technique,[3] and is called a decahedron. Our achievement of a nanodecahedron aluminum structure, proof of an oxide-layer-free surface (typically observed in noble metals such as Au or Pt rather than in base metals), is a first. The diameter of the Al nanoparticles ranged from 2 to 40 nm depending on electron irradiation conditions. Optimum diameter range for the twinned structure was 10 to 30 nm.

The mechanism of Al nanoparticle formation is considered to be based on release of oxygen atoms from the surface caused by the combination of electron-stimulated desorption (ESD) and the sputtering effect.

Derived Alumina Structure

The phases of the irradiated metastable alumina were mainly θ with a monoclinic structure: α and δ were minor phases. The α-alumina nanoparticles shown in Figure 1 were obtained through θ → α transformation. These particles have the hexagonal shape typical for trigonal crystallographic symmetry. In the range of low electron irradiation, 3×10^{19} e/cm^2 · sec, the transformation occurs from the surface layer of starting θ particles, as reported previously.[1] Another characteristic alumina was observed in this experiment. Figure 2 shows one directionally grown α-alumina which has c plane with a {100} faceted front and [110] growth direction.

The temperature of θ → α transformation depends on oxygen partial pressure (measured by DTA as 1150 to 1250°C at the atmospheric pressure), whereas the estimated value from our experiment was below 500°C in vacuum. The mechanism of the transformation and its growth to the preferred orientation is explained by the same process as for Al nanoparticles; that is, the ESD plus sputtering process. Free Al and oxygen atoms originated by ESD plus sputtering process recombine to form the stable alumina.

Bonding of Nanoparticles

Al nanoparticles formed close to alumina showed dynamic behavior by successive electron beam irradiation.[2] Al nanoparticles rotated and revolved around the irradiation center (as shown in Figure 3), which indicates that the rotation speed increased as the irradiation intensity became stronger. The nanoparticles migrated to coalesce with each other, as shown in Figure 4, in which a {111} plane rearranged to be parallel just before bonding.

The major driving force of such behavior is not the thermal activation, but the momentum transfer from spirally trajected electrons or ions to Al nanoparticles inside the pole piece of TEM.

CONCLUSION

In summary, the following findings were made:

- Oxide-free Al nanoparticles were obtained from metastable alumina by electron beam irradiation.
- Stable alumina having spherical and faceted shape grew from metastable alumina.
- Al nanodecahedra rotated, revolved, and bonded under electron beam without heating. The driving force of their behavior is the momentum transferred from spirally trajected electrons and ions inside the pole piece of TEM.

We propose a possible method of controlling/manipulating nanoscopic phases and their bonding behavior by an electron beam. This approach has the potential to be applied in the fabrication of new materials and new devices.

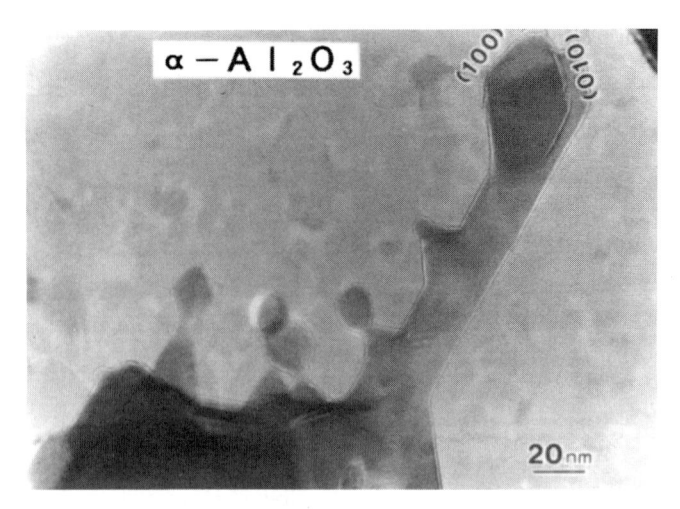

Figure 2. α-alumina single crystal plate grown from θ-alumina surface having a {100} front shape and [110] growth direction.

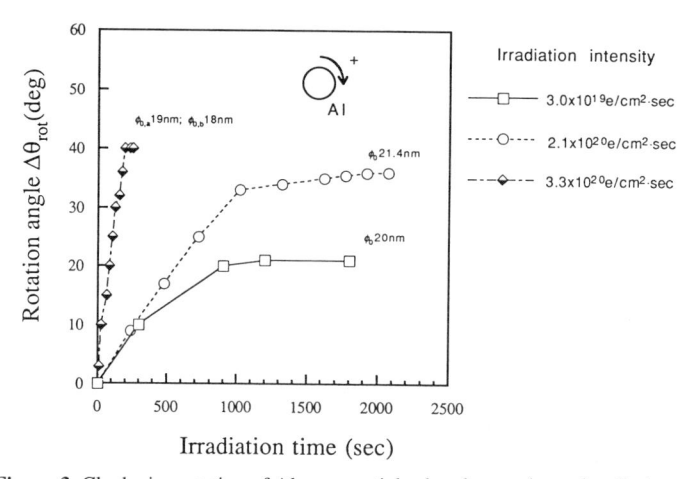

Figure 3. Clockwise rotation of Al nanoparticles by electron beam irradiation.

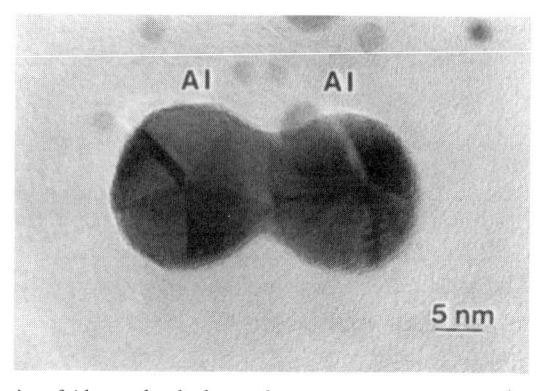

Figure 4. Bonding behavior of Al nanodecahedra on the room-temperature stage by electron beam irradiation.

ACKNOWLEDGEMENTS

The authors are indebted to Mr. S. Munekawa for his assistance with x-ray measurement and DTA of alumina.

REFERENCES

1. B. Xu and S. Tanaka, *Nano Structured Materials,* **6**:727 (1995).
2. B. Xu and S. Tanaka, *Materials Science Forum,* **137**:207 (1996).

MICROSTRUCTURE AND PROPERTIES OF SiC/SiC AND SiC/III-V NITRIDE THIN FILM HETEROSTRUCTURAL ASSEMBLIES

Robert F. Davis, S. Tanaka, S. Kern, M. Bremser, K. S. Ailey, W. Perry, and T. Zheleva

North Carolina State University
Department of Materials Science and Engineering
Box 7907
Raleigh, NC 27695-7907.

ABSTRACT

Monocrystalline thin films, multilayered heterostructures and solid solutions containing selected combinations of SiC, AlN and GaN have been grown on 6H-SiC(0001) substrates by gas-source molecular beam epitaxy (GSMBE) or metallorganic vapor phase epitaxy (MOVPE). Polytype control of the deposition of 3C(β)-SiC(111) and 6H-SiC(0001) was achieved via control of the substrate orientation, temperature and gas phase chemistry. Essentially atomically flat AlN films or island-like features were observed using on-axis or vicinal 6H-SiC substrates, respectively. The coalescence of the latter features at steps gave rise to incommensurate boundaries as a result of the misalignment of the Si/C bilayer steps with the Al/N bilayers in the growing film. The controlled growth of 3C-SiC films with low defect densities and atomically flat surfaces was achieved on the AlN films to form the first 6H-SiC/2H-AlN/3C-SiC multilayer heterostructures. Solid solutions of these two phases were also achieved. Monocrystalline GaN(0001) or $Al_xGa_{1-x}N$ ($0 \leq x \leq 1$) thin films were also grown on the AlN(0001) films or directly on the same SiC surface at 1100°C, respectively, via metallorganic vapor phase epitaxy (MOVPE). The stages of growth of each of the above films, their microstructure and selected other properties will be described.

INTRODUCTION

Silicon carbide exists in >250 polytypes or stacking arrangements of the Si/C bilayers along the direction of closest packing. There is a one cubic (zincblende) polytype referred to as 3C or β–SiC where the three indicates the number of bilayers in the unit cell. The hexagonal (wurtzite) form also exists. Both occur in more complex, intermixed forms to produce the wide range of ordered hexagonal or rhombohedral structures of which 6H and, more recently, 4H, are currently the most common. All of the non-cubic structures are known collectively as α–SiC.

Previous SiC film growth studies via MBE on 6H-SiC substrates have not resulted in homoepitaxy. Kaneda, et al.[1] deposited films of β-SiC(111) on on-axis 6H-SiC{0001} at 1150-1400 °C using electron-beam evaporated Si and C sources. Yoshinobu, et al.[2,3] deposited 3C films via MBE on 6H-SiC{0001} and 3C-SiC(001) substrates at 850-1160 °C using an alternating supply of C_2H_2 and Si_2H_6 controlled by continuous monitoring of the RHEED pattern reconstructions. Similar RHEED pattern monitoring was used by Fissel, et al.[4,5] to grow 3C-SiC at modest growth rates (to 1 nm min^{-1}) below 1000 °C on "Si-stabilized" 6H-SiC(0001) having a 3 × 3 RHEED pattern. Rowland et al.[6] reported the deposition of monocrystalline 3C on vicinal 6H-SiC(0001) at 1050°C using a simultaneous supply of C_2H_4 and Si_2H_6 and the very slow growth rate of ≈100Å h^{-1}; however, the films contained a high density of double positioning boundaries which formed as a result of β-SiC nucleation on terrace sites rather than step sites.

The potential applications of III-N materials for both optoelectronic and microelectronic applications has stimulated significant research[7-9]. Gallium nitride (GaN) in the wurtzite phase forms a continuous solid solution with AlN such that bandgap engineering from 3.4 eV to 6.2 eV is possible. Thus $Al_xGa_{1-x}N$ alloys are attractive for both ultraviolet emitters and detectors as well as high-power and high frequency microelectronic applications.

Single crystal GaN substrates are currently available in very limited quantities only via high pressure, high temperature processes. As such, heteroepitaxy is the dominant thin film growth process, and c-plane sapphire is the most commonly employed substrate. In order to accommodate the large lattice mismatch, nitridation of the sapphire followed by the deposition and annealing of a low temperature GaN or AlN buffer layer is conducted prior to GaN film growth[10,11]. Grain orientation competition occurs during the first 0.5 microns of GaN growth which results in the formation of low angle grain boundaries which persist throughout the entire epitaxial layer[12,13]. Substantial concentrations of threading dislocations exist, especially near the film/buffer layer interface. The growth of 3 to 4 μm of material is usually necessary in order achieve device quality epitaxy. In contrast, it has been shown that monocrystalline AlN can be deposited epitaxially on 6H-SiC(0001) at temperatures >1050°C[14,15]. Use of a 1000Å AlN buffer layer deposited 1100°C has been demonstrated to be an effective buffer layer. Subsequent GaN epitaxy contains only random threading dislocations and is free of low angle grain boundaries. Since conductive AlN has yet to be demonstrated, this buffer layer prohibits the effective use of the conductive SiC for the formation of backside contacts. Hence, the use of a conductive $Al_xGa_{1-x}N$ as a buffer layer appears to hold great promise.

In the following sections, recent research in the authors' laboratories regarding the growth and characterization of thin films of 3C- and 6H-SiC, GaN and $Al_xGa_{1-x}N$ alloys and heterostructures from these materials are described.

GAS SOURCE MBE GROWTH OF 3C- AND 6H-SiC, AlN, AND AlN/SiC HETROSTRUCTURES AND SOLID SOLUTIONS

The 3C- and 6H-SiC thin films were deposited on vicinal 6H-SiC(0001) substrates cut 3.5 ± 0.5° toward [11$\bar{2}$0]. The substrates were chemically cleaned *ex situ* in a 10% HF solution for 5 minutes, loaded immediately into the GSMBE growth chamber and further cleaned *in situ* using a silane exposure and UHV anneal [18].

The reactants of SiH_4 and C_2H_4 were used to grow the SiC epilayers. Deposition experiments were performed between 1000-1500 °C in either SiH_4-C_2H_4 or SiH_4-C_2H_4-H_2 environments. The flow rates of these gases were 0.75 sccm SiH_4, 0.75 sccm C_2H_4 and 5 sccm H_2. The base pressure of the system was ≈10^{-9} Torr and the operating pressure was varied between 3×10^{-6} and 2×10^{-4} Torr depending on the ambient processing conditions.

Films of AlN were subsequently deposited on the 6H-SiC(0001) films at 1100°C using evaporated Al from a Knudson cell at 1250°C and high purity ammonia. The 6H-SiC(0001) substrates were maintained at 1100°C for the deposition of the AlN/SiC heterostructures and the $(AlN)_x(SiC)_{1-x}$ solid solutions using the previously noted precursor sources. In the deposition of the former, gases species used to make one layer were exhausted from the MBE before the next layer was begun. For the solid solutions, variations regarding the relative amount of AlN and SiC were used with special consideration given to limit the formation of the Si-N bonding.

Epilayer growth using only SiH_4 and C_2H_4 always resulted in films of β-SiC(111). Figure 1 shows a plot of the log of growth rate (ln R_g) vs. T^{-1} for the β-SiC(111) on vicinal 6H-SiC(0001) for reactant input flows of 0.75 sccm for SiH_4 and C_2H_4. Analysis of this curve showed that it follows the general Arrhenius equation:

$$R_g = R_0 \exp(-\frac{\Delta H_a}{RT}) \qquad (1)$$

where R_g is the growth rate (Å hr^{-1}) perpendicular to the substrate surface, R_0 is a pre-exponential factor, ΔH_a is the effective activation energy (kcal•mol^{-1}), R is the ideal gas

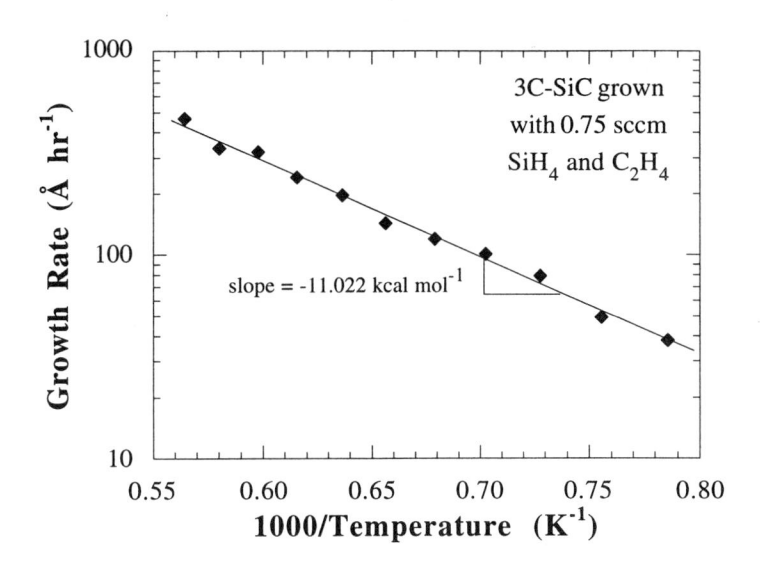

Figure 1. Arrhenius plot of growth rate for 3C-SiC(111) films grown between 1000-1500 °C with 0.75 sccm SiH4 and 0.75 sccm C2H4.

constant (1.987 cal mol^{-1} K^{-1}) and T is in °K. From the slope of the curve, the apparent activation energy, ΔH_a, was determined to be 21.9 kcal mol^{-1}. The rate expression describing this data is given by:

$$\ln R_g = 12.3 - \frac{11\,022}{T} \tag{2}$$

where R_g is in Å hr^{-1} and T is in K. From the shape (linear) of the fitted curve and the size of the activation barrier, the reaction appears to be surface reaction limited. The size of the activation barrier was determined to be independent of the partial pressures of the reactants. No change in growth rate was observed for SiH4 flow rates of 0.5, 0.75 and 1.0 sccm. Since the growth rate was constant for a given deposition temperature between 1000-1500 °C for all conditions where $0.33 \leq f_{SiH_4}/f_{C_2H_4} \leq 1.33$, the SiC deposition process was deemed to be independent of the SiH4 flow rate (f_{SiH_4}) within these regimes. This indicates that the deposition reaction (2SiH4 + C2H4 = 2SiC + 6H2) was zeroeth order with respect to the SiH4 flow rate. These results indicate that the deposition reaction is most likely governed by the decomposition of C2H4.

The flow rate of C2H4 was subsequently varied between 0.5 and 1.5 sccm with the SiH4 flow rate maintained at 0.75 sccm. Figure 1 shows a plot of $\ln R_g$ vs. T^{-1} obtained using a C2H4 flow rate of 0.75 sccm. Analogous curves, each representing a different input flow rate of C2H4, had identical slopes indicating that the reaction mechanism was unchanged as a function of the C2H4 flow rate.

Figure 2 is a series of plots of $\ln R_g$ vs. $\ln f_{C_2H_4}$ in the range of 0.5-1.5 sccm at 1100, 1200 and 1300 °C. In each case, the data appeared to be linear. Each curve had a slope of ≈0.63 which indicated that the deposition reaction was a two-thirds order power reaction which was limited by the decomposition of the C2H4 gas. The fractional order of the reaction may be indicative of a series of elementary step reaction(s) (i.e., the adsorption and decomposition of the
C2H4 into more reactive species), which were necessary for Equation 1 to proceed. The deposition rate as a function of temperature and C2H4 input flow rate was determined to be:

$$\ln R_g = 12.48 - \frac{11\,000}{T} + 0.63 \ln f_{C_2H_4} \tag{3}$$

where R_g is in Å hr^{-1}, T is in K and $f_{C_2H_4}$ is in sccm. Again, the apparent activation barrier for this process was ≈22 kcal mol^{-1}. Since the curves presented in Figure 2 were linear over the range studied and had a dependence on the C2H4 flow rate, the deposition process

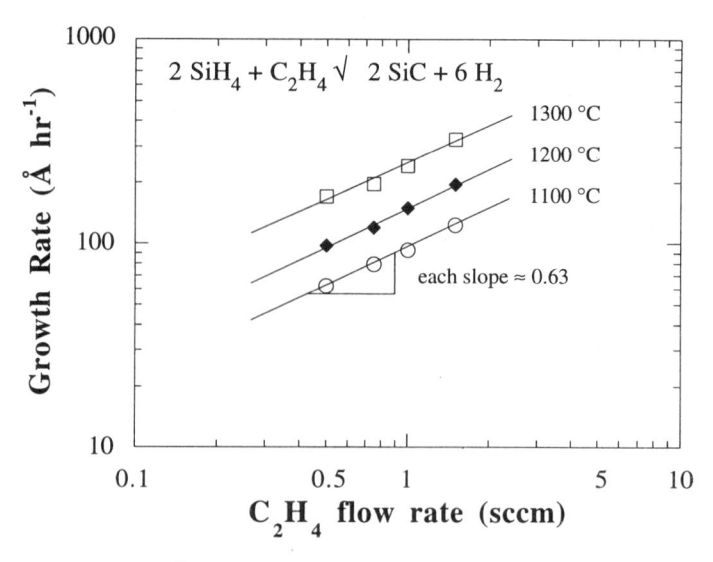

Figure 2. Plot of ln R_g vs. ln $f_{C_2H_4}$ for 3C-SiC(111) films grown at 1100, 1200 and 1300 °C with 0.75 sccm SiH$_4$ and 0.5-1.5 sccm C$_2$H$_4$. Note that the slope of each curve is ≈0.63 indicating that the reaction is a 2/3 order power reaction with respect to $f_{C_2H_4}$.

appeared to be controlled by surface reactions. The rate of species formation, reaction and desorption from C$_2$H$_4$ appeared to be the rate limiting step.

Combining H$_2$ with the SiH$_4$ and C$_2$H$_4$ during growth produced several advantageous changes in the resulting films. At temperatures below 1350 °C, the primary benefits of the H$_2$ addition were increased surface smoothness and a greatly enhanced growth rate of the β-SiC films. The activation energy for the films grown with 0.75 sccm SiH$_4$, 0.75 sccm C$_2$H$_4$ and 5 sccm H$_2$ at 1000-1300 °C was 21.6 kcal mol^{-1}. Thus the presence of H$_2$ did not change the magnitude of the barrier over this temperature range. However, since the slopes of the curves shown in Figures 1 and from this study were nearly identical, the reaction mechanism was both apparently unchanged and independent on the input flow of C$_2$H$_4$.

Subsequent increases in the $f_{C_2H_4}$ did not result in an increased growth rate. These observations are attributed to the presence of H$_2$ which apparently provided the impetus for at least one of the following processes: 1) adsorption of C$_2$H$_4$ onto the surface of the growing film, 2) sweeping of the unreacted source gas species or unwanted product species from the growth surface and/or 3) formation of a suitable reactant specie in tandem with the decomposition or reaction of C$_2$H$_4$ in the presence of H$_2$.

Further increases in the deposition temperature (≥1350 °C), resulted in the stabilization of the step flow growth mode, the consequent deposition of 6H-SiC(0001) epilayers, a marked increase in growth rate and an apparent decrease in the activation energy for the growth of these films were obtained with the addition of H$_2$ to the gas stream between 1350-1500 °C. The temperature dependence of the growth rate appeared to be linear on an Arrhenius plot which is shown in Figure 3. The activation barrier was determined to be 12.6 kcal mol^{-1}, which is in excellent agreement with values calculated from CVD research (13.0 kcal mol^{-1}) using the same reactant sources, substrate orientation and crystallographic face [20]. Although the growth rate was still strongly dependent on the deposition temperature, the decrease in the value of this activation barrier was an excellent indicator of the effect of the H$_2$ gas on the gas chemistry, growth kinetics and gas flow dynamics. Although the rate limiting step and the controlling factor could not be ascertained, limited studies showed that varying the C$_2$H$_4$ input from 0.375-1.0 sccm did not result in a change in the growth rate. This indicated that the rate controlling factor at these temperatures in the presence in H$_2$ had most likely changed. It is important to note that growth on both vicinal and on-axis substrates proceeded at approximately the same rate (within 5%) under similar growth conditions. However, β-SiC was always produced on the on-axis substrates,

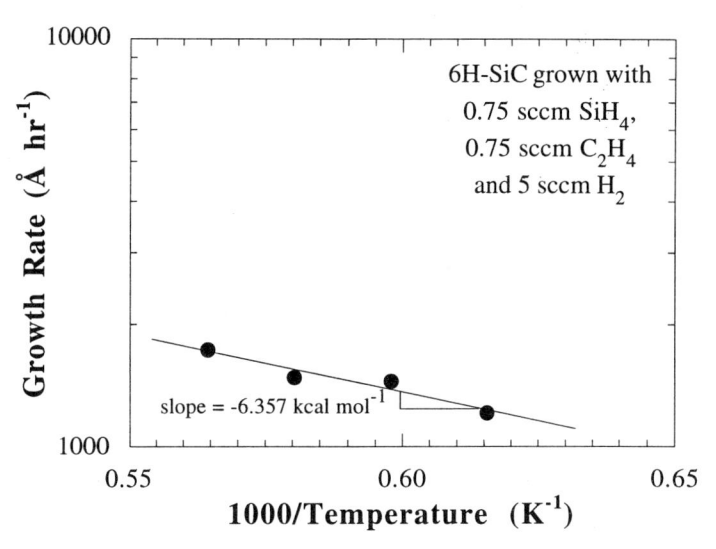

Figure 3. Arrhenius plot of growth rate for 6H-SiC(0001) films grown between 1350-1500 °C with 0.75 sccm SiH4, 0.75 sccm C2H4 and 5 sccm H2.

regardless of growth conditions because the smaller atomic diffusion distance relative to the distance between steps.

The first 3C-SiC/2H-AlN/6H-SiC multilayer ever produced has been grown in this research[19]. The interfaces between layers wre physically and chemically abrupt. The layers were oriented identically in all directions. The alignment of the atomic columns was continuous except in the vicinity of the surface steps which resulted in inversion domain boundaries in the AlN. Considerable <111> twinning and stacking faults were present in the 3C-SiC layer. Growth on the on-axis 6H-SiC substates resulted in much smoother sufaces for all layers and reduced defect densities in both the AlN and the SiC. For additional information regarding this research, the reader is referred to Ref. 19.

More recent growth kinetics research has determined the activation energy for 3C-SiC growth on the AlN to be 14.2 kcal mol^{-1}. This value is similar to that for the deposition of this film on 6H-SiC via CVD using the same sources and growth temperature[20] as well as for the homoepitaxial deposition of 6H-SiC described above. Changing the C_2H_4 flow rate (and thus the C/Si ratio) from 0.375 to 0.75 caused a change in the carrier concentration (Hall measurements) and electron mobility from 6.2×10^{15} cm^{-3} and 613 cm^2 V^{-1}s^{-1} to 1.2×10^{15} cm^{-3} and 721 cm^2 V^{-1}s^{-1}. Increasing the C concentration in the gas phase and at the surface of the growing film increased the competition between the n-type donor species of N and the C for the atomic site of the latter such that the latter was favored for occupation. This resulted in the decrease (increase) in the carrier concentration (Hall mobility) noted above.

Monocrystalline films of $(AlN)_x(SiC)_{1-x}$ solid solutions were achieved over the composition range of $0.2 \leq x \leq 0.8$. High-resolution transmission electron microscopy (HRTEM) revealed that monocrystalline films with $x \geq 0.25$ had the wurtzite (2H) crystal structure; however, films with $x < 0.25$ had the zincblende (3C) crystal structure. This cubic to hexagonal structural transition as a function of AlN concentration is the first ever reported for a monocrystalline film of these materials. No evidence of compositional nonuniformity or second phase formation was evident in the Auger depth profiles of each sample, an example of which is given in Figure 4 for the approximate composition of $(AlN)_{0.8}(SiC)_{0.2}$. However, the phase diagram of Zangville and Ruh[21] indicates that equilibrium solid solutions having the AlN concentrations deposited in the present research only exist above $\approx 2000°C$. Thus, using the nonequilibrium process route of MBE allows these non-equilibrium compositions to be produced.

GALLIUM NITRIDE AND $Al_xGa_{1-x}N$ SOLID SOLUTIONS

Monocrystalline GaN(0001) or $Al_xGa_{1-x}N$ ($0.05 \leq x \leq 0.96$) thin films were grown on either AlN(0001) buffer layers or directly on the same 6H-SiC(0001) surface, respectively, at 1100°C and 45 Torr in a cold-wall, vertical, pancake-style, RF inductively heated, OMVPE

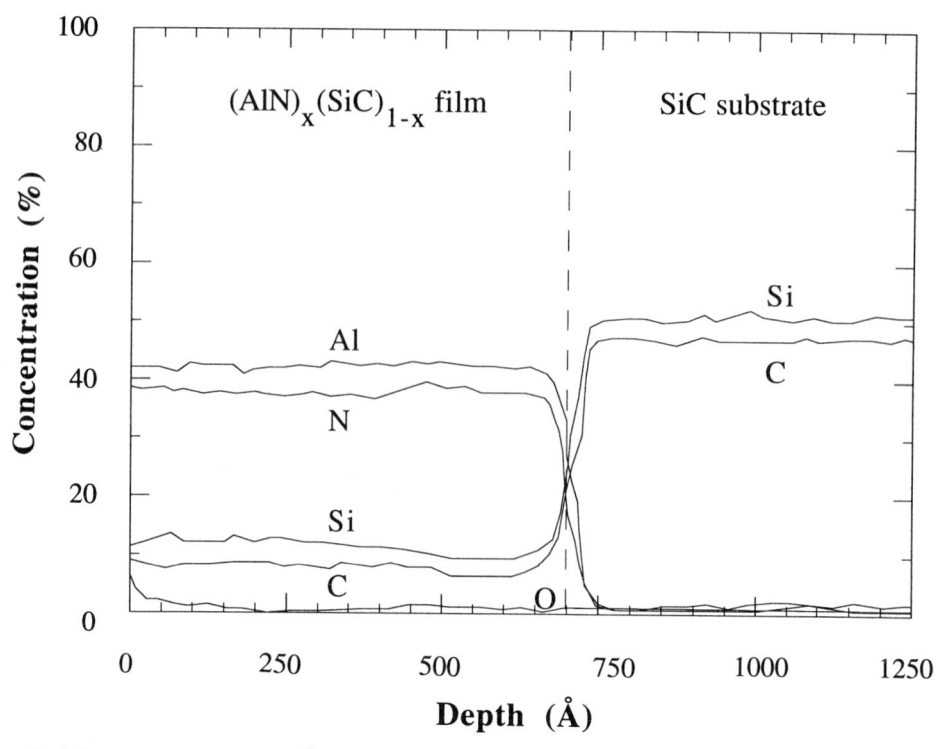

Figure 4. AES depth profile of a 700 Å thick $(AlN)_{0.8}(SiC)_{0.2}$ film deposited at 900°C via GSMBE.

system. Deposition of these films was initiated by flowing triethylgallium (TEG) or various ratios of triethylaluminum (TEA) and triethylgallium (TEG) in combination with 1.5 SLM of ammonia (NH$_3$) and 3 SLM of H$_2$ diluent. The total metalorganic precursor flow rate for both types of films was 32.8 μmol/min.

The GaN films deposited on AlN buffer layers previously grown on the on-axis 6H-SiC(0001)$_{Si}$ substrates had very smooth surfaces. A slightly mottled surface was observed for films deposited on vicinal 6H-SiC(0001)$_{Si}$ substrates and believed to be a result of inversion domain boundaries (IDBs) produced at selected steps on the growth surface due to the mismatch in the Si/C and Al/N bilayer stacking sequences [22]. The dislocation density within the first 0.5 μm of the GaN film on the vicinal 6H-SiC(0001)$_{Si}$ substrate was approximately 1×10^9 cm^{-2}. This density decreased rapidly as a function of thickness, as determined from plan view TEM. The on-axis wafers had less step and terrace features; thus, the AlN buffer layers on these substrates were of higher microstructural quality with smoother surfaces and fewer IDBs. Consequently, the microstructural quality of the GaN films were better for on-axis growth. Undoped, high quality GaN films grown on both vicinal and on-axis 6H-SiC(0001) substrates were too resistive for Hall-effect measurements.

Al$_x$Ga$_{1-x}$N (0.5≤x≤0.96) films were simultaneously deposited directly on vicinal and on-axis 6H-SiC(0001) substrates without the use of AlN buffer layers. However, the growth mechanism was found to be complex. Field-emission Auger electron spectroscopy revealed that the interface region was Al rich and that the desired bulk film composition was achieved at approximately 100 Å of film thickness. Therefore, the possibility exist that an extremely thin AlN or aluminum rich Al$_x$Ga$_{1-x}$N layer exist within the first few angstroms of the interface; thereby providing a graded buffer layer structure of less than 100Å in thickness. Coalescence of the resulting islands resulted in smooth films after a few thousands angstroms of growth. More specifically, films having x≤0.5 and deposited at 1100°C had smooth, featureless surfaces as revealed by SEM. Alloys with x>0.5 had smooth surfaces only if deposited at T=1150°C.

The cathodoluminescence (CL) emission peaks of the alloy compositions were compared with the bandgap as determined by spectroscopic ellipsometry. Using a parabolic model, the following relationship describe the CL peak emission (I_2-line emission) (Eq.(1)) as a function aluminum mole fraction for $0 < x < 0.96$.

$$E_{I_2}(x) = 3.50 + 0.64x + 1.78x^2 \qquad (1)$$

The measurements show a negative deviation from a linear fit. However, these films are highly strained due to incompletely relaxation of thermal expansion coefficient and lattice mismatch. The effect of strain is unquantified at this time. Comparison of CL spectra of $Al_{0.12}Ga_{0.88}N$ grown directly on 6H-SiC versus growth on a previously deposited 1000Å AlN which is used for GaN deposition revealed a 75 meV blue shift for films deposited with a buffer layer. This would indicate that the films deposited directly on SiC are in greater tension than the ones deposited using an AlN buffer layer. In addition, low temperature CL revealed near-bandedge emission located at 3.53 eV which indicates that considerable tensile strain is present in this film since thicker films of the same composition have emission located at 3.83 eV.

Undoped, high quality $Al_{0.05}Ga_{0.95}N$ films grown directly on vicinal 6H-SiC(0001) exhibited residual, ionized donor concentrations of 1×10^{18} cm^{-3}. The ionized donor concentration decreased rapidly with increasing Al content and was $<1 \times 10^{16}$ cm^{-3} for $Al_{0.35}Ga_{0.65}N$, as determined by CV measurements. This origin of this residual ionized donor concentration is under investigation, since concentrations of $<1 \times 10^{15}$ cm^{-3} have been measured for GaN films grown on AlN buffer layers in the same reactor. Moreover, AlN layers having $N_D - N_A$ of 8×10^{15} cm^{-3} has also been deposited. However, the controlled introduction of SiH_4 allowed the reproducible achievement of ionized donor concentrations within the range of 2×10^{17} cm^{-3} to 2×10^{19} cm^{-3} in films with $x \leq 0.52$. Silicon doping for higher aluminum compositions result in highly compensated material. The growth of p-type $Al_xGa_{1-x}N$ films for $x < 0.13$ via the introduction of Mg has been successful.

Several structures have been grown using GaN and $Al_xGa_{1-x}N$, e.g., a $Al_{0.2}Ga_{0.8}N$/GaN superlattice with periods of various thicknesses. Each superlattice period was repeated five times, and the structure was capped with 0.2 microns of GaN. Observation of the structure in plan-view TEM did not indicate a reduction in dislocation below that normally observed.

SUMMARY

Monocrystalline thin films of 3C(β)-SiC(111) were achieved via GSMBE on vicinal 6H-SiC(0001) substrates between 1000-1500°C using SiH_4 and C_2H_4 at a growth rate which was dependent on T and C_2H_4 flow rate. Similarly deposited films of 6H-SiC(0001) and an increase in growth rate were achieved between 1350-1500°C with the addition of H_2. The rate limiting step involved the adsorption and decomposition of C_2H_4. At the higher temperatures, the controlling growth rate mechanism changed, as indicated by a decrease in the apparent activation barrier. Monocrystalline, pseudomorphic 3C-SiC/2H-AlN/6H-SiC heterostructures and $(AlN)_x(SiC)_{1-x}$ films ($0.2 \leq x \leq 0.80$) were achieved. In the latter the structure changed from zincblende to wurtzite at $x \approx 0.25$. MOVPE has been used to grow monocrystalline GaN/AlN(0001), $Al_xGa_{1-x}N$ ($0.5 \leq x \leq 0.95$) films and superlattices of these materials on 6H-SiC(0001) substrates at 1100°C. All films possessed a smooth surface morphology and were free of low-angle grain boundaries and associated oriented domain microstructures.

ACKNOWLEDGEMENTS

The authors acknowledge the Office of Naval Research for the sponsorship of this research under Grants No. N00014-92-J-1500 and N00014-92-J-1477, Cree Research, Inc. for the SiC substrates and A. Reisman, H. H. Lamb, T. W. Weeks, and C. Wolden of NCSU and J. C. Angus of CWRU for helpful discussions.

REFERENCES

1. S. Kaneda, Y. Sakamoto, T. Mihara and T. Tanaka, J. Cryst. Growth 81:536 (1987).

2. T. Yoshinobu, M. Nakayama, H. Shiomi, T. Fuyuki and H. Matsunami, J. Cryst. Growth, 99:520 (1990).
3. T. Yoshinobu, H. Mitsui, I. Izumikawa, T. Fuyuki and H. Matsunami, Appl. Phys. Lett. 60:824 (1992).
4. A. Fissel, B. Schröter and W. Richter, Appl. Phys. Lett. 66:3182 (1995).
5. A. Fissel, U. Kaiser, E. Ducke, B. Schröter and W. Richter, J. Cryst. Growth 154:72 (1995).
6. L. B. Rowland, R. S. Kern, S. Tanaka and R. F. Davis, J. Mater. Res. 8:2753 (1993).
7. R. F. Davis, et al., J. Mater. Sci. Eng. B 1:77 (1988).
8. S. Strite and H. Morkoc, J. Vac. Sci. Technol. B 10:1237 (1992).
9. H. Morkoç, S. Strite, G. B. Gao, M. E. Lin, B. Sverdlov and M. Burns, J. Appl Phys. 76:1363 (1994).
10. Nichia Chemical Industries, Ltd., 491 Oka, Kaminaka, Anan, Tokushima 774, Japan
11. M.A. Khan, J.N. Kuznia, D.T. Olson and R. Kaplan, J. Appl. Phys., 73, 3108 (1993).
12. H. Amano, I. Akasaki, K. Hiramatsu, N. Koide and N. Sawaki, Thin Solid Films, 163, 415 (1988).
13. J.N. Kuznia, M.A. Khan, D.T. Olson, R. Kaplan and J. Freitas, J. Appl. Phys.,73, 4700 (1993).
14. S. Nakamura, Jpn. J. Appl. Phys., 30, L1705 (1991).
15. T.W. Weeks, Jr., M.D. Bremser, K.S. Ailey, E.P. Carlson, W.G. Perry, R.F. Davis, Appl. Phys. Lett., 67, 401 (1995).
16. L. B. Rowland, S. Tanaka, R. S. Kern and R. F. Davis, in *Amorphous and Crystalline Silicon Carbide IV*, C. Y. Yang, M. M. Rahman and G. L. Harris, eds., Springer-Verlag, Berlin, 1992, p. 84.
18. R. S. Kern, S. Tanaka, and R. F. Davis, in Silicon Carbide and Related Materials: M. G. Spencer, R. P Devaty, J. A. Edmond, M. A. Khan, R. Kaplan and M. Rahman, eds., Institute of Physics Conf. Series, No. 137, Institute of Physics, Bristol, United Kingdom: IOP Publishing (1994), p. 389.
19. L. B. Rowland, R. S. Kern, S. Tanaka, and R. F. Davis, Appl. Phys. Lett. 62, 2310 (1993).
20. H. S. Kong, J. T. Glass and R. F. Davis, J. Mater. Res. 4, 204 (1989).
21. R. Ruh and A. Zangville, J. Am. Ceram. Soc. 65, 260 (1988).
22. S. Tanaka, R. S. Kern, and R. F. Davis, Appl. Phys. Lett. 66, 37 (1995).

CORRELATING SIMULATION DEFECT ENERGY CALCULATIONS WITH HREM IMAGES OF PLANAR DEFECTS IN ELECTROCERAMICS.

W. E. Lee[1], M. A. McCoy[1,2] and R. W. Grimes[2],

[1]Department of Engineering Materials, University of Sheffield, UK
[2]Department of Materials, Imperial College, London, UK.

INTRODUCTION

The influence of microstructural features such as grain size, porosity and grain boundary structure on the characteristic (extrinsic) properties of electroceramics is reasonably well understood (Levinson and Hirano 1993). However, electroceramics frequently contain a variety of intragranular planar (extended two-dimensional) defects whose influence on properties is less well known. Such planar defects fall into three general categories (Amelinckx and van Landuyt 1978, Lee et al. 1996) defined by considering the macroscopic crystallographic symmetry (Figure 1).

Translation interfaces (Figure 1a) separate two parts of a crystal (variants or domains) that are related to each other by a constant displacement vector R, independent of the distance from the interface. Examples include stacking faults (SF's), antiphase domain boundaries (APB's) and crystallographic shear (CS) planes. **Orientation** interfaces (Figure 1b) separate variants which possess a displacement field in which R increases linearly with distance from the interface. These separate parts of a crystal at a different orientation e.g. twin variants related by a point symmetry element such as a mirror plane or rotation axis forming reflexion and rotation twins respectively. **Inversion** interfaces (180° boundaries or inversion twins) separate two domains related to one another by an inversion operation (Figure 1c) such as inversion domain boundaries (IDB's, McCoy et al. 1996a).

Such planar defects form in electroceramics by several mechanisms. They may arise as a **growth defect** produced accidentally as the crystal grows from the nucleus or from **deformation** as a means of relieving strain produced by an applied stress. Planar defects also form during **polymorphic transformations** when a higher symmetry crystal converts to a lower symmetry structure. Finally, the presence of a **dopant ion** or some **nonstoichiometry** may be accommodated in the crystal by the presence of planar boundaries. This paper is concerned with examples of planar defects formed to accommodate dopants or nonstoichiometry in electroceramic systems.

To achieve a full understanding of the mechanisms by which such defects form, a novel combination of atomistic simulation and High Resolution Electron Microscopy (HREM) is being used. The atomistic simulations are used to calculate (i) lattice energies of bulk phases; (ii) energies associated with point defects and defect clusters; and (iii) interfacial energies. The calculated energies indicate how easily the crystal accommodates dopants or nonstoichiometry either within the bulk or along a certain interface, as well as the

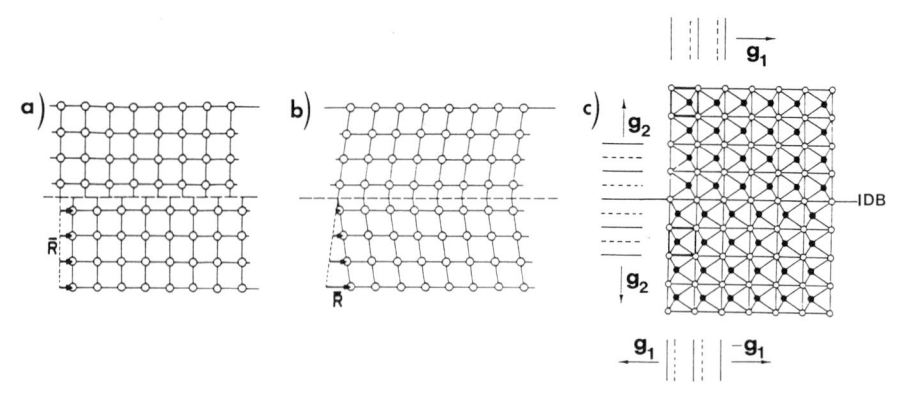

Figure 1. The three main types of planar interface, a) translation (e.g. SF's, APB's and CS planes), b) orientation (e.g. twins) and c) inversion (e.g. IDB's) interfaces.

effect of changes in interface chemistry on the interfacial energy; the HREM and associated image simulation calculations identify the location of dopants within the planar defect or the composition of defect phases. Based on this knowledge atomic level models can be derived for the planar defect formation mechanism.

EXPERIMENTAL

Atomistic simulation techniques are well developed for predicting defect structures in ceramics (*e.g.* Catlow and Stoneham, 1989). Lattice energies were calculated using the GULP code developed by J. D. Gale at Imperial College. The code is based on a description of an ionic crystal in which the lattice is defined in terms of effective two-body interionic potentials; **long-range** Coulombic interactions are summed to infinity by the Ewald technique and **short-range** interactions are modelled using a Buckingham potential of the form

$$V(r) = A \exp(-r/\rho) - C/r^6$$

where A, ρ and C are adjustable parameters. These parameters are fitted so that the calculated crystal properties, such as lattice parameter, elastic constants and dielectric constant, closely match the experimental data. The short-range forces, which account for electron overlap and van der Waals interactions, decrease sharply with distance and are effectively zero beyond a few atom spacings so that for computational economy these interactions are set to zero beyond about 2 nm. The effect of ionic polarisability is accommodated by a shell model (Dick and Overhauser, 1958) consisting of a massless shell of charge Y coupled to a core of charge X (total charge = X + Y) by a harmonic spring of force constant K.

Interfacial energies are calculated using the MARVIN code (Gay and Rohl 1995). In this treatment, the crystals on either side of the interface are divided into regions I and II. Ions in region I are relaxed explicitly to zero strain, while the ions in region II remain fixed and represent the electrostatic and short-range effects of the remaining bulk lattice on the interface. The optimum value for region I (typically 1-2 nm) is determined by increasing the region I size until there is no change in the calculated interfacial energy. The interfacial energy, ΔE_{int}, is defined by

$$\Delta E_{int} = (E_{interface} - \Sigma n^i E^i_{bulk})/Area_{(interface)}$$

for n unit cells of species i with relaxed bulk energy E_{bulk}, and where $E_{interface}$ is the total energy associated with the interface region (the combined energy of region I of crystal A and crystal B). In this technique, the effects of both the interfacial structure (e.g. orientation relationship and interfacial habit planes; termination surface of each crystal) and chemistry (introduction of local defects such as substitutional ions, vacancies and interstitials) on the interfacial energy can be calculated.

Mixed oxide pellets were uniaxially pressed and sintered to high density at various temperatures. Standard ceramographic grinding and polishing techniques were used along with ultrasonic drilling, dimpling and 6 kV Ar^+ ion beam milling of the TEM specimens. HREM imaging was done using a JEOL 3010 HREM with 0.18nm point to point resolution operating at 300keV. HREM image simulation calculations were done using the a multislice code in the EMS V3.30 software package developed by Stadelmann (1987). Supercells containing candidate defect structures were constructed and used to produce calculated HREM images using the appropriate microscope and specimen parameters.

RESULTS AND DISCUSSION

Chemically-Induced Planar Defects in In-doped ZnO.

Transparent, electrically-conducting, ZnO thin films are widely used in opto-electronic devices and solar cell technology. In such devices approximately 1 atom% In doping improves the electrical conductivity considerably. Unlike formation of IDB's in Sb-doped ZnO (McCoy et al., 1996a) no spinel phases are observed; instead, higher levels of In are incorporated in ZnO via formation of a series of homologous compounds, $Zn_mIn_2O_{3+m}$, composed of planar defects lying along (0001) of the wurtzite ZnO structure. The detailed structure of these defect compounds is unclear. The most recent model (Kimizuka et al., 1995) suggests that the defect phases are isostructural with $LuFeO_3(ZnO)_m$ with *single* In-O layers separating ZnO layers. Space groups of $R\bar{3}m$ and $P6_3/mmc$ were assigned based on XRD results for m odd and m even compounds, respectively. In contrast, previous studies by Kasper (1967) and Cannard and Tilley (1988) suggested that the defect phases are composed of wurtzite ZnO separated by *double* layers of In_2O_3 although there is disagreement over the configuration of the layer. Cannard and Tilley (1988) proposed that the indium-rich layers were thin layers of indium oxide in the bixbyite structure (room temperature cubic polymorph with space group Ia3) with $(0001)_{ZnO}$ // $\{111\}_{Bixbyite\ In2O3}$ so that the stacking of pseudo-close packed oxygen planes is preserved throughout. Kasper (1967), however, proposed that the indium oxide layers adopted a hexagonal form, isostructural with La_2O_3 (space group $C\bar{3}m$) with $(0001)_{ZnO}$ // $\{0001\}_{Hexagonal\ In2O3}$ in which the continuity of the oxygen sublattice is again maintained.

Calculated lattice energies for the three relaxed structures are shown in Table 1; note that a more negative lattice energy implies greater stability. The results suggest (McCoy et al. 1996b) that the most stable structures for the defect phases are based on the Cannard and Tilley model, having a double layer of indium oxide separating the slabs of wurtzite ZnO with an associated stacking fault of $1/3<10\bar{1}0>$ of one wurtzite layer relative to another across the indium oxide layer (Figure 2).

In the most stable model the indium oxide layers are relaxed considerably from the positions in the bixbyite structure, although the In-O and In-In bond distances are close to those in the parent bixbyite structure. The calculated space groups for the odd and even phases, R3m and P6₃mc, are consistent with the previously reported diffraction results in that the allowed reflexions for x-ray and electron diffraction are identical. Additionally, there is very good agreement between image simulation based on this model and HREM micrographs, and a generally better match than with the Kimizuka et al. model (Figure 3).

Crystallographic Shear Planes Accommodating Nonstoichiometry in $SrTiO_3$.

$SrTiO_3$-based ceramics are used in varistor applications being superior than ZnO in suppressing sharp pulse noise and electrostatic discharge pulses such as pulse interference

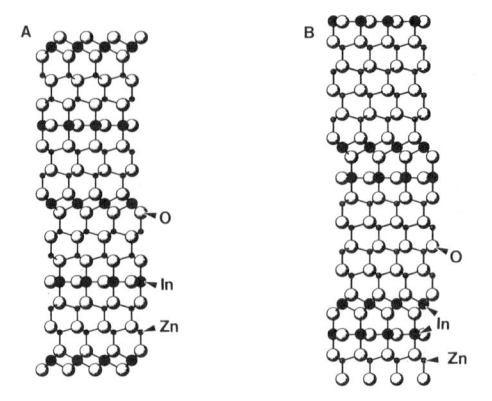

Figure 2. Schematic of relaxed structure based on (a) Kimizuka *et al.* model and (b) Cannard and Tilley model.

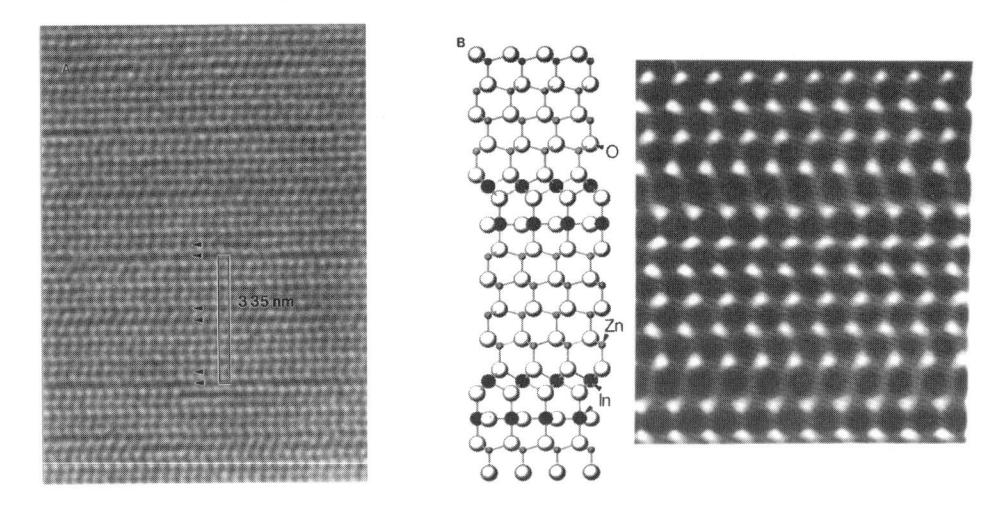

Figure 3. (a) HREM micrograph of $Zn_4In_2O_7$ in [1$\bar{2}$10] projection and (b) simulated images based on atomistic simulation refinement of the Cannard and Tilley model (imaging conditions near Scherzer defocus; specimen thickness 4 nm). Under these imaging conditions the black spots correspond to cation columns. Positions of In-rich layers are arrowed in (a).

Table 1. Calculated Lattice and Formation Energies for Defect Phases in the ZnO-In$_2$O$_3$ System (see McCoy et al. 1996b for details of potential and shell parameters).

Model	Lattice Energy (eV)	Formation Energy (eV)
Kimizuka et al. (1995)		
m-even, Zn$_4$In$_2$O$_7$	-297.2	+0.54
Kasper (1967)		
m-even, Zn$_4$In$_2$O$_7$	-296.77	+0.98
Cannard and Tilley (1988)		
m-even, Zn$_4$In$_2$O$_7$	-297.83	-0.09
m-odd, Zn$_5$In$_2$O$_8$	-337.54	-0.45

generated by electric micromotors. The cubic perovskite crystal structure of SrTiO$_3$ (Figure 4a), does not easily incorporate changes in A/B stoichiometry via isolated point defects (Balachandran and Eror, 1982) with excess solubility limits of only about 0.5 mol% TiO$_2$ and less than 0.2 mol% SrO. Excess TiO$_2$ is usually precipitated as a Ti$_x$O$_{2-x}$ compound (Fujimoto and Watanabe, 1985) while excess SrO is accommodated by formation of Ruddlesden-Popper defects (Ruddlesden and Popper, 1958), consisting of extra layers of strontium oxide between sheets of strontium titanate.

These Ruddlesden-Popper defects can be considered in terms of a crystallographic shear of the perovskite structure in which sheets of TiO$_6$ octahedra along (001) are removed and the remaining perovskite layer is sheared by 1/2[111] (McCoy et al. 1995). The excess SrO gives rise to a homologous series of defect phases denoted by the formula Sr$_{n+1}$Ti$_n$O$_{3n+1}$, where n is the number of perovskite layers separating the defects (Tilley, 1977) with a corresponding change in space group from perovskite Pm3m to I4/mmm. Sr$_2$TiO$_4$ is tetragonal and made up of interleaved single SrO (cubic rock salt) layers and SrTiO$_3$ (Figure 4b). Similarly, Sr$_3$Ti$_2$O$_7$ is comprised of double SrTiO$_3$ slabs and SrO (Figure 4c).

Figure 5 is an HREM micrograph for a sample of overall composition Sr$_{1.25}$TiO$_{3.25}$ (Sr$_5$Ti$_4$O$_{13}$) showing grains containing a high density of planar defects lying along (001). In this composition the most frequently observed SrO-rich phase was Sr$_3$Ti$_2$O$_7$, with occasional Sr$_2$TiO$_4$; the Sr$_3$Ti$_2$O$_7$ grains were almost always defective, containing regions of unfaulted SrTiO$_3$. Table 2 shows the calculated relaxed lattice energies for the Ruddlesden-Popper phases and a comparison between the calculated and published experimental lattice parameters.

The excellent agreement of the predicted lattice parameters of the compounds with experiment confirms the accuracy of the calculations. Since the lattice energies correspond to different compositions, the appropriate formation reactions must be considered e.g. SrO

Table 2. Calculated Lattice Energies and Parameters for the System SrO-TiO$_2$ (for details of potential and shell parameters see McCoy et al. 1996c).

Phase	Lattice Energy (eV)	Lattice Parameter (nm) Calculated	Experimental
SrO	-33.12	a = 0.5159	a = 0.5160
SrTiO$_3$	-155.52	a = 0.3903	a = 0.3903
Sr$_2$TiO$_4$	-188.79	a = 0.3892	a = 0.3890
		c = 1.27	c = 1.270
Sr$_3$Ti$_2$O$_7$	-344.37	a = 0.3901	a = 0.3900
		c = 2.054	c = 2.040
Sr$_4$Ti$_3$O$_{10}$	-499.87	a = 0.3903	a = 0.3900
		c = 2.830	c = 2.810
Sr$_5$Ti$_4$O$_{13}$	-655.38	a = 0.3903	
		c = 3.615	

+ 2SrTiO$_3$ = Sr$_3$Ti$_2$O$_7$. Summing the calculated lattice energies of SrO, SrTiO$_3$ and Sr$_3$Ti$_2$O$_7$ suggests that this reaction is energetically favourable with a reaction energy of -0.21eV. Similarly, the Sr$_3$Ti$_2$O$_7$ composition is more stable than the associated combination of Sr$_2$TiO$_4$ and SrTiO$_3$, since Sr$_2$TiO$_4$ + SrTiO$_3$ = Sr$_3$Ti$_2$O$_7$, has a favourable reaction energy of -0.06eV. However, the Sr$_3$Ti$_2$O$_7$ phase is only slightly more stable than other Ruddlesden-Popper phases with lower Sr/Ti ratios; for example, the reaction energy for Sr$_4$Ti$_3$O$_{10}$ = SrTiO$_3$ + Sr$_3$Ti$_2$O$_7$ is only -0.02 eV. Clearly, since the differences in reaction energies between the various Ruddlesden-Popper phases are small, a number of different metastable phases may coexist in the sintered ceramic. Alternatively, since the formation energy difference is relatively small, the various interfacial energies may also be a significant factor in discerning between the formation of the Ruddlesden-Popper phases.

For compositions richer in SrO than Sr$_3$Ti$_2$O$_7$, the reaction SrO + Sr$_3$Ti$_2$O$_7$ = 2Sr$_2$TiO$_4$ is energetically favourable (-0.09 eV) suggesting that equilibrium in this system may be based on a series of line compounds: Sr$_2$TiO$_4$, Sr$_3$Ti$_2$O$_7$, and SrTiO$_3$. For SrO-excess compositions greater than Sr$_3$Ti$_2$O$_7$ (60 mol% SrO, 20 mol % TiO$_2$), formation of both Sr$_2$TiO$_4$ and Sr$_3$Ti$_2$O$_7$ is energetically favourable. The experimental observation of the Sr$_2$TiO$_4$ phase may be due to local compositional inhomogeneities in which the local composition lies between Sr$_2$TiO$_4$ and Sr$_3$Ti$_2$O$_7$. For SrO-excess compositions less than Sr$_3$Ti$_2$O$_7$, formation of an intergrowth of Sr$_3$Ti$_2$O$_7$ and SrTiO$_3$ is energetically favourable.

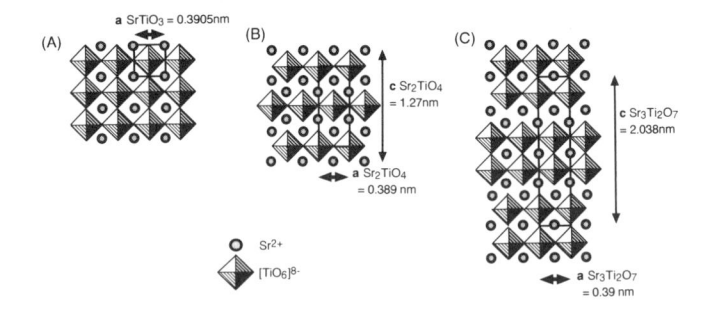

Figure 4. Schematic diagram of Ruddlesden-Popper defect structures, a) SrTiO$_3$, b) Sr$_2$TiO$_4$ and c) Sr$_3$Ti$_2$O$_7$ projected along [100] (after Tilley, 1977).

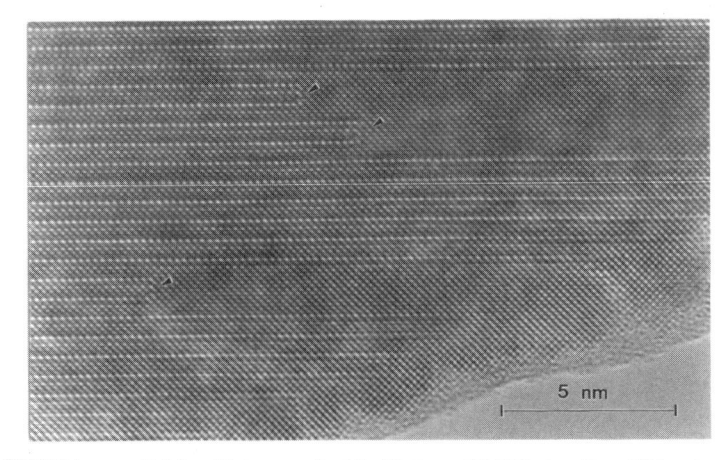

Figure 5. HREM image of slabs of intergrowth of Sr$_3$Ti$_2$O$_7$ and SrTiO$_3$ in a Sr$_{1.25}$TiO$_{3.25}$ composition (McCoy et al. 1996c) viewed in [100] projection. Cation columns appear as white spots. The steps showing the interface between the Sr$_3$Ti$_2$O$_7$ and SrTiO$_3$ phases are arrowed.

The Sr$_3$Ti$_2$O$_7$ phase and the SrTiO$_3$ intergrowths maintained a well-defined orientation relationship [100](001)$_{SrTiO3}$ // [100](001)$_{Sr3Ti2O7}$ with interfaces parallel to (001) and (010). The interface between the Sr$_3$Ti$_2$O$_7$ and the perovskite phase was structurally coherent along both (010) and (001), despite a 4% mismatch between the respective oxygen sublattices across the (010) boundary which is absorbed by a tetragonal [001] distortion of the SrTiO$_3$ layer.

The best fit between the HREM image simulation and the experimental micrographs suggests that both top and bottom (001) Sr$_3$Ti$_2$O$_7$/SrTiO$_3$ interfaces consist of continuous stacking of SrTiO$_3$ unit cells. The (010) interface is characterized by a change of intensity in the spot maxima and regions of diffuse intensity. Certain spot maxima (arrowed in Figure 6) are displaced away from the Sr$_3$Ti$_2$O$_7$ phase by approximately 0.02nm along [010]. In addition, these spots have a higher intensity than the adjacent corresponding spots to the left (Ti positions in the Sr$_3$Ti$_2$O$_7$ phase), suggesting a possible difference in local chemistry.

The interfacial models for the atomistic calculations were based on the experimentally-observed orientation relationship. The various ways of cutting the crystals are described in McCoy *et al.* (1996c). For each cut, the crystals were systematically translated in directions within the interfacial plane as well as perpendicular to the interface, and the minimum energy noted.

In the (001) interface the most stable interfaces maintained the strontium titanate stacking sequence across the interface with no defects on either the cation or anion sublattices. These interfaces had no excess interfacial energy, due to the negligible in-plane strain and good atomic matching across the interface.

In any configuration the (010) interface creates a mismatch for both cation and oxygen sublattices. The most stable (010) interface was the Ti-terminating SrTiO$_3$ configuration with an interfacial energy of 1.26 J/m^2, despite some Ti-Ti nearest neighbour configuration at the interface. The calculated interfacial energy is lowered by local substitution of Sr^{2+} for Ti^{4+} ions, and the creation of charge-compensating oxygen vacancies in the Sr$_3$Ti$_2$O$_7$ phase at the interface. A schematic of the (010) interface, containing details from both the atomistic simulation calculations and the HREM results is shown as Figure 6.

Total substitution of Sr^{2+} for interfacial Ti^{4+} ions gave the lowest calculated (010) interfacial energy (1.04 J/m^2). During the relaxation process, significant displacement parallel to and perpendicular to the interface plane was observed on both the cation and anion sublattice away from the interface. The maximum displacement occurred within the region containing the substitutional defects, consistent with the experimental HREM observations. These displacements were highly localized, and confined to ions within the first and second unit cell (ca. 0.4 - 0.8 nm) away from the interface.

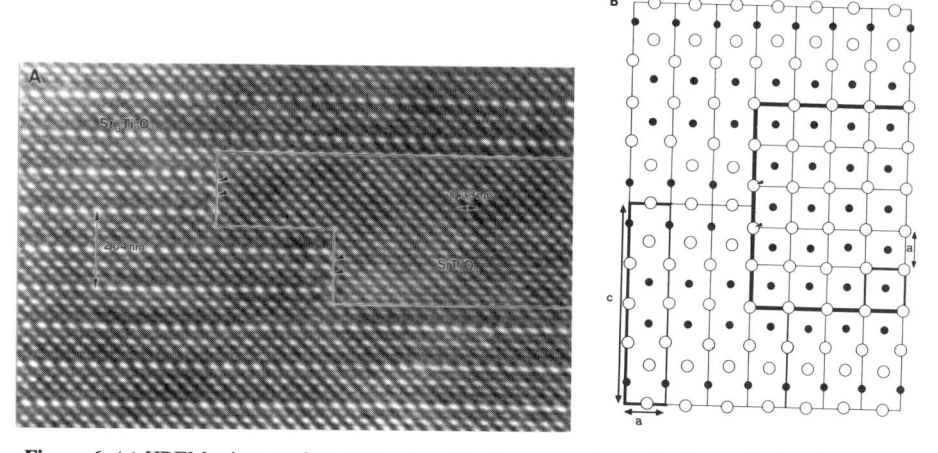

Figure 6. (a) HREM micrograph and (b) schematic diagram (cation sublattice only) showing proposed interface between the terminating Sr$_3$Ti$_2$O$_7$ layer and SrTiO$_3$. Positions of possible Sr^{2+} substitution are arrowed.

The experimental HREM results in the present study and from previous work (Tilley, 1977) support the results of the atomistic simulation calculations; the most frequently observed defect structure, $Sr_3Ti_2O_7$, was also found to be the most energetically stable. The calculated lattice energies are consistent with the hypothesis that the lattice would prefer to accommodate small concentrations of SrO in the form of isolated $Sr_3Ti_2O_7$ intergrowths in a $SrTiO_3$ matrix.

SUMMARY

The combined use of simulation defect energy calculations and HREM leads to an understanding of the atomic structure of planar defects and can be used to suggest models for their formation mechanisms.

ACKNOWLEDGEMENTS

This work was partly funded by the United Kingdom EPSRC under grants GR/J31629 and GR/K74302.

REFERENCES

S. Amelinckx and J. van Landuyt in *Diffraction and Imaging Techniques in Materials Science*, S. Amelinckx, R. Gevers and J. van Landuyt, Eds., N. Holland, Amsterdam, 1978, 107-151.

U. Balachandran and N. G. Eror, "On the defect structure of strontium titanate with excess SrO," *J. Mater. Sci.* **17** 2133-40 (1982).

P. J. Cannard and R. J. D. Tilley, "New intergrowth phases in the $ZnO-In_2O_3$ system," *J. Solid State Chem.* **73** 418-426 (1988).

C. R. A. Catlow and A. M. Stoneham (Eds.) *Computer Simulation of Defects in Polar Solids*, special issue of *J. Chem. Soc. Faraday Trans.* **85** [5] (1989).

B. G. Dick and A. W. Overhauser, "Theory of the Dielectric Constant of Alkali Halides", *Phys. Rev.* **112** (1) 90-103 (1958).

M. Fujimoto and M. Watanabe, "Ti_nO_{2n-1} Magneli phase formation in $SrTiO_3$ dielectrics," *J. Mater. Sci.* **20** 3683-90 (1985).

D. H. Gay and A. L. Rohl, "MARVIN-a new computer code for studying surfaces and interfaces and its application to calculating the crystal morphologies of corundum and zircon," *J. Chem. Soc. Far. Trans.* **91** [5] 925- (1995).

H. Kasper, "New phases with the wurtzite structure in the $ZnO-In_2O_3$ system," *Z. anorg. allg. Chem.* **349** [3-4] 113-123 (1967).

N. Kimizuka, M. Isobe and M. Nakamura, "Synthesis and single-crystal data of homologous compounds, $In_2O_3(ZnO)_m$ (m = 3, 4 and 5), $InGaO_3(ZnO)_{13}$, and $Ga_2O_3(ZnO)_m$ (m = 7, 8, 9, and 16) in the $In_2O_3-ZnGa_2O_4-ZnO$ system," *J. Solid State Chem.* **116**, 170-78 (1995).

W. E. Lee, I. M. Reaney and M. A. McCoy, "Planar defects in electroceramics," *British Ceramic Proceedings* **55** 199-212 (1996).

C. M. Levinson and S. Hirano (Eds.), *Grain Boundaries and Interfacial Phenomena in Electronic Ceramics* (Am. Ceram. Soc., Ohio, USA, 1993).

M. A. McCoy, R. W. Grimes and W. E. Lee, "Comparison of observed and calculated structures of Sr-rich planar defects in strontium titanate," *Proc. 4th Euro Ceramics*, **5** 451-458 (1995).

M. A. McCoy, R. W. Grimes and W. E. Lee, "Inversion domain boundaries in ZnO ceramics," To be published in J. Mater. Res. 1996a.

M. A. McCoy, R. W. Grimes and W. E. Lee, "Atomistic simulation and HREM of planar defect is indium-doped zinc oxide ceramics," in preparation for *Phil. Mag. A* 1996b.

M. A. McCoy, R. W. Grimes and W. E. Lee, "Phase stability and interfacial structures in the $SrO-SrTiO_3$ system," To be published in *Phil. Mag. A* 1996c.

S. N. Ruddlesden and P. Popper, "The compounds of $Sr_3Ti_2O_7$ and its structure," *Acta Cryst.* **11** 54-55 (1958).

P. Stadelmann, "EMS-A software package for electron diffraction analysis and HREM image simulation for materials science", *Ultramicroscopy* **21** 131-145 (1987).

R. J. D. Tilley, "An electron microscope study of perovskite-related oxides in the Sr-TiO system," *J. Solid State Chem.* **21** 293-301 (1977).

MICROSTRUCTURAL STUDIES OF STRONTIUM TITANATE DIELECTRIC CERAMICS

Zhigang Mao and Kevin M. Knowles

University of Cambridge
Department of Materials Science & Metallurgy
Pembroke Street
Cambridge CB2 3QZ, UK

INTRODUCTION

Perovskites are an important family of materials. Devices based on $SrTiO_3$ and $BaTiO_3$ are particularly important for a number of electronic applications. Depending on the way in which the materials are processed, perovskites can exhibit a number of phenomena, such as a positive temperature coefficient of resistance, ferroelectricity, varistor behaviour, dielectric behaviour, semiconductivity, superconductivity and infra-red sensitivity. $SrTiO_3$ is one of the few perovskites to have the ideal cubic *Pm3m* perovskite crystal structure at room temperature and is often used as substrate for ceramic superconducting thin films.

Suitably doped and processed, $SrTiO_3$ powders can be used to produce internal boundary layer capacitors (IBLCs) with high dielectric constants and low loss factors (Yamoaka, 1986). The microstructure of IBLCs consists of semiconducting grains and insulating grain boundaries. This microstructure can be achieved in two different ways. The first way is an interrupted two stage firing process in which the $SrTiO_3$ grains are first made semiconducting by the introduction of dopants such as Nb_2O_5 and firing the material in a reducing atmosphere. The grain boundaries are made insulating by painting low melting point metal oxides on the fired ceramics and then having a second firing stage where these oxides diffuse along the grain boundaries of the ceramic (Yamoaka, 1986). Transmission electron microscopy (TEM) of these devices show clear evidence for diffusion layers and second phase layers at the grain boundaries in which the low melting point metal oxides are concentrated (Franken et al., 1981; Fujimoto and Kingery, 1985).

The second way of producing the desired microstructure is to have an initial powder composition in which there is a carefully controlled level of acceptor dopant, which might be expected to segregate to grain boundaries (Chiang and Takagi, 1990) and donor dopant, which should be distributed reasonably homogeneously throughout the $SrTiO_3$ grains (Chiang and Takagi, 1990; Zhou et al., 1991; Mao and Knowles, 1995). The advantage of this second method is that there does not need to be an interruption in the firing schedule, making it a potentially attractive process for producing multilayer IBLCs. Recent microstructural studies of such IBLCs (Mao and Knowles 1995, 1996a, 1996b) show that they have a much smaller average grain size, thinner interphase layers at grain boundaries, and strontium-rich rather than titanium-rich second phases at grain boundaries when compared with IBLCs produced by the interrupted two stage firing process. However, in order to obtain a complete understanding of the microstructural development of these new IBLCs prepared without an interruption in the firing schedule and to establish the role of the acceptor dopants beyond reasonable doubt, it is necessary to make a comparison with 'control' samples of SrTiO₃. Here, we report observations on donor-doped SrTiO₃ and SrTiO₃ bicrystals and make a critical comparison of these observations with those we have made on the IBLCs produced without an interruption in the firing schedule.

EXPERIMENTAL DETAILS

The donor-doped SrTiO₃ ceramic was prepared from a powder mixture of 1.0 mol $SrCO_3$, 1.0 mol TiO_2, 0.007 mol Nb_2O_5, 0.003 mol Al_2O_3, and 0.004 mol SiO_2 wet milled for 48 hr. After drying, the powder was calcined at 1150 °C for 1 hr in air and pressed into 2 mm thick pellets under a pressure of 0.7 MPa. This was sintered at 1360 °C in a reducing atmosphere of methane and argon for four hours for consolidation and in order to make the SrTiO₃ grains semiconducting. Finally, the pellet was furnace cooled to room temperature. Procedures for producing IBLCs with acceptor dopants were similar, although they contained a second firing stage at a lower temperature. These procedures have been described elsewhere (Mao and Knowles, 1995, 1996a, 1996b, 1996c). Undoped SrTiO₃, 24° about [001] (i.e., near a $\Sigma = 13$ coincident site lattice orientation) symmetrical tilt bicrystals made by Shinkosha Co. Ltd. of Tokyo were purchased from Wako Bussan Co. Ltd. for TEM and HREM observation.

Specimens for TEM examination were prepared using standard ion beam thinning methods and examined uncoated in both a JEOL 2000FX at 200 kV and a JEOL 4000EX HREM at 300 kV. Impedance measurements were made on the donor-doped sample with a Solartron 1250/1286 FRA used at a nominal applied voltage of 50 mV, covering a frequency range f of 30 mHz –100 kHz.

EXPERIMENTAL OBSERVATIONS ON THE DONOR-DOPED SAMPLE

X-ray analysis of this material showed diffraction peaks attributable only to SrTiO₃, with no evidence of a second phase, as might be expected given the low levels of additives. General TEM observations showed that the average grain size was ≈ 20 μm, significantly larger than in samples in which either lithium or sodium had been introduced as an acceptor dopant (Mao and Knowles, 1995, 1996a, 1996b). A number of dislocations were observable in the grains, as in the two examples *A* and *B* shown in Figure 1 in dark field under three different **g**/4**g** condition weak beam imaging conditions. In Figure 1(a), the diffracting beam is $\mathbf{g} = 10\bar{1}$ and the electron beam direction **BD** is approximately $[\bar{3}13]$, in (b) $\mathbf{g} = 0\bar{2}0$ and $\mathbf{BD} = [\bar{1}03]$ and in (c) $\mathbf{g} = \bar{1}\bar{1}0$ and $\mathbf{BD} = [\bar{1}14]$. While it should be noted that these images are not taken in preferred weak beam imaging conditions, because the values of s_g are less than ideal and the choice of **g**/4**g** introduces dynamical scattering effects which serve to degrade the image resolution and complicate the image contrast (Thomas and Goringe, 1979), these imaging conditions are, however, very convenient for the determination of the Burgers vectors of the lattice dislocations using the method of Ishida et al. (1980), as we have shown elsewhere (Mao and Knowles, 1996c).

Since the **g** in Figure 1 are not coplanar, the Burgers vectors of these dislocations can be determined from the number of terminating thickness fringes at the exit of the dislocations from the wedge-shaped foil as a function of **g** (Ishida et al., 1980). The number of terminating fringes for dislocation *A* are 0, –2 and –1 respectively and for dislocation *B* are +1, –2 and –1 in Figure 1(a) – 1(c) respectively, with the sign of terminating fringes positive when they are on the left-hand side of the dislocations at the position arrowed and negative when they are on the right-hand

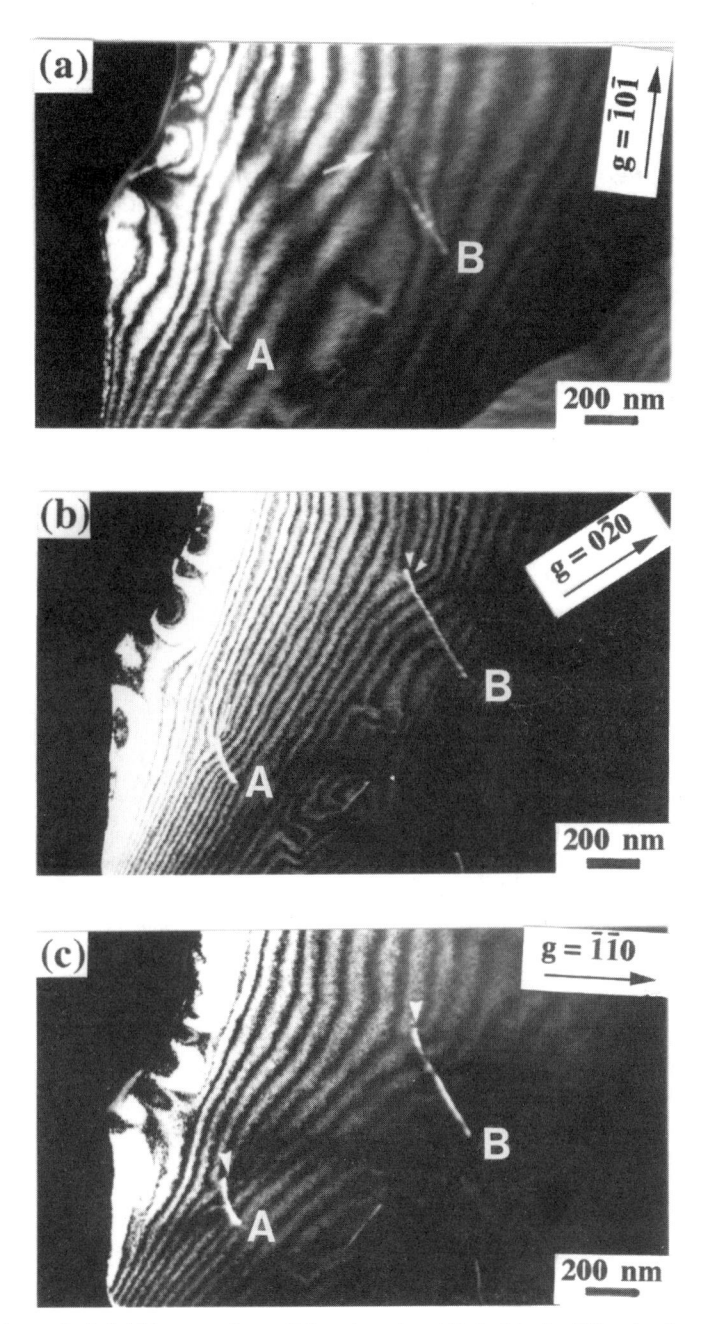

Figure 1. Weak beam dark field images of two dislocations A and B. In (a), the diffracting beam is $\mathbf{g} = \bar{1}0\bar{1}$ and the electron beam direction **BD** is approximately $\bar{3}13$, in (b) $\mathbf{g} = 0\bar{2}0$ and $\mathbf{BD} = \bar{1}03$ and in (c) $\mathbf{g} = \bar{1}\bar{1}0$ and $\mathbf{BD} = \bar{1}14$. The arrows indicate the terminating contours at the end of the dislocation.

side. Thus, as the number of terminating fringes $n = \mathbf{g} \cdot \mathbf{b}$ where $\mathbf{b}$ is the Burgers vector of the dislocation $\mathbf{g}$ (Ishida et al., 1980), the Burgers vector of dislocation A is $\mathbf{b} = [010]$, and for dislocation B, is $\mathbf{b} = [01\bar{1}]$. Trace analyses showed that the line directions of dislocation A and dislocation B were both $[1\bar{1}1]$, indicating a $(\bar{1}01)$ slip plane and a $(0\bar{1}\bar{1})$ slip plane respectively. In contrast to the clear evidence from weak beam imaging under these conditions of dislocation dissociation that we have obtained in $SrTiO_3$ samples doped with both niobium and sodium (Mao and Knowles, 1996c), no evidence for dislocation dissociation could be obtained in this material.

Examination of high angle grain boundaries showed very weak Fresnel contrast away from focus (Mao and Knowles, 1996b). The through-focal behaviour of the Fresnel fringes was indicative of ≈ 1 nm thick amorphous films at such boundaries. Occasionally, however, there was evidence of a crystalline phase formed at grain boundaries. An example is shown in Figure 2. Figure 2(a) is a dark field image of a second phase formed using the arrowed diffraction spot in the electron diffraction pattern in Figure 2(b). A comparison of the X-ray energy dispersive spectra in Figure 2(c) from one of the adjacent $SrTiO_3$ grains and in Figure 2(d) from the interface phase suggests that this phase is richer in titanium and deficient in oxygen relative to $SrTiO_3$. However, further analysis of this phase was complicated by its size and the difficulty in obtaining any recognisable electron diffraction patterns at all from low index zones.

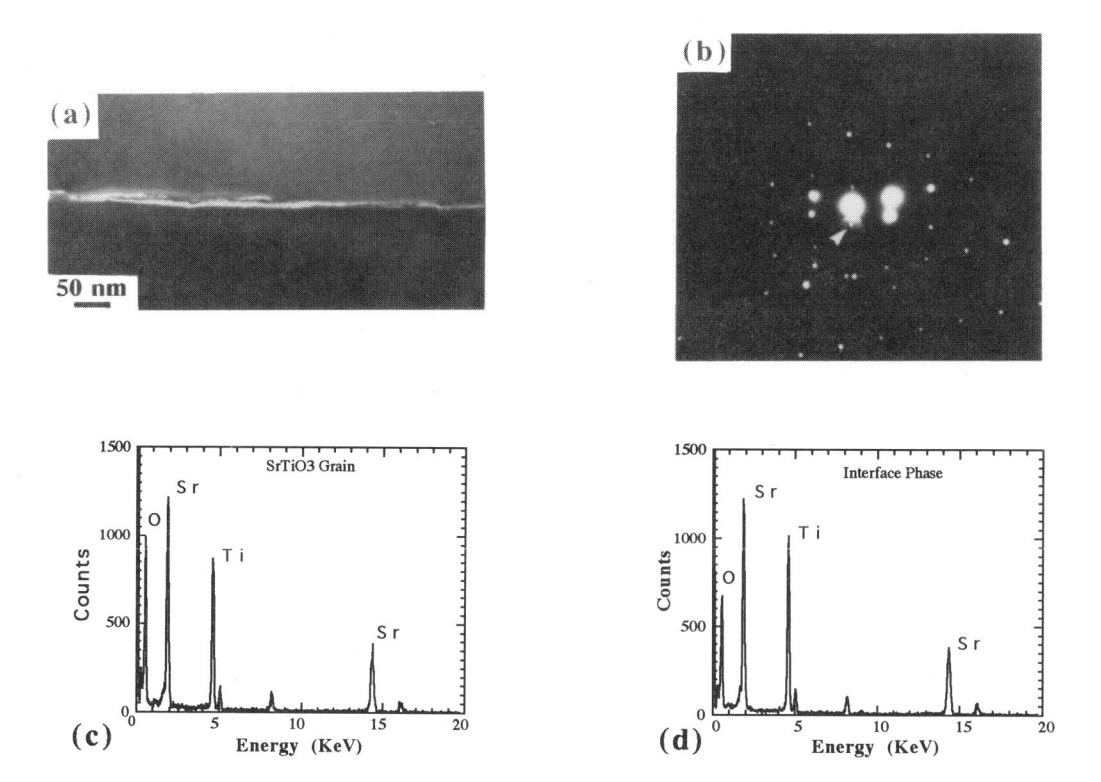

Figure 2. An example of a crystalline interface phase found in the donor-doped $SrTiO_3$ sample: (a) dark field image of the interface phase formed using the arrowed diffraction spot shown in the electron diffraction pattern in (b), (c) X-ray energy dispersive spectrum from the $SrTiO_3$ grains and (d) X-ray energy dispersive spectrum from the interface phase.

It is of interest to note that the orientation relationship between the two SrTiO$_3$ grains of ≈ 28° about the common (to within 1.5°) [311] direction is very close to Σ = 45, 28.62° about [311] (Grimmer et al., 1974) and that in addition the grain boundary in Figure 2 is noticeably faceted. Faceted boundaries were a common observation in this sample, most frequently without second phases. A further example of a faceted grain boundary in this sample is shown in the two dark field images in Figure 3. Here the two grains are oriented so that there is a near-common **g** = $\bar{2}$00, with the [001] zone in grain 1 ≈ 4.5° away from the [011] zone of grain 2, i.e. close to Σ = 29, 43.6° about [$\bar{1}$00]. The weak beam dark field micrograph in Figure 3(a) using the near-common **g** = $\bar{2}$00 and the weak beam micrograph from grain 2 formed in **g/4g** conditions with **g** = 01$\bar{1}$ show the faceting behaviour well, together with the dislocation-like contrast at the steps. No second phases are evident at this particular grain boundary.

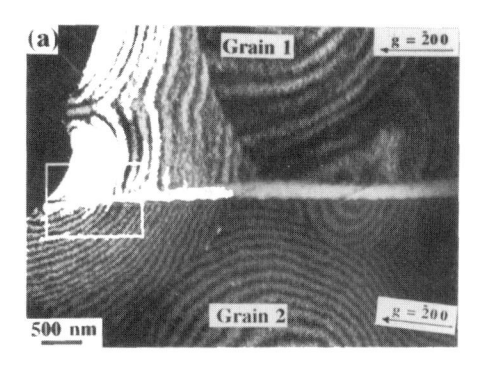

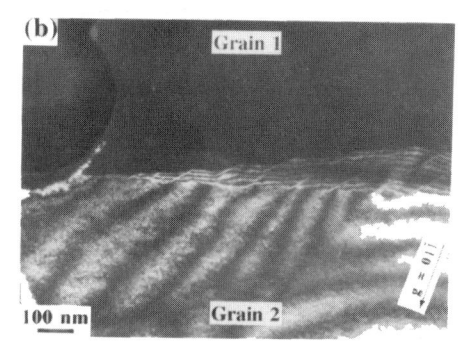

Figure 3. (a) Weak beam image taken using **g** = $\bar{2}$00 of a grain boundary free of second phase in donor-doped SrTiO$_3$. The **g** is a near-common diffraction vector in both grains. (b) Weak beam image of the boxed area in (a) formed in **g/4g** using **g** = 011 from the lower grain.

The semiconducting behaviour of this niobium-doped sample was confirmed from impedance measurements, data from which is shown in Figure 4. Within the frequency range under observation, there is a single dominant contribution to Z". The plot of log ε" against log f (where $f = 2\pi\omega$ and ω is the angular frequency) has a slope of –1 within this frequency range, indicative of the behaviour of a pure capacitor and resistor in parallel.

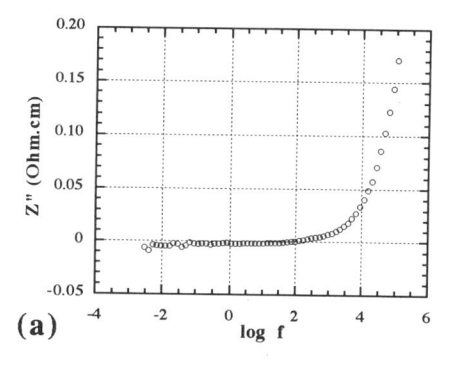

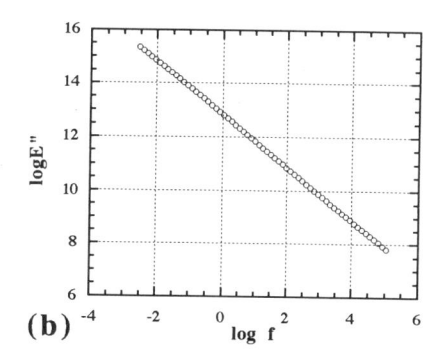

Figure 4. Impedance spectroscopy data from the niobium-doped sample. (a) is a plot of Z" against log f and (b) is a plot of log ε" against log f.

STRONTIUM TITANATE BICRYSTAL OBSERVATIONS

Fresnel fringe images and a high resolution transmission electron micrograph of the 24° bicrystal are shown in Figure 5, together with a plot of Fresnel fringe spacing as a function of defocus and the common [001] zone electron diffraction pattern from which the high resolution transmission electron micrograph was taken. There is no amorphous phase at this boundary and in comparison with Fresnel fringe profiles from the IBLCs we have investigated (Mao and Knowles, 1996a, 1996b), the contrast is very weak. Microfaceting of the boundary is evident on a scale of a few nanometres in the high resolution micrograph.

DISCUSSION

In comparison with the IBLCs prepared without interruption in the firing schedule, the donor-doped sample of SrTiO$_3$ had a significantly larger grain size. Given that the IBLCs had a

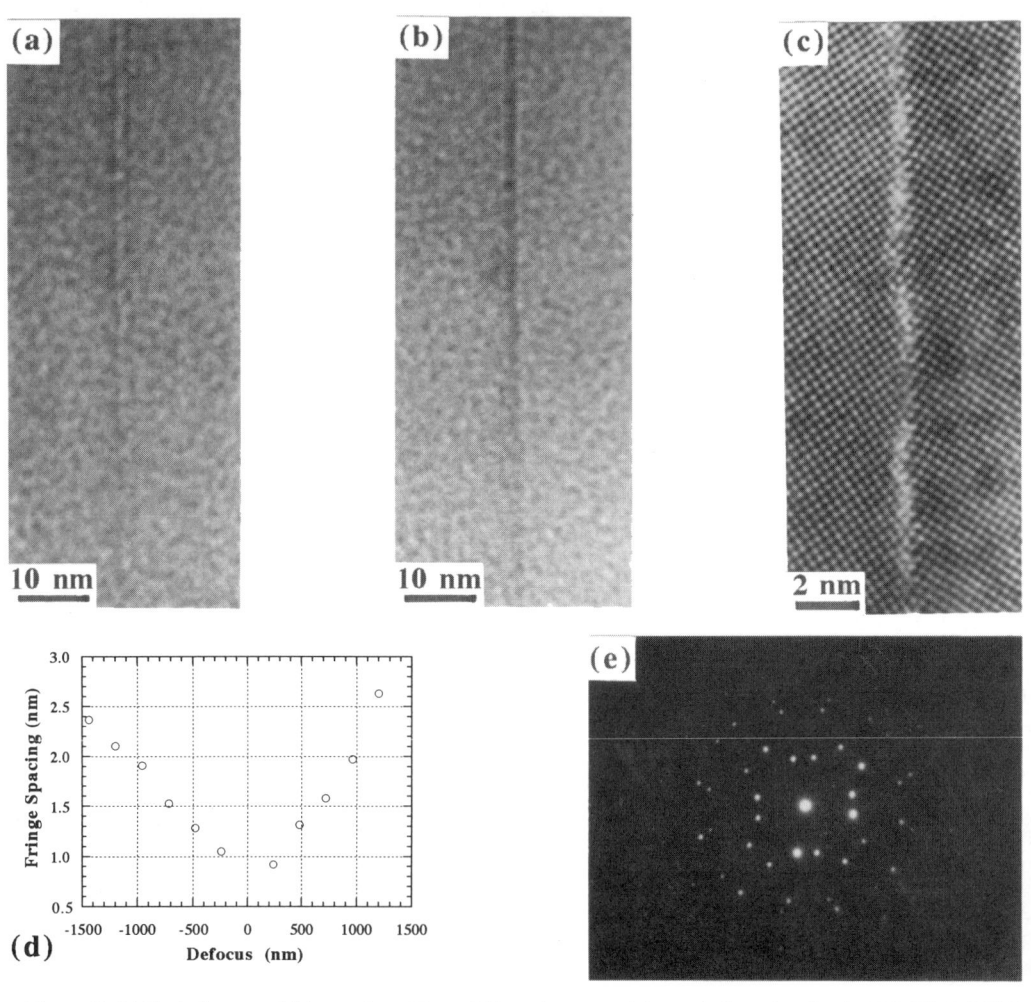

Figure 5. (a) Underfocus and (b) overfocus Fresnel fringe images of the grain boundary in the 24° [001] SrTiO$_3$ bicrystal, together with (c) a high resolution transmission electron micrograph of the boundary, (d) a plot of the Fresnel fringe spacing against defocus and (e) an electron diffraction pattern from the common [001] zone.

second firing, the clear implication is that the acceptor dopants of sodium and lithium that we used both act as grain refiners, preventing grain growth. Furthermore, the magnitude of the Fresnel contrast at grain boundaries when imaged out of focus in both the donor-doped sample and the SrTiO$_3$ bicrystal is significantly less than in samples where either sodium or lithium had been incorporated as an acceptor dopant (Mao and Knowles, 1995, 1996a, 1996b). In general, this difference could be attributable to (*a*) a thin amorphous layer at the grain boundary with a mean potential significantly lower than either grain or (*b*) an interfacial space charge layer (see for example, Dunin-Borkowski et al., 1996).

A comparison of the Fresnel fringe profiles as a function of defocus for samples prepared with and without acceptor dopant showed that in all cases, the boundary layer thicknesses were very thin, ranging from ≤ 1 nm to 2 nm. The dramatic difference in contrast of the Fresnel fringes between the IBLCs we have examined and the donor-doped sample can be rationalised in terms of segregation of lithium and sodium to the grain boundaries. This segregation produces the narrow negative interfacial space charge at the grain boundary and a wider region adjacent to the boundary of positive charge, the net result of which is a fall in mean specimen potential seen by high energy electrons (c.f. Dunin-Borkowski et al., 1996), consistent with the Fresnel contrast behaviour. The weak Fresnel contrast seen in Figure 5 is what would be expected in the absence of a significant space charge effect, as we would expect for the nominally pure SrTiO$_3$ bicrystal. In addition, of course, segregation of lithium and sodium to grain boundaries would also explain the grain refinement effect of these dopants.

Two further differences between the donor-doped sample and the IBLCs produced without an interruption in the firing schedule can be seen in the nature of the dislocations seen and in the crystalline second phases found between grains. The dissociated dislocations seen in the sodium-doped IBLC occurred in both climb and glide configurations with partial dislocation spacings of 6.5 ± 1 nm for the climb configuration and 11 ± 1 nm for the glide configuration (Mao and Knowles, 1996c), with parallel **b** = ½<011>. These spacings are large and in direct contrast to other observations on single crystals of SrTiO$_3$ by Takeuchi et al. (1989) and Nishigaki et al. (1991) in which dislocation dissociation was not observed. Our own observations in this paper on the donor-doped polycrystalline SrTiO$_3$ are in accord with those of Takeuchi et al. and Nishigaki et al. This therefore suggests that there is likely to be a further rôle for acceptor dopants in IBLCs: segregation to dislocations, with a lowering of the stacking fault energy. Future work in this area will examine the feasibility of establishing by parallel electron energy loss spectroscopy whether there is evidence for this.

The crystalline second phases found in the IBLCs produced without an interruption in the firing schedule were identifiable from EDX and their electron diffraction patterns as Sr$_2$TiO$_4$ (a = 3.9 Å, c = 12.6 Å (Tilley, 1977)), isostructural with K$_2$NiF$_4$. The second phase particles in the samples which were only doped with niobium were titanium-rich, rather than strontium-rich, as in the example shown in Figure 2. This difference is easily explicable. In both sets of samples, the Ti:Sr ratio was unity. The semiconducting behaviour of the grains evident from Figure 4 at the levels of Nb we have here can be rationalised in terms of an ionic conduction mechanism involving strontium vacancies (Balachandran and Eror 1982):

$$Nb_2O_5 + SrTiO_3 \longrightarrow 2Nb_{Ti}^{\cdot} + V_{Sr}^{''} + 6O_o + Sr_{Sr} + TiO_2$$

Thus there will be excess titanium in the sample, consistent with the existence of a small quantity of titanium-rich material, as we have observed. Lithium and sodium at the levels we have used for the IBLCs will segregate to grain boundaries and occupy strontium sites in the perovskite, giving rise to a space charge effect and a thin insulating, rather than semiconducting, diffusion layer, and causing the precipitation of a strontium-rich phase, rather than a titanium-rich phase between perovskite grains.

ACKNOWLEDGEMENTS

We would like to thank the Cambridge Commonwealth Trust for financial support during the course of this work.

REFERENCES

Balachandran, U. and Eror, N.G., 1982, Electrical conductivity in lanthanum-doped strontium titanate *J. Electrochem. Soc.* 129:1021.

Chiang, Y.M. and Takagi, T., 1990, Grain-Boundary chemistry of barium titanate and strontium titanate: I, high-temperature equilibrium space charge, *J. Am. Ceram. Soc.* 73:3278.

Dunin-Borkowski, R.E, Saxton, W.O. and Stobbs, W.M., 1996, The electrostatic contribution to the forward scattering potential at a space charge layer in high energy electron diffraction I. theory neglecting the effects of fringing fields, *Acta Crystallographica* A in the press.

Franken, P.E.C., Viegers M.P.A. and Gehring A.P., 1981, Microstructure of $SrTiO_3$ boundary-layer capacitor material, *J. Am. Ceram. Soc.* 64:687.

Fujimoto, M. and Kingery, W.D., 1985, Microstructures of $SrTiO_3$ internal boundary layer capacitors during and after processing and resultant electrical properties, *J. Am. Ceram. Soc.* 68:169.

Grimmer, H., Bollmann, W. and Warrington, D.H., 1974, Coincident-site lattices and complete pattern-shift lattices in cubic crystals, *Acta Crystallographica* A 30:197.

Ishida, Y., Ishida, H., Kohra, K. and Ichinose, H., 1980, Determination of the Burgers vector of a dislocation by weak-beam imaging in a HVEM, *Phil. Mag.* A 42:453.

Mao, Z. and Knowles, K. M., 1995, Characterisation of a novel $SrTiO_3$ internal boundary layer capacitor by transmission electron microscopy, *Inst. Phys. Conf. Ser.* 147:563.

Mao, Z. and Knowles, K. M., 1996a, Microstructure of novel lithium-doped $SrTiO_3$ internal boundary layer capacitors, submitted to *British Ceramic Transactions*.

Mao, Z. and Knowles, K. M., 1996b, Acceptor segregation to grain boundaries in strontium titanate internal boundary layer capacitors, submitted to *Electroceramics V*, Aveiro, Portugal, September 1996.

Mao, Z. and Knowles, K. M., 1996c, Dissociation of lattice dislocations in $SrTiO_3$, *Phil. Mag.* A 73:699.

Nishigaki, J., Kuroda, K. and Saka, H., 1991, Electron microscopy of dislocation structures in $SrTiO_3$ deformed at high temperatures, *phys. stat. sol.* (a) 128:319.

Takeuchi, S., Suzuki, K., Ichihara, M. and Suzuki, T., 1989, High resolution electron microscopy of core structure of dislocations in oxide ceramics, in: *Lattice Defects in Ceramics*, p. 17, Japanese Journal of Applied Physics Series No. 2, Japan Society of Applied Physics, Tokyo.

Thomas, G. and Goringe, M.J., 1979, *Transmission Electron Microscopy of Materials*, Wiley, New York.

Tilley, R. J. D., 1977, An electron microscope study of perovskite-related oxides in the Sr–Ti–O system, *J. Solid State Chem* 21:293.

Yamoaka, N., 1986, $SrTiO_3$-based boundary-layer capacitors, *Am. Ceram. Soc. Bull.* 65:1149.

Zhou, L., Jiang, Z., and Zhang, S., 1991, Electrical properties of $Sr_{0.7}Ba_{0.3}TiO_3$ ceramics doped with Nb_2O_5, $3Li_2O \cdot 2SiO_2$, and Bi_2O_3, *J. Am. Ceram. Soc.* 74:2925.

ELECTRICAL AND OPTICAL PROPERTIES AT THE INTERFACES OF n-BaTiO3 / p-Si

S. Sugihara, T. Ebina, S. Suzuki and T. Andoh

Department of Materials Science and Ceramic Technology,
Shonan Institute of Technology, 1-1-25 Tsujido Nishikaigan,
Fujisawa 251, Japan

INTRODUCTION

BaTiO$_3$ ceramic is a popularly studied ferroelectric with many studies of its semiconductor properties, too. The semiconductor BaTiO$_3$ can be easily fabricated by doping certain elements such as Y, Sb, La, Ce, Pr and Bi, and by oxidation or reduction of BaTiO$_3$. The additive type semiconducting BaTiO$_3$ is pasted by CeO$_2$ and Bi$_2$O$_3$ followed by a heat treatment, which leads to high resistivity. The electrical properties of semiconductive BaTiO$_3$ have been reported by Saburi et al.[1] They discussed the mechanisms associated with PTC characteristics. On the other hand, BaTiO$_3$ was one of the first crystals on which the photovoltaic effect was researched.[2] BaTiO$_3$ employed in this study was typically n-type semiconductor. Thin films of semiconductive BaTiO$_3$ have not been studied recently as much as the ferroelectric properties for potential use as non-volatile memory devices.

The semiconductive BaTiO$_3$ is easily fabricated and chemically stable. The purposes of the present study are to form thin films on a p-type Si single crystal, to structure the p-n junction with CO$_2$ laser heat treatments, to evaluate photovoltaic effects with a solar beam and, and also to study possible use in the infrared regions.

EXPERIMENTAL

The semiconductive BaTiO$_3$ powder (hereafter BT or BT powder) was pressed at 196 MPa, followed by sintering for 2 hrs at 1340°C. Then the sintered pellets were pulverized with ball-milling for 24 hrs. The milled powders were mixed with organic solvents to produce slurry. The slurry was spin-coated on the p-type Si single crystal (20 by 20 by 0.5 mm). After the organic solvents were vaporized, the system of BT/p-Si was set in a chamber and irradiated with CO$_2$ laser as shown in Fig. 1. The system was placed such that the laser beam hit the Si side instead of BT side, thereby protecting the film. Al$_2$O$_3$ powders absorb the laser energy effectively and the intermediate AlN plate transfers the heat to the BT/p-Si system. The joining of BT to the p-Si substrate required a laser power of 714

W/cm² for 2 min and 1143 W/cm² for 5 min. This method of CO_2 laser irradiation was successful to control the compounds that form at the interfaces between BT and Si. SEM, and EPMA were used to characterize reactions at the interface and film thicknesses. The photovoltaic evaluations were performed by solar beam and it's simulator; and the four probe method was provided for the electrical measurements. Optical properties of the system were measured by spectrophotometer from 200 to 900 nm.

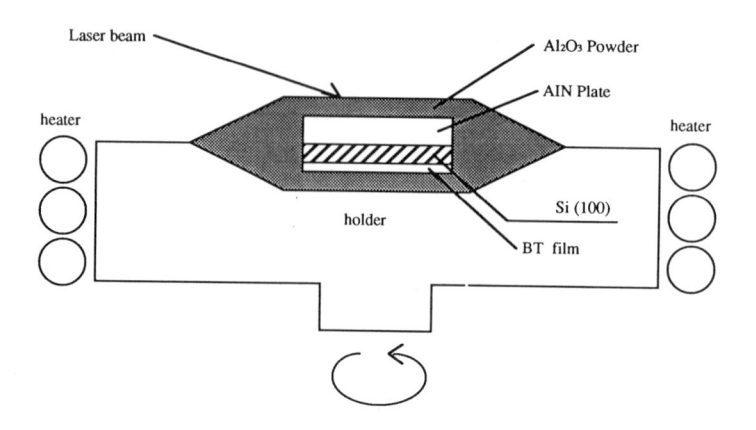

Figure 1. Specimen setting inside chamber.

RESULTS AND DISCUSSION

XRD ANALYSIS

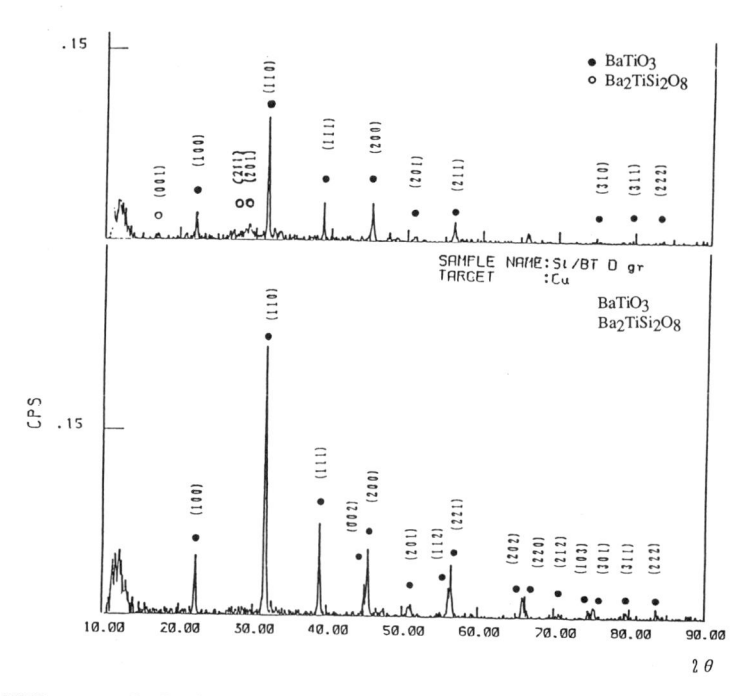

Figure 2. XRD patterns for $BaTiO_3$/Si system : lower; no heat treatment (green film), upper; 714 W/cm² CO_2 irradiation.

Evaluation of BaTiO3 films on p-Si by XRD was carried out for the system without irradiation (i.e. green film), and with heat-treatment at 714 W/cm^2 for 2 min and 1143 W/cm^2. The results were shown in the upper (714 W/cm^2) and lower (green) in Fig. 2. The green film of BT on p-Si showed the typical patterns of BT with cubic and tetragonal structures. However, the films irradiated at 714 W/cm^2 for 2 min mostly indicated a cubic structure. This is because a heat-treatment at 800°C corresponds to the values of this laser power where Ba$_2$TiSi$_2$O$_8$ was formed, based on the following reaction:

$$2\ BaTiO_3 + 2\ Si + 3\ O_2 \rightarrow Ba_2TiSi_2O_8 \tag{1}$$

At the higher power, 1143 W/cm^2, the XRD peaks (201) and (211) of Ba$_2$TiSi$_2$O$_8$ became more significant than the (110) peak of BT. The power of 1143 W/cm^2 corresponded to the about 1000°C.

MICROSTRUCTURES

The surfaces of BT and their interfaces at Si were influenced by the laser irradiation conditions. Figures 3(a) and (b) showed SEM photographs at the interfaces of BT/Si formed by CO$_2$ laser irradiation at power rates of 714 W/cm^2 for 2 min and 1143 W/cm^2 for 5 min, respectively. The grain size of the green BT film was only 0.5 to 1 μm, just like the powder. The surfaces of the film irradiated at the power rate 714 W/cm^2 for 2 min indicated a grain size of 1 to 3 μm with a tetragonal shape, and also large (5 mm long) needle-like grains. The needle-like grains exhibited a higher concentration of Ti than Ba, based on EPMA analysis. This may be due to the higher thermal conductivity of Ti, with heat taken away from the irregular small grains resulting in large needle shaped grains. Furthermore, shorter times of 30 sec irradiation produced sparse surfaces coverage while more dense surfaces were obtained for 2 min at the same laser power. SEM cross sectional analysis, Fig. 3(a), depicted a thin film of sintered BT grains (1 μm in thickness) and a smooth interface along Si, after irradiation at the lower laser energy of 714 W/cm^2 for 2 min. The film was thicker (5 μm) after higher energy irradiation (1143 W/cm^2 for 5 min) as shown in Fig. 3(b). In addition, a BT/Si system which had an intermediate layer of the ceramace G (consisting of SnO$_2$ and Sb$_2$O$_3$, mainly), was heat-treated at 900°C. The microstructures that developed at the interfaces between BT/the ceramace G/Si were analyzed by SEM in Fig. 4, and EPMA. The film thickness of BT was 3 to 4 μm including the Sn and Sb oxides which had diffused into the BT layer. The microstructures of the BT surface and the intermediate compounds were obtained by the appropriate conditions of the laser irradiation and annealing temperature. At this moment, the effects of the intermediate compounds were not clear, as to whether they act as an insulator to accumulate the charges or as photoconductive SnO$_2$ and Sb$_2$O$_3$.

ELECTRICAL PROPERTIES

The electrical properties of bulk BT were first evaluated as a function of temperature. A typical PTC phenomenon at about 120°C indicates a jump of about 10^4. Meanwhile, the electrical resistivity of BT/Si system decreased in the range from the room temperature up to about 200°C, with a negative temperature coefficient. The activation energy of the p-n junction was calculated to be about 0.12 eV. This reflected the temperature dependence of conductivity typical for an extrinsic Si semiconductor. The resistivity of Si increased only from 1 x 10^4 to 1.3 x 10^4 Ωcm as the temperature increased from 100 to 200°C. The current increased 0.5 to 3 nA as the temperature went from 23 to 100°C, and changed to 3.5 nA at 200°C. This fact suggests that the carriers from n-BT increase, resulting in a higher conductivity according to the following equation:

$$n_i \propto \exp(-E_g/2kT) \tag{2}$$

where n_i : electron density, E_g : activation energy, k : Boltzmann factor and T : temperature. Also the n-type BT stores the charges at the Curie temperature. Figure 5 shows the I-V curve for BT/Si in the forward and the reverse directions. The intermediate compounds of the ceramace G at the interfaces between BT and Si cause the increasing electrical current in the reverse direction, when the depletion layer forms at the interfaces.

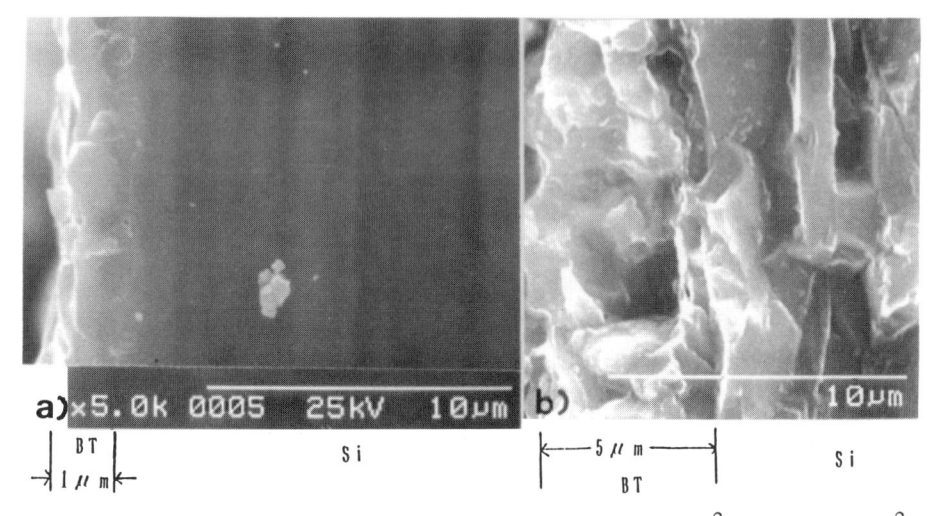

Figure 3. SEM photographs of the interfaces between BT and Si, a) 714 W/cm^2 and b) 1143 W/cm^2 CO_2 irradiation.

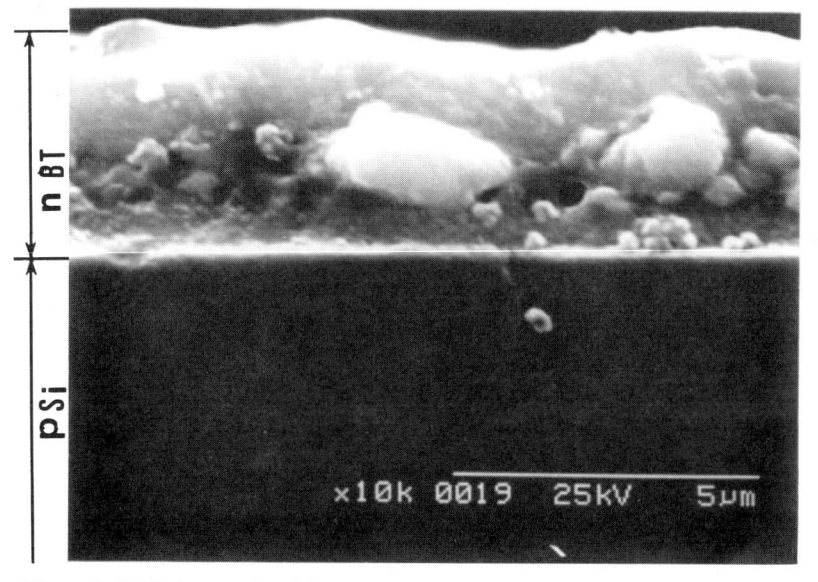

Figure 4. SEM photographs of the interfaces between BT and Si with intermediate layers.

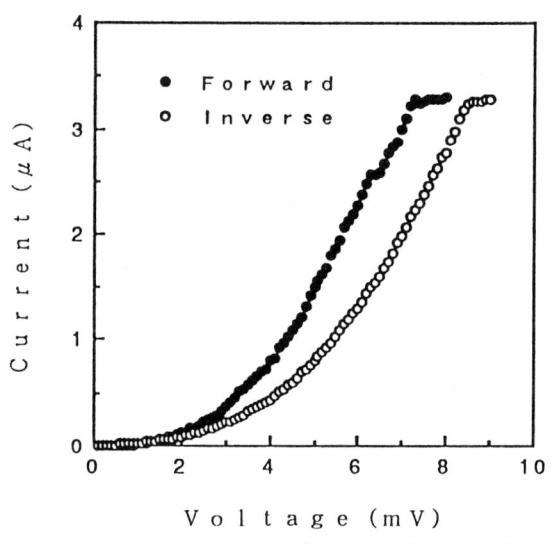

Figure 5. V-I characteristics for BaTiO$_3$/Si system in forward and reverse direction.

OPTICAL PROPERTIES

The photocurrent was 153 µA (voltage 68 mV) with the solar beam, for a dark current of 0.01 nA (0 mV) and a BT film thickness of 0.8 µm. A thicker film (2–3 µm) caused a photocurrent of 0.3 µA (voltage 1.4 mV).

Absorption changes cause the current sign to be altered. The current depends on polarization parallel to and perpendicular to the x-axis, with a change of the order of 10^{-13} A depending on the wavelength which has a maximum at around 400 nm.[3] Under monochromatic light (vector **q**, frequency ω), the electric field is given by the following equation,[4]

$$E(r,t) = \mathbf{e}\, a \exp(-i\omega t + i\mathbf{q}r) + \text{complex conjugate} \tag{3}$$

where **a** is the complex amplitude and **e** is polarization unit vector. The density of the constant current j is indicated in the next equation,

$$J = \Sigma^{d_{ij}} E_j + \Gamma_{ijl} E_j E_l + \Delta_{ijlm} E_j e_l e_m^* j \\ + X_{ijlm}\, q_j e_l e_m^* j + \beta_{ijl}\, e_l e_m^* j + \cdots \tag{4}$$

$\Sigma^{d_{ij}}$ is a tensor describing the intrinsic (dark) conductivity (Ohm's raw). Γ_{ijl} is the quadratic electrical conductivity and Δ_{ijlm} presents the anisotropic photoconductivity for our case, which considers a noncetrosymmetric system. β_{ijl} in the last term, is related to the photovoltaic effect. If the enough beam transmits and reaches the interface of the p and n junction, the last term in the equation be the primary cause for the current. According to the light absorbance as shown in Fig. 6, the BT/Si system exhibited the increasing transmittance with wavelength, so the infrared region was estimated to transmit the system to reach the p and n junction. Therefore, the light frequencies transmitting through the BaTiO$_3$ thin layer are ideally supposed to be wider ranges and include even the infrared regions.

The photocurrents around 13 nA, measured by the spectrophotometer from 200 to 900 nm did not change much as shown in Fig. 7. However, the photovoltaic effect was found to

Figure 6. Transmittance by BaTiO$_3$/Si system.

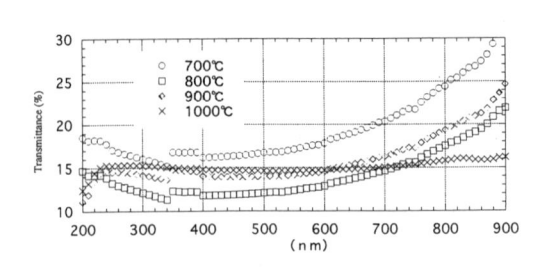

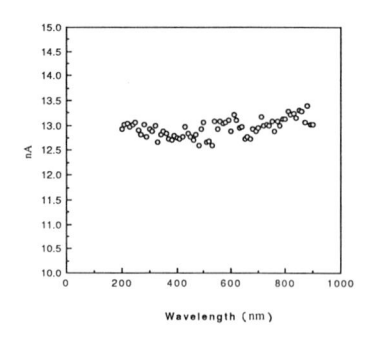

Figure 7. Photocurrent vs wavelength (200 to 900 nm).

show a larger photocurrent of 64 nA and 0.4 mV when 143 W/cm^2 of CO$_2$ laser beam (wavelength;10600 nm) was used for irradiation.

In a typical solar battery, the device is constructed by the layers of p-, n- and intrinsic (so called pin structure). The intrinsic layer plays an important roll between the p and n layers, with high resistivity leading to the efficient absorption of the beam. Under the simulator, the photocurrent and the photovoltage were 46.2 µA/cm^2 and 40.0 mV/cm^2, respectively. The solar beam generated more than twice the current and twice the voltage caused by the simulator.

CONCLUSIONS

1. The film formation of n-type BaTiO$_3$ on p-type Si was produced by a CO$_2$ laser beam with a rate of 715 W/cm^2 for 2 min, resulting in a film thickness of 0.8 to 1.0 µm.

2. The rectifying function was small and exhibited a negative temperature coefficient characteristic between 23 to 200°C. Furthermore, the photocurrent on the BaTiO$_3$/Si detected by the spectrophotometer was small (13 nA) and scarcely changed, for wavelength from 200 to 900 nm. The photovoltaic values of 64 nA and 0.4 mV reached into the infrared region, when the CO$_2$ laser beam was irradiated onto the BaTiO$_3$/Si system.

3. The p-n junction under a solar beam indicated 153 mA/cm^2 and 68 mV/cm^2 for the photocurrent and voltage, respectively.

REFERENCES

1. O. Saburi, Properties of Semiconductive Barium Titanate, J. Pys. Soc. Japan, **14**:1159 (1959) and **15**:733 (1960).
2. W.T.N. Koch, R. Munster, W. Ruppel and P. Wurful, Ferroelectrics, **13**:305 (1976).
3. W.T.N. Koch, R. Munster, W. Ruppel and P. Wurful, Solid State Comm., **17**:847 (1975).
4. I. Stuman and M. Fridkin, Ferroelectricity and related phenomena, pp. 128-129, vol. 8, Gordon & Breach Science Publisher, Philadelphia, USA (1972).

BULK AND THIN FILM PROPERTIES OF (Pb, La)TiO$_3$ FERROELECTRICS

Kiyoshi Okazaki
Okazaki Ceramic Laboratory, Yokohama, 236, Japan

Hiroshi Maiwa
Shonan Institute of Technology, Fujisawa, 251, Japan

INTRODUCTION

Dielectric, electrooptic, mechanical properties of ferroelectric ceramics including (Pb, La)TiO$_3$ are described. PbTiO$_3$ is a ferroelectric material with a higher Curie point (about 500°C) and a larger tetragonality (c/a = 1.06) compared to BaTiO$_3$ (Curie point of 120°C and tetragonality of c/a = 1.01). PbTiO$_3$ ceramics are very difficult to sinter because of this large tetragonality. When PbTiO$_3$ is doped with La, the tetragonality, Curie point, and internal stress incurred during the cubic-to-tetragonal transition all decrease.

(Pb, La)TiO$_3$ thin films are prepared using a multiple-cathode rf sputtering apparatus with four targets on platinum and ruthenium oxide electrodes. The ferroelectric properties are measured with D-E hysteresis loops, dielectric properties, and voltage-current characteristics. The dependences of dielectric permittivity on film thikness were also measured with depth profiles. In this paper, the ferroelectric properties of these thin films are compared to bulk properties.

DIELECTRIC, PIEZOELECTRIC, ELECTROOPTIC AND MECHANICAL PROPERTIES OF PLT CERAMICS

Pure PbTiO$_3$ is very difficult to sinter in bulk form because the crystal anisotropy is so large, and the ceramic body is typically broken at the Curie point during the cooling process after firing. Dielectric properties of La-doped PbTiO$_3$ [hereafter PLT] ceramics were first measured by Okazaki and Takahashi[1] in 1967. Ueda et al.[2] have investigated the piezoelectric properties of conventionally fired (0.995 PbTiO$_3$ + 0.045 La$_{2/3}$TiO$_3$) + 0.01 MnO$_2$. Materials with a 1.5-μm grain size have a high bend strength of 2000 kg/cm^2, which is three times larger than BaTiO$_3$ ceramics. Yamashita et al.[3] have examined (Pb$_{1-x}$Ca$_x$)[(Co$_{1/2}$W$_{1/2}$)$_\alpha$Ti$_{1-\alpha}$]O$_3$ and they found a very small k$_p$/k$_t$ ratio, less than 0.1 at a composition of x = 0.25. In ferroelectric ceramics, including PLT, two kinds of internal bias field, E$_i$ and E$_{-i}$, are generated by space charge effects. When the poled sample was aged after poling, an E$_i$ field was generated in the same direction as the poling field. It stabilized the remnant polarization.[4] The details of the space charge effects of ferroelectric ceramics are described in "Ceramic Engineering for Dielectrics."[5] The PbTiO$_3$ family of ceramics have special mechanical properties. Okazaki measured the internal stresses of pore-free PLZT ceramics using a microindentation technology as shown in Table 1.[6] Among the listed compositions, PLZT 10/0/100, 15/0/100, 22/0/100 and 30/0/100 are PLT ceramics. As can be seen, the internal stress decreases with the increase of La and the corresponding decrease in tetragonality. The internal stress of cubic samples is zero in all cases. This means that the internal stresses of tetragonal samples are induced at the Curie points during the cooling process after firing.

Anisotropic D-E loops,[5] anisotropic electrooptic properties,[7] and saturation phenomena of dc voltage-current characteristics in the ferroelectric temperature region[5] have been measured. Okazaki et al. have investigated electrical, optical, and acoustic properties of PLZT ceramics after two-stage processing.[8] This processing was applied in Latvia University. The first approach to PLZT ceramics defect structure appeared in '76 Ceramic Microstructure,[9] Proceedings of the 6th International Symposium, chaired by Pask and Fulrath. Theoretical density and PbO deficiency of pore-free transparent PLZT ceramics, investigated in 1978, are shown in Table 2.[10]

Table 1. Compositions, crystal structure, tetragonality (c/a), Curie point (T$_c$) and internal stress (σ_i) measured using microindentation for pore-free PLZT bulk ceramics.[6]

Composition (La/Zr/Ti)	Phase	c/a	T$_c$ (°C)	σ_i (MN/m^2)
10/0/100	Tet	1.028	340	150
15/0/100	Tet	1.02	283	90
22/0/100	Tet	1.010	85	30
30/0/100	Cub	1.000	below RT	0
2/50/50	Tet	1.025	376	40
2/54/46	Rhom	1.000	345	22
20/50/50	Cub	1.000	below RT	0

Table 2. Examples of equilibrium formula of PLZT x/65/35 and their PbO loss and calculated density.[10]

x	Equilibrium Formula	PbO loss(mol %)	Bulk density(g/cm^3)
0	UL Pb$_{100}$(Zr, Ti)$_{100}$O$_{300}$	0	8.024
0	LL Pb$_{98}$ □$_2$(Zr, Ti)$_{100}$O$_{298}\phi_2$	−2.0	7.916
4	UL Pb$_{95.2}$La$_4$ □$_{0.53}$(Zr, Ti)$_{99}$ □$_{0.73}$O$_{299.2}$	−0.8	7.960
4	LL Pb$_{92.6}$La$_4$ □$_{2.4}$(Zr, Ti)$_{99}$O$_{296.6}\phi_{2.6}$	−3.4	7.876
8	UL Pb$_{89.6}$La$_8$ □$_{1.6}$(Zr, Ti)$_{98}$ □$_{1.2}$O$_{297.6}$	−2.4	7.852
8	LL Pb$_{86}$La$_8$ □$_4$(Zr, Ti)$_{98}$O$_{294}$	−6.0	7.748
12	UL Pb$_{83.9}$La$_{11.57}$ □$_{3.13}$(ZT)$_{97}$La$_{0.43}$ □$_{1.17}$O$_{295.9}$	−4.1	7.738
12	LL Pb$_{80.3}$La$_{11.57}$□$_{5.36}$(ZT)$_{97}$La$_{0.43}$O$_{292.3}$	−7.7	7.632

Note: □ A or B, site vacancy; ϕ, oxygen vacancy.
       UL: upper limit, LL: lower limit

PREPARATION OF PbTiO$_3$ THIN FILMS

To prepare thin films in this experiment, a four-target-cathode rf sputtering apparatus was used.[11,12] The sputtering targets were three-inch disks of Pb(99.999%) or a combination of Pb(99.99%), La(99.9%) and Ti(99.99%) metals. Compared to the sputtering apparatus using a single compound target, this multiple-cathode sputtering can easily control the sputtering rate of each target over a larger input power regime. The chemical composition of sputtered thin films were characterized using ICP, EPMA and XPS; the crystallography was analyzed by XRD; the depth profile was obtained using the XPS and the film thickness was subsequently measured using a surface profilometer. Adachi et al.[13] investigated the phases of sputtered PbTiO$_3$ films as a function of substrate temperature. The results indicate that below 500°C, it is amorphous; above 500°C, it is crystalline pyrochlore; and above 600°C, it has the perovskite structure.

Maiwa et al.[11] investigated the crystalline structure of PbTiO$_3$ thin films using multiple-cathode sputtering. The sputtered Pb/Ti ratio was determined by XPS as a function of the input power for Pb target at room temperature under a constant input power condition for Ti target of 500 W on a substrate of SiO$_2$(1 μm)/Si. From their results the input power for Pb target should be 120 W, to get a thin film with an incident ratio of Pb/Ti = 6.1.

PbTiO$_3$ thin films were deposited at an incident Pb/Ti ratio of 6.1 and constant substrate temperature of 620°C on various substrates: Si(100), Pt/Ti/SiO$_2$/Si, SiO$_2$/Si, c-sapphire(0001), MgO(100), SrTiO$_3$(100). XRD exhibited the perovskite structure on all substrates except for Si(100).

Similar experiments were made for PbTiO$_3$ thin films on MgO and Si substrates at a constant substrate temperature of 500°C, as a function of the incident Pb/Ti ratio of 0.82–3.7. For the substrate temperature range of 460 to 500°C, a perovskite PbTiO$_3$ film can be formed on the Si substrate without a buffer layer.

Figure 1 is an example of an XPS depth profile of PLT thin film deposited on a RuO$_2$/Ru/SiO$_2$/Si substrate. The interdiffusion of elements or the formation of reaction layer seems to be minimal.

Figure 2 is the chemical composition of PLT thin film deposited on Pt/SiO$_2$/Si substrate as a function of deposition time. Deposition time corresponds to estimated thickness. It is noticed in this figure that a Pb deficiency is observed in a small thickness region less than 2 nm from the surface.

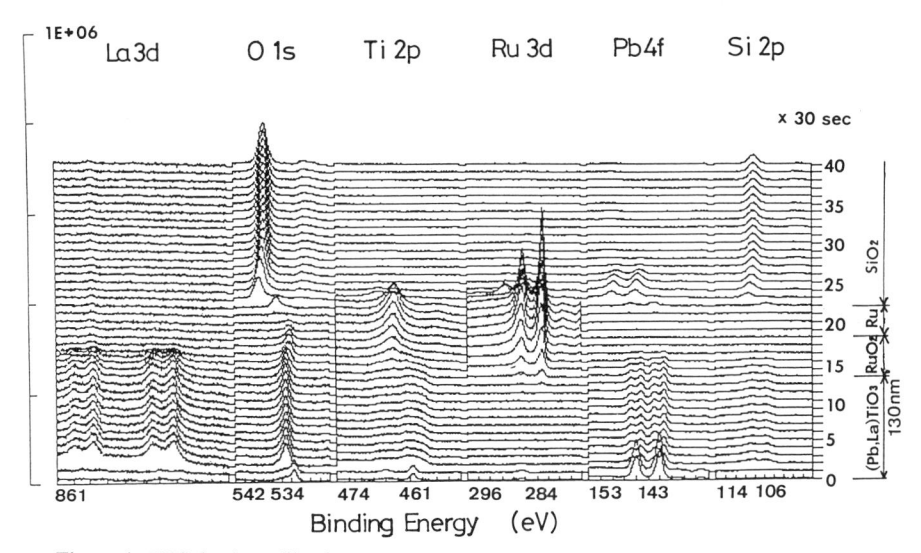

Figure 1. XPS depth profiles for PLT thin film deposited on RuO$_2$/Ru/SiO$_2$/Si substrate.

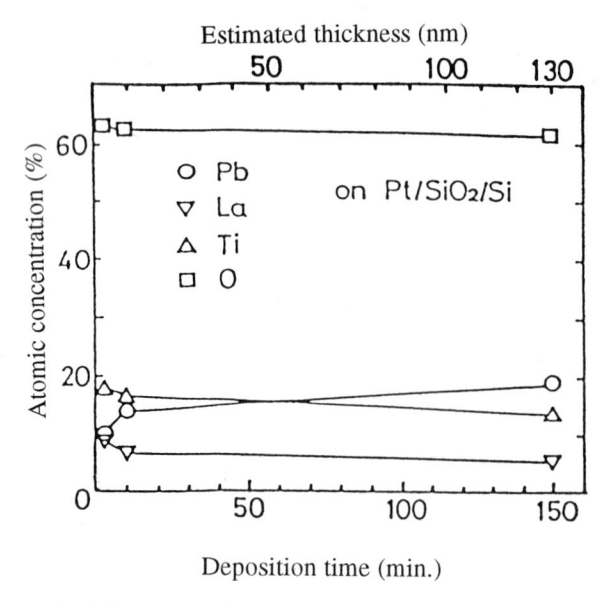

Figure 2. Surface analysis for PLT thin film deposited on Pt/SiO$_2$/Si substrate as a function of estimated thickness.

PREPARATION OF PLT THIN FILMS AND FERROELECTRIC PROPERTIES

PLT thin films were deposited on MgO, Pt/MgO, Pt/SiO$_2$/Si, RuO$_2$/Ru/MgO, RuO$_2$/Ru/SiO$_2$/Si at a substrate temperature of 540°C under rf input power conditions: 50 to 65 W for PbO, 450 W for Ti and 0 ~ 70 W for La target, respectively. The sputtering gas ratio of Ar/O$_2$ was 1:1. Figure 3 shows the lattice matching between the thin film and the substrate of MgO(100) and RuO$_2$(110).

The remanent polarization of 130 nm PLT thin film on Pt/SiO$_2$ substrate is P$_r$ = 6.3 μC/cm^2 and the coercive force is E$_c$ = 229 kV/cm. For the 270-nm-thick PLT, P$_r$ = 12.6 μC/cm^2 and E$_c$ = 92.5 kV/cm.

Figure 4(a) is a D-E loop for the 270 nm PLT thin film on Pt/SiO$_2$ substrate. Notice that this loop is somewhat asymmetrical by the internal bias field. Figure 4(b) is the loop for the 130-nm thin film on Pt/SiO$_2$ substrate.

Table 3 illustrates the dielectric permittivity, ε$_s$; loss tangent, tan δ; coercive field, E$_c$; and remanent polarization, P$_r$, of 270-nm-thick PLT thin films deposited on various substrates.

Figure 5 shows the leakage current/applied voltage characteristics of 270-nm PLT thin film deposited on a Pt/MgO substrate. For the BaTiO$_3$ family ceramics, the V-I curve shows a saturation phenomenon in the ferroelectric temperature region. In particular, the V-I curve of BaTiO$_3$ ceramics shows a linear ohmic relation before the breakdown voltage (BDV, V$_c$) which is above the Curie point. For BaTiO$_3$ ceramics, with a Curie point below room temperature, a linear ohmic relation exists at room temperature. The saturation behavior of PLT thin film in Figure 5 is very similar to that of BaTiO$_3$ ceramics. It is clear that the saturation phenomenon is closely related to the ferroelectric polarization. Also, notice that an asymmetric V-I curve is observed when a dc electric field is applied (plus or minus) on top of the thin film electrode.

Reduced fatigue is also important for lower residual internal stress in thin ferroelectric film. For this purpose, RuO$_2$(70 nm)/Ru(70 nm)/SiO$_2$(1000 nm)/Si and RuO$_2$/Ru/MgO substrates were prepared. The idea is to eliminate the lattice mismatch between the thin film

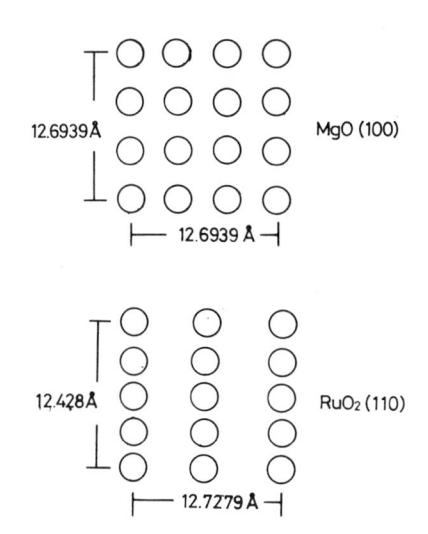

Figure 3. Lattice matching between the thin film and the substrate of MgO(100) and RuO$_2$(110).

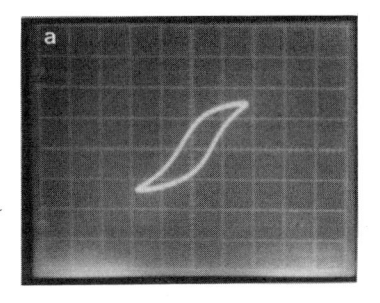

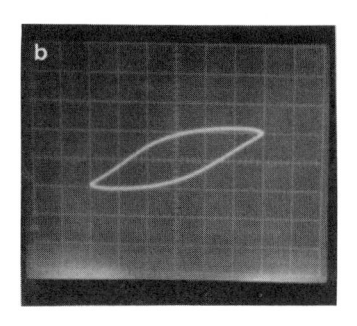

Figure 4. Ferroelectric D-E hysteresis loops of PLT thin film on Pt/SiO$_2$ substrate (a) 270 nm thick on Pt/SiO$_2$ substrate, X: 185/cm/div, Y: 15.7 µC/cm^2, P$_r$: 12.6 µC/cm^2, E$_c$: 92.5 kV/cm, (b) 130 nm thick on Pt/SiO$_2$ substrate, X: 153/cm/div, Y: 7.85 µC/cm^2, P$_r$: 6.3 µC/cm^2, E$_c$: 229 kV/cm.

and the substrate. The RuO$_2$ has lattice constants of a = 0.4500 nm and c = 0.3107 nm, and the √2a = 0.6364 nm and 1.5 a = 0.6320 nm and the 1.5 ∞ 2c = 0.9321 nm and 2√3 = 0.9373 nm as shown in Figure 3.

Okazaki used microindentation to investigate internal stresses - both anisotropic internal stress induced by cubic-to-tetragonal phase transition and anisotropic internal stress induced by the application of electric field.[4]

Table 3. Dielectric permittivity, ε_s, loss tangent, tan δ, coercive field, E_c, and remnant polarization, P_r, of 270 nm thickness (Pb,La)TiO$_3$ thin films deposited on various substrates.

Substrate	Pt/SiO$_2$/Si	RuO$_2$/Ru/SiO$_2$/Si	Pt/MgO	RuO
orientation	(111)	random	(001)	(001)
ε_s (at 100 kHz)	226	132	189	159
tan δ	0.04	0.08	0.05	0.05
E_c (kV/cm)	92.5	92.5	111	–
P_r (μC/cm^2)	12.6	10.2	18.8	–

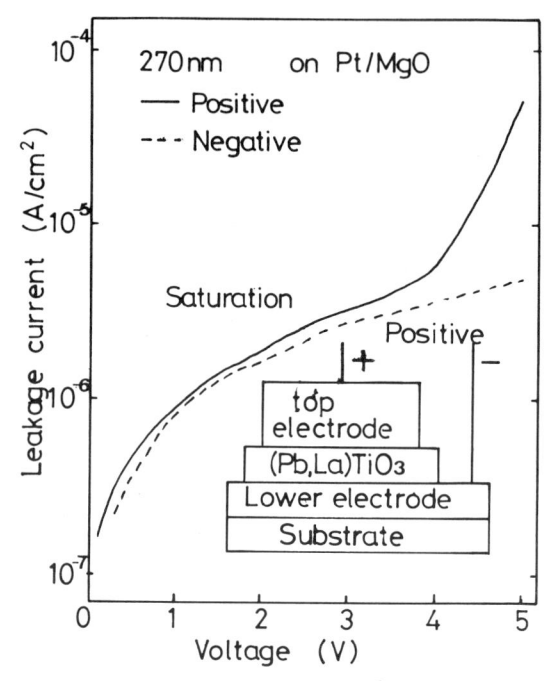

Figure 5. Leakage current-applied voltage characteristics of 270-nm-PLT thin film deposited on Pt/MgO substrate.

THICKNESS DEPENDENCIES

Figure 6 is an SEM Micrograph of the fractured cross section of a PLT thin film deposited on Ru/MgO. This shows relatively large grains growing up from the substrate.

Figure 7 is Log-Log plot of the dielectric permittivity and breakdown voltage (BDV, V_c) as a function of dielectric thickness, d, for bulk and thin film materials including recent results of PLT thin films less than 100 nm thick,[11] BaTiO$_3$ bulk ceramics with a dielectric thickness of 1 ~ 3 mm;[5] and commercial MLCs, including BaTiO$_3$ and relaxor families. As can be seen, the permittivity decreases with decreased thin film thickness, particularly in the region below 100 nm. This result suggests the importance of the surface effect on ferroelectric particles and ferroelectric thin films.

Dielectric bulk and thin film ferroelectric materials consist mainly of two parts. The first is the ferroelectric material inside ferroelectric grains or ferroelectric parts inside thin films with high dielectric permittivity. The second is the grain boundary layers with low dielectric permittivity because they have more defects, impurities, internal stresses induced by lattice mismatch, etc.

The equivalent circuits of C_F of ferroelectric part and C_N of non-ferroelectric parts are shown in Figure 7. When a dc or ac electric voltage is applied to the series circuit, most of the voltage is applied to C_N because in the C_F, ferroelectric polarization is induced. In other words, for an electric field in C_N, E_N is higher than that in C_F. Then, $E_F \gg E_N$. Therefore, the breakdown voltage should be dominated by the electrical properties of C_N. The dielectric permittivity of C_F is higher than that of C_N. The apparent permittivity should decrease with decreased dielectric thickness because the C_N thickness is almost constant. Loss tangent, tan δ, is also dominated by the dielectric properties of surface layer structure in the thinnest films.

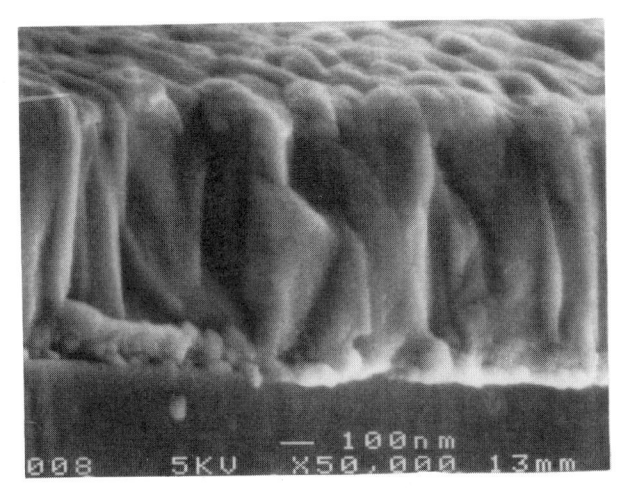

Figure 6. SEM micrograph of PLT thin film with large grains deposited on Ru/MgO.

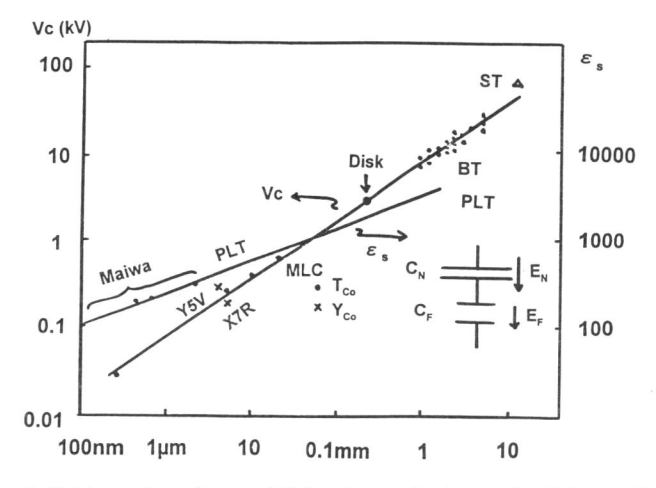

Figure 7. Thickness dependences of dielectric permittivity and breakdown voltage.

665

FATIGUE AND INTERNAL STRESSES

For an isotropic internal stresses induced during the cubic-to-tetragonal phase transition and anisotropic internal stress induced by the application of dc field, the aging effects of ferroelectric ceramics was reviewed by Okazaki.[2] In $(Pb_{1-x}, La_x)TiO_3$, the tetragonality, c/a; the Curie point, T_c; and the internal stress decrease with increasing La content in the transparent pore-free ceramic samples. Yamamoto et al. investigated the field-induced anisotropic internal stress[4] and found that after dc field application, tensile strength is induced in the direction of dc field application; and a compressive internal stress is also induced in the perpendicular direction to the electric field.

In 1953, Okazaki proposed the Space Charge Model in $BaTiO_3$ Ceramics.[14] Takahashi ascertained the space charge internal field in PZT ceramics[15] and revealed the relationship between the internal bias field and mechanical quality factor, Q_m.

Okazaki suggested the "space charge stabilized effect" for the piezoelectric applications, for the PZT ignitor k_{33} against mechanical compression,[16] for the high-Q_m and Low-Q_m PZT mechanical fatigue. This effect should also exist in ferroelectric thin films.

Figure 8 presents XRD of $PbTiO_3$ thin films for different deposition thicknesses.[16] The tetragonality decreases with decreased thin film thickness. Films less than 10 nm thick are essentially all cubic. This result strongly supports the existence of the Kaenzig's surface layer for non-ferroelectric with the cubic structure.[17]

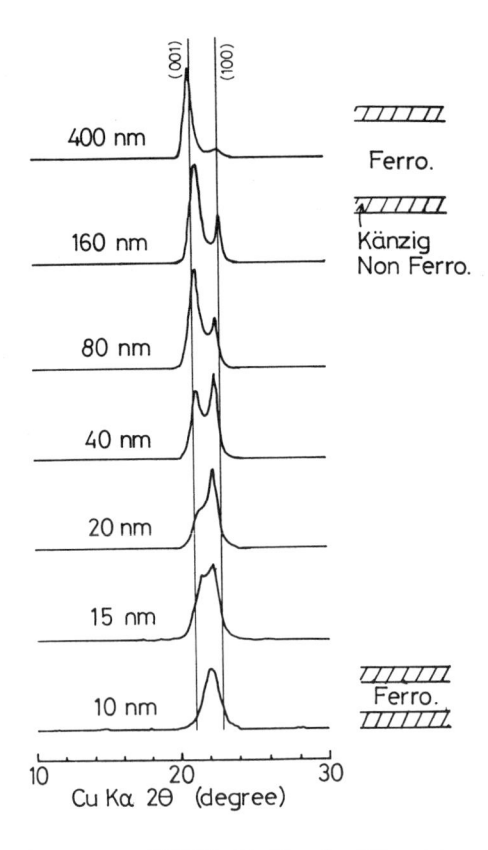

Figure 8. X-ray diffraction patterns of $PbTiO_3$ thin films for different deposited film thicknesses.[16]

CONCLUSION

PLT thin films were prepared using a four-target-cathode rf sputtering apparatus, ferroelectric properties were measured, and microstructures were analyzed. The film thickness dependence; fatigue and internal stresses; and the shape changes of D-E loops were also discussed in the comparison to other ferroelectric ceramics.

The ferroelectric properties of these thin films are very complicated because of the surface layer effects, the interface between substrates and thin films, the internal bias field effects, changes under ac field application,[18] defect structure analysis, mechanical internal stress,[19] aging effects, and other issues. Substrates were prepared which provide a smaller mismatch of lattice constants and less fatigue. The asymmetric polarization reversal is also solved for these practical applications. Isotropic internal stresses induced during the cubic-to-tetragonal phase transition and anisotropic internal stress induced by the application of dc field have already been reviewed by Okazaki.[2]

REFERENCES

1. K. Okazaki and K. Takahashi, Study on the High Permittivity Ceramics of $PbTiO_3$ Families, *Sci. Rep. Nat. Def. Acad.*, 5, **2**:183 (1967), in Japanese.
2. I. Ueda, M. Nishida, S. Kawashima, *Rep. IECE Jpn.*(1980, 8, 26) US80-25, in Japanese.
3. Y. Yamashita, K. Yokoyama, H. Honda and K. Okuma, 3-FMA, Kyoto, (1981) 21-F-14.
4. K. Okazaki, Normal Poling and High Poling of Ferroelectric Ceramics and Space-Charge Effects, *Jpn. J. Appl. Phys.*, **32**:4241 (1993).
5. K. Okazaki, Ceramic Engineering for Dielectrics (4th Edition), Gakkensha, Tokyo, 1992, in Japanese, translated into Russian (Energy Publishing Office in Moscow, (1976).
6. K. Okazaki, Mechanical Behavior of Ferroelectric Ceramics, *Am. Ceram. Soc. Bull.*, **63**:1150 (1984).
7. K. Okazaki, H. Igarashi, K. Nagata and A. Hasegawa, Effects of Grain Size on the Electrical Properties of PLZT Ceramics, Ferroelectrics (Proc. of 3rd Int'l. Meeting on Ferroelectricity, Edinburgh 1973), **7**:153 (1974).
8. K. Okazaki, I. Ohtsubo and K. Toda, Electrical, Optical and Acoustic Properties of PLZT Ceramics by Two-Stage Processing, *Ferroelectrics*, **10**:195 (1976).
9. K. Okazaki and H. Igarashi, Importance of Microstructure in Electronic Ceramics, in Ceramic Microstructure '76, Ed. R. M. Fulrath and J. A. Pask, Westview Press, Boulder, 1977.
10. K. Okazaki, Theoretical Density and PbO Deficiency of Pore-Free Transparent PLZT Ceramics, *J. Japan Soc. of Powder Metallurgy*, **25**:147 (1978).
11. H. Adachi et al. Proc. of 1990 IEEE 7the ISAF.
12. H. Maiwa, N. Ichinose and K. Okazaki, *Jpn. J. Appl. Phys.*, **31**:3029 (1992).
13. K. Okazaki, H. Maiwa and N. Ichinose, Preparation of $(Pb,La)TiO_3$ Thin Films by Multiple Cathode Sputtering, Ceram. Trans., Ferroic Materials: Design, Preparation, and Characteristics, pp. 15-26.
14. K. Okazaki; Space-Charge in Barium Titanate Ceramics, J. Inst. of Elect. Eng. of Japan, **80**:865 (1960).
15. S. Takahashi, *Jpn. J. Appl. Phys.*, **20**:95 (1981).
16. K. Okazaki and H. Maiwa, Space Charge Effects on Ferroelectric Ceramic Particle Surfaces, *Jpn. J. Appl. Phys.*, **31**:3113 (1992).
17. W. Keanzig, *Phys. Rev.*, **98**:549 (1955).
18. K. Okazaki, M. Masuda, S. Tashiro and S. Ishibashi, *Ferroelectrics*, **22**:631 (1978).
19. K. Okazaki and T. Tanimoto, Electro-Mechanical Strength and Fatigue of Ferroelectric Ceramics, *Ferroelectrics*, **131**:25 (1992).

PREPARATION AND PROPERTIES OF PZT/PbTiO$_3$ CERAMIC COMPOSITE

Hisao Banno and Noriyasu Sugimoto

NTK Technical Ceramics Div., NGK Spark Plug Co., Ltd.
14-18, Takatsuji-cho, Mizuho, Nagoya 467 Japan

Takashi Hayashi

Dept. of Materials and Ceramic Technology,
Shonan Institute of Technology
1-1-25, Tsujido-nishikaigan, Fujisawa, Kanagawa 251 Japan

ABSTRACT

An experiment was carried out in order to obtain a diphasic ceramic composite consisting of Pb(Zr,Ti)O$_3$ (PZT) and PbTiO$_3$. Starting raw materials were:

- Soft PZT powder with a composition near the morphotropic phase boundary and average particle size of 0.6 μm.
- Pure PbTiO$_3$ powder with average particle size of 5 μm.
- Pure PbTiO$_3$ powder with average particle size of 20 μm.

These powders were mixed at PZT/PbTiO$_3$ mass ratios of 75/25, 50/50 and 25/75, pressed into a disk and the compacts were fired at around 1000°C for 0.5 hours (short-time soak) or 12 hours (long-time soak). The short-time-soaked compact with PZT/PbTiO$_3$ (20 μm) mass ratio of 32.5/67.5 was hot-isostatically pressed (HIPed) at 1000°C and 200 MPa in O$_2$(5%)/Ar mixture gas atmosphere for 1 hour without encapsulating. PZT/PbTiO$_3$ diphasic ceramic composites were obtained at PZT/PbTiO$_3$ mass ratios of 50/50 and 25/75 for (a) a short-time soak (0.5 hour) with PbTiO$_3$ powders of 5 and 20 μm and (b) a long-time soak (12 hours) with PbTiO$_3$ of 20 μm, although they were not so dense. The HIPed one was of PZT/PbTiO$_3$ diphasic ceramic composite with comparatively high density. The composition of the PZT phase in the ceramic composite was shifted to the tetragonal (Ti-rich) side from that of starting PZT powder and its compositional fluctuation was observed, while that of PbTiO$_3$ was not detected by x-ray analysis.

INTRODUCTION

Extensive research has produced highly dense Pb(Zr,Ti)O$_3$ (PZT) ceramics in order to improve their piezoelectric performances[1-6] and highly porous (3-3 type) PZT ceramics have been fabricated in order to enhance hydrostatic piezoelectric constants of ceramic/polymer composites.[7,8] Hayashi *et al.* have attempted to make porous 3-3 PZT ceramics improve connectivity with a higher permittivity by using capsule-free oxygen hot isostatic pressing (HIP).[9]

In many cases, commercially available PZT ceramics have been produced by firing compacts or doctor-bladed sheets of a calcined powder prepared by solid-state reaction of a mixture of PbO, ZrO$_2$, TiO$_2$ and some additives. Reaction mechanisms in the formation of PZT solid solutions were investigated by Chandratreya *et al.*[10]

A method to determine compositional fluctuations in PZT solid solutions was developed by Kakegawa *et al.* in 1975.[11] They also reported a compositional fluctuation in PZT ceramics prepared by a solid-state reaction of a mixture of PbO, ZrO$_2$ and TiO$_2$, but not in those prepared from alkoxides.[12]

Piezoelectric properties of 0 to 3 composites consisting of a polymer and ceramic powder mixture of PZT and PbTiO$_3$ (PT) have been investigated in order to enhance the piezoelectric anisotropy d_{33}/d_{31} for hydrophone.[13,14]

Little work on fabricating PZT/PbTiO$_3$ ceramic composites, however, has been done because the PbZrO$_3$-PbTiO$_3$ system is solid-soluted into single-phase Pb(Zr,Ti)O$_3$ in full range by heat-treatment.

In this paper, the preparation and properties of PZT/PbTiO$_3$ ceramic composites are discussed with respect to processing/composite structure relations.

EXPERIMENTAL PROCEDURE

Starting raw materials were:

- Soft PZT powder with a composition of Bi oxide-added Pb(Zr,Ti,W$_{1/2}$Ni$_{1/2}$)O$_3$ in the tetragonal region near the morphotropic phase boundary and average particle size of 0.62 µm (abbreviated as 0.6 µm PZT powder, or pzt).

- Pure PbTiO$_3$ powder with average particle size of 4.6 µm (abbreviated as 5-µm PT powder, or pt).

- Pure PbTiO$_3$ powder with average particle size of 22.7 µm (abbreviated as 20-µm PT powder, or pT).

The particle size distributions of the powders are shown in Figure 1.

The powders were mixed at PZT/PbTiO$_3$ mass ratios of 75/25, 50/50 and 25/75, pressed into a disk (approximately 11 mm in diameter and 2 mm in thickness) at approximately 20 MPa; and the compacts were normally fired at around 1000°C for 0.5 hour (short-time soak, abbreviated as "S$_{1000}$" for 1000°C firing with 0.5-hour soak) or 12 hours (long-time soak, abbreviated as "L$_{960}$" for 960°C firing with 12-hour soak). Compacts with PZT/PbTiO$_3$ mass ratio of 32.5/67.5 were normally-fired at 1000°C for 0.5 hour and at 960, 1010 and 1060°C for 12 hours; some specimens that would have been normally fired at 1000°C for 0.5 hour were HIPed without encapsulating at 200 MPa at 1000°C for 1 hour in O$_2$(5%)/Ar mixture gas atmosphere (abbreviated as "HIP$_{1000}$"). The firing schedules are schematically shown in Figure 2.

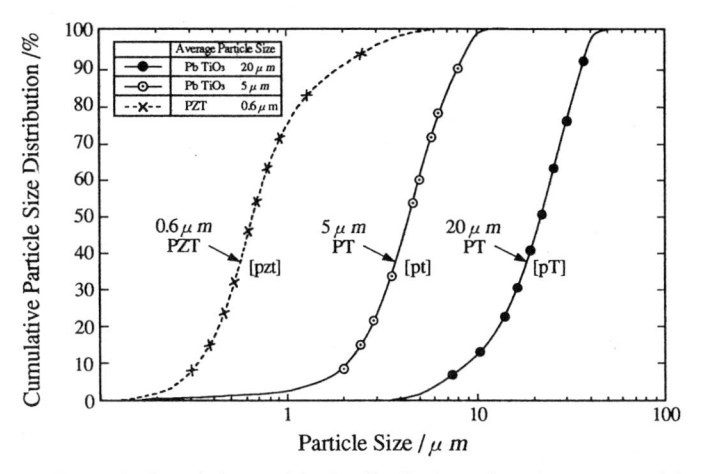

Figure 1. Cumulative particle size distributions of starting raw materials.

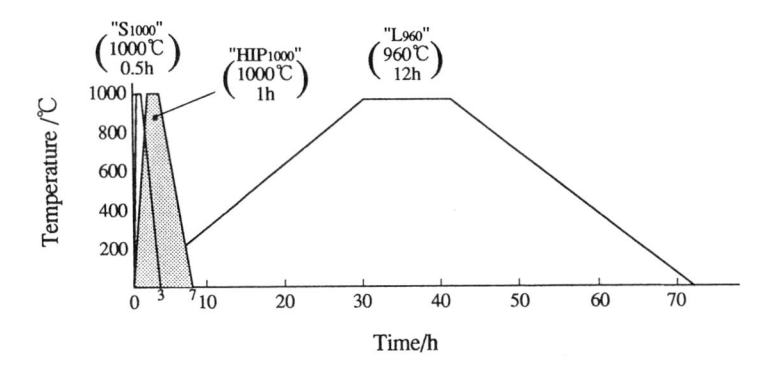

Figure 2. Firing schedules for the compacts of mixture of PZT and PbTiO$_3$.

Microstructures of the fired specimens were examined by scanning electron microscopy (SEM) and x-ray microanalysis (XMA). Crystal structures and compositional fluctuations of solid-state reacted compound of PZT and PT in the specimens were measured by powder x-ray diffraction (Cu K radiation through a nickel filter) and then determined by a method developed by Kakegawa et al.[11,12,15] Differential thermal analysis (DTA) of the specimens was done at a heating rate of 10/min. "HIP$_{1000}$" specimens (1.5 mm in thickness) were electroded with silver paste (4.5 mm in diameter) on surfaces, and then temperature dependencies of dielectric properties were measured at 10 kHz.

RESULTS AND DISCUSSION

Apparent densities, porosities and shrinkages of "S$_{1000}$" and "L$_{960}$" specimens prepared from 5 and 20-μm PT powders are shown as a function of starting composition of PZT/PT mixture in Figure 3.

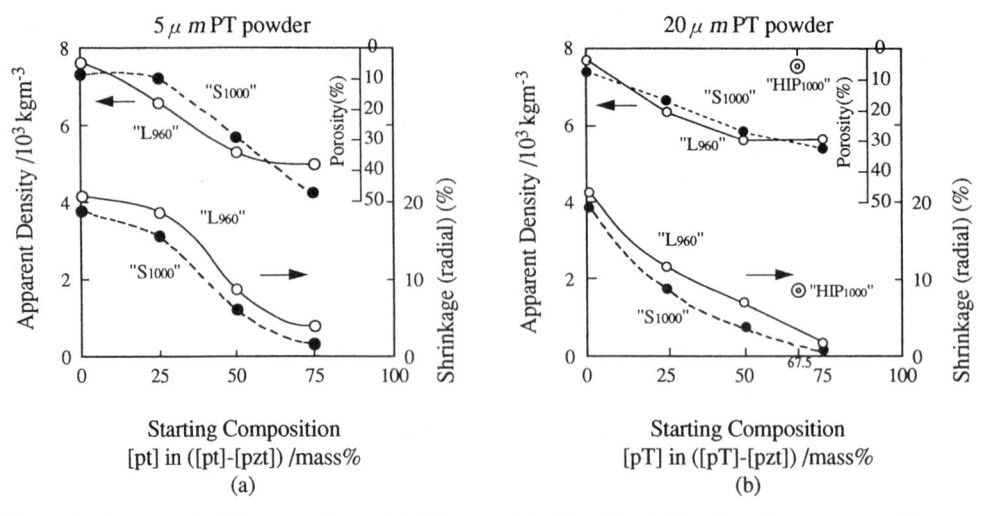

Figure 3. Apparent densities, porosities and shrinkages of the "S$_{1000}$" and "L$_{960}$" specimens prepared from PT powders as a function of starting composition of PZT/PT mixture. (a) Specimens prepared from 5-μm PT powder. (b) Specimens prepared from 20-μm PT powder. (Porosity was obtained assuming that densities of solid-state reacted compound of PZT and PT were 8.0×10^3 kgm^{-3}.)

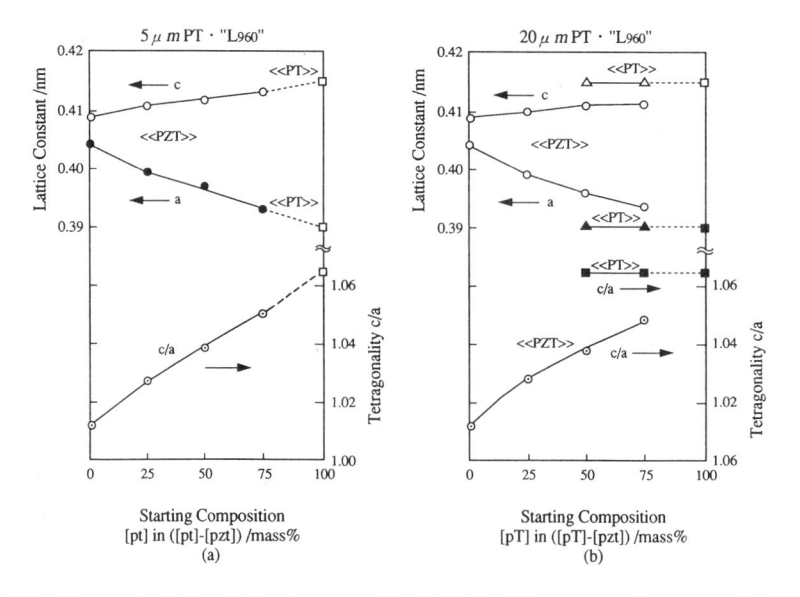

Figure 4. Lattice constants (a and c) and tetragonality (c/a) of solid-state reacted compounds of PT and PZT in the "L$_{960}$" specimens. (a) Specimens prepared from 5-μm PT powder. (b) Specimens prepared from 20-μm PT powder.

Lattice constants (a and c) and tetragonality (c/a) of solid-state reacted compounds of PT and PZT in the "L$_{960}$" specimens prepared from 5 and 20-μm PT powders are shown in Figure 4. The contents and crystalline phase of solid-state reacted compounds of PZT and PT in the specimens are illustrated as a ratio of intensity of PZT to (PT + PZT) for [(002) and (200)] peaks in x-ray diffraction patterns and their tetragonality as a function of the starting composition of the PZT/PT mixture in Figure 5.

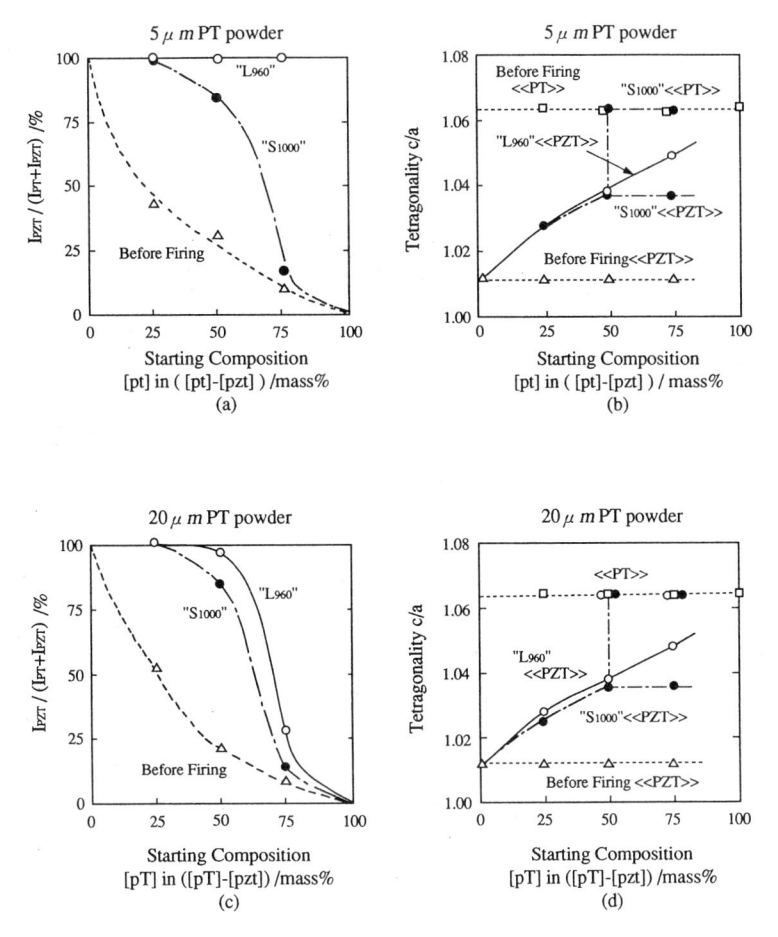

Figure 5. (a), (c) A ratio of intensity of PZT to (PT+ PZT) for [(002) and (200)] peaks in x-ray diffraction patterns and (b),(d) tetragonality of solid-state reacted compounds of PT and PZT as a function of starting composition of PT-PZT mixture (prepared from (a), (b) 5-µm, and (c),(d) 20 µm PT powders).

As illustrated in Figures 3 through 5, compacts of PZT/PT mixture were solid-soluted to a single phase of PZT at different starting composition PZT/PbTiO₃ mass ratios:

- Ratios of 75/25, 50/50, and 25/75 for long-time soak (12 hours at 960°C) with 5-µm PT powder.
- A ratio of 75/25 for short-time soak (0.5 hour, at 1000°C) with 20-µm PT powder.

PZT/PbTiO₃ diphasic ceramic composites were also obtained, although they were not so dense:

- Ratios of 50/50 and 25/75 for short-time soak (0.5 hour, at 1000°C) with 5-µm and 20-µm PT powders.
- Ratios of 50/50 and 25/75 for long time soak (12 hours, at 960°C) with 20-µm PT powder.

The composition of the PZT phase in the ceramic composite was shifted to the tetragonal (Ti-rich) side from that of the starting PZT powder.

By this heat-treatment, PZT (with c/a = 1.012, near the morphotropic phase boundary) and PT (with c/a = 1.065) have reacted with each other to form another PZT phase (with 1.012 < c/a < 1.065). For the short-time soak (0.5 hours, at 1000°C) with both 5- and 20-μm PT powders, the reacted PZT phase in the composite at the composition PZT/PbTiO$_3$ mass ratio of 25/75 had almost the same tetragonality as that of 50/50. A boundary between the single phase region and the diphasic ceramic composite existed around the composition PZT/PbTiO$_3$ mass ratio of 50/50 for this heat treatment.

A ratio of intensity of PZT to (PT + PZT) for [(002) and (200)] peaks and tetragonality of solid-state reacted compounds of PT and PZT in the normally-fired/long-time soaked and "HIP$_{1000}$" specimens prepared from a composition PZT/PT mass ratio of 32.5/67.5 were obtained as a function of firing condition as shown in Figure 6. Compositional fluctuations of solid-state reacted compound of PZT and PT were also obtained as shown in Figure 7. Scanning electron micrographs of as-fired and polished surfaces of the "HIP$_{1000}$" specimen are shown in Figure 8.

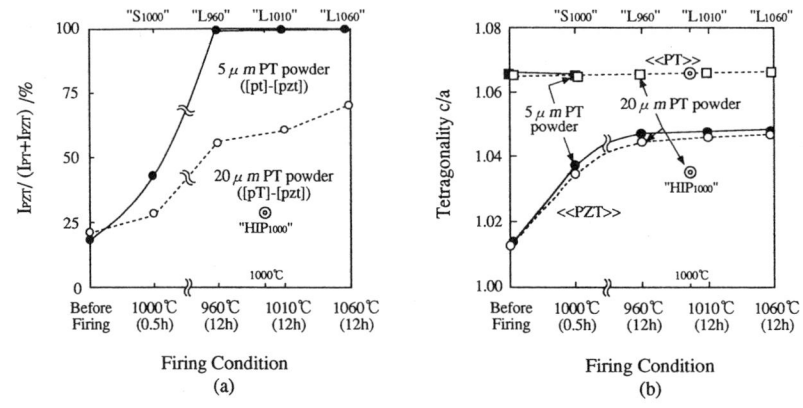

Figure 6. (a) A ratio of intensity of PZT to (PT + PZT) for [(002) and (200)] peaks in x-ray diffraction patterns. (b) Tetragonality of solid-state reacted compounds of PT and PZT in specimens prepared from a composition PZT/PT mass ratio of 32.5/67.5 as a function of firing condition.

Tetragonality of the PT phase in the diphasic ceramic composite was 1.06 to 1.065, or almost the same as that of the starting PT powders. Its compositional fluctuation was not detected by x-ray analysis.

Accordingly, the PT phase in the composite was considered to be nonreactive with a composition the same as that of the starting PT powder. When a part of PT and all of PZT in the compact of PZT/PT mixture were solid-soluted to another PZT and the rest of PT remained unreacted after the heat-treatment because of the large particle size of PT, a diphasic PZT/PT ceramic composite was obtained.

Behaviors of the composition and its fluctuating PZT reaction in the "HIP$_{1000}$" ceramic composite were remarkably different from those in the "long-time-soaked" one as shown in Figure 7. The composition of the reacted PZT in the "HIP$_{1000}$" ceramic composite was almost the same as that of the "S$_{1000}$" one (before HIPing), although its density was increased and compositional fluctuation was decreased by the capsule-free O$_2$ HIP. The HIPing was very effective in sintering the PZT region in the PZT/PT ceramic composite without causing a reaction in PZT and PT.

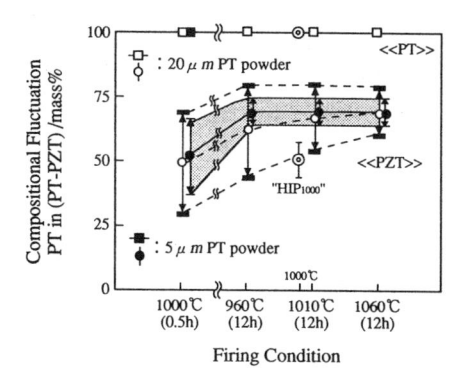

Figure 7. Compositional fluctuation of solid-state reacted compound of PZT and PT in specimens prepared from a composition PZT/PT mass ratio of 32.5/67.5 as a function of firing condition.

Microstructures of the normally-fired PZT/PbTiO$_3$ ceramic composites were structurally of the 3-0 type (PZT-PbTiO$_3$) in which PZT phase was of 3-3 type (PZT-pore) (i.e., highly porous). This highly porous PZT phase region was changed to be dense and with flat pores at the PZT/PT interface region by the HIPing, as shown in Figure 8. This was considered to be the same effect as observed in capsule-free O$_2$ HIPed 3-3 PZT ceramics by Hayashi et al.[9]

Figure 8. Scanning electron micrographs. (a) As-fired SEM. (b) Polished surfaces of "HIP$_{1000}$" specimen. (∇: flat pores)

DTA curve and temperature dependencies of capacitance and the dielectric loss tangent of the "HIP$_{1000}$" specimen are shown in Figure 9. Peaks and bends (maximum or minimum) observed in the figure were considered to appear at phase transformation temperatures corresponding to the reacted compounds of PZT and PT in the "HIP$_{1000}$" specimen. Two small bends at 410°C and 490°C were observed in the dielectric loss tangent. Also, a major peak centered at 410°C and an abrupt change at 490°C were observed in the capacitance. These abrupt changes may be caused by structure-properties relationships in the PZT/PbTiO$_3$ ceramic composites. A peak due to phase transformation temperature of PT phase was detected at 489°C, but that of PZT phase was not detected by the DTA. Further investigations are needed.

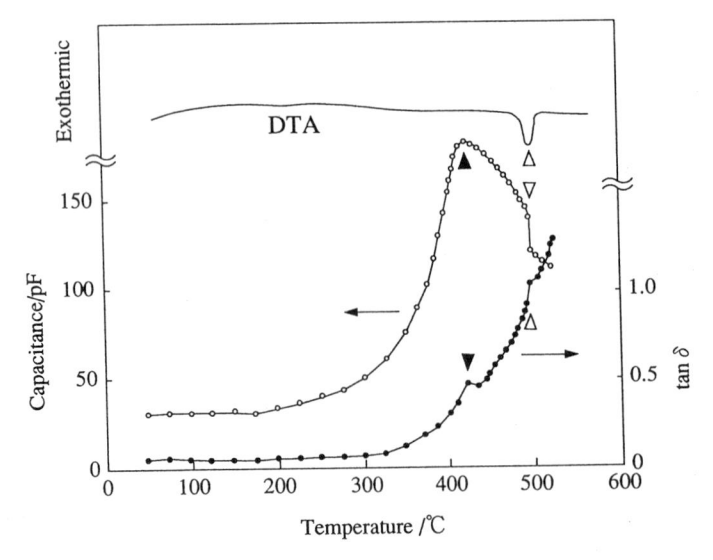

Figure 9. DTA curve and temperature dependencies of dielectric constant and dielectric loss tangent of "HIP$_{1000}$" specimen. (▲ and ▽: phase transformation temperatures corresponding to PZT and PT phases, respectively.)

As described above, a comparatively dense PZT/PT diphasic ceramic composite was obtained by normally firing and capsule-free O$_2$ HIPing compacts of a mixture of PZT powder with small (submicron) particle size and PT powder with large (more than 20-μm) particle size.

CONCLUSIONS

PZT/PbTiO$_3$ diphasic ceramic composites were obtained at PZT/PbTiO$_3$ mass ratios of 50/50 and 25/75 for two cases:

- Short-time soak (0.5 hour, at 1000°C) with 0.6-μm PZT and 520-μm PbTiO$_3$ powders.
- Long-time soak (12 hours, at 1000°C) with 0.6-μm PZT and 20-μm PbTiO$_3$ powders, although they were not so dense.

Comparatively dense PZT/PbTiO$_3$ diphasic ceramic composite was obtained at PZT/PbTiO$_3$ mass ratio of 32.5/67.5 with 0.6-μm PZT and 20-μm PbTiO$_3$ powder by hot-isostatically pressing a pre-fired (at 1000°C for 0.5 hours) specimen without encapsulating at 1000°C and 200 MPa in O$_2$ (5%)/Ar mixture gas atmosphere for 1 hour.

The composition of reacted PZT in the HIPed ceramic composite was almost the same as that of the pre-fired one, although its density was increased and compositional fluctuation was decreased by the capsule-free O$_2$ HIP. Tetragonality of PT in the diphasic ceramic composite was 1.06 to 1.065, or almost the same as that of the starting PT powders, and its compositional fluctuation was not observed.

REFERENCES

1. K. Okazaki and K. Nagata, *Trans. Inst. Elect. Commun. Eng.*, **70**/11, 53-C [11]:815 [in Japanese] (1970).
2. G. H. Haertling and C. E. Land, *J. Amer. Ceram. Soc.*, **54**:1 (1971).
3. K. Okazaki and K. Nagata, *J. Amer. Ceram. Soc.*, **56**:82 (1973).

4. K. Okazaki and H. Igarashi, Importance of Microstructure in Electronic Ceramics, in *Ceramic Microstructure '76*, R. M. Fulrath and J. A. Pask, eds., Westview Press, Boulder, Colorado (1976).
5. S. Tashiro, Y. Tsuji, and H. Igarashi, *Ferroelectrics,* **95**:157 (1989).
6. T. Yamamoto, R. Tanaka, K. Okazaki, and T. Ueyama, Proc. 7th Meeting on Ferroelectric Materials and Their Applications, Kyoto 1989, *Jpn. J. Appl. Phy.,* 28 Suppl. **28-2**:67 (1989).
7. T. R. Shrout, W. A. Schulze and Biggers, *Mat. Res. Bull.*, **14**:1553 (1979).
8. K. Nagata, H. Igarashi, K. Okazaki and R. C. Bradt, *Jpn. J. Appl. Phys.,* **19**:L37 (1980).
9. T. Hayashi, S. Sugihara, and K. Okazaki, *Jpn. J. Appl. Phys.,* **30**:2243 (1991).
10. S. S. Chandratreya, R. M. Fulrath and J. A. Pask, *J. Amer. Ceram. Soc.,* **64**:422 (1981).
11. K. Kakegawa, K. Watanabe, J. Mohri, H. Yamamura, and S. Shirasaki, *J. Chem. Soc. Jpn.*, **3**:413 [in Japanese] (1975).
12. K. Kakegawa, J. Mohri, T. Takahashi, H. Yamamura, and S. Shirasaki, *J. Chem. Soc. Jpn.,* **5**:717 [in Japanese] (1976).
13. H. Banno and K. Ogura, *Jpn. J. Appl. Phys.,* **30**:2050 (1991).
14. H. Banno and K. Ogura, *J. Ceram. Soc. Jpn.,* **100** [4]:551 (1992).
15. K. Kakegawa, J. Mohri, T. Takahashi, H. Yamamura, and S. Shirasaki, *Solid State Commun.*, **24** [11]:769 (1977).

GROWTH OF ORIENTED SnO$_2$ THIN FILMS FROM ORGANOTIN COMPOUNDS ON GLASS SUBSTRATES BY SPRAY PYROLYSIS

Shoji Kaneko,[1] Isao Yagi,[2] Tsuyoshi Kosugi,[2] and Kenji Murakami[3]

[1]Dept. Mater. Sci. Tech., [2]Grad. Schl Electron. Sci. Tech., and [3]Res. Inst. Electron., Shizuoka University, Johoku, Hamamatsu 432, Japan

INTRODUCTION

Tin(IV) oxide (SnO$_2$) is an oxygen-defect type semiconductor with a wide band gap and a large mobility. It is transparent in the visible region, and reflective in the infrared region.[1] Thus, SnO$_2$ thin films have been used as transparent electrodes in sophisticated electrochromic, electroluminescent, and photoconductive devices.[2] Recently, it has been also reported that gas sensitivity is high for (110) oriented SnO$_2$ thin flms[3] and a large electrical conductivity is obtained in (200) oriented SnO$_2$ thin films.[4]

Chemical techniques, including spray pyrolysis using simple apparatus with good productivity, for the preparation of thin films facilitate the design of materials on a molecular level. The influence of preparation conditions on the physical properties of sprayed SnO$_2$ thin films has been studied extensively; in contrast, the influence on film formation seldom has been examined in detail. Although source compounds play an important role in controlling the structure and morphology of thin films under spray pyrolysis, the effect of those compounds on the growth of thin films has not yet been investigated extensively. In the case of SnO$_2$ thin films also, other compounds except for tin(IV) chloride[1] have not been examined sufficiently.

For the present work, SnO$_2$ thin films were grown on glass substrates by a spray pyrolysis technique using a series of organotin compounds with butyl groups and/or acetate groups: tetra-n-butyltin(IV) [TBT],[5] tri-n-butyltin(IV) acetate [TBTA],[6] and di-n-butyltin(IV) diacetate [DBTDA].[7] This paper focuses on the effect of the molecular structures of the source compounds on the oriented growth of SnO$_2$ thin films.

EXPERIMENTAL PROCEDURE

(1) Film Formation

Tetra-n-butyltin(IV) (MERCK, >97% purity) and di-n-butyltin(IV) diacetate (NITTO KASEI, JAPAN, >95% purity) were dissolved into ethanol to use as source compounds. The apparatus used is illustrated schematically elsewhere.[5] The ethanol solution was atomized by a pneumatic spraying system at an air pressure of 0.1 MPa. The droplets were transported onto a heated Corning 7059 glass substrate (25x25x1 mm^3

in size). Since the substrate temperature was lowered by spraying the ethanol solution with the compressed air, the solution was atomized for 1 s intermittently at the interval of several tens of second. The substrate was mounted on a cordierite ceramic holder, and then heated.

(2) Characterization of films

The structural analysis of the deposited film was carried out by an X-ray diffraction (XRD) technique with CuKα radiation using a RIGAKU RINT-1100 diffractometer. The preferred orientation of the (hkl) plane of the film was expressed by a texture coefficient (TC) calculated from equation (1),[8]

$$TC_{hkl} = I_{hkl} / I_{ohkl} \tag{1}$$

where I and I_0 are the X-ray diffraction intensities of sample and standard non-oriented film measured under the constant thickness conditions of the same. A higher deviation of the texture coefficient from unity indicates a higher preferred orientation of the films. Crystallite size, D_{hkl}, was calculated using the Scherrer's equation. Film thickness was determined by a stylocontact method using a Sloan Dectak IIA.

The surface morphology of films was observed by a JEOL JSM-6300F field emission type scanning electron microscope (SEM). The crystal structure of the very thin films was determined by reflection high energy electron diffraction (RHEED).

RESULTS AND DISCUSSION

Tin(IV) chloride was mostly used as a source compound for forming SnO_2 thin films by spray pyrolysis.[4] Various growth conditions/parameters such as solvent, solution feed rate, solution concentration, film growth rate, film thickness, and annealing have been investigated in an effort to deposit oriented SnO_2 thin films onto glass substrates.[9] The relationship between the microstructure of thin films and the source organotin compounds has also been discussed.

(1) Oriented Growth of SnO_2 Films

A typical XRD profile of the SnO_2 thin film (0.1 μm thick) grown on a glass substrate from a 10% TBT solution under the following optimum conditions: substrate temperature, 340°C and film growth rate, 0.8 nm/s is shown in Fig. 1(A). A strong XRD peak attributed to the (110) plane of cassiterite and weak XRD peaks corresponding to the (211) and (220) planes were observed. Indeed, the TC_{110}, calculated from Eq. (1), showed a high value of 6, thus indicating a strong [110] preferred orientation of the SnO_2 film normal to the glass substrate.

The source compound was replaced with DBTDA in order to study the effect of substituents, since DBTDA has two butyl and acetoxy groups per molecule, whereas TBT has four butyl and no acetoxy groups. An XRD profile of the 0.3-μm-thick film grown on a glass substrate from a 1% DBTDA solution under the optimum conditions: substrate temperature, 480°C and film growth rate, 3.0 nm/s is shown in Fig. 1(B). Only a strong XRD peak attributed to the (200) plane of cassiterite and a very weak XRD peak of the (310) plane were observed. Indeed, the TC_{200} showed a very high value of 15. That result indicates that the film was formed along a [100] preferred direction normal to the glass substrate. The use of TBTA as a source compound, which has three butyl and one acetoxy groups per molecule, resulted in a drastic decrease of the TC_{110} and a gradual increase of the TC_{200}.[6] The spray pyrolysis of TBTA led to the deposition of a SnO_2 film with an intermediate orientation, as compared to those with the TBT and DBTDA source compounds. Therefore, the substitution of a butyl for an acetoxy group seemed to facilitate the [100] oriented growth of SnO_2 films.

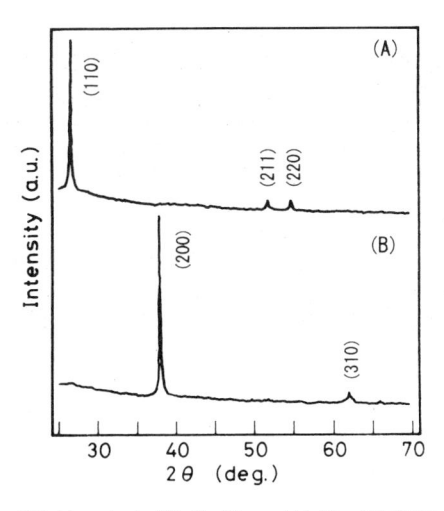

Figure 1. XRD profiles of highly oriented SnO_2 films. (A) 10 wt% TBT soln at 340°C; (B) 1 wt% DBTDA soln at 480°C.

(2) Initial Stage of Film Deposition

Tin(IV) oxide films with thicknesses ranging from 9 to 90 nm were prepared on glass substrates to investigate the relationship between oriented growth and film thickness, especially in the early growth region. Figure 2(A) shows a SEM micrograph of the SnO_2 film (90 nm thick) grown from a 1% DBTDA solution at a substrate temperature of 480°C. Clearly, the surface was densely packed with particles measuring ~30 to 60 nm in diameter, and they appeared to have columnar structures, suggesting full oriented growth around that film thickness. Although thinner films were deposited, no clear SEM images of the surface morphology for films of less than 30 nm thick were observed, because the films exhibited insufficient electrical conductivity on glass substrates.

To determine the initial stage of SnO_2 thin film formation, films were deposited onto Si(100) substrates and examined using SEM and RHEED. The surface morphologies of those films are shown in Figs. 2(B) and (C). The 9-nm-thick film in the initial stage of oriented growth displayed some recognizable grown nuclei (Fig. 2(B)). A corresponding RHEED pattern for that film (Fig. 3(A)) shows a few diffraction spots, which support the occurrence of crystallization, together with a number of diffused diffraction rings. As the film thickness increased, the number and size of nuclei also increased. Figure 2(C) shows the surface morphology of the 30-nm-thick film, in which the nuclei grew up to 30 to 60 nm in diameter and also appeared to grow in the direction normal to the surface.

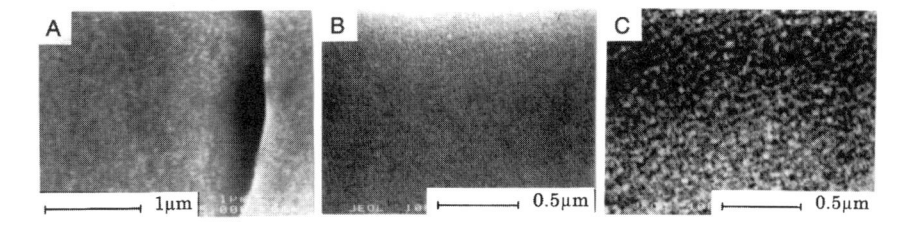

Figure 2. SEM photographs of SnO_2 films. (A) 90-nm-thick film on a glass substrate; (B) 9-nm-thick film on Si(100); (C) 30-nm-thick film on Si(100) (1 wt% DBTDA soln, 480°C).

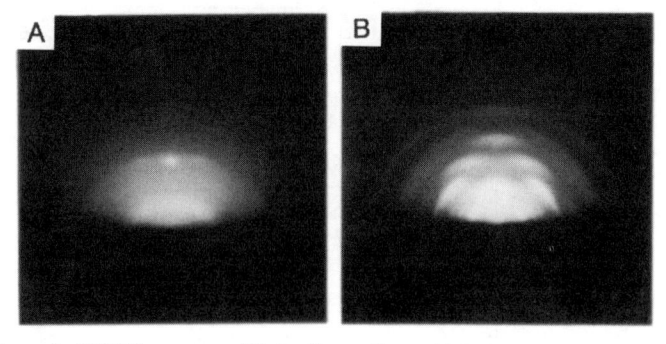

Figure 3. RHEED patterns of SnO_2 films of 9 nm thick (A) and 30 nm thick (B).

The diffraction spots of the (200) and (400) planes of SnO_2 were recognized in the corresponding RHEED pattern ((Fig. 3(B)), which suggests that the film had a [100] preferred orientation. Although the above results were obtained from films on Si substrates, SnO_2 films are believed to have grown on the Si substrates in the same manner as they would have grown on glass substrates, because the surface of the Si wafers was covered by an amorphous SiO_2 layer. The SnO_2 film, therefore, grows in the following manner (Fig. 4): (1) Nuclei form on the surface of a substrate (nucleation), (2) crystals with a preferred orientation and measuring up to 30 to 60 nm in diameter grow until most particles contact one another, and (3) further crystal growth occurs normal to the surface of the substrate.

(a) nucliation   (b) grain growth with preferred orientation   (c) oriented growth

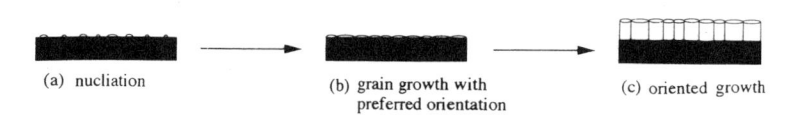

Figure 4. A growth model of oriented SnO_2 film.

Fujimoto et al.[10] have reported that the [100] preferentially oriented SnO_2 thin films were prepared by pyrohydrolytic decomposition onto a glass substrate along the direction of a flux flow. It has been suggested that the preferential nucleation occurred from an early stage in growth and that the shape of the nuclei initially formed on the substrate because of supersaturation of the flux influenced the initial stage of oriented growth. It has also been pointed out that when nuclei form without chemical influence from the substrate, oriented growth occurs from an early stage and results in a more highly oriented thin film. No reference, however, was made to the role of the source compound on oriented growth. Bruneaux et al.,[11] on the other hand, have reported that SnO_2 thin films prepared from concentrated $SnCl_4$ spray solutions showed a [100] preferred orientation whereas those prepared from diluted $SnCl_4$ solutions presented different structural characteristics. Those results indicate that oriented growth is affected not only by substrate temperature and supersaturation but also by the structure of the source compounds.

(3) Effect of Source Compound

Two kinds of oriented growth of SnO_2 thin films on glass substrates were realized by the spray pyrolysis technique, and the differences in the preferred orientations are attributed to the differences in the structures of the source compounds. Indeed, [100] and [110] preferentially oriented thin films were prepared from DBTDA and TBT, respectively. Possibly, then, the mechanism of oriented growth was deeply affected by the relationship between the structure of the source compounds and the atomic configuration of the growth surfaces.

Top views of the atomic configurations of the (110) and (200) planes of cassiterite are illustrated in Figs. 5(A)-(a) and 6(A)-(a), respectively. Two different tin atom sites exist in the (110) plane. One of those sites (designated Sn) is bonded with six oxygen atoms, which are divided into three groups: two oxygen atoms each belong to the second, the third, and the fourth layers. The other site (designated Sn*) is bonded with four of the six oxygen atoms in the third layer. The remaining bonding oxygen atoms belong one each to the second and fourth layers. The side views of the two different tin atom sites are illustrated in Figs. 5(A)-(b) and -(c), where the hatched second layers are for later discussion. In contrast, the (200) plane consists of only one kind of tin atom site (designated Sn). That site is bonded with three oxygen atoms in the second layer and with another three oxygen atoms in the fourth layer. The side view of that site is shown in Fig. 6(A)-(b).

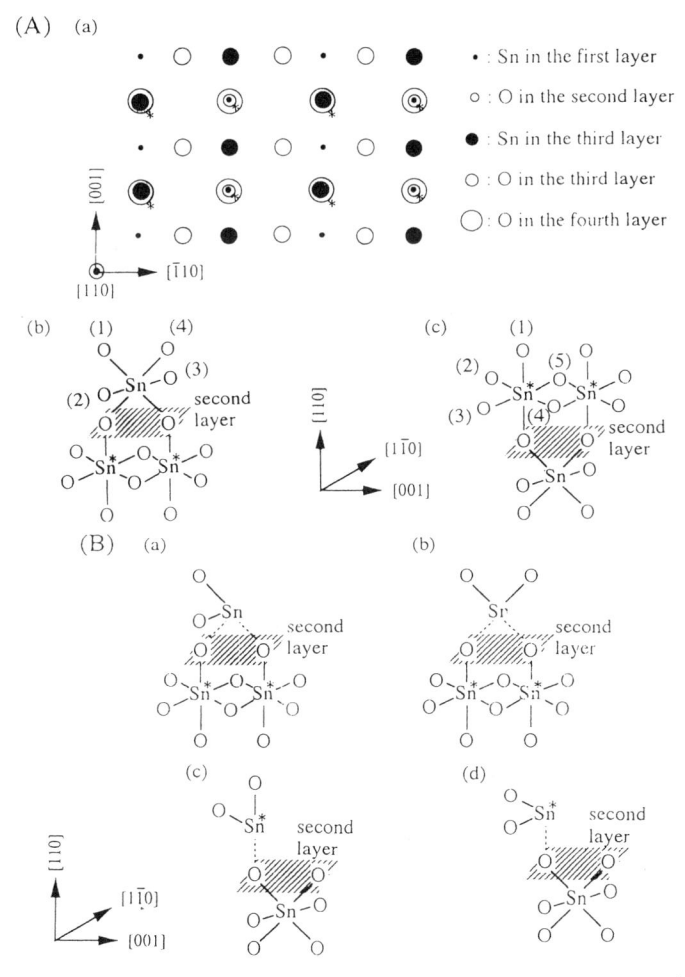

Figure 5. Atomic configuration of the (110) planes. (A)-(a) plane view, and -(b)cross-sectional views of Sn and -(c) Sn* sites; (B) [110] oriented growth model.

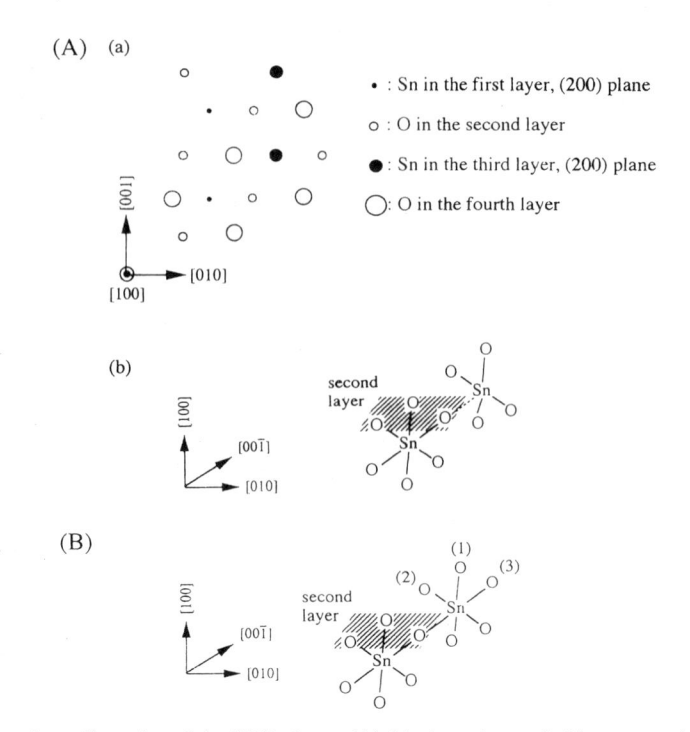

Figure 6. Atomic configuration of the (200) planes. (A)-(a) plane view and -(b) cross-sectional view of Sn site; (B) [100] oriented growth model

[100]-Oriented Growth. In the present study, DBTDA deposition led to [100] oriented growth as the substrate temperature increased and the film growth rate decreased. In a previous paper,[12] one of the authors investigated the saturated vapor pressure and pyrolytic properties of DBTDA with and without oxygen by means of vapor pressure measurement, mass spectroscopy, and gas chromatography. Pyrolysis of that compound was found to occur in two stages: the elimination of n-butyl groups in the temperature range between 280° and 310°C, and of acetoxy groups above 320°C. Moreover, tin, as well as SnO and SnO_2, were produced in a sample tube sealed under vacuum after pyrolysis of the DBTDA. In the presence of oxygen, however, only SnO and SnO_2 were produced. The pyrolysis of DBTDA with increasing temperature thus seems mainly to have produced a chemical species, O-Sn-O. Two different cases can be modeled for when the chemical species O-Sn-O approaches the growth front (the second layer in Figs. 5(A)-(b) and -(c)). In the first case, the tin atom of the O-Sn-O would occupy the Sn site in the third layer, the (110) plane (Fig. 5(A)-(b)). In that case, two oxygens of the O-Sn-O should occupy two oxygen sites among the (1)~(4). Occupation of the two oxygen sites of (1)-(2) (Fig. 5(B)-(a)), (1)-(3), and (2)-(3) is impossible because of the large electrostatic repulsive force among the oxygen atoms. Thus, only occupation of the two oxygen sites among (1)-(4) (Fig. 5(B)-(b)) is possible, since the electrostatic repulsive force there is small. In the second case, the tin atom of the O-Sn-O site would share the Sn* site in the third layer, the (110) plane (Fig. 5(A)-(c)). That case, however, is impossible because whenever two sites among (1)-(5) are occupied by two oxygens in the O-Sn-O, a large electrostatic repulsive force exists among the oxygen atoms (Fig. 5(B)-(c) and -(d)). A [110] oriented growth, therefore, is impossible from DBTDA because of mismatching with the Sn* site in the (110) plane.

On the other hand, when the growth front is approached in the third layer (the (200) plane in Fig. 6(A)-(b)), the occupation of two oxygen sites among the three oxygen sites of (1), (2), and (3) (Fig. 6(A)-(b)) is possible because of the small electrostatic repulsive force among the oxygen atoms (Fig. 6(B)). Moreover, two oxygens of the

O-Sn-O have three degrees of freedom when they occupy two oxygen sites among the three oxygen sites in the fourth layer, compared with only one degree of freedom when they occupy an Sn atom site in the third layer, the (110) plane (Fig. 5(B)-(b)). A [100] oriented growth, therefore, is possible from DBTDA because the atomic configuration of the (200) plane matches that of the O-Sn-O species.

[110]-Oriented Growth. TBT led to strong [110] oriented growth as the film growth rate decreased. It is possible that a chemical species of tin was produced by the pyrolysis of TBT.[13] According to the above discussion on [100] oriented growth, the chemical species of tin can occupy any sites, Sn and Sn* in the third layer, the (110) plane, and Sn in the third layer, the (200) plane, because no oxygen exists in the chemical species tin. A [110] oriented growth was more possible than [100] oriented growth, however, because the (110) plane was most closely packed and had the lowest potential energy of all the planes.

(4) Film Growth

In order to discuss the microstructure of the film deposited, we denote a θ_{hkl} as an angle between any (hkl) plane and the oriented plane of the constituent crystallite of the film, i.e., (200) for DBTDA and (110) for TBT in this study. When one (hkl) plane of the crystallite grown is parallel to the substrate, the axis of orientation of this crystallite is inclined by θ_{hkl} from the normal to the substrate. It is proved that TC_{hkl} is equal to the ratio of the volume of crystallites whose axis of orientation is inclined by θ_{hkl} to the normal of the substrate to the whole volume of the film. Figure 7 shows the relationship between TC_{hkl} and θ_{hkl} for SnO_2 films of different thicknesses. For small thickness the angle dependence of the volume ratio was weak, however, for large thickness the volume ratio became high only at the small angle range. The same tendency was also observed for the size of individual crystallite, D_{hkl}, whose (hkl) plane is parallel to the substrate. The results for TC_{hkl} suggest that the SnO_2 film is not oriented in the thin thickness range but oriented gradually toward the specific direction as the film thickness increased. Thus, the cross section of the oriented film is schematically presented in consideration of the result of D_{hkl} (Fig. 8).

Knuyt et al. have successfully explained the oriented growth of TiN film using a PVD method with the following principle; a growing film shows a tendency for having the lowest possible surface energy.[13] Their model could be applied to the present film growth after two modifications.[14] Firstly, the surface energy of the crystal is replaced by

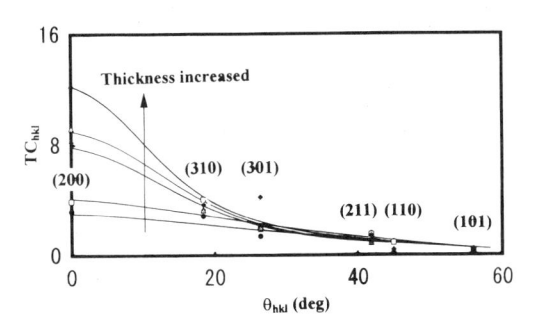

Figure 7. Texture coefficient (TC_{hkl}) as a function of the angle (θ_{hkl}) between oriented (200) and (hkl) planes.

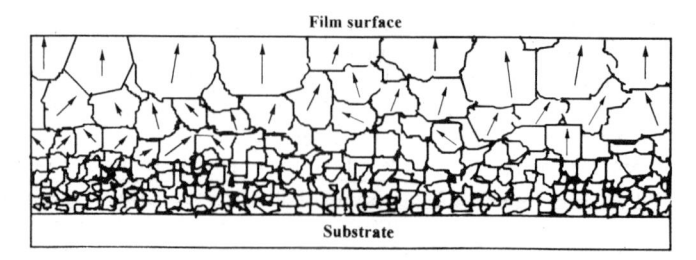

Figure 8. Schematic illustration of the cross section of oriented SnO_2 film.

electrostatic energy attributed to the interaction between thermally decomposed chemical species and surface atoms of the growing film, since the preferred orientation depends on the source compound. Secondly, it is needed to introduce consideration of the nucleation in the growing film in addition to on the substrate, because TC tends to saturate with increasing film thickness. If the nucleation occur only at the substrate surface according to the Knuyt's model, TC should increase infinitely with the film thickness.

REFERENCES

1. Z. M. Jarzebski and J. P. Marton, Physical properties of SnO_2 materials, *J. Electrochem. Soc.*, 123:199C, 299C, 333C (1976).
2. D. S. Albin and S. H. Risbud, Spray pyrolysis processing of optoelectronic materials, *Adv. Ceram. Mater.*, 2:243 (1987).
3. J. S. Ryu, Y. Watanabe, and M. Takata, Correlation between gas sensing properties and preferential orientations of sputtered tin oxide films, *J. Ceram. Soc. Jpn.*, 100:1165 (1992).
4. C. Agashe, M. G. Takwale, B. R. Marathe, and V. G. Bhide, Structural properties of SnO_2:F films deposited by spray pyrolysis, *Sol.Energy Mater.*, 17:99 (1988).
5. I. Yagi, Y. Hagiwara, K. Murakami, and S. Kaneko, Growth of highly oriented SnO_2 thin films on glass substrate from tetra-n-butyltin by the spray pyrolysis technique, *J. Mater. Res.*, 8:1481 (1993).
6. I. Yagi, K. Kakizawa, K. Murakami, and S. Kaneko, Preferred orientation of SnO_2 thin films growth from tri-n-butyltin acetate by spray pyrolysis technique, *J. Ceram. Soc. Jpn.*, 102:296 (1994).
7. I. Yagi and S. Kaneko, Growth of oriented tin oxide films from an organotin compound by spray pyrolysis, *Mat. Res. Soc. Symp. Proc.*, 271:407 (1992).
8. C. Barret and T. B. Massalshi, *Structure of Metals*, Pergamon, Oxford (1980).
9. I. Yagi, *Ph.D. Thesis*, Shizuoka University (1994).
10. M. Fujimoto, T. Urano, S. Murai, and Y. Nishi, Microstructure and x-ray study of preferentially oriented SnO_2 thin films prepared by pyrohydrolytic decomposition, *Jpn. J. Appl. Phys.*, 28:2587 (1989).
11. J. Bruneaux, H. Cachet, M. Froment, and A. Messad, Correlation between structure and electrical properties of sprayed tin oxide films with and without fluorine doping, *Thin Solid Films*, 197:129 (1991).
12. I. Yagi, E. Ikeda, and Y. Kuniya, Pyrolytic properties of di-n-butyltin(IV) diacetate as a precursor for sprayed SnO_2 thin films, *J. Mater. Res.*, 9:663 (1994).
13. G. Knuyt, C. Quaeyhaegens, J. D'Haen, and L.M. Stals; A quantitative model for the evolution from random orientation to a unique texture in PVD thin film growth, *Thin Solid Films*, 258:159 (1995).
14. T. Kosugi, N. Kamiya, I. Yagi, K. Murakami, and S. Kaneko, Spray pyrolysis of organotin compounds for the growth of oriented SnO_2 thin films on glass substrates, *Trans. Mater. Res. Soc. Jpn.*, 20:530 (1996).

THE ENERGY BARRIER AT THE INTERFACE OF ZnO CONTACT

Satoru Fujitsu, Koji Nakanishi, Shigeo Hidaka and Yoshio Awakura

Department of Materials Science and Ceramic Technology,
Shonan Institute of Technology,
1-1-25 Tsujido-Nishikaigan, Fujisawa, Kanagawa, 251 Japan

ABSTRACT

Large transparent specimens of polycrystalline zinc oxide with c-axis orientation were prepared by the vapor transport method. The obtained specimens were of n-type semiconductor with resistivity below 50 Ωcm^{-1} at room temperature. When mirror-finished samples obtained by mechanical polishing were contacted under mechanical pressure, the reversible large non-ohmic I-V behavior similar to that of a ZnO varistor was observed. This non-ohmic behavior disappeared as a result of heating the sample above 400°C or Ar ion etching. The results of XPS measurement suggested the presence of adsorbed OH$^-$ on the as-polished surface, which formed the Schottky-type energy barrier with a trapped conduction electron.

INTRODUCTION

Zinc oxide is a well-known n-type semiconducting oxide with an extra zinc ion at the interstitial site and/or oxygen vacancy as a donor.[1] Since the electronic mobility of ZnO is high ($\approx$100 cm^2/Vs) and its carrier concentration can be easily controlled by doping the aliovalent ion, we can obtain the sample with various resistivity from metallic to insulator (10^{-3}–10^{12} Ωcm at room temperature). The semiconducting properties of ZnO have been widely applied to various devices, including varistors, sensors for reducing gas, and transparent electrodes for solar cells.[2] In varistors and gas sensors, the origin of their characteristics is the Schottky-type energy barrier at the interface region. Usually, the energy barrier on the surface of n-type semiconductors such as ZnO is formed by the chemisorption of negative ions such as O^{2-}, O^-, O_2^-, OH^- or CO_3^{2-}.[3] When an atom or molecule of this kind chemisorbs on the surface of an n-type semiconducting oxide such as ZnO, it traps a conduction electron from the bulk to acquire a negatively charged ionic state. To compensate these negative charges on the surface, the electron depletion layer is generated with the energy barrier as shown in Figure 1.

Though the thickness of this depletion layer (λ) is usually very thin, the total electrical properties of the porous materials with narrow-neck connecting grains or thin films are controlled by this surface condition. The gas sensor is a typical example of this effect.

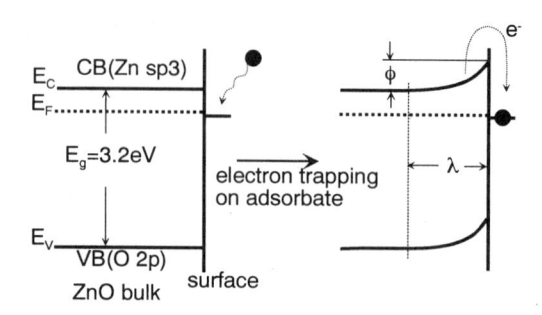

Figure 1. The formation of the surface energy barrier by chemisorption.

Varistor is a polycrystalline material with non-ohmic I-V behavior due to the Schottky-type energy barrier at the grain boundary. Since the non-ohmic characteristic of varistor depends strongly on the oxygen atmosphere during its heating process, it is obvious that oxygen has an important role to play in the formation of the energy barrier.[4,5] In previous work, we have examined the analogy between the formation of the energy barrier on the surface of porous ZnO and that in the grain boundary of the varistor,[6] and measured the diffusion of oxygen into the varistor in considerable detail.[7] Based on these results, it was pointed out that the excess oxygen at the grain boundary of the varistor should trap the electron from the ZnO grain and then generate the barrier.

The preparation of the single boundary is a useful technique to evaluate the energy barrier of the grain boundary. While the preparation of a large single crystal of ZnO is very difficult, we prepared large (1 cm cube) transparent specimens of zinc oxide with c-axis orientation by the vapor transport method.[8] In this sample, the a-axis oriented randomly, and the microstructure composed of pore-free columnar crystal (approximately 0.1 µm in diameter) growing normal to the substrate was observed by scanning electron microscopy (SEM). Such a microstructure is frequently observed in ZnO thin film prepared by the sputtering method. The optical transmittance of this sample with 2-mm thickness was >80% at 600 nm. X-ray diffraction patterns showed strong reflections only from the (*00l*) peaks with their c-axes aligned normally to the substrate. The use of these transparent ZnO specimens is thus a very convenient way to evaluate the bulk and interface properties of ZnO.

By applying this transparent material, we observed small non-ohmic behavior (α=2 at $-100°C$) in the junction system, where α is the non-ohmic exponent expressed as $I=(V/C)^\alpha$,[9] and large non-ohmic behavior (α=20 at 25°C) in the contacting system.[10] The key to developing a large non-ohmic property was mirror-finishing the contacting surface. Here, the mechanism of this large non-ohmic behavior, especially surface condition, will be discussed with respect to electrical measurements and surface characterization using x-ray photoelectron spectroscopy (XPS).

EXPERIMENTAL PROCEDURE

Sample Preparation

The apparatus used to obtain the transparent ZnO is shown in Figure 2(a). Sintered rodlike ceramics of ZnO were placed as a vapor source in the hottest zone (1200°C) of the alumina tube. The central area of this furnace was heated with a SiC heater to produce a temperature gradient extending from the hot zone to both sides of the tube. The maximum temperature gradient was 100 to 200°C/cm. A polished flat surface of ZnO ceramic was used as a substrate and placed in the lower temperature zone (≈800°C). This substrate was connected to the alumina tube by ceramic cement and was rotated at 0.3 rpm. Nitrogen gas with 6% hydrogen was used to maintain a reducing atmosphere in the tube. The growth rate using this condition was 0.3 mm/h.

The obtained crystal was cut into a rectangular bar (5 mm by 5 mm by 3 mm) and mirror-finished by a polishing machine (#8000 diamond wheel) with aqueous coolant, by hand with diamond paste (0.25 µm) in normal atmosphere (>60%RH), or by hand with diamond paste in dry atmosphere (<30%RH). The polished sample was cleaned in acetone using the ultrasonic cleaner for 5 minutes. The optical transmittance and x-ray diffraction were measured by the method reported in Noritake, et al.[8]

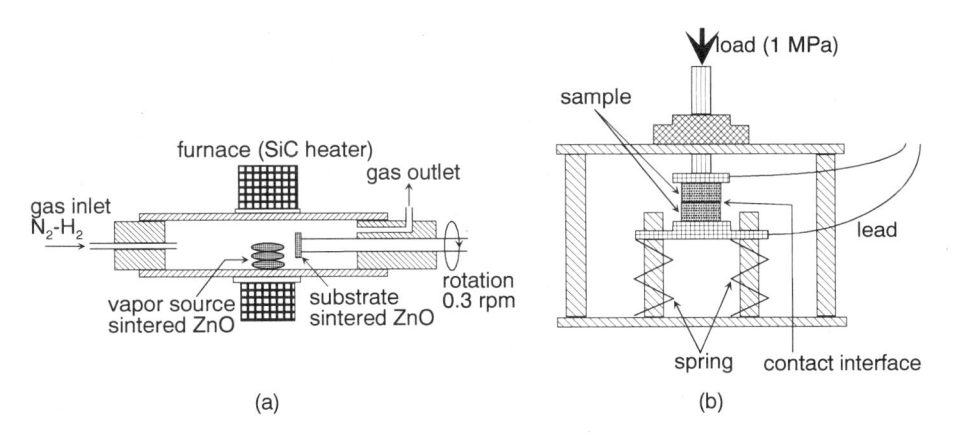

Figure 2. The schematic of the furnace used to make transparent ZnO (a) and I-V measurement assembly (b).

I-V Measurement

Two samples were contacted under the mechanical pressure of 1 MPa as shown in Figure 2(b). In-Ga liquid alloy was painted on the back side of the crystal faces as electrodes. The faces parallel and perpendicular to the c-axis of the crystal are the A and C plane, respectively. The contacts of the A faces will be called A-A . The data shown in this paper are the result of the measurement for the A-A contact. The I-V relation was measured by the Curve Tracer (Kikusui 5802) with 50 Hz at 25°C. The following treatments were applied to the sample after mirror finishing in order to change the surface condition:

- Heating at 100 to 500°C for 1 hour in air.
- Etching by Ar ion for 5 seconds to 1 hour in the XPS chamber.
- Roughening a surface using SiC paper of #400-1200.

The non-ohmic exponent, α, was estimated by the following equation:

$$\alpha = \frac{\log(0.4) - \log(0.1)}{\log(V_{0.4}) - \log(V_{0.1})} \qquad (1)$$

where $V_{0.4}$ and $V_{0.1}$ are applied voltage with 0.4 and 0.1 mA currents, respectively.

Capacitance under constant dc bias was measured to check the interface barrier using the impedance analyzer (HP4192A) at 25°C.

Characterization of Surface

The chemical state of the surface was characterized using XPS (Rigaku XPS-7000) with AlKα accelerated at 14 kV. The measured binding energy was corrected by using evaporated Au.

RESULTS

The transparent ZnO prepared in a strong reducing atmosphere turned a dark brown color and measured a low resistivity (<10 Ωcm). Once this sample was annealed at 800°C for >20 hours in air, its color changed to light yellow and the resistivity increased to 50 Ωcm. The I-V behavior was ohmic as shown in Figure 3.

When two of the samples with mirror-finished faces were contacted, a reversible large non-ohmic behavior similar to that of a varistor was observed. Though the sintered material contact system also showed non-ohmic behavior, its non-ohmic exponent was usually lower than that of a transparent material. The sintered material contact system had very unstable measurement, poor data reproducibility, and an occasional, abrupt, and irreversible dielectric failure. In the transparent ZnO system, dielectric failure occurred as a result of high current flow (>3 mA/cm^2).

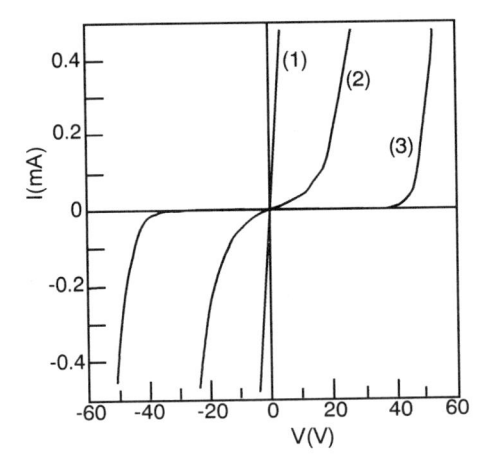

Figure 3. The I-V behavior for (1) ZnO with no contact showing ohmic property, (2) the contact system composed of the sintered ZnO (α=3, $V_{0.1}$=17 V) and (3) the contact system composed of the transparent ZnO (α=17, $V_{0.1}$=47 V). The reproducibility of the result and the stability of the measurement were very poor for the system composed of sintered material.

Figure 4 shows how roughening changes the α and $V_{0.1}$ of the contact system. The roughening decreased both α and $V_{0.1}$. Usually, α and failure voltage of the varistor depend on the grain boundary barrier height. The sample showing the small α has a small failure voltage. Normal varistor theory does not explain this result. Furthermore, the failure voltage

($V_{0.1}$, in the present case) of this system (>20 V) is considerably higher than that of the normal varistor (3 V per one grain boundary). Thus, the effect of the microstructure at the contacting area should be considered. From the result for the single grain boundary type of varistor,[11] the failure voltage depends on the Bi_2O_3 insulator layer thickness. In the present case, spreading the air gap and changing the contacting type from face-face to point-point by roughening should increase the $V_{0.1}$. Decrease in α should result from the decreased barrier height at the surface caused by scraping off the adsorbate.

The values (α and $V_{0.1}$) decreased by heating the sample. Figure 5(a) shows the α and $V_{0.1}$ for the A-A contact system composed of the samples kept in air for 1 hour at the temperature marked on the abscissa. The α decreased with increasing temperature and approached ≈1 by heating above 400°C. Since the $V_{0.1}$ is 0.01 V in the case of the bulk ZnO without contacting interface, the contacting resistance still remained.

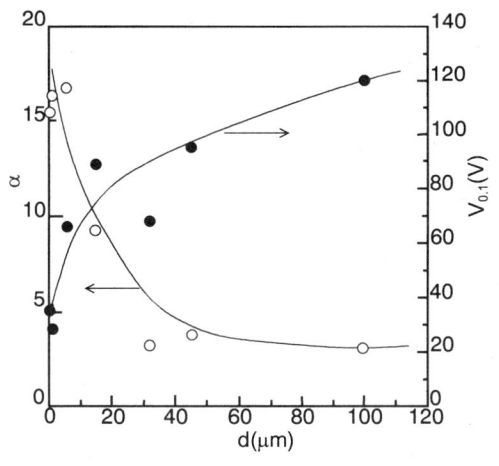

Figure 4. The α (O) and the $V_{0.1}$ (●) for contact systems with surface finished by various grinding particles with diameter, d. The mirror finishing condition was obtained for the samples polished by under 6 μm.

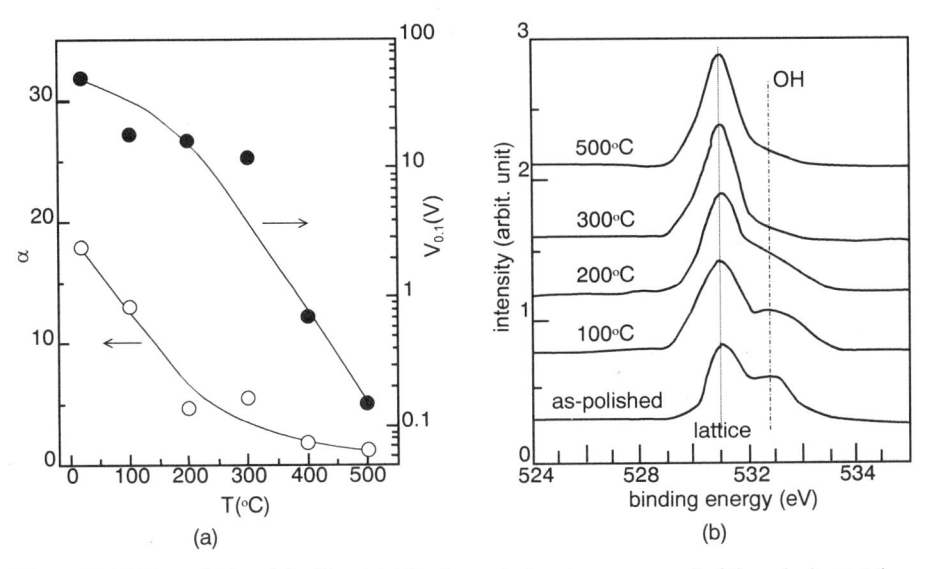

Figure 5. (a) The α (O) and the $V_{0.1}$ (●) for the contact systems composed of the pairs kept at the temperature marked on the abscissa in air for 1 hour. (b) The XPS spectra of O_{1s} for the surface of the measured samples in (a).

DISCUSSION

Formation of the Energy Barrier

To explain the non-ohmic behavior in this contact system, two types of surface conditions should be considered: (1) the presence of the insulating layer composed of the secondary phase, other than ZnO, and (2) the formation of the Schottky-type energy barrier with electron trapping by adsorbates or surface defects. The analysis by SEM and EPMA showed no phase other than ZnO. The multi-reflection spectrum of FT-IR showed the presence of adsorbed CO_2 and H_2O. From these observations, it could be deduced that the insulating layer based on (1) is not present on the polished surface. The capacitance decreased as the bias increased, suggesting the existence of a Schottky-type energy barrier.

The XPS spectrum for O_{1s} provided useful information for understanding the nature of the surface state. Figure 5(b) shows the XPS spectra for O_{1s} of the samples heated in air for 1 hour. From the as-polished sample, two peaks were observed at 531.2 and 532.8 eV. The peak at 531.2 eV was observed in all samples, while that at 532.8 eV disappeared when heated. The correlation between the presence of this peak and the non-ohmic behavior shown in Fig. 5(b) suggests that the peak at 532.8 eV is assigned to the adsorbed oxide on the surface, which is the key to forming the surface energy barrier. Gaggiotti et al.[12] reported that the XPS peaks for O_{1s} on SnO_2 at 530.5, 531.8 and 533.1 eV are assigned to the lattice oxygen of SnO_2, hydroxyl groups (OH^-), and adsorbed water (H_2O). Based on their report, the observed peaks in this work for ZnO at 531.2 and 532.8 eV should be assigned to the lattice oxygen of ZnO and hydroxyl groups, respectively. The chemisorption of this negative ion might form a surface energy barrier by trapping the electron from the ZnO bulk.

The effect of the polishing atmosphere on I-V behavior indicated the importance of water in forming an energy barrier. The large non-ohmic exponent was obtained in the contacting system composed of the samples finished by a polishing machine using a #8000 diamond wheel with aqueous coolant ($\alpha=17$), while the system composed of the samples polished in the dry atmosphere showed smaller non-ohmic exponent ($\alpha=5$).

Adsorption of OH^-

The possibility of forming a zinc hydroxide layer on the surface should be considered. The XPS peak assigned to OH^- disappeared easily after a short period of Ar ion etching (>5 seconds), and the etched sample shows the XPS spectrum as illustrated in Figure 6. It appears that the observed OH^- is an adsorbate and the thick hydroxide layer is not formed.

The adsorption of OH^- is an irreversible process. Readsorption does not take place after the sample is heated. The non-ohmic behavior lost by heating is not recovered by storing the specimens in air. One possible explanation is the effect of the inhomogeneity developed by polishing. It is well known that as-polished surfaces have such microscale inhomogeneities as steps, edges, kinks, and voids, even if they look like flat surfaces.[13] These inhomogeneities supply a stable point for chemisorption. During the polishing procedure, the active dangling bond prepared by scraping should be stabilized immediately by bonding or covered by chemisorbate. The chemisorbate (hydroxyl group in this case) bonding at steps, edges, kinks, and/or voids is stable on the surface. Such a surface is homogenized by a surface diffusion process at high temperature, causing the chemisorbate to lose its stable position. The XPS measurement using Ar ion etching in the instrument supported this model.

As mentioned above, the peak resulting from the OH^- at 532.8 eV, shown in Figure 6 (lines 2 and 3) disappeared after a brief period of etching (>5 seconds). When the sample was kept in air after etching for a short period (<20 seconds), its contacting system showed non-ohmic behavior and its α increased gradually during the storage in air. The peak at 532.8 eV was detected again after this storage (see Figure 6, line 5). Such a recovery for the non-ohmic behavior and the XPS peak was not observed for the sample etched for a long period (>1 minute) (see Figure 6, line 4). This result suggests that the OH⁻ disappears after a short period of etching, but the inhomogeneity still remains on the surface. The inhomogeneity disappears after a long period of etching.

Comparing the peak intensity at 532.8 eV for the A and C plane, the amount of adsorbed OH⁻ on the C plane is larger than that on the A plane. Since the C plane is a polar plane, a greater amount of OH⁻ should adsorb on the C plane rather than on the non-polar A plane.

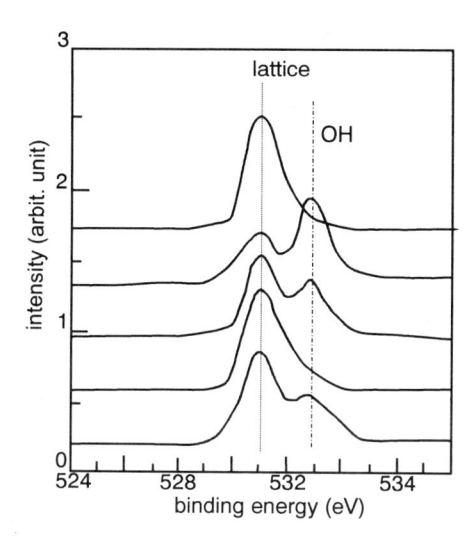

Figure 6. XPS spectra for the samples for Ar ion etched (1 minute) A plane; (line 1) as polished C plane (line 2), and A plane (line 3), kept in air for 1 hour at 25°C after Ar etched for 1 minute (line 4) and 10 seconds (line 5). Lines are in sequence top to bottom (line 1 is at top).

CONCLUSIONS

The contacting system composed of the mirror-finished transparent ZnO showed large non-ohmic I-V behavior resulting from the Schottky-type energy barrier at the contacting interface. This barrier is formed by OH⁻ chemisorbed on the surface of the sample during the polishing process.

ACKNOWLEDGMENTS

The authors wish to thank Mr. Akira Masuda of Shonan Institute of Technology for his assistance with the experimental studies.

This work was partially supported by a Grant-in-Aid for Scientific Research (c) from the Japanese Ministry of Education, Science, Sports and Culture (08650996)

REFERENCES

1. R. Littbarski, Carrier concentration and mobility, *Current Topics Materials Science*, Vol. 7, E. Kaldis ed., North Holland, Amsterdam (1981).
2. For example, H. Yanagida and M. Takata, Densizairyo seramikkusu, Gihodo, Tokyo (1988).
3. P. Bonasewicz, R. Lttbarski and M. Grunze, Adsorption phenomena, *Current Topics Materials Science*, Vol. 7, E. Kaldis ed., North Holland, Amsterdam (1981).
4. S. Fujitsu, H. Toyoda, K. Koumoto, H. Yanagida, M. Chikazawa and T. Kanazawa, Simultaneous measurement of electrical conductivity and the amount of adsorbed oxygen in porous ZnO, *Bull. Chem. Soc. Japan.*, **61**:1979, (1988).
5. E. Sonder, M. M. Austin and D. L. Kinser, Effect of oxidizing and reducing atmospheres at elevated temperatures on the electrical properties of zinc oxide varistors, *J. Appl. Phys.*, **54**:3566 (1983).
6. S. Fujitsu, Adsorbed oxygen and electronic materials, *Electronic Ceramics*, **91**:17 (1991) and S. Fujitsu, H. Toyoda and H. Yanagida, Origin of ZnO varistor, *J. Am. Ceram. Soc.*, **70**:C71 (1987).
7. S. Fujitsu, H. Toyoda and H. Yanagida, The enhanced diffusion of oxygen in ZnO varistor," *J. Ceram. Soc. Japan*, **96**:119 (1988).
8. F. Noritake, N. Yamamoto, Y. Horiguchi, S. Fujitsu, K. Koumoto and H. Yanagida, Vapor phase growth of transparent zinc oxide with high crystal-axis orientation, *J. Am. Ceram. Soc.*, **74**:232 (1991).
9. H. Ito, M. Kajita, S. Fujitsu and H. Yanagida, The measurement of electrical conduction of crystal axis oriented zinc oxide, *J. Ceram. Soc. Jpn.*, **100**:823 (1992).
10. S. Fujitsu, S. Hidaka, K. Kawamura and Y. Awakura, Formation of energy barrier by contact of ZnO, *Proc. Electroceramics IV*, Aschen **1**:591 (1994).
11. M. Ieda, Y. Suzuoki, N. Nakagawa and T. Mizutani, Electrical properties of ZnO/Bi_2O_3 two-layer composites, *IEEE Transactions on Electrical Insulation*, **25**:599 (1990).
12. G. Gaggiotti, A. Galdikas, S. Kaciulis, G. Mattogno and A. Setkus, *J. Appl. Phys.*, **78**:4467 (1994).
13. T. Osaka, Hyomen Koshi Kekkan, *Hyomen Kagaku no Kiso to Oyo*, Nihon Hyomen Kagakukai ed., NTS, Tokyo (1991).

FIBER COATING DESIGN PARAMETERS FOR CERAMIC COMPOSITES AS IMPLIED BY CONSIDERATIONS OF DEBOND CRACK ROUGHNESS

Ronald J. Kerans[1] and Triplicane A. Parthasarathy[2]

[1]Wright Laboratory Materials Directorate, Wright-Patterson AFB, OH 45433-7817
[2]UES Inc., 4401 Dayton-Xenia Rd, Dayton, OH 45432

INTRODUCTION

The key technological challenge to the use of ceramic composites in high temperature structural applications is oxidation resistant control of fiber/matrix interface properties. A number of approaches based on oxide coatings are being investigated (for a brief review and current references, see Refs. 1 and 2, respectively) and, while the principle role of these coatings is to promote debonding, the actual material and geometric requirements they must satisfy have been poorly understood. As the level of understanding of composite behavior has been increasing, so has appreciation for the potential complexities of fiber coating design. This has led to recognition of several potential design parameters; characteristics that must be controlled in the design of the coating or accommodated in the design of the overall composite. A good deal of our awareness of these coating design parameters stems from the study of the role and importance of surface topography.

The objective of this work was to summarize current understanding of the design requirements for fiber coatings, as derived from roughness related work, and to identify outstanding issues. It is based heavily upon references 3, 4 and 5.

ROUGHNESS

The potential of significant effects due to surface roughness was first discussed in the context of increased radial stresses due to the geometric misfits. Contradictions in measured sliding friction led to experimental confirmation of roughness effects with the "push back seating drop" of Jero, et al.[7,8] For a brief review of subsequent work, see Ref. 9.

Briefly, a fiber pulling out of a matrix under axial loading is subjected to radial stresses which differ along its length. The interfacial normal stress in the un-slipped region ahead of the crack tip (Region I of Fig. 1) is determined by the residual stresses, and to a minor degree, by differences in Poisson's ratios and the applied axial stress. In Region III, the normal stress is the sum of the residual stresses and the stresses resulting from the full effect of the topographical misfit. The roughness-induced stresses may be larger than the thermal stresses. Region II extends with increasing misfit from the crack tip to the beginning of

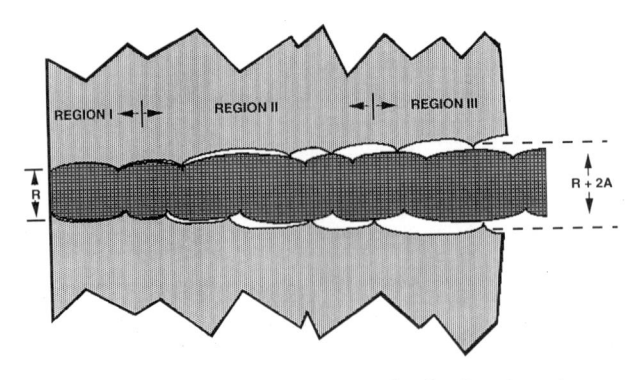

Figure 1. Schematic of misfit strain created by a rough fiber of radius R and roughness amplitude A sliding in a matching matrix hole. In practice, roughness is assumed to be asymmetric and small compared to fiber dimensions. (after [2])

Region III. This region both complicates analysis and gives rise to a number of interesting effects. A full solution of the problem has been accomplished for one simple form of roughness.[9] The works discussed in the following two sections derive from a simple consideration of the total misfit, i.e., Region III, and anticipated effects upon friction.[10] In fact, consideration of the full problem[4,9] has shown that the friction can actually be highest in Region II and that the effects can be very much larger than the Region III misfit alone, as evident in the plot of average interfacial friction versus debond length of Fig. 2. The properties used for these calculations were those of NICALON™/SiC, but no consideration was given to the elastic effects of the coating and the roughness parameters are speculative; hence, the plots explore possible effects of roughness and may not correspond well with experiment.

COMPLIANCE

Compliance in Conventional Coatings

Recognition of the potential for substantial effects of coatings upon interfacial stresses due to roughness misfit motivated consideration of evidence for such effects in the behavior of conventional composites.[10] One such possibility was the well known dependence of composite properties on coating thickness. Most ceramic composites that demonstrate good damage tolerance have at least one layer of C or BN between the fiber and matrix and it is widely held that the thickness of the layers are important parameters. This is based on abundant anecdotal evidence as well as more controlled studies.[11,12]

The implications of roughness related coating thickness/compliance effects were evaluated by estimating the maximum probable roughness of NICALON™ fibers, and calculating: 1. the effects on stress state after sliding of the fiber, 2. the implied maximum debond length and 3. the probable pullout length, as functions of applied coating thickness. Roughness was calculated to increase the compressive radial stress in a hypothetical uncoated-fiber composite from about 150 MPa before sliding to greater than 450 MPa after sliding. The roughness related stresses were calculated to be significantly alleviated by the coating with 0.5 mm of C reducing the stress by about one third. Calculated debond and pullout lengths are shown in Fig. 3 with coating thicknesses indicated. The exact relationship between stresses and thickness is dependent upon a number of assumptions and may not correlate precisely with experiment. Nevertheless, it is clear that in certain regions small changes in coating thickness can have a dramatic effect on debond length, and hence, on composite properties.

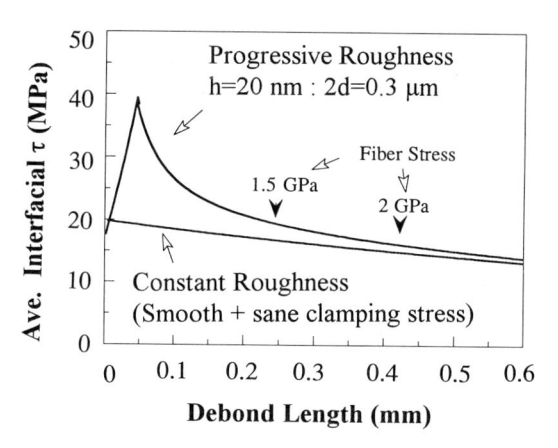

Figure 2. The progressive roughness model predicts that the average τ can be higher or lower than predicted by a constant roughness model depending on the roughness parameters amplitude = h and period = 2d. (after [4])

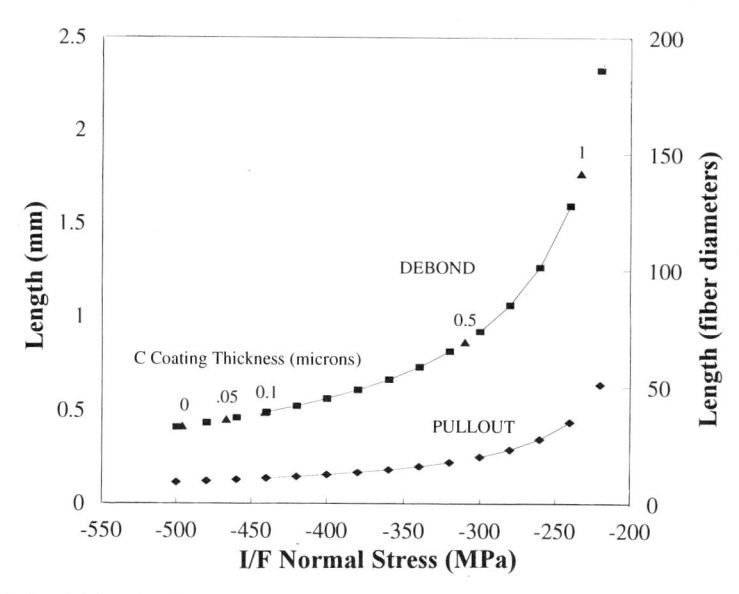

Figure 3. Calculated debond and pullout lengths vs. interfacial normal stress. C coating thicknesses indicated correspond to stresses at complete misfit. (after [3])

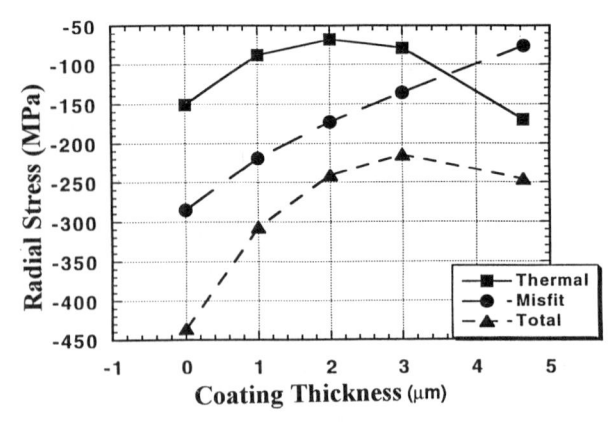

Figure 4. Maximum radial stresses in the coating vs. oxide coating thickness for Nicalon R = 8 mm and roughness amplitude A = 24 nm, f = 0.4, E_c = 70 GPa, σ_c = 9 ppm/°C, ΔT = -1000°C. (after [2])

Compliance and Design of Advanced Coatings

If compliance is a significant parameter for current coatings, it follows that it will be for advanced oxide-based coatings as well. However, in general, oxides are much less compliant materials than C and BN, which raises the question of the feasibility of oxide substitutes. This question led to an investigation of the viability of oxide coatings for accommodating misfit stresses analogous to the calculations summarized in the previous section.

The elastic moduli of crystalline oxides range from about 8 GPa for $MgO \cdot 2TiO_2$ to around 400 GPa for Al_2O_3.[13] A value of 70 GPa was somewhat arbitrarily taken as a value practically achievable with porous layers of one of several oxides, thus allowing some choice of chemistry and thermal expansion. An elastic modulus of 70 GPa would correspond, to, e.g., 74% dense yttria, 66% dense zirconia or 29% dense alumina. Sensitivities to coating elastic modulus and other parameters were also evaluated.

The radial misfit, thermal and total stresses as functions of coating thickness for a roughness amplitude of 24 nm on a fiber of 16 μm diameter are presented in Fig. 4. These plots are for fixed fiber volume fraction f=0.4 and coating ranging from no oxide coating to complete oxide matrix -- no SiC. The maximum in stress with increasing thickness is a result of the matrix transition from SiC with CTE = 4.5×10^{-6}/K to oxide with CTE = 9×10^{-6}/K. While oxides must be thicker than C, comparable effects can be achieved with what appear to be reasonable thicknesses of oxide coating, i.e., 2 μm or less.

Coatings that are such a large fraction of the composite as may be required for interfacial stress control can be expected to have a significant effect upon the properties of the composite. For example, a 2 μm thick 70 GPa coating imposes about 20% decrease in axial elastic modulus of the composite as compared to the no-coating case. While such a debit is in no way prohibitive in a general sense, it certainly merits consideration in composite or component design.

COATING DESIGN FOR CONTROL OF ROUGHNESS

Simple accommodation of interfacial roughness may not be sufficient in some circumstances and active control of the fracture path may be required. The following is based on a brief discussion of this possibility in reference 2. As an example of the role of the interfacial fracture path and the severity of the consequences, it is instructive to consider the work of Cinibulk and Hays[14,15,16] on magnetoplumbite coatings. There is good evidence that the coating fractures preferentially in a series of planar cleavage cracks joined by small "steps" ranging up to about 1/2 μm in height. Nevertheless, fibers cannot be pushed out in fiber pushout tests. The probable cause of the high resistance to pushout is the 1/2 μm steps in the coating fracture surface. For reasonably probable values of the modulus, the misfit stress resulting from such 1/2 μm steps is over 800 MPa. In this case of a 135 μm diameter fiber, a 2 μm coating is a very small fraction of the composite, and much thicker coatings could be considered from an elastic standpoint. However, there are at least four probable problems with significantly thicker coatings: the CTEs of coating and fiber must match very closely to avoid spalling; the fiber bend radius at which the coating fractures will increase with associated penalties in handling and processing constraints; there are direct coating processing and cost issues resulting from the probable need for multiple layer;[17] and finally, thicker coating would allow even greater meandering of the crack. Consequently, it will be necessary to control the fracture path in some way. The most obvious possibility is to make the weak coating very thin in order to confine the crack path.

The results of this study indicate that while oxide coatings must be significantly thicker than current C and BN coatings, similar control of misfit stresses is feasible. Significant but not prohibitive changes in composite properties can be expected. It also points up that there will be competing design requirements on such coatings.

COATING TOUGHNESS REQUIREMENTS

Experience with successful composites and theoretical treatments have led to the prevalent assumption that good composite behavior requires that the debond crack toughness and friction should both be quite low (see, for example, Cao et al.[18]). Reports of substantially improved properties of N/SiC composites resulting from the sole processing change of pretreating the surface of the fibers[19,20] have motivated reexamination of these assumptions regarding the upper limits of allowable values.[21,22,23] The behavior of the materials is not fundamentally different, but is sufficiently outside the assumed envelope of workable properties that they suggest new possibilities for interface coating properties. Recent preliminary work using rough interface formalism to analyze pushout tests on these composites[5] may provide a revision of the perceived requirements and is summarized in this section.

Both systems consist of pyrolytic carbon coated ceramic grade NICALON™ fibers in a matrix of CVI SiC. In one case, the fibers were subjected to a proprietary treatment to reduce oxygen levels at the surface of the fibers.[24] The composites made with treated fibers demonstrate 30% higher strength at the same strain-to-failure, much finer matrix crack spacing and significantly different stress-strain behavior.[19]

Interface property measurement has been difficult to rationalize by conventional means. Pushout tests have shown curves that differ somewhat from those expected from commonly used models and observed for conventional composites.[25,26] A combination of nanoindentor push-in tests and conventional pushout tests revealed that an initial crack grows with very little deflection with increasing load to what appears at first inspection to be the initial deviation from linearity, then quickly jumps to a greater length and grows with a load deflection trace that is concave upward for most of its length.

In this analysis, each of the two stages of cracking were described by the rough interface formalism with different roughness parameters describing the two types of cracks. The same fracture energy, G, was assumed to govern both cracks. The roughness parameters were adjusted to generate curves which fit the respective portions of the experimental curve. The elastic effects of the coating were introduced by adjusting the radial elastic modulus of the matrix to produce the correct interfacial normal stress.

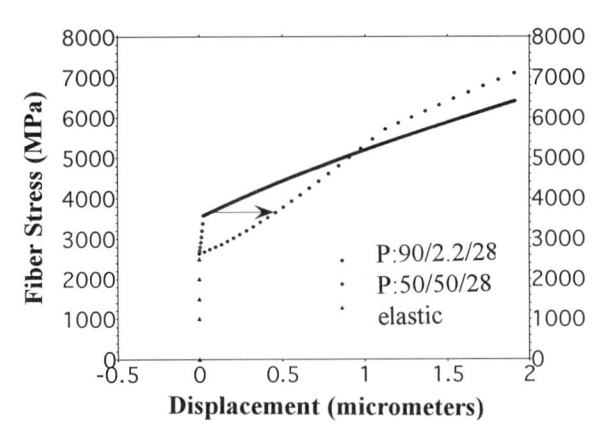

Figure 5. Pushout curves according to the rough interface model [30] with parameters chosen to fit a typical experimental curve assuming two modes of cracking. Transition between the two is indicated by the arrowed line. (after [4])

There is an inherent problem with the experimental data in that the displacement during the growth of the initial crack is comparable to the sensitivity of corrected pushout data. Therefore, the procedure used here should not be expected to yield precise parameters.

The pushout curves for the two fitted cracks are shown in Fig. 5 and the transition between the two is indicated by the arrowed line. The relevant parameters for the initial crack are roughness amplitude h = 50 nm, period d = 50 nm and G = 28 J/m^2. The parameters for the secondary crack are h = 90 nm, d = 2.2 µm, G = 28 J/m^2.

Predicted curves of bridging-fiber stress versus displacement for composite tension tests (Type II boundary conditions[27]) assuming the same crack characteristics are shown in Fig. 6. If transition to the secondary crack is taken (somewhat arbitrarily) to occur at the same crack progression -- slightly after the Mode II to III transition -- it will occur at about 2.2 GPa in the tension test, as indicated in Fig. 6. However, fiber strengths in composite tension tests have been found to typically be 1.8 to 2 GPa; hence composite failure can be expected to occur before transition to the secondary crack. Tensile specimens then should display only (or mostly) primary, short period cracking, which is, in fact, as observed.

The value of 28 J/m^2 for the fracture energy of pyrocarbon is in good agreement with literature mode I values.[28,29] Nevertheless, it is a good deal higher than the 0 to 3 J/m^2 typically inferred from composites demonstrating tough behavior, and often thought of as a necessary conditions. This, in turn, raises a question regarding the suitability of crack deflection criteria suggested by Cook and Gordon[30] and the more rigorous analyses of He et al.[31,32] (a planar interface between two semi-infinite solids) for application to a geometrically more complicated case such as this, a coated fiber in a matrix.

The average interfacial friction will rise very quickly to extremely high values in the primary, short-period crack regime then drop to much lower values in the secondary crack regime. Tests which probe the primary crack regime, such as matrix crack spacing or tensile hysteresis loops will return quite high values for friction. On the other hand, tests which probe the secondary crack regime, such as fiber pushout (in a conventional analysis), will return much lower values for friction.

The differences in behavior of the treated-fiber composite system considered here as compared to a more conventional untreated-fiber composite have been attributed to the different interfacial cracking behaviors.[19,20] The results of this work are consistent with that scenario. Specifically, the interfacial crack in untreated-fiber composites has been reported

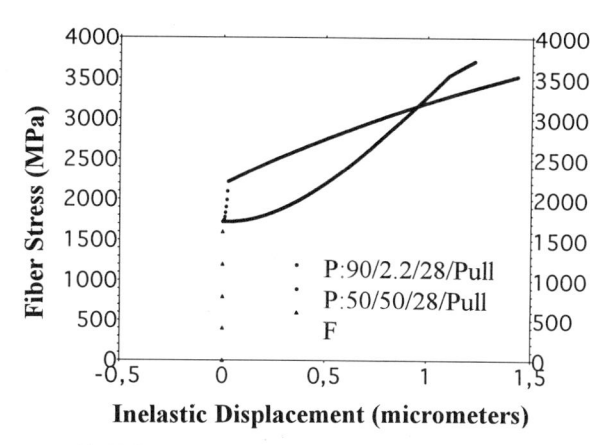

Figure 6. Predicted curves of bridging-fiber stress versus displacement for composite tension tests (Type II boundary conditions [26]) for the same two sets of crack characteristics as the pushout curves of Fig. 5. (after [4])

to be confined to the C/SiO$_2$ interface, or the C layers very near to the interface.[20,33] Such composites are measured to have interfacial fracture energies of no more than a few J/m^2. The fiber treatment has been inferred to strengthen the interface region to a level which is above the strength of the pyrocarbon coating itself thereby shifting the fracture to the pyrocarbon; the next-weakest link.[20,25] The measurements and interpretation of this work implies that both the fracture energy and the friction are very much higher than in the untreated-fiber composites, and, moreover, much higher than in any other reported composites showing crack deflection along fibers and good properties.

Perhaps the most important outcome of work related to the treated-fiber composites is the suggestion that the fracture energies of fiber coatings can be higher than often assumed and still perform the necessary function of crack deflection. This result suggests that the window of properties that can be possessed by an oxide alternative coating, for example, is larger than previously thought. This is an important result because the very low fracture energies previously thought necessary will likely be difficult to obtain with oxides or other alternative materials.

SUMMARY

The evidence implies that debonding crack surface roughness is generally a significant contributor to interfacial friction and, hence, to composite behavior. A simple description of roughness effects in terms of geometric misfit has been consistent with experimental results. Consideration of the implications of the role of misfit stresses has led to the identification or potential revision of several design parameters for fiber coatings; compliance, fracture surface roughness, and coating toughness.

It is evident that coatings can affect interfacial stresses due to roughness misfit, as well as thermal expansion mismatch. Coating thickness dependences most likely derive from these effects. These considerations led to the identification of compliance as a coating design parameters and raised the issue regarding alternative coatings.

Oxide coatings with sufficient compliance to accomplish stress reduction comparable to C coatings are feasible. However, the greater coating thickness required will affect composite properties and may be difficult to accomplish in processing.

The rough crack formalism appears to be useful in providing interpretation of the pushout behavior of high strength, treated-fiber NICALON™/SiC composites. It also offers a reasonable resolution of the wide variety of interface properties of such composites inferred from different test techniques and resolves macroscopic measurements of pyrocarbon properties with composite interface measurements. Finally, the analyzed tests indicate that the allowable toughness of coatings for deflecting cracks may be higher than previously believed.

Consideration of the implications of debond crack roughness has led to valuable insights regarding the possible roles of fiber coatings in influencing composite behavior. This has led in-turn to identification of several parameters that should be considered in the design and evaluation of alternative fiber coatings. If they are not considered, it is entirely possible that viable coating schemes will be discarded for invalid reasons. While these additional design parameters add to our appreciation of the complexity of the problem, they should not be considered as prohibitive. There are simple approaches to managing all of these parameters and, no doubt, more subtle ones yet to be identified.

REFERENCES

1. R. J. Kerans, Issues in the control of fiber/matrix interfaces in ceramic composites, *Scripta Metall. Mat.*, **31**[8]:1079-1085 (1994).
2. R. J. Kerans, Viability of oxide fiber coatings in ceramic composites for accommodation of misfit stresses, *J. Am. Ceram. Soc.*, **79**[6]:1664-68 (1996).
3. R. J. Kerans, Effects of coating thickness and compliance on fiber debonding and sliding in Nicalon/SiC composites, *Scripta Metall. et Mat.*, **32**[4]:505-509 (1994).
4. T. A. Parthasarathy and R. J. Kerans, Anticipated consequences of interfacial roughness effects on the behavior of selected ceramic composites, submitted to *J. Am. Ceram. Soc.*
5. R. J. Kerans, T. A. Parthasarathy, E. Rebillat and J. Lamon, Fiber debonding in high strength NICALON™/C/SiC composites as determined by rough surface analysis of fiber pushout tests, *J. Am. Ceram. Soc.*, in press.
6. R. J. Kerans and T. A. Parthasarathy, Theoretical analysis of the fiber pullout and pushout tests, *J. of Am. Ceram. Soc.*, **74**[7]:1585-1596 (1991).
7. P. D. Jero and R. J. Kerans, The contribution of interfacial roughness to sliding friction of ceramic fibers in a glass matrix, *Scripta Metall. et Mater.*, **24**:2315-2318 (1991).
8. P. D. Jero, R. J. Kerans and T. A. Parthasarathy, Effect of interfacial roughness on the frictional stress measured using pushout tests, *J. Am. Ceram. Soc.*, **74**[11]:2793-2801 (1991).
9. T. A. Parthasarathy, D. B. Marshall, and R. J. Kerans, Analysis of the effect of interfacial roughness on fiber debonding and sliding in brittle matrix composites, *Acta Metall. et Mater.*, **42**[11]:3773-3784 (1994).
10. R. J. Kerans, Effects of coating thickness and compliance on fiber debonding and sliding in Nicalon/SiC composites, *Scripta Metall. et Mat.*, **32**[4]:505-509 (1994).
11. R. A. Lowden, Fiber coatings and the mechanical properties of a fiber-reinforced ceramic composite, pp. 619-630 in *Ceramic Transactions, Vol. 19, Advanced Composite Materials,* ed. by Michael D. Sacks, The American Ceramic Society, Westerville, Ohio (1991).
12. E. Lara-Curzio, M.-K. Ferber and R. A. Lowden, The effect of fiber coating thickness on the interfacial properties of a continuous-fiber ceramic matrix composite, *Cer. Eng. Sci. Proc.*, **15**:5 (1994).
13. J. F. Lynch, C. G. Ruderer and W. H. Duckworth, Engineering properties of ceramics, *AFML-TR-66-52, Air Force Materials Laboratory, (1966).*
14. M. K. Cinibulk, Magnetoplumbite compounds as a fiber coating in oxide/oxide composites, *Ceram. Eng. Sci. Proc.*, **15**[5]:721-728 (1994).
15. M. K. Cinibulk, Microstructure and mechanical behavior of an hibonite interphase in alumina-based composites, *Cer. Eng. Sci. Proc.*, **16**[5]:633-641 (1995).
16. M. K. Cinibulk and R. S. Hay, Textured magnetoplumbite fiber-matrix interphase derived from solgel fiber coatings, *J. Am. Ceram. Soc.*, **79**[5]:1233-46 (1996).
17. R. S. Hay and E. E. Hermes, Sol-Gel coatings on continuous ceramic fibers, *Cer. Eng. Sci. Proc.*, **11** [9-10]:1526-1538 (1990).

18. H. C. Cao, E. Bischoff, O. Sbaizero, M. Ruhle, A. G. Evans, D. B. Marshall and J. J. Brennan, Effect of interfaces on the properties of fiber-reinforced ceramics, *J. Am. Ceram. Soc.,* **73**[6]:1691-99 (1990).
19. C. Droillard and J. Lamon, Fracture toughness of 2D woven SiC/SiC composites with multilayered interphases, *J. Am. Ceram. Soc.,* **79**[4]:849-58 (1996).
20. C. Droillard, J. Lamon and X. Bourrat, Strong interfaces in CMCs, condition for efficient multilayered interphases, Mat. Res. Soc. Symp. Proc., 1994, MRS Fall Meeting (Boston), *Materials Research Society,* Vol. 371-6, (1995).
21. R. Naslain, Fiber-matrix interphases and interfaces in ceramic matrix composites processed by CVI, *Composite Interfaces,* **1**[3]:253-86, (1993).
22. J. Lamon, Interfaces and Interfacial Mechanics: Influence on the mechanical behavior of ceramic matrix composites, *Journal de Physique IV,* Coloque C7, supplement au *Journal de Physique III,* Vol. 3, 1607-1616, (1993).
23. N. Lissart and J. Lamon, Damage and failure in ceramic matrix minicomposites: experimental study and model, *Acta Met.,* in press.
24. D. Cojean and M. Monthioux, Microtextures of interfaces related to mechanical properties in ceramic fiber reinforced ceramic matrix composites, *Proc. 5th Europ. Conf. Comp. Mater. (ECCM-5),* edited by A.R. Bunsell, J.F. Jamet, A. Massiah, EACM, Bordeaux, 729-734, (1992).
25. F. Rebillat, E. Lara-Curzio, J. Lamon, R. Naslain, M. K. Ferber and T. M. Besmann, Interfacial bond strength in Nicalon/C/SiC composite materials as studied by single-fiber push-out tests, submitted to *J. Am. Ceram. Soc.*
26. F. Rebillat, E. Lara-Curzio, J. Lamon, R. Naslain, M. K. Ferber and T. M. Besmann, Properties of multilayered interphases in SiC/SiC CVI-composites with "weak" and "strong" interfaces, submitted to *J. Am. Ceram. Soc.*
27. J. W. Hutchinson and H. M. Jensen, Models of fiber debonding and pullout in brittle composites with friction, *Mech. Mater.,* **9**[2]:139, (1990).
28. M. Sakai, R. C. Bradt, D. B. Fischbach, Fracture toughness anisotropy of a pyrolytic carbon, *J. Mat. Sci.* **21**:1491-1501 (1986)
29. R. O. Ritchie, R. H. Dauskardt, W. W. Gerberich, A. Strojny and E. Lilleodden, Fracture, fatigue and indentation behavior of pyrolytic carbon for biomedical applications, *Mat. Res. Soc. Symp. Proc.,* 383, Materials Research Society (1995).
30. J. Cook and J. E. Gordon, A mechanism for the control of crack propagation in all-brittle systems, *Proc. R. Soc. Lond.,* **282**:508-520 (1964).
31. M. Y. He and J. W. Hutchinson, Crack deflection at an interface between dissimilar elastic materials, *Int. J. Solids Structures,* **25**[9]:1053-1087 (1989).
32. M. Y. He, A. G. Evans and J. W. Hutchinson, Crack deflection at an interface between dissimilar elastic materials: role of residual stresses, *Int. J. Solids Structures,* **31**[24]:3443-3455 (1994).
33. C. Droillard, PH.D. Thesis, University of Bordeaux, France, (1993).

INTERFACIAL DESIGN AND PROPERTIES OF LAYERED BN(+C) COATED NICALON FIBER-REINFORCED GLASS-CERAMIC MATRIX COMPOSITES

John J. Brennan

United Technologies Research Center
East Hartford, CT 06108

ABSTRACT

Nicalon fiber-reinforced glass-ceramic matrix composites with controlled fiber/matrix interfaces based on a layered BN(+C) concept have been developed. These composites have been shown to possess high strength and good creep, stress-rupture, and fatigue resistance to temperatures of 1100°C. The unique microstructure of the BN(+C) interfacial coating has been shown to be optimum for minimizing fiber/matrix interdiffusion and maximizing the toughness and high temperature environmental durability of the composites.

INTRODUCTION

Fiber-reinforced glass-ceramic matrix composites are prospective materials for high-temperature, lightweight, structural applications.[1] In the last decade, research in this type of ceramic matrix composite has concentrated on the fiber/matrix interface and the relationship of the interfacial microstructure, chemistry, and bonding to the resultant composite mechanical properties and interface stability.[2-4] As was discussed at the Ceramic Microstructures '86, Role of Interfaces Symposium,[2] polymer derived SiC type fibers such as Nicalon (Nippon Carbon Co.), that contain excess carbon and oxygen over stoichiometric SiC, form a thin (20-50nm) carbon rich fiber/matrix interfacial layer when incorporated into glass-ceramic matrices at elevated temperatures. The formation of this weak interfacial carbon layer is responsible for the high toughness and strength observed in these composites, but is also responsible for composite embrittlement and concurrent strength and toughness degradation when either stressed at elevated temperatures in an oxidizing environment, or thermally aged in an unstressed condition in oxidizing environments. This embrittlement and strength degradation is a result of oxidation of the carbon layer and its replacement by a glassy oxide layer that is bonded strongly to both the fiber and matrix, thus inhibiting matrix crack deflection at the fiber/matrix interface.

It is imperative that the fiber/matrix interface in fiber-reinforced ceramic matrix composites be controlled, or "engineered", so that relatively weak interfacial bonding exists for matrix crack deflection while maintaining oxidative stability. One approach to accomplish this is to utilize coatings on the fiber surfaces that are applied before composite processing. Not only must these interfacial coatings be relatively weak and oxidatively stable, they must also be resistant to matrix and/or fiber interdiffusion so that interfacial reactions do not occur.

At UTRC, research into coated fiber-reinforced ceramic matrix composites has concentrated on BN and SiC/BN coated Nicalon SiC fibers in a barium-magnesium aluminosilicate (BMAS) glass-ceramic matrix. Previous research, both at UTRC and at other institutions, dealt primarily with either low temperature CVD BN deposited utilizing a BCl_3 precursor (usually containing excess B and O),[5] or a higher temperature deposited BN that yielded a uniform microstructure with a composition of ~42at%B, 42%N, and 15%C.[6-9] For these types of BN, a CVD SiC overcoat was found to be necessary to prevent matrix element diffusion into and crystallization of the turbostratic BN layer, as well as to prevent boron diffusion from the BN into the BMAS matrix. Recent studies at UTRC on nano-structured layered BN based coatings with carbon additions have shown improved resistance to matrix interdiffusion, potentially eliminating the need for the SiC overcoat. These composites have been shown to possess high strength and good fracture toughness to temperatures of 1200°C. From tensile creep, tensile fatigue, and long-time tensile stress-rupture tests, it was found that if the coating chemistry and microstructure are controlled, this composite system can withstand high temperature stressed oxidative exposure at stress levels significantly above the proportional limit, or matrix microcrack stress, which is key to the utilization of ceramic matrix composites for high temperature structural components.

EXPERIMENTAL PROCEDURE

The glass-ceramic matrix utilized for the present study was barium-magnesium aluminosilicate (BMAS), formulated to yield the barium osumilite phase ($BaMg_2Al_3(Si_9Al_3O_{30})$) on crystallization. The reinforcement was a polycarbosilane derived Si-C-O fiber (Nicalon NLM 202) with a BN or dual layer SiC over BN coating. The BN and SiC coatings discussed in this paper were deposited continuously on fiber tows by atmospheric chemical vapor deposition (CVD) at a temperature of ~1000°C, with BN and SiC coating thicknesses of ~300-400 and 200nm, respectively. The BN was deposited using a proprietary precursor chosen to yield a layered structure consisting of alternate layers of ~42at%B, 42%N, and 15%C and a very carbon rich (~50%C, 25%B, and 25%N) BN. From scanning Auger analysis, the oxygen content of both the BN and SiC layers was less than 2%. The measured tensile strength of the coated Nicalon fibers of >2 GPa indicated essentially no loss in fiber strength during coating.

Composites (10cm x 10cm) were fabricated by hot-pressing a layup of 0/90° oriented unidirectional fiber plus matrix powder plies at a maximum temperature of ~1450°C for 5 minutes under 6.9 MPa pressure, yielding an essentially fully dense matrix with a fiber volume of ~45%. After hot-pressing, the composite panels were machined into reduced cross-section tensile samples, most of which had an ~40mm gage length and 6.3mm gage width, and then heat-treated ("ceramed") in argon at 1200°C for 24 hrs to crystallize the BMAS matrix to the barium osumilite phase.

Although a variety of mechanical property measurements were done, including tensile creep and stress-rupture, only the results of uniaxial monotonic tensile and fatigue testing at temperatures of 550° and 1100°C, will be discussed here. All testing was conducted in air.

After composite testing, the fracture surfaces were examined using scanning electron microscopy (SEM), and the composite microstructures characterized by transmission electron microscopy (TEM) of polished composite cross-section replicas and ion beam thinned foil sections.

RESULTS AND DISCUSSION

As-Fabricated Composite Microstructure

A typical microstructure of a 0/90° ply layup BMAS matrix/SiC over nanolayered BN(+C) coated Nicalon fiber composite is shown in Figure 1. The light micrograph (Fig. 1A) shows the overall composite microstructure, while the TEM replica (Fig. 1B) shows the details of the SiC/BN interfacial coatings. The layered structure of the BN(+C) coating is quite evident. The particular composite shown in Fig. 1 is in the ceramed, or matrix crystallized condition, and from the TEM micrograph the lathe-like structure of the barium osumilite grains, along with a small number of residual mullite grains, can be seen in the matrix.

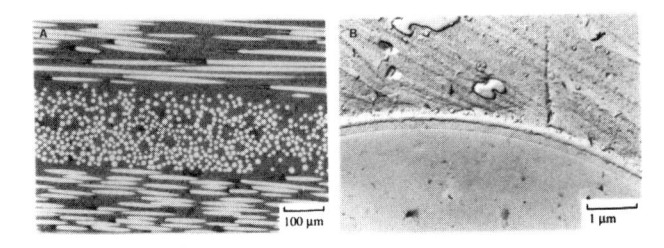

Fig. 1. Microstructure of 0/90° BMAS matrix/SiC/layered BN(+C) coated Nicalon fiber composite.

Figure 2 shows the interfacial microstructure of two BMAS matrix/BN coated Nicalon fiber composites without the SiC overcoat layer; one with the layered BN(+C) type of BN (Fig. 2A), and one with a standard BN coating that was deposited at relatively low temperature (~850°C) utilizing a BCl_3 plus ammonia precursor (Fig. 2B). The latter BN layer has crystallized throughout its total thickness, while the layered BN(+C) has exhibited only minimal crystallization. From previous studies,[7,8] it was found that crystallization of the BN layer is caused by matrix element (Ba, Mg, Al, Si, O) diffusion into the BN during composite fabrication, and results in a composite with lower mechanical properties and reduced toughness. For example, the composite in Fig. 2A exhibited a RT flexural strength of 640 MPa and a strain-to-failure of 0.85%, while the composite of Fig. 2B exhibited a RT flexural strength of 349 MPa and a strain-to-failure of 0.42%. The relative tensile properties at elevated temperature of the two composites also reflected the difference in interfacial chemistry and microstructure. At 1100°C, the ultimate tensile strength of the composite in Fig. 2A was 370 MPa, with a strain-to-failure of 0.98%, while the 1100°C UTS for the composite in Fig. 2B was 220 MPa, with a strain-to-failure of 0.55%.

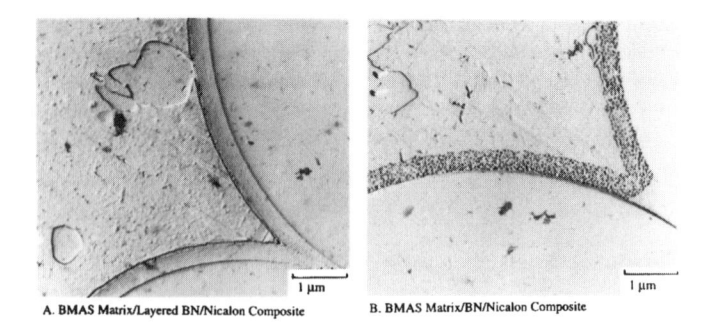

Fig. 2. Interfacial microstructure of (A) BMAS matrix/layered BN(+C) and (B) BMAS/low temperature BN/Nicalon fiber composites.

TEM thin foil analysis of the BN interfacial layer of these two types of composites is shown in Fig. 3. For a layered BN(+C) thickness of ~275nm, as shown in Fig. 3A, the BN is coarsened (crystallized) only through the first BN layer (~40nm), and there is very little evidence of any interaction between the BN and the Nicalon fiber. The first carbon rich layer nearest the matrix appears to act as a matrix element diffusion barrier. However, for a low temperature deposited BN layer (Fig. 3B), the total BN layer thickness (~200nm) is coarsened (crystallized), with subsequent carbon plus silica sublayers formed at the BN/Nicalon fiber interface. From EDS and scanning Auger analyses, a significant amount of matrix element (Si, Al, Mg, Ba, O) diffusion was found in the coarsened BN. These layers are formed by the carbon condensed oxidation of the Nicalon fibers, as initially described by Cooper and Chyung[3]:

$$SiC_{(s)} + O_{2(g)} \rightarrow SiO_{2(s)} + C_{(s)}$$

This reaction is enhanced by the diffusion of oxygen from the BMAS matrix through the low temperature deposited BN layer during composite fabrication.

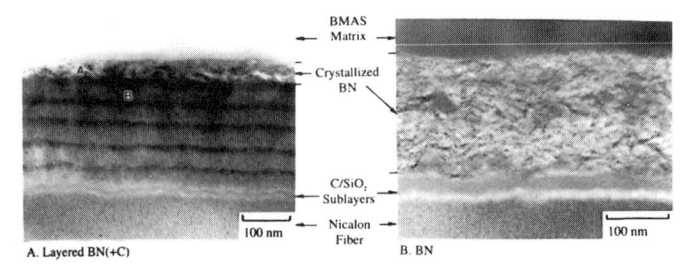

Fig. 3. Interfacial TEM analysis of (A) BMAS matrix/layered BN(+C) and (B) BMAS/low temperature BN/Nicalon fiber composites.

From mechanical property measurements, as mentioned previously, the crystallization of the BN was found to result in lower strength and toughness composites, due to the crystallized BN being a less effective crack deflecting interfacial layer than the layered turbostratic BN. As shown in Fig. 4, the basal planes of the hexagonal BN in the crystallized layer (Fig. 4A) tend to be randomly oriented, while those of the layered BN(+C) are essentially aligned parallel to the fiber surface, thus enhancing crack deflection within the BN(+C) layer.

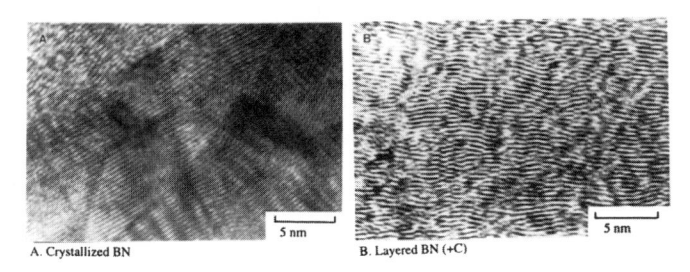

Fig. 4. HRTEM of (A) crystallized and (B) uncrystallized layered BN(+C) from Fig. 3A.

Monotonic Tensile Test Results

The BMAS matrix composites with layered BN(+C) and SiC/BN(+C) fiber coatings were tested in monotonic tension at 550° and 1100°C, as shown in Table 1. The loading rate was 20 MPa/s, and the proportional limit (PL) stress, (deviation from linearity) was measured utilizing a 0.005% offset strain. In general, the BN(+C) coated Nicalon fiber composites exhibited higher ultimate strengths and strain-to-failures, and lower proportional limit stresses and elastic moduli at each temperature, compared to the SiC/BN(+C) coated Nicalon composites. The elastic modulus of the composites was lower at 1100°C than at 550°C, reflecting the softening of the residual glassy phase in the BMAS matrix. All composites exhibited very fibrous fracture surfaces.

Table 1. Tensile Properties for BMAS Matrix/BN & SiC/BN/Nicalon Fiber Composites (0/90°).

Interface	Temp-°C	UTS-MPa	PL-MPa	E-GPa	ε_f%
BN	550	371	65	119	0.83
SiC/BN	550	336	89	127	0.73
BN	1100	370	74	66	0.98
SiC/BN	1100	301	105	73	0.79

Tensile Fatigue Testing

Tensile fatigue (tension-tension, R=0.1) experiments were conducted on the two types of coated fiber composites at 550°C under a maximum stress of 104 MPa, and at 1100°C under maximum stresses of both 104 MPa and 138 MPa. Both of these stresses are at or above the proportional limit stress of the composites, as shown in Table 1. The frequency was 3 Hz, which resulted in a 10^6 cycle runout time of ~93 hrs. All of the samples fatigued at both stress levels survived 10^6 cycles without failure, and were then tested to failure in monotonic tension at temperature (20 MPa/s loading rate). The temperature of 550°C was chosen because previous studies[2] indicated that this temperature was particularly severe for the oxidation of carbon interfaces, and thus there was concern that the carbon rich BN layered interfaces might suffer enhanced oxidation at this temperature. The temperature of 1100°C was chosen since this is near the planned use temperature for composites of this type. The residual monotonic tensile properties of the composite samples after fatigue runout are shown in Table 2.

Table 2. Tensile Properties for Fatigued BMAS Matrix/BN & SiC/BN/Nicalon Fiber Composites (0/90°).

Interface	Temp-°C	Fatigue σ-MPa	UTS-MPa	PL-MPa	E-GPa	ε_f-%
BN	550	104	457	110	103	0.99
SiC/BN	550	104	384	107	124	0.98
BN	1100	104	398	105	68	0.98
SiC/BN	1100	104	372	88	78	1.06
BN	1100	138	323	62	60	1.00
SiC/BN	1100	138	284	55	79	0.56

From the data shown in Table 2, it can be seen that the residual tensile strengths of the samples fatigued at 104 MPa at both 550° and 1100°C, and with both BN and SiC/BN coatings, increased over that measured on as-fabricated composites, as was shown in Table 1. This increase may be due to straightening of the 0° plies during the fatigue testing. The strain-to-failure of the composites also tended to increase. At 1100°C and 138 MPa fatigue stress, the residual ultimate strength of the composites tended to decrease, as did the proportional limit stress. In the case of the SiC/BN coated Nicalon fiber composite fatigued at 138 MPa, the strain-to-failure also decreased substantially. The stress-strain behavior of the BN and SiC/BN coated Nicalon fiber composites at 1100°C, as-processed and after 104 MPa and 138 MPa fatigue, are shown in Fig. 5. From this figure, the increase in ultimate strength after 104 MPa fatigue is evident, as is the degradation of the SiC/BN coated Nicalon fiber composite after 138 MPa fatigue testing.

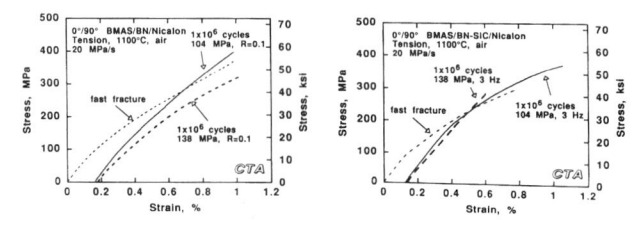

Fig. 5. Stress-strain behavior of BN and SiC/BN coated Nicalon fiber composites before and after fatigue.

The fracture surfaces of all fatigued composites were very fibrous in nature, except for a region of the SiC/BN coated Nicalon fiber composite that had experienced the 1100°C, 138 MPa fatigue, as shown in Fig. 6. From this figure, it can be seen that one of the edge regions on the fracture surface exhibited a very brittle and glazed appearance, indicating that a crack had formed during the fatigue test and propagated partially through the composite. Even though this crack formation was not enough to fail the composite during fatigue, it did cause a significant reduction in the residual 1100°C tensile strength. Interestingly, no such brittle region was evident on the fracture surface of the composite fatigued at 1100°C, 138 MPa that contained a BN coating only on the Nicalon fibers, even though this stress level is significantly higher than the proportional limit, or matrix microcrack stress.

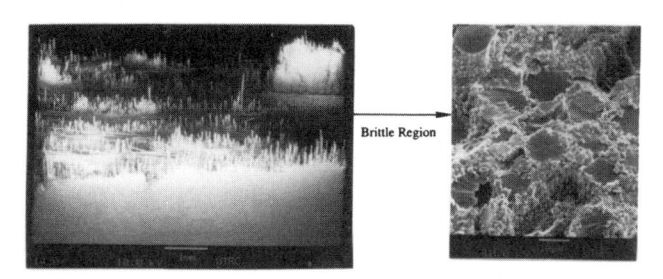

Fig. 6. Fracture surface of BMAS matrix/SiC/BN coated Nicalon fiber composite after 1100°C, 138 MPa, 10^6 cycle tensile fatigue.

The difference in behavior of the BN(+C) vs SiC/BN(+C) coated Nicalon fiber composites in 1100°C, 138 MPa fatigue, may be associated with the somewhat different fracture path that matrix cracks take when propagating into the fiber coatings, as shown in Fig. 7. The crack path in the dual layer composite with the SiC overcoat is usually found to be in the carbon rich layer nearest to the Nicalon fiber surface (Fig. 7A), whereas that in the BN(+C) coating is invariably between the crystallized BN layer and the first carbon rich layer nearest the matrix (Fig. 7B).

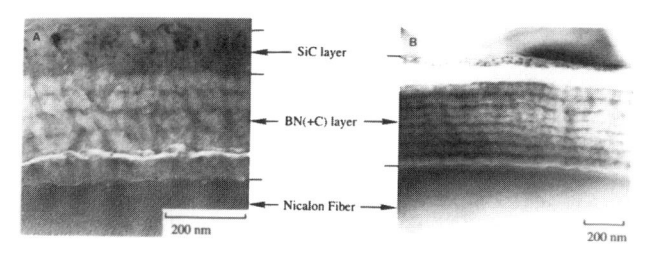

Fig. 7. Crack paths in layered BN(+C) coatings in composites with (A) and without (B) a SiC overcoat.

Oxidation of the fibers during fatigue at stress levels significantly above the proportional limit stress, as was seen in Fig. 6, may be more severe when the crack path is closer to the fiber surface. Previous studies[8,9] with uniform turbostratic SiC/BN fiber coatings showed that when the interfacial cracks followed the BN/Nicalon fiber interface, limited oxygen access through the matrix cracks during 1100°C tensile fatigue led to oxidation of the surface of the Nicalon fibers to silica before significant oxidation of the BN,[9] resulting in increased bonding of the fiber/matrix interfacial region.

CONCLUSIONS

From the mechanical property data and microstructural information presented in this paper, it has been demonstrated that nanolayered BN(+C) fiber coatings in BMAS glass-ceramic matrix composites reinforced with Nicalon fibers, result in optimized modes of matrix crack deflection and improved high temperature properties in oxidizing environments compared to previous uniform microstructure BN fiber coatings. In particular, it has been shown that this composite system can withstand fatigue stress levels at elevated temperatures for many cycles that are significantly *higher* than the proportional limit, or matrix microcracking stress of the composite. The diffusional characteristics of the carbon rich layers within these BN(+C) fiber coatings also obviate the need for a SiC diffusion barrier overcoating, thus resulting in a simpler, lower cost composite system.

ACKNOWLEDGEMENTS

The author would like to thank Dr. Chris Griffin of 3M for the CVD BN(+C) fiber coatings, Mr. Gerald McCarthy of UTRC for the TEM analyses, Dr. John Holmes of CTA for the composite mechanical testing, and to Dr. Alexander Pechenik of the Air Force Office of Scientific Research (AFOSR), Bolling AFB DC, for his sponsorship of this program.

REFERENCES

1. K.M. Prewo, J.J. Brennan, and G.K. Layden, "Fiber reinforced glasses and glass-ceramics for high performance applications," *Am. Cer. Soc. Bull.*, 65 [2] 305 (1986)
2. J.J. Brennan, "Interfacial characterization of glass and glass-ceramic matrix/Nicalon SiC fiber composites," *Materials Science Research*, Vol. 20, Plenum Press, New York, 549 (1986).
3. R.F. Cooper and K. Chyung, "Structure and chemistry of fiber-matrix interfaces in SiC fibre-reinforced glass-ceramic composites: An electron microscopy study," *J. Mater. Sci.*, 22, 3148 (1987).
4. C. Ponthieu, M. Lancin, J. Thibault-Desseaux, and S. Vignesoult, "Microstructure of interfaces in SiC/glass composites of different tenacity", *Journal de Physique,* 51, C1-1021, (1990).
5. R.W. Rice: "BN coating of ceramic fibers for ceramic fiber composites", US Patent 4,642,271, Feb. 10, 1987.
6. J. Brennan: "Interfacial studies of fiber reinforced glass-ceramic matrix composites," *High Temperature Ceramic Matrix Composites, HT-CMC-1*, Editors: R. Naslain, J. Lamon, and D. Doumeingts, Woodhead Publishing, Ltd, Abington Cambridge, England, 269 (1993).
7. E. Sun, S. Nutt, and J. Brennan: "Interfacial microstructure and chemistry of SiC/BN dual coated Nicalon fiber reinforced glass-ceramic matrix composites", *J. Am. Ceram. Soc.* 77, [5] 1329 (1994)
8. J. Brennan, S Nutt, and E. Sun: "Interfacial microstructure and stability of BN coated Nicalon fiber/glass-ceramic matrix composites," *High Temperature Ceramic Matrix Composites II: Manufacturing and Materials Development*, Editors: A. Evans and R. Naslain, *Ceramic Transactions,*[58] 53 (1995).
9. B.W. Sheldon, E.Y. Sun, S.R. Nutt, and J.J. Brennan: "Oxidation of BN-coated SiC fibers in ceramic matrix composites", *J. Am. Ceram. Soc.* 79, [2] 539 (1996)

STABILITY OF CARBON INTERPHASES AND THEIR EFFECT ON THE DEFORMATION OF FIBER REINFORCED CERAMIC MATRIX COMPOSITES

Margaret M. Stackpoole[1], Rajendra K. Bordia[1],
Charles H. Henager, Jr.[2], Charles F. Windisch[2], Russell H. Jones[2],

[1]Department of Material Science and Engineering, University of Washington, Seattle, WA,
[2]Pacific Northwest National Laboratories, Richland, WA

ABSTRACT

Time-dependent deformation behavior, in bending, of notched ceramic matrix composite samples was investigated at elevated temperatures under varying pO_2. Materials studied consisted of a chemical vapor infiltrated (CVI) β-SiC matrix reinforced with Nicalon® SiC fibers with a 1μm C interphase. From these results, activation energies for composite deformation were obtained as a function of oxygen partial pressure. Finally Thermo-Gravimetric experiments were conducted to obtain the oxidation kinetics of the C interphase. The activation energy for oxidation was correlated to the activation energy for damage evolution.

INTRODUCTION

Many of the most promising ceramic matrix composites have been optimized for high toughness at relatively low temperatures. However, the most promising applications of continuous fiber reinforced composites are at high temperatures, e.g. boilers, turbine combustors and heat exchangers. It is realized that superior composite properties are dominated by the interphase and for these composites high ambient temperature toughness is obtained by coating the fibers with a material that allows weak bonding between the fiber and the matrix. This leads to crack deflection, debonding at the fiber matrix interface, crack wake fiber bridging and fiber pull out which contribute to high toughness and damage tolerance[1]. Since these materials have been optimized for high toughness at relatively low temperatures the long term, high temperature performance of these composites needs further study. It is expected that time dependent crack growth and deformation will control the lifetime of these materials at elevated temperatures [2]. The interphase of these composite materials has been specifically tailored so that crack wake processes lead to debonding and fiber pull-out giving a damage tolerant material, therefore, crack wake bridging by the fibers dictates their desirable mechanical behavior. If the temperature is high enough, then the bridging fibers could creep and relax the bridging forces. Thus, at elevated temperatures the time dependent response of the fibers in the bridging zone greatly influences the deformation of these composites. The crack velocity-stress intensity factor relationship of these composites has been previously investigated and higher crack velocities were observed in Ar with up to 20% oxygen than in pure Ar. Properties of the C interphase deteriorated when oxygen was present in the environment and this leads to a further decrease in the effectiveness of bridging fibers[3].

Fiber creep reasonably predicts the correct crack velocity and time dependence in pure Ar, however, other factors need to be considered when oxygen is introduced into the environment. In these environments interphase removal or reaction must also be considered. Recent investigations [4] addressing the kinetics of C - interphase removal from a SiC/SiC system resulted in a mass gain due to SiO_2 formation. However, the interphases present were significantly thinner than the interphases in this investigation and the oxygen partial pressure was much greater. The same investigators also observed a mass loss for thicker interfaces which followed parabolic kinetics in the same environment. The parabolic behavior suggests that the transport of oxygen along the interface region is rate controlling and after a certain time the reaction front will be deprived of oxygen. Other investigators [5] have observed linear kinetics in an oxidation study of a C interface SiC material. Recent work examines the effect of SiO_2 formation at the interface and at matrix cracks [4,6] and a model has been developed to explain the results [7]. If sufficient SiO_2 is formed and the interface is thin enough then the interface channel is pinched off and further reaction prevented. From this brief review of the literature, it is apparent that the effect of pO_2 on the oxidation of the interphase and the mechanical properties of composites is rather complex and depends on many factors. These include sample geometry, interphase layer thickness, pO_2 and temperature. The present work will investigate the effect of pO_2 and temperature on the time dependent deformation of the composite and compare these results with data taken from an inert environment. The effect of pO_2 on interface reaction is also investigated on the same materials and an attempt made to correlate interface reaction with composite deformation.

MATERIALS USED AND EXPERIMENTAL PROCEDURE

The composite materials in this investigation consisted of a woven Nicalon fiber cloth ($0°/90°$) which was 8 plies in thickness. The matrix material, β-SiC, was applied by chemical vapor infiltration. The composite had a residual porosity of between 10% and 15%. The interface present was a 1μm Carbon (C) layer which was deposited on the fibers before the CVI step. The as received composites were 4mm x 13mm x 50mm in dimensions and were fabricated by RCI of Whittier, CA. Single-edge-notched bend bars (SENB) with dimensions 3.5 mm x 5mm x 50mm were prepared from the as received bars. The bars had a blunt U notch and an initial a/w ratio of ~ 0.25. These specimens were loaded in four-point bending using a fully articulated SiC bend fixture. All tests were performed inside a sealed mullite tube and a split clamshell furnace [8].

All the experiments were conducted under static loading for times up to 1×10^5 seconds giving long term displacement/time data. Temperature of tests varied between $800°C$ to $1100°C$. Samples were tested in pure Ar and in Ar with varying oxygen partial pressures. The specimens were loaded such that the outer fiber stress was slightly higher than the matrix cracking stress. Testing continued until sample failure. Finally, samples that were tested in an oxygen environment were brought up to temperature in pure Ar and the gas mixture was switched over just prior to testing. The gas flow rate through the mullite tube holding the specimen was set to ~ 1.6×10^{-6} m^3/s. Time, load, displacement, oxygen partial pressure and temperature were recorded for each experiment. All the load/displacement data was corrected for the compliance of the SiC four point bend fixture used in the experiments.

Thermo-Gravimetric Analysis (TGA) was used to monitor the oxidation behavior of the composite at temperatures between $800°C$ and $1100°C$ in varying oxygen environments for periods up to 12 hours. All studies were performed on a thermogravimetric analyzer (Netzsch Instrument STA409). Gas flow was maintained at 1 SCFH throughout the experiment. Before each test the sample was placed in an Al_2O_3 pan in the furnace. The furnace was then sealed and the gas flow was allowed to equilibrate before heat up. The furnace was then ramped up to the required temperature and held for the specified time. Mass change was recorded as a function of time. A sample was also run in pure Ar and this showed little mass change indicating that all mass changes are attributed to reactions requiring oxygen. All TGA samples had dimensions of 4mm x 4mm x 8mm. One of the faces was free from SiC coating applied during processing while all the other surfaces had a 5 μm coating. During the experiment the surface without the coating was the only one that underwent a reaction. Finally all the rates obtained for each sample were normalized with respect to the area of the uncoated surface as these varied slightly between samples.

RESULTS AND DISCUSSION

Displacement/Time Data

From the four point bend experimental data, plots of deflection as a function of time were constructed. Fig. 1 shows an example of the raw data for the C interphase composite material held at a constant load at 1100°C for different oxygen partial pressures. Note that the initial deflection is different for the different samples. In order to account for the slight differences between samples (geometry, load) the deflection needs to be normalized.

The creep rate of a material is influenced by many variables including temperature, T, applied stress, σ, creep strain, ε, elastic modulus, G and grain size, d. In general the creep rate may be given by:

$$\dot{\varepsilon} = f(T, \sigma, \varepsilon, G, d) \tag{1}$$

Di Carlo[9] investigated the creep behavior of several ceramic fibers and found that the fiber creep followed the standard expression for transient creep:

$$\varepsilon = A\sigma^n \left[t \cdot \exp\left(\frac{-Q}{RT}\right) \right]^P \tag{2}$$

where A, P and n are empirically determined creep parameters. The creep parameter, P, is obtained by plotting $\ln(\varepsilon)$ vs $\ln(t)$ at a constant temperature and the activation energy is obtained from a plot of $\ln(\varepsilon)$ vs 1/T.

In this study the creep deformation of composites was investigated using this approach on normalized deflection. The deflection was normalized using the following approach. Let the mid point deflection = δ, and the initial mid point deflection = δ_0 (at time t = 0). Then we define normalized deflection as follows:

$$\delta^* = \frac{\delta - \delta_0}{\delta_0} \tag{3}$$

Using this normalization, small variations in load and sample dimensions are accounted for allowing different samples to be compared. Equation 4 relates creep strain to midpoint displacement for the geometry used in this study:

$$\varepsilon = \frac{4\delta h}{s^2} \tag{4}$$

where ε is the outer fiber bending strain, s is the inner span, h is the sample height. Thus:

$$\varepsilon_{norm} = \frac{\varepsilon - \varepsilon_0}{\varepsilon_0} = \frac{4h\delta - 4h\delta_0}{4h\delta_0} = \frac{\delta - \delta_0}{\delta_0} = \delta^* \tag{5}$$

Eqn. 2 describes the creep behavior in inert environments, however in this study the pO_2 has also been varied. In general the pO_2 dependence is a power law for an oxidation reaction. We therefore use Eqn. 6a as a modification for Eqn. 2 to account for oxidizing environments.

$$\varepsilon = A_o pO_2^m \sigma^n \left[t \cdot \exp\left(\frac{-Q}{RT}\right) \right]^P \tag{6a}$$

therefore

$$\delta^* = \frac{\varepsilon - \varepsilon_0}{\varepsilon_0} = A_1 pO_2^m \left[t \cdot \exp\left(\frac{-Q}{RT}\right) \right]^P - 1 \tag{6b}$$

Eqn. 6b assumes that the stress dependence of deflection at zero time and as a function of time is the same. It should be noted that in this study, all the samples were tested at approximately the same outer fiber stress. Thus, even if this assumption is not correct, its effect on the result is small.

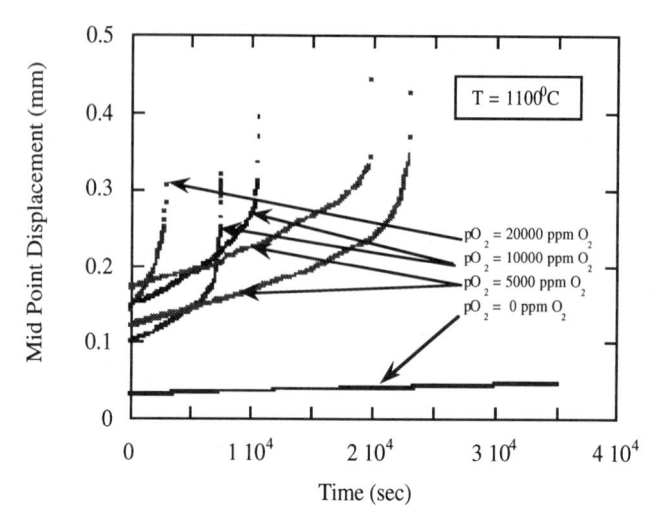

Fig.1: The Deflection behavior of the Carbon Interphase Composite for varying oxygen partial pressures at 1100°C. This data is not normalized.

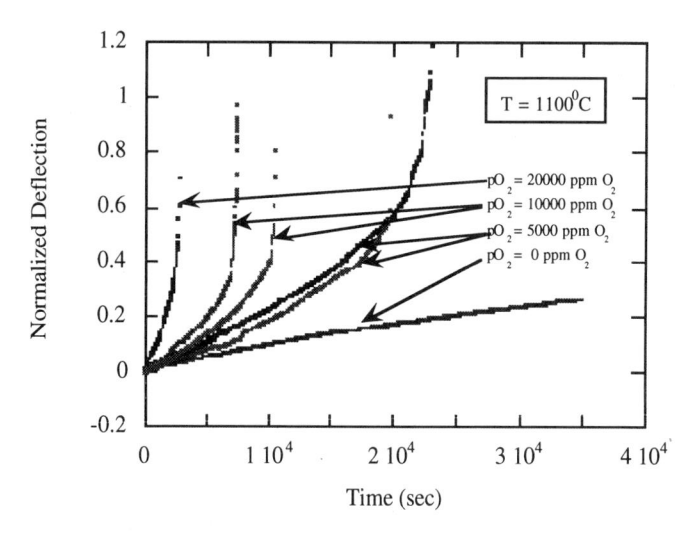

Figure 2: The Normalized Deflection vs Time of the C Interphase Composite for varying oxygen partial pressures at 1100°C.

An example of normalized deflection vs time is given in Fig 2. From this plot it was observed that the deflection increased with increasing oxygen partial pressure at a fixed temperature. At a fixed pO_2, the deflection also increased as temperature increased. Taking the natural log of both sides of Equation 6b the creep parameter, P, is obtained by plotting ln (δ) vs ln (t) at a fixed temperature and pO_2, and the activation energy is obtained from a plot of ln (δ) vs 1/T at a constant pO_2 and time.

The empirical creep parameter, P, was obtained for the different oxygen partial pressures and temperatures. P was relatively constant at about 0.4 in the inert environment and this is the same as the P value obtained by DiCarlo[9] for Nicalon fibers tested in the same temperature range. P approached 1 for most oxygen partial pressures for the C interphase composite. There is no clear physical significance of P. It can be viewed as a measure of deviation from steady state. Note P = 1 for steady state creep and less than 1 for primary creep. Once the creep parameter P has been obtained the activation energy Q can be obtained in the following manner. Normalized deflection is plotted as a function of time for different temperatures at a fixed pO_2. The normalized deflection at a fixed time is obtained and inserted into Eqn. 6b. For C interphase material in an Ar atmosphere the activation energy for composite deformation, Q, was observed to be constant at 550 ± 20 kJ/mol. (see Fig. 3) The activation energy for Nicalon fiber creep is estimated at 600 kJ/mol[9] and this value is very similar to Q obtained for deformation in these experiments which suggests that in an inert environment fiber creep controls composite deformation. In addition, for testing in Ar environments very long times to failure were observed.

Oxidizing environments of various partial pressures of O_2 were chosen to investigate oxidation effects on the composite material. The activation energy, Q, was observed to decrease in all cases when O_2 was present in the environment. The oxidation effect on the C interphase was very rapid, especially at higher oxygen partial pressures and Q was calculated to be 50 ± 10 kJ/mol (see Fig. 3). The activation energy was also independent of pO_2 However, the deformation rate increased as the pO_2 increased (see Fig. 2). SEM micrographs (Fig. 4) from the composite tested in controlled oxygen environment show the removal of the interphase due to oxidation. Similar effects were observed at all oxygen partial pressures. As already mentioned the presence of the interphase is critical for the superior properties displayed by these composites. Removal of the interphase leads to severe degradation of mechanical properties. Even though very long pull out lengths are observed in oxidizing environments for C there is no load transfer because there is no physical contact between the fibers and matrix due to interphase removal.

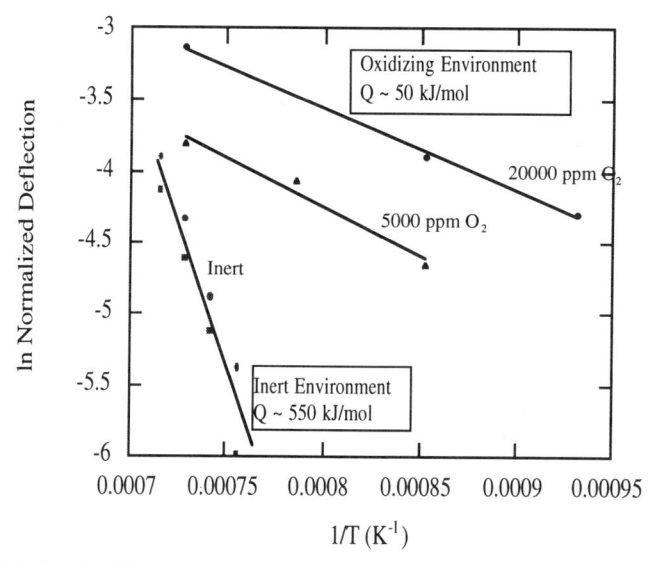

Figure 3. Activation Energies obtained for Composite Deformation at different pO_2's.

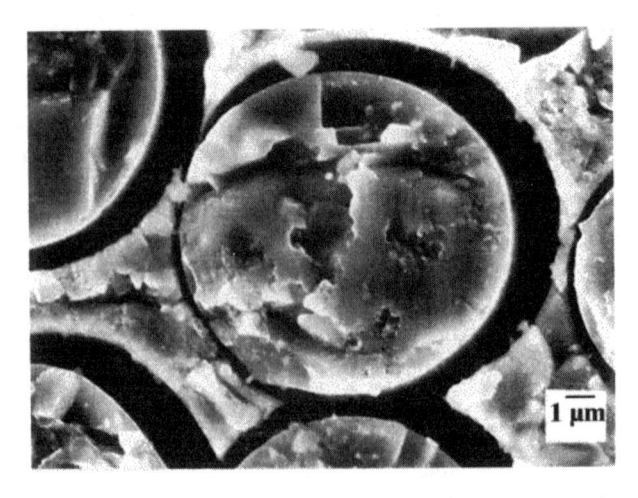

Figure 4. SEM Micrograph of C Interphase material after testing at 1100°C with 20,000 ppm oxygen environment.

TGA DATA

From the TGA data, plots of weight loss as a function of time were obtained (Fig. 5.). Initially all the mass loss vs time data is normalized with respect to the exact exposed area as this varied slightly from sample to sample. Since the plots give a straight line linear kinetics is predominant.

The oxidation of carbon along the interface has been investigated. From Cawley[11], for the carbon interphase, the following linear rate constant is obtained.

$$k_l = \left(\frac{1}{N}\right)\frac{xP}{RT}k_o \exp\left(\frac{-Q}{RT}\right) \quad (7)$$

where N is the molar density of carbon = 1.5×10^5 mol/m^3, P is the total gas pressure in Pa, k_o is a constant (m/s), R is the gas constant (J mol^{-1} K^{-1}), T is the temperature (K), Q is the activation energy (J/mol) and x is the fraction defining the partial pressure of oxygen. From Windisch *et al*[5] the recession of the interphase results in weight loss and the rate in is related to the weight change as follows:

$$k_l = \frac{\Delta M}{t}\frac{1}{\rho A} \quad (8)$$

where ΔM is the mass loss (g), t is time (sec), ρ is the density of Carbon = 1.8g/cm^3, A is the calculated area of exposed surface that is interphase ($1.7 \pm 0.5 \times 10^{-2}$ cm^2). Combining equations 7 and 8 it is possible to get an activation energy for the recession rate from mass loss.

$$k_l = \frac{\Delta M}{t}\frac{1}{\rho A} = \left(\frac{1}{N}\right)\frac{xP}{RT}k_o \exp\left(\frac{-Q}{RT}\right) \quad (9)$$

If pO_2 is held constant then equation 9 has the following form:

$$k_l T = Const. \exp\left(\frac{-Q}{RT}\right) \quad (10)$$

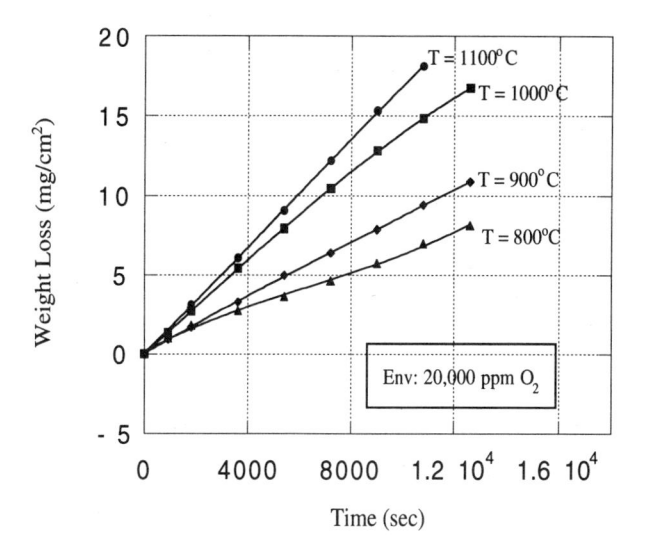

Figure 5. Weight Loss vs Time for Composite material in an Ar. env. with 20,000 ppm O_2

A plot of ln (k_1T) vs 1/T will yield a straight line with a slope of -Q/R. From this it is possible to obtain the activation energy for recession of the Carbon interphase. This plot is presented in Fig. 6 and from this the activation energy has been calculated to be 45 ± 5 kJ/mol independent of the partial pressure of oxygen. This is similar to the activation energy obtained for composite deformation in the previous sub-section (50 ± 10 kJ/mol) and also similar to the activation energy for C oxidation from literature[10].

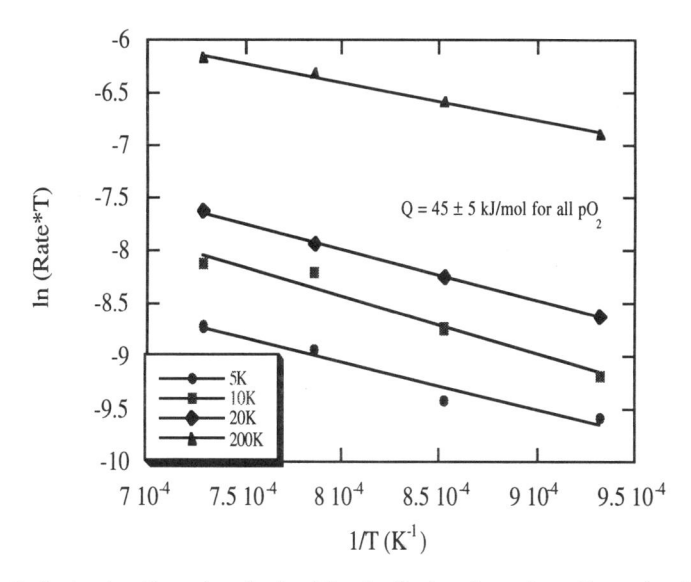

Fig. 6: Activation Energies obtained for the Carbon Interphase Recession Rates

SUMMARY AND CONCLUSIONS

The time dependent deformation under constant flexural load of SiC/SiC composites with a C interphase has been obtained as a function of pO_2 and temperature. The samples tested were notched beams. It is shown that the normalized deformation behavior can be analyzed in terms of the formalism developed for fiber creep. From these results activation energies for deformation were obtained in both inert and oxidizing environments. In inert environment the activation energy observed for time dependent deformation (550 ± 20 kJ/mol) is similar to the activation energy for Nicalon fiber creep. This suggests that composite deformation is controlled by fiber creep in this environment. In oxidizing environments the activation energy for time dependent deformation decreases significantly to 50 kJ/mol. It was also shown that the activation energy for time dependent deformation in oxidizing environments and the activation energy obtained for interphase oxidation from the TGA data are similar suggesting that the reaction of oxygen with the interface controls time dependent deformation in oxidizing environments. Thus, based on the approximately equal values of the activation energy, it can be concluded that the kinetics of deformation in an inert environment is controlled by the creep of fibers. Further, in an oxidizing environment the kinetics of deformation is controlled by the rate of interface oxidation.

REFERENCES

1. A.G. Evans and D.B. Marshall,. *Acta Metall.*, **37**, [10], 2567, (1989)
2. C.H. Henager Jr. and R.H. Jones, *Ceramic Eng. Sci. Proc.*, **13**, 411 (1992)
3. C.H. Henager Jr. and R.H. Jones, *Ceramic Eng. Sci. Proc.*, **14**, 7 (1993)
4. L Filipuzzi, G Cumas, R Naslin and J Thebault, *J. Am. Ceram. Soc.*, **77**, [2], 459, (1994)
5. R.H. Jones, C.H. Henager Jr., and C.F. Windisch Jr., *Mat. Sci. and Eng.*, **A198**, [1-2], 103, (1995)
6. F. Lamouroux, G. Cumas and J. Thebault, *J. Am. Ceram. Soc.*, **77**, [8], 2079, (1994)
7. L. Filipuzzi and R. Naslin, *J. Am. Ceram. Soc.,* **77**, [2], 467, (1994)
8. C.H. Henager Jr. and R.H. Jones, *Mat. Sci. and Eng.*, **A166**, [1-2], 211, (1993)
9. J.A. DiCarlo, *Comp. Sci. and Tech.* **51**, [2], 213, (1994)
10. B.R. Stanmore, *Combustion and Flame*, **83**, [221], (1991)
11. A.J. Eckel, J.D. Cawley, and T.A. Parthasarathy, *J Amer. Ceram. Soc.*, **78**, [4], 972, (1995)

ACKNOWLEDGMENTS

PNNL is operated for the U.S. Department of Energy by Battelle Memorial Institute under contract DE-AC06-76RLO 1830. The research at The University of Washington was supported by the National Science Foundation under Grant Number DMR-NYI-9257827 and by the Du Pont Young Professor Award.

THERMAL STABILITY OF NEXTEL 720 ALUMINO SILICATE FIBERS

H. Schneider, J. Göring, M. Schmücker, and F. Flucht

Institute for Materials Research
German Aerospace Research Establishment
51140 Köln, Germany

ABSTRACT

Polycrystalline 3M Nextel 720 alumino silicate fibers ($\approx$ 85 wt.% Al_2O_3, = 15 wt.% SiO_2, diameter: $\approx$ 12 µm) consisting of 70 wt.% mullite and 30 wt.% α-Al_2O_3 were characterized mechanically and microstructurally.

The fibers were annealed at temperatures up to 1500°C in air, either in a laboratory furnace or directly in the furnace of the tensile strength testing equipment. After this heat-treatment, tensile strength of the fibers was measured at room temperature with two different fiber gauge lengths, 5 mm and 100 mm (designated as F5 and F100 fibers), respectively.

Heat treatment at 1200°C leads to a fiber strength decrease to $\approx$ 85% of the starting value, probably due to locally starting α-Al_2O_3 and mullite grain growth processes. Annealing the fibers at 1200 and up to 1350°C shows that strength decrease to $\approx$ 35% can be directly correlated with gradual grain growth and recrystallization effects as determined by transmission electron microscopy.

Above 1350°C, the short (F5) and long (F100) fibers behave differently: While F5 fiber still displays grain growth controlled strength reduction ($\approx$ 40% of the starting value at 1500°C), the F100 exhibits drastic fiber strength loss between 1350 and 1400°C. Obviously, degradation of the F100 fibers is caused by the occurrence of critical defects in an average distance far beyond the length of F5 fibers (>> 5 mm). This is possibly due to critical defects formed at processing-induced locally inhomogeneous fiber areas.

INTRODUCTION

An important potential application of continuous, fiber-reinforced ceramic matrix composites is the production of shingles for thermal protection of combustion chamber metal walls in aircraft turbine engines. The main aim of thermal protection systems is to reduce the cooling air flow in the combustion chamber, so that burning process efficiency is improved. The application of composites requires long-term high temperature stability

(20,000 hours at 1000°C with temperature maxima up to 1300°C) in oxidizing atmosphere, thermal shock resistance, damage tolerance, and chemical stability. Many nonoxide/nonoxide ceramic composites with Si_3N_4, SiC, and C matrices and fibers have excellent mechanical properties. The high temperature stability of these materials is due to their tendency to display grain coarsening at such temperatures from the strong covalent bonding of these compounds and correlated low diffusion coefficients. A major limitation of nonoxide/nonoxide composites, however, is their low oxidation resistance, especially under cyclic, long-term conditions. At that point, oxide/oxide composites come into play because of their inherent stability in air. Oxide composite materials that best fit the above-described requirement are continuous alumino silicate or alumina fiber reinforced mullite matrix composites.[1-5]

Different types of commercial polycrystalline fibers with near-mullite composition have been used to reinforce mullite matrices. Important examples are 3M Nextel 440, 480, 550 and Sumitomo Altex 2K fibers.[6-9] Currently, the most promising commercial fibers for long-term high temperature application under oxidizing conditions are 3M Nextel 720 fibers. All commercial polycrystalline alumino silicate fibers with near-mullite composition consist of submicron-sized grains. At elevated temperatures, grain coarsening occurs. This is caused by the high grain surface and the high cation and oxygen-diffusion coefficients in oxides (these are much higher than those of most nonoxide materials). In an earlier study on Nextel 440 and Sumitomo Altex 2K alumino silicate fibers, it has been shown that the reduction of the mechanical properties of fibers annealed above 1000°C can be directly attributed to grain coarsening.[7]

The objective of this paper is to correlate the mechanical strength data of Nextel 720 fibers after annealing at temperatures up to 1500°C to the temperature-induced development of the fibers' microstructures (especially grain growth and recrystallization processes).

EXPERIMENTAL

Fiber Material

Commercially produced 3M Nextel 720 fibers with mean diameters of ≈ 12 µm have been used for the study. In the as-received state, the fibers consist of very small α-Al_2O_3 crystals (mean grain size: ≈ 70 nm), and larger mosaic-type mullite (mean grain size: ≈ 300 nm). The chemical composition of the fibers is ≈ 85 wt.% Al_2O_3 and ≈15 wt.% SiO_2.

Annealing Experiments

In a first series of annealing experiments, Nextel 720 fiber bundles were heat-treated in a laboratory furnace in air at 1000, 1100, 1200, 1300, 1400, 1500, and 1600°C for 2 hours. After annealing, single fibers were pulled out of the bundles and were glued on a paper frame for strength measurements. Gauge lengths were variable, though most experimental runs were carried out on 5-mm-long fibers (the F5 samples, so called from their fiber length). The paper frame was installed into a screw-driven universal testing machine with a 10 N load zone.

In a second series of annealing experiments, single fibers (200 mm long) were directly installed into the tensile strength test machine and were heat-treated there for 2 hours in air at 1300, 1350, and 1400°C by an induction-heating furnace (150 mm heating zone) having a 100-mm zone of constant temperature (T = T_c ± 10°C). After being cooled to room temperature, the fibers were tested for tensile strength in a screw-driven testing machine without further handling. Fibers from this test procedure have been designated as the F100 samples from the 100-mm zone of constant temperature.

Description of Microstructures and Phase Contents

Microstructural investigations were carried out on F5 fibers heat-treated for 2 hours in temperatures of between 1200 and 1600°C. Due to the small grain size of fibers, microstructures were examined by scanning electron microscopy (SEM) (Philips, 250M) and by transmission electron microscopy (TEM) (Philips EM 430) with an accelerating voltage of 300 kV. Electron transparent areas of long sections of fibers were achieved by argon ion beam milling (Gatan Duomill). Grain size developments of the fibers were determined by measuring about 100 grains in each sample on the TEM monitor. Intersectional effects were neglected.

X-ray diffraction studies were performed with a Siemens D5000 diffractometer using Cu-K$_\alpha$ radiation. Diffraction patterns were recorded in the step mode (0.01° step size, 10 s counting time). Lattice constants of mullite solid solutions were calculated by means of a PC least square program (LCLSQ) using the α-Al$_2$O$_3$ of Nextel 720 fibers as an internal standard. Changes of fiber phase content were deduced from the development of $I_{mullite}/I_{corundum}$, intensity ratios, where $I_{mullite}$ corresponds to the sum of 4 mullite x-ray diffraction peaks ((110) + (001) + (220) + (121)) and $I_{corundum}$ to the sum of 5 corundum peaks ((110) + (113) + (024)+ (124)+ (030)), respectively.

TEMPERATURE-DEPENDENT DEVELOPMENTS OF MECHANICAL STRENGTH AND OF MICROSTRUCTURE

Strength values of F5 fibers (gauge length: 5 mm) and of F100 fibers (gauge length: 100 mm) annealed for 2 hours at high temperatures are given in Table 1 and Figure 1. The absolute strength values obtained from F5 and F100 fiber measurements are different, due to a pure statistical volume effect, leading typically to higher strength values for shorter fibers. Therefore, relative values are compared instead of absolute data. At up to 1100°C, the short F5 fibers show no significant change of strength, while above this temperature limit, strength gradually decreases to reach a residual strength of ≈ 41% at 1500°C. At 1600°C complete fiber degradation takes place. The mechanical temperature stability of Nextel 720 fibers is significantly higher than those of 3M Nextel 440 and Sumitomo Altex 2 K alumino silicate fibers (Figure 2; see also Schmücker et al.[7]). The long F100 fibers that were heat-treated directly in the tensile strength testing furnace display behavior similar to the short F5 fibers up to 1350°C (Table 1, Figure 1). However, after annealing at 1400°C, F5 and F100 fibers behave quite differently. Five out of ten tested F100 fibers show a residual strength of ≈ 65%, which is similar to that of F5 fibers (≈ 50%). Three fibers display a dramatic strength decrease to a value of < 30%, while two other fibers totally degraded. Obviously, major degradation processes take place in a very narrow temperature interval of only 50°C.

TEM microstructural investigations on heat-treated, short F5 fibers show that grain growth and recrystallization processes take place at temperatures • 1200°C. Starting fibers consist of an admixture of very small α-Al$_2$O$_3$ crystals (mean grain size ♠ 70 nm) and of larger mullite grains (mean grain size ♠ 300 nm), the latter having a mosaic-type appearance. At up to 1200°C, this microstructure remains unchanged. At temperatures higher than 1200°C and up to 1400°C, grain growth is observed, which is stronger in the case of α-Al$_2$O$_3$ than in mullite. At temperatures above 1400°C, recrystallization of the mullite mosaic crystals, with disappearance of the small angle grain boundaries and polygonization of crystals, occurs. After heat-treatment at 1500°C, morphologies and grain sizes of mullite and α-Al$_2$O$_3$ are very similar (Figure 3). The increase of grain sizes of mullite and α-Al$_2$O$_3$ in Nextel 720 fibers as a function of temperature is shown in Figure 4.

Table 1. Tensile strength of annealed Nextel 720 alumino silicate fibers.

Annealing temperature [°]	20	1200	1300	1350	1400	1500
F5 fibers						
No. of measurements	10	9	10	/	10	10
Strength [MPa]	2585	2220	1943	/	1281	1050
Residual strength [%]	100	85	75	/	50	41
F100 fibers						
No. of measurements	30	/	10	10	9	/
Strength [MPa]	1510	/	1230	1197	980 (5) / 450 (3) / 0 (2)	/
Residual strength	100	/	81	79	62 / 21 / 0	/

<u>F5 fibers</u> were annealed in laboratory furnace. After heat-treatments, single fibers with testing lengths of 5 mm were mounted on the testing device.
<u>F100 fibers</u> with testing length of 100 mm were mounted into the testing machine, annealed, and subsequently cooled down to room temperature.
All strength measurements were carried out at room temperature.

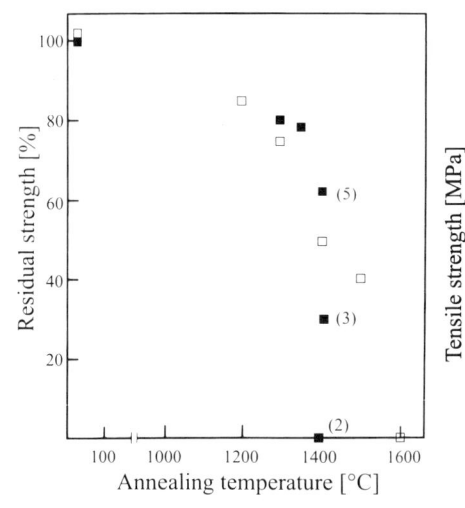

Figure 1. Residual room temperature strength of long (■ F 100) and short (□ F 5) fibers after two-hour heat treatments at various temperatures.

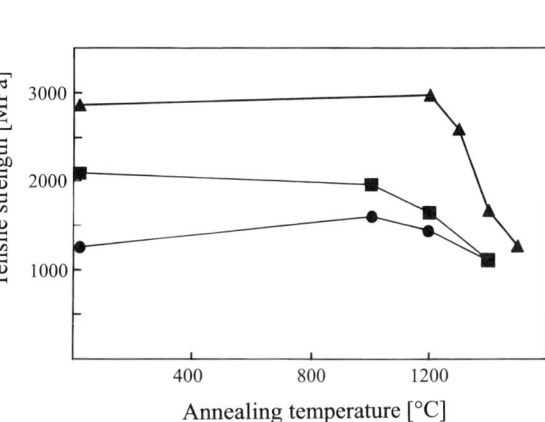

Figure 2. Residual room temperature strength of Nextel 720 (▲), Nextel 440 (■) and Altex 2 K fibers (○) after two-hour treatments at various temperatures.

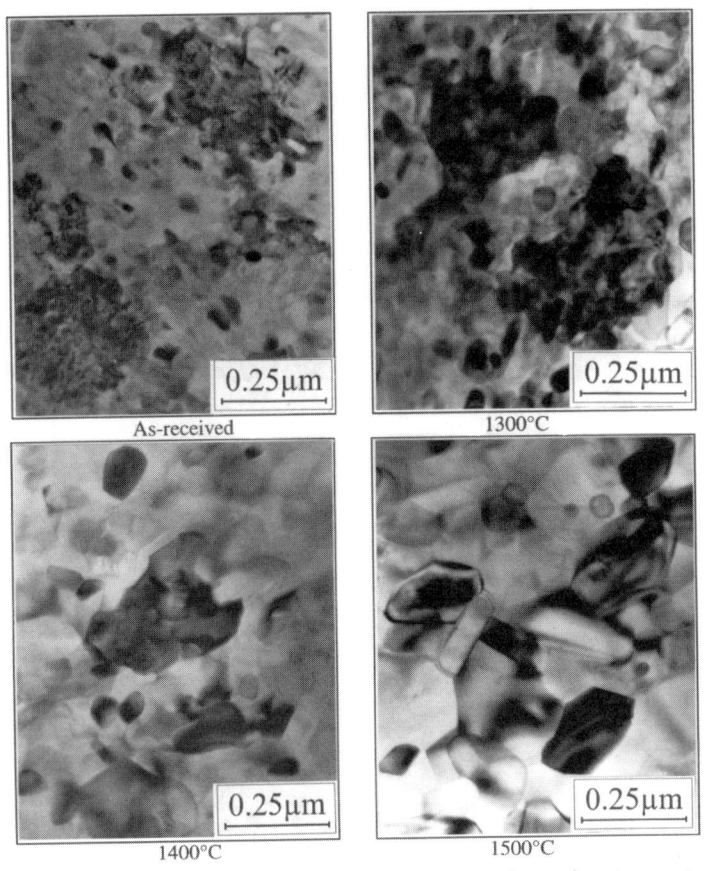

Figure 3. TEM-micrographs of Nextel 720 fibers heat-treated at various temperatures.

When heat-treated up to 1350°C, the Nextel 720 fibers typically fracture in a transgranular mode. The same is true of those F100 fibers surviving 1400°C annealing runs with little loss of strength. But F100 fibers that exhibit a catastrophic drop in residual strength at 1400°C have totally different fracture surfaces. They frequently show recrystallization with exaggerated grain growth directly at the fracture planes or on the fiber surfaces near to fracture planes (Figure 5).

The chemical composition of mullite at the different annealing stages of Nextel 720 fibers was determined using the correlation between Al_2O_3-content and a-lattice constants published by Cameron[10] and Fischer et al.[11] The a-lattice constant determination of the as-received fibers yields an Al_2O_3 mullite content of ≈ 66 mole %, which is typical for sol-gel-derived mullite crystallization below 1200°C.[11] At temperatures greater than 1200°C, a-lattice constants and corresponding Al_2O_3 contents gradually decrease, approaching the value of stoichiometric mullite (60 mole % Al_2O_3) at ≈ 1500°C. The exsolution of Al_2O_3 out of mullite along with temperature increase produces excess "free" Al_2O_3 leading to an increase of the α-Al_2O_3 fraction. Calculations on the basis of the bulk chemical composition of the fibers (85 wt.% Al_2O_3, 15 wt.% SiO_2), and on the Al_2O_3 contents of mullite, indicate that the α-Al_2O_3 fraction increases from ≈ 30 wt.% in the as-received state to ≈ 45 wt% at 1500°C. This increase of the α-Al_2O_3 fraction is expected to cause a volume contraction of ≈ 3.5% or a linear shrinkage of ≈ 1.0%. Actually, after the 1300°C heat-treatment, shrinkage values of ≈ 0.5% were determined, while firing at 1350°C and 1400°C caused length reductions of ≈ 0.7% and ≈ 0.8%, respectively. According to these data, ≈ 40% of the overall reaction takes place up to 1300°C, ≈ 60% up to 1350°C, and ≈ 70% up

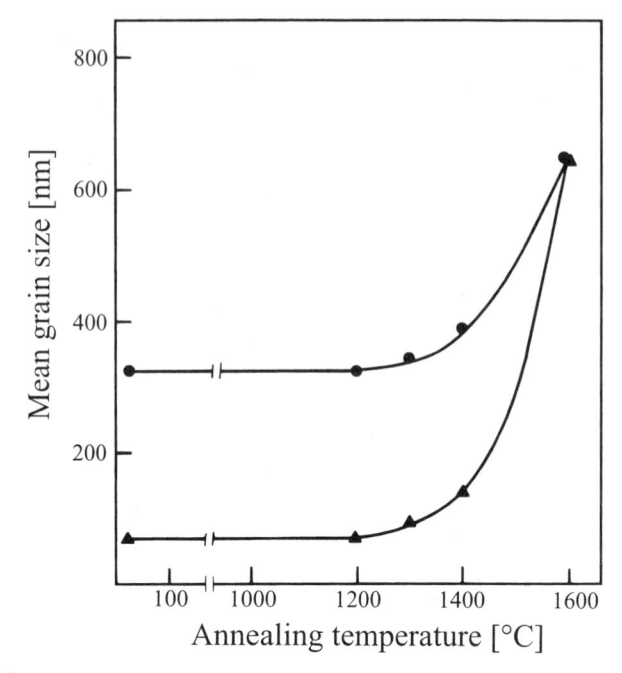

Figure 4. Temperature-dependent grain size development of mullite (●) and α-Al$_2$O$_3$ (▲) in Nextel 720 fibers.

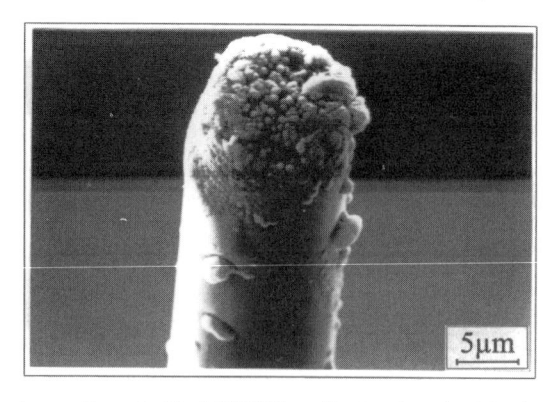

Figure 5. Fracture surface of a Nextel 720 fiber after two-hour heat-treatment at 1400°C.

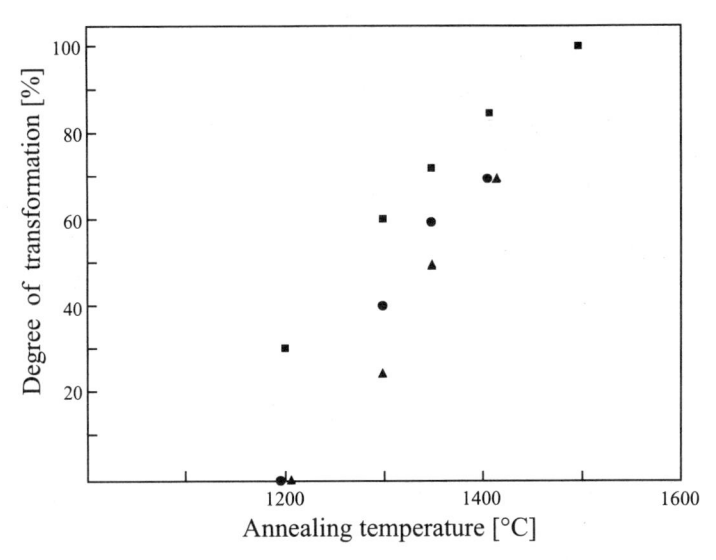

Figure 6. Temperature-dependent transformation of Al_2O_3-rich to stoichiometric mullite in Nextel 720 fibers as determined by lattice constants (▲), semi-quantitative phase analysis (■), and fiber shrinkage (○).

to 1400°C. These data are in good agreement with reaction rates determined on the basis of a-lattice constants and semi-quantitative x-ray diffraction phase analyses (Figure 6). The temperature-induced fiber shrinkage must be taken into account in considering the mechanical behavior of Nextel 720 fiber reinforced ceramic composites.

DISCUSSION

According to strength measurements and microstructural observations, thermal stability of Nextel 720 fibers is essentially controlled by grain coarsening and activation of critical defects. Our studies showed that 3M Nextel 720 fibers display higher starting strengths and also higher temperature stability than 3M Nextel 440 and Sumitomo Altex 2 K alumino silicate fibers (Figure 2).

In fibers annealed at 1200°C, a strength decrease of ≈ 15% is observed with respect to the starting value though no change of microstructure is determined. A possible explanation is locally starting grain growth: While strength is directly influenced by these local effects, it is difficult to detect areas with beginning grain coarsening with the sub-micrometer scale TEM probe. Above ≈1200°C and up to 1350°C, the residual strength in both F5 and F100 testing procedures are controlled by grain growth and recrystallization effects in the fibers. At temperatures above 1350°C, different strengths can be seen depending on which tensile strength testing method is used. While residual strength reduction of the short, F5 fibers at above 1350 and up to 1500°C is controlled by grain growth and recrystallization, most of the long, F100 fibers exhibit a dramatic residual strength loss between 1350 and 1400°C. The probable reason for the latter is the formation of critical defects in the very narrow temperature field between 1350 and 1400°C. At 1400°C, five critical defects occur in a total test length of 1000 mm (ten 100-mm test-length fibers). This could mean that a critical defect occurs every 200 mm in the average, assuming a homogeneous distribution of critical defects along the fiber axis with no defect accumulation. Starting from this suggestion, the explanation of the higher temperature stability of the F5 fibers is very easy: These fibers were annealed in a laboratory furnace and were subsequently mounted into the strength test equipment. This required further handling of the annealed fibers prior to testing, by which most fibers broke at critical defects. Since the critical distance is ≈ 200 mm (i.e. >> 5 mm) this handling automatically produced a selection: only intact volumes between the defects were taken for the F5 fiber tests.

There is little information on the mode of critical defects producing Nextel 720 fiber failure after high temperature treatment. In the optical microscope, formerly clear and transparent fibers become opacified after heat-treatment at 1400°C. Sometimes local increase of the fiber diameter with flowery patterns due to extreme grain coarsening is observed near to the defect areas (Figure 5). Possibly these effects may be associated with locally enriched Fe_2O_3 contents.* Due to the flux character of Fe_2O_3 in silicate matrices, higher amounts of low viscous liquids can be formed in these fiber areas, accelerating fast local recrystallization and crystal growth.

SUMMARY

A main result of our study is the finding that 3M Nextel 720 fibers have favorable high temperature stability. One key factor limiting this high temperature fiber stability is certainly grain growth. We believe that grain growth rates can be reduced by lowering α-Al_2O_3 contents of the fibers. This idea is derived from the fact that diffusion coefficients are higher in α-Al_2O_3 than in mullite[12] and, therefore, decrease of the amount of α-Al_2O_3 may reduce the mass transport necessary for grain growth. This is an important point, considering the increase of the α-Al_2O_3 contents in the heat-treated fibers, due to Al_2O_3 exsolution from mullite. In spite of this general tendency, some slight excess of α-Al_2O_3 may be favorable, in preventing the occurrence of glassy silicate grain boundary phases in the fibers. Mullite grain growth may be further reduced by controlled decoration of the grain boundaries of the larger mullite mosaic crystals by small α-Al_2O_3 crystallites or other suitable phases: Grain boundary decoration does reduce mullite grain boundary mobility and therefore decreases grain growth rates. A further important way of fiber improvement is to decrease the number of critical defects in the fibers. However, the latter will only be possible if the character of these defects is better understood.

ACKNOWLEDGMENTS

We thank Mrs. S. Schneider and Mr. H. Hermanns for writing the manuscript and for their help in preparing the figures.

REFERENCES

1. H. Schneider, H. Okada, and J. A. Pask, *Mullite and Mullite Ceramics*, Wiley & Sons, Chichester (1994).
2. D. Holz, S. Pagel, C. Bowen, S. Wu, and N. Claussen, Fabrication of low-to-zero shrinkage reaction-bonded mullite composites, *J. Europ. Ceram. Soc.,* **16**:255 (1996).
3. J. Brandt and R. Lundberg, Processing of mullite-based long-fiber composites via slurry route and by oxidation of an Al-Si alloy powder, *J. Europ. Ceram. Soc.,* **16**:261 (1996).
4. B. Saruhan, W. Luxem, and H. Schneider, Preliminary results on a novel fabrication route for α-Al_2O_3 single crystal monofilament-reinforced reaction-bonded mullite (RBM), *J. Europ. Ceram. Soc.,* **16**:269 (1996).

---

*According to analytical transmission electron microscopy (ATEM), the bulk Fe_2O_3-content of the Nextel 720 fibers is < 0.2 wt%.

5. K. K. Chawla, H. Schneider, M. Schmücker, and Z. R. Xu, Oxide fiber/ oxide matrix composites, in: *Proc. Processing and Design Considerations in High Temperature Materials, Davos, Switzerland*, TMS, Warrendale, PA, in press (1996).
6. J. A. Fernando, K. K. Chawla, M. K. Ferber, and D. Coffey, Effect of boron nitride coating on the tensile strength of Nextel 480 fiber, *Mater. Sci. & Eng.*, **A154**:103 (1992).
7. M. Schmücker, F. Flucht, and H. Schneider, High temperature behaviour of polycrystalline alumosilicate fibers with mullite bulk composition, Part I: Microstructural development and strength, *J. Europ. Ceram. Soc.*, **16**:281 (1996).
8. M. Schmücker, H. Schneider, K. K. Chawla, Z. R. Xu, and J.-S. Ha, Thermal degradation of fiber coatings in mullite-fiber-reinforced mullite composites, *J. Amer. Ceram. Soc.*, in press (1996).
9. K. K. Chawla, *Ceramic Matrix Composites*, Chapman & Hall, London (1993).
10. W. E. Cameron, Composition and cell dimensions of mullite, *Amer. Ceram. Soc. Bull.*, **56**:1003 (1977).
11. R. X. Fischer, H. Schneider, and M. Schmücker, Crystal structure of Al-rich mullite, *Amer. Min.*, **79**:983 (1994).
12. I. A. Aksay, Private communication.

TEXTURED CALCIUM HEXALUMINATE FIBER–MATRIX INTERPHASE FOR CERAMIC-MATRIX COMPOSITES

M. K. Cinibulk, R. S. Hay, and R. E. Dutton

Wright Laboratory Materials Directorate
Wright Patterson Air Force Base, OH 45433

ABSTRACT

The concept of an oxidation resistant fiber–matrix interphase that deflects cracks by cleavage is explored. Sols with the nominal composition $CaAl_{12}O_{19}$ were used to coat single-crystal alumina fiber and plates and yttrium-aluminum garnet fiber. Subsequent annealing, or consolidation in a matrix, results in strong texturing of the hibonite coating with basal planes parallel to the interface. Texture development within the hibonite interphase occurs by a complex series of phase transformations and reactions. Fiber pushout tests indicate high sliding resistance; however, strain-energy release rates of hibonite-bonded alumina laminates were found to be adequate to provide for crack deflection within the interphase via cleavage. Model composites containing hibonite-coated fibers were tested and showed evidence of crack deflection within the interphase.

INTRODUCTION

Most high-temperature applications for ceramic-matrix composites require that they be resistant to long-term oxidation. The feature of most current fiber-reinforced composites that is most susceptible to intermediate- and high-temperature degradation is the carbon or boron nitride fiber coating that is required to protect the fibers from matrix cracks.[1-3] With the need for increased chemical compatibility and stability in high-temperature oxidizing environments, alternative fiber coating concepts for oxide-based composites are being explored. One coating concept under consideration is the use of hexagonal aluminates with the β-alumina or magnetoplumbite structures, having high fracture-toughness anisotropy, with easy cleaving crystallographic planes oriented along the fiber axis, to encourage matrix-crack deflection away from the fiber.[4-7] A large number of alkali, alkaline-earth, and rare-earth cations can be used to stabilize the β-alumina and magnetoplumbite structures,[8] however, not all are stable at elevated temperatures. Therefore, we have limited our work to the alkaline-earth and rare-earth stabilized hexaluminates to avoid the incongruent vaporization of alkali oxides.[9] The β-alumina and magnetoplumbite structures are very similar, both consisting of layered spinel blocks $[Al_{11}O_{16}]^+$ with stabilizing cations and oxygen anions situated in mirror planes between the spinel layers, as shown in Fig. 1(a). The two structures differ in the arrangement of the stabilizing cations and oxygen anions in the mirror planes. The structure that is adopted is determined by the charge and radius of the cation in the mirror plane (Fig. 1(b)).[8]

Single crystals of these materials exhibit high fracture-energy anisotropy with G_b/G_p ~ 0.01, where the subscripts b and p indicate planes parallel and normal to the basal plane, respectively.[4, 10] Therefore, basal planes should deflect cracks since the fracture-energy ratio is less than 0.25.[11] The growth-rate anisotropy of the hexaluminates gives rise to an elongated or plate-like morphology, which is beneficial in obtaining a textured coating or interphase.

The phase development and reactions occurring in a calcium hexaluminate ($CaAl_{12}O_{19}$, the mineral hibonite) precursor that results in a highly textured film or fiber coating have been recently reported.[7] Here, we present a brief review of texture development and highlight some of the issues that are critical in forming the textured interphase *in situ* in alumina-matrix composites. The mechanical behavior of the interphase in model fiber-reinforced composites and laminates is also discussed.

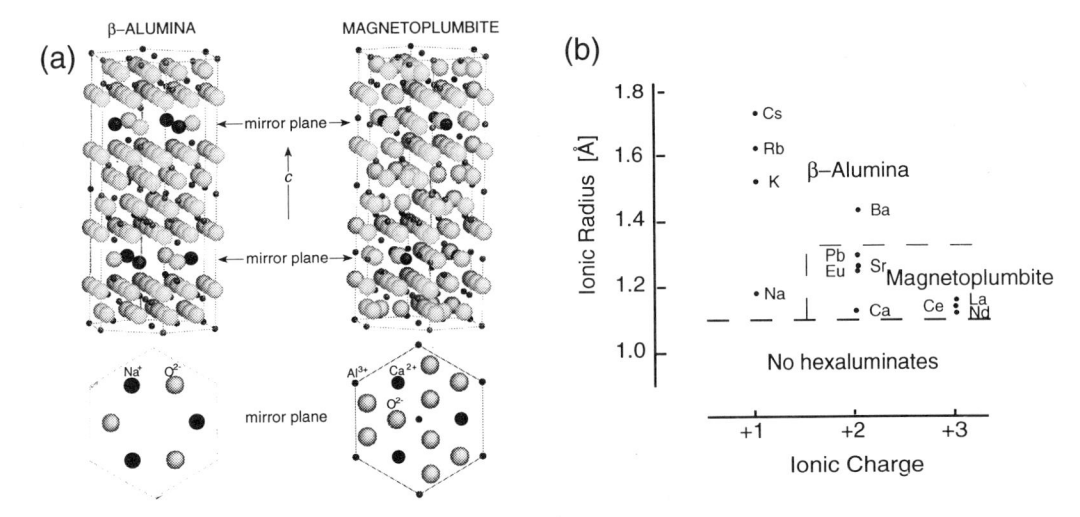

Figure 1. (a) Hexagonal β-alumina ($NaAl_{11}O_{17}$) and magnetoplumbite ($CaAl_{12}O_{19}$) structures. Below each structure is the interspinel mirror plane viewed down the c-axis. (b) Schematic depicting the stability of the hexaluminate structures as a function of ionic radius and charge of the stabilizing cation.

EXPERIMENTAL

Coating of Fibers and Plates

Colloidal $CaAl_{12}O_{19}$ sols were prepared by doping a boehmite (AlOOH) sol (DISPERAL Sol 10/2, CONDEA Chemie) with calcium acetate. Characterization of phase development in the bulk sol has been presented elsewhere.[7] Single-crystal *a*-axis-oriented alumina plates (Saphikon, Inc.) were coated with 50 g/L sol by spin coating at 2000 rpm to obtain an ~1 μm thick film.[6] The coated plates were pyrolyzed at 800°C prior to subsequent coating or bonding. Coated plates were bonded together at 1500°C for 1 h under a uniaxial pressure of 1 MPa. Single-crystal, *c*-axis-oriented alumina fiber and YAG (nominally [111]-

oriented) fiber (Saphikon, Inc.) were desized and coated with the sol in a continuous-fiber coating apparatus.[7, 12] The fibers were passed through a 60 g/L sol at ~40 mm/s and into the tube furnace at ~1450°C. Multiple passes were necessary to obtain a 1–2-μm thick coating. Some coated alumina fiber was passed just through the furnace for a second heat treatment at ~1 mm/s and a temperature of 1650°C. Some of the fibers were heat treated at 1300°–1500°C for times of 1 min to 60 min.

Composite Processing and Characterization

The coated alumina fibers were hot-pressed in a matrix of either alumina or YAG powder at 1500°C for 20 min under 25 MPa axial stress in vacuum to obtain unidirectional composites having a fiber volume of 1–5 %. Doping of the matrix with 0.5 wt% CaO was necessary to avoid loss of CaO from the coating. These composites were characterized for interphase microstructure and constituent compatibility. Sections were also prepared for pushout tests.

A second type of composite, for tensile testing, was prepared by tape casting.[13] A single row of hibonite-coated alumina fibers, first heat-treated at 1400°C for 1 h, were laminated within tapes of a potassium-borosilicate glass at 90°C under 0.1 MPa uniaxial pressure. The composite was sintered at 710°C in vacuum for 1 h, followed by 20 min at 710°C under 0.1 MPa argon. The composite was then hot-isostatically pressed at 650°C and 35 MPa argon for 30 min to remove any residual porosity. The fiber fraction was less than 0.01 in the composites, having dimensions 50 mm parallel to the fiber axes, 20 mm wide and 1 mm thick. The composites were tested in tension.

Coated fibers, composite cross-sections and fracture surfaces were characterized by light microscopy (LM) and scanning electron microscopy (SEM). Petrographic thin sections of composites were prepared by mechanically thinning a section of the composite normal to the fiber axes, using diamond abrasives, and then ion-beam thinning. Transmission electron microscopy (TEM) and energy-dispersive X-ray spectroscopy (EDS) were carried out on both coated fibers and composites

OBTAINING AN HIBONITE INTERPHASE IN COMPOSITES

The as-coated fibers had a polyphasic coating, consisting of α-Al_2O_3, $CaAl_{12}O_{19}$, and other more-CaO-rich phases. Various combinations of coated and uncoated fibers and matrices were used to study the formation and/or retention of a calcium hexaluminate interphase during hot-pressing. Table 1 summarizes the constituents and interphase microstructures formed. A few of the important issues regarding formation and/or retention of a textured hibonite interphase are discussed below.

Early attempts to form an hibonite interphase in situ by reacting excess CaO in a YAG matrix with a single-crystal alumina fiber did not yield the desired microstructure.[7] Since the coupled diffusion of Ca^{2+} and O^{2-} from calcia-rich to alumina-rich phases controls the formation of hibonite,[14, 15] the reaction front consumes the fiber in a radial direction. Rapid lattice diffusion down the basal planes of hibonite produces oriented grains with basal planes normal to the reaction front and aligned with the fiber radius. This texture is perpendicular to the desired texture of hibonite basal planes aligned with the plane of the interface.

In the composites that contained as-coated fibers with the polyphasic coatings, it was necessary to dope the matrix with CaO to prevent solution and segregation of CaO from the coating to the matrix grain boundaries. The coating was lost in the composite made with undoped alumina as the matrix, shown in Fig. 2(a). Consequently the fiber seeded abnormal

Table 1. Constituents and interphase microstructure obtained by hot-pressing

Fiber	Matrix	Fiber coating	Interphase microstructure
Alumina	4-wt% calcia-doped YAG	—	Growth of hibonite grains into fiber
Alumina	alumina	Polyphase, as-coated	No interphase, fiber growth
Alumina	0.5-wt%-calcia-doped alumina	Polyphase, as-coated	Textured interphase, fiber growth
Alumina	0.5-wt%-calcia-doped alumina	Single-phase hibonite	Textured interphase
Alumina	0.5-wt%-calcia-doped YAG	Polyphase, as-coated	Textured interphase
YAG	0.5-wt% calcia-doped alumina	Polyphase, as-coated	Textured interphase

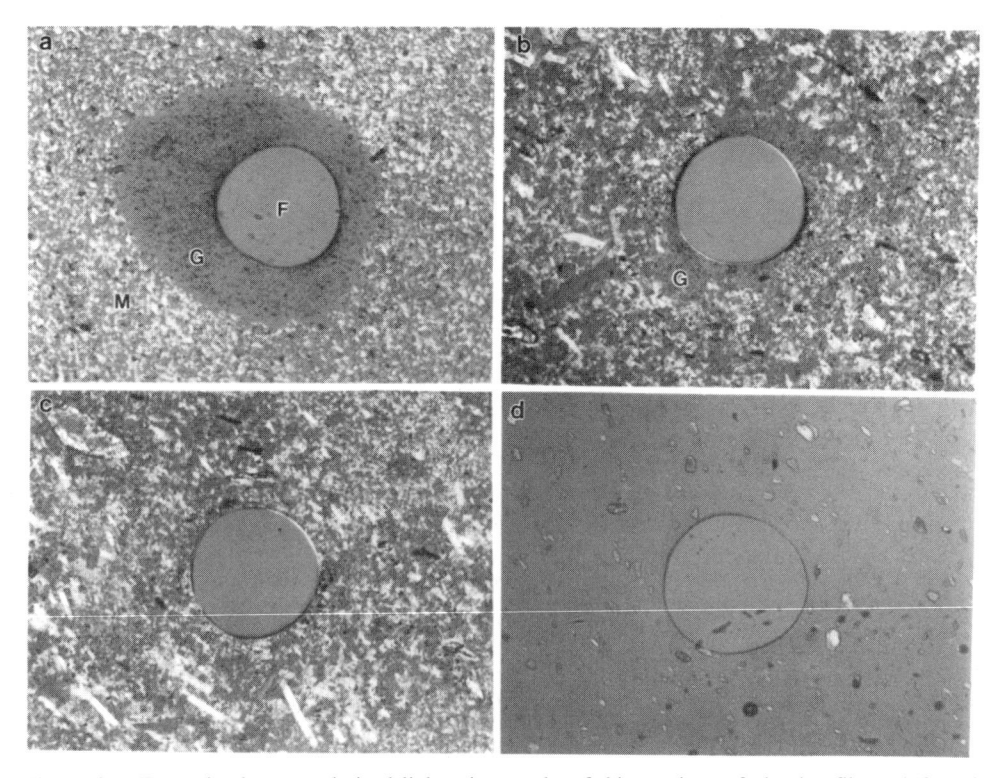

Figure 2. Transmitted cross-polarized light micrographs of thin sections of alumina-fiber reinforced composites. (a) Coated fiber (F) in alumina matrix (M) showing loss of coating and seeding of matrix, resulting in growth (G) of fiber. (b) Coated fiber in CaO-doped alumina matrix with bright contrast around fiber indicating textured hibonite interphase. (c) Heat-treated fiber in CaO-doped alumina. (d) Coated fiber in YAG matrix. Fiber diameter is ~135 μm.

grain growth in the matrix, with the original fiber–matrix interface delineated by porosity with trace levels of Ca on the pore surfaces, as measured by EDS. Doping of the matrix with 0.5-wt% CaO countered the loss of CaO from the coating and reduced, though did not eliminate, seeding of the matrix by the fiber (Fig. 2(b)).

To eliminate seeding of the matrix by the fiber, the coating had to be first fully converted to hibonite. Heat treating the fiber prior to incorporating it into an alumina matrix allowed a textured coating to form (Fig. 2(c)). Hot-pressing did not contribute to interphase development and, therefore, seeding of the matrix did not occur. Hot-pressing either the as-coated or heat-treated fibers in a YAG matrix also resulted in a textured interphase with no abnormal grain-growth in the matrix by fiber seeding (Figs. 2(d) and 3). Coated YAG fiber hot-pressed in alumina also produced the desired interface microstructure.

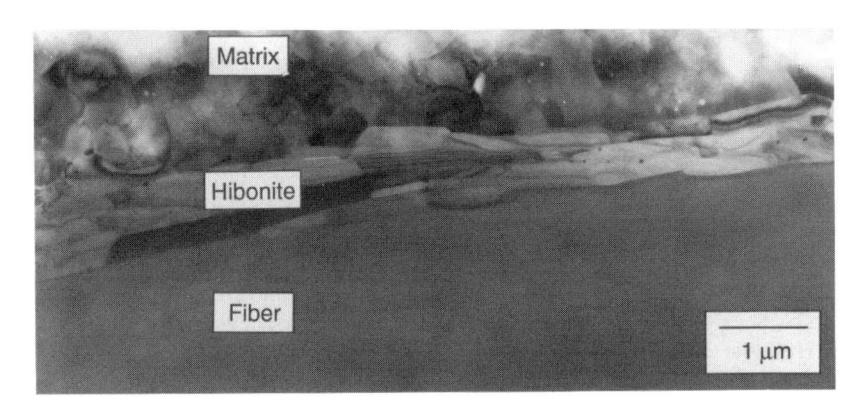

Figure 3. TEM image of the textured interphase in alumina-fiber-reinforced YAG-matrix.

DEVELOPMENT OF TEXTURE IN HIBONITE FIBER COATINGS

The development of texture in hibonite coatings and interphases is depicted schematically in Fig. 4. Supporting evidence for the proposed reaction sequence and texture mechanism was observed by TEM in coatings and interphases at various stages of evolution and was recently reported in detail.[7,16] A brief summary is given below.

Hibonite coatings are deposited as a sol, which gels to a mechanical mixture of colloidal alumina and calcia. In the case where the film is either freestanding or deposited on an inert substrate (such as YAG), reaction to and nucleation and growth of hibonite grains in random and/or preferred orientations (along basal planes) occurs through $CaAl_4O_7$, an intermediate phase. As the reaction to hibonite proceeds and reaches completion, grains with basal planes parallel to the substrate continue to grow along the rapid-growth basal planes at the expense of grains with other orientations. Subsequent abnormal grain growth of hibonite is essentially confined to within the plane of the thin film, possibly enhanced by interface-energy anisotropy.

If the substrate is single-crystal alumina, either a fiber or a plate, a more complex series of steps occurs in the development of basal texture. As the colloidal alumina and calcia coating is fired, e.g., in the fiber coater furnace, syntactic seeding of α-Al_2O_3 in the coating by the fiber appeared to be accompanied by a simultaneous diffusion of CaO away from the phase transformation front, forming α-Al_2O_3 tooth-like extensions of the single-crystal fiber ~100–200 nm wide. These teeth are separated by CaO-rich phases, including $CaAl_{12}O_{19}$,

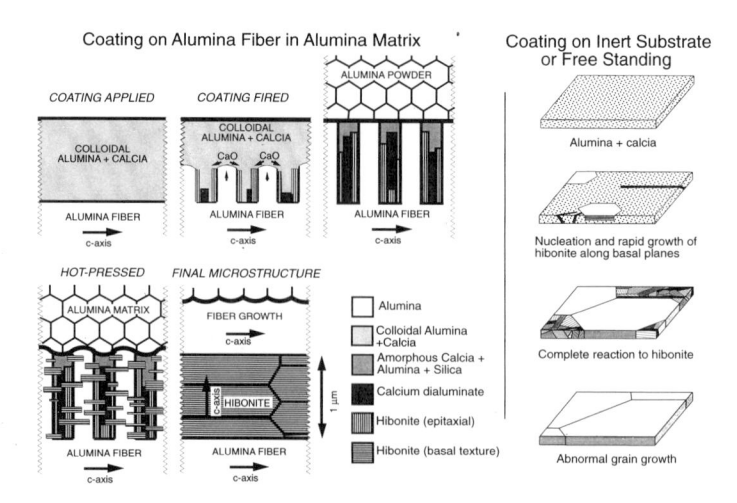

Figure 4. Schematic of texture development in hexaluminate films on an alumina fiber in an alumina matrix or as a free standing film.

which forms epitaxially on the (0001) sides of the α-Al_2O_3 extensions. The fiber leaves the coater with a polyphase coating consisting of alumina teeth spanning the thickness of the coating, hibonite epitaxially formed on the alumina, with basal planes extending radially outward, and other calcia-rich phases to maintain mass balance. Upon further heating the calcia-rich phases not compatible with alumina react with the syntactic alumina teeth to form hibonite. Hibonite grains with the fast-growing basal planes parallel to the reaction path (axial) dominate. Interface-energy anisotropy eventually favors those grains with basal planes oriented parallel to both the reaction direction and the fiber surface. Abnormal grain growth continues in the coating in a manner analogous to that discussed above, giving the basal-textured hibonite film.

When fiber with the incompletely reacted polyphase coating is hot-pressed in calcia-doped alumina, the syntactic alumina extensions seed abnormal grain-growth of the matrix prior to complete formation of an hibonite interphase. The resulting microstructure is that of a tubular polycrystalline inclusion of basal-textured hibonite inside an enlarged single-crystal alumina fiber. When the fiber is YAG, texture formation in the coating does not involve the fiber; therefore, seeding of the matrix does not occur.

MECHANICAL BEHAVIOR OF INTERPHASE

During TEM specimen preparation, cracking of the foil invariably occurred. Cracks were frequently observed during TEM characterization in the vicinity of the fibers. Upon interaction with the interphase, these cracks always were deflected within the textured hibonite and a single transgranular crack usually ran along the basal planes around the fiber. In Fig. 5, a crack can be seen cleaving grains in the interphase, leaving a smooth debond interface. The formation of a rough debond interface can also occur, due to intergranular crack propagation at grain boundaries with steps of up to 0.5 µm in size. Significant compressive stresses can develop upon attempting to push or pull a fiber out of the matrix with this degree of interfacial roughness, preventing fiber sliding.[6,17]

The tape-cast composites contained a volume fraction of fibers that was too small to influence the stress–strain behavior of the matrix; i.e., the modulus of the composite was identical to that of the matrix alone. Nevertheless, characterization of the fracture surface

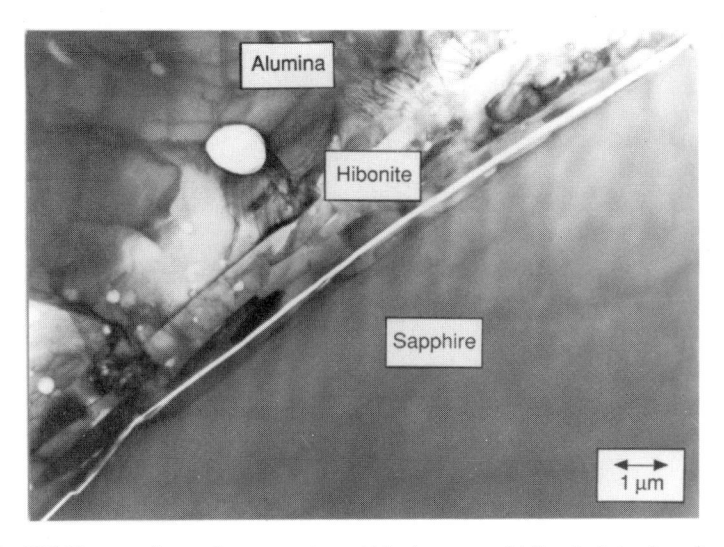

Figure 5. TEM image of a crack propagating within the textured hibonite interphase by cleavage.

revealed information on the behavior of the textured hibonite interphase. The crack plane was slightly inclined from the normal of the axis of the fibers. Fiber–matrix debonding was observed to have occurred by deflection of the crack within the hibonite interphase. Usually, only the side of the fiber facing the direction of the crack was exposed, as shown in Fig. 6(a). Figure 6(c) shows the matching fracture surface of the one in Fig. 6(a), but at a different angle. Figures 6(b) and (d) are higher magnifications of the surfaces of the fiber and of the trough from which the fiber was separated, respectively. The presence of the hibonite interphase can be seen attached to both the fiber and trough at the same location. Therefore, the crack must have propagated within the interphase. The flat regions on the debonded fiber and matrix are cleaved grains. The roughness of the surface arises from the stepping of the crack at grain boundaries, which is probably the reason for the limited fiber pullout observed in this composite. Attempts to determine the interfacial shear stress were complicated by the radial compressive stress arising from clamping of the fiber by the steps.

Strain-energy release rates of hibonite-bonded single-crystal alumina plates were measured by loading a precracked specimen in four-point bending.[18] Strain-energy release rates of the order of 2.2 J/m^2 were obtained (Fig. 7(a)), confirming the low fracture-energy suggested by the intragranular propagation of cracks observed in the interphase of TEM specimens and the tape-cast composites. X-ray diffraction and SEM of the debonded surfaces revealed cleavage of hibonite as the mode of crack propagation through the textured hibonite interlayer (Fig. 7(b)).

CONCLUSIONS

Fibers and plates were coated with precursor sols, which transformed to a textured hibonite film by a complex series of phase transformations and reactions. The textured fiber coating could be retained after hot-pressing in an alumina or YAG matrix, or formed *in situ*, under the right processing conditions. The current temperature of 1400°C required for complete reaction and abnormal grain growth for texturing to occur is too high for commercially available oxide fibers, such as Nextel 610 and 720, which are generally limited to a maximum temperature of 1300°C. Alternate precursors to hibonite or other hexaluminates may allow for the coating, and subsequent processing, of these fibers at lower temperatures.

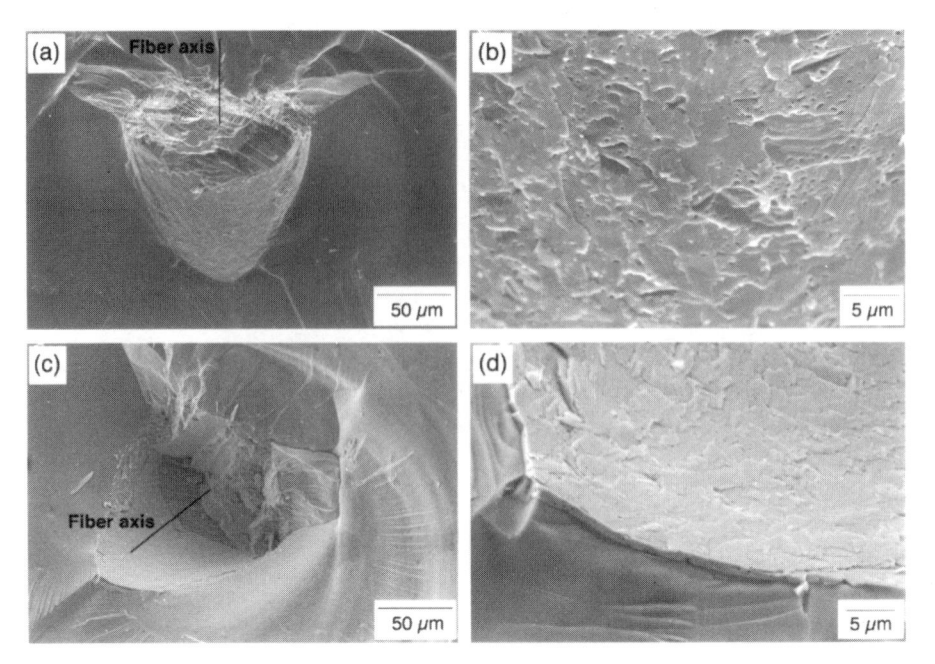

Figure 6. SEM images of two mating fracture surfaces in a fractured tape-cast composite. (a) and (c) show the two halves of the fiber, which has partially pulled out of the matrix. The higher magnification images are of opposing surfaces showing the hibonite interphase bonded to (b) the fiber and (d) the matrix, indicating that the crack deflected within the interphase and propagated by cleavage.

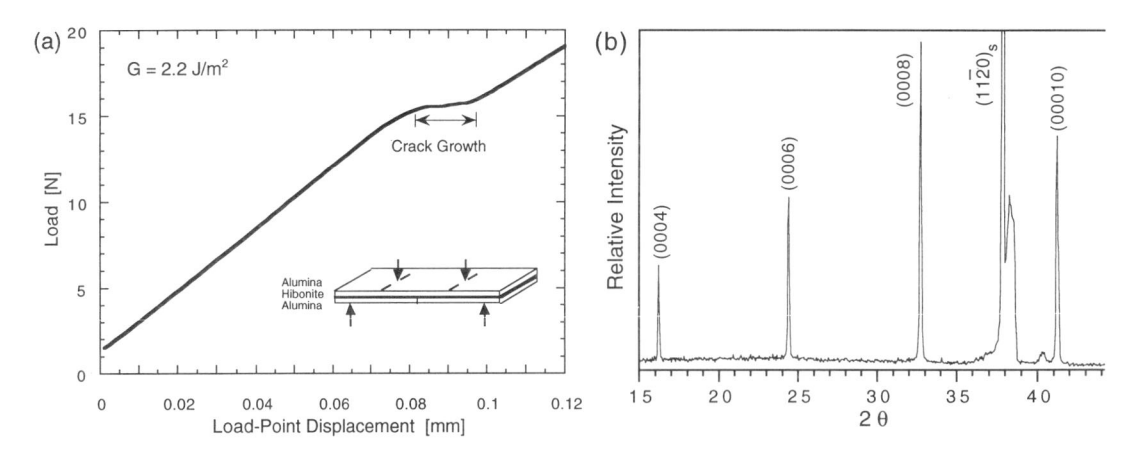

Figure 7. (a) Plot of load vs. displacement of the laminated sample shown inset. Plateau in the curve corresponds to steady-state propagation of the crack through the hibonite interphase, between the upper load points. (b) X-ray diffraction pattern of one face of the debonded laminate, showing only hibonite diffraction peaks of the type (000l), indicating a high degree of basal texture and delamination by cleavage through the interphase. The (11$\bar{2}$0) peak is from the alumina substrate. The other debonded face gave indentical results.

The fracture energy of cleavage of a textured polycrystalline hibonite interlayer has been determined to be about 2.2 J/m^2. Observations of fracture surfaces in model composites indicate crack deflection within the hibonite coating, with subsequent crack propagation occurring by cleavage. These observations indicate that a textured, cleavable oxide interphase should be a viable fiber coating for alumina-based composites. However, the interfacial shear stresses developed after debonding during fiber pullout are likely to be very high in composites containing a thick interphase, but thinner fiber coatings may lower these stress to an acceptable range.

ACKNOWLEDGMENTS

We thank J. L. Schuck for processing the tape-cast composites. MKC was supported by AFOSR under Contract No. F336150-91-C-5663 with UES, Inc.; RED was supported by NIST.

REFERENCES

1. T. Mah, M.G. Mendiratta, A.P. Katz, R. Ruh, and K.S. Mazdiyasni, High-temperature mechanical behavior of fiber reinforced glass-ceramic-matrix composites, *J. Am. Ceram. Soc.* 68:C-248 (1985).
2. E. Bischoff, M. Rühle, O. Sbaizero, and A.G. Evans, Microstructural studies of the interfacial zone of a SiC-fiber-reinforced lithium aluminum silicate glass-ceramic, *J. Am. Ceram. Soc.* 72:741 (1989).
3. S.S. Lee, L.P. Zawada, J.M. Staehler, and C.A. Folsom, Mechanical behavior and high-temperature performance of a woven Nicalon/SiNC ceramic matrix composite, *J. Am. Ceram. Soc.* in review.
4. P.E.D. Morgan and D.B. Marshall, Functional interfaces for oxide/oxide composites, *Mater. Sci. Eng.* A162:15 (1993).
5. M.K. Cinibulk, Magnetoplumbite compounds as a fiber coating in oxide/oxide composites, *Ceram. Eng. Sci. Proc.* 15:721 (1994).
6. M.K. Cinibulk, Microstructure and mechanical behavior of an hibonite interphase in alumina-based composites, *Ceram. Eng. Sci. Proc.* 16:633 (1995).
7. M.K. Cinibulk and R.S. Hay, Textured magnetoplumbite fiber–matrix interphase derived from sol–gel fiber coatings, *J. Am. Ceram. Soc.* 79:1233 (1996).
8. N. Iyi, S. Takekawa, and S. Kimura, The crystal chemistry of hexaluminates: β-alumina and magnetoplumbite structures, *J. Solid State Chem.* 83:8 (1989).
9. M.K. Cinibulk, Thermal stability of some hexaluminates at 1400°C, *J. Mater. Sci. Lett.* 14:651 (1995).
10. D.C. Hitchcock and L.C. De Jonghe, Fracture toughness anisotropy of sodium-β-alumina, *J. Am. Ceram. Soc.* 66:C-204 (1983).
11. M.Y. He and J.W. Hutchinson, Crack deflection at the interface between dissimilar materials, *Int. J. Solids Struct.* 25:1053 (1989).
12. R.S. Hay and E.E. Hermes, Sol-gel coatings on continuous ceramic fibers, *Ceram. Eng. Sci. Proc.* 11:1526 (1990).
13. C.M. Gustafson, R.E. Dutton, and R.J. Kerans, Fabrication of glass matrix composites by tape casting, *J. Am. Ceram. Soc.* 78:1423 (1995).
14. I. Kohatsu and G.W. Brindley, Solid state reactions between CaO and α-Al$_2$O$_3$, *Z. Physik. Chemie N. F.* 60:79 (1968).
15. R.C. Ropp and B. Carroll, Solid-state kinetics of LaAl$_{11}$O$_{18}$, *J. Am. Ceram. Soc.* 63:416 (1980).
16. M.K. Cinibulk, Synthesis and characterization of sol–gel derived lanthanum hexaluminate powders and films, *J. Mater. Res.* 10:71 (1995).
17. R.J. Kerans, The role of coating compliance and fiber/matrix interfacial topography on debonding in ceramic composites, *Scripta Metall. Mater.* 32:505 (1995).
18. P.G. Charalambides, J. Lund, A.G. Evans, and R.M. McMeeking, A test specimen for determining the fracture resistance of bimaterial interfaces, *J. Appl. Mech.* 56:77 (1989).

MICROSTRUCTURE OF THE PARTICULATE COMPOSITES IN THE (Y) TZP - WC SYSTEM

K. Haberko and Z. Pedzich
University of Mining and Metallurgy, Department of Special Ceramics,
al. Mickiewicza 30, 30-059 Krakow, Poland

J. Dutkiewicz and M. Faryna
Institute of Metallurgy and Material Science of the Polish Academy of Sciences, Krakow

A. Kowal
Institute of Catalysis of the Polish Academy of Sciences, Krakow

INTRODUCTION

Tetragonal zirconia polycrystals (TZP) show high fracture toughness and strength. But the relatively low hardness of these materials limits their potential applications. Hardness, wear resistance, and fracture toughness could be improved by incorporating hard carbide inclusions into the TZP matrix. Earlier investigations have shown that the additives of the silicon carbide whiskers,[1,2] platelets,[3] and isometric particles[4] lead to improved body properties. Unfortunately, densification of such materials under pressureless conditions was not sufficiently high. Thus, hot uniaxial or isostatic pressing had to be applied.

Our previous work[5-7] has pointed out that incorporating tungsten carbide particles into the yttria (Y)TZP matrix improved its mechanical and useful properties, especially wear resistance. It is worth noticing that such composites could be sufficiently densified under pressureless sintering conditions at relatively low temperatures. Because of the carbide component, oxygen free (argon) atmosphere had to be used. The samples were fired in a carbon bed.

Thermodynamic calculations[5] allowed us to state that under the applied sintering conditions, the reaction between ZrO_2 and WC could result in the formation of several tungsten oxides. But

$$ZrO_2 + 6\ WC = ZrC + 3\ W_2C + 2\ CO \qquad (1)$$

could proceed towards the right-hand side, providing that the CO partial pressure is lower than the limiting values indicated in Table 1.

Table 1. Limiting CO partial pressure (atm)

Temperature, °C	1400	1500	1600	1700
Reaction 1	0.95	5.20	23.8	93.0

The microstructure of the materials under discussion influences greatly their properties. The aim of the present work was to recognize the mutual arrangement of the phases within the TZP-WC composites and changes to this arrangement in relation to sintering temperatures.

EXPERIMENTAL

The 2.8 mole % Y_2O_3 - 97.2 mole % ZrO_2 solid solution was selected as a matrix material. The starting powder was prepared by the coprecipitation-calcination technique.[8] Commercially available tungsten carbide powder (Baildon, Poland) was applied. Table 2 shows the characteristics of the starting powders.

Table 2. Characteristics of the starting powders.

Material	Phase composition	S* [m²/g]	D_{BET}† [nm]	D_{111}** [nm]
2.8% Y_2O_3-ZrO_2	91 vol.% tetragonal 9 vol.% monoclinic	51.9	19.1 ± 0.2	18.4 ± 0.2
WC	WC, ~6 wt.% C, traces W_2C and W	0.23	1600 ± 600	–

*S = specific surface area measured by the BET method
†D_{BET} = 6/S.r, where r is the physical powder density
**D_{111} = crystallite size determined by the X-ray line broadening

One-hour attrition milling in ethanol using 2 mm Y-TZP balls secured good homogenization of the systems. Carbide additives ranged from 10 to 30 vol.%. Dry composite powders with no lubricants were compacted uniaxially under 50 MPa and then repressed isostatically under 350 MPa. By this method, cylindrical samples 23 mm in diameter and 10 mm thick and bars 4 mm × 3 mm in cross section and 45 mm long (after sintering) were prepared. Sintering was performed at temperatures ranging from 1400 to 1700°C with 2 hour soaking at the peak temperature.

Bulk densities of the sintered samples were determined by hydrostatic weighing and recalculated to relative densities. Elastic properties were measured ultrasonically.[9] Vickers indentation allowed us to measure hardness and fracture toughness. In the latter case, Palmqvist's crack model was applied.[10] Flexural strength was determined by four point bending. Wear resistance was found by a method similar to the "Dry Sand Test," using SiC grains (0.4 to 0.5 mm) as an abrasive medium.

RESULTS AND DISCUSSION

Mechanical Properties

Table 3 shows mechanical properties of the composites and the "pure" matrix and volume of the material removed during the wear test.

Tungsten carbide inclusions lead to the decreased densification of the sintered samples. The effect is stronger as the WC volume fraction increases. This phenomenon can be

Table 3. Properties of the matrix and composites.

Material / / Sintering temp.		Density* [% theo.]	Young modulus E, [GPa]	Vickers hardness HV [GPa]	Fracture toughness K_{ic} [MPam$^{0.5}$]	Flexural strength s, [MPa]	Wear [mm^3]
matrix	1400	99.9	217±2	14.7±0.5	6.8±0.6	-	11.77
	1500	99.6	221±2	14.4±0.5	6.8±0.2	695±90	12.05
10% WC	1400	98.7	233±3	16.2±1.3	7.3±0.8	-	5.29
	1500	98.7	238±3	16.5±1.5	8.6±0.7	570±100	6.24
20% WC	1400	97.4	255±3	15.0±0.5	7.8±0.8	-	5.61
	1500	97.1	259±3	16.8±1.6	8.6±0.7	675±115	6.92
	1600	96.8	255±2	14.7±0.4	8.8±0.3	-	8.83
	1700	93.6	173±1	12.7±2.0	5.8±0.6	-	8.99
30% WC	1400	94.6	263±2	11.8±0.8	8.6±1.1	-	9.49
	1500	95.7	274±4	12.3±0.6	6.7±0.7	650±100	7.77

*Relative densities were calculated on the basis of the WC (15.7 g/cm^3) and Y-TZP (6.1 g/cm^3) densities and their proportions in the mixtures.

plausibly attributed to the hindering influence of the inclusions of volume concentration below the percolation threshold on the densification of a composite.[11] Moreover, at the highest applied sintering temperatures, samples of the decreased density occur. It will be shown further that this effect is related to the reaction between ZrO_2 and WC shown in Equation (1), resulting in gaseous product release, which results in the secondary porosity formation.

The WC - ZrO_2 composites show increased Young's modulus, fracture toughness and hardness compared to the pure matrix. Flexural strength remains practically unchanged. An essential improvement of the wear resistance is observed.

X-Ray Diffraction

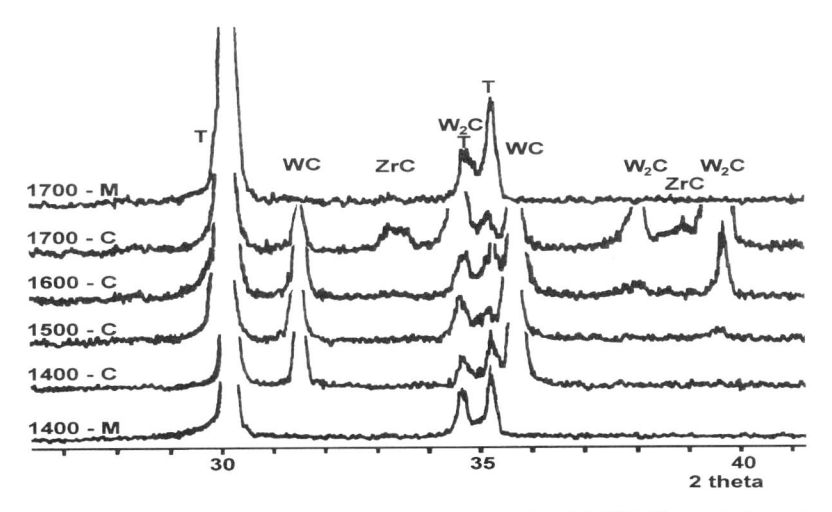

Figure 1. X-ray diffraction patterns of the composites containing 20 vol.% WC (C) and the "pure" matrix (M) sintered at indicated temperatures.

Figure 1 demonstrates x-ray diffraction patterns of composites containing 20 vol.% WC and the pure matrix sintered at indicated temperatures. The matrix remains mainly tetragonal. A certain content of the cubic phase is also observed because of the deoxidation of the yttria - zirconia solid solution.[5,12] In the composite sintered at 1400°C, only WC and zirconia solid solution have been identified. At higher temperatures the reaction between composite components result in the W_2C and ZrC formation. W_2C becomes the dominating tungsten carbide phase in the samples sintered at 1700°C. Since the CO partial pressure, even at 1400°C, is lower than indicated in Table 1, the reaction of Equation (1) is also expected to proceed at this temperature. Therefore the lack of its products in the X-ray diffraction pattern suggests the low advance of this reaction. This conclusion was confirmed by the TEM investigations: W_2C and ZrC particles were observed under the transmission electron microscope in the samples sintered at 1400°C.

SEM, AFM, AND TEM OBSERVATIONS

The microstructures of the composites sintered at different temperatures are shown in Figure 2. They prove the zirconia matrix grain growth. This is illustrated quantitatively by the data of Figure 3. The plots show the mean grain size of the matrix calculated by the Saltykov reverse diameter method.[13] According to the Zener rule the matrix grain growth is restrained by the carbide inclusions.

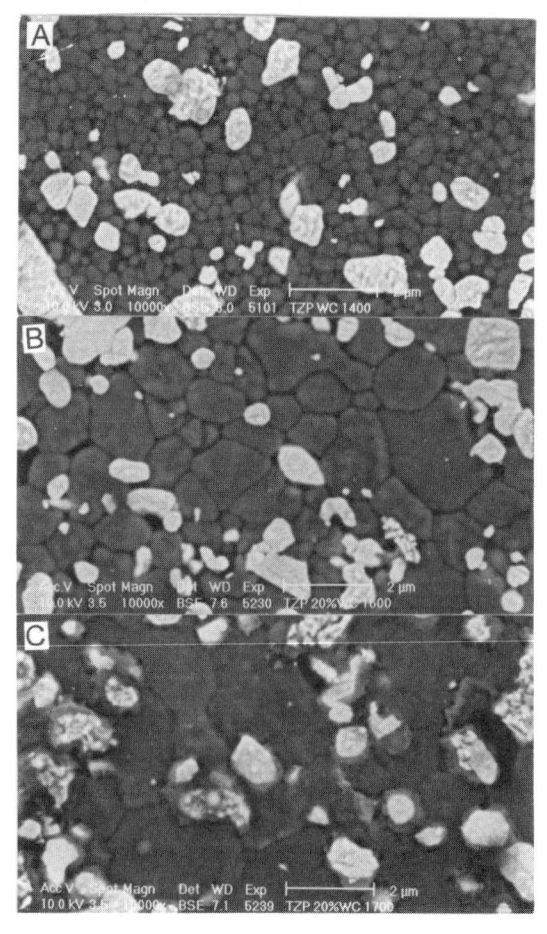

Figure 2. SEM micrographs of the composites containing 20 vol.% WC: (a) Sintering temperature 1400°C. (b) Sintering temperature 1600°C. (c) Sintering temperature 1700°C.

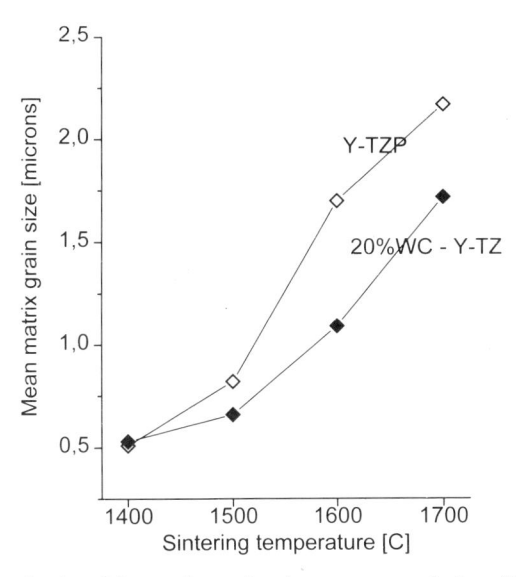

Figure 3. The mean grain size of the matrix vs. sintering temperature in "pure" matrix and composite containing 20 vol.% WC.

In the samples sintered at 1700°C, the secondary porosity appears close to the carbide grains. It is visible both in the SEM micrograph (Figure 2(c)) and the AFM one (Figure 6(b), (c)). As has been pointed out, this porosity should be related to the carbon monoxide release caused by the reaction between WC and ZrO_2. Rims surrounding carbide inclusions are visible. The EDS analysis (Figure 4) shows that both tungsten and zircon are present within the rim.

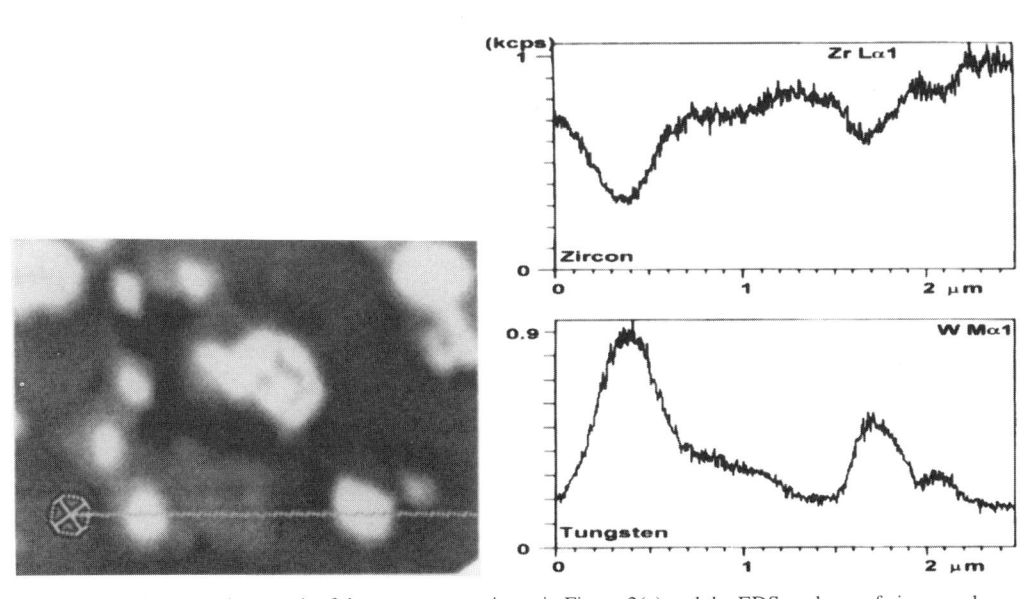

Figure 4. SEM micrograph of the same composite as in Figure 2(c) and the EDS analyses of zircon and tungsten. The line along which analyses have been made is indicated in the micrograph.

The AFM investigations of the polished and thermally etched samples show that during polishing, the rims around the WC grains in the material sintered at 1700°C behave in a intermediate way between the tungsten carbide core and the zirconia matrix (Figures 5 and 6). This is demonstrated by the step in the surface profile (Figure 6(a)). No phenomenon like this occurs in the sample sintered at the lower temperature of 1400°C (Figure 5(a)). The data presented confirm the conclusions coming from the x-ray diffraction on the extensive reaction proceeding between zirconia and tungsten carbide at high temperatures. Plausibly, the rims around carbide inclusions contain the products of the reaction between ZrO_2 and WC.

Close adhering of the matrix and carbide grains with no discontinuities occurs in the composites. Figure 7(a) demonstrates the TEM micrograph of the system sintered at 1400°C. The hexagonal WC and tetragonal ZrO_2 grains are visible. The identification of the grains is based on indexing of the electron diffraction patterns (Figure 7(b) and 7(c)). The diffraction pattern taken from the border zone between these grains (Figure 7(d)) indicates the existence of the following crystallographic relationship of the WC and tetragonal ZrO_2 particles:

$$[0001]\ WC \quad | \quad [001]\ t\text{-}ZrO_2$$
$$[\bar{1}010]\ WC \quad | \quad [010]\ t\text{-}ZrO_2$$

This correlation was observed in several cases.

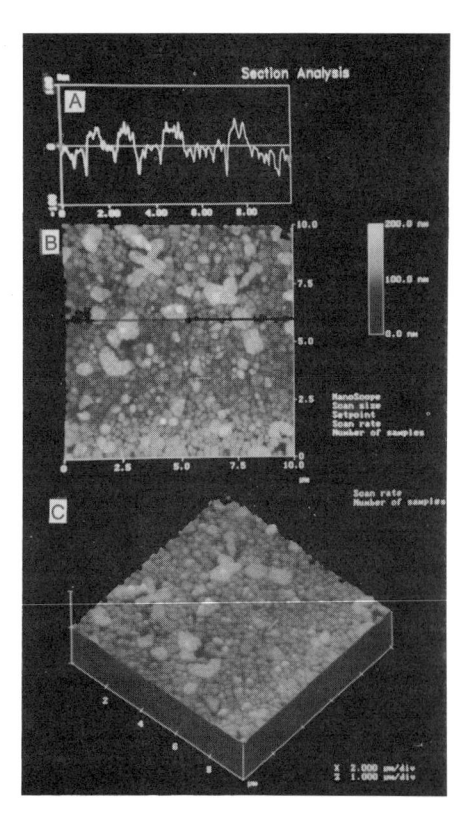

Figure 5. The surface of the composite containing 20 vol.% WC inclusions sintered at 1400°C visualized by the AFM technique. (a) Surface profile along the line indicated in B, C- three dimensional projection of the B.

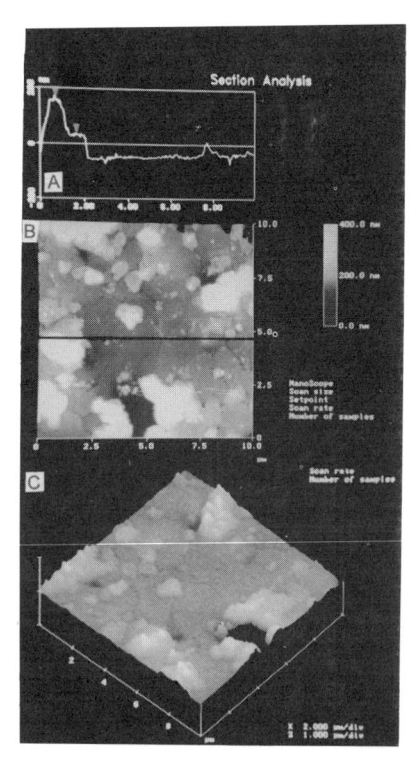

Figure 6. The surface of the composite containing 20 vol.% WC inclusions sintered at 1700°C. The same technique of visualization as applied in Figure 5.

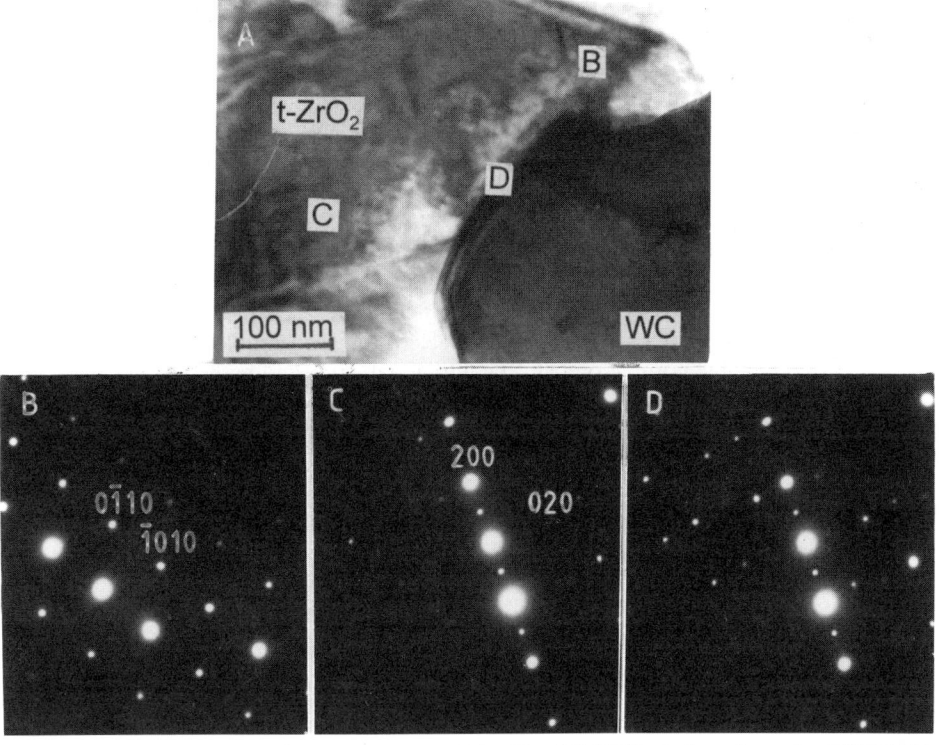

Figure 7. TEM micrographs of the composite containing 20 vol.% WC inclusions sintered at 1400°C. (a) Diffraction patterns taken from the indicated spots. t-ZrO$_2$ - B (zone axis [001]), WC - C (zone axis [0001]), grains boundary-D

In the case of the samples sintered at 1700°C, W$_2$C is the dominating tungsten carbide phase. Figure 8(a) shows the TEM micrograph in which the W$_2$C, tetragonal, and monoclinic ZrO$_2$ grains coexist. Again the crystallographic correlation between tungsten carbide (W$_2$C) and zirconia grains was observed in several cases:

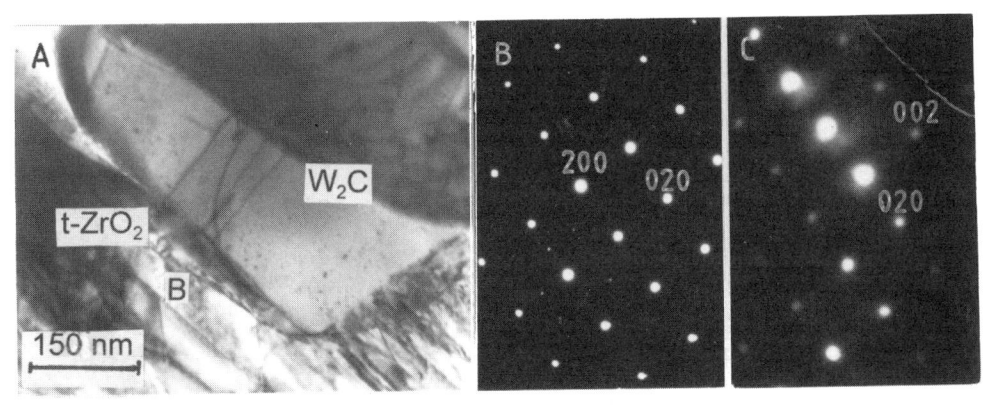

$$[100]\ W_2C\ \mid\ [001]\ t\text{-}ZrO_2$$
$$[010]\ W_2C\ \mid\ [010]\ T\text{-}ZrO_2$$

Figure 8. TEM micrographs of the composite containing 20 vol.% WC inclusions sintered at 1700°C - A. Diffraction patterns taken from the indicated spots: t-ZrO$_2$ - B (zone axis [001]), W$_2$C - C (zone axis [100])

CONCLUSIONS

Incorporation of WC grains into the Y-TZP matrix leads to a material characterized by the better properties than the "pure" matrix. The composites show increased fracture toughness, hardness, and Young's modulus. Flexural strength of the material remains unchanged. Increased wear resistance depends mainly on sample densification. Carbide inclusions restrain densification of the composites. That is why the highest densities by pressureless sintering could be obtained with the smallest WC content (10 vol.%). A sintering temperature that is too high causes the extensive reaction between the components, resulting in secondary porosity and formation of new phases.

Because of the Zener effect, the carbide inclusions limit the zirconia matrix grain growth. Transmission electron microscopy revealed that the carbide / matrix contact was tight. A crystallographic correlation between tetragonal ZrO_2 and both WC and W_2C grains was observed. It indicates that under sintering conditions, the carbide and oxide grains tend to arrange in such a way that a minimum interface energy orientation is achieved.

ACKNOWLEDGEMENT

This work was financially supported by the Polish Committee of Scientific Research under Grants 7 T08A 052 10 and 7 T08D 031 08.

REFERENCES

1. N. Claussen, K.L. Weisskopf, M. Ruhle, Tetragonal zirconia polycrystals reinforced with SiC whiskers, *J. Am. Ceram. Soc.*, **68**:288 (1986).
2. G.Y. Lin, T.C. Lei, S.X. Wang, Y. Zhou, Microstructure and mechanical properties of SiC whisker reinforced ZrO_2 (2 mol.% Y_2O_3) based composites, *Ceramics International*, **22**:199 (1996).
3. M. Poorteman, P. Descamps, F. Cambier, J. Leriche, B. Thieny, Hot isostatic pressing of SiC - platelets / / Y-TZP, *J. Europ. Ceram. Soc.*, **12**:103 (1993).
4. Zh. Dingh, R. Oberacker, F. Thummler, Microstructure and mechanical properties of yttria stabilized tetragonal zirconia polycrystals (Y-TZP) containing dispersed silicon carbide particles, *J. Europ. Ceram. Soc.*, **12**:377 (1993).
5. K. Haberko, Z. Pedzich, G. Róg, M. M. Bucko, M. Faryna, The TZP matrix - WC particulate composites, *Europ. J. Solid State Inorg. Chem.*, **32**:593 (1995).
6. K. Haberko, Z. Pedzich, J. Piekarczyk, G. Róg, Zirconia - tungsten carbide particulate composites. Part I: Manufacturing and physical properties, in: *Fourth Euro-Ceramics, vol. 4, Basic Science - Trends in: Emerging Materials and Applications,* A. Bellosi, ed., Gruppo Editoriale Faenza Editrice S. p. A. (1995).
7. Z. Pedzich, K. Haberko, M. Faryna, Fourth Zirconia - tungsten carbide particulate composites. Part II: Microstructure and wear resistance, Fourth Euro-Ceramics, vol. 3, *Basic Science - Optimisation of Properties and Performance by Improved Design and Microstructura Control,* S. Meriani, V. Sergo, eds., Gruppo Editoriale Faenza Editrice S. p. A., (1995).
8. Z. Pedzich, K. Haberko, Coprecipitation conditions and compaction behaviour of Y-TZP nanometric powders, *Ceramics International*, **20**:85 (1994).
9. J. Piekarczyk, H.W. Hoennike, R. Pampuch, On determining the elastic constants of porous zinc ferrite materials, *Ceramic Forum International, (Ber. Dtsch. Keram. Ges.)*, **59**:227 (1982).
10. K.A. Niihara, A fracture mechanics analysis of indentation, *J. Mater. Sci. Lett.*, **2**:221 (1983).
11. M.W. Weiser, L.C. De Jonghe, Inclusion size and sintering of composite powders, *J. Am. Ceram. Soc* - **71**:C125 (1988).
12. K. Haberko, Z. Pedzich, J. Piekarczyk, M.M. Bucko, W. Suchanek, Zirconia-based composites, in: *Third Euro-Ceramics, vol. 3, Engineering Ceramics,* P. Duran, J.F. Fernandez, eds., Faenza Editrice Iberica S. L., Madrid, (1993), 629.
13. S. A. Saltykov, *Stereological Metallurgy,* Ed. Metallurgia, 3rd edition, Moscow, (1970).

REACTION MECHANISMS AND MICROSTRUCTURES OF CERAMIC-METAL COMPOSITES MADE BY REACTIVE METAL PENETRATION

William G. Fahrenholtz[2], Kevin G. Ewsuk[1], Antoni P. Tomsia[3], and Ronald E. Loehman[1]

[1]Advanced Materials Laboratory, Sandia National Laboratories, Albuquerque, NM 87106
[2]Advanced Materials Laboratory, University of New Mexico, Albuquerque, NM 87106
[3]Lawrence Berkeley Laboratory, Berkeley, CA 94703

ABSTRACT

Ceramic-metal composites can be made by reactive penetration of molten metals into dense ceramic preforms. The metal penetration is driven by a large negative Gibbs energy for reaction, which is different from the more common physical infiltration of porous media. Reactions involving Al can be written generally as $(x+2)Al + (3/y)MO_y \rightarrow Al_2O_3 + M_{3/y}Al_x$, where MO_y is an oxide, such as mullite, that is wet by molten Al. In low P_{O_2} atmospheres and at temperatures above about 900°C, molten Al reduces mullite to produce Al_2O_3 and Si. The Al/mullite reaction has a $\Delta G_r°(1200K)$ of -1014 kJ/mol and, if the mullite is fully dense, the theoretical volume change on reaction is less than 1%. A microstructure of mutually-interpenetrating metal and ceramic phases generally is obtained.

Penetration rate increases with increasing reaction temperature from 900 to 1150°C, and the reaction layer thickness increases linearly with time. Reaction rate is a maximum at 1150°C; above that temperature the reaction slows and stops after a relatively short period of linear growth. At 1300°C and above, no reaction layer is detected by optical microscopy. Observations of the reaction front by analytical transmission electron microscopy show only Al and Al_2O_3 after reaction at 900°C, but Si is present in increasing amounts as the reaction temperature increases to 1100°C and above. The kinetic and microstructural data suggest that the deviation from linear growth kinetics at higher reaction temperatures and longer times is due to Si build-up and saturation at the reaction front. The activation energy for short reaction times at 900 to 1150°C varies from ~90 to ~200 kJ/mole, depending on the type of mullite precursor.

INTRODUCTION

Ceramic-metal composites are being developed because they exhibit combinations of properties that are unobtainable in single phase materials, for example, high stiffness, high toughness and oxidation resistance.[1,2] A number of methods for processing composites are being studied, including physical mixing of ceramic powders or whiskers in molten metals,[2] metal infiltration of fibrous or porous ceramic preforms,[3,4] and reaction routes such as directed oxidation of molten metals,[5-9] in-situ displacement reactions,[10-13] and reactive metal penetration of dense ceramics.[14-18] Reactive techniques have several advantages, including the prospect of near-net shape forming. In this paper we review and summarize work we have done over the past several years to understand and control the synthesis and properties of composite formation by reactive metal penetration (RMP). Key to this control is our emerging understanding of reaction mechanisms, which is also presented.

BACKGROUND ON REACTIVE METAL PENETRATION

Evaluation of different ceramic and metal combinations has shown that the RMP process is applicable to a variety of systems, and in many of them composites are produced to near net shape.[14-18] Two requirements for RMP are that the molten metal must wet the ceramic (i.e., the contact angle $\theta < 90°$) and that the Gibbs energy for reaction must be negative.[14] For composites made from Al and oxide preforms reactions can be written in general as,

$$(2 + x)\,Al + (3/y)\,MO_y \rightarrow Al_2O_3 + Al_xM_{3/y} \qquad (1)$$

where MO_y is an oxide and $Al_xM_{3/y}$ is the residual metal phase.

Most of our work has been on the Al/mullite system in which molten Al reduces mullite to produce alumina and elemental Si according to the reaction,

$$3\,Al_6Si_2O_{13} + (8 + x)\,Al \rightarrow 13\,Al_2O_3 + 6\,Si + x\,Al \qquad (2)$$

For the reaction with x=0, $\Delta G_r(1200K) = -1014$ kJ and $\Delta G_r(1600K) = -828$ kJ.[19-21] For excess Al (i.e., x > 0), a composite of alumina, Si, and Al is formed. When the mullite preform is in contact with a large external source of molten Al, the Si produced from the reaction diffuses out of the ceramic and into the Al source, as shown in Figure 1, leaving only a small amount in the composite. Thus, the process naturally avoids leaving significant concentrations of Si that could embrittle the composite.

The change in volume, ΔV, on reaction can be calculated from the molar volumes, V_m, of reactants and products in Reaction 2.[16] In the Al/mullite system, the relevant values in cm^3/mol are: $Al_6Si_2O_{13}$ (135.26), Al (9.99), Al_2O_3 (25.62), and Si (12.00). For Reaction 2, assuming x = 0 and a dense mullite preform, the predicted volume change, ΔV, after Al penetration is -0.72 cm^3/mole. The calculated fractional volume change on reaction is -0.0018 as shown by:

$$\frac{\Delta V}{3\,V_m(Al_6Si_2O_{13})} = \frac{13\,V_m(Al_2O_3) + 6\,V_m(Si) - 3\,V_m(Al_6Si_2O_{13})}{3\,V_m(Al_6Si_2O_{13})} = -0.0018$$

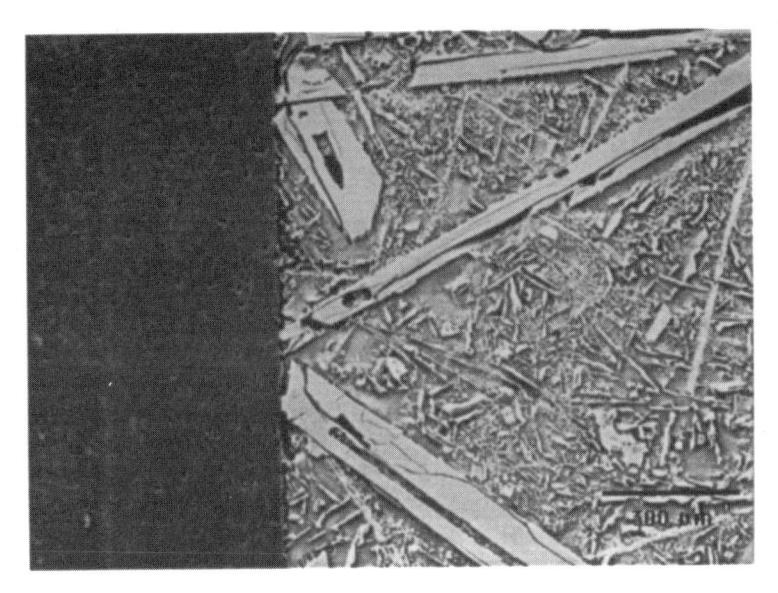

Figure 1. The interface between a reacted RMP composite (left) and contacting aluminum source (right) showing Si precipitates in the Al after cooling to room temperature. The composite was reacted for 60 min at 1100°C.

The volume change on reacting Al with commercial mullite-SiO$_2$ ceramics can be calculated in a similar way by assuming that the molar volumes of the mullite and silica phases in the preform are additive.

With excess Al (i.e., x >0), reactive metal penetration produces a composite of Al$_2$O$_3$, Al, and Si whose molar volume depends on the x value in Reaction 2. Net-shape composites also can be obtained with x > 0 compositions by adjusting the porosity of the ceramic preform to accommodate the increased volume of Al present in the composite after reaction.[14]

Understanding and controlling RMP allows physical properties to be optimized for particular applications. For composites in the Al$_2$O$_3$/Al/Si system, improvements in Young's modulus, fracture strength, density, hardness, fracture toughness, and specific modulus are apparent, compared with the unreacted mullite preform, as shown[14] in Table 1. The fracture toughness of reactively-formed ceramic-metal composites increases with increasing Al concentration, while modulus and density remain relatively constant, making it possible to make materials that are both tough and have a high specific modulus (i.e., high modulus/density).

Measurements of physical properties and microstructures show that a three-dimensional skeleton of alumina is formed during reactive penetration.[14] If the Al concentration is high enough, that phase also is continuous, resulting in a composite with mutually-interpenetrating metal and ceramic phases. Our measurements show that, relative to the ceramic preform, the composites are more flaw tolerant; that is, the strength degradation from crack formation and propagation appears to be less severe in the ceramic-metal composite than in the original mullite preform.[14]

EXPERIMENTAL

Ceramic preforms for most of our studies have been made by sintering mullite powders (Chichibu MP40, Scimarec Ltd. Tokyo, Japan or MULCR, Baikowski

Table 1. Properties of Ceramic-Metal Composites Formed by Reactive Metal Penetration Compared to Those of the Ceramic Preform[14]

Property	Mullite	Al$_2$O$_3$-Al-Si Composite
Bend Strength(MPa)	180	250 - 320
Hardness (GPa)	11.2	10 - 11
Toughness (MPa • √m)	1.7	6 - 10.5
Density (g/cc)	3.16	3.45 - 3.75
Young's Modulus (GPa)	200	240 - 394
Specific Modulus(GPa•cm^3/g)	63	71 - 105

International Corp., Charlotte, NC), kaolin (EPK, Feldspar Corp., Edgar, FL), or mixtures of mullite and SiO$_2$ (Corning Inc., Corning, NY) to >90% theoretical density by heating in air at 1200 - 1700°C. Some porous ceramic specimens were made by sintering at lower temperatures and for shorter times. In some experiments commercial mullite tubes (MV 30, Vesuvius McDanel, Beaver Falls, PA) or blocks (Ceraten, SA, Getafe, Spain) were used as preforms. Both pure Al (99.9%, Johnson Matthey, Ward Hill, MA), commercial Al alloys (6061 and 5052, Alcoa, Pittsburgh, PA), and synthesized alloys of Al and varying concentrations of Si have been used as the reactive metal.

Wetting and reactive penetration of Al with different ceramics were studied by melting small cubes of metal on polished ceramic disks at fixed temperatures between 900 and 1500°C in a resistance-heated furnace in an atmosphere of flowing, gettered Ar.[14] The furnace has a viewing port that allowed the contact angle of the molten metal to be measured in real time. Kinetics and mechanisms of composite formation were studied on samples made in another furnace[16] in which ceramic preforms were dipped in molten Al at a fixed temperature from 900 to 1400°C for up to 250 min. and then removed. The thickness of composite layers was determined by optical microscopic measurements on polished cross sections of reacted specimens. The rate of composite formation was determined from the measured reaction thicknesses. Other specimens were prepared for examination by scanning and transmission electron microscopy using standard techniques. Still others were sectioned into bars that were used to measure mechanical properties. Some specimens were made by hot pressing reactive mixtures of ceramic and metal powders in graphite dies in the processing range: 900 - 1450°C, 5 - 60 min., 1000 - 5000 psi. Reaction to produce the composite occurred in the graphite die, which also prevented the Si from diffusing out of the composite.

RESULTS

Contact angles of Al on aluminosilicate ceramics such as mullite[14] or fired kaolin decrease with increasing temperature and with time at fixed temperatures of 900°C and above. The steady decrease in Al contact angle with time at temperature is evidence of an on-going chemical reaction. The molten Al drop decreases in volume during a typical experiment as it reacts with and penetrates the ceramic. Wetting of mullite by Al generally is not observed at 900°C (i.e., θ > 90°) and Al wets mullite only after ~30 minutes at 1000°C. Sessile drop experiments conducted on nominally 100%, 99.5%, and 99% pure, dense mullite (with the remainder SiO$_2$) at 1100°C reveal no significant differences in wetting behavior with minor variations in silica content of the ceramic preform[14]. Aluminum-magnesium alloys readily wet mullite but do not react with it. Al alloys saturated with Si at the reaction temperature also do not react with mullite or mullite-SiO$_2$

ceramics. X-ray diffraction analysis shows reacted specimens contain Al_2O_3, Al, and Si, in accord with Reaction 2. Both Al and Al_2O_3 are present in specimens formed by reactive hot pressing of mixed powders or by penetration of Al into dense ceramics. However, little or no Si is found in samples made by ceramic-molten Al reaction. In them, much of the Si produced in Reaction 2 is found as precipitates in the Al source after cooling to room temperature.

In addition to mullite, the RMP reaction forms composites with other aluminosilicate ceramics as well. One example is kaolin, $Al_2Si_2O_5(OH)_4 \cdot 2H_2O$, which dehydrates to meta-kaolin, $Al_2Si_2O_7$, when heated to 800°C. At typical kaolin processing temperatures of 1300-1450°C, crystalline mullite, $Al_6Si_2O_{13}$, nucleates and grows within a glassy matrix. Thus, fired kaolin is similar to mullite but with a continuous glassy grain boundary phase. Reactions 3 and 4 illustrate the sequential conversion of kaolin to meta-kaolin and then to mullite/silica ceramics.

$$Al_2Si_2O_5(OH)_4 \cdot 2H_2O \rightarrow Al_2Si_2O_7 + 4\ H_2O \qquad (3)$$

$$3\ Al_2Si_2O_7 \rightarrow Al_6Si_2O_{13} + 4\ SiO_2 \qquad (4)$$

As with the Al/mullite reaction, initiating the Al/kaolin reaction depends on first forming a ceramic-metal interface. Results in Figure 2 show that Al wets dense kaolin, which should produce a uniform ceramic-metal interface and, therefore, a stable reaction front as the metal reacts with and penetrates into the preform.

Aluminum reacts with fired kaolin preforms in much the same way as it does with phase-pure mullite. The reaction is near-net-shape and Si diffuses away from the reaction front and out of the developing composite, to produce an Al_2O_3/Al composite. Due to the higher silica content of the preform, composites prepared from kaolin preforms contain more Al than those prepared from mullite. The Al/kaolin reaction can be represented as the reaction of Al separately with mullite and silica:

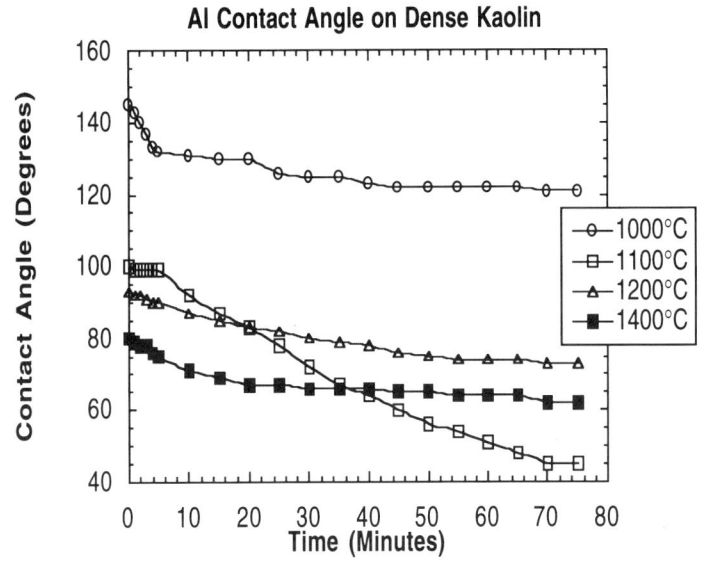

Figure 2. The contact angle of molten Al on fired kaolin substrates as a function of time at the indicated temperatures.

$$(8 + x) Al + Al_6Si_2O_{13} + 4 SiO_2 \rightarrow 7 Al_2O_3 + 6 Si + x Al \qquad (5)$$

Mass and volume balance calculations show that stoichiometric reaction of dense kaolin with Al produces a composite that contains approximately 30 vol.% Al and 70 vol.% Al_2O_3. This corresponds to an x value of ~ 7 in Reaction 5. The predicted composite density is 3.62 g/cm^3. The Al/kaolin reaction seems to be more favorable than Al reaction with mullite as suggested by the observation that complex shapes are more easily formed from kaolin preforms than from mullite.

Samples made by immersing dense mullite preforms in molten Al or Al-Si alloys with Si concentrations below the saturation level exhibited identical penetration rates for temperatures of 1000-1150°C. There is an induction period of about 5 min. where no reaction occurs, after which the rate increases with temperature and is linearly proportional to time. Composite formation rate reaches a maximum of ~6 mm/hour at 1150°C and then decreases with increasing temperature above 1150°C. Closer examination of growth kinetics at 1150°C and above shows that reaction layer formation slows or stops after a relatively short period of linear growth. The duration of linear growth decreases from 25 minutes at 1150°C to less than a minute at 1250°C. At 1300°C and above, no reaction layer is detected by optical microscopy.

Transmission electron microscopic (TEM) analysis of specimens quenched from the growth temperature shows that the reaction front contains little or no silicon after reaction at 900°C, but that silicon is present at the Al-mullite interface after reaction at 1100°C. The amount of Si at the interface increases with reaction temperature. As shown in Figure 5, there is a lamellar microstructure of alternating layers of Al_2O_3 and Al at the growth front. Si precipitates form at the end of the Al_2O_3 lamella for the higher reaction temperatures. The kinetic data and microstructural information suggest that the deviation from linear growth kinetics at higher reaction temperatures and longer times is due to silicon build-up and saturation at the reaction front.

Composite formation rates exhibit Arrhenius behavior up to 1150°C, with activation energies between 90 and 200 kJ/mol depending on the source of the mullite preform.[16] The reason for the variation in activation energy is still under investigation, but it may be related to the fact that MP40 mullite contains a small amount (<<1%) of a finely dispersed phase that is high in zirconium.

DISCUSSION

We have proposed a mechanism for reactive metal penetration of Al into mullite that accounts for the experimental observations described above:
(1) Al does not wet mullite below about 1000°C, which inhibits the reaction;
(2) when molten Al contacts the mullite surface there is an induction period of about 5 min. before reaction starts;
(3) the kinetics of composite layer formation are linear from 1000 to 1150°C;
(4) the reaction rate increases with temperature to a maximum at about 1150°C and decreases at higher temperatures; and
(5) the reaction front shows a buildup of Si for temperatures above 1150°C.

The lack of wetting by Al is explained by its strong tendency to form a thin surface oxide coating when melted, even in low P_{O_2} atmospheres. Although Al melts at 660°C, it does not initially react or spread readily on surfaces because the oxide coating forms a cohesive, passivating layer. At higher temperatures (≈ 1000°C) the oxide tends to be

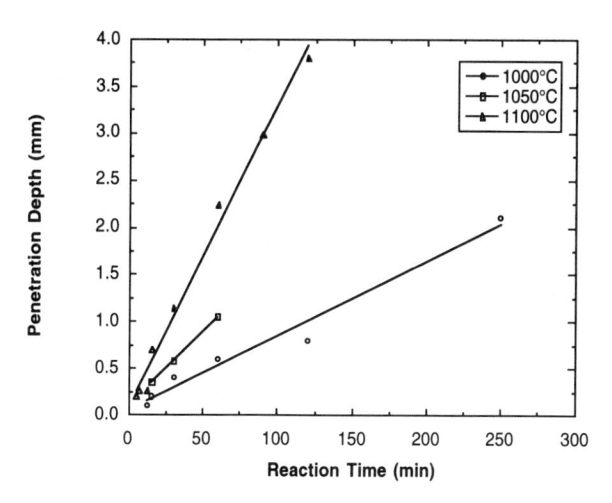

Figure 3. Aluminum penetration depth as a function of time for the reaction of aluminum and Ceraten mullite at 1000, 1050, and 1100°C.

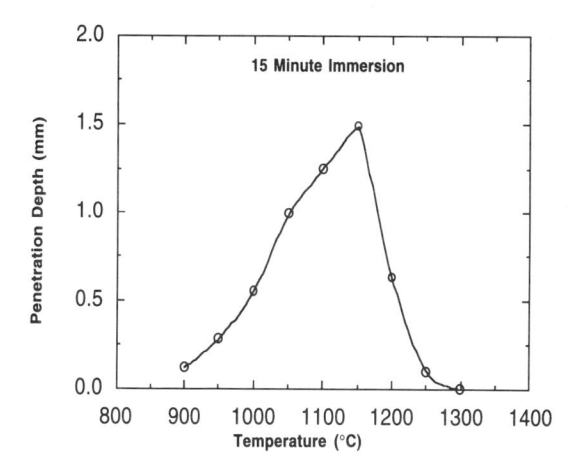

Figure 4. The penetration depth of aluminum into MP40 mullite as a function of temperature measured after a 15 minute reaction.

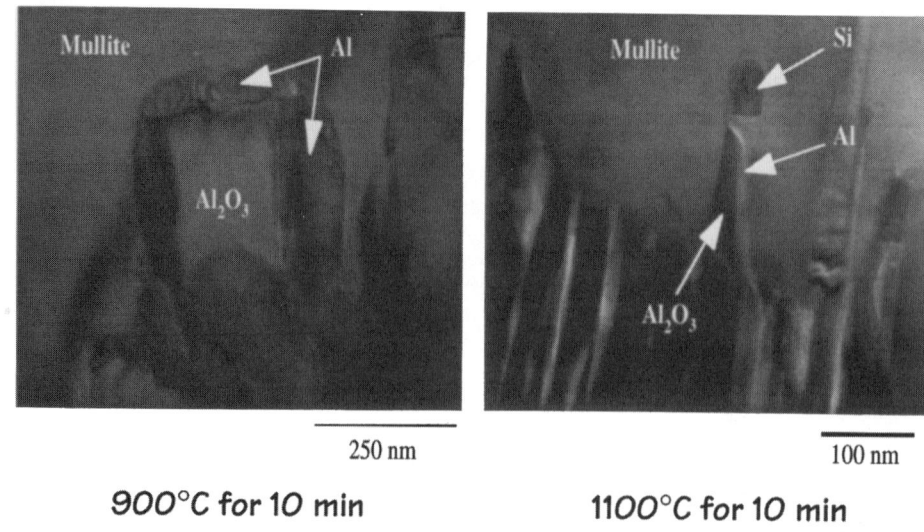

Figure 5. Transmission electron micrographs of the reaction interface for MP40 mullite reacted with pure aluminum at 900°C and 1100°C for 10 min.

removed by vaporization so Al is able to wet surfaces such as mullite, and reaction can proceed.

There is a strong thermodynamic driving force for Reaction 2, for which ΔG_r = -1014 kJ at 1200K. The Gibbs energy for reaction between Al and atmospheric oxygen to form Al_2O_3 is also strongly negative with ΔG_r = -1163 kJ/mole at 1200K.[19-21] The Ellingham diagram for Al/Al_2O_3 indicates that Al and alumina are in equilibrium at a P_{O_2} of $10^{-28.5}$ atm. at 1200K. Thus, to promote Al-mullite reaction and to reduce competition from atmospheric oxidation of Al, the system P_{O_2} should be as low as possible, although

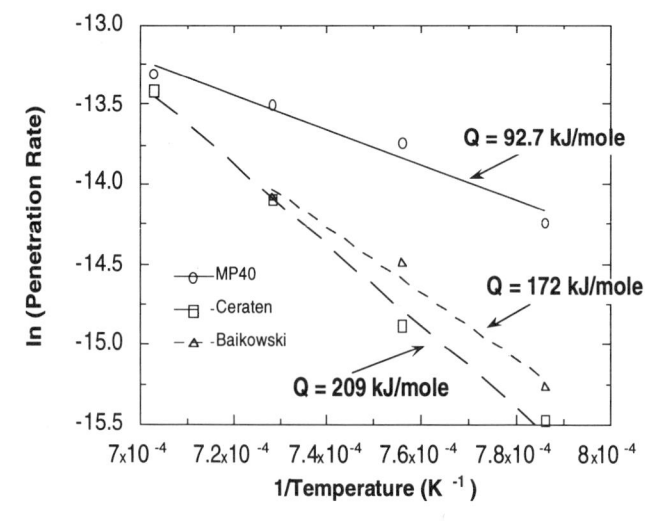

Figure 6. The activation energy for the reaction of Ceraten, Baikowski, and MP40 mullite preforms with aluminum

obtaining a P_{O_2} below $10^{-28.5}$ atm. in the overall reaction atmosphere is extremely difficult. The induction period prior to reaction can be attributed to the time it takes for the Al-mullite interface that is isolated from the furnace atmosphere to reach the local equilibrium P_{O_2} so conditions are favorable for wetting and reaction.

The linear growth kinetics below 1150°C for the reactive penetration of mullite by Al is similar to the linear growth behavior found for directed metal oxidation of Al and other liquid-solid displacement reactions.[5-9] Rather than atomic-scale diffusion through narrow grain boundaries, the metal layers at the reaction front of RMP composites, seen in Figure 5, form channels that allow macroscopic flow of molten Al to the reaction front. The observed linear growth kinetics suggest that transport of Al from the molten source to the reaction front is not rate-limiting below 1150°C. If it were, the increase in thickness of the reaction layer would be proportion to the square root of time. Below 1150°C silicon diffusion away from the reaction front also cannot be rate-limiting because that would allow the concentration of Si at the reaction front to steadily increase, which would shut down the reaction as it does at higher temperatures. However, the rate-limiting step in the linear growth regime has not been determined. Potential candidates are the rate of Reaction 2 itself or, possibly, the rate of Al transport through a transitional zone of constant thickness near the reaction front.

Because RMP is thermodynamically driven, the decrease in reaction rate with temperature above 1150°C is unexpected. The reaction rate maximum at 1150°C suggests two different mechanisms with different temperature dependences are operative simultaneously, one being dominant below 1150°C and the other dominant above. TEM analysis of reaction couples heated at different temperatures (Figure 5) show Si precipitates in the reaction zone at 1100°C, with larger concentrations of Si present at higher temperatures. These results suggest a mechanism where Si production in Reaction 2 increases faster with temperature than the rate of Si transport away from the reaction front. At reaction temperatures of 1150°C and above, the buildup of Si progressively slows the reaction and eventually stops it when the interface becomes saturated with Si. Because silicon does not react with aluminum, it forms a stable diffusion barrier that prevents further reaction. This proposed mechanism is shown schematically in Figure 7.

In the Reaction Zone:

$3Al_6Si_2O_{13} + (8+x)Al \rightarrow 13Al_2O_3 + 6Si + xAl$

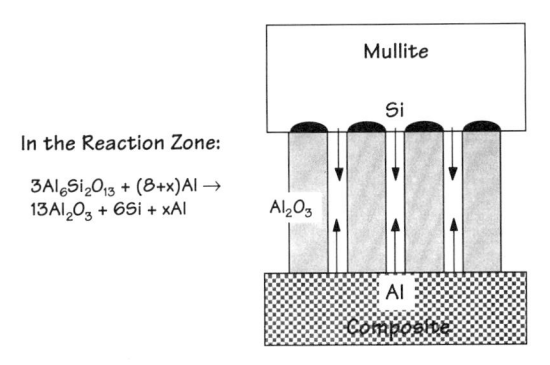

Figure 7. Schematic of the reaction mechanism for formation of composites by reactive metal penetration.

CONCLUSIONS

Al reacts with dense mullite to form ceramic-metal composites according to the reaction: $3Al_6Si_2O_{13} + (8 + x)Al \rightarrow 13Al_2O_3 + 6Si + x\,Al$. Reaction between Al and fired kaolin can be represented as separate reactions with mullite and silica according to $(8 + x)Al + Al_6Si_2O_{13} + 4\,SiO_2 \rightarrow 7\,Al_2O_3 + 6\,Si + x\,Al$. The resulting composites are stronger and tougher than the ceramics from which they are made. For reaction between Al and mullite or mullite-SiO_2 ceramics, the volumes of reactants and products differ by less than 2%, resulting in a net-shape process. Below 900°C reaction is inhibited by a passivating oxide layer on the molten Al. The penetration rate increases with temperature from 900 to 1150°C, reaching a maximum of ~6 mm/hour at 1150°C. Above 1150°C, the penetration rate slows or stops after a relatively short period of linear growth. The reaction produces Si as a by-product, which must diffuse away from the reaction front for reaction to continue. The results suggest that the rate of Si production has a steeper temperature dependence than the rate of Si diffusion away from the reaction front. As the temperature increases above 1150°C, Si formation is faster than its transport, and Si saturates the reaction front, which causes the reaction to slow drastically. At 1300°C and above, no noticeable reaction occurs.

ACKNOWLEDGMENT

The authors express their thanks to Professor Ping Lu of the New Mexico Institute of Mining and Technology for providing the transmission electron micrograph of Figure 5.

REFERENCES

1. R. W. Rice, Processing, microstructures, performance, and economics of ceramic composites, in: *Metal and Ceramic Matrix Composites: Processing, Modeling, and Mechanical Properties*, R.B. Bhagat, A.H. Clauer, P. Kumar, and A.M. Ritter, eds., The Minerals, Metals and Materials Society(1990).
2. T. W. Clyne and P. J. Withers *An Introduction to Metal-Matrix Composites* Cambridge University Press, Cambridge(1993).
3. W. B. Hillig, Making ceramic composites by melt infiltration, *Am. Ceram. Soc. Bull.* 73(4):56(1994).
4. M. J. Koczak and M. K. Premkumar, Emerging technologies for the in-situ production of MMCs, *J. Met.* 44(1993).
5. A. S. Nagelberg, S. Antolin, and A.W. Urquhart, Formation of Al_2O_3/metal composites by the directed oxidation of molten aluminum-magnesium-silicon alloys: Part II, growth kinetics, *J. Am. Ceram. Soc.*, 75(2):455(1992).
6. E. Breval, M. K. Aghajanian, and S. J. Luszcz, Microstructure and composition of alumina/aluminum composites made by directed oxidation of aluminum, *J. Am. Ceram. Soc.* 73(9):2610(1990).
7. M. Sindel, N. A. Travitzky, and N. Claussen, "Influence of magnesium-aluminum spinel on the directed oxidation of molten aluminum alloys, *J. Am. Ceram. Soc.*, 73(9):2615(1990).
8. M. S. Newkirk, A. W. Urquhart, and H. R. Zwicker, Formation of Lanxide™ ceramic composite materials, *J. Mater. Res.*(1):81(1986).
9. A. S. Nagelberg, Growth kinetics of Al_2O_3/metal composites from a complex aluminum alloy, *Solid State Ionics*, 32/33:783(1989).
10. A. E. Standage and M. S. Gani, Reaction between vitreous silica and molten aluminum," *J. Am. Ceram. Soc.* 50(2):101(1967).
11. K. Prabriputaloong, and M. R. Piggott, Reduction of SiO_2 by molten Al, *J. Am. Ceram. Soc.* 56(4):184(1973).
12. S. Matsuo and T. Inabe, Fabrication of Al-Al_2O_3 composites by substitutional reaction in fused aluminum, *Tokyo Ceramics*, 222(1991).

13. M. C. Breslin, J. Ringnalda, J. Seeger, A. L. Marasco, G. L. Daehn, and H. L. Fraser, Alumina/aluminum co-continuous ceramic composite (C^4) materials produced by solid/liquid displacement reactions: processing kinetics and microstructures, *Ceramic Engineering and Science Proceedings*, 15(4):104(1994).
14. R. E. Loehman, K. G. Ewsuk, and A. P. Tomsia, Synthesis of Al_2O_3-composites by reactive metal penetration, *J. Am. Ceram. Soc.* 79(1):27(1996).
15. W. G. Fahrenholtz, K. G. Ewsuk, D. T. Ellerby, and R .E. Loehman, Near-net-shape processing of metal-ceramic composites by reactive metal penetration, *J. Am. Ceram. Soc.* 79(9):2497(1996).
16. W. G. Fahrenholtz, K. G. Ewsuk, R. E. Loehman, and A. P. Tomsia, Synthesis and processing of Al_2O_3/Al composites by in-situ reaction of aluminum and mullite, in: *In-Situ Reactions for Synthesis of Composites, Ceramics, and Intermetallics*, E.V. Barrera, et. al., eds. The Minerals, Metals, and Materials Society, Warrendale, PA(1995).
17. Y. Gao, J. Jia, R.E. Loehman, and K.G. Ewsuk, TEM studies of metal-ceramic composite formation by reactive metal infiltration, *J. Mater. Res.* 10(5):1216(1995).
18. Y. Gao, J. Jia, R. E. Loehman, and K. G. Ewsuk, TEM characterization of Al_2O_3-Al composites fabricated by reactive metal infiltration, in: *Ceramic Matrix Composites - Advanced High-Temperature Structural Materials, Mater. Res. Soc. Proc.*, 365:209(1995).
19. R. A. Robie and D. R. Waldbaum, Thermodynamic properties of minerals and related substances at 298.15 K and one atmosphere pressure and at higher temperatures, *Geological Survey Bulletin*, 1259, U. S. Department of the Interior, U. S. Government Printing Office, Washington, D. C.(1968).
20. M. W. Chase, Jr., C. A. Davies, J. R. Downey, Jr., D. J. Fruip, R. A. McDonald, and A. N. Syverud, JANAF Thermochemical tables, third edition, *J. Phys. Chem. Ref. Data*, 14(1):1(1985).
21. *Thermochemical Properties of Inorganic Substances, Second Edition*, O. Knacke, O. Kubaschewski, and K. Hesselmann, eds., Springer-Verlag, Berlin(1991).

EUTECTIC STRUCTURES THAT MIMIC POROUS HUMAN BONE

P. N. De Aza, F. Guitian and S. De Aza*

Instituto de Ceramica, Universidad de Santiago, Santiago, Spain
*Instituto de Ceramica y Vidrio, CSIC. Arganda del Rey, Madrid, Spain

INTRODUCTION

Ceramic biomaterials for bone replacement are becoming increasingly important[1]. With the exception of the phosphate ceramics,[2] all the bioactive materials developed since the discovery of a bioactive glass by Hench in 1970,[3] have been glasses or glass-ceramics.[4,5] Only recently, two polycrystalline bioactive ceramic materials belonging to the chain silicates have been developed: wollastonite[6-8] and diopside.[9-10]

A common characteristic of all these materials is that they form a hydroxyapatite (HA) layer on their surfaces when are soaked in acellular simulated body fluid (SBF),[4-11] while the bulk of the material remains without any modification.

Similarly, when bioactive ceramic materials are implanted in a living body the interaction between the bone and these materials also takes place only on their surface. To improve the ingrowth of new bone into the implant, the use of materials with an appropriate porous structure is recommended.[12,13] However porous materials have very poor mechanical properties.

A new approach to overcoming this problem is to design dense bioactive ceramic materials capable of developing porous structure *in situ* when they are implanted. To fulfill this goal, the material should contain at least two phases, one bioactive and the other resorbable. At the same time the microstructure should be as homogeneous as possible and easily controlled. This can be achieved if binary eutectic structures can be obtained, whose morphologies have very well known intrinsic properties.[14-16]

Taking into account all these considerations, the binary system wollastonite (W) - tricalcium phosphate (TCP) was selected.[17] W is bioactive[6-8] and TCP is resorbable[2,4] and eutectic point exists at 1402 ±3°C for the composition 60 wt% W and 40 wt% TCP (Figure 1). Due to the volume fractions of the two phases at the eutectic point ($V_{fW} = 0.61$ and $V_{fTCP} = 0.38$), the microstructure of the eutectic has to be of the lamellar type,[14-16] characterized by an irregular morphology as a consequence of the high dimensionless entropy of fusion ($\alpha = [S_f/R] > 2$) of both phases ($\alpha_W = 3.71$ and $\alpha_{TCP} = 9.66$).[14-16]

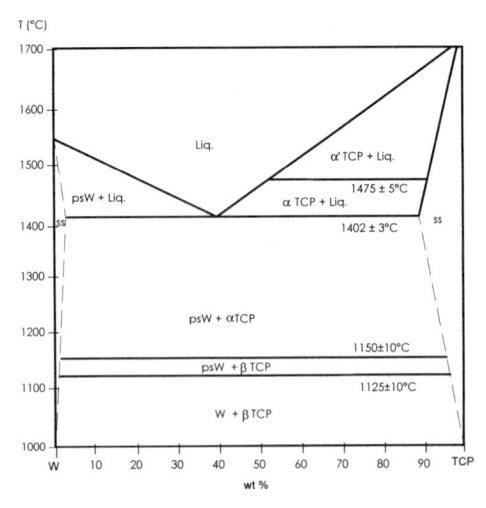

Figure 1. Phase diagram of the system wollastonite-tricalcium phosphate (W= wollastonite; psW= pseudowollastonite: TCP= tricalcium phosphate).

EXPERIMENTAL

To obtain a eutectic structure, heat must be extracted from the eutectic liquid. Of all the methods of heat extraction available,[14-16] radial heat flux was selected in order to obtain microstructures formed by rounded colonies (rosettes).

For this purpose, pellets of the eutectic composition, wrapped in platinum foil crucibles, were heated up to 1500°C in 2 h to obtain homogeneous liquids which were then cooled to 1410°C at a rate of 3°C/min. From this temperature, the samples were further cooled inside the furnace down to 1390°C, 12°C below the eutectic temperature (1402 ± 3°C), at rates of 2°, 1° and 0.5°C/h.

Plates of the eutectic material obtained at a cooling rate of 0.5°C/h, 3 mm in diameter and 2 mm thick, were immersed in 100 ml of simulated body fluid (SBF) which had an ion concentration almost equal to that of human blood plasma.[11] The plates and SBF were contained in polyethylene bottles held at human body temperature (36.5°C). The immersion periods varied from 1 day to 1 week.

At intervals, the specimens were removed from the fluid and analyzed by scanning electron microscopy (SEM) using a JEOL 6400 operated at 20 kV and fitted with energy-dispersive and wavelength-dispersive spectrometers (EDS and WDS). In addition, selected area diffraction (SAD) in a transmission mode (TEM) on the electron beam transparent specimens was performed using a JEOL Jem 2010 operated at 200 kV. The samples were also studied by thin film X-ray diffraction and Raman spectroscopy fitted with an optical microscope which allows a spatial resolution on the sample close to 1 µm.

RESULTS AND DISCUSSION

The SEM studies show (Figure 2) that all the samples obtained consists of quasi-spherical colonies formed by alternating lamellae of W and TCP with morphologies which correspond to irregular eutectic structures as was previously suggested.

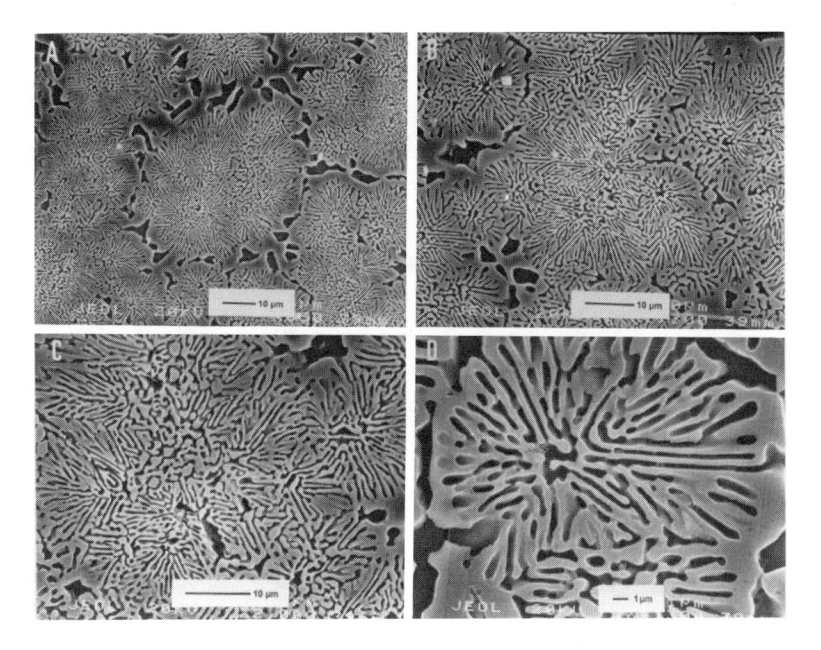

Figure 2. SEM images of the eutectic materials obtained at cooling rates of: (A) 2°C/h; (B) 1°C/h; (C) 0.5°C/h. (D) Detail of a colony belonging to sample (C).

EDS microanalysis showed that the white phase is W and the darker phase is TCP. The only difference that can be observed between the samples cooled at the different rates, is the greater volume of the colonies in samples with lower cooling rate, and consequently, the smaller number of colonies per unit volume (Figure 3).

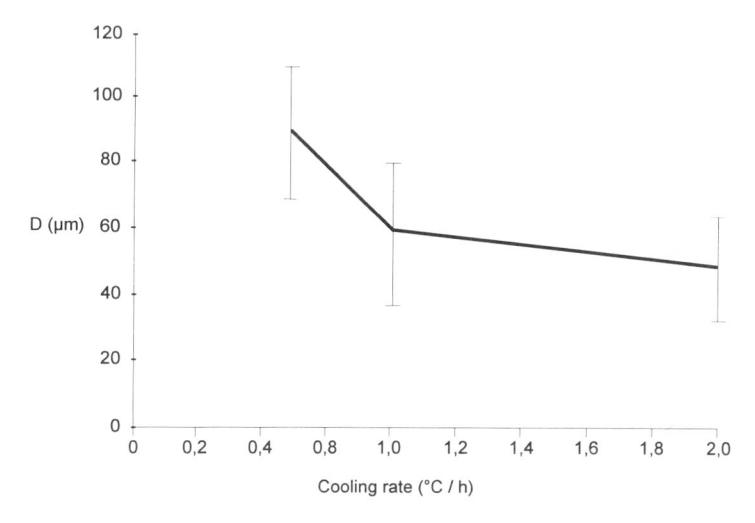

Figure 3. Size of the colonies versus cooling rate.

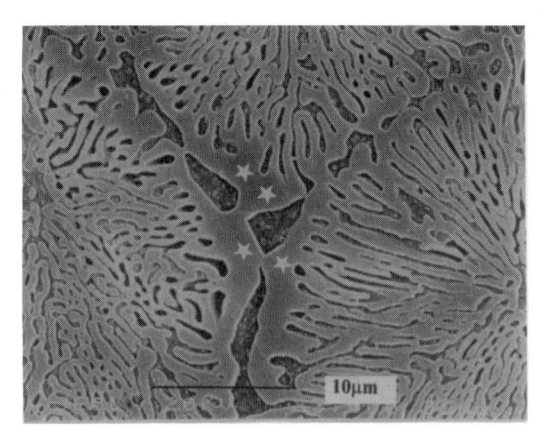

Figure 4. Triple point between colonies. Sample cooled at a rate of 0.5°C/h.

Table 1. WDS mean area microanalysis (wt%) of the triple point in Figure 4.

SiO_2	CaO	Al_2O_3	P_2O_5
54.1	40.2	3.9	1.8

However, the width of the lamellae of W and TCP, ~1.3 and ~0.9 µm, respectively, seems to remain essentially constant. As can be seen in all the micrographs, the phase which bounds or envelopes the colonies is always W, which is consistent with its greater volume fraction in the eutectic composition. It is this phase which bonds or connects the colonies to one another.

Small quantities of impurities affect drastically the way the colonies are joined together. Impurities segregate to the periphery of the colonies as they grow, and they are mainly concentrated at triple points. This fact breaks the continuous structure of alternating lamellae of W and TCP to form a continuous matrix of W joining the colonies with dispersed TCP inclusions. Figure 4 exhibits clearly this fact and Table 1 shows the mean area microanalysis of the triple point obtained by WDS showing the presence of impurities of alumina.

The TEM studies confirm the presence of W and TCP phases as shown in Figure 5. The dark phase is TCP and the clear phase is W. In the same figure, the SAD of the two phases are shown, indicating good crystallization in both cases. The material appears to be compact and without any cracks, although some mis-match dislocations were found due to the different crystallographic orientation of the two phases, as well as loop dislocations in the wollastonite due to the electron beam incidence.

A fresh fracture surface of the eutectic microstructure, belonging to the sample obtained at a cooling rate of 0.5°C/h, etched with diluted HCl acid to eliminate the TCP phase, is shown in Figure 6. The obtained microstructure bears great similarity to that of porous human bone.

The microstructure of a polished cross-section of the eutectic material after soaking in SBF for one week as well as elemental X-ray maps of silicon, calcium and phosphorus are shown in Figure 7. We can clearly observe the selective solution of one of the phases which forms the colonies of the eutectic material to a depth of 50 µm. The dissolved phase is the wollastonite which enosed the colonies and joined them in the original microstructure.

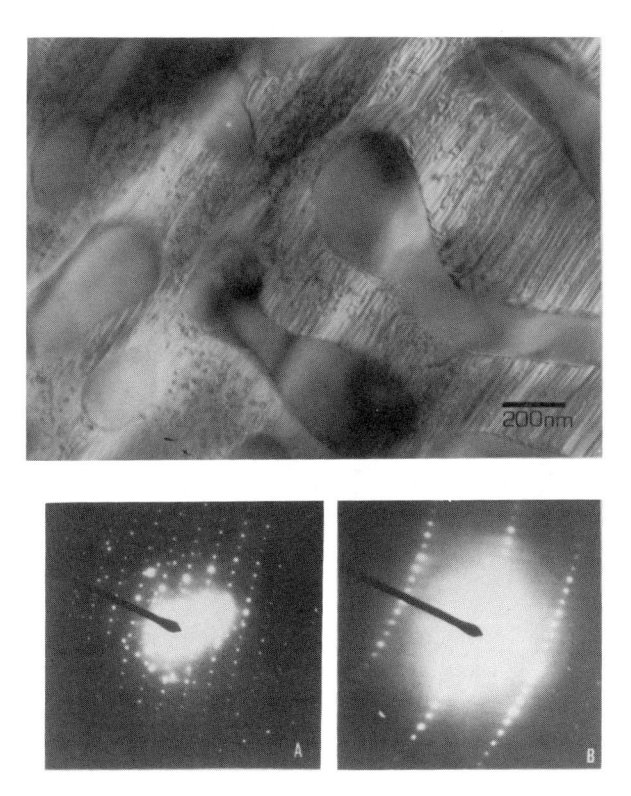

Figure 5. High-resolution TEM image showing mis-match and loop dislocations and SAD of the wollastonite (A) and tricalcium phosphate (B)

    This is confirmed by the elemental X-ray maps and EDS microanalyses. These showed that the remaining discontinuous phase is a calcium phosphate. Raman microscopy of the reaction zone (Figure 8) confirms that the calcium phosphate phase corresponds to a hydroxyapatite of low crystallinity well differentiated from the original TCP. Consequently, a selective solution of the wollastonite in the SBF takes place, leaving a residual phase of HA pseudomorphic with the original TCP.

    The results obtained can be explained through the following mechanism (Figure 9): at the beginning of the reaction the wollastonite phase in contact with the SBF starts to react with it through an ionic exchange of two protons from the SBF for one calcium ion from the wollastonite network. This exchange transforms the W crystals into an amorphous silica phase as has been demonstrated in previous works.[6-8] This reaction starts at the surface of the material, increasing the pH up to a value of 8.5 at the material-SBF interface, as was measured by using an ISFET-meter,[18] and releasing calcium and silica ions into the SBF media. The reaction progresses into the material through the W lamellae in the confined media between the TCP lamellae as wollastonite is dissolved. This reaction triggers the process.

    Simultaneously. the TCP starts to react with the Ca^{++} and OH^- ions present in the confined media at pH $\approx$ 10.5[18] giving rise to the pseudomorphic transformation of TCP into HA according to the reaction shown in Figure 9.

Presently, we are carrying out much longer experiments in SBF and also examining *in vivo* implants in rats. The results will be the subject of future publications

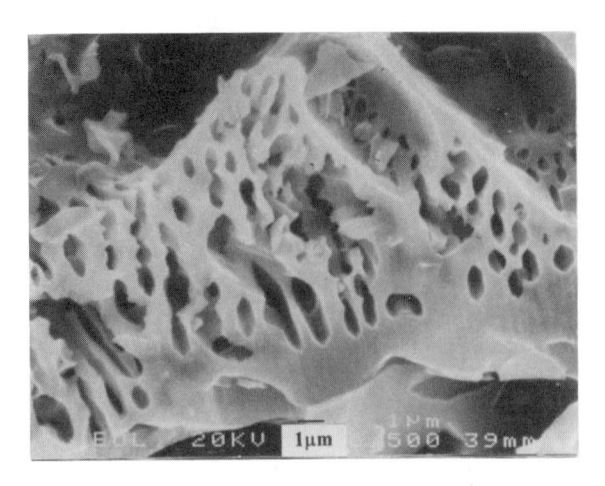

Figure 6. SEM image of a fresh fracture surfaces of the eutectic material after etching with dilute HCl acid.

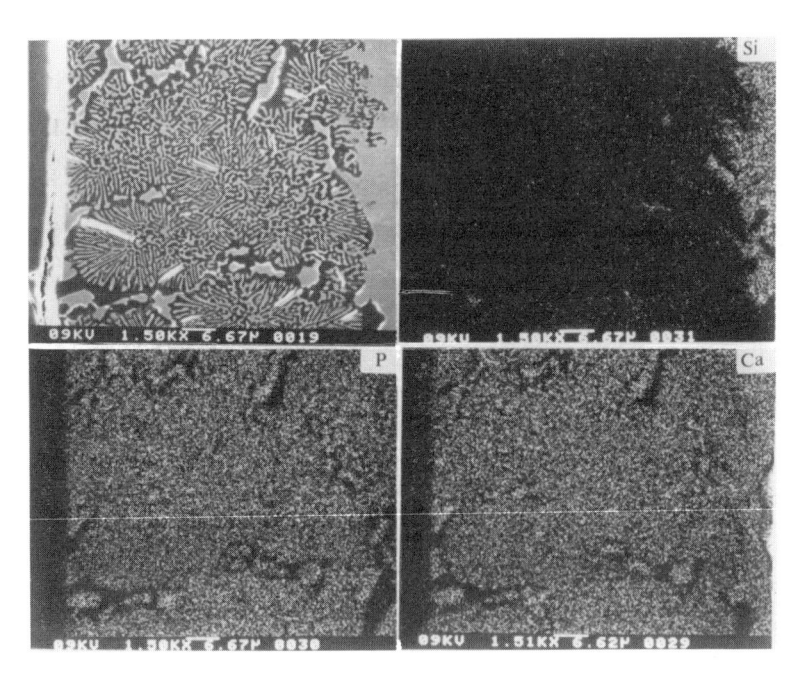

Figure 7. SEM image of a cross-section of the eutectic plate after one week soaking in SBF and Si, P and Ca x-ray maps.

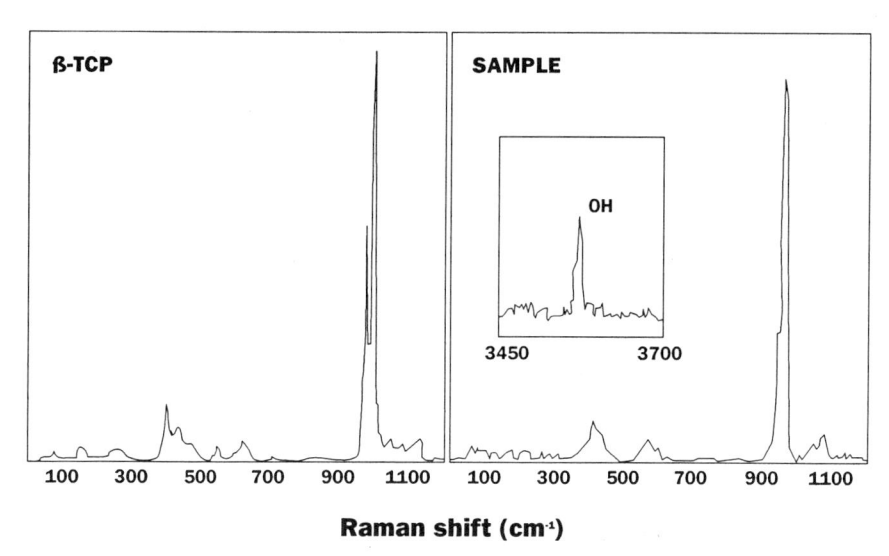

Figure 8. Raman spectra of standard TCP and of the reaction zone in Figure 7.

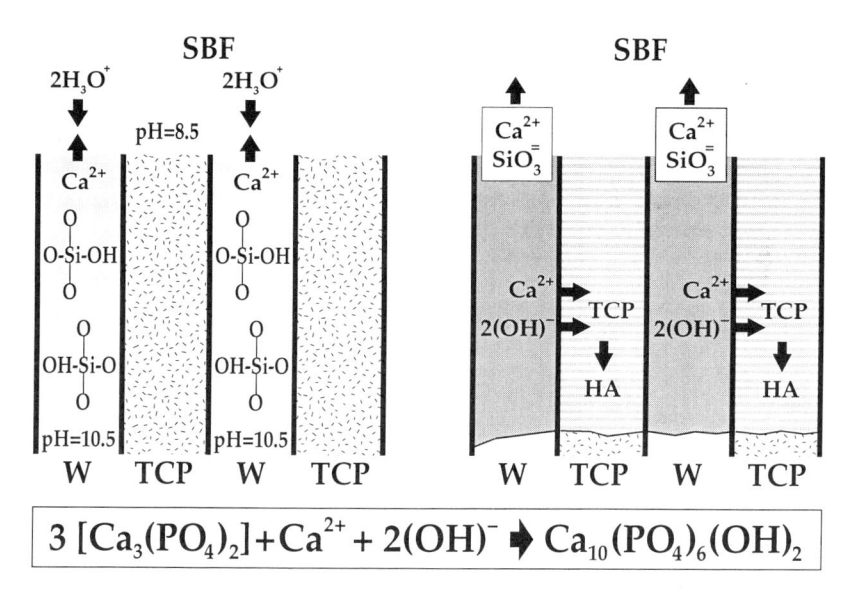

$$3\,[Ca_3(PO_4)_2] + Ca^{2+} + 2(OH)^- \rightarrow Ca_{10}(PO_4)_6(OH)_2$$

Figure 9. Schematic representation of the mechanism of HA formation in the eutectic sample after immersion in SBF for one week.

CONCLUSIONS

By making use of the eutectic in the wollastonite-tricalcium phophate system, eutectic microstructures with a colonies morphology can be obtained.

These materials provide a high bioactivity *"in vitro"*, forming after soaking in SBF a porous structure of HA that mimic porous bone.

We can expect that the eutectic material is bioactive *"in vivo"* because both phases individually behave as such.[4,6-8]

We believe we have provided enough evidence to show that in the *in vitro* experiments the reaction starts on the surface of the material and progresses deeply into it. Consequently we expect it will behave similarly in the *in vivo* experiments, facilitating the bone colonization of the implant.

Finally the method used opens the opportunity to develop a new family of bioactive materials with different constituents, binary or ternary, for which the authors propose the general name of ***bioeutectics®***.

Acknowledgements

This work was supported by CICYT under Project MAT95-0385. In addition P.N. De Aza thanks the Ministry of Education and Science of Spain for the Fellowship given to her. The authors wish to thank Dr. A. P. Tomsia and Dr. E. Saiz and also Dr. Z. B. Luklinska and Dr. Anseau for the facilities given to P. N. De Aza during her respective stays at Lawrence Berkeley Laboratory (CA. USA) and at IRC in Biomedical Materials, The London Hospital, Medical College, University of London.

REFERENCES

1. S. F. Hulbert, J. C. Bohros, L. L. Hench, J. Wilson and G. Heimke, *Ceramic in Clinical Application: Past, Present and Future, High Tech Ceramics,* P. Vicenzini, ed., Elsevier Amsterdam, The Netherlands, (1987), pp. 189-213.
2. K. De Groot, *Bioceramics of Calcium Phosphate,* K. De Groot, ed., CRC Press, Boca Raton, FL. (1983).
3. L. L. Hench, R. J. Splinter, T. K. Greenle and W. C. Allen, Bonding mechanism at the interface of ceramic prosthetic materials, *J. Biomed. Mater. Res. Symp.,* 117-141 (1971).
4. L. L. Hench, Bioceramics: from concept to clinic, *J. Am. Ceram. Soc.,* **74**:1487-1510 (1991).
5. T. Kokubo,. Novel biomedical materials derived from glasses. *Ceramics: Toward The 21st. Century.* N. Soga, ed., *The Ceram. Soc. of Japan.,* Tokyo, (1991), pp. 500-518.
6. P. N. De Aza, F. Guitian and S. De Aza, Bioactivity of wollastonite ceramics: in vitro evaluation, *Scripta Metallurgica et Materialia,* **31**:1001 (1994).
7. P. N. De Aza, F. Guitian and S. De Aza, Polycrystalline wollastonite ceramic biomaterials free of P_2O_5, *Advances in Science and Technology, 12. Materials in Clinical Application,* P. Vincenzini, ed., Techna Srl., Florenze, (1995), pp.19-27
8. P. N. De Aza, Z. B. Luklinska, M. Anseau, F. Guitian and S. De Aza, Morphological studies of pseudowollastonite for biomedical application, *Journal of Microscopy,* **182**:24 (1996).

9. T. Nonami, Developmental study of diopside for use as implant material, *Materials Research Society Symposium. Proceedings, Vol. 252, Tissue-Inducing Biomaterials.* L. Cima and E. Ron, ed., Mat Res. Soc., (1992), pp. 87-92.
10. T. Nonami, In vivo and in vitro testing of diopside for biomaterial, *J. of the Society of Materials Engineering for Resources of Japan,* **8**:12 (1995).
11. J. Gamble, *Chemical Anatomy, Physiology and Patology of Extracellular Fluid, Cambridge,* Harvard University Press, Cambridge, (1967).
12. K. De Groot and R. Le Geros, Significance of porosity and physical chemistry of calcium phosphate ceramics. *Bioceramics Material Characteristics Versus In Vivo Behaviour, P.* Ducheyne, ed., J. Ann. N.Y. Acad. Sci., New York, (1988), pp. 268-277.
13. K. De Groot. Effect of porosity and physicochemical properties on the stability, resorption and strength of calcium phosphate bioceramics, *Bioceramics Material Characteristics Versus In Vivo Behaviour,* P. Ducheyne, ed.,. J. Ann. N.Y. Acad. Sci.,New York, (1988), pp. 227-235.
15. R. Elliott, *Eutectic solidification*, Int. Metals Review, 161-186 (1977).
16. R. L. Ashbrook, Directionally solidified ceramic eutectics, *J. Am. Ceram. Soc.,* **60**:428 (1977).
17. P. N. De Aza, F. Guitian and S. De Aza, Phase Diagram of Wollastonite-Tricalcium Phosphate, *J. Am. Ceram. Soc.,* **78**:1653 (1995).
18. P. N. De Aza, F. Guitian, A. Merlos, E. Lora-Tamayo and S. De Aza, Bioceramics-simulated body fluid interface: pH and its influence on hydroxyapatite formation, *J. Mat. Science: Mat. In Medicine,* **7**:399 (1996).

MULLITE - A MODERN CERAMIC MATERIAL

S. Sōmiya,[1] H. Ohira[2] and T. Akiba[2]

[1]Teikyo University of Science and Technology, Uenohara,
 Yamanashi 409-01, Japan
[2]Chichibu Onoda Cement Corp., Mikajiri, Kumagaya, Saitama, 360 Japan

ABSTRACT

There are many ways to produce mullite powder. One of the methods to produce mullite powder is sol-gel processing. This paper describes properties and processing of mullite powder and also properties of sintered mullite.

1. Introduction

Mullite is one of the ceramic materials, such as alumina, silica, magnesia, calcia, zirconia, SiC, Si_3N_4, etc. Among these materials, mullite is one of the most common materials because whenever clay bodies were fired, mullite appears in the fired bodies. They, however, contain mullite and also impurities. The impurities result in undesirable properties such as liquid formations at high temperatures. Therefore a new and modern mullite powder was developed.

There are numerous publications related to mullite powder and ceramic technology in the form of review and overview papers. The interested reader is referred to references 1-9.

2. Characteristics of a new mullite ceramic

Mullite has many attractive features. These include: 1) good thermal shock resistance, 2) excellent high temperature strength, 3) good creep resistance, 4) little change of strength with increasing temperature, 5) no oxidation, 6) isotropic thermal expansion, 7) good resistance to corrosion by iron oxides, slags, glasses, etc., 8) limited solubility of gas even at 1400°C, 9) low thermal conductivity, 10) the ability to make high toughness ceramic, and 11) desirable dielectric properties.
These are good reasons why mullite is a good material among the ceramic materials.

3. Preparation of a new mullite powder

As for preparation of mullite, a good review paper is that by M.D. Sacks, H. Lee and J.A. Pask in ref.3. And also many papers on this topic have appeared in 3) and 8) as well as in reports[6,7].

Some of the available methods that have been described are as follows: diphasic nano-composite gels,[10] colloidal mullite sol-gel[11], sol-gel method [12,13], alkoxide mixtures

[14,15,16,17,18], pyrolysis process[19,20], coprecipitation method[21], hydrothermal processing[22], chemical vaper deposition[23,24], spray pyrolysis[25], dip-coating[26], directional solidification[27], rapid quenching[28], refined bauxite processing[29], and reaction sintering[30].

Other methods are also available, and thus, there are many methods to produce mullite powder.

4. Preparation of a new mullite powder by Chichibu Onoda Cement Corp.[3,12]

4. 1. Starting materials

γ-Aluminium oxide (UA-5605, Showa Keikinzoku Co. Ltd. Japan) was used as the source of aluminum oxide. Colloidal silicon dioxide (Nipsil E220A, Nihon Silica Kogyo Co. Ltd. Japan) was the source of silicon dioxide. The specific surface areas (BET) of the γ-aluminum oxide and the colloidal silicon dioxide were 68 m^2g^{-1} and 130 m^2g^{-1}, respectively. The impurities present in γ-aluminum oxide were Na (16 ppm), K (5 ppm) and Si (12 ppm). The chemical composition of the colloidal silicon dioxide used was SiO_2 (95.3%), Al_2O_3 (0.55%), Na_2O (0.30%), Fe_2O_3 (0.09%), TiO_2 (0.09%) and CaO (0.05%); loss on ignition, 3.8% by weight.

4. 2. Preparation of boehmite and silicon dioxide sols

Hydrolysis of γ-aluminum oxide was carried out at temperatures above 90°C under vigorous agitation to form γ-aluminum hydrate (γ-$Al_2O_3·H_2O$), which was peptised using concentrated nitric acid, maintaining a γ-aluminum oxide to concentrated nitric acid molar ratio of 1:0.06. The temperature of peptisation was 95°C and it was carried out under refluxing conditions. A cloudy sol was formed after 3 h of peptization at 95°C.

Colloidal silicon dioxide was dispersed in distilled water under acidic conditions in a mechanical blender. The dispersion formed was slightly cloudy. No settling of silicon dioxide particles was observed even after standing for 12 h.

4. 3. Preparation of mullite gel

The boehmite sol formed by precipitation of γ-aluminium oxide was mixed in a blender with the silicon dioxide sol for 30 min. The mullite sol thus formed was gelled to a solid mass by evaporation of the excess water. The pH of the mullite sol before gelation was 1·8.

At present time, Japan is not in a good economic condition. Therefore, little effort is being made to process mullite powder by this method. Coprecipitation and sol-gel methods are still active for preparation of mullite powder.

4. 4. Processing of mullite powders

A Schematic illustration of the processing steps used in mullite powder preparation is shown in Fig. 1.

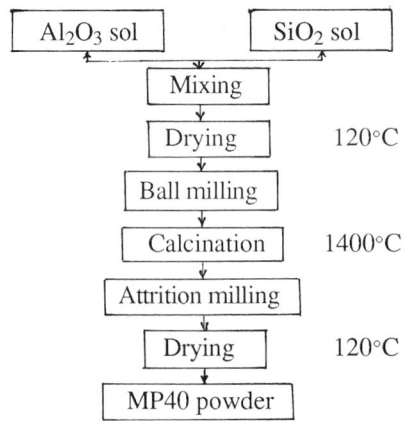

```
    Al2O3 sol        SiO2 sol
         ↓              ↓
            Mixing
              ↓
            Drying          120°C
              ↓
          Ball milling
              ↓
          Calcination       1400°C
              ↓
        Attrition milling
              ↓
            Drying          120°C
              ↓
          MP40 powder
```

Fig. 1 Powder synthesis flow chart of mullite powder

4. 5. Formation of mullite

A DTA curve of mullite gel is shown in Fig. 2. Endothermic reactions at 96°C and 420°C are drying of water and dehydration of water and formation of γ-Al_2O_3, respectively. The broad peak extending from 540°C to 1296°C is due to formation of spinel, and the peak at 1290°C is due to crystallization of mullite.

X-ray diffractions patterns of mullite gel, fired at 1000°C, 1300°C and 1400°C for 1hr. are shown in Fig. 3. Only mullite patterns appeared in the specimen D fired at 1400°C for 1 h. At 1000°C patterns show δ-Al_2O_3 and silica. In Fig 4, experiment flow chart is shown.

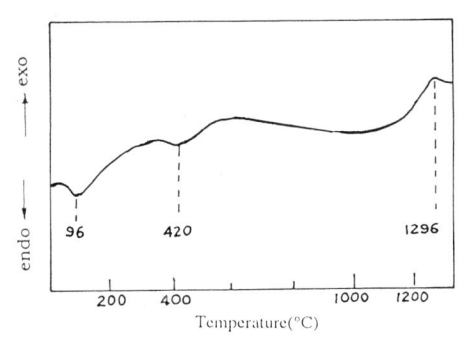

Fig. 2 DTA heating curve of gelled mullite powder.

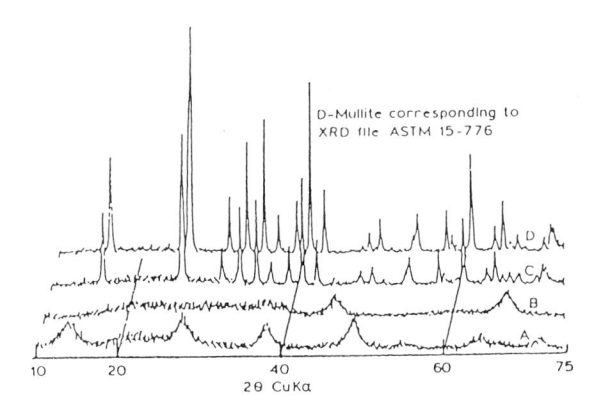

Fig. 3 X-ray diffraction patterns of the gel calcined at different temperatures.
A, gel; B, gel at 1000°C; C, gel at 1300°C for 1h; D, gel at 1400°C for 1h.

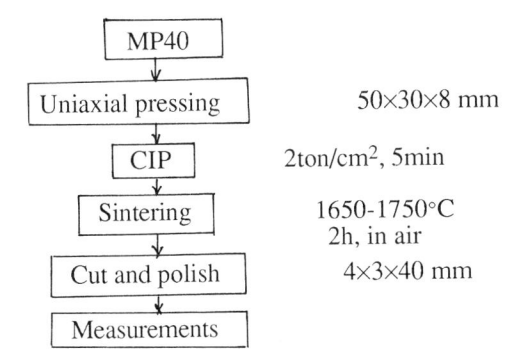

Fig. 4 Schematic Flow Chart of Mullite Body by Chichibu Onoda Cement Corp.

5. Properties of powder [31) 32) 33) 34)]

Characteristics of mullite powder are shown in Table 1. Microstructures of produced mullite powder are shown in Fig. 5 and Fig. 6. Pure stoichiometric mullite products are specially suitable for high temperature applications. Spray dried powder, MP41, is added with a binder and suitable for dry pressing. Spherical pure mullite powder is now available. This powder could be utilized as a filler where mullite characteristics are required. The low surface area and uniform shrinkage along three axes make this powder attractive for manufacture of electronics packages.

6. Properties of Sintered Bodies

Green density of the body is about 55-57% and the density of sintered bodies is shown in Fig. 7. The thermal shrinkage curve of a mullite pressed body is shown in Fig. 8. The modulus of rupture data of mullite sintered bodies is shown in Fig. 9. Even at 1300°C, the strength is over 350MPa.
The thermal shock resistance of mullite sintered body is shown in Fig. 10.

Fracture toughness, Vickers hardness, Young's modulus are 2.4 MPa·m$^{1/2}$, 12.5GPa, 246GPa respectively. Microstructue of sintered body are shown in Fig. 11 and Fig. 12.

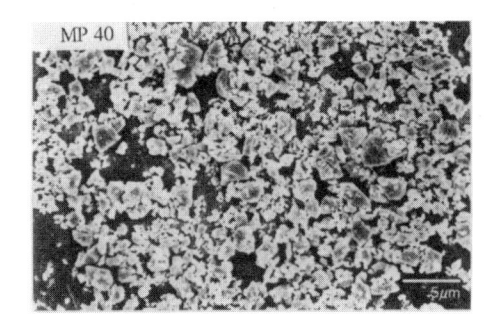

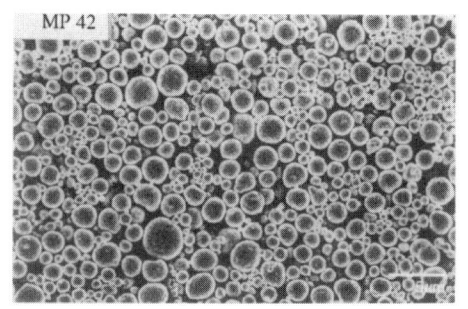

Fig. 5 High purity mullite powder/MP40·MP41

Fig. 6 Spherical pure mullite powder/MP42

Table 1 Characteristics of mullite powder

High purity mullite powder			MP40	MP42
Composition	Al$_2$O$_3$	wt%	72	72
	SiO$_2$	wt%	28	28
	TiO$_2$, Na$_2$O, Fe$_2$O$_3$	wt%	<0.15	<0.15
Average particle size		μm	14	17
Specific surface area		m^2/g	8	2
Alpha radiation		count/cm^2/h	<0.05	<0.05

Table 2 Properties of the sintered mullite ceramics

			(1650°C×2hrs)
Bulk density		g/cm³	3.12
Flexural strength[1]	(20°C)	MPa	350
	(1400°C)	MPa	350
Fracture toughness[2]		MPa·m^{1/2}	2.4
Vicker's hardness		GPa	12.5
Thermal conductivity		W/m·K	5.2
Thermal expansion		×10⁻⁷/°C	45
Electrical resistivity		Ω·cm	10¹⁴
Dielectric constant	(1MHz, 20°C)		7.2
Thermal shock resistance	(ΔT)	°C	300

1. 3point bending   2. SEPB method

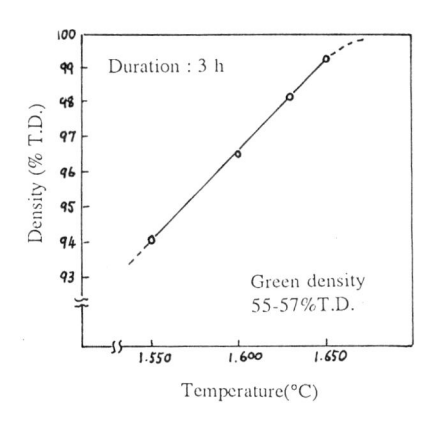

Fig. 7 Sinterability of synthesized mullite powder.

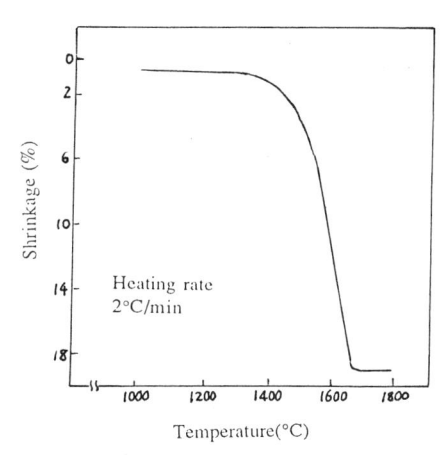

Fig. 8 Thermal shrinkage curve of mullite green body (CIP, 200MPa).

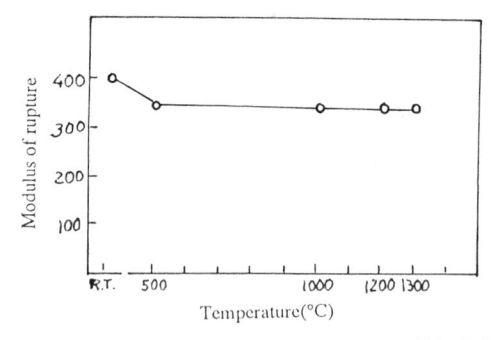

Fig. 9 Modulus of rupture of mullite sintered at 1650°C for 1.5 hours.

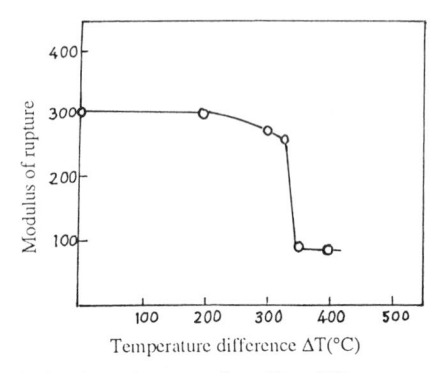

Fig. 10 Thermal shock resistance of mullite (Water quenching method).

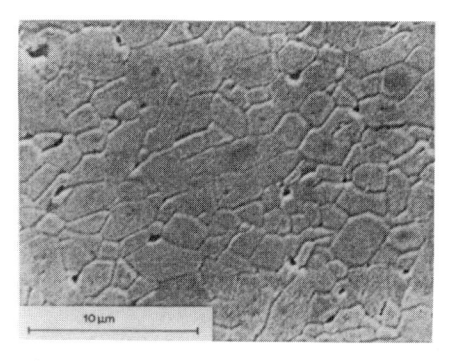

Fig. 11 SEM microphotograph of mullite sintered at 1650°C for 1.5h (bar=1μm).

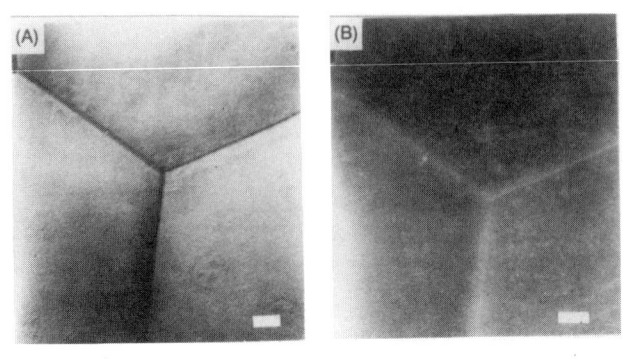

Fig. 12 TEM microphotographs of the grain boundary of mullite sintered at 1650°C for 3h: (A) bright field (bar=50nm); (B) dark field (bar=100nm).

7. SUMMARY

This paper describes the powder of mullite and sintered mullite bodies which are produced by Chichibu Onoda Cement Corp.

REFERENCES

1) J. Grofcsik, Mullite, Its Structure, Formation and Significance. pp.163 Akademiai Klado, Budapest, 1961
2) R.F. Davis and J.A. Pask, Mullite, 37-76, High Temperature Oxides Part IV. ed. by A. M. Alper. Academic Press (1971)
3) S. Sōmiya, R.F. Davis, J.A. Pask, Mullite and Mullite Matrix Composites Ceramic Transactions Vol.6 pp.649 Am. Ceram. Soc. 1990
4) H. Schneider, K. Okada, and J.A. Pask, Mullite and Mullite Ceramics, pp. 251 John Wiley & Sons, (1994)
5) H. Schneider, Abstracts, Mullite '94. Spt. 7-9 1994 Irsee, Germany; J. European Ceramic Soc. **16** [2] 101-320 (1996) Special Issue, Mullite '94
6) Shigeyuki Sōmiya, Yoshihiro Hirata, Mullite Powder Technology and Applications in Japan. Am. Ceram. Soc. Bull. **70** [10] 1624-1632 (1991)
7) Kiyoshi Okada, Nozomu Otsuka, Shigeyuki Sōmiya, Review of Mullite Synthesis Route in Japan. Am. Ceram. Soc. Bull, **70** [10] 1633-1640 (1991)
8) S. Sōmiya,(Ed) mullite I, and II Uchida Rokakuho Publishing Co. Ltd. Tokyo In Japanese.
9) H. Ohira, M.G.M.U. Ismail, Y. Yamamoto, T. Akiba, S. Sōmiya; Mechanical Properties of High Purity Mullite at Elevated Temperatures, J. Europ. Ceram. Soc. **16** [2] 225-229 (1996)
10) S. Komarneri, R. Roy, Mullite Derived from Diphase Nano-composites Gels, Ceram. Trans. **6** 209-219 (1990) Ed. S. Somiya, R.F. Davis, J.A. Pask, Am. Ceram. Soc.
11) J.C. Huling, and G.L. Messing Surface Chemistry Effects on Homogeneity and Crystallization of Colloidal Mullite Sol-Gels in Ref. 3, 221-229 (1990)
12) M.G.M.U. Ismail, Z. Nakai, S. Sōmiya Sintering of Mullite Prepared by Sol-Gel Method in Ref.3, 231-241 (1990)
13) T. Hiraiwa: Synthesis of High Purity Mullite Powder, Mullite 2, 13-19 (1987), Ed. S. Sōmiya, Uchida Rokakuho Publishing Co. Ltd.
14) K.S. Mazdiyasni: Preparation and Characterization of Mullite Powders from Alkoxides and Other Chemical Routes in Ref.3, 243-253 (1990)
15) B.E. Yoldas: Mullite Formation from Aluminum and Silicon Alkoxides in Ref. 3, 255-261 (1990)
16) H. Suzuki, H. Saito, Y. Tomokiyo, Y. Suyama: Processing of Ultrafine Mullite Powder through Alkoxide Route in Ref. 3, 263-274 (1990)
17) S. Mitachi, M. Matsuzawa, K. Kaneko, S. Kanzaki and Y. Tabata: Characterization of SiO_2-Al_2O_3 Powders Prepared from Metal Alkoxides, in Ref. 3. 275-
18) S. Mitachi, N. Tanaka, M. Matsuzawa, K. Kaneko: Characterization of Mullite Powders Prepared from Metal Alkoxides. Mullite 2. 21-29 (1987). Ed. S. Sōmiya, Uchida Rokakuho Publishing Co. Ltd.
19) M. Mizuno, H. Saito: High-purity Mullite Powder Prepared by Pyrolysis and Its Properties Mullite 2, 31-41 (1987), Ed. S. Sōmiya, Uchida Rokakuho Publishing Co. Ltd.
20) C. Yamaishi, H. Kamiaki, J. Asaumi: High Purity Mullite Ceramics. Mullite 2 81-92 (1987), Ed. S. Sōmiya, Uchida Rokakuho Publishing Co. Ltd.
21) Y. Kubota, H. Takagi: Preparation and Mechanical Properties of Mullites and Mullite-Zirconia Composites, Mullite 2, 105-118 (1987), Ed. S. Sōmiya, Uchida Rokakuho Publishing Co. Ltd.
22) S. Sōmiya, M. Yoshimura, M. Suzuki, T. Yamaguchi: Mullite Powder from Hydrothermal Processing in Ref. 8, 287-310 (1990)
23) S. Hori, R. Kurita: Characterization and Sintering of Al_2O_3-SiO_2 Powders Formed by Chemical Vapor Deposition in Ref. 8, 311-322 (1990)
24) Y. Hirata and I.A. Aksay: Processing of Mullite with Powders Processed by Chemical Vapor Deposition in Ref. 3, 323-338 (1990)
25) S. Kanzaki, H. Tabata, and T. Kumazawa: Sintering and Mechanical Properties of Mullite Derived Via Spray Pyrolysis, in Ref. 3, 339-351 (1990)

26) K. Okada, N. Otsuka: Preparation of Transparent Mullite Films by Dip-Coating Method, in Ref. 3, 425-434 (1990)
27) D. Michel, L. Mazerolles, R. Portier: Directional Solidification in the Alumina-Silica System Microstructures and Interfaces in Ref. 3, 435-447 (1990)
28) M. Yoshimura, Y. Hanaue, S. Sōmiya: Non-Stoichiometric Mullite From Al_2O_3-SiO_2-ZrO_2 Amorphous Materials by Rapid Quenching in Ref. 3, 449-459 (1990)
29) Li Nan, W.Xitang, Investigation on Mullite prepared from Refined Bauxite, in Ref. 3, 457-462 (1990)
30) P. Boch, T. Chartier, P.D.D. Rodrigo: High-Purity Mullite Ceramics by Reaction-Sintering in Ref. 3, 353-374 (1990)
31) M.G.M.U. Ismail, Zenjiro Nakai, Shigeyuki Sōmiya: Sintering Mullite Prepared by Sol-Gel Method, 231-241 (1990) Mullite Ed. S. Sōmiya R.F. Davis, & J.A. Pask, Am. Ceram. Soc.
32) M.G.M.U. Ismail, Z. Nakai, H. Ohira, S. Sōmiya: Preparation and Characterization of Containing Materials, Vol 1, Part B Ceramic Powder Science, 1108-1114 (1988) Am. Ceram. Soc.
33) M.G.M.U. Ismail, Zenjiro Nakai, Shigeyuki Sōmiya: Microstructure and Mechanical Properties of Mullite Prepared by Sol-Gel Method, Communication Am. Ceram. Soc. **70** [1] c-7, c-8 (1987)
34) M.G.M.U. Ismail, Zenjiro Nakai, Keichi Minegishi, Shigeyuki Sōmiya: Synthesis of Mullite Powder and Its Characteristics, Int. J. High Technology Ceramics, **2**, 123-134 (1986)

CONTROL OF INTERFACE FRACTURE IN SILICON NITRIDE CERAMICS: INFLUENCE OF DIFFERENT RARE EARTH ELEMENTS

Ellen Y. Sun,[1] Paul F. Becher,[1] Shirley B. Waters,[1] Chun-Hway Hsueh,[1] Kevin P. Plucknett[1] and Michael J. Hoffmann[2]

[1]Metals and Ceramics Division, Oak Ridge National Laboratory, Oak Ridge, Tennessee 37831-6068
[2]University of Karlsruhe, Institute for Ceramics in Mechanical Engineering, Karlsruhe, Germany

INTRODUCTION

The toughness of self-reinforced silicon nitride ceramics can be improved by enhancing crack deflection and crack bridging mechanisms.[1-3] Both mechanisms rely on the interfacial debonding process between the elongated β-Si_3N_4 grains and the intergranular amorphous phases. The various sintering additives used for densification may influence the interfacial debonding process by modifying (1) the thermal and mechanical properties of the intergranular glasses, which will result in different residual thermal expansion mismatch stresses;[4] and (2) the atomic bonding structure across the β-Si_3N_4/glass interface.[5] Earlier studies indicated that self-reinforced silicon nitrides sintered with different rare earth additives and/or different Y_2O_3:Al_2O_3 ratios could exhibit different fracture behavior that varied from intergranular to transgranular fracture.[6-8] However, no systematically studies have been conducted to investigate the influence of sintering additives on the interfacial fracture in silicon nitride ceramics. Because of the complexity of the material system and the extremely small scale, it is difficult to conduct quantitative analyses on the chemistry and stress states of the intergranular glass phases and to relate the results to the bulk properties.

In the current study, the influence of different sintering additives on the interfacial fracture behavior is assessed using model systems in which β-Si_3N_4 whiskers are embedded in SiAlRE (RE: rare-earth) oxynitride glasses. By systematically varying the glass composition, the role of various rare-earth additives on interfacial fracture has been examined. Specifically, four different additives were investigated: Al_2O_3, Y_2O_3, La_2O_3, and Yb_2O_3. In addition, applying the results from the model systems, the R-curve behavior of self-reinforced silicon nitride ceramics sintered with different Y_2O_3:Al_2O_3 ratios was characterized.

EXPERIMENTAL PROCEDURE

In the model system, 5 vol.% β-Si₃N₄ whiskers were embedded in oxynitride-glasses. The processing parameters and the compositions of the glasses are listed in Table 1. The processing procedures are described in detail in Ref. 9. For each sample, glass formation, complete dissolution of the starting powders, and retention of β-Si₃N₄ whiskers were confirmed by x-ray diffraction analyses. The linear thermal expansion coefficients (α) and the glass transition temperatures (T_g) were measured using a dual rod dilatometer, following the procedures described in Ref. 10. Microstructural and compositional analyses were carried out using scanning electron microscopy (Hitachi S4100) equipped with energy dispersive spectrometry capable of light element detection.

Table 1. Compositions and processing conditions of the β-Si₃N₄(whisker)/oxynitride-glass model systems.

Sample	Composition (eq.%)					Temp.	Time at Temp.
	Si	Al	Y or RE	O	N	(°C)	(minute)
AlY10	55	25	20	90	10	1700	1
AlY20-I	55	25	20	80	20	1700	6
YAl10	55	10	35	90	10	1650	1
YAl20-I	55	10	35	80	20	1700	8
AlY20-II	55	25	20	80	20	1600*	60*
YAl20-II	55	10	35	80	20	1600*	60*
LaAl	50	25	25	67	33	1680	30
YbAl	50	25	25	67	33	1680	30
La	57	0	43	79	21	1700	4

* AlY20-II and YAl20-II were obtained by annealing AlY20-I and YAl20-I under these conditions.

The debonding response of the whisker/glass interface in the different systems was evaluated by an indentation-induced crack-deflection method, as illustrated by the schematical diagram in Figure 1(a). A cube-corner diamond indenter with a 30-35 gram applied load was used to generate cracks in the glass. When the indentation crack plane intersects the longitudinal axis of the whisker, the crack will either deflect at the whisker/glass interface or penetrate the whisker, depending on the angle of incidence (θ). For a specific interface, it becomes increasingly more difficult for a crack that is propagating in the matrix to deflect at and travel along the interface as θ is increased towards 90°. By characterizing the interface debond length, l_{db}, versus θ, the maximum angle of incidence for the onset of interfacial debonding (θ_{crit}) can be determined, as shown in Figure 1(b). By comparing the θ_{crit} and l_{db} values, the interfacial debonding energy in different systems can be assessed.

Self-reinforced silicon nitride ceramics sintered with different Y₂O₃:Al₂O₃ additive ratios (but same total amounts) were studied in conjunction with the Si₃N₄(whisker)/oxynitride-glass model systems. Three different Y₂O₃:Al₂O₃ ratios were employed: 1:1, 2:3 and 3:1

(ratio in eq.%). The fraction of large elongated grains in these samples was controlled by incorporating 2 wt.% elongated β-Si₃N₄ seeds into the ceramics, following the procedures described in Ref. 11. The R-curve behavior of the ceramics was characterized *in-situ* using an applied moment DCB testing stage operated either under an optical microscope (Nikon MM-11) or in the chamber of an SEM (Hitachi S4100).[12]

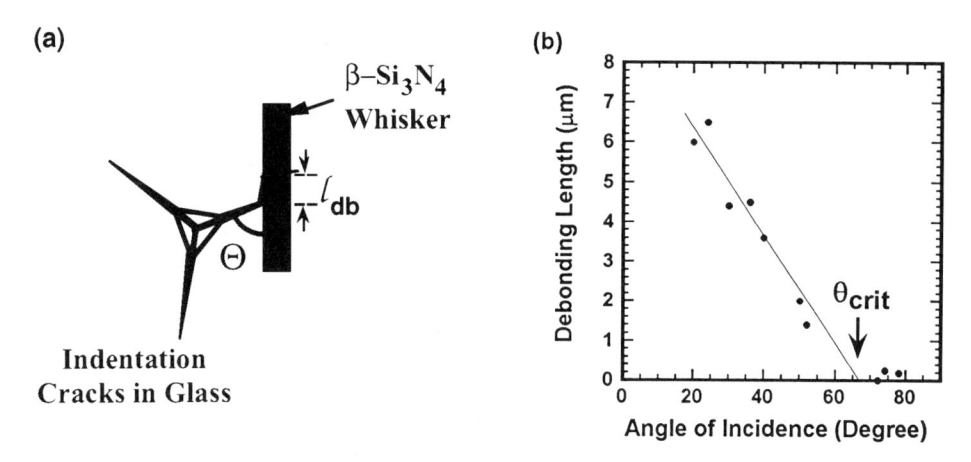

Figure 1. (a) Schematic diagram of the debonding experiment; and (b) data analyses of the debonding experiment. θ_{crit} can be determined by plotting l_{db} versus θ.

RESULTS AND DISCUSSION

Interfacial Debonding Behavior in the Si₃N₄(whisker)/Oxynitride-Glass Systems

The interfacial debonding behavior in the Si₃N₄/Si-Al-Y glass systems processed at high temperatures for a short period of time (AlY10, AlY20-I, YAl10 and YAl20-I)[9] is briefly summarized here. As shown in Figure 2, systems AlY10, YAl10 and YAl20-I showed similar debonding behaviors, while system AlY20-I exhibited much lower θ_{crit} and l_{db} values compared to the other three systems, indicating a higher interfacial debonding energy. Microstructural characterization revealed formation of a β'-SiAlON layer at the Si₃N₄/glass interface in system AlY20-I,[5] which was absent in the other systems. These results indicate that the θ_{crit} and l_{db} values are decreased when an interfacial SiAlON layer forms.

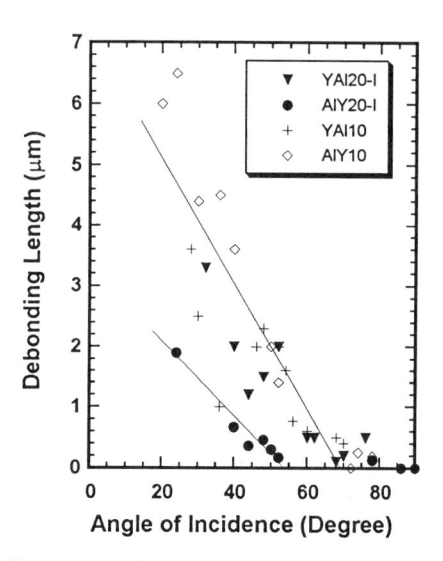

Figure 2. Debonding behavior in the AlY and YAl systems.

781

Phase equilibrium indicates that the formation of β'-SiAlON phase from the Si–Al–Y oxynitride glasses is thermodynamically favorable when the nitrogen content is greater than 16 eq.%.[13] However, the kinetics of the interfacial phase formation depends upon the specific glass composition and processing conditions. Among the four systems discussed above, SiAlON-formation was observed only in system AlY20-I under the processing conditions employed (1600°–1700°C for several minutes). It is possible for SiAlON-formation to occur in the other high-nitrogen system (YAl20) with extended holding times at elevated temperatures.

The formation of SiAlON layers and its influence on the interfacial debonding strength were studied by examining the microstructure evolution and interfacial debonding behavior of systems AlY20-II and YAl20-II, which were obtained by annealing AlY20-I and YAl20-I respectively. β'-SiAlON growth on the β-Si$_3$N$_4$ whisker indeed occurred in the YAl20 system, as shown in Figure 3(a). Furthermore, the interfacial debonding behavior of system YAl20 changed dramatically after the annealing treatment (Figure 3(b)). On the other hand, the debonding behaviors of system AlY20 remained the same after the annealing treatment. Compared with the data in Figure 2, it is noted that the θ_{crit} values are significantly lower in all the systems with SiAlON-formation (AlY20-I&II and YAl20-II) — ~50° in systems with SiAlON versus ~70° without SiAlON. These results appear to confirm that SiAlON growth on the β-Si$_3$N$_4$ grains induces in a high interfacial debonding energy.

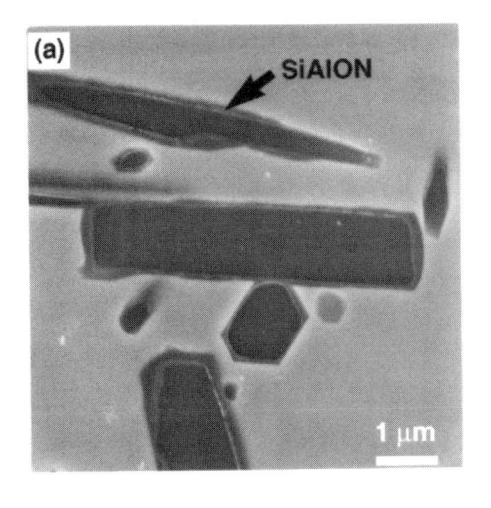

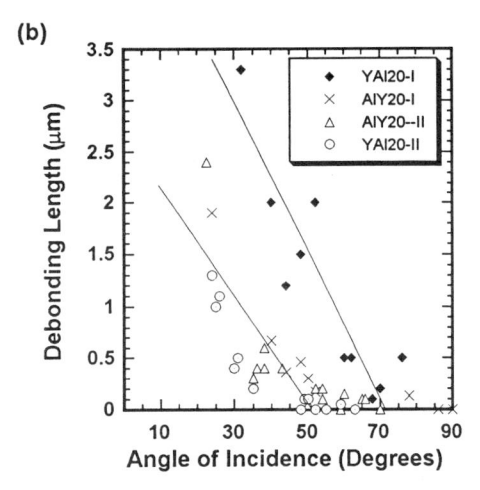

Figure 3. (a) A SiAlON layer formed on surface of the β-Si$_3$N$_4$ whiskers in system YAl20 after the annealing treatment; and (b) debonding behavior in systems AlY20 and YAl20 before and after the annealing treatments.

The SiAlON formation has a similar influence on the interfacial debonding energy in other Si-Al-RE-O-N (RE: rare earth) glass systems. SiAlON growth occurred in the LaAl and YbAl systems because the materials were prepared at high temperatures for 30 minutes. The SiAlON growth band exhibited a similar structure as that shown in Figure 3(a). Comparing with the AlY and YAl systems, the θ_{crit} and l_{db} values in the LaAl and YbAl systems are comparable to those of the AlY and YAl series with SiAlON formation, as illustrated in Figure 4.

Systems not containing Al were also studied, where the SiAlON formation would not be an influence on the interfacial debonding behavior. Specifically, systems with Si-La-O-N glasses were examined. The θ_{crit} and l_{db} values in the La-system were compared with those in the AlY and YAl systems without SiAlON formation, as shown in Figure 5. Compared with the LaAl system with SiAlON formation, interfacial debonding was enhanced in the Al-free La system.

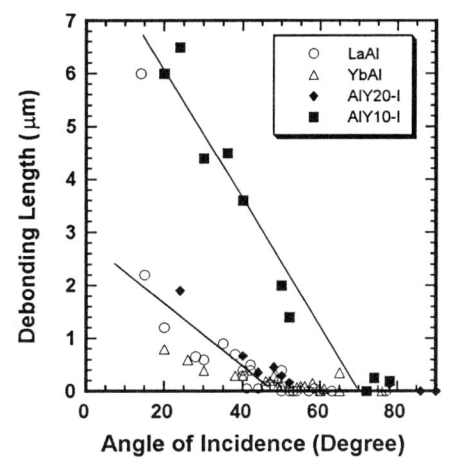

Figure 4. Debonding behavior in the LaAl and YbAl systems.

Figure 5. Debonding behavior in the La based systems.

Residual Thermal Mismatch Stresses

The residual thermal mismatch stresses in these $Si_3N_{4(whisker)}$/oxynitride-glass systems were analyzed using a modified Eshelby model, in which the whiskers were simulated as ellipsoidal inclusions with an aspect ratio of 10:1.[14, 15] The thermal and mechanical properties of the glasses and the β-Si_3N_4 crystal used in the predictions were measured (Table 2). Previous studies found that the elastic modulus of oxynitride glasses does not vary significantly with composition[9] and the residual thermal mismatch stresses were more sensitive to the thermal properties than the mechanical properties. Therefore, an average elastic modulus value of 145 GPa was used in the current calculations. Poisson ratios of the whiskers and glasses were assumed to be 0.29 and 0.26 respectively.

Table 2. Measured thermal and mechanical properties of the oxynitride glasses and the β-Si_3N_4 crystal.

Sample	AlY10	AlY20	YAl10	YAl20	LaAl	YbAl	La	β-Si_3N_4	
α (10^{-6}/°C)	5.25	5.17	6.66	6.38	6.5	5.9	7.2	2.01[a]	2.84[b]
T_g (°C)	915	950	970	1005	1030	990	1010	---	
E (GPa)	145							380	

[a]a-axis, [b]c-axis, Ref. 16

Stress analyses revealed that the resultant radial and axial thermal expansion mismatch stresses within a rod embedded in oxynitride glass were compressive due to the lower thermal expansion coefficient of the silicon nitride. The relationship between the compressive radial residual stresses and the θ_{crit} and l_{db} values are shown in Figures 6(a) and 6(b). (The axial residual stresses show a similar trend. The stress levels only change ~2% and the ranking of the stresses remains the same when the SiAlON layer is considered.[9]) The results indicate that SiAlON formation determines the θ_{crit} values while the influence of the residual thermal mismatch stresses appears to be negligible (Figure 6(a)). On the other hand, it is noticed that among systems without the SiAlON formation, the debonding length at a fixed angle of incidence generally increases with decreasing residual stresses (Figure 6(b)). However, no such relationship was observed in systems with the SiAlON formation.

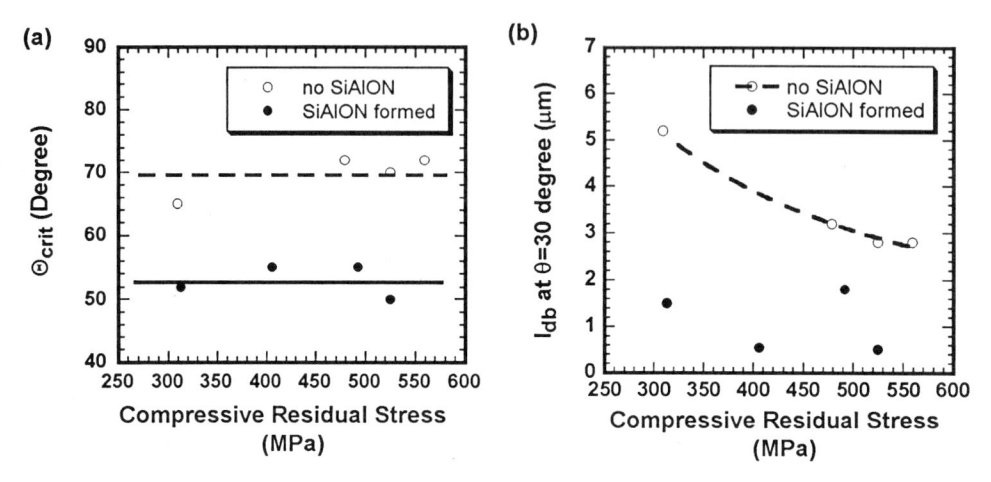

Figure 6. Relationship between the compressive radial residual stresses and the (a) θ_{crit} and (b) l_{db} values.

R-Curve Behavior of Seeded Silicon Nitrides with Different Sintering Additives

The seeded silicon nitride ceramics exhibited *R*-curve responses that were dependent on the ratio of yttria to alumina sintering additives. As shown in Figure 7, the materials sintered with the highest $Y_2O_3:Al_2O_3$ ratio exhibit the highest steady-state toughness and a steeply rising *R*-curve, while the materials sintered with the lowest $Y_2O_3:Al_2O_3$ ratio have the lowest steady-state toughness. *In-situ* observation of crack propagation and interaction with microstructural features indicated that crack-deflection and bridging occurred more readily in the higher yttria-content samples (Figure 8). However, the main cause for the different interface fracture behavior in these three ceramics was residual stresses, instead of interfacial phase formation as shown in the whisker/glass model systems, because SiAlON growth was present in all the three ceramics studied due to the long processing time at elevated temperatures. Also, it is possible that the influence of residual stresses on the interface fracture is greater in the ceramics than in the whisker/glass systems due to the significantly different volume fractions of the glassy phases. Ongoing research is focusing on the measurement and analytical modeling of residual stresses in silicon nitride ceramics.

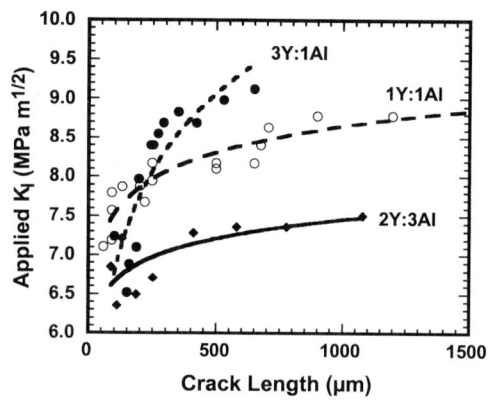

Figure 7. *R*-curve response of self-reinforced silicon nitrides sintered with different yttria to alumina additives.

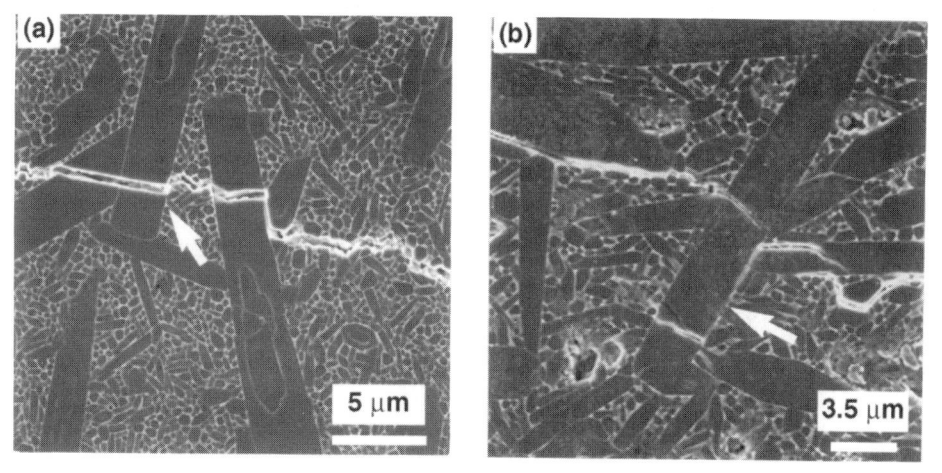

Figure 8. Crack deflection and bridging by the elongated grains in seeded silicon nitride sintered with different $Y_2O_3:Al_2O_3$ additive ratios, (a) 2Y:3Al and (b) 3Y:1Al.

CONCLUSION

In $Si_3N_{4(whisker)}$/oxynitride-glass model systems, interfacial debonding behavior is determined by the interfacial microstructure and chemistry. In Si-Al-RE(Y)-O-N glasses, the interfacial debonding energy increases significantly with SiAlON formation. Al-free glasses enhances interfacial debonding by inhibiting SiAlON formation. Compared to the interfacial microstructure/chemistry, the residual thermal mismatch stresses is a secondary influence on the debonding behavior. In systems without SiAlON formation, the residual stresses modify the debonding length. In self-reinforced silicon nitride ceramics, a higher yttria to alumina additive ratios resulted in a higher steady state toughness. Sophisticated experimental and analytical-modeling work are required to understand the influence of the residual stresses on the interfacial fracture behavior in self-reinforced silicon nitride ceramics.

ACKNOWLEDGMENTS

The authors thank Drs. K. Hirao and M. Brito of the National Industrial Research Institute-Nagoya for their assistance and the MITI Agency for International Science and Technology Fellowship (Japan) for supporting KPP producing the silicon nitride ceramics at NIRI-Nagoya. Drs. H. T. Lin and E. Lara-Curzio are thanked for reviewing the manuscript. Research is sponsored by the U.S. Department of Energy, Division of Materials Sciences, Office of Basic Energy Sciences, under contract DE-AC05-96OR22464 with Lockheed Martin Energy Research Corp. and by appointments of EYS and KPP to the Oak Ridge National Laboratory Postdoctoral Research Associates Program, which is administered jointly by the Oak Ridge Institute for Science and Education and Oak Ridge National Laboratory.

REFERENCES

1. P. F. Becher, S. L. Hwang, and Chun-Hway Hsueh, Using microstructure to attack the brittle nature of silicon nitride ceramics, *MRS Bull.* 20[2]:21 (1995).
2. T. Kawashima, H. Okamoto, H. Yamamoto, and A. Kitamura, Grain size dependence of the fracture toughness of silicon nitride ceramics, *J. Am. Ceram. Soc. Japan,* 99:1 (1991).
3. P. Sajgalik, J. Dusza, and M. J. Hoffmann, Relationship between microstructure, toughening mechanism, and fracture toughness of reinforced silicon nitride ceramics, *J. Am. Ceram. Soc.,* 78[10]:2619 (1995).
4. I. M. Peterson and T. Y. Tien, Effect of grain boundary thermal expansion coefficient on the fracture toughness in silicon nitride, *J. Am. Ceram. Soc.,* 78[9]:2345 (1995).
5. E. Y. Sun, K. B. Alexander, P. F. Becher, S. L. Hwang, β-Si_3N_4 whiskers embedded in oxynitride-glasses: interfacial microstructure, *J. Am. Ceram. Soc.,* in print.
6. Y. Tajima, K. Urashima, M. Watanabe, and Y. Matsuo, Fracture toughness and microstructure evaluation of silicon nitride ceramics, in *Ceramic Transactions, Vol. 1*, E. R. Fuller and H. Hausner, ed., Am. Ceram. Soc., Westerville (1988).
7. Y. Tajima, Development of high performance silicon nitride ceramics and their application, in *MRS Proc., Vol. 287*, I. W. Chen, P. F. Becher, M. Mitomo, G. Petzow and T. S. Yen, ed., MRS, Pittsburgh (1993).
8. G. Wotting and G. Ziegler, Influence of powder properties and processing conditions on microstructure and mechanical properties of sintered Si_3N_4, *Ceramics Intl.,* 10[1]:18 (1984).
9. P. F. Becher, E. Y. Sun, C. H. Hsueh, K. B. Alexander, S. L. Hwang, S. B. Waters, and C. G. Westmoreland, Debonding of interfaces between beta-silicon nitride whiskers and Si-Al-Y oxynitride glasses, *Acta Metall.,* in print.
10. E. Y. Sun, P. F. Becher, S. L. Hwang, S. B. Waters, G. M. Pharr, and T. Y. Tsui, Properties of silicon-aluminum-yttrium oxynitride glasses, *J. Non-Crystal. Solids,* in print.
11. K. Hirao, M. Ohashi, M. E. Brito, and S. Kanzaki, Processing strategy for producing highly anisotropic silicon nitride, *J. Am. Ceram. Soc.,* 78[6]:1687 (1995).
12. P. F. Becher, C. H. Hsueh, K. B. Alexander, and E. Y. Sun, Influence of reinforcement content and diameter on the R-curve response in SiC-whisker-reinforced alumina, *J. Am. Ceram. Soc.,* 79[2]:298 (1995).
13. A. Drew, *Nitrogen Glass*, P. Evans, ed., the Pathenon Press, Casterton Hall, U.K. (1986).
14. C. H. Hsueh and P. F. Becher, Residual thermal stresses in ceramic composites, part I with ellipsoidal inclusions, *Mater. Sci. Eng.* A212:22 (1996).
15. C. H. Hsueh and P. F. Becher, Residual thermal stresses in ceramic composites, part II with short fibers inclusions, *Mater. Sci. Eng.* A212:29 (1996).
16. C. M. B. Henderson and D. Taylor, Thermal expansion of the nitrides and oxynitride of silicon in relation to their Structure, *Trans. Brit. Ceram. Soc.* 74[2]:49 (1975).

MICROSTRUCTURAL EVALUATION OF DEFORMATION MECHANISMS IN SILICON NITRIDE CERAMICS

J. A. Schneider and A. K. Mukherjee

Department of Chemical Engineering and Material Science
University of California, Davis

ABSTRACT

Changes of phase composition and morphology were investigated in silicon nitride (Si_3N_4) before and after compressive deformation testing. Using different viscosity additives, rapid consolidation techniques were used to achieve specimens with different α-phase content and grain morphology. Low viscosity additive systems in which the phase change occurred before densification produced elongated microstructure with preferred orientation noted in the prismatic [210] direction. Limited deformation was accommodated with this grain morphology which indicated a flow stress dependency of $n=1$. High viscosity additive systems in which the densification occurred before the phase change produced equiaxed microstructures with virtually no preferred orientation. Enhanced deformation was accommodated in this grain morphology with an indicated flow stress dependency of $n=2$. Transmission electron microscope (TEM) photographs corresponded with x-ray diffraction (XRD) analysis for the grain morphology. Selected area diffraction (SAD) was used to verify phases.

INTRODUCTION

Silicon nitride (Si_3N_4) based materials are of interest in both high temperature structural applications as well as chemically corrosive environments. However, the mechanical properties that make this ceramic desirable in various applications also make it very difficult and costly to conventionally machine into complex shapes. Thus, methods to produce net shape components are of interest. One method that is commercially attractive in metal-based systems is the use of enhanced or superplastic (SP) deformation. These techniques of metal-based systems provided empirical guidelines which correlate enhanced deformation with the presence of a fine, equiaxed grain morphology. This requirement is counter to that of the typical Si_3N_4 microstructure in which elongated β-phase grains are observed. Thus, this study considered the effect of microstructural morphology on the flow stress dependency observed in enhanced plasticity compression tests. Various researchers [1-8] have reported flow stress dependencies in superplasticity studies on Si_3N_4 that range from 0.5 [3,5] to > 2 [2]. Based on various models for deformation of a solid with a viscous grain boundary phase, a flow stress dependency of *n=1* is expected for viscous flow of the grain boundary phase [9]. Deformation in this case is expected to be limited. A value of *n=2* is expected for superplasticity in which deformation is accommodated by grain rearrangement. Enhanced deformation is associated with this mechanism [9].

Silicon nitride is noted to co-exist in two crystalline phases, a metastable α-phase and a stable β-phase [10-12]. If these phases are formed unconstrained, the α-phase displays an equiaxed morphology and the β-phase an elongated morphology [13,14]. The phase change is generally considered to result from the α-phase going into solution with the grain boundary phase and either homogeneously precipitating out as β-grains or heterogeneously precipitating out on existing β-grains [10-12,15]. The grain boundary phase forms in Si_3N_4 from the inherent SiO_2 coating on the starting powders along with the oxide additives intentionally added to assist in sintering. Use of low viscosity additives has been shown to result in densification prior to phase change with the retention of the metastable α-phase [13]. As the additives increase in viscosity, phase change can precede the densification resulting in complete conversion to the stable β-phase [13].

EXPERIMENTAL PROCEDURE

Material Preparation

Ube Industry SN-E10 Si_3N_4 powders with an average particle size of 500 nm and >95% α-phase were used in this study. Two additive systems were used in this study to produce specimens with either elongated and equiaxed grains. Additives used singularly and in combination were Y_2O_3 and $MgAl_2O_4$. The powders were packed in a graphite die and rapidly consolidated using Plasma Assisted Sintering (PAS) equipment by Sodick, Co. Details of the PAS process, which uses pulsed and applied resistance heating, are described elsewhere [16,17]. Processing parameters for the consolidate of the specimens ranged from 2023 to 2173 K for 2 to 5 minutes under an applied load of 66 MPa, followed by rapid cooling in air. Compression specimens of approximately 6 mm in length and 3 x 3 mm in cross sectional area were machined from the consolidated specimens with the length perpendicular to the hot pressing direction.

Compression Tests

Compression tests were performed in an air furnace at 1673-1773 K with molydisilicide heating elements. Temperature was measured with a Pt-Rh thermocouple and controlled by a programmable temperature controller to within 2° of the set temperature. The push rods were made of alumina and were 50 mm in diameter. To prevent possible chemical reaction, SiC spacers coated with BN spray were placed under the test specimen and graphfoil™ was used on the top of the test specimen. Heating and cooling times were held constant at 7 hours to avoid thermally shocking the alumina push rods and ensuring thermal equilibrium was reached. To evaluate the effects of atmosphere, a nitrogen purge was installed around the test specimen for several tests.

Compression tests were conducted over a range of temperatures and strain rates to determine the stress sensitivity and activation energy. Steady state flow behavior was observed at constant strain rates ranging from 5×10^{-6} to 1×10^{-4} s^{-1} at a constant temperature of 1723 and 1773 K. A MTS servohydraulic mechanical tester was used which has both stroke and load controlling functions with an automated data acquisition system. Compression tests were terminated at predetermined deformation stages of 70%, 40% and 10% true strain.

Microstructural Evaluation

Phase content of specimens before and after compression testing were determined from x-ray diffraction (XRD) analysis using Cu-Kα radiation. The method of Gazzara and Messier [18] was used to determine the α and β-phase contents. Densities of the consolidated specimens were measured using the Archimedes' immersion method. Specimens were prepared for transmission electron microscopy (TEM) by ion beam thinning for electron transparency. A thin layer of carbon was deposited on the surface to minimize charging under the electron beam.

RESULTS

Flow stress versus strain data obtained in constant strain rate tests at 1773 K for specimens of 12% vs 66%, and 89% initial α-phase are plotted in Figures 1 and 2, respectively. The plots are shown intentionally truncated at 10% true strain for the 12% α-phase as specimen breakage was observed at higher strains. The plots of the 66% and 89% α-phase specimens are shown truncated at 40% true strain where steady state flow behavior was obtained in earlier tests [19].

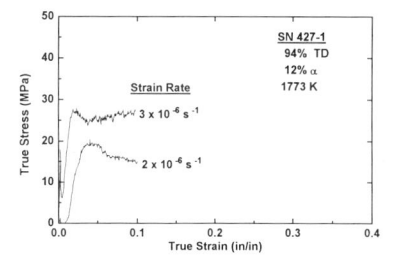

Fig. 1. Stress vs. Strain Plot of SN 427-1 (12% α-Phase).

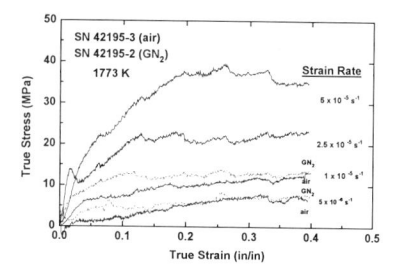

Fig. 2. Stress vs. Strain Plot of SN 042195-2 (89% α -Phase) and SN 042195-3 (69% α -Phase).

TEM micrographs of the starting specimen morphology are shown in Figure 3 for the 12% α-phase and Figure 4 for the 89% α-phase specimen. XRD analyses of the respective phase content before compression testing are shown in Figures 5 and 6. Control specimens were also analyzed which were subjected to the same heat cycle without deformation. Figures 7 and 8 show a comparison of representative control and compressed specimen XRD data. Some preferred orientation is noted as summarized by the ratio of β- [210] to β- [101] peaks in Figure 9.

The flow stress dependency was evaluated for the three specimens from the slope of the plot in Figure 10 using the constitutive relationship of $\varepsilon = \sigma^n$. A value approaching $n=2$ is noted for the equiaxed specimen. Data from the 1723 K tests fall along the line with slope n=2 whereas data from the 1773 K tests are shifted to the left. This has previously been correlated with a tendency for specimens with this additive system to dissociate at 1773 K which results in an underprediction of the flow stress [19]. This dissociation is not observed at 1723 K or with only the Y_2O_3 additive. A value approaching $n=1$ is noted for the elongated specimen in which no indication of dissociation was noted.

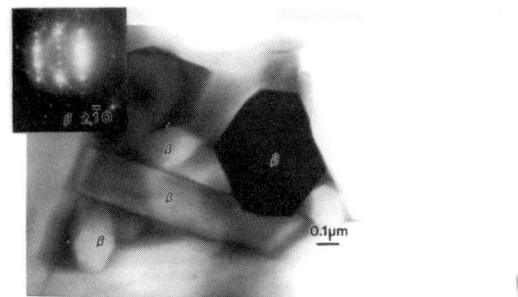

Fig. 3. TEM of SN 427-1 Elongated Morphology (12% α-Phase).

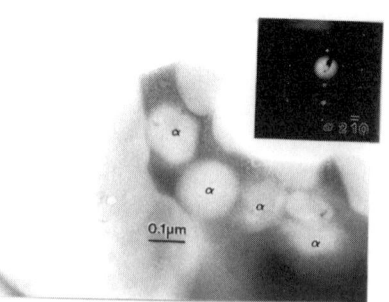

Fig. 4 TEM of SN 011695-1 Equiaxed Morphology (89% α-phase).

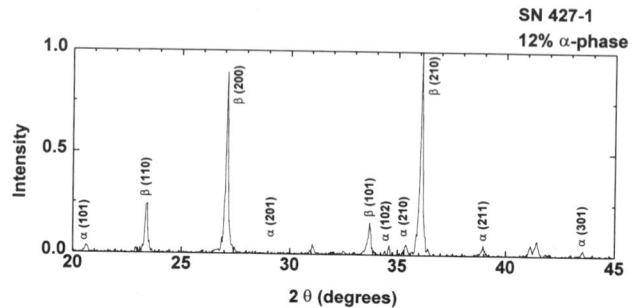

Fig. 5. Initial XRD of Elongated Morphology Specimen SN 427-1 (12% α-Phase).

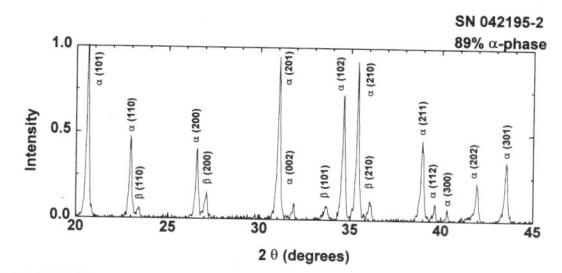

Fig. 6. Initial XRD of Equiaxed Morphology Specimen SN 042195-2 (89% α-Phase).

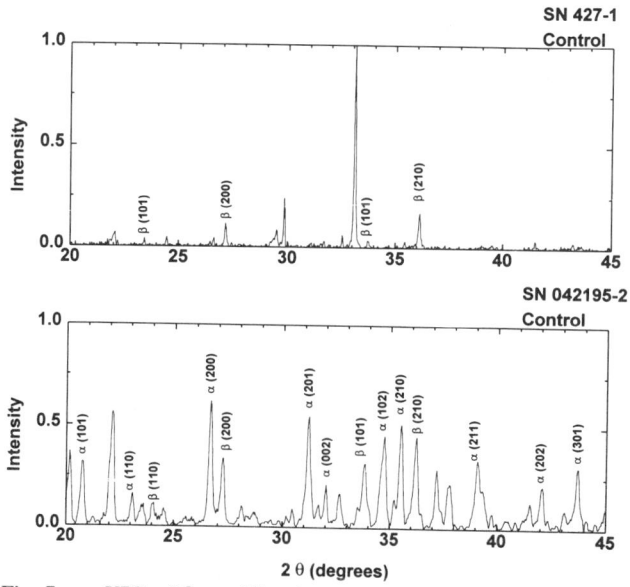

Fig. 7. XRD of Control Specimens after Heat Treatment at 1773 K.

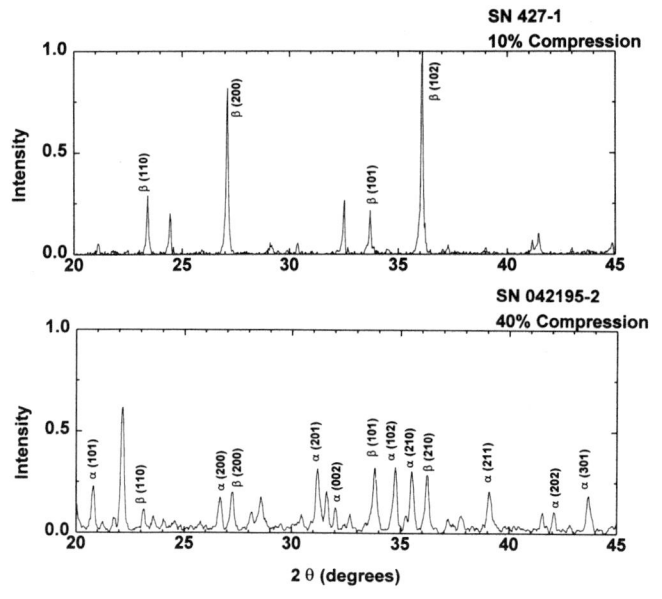

Fig. 8. XRD of Specimens after Compression Testing at 1773 K.

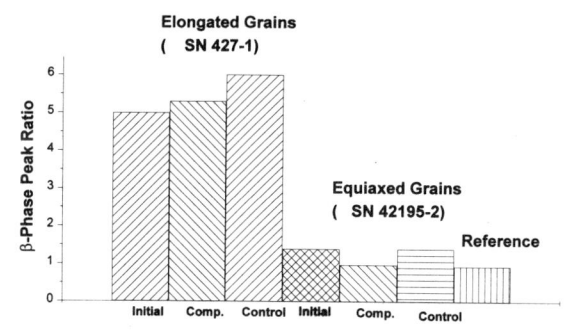

Fig. 9. Comparison of β-(210) to β-(101) Peak Ratio from XRD Analysis.

DISCUSSION

The microstructure in all specimens tested was that of Si_3N_4 grains embedded in a glassy grain boundary phase. Various models have been proposed for deformation of a solid with a liquid or glass at the interface in which a stress dependency of *n=1* is predicted [20-23]. These models consider the rate controlling deformation mechanisms to be either viscous flow of the glassy phase or mass transport of material from one grain to another. If the rate controlling mechanism is viscous flow, the accommodated strain is expected to be limited as the grains would eventually come in intimate contact under the applied load. If mass transport of material through the grain boundary phase is rate controlling, changes in grain morphology or elongation would be noted. Again the accommodated strain would be limited as the diffusional paths increased with increasing grain size. These arguments suggest that accommodation of a large amount of strain is expected only if the grain size remains fine and equiaxed. A model has recently been proposed in which a flow stress dependency of n=2 is predicted for materials with a glassy grain boundary phase. In this model, accommodation of generated stresses is controlled by mass transfer at the interface rather than through the grain boundary diffusional path [24].

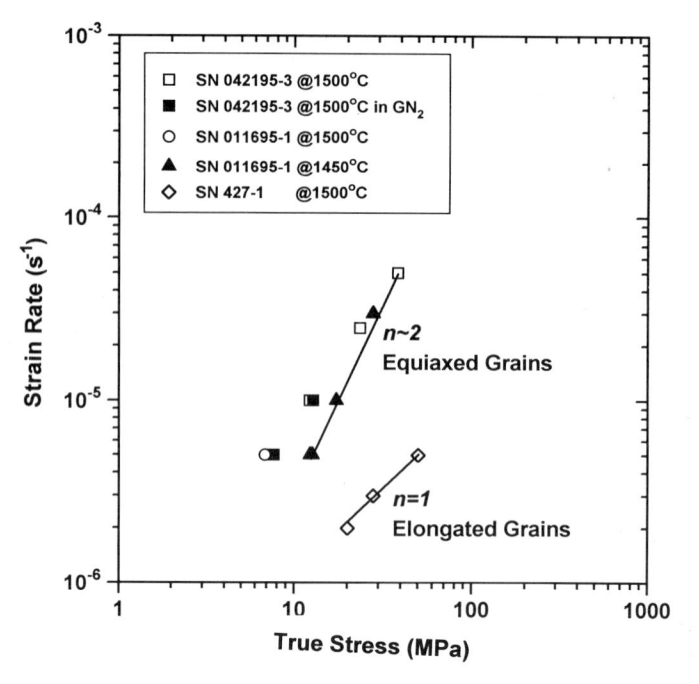

Fig. 10. Flow Stress Dependency.

Specimens with high viscosity grain boundary phases displayed a preferred orientation relationship that increased following time at temperature to 10% true strain both with and without compressive load applied. The preferred orientation was noted to be perpendicular to the hot press direction and parallel to the compressive load applied, as illustrated in Figure 11. In this microstructure stresses generated by the applied load would be expected to be accommodated by the sliding of the elongated grains past one another. Ultimately failure would be expected as grain impingement prevented further sliding.

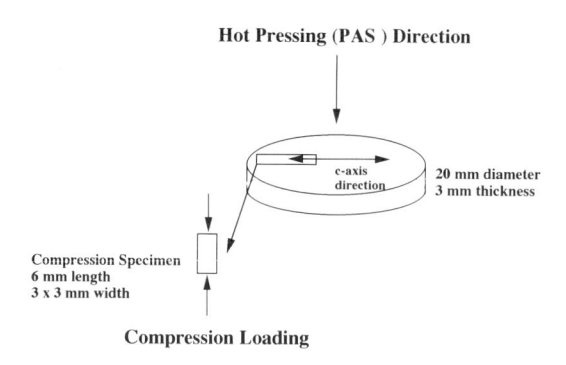

Fig. 11. Specimen Orientation

Specimens with low viscosity grain boundary phases displayed lesser amounts of preferred orientation. Data from the peak ratio in the reference JCPDS [25] files were taken as a reference. Specimens which were not compressed showed a slight increase in the ratio of the β-phase peaks. Whereas specimens which were compressed showed a slight decrease in the ratio of the β-phase peaks. Since the starting morphology of these specimens were equiaxed α-phase and the final morphology was equiaxed β-phase, the 500 nm grains may rotate in response to the applied load. Flow stress dependency values approaching n=2 would be in agreement with recent models which propose a rate controlling mechanism of interface control. The mass transport under applied load would still promote dissolution of the α-phase grains and small β-phase grains. As the grains rotate, new sites would continually be exposed for attachment sites. Deposition can now occur uniformly on all sides of the grain resulting in the equiaxed morphology exhibited.

SUMMARY

Si_3N_4 specimens have been consolidated that retained the equiaxed, metastable α-phase. Use of additives such as $MgAl_2O_4$ promote formation of a low viscosity additive system which rapidly pulls the grains together by capillary forces during the liquid phase sintering. If additives are used such as Y_2O_3 which promote formation of a high viscosity additive phase, the phase transformation from α to β-phase occurs prior to completion of the liquid phase sintering. Equiaxed initial structures of the α-phase are noted to transform to an equiaxed β-phase microstructure following the elevated temperature compression tests. Compressive deformation tests indicate different flow stress dependencies for the equiaxed and elongated morphology. Limited deformation is noted in the elongated morphology with a flow stress of n=1 indicating viscous flow of the grain boundary phase. Enhanced deformation is noted in the equiaxed morphology with a flow stress tending toward n=2 indicating grain rearrangement. Recent models [2-4] have proposed an interface control mechanism that may be correlated with enhanced deformation in Si_3N_4. These studies are in agreement with this model and indicate an equiaxed morphology is required to permit grain rearrangement in response to applied stress.

ACKNOWLEDGMENTS

The authors wish to thank Mr. C.J. Echer of the National Center for Electron Microscopy, Lawrence Berkeley Laboratory, for assistance with the analytical electron microscope. This work has been supported in part by the DP Materials Initiative, with Dr. Wendell Kawahara as the Sandia National Laboratory collaborator, and by the Ceramics Program of the Division of Materials Research Grant #NSF DMR-9314825 monitored by Lisolette J. Schioler.

REFERENCES

1. Wakai, F., Kodama, Y., Sakaguchi, S., Murayama, N., Izaki, K., Niihara, K., *Nature*, **344**, 421 (1990)
2. Rouxel, T., Wakai, F., *J. Am. Ceram. Soc.*, **75**, 2363 (1992).
3. Chen, I.-W., Hwang, S.-L., *J. Am. Ceram. Soc.*, **75**, [5], 1073 (1992).
4. Wu, X., Chen, I.-W., *J. Am. Ceram. Soc.*, **75**, [10], 2733 (1992).
5. Chen, I.-W., Hwang, S.-L., Mat. Res. Soc. Symp. Proc., **287**, 209 (1993).
6. Hwang, S.-L., Chen, I.-W., *J. Am. Ceram. Soc.*, **77**, [10], 2575 (1994).

7. Mitomo, M., Hirotsuru, H., Suematsu, H., Nishimura, T., *J. Am. Ceram. Soc.*, 78, [1], 211 (1995).
8. Rossignol, F., Rouxel, T., Besson, J.-L., Goursat, P., Lespade, P., *J. Phys. III France*, 127 (1995).
9. Arzt, E., Ashby, M.F., Verral, R.A., *Acta. Metall.*, 31, 1977 (1983).
10. Jack, K.H., *J. Mater. Sci.*, 11, [6], 1135 (1976).
11. Lange, L.L., *Int'l Metals Reviews*, [1], 1, (1991).
12. Lange, H., Wotting, G., Winter, G., *Angew. Chem. Int. Ed. Engl.*, 30, 1579 (1991).
13. Hampshire, S., **Materials Science and Technology**, ed. R.W. Cahn, P. Haasen, E.J. Kramer, 11, 119 (1991).
14. Chen, I.-W., Mat. Sci. Engr., A166, 51 (1993).
15. Sajgalik, P., Galusek, D., *J. Mat. Sci. Letters*, 12, 1937 (1993).
16. Schneider, J.A., Risbud, S.H., Mukherjee, A.K., *J. Mat. Res.*, 11, [2], 358 (1996).
17. Schneider, J.A., Mishra, R.S., Mukherjee, A.K, Proc., Second Int'l Symp. on Advanced Synthesis and Processing (1996).
18. Gazzara, C.P., Messier, D.R., *Am. Cer. Soc. Bull.*, 56, [9], 777 (1977).
19. Schneider, J.A., Mukherjee, A.K., Proc. 20th Annual Ceramic Society Conf., (1996).
20. Stocker, R.L., Ashby, M.F., *Reviews of Geophysics and Space Physics*, 11, [2], 391 (1973).
21. Raj, R., Chyung, C.K., *Acta. Metall.*, 29, 159 (1981).
22. Pharr, G.M., Ashby, M.F., *Acta. Metall*, 31, 129 (1983).
23. Dryden, J.R., Kucerovsky, D., Wilkinson, D.S., Watt, D.F., *Acta. Metall.*, 37, [7], 2007 (1989).
24. Wakai, F., *Acta. Metall.*, 42, [4], 1163 (1994).
25. JCPDS-ICDD, Powder Diffraction Files No. 33-1160 and 41-0360, (1991).

MICROSTRUCTURALLY INDUCED INTERNAL STRESSES IN THE SILICON NITRIDE LAYERED COMPOSITES

Pavol Šajgalík, and Zoltán Lenčéš

Institute of Inorganic Chemistry, Slovak Academy of Sciences, Dubravská cesta 9, SK-842 36 Bratislava, Slovakia

INTRODUCTION

A design of composites with layered structure seems to be a way of preparation of ceramic materials with decreased sensitivity to the defects[1,2] Recently, this approach was applied also for silicon nitride based composites[3-7]. The decreased flaw sensitivity of these materials is probably reached by presence of residual stresses which are conserved in these materials after cooling. These stresses can diminish the quantity of the stresses concentrated on the largest defect and so decrease the sensitivity of layered materials to the flaw size[6,7]. The residual stresses are a consequence of different material constants of individual layers (thermal expansion coefficients and Young's modulus)[6]. An increase of the bending strength of layered materials was achieved, however the size of the flaws was the same as in the relative monoliths[3-7]. Additionally, the stress status of the layer can be modified by the layer thickness, green density of the layer, sinterability of the used powder and used sintering method[7]. Layered ceramic materials pay for the increased quality by loosing of isotropy of their mechanical properties.

Present paper deals with the preparation of Si_3N_4 composites with the layers of different microstructure but the same chemistry. Paper attempts to understand the physical background and principles of their behaviour. Their room temperature properties are reported.

EXPERIMENTAL

The starting mixtures were prepared by attrition milling of Si_3N_4 powder with sintering aids and microstructure forming additions in dry isopropanol for 4 h in the weight ratios listed in Table 1. The layers of the dried starting powders were poured into the die in the desired succession and then pressed into the cylindrical samples with uniaxial pressure of 100 MPa. Green bodies were hot pressed (HP) at temperatures ranging from 1750°C to 1850°C and mechanical pressure of 30 MPa for 3 h under nitrogen atmosphere. The samples for gas pressure sintering (GPS) were CIP-ed at 250 MPa in a rubber mould. The layers of starting powder were sequenced in the same way as in case of green bodies for hot pressing. Rectangular bars were sintered at 1900°C for 3 h under nitrogen overpressure of 10 MPa. All the samples were 99% dense after either HP or GPS.

Table 1. Starting powders composition.

Starting Powder	Preparation Method	Si$_3$N$_4$ wt%	Sintering Aids wt%		Microstructure Forming Powders /wt%
Mixture			Al$_2$O$_3$	Y$_2$O$_3$	amorph. Si$_3$N$_4$
A	HP	72[+]	3.4[⋄]	4.6[*]	-
B	HP	72[+]	3.4[⋄]	4.6[*]	20[h]
E	GPS	92[*]	3.4[⋄]	4.6[*]	-
F	GPS	92[#]	3.4[⋄]	4.6[*]	-

[+] LC-12-S, H.C. Starck, Germany; [*] UBE SNE10, Japan; [#] UBE SN03, Japan; [⋄] Fluka AG, Germany; *Techsnabexport, Russia; [h] prepared in house.

For ceramographic and mechanical testing the bars of dimensions (3x4x45) mm were cut from the dense ceramic bodies with the tensile face (15 µm finish) parallel to the layer direction. Bending strength was evaluated using a four point bending fixture with inner/outer span 20/40 mm and a cross head rate 0.5 mm/min. The microstructures were observed by SEM on polished and plasma etched surfaces. The internal stresses in the individual layers were measured by indentation method from the different length of indentation cracks in the presence of residual stresses when compared with the stress-free state.

RESULTS AND DISCUSSION

Figure 1 shows the deflection of crack crossing the layer boundary between two Si$_3$N$_4$ microstructures prepared by HP. The crack deflection is caused by the stress status of the particular layers. The residual stresses are step-wise changed at the layer boundary. The stresses of the adjacent two dimensional layers with different elastic constants and thermal expansion coefficients are described by Chartier et al.[8] Their difference at the boundary can be expressed by the following equation:

$$\Delta\sigma = \frac{E_1 E_2 (\alpha_2 - \alpha_1)(d_1 + 2d_2)\Delta T}{(1 - \nu_1)E_2 d_2 + 2(1 - \nu_2)E_1 d_1} \quad (1)$$

where ν is the Poisson's ratio, E the Young's modulus, d the layer thickness, $\Delta T = T_j - T_o$, the difference between temperature of rigid joints formation between the layers (T_j) and room temperature (T_o), α_1 and α_2 are the thermal expansion coefficients of adjacent layers. According to the Eq. (1) the difference in the stresses are influenced by the difference in elastic constants (Young's modulus and Poisson ratio), thermal expansion coefficients and the layer thickness. The role of these factors are studied in our previous work[7]. For the present case, the layered composite consisting of the silicon nitride layers differing only by microstructure, not the chemistry, should be stress free because of $\alpha_1 = \alpha_2$ as follows from the Eq. (1). In fact, the measurement of residual stress by indentation on the three layer

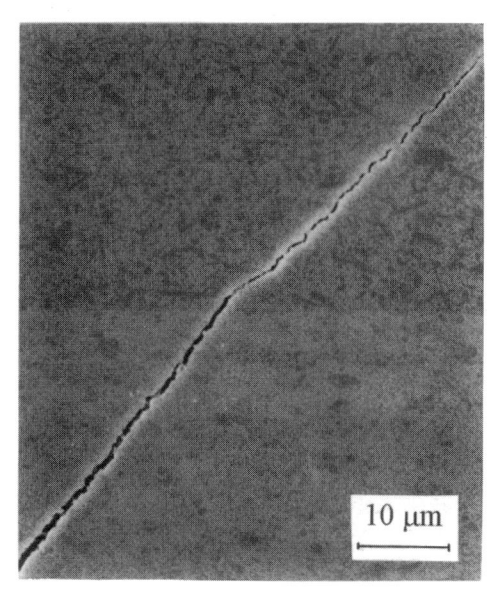

Figure 1. SEM micrograph of the crack crossing the layer boundary.

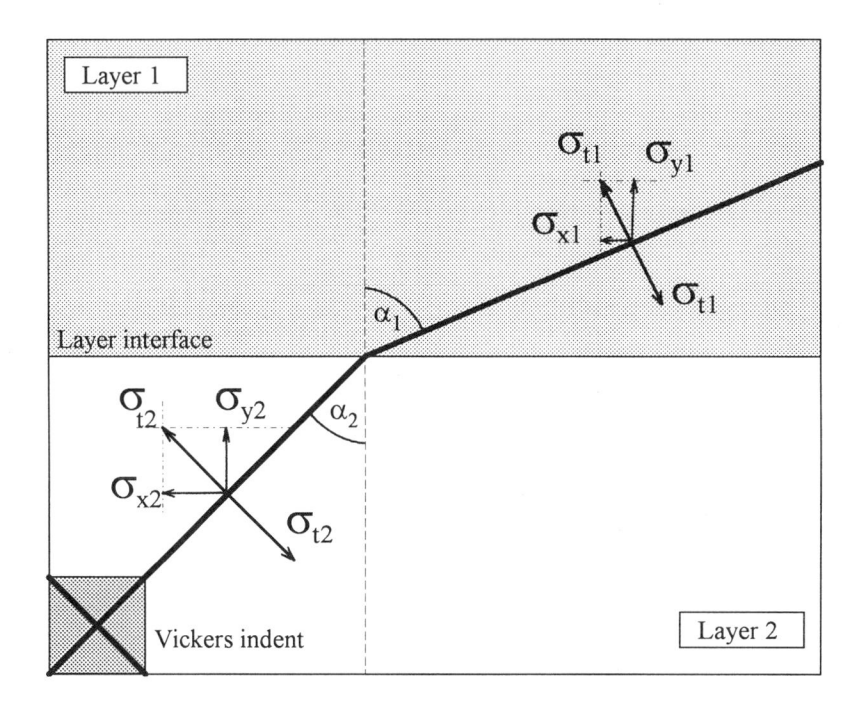

Figure 2. Schematic of the crack deflection on the layer boundary.

composite confirmed this statement for samples prepared by GPS. No residual stresses were detected. For samples prepared by HP the residual stress was observed, which is demonstrated by the crack deflection on the layer boundary, Fig. 1. The deflection of the crack can be described by the following equation:

$$\sin \Delta\alpha = \sin(\alpha_1 - \alpha_2) = \frac{\sigma_{y1}\sigma_{x2} - \sigma_{x1}\sigma_{y2}}{\sigma_{t1}\sigma_{t2}} \qquad (2)$$

where σ_{t1} and σ_{t2} are the tensile stresses responsible for the crack propagation in the layer 1 and 2, respectively. σ_x and σ_y are their components, Fig. 2. The indentation cracks length is a function of the stresses in the individual layers. The deflection angle of the crack crossing the layer boundary can be estimated by Eq. (2). The stresses σ_x and σ_y in the Eq. (2) can be replaced by K_{IC} values from Fig. 3 using the Griffith equation, i.e. $\sigma_x = Y \cdot a^{-1/2} \cdot K_{IC}^{perp}$, $\sigma_y = Y \cdot a^{-1/2} \cdot K_{IC}^{par}$ at the assumption that the flaw size and shape are the same in the both adjacent layers. The $\Delta\alpha_{calc}$ calculated in this way from the Eq. (2) is 6°. The deflection angle can be also directly measured from the Fig. 1, $\Delta\alpha_{meas} = 8°$. Both values of crack deflection, the calculated and measured ones are close each other. This confirmes a relative good reliability of the indentation measurements.

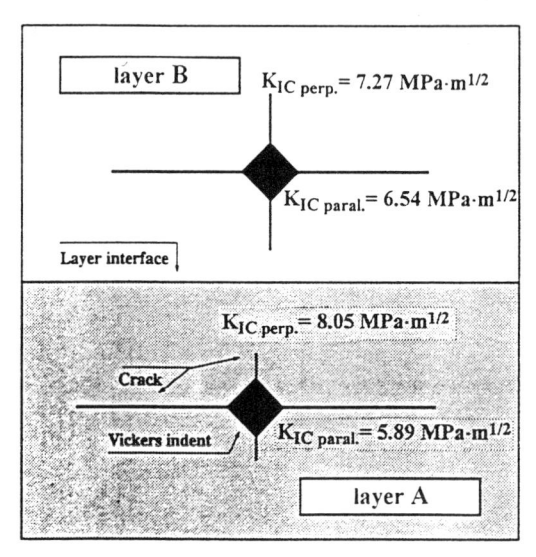

Figure 3. Fracture toughness anisotropy. Layer A - coarse Si_3N_4, layer B - fine Si_3N_4.

The next two paragraphs will be devoted to the possible reasons for arrising of the residual stresses in the samples prepared by HP with the layers of different microstructure.

Role of the Densification Method

Deeper study showed that hot pressing itself introduced some residual stress anisotropy into the samples prepared by this method. The conservation of the stress anisotropy in the HP samples can be explained by the oriented growth of the β-Si_3N_4 grains under the pressure. Figures 4 a,b shows both sections perpendicular and parallel to the HP direction of the same sample and Fig. 4c shows the pole diagram of the surface perpendicular to the HP direction. From Fig. 4 it is visible that β-Si_3N_4 grains are aligned with the c-axis in the direction perpendicular to the applied pressure. Taking into account this fact and the thermal expansion anisotropy of β-Si_3N_4, $\alpha_a = 3.3 \cdot 10^{-6}$ K^{-1}, $\alpha_c = 3.8 \cdot 10^{-6}$ K^{-1} [calculated on the data from Ref. 9] the residual compressive stresses after HP will arise in the direction of the higher thermal expansion. If the majority of grains is oriented in the direction perpendicular

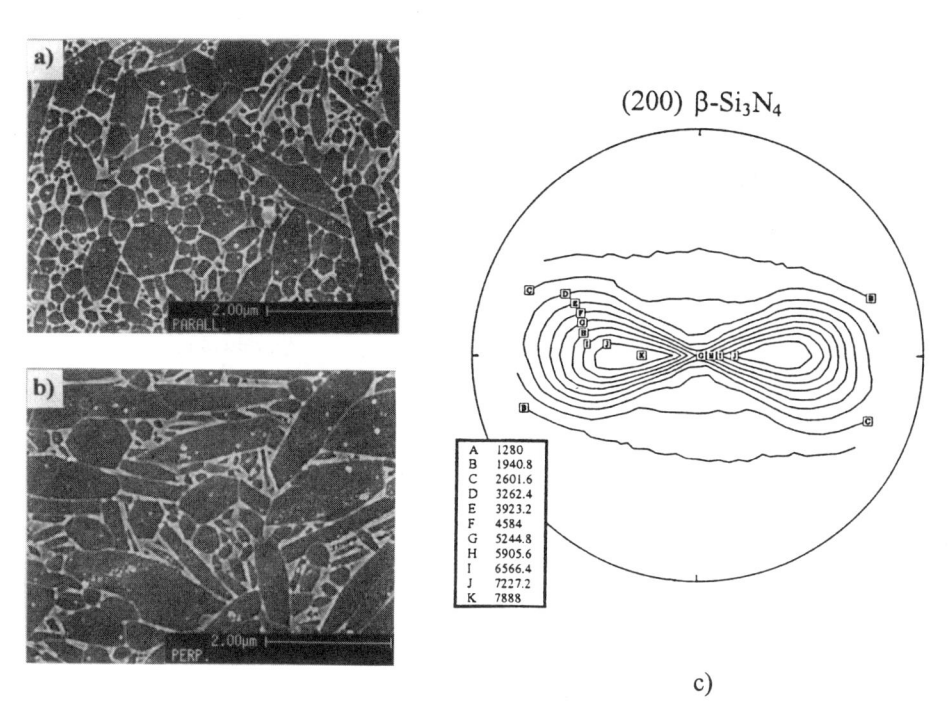

Figure 4. SEM micrographs of plasma etched surfaces parallel (a) and perpendicular (b) to the hot pressing axis. Pole diagram of the perpendicular surface (c) shows high grain alignement.

to the HP, the compressive stresses in this direction will occur. These stresses from HP are superimposing on the stresses having other origin (e.g. indentation). For the two adjacent layers with the same degree of alignment, the layer with coarser needle-like particles should have higher length change than with the finer ones. This statement is based on the fact that

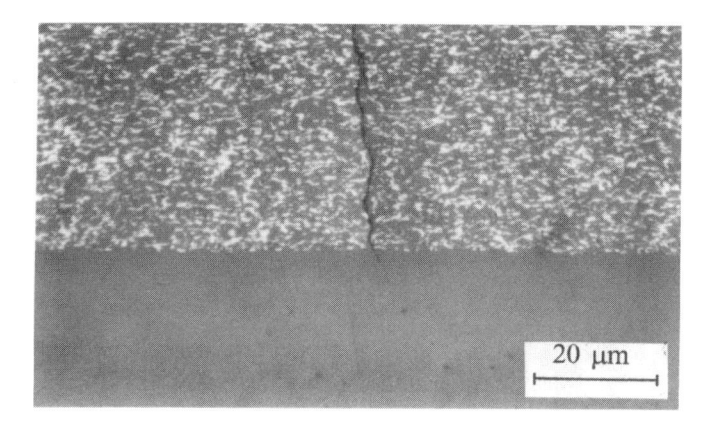

Figure 5. Tensile stress induced crack propagation in the Si_3N_4 + 20% TiN layer.

the grain boundaries have the same thickness at the same chemistry of the sintering additives without respect to the grain size. The finer microstructure has much higher number of these boundaries which is supposed to have higher thermal expansion coefficient as the silicon nitride. Thus the layer with coarse grains should be under compresive stress (see Fig. 1). The role of the expansion of the tripple points (different size for the fine and coarse microstructure) is negligible because of high degree of alignment in both layers.

Tensile stress was detected also in the Si_3N_4 + 20% TiN layer which was in a neighbourhood of Si_3N_4 layer after HP (the crack shown in Fig. 5 arrised as a consequence of these stresses). Due to the much higher thermal expansion coefficient of TiN the Si_3N_4 matrix is under tensile stress. The alignement of TiN agglomerates increases the tensile stress state in prependicular direction to the HP axis.

Role of the Green Density and Shrinkage Rate of Individual Layers

In case that instead of the thermal expansion coefficients in the Eq. (1) linear shrinkage of the E and F monoliths was used, the calculated $\Delta\sigma$ was 14 GPa. It results from the different green density of individual layers and substantially slower sintering rate of the composition containing the coarse Si_3N_4 starting powder. This extreme negative value is the theoretical one without considering the relaxation process which proceeds during the sintering as a result of present liquid phase. The relaxation will release majority of these stresses, but if only 1% of residual stresses from difference of shrinkage rate of adjacent layer are not relaxed, the stress of 140 MPa is still present. This phenomenon can change the stress status of the sample and so makes the evaluation of the actual role of particular factors difficult. The samples sintered for short time can be influenced by shrinkage rate phenomena.

Relationship of the Residual Stresses and Mechanical Properties

Fracture Toughness. As it is shown in previous paragraphs the layer boundary significantly influences the fracture behaviour of the composite. The difference in kinetic energy of the crack in the two adjacent layers can be expressed as follows,[6]

$$\Delta W_k = W_{k2} \{1 - \sigma_{t2}^2 E_2 / \sigma_{t1}^2 E_1\} = \chi W_{k2}. \qquad (3)$$

where W_{k2} is the kinetic energy in the layer 2, σ_{t1} and σ_{t2} are the tensile stresses responsible for the crack propagation in the layers 1 and 2, respectively and E_1 and E_2 are the Young's modulus of relating layers.

From Eq. (3) can be drawn the conclusion: the energy of the crack tip (propagating from the layer 2 to the layer 1) is consumed by the boundary in case the term in the brackets, marked as χ, is positive, $\chi > 0$. On the other hand, the crack is supplied by energy released from the boundary in case $\chi < 0$ (crack propagates from the layer 1 to the layer 2) and finally in case $\chi = 0$, the boundary does not influence the propagating crack. The ratio of Young's modulus and the ratio of internal stresses of both adjacent layers determine the energetic status of the boundary. The direction of the crack propagating through the boundary between layer 1 and 2 or vice versa determines the energy consumption/release mode and so the contribution of the layer boundaries to the fracture toughness of the layered composite. The measured fracture toughness by ISB method showed that K_{IC} for the 4 layer composite A-B(4), Table 1 consisting of two different silicon nitride-based microstructures is 7.3 $MPa.m^{1/2}$, while for the relating monoliths the values 6.1 and 6.9 $MPa.m^{1/2}$ were measured. This composite was prepared by HP method. The K_{IC} by ISB for three layered composite E-

F (3), Table 1, prepared by GPS method is 6.7 MPa.m$^{1/2}$ which is almost the same value as for the relating monoliths.

Strength. The strength of the layered composite which is stress free can be described according to the rule of mixtures, first term in the Eq. (4), where V_i is the volume fraction of the particular layer and σ_i is its strength. For the composite containing residual stresses at the layer boundaries the rule of mixtures should be modified:

$$\sigma = \sum_{i}^{N} V_i \sigma_i + \sum_{i}^{N-1} \Delta \sigma_i \qquad (4).$$

The sign of $\Delta\sigma$ dictate the contribution to the strength, if this is positive or negative, i.e. if the fracture energy necessary for the composite fracture will be increased or decreased. This depends on the sequencing of the layers. For 4 layer composite A-B(4) consisting of 2-2 silicon nitride layers with different microstructures the four point bending strength with the tensile surface of coarse layer was 785 ± 83 MPa. In case, the fine Si_3N_4 was on the tensile surface, this value was only 630 ± 107 MPa. These composites were prepared by HP method.

This fact can be qualitatively explained by previous approach. In case of higher bending strength two of three layer boundaries contribute positively ($\Delta\sigma_1+\Delta\sigma_3$) to the composite strength ($\sigma_{comp.}$) and one negatively ($-\Delta\sigma_2$). In case of lower bending strength, two boundaries contribute negatively ($-\Delta\sigma_1 -\Delta\sigma_3$) and only one positively ($+\Delta\sigma_2$). When the composite consists of 7 layers, i.e. with 3 boundaries positively contributing to the fracture energy and 3 boundaries contributes negatively, the four point bending strength of such composed composite was 730 ± 71 MPa, independently on the orientation of the sample with respect to the layer sequence during the bending test.

For the three layer composite E-F(3) prepared by GPS method the 4-point bending strength was 870 ± 154 MPa and for the relating monoliths 921 ± 214 MPa and 893 ± 124 MPa, respectively. The differences among these three mean strength values are smaller as the measured strength scatter.

Generally, the values of the strength for the GPS samples are higher comparing to HP samples. The reason is that the starting powders for the GPS were get rid of the agglomerates by sieving through 25 μm sieve.

CONCLUSIONS

The layeres of different microstructure in the composite prepared by HP conserved the residual stresses after cooling.

The layered composite prepared by hot pressing had in all cases higher strength and fracture toughness than the comparable monoliths. The layers differed only by the microstructure, not by the chemical composition.

Shrinkage rate and green density of the particular layer are the factors seriously influencing final stress state of the layered composite. The composites should be annealed for longer time at the temperature above T_j (temp. of rigid joint formation) in order to avoid formation of stresses acting opposite to the stresses arising from the other factors as Young's modulus and thermal expansion coefficients. This is valid for both HP and GPS.

Hot pressing introduces the microstructure anisotropy because of oriented growth of Si_3N_4 needle-like particles in the direction perpendicular to the HP-direction. In the case of the Si_3N_4 ceramics, the residual stresses are compressive in the direction perpendicular to the HP-direction.

ACKNOWLEDGEMENT

The investigation was supported by the Slovak Grant Agency for Science No. 2/1169/95 and partially by the United States - Slovak Science and Technology Program, project No. 94-039.

REFERENCES

1. M.P. Harmer, H.M. Chan, G.A. Miller, Unique opportunities for microstructural engineering with duplex and laminar ceramic composites, *J. Am. Ceram. Soc.* **75**, 1715-1728 (1992).
2. D.B. Marshall, J.J. Ratto, F.F. Lange, Enhanced fracture toughness in layered microcomposites of Ce-ZrO$_2$ and Al$_2$O$_3$, *J. Am. Ceram. Soc.* **74,** 2979-87 (1991).
3. P. Šajgalík, Z. Lenčéš, J. Dusza, Si$_3$N$_4$ based composite with layered microstructure, pp. 198-201, in: *Ceramic Materials and Composites for Engines*, D.S. Yan, X.R. Fu and S.X. Shi, eds., World Scientific, Singapore-New Jersey-London-Hong Kong (1995).
4. P. Šajgalík, J. Dusza, Z. Lenčéš, Layered Si$_3$N$_4$ composites, pp. 603-608, in: Ceramic Processing Science and Technology, *Ceramics Transactions* **51**, H. Hausner, G.L. Messing and S. Hirano, eds.,The Am. Ceram. Soc., Westerville, Ohio (1995).
5. P. Šajgalík, Z. Lenčéš, J. Dusza, Tailoring of microstructure and room-temperature properties of Si$_3$N$_4$ based composites: from the nanocomposites to the layered millicomposites, pp. 5-12, in: "Fourth Euro-Ceramics", Vol. 4, Basic Science: *Trends in Emerging Materials and Applications*, A. Bellosi ed., Gruppo Editoriale Faenza Editrice, (1995).
6. P. Šajgalík, Z. Lenčéš, J. Dusza, Layered Si$_3$N$_4$ composites with enhanced room temperature properties, *J. Mater. Sci.*, to be published.
7. P. Šajgalík, Z. Lenčéš, J. Dusza, Residual stresses in layered silicon nitride-based composites, NATO ARW Proceedings, *Engineering Ceramics '96*, Kluwer Publ., to be published.
8. T. Chartier, D. Merle, J.L. Besson, Laminar ceramic composites, *J. Eur. Ceram. Soc.* **15**, 101-107 (1995).
9. D.R. Messier, W.J. Croft, Silicon nitride, pp. 178-179, in: *Preparation and Properties of Solid State Materials*, Vol. 7, W.R. Wilcox ed., Marcel Dekker Inc., New York - Basel, (1982).

CHARACTERIZATION OF MICROSTRUCTURE AND CRACK PROPAGATION IN ALUMINA USING ORIENTATION IMAGING MICROSCOPY (OIM)

S. Jill Glass,[1] Joseph R. Michael,[1] Michael J. Readey,[2] Stuart I. Wright,[3] and David P. Field[3]

[1]Sandia National Labs, Albuquerque, NM 87185
[2]Caterpillar Inc., Peoria, IL 61656
[3]TSL Inc., Provo, UT 84604

ABSTRACT

Conventional studies of structure-property relationships for polycrystalline materials have focused either on descriptions of the morphological aspects of the microstructure, such as grain size and shape, or on the chemistry and structure of individual grain boundaries using transmission electron microscopy (TEM). TEM, while capable of determining the misorientation of adjacent grains, can provide information only for a small number of boundaries. A more complete description of a polycrystal requires the lattice orientations of a statistically significant number of grains, coupled with the morphological aspects, such as grain size and shape. This description can be obtained using a relatively new technique known as orientation imaging microscopy (OIM), which utilizes crystallographic orientation data obtained from Backscattered Electron Kikuchi patterns (BEKP) collected using a scanning electron microscope. This paper describes the OIM results for Al_2O_3. The results include image quality maps, grain boundary maps, pole figures, and lattice misorientations depicted on MacKenzie plots and in Rodrigues space. High quality BEKP were obtained and the images and data readily reveal the grain morphology, texture, and grain boundary misorientations, including those for cracked boundaries.

INTRODUCTION

The mechanical and electrical behavior of a polycrystalline material depends upon the properties of the individual grains, their spatial orientation, and the properties and orientations of the grain boundaries. While relationships between grain size and shape and material properties have been studied extensively, the effects of lattice orientations (microtexture) and misorientations between grains (mesotexture or grain boundary texture) have been difficult to identify because a measurement technique that accounted for a sufficient number of grains to be representative of a complete microstructure was not available. Many studies have reported that the properties of a polycrystal vary as a function of grain size. For example, in many ceramics a transition from intergranular to transgranular fracture occurs as the grain size increases. Explanations for this include grain boundary impurity segregation and microcracking; however, these may be symptoms of the cause of the transition rather than the cause itself. The true cause may be that the distributions of lattice orientations and misorientations are changing as the microstructure evolves. Certain boundaries have a higher mobility than others,[1,2] which may lead to the preferential removal of these boundaries, leaving a microstructure that is less random than it was. If high energy boundaries, which generally correspond to high angle boundaries, are preferentially eliminated, this may lead to a microstructure that has more fracture resistant boundaries with a concomitant increase in the amount of transgranular fracture.[3] There is mounting evidence that lattice orientations and boundary types in polycrystals play a crucial role in determining the intrinsic response of the material and its overall properties.[4] Special boundaries may dominate the behavior of a material. For example, in Ni_3Al, low-angle and Σ3 boundaries are strong, whereas high-angle boundaries are prone to cracking.[5] Other properties that are different for special boundaries compared to the general population include impurity segregation,[6] diffusion, mobility, energy, resistivity, and corrosion resistance. The dramatic differences in properties as a function of the misorientations are exemplified by Σ3 boundaries on the 110 zone that have energies of 0.01-0.61 J/m^2, compared to values of ~ 1 J/m^2 for a totally disordered, general boundary.[7] The implication is that if a polycrystal could be engineered with a high percentage of special boundaries, then there would be an opportunity to enhance the overall properties.[3,8]

The misorientation between two grains is completely described by five parameters, three for the lattice misorientation and two for the boundary normal. It is typically the lattice misorientation that is related to the properties of the grain boundaries, although it is clear that the boundary inclination is also important, especially for special boundaries.[9] A convenient framework for describing the crystallography of the grain boundary is the coincidence site lattice (CSL) model.

Although there is very little information regarding microtexture[10] and mesotexture[4,11] and how they are influenced by processing and microstructural evolution, it is well known that preferred orientation or macrotexture (non random distributions of lattice orientations with respect to the specimen axes) can produce an anisotropic response in polycrystalline materials. In structural materials the anisotropy can be in the elastic properties, fracture toughness, and strength. Macrotexture measurements indicate that directionally-dependent processing techniques, such as hot-pressing and forging, lead to preferred orientations. During hot-pressing, which is the approach used to densify many ceramics, both grain rotation and preferred grain growth can contribute to the development of macrotexture.[12] Strong basal textures are produced in hot-pressed or forged alumina.[13]

Macrotexture measurements are usually made using X-Ray or neutron diffraction techniques, and the results are expressed using the orientation distribution function (ODF). The ODF provides the volume fraction of crystals with a given orientation. Although X-Ray and neutron diffraction macrotexture measurements suggest the presence of a higher than random distribution of special boundaries in many materials, these characterization techniques are unable to provide specific information on the orientation relationship between individual grains (mesotexture or grain boundary texture) or on the relationship between the distribution of lattice orientations and microstructural features, such as grain size and grain shape (microtexture). For example, are specific orientations associated with the large or small grains in the microstructures? Also, small components of macrotexture may be missed, or complementary texture components with similar volume fractions may cancel each other out, producing the effect that no texture is detected. Thus, models based on ODF macrotexture data may not be detailed enough to provide valid structure-property relationships.

Even in the absence of directionally dependent processing and when there is no evidence of macrotexture, microtexture and mesotexture may exist because they are the natural outcome of microstructural evolution. As mentioned earlier, a higher than random frequency of special boundaries may be present after grain growth because certain boundaries are likely to be preferentially eliminated. Certain boundaries have a higher velocity because of their higher mobility and/or the higher driving forces for their migration. This is especially true for materials with a plate-like morphology.[14] The importance of the mesotexture has been noted for both the development of anisotropic microstructures and abnormal grain growth.[15] Similar to the ODF, the misorientation distribution function or MDF has been developed to help characterize the statistical distribution of grain boundary misorientations in a polycrystal.

Techniques for obtaining the lattice orientation data include optical mineralogical techniques, etch pits, back-reflection Laue patterns, electron diffraction in the transmission electron microscope (TEM), electron channeling, X-ray diffraction using a conventional laboratory diffractometer or a synchotron radiation source, Kossel X-Ray diffraction, and electron backscatter diffraction in the SEM. These techniques are reviewed by Wright.[16]

Except for electron backscatter diffraction in the SEM, each technique has limitations that prevent the acquisition of data regarding the effects of lattice orientation and grain misorientation on properties. These limitations include low resolution, unavailability of an appropriate radiation source, sample preparation difficulties, and the inability to study large areas of the specimen.

A fairly recent technique for characterizing lattice orientations makes use of backscattered electron Kikuchi patterns (BEKP). The first observations of these patterns were reported in 1954 and called High-angle Kikuchi patterns by Alam et al.[17] The first use of BEKP in an SEM for lattice orientation determination was in 1973 by Venables and Harland.[18] In order to obtain a BEKP, the incident electron beam is focused and held stationary on a feature of interest on a specimen that is tilted about 70° toward the BEKP detector. Inside the specimen, the electrons are inelastically and elastically scattered; some are scattered at high angles and exit the specimen. Some of these backscattered electrons satisfy the Bragg condition and are diffracted into pairs of cones that can be detected using a suitably placed phosphor screen or photographic film. The cones are imaged as conic sections but appear as parallel sets of straight lines due to the large apex angle of the cones. Interpretation of the pairs of lines, known as Kikuchi bands, allows the specific orientation of the crystal to be determined.[19]

Pioneering work by Dingley, including the introduction of the use of a low light level video camera and on-line computer analysis to aid in the identification of patterns, revealed the potential power of the technique.[20] Substantial progress in the last decade in hardware and software has provided a unique opportunity to interrogate microstructures. Completely automated systems are now available for determining lattice orientations from backscattered electron Kikuchi patterns (BEKPs). More detailed descriptions of the hardware and software routines can be obtained in recent references.[16,21] Presently, a spatial resolution of 200 nm and a precision of 1° can be obtained.[16] Approximately thirty orientations can be determined per minute, which allows hundreds to thousands of grains to be analyzed in a day, depending on the grain size.

Not only does the BEKP technique allow the determination of orientations in a statistically meaningful manner, but it can also be coupled with information about morphological parameters such as grain size and shape. The term that Adams et al. coined for this coupled information is Orientation Imaging Microscopy (OIM).[22] In OIM, the crystallographic orientation is obtained from automatic indexing of BEKPs. The computer controls the electron beam in the SEM so that BEKPs and the corresponding lattice orientation can be obtained at many points on the sample on a user defined grid. In addition to the orientation, the computer records the x,y coordinates, a parameter characterizing the image quality (IQ) of the corresponding BEKP, and a confidence index (CI) describing the confidence the computer has that the indexing algorithm has correctly identified the lattice orientation. An image can then be generated by mapping any of these parameters onto a color or gray scale and shading each point on the data measurement grid accordingly. Such images enable the spatial arrangement of orientation to be graphically displayed providing visual cues to the connection between morphological features of the microstructure and lattice orientation.

To fully utilize the range of information from BEKP data requires statistical measures of the microstructure, such as the ODF and MDF. Plotting both the distribution functions and discrete orientation and misorientation data using a variety of representations can help identify the salient crystallographic elements of the microstructure. In this study we have used several representations of the crystallographic data including pole figures recalculated from ODFs, MacKenzie plots, and discrete plots of misorientation in Rodrigues space.[23] The MacKenzie plot shows only the distribution of the misorientation angle and does not include any information about the axis of rotation.

Quantitative, statistical information on the orientations of thousands of grains in a polycrystal is expected to reveal previously unknown characteristics of materials and to provide a better understanding of structure-property relationships. OIM is still in its infancy in terms of an understanding of its powers and capabilities, the number of materials that remain to be analyzed, and identifying which crystallographic features are most relevant. Once we have gained a better understanding of the range of information available and its relevance, some long-standing questions regarding structure-property relationships may finally be answered.

The objectives of this study were to determine whether useable BEKPs could be obtained for polycrystalline alumina materials and to use OIM to examine microtexture and mesotexture as a function of purity (99.7 and 99.99%) and grain size. We were also interested in examining the distribution of boundary misorientations along cracks in alumina and comparing them to the distribution of boundary misorientations in the bulk. Details regarding the fracture behavior of these alumina materials can be found in recent references.[24,25]

EXPERIMENTAL PROCEDURE

Sample Processing

Two commercially available alumina powders (99.7%[i] and >99.99%[ii]) were uniaxially pressed at 28 MPa into disks 25 mm in diameter and approximately 3 mm thick. The disks were subsequently isostatically pressed at 280 MPa. The isopressed density was ~ 57% of the theoretical density, 3.98 g/cm^3. The disks were buried in a bed of identical powder in high purity alumina crucibles and fired at 1600°C for 5 hr at a heating rate of 5°C/min and a cooling rate of 10°C/min. To increase grain size, specimens were subsequently fired at 1720°C for times up to 48 hr. Quantitative stereology on SEM micrographs provided mean grain sizes (d_{avg}) of 5, 10, and 27 μm for the 99.99% Al_2O_3, and 4 and 13 μm for the two 99.7% Al_2O_3 materials. The grain size distributions appear to be self-similar as grain size increases, but the distribution is broader for the lower purity material. The mean aspect ratio was ~1.5 for the 99.99% Al_2O_3, representing a grain shape close to equiaxed for all grain sizes and ~2 for the lower purity alumina, representing a more elongated shape. Densities ranged from 98.6% theoretical density (TD) for the fine-grained 99.99% material to 99.2% TD for the coarse grained 99.99% material and from 98.7 to 98.29% TD for the 99.7% Al_2O_3 materials.

Sample Preparation for OIM

Conventional metallographic polishing techniques, which consisted of grinding flat with a 9 μm fixed diamond wheel at an applied load of 150 N, followed by 9, 6, and 3 μm diamond polishing at 150 N, were used to prepare the samples. The final polish was done using colloidal silica for three minutes at an applied load of 100 N. Samples were lightly coated with carbon to prevent charging effects. Coating does not noticeably degrade the backscattered electron Kikuchi patterns.[26]

Pattern Collection and Data Analysis

The BEKP analyses were performed using a Philips XL30 tungsten source SEM at a voltage of 30 KeV. The beam current was approximately 5 nA. The SEM was equipped with a low-light, silicon-intensified tube (SIT) camera capable of capturing BEKP images at a light level of 5×10^{-5} lux. BEKP data were obtained over a regular hexagonal grid on the surface of each specimen. The scan on the coarse grained alumina (d_{avg}=27 μm) covered an area of 400μm x 400μm with a step size of 2 μm. The fine grained alumina (d_{avg}=5 μm) was scanned over an area of 200μm x 200μm with a step size of 1 μm. Alumina was indexed using trigonal crystal symmetry with lattice parameters of a=b=4.76 Å and c=12.99 Å. The alumina data sets consisted of approximately 46,000 orientation measurements each. BEKP images were transferred to a Silicon Graphics Indy workstation, where TSL's Orientation Imaging Microsocopy™ software processed the desired data sets of x-y coordinates, Euler angles, image quality, and confidence index measures.

RESULTS AND DISCUSSION

Useable backscattered electron Kikuchi patterns (BEKP) were obtained from all five alumina samples. Figure 1 shows the OIM image produced for the 27 μm 99.99% alumina

---

[i] A16 SG Alumina, Alcoa Industrial Chemicals, Pittsburgh, PA.
[ii] AKP-50, Sumitomo Chemical Company, New York, NY.

a)   b)

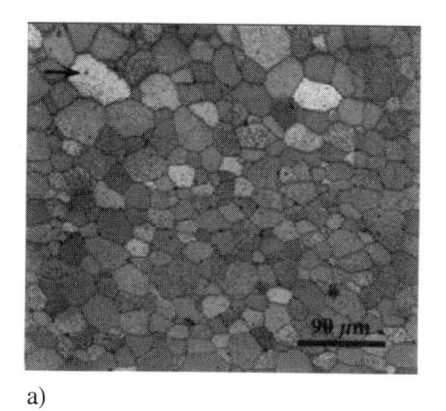

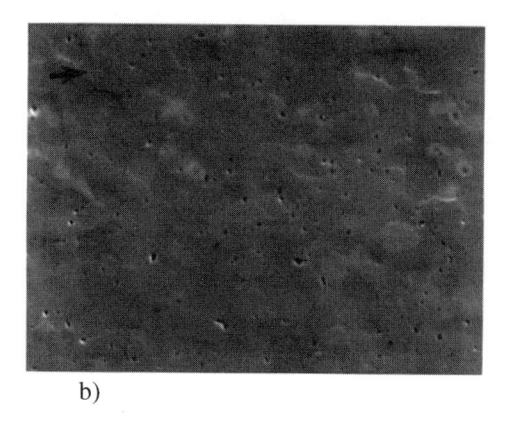

Figure 1. a) OIM image for the 27 μm, 99.99% alumina sample together with b) a conventional scanning electron microscope micrograph of the same region. The arrow highlights the same grain in both images.

sample together with a conventional scanning electron microscope micrograph of the same area. One grain is highlighted with an arrow to indicate its position in both images. Note that the sample was not deliberately etched, but grains are apparent in the conventional SEM micrograph because of chemical etching that occurred during the colloidal silica polishing step. One of the advantages of OIM is that no etching of the samples is required to delineate the grains. Grain sizes were calculated from the OIM images and compared to the results obtained in a previous study where standard stereological analyses were performed on conventional SEM micrographs.[24] The average grain size from the OIM image of the largest grain alumina sample was 28 μm compared to 27 μm from the conventional SEM measurements.

Figure 2 is an image quality map for the 10 μm, 99.99% alumina sample with darker pixels representing a lower image quality. The image quality map is determined from the confidence index measured for each pattern, which is a function of diffuseness of the BEKP image. The diffuseness of the image is related to both the orientation and to the perfection of the material. High dislocation densities and surface imperfections, including roughness and contamination, produce more diffuse images. The image quality measure is not normalized for orientation, so a distinct difference in this parameter is seen from grain to grain. Grain boundaries produce low image quality because measurement points near the grain boundaries are generally affected by the superposed diffraction patterns from the two grains separated by the grain boundary, resulting in a transition area of random noise between the true orientation measurements in the grain interiors. A crack, which also produces a low image quality, runs from the middle left of the image towards the upper right (indicated with arrows). Figure 3 shows the same image with the addition of grain boundaries, which are drawn for misorientations between neighboring measurements of greater than 15°. Some regions of the image appear as if each measurement is a different orientation and give a chicken wire appearance. These are regions of low confidence that are assumed to have been indexed improperly. All such data can be disregarded in the analysis or corrected for by using a clean-up routine that uses a voting procedure to determine the most likely orientation for a given diffraction pattern (described below).

A more detailed analysis of the points contained within the box in Fig. 3, which appear to be part of a single grain, indicate that two orientations are obtained consistently. The fact that the orientations measured within this grain are not the same and that the average confidence index for the grain is low, along with the fact that there are similarly oriented, but scattered measurement points within the grain, suggests that the indexing algorithm may have difficulty properly indexing the diffraction pattern associated with this grain and others that display the same chicken wire appearance. This ambiguity in pattern recognition for alumina needs to be investigated in more detail. While the orientations from these grains seem to be random, there is a possibility that certain orientations are more likely to produce diffraction patterns that produce non-unique indexing solutions.

Figure 2. Image quality map for the 10 μm, 99.99% alumina sample with darker pixels representing a lower image quality. Arrows highlight a crack that runs from the middle left of the image to the upper right.

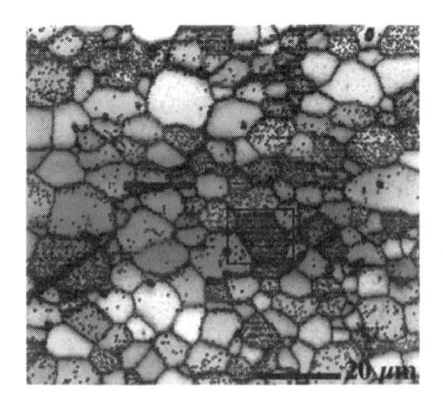

Figure 3. The OIM image of the 10 μm, 99.99% alumina with the addition of grain boundaries, which are drawn for misorientations between neighboring measurements of greater than 15°. Dots within grains, which produce a chicken wire like appearance, indicate regions whose orientation has been determined with low confidence.

This would bias the texture results away from these types of orientations. The data set was processed to ignore data from these types of grains using the following clean-up procedure. When a point is surrounded by three points with the same orientation, its orientation is changed to match the orientation of those three points. Then the data is grouped into grains by grouping neighboring points whose orientation does not differ more than 5 degrees. Finally, grains with less than 25 measurement points per grain were neglected. This procedure results in the image shown in Fig. 4. The shades in this image are not related to the grain orientations in any way but are used to help delineate the grains. The clean-up procedure is not entirely effective in eliminating the unreliable data, but it allows the noise associated with points near grain boundaries to be eliminated, producing a better structure for generating accurate orientation and misorientation distributions.

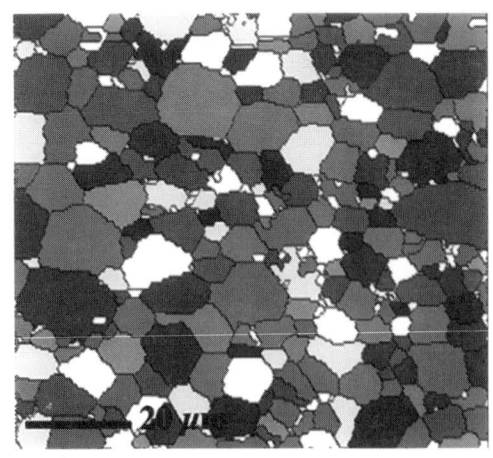

Figure 4. The OIM image of the 10 μm, 99.99% alumina sample after a clean-up procedure was used to minimize the presence of ambiguous patterns. The shades in this image are used only to help delineate the grains.

Figure 5 shows the 002 (using the 3 index notation for hexagonal symmetry) intensity pole figures for each of the alumina samples. The rolling direction (RD) and transverse direction (TD) represent the two directions in the plane of the sample, which is perpendicular to the original uniaxial pressing direction.

Figure 6 shows the misorientation distributions as MacKenzie plots, which show the frequencies of given misorientations. The predicted random misorientations for crystals with n-fold dihedral symmetry (where trigonal is represented by n=3) are shown in Fig. 7.[27] Misorientation data can also be represented in Rodrigues space, which is particularly useful for the display of mesotexture because special boundaries can be identified.[19] Misorientations for the 13 μm, 99.7% alumina sample, which appears to exhibit the greatest degree of mesotexture, are shown in Rodrigues space in Fig. 8. Rodrigues space plots for the other samples are not included because of space considerations, but the results are discussed below.

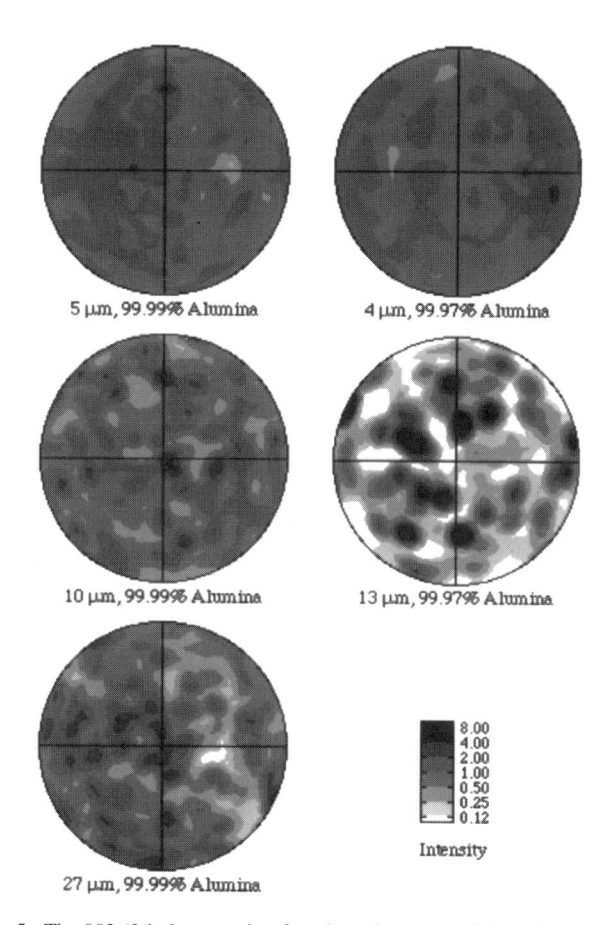

Figure 5. The 002 (3 index notation for trigonal symmetry) intensity pole figures.

Comparison of the MacKenzie plots in Fig. 6 shows that all samples appear to have a greater frequency of misorientations at angles between 55 and 60° than are indicated in the predicted random distribution in Fig. 7. As discussed earlier, in the alumina crystal structure (trigonal - di-pyramidal), there exist certain orientations for which the solution of the BEKP is ambiguous. We believe that the 60° peak in the plots is an artifact of this indexing difficulty. Other than this anomaly, the data in the MacKenzie plots appear to be close to random. The only distribution that is somewhat different than the rest is that for the 13 μm, 99.7% alumina sample, but it represents a smaller number of grains. Its distribution has relatively more misorientations at angles between 5 and 10°, fewer at angles between 10 and 20°, and more variability in the frequency for angles between 60 and 95°. Its pole figure in Fig. 5 also has the least random texture.

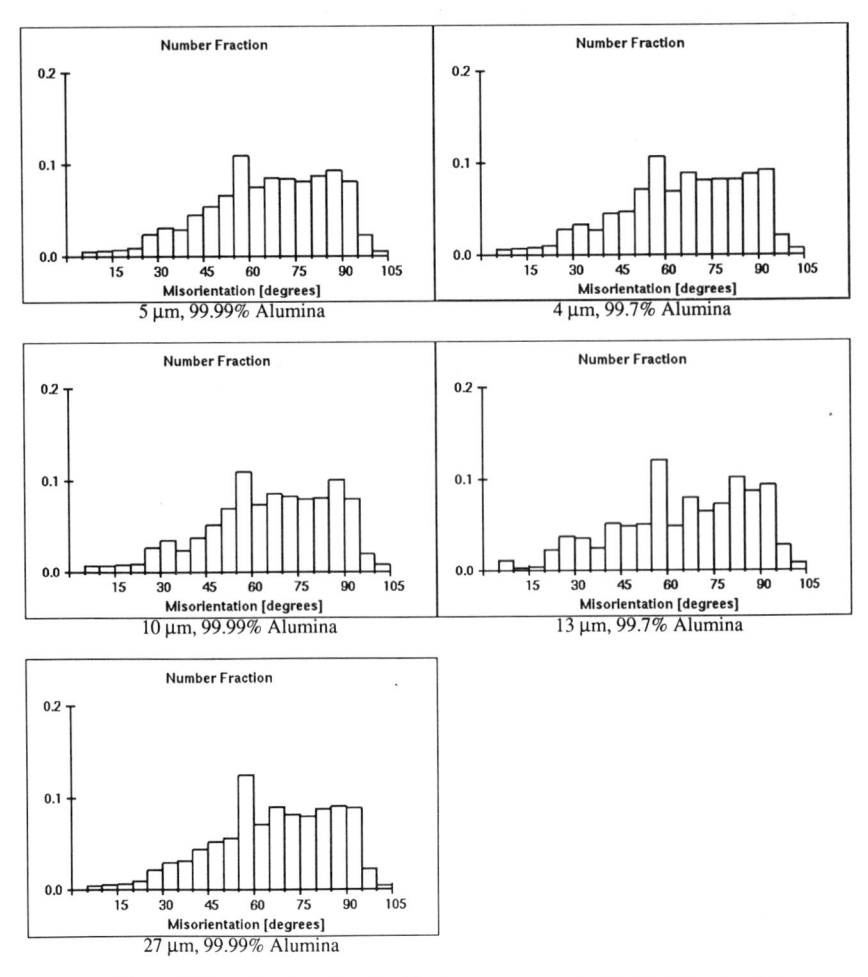

Figure 6. The misorientation distributions as represented in MacKenzie plots.

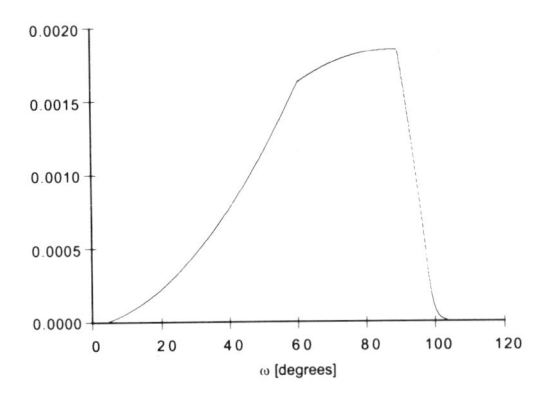

Figure 7. The predicted MacKenzie plot of random misorientations for crystals with 3-fold dihedral symmetry (trigonal).

General Features of the Intensity Pole Figure Microtexture Representations (Fig. 5) and Rodrigues Space Mesotexture Representations

5 μm, 99.99% alumina. The texture is quite random as seen by the intensity pole figure for the c-axis. The MacKenzie plot is also fairly random. The misorientations plotted in Rodrigues space are somewhat random but shifted towards c-axis rotations in the distribution (left-hand vertex of each triangle in the space). There are also a number of low angle boundaries (upper-left hand triangle c-axis position).

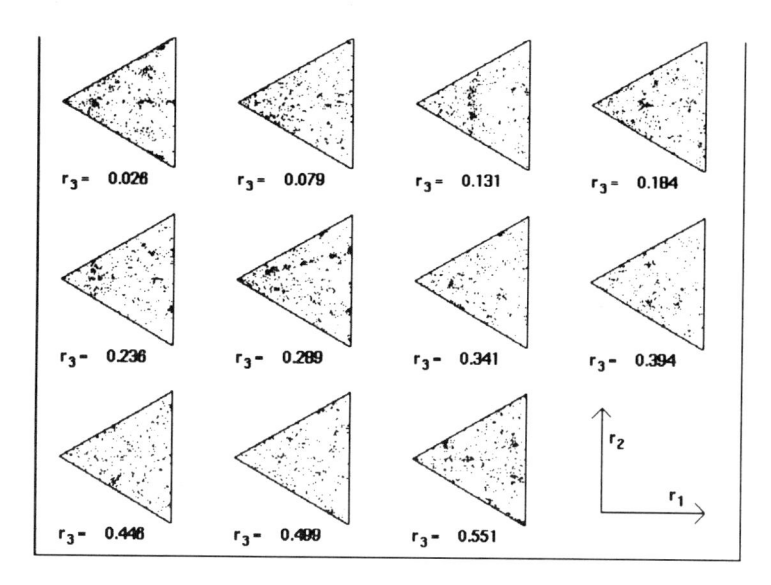

Figure 8. Misorientations displayed in Rodrigues space for the 13 μm, 99.7% alumina sample.

10 μm, 99.99% alumina. The texture is somewhat random, but some clusters appear to be forming. The pole figure shows the c-axis has a weak component as well as a few positions rotated 50-60° off the c-axis. The Rodrigues space plot of misorientations indicates the presence of some low angle and c-axis misorientations but also contains a small cluster at a position of 50° about an a-axis.

27 μm, 99.99% alumina. The texture shown in the pole figure is again quite random, and the Rodrigues space plot is similar to that of the 5 μm, 99.99% alumina.

4 μm, 99.7% alumina. The texture is weak but has a different character than any of the others. The c-axes are aligning about 70° off the specimen surface normal but only in one general direction. There are some significant features near the rolling direction (RD), and there appears to be some in-plane near c-axis texture.

13 μm, 99.7% alumina. The texture in this sample is the strongest of all of the aluminas, but there are also fewer grains in the data set. The distribution tends to be off the c-axis by about 40° in a random direction. The misorientations show a peak in the same position as that described for the 10 μm, 99.99% alumina. Additional components exist, each of which lie near the boundary of the fundamental region, indicating some special but not well defined symmetries.

Misorientations for Cracked Boundaries

Figure 9 shows the OIM image and the points that were used to determine the distribution of misorientations along a crack in the 10 μm, 99.99% alumina sample. Figure 10 is the MacKenzie plot for these misorientations. Low angle misorientations represent transgranular fracture. Although there appears to be a higher percentage of fracture between grains with large misorientations, relatively few data were used to construct the histogram. Larger numbers of misorientations will need to be measured to identify the true distribution and to determine whether special boundaries, such as low angles and twin boundaries, are absent from the distribution compared to the bulk population. This would provide support for the hypothesis that special boundaries are more fracture resistant. Statistical analyses of misorientation distributions for cracked materials as a function of grain size will also provide information on how the structure of the boundaries is changing as a microstructure evolves.

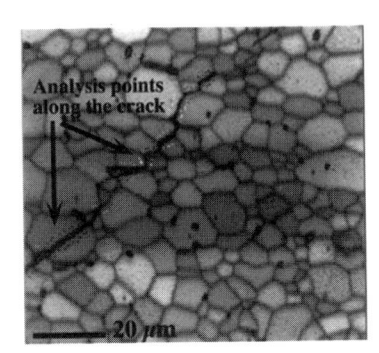

Figure 9. The OIM image of the 10 μm, 99.99% alumina sample and the points that were used to determine the distribution of misorientations along the crack it contains.

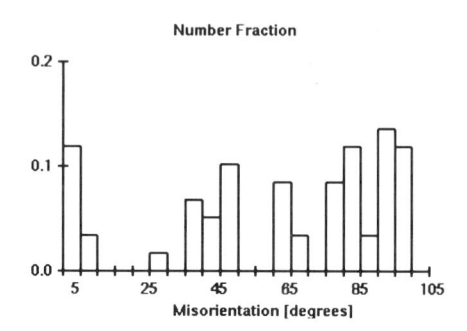

Figure 10. The MacKenzie plot for the misorientations of the crack in Fig. 9. Low angle misorientations represent transgranular fracture.

CONCLUSIONS

Polycrystalline alumina samples were characterized using Orientation Imaging Microscopy (OIM). Good BEKP data were obtained and were used to produce pole figures showing microtexture, and MacKenzie plots and Rodrigues space representations of mesotexture. For the alumina crystal structure, some of the data are unreliable because of the ambiguity in indexing certain orientations. This ambiguity needs to be investigated further. Differences in microtexture and mesotexture in alumina as a function of purity and grain size were subtle but measurable; however, additional measurements will need to be made to provide a more complete picture. Misorientations of cracked boundaries can be measured and compared to the bulk distribution, but larger numbers of misorientations will also need to be obtained to make comparisons. A powerful tool now exists for obtaining meaningful numbers of orientations and misorientations in ceramic materials. This information is expected to provide a better understanding of how these characteristics relate to macroscopic properties such as strength and toughness.

Acknowledgments

The authors thank Desi Kovar for providing the alumina samples. This work was supported by the U.S. Department of Energy under contract No. DE-AC04-94AL85000 at Sandia National Laboratories.

REFERENCES

1. V. Randle, B. Ralph, and D. Dingley, The relationship between microtexture and grain boundary parameters, *Acta metall. mater.* 36:267 (1988).
2. G. S. Grest, D. J. Srolovitz, and M. P. Anderson, Computer simulation of grain growth - IV. anisotropic grain boundary energies, *Acta metall. mater.* 33:509 (1985).
3. T. Watanabe, Grain boundary design and control for high temperature materials, *Mater. Sci. Eng.* A166: 11 (1993).
4. V. Randle, Grain assemblage in polycrystals - overview no. 115, *Acta metall. mater.* 42:1769 (1994).
5. H. Lin and D. P. Pope, Weak grain boundaries in Ni_3Al, *Mater. Sci. Eng.* A192/193:394 (1995).
6. W. Swiatnicki, S. Lartigue-Korinek, and J. Y. Laval, Grain boundary structure and intergranular segregation in Al_2O_3, *Acta metall. mater.* 43:795 (1995).
7. V. Randle, An investigation of grain-boundary plane crystallography in polycrystalline nickel, *J. Mater. Sci.* 30:3983 (1995).
8. T. Watanabe, The impact of grain boundary character distribution on fracture in polycrystals, *Mater. Sci. Eng.* A176:39 (1994).
9. A. Garbacz, B. Ralph, and K. J. Kurzydlowski, On the possible correlation between grain size distribution and distribution of CSL boundaries in polycrystals, *Acta metall. mater.* 43:1547 (1995).
10. T. T. Wang, B. L. Adams, and P. R. Morris, Development of orientation coherence in plane-strain deformation, *Met. Trans.* 21A:2223 (1990).
11. V. Randle, Origins of misorientation texture (mesotexture), in: *Textures and Microstructures*, Vol. 14-18, H. J. Bunge, ed., Gordon and Breach Science Publishers SA, United Kingdom (1991).
12. F. Lee and K. J. Bowman, Texture and anisotropy in silicon nitride, *J. Am. Ceram. Soc.* 75:1748 (1992).
13. Y. Ma and K. J. Bowman, Texture in hot-pressed or forged alumina, *J. Am. Ceram. Soc.* 74:2941 (1991).
14. M. S. Sandlin, C. R. Peterson, and K. J. Bowman, Texture measurement on materials containing platelets using stereology, *J. Am. Ceram. Soc.* 77:2127 (1994).
15. J. Rodel and A. M. Glaeser, Anisotropy of grain growth in alumina, *J. Am. Ceram. Soc.* 73:3292 (1990).
16. S. I. Wright, A review of automated orientation imaging microscopy, *J. Comput.-Assist. Microsc.* 5:207 (1993).
17. M. N. Alam, M. Blackman, and D. W. Pashley, High angle Kikuchi patterns, *Proc. Roy. Soc.* 221A:224 (1954).
18. J. A. Venables and C. J. Harland, Electron back-scattering patterns - a new technique for obtaining crystallographic information in the scanning electron microscope, *Phil. Mag.* L2:1193 (1973).
19. V. Randle, *Microtexture Determination and its Applications*, The Institute of Materials, London (1992).
20. D. J. Dingley, A comparison of diffraction techniques for the SEM, *Scanning Electron Microsc.* IV:273 (1981).
21. K. Kunze, S. I. Wright, B. L. Adams, and D. J Dingley, Advances in automatic EBSP single orientation measurements, in: *Textures and Microstructures*, B. L. Adams and H. Weiland, ed., Gordon and Breach Science Publishers SA, United Kingdom (1993).
22. B. L. Adams, S. I. Wright, and K. Kunze, Orientation imaging: the emergence of a new microscopy, *Metall. Trans. A*, 24A:819 (1993).
23. F. C. Frank, Orientation mapping, *Metall. Trans. A*, 19A:403 (1988).
24. D. Kovar and M. J. Readey, Role of grain size in strength variability of alumina, *J. Am. Ceram. Soc.* 77:1928 (1994).
25. D. Kovar, *The Role of Microstructure on the Mechanical Reliability of Alumina Ceramics*, Ph.D. Thesis, Carnegie Mellon University (1995).
26. J. R. Michael and R. P. Goehner, Advances in backscattered-electron Kikuchi patterns for crystallographic phase identification, in: *Proc. 52nd Annual Meeting*, G. W. Bailey and A. J. Garratt-Reed, ed., Microscopy Society of America (1994).
27. A. Morawiecz and D. Field, Misorientation angle distribution of randomly oriented symmetric objects, submitted to *J. Appl. Cryst.* (1996).

GRAIN BOUNDARY CHEMISTRY AND CREEP RESISTANCE OF ALUMINA

Yan Z. Li, Martin P. Harmer, Helen M. Chan, and Jeffrey M. Rickman

Department of Materials Science and Materials Research Center,
Lehigh University, Bethlehem, PA 18015

INTRODUCTION

Recently, it has been reported that a dramatic improvement in creep resistance of alumina ceramics is obtained by the addition of yttrium (Y)/lanthanum (La) ions [1,2]. The original discovery of the beneficial effect of Y_2O_3 doping was made by French et al. [1] in an investigation of the creep behavior of equi-volume duplex structures, Al_2O_3 : c-ZrO_2(Y) and Al_2O_3 : YAG systems. These investigators found that the creep rate of the duplex structures was significantly lower than their single phase constituents, and that the result could not be accounted for by simple composite theory. They suspected that the reduced creep rate of the duplex structures was a result of Y segregation to the grain boundaries, where the larger ionic size of the segregants obstructed the grain boundary mass transport process, the dominant creep mechanism for fine grained alumina. To test this concept, French et al. [1] added 1000 ppm Y (Y/Al atomic ratio) to pure Al_2O_3, and demonstrated that the creep rate was further reduced relative to that of the duplex structures. The tensile creep rate of the 1000 ppm Y-doped alumina was reduced two orders of magnitude as compared with undoped alumina. Subsequent work showed that the addition of 500 ppm La to alumina resulted in about the same amount of reduction in the creep rate as with Y addition [2]. Co-doping alumina with Mg and Y has also been reported by other investigators [3-5] to reduce the creep rate, but the effect was not as dramatic. Wakai et al. [6] reported that 1000 ppm Zr addition to alumina enhanced the creep resistance by a factor of about 15.

The steady-state creep rate predicted by diffusional mechanisms is commonly expressed in the form,

$$\dot{\varepsilon} = A\,(d)^{-p}\,(\sigma)^n\,\exp(-Q/RT)$$

where A is a constant, d is the grain size, σ is the applied stress, p is the inverse grain size exponent, n is the stress exponent, Q is the apparent activation energy for creep, R is the gas constant, and T is the absolute temperature. For lattice diffusion controlled (Nabarro-Herring) creep [7,8] p=2 and n=1, while p=3 and n=1 for grain boundary diffusion controlled (Coble) creep [9]. If the grain boundaries do not behave as perfect sources and sinks of vacancies, the interface reaction process may be rate-limiting [10,11], then p=1 and n=2. Details of creep characteristics and other mechanisms can be found in a number of review articles [12,13].

In the previous studies of Y [1], La [2], and Zr [6] doped alumina, the doping level exceeded the solubility limits of these elements in alumina, and thus resulted in dopant-rich second phase precipitates at the grain boundaries. As a result, the role (if any) of these second phase particles on the creep behavior of these materials is not clear. To avoid this complication, in the present study, alumina was doped with Zr and Nd ions at levels below their solubility limits in alumina. The effect of the dopants on the creep behavior of alumina may thus be attributed solely to solid solution/segregation effects. In this paper, we report the preliminary results of this investigation. The reason that Nd and Zr ions were chosen for this study was their ionic size. The size of Zr^{+4} (0.73Å) and Nd^{3+} (0.995Å), together with that of Y^{3+} (0.89 Å) and La^{3+} (1.06Å) makes up a range of ionic size, larger than that of Al^{3+} (0.53Å). The effect of the ionic size of the dopants on the creep resistance is currently under investigation.

EXPERIMENTAL PROCEDURE

Ultra pure α-Al_2O_3 powders (AKP-53, 99.995%, a mean particle size of 0.35 μm, Sumitomo Chemical America, New York, NY) and neodymium nitrate ($Nd(NO_3)_3 \cdot 6H_2O$, 99.99%, Alfa AESAR, Ward Hill, MA) were used as the starting powders. Alumina slurries of roughly equal mass of alumina powders and DDI (deionized + distilled) water were first made. Approximately 100 ppm Zr doping was achieved by ball-milling the alumina slurry with ZrO_2 grinding media (Tosoh U.S.A. Inc., Bridgewater, NJ) for 20 hours. Nd doping at 100 ppm level was achieved by adding the appropriate amount of neodymium nitrate aqueous solution to the alumina slurry. Nd-doped slurry was also ball-milled with ZrO_2 grinding media for 20 hours to achieve uniform mixing. Chemical analysis by ICP-OES (inductively coupled plasma-optical emission spectroscopy) technique revealed that the Zr-doped material had a doping level of 110 ppm Zr, and the Nd/Zr- codoped material contained a 40 ppm Nd and 70 ppm Zr dopant. After drying and crushing, the powders were then calcined at 850°C for 10 hours in high purity alumina crucibles (99.8%, Vesuvius McDanel Ceramic Company, Beaver Falls, PA). A new set of labware was used for each composition to avoid cross contamination. All powder preparation was carried out in a class 100 clean environment using acid-washed containers. Typical powder batch sizes were 200-250 grams each.

The processed powders were hot-pressed under vacuum in graphite dies at NIST (National Institute of Standards and Technology, Gaithersburg, MD) for 30 min with an applied pressure of 45 MPa. The Zr-doped Al_2O_3 was hot-pressed at 1320°C and the Nd/Zr-codoped Al_2O_3 at 1400°C. The grain size was measured by the linear intercept technique (1.5 × average intercept) with at least 600 intercepts counted for each measurement. The above hot-pressing conditions resulted in an average grain size of 0.48 and 0.72 μm, and a density of 97.1% and 98.0%, for the Zr-doped and Nd/Zr-codoped Al_2O_3, respectively.

The hot-pressed billets were machined commercially (Bomas Machine Specialties, Inc., Somerville, MA) into tensile specimens of the 2-inch specimen design developed at NIST by French and Wiederhorn [14]. SiC flags were attached to the specimen with carbon cement to establish the gauge length (the flags remain attached by friction at high temperatures). A laser extensometer (LaserMike, Inc., Dayton, OH) was used to measure the gauge length as a function of time under load with a precision of ±1 μm. Stress was applied to the specimen at constant load using either a lever-arm or hydraulic testing machine (Applied Test Systems, Inc., Butler, PA). Since all the experimental data were collected under a total strain < 3%, the tests could be considered to be constant stress. The maximum strain at rupture was about 10-12% for both materials. The temperature and stress range tested were respectively 1200-1350°C and 20-100 MPa. For each specimen, incremental temperature changes at constant stress, or incremental stress changes at constant temperature during the steady state creep were used to extract the maximum number of data points. For most specimens, the creep rate under each set of conditions was measured two or three times on a given specimen to ensure that the strain rate was not strain-history dependent. Typically, the strain rates under the same conditions were within about 10%, and the average value was taken as the strain rate under that set of conditions. No significant concurrent grain growth during creep testing was observed in these materials. A maximum of 20% grain growth was observed for the Zr-doped alumina after being crept 12% at rupture.

RESULTS AND DISCUSSIONS

The as-fired microstructures of Zr- and Nd/Zr-codoped alumina are shown in Fig. 1. The microstructures of both materials were very uniform. The grains were mostly equiaxed, though some grains showed irregular shapes. The range of the grain size for both materials was rather broad, 0.1-1.5 µm for Zr-doped alumina with an average size of 0.48 µm, and 0.1-3.0 µm for Nd/Zr-codoped alumina with an average of 0.72 µm. The large ratio of the largest grain size to the smallest grain size may be an indication of microscopic inhomogeous distribution of the dopants.

The creep curves from tensile creep rupture tests at a fixed temperature and under a

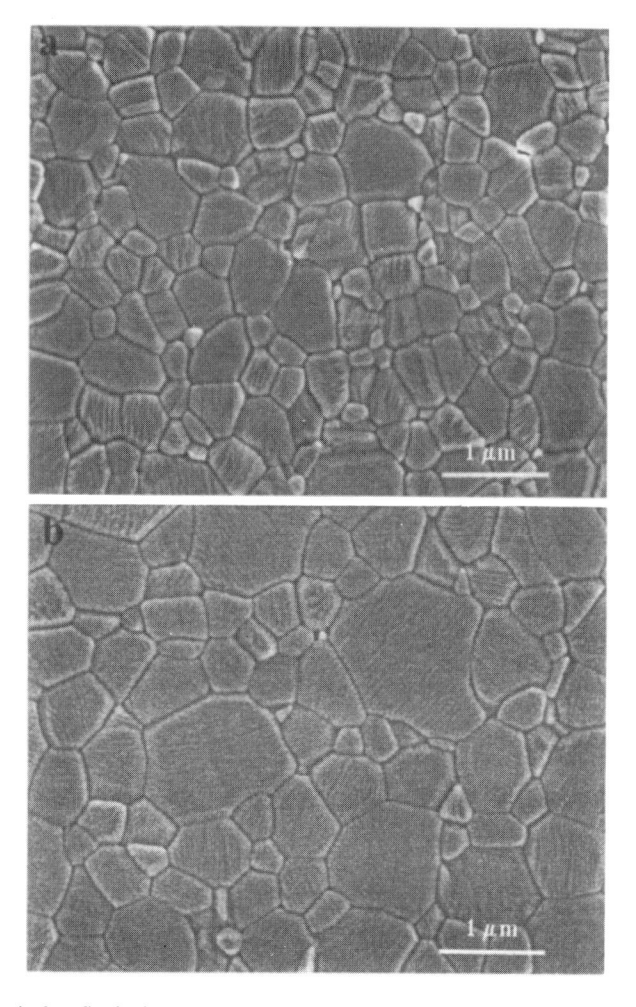

Fig. 1 Typical as-fired microstructures of (a) Zr-doped and (b) Nd/Zr-codoped alumina

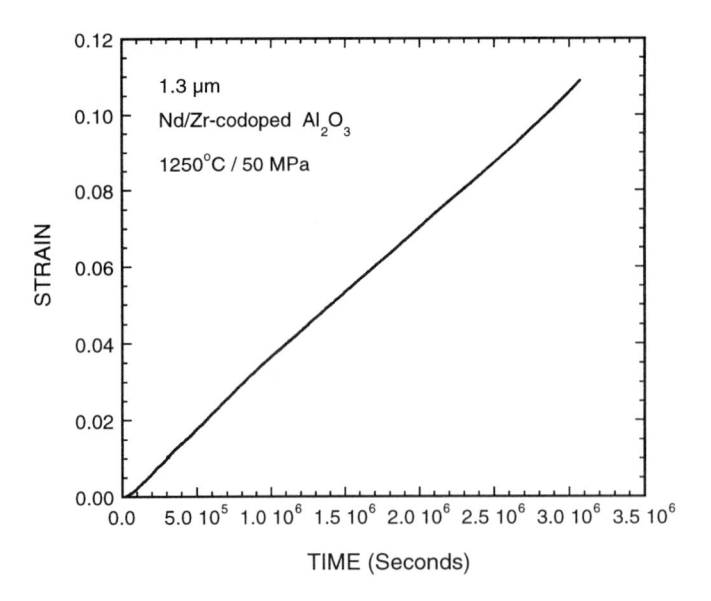

Fig. 2 A typical strain-rate relationship for a creep rupture test.

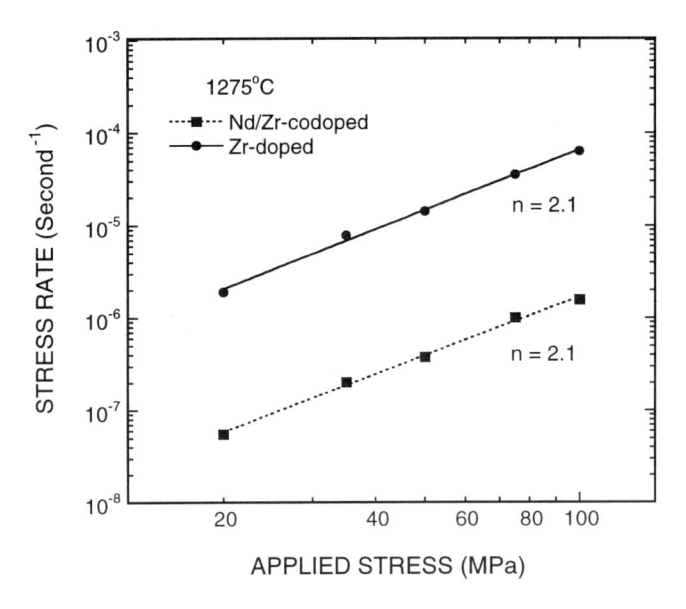

Fig. 3 The effect of applied stress on the creep rates at 1275°C.

constant load for both Zr- and Nd/Zr-codoped materials were very similar. Fig. 2 shows a typical strain-time curve for the Nd/Zr-codoped material. The specimen was annealed at 1200°C for 24 hours in the creep testing machine with a stress <5 MPa before the creep test was carried out. This annealing was found to significantly shorten the primary stage of creep, leaving an extensive steady state creep stage. The creep curve in Fig. 2 shows all the three stages of creep. However, since the creep rate difference among the different creep stages was less than 10%, (which was within the experimental reproducibility for multiple measurements under the same set of conditions), the primary and tertiary creep can be neglected, and the entire creep can be considered to be steady state creep for the purpose of creep rate measurements.

The effect of stress on the steady state creep rate at 1275°C is shown in Fig. 3. The stress exponent was measured to be close to 2 for both materials, which is consistent with the results of French et al. [1] and Robertson et al. [4]. This value reflects a possible role of interface-reaction in the creep process [10,11].

The effect of temperature on the creep rate under an applied stress of 50 MPa is shown in Fig. 4. Due to the grain size differences of these materials, however, the effect of dopant

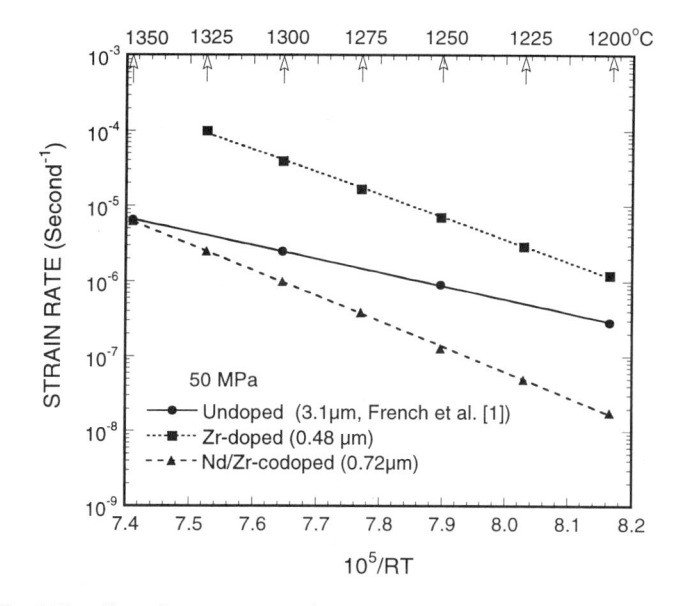

Fig. 4 The effect of temperature on the creep rate at an applied stress of 50 MPa.

on the creep rate is not clear. Sintering studies, carried out at Lehigh on samples processed from the same powder batches as the creep specimens, have shown that grain boundary diffusion was the dominant mechanism in the densification of Y-doped and La-doped alumina with similar dopants [15]. If, however, grain boundary diffusion is assumed to be the dominant creep mechanism [16-22], then one can assume an inverse grain size exponent of 3. The creep rates of the pure alumina and Nd/Zr-codoped alumina corrected to a grain size of 0.72 μm, (the grain size of Nd/Zr-codoped alumina), are shown in Fig. 5. As seen in this figure, 100 ppm Zr addition reduces the creep rate by a factor of about 15 as compared with undoped alumina. The addition of 40 ppm Nd further reduces the creep rate by another factor of about 20, which represents a combined effect of two and a half orders of magnitude as compared with the undoped alumina.

The activation energy was 470, 690, and 780 kJ/mol for undoped, Zr-doped, and Nd/Zr-codoped Al_2O_3, respectively. Clearly, the introduction of Nd/Zr- dopant ions resulted in an increase in the activation energy. This is consistent with the obstruction of grain boundary transport processes, by the oversize dopant ions segregated to the grain boundaries. Higher activation energies caused by dopants were also confirmed by Cho et al. [2] in Y and La doped alumina, by Sato and Carry [23] in Y and Mg codoped alumina, and by Wakai et al. [6] in Zr

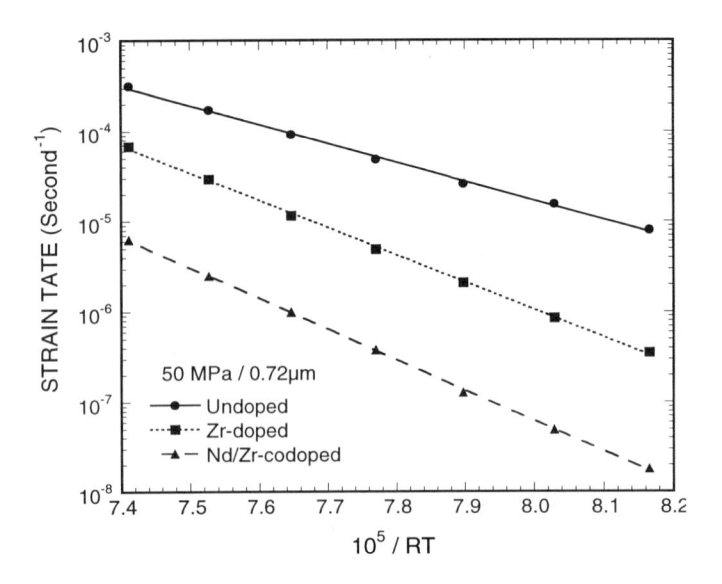

Fig. 5 The effect of temperature on the creep rate and corrected to a grain size of 0.72 μm according to the grain boundary diffusion controlled creep.

doped alumina.

From microstructural observation in the SEM, all the samples appeared to be single phase (Fig. 1); no second phase particles were observed in either material. This result clearly suggests that the solubility limit of Zr in alumina is >110 ppm, and that of Nd is >40 ppm. Moreover, the data demonstrate that the enhanced creep resistance is primarily a solid solution effect. This contrasts with the work of Wakai et al. [6], where the creep rate reduction by the Zr addition was attributed to the ZrO_2 particles at grain boundaries, reducing the interface-reaction rate and thus the creep rate. Interestingly however, the extent of creep rate reduction reported by Wakai et al. and observed in the present study agree closely (factor of 15).

To date, the codoping combination of Nd and Zr has produced the greatest benefit in terms of reduction of creep rate (two and half orders of magnitude versus two orders of magnitude for Y- and La-doped alumina). From the present results alone, it is not possible to conclude whether the Nd ions are having a greater effect than the Zr ions, or whether there is some beneficial interaction between the influences of the two dopant ions. The relative contribution of Nd ions alone to the creep rate is under current investigation.

As stated previously, our belief is that the improvement in creep results from the lowering of the grain boundary diffusivity due to the presence of oversize dopant ions. Lartigue and coworkers have proposed an alternative explanation based on their observation [24,25] that Mg doping and Mg and Y codoping in alumina resulted in a large proportion (~30%) of special grain boundaries (near coincidence boundaries), compared with undoped alumina. Lartigue et al. [26] further observed that Mg and Y codoping also lead to an increase in dislocation density both in the grains and in the grain boundaries. These investigators postulate that because the special grain boundaries are less able to accommodate intergranular dislocations, the grain boundary sliding process was limited. However, in the investigations of Y- and La-doped aluminas, neither grain boundary nor lattice dislocations were observed by the Lehigh researchers [1,2]. One possible explanation for the different observations is that the creep tests were conducted at different temperatures. The Y- and La-doped materials were tested in the range of 1200-1350°C by the Lehigh group, while the creep tests by the above investigators were carried out at or above 1450°C. At higher temperature, dislocations can be created more easily and dislocation mobility is higher, hence the contribution of dislocation processes to creep may be greater. Given the absence of any detected dislocation activity to date, it is believed that the segregation argument provides the explanation which is the most consistent with the results. The relationship between the creep rate (grain boundary diffusivity) and the ionic size of the dopant is currently under our investigation both experimentally and by means of computer simulation.

SUMMARY

The tensile creep behavior of 100 ppm Zr-doped and Nd/Zr-codoped alumina ceramics was studied. The addition of 100ppm Zr alone resulted in a reduction of the creep rate by a factor of 10, whereas the presence of both Zr and Nd ions, reduced the creep rate by two and a half orders of magnitude. The activation energy was increased from the 470 kJ/mol for undoped alumina to 690 and 780 kJ/mol for 100 ppm Zr-doped and Nd/Zr-codoped alumina, respectively, consistent with impeded grain boundary transport process by the segregated dopant ions. Given that the dopant concentration of both the Nd and Zr ions was below the solubility limit, the present study suggests that the enhanced creep resistance is primarily a solid solution effect.

ACKNOWLEDGMENTS

The authors would like to thank R. Krause and E. R. Fuller, Jr. at NIST for help with hot-pressing the 3-inch billets, and W. Luecke at NIST for providing the engineering graph of the 2-inch tensile specimen design. This work was supported by AFOSR under grant # F49620-94-1-0284.

REFERENCES

[1]. J.D. French, J. Zhao, M.P. Harmer, H.M. Chan, and G.A. Miller, "Creep of Duplex Microstructures," J. Am. Ceram. Soc. 77:2857 (1994).
[2]. J. Cho, M.P. Harmer, H.M. Chan, J.M. Rickman, and A.M. Thompson, The effect of Y and La on tensile creep behavior of aluminum oxide, to be submitted to J. Am. Ceram. Soc. (1996).
[3]. S. Lartigue, C. Carry, and L. Priester, Grain boundaries in high temperature deformation of yttria and magnesia co-doped alumina, J. Phys. (Paris), C1-51:985 (1990).
[4]. A.G. Robertson, D.S. Wilkinson, and C.H. Cáceres, Creep and creep fracture in hot-pressed alumina, J. Am. Ceram. Soc. 74:915 (1991).
[5]. P. Gruffel and C. Carry, Strain-rate plateau in creep of yttria-doped fine grained alumina, pp.305-11 in Proceedings of the 11th RISø International Symposium on Metallurgy and Materials Science: Structural Ceramics-Microstructure and Properties. J.J. Bentzen, J.B. Bilde-Sorensen, N. Christiansen, A. Horsewell, and B. Ralph, ed. Risø National Lab. Roskilde, Denmark, 1990.
[6]. F. Wakai, T. Iga, and T. Nagano, Effect of dispersion of ZrO_2 particles on creep of fine-grained Al_2O_3, J. Ceramic Soc. Jpn. 96:1206 (1988).
[7]. F.R.N. Nabarro, Deformation of crystals by the motion of single ions, in: Report of a Conference on the Strength of Solids, p. 231, Phys. Soc., London (1948).
[8]. C. Herring, Diffusional viscosity of a polycrystalline solid, J. Appl. Phys. 21:437 (1950).
[9]. R.L. Coble, A model for boundary diffusion controlled creep in polycrystalline materials, J. Appl. Phys. 34:1679 (1963).
[10]. E. Arzt, M.F. Ashby, and R.A. Verrall, Interface controlled diffusional creep, Acta Metall. 31:1977 (1983).
[11]. G.W. Greenwood, The possible effects on diffusional creep of some limitation of grain boundaries as vacancy sources and sinks, Scripta Metall. 4:171 (1970).
[12]. A.G. Evans and T.G. Langdon, Structural ceramics, Prog. Mater. Sci. 21:171 (1976).
[13]. W.R. Cannon and T.G. Langdon, Review creep of ceramics, Part I, J. Mater. Sci., 18:1 (1983); Part II, J. Mater. Sci., 23:1 (1988).

[14]. J.D. French and S.M. Wiederhorn, Tensile specimens from ceramic components, J. Am. Ceram. Soc. 79:550 (1996).
[15]. J. Fang, A.M. Thompson, H.M. Chan, and M.P. Harmer, Effect of Y and La on the sintering behavior of ultra-high-purity Al2O3, in press, J. Am. Ceram. Soc. (1996)
[16]. P. Gruffel and C. Carry, Effect of grain size on yttrium grain boundary segregation in fine grained alumina, J. Eur. Ceram. Soc., 11:189 (1993).
[17]. B.A. Pint, Experimental observations in support of the dynamic-segregation theory to explain the reactive-element effect, Oxidation of Metals, 45:1 (1996).
[18]. M. Le Gall, A.M. Huntz, and B. Lesage, Self-diffusion in α-Al_2O_3 and growth rate of alumina scales formed by oxidation: effect of Y_2O_3 doping, J. Mater. Sci. 30:201 (1995).
[19]. S. Lartige-Korinek, C. Carry, F. Dupau, and L. Priester, Transmission electron microscopy analysis of grain boundary behavior in superplastic doped aluminas, Mater. Sci. Forum, 170-172:409 (1994).
[20]. A.M. Thompson, K.K. Soni, H.M. Chan, M.P. Harmer, D.B. Williams, J.M. Chabala, and R. Levi-Setti, Dopant distributions in rare earth-doped Al2O3, in press, J.Am. Ceram. Soc. (1996).
[21]. J. Bruley, J. Cho, H. Chan, M.P. Harmer, and J.M. Rickman, STEM analysis of grain boundaries of creep resistant Y and La doped alumina, to be submitted to J. Mater. Res. (1996).
[22]. T.G. Langdon, The significance of diffusion creep in simple and multicomponent ceramics, Defect and Diffusion Forum, 75:89 (1991).
[23]. E. Sato and C. Carry, Yttria doping and sintering of submicrometer-grained α-alumina, J. Am. Ceram. Soc. 79:2156 (1996).
[24]. S. Lartigue and L. Priester, Influence of doping elements on the grain boundary characteristics in alumina, J. Phys. C5,49:451 (1988).
[25]. H. Grimmer, R. Bonnet, S. Lartigue, and L. Priester, Theoretical and experimental descriptions of grain boundaries in rhombohedral alpha-Al_2O_3, Phil. Mag. A6:493 (1990).
[26]. S. Lartigue, C. Carry, and L. Priester, Grain boundary in high temperature deformation of yttria and magnesia co-doped alumina, J. Phys. C1, 51:985 (1990).

FRACTURE OF COPPER/ALUMINA INTERFACES: THE ROLE OF MICROSTRUCTURE AND CHEMISTRY

Ivar E. Reimanis[1] and Kevin P. Trumble[2]

[1]Metallurgical and Materials Engineering
Colorado School of Mines
Golden, CO 80403

[2]School of Materials Engineering
Purdue University
W. Lafayette, IN 47907

ABSTRACT

Crack propagation at metal/ceramic interfaces may depend on the presence of chemical interphases. A particularly convenient model system for examining this dependence is copper/alumina. The recently developed ability to control the interface chemistry has made possible systematic experiments. In particular, it has been observed that the presence of discontinuous Cu_2O at the Cu/Al_2O_3 interface accelerates crack propagation, while the presence of discontinuous $CuAlO_2$ impedes it. The reason for these effects are not well-known, and are explored in this paper, in the context of microstructural observations obtained using electron microscopy. The nature of the chemical and structural bonding likely plays a role in governing the adhesive strength, but other important variables include the degree of growth of the interphase into the oxide and the metal, and the thermal expansion coefficient of the interphase. The relative importance of these variables is discussed.

INTRODUCTION

A number of workers have studied the mechanical properties of interfaces between Cu and Al_2O_3 [1-8]. Most of the interfaces in these studies [1-6] were synthesized by the eutectic bonding process; some [7,8] were synthesized by solid state bonding. Attempts have been made in these studies to understand the influence of either Cu_2O, $CuAlO_2$ or $CuAl_2O_4$ at the interface on the strength and fracture behavior. Unfortunately, due to a lack of complete control over the formation of these interphases, an understanding of their influence on interface fracture is incomplete. However, two observations common in these studies stand out. First, the presence of Cu_2O is associated with strong interfaces; this observation is consistent with the fact that Cu_2O forms from the eutectic liquid film, which assists in interface pore removal. While it is logical that pore removal results in higher interface strengths, the effect of the Cu_2O phase on crack propagation has not been directly studied prior to the present work. The second common observation on Cu/Al_2O_3 interfaces is that the presence of $CuAlO_2$ usually accompanies high bond strengths. The reasons $CuAlO_2$ affects the bond strength are not known.

Recent in-situ studies [9,10] have revealed how isolated Cu_2O and $CuAlO_2$ interphases influence crack propagation at Cu/Al_2O_3 interfaces. In short, the presence of Cu_2O accelerates crack growth while the presence of $CuAlO_2$ impedes it. This paper reviews these results and then focuses on the influence of local residual stresses associated with interphases.

EXPERIMENTAL

Bond Synthesis and Heat Treatments

The solid state diffusion bonding technique used to produce Cu/Al_2O_3 interfaces has been described in detail elsewhere [11]. The bonding conditions are briefly summarized here. Two grades of Cu foil 130 µm thick (99.95 wt. % and 99.998 wt. %, Johnson Matthey, Ward Hill, MA) were used, representing two different initial oxygen levels. The 99.95 wt. % Cu was found to contain 0.021 wt.% oxygen (~5 times saturation), whereas the 99.998 wt. % Cu was pre-annealed in hydrogen to remove all oxygen. The Cu foils were cleaned and diffusion bonded between basal-oriented sapphire plates (Crystal Systems, Salem, MA) 50 mm in diameter X 1.2 mm and 3.0 mm thick. Bonding was conducted at 1040°C for 24 h under a high vacuum of ~1.3×10^{-4} Pa with 3 MPa applied pressure.[†] Samples with the higher oxygen content contained isolated needles of Cu_2O throughout the interface, as shown in figure 1a. Since the needles covered less than 5 area percent of the interface, the majority of the interface regions were "clean" at this microstructural scale. The needles typically grow approximately 1 µm into the Cu [10]. They exhibit a (110) texture, but are not crystallographically aligned with the basal sapphire. These bonds were used for the fracture experiments described below.

Both higher and lower-oxygen content samples were heat treated to 1000°C for 100 h in a CO/CO_2 buffer-controlled furnace at $PO_2 = 1 \times 10^{-2}$ Pa. This heat treatment resulted in a uniform distribution of $CuAlO_2$ needles [11], as shown in figure 1b. As evident in figure 1b, the $CuAlO_2$ needles are much finer than Cu_2O needles. The needles exhibit a growth orientation of $(001)CuAlO_2//(001)Al_2O_3$ with preferred needle axes of $[1\bar{1}0]CuAlO_2//[0\bar{1}0]Al_2O_3$ and $[110]CuAlO_2//[210]Al_2O_3$ [12]. The needles are typically less than 0.5 µm in thickness, and grow into the sapphire as well as the Cu.

[†] The authors acknowledge Dr. B. J. Dalgleish for synthesizing all of the Cu/Al_2O_3 interfaces in this work.

Crack Propagation Experiments

Cracks were propagated at the interface in ambient air in a controlled manner using a delamination specimen geometry; this provided a measurement of the fracture energy [13]. Beams (approximately 4 mm X 4 mm X 40 mm) were cut from the bonds by low speed diamond sawing and tested in the geometry shown in Figure 2. The phase angle of loading for this geometry in the present configuration is approximately 45° [10]. Once sharp precracks were produced, as described in detail in ref. 10, interface cracks

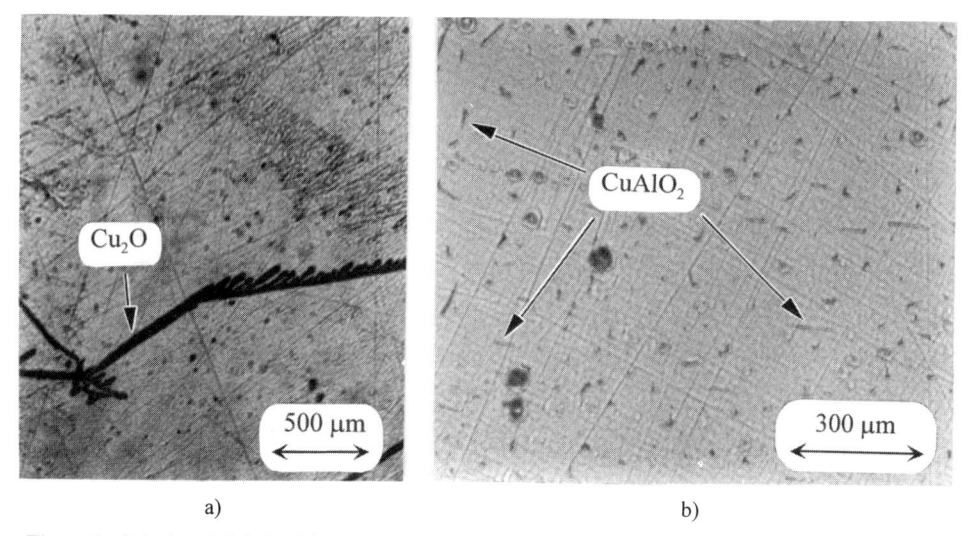

Figure 1. a) As-bonded, 'clean' interface between Cu and Al_2O_3, showing the occasional Cu_2O needle; b) heat treated interface showing $CuAlO_2$ needles on approximately 20 % of the interface area.

were propagated by loading the sample in four point bending. Crack growth was monitored by viewing through the optically transparent layer of sapphire onto the interface plane. Crack growth was stable, suggestive of R-curve behavior for these interfaces. The crack propagation experiment was usually terminated by fracture in the lower sapphire beam (indicated in Figure 2), an event governed by the strength of the sapphire. The fracture strength of the as-bonded specimens, based on the load required to fracture the lower sapphire layer, was 632 MPa ± 39 MPa for four specimens tested. The heat treated specimens failed at lower fracture stresses; one of the two specimens failed at 473 MPa, and the other failed at 341 MPa. As discussed below, the variation in strength in the sapphire is attributed to differing amounts of local interface residual stress.

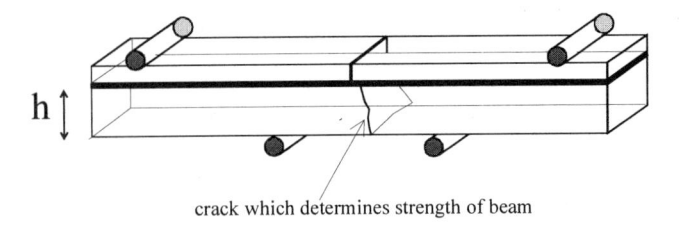

Figure 2. a) Delamination specimen geometry used for in-situ crack propagation experiments.

EXPERIMENTAL RESULTS

Generally, interface crack fronts for all specimens were straight except when they encountered a second phase (Cu_2O or $CuAlO_2$) at the interface. These interactions are discussed below. The interfaces in both kinds of bonds (figure 1a and 1b) failed by decohesion at the crack tip. That is, fracture did not occur by cracking or void growth ahead of the main crack, as has been observed to occur for several other fcc metal/Al_2O_3 interfaces [14,15]. In further contrast to these other systems, subcritical crack growth was not observed in this system, even though the interface was loaded for up to two hours at 80% of the load required to cause crack propagation. This result suggests that the Cu/Al_2O_3 interfaces does not exhibit H_2O-induced stress corrosion, in contrast with several other metal/ceramic interfaces [15-17]. It is noted that these experiments cannot totally rule out the possibility of stress corrosion in this system; it is possible that the velocity-stress intensity factor curve is steep enough that 80% of the critical load is not high enough to initiate subcritical crack growth [17].

The fracture energy at which cracks start to propagate at the 'clean' interfaces is approximately 125 J/m^2; this is termed the initiation fracture energy. Higher energies were required to cause continued crack propagation, indicating that the interface exhibits R-curve behavior. The highest fracture energies for the 'clean' interfaces, near the plateau of the R-curve, are approximately 200 J/m^2. Beyond this load, the lower sapphire beam fractures.

The initiation fracture energy of the $CuAlO_2$-containing interfaces is 190 J/m^2, significantly higher than that for the 'clean' interfaces (125 J/m^2). The extent of R-curve behavior in these interfaces was not determined because the lower sapphire beam fractured when the load was increased beyond that required to sustain a crack tip energy release rate greater than 190 J/m^2.

The effect of the Cu_2O and $CuAlO_2$ interphases on crack propagation is apparent in the in-situ optical micrographs of Figure 3. As figure 3 shows, the presence of Cu_2O at the interface has a very different effect from that observed when $CuAlO_2$ is present. Because of the small area fraction of Cu_2O at the interface, it is believed that its presence did not significantly affect the fracture energy of the 'clean' interface regions. However, the in-situ observations reveal that the Cu_2O needles tend to accelerate crack growth in their vicinity. Specifically, once the crack was within about 20 µm of a Cu_2O needle, it

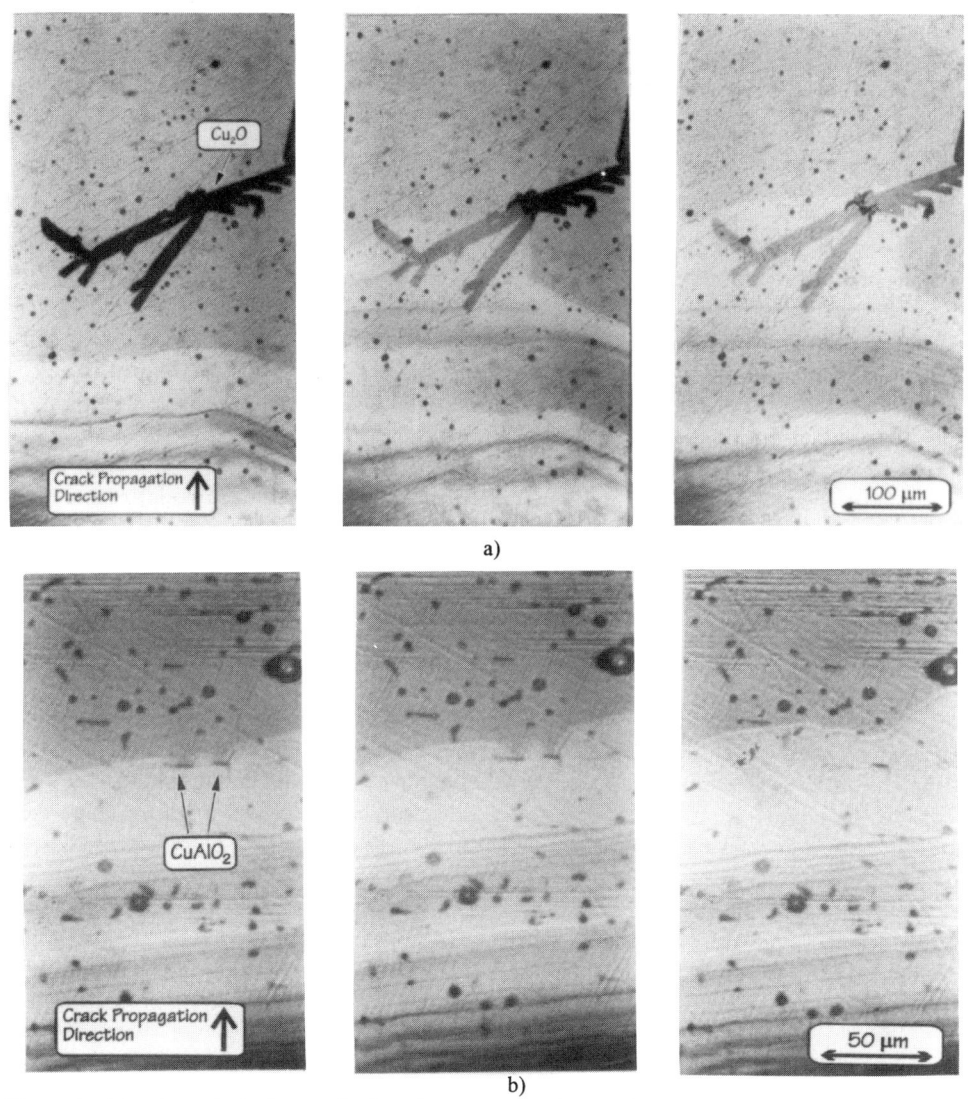

Figure 3. a) In-situ optical micrograph of crack propagating along Cu/Al$_2$O$_3$ interface with Cu$_2$O present. The crack is drawn into the Cu$_2$O phase. b) Similar to a) for heat treated sample showing crack front being pinned by the CuAlO$_2$ phase.

accelerated rapidly towards the particle and encompassed the particle, as is seen in figure 3a. From this observation, it was clear that within about 20 μm of the needle, either the driving force for crack growth increased or the interface fracture energy decreased.

CuAlO$_2$ particles were directly observed to impede crack growth as shown in figure 3b. It may be seen that the crack is pinned at the particles sites and only 'breaks through' after most of the crack has passed the particles. Because the particles represent a significant area fraction of the interface (approximately 20%), it is believed that their presence is the reason that the fracture energy is higher than that for the 'clean' interfaces.

DISCUSSION

The fracture energy of metal/ceramic interfaces is known to depend on the work of adhesion, the plasticity in the metal, roughness, and the residual stresses [18]. However, predictive physically-based laws have not yet been developed. The presence of interfacial phases may alter the contribution of each of these factors, but the discussion here concentrates on the effect of local residual stresses associated with the interfacial phases. For most metal/ceramic interfaces processed at elevated temperatures, thermally induced residual stresses develop during cooling; the metal is typically in a state of biaxial tension, the magnitude of which does not exceed the metal flow stress. The presence of an interfacial phase may alter the local thermally induced residual stresses which may in turn alter the local driving force for crack propagation.

Three potential sources of residual stress with respect to interphases have been identified: 1) the volume change associated with the formation of the interphase; 2) the lattice mismatch in the case of a coherent interface between the interphase and either Cu or Al_2O_3; 3) the coefficient of thermal expansion mismatch between the interphase and Cu or Al_2O_3.

The formation of Cu_2O within Cu results in an increase in molar volume of approximately 56%. Thus, it is predicted that the Cu_2O phase experiences residual axial compression. It is noted however, that the reaction takes place at 1000°C where the yield stress of the copper is very low[†] and can readily accommodate the stresses. Thus, the stresses associated with the volume expansion in growth of the Cu_2O phase are predicted to be negligible.

The formation of $CuAlO_2$ at a Cu/Al_2O_3 interface results in an approximately 20% molar volume increase. Because the $CuAlO_2$ consumes the Al_2O_3, it is partially surrounded by Al_2O_3. At 1000°C the stresses in Al_2O_3 would not be relieved; indeed, recent transmission electron microscopy studies have shown that dislocations exist in the Al_2O_3 adjacent to $CuAlO_2$ [12], evidence of the presence of high residual stress. Thus, volume expansion is expected to induce significant residual stresses near the interface (axial compressive in the $CuAlO_2$ and axial tensile in the Al_2O_3).

Based on the observation [12] that interfaces between Al_2O_3 and Cu_2O or $CuAlO_2$ are not coherent, it is not likely that significant coherency strains exist.

Thermally induced residual stresses are considered by examining the coefficients of thermal expansion (CTE) as shown in Table 1. The CTE for Cu_2O is well below the CTE of Cu and that of Al_2O_3. The residual stress state in the vicinity of the Cu_2O needle may be predicted using a model similar to that developed for second phase particles in a matrix [23]. Figure 4 indicates schematically the sign of the stress; the magnitudes are

---

[†] The melting point of copper is 1084°C [19].

difficult to predict because it is not known exactly how these stresses are superposed on the thermally induced residual stresses between Al$_2$O$_3$ and Cu. However, the sign of the

Table 1. Coefficients of thermal expansion for the phases relevant in this work. Values taken for temperatures between 25°C and 1000°C, except for Cu$_2$O which is taken for temperatures between -150°C and 200°C.

	$\alpha_{\text{a-axis}}$ (x 10^{-6}/°C)	$\alpha_{\text{c-axis}}$ (x 10^{-6}/°C)
Cu [19]	20	--
CuAlO$_2$ [20]	11.2	4.1
Cu$_2$O [21]	1.9	--
Al$_2$O$_3$ [22]	8	9

stress affects the crack propagation. Specifically, if a crack encounters a tensile axial stress prior to intercepting the needle, the crack driving force is increased, and the crack should be drawn preferentially towards the particle. The experimental observations discussed above indicate that the crack is indeed drawn preferentially towards the particle, once the crack is within about 20 µm (Figure 3a). The observation that the crack preferentially debonds the particle (Figure 3a) (in contrast to being pinned, as Figure 4 suggests would occur) may be due to several factors. First, the compressive axial stress

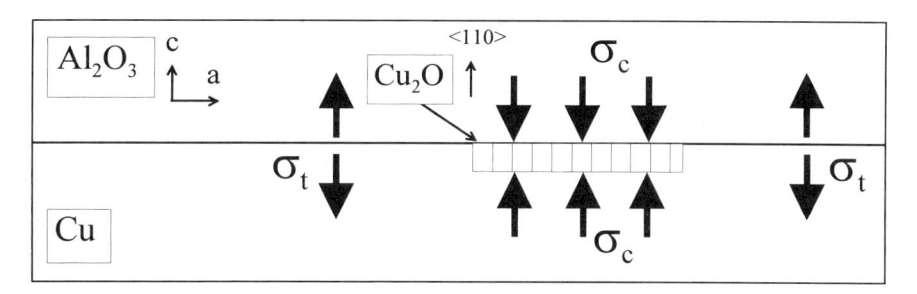

Figure 4. Schematic showing the predicted stress state associated with a Cu$_2$O needle at a Cu/Al$_2$O$_3$ interface based on the thermal expansion mismatch.

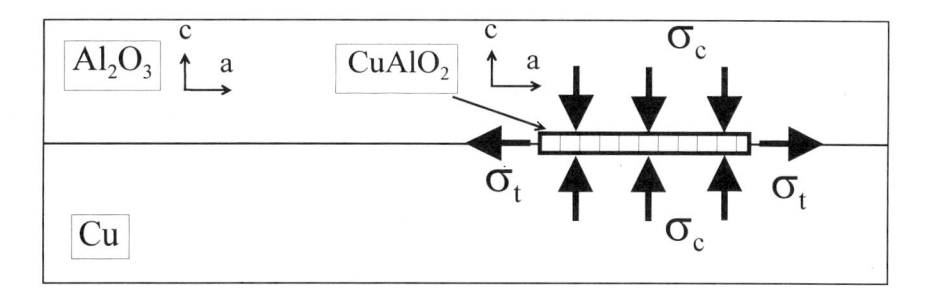

Figure 5. Schematic showing the predicted stress state associated with a CuAlO$_2$ needle at a Cu/Al$_2$O$_3$ interface based on the thermal expansion mismatch.

exists within the particle, not at the interface where the crack propagates. Second, the inherent bonding between Cu_2O and Al_2O_3 may be weak. The last factor contributing to preferential debonding between Cu_2O and Al_2O_3 is that the plasticity in the Cu may be limited due to the presence of the Cu_2O.

The $CuAlO_2$ phase has a delafossite structure with a large anisotropy [20]. Based on the observed orientation relationships between $CuAlO_2$ and Al_2O_3, the signs of the thermally induced residual stresses were evaluated as shown in the schematic in Figure 4. This stress state is quite different from that predicted when Cu_2O is present. It is apparent from Figure 5 that a crack would not be preferentially drawn into the $CuAlO_2$ phase. Furthermore, if the crack remains in the interface plane (and therefore propagates through the $CuAlO_2$ phase), the crack would experience a state of compression, thereby lowering the crack growth driving force. On the other hand, if the crack deviates into the Al_2O_3 or the Cu, the crack driving force would also be expected to decrease due to crack tip shielding, similar to that predicted for 'rough' interface[st][24].

Further evidence that suggests the schematic in Figure 5 represents the stresses appropriately is the observation that the heat treated samples exhibit lower strengths to failure than the as-bonded 'clean' specimens (data in Figure 2b). The tensile stresses indicated in Figure 5 would lower the strength in bending of the delamination specimen geometry (Figure 2). Even though the $CuAlO_2$ grows into the Al_2O_3, the thickness of the $CuAlO_2$ phase (<0.5 µm) is too small for it to act as a strength-reducing flaw itself; thus, it appears that increased thermally induced residual stress result in the lower strengths. It is noted that the source of those stresses has not been unequivocally identified with the presence of the $CuAlO_2$ phase. Because the Cu foil in the $CuAlO_2$-containing specimens has experienced a heat treatment, its flow strength may differ from the Cu in the as-bonded specimens; the Cu flow stress is directly related to the interface fracture energy as well as the state of thermally induced residual stress associated with the interface. Thus, it is difficult to directly compare the two kinds of specimens. Future work will focus on evaluating the differences in Cu flow strength between the as-bonded and heat-treated specimens.

CONCLUSIONS

Local residual stresses associated with interphases play an important role in determining how a particular interphase affects crack propagation at metal/ceramic interfaces. In the case of Cu/Al_2O_3 interfaces, the presence of Cu_2O increases the crack driving force while the presence of $CuAlO_2$ decreases it. Provided it is possible to control the presence of particular interfacial phases, the interface fracture energy may be tailored; however, a clear understanding on how the residual stress develops and how it affects crack propagation does not yet exist.

REFERENCES

1. J. F. Burgess, C. A. Neugebauer, G. Flanagan, The direct bonding of metals to ceramics by the gas-metal eutectic method, *J. Electrochem. Soc.*, 122[5], 688-690 (1972).
2. Y. S. Sun and J. C. Driscoll, A new hybrid power technique utilizing a direct copper to ceramic bond, *IEEE Transactions on Electron Devices*, Vol. ED-23, No. 8, 961-967 (1976).
3. Y. Yoshino, "Role of oxygen in bonding copper to alumina, *J. Am. Ceram. Soc.*, 72[8], 1322-1327 (1989).
4. S. T. Kim and C. H. Kim, Interfacial reaction product and its effect on the strength of copper to alumina eutectic bonding, *J. Mater. Sci.* 27, 2067-2066 (1992).
5. Y. Yoshino and T. Shibata, Structure and bond strength of a copper-alumina interface, *J. Am. Ceram. Soc.* 75[10] 2756-2760 (1992).
6. W. L. Chiang, V. A. Greenhut, D. J. Shanefield, L. A. Johnson and R. L. Moore Gas-metal eutectic bonded Cu to Al_2O_3 substrate-mechanism and substrate additives effect study, *Ceram. Eng. Sci. Proc.* 14 [9-10], 802 (1993).
7. C. Beraud, M. Courbiere, C. Esnouf, D. Juve and D. Treheux, Study of copper-alumina bonding, *J. Mater. Sci.*, 24, 4545-4554 (1989).
8. B. J. Dalgleish, E. Saiz, A. P. Tomsia, R. M. Cannon and R. O. Ritchie, Interface formation and strength in ceramic-metal systems, *Scripta Metall. et Mater.*, 31, 1109-1114 (1994).
9. I. E. Reimanis, B. J. Dalgleish and K. P. Trumble, Fracture at Cu/sapphire interfaces, *Ceramic Transactions*, Vol. 35, Structural Ceramics Joining II, eds. A. J. Moorhead, R. E. Loehman and S. M. Johnson 219-228 (1993).
10. I. E. Reimanis, K. P. Trumble, K. A. Rogers, and B. J. Dalgleish, The influence of Cu_2O and $CuAlO_2$ interphases on crack propagation at $Cu/\alpha\text{-}Al_2O_3$ interfaces, accepted for publication *in J. Amer. Ceram. Soc.*, (1996).
11. K. A. Rogers, K. P. Trumble, B. J. Dalgleish and I. E. Reimanis, The role of oxygen in microstructure development at solid state diffusion-bonded $Cu/\alpha\text{-}Al_2O_3$ interfaces, *J. Amer. Ceram. Soc.*, 77 [8] 2036 (1994).
12. K. A. Rogers, K. P. Trumble and B. J. Dalgleish, The morphology and mechanism of $CuAlO_2$ formation at solid $Cu/\alpha\text{-}Al_2O_3$ interfaces, submitted to *J. Am. Ceram. Soc.*, May 1996.
13. P. G. Charalambides, J. Lund, A. G. Evans and R. M. McMeeking, A test specimen for determining fracture resistance of bimaterial interfaces, *J. Appl. Mech.* 56, 77 (1989).
14. B. J. Dalgleish, K. P. Trumble, and A. G. Evans, *Acta Metall.* 37, 1923 (1989).
15. I. E. Reimanis, B. J. Dalgleish and A. G. Evans, The fracture resistance of a model metal/ceramic interface, *Acta metall. mater.* 39, 3133 (1991).
16. J. R. Stolken, University of California, Santa Barbara, personal communication (1995).
17. T. S. Oh, R. M. Cannon and R. O. Ritchie, *J. Am. Ceram. Soc.* 70, C352 (1987)..
18. A. G. Evans and B. J. Dalgleish, The fracture resistance of metal-ceramic interfaces, *Acta Metall. et Mater.*, 40, S295-S306 (1992).
19. J. F. Shackelford, W. Alexander and J. S. Park editors of *Materials Science and Engineering, CRC Handbook*, edited, second edition, CRC Press, Boca Raton (1994).
20. T. Ishiguro, N. Ishizawa, N. Mizutani and M. Kato, High temperature structural investigation of the delafossite type compound $CuAlO_2$, *J. Solid State Chem.*, 41, 132-137 (1982).
21. T. Suzuki, X-ray study on the binding properties of Cu_2O and Ag_2O crystals, *J. Phys. Soc. Japan*, 15, 2018-2024 (1960).

22. N. Ishizawa, T. Miyata I. Minato, F. Marumo and S. Iwai, A structural investigation of α-Al$_2$O$_3$ at 2170K, *Acta Cryst.* B36, 228-230 (1980).
23. R. W. Davidge and T. J. Green, The strength of two-phase ceramic/glass materials", *J. Mater. Sci.* 3 629-634 (1968)..
24. A. G. Evans and J. W. Hutchinson, *Acta Metall.* 37, 909-915 (1989).

EFFECT OF MICROSTRUCTURE AND INTERNAL STRESS ON MECHANICAL PROPERTIES OF WC/Co DOPED Al$_2$O$_3$/TiC/Ni FGMs

J. Lin, Y. Miyamoto, and K. Tanihata

The Institute of Scientific & Industrial Research, Osaka University
Ibaraki, Osaka 567, Japan

M. Yamamoto and R. Tanaka

Japan Ultra High Temperature Material Research Institute
Okiube 573-3, Ube 755, Japan

INTRODUCTION

The intrinsic low fracture toughness of ceramics limits their potential applications. Various processes have been developed to improve the fracture toughness of ceramics. One is to incorporate reinforcements, such as crack bridging, crack deflection and whisker or fiber pull-out, into the ceramic matrix to activate toughening mechanisms.[1-3] Another well-recognized method to enhance crack propagation resistance is to introduce a compressive stress in the surface of ceramics. Lange[4] introduced a compressive surface stress into Si$_3$N$_4$/ZrO$_2$ ceramics by utilizing the volume expansion of the Zr oxynitride to monoclinic ZrO$_2$ by oxidation. He found that the apparent surface fracture toughness was directly related to the oxidation kinetics and the initial volume content of ZrO$_2$.

Green also developed a technique to create a ZrO$_2$ ceramic surface with compressive stresses by a special heat treatment method.[5] As indicated by Virkar, however, it is difficult to build a deep compression layer by using the above processes. In order to overcome this shortcoming, Virkar and co-workers[5-8] fabricated an Al$_2$O$_3$/ZrO$_2$ layered material. Compressive stress was introduced in the outer layers by the transformation of unstabilized ZrO$_2$ upon cooling. The results demonstrated the beneficial effects of deep compressive surface stress on strength and indentation fracture toughness.

In the case of ceramics without phase transformation, surface compressive stresses can be introduced by the mismatch of thermal expansion between the inner and outer layers. For example, SiC/AlN composites were fabricated in three layers by Sathyamoorthy.[9] Miyamoto developed a self-propagating high temperature synthesis aided hot isostatic pressing (SHS/HIP) to fabricate Al$_2$O$_3$/TiC/Ni graded composites. The materials produced showed simultaneous improvements of bending strength and indentation fracture toughness.[10] Recently, Miyamoto and co-workers applied this concept to fabricate

(Al₂O₃TiC)/TiC/Ni cutting tools for cast iron. The cutting life was elongated by four times as compared with the conventional ceramic tools of the monolithic Al₂O₃/TiC material.[11]

In this study, we fabricated a symmetrically graded composite in the Al₂O₃/TiC/Ni system with the SHS/HIP process, and evaluated the effect of surface thermal residual stress on bending strength and indentation fracture toughness. In order to control the thermal expansion coefficient of the outer layer, WC/Co particles were doped into the Al₂O₃ matrix. X-ray diffraction and finite element method (FEM) were employed to determine the thermal residual stress experimentally and theoretically. The microstructure of the surface layer was analyzed by scanning electron microscopy (SEM) and scanning electron transmission (TEM) observation.

EXPERIMENTAL PROCEDURE

Material Fabrication

The raw materials used for making (Al₂O₃-WC/Co)/TiC/Ni graded materials were Al₂O₃, WC/Co, TiC, and Ni powders. As-received Al₂O₃ powder has an average particle size of 0.4 µm; TiC one of 1.4 µm; and WC/Co and Ni, 1 µm. The powders in pre-determined compositions were wet-mixed by ball milling for over 48 hours, and then dried in a vacuum furnace. The structure and composition of the green compact are shown in Figure 1. For all samples, the intermediate (I) and central (C) layers have the same compositions, respectively. The outer (O) layers were controlled to have different WC/Co volume fractions in order to give a different thermal expansion coefficient and Young's modulus. The green compact was sealed into a borosilicate glass container with a BN powder bed in vacuum, then placed into a graphite chemical oven. The glass container was embedded in the low-cost silicon powder, which was used as a fuel for instantaneous heating. When the sample was 30 mm in diameter and 6 mm thick, 40 g silicon was added. The ignition pellets were made of thermit agents.

In HIP, after the chemical oven was heated to 780°C, the nitrogen pressure was applied up to 93 MPa, and then the chemical oven was heated to 1150°C. At about 1030°C, the silicon powder was ignited by the exothermic reaction of the thermit pellets. Internal temperature of the chemical oven can reach as high as 2500°C because of the reaction heat of silicon nitriding under the nitrogen pressure of 100 MPa. The samples were successively kept at 1150°C for 30 minutes and then cooled in HIP. The details of the SHS/HIP process are given in Miyamoto et al.[12]

Surface Fracture Toughness, Bending Strength, and Residual Stress

Hardness and indentation fracture toughness were measured by using a Vickers hardness testing machine. The specimen was polished to a 3-µm diamond-surface finish before the test. The load used was 20 kg, with a loading time of 15 seconds. The indention-induced crack lengths, $2c$, were measured with an optical microscope, and the indentation fracture toughness, K_c, was calculated using the following expression[13]:

$$(K_c \phi H_v a^{1/2})(H_v / E\phi)^{2/5} = 0.129(c/a)^{-3/2} \qquad (1)$$

where ϕ is a material-independent constant. H_v, E and a are Vickers hardness, Young's modulus and half-diagonal length of indentation, respectively.

The strength of the 2-mm×6-mm×25-mm bar specimens was determined by a three point bending test with an 18 mm span. The loading rate was 0.5 mm/min.

The finite element method was used to analyze the macrothermal residual stress produced by the graded structure, and the microthermal residual stress produced by the mismatch of the thermal expansions of Al_2O_3, WC, and Co. The effect of WC/Co volume fractions on the surface residual stress in Al_2O_3 matrix, and the volume ratio of Co to WC on micro residual stress were calculated using FEM software MENTAT II on a Hewlett Packard workstation.

Macro residual stress on the surface of samples was determined by the x-ray diffraction method. The measured stresses were compared with FEM results.

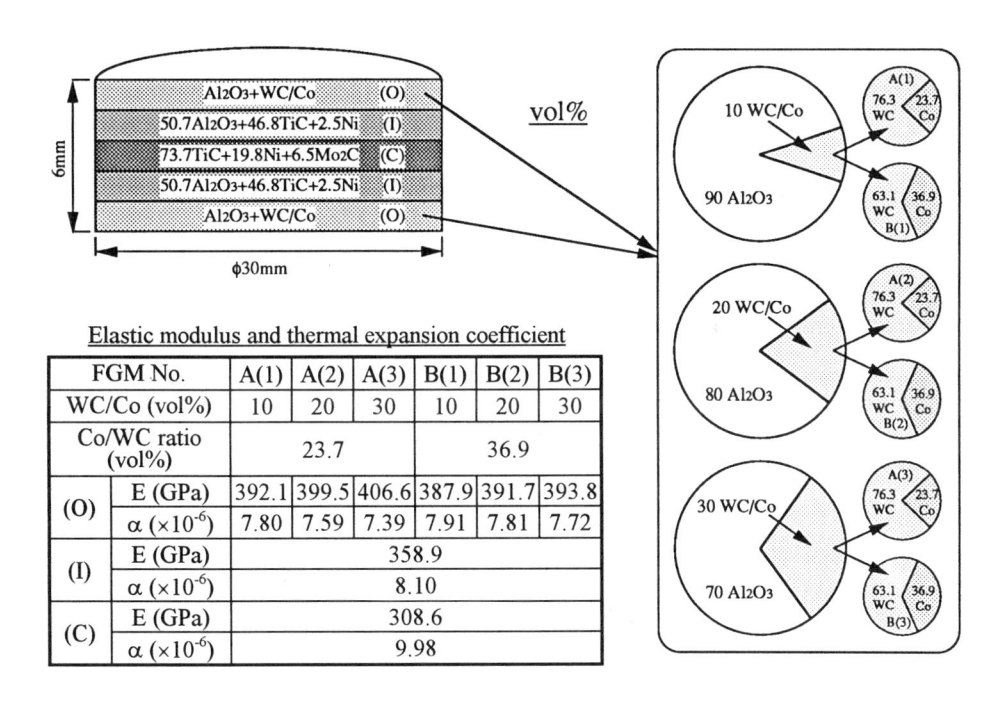

Elastic modulus and thermal expansion coefficient

FGM No.		A(1)	A(2)	A(3)	B(1)	B(2)	B(3)
WC/Co (vol%)		10	20	30	10	20	30
Co/WC ratio (vol%)		23.7			36.9		
(O)	E (GPa)	392.1	399.5	406.6	387.9	391.7	393.8
	α (×10⁻⁶)	7.80	7.59	7.39	7.91	7.81	7.72
(I)	E (GPa)	358.9					
	α (×10⁻⁶)	8.10					
(C)	E (GPa)	308.6					
	α (×10⁻⁶)	9.98					

Figure 1. Composition and structure of the green compact.

Characterization of Microstructure

Microstructural observations were performed on the polished cross section using SEM. TEM observations were carried out on the microstructure in the outer layer of the sample functionally graded material (FGM) A(2). TEM specimens were prepared by cutting a thin plate, about 500 μm thick, from the surface layer of the bulk sample; polishing the plate to a thickness of about 100 μm; drilling 3-mm-diameter discs from the plate using an ultrasonic drilling machine; dimpling the discs to a thickness of about 20 to 30 μm on a dimpler; and then further thinning the specimens with a Gatan ion-beam thinning machine. The discs were observed on HITACHI 600 TEM with energy dispersive spectroscopy (EDS).

RESULTS AND DISCUSSION

Microstructure

The obtained materials were well densified and showed a uniform microstructure at each layer. Figure 2 is SEM photograph of FGM A(2). In the central layer, Figure 2(c), the TiC particles (black phase) are surrounded by a Ti-Mo-C phase. The white phase is nickel. In the intermediate layer, the TiC particles are uniformly distributed in the Al_2O_3 matrix with a little of the Ni phase, as seen in Figure 2(b). In the outer layer, some WC particles, taking the metal Co as a cementing agent, tend to be clustered due to the poor wettability between Co and Al_2O_3, as seen in Figure 2(a). In observation of indention-induced cracks, we found that clustered WC/Co islands can effectively restrain crack propagation.

Figure 3 is a TEM photograph in the outer layer of FGM A(2). Isolated WC particles predominate at the grain boundary and grain junctions of the Al_2O_3 matrix. For these isolated particles, cobalt is sometimes present at the interface between the WC and Al_2O_3, as seen in Figure 4(a). When WC particles are clustered, cobalt is always found in the area between WC particles or WC and Al_2O_3, as seen in Figure 4(b).

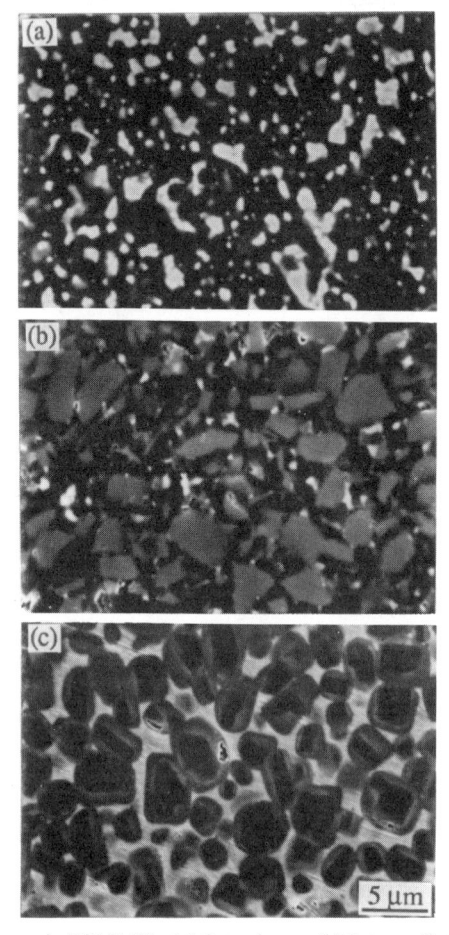

Figure 2. SEM microstructures in FGMA(2). (a) Outer layer. (b) Intermediate layer. (c) Central layer.

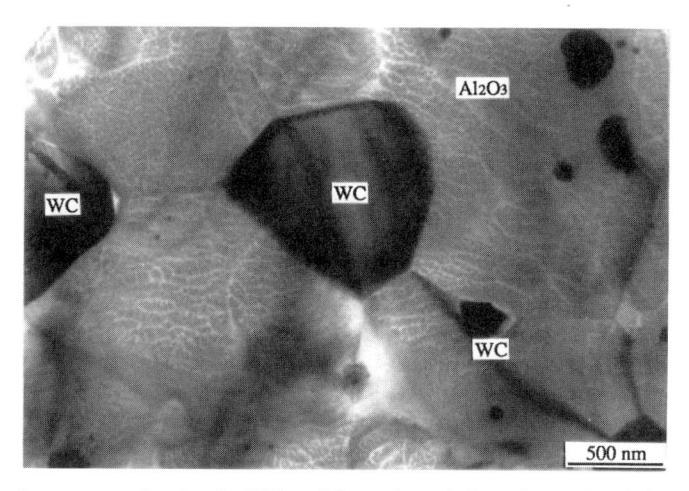

Figure 3. TEM microstructure showing the WC particles at the grain boundary and grain junctions of the Al$_2$O$_3$ matrix.

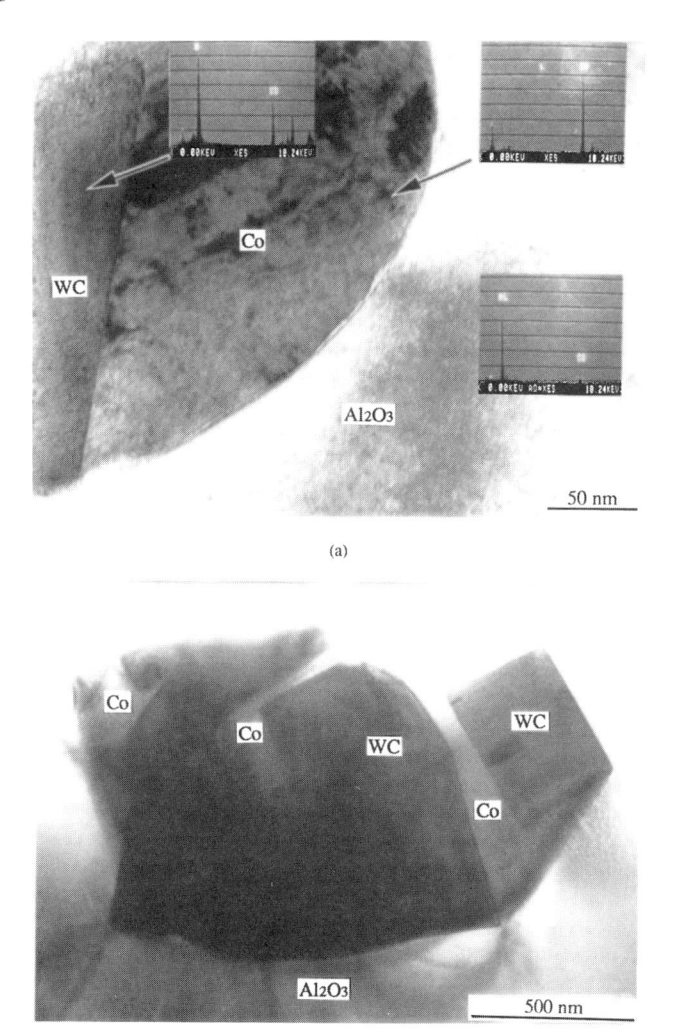

(a)

Figure 4. TEM microstructures showing cobalt presence. (top) At the interface between the isolated WC and the Al$_2$O$_3$ matrix. (bottom) In the area of the clustered WC particles.

837

Mechanical Properties

The flexural strength and indentation fracture toughness are plotted as a function of WC/Co volume fraction in Figures 5 and 6, respectively. With WC/Co content increasing, the strength of the FGMA slightly increases, while the hardness slightly decreases. The Co metal is responsible for the decrease of hardness. However, all samples in this study showed higher hardness than the monolithic Al_2O_3.

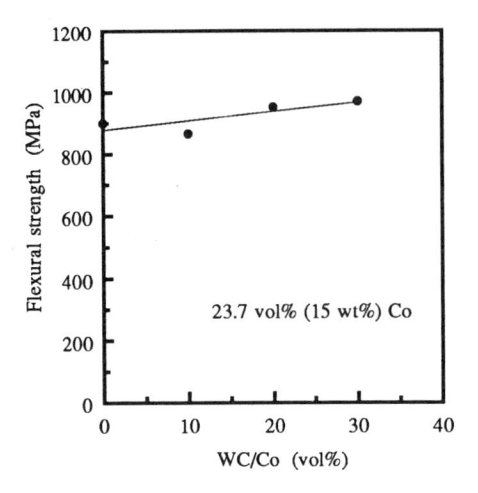

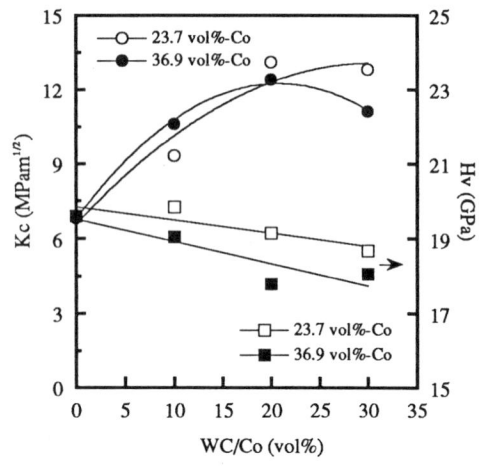

Figure 5. Flexural strength of FGMs as a function of WC/Co volume fraction.

Figure 6. Hardness and toughness of FGMs as a function of WC/Co volume fraction.

With increase of the WC/Co content, the surface toughness increases and tends to be saturated at about 20 to 30 vol% of WC/Co. The FGM A(2) shows a peak hardness of 12.8 MPam$^{1/2}$; that is, 88% higher than the toughness of the Al_2O_3/TiC/Ni FGM without WC/Co (6.8 MPam$^{1/2}$).

The addition of WC/Co into the outer Al_2O_3 layer can produce the dual effects of enhancing the compressive stress and activating the toughening mechanism due to the dispersion of second phase particles. The thermal residual stress present in the surface layer of Al_2O_3-(WC/Co) can be divided into two types as shown in Figure 7(a). One is the macro-thermal residual stress induced by the symmetrically graded structure. The other is micro-thermal residual stress induced by the thermal expansion mismatch of WC, Co, and Al_2O_3. FEM analyses were carried out to calculate the macro- and microthermal residual stresses at a temperature drop of 1000°C. Figures 7(b) and (c) show the FEM meshes for the analyses. Because of the symmetry of the FEM structure, element meshes were divided on 1/4 sample. Four-node elastic axisymmetric element and plane stress element were employed to analyze macrostress and microstress, respectively. In the case of microstress analysis, an elastic-plastic model was used for Co metal. The reported yield stress, 250 MPa, was referred.

Figure 8 shows the distribution of the residual stress in the sample of FGM A(2). In the outer layer, uniform compressive stress was developed except in the edge region. In every layer, the residual stress is generally constant. The outer and intermediate layers were subjected to a compressive stress as shown in Figure 8(b). The central layer sustained a high tensile stress. With increase of WC/Co content, the measured stress in the outer layer increases. The FEM calculation showed that the macrostress increases linearly with WC/Co content, as shown in Figure 9.

As illustrated in Figure 7(a), the residual stress in the outer layer should be a synthetic result of the macro- and microstresses. The thermal expansion mismatch of WC, Co, and Al$_2$O$_3$ will lead to a local stress field. Figure 10 shows the distribution of principal stress as a function of the distance from the center of the WC/Co particle (along OC line in Figure 7(c)). The direction of the maximum principal stress is parallel to the Co-Al$_2$O$_3$ interface, while the minimum stress is perpendicular to the interface. As seen in Figure 10, a compressive stress exists in the direction perpendicular to the interface, but a strong tensile stress, reaching to 400 MPa, occurs in the direction along the interface in the Al$_2$O$_3$ matrix. When the macro compressive stress exceeds the micro tensile stress, a compressive stress can remain in the matrix. Otherwise, a tensile stress will appear in the local area. The microstress is expected to be dependent on the volume fraction of WC/Co and the volume ratio of Co to WC. For the present materials, the correlation of the macro compressive stress, micro tensile stress, and WC/Co volume fraction are plotted in Figure 11. It can be found that the micro tensile stress tends to exceed the macro compressive stress when the WC/Co content is over 22 vol% for FGM B samples and 27% for FGM A samples. This result suggests that a local tensile stress will appear in the Al$_2$O$_3$ matrix near the Co-Al$_2$O$_3$ interface. This stress will result in reduced crack propagation resistance, which may be responsible for the increased toughness at 20 to 30 vol% of WC/Co, as seen in Figure 6.

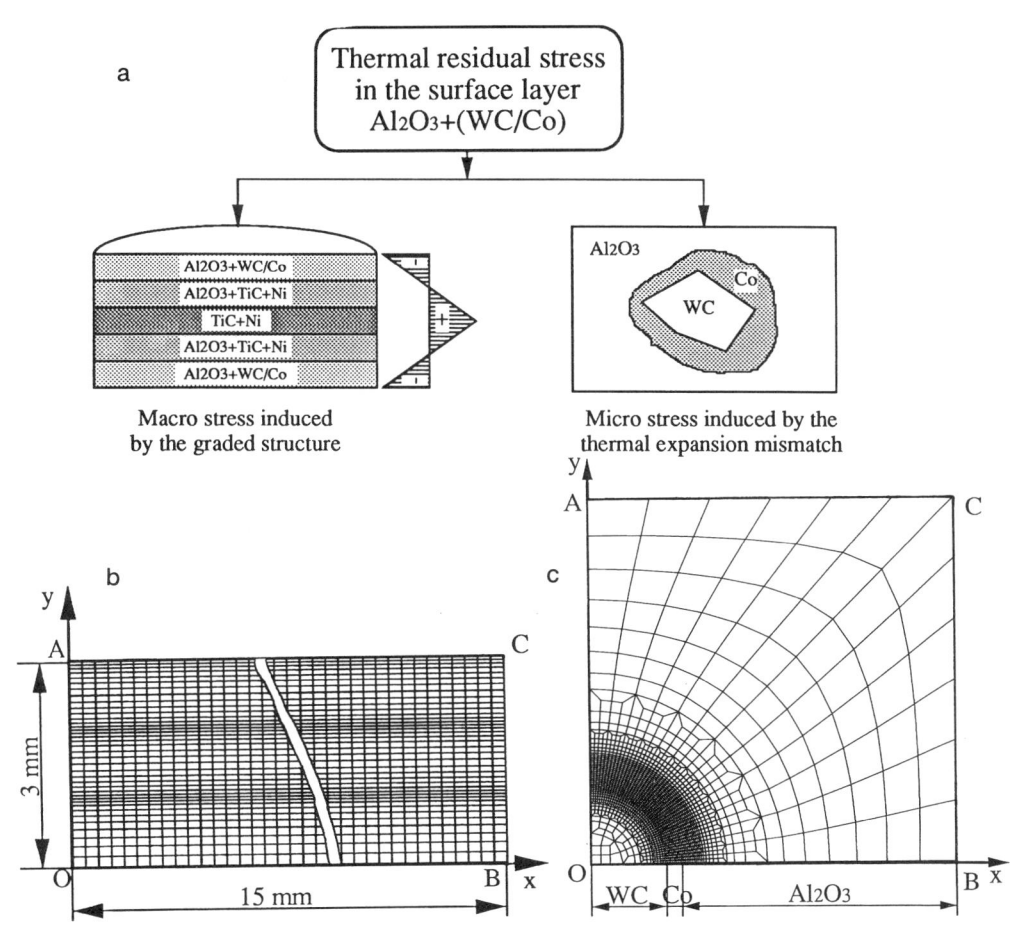

Figure 7. (a) Schematic diagram of residual stress. (b) FEM model for macrostress analysis. (c) FEM model for microstress analyses. In calculation, boundaries OA and OB were fixed in the direction of x and y, respectively.

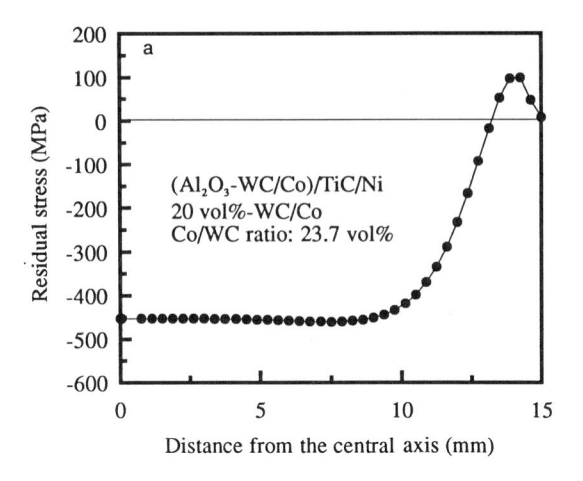

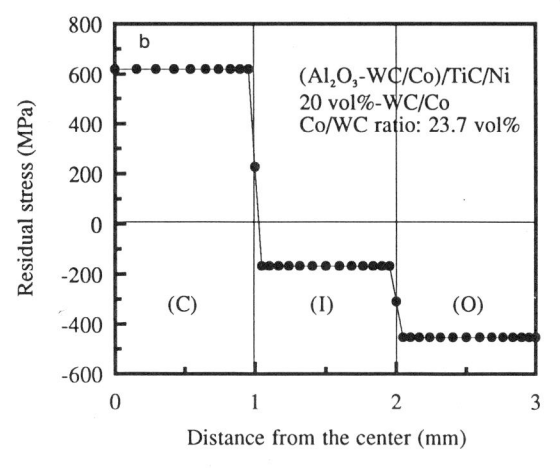

Figure 8. Macrostress distribution in the sample of FGM A(2). (a) Radial stress along AC line. (b) Radial stress along OA line.

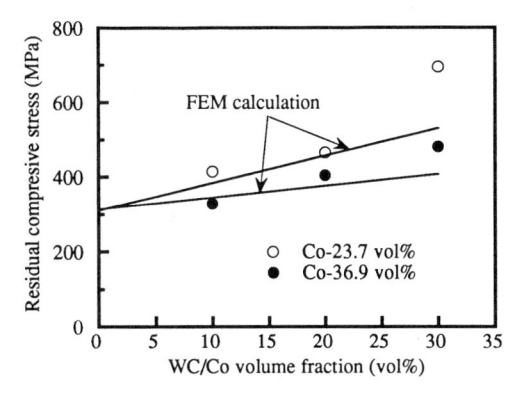

Figure 9. Relation between the surface compressive stress and WC/Co volume fraction. Open and solid circles denote the measured stress by x-ray diffraction.

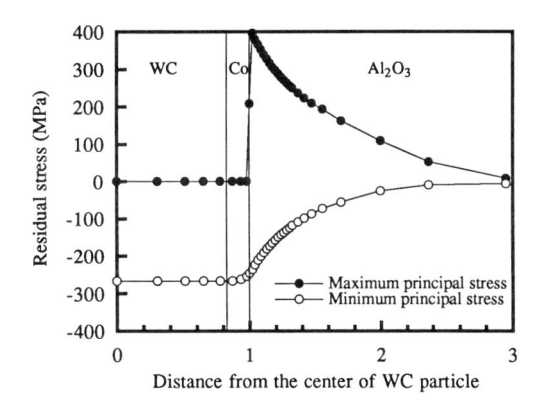

Figure 10. Residual microstress distribution along the OC line shown in Figure 7(c).

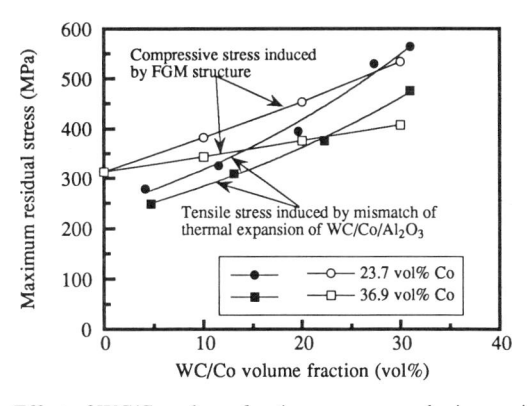

Figure 11. Effect of WC/Co volume fraction on macro and micro residual stress.

CONCLUSIONS

Dense WC/Co doped Al$_2$O$_3$/TiC/Ni FGMs were prepared by the SHS/HIP process. The WC particles tended to form clusters with Co in the Al$_2$O$_3$ matrix. Both the strength and toughness of WC/Co doped FGMs were remarkably improved due to the dual effects of surface compressive stress induced by the symmetrically graded structure and secondary phase dispersion. The FEM analysis showed that the addition of WC/Co to the outer Al$_2$O$_3$ layer will produce the micro tensile stress at the Co-Al$_2$O$_3$ interface in the Al$_2$O$_3$ matrix when the WC/Co content increases to 20 to 30 vol%. Increasing the volume of WC/Co reinforcements would reduce the crack propagation resistance.

REFERENCES

1. P. F. Becher, Microstructural design of toughened ceramics, *J. Am. Ceram. Soc.*, **74**:255 (1991)
2. V. D. Krstic and P. S. Nicholson, Toughening of glasses by metallic particles, *J. Am. Ceram. Soc.*, **64**:499 (1981).
3. W. H. Tuan, H. H. Wu and T. J. Yang, Preparation of Al$_2$O$_3$/Ni composites by a powder coating technique, *J. Mater. Sci.*, **30**:855 (1995).
4. F. F. Lange, Compressive surface stresses developed in ceramic by an oxidation-induced phase change, *J. Am. Ceram. Soc.*, **63**:38 (1980).
5. A. V. Virkar, J. L. Huang and R. A. Cutler, Strengthening oxide ceramics by transformation-induced stresses, *J. Am. Ceram. Soc.*, **70**:164 (1987).
6. R. A. Culter, J. D. Bright, A. V. Virkar and D. K. Shetty, Strength improvement in transformation-toughened alumina by selective phase transformation, *J. Am. Ceram. Soc.*, **70**:714 (1987).
7. A. V. Virkar, J. F. Jue, J. J. Hansen and R. A. Culter, Measurement of residual stresses in oxide-ZrO$_2$ three-layer composites, *J. Am. Ceram. Soc.*, **71**:C148 (1988).
8. J. J. Hansen, R. A. Cutler, D. K. Shetty and A. V. Virkar, Indentation fracture response and damage resistance of Al$_2$O$_3$-ZrO$_2$ composites strengthened by transformation-induced residual stresses, *J. Am. Ceram. Soc.*, **71**:C501 (1988).
9. R. Sathyamoorthy, A. V. Virkar and R. A. Cutler, Damage-resistant SiC-AlN layered composites with surface compressive stresses, *J. Am. Ceram. Soc.*, **75**:1136 (1992).
10. Y. Miyamoto, Z. Li and K. Tanihata, Recycling processes of Si waste to advanced ceramics using SHS reaction.
11. H. Morignchi, A. Ikegaya, T. Nomura, Y. Miyamoto, Z. Li and K. Tanihata, Cutting performance of hyperfunctional ceramics, *Powder & Powder Metall.*, **42**:1389 (1995) (in Japanese).
12. Y. Miyamoto, K. Tanihata, Z. Li, Y. S. Kang, K. Nishida and T. Kawai, SHS/HIP compaction of functionally graded materials, Proc. 8th CIMTEC Conf. Advanced Science and Technology 4, 387 (1995).
13. K. Niihara, R. Morena and D. P. H. Hasselman, Further reply to "Comment on 'Elastic-plastic indentation damage in ceramics: the Median/radial crack system,'" *J. Am. Ceram. Soc.*, **65**:C116 (1982).

INDEX

$3Al_2O_3 \cdot 2SiO_2$, mullite, *see also* Mullite ceramics, mullite fiber and mullite formation
  anisotropic grain growth in, 304–307
  coatings for SiC_f/LAS composites, 514, 516–524
  crystallization of glass fibers, 170, 174, 175
  formation from mullite gels, 773–774
  glass fiber fabrication, 169–170, 175
  growth of, 285–290, 294, 296, 299
  incubation period for formation, 286–290, 293–294, 299
  phase stability/formation, 255–261, 285–290, 292–300, 516–518
  powder characteristics, 774
  powder synthesis, 771–774
  processing, 287–288, 772–775
  properties, 771, 774–777
  reactive metal penetration of, 749–759
  Ti-doped, 304–307
Acoustic microscopy, 552–554, 556–557
Adhesion strength
  Al_2O_3/Cu, 823–830
  AlN/W, 392, 397
Adsorption
  Gibbs adsorption isotherm, 67–68, 73, 351–354
  Langmuir adsorption isotherm, 68, 79
AFM: *see* Atomic force microscopy
Al, aluminum
  bonding *via* electron beam, 623, 625, 627
  nanoparticle morphology, 624–625
  reactive metal penetration into mullite, 749–759
Al_2O_3, alumina
  anisotropic grain growth, 304–306, 308–310; *see also* Alumina, grain growth/boundary mobility
  brazing, 359–360, 371, 380–382
  $CaAl_{12}O_{19}$ (calcium hexaluminate) coatings on, 447–451
  crack healing, 229
  creep resistance, 815–821
  crystal structure calculations, 4
  densification, 241–244
  grain boundaries, 1, 5, 199, 201–203, 323–325, 328–329, 815–816, 819–821

Al_2O_3, alumina (*cont.*)
  grain boundary chemistry, 815–816, 819–821
  grain growth/grain boundary mobility, 199, 201, 234–237, 240, 242–245, 304–306, 308–310, 323–329, 817
  green microstructure, 239–245
  impurities
    Ca, 6, 323–326, 733–735
    La, 815–816, 819–820
    Mg, 6, 311–320, 323–329, 815, 819–820
    Nd, 816–821
    Si, 323–329
    Ti, 232–237
    Y, 5, 6, 311–312, 314, 316–317, 319–321, 815–816, 819–820
    Zr, 815–821
  La-doped, 815–816, 819–820
  liquid metals on, 66, 75, 350–357, 360, 407–411, 413–414
  liquid-phase sintered, 304–306, 308–310, 314, 317, 320
  Mg-doped, 311–320, 323–329
  $MgAl_2O_4$ precipitates in, 311–320
  Nd-doped, 816–821
  Nd/Zr co-doped, 816–821
  plasma activated sintering, 201–203
  pore channel instability, 229–232
  powders, 239–240, 243
  relative surface energies, 229, 232–234
  sintering aids, 311–321
  surface diffusion, 229–232
  tensile creep of Y/La co-doped Al_2O_3, 815–816, 819–820
  tensile creep of Zr-doped and Nd/Zr co-doped Al_2O_3, 817–821
  texture characterization by orientation imaging microscopy (OIM), 803–812
  Ti-doped, 232–237
  transient liquid phase bonding of, 407–414
  Y-doped, 5, 6, 311–312, 314, 316–317, 319–321, 815–816, 819–820
  Zr-doped, 815–821
Al_2O_3-(Al-Si) (alumina–aluminum-silicon alloy) composites, 749–759

843

Al$_2$O$_3$-Al$_2$O$_{3(f)}$ (Alumina-sapphire fiber) composite, 732–739
Al$_2$O$_3$-SiC nanocomposites, 551–552, 554–557
Al$_2$O$_3$-SiO$_2$, alumina-silica/aluminosilicate
   characterization of fibers, 170–173, 175, 721–728
   fibers, 169–175, 721–728
   grain growth in fibers, 721–728
   mechanical properties of fibers, 170–171, 174–175, 721–728
   metastable equilibria, 254, 261
   phase equilibria, 255–261
   reaction kinetics, 255–261, 285–290, 292–300
   reaction thermodynamics, 257–261
   thermal stability of fibers, 170, 174–175, 721–728
Al$_2$O$_3$-WC-Co/TiC/Ni (alumina-tungsten carbide-cobalt/titanium carbide/nickel) graded composites, 834–842
   finite element analysis, 835, 838–840
   mechanical properties, 834, 838–842
   microstructure, 835–837
   processing, 834–835
Al$_2$O$_3$-Y$_3$Al$_5$O$_{12}$ (alumina–yttrium aluminum garnet fiber) composite, 734–736
Al$_2$O$_3$-ZrO$_2$ composites, 613–617, 833
Al$_2$O$_3$-ZrO$_2$ composites, processing, 613–617
Al$_2$O$_3$/Cu interfaces, 823–830
   fabrication by diffusion bonding, 824
   fracture, 823–830
   interphases, 823–830
Al$_2$O$_3$/glass interfaces, 535–542
Al$_2$O$_3$/MgO diffusion couples, 578–582
Al$_2$O$_3$/SiO$_2$ diffusion couples, 257
Al$_2$O$_3$/TiC/Ni graded composites, 833–834
Al$_2$OC and Al$_4$O$_4$C, aluminum oxycarbides, in SiC, 103, 183
AlN
   on Si, epitaxy, 192
   on Si, thin film characterization, 193–198
AlN thin films, oxygen content, 194
AlN, aluminum nitride
   film growth mechanism, 192–193
   liquid phase sintering, 391, 394–397, 399
   for packaging, 391–397, 399–405
   plasma activated sintering, 202
   sintering aids, 391–397, 399–405
   thin film deposition, 191–192, 629–630, 633–635
   W metallization of, 391–397, 399–405
Al$_x$Ga$_{1-x}$N, aluminum gallium nitride, thin film deposition, 629–630, 633–635
Amorphous grain boundary films, see also Grain boundaries
   in Al$_2$O$_3$ ceramics, 202–203, 323–325, 328–329
   in Bi$_2$O$_3$-doped ZnO, 131–145
   in olivine, 104–105
   in SiC ceramics, 102–103, 177, 183–184, 186–188, 492–493
   at SiC$_w$/Si$_3$N$_4$ interfaces, 97
   in Si$_3$N$_4$ ceramics, 1, 23–27, 31–33, 95–97, 100–102, 104, 107–111, 113–114, 116, 118, 120–121, 123–130, 779–780, 788, 791–792

Amorphous grain boundary films (cont.)
   in SrTiO$_3$ ceramics, 645–646, 648–651
   thermodynamic models of, 150–156, 324
   in TiO$_2$-SiO$_2$ system, 149–160
   in ZnO ceramics, 131–145, 344–346
Analytical transmission electron microscopy AEM: see Transmission electron microscopy
Atomic force microscopy, AFM, 13–15, 199, 227–228, 232–233, 441, 619, 744–746
   mesoscopic-scale interactions, 15
   stick-slip behavior, 16
   tip-ionic surface interactions, 14, 15
   tip-surface defect interactions, 15
Atomistic simulation: see Modelling, atomistic
Auger electron spectroscopy, 179, 193, 633–634, 706, 708

Basicity, importance in sealing glass design, 537–542
BaTiO$_3$, barium titanate
   abnormal grain growth in, 331–337
   coatings on TiO$_2$ (rutile), 447, 455–457
   cubic-to-hexagonal transformation, 331–337
   electrical properties, 655–658
   liquid-phase sintered, 331–337
   optical properties, 653–654, 657–658
   photovoltaics, 653–658
   processing, 331–337
   semiconductive films on Si, 653–658
   sintering atmosphere effects, 331–337
Bioactive ceramics, 543–550, 761–768
Bioactive glasses, for coatings, 543–550
Biocompatibility, 543, 545, 549–550
Bioeutectic materials, characterization, 762–767
Bioeutectic materials, processing, 761–763
Biomaterials
   bioactive ceramics, 543–550, 761–768
   for bone replacement, 761–768
   resorbable ceramics, 761–768
Biomimetic processing, 437–438
Biomineralization, 437–438
BN, boron nitride, coatings on SiC$_f$ (Nicalon SiC fibers), 705–712
Bonding, at grain boundaries in Si$_3$N$_4$, 31–33, 112–114, 779, 781–785
Bonding; type via analytical TEM, 3
Brazing
   of Al$_2$O$_3$, 359–360, 371, 380–382
   of Si$_3$N$_4$, 359, 361–364, 366–367, 373, 415–419
   of ZrO$_2$, 359–360, 365

Ca$_5$(PO$_4$)$_3$OH, hydroxyapatite
   as a bioceramic, 761, 765–767
   coatings, 543–550
   microwave heating of, 472, 475–478
   powder synthesis, 429, 431, 433–434
   synthesis on Langmuir monolayers, 438–440, 444
CaAl$_{12}$O$_{19}$, calcium hexaluminate
   coatings on sapphire and YAG fibers, 731–739
   in situ texturing of coatings, 731–736
Calcium phosphates, powder synthesis, 429–434
Capacitors, SrTiO$_3$-based, 645–652

Capillary pressure: *see* Colloidal phenomena
Carbon, coatings on SiC$_f$ (Nicalon SiC fibers), 713–720
Cathodoluminescence, of Al$_x$Ga$_{1-x}$N films, 634–635
Ceramic/ceramic interfaces
  Al$_2$O$_3$/CeO$_2$, 17
  Al$_2$O$_3$/MgO, 577–587
  Al$_2$O$_3$/ZnO, 577, 584
  BaO/MgO, 17
  Y-TZP/WC, yttria-stabilized tetragonal zirconia/tungsten carbide, 741–748
Ceramic/glass interfaces
  Al$_2$O$_3$/glass, 535–542
  bonding morphologies, 535–539
  reactions at, 535–542, 781–785
  Si$_3$N$_4$/oxynitride glass, 779–785
Ceramic/glass-ceramic interfaces, SiC$_f$/BMAS (Nicalon SiC fibers/barium-magnesium aluminosilicate), 705–712
Ceramic/metal interfaces
  Al$_2$O$_3$/Ag, 350–357
  Al$_2$O$_3$/Al, 75–77, 371
  Al$_2$O$_3$/Al-Si, 749–759
  Al$_2$O$_3$/Au, 350–356
  Al$_2$O$_3$/Au-Cu, 350–357
  Al$_2$O$_3$/Cu, 1, 6–10, 71–81, 350–357, 408–413, 823–830
  Al$_2$O$_3$/Cu-Ag-Ti, 380–382
  Al$_2$O$_3$/Cu-Ni, 408–413
  Al$_2$O$_3$/Cu-Ni-Cr, 408–414
  Al$_2$O$_3$/Ni, 78–80, 408–413
  Al$_2$O$_3$/Sn, 78
  Al$_2$O$_3$/Sn-Ti, 380, 382–383
  AlN/Fe, 605–608, 611
  AlN/W, 391–397, 399–405
  atomic structure, 8, 415–419
  fabrication of: *see* Joining, brazing, diffusion bonding, reactive metal penetration
  geometric structure, 6, 7
  heteroepitaxial relationships, 7, 415–419
  metal oxidation state, 8
  neutron reflection analysis of, 377–384
  oxygen segregation at, 349–354, 356–357
  reaction product formation at, 386–387, 391, 393, 396–397, 409–410, 415–419, 543–544, 546–549, 605, 608, 611, 823–830
  reactions during brazing, 360–362, 364–367, 371, 377, 379–382, 415–419
  Si$_3$N$_4$/Al, 373–374, 386, 388
  Si$_3$N$_4$/Cu-Ag-Ti, 415–419
  strength, 366, 372–373, 407–409, 411–414, 823–836, 830
Coarsening, in liquid-phase sintered materials, 277–283, 331–337, 339–340, 343–347, 399, 402–405
Coatings, *see also* Fiber coatings
  adherence of, 543–550
  BaTiO$_3$ on Si, 653–658
  BaTiO$_3$ onTiO$_2$ (rutile), 447, 455–457
  bioactive, 543–550
  BN/C on SiC$_f$ (Nicalon SiC fibers), 705–712

Coatings (*cont.*)
  C on SiC$_f$ (Nicalon SiC fibers), 713–720
  Ca$_5$(PO$_4$)$_3$OH (hydroxyapatite) on Ti,Ti6Al4V alloy, 543–550
  CaAl$_{12}$O$_{19}$ on Al$_2$O$_3$, 447–451, 731–739
  diamond, 221–226
  oxide on LAS/SiC$_f$ (lithium aluminosilicate-Nicalon silicon carbide fiber) composites, 513–524
  on particles, 559–565
  plasma sprayed, 19, 520
  reactive, 447–461
  texture development in, 731–736
  ZrTiO$_4$ on TiO$_2$, 447, 452–455
Coercivity, Fe/AlN multilayers, 606
Coincident site lattice theory CSL, 17, 163, 167, 192, 197, 804
Colloidal phenomena
  capillary pressure, 131–145
  colloidal stability, 495–500
  depletion force, 495–500
  double layer compression, 223
  electrostatic forces, 495, 496, 498, 500
  Hamaker constants, 132, 134, 136, 139, 150, 154
  high temperature, 503–511
  interparticle potentials, 495–500, 505–507
  polymer effects, 495–500
  solvation forces, 496–498, 500
  steric repulsion forces, 129, 150–151, 154, 495–496, 499, 500
  Van der Waals forces, 14, 27, 128–129, 132–135, 140, 145, 150, 153–154, 157, 495–496, 499, 500
  zeta potential, 222–225, 495–496, 498, 500
Colloidal processing, of functionally graded materials, 503–511
Combustion synthesis: *see* Self-propagating high-temperature synthesis and powder synthesis
Composites
  3Al$_2$O$_3$·2SiO$_2$ (mullite) fibers as reinforcement in, 170, 172–173, 175, 721–728
  α-β-Si$_3$N$_4$-β-SiC, strength/fracture, 468
  Al$_2$O$_3$-(Al-Si) alloy, 749–759
  Al$_2$O$_3$-(Al-Si) alloy, interfacial reactions, 749–751, 753, 756–757, 759
  Al$_2$O$_3$-(Al-Si) alloy, processing of, 749–759
  Al$_2$O$_3$-Al$_2$O$_{3(f)}$ (alumina-sapphire fiber), 83, 85, 88, 732–739
  Al$_2$O$_3$/MgO multilayers, 578–583, 586
  Al$_2$O$_3$-SiC$_f$ (alumina-Nicalon silicon carbide fiber), 95, 96, 101
  Al$_2$O$_3$-SiC$_p$ nanocomposites, 551–552, 554–557
  Al$_2$O$_3$-SiO$_2$ fibers, 169–175, 721–728
  Al$_2$O$_3$-SiO$_2$ microcomposite powders, 285, 287–300
  Al$_2$O$_3$/TiC/Ni graded composites, 833, 34
  Al$_2$O$_3$-Y$_3$Al$_5$O$_{12(f)}$ (alumina-YAG fiber), 734–736
  Al$_2$O$_3$-ZrO$_2$, 83, 88, 833
  AlN/Fe multilayer films, 605–608
  β-SiC-B$_4$C, strength/fracture, 464–466, 468

Composites (*cont.*)
  BMAS-SiC$_f$ (barium-magnesium alumino-silicate-Nicalon silicon carbide fiber), 706–712
    processing, 706–707
    strength/fracture, 706–707, 709–712
  CaO·SiO$_2$-Ca$_3$(PO$_4$)$_2$ (wollastonite-tricalcium phosphate) for bone replacement, 761–768
  ceramic matrix, CMCs, 36, 50, 53, 54, 56–60, 62, 95–98, 100, 101, 695–702, 731–739
  fiber-reinforced, 83, 95, 695–702, 705–712, 721–728, 731–739
  fiber/matrix interface, 83, 705–712, 731–739
  fiber/matrix roughness misfit effects, 695–702, 737
  functionally graded materials, 503–511, 833–842
  interphase materials/coatings, 83, 84, 170, 172, 175, 448, 451, 695–702, 705–712, 731–739
  LAS-SiC$_f$ (lithium aluminosilicate-Nicalon silicon carbide fiber), 513–515, 519–524
  MgO-SiC system, 589–596
  misfit strains in, 695–699
  nanocrystal CdS or ZnSe in glass, 199–200
  oxide-based, 83, 91, 722
  Pb(ZrTi)O$_3$-PbTiO$_3$ (PZT–lead titanate)
    chemical homogenization, 669, 672–676
    densification, 669, 671–673, 675–676
    electric properties, 669–671, 675–676
    powders, 669–676
    processing, 669–676
  powder synthesis, 559–565
  via reactive metal penetration, 749–759
    interfacial reactions, 741, 743–748
    mechanical properties, 741–743, 748
    processing, 741–747
    Y-TZP-Al$_2$O$_3$ (yttria-stabilized zirconia-alumina) system, 613–618
    Y-TZP–WC (yttria-stabilized tetragonal zirconia–tungsten carbide), 741–748
  Si$_3$N$_4$-oxynitride glass, 779–785
  Si$_3$N$_4$/Si$_3$N$_4$ multilayer laminates, 795–801
    fracture behavior, 796–801
    processing, 795–796, 800–801
    residual stresses in, 795–801
  Si$_3$N$_4$/SiC, 95, 97–100, 263, 270, 272, 274, 567–569, 571–576
  Si$_3$N$_4$/ZrO$_2$, 833
  SiC-C, 487–493
  SiC-SiC$_f$ (Nicalon)
    mechanical properties, 713–717, 720
    oxidation, 713–720
    processing of, 714
    thermal stability, 713–720
  SiC/AlN, 833
  SiC/SiC and SiC/III-V nitride heterostructures, 629–635
  in situ via templated grain growth, 303–310
  ZrSiO$_4$/glass-ceramic multilayer, 447, 458–461

Contact angle
  advancing/receding, 76
  Al on 3Al$_2$O$_3$·2SiO$_2$ (mullite), 749–750, 752–753, 757–758
  effect of P$_{O_2}$, 65–81, 349–356, 410–411, 505, 508–509
  effect of ridge formation, 66, 74–76, 81, 509
  effects of adsorption, 65–81, 349–356, 506–507
  glass on solid, 503–505, 508–509, 511, 539–540, 543, 547–549
  liquid in Bi$_2$O$_3$-doped ZnO on solid, 136, 139
  liquid metals on Al$_2$O$_3$, 66, 350–351, 355–356, 407–411, 413–414
  reactive metal brazes, 360–361
  Young-Dupré equation, 65–66, 349–351, 356, 410–411, 509
CoO, cobalt oxide, nonstoichiometry of, 421–427
Coordination number; *via* analytical TEM, 3
Cordierite, Mg$_2$Al$_4$Si$_5$O$_{18}$, coatings for SiC$_f$/LAS composites, 514, 518–524
Crack healing
  in Al$_2$O$_3$, 229
  in Al$_2$O$_3$-SiC$_p$ nanocomposites, 557
Creep
  diffusion controlled, 815, 819–821
  interfacial reaction controlled, 815, 819–820
  of Mg-doped and Mg/Y co-doped Al$_2$O$_3$, 815, 819–820
  of Y/La co-doped Al$_2$O$_3$, 815–816, 819–820
  of Zr-doped and Nd/Zr co-doped Al$_2$O$_3$, 817–821
Cu/Al$_2$O$_3$: *see* Ceramic/metal interfaces

Damage tolerance: *see* Fracture
Debonding: *see* Fracture
Defect chemistry, bulk, 214–218, 421–424, 426–427, 584–587, 589, 645–646, 651
Defect chemistry, surface/interfacial region, 421–427
Defects
  in Al$_2$O$_3$-SiO$_2$ (aluminosilicate) fibers, 721–722, 727
  in green bodies, 239–245
  intragranular planar in electroceramics, 637–644
  via lithographic processing, 229–232
Diamond
  applications, 221
  electrostatic deposition, 222–228
  processing/synthesis, 221–226
  thin films, 221–226
Differential densification, 247–252, 254, 800
Diffusion bonding, 162, 164, 167, 359–360, 364–367, 373, 824
Diffusion couples
  Al$_2$O$_3$/MgO, 578–582
  Al$_2$O$_3$/SiO$_2$, 257
Diffusion, grain boundary, 145, 161, 613, 619–620, 819–821
Diffusion, surface, 229–232
Dihedral angle
  effect on sintering, 248–249, 503–511
  effects of adsorption, 506–507, 510–511
  grain boundary-liquid, 131, 136–139, 145, 503–507, 509–511

846

Dislocations
   misfit, 7, 16, 18
   plasticity, 35
Ductility: *see* Fracture

EELS: *see* Electron energy loss spectrometry
Electroceramics, *see also* Specific ceramic of interest
   dopant-induced defects in, 637–641
   intragranular planar defects in, 637–644
   nonstoichiometry in, 637–638, 641–644
Electromagnetic fields, effects on particle growth, 429–436
Electron beam processing, 623–627
Electron beam, for nanoparticle fabrication, 623–626
Electron energy loss spectrometry, EELS, of
   Al_2O_3/Cu interfaces, 8, 71
   AlN/Si interfaces, 192–193, 196–197
   grain boundaries
      in Al_2O_3 prepared by PAS, 199, 201–203
      in SiC, 103, 179, 184, 187
      in Si_3N_4, 24–25, 95–98, 111, 113, 128
   interfaces in Si_3N_4-SiC_f composites, 95–98, 100–101
Electron microprobe analysis, EPMA, of
   Al_2O_3/Cu(-Ni-Cr) interfaces, 409, 412–413
   $BaTiO_3$/Si interfaces, 654–655
   W-Mn/AlN interfaces in cofired multilayer substrates, 399, 401, 403–404
   ZnO ceramics, 692
Electronic structure *via* ELNES analysis, 3
ELNES: *see* Energy-loss near-edge structure
Energy dispersive x-ray spectrometry, EDS, (of)
   $3Al_2O_3 \cdot 2SiO_2$ (mullite) fibers, 170
   Al_2O_3-$MgAl_2O_4$ precipitates, 313, 321
   Al_2O_3/TiC/Ni graded composites, 835
   Bi_2O_3-doped ZnO, 135, 137–145
   $Ca_5(PO_4)_3OH$/Ti;Ti alloy interfaces, 545
   $CaAl_{12}O_{19}$/sapphire or YAG fiber interfaces, 733, 735
   $CaO \cdot SiO_2$-$Ca_3(PO_4)_2$ (wollastonite-tricalcium phosphate), 762–763, 765–766
   detectors, 3
   Fe/AlN thin films, 608–609
   grain boundaries in Si_3N_4, 109, 111, 113
   interfaces
      in BMAS/Nicalon SiC composites, 708
      in LAS-SiC_f composites, 520–521
      in Si_3N_4-SiC_f composites, 95, 99
   MgO-SiC composites, 590, 593
   olivine, 104
   PZT-$PbTiO_3$ composites, 671
   Si_3N_4 ceramics, 109, 111, 113, 264
   SiC ceramics, 179, 187
   SiC powder prepared by combustion synthesis, 467
   $SrTiO_3$ ceramics, 648, 651
   TiO_2-SiO_2 system, 157–158
   W-Mn/AlN interfaces in cofired multilayer substrates, 392–395
   Y-TZP-WC composites, 745
   Y-TZP/Al_2O_3 powders, 616
   ZnSe nanoparticles, 206

Energy-loss near-edge structure, ELNES, 3, 8, 9, 24–26
EPMA: *see* Electron microprobe analysis
Eutectic structures, as biomaterials, 761–768
EXAFS, 608–609; *see also* Extended x-ray absorption fine structure
EXELFS: *see* Extended energy-loss fine structure
Extended energy-loss fine structure (EXELFS), 3
Extended x-ray absorption fine structure, 608–609

Fatigue, 36, 55, 59, 613, 618–619, 706, 709–712
   BMAS-SiC_f (barium-magnesium aluminosilicate-Nicalon silicon carbide fiber) composites, 706, 709–712
   tensile, 613, 618–619, 706, 709–712
   Y-TZP, 613, 618–619
Fiber coatings
   BN/C on SiC_f (Nicalon), 705–712
   C on SiC_f (Nicalon), 713–720
   $CaAl_{12}O_{19}$ (hibonite) on sapphire and YAG, 731–739
   coating compliance as design parameter, 698–700, 702
   design parameters for composites, 695–702, 705–712
   effect on debond length, 695–697
   $LaPO_4$, monazite, 83–91
   layered structures, 84, 91, 705–712
   oxidation of, 695, 713–720
   oxide, 695, 698, 701–702, 731–739
   porous, 83, 85, 91
Fibers, *see also* Composites and fiber coatings
   $3Al_2O_3 \cdot 2SiO_2$ (mullite) fibers, 169–176
   effect of fiber/matrix surface roughness misfit, 695–702, 737
   mechanical properties of Nextel 720 fibers, 721–728
   tensile testing, 170–171, 174–175, 721–728
Fracture
   Al_2O_3-SiC_p nanocomposites, 551–552, 554–557
   Al_2O_3/Cu interfaces, 823–830
   crack bridging, 43–45, 59, 177, 187, 270, 272, 779, 784–785, 833
   crack deflection, 177–178, 270, 272, 551, 731–732, 736–39779–39780, 782–783, 785, 796–798, 833
   damage tolerance, 35, 36, 83, 91, 696, 713
   debonding, 44–46, 48, 51, 52, 54, 84–86, 90, 91, 270, 695–697, 699, 713, 731–732, 736–739, 779–785, 825–828
   ductile phase toughening, 44–46, 50
   ductility, 35, 36, 46, 61, 62
   effect of fiber/matrix roughness misfit on, 695–702
   energy release rate, 41, 59
   frictional effects/toughening, 46–48, 52, 55, 61, 62, 187, 695–702, 705–707, 709, 713
   inelastic strain/deformation, 36, 38, 43, 51, 52, 54, 55, 62
   initiation toughness, 40, 826

Fracture (*cont.*)
  interface fracture energy/resistance, 44, 48, 65, 87, 186, 699–702, 705–707, 709, 731–732, 736–739, 779–785, 826–828
  life prediction, 35, 36, 55, 59, 60–62
  notch insensitivity, 36, 38, 39, 40
  notch sensitivity, 35, 48
  plastic blunting/plasticity, 35, 828, 830
  process zone toughening, 43–45
  residual stress effects, 795–801, 824–830, 833–835, 838–842
  SiC, 179, 184, 186–188
  Si_3N_4, 32, 33, 112, 263, 265, 272–274, 779–785
  Si_3N_4/Si_3N_4 multilayer composites, 796–801
  stress concentrations, 36, 38–40
  stress corrosion, 36, 60, 61, 826
  stress redistribution mechanisms, 38, 50, 54, 62, 274
  survival probability, 36, 37, 60, 61
  tearing index, 40–43, 46, 47, 49, 50, 61
  transformation toughening, 44, 88
  weak link scaling, 36, 40, 61
  Weibull distributions/parameter, 37, 42, 48, 55, 184, 186, 372, 554
Fracture origins, 241, 272, 274, 721, 722, 727
Fracture resistance, R-curve, 38, 41, 184, 187, 551, 556–557, 779, 781, 784–785, 825–826
Fracture surfaces, characterization, 179, 187, 367, 392–394, 399, 401, 403, 554, 556–557, 711, 737–738; *see also* Topics under fracture
Fracture toughness, 35, 36, 38, 40, 42, 51, 61, 62, 85, 87, 177–179, 184–188, 263, 265–266, 270, 551–552, 554–557, 618, 743, 750–752, 779, 781, 784–785, 798, 800–801, 833–834, 838–839; *see also* Fracture
Functionally graded materials
  Al_2O_3 *via* dopant valence gradients, 234–235
  via high temperature colloidal processing, 503–511

GaN, thin film deposition, 629–630, 633–635
Gas-solid reactions, 421–423, 425–427, 589–596
Glass
  bioactive for coatings, 543–550
  compositional design of sealing glasses, 537–542
  interstitial structure, 527–533
  metallic, 528–530
  modification of thermal expansion coefficient, 544–547, 550
  ordering in, 527
  penetration along grain boundaries, 535–542
  prediction of thermal expansion coefficient in glass, 541–542
  structure of, 527–533
Grain boundaries
  in Al_2O_3, 1, 5, 199, 201–203, 323–325, 328–329, 803, 805–812, 815–816, 819–821

Grain boundaries (*cont.*)
  amorphous films
    in Al_2O_3 ceramics, 202–203, 323–325, 328–329
    in Bi_2O_3-doped ZnO, 131–145
    in olivine, 104–105
    in SiC ceramics, 102–103, 177, 183–184, 186–188, 492–493
    at SiC_w/Si_3N_4 interfaces, 97
    in Si_3N_4 ceramics, 1, 23–27, 31–33, 95–97, 100–102, 104, 107–111, 113–114, 116, 118, 120–121, 123–130, 779–780, 788, 791–792
    in $SrTiO_3$ ceramics, 645–646, 648–651
    thermodynamic models of, 150–156, 324
    in TiO_2-SiO_2 system, 149–160
    in ZnO ceramics, 131–145, 344–346
  Bi_2O_3-doped ZnO, 131–145
  ceramic/metal: *see* Ceramic/metal interfaces
  misorientation characterized by orientation imaging microscopy, 803–812
  orientation relationships in Y-TZP–WC (yttria-stabilized tetragonal zirconia–tungsten carbide) composite system, 746–748
  penetration by liquid phase, 535–542
  $\Sigma 5$ in $Y_3Al_5O_{12}$ (yttrium aluminum garnet), 161–167
  in SiC, 102–104, 177–188, 487, 492–493
  in SiC-C system, 487, 490–493
  in Si_3N_4, 1, 23–27, 31–33, 97–100, 107, 109–110, 113–114, 116–121, 123–130, 149–150, 779–785
  special misorientation boundaries, 804, 809, 811–812, 820
  in $SrTiO_3$, 645–652
  structure calculations, 5, 17, 161, 163–167
  in TiO_2-SiO_2 system, 149–160
Grain boundary, diffusion, 145, 165; *see also* Diffusion, grain boundary
Grain boundary, penetration by liquid phase, 503–511
Grain boundary segregation
  additives in ZnO, 137–139, 141–144, 340, 344
  anion impurities in Si_3N_4, 107–112
  impurities in Al_2O_3, 5, 6, 733–736, 815–816, 819, 821
  impurities in $SrTiO_3$, 645–646, 651–652
  sintering aids in AlN, 399, 403
  sintering aids in Si_3N_4, 98–100
Grain boundary structure, modelling of, 527–530, 532–533
Grain growth
  in Al_2O_3-SiO_2 (aluminosilicate) fibers, 721–728
  α grains on β SiC seeds, 177–188
  anisotropy of, 181, 234–237, 277–283, 303–310, 331–337, 732–736, 798, 803–804
  during densification, 103
  effect on texture, 803–806, 811–812
  exaggerated/abnormal, 89, 201, 240, 242–245, 306–307, 323–329, 331–337, 733–736
  in fiber coatings, 89, 732, 735–736
  during intermediate stage sintering, 247, 249–250, 252, 254

Grain growth (*cont.*)
  in liquid-phase sintered materials, 277–283, 331–337, 339–341, 343–347, 399, 402–405
  precipitate pinning, 345–347, 744, 748
  role of twinning, 337, 345
  in Si_3N_4 ceramics, 268–269, 272, 282
  in SiC, 180–182, 184–185
  seeded, 303–310, 733–736
  in 3Y-TZP, 479–481, 482
  in Y-TZP–WC (yttria-stabilized tetragonal zirconia–tungsten carbide) composites, 744–745, 748
  in ZnO ceramics, 339–347
Grain growth/grain boundary mobility, in Al_2O_3, 199, 201, 234–237, 304–306, 308–310
Green body, packing defects, 239–245

Heteroepitaxy
  interfacial energy, 16–18, 735–736
  misfit, 17, 18
  stress relaxation, 18
  thin film texture, 597, 599–600, 603–604, 731–736
  thin films on SiC, 629–635
Heterostructures
  2H-AlN/Al_xGa_{1-x}N, aluminum nitride on aluminum-gallium nitride, 629–630, 633–635
  6H-SiC/2H-AlN/3C-SiC, silicon carbide (6H) on aluminum nitride on silicon carbide (3C), 629–630, 633, 635
  epitaxial growth on SiC, 629–635
High resolution transmission electron microscopy, HRTEM: *see* Transmission electron microscopy
HRTEM: *see* Transmission electron microscopy
Hydroxyapatite, $Ca_5(PO_4)_3OH$
  as a bioceramic, 761, 765–767
  coatings, 543–550
  microwave heating of, 472, 475–478
  powder synthesis, 429, 431, 433–434
  synthesis on Langmuir monolayers, 438–440, 444

Image simulation: *see* Transmission electron microscopy
Immersion liquid technique, 239–244
Impedance spectroscopy, of $SrTiO_3$, 649
Interface strength, effect on fracture path, 32, 33, 65, 84, 95, 701–702, 713, 731–732, 737–739, 779–785, 823–824, 826–828, 830
Interfaces: *see* Ceramic/metal, ceramic/glass, and ceramic/ceramic interfaces
Interfacial energy, Al_2O_3 solid-vapor, 67
Interfacial energy, heteroepitaxial systems, 18, 735–736
Intergranular film thickness: *see* Grain boundaries, amorphous films
Intermediate stage sintering, grain growth, 247–250, 252, 254
Intermediate stage sintering, models, 247–254
Internal friction measurements, 107–109, 112–113
Interstitial structure models, 527–533

Joining, 359–367, 369–376, 407–414, 415–419, 458–461; *see also* Brazing, diffusion bonding, and specific materials combinations of interest
  glass-to-ceramic bonding, 535–542
  by partial transient liquid phase bonding, 407–414
  by reactive coatings, 458–461
  by squeeze casting process, 369–376
  by surface activated bonding, 385–389

Kirkendall effect, during $3Al_2O_3 \cdot 2SiO_2$ (mullite) formation, 294, 296, 299
Kirkendall plane, ceramic/metal joining, 367

LaMer diagram, 559–560
Langmuir monolayers, for templated growth, 437–440, 444
Langmuir-Blodgett films, for templated growth, 437–438, 441–444
$LaPO_4$, monazite, as a fiber coating, 83–91
Laser driven powder synthesis, 567–576
Laser, effect on $BaTiO_3$ thin films on Si, 653–655, 658
Lead oxide, PbO, effect on interfacial reactions in glass/ceramic systems, 535–542
Lead zirconium titanate (PZT)-lead titanate, $Pb(Ti,Zr)O_3$-$PbTiO_3$
  chemical homogenization, 669, 672–676
  composite, 669–676
  densification, 669–673, 675–676
  electric properties, 669–671, 675–676
  processing, 669–676
Life prediction: *see* Fracture
Liquid phase sintering of
  AlN, 391, 394–397, 399–400
  $BaTiO_3$, 331–337
  functionally graded materials, FGMs, 503–511
  model system, 131, 149
  SiC, 184–185
  Si_3N_4, 266–269
  tungsten conductor in aluminum nitride, W-AlN, 399–400, 402–405
Lithium aluminosilicate matrix composites: *see* Composites, SiC-LAS
Lithography, for producing defect structures, 229–232
Low energy electron diffraction, LEED, 192–194
Luminescence, in semiconductor/glass nanocomposites, 199–200

Magnetic behavior
  Fe-Al-N system, 605–611
  $La_{1-x}Sr_xMnO_3$ thin films, 600–604
Magnetic materials
  Fe-Al-N films, 605–611
  $La_{1-x}Sr_xMnO_3$ thin films, 597–604
Magnetoresistance, of $La_{1-x}Sr_xMnO_3$ thin films, 597, 602–604
Metallization, of AlN, 391–397

849

MgAl$_2$O$_4$, magnesium aluminate spinel
   formation in Al$_2$O$_3$/MgO system, 577–587
   lattice parameter in Al$_2$O$_3$ matrix, 311–315, 317–318
   precipitates in Al$_2$O$_3$, 311–320
Mg$_2$Al$_4$Si$_5$O$_{18}$, cordierite, coatings for SiC$_f$/LAS composites, 514, 518–524
Microscopy, acoustic, 552–554, 556–557
Microwave effect in sintering, 471, 481, 483, 484
Microwave heating
   of Ca$_5$(PO$_4$)$_3$OH (hydroxyapatite), 472, 475–478
   of SiC, 472, 476–477
Microwave sintering, 471–484, 567–568, 573–575
   susceptor design, 477–478, 482, 484
   temperature measurement, 471–477, 483, 568
Mixed oxides, internal reduction, 209–219
Modelling
   AFM tip-ionic surface interactions, 14, 15
   amorphous thin films, 27, 31–33
   atomistic, 13, 17, 637–644
   bulk properties, 13, 19
   coarsening of solid-liquid systems, 277–283
   defect structure in electroceramics, 637–644
   density functional calculations, 17
   effect of P_{O_2} on wetting, 349–356
   glass penetration into Al$_2$O$_3$, 539–541
   glass structure, 527–533
   grain boundary strength in Si$_3$N$_4$, 779, 781, 783–785
   grain boundary structure, 17, 25–27, 31–33, 527–530, 532–533
   grain boundary structure in heteroepitaxial systems, 17
   grain boundary structure in Si$_3$N$_4$, 25–27, 31–33
   grain growth in solid-liquid systems, 277–283, 323–326, 328–329
   heteroepitaxial systems, 16–18
   intermediate stage sintering, 247–254
   mechanical behavior of SiC-SiC$_f$ composites, 714–716, 720
   mesoscopic, 13, 15
   molecular orbital method, 24–26
   oxidation of fiber coatings, 718–720
   pair-potential models, 14, 17
   plasma sprayed coating properties, 19
   rare earth ion solubility in Si$_3$N$_4$, 28–30
   spinel formation in Al$_2$O$_3$/MgO system, 584–587
   stress relaxation in heteroepitaxial systems, 16
Molecular beam epitaxy, 629–633, 635
Monazite, LaPO$_4$, as a fiber coating, 83–91
Mullite ceramics, via Al$_2$O$_3$-SiO$_2$ composites, 285–300
Mullite fiber, characterization of, 170–173, 175, 721–728
Mullite fiber, mechanical properties, 170–171, 174–175, 721–728
Mullite formation, Kirkendall effect, 294, 296, 299
Mullite, 3Al$_2$O$_3$·2SiO$_2$
   anisotropic grain growth, 304–307
   coatings for SiC$_f$/LAS composites, 514, 516–524
   crystallization of glass fibers, 170, 174, 175

Mullite, 3Al$_2$O$_3$·2SiO$_2$ (*cont.*)
   formation from mullite gels, 773–774
   glass fiber fabrication, 169–170, 175
   growth of, 285–290, 294, 296, 299
   incubation period for formation, 286–290, 293–294, 299
   phase stability/formation, 255–261, 285–290, 292–300, 516–518
   powder characteristics, 774
   powder synthesis, 771–774
   processing, 287–288, 772–775
   properties, 771, 774–777
   reactive metal penetration of, 749–759
   Ti-doped, 304–307

Nanoparticles
   Al$_2$O$_3$, 623–626
   Al, 623–627
   C fullerenes, 623
Neutron reflection analysis, of ceramic/metal interfaces, 377–384
NiO, nickel oxide, nonstoichiometry of, 421–427
Nonstoichiometry
   effect of P_{O_2}, 422–424, 426
   in electroceramics, 637–638, 641–644
   oxides, 421–427, 637–638, 641–644

Optical absorption, in semiconductor/glass nanocomposites, 199–200
Orientation imaging microscopy, OIM, 803–812
Ostwald ripening: see Coarsening
Oxidation
   of BMAS-SiC$_f$ (barium-magnesium aluminosilicate-Nicalon silicon carbide fiber) composites, 705–706, 710, 712
   of fiber coatings, 695, 705–706, 710, 712, 718–720, 731
   of LAS-SiC$_f$ (lithium aluminosilicate-Nicalon silicon carbide fiber) composites, 520–522
   of MgO-SiC composites, 589–596
   of SiC-SiC$_f$ (Nicalon fiber) composites, 713–720
Oxides, nonstoichiometry of, 421–427
Oxynitride glasses, thermomechanical properties, 783

Parallel electron energy loss spectrometry, PEELS, 8, 109, 192–195, 197–198, 204
Partial transient liquid phase bonding, PTLP, 407–414
PbO, effect on interfacial reactions in glass/ceramic systems, 535–542
Pb(Ti,Zr)O$_3$-PbTiO$_3$, PZT-PbTiO$_3$
   chemical homogenization, 669, 672–676
   composite, 669–676
   densification, 669–673, 675–676
   electric properties, 669–671, 675–676
   processing, 669–676
PEELS: see Parallel electron energy loss spectrometry

Phase diagram(s)
 Al$_2$O$_3$-CaO system, 396, 448, 450
 Al$_2$O$_3$-CaO-MgO system, 448, 450
 Al$_2$O$_3$-CaO-SiO$_2$-ZrO$_2$ system, 458, 460–461
 Al$_2$O$_3$-MgO system, 312, 314, 316, 586
 Al$_2$O$_3$-MgO-Y$_2$O$_3$ system, 312, 314, 316
 Al$_2$O$_3$-SiO$_2$ system, 255–261, 287, 296, 725–727
 Al$_2$O$_3$-WO$_3$-CaO system, 397
 BaO-TiO$_2$ system, 331–337
 BaTiO$_3$-TiO$_2$ system, 456–457
 CaO·SiO$_2$-Ca$_3$(PO$_4$)$_2$ (wollastonite-tricalcium phosphate), 761–762, 768
 SiO$_2$-TiO$_2$-ZrO$_2$ system, 452, 454–455
 use in joining, 362–363, 381, 412
 use in processing of FGMs, 503–504, 508, 510–511
 W-Mn system, 404–405
Phase transformation(s)
 in CaAl$_{12}$O$_{19}$-based (hibonite-based) fiber coatings, 731–736
 liquid-aided cubic to hexagonal transformation in BaTiO$_3$, 331–337
 seeding of β to α transformation in SiC, 177–188
Plasma activated sintering, PAS
 of Al$_2$O$_3$, 201–203
 processing variables, 200–201
 of SiC-C composites, 487–489
Plastic blunting: see Fracture
Point defects, 214–218, 421–427, 584–587, 589
Polyhedra, as structural building blocks, 527–533
Pore-boundary separation
 in Al$_2$O$_3$, 229
 in 3Y-TZP (yttria-stabilized zirconia), 480–481
Potts model, application to grain growth, 277–283
Powder synthesis
 α-β-Si$_3$N$_4$-β-SiC by SHS, 463, 466–470
 β-SiC-B$_4$C by SHS, 463, 466–469
 by combustion synthesis, 463–464, 466–470
 composite/coated powders, 559–565
 laser-driven, 567–576
 SiC by CO$_2$ and excimer laser, 567–571, 575–576
 Si$_3$N$_4$/SiC, 567–569, 571–576
 from solutions, 429–435, 437–440
Precipitation
 internal stresses, 209–210, 214–219
 in oxides, 209–219
 stress relaxation mechanisms, 209–210, 214–215, 218–219
 volume changes of metals, 209–219

Quantum dots, in semiconductor/glass composites, 204–206

Rayleigh instabilities, in Al$_2$O$_3$, 229–232
Reactive metal penetration, kinetics, 749, 752, 754–759
Reactive metal penetration, thermodynamics, 749–750, 756–758
Reduction reactions
 in Cr$_2$O$_3$-doped Al$_2$O$_3$, 209–213, 216–219
 in NiO-doped MgO, 209–211, 214–215

Resorbable ceramics, for bone replacement, 761–768
RHEED: see Transmission electron microscopy
Rheological measurements, Al$_2$O$_3$-PMMA-toluene system, 498–499

Saturation magnetization
 of Fe-Al-N granular films, 608–610
 of Fe/AlN multilayers, 606–608
Scanning transmission electron microscopy, STEM: see Transmission electron microscopy
Scanning tunneling microscopy, STM, 14, 199
Secondary ion mass spectroscopy, SIMS, 590, 592, 596
Seed particle(s)
 orienting of, 304–305
 seed/matrix size ratio effects, 306, 308, 310
 volume fraction effects, 305–307
Seeding of phase transformations,: see Phase transformations and grain growth
Seeding
 abnormal grain growth, 733–736
 3Al$_2$O$_3$·2SiO$_2$ (mullite) formation, 296–297, 300
 alumina precipitation from aluminosilicate liquids, 255, 258–260
 anisotropic grain growth, 304–310
 diamond thin film growth, 222–226
 textured coatings, 731–736
Segregation, in space-charge layer, 421–423, 426–427
Self-propagating high-temperature synthesis, SHS, 463–464, 466–470, 833–834
Semiconductor/glass composites, 199–200, 204–206
 characterization, 204–206
 fabrication, 204
Sessile drop: see Contact angle
SHS: see Self-propagating high-temperature synthesis
SiC, silicon carbide
 α-SiC, 95, 96, 102–104, 177–188
 β to α transformation, 177–188, 492
 devitrification of grain boundary phase, 183
 grain boundaries, 102–104, 177–188, 487, 492–493
 grain boundary chemistry, 102–103, 183–184, 186–187
 microwave heating of, 472, 476–477
 oxidation, 520–521
 polytype control during thin film growth, 629–635
 powder synthesis, 567–571, 575
 processing, 179–180
 sintering aids, 96, 102–104, 177, 179, 183–185, 188, 468–469, 568, 573, 575
 in situ toughened, 177–188
 as substrate for thin film growth, 629–635
SiC-AlN composites, 833
SiC$_f$-BMAS (Nicalon silicon carbide fiber-reinforced barium-magnesium aluminosilicate) composites
 fatigue, 706, 709–712
 oxidation, 705–706, 710, 712
 processing, 706–707
 strength/fracture, 706–707, 709–712

851

SiC-C composites
    microstructures, 488–493
    plasma sintering, 487–489
    processing, 487–489
    sintering aids, 487–488
SiC$_f$-LAS (silicon carbide fiber-reinforced lithium aluminosilicate) composites
    abrasion testing of, 522–524
    oxidation of, 520–522
    oxide coatings for, 513–524
SiC-MgO composites, oxidation of, 589–596
SiC-SiC$_f$ composites
    mechanical properties, 713–717, 720
    oxidation, 713–720
    processing, 714
    thermal stability, 713–720
Si$_3$N$_4$, silicon nitride
    α to β transformation, 787–790, 793
    brazing of, 359, 361–367, 373, 415–419
    compressive deformation behavior, 787–793
    devitrification of amorphous phase, 116–118, 270
    diffusion bonding of, 364–367
    ELNES interpretation, 24–26
    fracture, 32, 33, 263, 265, 272–274, 779–785
    fracture toughness, 265–266, 270, 779, 784–785
    grain boundaries, 1, 23–27, 31–33, 97–100, 107, 109, 113–114, 116–121, 123–130, 149–150, 779–785, 787–788, 791–793
    grain boundary chemistry, 95, 97–100, 104, 107–112, 116–121, 779–785, 787–788, 792–793
    grain boundary strength, 32, 33, 779–785
    grain morphology, 787–793
    in situ toughened, 263, 267, 270–272, 274
    modelling of rare earth ion solubility, 28–30
    NC132, 116, 119–121, 123–130
    processing, 263–264, 272, 780–785, 787–788
    sintering aids, 28–33, 96, 98, 99, 104, 107–110, 113–114, 116–121, 263–264, 268, 270, 272, 274, 468–469, 568, 573, 575, 779–785, 787–789, 792–793, 800
    superplastic deformation, 787–788
    texture development in, 787, 789–793, 798–799
    thermal expansion coefficient, 265–266, 270, 779–780, 783, 795–796, 798
    Young's modulus, 265–266, 270, 779–780, 783, 795, 800
Si-C-N powder synthesis, 567–569, 571–576
Si$_3$N$_4$-SiC composites
    fracture, 573–574
    hot pressing, 567–568, 573–575
    microwave sintering, 567–568, 573–575
Si$_3$N$_4$/Si$_3$N$_4$ multilayer composites
    fracture behavior, 796–801
    processing, 795–796, 800–801
    residual stresses in, 795–801
Si$_3$N$_4$/ZrO$_2$ composites, 833
Sintering
    constraining forces, 247–253
    differential densification, 247–252, 254, 800

Sintering (*cont.*)
    dihedral angle effects, 248–249
    glass/metal and glass/ceramic systems, 503–511
    intermediate stage, 247–254
    interparticle rupture, 247–250, 252, 254
    microwave, 471–484, 567–568, 573–575
    models: *see* Modelling
    Pb(Ti,Zr)O$_3$-PbTiO$_3$ composite powders, 669–676
    pore-boundary separation, 229
    SrTiO$_3$, 645–646, 651
    temperature gradients, 472–473, 475–477, 482–484
    Y-TZP–WC composites (yttria-stabilized tetragonal zirconia–tungsten carbide), 741–748
SnO$_2$, tin oxide
    oriented growth of thin films, 680–686
    thin films growth model, 681–686
    thin films *via* spray pyrolysis, 679–686
Sol-gel methods
    for fabrication of mullite powders, 772–774
    for producing oxide coatings/films, 513–520, 524, 597–603, 732–735
Solution coating of powders, 559–565
Space-charge/Helmholtz layer, effect on powder growth, 429–436
Space-charge layer
    effect on defect concentrations, 421–423, 426–427
    in strontium titanate ceramics, 651–652
Spinel formation
    in Al$_2$O$_3$/MgO system, 311–320, 577–587
    in Al$_2$O$_3$/ZnO system, 577, 584
    modelling, 584–587
    morphology, 577–587
Spray pyrolysis, for growth of tin oxide thin films, 679–686
Strain hardening, 38
Strain softening, 38
Strength
    bend testing, 179, 184, 408, 411–414, 552, 554–556, 618, 714–717, 743, 748, 796, 801, 834, 838
    ultimate tensile (UTS), 38, 41, 42, 54, 55, 59, 62, 709–710, 722–724, 728
Stress corrosion: *see* Fracture
Stress redistribution: *see* Fracture
SrTiO$_3$, strontium titanate
    bicrystals, 650–651
    as boundary layer capacitors, 645–652
    dislocation structure, 646–648, 651
    impedance spectroscopy, 649
    nonstoichiometry in, 641–644
    planar defects in, 641–644
    processing of, 645–646, 651
Superplastic deformation, Si$_3$N$_4$, 787–788
Superplastic deformation, Y-TZP, 613, 618–620
Surface activated bonding, 385–389
Surface diffusion, Al$_2$O$_3$, 229–232
Surface energies, in Al$_2$O$_3$, 229–230, 232–234
Surface/interfacial energy, anisotropy of, 229–230, 232–234, 277–283, 329, 735–736

TEM: *see* Transmission electron microscopy
Templated growth, 437–444; *see also* Seeding and seed crystals
Texture, determination by orientation imaging microscopy, 803–812
Thermal expansion coefficient, prediction of in glass systems, 541–542
Thin films
    BaTiO$_3$ on Si, 653–658
    diamond *via* CVD processing, 221–228
    effect of source compound on film characteristics, 679–686
    Fe-Al-N granular films, 605–611
    Fe/AlN multilayers, 605–608, 611
    growth by gas source MBE, 629–633, 635
    growth by metallorganic vapor phase epitaxy, 629, 633–635
    growth by spray pyrolysis, 679–686
    La$_{1-x}$Sr$_x$MnO$_3$, 597–604
    oriented growth on Langmuir-Blodgett films, 437–438, 441–444
    by polymeric sol-gel methods, 597–603, 732–735
    by pulsed laser deposition, 597–604
    via rf sputter deposition, 605–611
    SnO$_2$ on Corning 7059, 679–686
    stress during growth, 18
Tin oxide, SnO$_2$
    oriented growth of thin films, 680–686
    thin films growth model, 681–686
    thin films *via* spray pyrolysis, 679–686
TiO$_2$-SiO$_2$, intergranular films, 149–160
Transient liquid phase bonding, 407–414
Transmission electron microscope, electron beam processing in, 623–627
Transmission electron microscopy TEM
    analytical, AEM, *see also* scanning, STEM, and other specific techniques under this heading
    Al$_2$O$_3$-(Al-Si) alloy composites, 756
    Al$_2$O$_3$-MgAl$_2$O$_4$ precipitates, 311–314, 317, 319–321
    Al$_2$O$_3$-SiC$_f$ (alumina-Nicalon silicon carbide fiber), 101, 104–105
    Al$_2$O$_3$-SiO$_2$ (aluminosilicate) fibers, 728
    Al$_2$O$_3$/Cu interfaces, 8, 9
    CdS, ZnSe nanocrystals in glass matrix, 202, 204–205
    Cr$_2$O$_3$-doped Al$_2$O$_3$, 210
    general features, 2
    of grain boundaries in Al$_2$O$_3$, 5
    grain boundaries in Si$_3$N$_4$ ceramics, 95, 107, 111, 115, 117, 124
    interfaces in Si$_3$N$_4$-SiC$_f$ composites, 95–101, 104–105
    NiO-doped MgO, 210
    SiC ceramics, 102–105
    W-Mn/AlN interfaces in cofired multilayer substrates, 392–395
    conventional, CTEM
        3Al$_2$O$_3$·2SiO$_2$ (mullite) ceramics, 291, 293, 776
        3Al$_2$O$_3$·2SiO$_2$ (mullite) fibers, 170, 172–174
        Al$_2$O$_3$-(Al-Si) alloy composites, 756

Transmission electron microscopy TEM (*cont.*)
    conventional, CTEM (*cont.*)
        Al$_2$O$_3$-SiO$_2$ (aluminosilicate) fibers, 721, 723–725
        Al$_2$O$_3$/TiC/Ni graded composites, 835–837
        AlN/Si interfaces, 192–193, 195–197
        BMAS/BN/Nicalon interfaces, 708
        BMAS/Nicalon SiC composites, 707–708
        CaAl$_{12}$O$_{19}$/sapphire or YAG fiber interfaces, 733, 735–737
        CdS, ZnSe nanocrystals in glass matrix, 203, 205
        Cr$_2$O$_3$-doped Al$_2$O$_3$; Cr/Al$_2$O$_3$, 210–213
        general features, 2
        La$_{1-x}$Sr$_x$MnO$_3$ thin films, 599–600
        NiO-doped MgO; Ni/MgO, 210–211
        Si-C-N powders, 572
        Si$_3$N$_4$ ceramics, 110, 124, 789–790
        SiC ceramics, 179–182
        SiC-C composites, 487–490
        SiC-MgO composites, 590, 593–594, 596
        SiC/SiC and SiC/III-V nitride heterostructures, 633–634
        SrTiO$_3$ ceramics, 645–649
        TiO$_2$-SiO$_2$ system, 157
        W-Mn/AlN interfaces in cofired multilayer substrates, 392–395
        Y-TZP-Al$_2$O$_3$ composite powders, 616
        Y-TZP–WC composites, 744, 746–747
    convergent beam electron diffraction, CBED, 4, 118, 311–315, 317–318, 320
    energy filtering, EFTEM, 2, 4, 96, 97
    high resolution, HRTEM
        Al,Al$_2$O$_3$ nanoparticles, 623–627
        Al/Al interface, 387
        Al$_2$O$_3$-SiC$_f$ (alumina-Nicalon silicon carbide fiber), 101
        Al$_2$O$_3$/Cu interfaces, 6–9
        AlN/Si interfaces, 192, 195–196
        Bi$_2$O$_3$-doped ZnO, 131, 133, 139–144
        BN/C multilayer coatings on Nicalon fibers in BMAS matrix, 709
        CaO·SiO$_2$-Ca$_3$(PO$_4$)$_2$ bioeutectics, 764–765
        CdS, ZnSe nanocrystals in glass matrix, 199, 201–206
        Cr$_2$O$_3$-doped Al$_2$O$_3$, 210, 212
        general features, 1
        of grain boundaries in Al$_2$O$_3$, 4–6
        interface in SiC-MgO composite, 595
        interfaces in SiC-C composites, 491–493
        interfaces in Si$_3$N$_4$-SiC$_f$ composites, 97, 99
        interfaces in TiO$_2$-SiO$_2$ system, 157–159
        La$_{1-x}$Sr$_x$MnO$_3$ thin films, 599–600
        NiO-doped MgO, 210
        planar defects in electroceramics, 640–644
        SiC ceramics, 177, 179, 181, 183, 186
        SiC/SiC and SiC/III-V nitride heterostructures, 633
        SiC/Si$_3$N$_{4(f)}$ composites, 104
        Σ5 in Y$_3$Al$_5$O$_{12}$ (yttrium aluminum garnet), 161–167

Transmission electron microscopy TEM (*cont.*)
 high resolution, HRTEM (*cont.*)
  Si$_3$N$_4$/Al interface, 374, 388
  Si$_3$N$_4$ ceramics, intergranular films, 25, 107, 109–111, 115–116, 118, 124–127, 131
  Si$_3$N$_4$/Cu-Ag-Ti interfaces, 415–419
  SrTiO$_3$ ceramics, 646, 650
 HOLZ line analysis, 311–318
 image simulation, 4, 5, 9, 10, 163–167, 197–198, 637–644
 reflection high-energy electron diffraction RHEED, 6, 680–682
 scanning, STEM, 2, 8, 95, 96, 98, 99, 131–133, 137–144, 157, 192, 204, 590
 techniques, 2
 Z contrast imaging, 95–99

Vapor phase epitaxy, 629–630, 633–635
Varistors, ZnO, 339–341, 343–347

Wear, alumina-silicon carbide (Al$_2$O$_3$-SiC$_p$) nanocomposites, 552, 554–555, 557
Weibull modulus: *see* Fracture
Wetting behavior, 65–66, 75–81, 135–145, 349–351, 353, 355–357, 407–411, 413–414, 503–511
Wetting, ceramic/metal systems, 349–351, 353, 355–357, 360–361, 407–411, 413–414, 503–511, 749–750, 752–753, 756–757
Wetting, glass/metal systems, 503–506, 508–511, 543, 547–549
Work of adhesion, 65–66, 76–81, 349–351, 356–357, 385, 410–411
Wulff shape
 of alumina (Al$_2$O$_3$), 229, 232–234
 effect of temperature on, 232–234
 of solid in liquid, 135, 277

X-ray photoelectron spectroscopy (XPS) of
 Si/C/N composite powders, 572–573
 thin films grown on Langmuir-Blodgett films, 442–443
 ZnO ceramics, 688, 690–693
 Fe/AlN thin films, 608–611
 AlN/Si interfaces, 192–193, 196–197

Yttrium aluminum garnet, Y$_3$Al$_5$O$_{12}$, grain boundaries, 161–167

Yttrium aluminum garnet, Y$_3$Al$_5$O$_{12}$, precipitates in alumina, 311–312, 314, 317, 319–321
Yttrium aluminum garnet–sapphire fiber (Y$_3$Al$_5$O$_{12}$-Al$_2$O$_{3(f)}$) composites, 733–735

Zeta potential, diamond, 222–225; *see also* Colloidal phenomena
Zinc oxide, ZnO
 additives, 339–341, 343–347, 639–641
 Bi$_2$O$_3$-doped, 131–145, 344–346
 devitrification of intergranular film, 145
 grain boundaries, 131–145, 339–342, 344–347
 grain growth/boundary mobility, 339–347
 growth from vapor phase, 687–690
 I-V characteristics, 687–693
 intergranular films, 131–145, 344–346
 liquid phase sintered, 340, 343–346
 non-ohmic behavior, 687–693
 oriented polycrystals, 687–693
 planar defects in In$_2$O$_3$-doped ZnO, 639–641
 processing, 132–133
 spinel formation in, 340, 345–347
 surface chemisorption as Shottky barrier source, 687–688, 691–693
Zircon, ZrSiO$_4$
 coatings for SiC$_f$/LAS composites, 514, 518–521, 523–524
 joining *via* reactive coatings, 458–461
Zirconia, ZrO$_2$
 3 mole % yttria-stabilized tetragonal zirconia polycrystals (3Y-TZP); yttria-stabilized tetragonal zirconia (Y-TZP)
  conventional sintering of, 478–482
  grain growth/boundary mobility, 479–481, 482
  microwave sintering of, 472, 474, 478–483
 brazing of, 359–360, 365
 nanoparticle synthesis, 613–615
Zirconia (yttria-stabilized tetragonal)–tungsten carbide composites, Y-TZP–WC, 741–748
 grain growth in, 744–745, 748
 interfacial reactions, 741, 743–748
 mechanical properties, 741–743, 748
 processing, 741–748
Zirconium titanate, ZrTiO$_4$, coatings for SiC$_f$/LAS composites, 514, 518–521, 523–524